# Ore Deposits of the United States, 1933-1967

## 1933-1967

### *The Graton-Sales Volume*

*Other Volumes in the Rocky Mountain Fund Series*

# Ore Deposits

## of the

# United States, 1933-1967

### THE GRATON-SALES VOLUME

### John D. Ridge, editor

FIRST EDITION
*Sponsored by The Rocky Mountain Fund*

*Volume I*

*Published by*
**The American Institute of Mining, Metallurgical, and Petroleum Engineers, Inc.**
NEW YORK    1968
Reprinted 1970

Reprinted 1970
*Copyright © 1968 by the*

American Institute of Mining, Metallurgical, and Petroleum Engineers, Inc.
*Printed in the United States of America*
by THE MAPLE PRESS COMPANY, York, Pennsylvania

Library of Congress Catalog Card Number 68-24170.

# A. I. M. E. Committees for the Volume

# Contributors to the Graton-Sales Volume

Charles A. Anderson
Gerald J. Anderson
Roy A. Anderson
Richard J. Bagan
Arthur Baker, III
Marvin P. Barnes
Richard W. Bayley
Edward L. Beutner
Arthur A. Bookstrom
James C. Bradbury
Douglas C. Brockie
J. M. Brown
John S. Brown
W. Horatio Brown*
Gordon B. Brox
Richard W. Brummett
Wilbur S. Burbank
George I. Burns
William H. Callahan
Robert H. Carpenter
Frederic M. Chace
Robert L. Clayton
D. M. Clippinger
Robert R. Coats
Douglas R. Cook
Manning W. Cox
Johnson Crawford
Robert M. Crump
John T. Cumberlidge
Frank W. Dickson
David B. Dill, Jr.
Paul R. Dingess
Robert L. DuBois
Carl E. Dutton
John T. Eastlick
J. James Eidel
James W. Emanuelson
John A. Emery
Chester O. Ensign, Jr.
Richard P. Fischer
Elwin L. Fisk
John J. Fritts
J. E. Frost
Verne C. Fryklund, Jr.
Roy P. Full
W. J. Garmoe
Paul Gemmill

Paul E. Gerdemann
Paul Gilmour
Charles C. Goddard, Jr.
Robert M. Grantham
L. C. Graton*
Raymond F. Gray
Tunstall R. Gray
Robert M. Grogan
Stanford O. Gross
John M. Guilbert
Arthur F. Hagner
Donald F. Hammer
Edward H. Hare, Jr.
E. N. Harshman
Olin M. Hart
Owen J. Hart
D. J. Hathaway
Donald M. Hausen
Robert M. Hernon*
William P. Hewitt
Allen V. Heyl
Alan D. Hoagland
S. Warren Hobbs
Victor J. Hoffman
Richard N. Hunt
Robert G. Ingersoll, Jr.
Everett D. Jackson
William R. Jones*
David C. Jonson
Vincent C. Kelley
Paul F. Kerr
Dale F. Kittel
Davis M. Lapham
Edgar R. Lea
Robert A. Laurence
R. J. Leone
Robert K. Linn
Robert G. Luedke
Joseph F. McAleer
Roger H. McConnel
Harold L. McKinley
W. Bruce Mackenzie
J. Hoover Mackin*
Maurice Magee
Roger C. Malan
Ralph W. Marsden
Paul E. Melancon

J. S. Merchant
Charles Meyer
Richard N. Miller
Hal T. Morris
John E. Motica
Vernon A. Mrak
Neil K. Muncaster
John E. Murphy
Thomas B. Nolan
Ernest L. Ohle
John S. Owens
Charles F. Park, Jr.
J. L. Patrick
Vincent D. Perry
Donald W. Peterson
Walden P. Pratt
Robert E. Radabaugh
John D. Ridge
Richard D. Rubright
C. L. Sainsbury
John H. Schilling
Daniel R. Shawe
Edward P. Shea
William M. Shepard
John G. Simos
Thomas A. Simpson
Samuel J. Sims
A. L. Slaughter
Frank G. Snyder
Edward C. Stephens
Thomas A. Steven
Arthur R. Still
Vaughn E. Surface
John W. Trammell
George Tunell
Ogden Tweto
Neal E. Walker
Stewart R. Wallace
Edgar L. Weinberg
Robert F. Werner
Walter S. White
Thomas Wigal
Hiram B. Wood
J. C. Wright
Thomas L. Wright
Robert S. Young
Lester G. Zeihen
Paul W. Zimmer

* Deceased

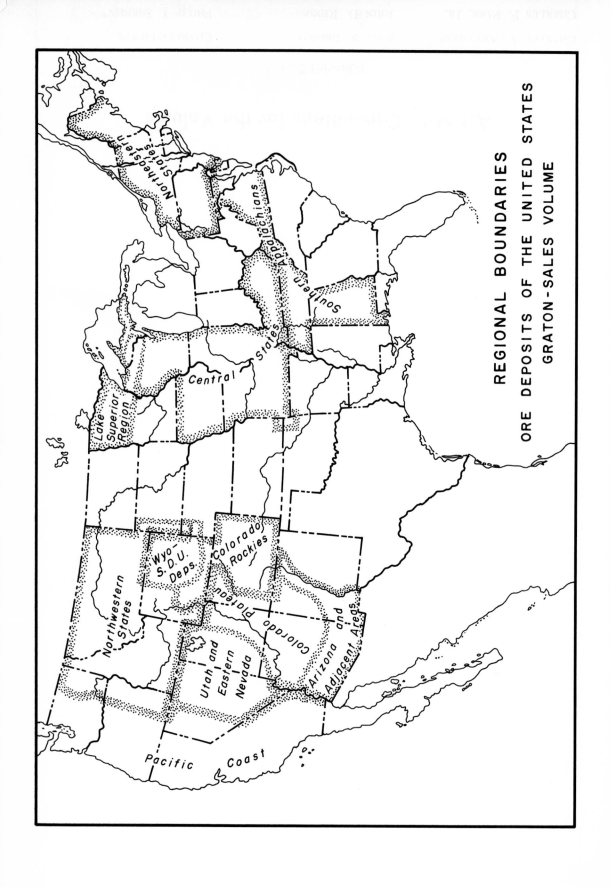

REGIONAL BOUNDARIES

ORE DEPOSITS OF THE UNITED STATES

GRATON-SALES VOLUME

# Editor's Preface

The oldest material I have in my files of the Graton-Sales Volume is a letter from Roland Mulchay dated July 20, 1962, in which he expresses the hope that the successor to the Lindgren Volume might be published in from 3 to 5 years. Actually, the Graton-Sales Volume will appear almost exactly 6 years from the date of Mulchay's letter. It was his hope that the Volume might include descriptions of new mines developed in the entire United States since the Lindgren Volume [Ore Deposits of the Western States, (1)] was published in 1933 and that it might contain papers on modern developments in [economic] geologic theory and research. The present Volume meets well the first of Mulchay's goals; the only major omissions being descriptions of the porphyry copper deposits. These latter ore bodies were largely covered in the 1966 Wilson Volume [Geology of the Porphyry Copper Deposits—Southwestern United States, (2)] as were some of the modern developments in geologic theory, exploration, and research. If a reader cannot understand why any deposit in the United States in which he is interested is not included in the Graton-Sales Volume, the most probable reason, in addition to that given for the omission of porphyry coppers, is that permission to publish a paper on that deposit could not be obtained. The publication in 1967 of the volume edited by H. L. Barnes, Geochemistry of Hydrothermal Ore Deposits (3) provided a synthesis of experimental and theoretical work in the field of ore genesis.

Because the plans for the publication of these two volumes were known to the Advisory Committee at its first session at the A.I.M.E. meeting in Dallas in February 1963, it was decided to limit the Graton-Sales Volume to descriptions of ore deposits anywhere in the fifty States that had been discovered since 1933 or in which new geologic developments had taken place since that date, less those described in the Wilson Volume. It also was agreed that a summary of the changes and developments in the concepts of ore genesis since 1933 should be included.

As the new Volume was to be the successor to the Lindgren Volume, it was decided that it should bear the name (or names) of one or two noted American economic geologists, all or much of whose work had been done in the period since the publication of the Lindgren Volume. After much discussion, it was decided to dedicate the Volume to Louis Caryl Graton and Reno Haber Sales.

The Advisory Committee concluded that the actual work on the Volume would be best left to an Editorial Board which the Committee selected. The Board was charged with choosing an Editor for the Volume. This board, as it originally was appointed, consisted of C. A. Anderson, R. M. Garrels, Charles Meyer, C. F. Park, Jr., J. D. Ridge, P. J. Shenon, and F. G. Snyder. When Garrels left Harvard early in 1965, he found it necessary to resign from the Board, and he was replaced by J. S. Brown. Phil Shenon died on December 1, 1966, and was not replaced on the Board. In December 1964, the Editorial Board selected John D. Ridge of The Pennsylvania State University as the Editor of the Graton-Sales Volume.

The all-important matter of funds to make publication of the Volume possible was solved by a contribution of $50,000 from the Rocky Mountain Fund. This money is to be repaid to the Fund through sales of the Volume.

If this Volume does nothing else, it shows the wide diversity of the geology of ore deposits in the United States and in the opinions held concerning their genesis. Each author has expressed his ideas as to how the ores of his deposit were formed, and anyone expecting to derive a uniform theory of ore genesis from the district papers in this Volume will be disappointed. He will, however, receive something even more important than such a theory and that is the realization of how far we are from a complete understanding of the processes of ore genesis and how much work, both in the field and in the laboratory, needs to be done if one eventually is to be produced.

The organization of the Volume can readily be ascertained by consulting the Contents pages and, except for the two general papers located at the end of the Volume, is entirely geographic. The boundaries of the 11 geographical regions to which the various district papers are assigned are shown in the Frontispiece.

\* \* \*

In memory of the members of the Graton-Sales family who died during the course of the writing of the Volume—W. Horatio Brown, Robert M. Hernon, William R. Jones, Harrison A. Schmitt, and Philip J. Shenon—copies of the Volume will be given, in their names, to academic institutions outside the American continent, institutions that might have difficulty in obtaining American currency with which to purchase the Volume.

\* \* \*

It is so obvious as hardly to need saying that the Graton-Sales Volume has been produced by the cooperative effort of the authors of regional, district, and general papers and by the help of the Editorial Board and the Advisory Committee. My special thanks must

go to Charles P. Thornton for his preparation of valuable indices at the end of the Volume and to Arthur W. Rose for having read and commented on all the papers in the Volume. The authors' facts and opinions are, of course, their own; such mechanical erors as yet remain are my responsibility.

*University Park,*                    JOHN D. RIDGE
*Pennsylvania*

## REFERENCES CITED

1. Finch, J. W., *Chairman of the Committee on the Lindgren Volume,* 1933, Ore deposits of the western states: A.I.M.E. (Rocky Mountain Fund), N.Y., 797 p.
2. Titley, S. R. and Hicks, C. L., *Editors,* 1966, Geology of the porphyry copper deposits—southwestern United States: Univ. Ariz. Press, Tucson, 287 p.
3. Barnes, H. L., *Editor,* 1967, Geochemistry of hydrothermal ore deposits: Wiley, N.Y., 670 p.

# Louis Caryl Graton[*]

Louis Caryl Graton was born in Parma, Monroe County, New York, on June 10, 1880, the son of Louis Graton, Sr. and A. Ella Gould Graton. His father was a native of a small village of St. Martin, just north of Montreal, who had emigrated at the age of 17 to New York where he had learned both the art of glass embossing and the English language. Disliking the great city, he decided to return northward. The use of embossing was rapidly declining, however, and he found work, as he passed from town to town, as an interior decorator. In Parma, he met Miss Gould, the village school teacher, and they were married in 1877. Since the family moved at frequent intervals, Caryl Graton's parents decided to teach the child at home. A pattern developed whereby, after supper, his father would read aloud to the boy from such works as Darwin's "Origin of Species," Tyndall's "Fragments of Science," and Tyndall's "Lectures in America." These books still are retained in the Graton family as valued heirlooms.

At the age of seven, Caryl Graton's mother thought he might be interested in stamps and brought from the attic a case filled with letters to and from her long deceased father. Among these, he found an age-yellowed newspaper clipping that told of a keen young chap from the East who went West in search of adventure and discovered and mined silver—lots of it. Then and there, at the age of seven, Caryl Graton committed his life to ore-hunting.

His formal schooling began at the age of nine when he entered fifth grade in the Friendship, New York, school. From Friendship, his family moved to Hornell where he found and profitted greatly from excellent teachers in mathematics, chemistry, and physics. He graduated from high school at 16 and, as Class Valedictorian, chose as his subject "The Life of Louis Pasteur." Immediately after his graduation, Caryl Graton took the annual State Regents' Examination for a four-year scholarship at Cornell and won. His family then moved to Ithaca, and the young man entered Cornell in the autumn of 1896. He elected courses in chemistry, physics, and mathematics under such well-known professors as Emile Monnin Chamot in microchemistry, A. G. Gill in mineralogy and petrography, and R. S. Tarr and Heinrich Ries in geology. During Graton's sophomore year, a distinguished chemist, visiting at Cornell, noted the care with which two students (Norman Dodge and Graton) carried out a common assignment and challenged them to undertake a more exacting program. The result of their work was Graton's first published paper (1) on the system water, alcohol, and potassium nitrate in Volume 2 of the Journal of Physical Chemistry.

The day after he received his B.S. degree, Caryl Graton left for his first job in the mining industry—as assayer for the Ledyard Gold Mines, Ltd., near Rockdale, Ontario. The deposit, as shown by Graton's assays, was worthless, and in 1900 he quit to work for the Canadian Goldfields, Ltd., at nearby Deloro, where the company was profitably producing gold and arsenic. That same autumn, he became a demonstrator in chemistry at McGill University and had the opportunity to take courses under such outstanding earth scientists as B. J. Harrington in mineralogy and Frank D. Adams in geology. He spent the summer of 1901 investigating mines in such areas as Timascaming, Timagami, Porcupine, Kirkland Lake, and Haliburton and returned that winter to McGill. Early in 1902, an expedition was organized by Dr. Thomas W. Gibson, Director of the Ontario Department of Mines, a friend of Caryl Graton, to explore the virgin southwest portion of the province along the Mississagi River, and he offered Graton the position of geologist, which he accepted. On his way to his field area, he delayed at Sudbury and took the opportunity to spend a shift underground at the famous Frood Mine of Interna-

* Prepared by John Ridge from material supplied by L. C. Graton and Philip W. Chase, Bend, Oregon.

tional Nickel. Before leaving Sudbury, Caryl Graton learned that he had been awarded the Schuyler Scholarship at Cornell for 1902–1903. During this period, he carried on further study in his field of specialization and became a life member of the Society of Sigma Xi. His summer's work on the Mississagi was later published by the Ontario Department of Mines (2).

In 1903, Waldemar Lindgren of the U.S. Geological Survey hired Graton as his assistant (on the recommendation of Professor Gill) in his restudy of the geology of the Cripple Creek district in Colorado. Thus began an association that continued until Lindgren's death in 1939.

During the following six years, Caryl Graton worked with the Survey, studying not only the Cripple Creek ores but also those of gold and tin in the southern Appalachians, various mineral deposits in New Mexico, and copper in California (4,5,6,7,8,9,10,11). During this period, he was named to write the annual reviews of copper in the U.S. Geological Survey's Mineral Resources of the United States. For the first time in the history of this publication, it stressed the geological occurrences of the metal, as well as provided information on such things as stocks, production, and prices. As a consequence of his work in this field, he was named to make the first estimate of copper reserves in the United States.

In 1909, Caryl Graton resigned from the Survey to become assistant professor of Mining Geology at Harvard; in 1912, he was advanced to the rank of full professor, a post he held for the next 37 years. His connection with Harvard, however, did not prevent him from continuing his interests outside the academic world. In 1913, the Geophysical Laboratory, a number of American copper companies, and the Harvard Mining School jointly undertook the Secondary Enrichment Investigation. Caryl Graton was named the Director of this project and enlisted the assistance of such eminent geochemists as E. T. Allen, Eugene Posnjak, and E. G. Zies, mineralogist H. E. Merwin, and geologists, Augustus Locke, Alan M. Bateman, and E. H. Perry. All the important copper districts in the United States were studied, and no less than 21 papers resulted from the work and are listed in the Survey's Bibliography of North American Geology. The investigation not only placed the chemistry and geology of the processes of secondary enrichment on a sound scientific basis, processes that have been such an important factor in making economically valuable so many of the world's

major copper deposits, but also established methods of geologic investigation that subsequently were used to solve other complicated geological problems.

On the U.S. entry into World War I (1917), Graton was asked to serve as Secretary of the Copper Producers' Committee for War Service, formed at the instigation of the War Industries Board. When he concluded this important assignment in 1919, he was asked by the Internal Revenue Service to assist that organization in developing a rational method of mine valuation, one acceptable to both the IRS and the mining industry. As part of his work on this assignment, Graton was influential in gaining acceptance of the principle of applying percentage depletion to mining operations.

When he had finished his work with the Internal Revenue Service (1920), he returned enthusiastically to the teaching and practice of geology. His keen desire to obtain the fundamental facts of ore occurrence at first hand led to extensive travel, in the course of which he visited at least 77 countries, in 50 of which he made geological observations of basic importance. As a result of his ever-active curiosity on the problems of ore genesis, there were few notable mining districts that he did not visit and study.

During his tenure at Harvard (1909–1949), Graton, with the collaboration of some of his gifted students, produced several items of equipment that have notably improved the instrumention for laboratory work in economic geology. Notable among these is the Graton-Vanderwilt Polishing Machine for the preparation of polished surfaces on opaque minerals. This machine permits the preparator to maintain an ideal plane across the boundaries between hard and soft minerals, a condition maintained even at 6000 times magnification. He also perfected a micro-drill of the Haycock type that permitted taking an uncontaminated sample from a grain with a surface exposure of as little as 20 microns in diameter. The Graton-Dana Microcamera is designed for use on both polished and thin sections, and, with first-class polished sections, it permits sharp negatives and fine resolution in the magnification range of 4000 to 6000 diameters. The original Graton-Dana microcamera was installed on the base originally used by the Harvard Seismograph (located in the basement of the Geology wing); this base provided both sufficient length and strength for the practical operation of the camera. Still another valuable instrument is the Graton-Talmadge Hardness

Tester that holds the mounted specimen in a clamp under a sharp slightly down-curving diamond blade and, by lowering the blade only a tiny distance, makes a minute elongate scratch on the surface of the specimen. The stage may then be rotated through any required set of angles and additional scratches made, showing whether the mineral has the same or different hardness in various orientations. The scratch is of such shallow depth that there is only the slightest chance of its penetrating an underlying mineral of different composition and hardness.

In 1912, Caryl Graton was promoted to full professor of Mining Geology at Harvard; in 1942, he was raised to the title of Sturgis Hopper Professor of Mining Geology and held that post until 1949 when he was granted emeritus rank.

From 1920 to 1928, Graton was principally engaged in the investigation of deep mines and tunnels in the U.S., Canada, Brazil, France, Italy, Switzerland, South Africa, and India. In the period 1936 to 1949, he conducted a study of volcanoes, hot springs, and geysers that took him to locations in the U.S., Mexico, Guatemala, Salvador, Nicaragua, Costa Rica, Peru, Hawaii, New Zealand, Java, Bali, and Iceland. This work resulted, along with other publications, in the printing of his major contribution in this area (45) in the Daly Volume in 1945.

His connection with the Cerro Corporation (then the Cerro de Pasco Copper Corporation) began when he was retained as a consultant in 1920, and although his actual consulting stopped in 1950, he was appointed to the Cerro Board of Directors in 1945 and served with that body until 1967. On the occasion of his retirement from that Board, he was presented with an impressively engraved silver tray that records his many services to that Corporation. A number of papers on various aspects of the geology of the several Cerro deposits (34,39,51) show his broad understanding of their geology and are notable contributions to the understanding of ore genesis. Graton's other clients included Calumet and Hecla, Noranda Copper, U.S. Bureau of Mines, Hercules Mining Company, Hollinger Consolidated Gold Mines, and International Nickel Company.

Probably Caryl Graton's outstanding contribution to ore geology was his paper (43), published in 1940, on the nature of the ore-forming fluid. Graton devoted the first part of this paper (summarized in Chapter 82 of this Volume, reference number 8), to the refutation of the Fenner-Bowen concept that ore elements leave the magma as gaseous compounds and then makes his two principal points: (1) that the hydrothermal fluid almost certainly does not develop in the magma chamber until late in the crystallization cycle of that silicate-water system and, therefore, cannot leave the chamber until late in that cycle and (2) that the ore fluid was segregated from the body of the magmatic (silicate) material as a separate and independent phase that differed in density to no great degree from a liquid of similar composition under standard conditions. It was his contention, as was further shown in his paper on zoning in the Cerro deposit (39), that the ore fluid began its upward journey on the alkaline side of neutrality and ended it probably on the acid side, a concept since confirmed by much theoretical and experimental work.

No study of Caryl Graton's life can be complete without mention of his controversial paper on the genesis of the Witwatersrand deposits (28), published before the period covered by this Volume. In that paper, Graton put forward arguments for the hydrothermal character of the Rand mineralization that have been much contended against in the years since 1930 but never have been convincingly refuted. He emphasized that the outstanding manifestations of hydrothermal activity exhibited by the Rand ores, in other areas, would have been universally recognized as definitive of that process.

Among his other contributions to the science of geology, the following are of particular importance:

1. Authorship of chapter on "Ore Deposits" in the 50th Anniversary Volume of the Geological Society of America in 1941, and the chapter on "Mining Geology" in the 75th Anniversary Volume of the American Institute of Mining and Metallurgical Engineers in 1947.

2. Studies on the behavior of rocks under increasing loads at increasing depth; and the necessity for integrating actual underground conditions with findings of laboratory experiments if true significance is to be had. These studies were based on actual observations carried out in many deep mines and tunnels throughout the world.

3. Study of volcanic heat, including the source of volcanic material reaching the surface; the eruptive mechanism and causative agent; fumaroles and hot springs in relation to active magma and to volcanoes; erupting hot springs (geysers) and the geyser mecha-

nism; evidence that in areas of very hot water this represents the mineral-depleted liquid waste from mineral-forming, perhaps ore-forming, processes along the channelways at greater depth. During the course of these studies Graton visited most of the world's active volcanic areas.

4. Amplification and partial modification of Lindgren's genetic classification of Depth Zones in Ore Deposition by proposing that while *mineralization* from a magmatic source along discovered upward-leading channelways may be continuous from the deep beginning of super-saturation all the way to surface outlet (which may there constitute a hot spring), that continuous mineralization of *economic tenor* is probably rare. Many a mine is abandoned at depth, not because the system dies, but only because a gap in adequate tenor has been reached.

From his earliest work with the United States Geological Survey, Graton became convinced that in seeking an interpretation of ore deposits and in the analysis of the more difficult geological problems of ore genesis, field relations were of critical importance. At the same time, he was among the first to recognize the importance of laboratory study of ores in contributing information not available in the field. This insistence that field and laboratory observations go hand in hand, his acute observation of natural phenomena, his scientific inquisitiveness and analytical approach were at the root of his success.

Graton also made important contributions in the application of the scientific principles of geology to the practical job of ore finding. He was notably successful in "selling" geology to the mining industry and imparting to management his own conviction of the benefits that scientific geology could bring to the practical task of finding more ore. As a consultant, Graton influenced numerous important mining companies to set up modern geological departments and to establish high standards of scientific performance within these departments. He instilled into the staff geologists his own insistence on arriving at the fundamental truths of geological processes and discarding all theories that did not fit the evidence, both field and laboratory. He inspired his associates to engage in the use of imagination and resourcefulness for the solution of varied geological problems; and through his wide knowledge of ore deposits and their genesis, he was able to aid the resident geologist to avoid allowing details to obscure the whole.

Graton's accomplishments in applied geology were immense, even though they are not capable of measurement in terms of tons of ore developed or mines discovered. His success can best be measured by considering the success of those directly charged with finding ore for the many mining companies for which Graton acted as consultant.

Another major area in which Graton achieved distinction was that of education. From 1910 to 1949 he was an important figure in the Geological Department at Harvard University. He had a remarkable faculty for sparking interest among his students in the genesis of ore bodies and then nurturing and encouraging this interest. He not only gave his students an enthusiasm for geology but likewise a firm foundation in its scientific facts and a deep sense of scientific honesty and professional ethics. He impressed upon them his own conviction that field observation and laboratory investigation must go hand in hand. Perhaps his accomplishment in the field of education is best evidenced by the number of his students who have attained high stature in professional life.

As a scientist and as a teacher, Graton has occupied a pre-eminent position in economic geology for over half a century. Through his students and colleagues his wholesome influence on the science of geology will continue for many years into the future. The bibliography of his publications and the list of his memberships in professional societies in numerous different countries (as set forth at the end of this vita) give evidence of the breadth and depth of his scientific interests and the extent of his contributions to the science of geology, while the positions he has held in the National Resources Council, the National Science Foundation, the American Academy of Arts and Sciences, the Society of Economic Geologists—President 1931 and Penrose Medalist 1950 ("for unusual original work in the earth sciences"), and his position as Honorary Fellow in Geology at Yale University attest to the esteem in which he is held by his colleagues. His membership in professional societies—Fellow—American Association for Advancement of Science, Geological Society of America, Mineralogical Society of America, American Academy of Arts & Sciences, Geological Association of Canada, Geological Society of London; Honorary Member—Geological Society of Belgium; Member—Mining & Metallurgical Society, Geochemical Society, Society of Economic Geologists, Geophysical

Union, Canadian Institute of Mining & Metallurgy, Mexican National Academy, Geological Society of South Africa, South African Institute of Mining & Metallurgy, and Geological Society of Peru—show his wide breadth of interest and accomplishment geographically and topically.

## SELECTED BIBLIOGRAPHY OF LOUIS CARYL GRATON

1. 1898 (with Dodge, N.), Alcohol water and potassium nitrate: Jour. Phys. Chem., v. 2, p. 498–501.
2. 1903, Up and down the Mississaga: Ont. Bur. Mines 12th Ann. Rept., v. 12, p. 157–172.
3. 1902, On the petrographical relations of the Laurentian limestones and the granite in the township of Glamorgan, Haliburton County, Ontario: Canadian Rec. Sci., v. 9, p. 1–38.
4. 1905, The Carolina tin belt: U.S. Geol. Surv. Bull. 260, p. 188–195.
5. 1905 (with Schaller, W. T.), Purpurite, a new mineral: Amer. Jour. Sci., 4th ser., v. 20, p. 146–151; also in Zeitsch. Kryst., v. 41, p. 433–438.
6. 1905, Consanguinity in the eruptive rocks of Cripple Creek (Colo.) (abs.): Science, n.s., v. 21, p. 391.
7. 1905 (with Hess, F. L.), The occurrence and distribution of tin: U.S. Geol. Surv. Bull. 260, p. 161–187.
8. 1906, Description and petrology of the metamorphic and igneous rocks (Cripple Creek district: U.S. Geol. Surv. Prof. Paper 54, p. 41–113.
9. 1906, Reconnaissance of some gold and tin deposits of the southern Appalachians: U.S. Geol. Surv. Bull. 293, p. 9–118.
10. 1906 (with Gordon, C. H.), Lower Paleozoic formations in New Mexico: Amer. Jour. Sci., 4th ser., v. 21, p. 390–395; also in Science, n.s., v. 23, p. 590–591.
11. 1906 (with Lindgren, W.), A reconnaissance of the mineral deposits of New Mexico: U.S. Geol. Surv. Bull. 285, p. 74–86.
12. 1907, Copper: Min., Res. of the U.S., 1906, U.S. Geol. Surv., p. 373–438.
13. 1908, Copper: Min., Res. of the U.S., 1907, U.S. Geol. Surv., pt. 1, p. 571–644.
14. 1908 (with Siebenthal, C. E.), Silver, copper, lead, and zinc in Central States: Min., Res. of the U.S., 1907, U.S. Geol. Surv., pt. 1, p. 483–549.
15. 1910, The occurrence of copper in Shasta County, California: U.S. Geol. Surv. Bull. 430, p. 71–111.
16. 1910 (with Lindgren, W.), The ore deposits of New Mexico: U.S. Geol. Surv. Prof. Paper 68, 361 p.
17. 1913, Investigation of copper enrichment: Eng. Min. Jour., v. 96, p. 885–887.
18. 1913, Ore deposits at Butte, Montana: (disc.), A.I.M.E. Bull. 83, p. 2735–2736.
19. 1914 (with Murdoch, J.), The sulphide ores of copper; some results of microscopic study: A.I.M.E. Tr. 45, p. 26–93, 529–530.
20. 1913, Notes on rocks from the coppermine River region, Canada: Canadian Min. Inst. Tr. 16, p. 102–114.
21. 1915 (with others), To what extent is chalcocite a primary and to what extent a secondary mineral in ore deposits?: (disc.), A.I.M.E. Tr. 48, p. 194–200.
22. 1917 (with McLaughlin, D. H.), Ore deposition and enrichment at Engels, California: Econ. Geol., v. 12, p. 1–38.
23. 1918 (with McLaughlin, D. H.), Further remarks on the ores of Engels, California: Econ. Geol., v. 13, p. 81–99.
24. 1918, The relation of sphalerite to other sulphides in ores: (disc.), A.I.M.E. Bull. 136, p. 844–845.
25. 1928 (with Davidson, S. C.), Microscopical interpretations of folded structures: Econ. Geol., v. 23, p. 158–184.
26. 1928, Notes on deep mines of Michigan and Brazil: Chem. Met. Min. Soc. Africa, 9th ser., v. 28, (March), p. 209.
27. 1929 (with Butler, B. S. and Burbank, W. S. plus T. M. Broderick, C. D. Hohl, C. Palache, M. J. Scholz, A. Wandke, and R. C. Wells), The copper deposits of Michigan: U.S. Geol. Surv. Prof. Paper 144, 238 p.
28. 1930, Hydrothermal origin of the Rand gold deposits—Part I. Testimony of the conglomerates: Econ. Geol., v. 25, 185 p.
29. 1930, Some economic aspects of the copper industry: Min. and Met. Soc. Amer. Bull. 208, (January), p. 5–35.
30. 1931, Future gold production—The geological outlook: A.I.M.E. Tr., 1931, p. 534–557.
31. 1931 (with Bastin, E. S., Lindgren, W., Newhouse, W. H., Schwartz, G. M., and Short, M. N.), Criteria of age relations of minerals: Econ. Geol., v. 26, p. 561–610.
32. 1933, Life anl scientific work of Waldemar Lindgren: Ore deposits of the western states: A T.M.E. (Lindgren volume), p. 13–32.
33. 1933, The hydrothermal depth zones: Ore deposits of the western states: A.I.M.E. (Lindgren volume), p. 181–197.
34. 1933 (with McLaughlin, D. H. and others), Copper in the Cerro de Pasco and Morococha Districts, Department of Junin, Peru: in Copper Resources of the World, 16th Int. Geol. Cong., v. 2, p. 513–544.
35. 1933 (with McKinstry, H. E. and others), Outstanding features of Hollinger geology: Canadian Inst. Min. and Met. Tr., v. 36, p. 1–20, disc., p. 606–618.

36. 1933, The depth zones in ore deposition: Econ. Geol., v. 28, p. 513–555.
37. 1935 (with Harcourt, G. A.), Spectrographic evidence on origin of ores of the Mississippi Valley Type: Econ. Geol., v. 30, p. 800–824.
38. 1935 (with Fraser, H. J.), Systematic packing of spheres, with particular relation to porosity and permeability: Amer. Assoc. Petrol. Geols. Bull., v. 43, p. 785–909.
39. 1936 (with Bowditch, S. I), Alkaline and acid solutions in hypogene zoning at Cerro de Pasco: Econ. Geol., v. 31, p. 652–698.
40. 1937, Technique in mineralography at Harvard: Amer. Mineral., v. 22, p. 491–516.
41. 1937 (with Dane, E. B., Jr.), A precision, all-purpose microcamera: Jour. Opt. Soc. Amer., v. 27, p. 355–376.
42. 1938, Ores: from magmas or deeper? A reply to Arthur Holmes: Econ. Geol., v. 33, p. 251–286.
43. 1940, Nature of the ore-forming fluid: Econ. Geol., v. 35, p. 197–358.
44. 1941, Ore deposits: in Geology, 1888–1938, Geo. Soc. Amer., 50th Anniv. Vol., p. 473–509.
45. 1945, Conjectives regarding volcanic heat: Amer. Jour. Sci., v. 243A (Daly Volume), p. 135–259.
46. 1945, Ciertos aspectos geneticos lel Parícutin, neuvo volcan de Michoacan: in El Parícutin, Univ. Mexico, Inst. Geología, Estudios volcanologicos, p. 59–91.
47. 1945, The genetic significance of Parícutin: Amer. Geophys. Union Tr., v. 26, p. 249–254.
48. 1947, Seventy-five years of progress in mining geology: in Seventy-five years of progress in the mineral industry, A.I.M.E. Anniv. Volume, p. 1–39.
49. 1947, Nature of certain ore-forming solutions (abs.): N. Y. Acad. Sci., ser. II, v. 9, no. 8, p. 285–286.
50. 1947, Causes for downward limits of hypogene deposits: Report on ore deposits, Committee on Research, Part VIII: Econ. Geol., v. 42, p. 547–556.
51. 1950 (with Cerro Geological Staff), Lead and zinc deposits of the Cerro de Pasco Copper Corporation in Central Peru: 18th Int. Geol. Cong., pt 7, p. 154–186.
52. 1960, If Lindgren were here: Econ. Geol., v. 55, p. 192–200.
53. 1961, Comments on program for the symposium on ore deposition, Prague, 1963: Geochem. News, no. 28, p. 4–5.
54. 1968, Lindgren's ore classification after fifty years: This Volume, p. 1703.

## REVIEWS

1. 1933, Ore deposits of the western states (Lindgren volume): Econ. Geol., v. 31, p. 222–226.
2. 1936, Economic geology of mineral deposits (Lilley, E. R.): Econ. Geol., v. 31, p. 882–884.
3. 1937, Gold deposits of the world (Emmons, W. H.): Econ. Geol., v. 34, p. 116–120.

# Reno H. Sales

VINCENT D. PERRY,* CHARLES MEYER†

Reno Haber Sales was born at Storm Lake, Iowa, September 10, 1876. In 1881, his family moved to Bozeman, Montana, where his father was an early day rancher in Montana's newly developing Gallatin Valley. Salesville, one of the Valley's first settlements, and now known as the Gallatin Gateway to Yellowstone Park, was named for the pioneer family. Stimulated by the magnificent mountain environment of his early childhood, young Sales had ample opportunities to develop an interest in geology. He attended Montana State College at Bozeman, graduating in the Class of 1898 with a Bachelor of Science degree. This was followed by two years of graduate work at Columbia School of Mines in New York City, from which school he received the degree of Engineer of Mines in 1900.

Reno Sales began his professional career in 1900 as an engineer-surveyor for the Boston and Montana Consolidated Copper and Silver Company at Butte, Montana. In that year, a start was being made on the organization of Anaconda's Geological Department by Horace V. Winchell, and in 1901 Sales became his assistant. Winchell and Sales, in conjunction with D. M. Brunton, worked out the essential methods of procedure for surface and underground mapping of Butte's geology and, in 1906, Sales succeeded Winchell as Chief Geologist of The Anaconda Copper Mining Company.

The mines of Butte at that period were divided among various ownerships. Problems of litigation, involving extra lateral extensions and faulted segments of high-grade veins, and the protection of Company-owned mineral rights against illegal trespass by outsiders required constant vigilance and defensive action. Reno Sales developed and applied a system of precise geologic field mapping that proved highly successful in the legal defense and protection of his Company's rich ore bodies. The high

esteem of the courts for Reno Sales' testimony as an expert witness is a monument to his uncompromising integrity as well as to his technical competence.

From his early work, there evolved a standardization of mine geological mapping that has been adopted throughout the mining world. His emphasis on engineering accuracy in realistically recording geological observations and his system of plotting these notes on oriented level plans and sections gave a simple, uniform, and concise style to graphic geological illustrations. Just as good English composition promotes clear thinking, Sales' geological mapping methods, based on accurate recording and interpretation of factual observations, stimulates sound geological reasoning.

By 1910, many of Butte's mining operations were consolidated under Anaconda, and Sales then had the opportunity to coordinate the geological patterns and details of the district. The result was his monumental work entitled "Ore Deposits at Butte, Montana" presented at a Butte meeting of the A.I.M.E. in 1912 and published in the A.I.M.E. Transactions in 1913. Commenting on this work, the late Professor C. F. Tolman of Stanford University said in 1939—"In my opinion Sales' description of the Butte ore deposits was the most important contribution to economic geology to date." Probably few mining geoolgists would care to dispute that statement as applied in 1967.

Sales' abilities contributed significantly to Anaconda's growth. Assisted through the creative years by Frank A. Linforth, Murl H. Gidel and Chester H. Steele, the Geological Department at Butte met the growing needs for maintaining a healthy backlog of ore reserves by a practical, integrated system of written geological recommendations for extending underground workings to develop new ore, both laterally and in depth. The detailed geo-

* The Anaconda Company, New York, N.Y.
† University of California, Berkeley, California.

logical records of old and presumably worked-out upper levels served repeatedly as guides to neglected prospects. More important, the records have become a basis for the most significant event in recent Butte history when their further evaluation pointed up the possibilities for large tonnages of low grade ore, mineable by low cost underground and open pit methods that contribute in such an important way to Butte's current production and future ore reserves.

As Anaconda's stature increased and as the need grew for additional metal supplies, the Company turned to foreign fields, and mine examinations took Reno Sales to Europe, Africa, and nearly every part of North and South America. Among other properties Anaconda acquired control of Cananea in Mexico, and Chuquicamata and Potrerillos in Chile. Again, Sales performed two major services; first by application of detailed mapping methods so successful in Butte, he pointed the way to the important breccia-pipe development at Cananea and to the great sulfide ore development at Chuquicamata; second, with the co-operation of mine management, he started new geological departments at various Anaconda subsidiaries, patterned after Butte and staffed with geologists trained there. Still later, exploration offices were opened at various locations throughout the Western Hemisphere initially staffed with geologists trained under Sales at Butte. In fact, most American and many foreign mining companies have key staff members who received their early experience and training as mine geologists under Sales at Butte, and his department long ago earned the reputation of "the best graduate school of mining geology in the country."

Sales' alma mater, Montana State College, began the series of honors which have accrued to him in succeeding years. It awarded him the honorary Doctor of Science degree in 1935. In 1937, Sales was president of the Society of Economic Geologists, and, in 1939, he was given the highest award in the field of economic geology by that Society, the Penrose Gold Medal, given for outstanding accomplishments; Sales is one of a select few who have been so recognized. In 1942, Reno Sales was honored by the Columbia Engineering School Alumni Association when he was awarded the Egleston Medal for distinguished engineering achievement.

The Mining Geology and Geophysics Division of the American Institute of Mining and Metallurgical Engineers established the Daniel C. Jackling Lecture in 1953 and selected Reno Sales as the first lecturer and recipient of this distinguished honor. In the words of Fred Searles—"He stands alongside of Mr. Jackling in his own right as another of the giants of United States mining, past and present." More recently, in 1960, the Montana School of Mines awarded Reno Sales "special recognition" for his lifetime of accomplishment. The editor of the Montana Standard said, "Perhaps more than any other man he has contributed to the welfare of Butte and Butte people. In honoring Mr. Sales, Montana School of Mines has honored itself, this community and the entire state." We may extend this statement— the nation owes him a similar debt of gratitude for his contributions to its economic and scientific stature.

In 1948, Reno H. Sales retired as Chief Geologist of The Anaconda Company to become its Consulting Geologist. In 14 years of "retirement" he has produced at least four major papers, each rated as an outstanding scientific contribution to geological literature. His advice and guidance are still sought by his company and by the entire profession.

Sales' contributions to mining geology are basic and may be outlined under three broad categories:

First, he provided in the early years of mining geology a superb standard of technical competence and intellectual integrity at a time when others were confusing observations and "authority." Precise geological mapping led Sales to delineate the geological history of the Butte ore deposit in 1912. Few important additions have been made since then. Many of Sales' conclusions in that paper were slow of acceptance; for example, his advocacy of primary chalcocite. But practically none of his conclusions is questioned now.

Second, few men within the universities or outside have directly and personally influenced as many students of mining geology as has Reno Sales. His "students" have been, and are now, in many of the principal mining and oil companies and in several of the great universities throughout the world. They number nearly 100 men.

Third, and perhaps most importantly, Reno Sales has left us with a basic philosophy. He has endeavored throughout his life as a scientist and engineer to minimize the distinction between pure and applied geology. In his Jackling Lecture, Sales insisted once more that "every mine operator, geologist, and even prospector appraises the future of mining properties or prospects on the basis of a theory of ore deposition whether he realizes the fact or not, and wherever the ultimate value of a mine or prospect is dependent on undeveloped ore, any estimate by an engineer or a geologist necessarily involves a theory of the origin of the deposit. Thus, the importance of a thorough understanding of the genesis of ore deposits cannot be overestimated." Under Reno Sales, Anaconda implemented this philosophy by attaching Geological Research Laboratories to its Geological Departments in 1940 (Butte), in 1960 (El Salvador, Chile), in 1965 (Britannia Beach, British Columbia), and in 1967 (Salt Lake City, Utah). Sales' directives to his various staffs have always stressed the urgency of achieving a better knowledge of how ore-forming processes take place. He has repeatedly demonstrated that ore-finding capabilities are improved through better understanding of ore genesis.

The well ordered accomplishments of this illustrious engineer and scientist are many and varied; basically he has contributed a system for mapping mine geology recognized and practiced throughout the mining world; he has trained a school of geologists in that system, and he has given mining geology a philosophy and purpose by which the science of ore genesis is applied to the art of ore finding.

## SELECTED BIBLIOGRAPHY OF RENO SALES

1. 1908, The localization of values in ore bodies and the occurrence of shoots in metalliferous deposits: (disc.), Econ. Geol., v. 3, p. 326–331.
2. 1910, Superficial alteration of the Butte veins: (disc.), Econ. Geol., v. 5, p. 15–21; 681–682.
3. 1913, Ore deposits at Butte, Montana: A.I.M.E. Tr., v. 46, p. 1523–1626.
4. 1933, Ore deposits of the Tri-State district: (disc.), Econ. Geol., v. 28, p. 780–786.
5. 1935, Chertification in the Tri-State (Oklahoma-Kansas-Missouri) mining district: (disc.), A.I.M.E. Tr., v. 115, p. 151–152.
6. 1935, Government surveys and the mining industry from the viewpoint of the mining geologist: A.I.M.E. Tr., v. 115, p. 393–406.
7. 1938, More intensive field studies for laboratory investigations of ore deposits: Econ. Geol., v. 33, p. 239–250.
8. 1941, Some observations in ore search (Symposium): A.I.M.E. Tr., v. 144, p. 116–117, 129–130, 134–135 (T.P. 1209).
9. 1942, The mining geologist's service to the mineral industry: Min. and Met., v. 23, no. 427, p. 381.
10. 1942, Factors in ore discovery: Min. Cong. Jour., v. 28, no. 1, p. 32.
11. 1946, Ore reserves and future exploration: Min. Cong. Jour., v. 32, p. 30–34.
12. 1948 (with Meyer, C.), Wall-rock alteration at Butte, Montana: A.I.M.E. Tr., v. 178, p. 9–35 (T.P. 2400).
13. 1949 (with Meyer, C.), Results from preliminary studies of vein formation at Butte, Montana: Econ. Geol., v. 44, p. 465–484.
14. 1950 (with Meyer, C.), Interpretation of wall-rock alteration at Butte, Montana: Colorado School of Mines Quarterly, v. 45, no. 1B, p. 261–273.
15. 1951 (with Meyer, C.), Effect of post-ore dikes intrusion on Butte ore minerals: Econ. Geol., v. 46, p. 813–820.
16. 1954, Genetic relations between granites, porphyries and associated copper deposits: (Daniel C. Jackling Award Lecture), Min. Eng., v. 6, no. 5, p. 499–505; also in A.I.M.E. Tr., v. 199 (T.P. 37861).
17. 1960, Critical remarks on the genesis of ores as applied to future mineral exploration: Econ. Geol., v. 55, p. 805–817.
18. 1962, Hydrothermal versus syngenetic theories of ore deposition: Econ. Geol., v. 57, p. 721–734.
19. 1964, Hydrothermal versus syngenetic theories of ore deposition: (disc.), Econ. Geol., v. 59, p. 162–167.

# Contents

## Volume I

## Part V. COLORADO ROCKIES

## Part VI. COLORADO PLATEAU URANIUM DEPOSITS

## Part IX.   ARIZONA AND ADJACENT AREAS

## Part X.   NORTHWESTERN STATES

\* Chapter 82 is indexed separately from the rest of the Volume, and the index for this chapter follows it, beginning in page 1833.

# Ore Deposits of the United States, 1933-1967

## 1933-1967

*The Graton-Sales Volume*

# 1. Ore Deposits of the Northeastern United States

## JOHN S. BROWN*

## Contents

## Illustrations

## Tables

* Towson, Maryland.

## ABSTRACT

The northeastern United States embraces that area of the Appalachian Mountains, and adjacent territory, beginning on the south at the Potomac River. It thus extends from the flat-lying Paleozoic terrane of the Appalachian plateau on the west, across the folded Paleozoics of the Great Valley and adjacent ridges (Paleozoic and Precambrian), over the Piedmont and New England metamorphic complex (mixed Paleozoic and igenous rocks), broken by the great Triassic rift remnants, onto the subdued coastal plain of Cretaceous and Tertiary sediments. Although not noted as a metal mining region, the area does contain numerous substantial and several outstanding mines, chiefly of the more prosaic metals, particularly iron, zinc, and copper, as well as many minor occurrences of these and a few additional items. These deposits occur to some extent in every subdivision of the area.

The region to date (1966) has produced almost 500,000,000 tons of medium to low-grade iron ore, especially magnetite, including ilmenitic magnetite; 50,000,000 tons of high-grade zinc ore; and 5,000,000 tons of low-grade copper. In the past, it has yielded small amounts of chromite and nickel, and at present is a large producer of titanium. Precious metals are conspicuously absent, and so also are the Tertiary or Mesozoic igneous phenomena so commonly associated with mines of the West. Hence, a survey of the extensive mineralization of this region should be instructive.

## INTRODUCTION

The northeastern United States, as here defined, includes the states of Maryland and Pennsylvania and all United States territory to the northeast, thus embracing Delaware, New Jersey, and all of New England as well as New York. The writer served for eighteen years as resident geologist for the Edwards-Balmat zinc mines of the St. Joseph Lead Company in New York and for twelve years thereafter as Chief Geologist in close contact with their operation. This afforded an opportunity to become acquainted with most of the types of mineralization and many of the particular deposits in the area, although only incidental attention ever was given to the important iron ores of the region. The historical review in the next section is drawn partly from personal observation and investigation but is based mainly on data in the geological references cited later at more appropriate places.

The conclusions as to ore genesis are the writer's matured opinion offered merely as that and with the expectation that many competent authorities may have different ideas.

## HISTORICAL INFORMATION

### Mining Operations in the Area

Precious metals never have been found in paying quantity in this region, and, since the early settlers were not highly mineral-conscious, there was none of the feverish search for mineral wealth which characterized the colonization of western North and South America. The utilitarian needs of their agricultural economy led the colonists to seek iron ore fairly early in their history, and this they found in adequate amount in many small bodies of limonitic bog iron ore and residual limonite or hematite. Crudely smelted in charcoal furnaces (bloomeries), refined in water powered forges, and finished in innumerable smithies, they supplied the simpler needs of the day, but the mines have been abandoned so long that many are all but lost and forgotten. Saugus, Massachusetts claims the honor of being the "birthplace" of the iron industry in the United States, in 1650 (36, p. 35), but, generally speaking, the iron ores of New England have been of little importance. However, some iron has been produced from local ores in every state within the region, even including Rhode Island and Delaware.

The less easily worked but much more extensive deposits of magnetite in southeastern New York, northern New Jersey, and eastern Pennsylvania attracted attention about 1740 and became important during the American Revolution. The bedded hematite ores of the Clinton formation were worked first in western New York and Pennsylvania in the decade beginning with 1800. Substantial production from these and more especially from the magnetite ores of New York, New Jersey, and Pennsylvania began about 1840 with the introduction of coal, and shortly thereafter coke, in place of the dwindling supply of charcoal (19). The blast furnace, which already had supplanted bloomeries in the more productive areas, became standard practice in this era. In spite of many cyclical adjustments and local interruptions, iron mining has been the most important, sustained segment of the metal mining industry in the region.

Lead for bullets was sought with minor success during the American Revolution as far west as Blair County, near Altoona, Pennsylvania. In the early 1800's, numerous lead deposits on the Maine coast, in Massachusetts, in southern New York, in Saint Lawrence County, New York, and in eastern Pennsylvania were discovered, and, in the expansionist era of the 1840's, some were exploited energetically, only to be eclipsed shortly by cheaper and richer deposits found in the Mississippi Valley and the far west. Some of these lead mines later became zinc mines, usually small. Lead itself never has achieved more than minor importance in the region, chiefly as a by-product of the Balmat zinc mines.

The mining of zinc became established at Friedensville, Pennsylvania and Franklin, New

*TABLE I.   Iron Ore Production, Northeastern United States in 1000 Long Tons*

**A. Magnetic Ore, Including Minor Hematite**

| Years | New York | New Jersey | Pennsylvania | Est. Grade |
|---|---|---|---|---|
| Prior 1881 | 20,000 | 10,000 | 20,000 | 50% Fe. |
| 1881–1890 | 9,000 | 5,424 | 7,000 | 45 |
| 1891–1900 | 3,801 | 3,374 | 5,879 | 45 |
| 1901–1910 | 6,776 | 4,863 | 5,749 | 45 |
| 1911–1920 | 10,114 | 3,000 | 5,056 | 40 |
| 1921–1930 | 5,600 | 3,000 | 9,000 | 40 |
| 1931–1940 | 6,400 | 2,300 | 8,724 | 38 |
| 1941–1950 | 48,000 | 8,000 | 20,000 | 35 |
| 1951–1960 | 76,000 | 14,000 | 16,000 | 30 |
| Total | 185,691 | 53,961 | 97,408 | 40 |

**B. Limonites Including Bog Ore and Brown Hematite**

| | New York | Maryland | Pennsylvania | Estimated |
|---|---|---|---|---|
| Prior 1881 | 1,000 | 1,800 | 20,000 | Average |
| 1881–1890 | 1,000 | 150 | 5,000 | Grade |
| 1890–1901 | 317 | 50 | 2,119 | |
| 1901–1910 | 60 | 75 | 1,353 | |
| Since 1910 | Nil | 25 | 217 | |
| Total | 2,377 | 2,100 | 28,689 | 40–45% |

**C. Hematite; Clinton Ore, Plus St. Lawrence Co., New York**

| | New York | Maryland | Pennsylvania | Estimated |
|---|---|---|---|---|
| Prior 1881 | 2,000 | 200 | 5,000 | Average |
| 1881–1890 | 2,000 | 50 | 1,600 | Grade |
| 1891–1900 | 426 | 25 | 579 | |
| 1901–1910 | 896 | 15 | 148 | |
| 1911–1920 | 600 | 10 | Nil | |
| Since 1920 | 350 | Nil | Nil | |
| Total | 5,272 | 300 | 7,327 | 35–40% |

**D. Carbonate Ore (Siderite)**

| | New York | Maryland | Pennsylvania | Estimated |
|---|---|---|---|---|
| Prior 1881 | 300 | 2,200 | 2,000 | Average |
| 1881–1890 | 400 | 300 | 600 | Grade |
| 1891–1900 | 285 | 75 | 42 | |
| Since 1900 | 3 | 75 | 50 | |
| Total | 988 | 2,650 | 2,692 | 40% Fe. |

| Grand Totals | New York | New Jersey | Pennsylvania | Maryland |
|---|---|---|---|---|
| | 199,328 | 53,961 | 130,116 | 5,050 |

Jersey about 1850. With the addition of the Edwards-Balmat area of northern New York, in 1915, zinc has proved to be easily the second most important metal product of the region, with a much more stable record of production, and doubtless of profit, than for iron.

Copper ores attracted attention in New England and elsewhere early in the 19th century, concurrently with the interest in lead. Production attained modest importance in Vermont in this early period and has had several revivals, usually of short duration. Copper was produced near Summitville, New Jersey, in 1824 and erratically thereafter until early 1900's. A lead mine at Phoenixville, Pennsylvania became a fairly important copper mine, briefly, around 1850, and the Bristol copper mine in Connecticut also enlivened this period. Copper was mined in small amounts in Maryland for many years.

Pyrite ores have been mined for their sulfur content or produced as a by-product at various times and places, from 1860 to 1950, particularly at the Davis mine in Massachusetts, the Milan mine, New Hampshire, and in Saint Lawrence County, New York.

Chrome mining had an auspicious begining near Baltimore, Maryland, in 1827 and was extended into nearby Pennsylvania but suc-

cumbed to cheaper foreign sources. The nickel ores of Lancaster Gap, Pennsylvania, developed in 1863, had a similar history.

One minor metal, at least with respect to consumption, in recent years has been more successful. Titanium, in substantial volume, has been added to the list of area products as detailed elsewhere in this volume, in the description of the Sanford Lake deposits of New York, where it occurs as large bodies of ilmenite-bearing magnetite. Some titanium also is being produced from black sands discovered recently in New Jersey (53).

In summary, mining in this area to date has been mainly of three basic commodities, iron, zinc, and copper. Lead has been a small contributor. Minor accessories include manganese with New Jersey zinc, silver with Balmat lead, and cobalt, associated with by-product pyrite, at the Cornwall iron mines. Chrome and nickel have had fleeting importance, and uranium has achieved publicity recently but without production.

## Statistics of Mine Production

The major statistics of output are summarized imperfectly, but approximately, in the following tables. Table I is based initially on

TABLE II.   *Zinc-Lead-Copper Production in Principal Districts of Northeastern United States*
*Short Tons Zinc*

| Franklin, New Jersey | | | Friedensville, Pennsylvania | | |
|---|---|---|---|---|---|
| Years | Tons Ore Mined | % Zn | Years | Tons Ore Mined | % Zn |
| 1850–1867 | Est. 100,000 | 19.00 | 1850–1880 | Est. 300,000 | 33.00 |
| 1868–1900 | 1,485,019 | 19.00 | 1958–1964 | 2,859,000 | 6.25 |
| 1901–1910 | 3,433,169 | 19.00 | Balmat-Edwards | | |
| 1911–1920 | 5,424,574 | 19.50 | District, New York | | |
| 1921–1930 | 4,710,518 | 19.24 | 1915–1964 | 17,432,184 | 10.11 |
| 1931–1940 | 3,804,506 | 20.67 | (Divided approximately 60% to Balmat | | |
| 1941–1950 | 3,163,438 | 20.11 | (1930–1964); 39% Edwards | | |
| 1951–1954 | 892,130 | 18.50 | (1915–1964); 1% Hyatt (1940–1949)). | | |
| Total | 23,053,394 | 19.50 | | | |

Lead, Balmat, New York

Recovered metal in ore (Balmat No. 2): 49,174 Tons (1930–1964)

Copper, Vermont

| Mine | Years | Tons Mined | % Cu. | Tons Metal Recovered |
|---|---|---|---|---|
| Pike Hill | — | 150,000 | — | 2,800 |
| Ely | 1800–1890 | 500,000 | 3.5 | 17,500 |
| Elizabeth | 1830–1900 | 250,000 | 2.2 | 5,250 |
| | 1943–1957 | 2,967,000 | 1.706 | 45,456 |
| | Total | 3,867,000 | — | 70,206 |

references (6) to (8) supplemented for later years by data in Minerals Yearbook, U.S. Bureau of Mines, which it has been necessary, in some cases, to apportion between states on the author's personal judgment. Bethlehem Steel Company has kindly supplied the figure of 106,000,000 tons for the total output of Cornwall deposits. This figure exceeds somewhat the total amount of magnetite for which Pennsylvania is accountable in Table I, a discrepancy for which no explanation has been found. The New Jersey Zinc Company has obliged with complete yearly figures for the Franklin Mine from 1897 to 1954 (Table II), and Saint Joseph Lead Company has provided overall figures for Edwards and Balmat. Early Vermont copper data are from geological references cited later, but only the 1943–1957 figures for metal (Minerals Yearbook) can be regarded as accurate. All figures prior to 1881 are mere "guesstimates."

## PHYSIOGRAPHY AND TOPOGRAPHY

The physiography and topography of this region, and their development, scarcely are separable from the geology and perhaps can be described best by reference to the accompanying geological outline map, Figure 1.

The axis and unifying feature of the region is the Appalachian mountain system, closely related to belt C, Figure 1. The Paleozoic strata of this belt were folded and faulted strongly at intervals during Paleozoic time, commonly referred to as the Taconic and Acadian revolutions or orogenies, and culminating at the close of the Paleozoic with the Appalachian orogeny. Following this, areas A, B, and C were uplifted and have been deeply dissected. In the folded area of belt C, this resulted in the etching of deep valleys along the softer or more soluble beds of tilted strata, leaving the harder and more resistant

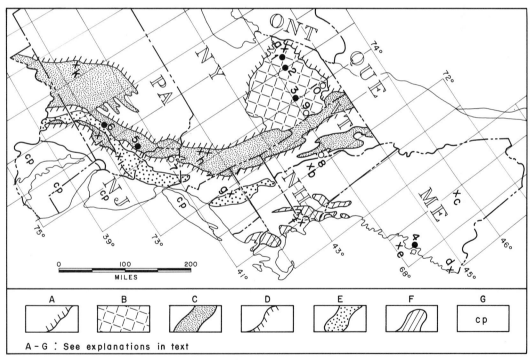

FIG. 1. *Geological Outline and Index Map of Northeastern United States. A–G, see explanations in text, under Structure, pages 7–9; 1 = Balmat-Edwards; 2 = Benson Mines; 3 = Sanford Lake; 4 = Blue Hill (Black Hawk); 5 = Friedensville; 6 = Cornwall; 7 = Franklin-Sterling; 8 = Elizabeth; 9 = Mineville; 10 = Lyon Mountain; a = Rossie Lead Mine, N.Y.; b = Warren Zinc Mine, New Hampshire; c = Katahdin Pyrrhotite Deposit, Maine; d = Dolson Lead-Zinc Discovery, Maine; e = Cape Rosier Zinc Copper Deposit, Maine; f = Chipman Lead Mine, Newburyport, Massachusetts; g = Bristol Copper Mine, Connecticut; h = Ellenville-Summitville Zinc Deposits, New York; i = Perkiomen Mine, Pennsylvania; j = Bamford Zinc Mine, Pennsylvania; k = Blair County Lead Mines, Pennsylvania.*

beds as ridges or "mountains," with all elements elongated along the strike of the fold belt in northeast-southwest trends.

Since the immediately adjacent area to the southeast of belt C at many places consists of Precambrian and Paleozoic igneous or metamorphic rocks that are highly resistant to erosion, these constitute attached or related mountains that are difficult of physiographic distinction and so are included in the term Appalachians. Prominent elements of this ridge and valley system are the Catoctin Mountains in Maryland, Kittatiny Mountain and South Mountain in Pennsylvania, the Delaware Water Gap area of Pennsylvania-New Jersey, the Nevesink-Wallkill and upper Hudson valleys of New York, the Champlain valley separating New York and Vermont, the Shawangunk mountains of southeastern New York, the Berkshires of western Massachusetts, and the Green Mountains of Vermont. Elevations of the valleys are low, generally well under 1000 feet above sea-level, whereas the mountains may rise to 4000 feet above sea-level at their crests. The Appalachian belt crosses from Vermont into eastern Quebec, curves eastward skirting Maine, and constitutes the axis of the Gaspé peninsula.

Westward of the Appalachian belt (A, Figure 1) lies a broad area where the rocks are nearly flat-lying, or at most only moderately folded, giving rise to broad but dissected uplands generally known as the Appalachian (or Allegheny) plateau region in western Pennsylvania. This continues into southern New York and in its higher eastern portion is known as the Catskill Mountains, which rise abruptly above the Hudson valley. The drainage pattern in all the plateau region is dendritic rather than lineal as in belt C. Maximum elevation in area A gradually diminishes from 3000 or 4000 feet in the east to less than 2000 in the west.

In northern New York, the plateau region is replaced by the Adirondack uplift (B, Figure 1), a sharp elevation of the Precambrian basement. This has a nucleus of impressive, randomly arranged mountains that rise to nearly 5000 feet in a few places. Between them and the Catskill upland lies the east-west lowland of the Mohawk valley connecting the Hudson valley with the Great Lakes basin.

Southeastward from Appalachia, as previously defined, lies a region of scattered mountains, or more commonly rolling hills, that embraces nearly all New England with the exception of Vermont, plus the Piedmont belt of Pennsylvania, Delaware, and Maryland (D &

E, Figure 1). Much of its surface approaches the concept of a gently sloping peneplain, broken by occasional prominent elevations, such as Mount Monadnock, New Hampshire, Mount Katahdin, Maine, and most prominent of all, the White Mountains of New Hampshire where Mount Washington (el. 6288 feet) is actually the highest point in the northeastern United States. Quite understandably, the White Mountains often are grouped with the Appalachians in popular terminology.

Finally, Long Island, New York, the southern half of New Jersey, Delaware, and the Chesapeake Bay area of Maryland constitute a low and nearly flat coastal plain (G, Figure 1) underlain by semi-consolidated, unfolded sediments of Cretaceous to Recent age. Elevations rise at most only a few hundred feet above sea-level, usually much less.

## GEOLOGIC HISTORY

### Stratigraphy

The range of geological features in the northeastern United States is wide and complex. Approximately 10 per cent of the region, in area B, Figure 1, constituting the Adirondack massif, is underlain by Precambrian igneous and metamorphic rocks of much complexity. Similar Precambrian rocks, in scattered patches along the Appalachian margin and eastward in the Piedmont and seaboard slopes, outcrop over nearly as much more territory. At the opposite end of geologic history, largely unconsolidated sediments of Cretaceous to Recent age in the coastal plain make up nearly 10 per cent more of the area in Maryland, Delaware, New Jersey, and Long Island, New York, Figure 1, G.

South of the Adirondacks and west of the Appalachian fold belt, approximately one-fourth of the region, is underlain by comparatively flat-lying strata of late Paleozoic age, mainly Devonian and Carboniferous, but extending downward in the Mohawk valley and Adirondack perimeter to the Cambrian (Figure 1, A).

The Appalachian fold belt, Figure 1 C, amounting to possibly 12 per cent of the area, displays the full range of Paleozoic stratigraphy from Cambrian to Mississippian in structural patterns grading from simple to complex, but without significant metamorphism or igneous association.

Most of the remainder, Figure 1, D, nearly 30 per cent by area, comprising most of New

England and the Piedmont section of New Jersey, Pennsylvania and Maryland is, basically, a metamorphic and igneous complex predominantly, but by no means wholly, of Paleozoic age, equivalent largely to the other rocks of the Appalachian fold belt but with the addition of extensive mid-Paleozoic intrusive and extrusive rocks. A minor portion, in the Narraganset basin of Rhode Island and the Boston basin, is late Paleozoic Carboniferous, which suffered less metamorphism. Throughout are small bodies of Precambrian, by no means easily identified. The tendency recently seems to be in the direction of extending the Paleozoic at the expense of the Precambrian. Striking lengthwise through the metamorphic complex of area D is a belt of Triassic rocks, Figure 1, E, that underlies possibly 5 per cent of the region as a whole. This is renowned for its red, mainly terrigenous, shales and sandstones, and their closely associated dikes and intrusive and extrusive sheets of dark diabase or "trap." These rocks occupy the Connecticut valley across Massachusetts and Connecticut, reappear in the Palisades of the Hudson west of New York city, and extend across New Jersey, Pennsylvania, and Maryland, southwest into Virginia.

Complicating this already over-complex situation, and not indicated in Figure 1, are the superficial effects of recent glaciation, which has spread a thin mantle of drift over all of New England, practically all of New York, and much of northern New Jersey and Pennsylvania. This renders the differentiation of Paleozoic and Precambrian more difficult in New England. Even to the south, beyond the limits of glaciation, there is still much uncertainty as to whether the metamorphic rocks of the Maryland-Pennsylvania Piedmont are predominantly Paleozoic or Precambrian. The current tendency in Maryland seems to favor the latter interpretation, but the writer personally is very skeptical regarding this.

The stratigraphy of the entire region, in generalized form, is summarized in Table III.

## Structure

The structure of the northeastern United States, as previously suggested, differs widely and characteristically with each geographic-geologic subdivision of Figure 1. Taking these in turn we may describe them briefly as follows:

(A) The Mohawk valley and Catskill-Allegheny plateaus are underlain by Paleozoic formations that, at surface, are comparatively flat-lying everywhere except in the Pennsylvania anthracite fields, which are fairly sharply folded and have suffered incipient metamorphism sufficient to convert bituminous coal to anthracite. Some would include this section in the Appalachian fold belt (C), but with respect to a mineral classification it seems to fit better with area A, the region of non-metallic mineral wealth, especially coal, oil and gas, and salt. If included in C, it would extend the wide portion of that belt almost all the way across Pennsylvania. Outside the anthracite district, dips in area A generally do not exceed 50 to 100 feet per mile, with occasional minor folds of oil-field type.

(B) The Precambrian massif of the Adirondacks (20), properly an outlier of the Canadian shield, has a framework of metasedimentary gneisses and limestones isoclinally folded and probably refolded (38) interlayered with many sheet-like masses of syenite and granite that commonly, though not always, conform to the fold patterns. Strikes are prevailingly northeast-southwest, with moderate to steep northwesterly dips. A large central mass of anorthosite of ill-defined structure, is definitely of igneous nature, but its exact age and origin still are a matter of lively debate.

(C) The Appalachian fold belt comprises a great series of sharp but generally open, usually asymmetric folds, which, to the southeast, may become overturned, recumbent, and perhaps merged into thrust sheets, the interpretation of which may vary widely according to the times and persons involved in their study (32).

(D) The western edge of this metamorphic complex, predominantly Precambrian in New Jersey and southeastern New York, closely resembles the Adirondacks, with similar lithology and northeast-southwest strikes, but with a tendency to dip southeastward. The situation in the Green Mountains of Vermont is more complex, with central cores of Precambrian flanked by metamorphosed Paleozoic, intensely faulted. In the Piedmont of Maryland and Pennsylvania, which may well be largely Paleozoic rather than Precambrian, northwestward dips again prevail.

Throughout New England a great variety of structures is encountered with many reversals of dip, but there is still some tendency to northeast-southwest strike trends. Both this area and the Piedmont contain many stock-like or irregular intrusive masses, isotopically datable as representing a considerable range of Paleozoic time (54), and undoubtedly related to the mid-Paleozoic orogenies and meta-

*TABLE III.   Stratigraphic Section, Northeastern United States*

| Age | Pennsylvania | New York | New England |
|---|---|---|---|
| Tert.<br>Cret. | Tertiary silt, sand, gravel<br>Cretaceous clay, marl, conglomerate<br>0–3000'+ | | Absent. |
| Triass. | Red shales, sandstone, conglomerate; thick basalt sills, sheets, dikes.<br>10,000'–20,000' | | |
| Pennsylv. | Conemaugh,<br>Allegheny,<br>Pottsville.   2500' | Coal measures   Absent in<br>New York | Boston basin series.<br>Narraganset series.<br>Cgl, slate, coal, volcanics.<br>12,000' |
| Miss. | Mauch Chunk shale<br>Pocono sandstone 1500' | Absent | Absent? |
| Devonian | Catskill—Chemung shale, sandstone<br>Hamilton shales<br>Onondage limestone          5000'<br>Schoharie grit<br>Oriskany sandstone          Helderberg limestone | | Littleton shale and minor<br>volcanics<br>5000' |
| Silurian | Cayuga limestone—Lockport dolomite<br>Salina salt          2000'<br>Clinton shale, sandstone, iron ore<br>Tuscarora quartzite          Shawangunk cgl | | Fitch limestone, dolomite,<br>sandstone.<br>Clough cgl<br>Waits river? |
| Ordovician | Juniata shale<br>Martinsburg shale<br>Beekmantown limestones<br>6000' | Utica shale =<br>Hudson river slate<br>Trenton dolomite<br>Beekmantown dolomite<br>2000' | Partridge shale, slate<br>Ammonoosuc volcanics<br>Albee quartzite, slate<br>Waits River?          7000'<br>Stockbridge limestone |
| Cambrian | Conococheague limestone<br>Elbrook limestone<br>Tomstown dolomite<br>Antietam sandstone<br>Harpers phyllite<br>6000' | Little Falls dolomite<br>Potsdam sandstone<br>—          2000'?<br>—<br>Schodack shale, ls.<br>Bomoseen grit | Castine volcanics<br><br>Ellsworth schists (Maine)<br><br>Age uncertain? |
| Pre-Cambrian | Pickering gneiss | Granite, syenite<br>Anorthosite<br>Grenville limestone<br>and gneiss<br>(Adirondacks) | Byram granite gneiss<br>(Manhattan schist?)<br>Franklin ls. (Inwood ls.?)<br>Pochuck gneiss<br>(New Jersey; S.E. New York) |

morphism. Here also should be mentioned a band, or in places, a double line of ultra-basic intrusives, now largely serpentinized, which traverse the Piedmont from just west of Baltimore, Maryland to Staten Island, New York (44). These also are regarded as of Paleozoic age but prior to most of the granitic intrusives of the area.

(E) The Triassic belt differs markedly from any of the preceding, or succeeding, areas. It seems likely that Triassic sediments were deposited originally over a considerably wider area than they now occupy, confined, however, to a zone parallel to, and easterly of, the Appalachian axis. Within this belt, rift faulting developed during sedimentation resulting in the dropping of narrow graben-like or mono-clinally tilted blocks of Triassic below the level of their metamorphic basement. In the subsequent base-levelling of this region, only these remnants of the Triassic have been preserved. Concurrently with the rifting and accompanying the later phases of sedimentation, great masses of deep-seated basic lava were injected along dikes, large and small, and spread out as intrusive sills and extrusive sheets interbedded with the sediments. The Palisades of the Hudson and the great trap sheets of New Jersey are expressions of this igneous activity and have counterparts elsewhere, especially in the Connecticut valley.

(F) The Carboniferous of the Narraganset and Boston basins, deposited unconformably on the previously folded and metamorphosed Precambrian and Paleozoic terrane, itself suffered strong folding and mild metamorphism in the Appalachian orogeny. Doubtless it once covered a considerably wider area.

(G) Following the post-Triassic erosion and base-levelling of areas D, E, and F, much of this seaboard slope was submerged and buried beneath coastal plain sediments, beginning with the upper Cretaceous and continuing through Tertiary and Quaternary time. The sediments are mainly sands, grits, and clays, poorly consolidated. How far inland beyond their present

TABLE IV.   *Lead Isotopes of Northeast U.S. and Adjacent Canada*

**Group I. Anomalous Ores of Mississippi Valley Type Age Uncertain**

| Locality | % 204 | Ratios to 204 206 | 207 | 208 | Other Ratios '06/'07 | '08/'06 | '08/'07 | 208 '06 + '07 | Occurrence |
|---|---|---|---|---|---|---|---|---|---|
| N'b'port, Mass. | 1.210 | 22.93 | 16.30 | 42.42 | 1.407 | 1.850 | 2.602 | 1.081 | Pb. Veins in Gran. and gneiss. |
| Friedensville, Pa. | 1.323 | 19.24 | 15.68 | 39.66 | 1.228 | 2.061 | 2.529 | 1.136 | Zn. Bedded Repl. in Ord. Ls. |
| Rossie, N.Y. | 1.324 | 19.28 | 15.63 | 39.61 | 1.233 | 2.054 | 2.534 | 1.135 | Pb.-Zn. Late Vein in Precamb. gneiss. |
| Galetta, Ont. | 1.324 | 18.97 | 15.79 | 39.77 | 1.201 | 2.096 | 2.519 | 1.144 | Pb. Vein in Ord. Ls. |

**Group II. Normal Ores of Mesozoic (Triassic?) Period**

| Guymard, N.Y. | 1.332 | 18.90 | 15.82 | 39.35 | 1.195 | 2.082 | 2.487 | 1.133 | Pb. Vein in Silurian Quartzite. |
| Walkill, N.Y. | 1.340 | 18.74 | 15.69 | 39.19 | 1.194 | 2.091 | 2.498 | 1.138 | Zn.-Pb. In fractured Silurian Quartzite. |
| Ellenville, N.Y. | 1.345 | 18.79 | 15.64 | 38.92 | 1.201 | 2.071 | 2.488 | 1.130 | Zn.-Pb. In fractured Silurian Quartzite. |
| Wheatley, Pa. | 1.343 | 18.79 | 15.74 | 38.94 | 1.194 | 2.072 | 2.474 | 1.128 | Cu.-Pb. Vein in Triassic Shl. and SS. |
| Perkiomen, Pa. | 1.346 | 18.71 | 15.61 | 38.97 | 1.199 | 2.083 | 2.496 | 1.135 | Pb. Veins in gneiss. |

**Group III. Normal Ores of Paleozoic Age**

| Blair Co., Pa. | 1.349 | 18.61 | 15.71 | 38.80 | 1.184 | 2.085 | 2.470 | 1.130 | Zn.-Pb. Veins in Ord. ls. |
| Warren, N.H. | 1.352 | 18.54 | 15.87 | 38.56 | 1.168 | 2.080 | 2.430 | 1.121 | Zn.-Pb. Bedded in Met. Ord. Volc. |
| Denbow, Me. | 1.360 | 18.38 | 15.63 | 38.51 | 1.176 | 2.095 | 2.464 | 1.132 | Pb. Veins in Ord. Sil. tuffs. |
| Deer I., Me. | 1.363 | 18.19 | 15.74 | 38.44 | 1.156 | 2.113 | 2.442 | 1.133 | Zn. Bedded Repl. in Met. Ord. Volc. |
| Brunsw. 12, N.B. | 1.355 | 18.39 | 15.85 | 38.55 | 1.160 | 2.096 | 2.432 | 1.126 | Zn.-Pb. Bedded Repl. in Met. Ord. Sed. & Volc. |
| N. Larder, N.B. | 1.358 | 18.37 | 15.85 | 38.42 | 1.159 | 2.091 | 2.424 | 1.123 | Zn.-Pb. Bedded Repl. in Met. Ord. Sed. & Volc. |
| Keymet, N.B. | 1.362 | 18.30 | 15.79 | 38.33 | 1.159 | 2.095 | 2.427 | 1.124 | Zn.-Pb. Vein in Met. Silurian ? Sediments. |
| Fed. Met. Que. | 1.370 | 18.16 | 15.72 | 38.11 | 1.155 | 2.100 | 2.424 | 1.125 | Zn. Veins in Ord. ls. |
| Brunsw. 6, N.B. | 1.370 | 18.22 | 15.72 | 38.06 | 1.159 | 2.089 | 2.421 | 1.121 | Zn.-Pb. Bedded Repl. in Ord. Sed. & Volc. |
| Candego, Que. | 1.375 | 18.02 | 15.64 | 38.07 | 1.152 | 2.113 | 2.434 | 1.131 | Pb. Vein in Ordov. Slates. |

**Group IV. Precambrian Ores**

| Franklin, N.J. | 1.420 | 17.14 | 15.60 | 36.68 | 1.099 | 2.140 | 2.351 | 1.120 | Zn. Bedded Repl. in Grenville ? ls. |
| Balmat 2, N.Y. | 1.426 | 16.96 | 15.54 | 36.62 | 1.091 | 2.159 | 2.356 | 1.127 | Zn.-Pb. Bedded Repl. in Grenville ls. |
| Edwards, N.Y. | 1.422 | 16.80 | 15.60 | 36.92 | 1.077 | 2.198 | 2.367 | 1.139 | Zn. Bedded Repl. in Grenville ls. |
| Calumet I., Que. | 1.434 | 16.57 | 15.75 | 36.42 | 1.052 | 2.198 | 2.312 | 1.127 | Zn.-Pb. Bedded Repl. in Grenville ls. |
| Anacon, Que. | 1.446 | 16.46 | 15.50 | 36.20 | 1.060 | 2.200 | 2.330 | 1.130 | Zn.-Pb. Bedded Repl. in Grenville ls. |

limits they may have spread is unknown, but almost certainly some marginal areas have been removed by erosion. The latest event has involved mild seaward tilting and resubmergence of the coast, thus accounting for its drowned appearance and many embayments. In New England, this submergence has obliterated the entire coastal plain and encroached on areas D, E, and F.

## Mineralization

Metallic ores of the region as exploited to date are essentially of three classes.

(1) Iron ores. These include minor limonites and considerable hematite but are predominantly magnetite, including here the ilmenitic magnetites utilized primarily for titanium products. Closely related are some pyritic deposits that have been mined chiefly for their sulfur content.

(2) Zinc ores, some of which contain recoverable lead.

(3) Copper ores. These are chiefly pyritic masses, usually mixtures of pyrite and pyrrhotite, in which minor chalcopyrite provides the principal economic worth; copper is associated, in some places, with recoverable zinc and possibly lead. There is also copper mineralization that lacks the pyritic association, but it has not proved to be of commercial value.

These three classes of mineralization are separable, likewise, into three distinct time periods: (1) Precambrian, (2) Paleozoic, and (3) Triassic, or later. The Precambrian comprises most, but not all, of the magnetite iron ores, and similarly the most important but not all zinc ores. It includes pyrite, but without copper. The Paleozoic embraces all the pyritic copper ores, several zinc deposits thus far of no great value, with some lead, and beds of hematite that are extensive but presently uneconomic, as well as minor siderite beds.

The Triassic has important magnetite and minor copper ore bodies, plus minor ones of lead and zinc, some of which last are associated with those of copper. Minor post-Triassic beds of siderite are in the Coastal Plain sediments, and fossil beach sands are exploitable for ilmenite.

Of questionable date are the important Friedensville zinc deposits and certain minor lead-zinc veins.

These age classifications, based mainly on sound geological evidence, are supported, confirmed, and in some cases may be extended with reasonable probability, by the aid of lead isotope data obtained from sulfide ores in which lead is associated, as shown in the following Table, IV.*

The interpretation of lead isotopes is a complex and, to some degree, unsettled problem (45,46,56), but it can be seen that the several groupings constitute well-defined subdivisions, which conform closely to the geological evidence to be developed later. Several Canadian deposits have been included because they resemble various deposits in the area under consideration and, in some cases, as in New Brunswick, have been demonstrated to be of major economic importance. This makes the search for similar deposits under like conditions in the United States more hopeful.

Measured by value and volume, the Precambrian mineralization has been much the most important thus far. In contrast to much of the United States, the virtual absence of gold and silver, even in accessory amounts, is remarkable. So also is the total lack of Tertiary igneous activity and associations so characteristic of the western United States.

## ECONOMIC GEOLOGY

### Forms of Ore Bodies

It is evident from Table IV that the ores of this region fall into two main structural types, bedded ores and veins. Moreover, both classes are represented in most of the groups in the table, and, if deposits lacking isotopic data are considered, can be found in all groups. However, since the true vein deposits usually are small in volume and of little value, attention can be concentrated mainly on the bedded type. It will be necessary to expand this slightly beyond the strict limits of bedded replacements to include certain bedded ores that are probably or possibly syngenetic, such as the Clinton iron ores and the Precambrian pyrite beds, and also certain irregular skarn-like replacements among the iron ores.

Since the most important ores, both Precambrian and Paleozoic, occur generally in rocks that have suffered intense or at least moderate folding and metamorphism, the bedded ores commonly reflect the deformational shapes of the enveloping strata. They occur especially in synclines, less commonly in anticlines, in plunging folds, in drag folds enlongated down dip, and in many cases in forms described

* Data for Blair County, Pa., from Heyl, A. V., *et al.,* 1966, Econ. Geol., v. 61, p. 933–961.

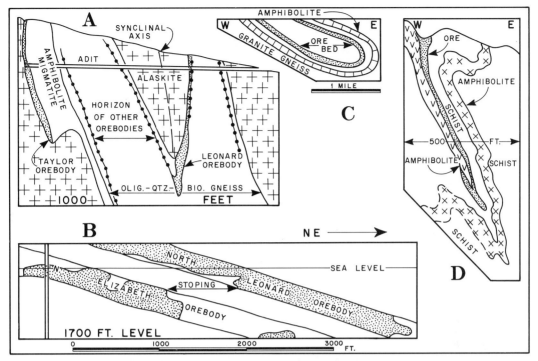

FIG. 2. *Typical Structural Patterns of Ore Bodies. Characteristic Structures of Ore Bodies in the Northeastern United States, showing synclinal preferences. (A) Cross Section, Mount Hope Magnetite Mine, New Jersey. From Sims (42). (B) Vertical Longitudinal Projection, showing lenticular "pencil like" habits of ore bodies. Mount Hope Mine. After Sims (42). (C) Cross Section, Russia Station Magnetite Ore Band, Clinton County, New York. From Postel (35). (D) Composite Section, Elizabeth Copper Mine, Vermont. From McKinstry and Mikkola (37) and Howard (38). Note acute synform structure even though beds, supposedly, are upside down; the ore horizon is older than enclosing schists.*

as lenticularly pencil-like (34. Figure 1).

### Stratigraphic Relations of Ore Bodies

The Precambrian ores of zinc and iron deserve first attention. In both the Franklin, New Jersey, and Edwards-Balmat, New York, zinc districts the ores occur in highly metamorphosed limestones, or dolomites, generally classified as Grenville and representing a relatively late phase of sedimentation in the enormous span of Precambrian time. Pegmatites do occur to some extent in all these ores, though very uncommonly at Balmat, but there is no close or obvious association with granite or other intrusive matter in any of the deposits.

The iron ores of the Adirondacks and the New York-New Jersey area generally occur in metasedimentary gneisses that appear to be sequential with the Grenville limestone, and hence are assumed to be of reasonably similar age, that is late Precambrian. All the iron ores, however, are associated closely with masses of granitic rock that have intruded the gneisses,

usually but not always in the manner of conformable sills or phacoliths. Indeed, in some cases, as at the Scrub Oaks mine in the Dover district, New Jersey, the ore is in rock described as granite or granitic gneiss (42). Hence the iron ores have a clear association with granitization. In the case of the iron-titanium ores, it is necessary only to substitute anorthosite or gabbro for granite.

The next most important grouping is that of the Paleozoic pyritic ores containing economic amounts of copper or zinc. Prime examples are the Elizabeth and Ely copper mines of Vermont and the Black Hawk mine currently being developed in Maine. The type is better known at present in adjacent portions of Quebec northeastward from Vermont and in New Brunswick east of Maine. In all instances, the ores occur in certain favored horizons within a thick series of clastic marine sediments and alternating volcanic lava and tuffs; the age of these rocks is not known very exactly but ranges from Ordovician through Silurian to Devonian with the Ordovician probably the most productive portion. These rocks suffered strong folding and metamorphism, accompanied by numerous intrusions, in the mid-Paleozoic orogenies. The ore bodies, however, are not associated clearly with the intrusives, in most localities, but do conform closely to the host structures and, in many places, seem to favor areas of excessive plication (41) so that the resulting forms rather closely resemble those of the Precambrian, as indicated in Figure 2.

Primary iron ores occur in the Paleozoic of the northeastern United States chiefly as the Clinton type of hematite, restricted to a very limited horizon in the Silurian of western New York (3,24) and Pennsylvania, all within or bordering area A, Figure 1. Though widely distributed and of a grade close to that of the valuable Alabama ores, about 35 per cent iron, they are so thin that, in spite of representing large tonnages, they have little economic value now or in the foreseeable future, even though once they were worked fairly extensively. They are regarded as original syngenetic deposits, and one might expect metamorphic equivalents in New England, but none is known. However, in the Bathurst area of New Brunswick, an extensive band of magnetite of similar grade is associated closely with the sulfide deposits of that district. Beds of siderite have yielded some iron, chiefly from the coal measures of Pennsylvania and the Cretaceous of Maryland.

Mention should also be made of the bedded manganese deposits of eastern Maine and adjacent New Brunswick, which are extensive but never have proved economic (27,50).

The great Friedensville zinc deposit, described (p. 95), occurs in Ordovician strata of the Appalachian fold belt. Only minor occurrences of related types are known elsewhere in the region, but closely similar deposits are of great importance in the southern Appalachians (p. 154 *et seq.*).

Mainly for completeness, it is necessary to mention the bedded copper ores of the Triassic, once worked fairly extensively though probably unprofitably, in New Jersey (23) and known also in Pennsylvania (4). These occur as lean impregnations in shales and sandstones, usually where these beds are directly in contact with a major trap sheet. Although of low grade, they are extensive enough to have conceivable future value were it not for the fantastic spread of urbanization and skyrocketing of land values in their chief known area of occurrence within sight of New York City.

A few replacement deposits will not fit comfortably into any stratigraphic classification. Most important are the Cornwall type magnetite mines of Pennsylvania (p. 72). These deposits actually are replacements of favorable limestone beds in the Paleozoic sequence that were conditioned by the proximity of huge diabase dikes or sheets along the faulted border of the Triassic sediments. The ores were formed, obviously, as some effect of the intrusive process. Closely related in type is the famous Bristol copper mine of Connecticut (12). Here copper impregnates the Triassic sandstones along the marginal fault and extends outward as minor veins in the bordering gneisses. The trap association here is not evident but is assumed.

## Mineralogy of the Ores

The mineralogy of ores of the northeastern United States is best treated by subdivision into the three principal classes of metal values: (1) iron, (2) zinc, with or without lead, and (3) copper.

IRON ORES The Clinton type of sedimentary iron may be dismissed as essentially hematite, with some chamosite, oolitic or otherwise, fine-grained, in a mixed siliceous, shaly, and calcareous, often fossiliferous, gangue.

Precambrian magnetite ores are predominantly magnetite, with hematite very subordinate, except in the lower grade disseminated ores such as at Scrub Oaks, New Jersey (42) and Benson mines, New York (55) where it is substantial. Sulfides, chiefly of iron, are accessories only, most conspicuously in the ores of New Jersey and southeastern New York (34). Accessories associated with mineralization include apatite, fairly abundant in the "old bed" ores of Port Henry in the Adirondacks (2), and fluorite, common in most ores of the eastern Adirondacks. Ilmenite in magnetite is abundant enough at Sanford Lake (p. 140) to make titanium the prime object of recovery. Gangue minerals in all these ores include the common constituents of their gneissic host rocks, quartz, feldspar, mica, and pyroxene. In the skarn type ores, calcite and silicates such as pyroxene and garnet are present and occasionally barite.

The Triassic magnetite of the Cornwall type (p. 72) differs mainly in having a gangue predominantly of calcite or calc-silicates and more abundant accessory sulfides, chiefly pyrite and chalcopyrite, which carry appreciable cobalt, silver, and rare metals.

In the magnetite iron ores, supergene minerals are inconsequential, though superficial limonite or hematite may occur near the surface. Secondary limonites of various origins, mainly derived from limestones or pyritic beds, were important in the early days of mining, particularly in eastern Pennsylvania, and the Salisbury district of western Massachusetts, Connecticut, and eastern New York (1). Hematite derived from pyrite was important in Saint Lawrence County, New York throughout the 19th century (10, p. 101–140).

ZINC-LEAD ORES Primary zinc ores are of two kinds, the Franklin, New Jersey type, and the normal sulfide ores. The Franklin ores, predominantly franklinite, willemite, and zincite, have no known counterpart anywhere. They are accompanied by numerous accessories, both zinc-bearing and otherwise. The literature concerning these is voluminous (15). The basic gangue is crystalline limestone with a wide variety of silicate gangue minerals. Sulfides, including sphalerite and galena, are present in the merest traces and may have no direct relation to the major ore minerals. Manganese, in the franklinite, however, yielded a byproduct of some value.

Balmat-Edwards ores (p. 20) consist, economically, entirely of sulfides, predominantly sphalerite fairly high in iron, with abundant pyrite, and accessory galena, the latter sufficient to be recoverable only in the Balmat No. 2 mine and restricted, even there, to certain ore shoots only. The gangue is dolomitic crystalline limestone, highly silicated, so that the principal gangue minerals are diopside, tremolite, quartz, serpentine, and talc. The large volume of pyrite at Balmat was recovered as a byproduct until about 1950 when the operation ceased to be profitable. Both the Balmat and Franklin assemblages imply subjection to high temperatures at some point in the history of their genesis.

Friedensville ore (p. 95), in contrast, consists of simple zinc sulfide, and pyrite, fine-grained, in a dolomitic limestone gangue that has suffered no visible thermal metamorphism.

Several minor bedded replacement deposits of zinc and lead range from types suggestive of fairly high-temperature conditions, as at Warren, New Hampshire (30), and Cape Rosier, Maine, as well as the Black Hawk mine in Maine, (p. 130) to others that are low in iron and unaccompanied by significant metamorphism as at Brandon, Vermont (21).

The quite numerous, but economically insignificant, veins also deserve some mention. Those apparently of at least moderately elevated temperature types would include the Chipman mine at Newburyport, Massachusetts

(5,59), the Perkiomen mine, the nearby Wheatley mine, and others near Phoenixville, Pennsylvania (13, pp. 21–46), possibly including some related veins in nearby Bucks County. The Chipman mine produced galena associated with pyrite in a carbonate (ankerite?) gangue with walls of granite or gneiss. The Perkiomen mine, in Triassic sandstone, contained some galena with more chalcopyrite, but the Wheatley and related mines, in granite or gneiss, contained mainly galena and sphalerite in quartz and carbonate, and the Bucks County veins, in Triassic shales, contained galena, sphalerite, and barite with carbonates. The veins of Orange, Sullivan and Ulster counties, southeastern New York, (Guymard, Ellenville, etc.) in Silurian quartzite are questionable as to classification. The principal ore minerals are galena and sphalerite, with quartz gangue.

Clearly of low-temperature type are numerous simple calcite veins carrying subordinate galena and sphalerite. Possibly the most important group in the northeastern United States are the veins of Blair County, Pennsylvania (13, pp. 13–16; 31). Others of less importance occur elsewhere in the state. The wall rock usually is limestone of the earlier Paleozoic. A second series, fairly common in New York west of the Adirondacks and even more abundant in adjacent Ontario, often is referred to as the Rossie group or type, present in western Saint Lawrence County, New York. These veins occur in various formations including the Precambrian gneisses and Grenville limestones upward through Cambrian Potsdam sandstone into the Ordovician limestones (10, pp. 296–307, 14). Fluorite is an occasional accessory; some pyrite is present.

Numerous minor veins of uncertain classification occur in New England and New Jersey.

In most of these deposits, whether bedded or vein type, changes in mineral composition of wall rock are not conspicuous. Only in the major bedded deposits has exploration in depth been sufficient to be significant. The Franklin district of New Jersey has revealed no startling change of character to the limit of depth at Franklin (about 1100 feet), or even to 3100 feet at Sterling Hill. Balmat and Edwards, explored to more than 3000 feet (5000 feet on dip), reveal no notable changes in the nature of primary mineralization, nor does Friedensville (p. 95) down to limits of 1500 feet. Vertical zoning appears to be almost nonexistent.

Supergene alteration of the zinc ores is inconsequential at Franklin but extensive at

Friedensville, where zinc carbonate fostered the entire early period of development and yielded some 100,000 tons of zinc. The carbonate extended to a maximum depth of only a few 100 feet, however, and all present production is based on unaltered sulfide ore. Friedensville is south of the line of effective glaciation, in the zone of lateritic weathering.

At Balmat, oxidation definitely has affected some ore bodies to considerable depth, but its extent is insufficient to have generated workable oxidized ores. Considerable willemite is present locally and has been described as supergene (17). It is accompanied by substantial amounts of earthy hematite and magnetite derived from the pyrite and by minor secondary sphalerite and galena. The disappearance of the willemite and most of the earthy hematite at 700 to 1000 feet depth seems to support the supergene interpretation, but the persistence of appreciable magnetite, and some hematite to much greater depth (2000 feet) may justify review of this problem. However, the alteration is confined to a small area at one end of the main ore body where a zone of definite post-ore shattering exists so that the interpretation may still hold.

Copper Ores   The Black Hawk mine, Maine, is described on p. 130. The important pyritic ores of Vermont have been treated in detail by McKinstry and Mikkola (37) and by Howard (43). The prime mineral is pyrrhotite with minor pyrite and appreciable chalcopyrite. Howard concludes that "The wall rocks of the ore bodies show considerable metasomatic reaction involving the addition of potassium, copper, sulphur and water, and the removal of sodium, calcium, magnesium, carbonaceous matter and $CO_2$." McKinstry and Mikkola do not emphasize this, however. Depth changes, observable only to about 1300 feet, the maximum depth at the Elizabeth, but to 4500 feet down slope at the Ely, appear to have been inconspicuous, and a similar condition persists for the most part to the 7000-foot slope distance at the famous Eustis pyrite mine in Quebec (52), although the lower of four lenses of ore was somewhat richer in chalcopyrite. No significant depth changes have been reported thus far in the sulfide ores of New Brunswick (41).

These statements by no means apply to the Triassic copper ores of New Jersey where primary chalcocite was altered to chrysocolla, malachite, cuprite, and native copper to slope depths of several 100 feet (23).

At the Perkiomen mine in Pennsylvania,

chalcopyrite was altered extensively to malachite, cuprite, and azurite and even galena to anglesite, cerussite, and pyromorphite (13, p. 28). At the Bristol copper mine, Connecticut, the abundant chalcocite is considered primary (12), but secondary carbonate and silicate also occur.

## Factors Controlling Form and Location of Ore Bodies

Detailed Structural Patterns   As previously suggested, (p. 7) the major deposits, both of the Precambrian and the Paleozoic, commonly reflect the deformational patterns of their host rocks. The strong preference for synclines in magnetite deposits is illustrated nicely in the description of the Saint Lawrence County area (e.g., the Benson Mines) by Leonard and Buddington's Figure 3, (55). The Forest of Dean mine in the New York Highlands is another example (9). However, in the New York-New Jersey area, descriptions more commonly emphasize lenticular, tabular shapes described as "pencil-like," frequently associated with plunging drag folds (34,42). The great plunging syncline at Franklin, New Jersey, is world-famous, and the major Balmat structure currently is interpreted as a steeply plunging syncline (38). Many of these structures are overturned. Figure 2 illustrates some of the typical forms. There are, of course, a good many deposits that appear monoclinal. This intimate relationship to structure has an important bearing on genesis, as will be discussed hereafter.

Vein deposits, for the most part, consist of simple fracture fillings, usually of steep dip (e.g., ores of the Rossie type). However, the zinc deposits at Ellenville and Summitville, New York, appear to be in shear zones, possibly thrust planes, of moderate dip (33), though the nearby Guymard mine occupies a simple fracture. The veins near Altoona, Blair County, Pennsylvania, probably occupy tension fractures related to strong anticlinal folding (24,31). Some of the veins along the Maine coast may well have been modified by folding attendant on, or subsequent to, their formation.

Favorable Beds   A prime factor in the location of most of the important and characteristic types of deposits seems to be the presence of certain favorable beds or rock layers. Thus, no zinc deposits of any consequence are known in the Precambrian except as bedded replacements of what commonly is called Grenville limestone, though this does not imply at all that these are of identical age. The host rocks differ widely in composition from pure limestone through dolomite to highly silicated phases with much diopside, other silicates, and quartz. The Friedensville zinc deposit, clearly akin to the extensive ores of Virginia and eastern Tennessee in nature and stratigraphic position, indicates that this affinity for carbonate hosts persisted into the Paleozoic.

The pyritic copper ores of New England again occur in favorable beds of Ordovician to Devonian age, similar to their location in Quebec and New Brunswick. Here, a host rock apparently much more siliceous and with only minor carbonate content seems to have been preferred, and volcanics are almost always nearby if, indeed, they may not provide the host in some cases.

The Precambrian iron ores are found generally in areas so highly granitized and metamorphosed that less can be said regarding preferred horizons, but Leonard and Buddington (55) emphasize the importance of skarn layers, implying an original calcareous content, as a conditioning factor for many of these deposits in the western Adirondacks. Skarns likewise are abundant, but usually as deposits of smaller size, subordinate to those in the gneisses, in the New York-New Jersey magnetite area (34, 42).

Triassic mineralization, both with respect to the important magnetite ores and the many unimportant bedded copper deposits, seems clearly related to the great trap dikes and sheets.

PATHS OF ORE AND GANGUE MOVEMENT   This topic seems to the writer best reserved for discussion hereafter under the heading of ore genesis.

## ORE GENESIS

Practically all authorities of the present century from Newland and Kemp (2) through Colony (9), Gallagher (18), Postel (35), Hotz (34), Gillson (39) and Sims (42), to Leonard and Buddington (55) have agreed on a magmatic origin from hydrothermal solutions, or gases, for the magnetic iron ores of New York and New Jersey. This results very naturally from the prevailing mood of the period, supported by the obvious close association of the ores with granitization. However, in earlier years there was some dissent, notably by Nason (11), who advocated a sedimentary origin, modified by subsequent metamorphism,

and supported this by strong evidence of the stratiform nature of the ores. Recently, a variation, involving metasomatic migration of iron out of the ferromagnesian phases of certain gneisses under the influence of metamorphic stress and its concentration in adjacent rocks as ore, has been advanced (51) with some supporting geochemical evidence. The writer has never studied closely any of these iron deposits and will reserve comment until later.

The writer has studied intimately and for many years the zinc-lead ores of the Balmat-Edwards district and numerous other important, and unimportant, lead-zinc deposits, within and outside the area. Indeed, this subject has been his prime life-time concern. Through the years, he has experienced a considerable metamorphosis of opinion, or prejudice, as the reader may prefer. Originally, inculcated with the lateral secretion theories of Buehler and Siebenthal, in Missouri, he was converted early in his career to the magmatic views of Kemp and Colony, so that he first accepted a magmatic source for Balmat ores as inevitable (16) even though not obvious. Later, after extensive observation and considerable microscopic and laboratory investigation, he concluded that the porosity of the ores as they now exist was so low as to preclude the possibility of their introduction by hydrothermal solutions (28). This led him to evolve a whole theory of ore genesis (29) based on vapor transport at high temperature. Most of this the author now is willing to repudiate as having been disproved by geochemical advances and his own later experience, though the basic objections to the universal application of the hydrothermal magmatic theory he still considers valid.

Finally, after extended experience with quite different types of ores, and with the development of the compelling evidence of lead isotopes in relation to ore genesis, he has concluded that these ores were emplaced as normal bedded deposits, such as those of North Africa, or Rammelsberg, or, with certain reservations, Friedensville, and that they have participated in all the metamorphism to which their host rocks have been subjected, thus being converted to high-temperature forms and robbed of their original porosity. This avoids completely the necessity of introducing the ore into these rocks as they now are constituted, though it implies considerable probable local migration and redistribution. These conclusions were expressed recently as follows: The chief Precambrian lead-zinc ores (Broken Hill,

Mount Isa, Sullivan, Balmat) "are essentially conformable deposits in metamorphosed sedimentary strata, and must have formed under conditions similar to those that produced the bedded deposits in Paleozoic rocks. This means that all have been subjected to whatever degree of metamorphism has affected their host rock, . . . . Their high temperature characteristics, therefore, are due mainly, or entirely, to their subsequent history and are not genetic" (56, p. 66).

The chief objection to this interpretation in recent years has been that, microscopically, the ores show little or no evidence of metamorphic reorganization. They appear to fit naturally into their environment as the latest phase of the paragenetic sequence and they clearly are of high temperature forms. It was natural to infer that they represent the latest phase of the history of their setting. However, it is not reasonable to believe that this can be true, as it seems to be superficially, of all bedded ore bodies in rocks as far back as those of central Canada, with ages of two billion years or more. A salutary change in attitude and awareness of possible significant evidence is set forth in a recent article by Kalliokoski (58) advocating this principle, not only with respect to the Precambrian ores, but for those of the metamorphosed Paleozoic as well. This will be discussed hereafter.

Returning to the isotopic evidence, the argument, briefly, is this: the lead isotopes of the ores at Balmat are essentially identical with those of the dispersed lead in their associated host rocks, if one allows for the slight effect of radiogenic decay of associated uranium minerals in the period of roughly one billion years that has elapsed since the host rocks were metamorphosed. Lead ore itself, once formed, is unaffected isotopically by time. The basic data for this conclusion are derived from the research of Doe (47,48) on the lead isotopes of Balmat ore and rocks, and on the writer's interpretation of these and similar data from numerous other districts (57). This conclusion, admittedly, may not be fully acceptable to Doe and others of the geochemical fraternity.

The Franklin, New Jersey, zinc ores pose a variation of the same problem, since they are unique mineralogically. Franklinite, zincite, and willemite are unknown in economic quantity elsewhere, but here constitute perhaps the world's largest single concentration of zinc. Lead is a mere trace accessory, as galena. The available isotope data show it to be very similar to that at Balmat (49), and the implication

as to age and origin should be the same. There has been endless speculation regarding the origin of Franklin ores that it seems useless to review here (29, p. 174–180). Most current opinion in North America probably still favors some magmatic process of generation, though derivation by metamorphism of a previously oxidized sulfide ore body has received serious consideration. Presently, and in the light of the isotopic evidence, the writer would conclude, as at Balmat, that the ores are late Precambrian but antedate the last Precambrian metamorphism, and he would welcome any geochemical explanation that could account for the complete desulfurization of so great an ore body. The large amount of manganese present may have some bearing on this, since manganese is a strong desulfurizer. But where did the sulfur go?

If one accepts the process of metamorphism as an essential post-genetic phase in the history of the great Precambrian lead-zinc deposits of this region, is it any less reasonable to apply it likewise to those magnetite ore bodies that show definite stratiform characteristics and occur in rocks believed to be paragneisses, even though extensively affected by granitization (60)?

With the bedded replacements of metamorphosed Paleozoic strata, we face the same problem. Here there has been considerable diversity of opinion lately. Hydrothermalists almost invariably find a Devonian or Carboniferous granite somewhere in the vicinity to which the ores can be attributed, though these seldom are obviously associated. A considerable school of thought advocates syngenesis—deposition concurrent with sedimentation. Others advocate lateral secretion from favored "source beds." Certainly there is a strong preference for certain favorable horizons. Kalliokoski (58) argues cogently that the ores have been metamorphosed with their host rocks, though he does not openly advocate syngenesis.

The writer has a great deal of sympathy for Kalliokoski's interpretation but based more on isotopic than microscopic evidence. Personal microscopic study of Brunswick ores in 1955, seemed to him to indicate virtually perfect adjustment of the sulfides to their environment, without any clear evidence of deformation, though the evidence of deformation of the host rock was most impressive. The isotopes, especially at Brunswick No. 6, however, appear to be older than any granitic source to which they might be related, essentially as old as the host rock itself. Hence, the writer concluded that these ores had, indeed, been folded and metamorphosed with their hosts (49, Table III). This discussion, centered around New Brunswick ores, is intended to apply, by extrapolation, to the bedded ores in the metamorphosed Paleozoic of New England (e.g., Black Hawk, Vermont copper).

There remain only the ores which are post-Paleozoic (Triassic, or later) and those that appear to be isotopically anomalous. The large Cornwall magnetite deposits obviously have some genetic relation to the great diabase dike with which they are associated. The problem is whether the iron was supplied mainly by the diabase or by the sediments that it intruded. The Triassic sediments suggest an obvious source of iron which could, conceivably, have been dissolved and reprecipitated in the Paleozoic limestones. The diabase itself more likely supplied the accessory copper and cobalt, since these are unusual associates of magnetite ores. Many doubtless will continue to regard the diabase as the chief source of the iron as well.

The minor copper ores of the Triassic, in New Jersey, at Bristol, Connecticut, and elsewhere seem to be related closely to the trap sheets, or dikes, and again suggest that the trap was the chief source of the copper, probably mainly at the time of intrusion but in some cases possibly considerably later by supergene leaching.

The small, but significant, zinc-lead district of southeastern New York (Guymard to Ellenville), in the Silurian Shawangunk conglomerate, the writer would place in the Triassic or post-Triassic period on the basis of its well-established lead isotopes (Table IV). These appear to be essentially normal (not anomalous) yet as near modern as normal leads can well be and not far removed from the Triassic veins in isotopic character. Surely no hydrothermalist can find any Tertiary igneous source for this mineralization. Derivation by downward solutions from a former extension of the Triassic sediments and volcanics is suggested as a possibility.

The lead isotopes of Rossie and Friedensville clearly transcend the limits of normal Phanerozoic lead (56) and verge on the Mississippi valley type of anomaly. The validity of the Friedensville analysis may be questionable, since galena seems to be virtually unknown in the deposit at present, but its similarity to the deposits of eastern Tennessee render it plausible.

The Newburyport, Massachusetts, lead (Chipman mine) is highly anomalous, implying to the writer that the host rock here, at least the attendant gneisses, not the Paleozoic granite, is Precambrian. Only from Precambrian rocks could such a high degree of anomaly have been developed (56).

At Phoenixville, Pennsylvania, on the contrary, the Perkiomen lead has an isotope composition normal for the Triassic sediments in which it occurs, but the lead in the gneisses at the Wheatley mine is virtually identical in composition, indicating that these rocks are not Precambrian. Otherwise, the normal Triassic lead would have been seriously contaminated by the radiogenic lead inevitably generated in the older basement.

There is some question whether the fairly extensive pyrite ores, with some pyrrhotite, that have been mined at various times in the past for their sulfur content alone should be included in this review, but since they are assumed by some to be related to the other metallic ores in origin, a brief discussion seems in order. Most important are the Precambrian pyrite ores of Saint Lawrence County, New York, last mined in the 1920's. These occur in well-defined stratigraphic units of broad extent interbedded in the Grenville limestone series. It seems entirely possible that the many isolated occurrences may constitute a single horizon disrupted by metamorphism and repeated by folding. Certainly not more than two or three horizons can be represented. The sulfur content generally ranges from 10 to 20 per cent or, exceptionally, 25 per cent. Visible lead, zinc, or copper minerals are virtually unknown, even in exposures only a mile or less distant from the Balmat-Edwards zinc mineralization. Graphite is an invariable and conspicuous associate. The writer endorses the conclusion of Prucha (40), and others, that these are original sedimentary syngenetic deposits, substantially modified by later metamorphism.

Pyrite ores of the Paleozoic metamorphic terrane in New England, such as the Davis mine in Massachusetts (59) and the Milan mine, New Hampshire (7) are mere variants of the Vermont type of pyrrhotite, pyrite, chalcopyrite ores that simply carry less, but still appreciable, amounts of copper. A more celebrated example is the Eustis pyrite mine in nearby Quebec (52). Admittedly problematic is the enormous pyrrhotite deposit at Katahdin, Maine, of undetermined but evidently substantial depth, estimated to contain 6.5 million tons of 44 per cent iron and 27 per cent sulfur per 100 feet vertically (26). Wholly enclosed in a gabbro mass of moderate size intrusive into the surrounding Paleozoic slates, most present opinion would relate it to the gabbro in origin. Much better exposure and more evidence are needed to justify speculation on this situation.

## CONCLUSION

Metal mining in the northeastern United States has made a substantial, though generally prosaic, contribution to the economy of the region, and prospects seem good for this situation to continue as far as one may foresee the future. Recent developments in geophysics and geochemistry should assist materially, as they have already, in the discovery of new sources of ore to replenish those that have been depleted, especially in areas where favorable horizons are concealed by a thin cover of later strata or glacial deposits. It seems obvious from this review that the Precambrian constitutes "elephant country" for this region, at least with respect to ores of iron and zinc. Strangely, the fairly extensive Precambrian exposures in Vermont thus far have made no contribution. This situation may deserve some investigation. The possibilities for additional Friedensville-type zinc ores in the Appalachian belt still are intriguing, as also the likelihood of more Cornwall-type magnetite or bodies bordering the Triassic. In New England, recent developments afford hope for a fair number of discoveries, but past history suggests that they may not be of major proportions individually.

## REFERENCE CITED

1. Hobbs, W. H., 1907, The iron ores of the Salisbury district, Connecticut, New York and Massachusetts: Econ. Geol. v. 2, p. 153–181.
2. Newland, D. H., 1908, Geology of Adirondack magnetite ores, with a report on the Mineville—Port Henry group by J. F. Kemp: N.Y. State Mus. Bull. no. 119, p. 5–182.
3. Newland, D. H. and Hartnagel, C. A., 1908, Iron ores of the Clinton formation in New York State: N.Y. State Mus. Bull. no. 123, 76 p.
4. Wherry, E. T., 1908, The Newark copper deposits of Southeastern Pennsylvania: Econ. Geol. v. 3, p. 723–738.
5. Clapp, C. H. and Ball, W. A., 1909, The lead-silver deposits at Newburyport, Massachusetts and their accompanying contact zones: Econ. Geol. v. 4, p. 239–250.
6. Bayley, W. S., 1910, Iron mines and mining in New Jersey: N.J. Geol. Surv. v. 7, 512 p.
7. Emmons, W. H., 1910, Some ore deposits in Maine and the Milan mine, New Hampshire: U.S. Geol. Surv. Bull. 432, 62 p.
8. Singewald, J. T., Jr., 1911, Report on the iron ores of Maryland, with an account of the iron industry: Md. Geol. Surv. v. 9, pt. 2, p. 121–327.
9. Colony, R. J., 1921, The magnetite deposits of southeastern New York: N.Y. State Mus. Bull. nos. 249, 250, 161 p.
10. Newland, D. H., 1921, Mineral resources of the State of New York: N.Y. State Mus. Bull. nos. 223, 224, 315 p.
11. Nason, F. L., 1922, The sedimentary phases of the Adirondack magnetite iron ores: Econ. Geol. v. 17, p. 633–654.
12. Bateman, A. M., 1923, Primary chalcocite, Bristol copper mine, Connecticut: Econ. Geol. v. 18, p. 122–166.
13. Miller, B. L., 1924, Lead and zinc ores of Pennsylvania: Pa. Geol. Surv., 4th Ser., Bull. M5, 91 p.
14. Buddington, A. F., 1934, Geology and mineral resources of the Hammond, Antwerp and Lowville quadrangles: N.Y. State Mus. Bull. no. 296, p. 202–209.
15. Palache, C., 1935, The minerals of Franklin and Sterling Hill, Sussex county, New Jersey: U.S. Geol. Surv. Prof. Paper 180, 135 p.
16. Brown, J. S., 1936, Structure and primary mineralization of the zinc mine at Balmat, New York: Econ. Geol., v. 31, p. 233–258.
17. ——— 1936, Supergene sphalerite, galena and willemite at Balmat, New York: Econ. Geol. v. 31, p. 331–354.
18. Gallagher, D., 1937, Origin of magnetite deposits at Lyon Mountain, New York: N.Y. State Mus. Bull. no. 311, 84 p.
19. Pennsylvania Historical Commission, 1938, Pennsylvania iron manufacture in the eighteenth century: v. 4, Harrisburg, 227 p.
20. Buddington, A. F., 1939, Adirondack igneous rocks and their metamorphism: Geol. Soc. Amer. Mem. 7, 354 p.
21. Brown, J. S., 1941, Private reports to Saint Joseph Lead Company.
22. Pennsylvania Bureau of Statistics and Pennsylvania State College, 1944, Pennsylvania's mineral hertiage: Harrisburg, 248 p.
23. Woodward, H. P., 1944, Copper mines and mining in New Jersey: N.J. Geol. Surv. Bull. 57, 156 p.
24. Butts, C., 1945, Description of the Hollidaysburg and Huntington quadrangles, Pennsylvania: U.S. Geol. Surv. Atlas Folio 227, 20 p.
25. Quinn, A. W., 1945, Geology of the Charlemont-Heath area (with special reference to pyrite and copper deposits): U.S. Geol. Surv. Open File Rept., 28 p.
26. Miller, R. L., 1945, Geology of the Katahdin pyrrhotite deposit and vicinity, Piscataquis county, Maine: Me. Geol. Surv., Bull. 2, 21 p.

27. —— 1947, Manganese deposits of Aroostook County, Maine: Me. Geol. Surv. Bull. 4, 77 p.

28. Brown, J. S., 1947, Porosity and ore deposition at Edwards and Balmat, New York: Geol. Soc. Amer. Bull., v. 58, p. 505–545.

29. —— 1948, Ore genesis; an alternative to the hydrothermal theory: Hopwell Press, N.J., 204 p.

30. Hermance, H. P. and McHenry, M., 1948, Investigation of Ore Hill zinc-lead mine, Grafton county, New Hampshire: U.S. Bur. Mines R.I. 4328, 13 p.

31. Reed, D. F., 1949, Investigation of the Allbright farm lead-zinc deposit, Blair county, Pennsylvania: U.S. Bur. Mines R.I. 4422, 7 p.

32. Eardley, A. J., 1951, Structural geology of North America: 1st ed., Harper and Bros., N.Y., 654 p.; 2d ed., 1962, 743 p.

33. Sims, P. K. and Hotz, P. E., 1951, Zinc-lead deposit at Shawangunk Mine, Sullivan County, New York: U.S. Geol. Surv. Bull. 978-D, p. 101–121.

34. Hotz, P. E., 1953, Magnetite deposits of the Sterling Lake, New York-Ringwood, New Jersey area: U.S. Geol. Surv. Bull. 982-F, p. 153–244.

35. Postel, A. W., 1952, Geology of the Clinton county magnetite district, New York: U.S. Geol. Surv. Prof. Paper 237, 88 p.

36. New England-New York Inter-Agency Report, 1954, U.S. Geol. Surv. Open file
    a. Nicholson, H. P. and Mitchell, D. W., History
    b. Thayer, T. P., *et al.,* Iron Ore

37. McKinstry, H. E. and Mikkola, A. E., 1954, The Elizabeth copper mine, Vermont: Econ. Geol. v. 49, p. 1–31.

38. Brown, J. S. and Engel, A. E. J., 1956, Revision of Grenville stratigraphy and structure in the Balmat-Edwards district, northwest Adirondacks, New York: Geol. Soc. Amer. Bull., v. 67, p. 1599–1622.

39. Gillson, J. L., 1956, Genesis of the titaniferous magnetites and associated rocks of the Lake Sanford district, New York: A.I.M.E. Tr., v. 205, p. 296–301 (in Min. Eng., v. 8, no. 3).

40. Prucha, J. J., 1956, Nature and origin of the pyrite deposits of Saint Lawrence and Jefferson counties, New York: Econ. Geol., v. 56, p. 333–353.

41. Lea, E. R. and Rancourt, C., 1958, Geology of the Brunswick mining and smelting ore bodies, Gloucester County, N.B.: Canadian Inst. Min. and Met. Tr., v. 61 (Bull. no. 551), p. 95–105.

42. Sims, P. K., 1958, Geology and magnetite deposits of the Dover district, Morris county, New Jersey: U.S. Geol. Surv. Prof. Paper 287, 162 p.

43. Howard, P. F., 1959, Structure and rock alteration at the Elizabeth mine, Vermont: Econ. Geol. v. 54, p. 1214–1249, 1414–1443.

44. Pearre, N. C. and Heyl, A. V., Jr., 1960, Chromite and other mineral deposits in serpentine rocks of the Piedmont Upland—Maryland, Pennsylvania and Delaware: U.S. Geol. Surv. Bull. 1082-K, p. 707–833.

45. Russell, R. D. and Farquhar, R. M., 1960, Lead isotopes in geology: Interscience Publishers, N.Y., 243 p.

46. Cannon, R. S., Jr., et al., 1961, The data of lead isotope geology related to ore genesis: Econ. Geol. v. 56, p. 1–38.

47. Doe, B. R., 1961, Distribution and composition of sulfide minerals at Balmat, New York: Geol. Soc. Amer. Bull., v. 73, p. 833–854.

48. —— 1961, Relationships of lead isotopes among granites, pegmatites, and sulfide ores at Balmat, New York: Jour. Geophys. Res., v. 67, p. 2895–2906.

49. Brown, J. S., 1962, Ore leads and isotopes: Econ. Geol. v. 57, p. 673–720.

50. Pavlides, L., 1962, Geology and manganese deposits of the Maple and Hovey mountains area, Aroostook County, Maine: U.S. Geol. Surv. Prof. Paper 362, 116 p.

51. Hagner, A. F., et al., 1963, Host rock as a source of magnetite ore, Scott mine, Sterling Lake, New York: Econ. Geol. v. 58, p. 730–808.

52. Stockwell, C. H., *Editor,* 1963, Geology and economic minerals of Canada: Econ. Geol. Ser. no. 1, 5th ed., p. 188–189.

53. Yeloushan, C. C., 1963, Mineral industry of New Jersey: U.S. Bur. Mines, Minerals Year Book, v. 3, p. 725–738.

54. Hadley, J. B., 1964, Correlation of isotopic ages, crustal heating and sedimentation in the Appalachian region: p. 33–46 *in* Lowry, W. D., Editor, Tectonics of the southern Appalachians, V.P.I. Mem. 1, 114 p.

55. Leonard, B. F. and Buddington, A. F., 1964, Ore deposits of the Saint Lawrence county magnetite district, northwest Adirondacks, New York: U.S. Geol. Surv. Prof. Paper 377, 259 p.

56. Brown, J. S., 1965, Oceanic lead isotopes and ore genesis: Econ. Geol. v. 60, p. 47–68.

57. —— 1965, Lead isotopes of pegmatites, granites and ores: Econ. Geol. v. 60, p. 1167–1184.

58. Kalliokoski, J., 1965, Metamorphic features in North American massive sulfide deposits: Econ. Geol. v. 60, p. 485–505.

59. Brown, J. S., various dates, Private investigations for Saint Joseph Lead Company.

60. Palmer, D. F., 1970, Geology and ore deposits near Benson Mines, New York: Econ. Geol. v. 65, p. 31–39.

# 2. Zinc Deposits of the Balmat-Edwards District, New York

EDGAR R. LEA,* DAVID B. DILL, JR.*

## Contents

## Illustrations

* St. Joseph Lead Company, Edwards Division, Balmat, New York.

## ABSTRACT

The zinc deposits of the Balmat-Edwards Division of the St. Joseph Lead Company in northern New York State provide some 10 per cent of the domestic zinc produced annually within the United States. These complex ore deposits are contained within marbles of the Precambrian Grenville series, in a repetitive sequence of dolomites and silicated units. The Balmat ore bodies are localized at various intervals throughout some 1800 feet of the stratigraphic column on the northwest side of a band of gneiss that has continuity to the Edwards area, some 10 miles to the northeast. The Edwards ore bodies are located to the southeast of this gneiss. The marble formations are thought to have included repetitive intervals of marine sedimentation that provided mineralized source "beds" and were subsequently folded, flooded, and injected with melts at great depth and were sheared, recrystallized, and reconstituted.

Detailed structural and stratigraphic studies throughout the northwestern Adirondacks, in the Balmat-Edwards district, and at each mine in particular indicate the close dependence of the ore deposits on the enveloping structure whether it be local, district-wide, or regional in scope. Primary folding has developed a northeast regional trend with northwest dips. Secondary folding has been developed by an almost horizontal northeast-southwest-oriented shear couple in the plane of the regional trend. The secondary folding has produced crossfolds that intersect the regional northeast trend and plunge north to northwest at various angles.

The ore bodies are found within certain stratigraphic horizons where repeated alternation of competent and incompetent beds encouraged unusually complex folding. Analysis of individual mine structures indicates that the location of each deposit is controlled by structural crossing of primary and secondary folding. The ore bodies, which are tabular, lenticular, or pod-like in form, have features that suggest parallelism to true bedding or modified flow banding, developed by northeast-southwest shearing stresses related to the development of the crossfolds. The ore deposits in general follow both structural and recrystallization trends within the wall rocks.

Mobilization of the pre-metamorphic ore minerals is believed to have involved a combination of plastic flow under stress and solution of the sphalerite, minor galena, and pyrite. The ore fluids are thought to have been guided by secondary shearing and microbrecciation into structurally prepared areas where reprecipitation took place. The presumed premetamorphic mineralization now exhibits epigenetic high-temperature features related to late-stage metamorphism.

## INTRODUCTION

The Balmat-Edwards zinc district is situated a few miles southeast of Gouverneur, St. Lawrence County, New York, within the St. Lawrence River Valley and adjacent to the northwestern slopes of the Adirondack Mountains (Figure 1). The district extends for 10 miles from Sylvia Lake and Balmat northeasterly to the village of Edwards. Active metal mining is limited to the operations of the Edwards Division of the St. Joseph Lead Company.

The writers have conducted joint studies within the Balmat-Edwards district since 1957.

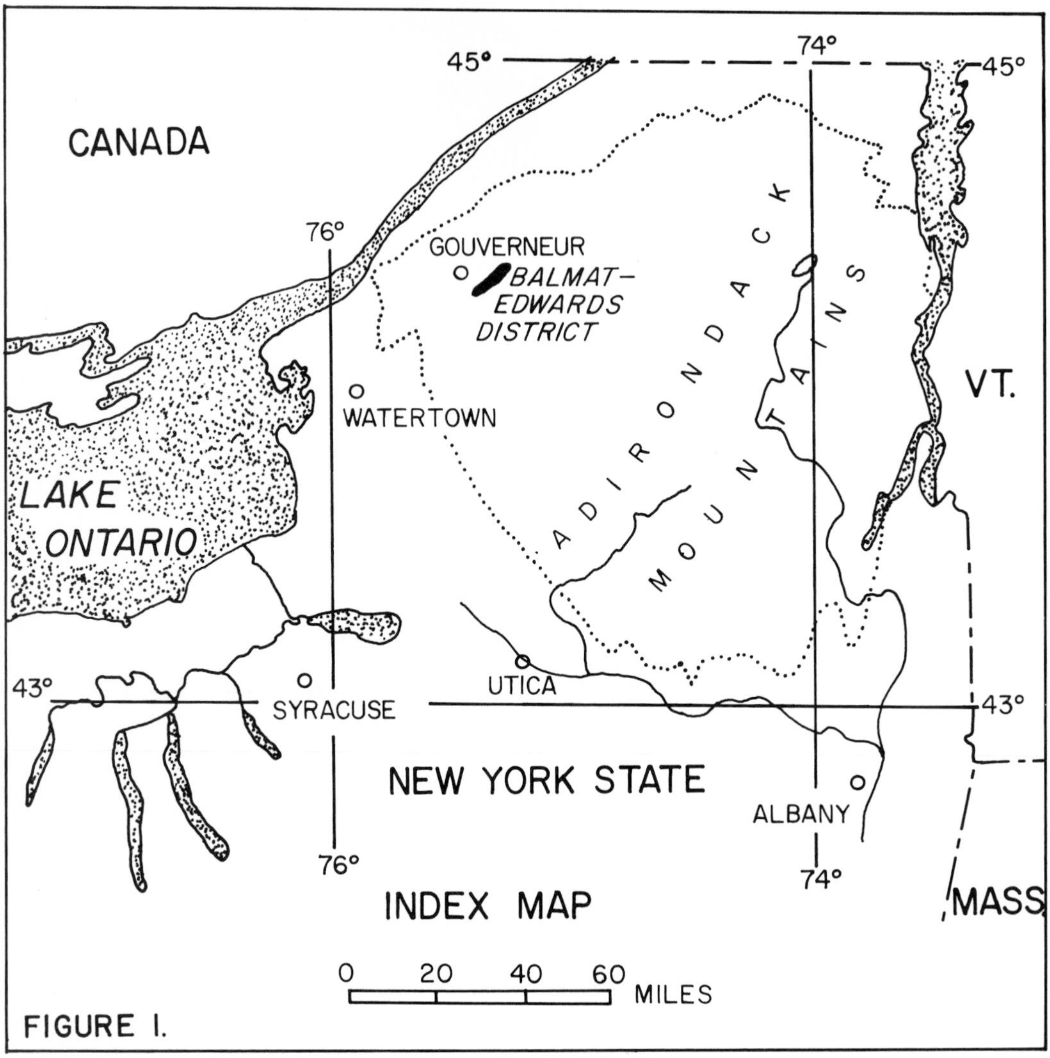

FIG. 1. *Index Map, showing location of the Balmat-Edwards District, New York.*

Emphasis has been on gathering detailed geological data from underground, in plan and section, and the correlation of these data with closer mapping of stratigraphy and structure of surface formations surrounding particular mine areas; new photogrammetric maps have been utilized.

The St. Joseph Lead Company has kindly permitted the authors to use company information on the geology of individual mine areas. The writers are indebted to all earlier workers in the northwestern Adirondacks, and particularly to A. F. Buddington and B. F. Leonard, A. E. J. Engel and Celeste G. Engel, J. S. Brown, and N. H. Donald, Jr., whose basic studies have contributed much to present understanding of regional, district, and mine geol-

ogy. More recent contributions by S. P. Brown, W. J. Cropper, and C. M. Grout and assistance by M. W. Hurley are gratefully acknowledged. Illustrations were drafted by J. Ward.

## HISTORICAL INFORMATION

### Mining Operations in the Area

The occurrence of zinc in the Balmat-Edwards district was recognized as early as 1838, when Ebenezer Emmons (1, p. 213) in a state geological report mentioned the presence of sphalerite and galena in shallow prospect pits near the Balmat farmhouse. No further interest in zinc prospecting was shown until 1903, when workmen uncovered part of the Edwards ore zone while quarrying for road material. Legal involvement and metallurgical problems delayed initiation of regular production at Edwards until 1915. The mine was operated by the Northern Ore Company and the New York Zinc Company until June 1926, when the property was purchased by the St. Joseph Lead Company. The Edwards mine and mill have operated continuously since 1915, and production has increased in stages to the present rate of 600 tons per day.

The purchase agreement of 1926 included an option covering the Balmat property. While diamond drilling in 1927 for depth possibilities of the meager sulfide showings described by Emmons, structurally distinct and much more important zinc mineralization was encountered by the St. Joseph Lead Company. The Balmat No. 2 mine and the Balmat concentrator began regular operations in 1930. Later surface drilling, initiated in 1946, resulted in the discovery of additional ore bodies to the north. These are now in production and serviced by the No. 3 shaft. Production of the Balmat mill has increased gradually from an initial 500 to the present 2100 tons per day.

The small Hyatt zinc mine, initially explored by the Grasselli Chemical Company in 1917, was operated by J. H. McLear and lessees from 1918 to 1922, resulting in the production of several thousand tons. Following additional diamond drilling, the Universal Exploration Company operated the mine from 1940 to 1949. The property has been idle since that time.

Mining of tremolite-talc has been an important industry in the district since 1875. Present production is shared by the Gouverneur Talc Company and the International Talc Company, at some five mines located in Balmat, Fowler, and Talcville. Other ore deposits within the northwest Adirondacks consist of magnetite (29), pyrite (14) and lead fissure veins (8) (Figure 2).

*TABLE I.   Statistics of Mine Production*

| Mine | Period | Short Tons Crude Ore | % Zn Content | % Pb Content |
|------|--------|------|------|------|
| Edwards | 1915–1964 | 4,505,925 | 11.25 | |
| Hyatt | 1940–1949 | 223,000 | 7.00 | |
| Balmat No. 2 | 1930–1964 | 10,656,730 | 9.65 | 0.73 |
| Balmat No. 3 | 1951–1964 | 2,046,529 | 10.35 | |
| Totals | | 17,432,184 | 10.11 | |

Source of Data: St. Joseph Lead Company.

## PHYSIOGRAPHIC HISTORY AND PRESENT TOPOGRAPHY

The district lies within the northeast-striking belt of Grenville lowlands, underlain largely by highly deformed Precambrian metasediments. It is bounded on the northwest by the St. Lawrence lowlands of quite undisturbed Paleozoic sediments and on the southeast by predominantly igneous gneisses of the Adirondacks.

By Cambrian time, the highly contorted Precambrian rocks of the Balmat-Edwards district had been levelled to a surface of gentle relief that included minor irregularities such as low ridges, valleys, and basins (2, p. 14). Continental deposition of siliceous sand, pebbles, and breccia fragments filled karst-like basins and solution cavities in the old Grenville-marble terrain. Such material is generally referred to as Potsdam but could be considerably older than the Potsdam marine sandstone of Upper Cambrian age that occurs elsewhere in the region. In contradiction to Bloomer's (31) conclusions, which were based on the Dekalb area, the writers believe the evidence clearly indicates that these sandstone residuals are unconformable with the Grenville series and have not been subjected to the intense deformation that affected the latter in Precambrian times.

The Grenville surface, modified by earlier Potsdam continental deposition, apparently was downwarped and then invaded by the westward-advancing Potsdam sea, eventually being covered by as much as 2000 feet of Cambro-Ordovician sediments. Following emergence in Upper Ordovician time, the land surface never again was submerged until the waning stages of Pleistocene glaciation. Probably more than one period of domal uplift and erosion resulted in the complete removal of marine Cambro-Ordovician sediments and the restoration of the original Grenville surface on which a few Potsdam outliers remained.

Pleistocene glaciation modified the terrain to a considerable extent. Glacial Lake Iroquois, the ancestor of Lake Ontario, submerged much of the area as a result of the impounding of the St. Lawrence River by the retreating Wisconsin ice sheet. As Cushing and Newland write (7, p. 9–10), the shore lines of the lake were extremely irregular, with long narrow bays and promontories. The lake waters washed the ridges clean of glacial debris, drowning the valleys and slopes with fine sand and clay. Extensive sand plains were formed north of Fullerville and at Edwards, probably due to pondings of the Oswegatchie River dur-

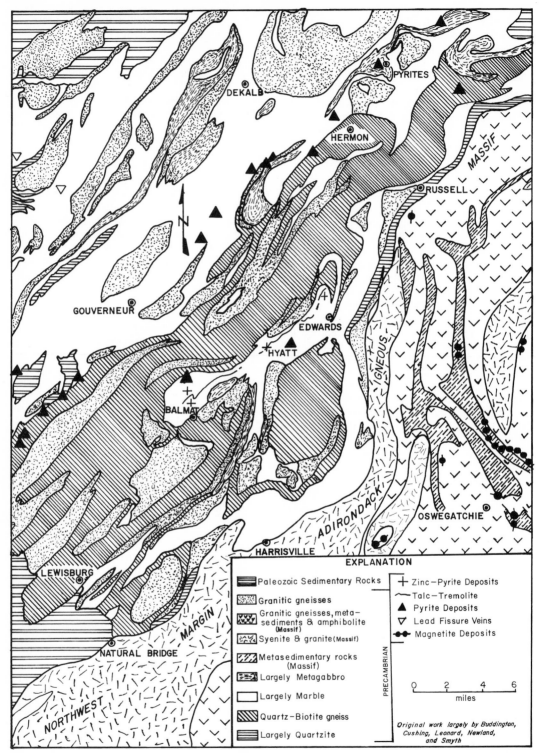

FIG. 2. *Generalized Geologic Map of the Balmat-Edwards District within the Northwestern Adirondacks, showing the distribution of major rock types and mineral deposits.*

ing the cutting of the gorges at Hyatt and Emeryville (6, p. 60). The subsequent marine submergence did not affect the area. The marine shoreline is found at least 5 miles to the northwest. Post-glaciation uplift has amounted to about 500 feet.

Present topography is not unlike that of pre-Potsdam time, but it is considerably masked by glacial deposits. Elevations above sea level range from a low of 580 feet at Fowler to 900 feet on a hill southeast of Talcville, but the local relief does not exceed 200 feet. There is generally a close correlation of topography with structure and lithology. Gneisses outcrop well and form long glaciated ridges, the trend of which is invariably parallel to the foliation. Siliceous and silicated marbles form hogbacks and ragged ridges, the previously polished glaciated surfaces of which have been roughened by post-glacial weathering. The purer and more deeply eroded dolomitic marbles occur either on level hummocky plains or beneath sand-choked depressions between ridges. Potsdam sandstone is erratically distributed in the lowlands as unconformable patches up to 0.5 mile in width and up to 2 miles in length, with an expressionless relief not representative of its very considerable depth (up to 400 feet as indicated by diamond drilling). The presence of Potsdam rock as an aquifer over part of the Balmat No. 3 mine has created a groundwater problem not encountered at the other mines. About half of the surface area is covered by sand plains, swamp deposits, and glacial debris.

The Oswegatchie River, together with its west branch, roughly parallels the marble belt but transects both the marble and gneiss. Tributaries are few because of the narrowness of the basin. Sylvia Lake, a prominent feature near Balmat, is spring-fed and deep (over 100 feet); it may have been formed either by glacial scour or solution activity (15, p. 1607).

## GEOLOGIC HISTORY

### Stratigraphic Column (See Table II)

For comprehensive descriptions of the entire stratigraphic column in the Balmat-Edwards district, the reader is referred to the basic studies by Engel and Engel (12,13), Brown and Engel (15) and Engel (24). Emphasis here is placed on the lithologic units containing ore zones in the Balmat-Sylvia Lake area, where mapping, diamond drilling, and mine workings have provided much, although far

from conclusive, information. The known ore-bearing horizons, units No. 6 through No. 14, have an estimated stratigraphic range of about 1800 feet.

The series is characterized by an alternation of almost pure dolomites and strongly silicated units. Thus, units 1, 3, 5, 7, 9, and 12 are dolomitic marbles with very minor quartz and silicates, and units 2, 4, 6, 8, 10, 11, and 13 through 15 are predominantly silicated members of various types. Many of the silicated units, in turn, can be subdivided into subunits that include alternating dolomitic or calcitic marbles, serpentinous and silicated marbles, quartzose and micaceous rocks. The repeated alternation of competent and incompetent beds on all scales from hundreds of feet down to inches is undoubtedly a prime factor in the formation of the usually complex folding in the district.

Units 1 and 3 at the base of the series are coarsely crystalline, white to buff dolomitic marbles, indistinguishable from each other. Thin lenses of quartz and diopside are sparsely distributed through these two units. They are separated by unit 2, which is a rather non-persistent pyritic schist composed of chlorite, feldspar, mica, and pyrite in which a little skarn has been developed. The heavier pyritic concentrations within unit 2 have been the loci of a few attempts at pyrite mining in the district.

Unit 4 is an alternating sequence of white dolomite, serpentinous diopside rock, and quartz lenses. The dolomite is finer-grained than in the major dolomitic units. There are excellent examples of retrograde metamorphism (diopside-serpentine-talc), in which diopside has been created as a reaction between quartz and dolomite, with serpentine and talc developing as rims on the diopside (15, p. 1605). Some sphalerite occurs as thin streaks and disseminations.

Unit 5, a white to gray coarse-grained dolomite, is somewhat less pure than units 1 and 3, and contains rather frequent lenses of quartz and serpentine-diopside. Occasionally, the grayer dolomite is fetid and very similar to unit 7.

The Gleason ore bodies occur in unit 6. Originally some 14 subunits of unit 6 had been identified, but more detailed underground work has shown that many of the subdivisions lack persistence as mappable units. Unit 6 consists of alternating layers and lenses of white and gray dolomite, quartz, and quartz-diopside. The layers range in thickness from inches to scores of feet. More massive quartz occurs

TABLE II.   Stratigraphic Column

| System | Series | Formation | Lithology | Typical Thickness (feet) | Ore Zones |
|---|---|---|---|---|---|
| Quaternary | Recent | | Soil mantle | 0–3 | |
| | Pleistocene | | Sand, clay, boulders | 0–375 | |
| | | | UNCONFORMITY | | |
| Cambrian | Upper Cambrian | | Potsdam sandstone, siliceous breccia | 0–400 | |
| | | | UNCONFORMITY | | |
| Precambrian | Upper Precambrian | Hermon granite California granite Amphibolite | Gneissic biotitic granite Alaskitic granite Granitized gabbro | | |
| | | | INTRUSIVE CONTACT | | |
| | Grenville | udf | Serpentinized dolomitic marble, minor quartz–diopside rock, calcitic marble, tremolite–anhydrite | 350 | Edwards |
| | | 16 | Median gneiss: quartz–mica–feldspar | 200 | |
| | | 15 | Phlogopitic silicated calcitic marble | 30 | |
| | | 14a | Serpentinized dolomitic and calcitic marbles, diopside, siliceous diopside | 120 | Main, Streeter, Hyatt |
| | | b | Calcitic marble, quartz augen | 130 | |
| | | c | Quartz–diopside rock, minor dolomite | 110 | Hanging Wall |
| | | 13 | Tremolite schist, anhydrite | 80 | |
| | | 12 | Dolomitic marble, locally anhydrite | 150 | Sylvia Lake |
| | | 11 | Diopside, calcitic and dolomitic marble, quartz, locally anhydrite | 300 | Sylvia Lake |
| | | 10 | Anhydrite, gypsum, talcose diopside | 50 | |
| | | 9 | Dolomitic marble | 60 | Loomis "B" |
| | | 8 | Diopside, quartz, dolomitic marble | 130 | Loomis "A" |
| | | 7 | Gray fetid dolomite | 120 | Loomis "A" |
| | | 6 | Siliceous diopside, quartz, dolomitic marble | 550 | Gleason |
| | | 5 | Dolomitic marble with diopside | 170 | |
| | | 4 | Diopside, quartz, dolomitic marble | 300 | |
| | | 3 | Dolomitic marble | 400 | |
| | | 2 | Pyritic schist | 100 | |
| | | 1 | Dolomitic marble | 400 | |
| | | Paragneiss | Quartz–biotite–oligoclase gneiss | 2500 | |

in the basal part of this unit than in any member of the Balmat series; much of the quartz is believed to have been $SiO_2$ introduced into the sediments during metamorphism. Noteworthy horizons are an intermediate gray dolomite subunit, a thin-banded micaceous layer known as the "E" bed, and a dark banded quartzite in contact with the No. 7 unit. Serpentine and talc alteration of diopside is not prominent.

Unit 7 is a remarkably persistent and uniformly dark gray, fetid, and graphitic dolomite, serving as an excellent marker bed not only near Balmat but probably as far northeast as Hyatt. At Balmat No. 3 mine, zinc ore occurs in thin lenses within the No. 7 dolomite and in important Loomis ore bodies at the contact with the No. 8 unit. Loomis ore is spread out through the 7, 8, and 9 units.

Unit 8 is another member composed of

layered diopside, quartz-diopside, and white dolomite, and some diopside has been altered to serpentine and talc.

Following in the sequence is unit 9, a white to light-gray, coarsely crystalline dolomite. Unit 10 is known as the Marker Bed because of its persistence and characteristic pea-green color. Typically, it consists of fine-grained talcy diopside and minor calcite. Anhydrite and near surface gypsum frequently occur as massive beds within this unit. At No. 2 mine, the typical pea-green diopside encloses a thin amphibolite member. The unit generally is unmineralized.

Unit 11 differs from the previously described silicated units in that brownish calcite is fairly conspicuous. Otherwise, it is a similar assemblage of dolomite, quartz, and diopside, with some diopside altered to serpentine and talc. The Sylvia Lake ore zone occurs within this unit and extends into Unit 12, which is another relatively pure dolomitic marble. Usually coarse-grained, Unit 12 ranges in color from dead white to light gray. Yellowish-green serpentine clots are common in the white marble. Diopside and quartz lenses are rare but characteristic. Massive anhydrite occurs as non-persistent beds in both Unit 11 and Unit 12.

Tremolite schist, the major division of unit 13, can be traced from Balmat to Edwards and is the source of all "talc" mined in the district. For a comprehensive treatment of the unit, see Engel (24). The schist is basically tremolite in which retrograde metamorphism has taken place in the series tremolite-anthophyllite-serpentine-talc to differing extents. Engel (24, p. 265–268) proposes that the tremolite was formed during metamorphism by dedolomitization of dolomite or siliceous dolomite, a process that has almost completely obliterated the host rock. It is remarkable, however, that the occurrence of tremolite is almost entirely restricted to Unit 13, in contrast with diopside, and that the tremolite schist is frequently in sharp contact with nearly pure dolomites of units 12 and 14. The origin of the tremolite is still very much of an enigma.

Purple, coarsely crystalline anhydrite with minor phlogopite and calcite occurs erratically as part of unit 13 in upper levels of Balmat No. 2 mine, usually as layers around the margins of the tremolite schist. With increasing depth, anhydrite gradually becomes the dominant mineral in many areas. On the 2100 level northwest, almost pure anhydrite occupies the entire horizon in a bed 50 feet thick. The problem of the tremolite-anhydrite relation is

under study. It is the present belief that the anhydrite is of evaporite origin, particularly since halite and natural gas are present in cavities and fissures nearby.

Unit 14, in which occur the Balmat No. 2 mine ore bodies, generally has been described as a quartzose calcitic marble (15, p. 1608–1609), because surface exposures are very largely of this type. Mapping of underground workings, however, has shown the possibility of as many as fourteen subunits, several of which may be repetitions by folding. Adjacent to the tremolite is a white to light-gray dolomite that often is missing. Next is a banded quartz-diopside rock, locally graphitic, with thin bands of a light-gray serpentinous dolomite. Following the quartz-diopside is the coarse-grained, brownish, calcitic marble with wisps of quartz, for which Engel (24, p. 140–141) makes a strong case of dedolomitization and fixing of the freed magnesium in the tremolite of the tremolite schist. Underground relationships, such as the usual lack of direct contact of tremolite and calcitic marble, do not entirely support this theory.

The remaining third of unit 14 is a surprisingly varied assemblage of serpentinous dolomitic and calcitic marbles, non-banded diopside and quartzose diopsidic marbles. Diagnostic minerals, color, and percentage of serpentine and relative proportions of calcitic to dolomitic matrix have made possible lithologic correlation of subunits that repeatedly are squeezed out and then reappear. Serpentine is present as discrete needle-like grains, solid patches, and as rims on diopside boudins. Diopside is usually white to greenish and granular in texture. Quartz is not nearly as abundant as in units, 4, 6, 8, and 11, being restricted to less than 50 per cent of the quartzose diopsides. Thin sill-like lamprophyres are non-persistent and have been disrupted by folding.

Forming the hanging wall of the Streeter ore zone is unit 15, known as the "rusty marble" or "hanging wall mica." It is highly phlogopitic marble, with a matrix of calcite, diopside, and feldspar, and is notably inhospitable to zinc mineralization. Completing the Balmat sequence is unit 16, the "median gneiss," a migmatitic paragneiss composed of mica, feldspar, and quartz.

The continuity of recognizable unit northeasterly from Balmat toward Edwards is imperfectly known. Both the tremolite schist and the median gneiss are essentially continuous throughout the marble belt, but none of the other units known at Balmat can definitely be recognized at Edwards. The wall-rock at the Edwards mine is predominantly a coarse-grained, light gray serpentinous dolomite, with non-persistent layers of quartz-diopside, calcitic marble, and tremolite-anhydrite. These rocks underlie the median gneiss.

## Structure

Present geological understanding of the structures enveloping the Balmat-Edwards district and the ore bodies owes much to the early work of Cushing and Newland (7), Newland (3), and Smyth (5). Subsequent to the initial comprehensive work in the district by Newland in 1916, detailed geological studies have continued within the district and throughout the northwestern Adirondacks. Later studies have been largely summarized by Brown (9,11), Brown and Engel (15), Buddington and Leonard (21), Engel and Engel (18), and Engel (24).

It is suggested that a considerable part of the present geological understanding of the Balmat-Edwards district ore deposits has evolved through recognition of geologic data indicating the close dependence of the ore bodies on the enveloping structure, whether it be local, district-wide, or regional in scope. These ore deposits have been formed within sediments that have been folded, flooded, and injected with silicate melts at great depth, sheared, recrystallized, and reconstituted. These events were regional in extent, having occurred throughout the northwestern Adirondacks.

The structure of the northwest Adirondacks has been studied in detail by Buddington and Leonard (21, p. 108, 109, Fig. 13, Fig. 14). They divide the region into three major structural units. These consist of: (1) a unit composed largely of granite gneiss and subordinate metasedimentary rocks in the extreme northwest; (2) adjoining these rocks to the southeast is the Grenville lowlands unit which is a broad belt some 25 to 30 miles wide consisting of about 70 per cent metasedimentary rocks with subordinate igneous rocks, largely granitic; and (3) the Grenville lowlands are, in turn, adjoined on the southeast by the main igneous complex unit of the northwest Adirondacks. (Figure 2)

The broad form of the Grenville series throughout the Grenville lowlands has been interpreted to be anticlinorial (18, p. 1373–1374, Fig. 2, p. 1373). The southeastern flank of the anticlinorium that occurs within

the vicinity of the Balmat-Edwards district has been overturned and dips at moderate angles toward the northwest. The opposite flank of this structure is thought to be just northwest of the St. Lawrence River in the Gananoque-Brockville-Mallorytown area, Ontario, where the beds face northwest (28,25). This northwest area lies astride the Frontenac axis, which is a narrow neck of the Canadian Shield connecting the Grenville Series of the Laurentian Plateau with the Adirondacks of New York State. Northeast of the crest of the Frontenac axis, folds plunge to the northeast, and southwest of the axis, folds most commonly plunge to the southwest. This reversal of plunge is thought to be caused by uplift of the Frontenac axis itself (25) and is a feature common to both flanks of the major anticlinorial structure.

Engel (24, p. 221–225) gives a detailed description of the structure and folding developed within the Grenville lowlands. Briefly, the highlights of this description follow.

". . . The Adirondack massif dips at a moderate angle (25° to 65°) northwest under the Grenville lowlands where metasedimentary rocks predominate.

". . . Data from many parts of the lowlands indicate that most of the early rock movements seem to have involved rolling and thrusting of the metasedimentary rocks over a synclinorial structure along the perimeter of the massif. Major structural elements formed at this time were folds overturned to the southeast with subhorizontal axes.

". . . Later rock movements along the outer parts of the massif and in the lowlands seem to have been dominantly strike-slip. Many major structural features indicate that rocks of the lowlands were dragged northeast relative to those of the massif and great refolds having a common direction sense (asymmetry) were superimposed upon the already folded complex in both lowlands and massif.

". . . Two (belts of refolding) of particular interest lie on opposite sides of the Grenville lowlands. The belt on the northwest plunges steeply (65° to 90°) downward along the nearly vertical contact with the Alexandria batholith. The belt on the southeast side varies in width from 5 to 15 miles or more and includes both metasediments of the lowlands and syenite, gabbros and granites on the northwest margin of the massif.

". . . Because axes of the folded folds along the southeastern side of the lowlands plunge northwestward athwart the northeast regional lithologic trends imparted by the primary deformation, these folds are referred to as crossfolds.

". . . The two belts of crossfolding in the lowlands are separated by a medial belt in which crossfolding is only locally or incipiently developed. Most folds in this medial zone retain their northeast trend and gentle northeast to southwest

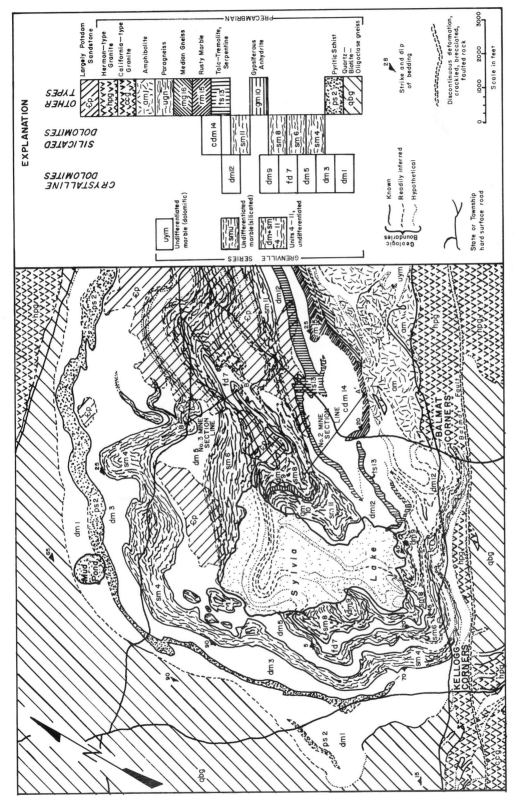

Fig. 3. *Geologic Map of the Balmat-Edwards District: Balmat Area.*

plunges characteristic of the features of the primary deformation of the area."

It is inferred (24, p. 223) that the earlier of the two major deformations occurred slightly before and during primary emplacement of the granites. The second major deformation is thought to have taken place during the culmination of the granitic intrusions.

For those interested in a more complete discussion of the structure of the Balmat-Edwards district than can be attempted here, the reader is referred to Brown and Engel (15).

The dominant controlling structure of the Balmat-Edwards district is the northeast-trending, northwest-dipping, overturned limb of the anticlinorial structure previously discussed. Thus, in Figure 3 although the quartz-biotite-oligoclase gneiss is the oldest formation in the area, it is seen to overlie the marbles to the northwest, because of this overturning. The Balmat-Edwards marbles that envelop the zinc deposits of the district lie conformably upon the gneiss, which is of uncertain thickness.

The Balmat fault is a northeast-trending zone of sheared and crackled rocks between primary folds called the Sylvia Lake syncline and the California anticline, structures that have developed at the southwest end of the district. (Figure 3) Brown (9, p. 243) describes this fault zone as 200 or 300 feet wide, with either minute shattering and jointing or strong flow banding and micro-granulation of the rocks involved. Brown and Engel (15, p. 1620) consider this cataclastic deformation as younger than the California phacolith and the Hermon-type granite, which it cuts off and brecciates. Because this fault zone parallels the direction of northeast-southwest shear couples related to the development of crossfolds in the district, they are inclined to relate its origin to a final stage of the crossfolding. The ruptures seem to form a thrust that dips northwest at moderate angles. Brown and Engel are of the opinion that thrusting up dip, or strike-slip motion of the northwest side to the northeast, could account for the lithologic discontinuities caused by this fault zone.

The median gneiss, so designated by Brown (9, p. 238, Fig. 1), provides an important horizon marker between Balmat and Edwards to which all stratigraphic studies have been tied. This characteristic layer of granitized quartzites separates the so-called Sylvia Lake-Cedar Lake marble belt northwest of the gneiss from the Fullerville-Edwards marble belt to the southeast.

The overturned, northwest-dipping flank of the Sylvia Lake syncline appears to have continuity to the Cedar Lake area, northeast of the Edwards mine area (Figure 4). Geologic mapping of outcrops of median gneiss between the Sylvia Lake area and Edwards would suggest at least a well-developed northwest-dipping flank of a possible fold between these two areas.

The moderate northeast plunge of the primary Sylvia Lake syncline suggests a trend toward the Edwards area, with the northwest dipping under-limb of the isoclinal syncline more or less paralleling the Balmat fault on its northwest side, between Kellogg and Balmat Corners (Figure 3). Northeast of Balmat, continuity of this limb is obscured by Hermon-type granite and amphibolite. A thick section of ambiguous amphibolite is mapped, but neither the original or transitional rock types are present. There is little wonder that stratigraphic and structural interpretation become difficult and open to question in this area.

At least some features of stratigraphy and structure suggest that the northeastern continuation of the Sylvia Lake syncline has been so tightly compressed that its trough (assumed here to be near horizontal in plunge) has been almost completely obliterated by isoclinal folding within the median gneiss. Brown and Engel (15, p. 1612) conclude that,

". . . the data suggest that an accordant fold, the Sylvia Lake-Edwards syncline has been mashed tightly together in its northeastern portion, and subsequently refolded into a northwest plunging crossfold."

As illustrated in Figure 4, the so-called Edwards "fish hook" fold, 'Z'-shaped in plan as outlined by the median gneiss, has a sudden northeast termination. Opposed units of the rusty marble within the median gneiss appear to converge and possibly to join at this point of termination of the gneiss. If this interpretation by Brown and Engel (15, p. 1614) is correct, the point of termination of the gneiss could well be the opposite apex of the Sylvia Lake syncline.

The thickened median gneiss, whether it be part of the overturned, northwest-dipping flank of a fold or a synclinal crease caused by tremendous compression and resultant isoclinal folding within the gneiss, has been refolded into a northwest-plunging crossfold. Development of Edwards ore bodies, found within underlying marbles around which the median gneiss has been rotated (see Figure 4), has proceeded down the plunge of this fold to the northwest in the plane of the regional northwest dip. This crossfold development is logically the result of an almost horizontal

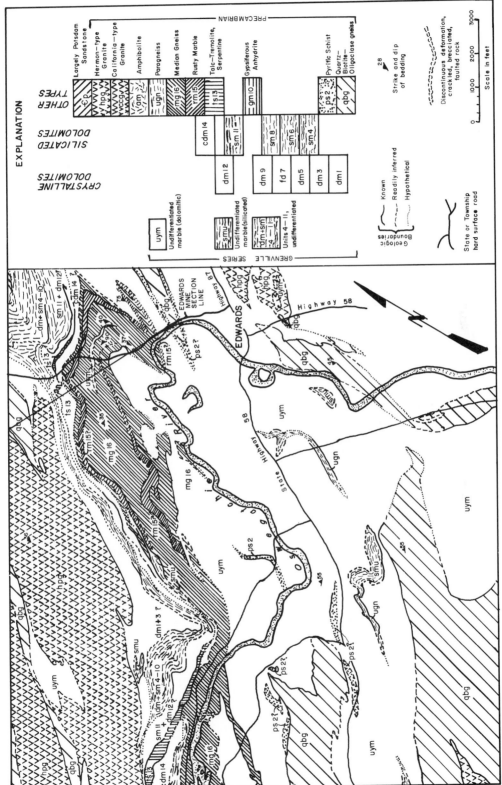

FIG. 4. *Geologic Map of the Balmat-Edwards District: Edwards Area.*

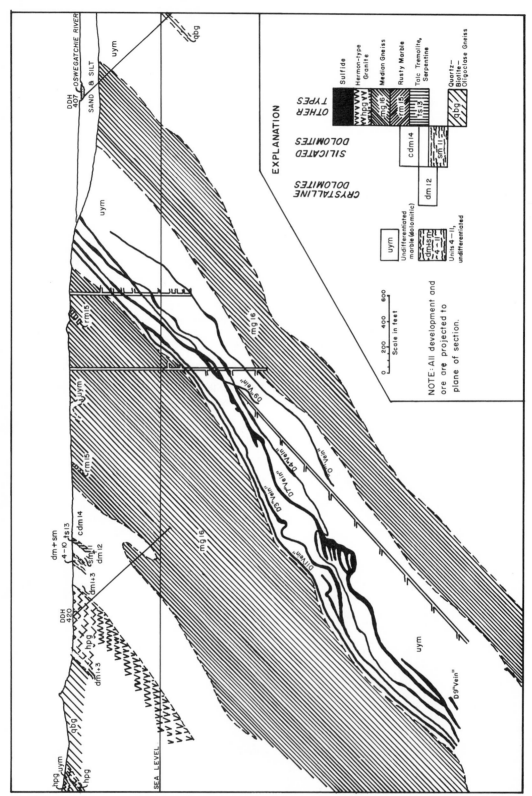

Fig. 5. *Edwards Mine: Northwest-Southeast Section.*

northeast-southwest oriented shear couple, consistent with the previously noted northeastward movement of the lowlands relative to the Adirondack massif. (Figure 2)

The Edwards mine thus is located within a structural intersection formed by northeast-striking and northwest-dipping elements of the primary folding and by the controlling northwest-trending secondary crossfold, plunging at about 40° between surface and the 3100-foot level. (Figure 5)

Brown and Engel (15, p. 1601–1602) observe that the Balmat-Edwards district lies largely within the belt of crossfolds, but its western-most part (Sylvia Lake area) coincides with the zone of transition between the belt of crossfolds to the southeast and central belt of accordant folds to the northwest. This central belt has crossfolding that is only locally or incipiently developed.

To illustrate, the Edwards crossfold plunges more or less at right angles to the regional northeast trend, in the plane of the regional dip. Development has proceeded within this structure for almost a mile down an average plunge of 40°. The direction of plunge of the Edwards crossfold averages approximately N60°W, whereas the average regional trend is N30°E. In comparison, crossfolds developed in the Sylvia Lake area have an average plunge of about 25°. The directions of plunge average approximately N10°W, whereas the average regional trend in the area is N30°E, similar to that of Edwards. Thus, the crossfolds in the Balmat area have a trend that is intermediate between the trend of the accordant folds in the district and that of the Edwards crossfold.

Accordant drag folds with a northeast-trend have been shown by underground mapping in the Balmat No. 3 mine and by exploration drilling from the surface to be drag folds related to, and dependent upon, the development of the primary Sylvia Lake syncline. Underground at certain locations, detailed mapping shows these folds to reverse their plunge at gentle angles from northeast and southwest and then plunge again to the northeast. These reversals are possibly related to the effects of the north-northwest plunging crossfolds.

Shearing stresses related to the development of crossfolds in the area tend to rotate and compress these folds in a clockwise fashion, causing smearing of their outlines in certain cases. In a similar manner, elements of the Sylvia Lake syncline, now exhibiting a composite and complex structure, have been rotated toward the northeast and compressed. (Figure 3)

The Balmat No. 2 mine is located just northwest of Balmat Corners, within a structural crossroads formed by northeast-trending, northwest-dipping overturned components of the Sylvia Lake syncline and by a north-northwest plunging crossfold. This refold is the dominating mine structure and takes the form of an isoclinal syncline or trough, overturned toward the southeast and plunging to the north-northwest at about 28°, between the 300- and 1900-foot levels.

Whereas the Edwards mine is located within a refold of calcareous sediments lying below the median gneiss, the Balmat No. 2 mine is located within a refold of calcareous sediments lying above the median gneiss. (Figures 3, 6)

The Balmat No. 3 mine is located within a structural crossing formed by northeast-trending drag folds, dependent upon the overturned northwest-dipping limb of the primary Sylvia Lake syncline, and by elements of secondary crossfolds plunging at about 25° to the north-northwest between the surface and the 900-foot level. The dominating mine structure is that formed by accordant drag folds gently plunging to the northeast that reverse themselves within the mine area and then return to a normal northeast plunge. (Figure 7)

## Age of Mineralization

Newland (3, p. 625) in 1916 assigned the ore deposits of the Balmat-Edwards district to the Precambrian. The introduction of sulfides as replacements of the Grenville Series marbles was thought to follow the metamorphic processes which led to the silication of the limestones. Newland was supported in his conclusions by Smyth (5, p. 11).

It is of interest here to note that Uglow (4), also in 1916, made a study of the Long Lake zinc mine in Frontenac County, Ontario, situated northwest of the Balmat-Edwards district. The deposit has similar chemical and physical characteristics to the ore bodies of this district and also occurs within the Grenville Series limestones. Uglow (4, p. 231) suggested that the metamorphic zone including the ores might be the result of recrystallization under conditions of high temperature and pressure of materials previously existing within the limestones.

Brown (9, p. 254) suggested that the Balmat-Edwards sulfides were emplaced during the later stages of metamorphism as evidenced by certain micro-textures of talc relative to those of the ore sulfides.

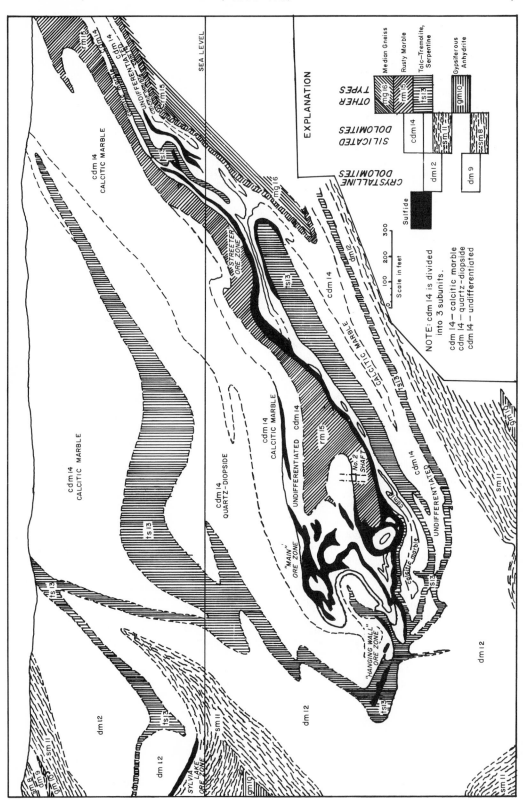

FIG. 6. *Balmat No. 2 Mine: East-West Section.*

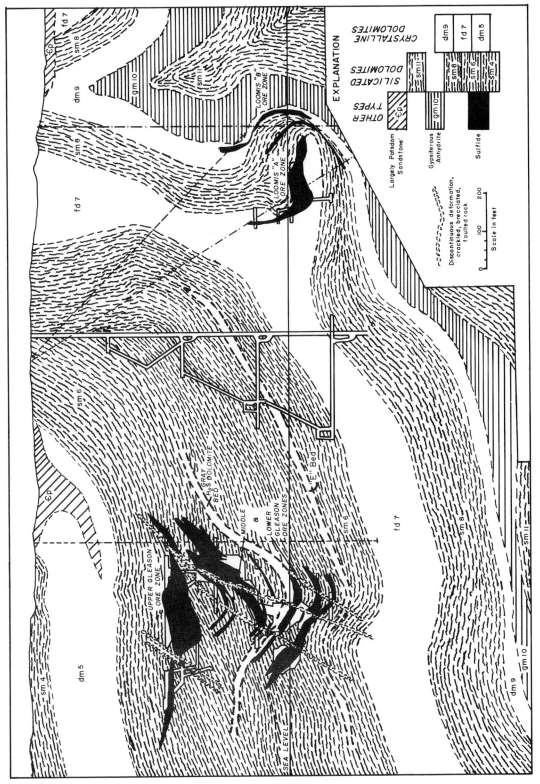

FIG. 7. *Balmat No. 3 Mine: Northwest-Southeast Section.*

Engel and Engel (18, p. 1385) confirmed Brown's (9, p. 254) reasoning by concluding that the paragenesis of the metamorphic minerals in the marbles was diopside, tremolite, anthophyllite, serpentine, and then talc, with the sulfides introduced during the anthophyllite stage of retrograde metamorphism.

Brown and Kulp (19, p. 138) on the basis of lead-isotope studies calculated the age of mineralization to be 1050 ± 100 m.y. This gave good agreement with the metamorphism of the Grenville Series as determined by Aldrich, *et. al.* (16, p. 105–108) for K/Ar, Rb/Sr, and Pb isotopes.

Doe (20, p. 37) derived an age of 1050 ± 100 m.y. both for the most recent metamorphism within the Grenville series and for the introduction of the sulfides.

Brown (26, p. 45) questioned the validity of an age of 1050 ± 100 m.y. for both metamorphism around Balmat and the assumed late stage introduction of sulfides, on the basis of his study of isotopic data on the Balmat-Edwards district lead mineralization and elsewhere, indicating that regional metamorphism resulted in homogenization of the originally varied isotopes. On the basis of this conclusion Brown (30, p. 66) has revised his earlier concepts and now believes that the Balmat-Edwards ores were not only present before, but took a complete part in, the Grenville metamorphism, thus explaining their present high-temperature characteristics.

Solomon (27, p. 154) advances the coarseness of the sulfides and gangue textures, the intense folding and microbrecciation of the Balmat ore body, and the confused paragenesis, together with the unique type of distribution of the sulfur isotopes, in support of premetamorphic sulfide mineralization.

Engel and Engel (12, p. 1044) assume an average age of 1100 m.y. for the culmination of the Grenville orogeny. They speculate that the inception of the Grenville sedimentation might have been as early as 1500 m.y. ago and probably exceeded 1200 m.y. If this speculation is valid, the age of the assumed premetamorphic sulfide mineralization would fall somewhere within these limits.

## ECONOMIC GEOLOGY-PRIMARY ORE

### Forms of the Ore Bodies

In general, the ore bodies can be defined as usually conformable, with cross-sectional areas of rather moderate size but characteristically with a large, often extremely large, down-plunge dimension. It is somewhat difficult to give dimensions to particular ore shoots because of their general habit of continually changing in details of form as they proceed down plunge. This feature, of course, is a direct result of the intimate relationship of the complexly folded enclosing rock formations with the ore shoots. Because of these varied characteristics of ore shoots in dimension and form, the reader is referred to particular plans (Figures 8, 9) and sections (Figures 5, 6, 7) for details within the particular mines.

The dimensions of individual mineable ore shoots for all mines within the district range as follows:

Thickness: 2 to 50 feet;

Horizontally along the strike: 50 to 800 feet;

Down plunge: Various, often according to the size of the particular ore shoot, but may be up to thousands of feet with downward limits now unknown;

Shape: Lenticular, pod-shaped, or tabular.

There are long, thin, tabular bodies in the less-folded areas. These usually strike northeast-southwest, consistent with the trend of primary folding, and progressively become thinner in their extensions until the mineralization is only faintly visible.

Elsewhere the ore bodies may become lenticular or pod-like as they pass into more intensely refolded areas or bulge out into greatly thickened masses reflecting the form of the folds. Some of these fold forms may be recumbent, doubled, and mashed together synclines and anticlines or troughs and crests. All orientations of strike and dip occur within the ore shoots, but, in all places, they faithfully conform to the direction and angle of plunge of the controlling crossfold structure. Within some ore shoots, remnants of diopside and in some cases carbonate show flexures and rock type characteristics similar to those present in the enveloping country rock. These remnants are preserved by surrounding high-grade sphalerite at many locations.

### Mineralogy of the Deposits

The assemblage of ore minerals is simple. Pyrite and sphalerite are abundant, and galena, pyrrhotite, and chalcoprite are minor in amount. Marcasite has been reported but is not conspicuous. The arsenic minerals, jordanite and realgar, occur extremely rarely and are mere curiosities. Supergene minerals, present largely in Balmat No. 2 mine, consist of magnetite, hematite, and secondary forms

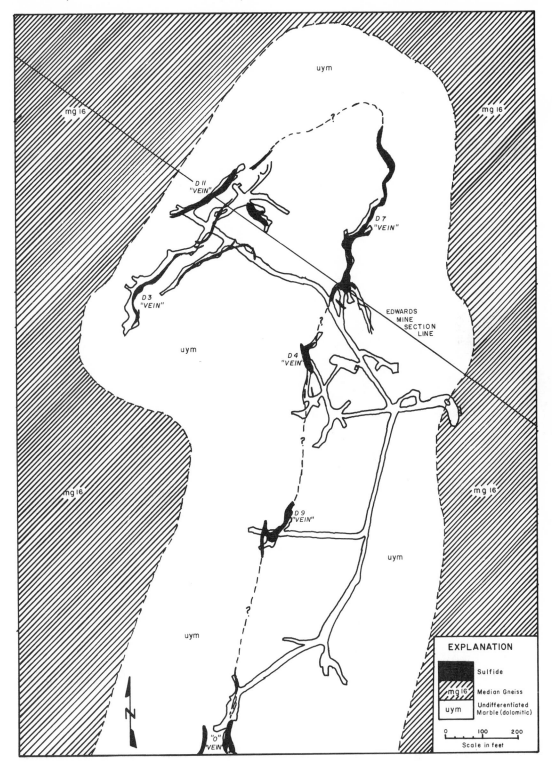

FIG. 8. *Geologic Map of the 1500-Foot Mining Level: Edwards Mine.*

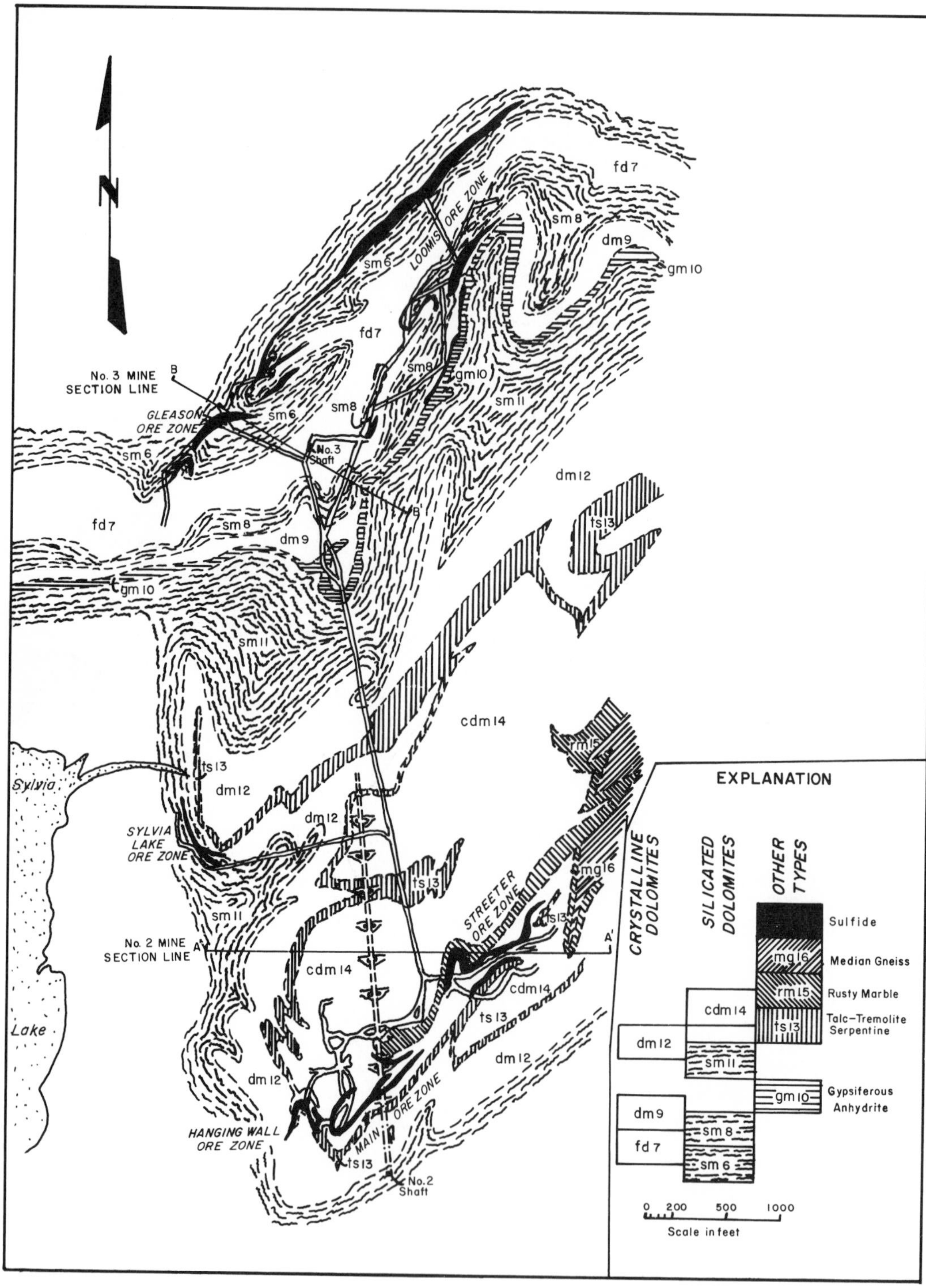

FIG. 9. *Geologic Map of the 900-Foot Mining Level: Balmat No. 2 and No. 3 Mines.*

of sphalerite, galena, chalcopyrite, bornite, and willemite.

Gangue minerals include quartz, diopside, tremolite, serpentine, talc, carbonate, barite, and anhydrite. In the supergene ores, chlorite is abundant, with minor ilvaite and garnet (grossularite). In most, if not all cases, the nonmetallic minerals found within the ores are either reworked fragments of adjacent wall rock or late products of retrograde metamorphism in the surrounding calcsilicate marbles.

Aside from a gradual diminishing of supergene effects with depth, there is no general change in mineralogy either laterally or in depth, in the sense of metal zoning or impoverishment in zinc. The tendency for pyrite to be segregated and for sphalerite to be disseminated leanly around the rims and extensions of individual ore bodies does not apply as a generalization to groups of ore zones. In comparing groups of ore zones, and also individual ore bodies within ore zones, fairly wide deviations in ore grade exist from high-grade, low-pyrite shoots to lean, weakly mineralized or highly pyritic bodies. These variations, however, occur in about the same proportion throughout the strike and plunge of the ore bodies and are a measure of structural control. The only possible exception to the foregoing statements is the habit of galena at Balmat No. 2 mine to accumulate, for reasons not yet understood, in footwall portions of the structure. In general, for each mine's composite production, there is a remarkable consistency in zinc-ore grade. For example, the ore grade at Edwards is essentially the same at the surface as at the 3100-foot level, nearly a mile down the plunge.

Wall rock alteration is virtually absent in the Balmat-Edwards district. Occasionally bleached rims measured in a few inches, may be observed in the more pure dolomites at ore contacts in the Balmat No. 2 and No. 3 mines, but these are exceptional.

There is a wide range of mineral association and texture. Macroscopically, most of the ores are massive to disseminated coarsely crystalline aggregates of sphalerite and pyrite in gangue of quartz, carbonate, and diopside. Where inclusions of wall-rock are abundant, the sulfide bodies have the aspect of ore breccias. Pyrite occurs with sphalerite either as cubes and pyritohedra in discrete subhedral or anhedral grains up to 10 cm, shattered and microbrecciated or in massive aggregates almost without sphalerite or gangue. Sphalerite usually is coarse-grained but ranges from .01 mm grains

to subhedral crystals over 8 cm in length and has well-developed polysynthetic twinning. Color ranges from a light chocolate-brown at Balmat No. 3 mine to dark chocolate-brown at Edwards and Balmat No. 2. The color is an index of the FeS content in the sphalerite. Doe (20, p. 75–82), in a study of the sphalerite geothermometer, determined the FeS content in pure sphalerite for 170 specimens from Balmat No. 2 mine and for 25 specimens from No. 3 mine. No valid temperature data could be determined, but the arithmetical average of analyses showed 9.6 weight per cent FeS for Balmat No. 2 and 3.1 weight per cent FeS for Balmat No. 3. Available data give about the same percentages for Edwards as for Balmat No. 2.

Galena very rarely appears within the pyrite-sphalerite ore except in footwall areas of Balmat No. 2 ore zones, where it occurs as discrete grains and fracture fillings in pyrite, sphalerite, and silicates. Commonly, at Edwards and Balmat No. 2 mines, galena is conspicuous as veinlets up to 3 cm in diopside rock. A little lead is recovered as a by-product at Balmat.

Chalcopyrite and pyrrhotite, as is true of galena, are most conspicuous as gash veins and disseminated blebs, often with minor sphalerite, in diopside rock at some distance from zinc ore bodies. Occurrences of chalcopyrite and pyrrhotite within ore bodies can be observed only in polished section. Chalcopyrite appears as fine blebs and chains within sphalerite and pyrrhotite, and pyrrhotite has complex relations with pyrite and sphalerite. There is no agreement on whether pyrrhotite is exsolved from sphalerite or replaces it.

Gangue minerals may be nearly absent in some of the high-grade ores or may predominate in disseminated or weakly mineralized zones. Quartz is much more common than the microscopical studies would indicate. It is particularly abundant in the Upper Gleason, Streeter, and Main ore zones as partly rounded, semi-replaced inclusions. Rounded and subangular fragmental inclusions of diopside and tremolite are frequent, very often largely altered to serpentine and talc. Carbonate, usually calcite, may occur either as granular gangue or as unreplaced bands of wall-rock that are often distorted in complex folds. Barite is a rather common gangue mineral at Edwards and in the southwest portion of Balmat No. 2 mine; since it is not found in the wall-rock, it is presumed to be closely related to the ore minerals. Anhydrite previously reported as a gangue mineral and related

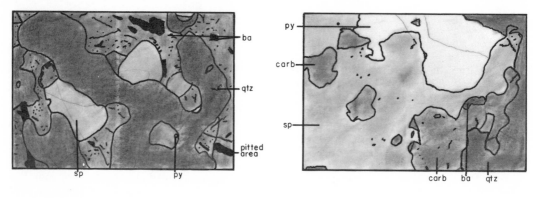

a.  Edwards Ore

b.  Edwards Ore

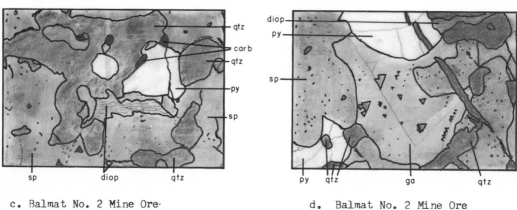

c.  Balmat No. 2 Mine Ore·

d.  Balmat No. 2 Mine Ore

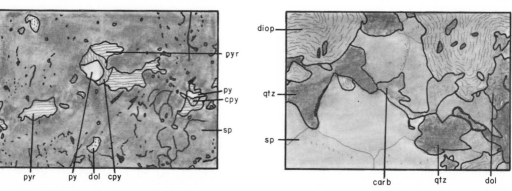

e.  Balmat No. 3 Mine Ore

f.  Balmat No. 3 Mine Ore

FIG. 10. *Photomicrographs of the Balmat-Edwards District Ores. qtz = quartz, diop = diopside, ba = barite, dol = dolomite, cpy = chalcopyrite, carb = carbonate, sp = sphalerite, py = pyrite, ga = galena, pyr = pyrrhotite. Magnification, ×140.*

to barite in some way, is probably unreplaced wall-rock.

Pyrite and sphalerite in some ore bodies may be evenly mixed, in others, they are nearly mutually exclusive, and in some cases may be segregated into fine-grained and coarse-grained alternating layers of pyrite-rich and sphalerite-rich sulfides that often exhibit flow banding. District-wide average composition of the ores is as follows: 10 to 30 per cent sphalerite, 0 to 80 per cent pyrite, and 5 to 80 per cent gangue.

Mineral paragenesis and texture has been studied in detail by Brown (9,10), Doe (20,22,23), and Solomon (27), and their reports have been freely used in the foregoing descriptions. Drawings of some recent polished-section work on ores from the various mines by G. Aletan are represented in Figure 10. Many conflicting observations in the published work emphasize the complexity of the mineral relationships, and, although many hundreds of polished sections have been examined, the interpretations unavoidably have been subjective and incomplete.

Only general statements can be made within the scope of this article. Mutual boundaries without obvious order of crystallization occur in the pairs sphalerite-pyrrhotite, sphalerite-carbonate, galena-sphalerite, galena-pyrite and galena-chalcopyrite, but many cross-cutting and reversible relationships also occur in these minerals. Pyrite has especially complex relations, but much of it is intensely microbrecciated and all other sulfides fill fractures in the pyrite and occur as grains within the pyrite crystals. Barite and quartz may be the only exotic nonmetallic minerals in the sulfides and appear to be closely related to the deposition of sulfides. Considering the ores in their present environment, serpentine and talc are definitely post-sulfide, and diopside-tremolite are pre-sulfide; these relationships assist in timing sulfide crystallization within Engel's series of retrograde metamorphism: Diopside-tremolite-anthophyllite-serpentine-talc. The preceding evidence strongly suggests to us that the primary ores in their present form were metamorphosed from previously existing sulfides.

## Factors Controlling Form and Location of Ore Bodies

It has been indicated previously in this discussion that the Balmat-Edwards mines are located in silicated and dolomitic marbles within areas forming structural crossroads between northeast-trending, northwest-dipping

(overturned) components of the primary or accordant folding and later development of north-northwest-plunging secondary cross folds.

Figure 8 of the Edwards mine at the 1500-foot mining level illustrates the relation of the siliceous, generally non-calcareous, and granitized median gneiss, as the more competent rock, to the less competent dolomitic marble series. The dolomites contain some silica as distinctly siliceous or quartzitic horizons, as banded quartz diopside, or as irregular disseminations within the dolomite. Some areas contain tremolite and diopside in association with their alteration products of talc and serpentine, and anhydrite, of probable sedimentary origin, is a mappable unit in certain areas of development.

Toward the close of all but the very last stage of major deformation and subsequent to silicate alteration within the dolomite, pegmatite dikes cut through the folded formations, and pegmatitic material soaked both the dolomites and the gneiss. Today, this granitic material shows up clearly within the gneiss as dikes, clots, and dissemination but loses its identity in the dolomites except where clear-cut pegmatite dikes have been intruded. The Edwards ore emplacement, at least in its present form of high-grade ore shoots, is post-pegmatite. At several locations, high-grade, coarse-grained sphalerite fills fractures and surrounds brecciated knots of pegmatite. In others, where pegmatite has become involved in a northeast-southwest-trending ore lens, brecciation of the pegmatite dike occurs along the strike of the ore lens in plan, and high-grade sphalerite ore surrounds and partially replaces the irregular pegmatite fragments. Definitely, some northeast-southwest shearing stresses are indicated as having been post-pegmatite but prior to the emplacement of ore shoots in their present form.

The effect of this shear couple acting on northeast-trending, northwest-dipping elements of the primary folding has been to rotate the gneisses around the dolomites to form Z-shaped folds, plunging to the northwest more or less down the regional dip. As seen in section and plan (Figures 5, 8) the rotation of the gneiss, enclosed and isoclinally rotated the marble and the mineralized horizon along which the major high-grade ore shoots were localized. Detailed mapping of these ore zones in most cases will give evidence of the same shearing stresses that developed the major controlling crossfold. For example, the D-9 'vein' (Figure 8) shows a well-developed Z-shaped fold, indicating that northeast-southwest shearing stresses have been important in controlling the form and location of the ore bodies. Elsewhere the folding may be more obscure, but generally in detail, even in tabular ore bodies, microbrecciation and Z-shaped minor folds give evidence of northeast-southwest shearing stresses along the strike of the ore shoot. The ore shoots were developed along the strike to limits determined by the width and richness of ore considered as economic for mining. The sulfides do not terminate completely but continue beyond the limits of economic development, usually as pinched stringers of pyrite and minor sphalerite that, together with information from diamond drilling between ore shoot areas, suggest weak continuity of the mineralized beds. The form suggested is that of irregular or elongated beads on a string.

The lenticular ore shoots or the complexly folded irregular pod-like ore shoots take on the plunge of the major crossfold in the rotated gneiss. In vertical cross section, the dips change from steep to flat to steep. This feature has been shown to be indicative of changing plunges not only for the major cross fold, as indicated by the rotated gneiss, but also for individual ore shoots. Where the beds are steeper, ore shoots are more tabular and tend to straighten out, as is shown at the 1500-foot level and above. Where the beds dip more gently, generally there is a major development of Z-shaped minor crossfolds that, within the mineralized horizon, control the shape of fantastically complex pod-like ore shoots. These fold structures apparently were related in their development to shearing stresses along the fabric of bedded formations that had different degrees of competency within the zone of plastic flowage. The angle of plunge, whether steep or gentle in the ore bodies, is dependent upon, and consistent with, the plunge of the major crossfold in the gneiss for particular elevations within the mine.

The factors controlling the form and location of the Balmat No. 2 and No. 3 mine can best be considered by reference to Figure 9 of the 900-foot mining level and Figure 6 and Figure 7 illustrating vertical sections through each mine.

The geologic map of the 900-foot mining level, with its accompanying vertical section, illustrates the localization of the Balmat No. 2 mine with reference to the structural crossing previously indicated between the northeast-trending, northwest-dipping beds of the primary folding and the north-northwest plunging

elements of the secondary cross folding. As seen in east-west section (Figure 6), the dominant structure within the No. 2 mine is the north-northwest-plunging crossfold that caused extreme folding of the dolomites, silicated dolomites, tremolite, and other formations into an isoclinal syncline, overturned toward the southeast. The "rusty marble" formation (unit 15) is completely doubled by northeast-southwest shearing stresses that generated the secondary fold. The subunits of the No. 14 formation, which contains the ore bodies of the No. 2 mine, generally strike northeast and dip northwest in the Streeter ore zone on the footwall side of the Balmat trough, consistent with the trend and dip of the primary folding. At the keel of the synclinal crease formed by the rusty marble in vertical east-west section may be seen the complex configurations caused by plastic flow of both the enveloping rock formations and the enclosed ore horizon. The subunits of the No. 14 formation, in plan and section, indicate clockwise isoclinal rotation around the squeezed keel of rusty marble, thus causing footwall No. 14 formation to repeat itself within the Balmat No. 2 trough. The shearing has operated along the bedding planes of the northwest-dipping overturned limb of the accordant Sylvia Lake syncline. The Hanging Wall ore zone is composed essentially of high-grade coarse-grained sphalerite and very minor pyrite. The Main and Streeter ore zones are more pyritic, containing variable amounts of sphalerite, resulting in medium grades generally. Some of the northeast extensions of the Streeter ore zone may be unusually pyritic.

In contrast to the No. 2 mine, the dominating structure of the No. 3 mine is that formed by the accordant drag folds that plunge gently to the northeast and reverse themselves in the mine area.

In detail, the No. 3 mine may be separated into two parts, the Gleason area to the northwest and the Loomis area to the southeast. The Gleason ore bodies have been called the Upper, Middle, and Lower Gleason ore zones (Figure 7). These ore bodies are localized in the general area of intersection between northeast-trending, northwest-dipping beds and what is believed to be north-northwest extensions of the axis of the same crossfold structure which dominates the No. 2 mine some 5000 feet to the south-southeast. The Gleason ore bodies enlarge in the general area of the section used for illustration. The Upper Gleason ore body is localized within a crossfold structure plunging to the north-northwest at about

25° with a limited down-plunge extension to the northwest and south-southeast extensions up the plunge to surface. The ore horizon is a coarsely crystallized high-grade sphalerite and pyrite replacement of a brecciated gray-quartz lens, with minor diopside, that is enclosed within the No. 6 formation. The Middle and Lower Gleason ore zones lie within northeast-trending structures. The very irregular high-grade, coarse-grained sphalerite and generally minor pyrite replacements of fractured or brecciated gray quartz and silicated dolomite lend themselves to longhole-stope extraction.

It is interesting to note the presence of northeast-southwest-trending shear planes that dip steeply to the northwest and have large horizontal displacements and small vertical throws. These shears are believed to be expressions of the northeast-southwest-acting shearing stresses related to the development of crossfolds. Some slight post-ore movement may have occurred along these fault planes in the more competent silicated and quartzitic dolomites of the No. 6 formation. Within the dolomitic horizons, the shear planes have since healed or are difficult to see in these more plastic formations.

The northeast-southwest section shows Loomis A ore zone development on a small drag fold structure related to the overturned limb of the Sylvia Lake syncline. The ore body forms at the sheared contact between the plastically deformed No. 7 dolomite formation and the more competent No. 8 silicated dolomite. The Loomis B ore zone is weak on this section but blossoms within the No. 8 silicated dolomite and No. 9 dolomite to the northeast (Figures, 7, 9), where a north-northwest-plunging crossfold intersects the northeast trends of the Loomis A ore zone. Both the Loomis A and B horizons are characterized by ore formed of light-brown coarse-grained sphalerite containing less than 1 percent Fe and accompanied by very minor pyrite.

Lithology frequently controls the shape of the ore bodies, particularly with reference to their contacts. In the No. 2 mine for example, sharp contacts of massive sulfides are usually seen with the rusty marble, tremolite, calcitic marble, or dolomite. On the other hand, contacts with serpentinous or diopsidic marbles usually are vague and poorly defined, and disseminations of sulfides, discontinuous pods, bunches, horsetails, and stringers outward from the more massive ore suggest irregular migration of ore minerals into slightly crushed foot

wall or hanging wall rocks. Ore bodies in diopside, quartz diopside, or quartz show contacts that are highly irregular and erratic, have complex patterns of penetration into the wall-rocks, and contain sulfide-filled gash veins.

Generally the thickest ore concentrations are found in synclinal or trough structures. Lesser thickening for some reason occurs at their crests. There is thinning along all flanks almost without exception. The favorable horizons are considered to be those in which base-metal sulfides were disseminated during the original deposition of the sediments, a process that was repeated at intervals during the development of the stratigraphic column.

Certainly, with ore bodies spread throughout many different horizons within the some 1800 feet of the stratigraphic column in the Balmat area, it would be difficult to attribute the source of present ore bodies to migration from one single source bed as advocated by Knight (17). Wisps and disseminations of pyrite, sphalerite, and very minor galena are found extending beyond and between known ore-bearing areas underground. This provides forceful evidence, as do the mineral textures, lack of wall-rock alteration, largely poor definition of mineral contacts, and general lack of obvious channelways or conduits, for considering the possibility of pre-metamorphic mineralization.

Intense folding and metamorphism have caused almost total destruction of primary sedimentary structures. This prevents a complete understanding of the original petrology of the Grenville marbles. We can but surmise on how the ore bearing source beds acquired their sulfide minerals.

Solomon (27, p. 154–155) proposes the following sequence of events to account for the known concentrations.

"a) The periodic formation in a rather shallow carbonate-precipitating basin of localized bottom depressions in which there developed a physico-chemical environment amenable to the reduction of $SO_4^=$ to $S^=$ and its contemporaneous interaction with an abundant supply of $Zn^{++}$, and $Fe^{++}$ and other metallic ions to form fine grained sulfide accumulations.

b) Subsequent compaction and incorporation of these sulfide concentrations into the sedimentary sequence.

c) Folding and metamorphism during which the ores were recrystallized, isotopically homogenized and locally concentrated to give their present distribution."

## Effects of Metamorphism on the Ores and Their Immediate Environment

It is assumed, for purposes of discussion, that the so-called source beds containing disseminated pyrite, sphalerite, and minor galena were pre-metamorphic and took a complete part in the Grenville metamorphism. It is thought that this metamorphism took place at great depth within the zone of plastic flowage, resulting in recrystallization and reconstitution of both source beds and enclosing rock formations.

The dolomites, silicated dolomites, and gneisses became thoroughly recrystallized under pressure and heat. Engel and Engel (18, p. 1385) indicate that the regional metamorphism, which extends throughout and beyond the Balmat-Edwards district, evolved the assemblage diopside, dolomite, calcite, and quartz in the marble and that of quartz, biotite, plagioclase, and muscovite in the gneiss. Their paragenesis in the marble was diopside, tremolite, anthophyllite, serpentine, and talc with the appearance of clinozoisite, chlorite, sericite, and albite in parts of the gneiss. Two major deformations affected the marbles. It has been suggested that the earlier of the two occurred slightly before, and during, primary emplacement of the granites. The second major deformation is thought to have taken place during the culmination of the granitic intrusions (24, p. 223). During the latter stage, the gneiss was subject to a drenching by melts and injected with granite. Late stages of the granite intrusion have resulted in pegmatite dikes cutting the marbles.

The early major deformation developed northeast-striking, northwest-dipping structures. The later deformation resulted in northeast movements of formations relative to those to the southeast, in the plane of regional trend. These shearing stresses developed the cross folds that intersected the primary regional trend at different angles to form the major structural control within particular mine areas.

At certain locations, possibly dependent on the competency of wall-rocks relative to each other and on the distribution of the shearing stresses, minor shear zones or fold structures developed along the mineralized beds. The plunge of both shear zones and Z-type minor fold structures, were dependent upon and followed the major controlling crossfolds in both direction and angle of plunge. The disseminated pyrite, sphalerite, and minor galena at certain locations within the source beds flowed plastically, developing pinched limbs and thickened concentrations of sulfides at the crests and troughs of minor folds. Shearing stresses along the fabric of the bedding were important in the development of local microbrecciation and shearing or of well-developed minor crossfold structures, involving rocks of different competencies. In these areas, the disseminated sphalerite and pyrite are believed to have acted quite differently. It is suggested that the pyrite, under these particular conditions of directed stress causing shearing between beds of different competency, was largely segregated and stayed in place. Mobilization of the disseminated sphalerite and minor galena and pyrite may have resulted by these minerals having gone into solution and having been reprecipitated elsewhere in the system (27, p. 74). These solutions are believed to have flowed along planes of secondary shear that followed the sedimentary fabric of the previously mineralized beds and redeposited their load in structurally prepared areas, thus forming the present high-grade ore shoots.

Within particular ore shoots and surrounded by ore, remnants and fragments of silicate bands are found. These have been stretched and torn apart by the shearing stresses described. Similar silicate remnants and fragments enclosed by dolomites and serpentinous dolomites may be seen in the surrounding wall-rocks. They suggest an analogous origin, i.e., previous brecciation under conditions of plastic flow. As a result of this plastic flow, nearly all evidence of shearing, brecciation, or fracturing was healed, leaving a general lack of obvious channelways or conduits for transport of ore solutions.

The introduced high-grade sphalerite is very coarse, and this suggests that it filled open spaces within originally brecciated zones that were developed in rocks of differing competency such as dolomite and diopside or dolomite and quartz. The high-grade, light to dark chocolate-brown sphalerite exhibits cleavage faces of up to 8 cm in width. These cleavage planes provide mirror surfaces and return bright flashes of light from mine lamps thus giving the name "spangle" ore in mining parlance. Spangle ore is a characteristic of all the Balmat and Edwards mines and provides necessary "sweetener stopes," or permits the mining in certain stoping areas of lower-grade disseminated ores through which it cuts with sharp contacts.

The introduction of sulfides at a late stage in the metamorphism into structurally prepared areas, following the formation of silicates, would be consistent with Engel and

Engel's (18, p. 1385) paragenesis of the metamorphic minerals in the marbles of diopside, tremolite, anthophyllite, serpentine, and then talc, with the sulfides introduced during the anthophyllite stage of retrograde metamorphism.

## Summary

In general, the authors would be sympathetic with Brown (30, p. 66) and Solomon (27, p. 154–155) in their interpretation of premetamorphic sulfide mineralization in explaining the sequence of geologic events required for the formation of primary ores.

Fetid dolomitic marbles, containing $H_2S$, and sedimentary anhydrite occur at a number of different horizons within the stratigraphic sequence. In addition, a repetitive sequence is indicated by alternating formations of dolomite and silicated dolomites of almost identical composition. This would suggest that similar conditions prevailed in the sedimentary environment at repeated intervals during deposition of the Grenville limestones in the Balmat-Edwards district.

It is conceivable that these repeated occurrences of similar conditions in the sedimentary environment explain the recurrence of disseminated base-metal sulfides at numerous calcareous horizons throughout much of the stratigraphic column.

It is thought that these mineralized beds were deeply buried, folded, and metamorphosed, during which events the disseminated base-metal sulfides were recrystallized and mobilized. The sphalerite, minor galena, and pyrite are presumed to have flowed either plastically under directed stress or in solution along planes of secondary shear; these solutions followed the sedimentary fabric of the previously mineralized beds and reprecipitated and concentrated the sulfides in structurally prepared areas. The pyrite was partially segregated. The sphalerite, minor galena, and pyrite may have behaved much like a distillate or as if they were filter-pressed from the disseminated sulfides within the carbonate rocks. Mobilization of the zinc and lead sulfides along the strike and dip of the mineralized beds has resulted in zones of concentration similar to irregular beads along a contorted string in plan and with continuity up and down the dip by virtue of their plunge (see Figures 8, 5). These concentrations of high-grade sphalerite formed as replacements of microbrecciated zones between rocks of different competency along planes of secondary shear or as replacements of all

manner of complex crumple structures, related in their origin to secondary shearing and to the development of plunging crossfolds.

## ECONOMIC GEOLOGY-SECONDARY ORE

To generalize, supergene processes have had but little effect upon the sulfide ores of the district. Thin limonitic gossan, often measured only in inches, caps the rather meager outcrops of the Edwards and many other ore bodies, giving way immediately below to fresh, unaltered sulfide. Cavities and vugs, lined with specularite, calcite, and other secondary minerals, are common in upper levels of the mines, but more often than not they are in the wall-rock rather than within the ore bodies. Hematite is widespread in the Balmat area, occurring occasionally as earthy and specular varieties filling fractures in both ore and wall-rock. In these instances of hematitization, the sulfides are unaffected, and the hematite, presumably derived from pyrite, has been transported. The notable exceptions are in the Upper Gleason ore body of Balmat No. 3 mine and the Hanging Wall ore zone of Balmat No. 2 mine. In the first instance, a very strong hematitization occurs only in the upper 50 feet of a thick massive sulfide lens. In the second, however, an intricately folded ore structure shows the effect of supergene activity to unusual depths.

The Hanging Wall zone is an uninterrupted complex of individual ore bodies following extremely tight, plunging keels, crests, and troughs. Concomitant with the tight folding, much brecciation and shearing were developed along contacts of rocks of dissimilar competency. It is in this environment that an unusual assemblage of supergene minerals has been developed; they are lacking in all other ore zones in the district.

The occurrence of secondary sphalerite, galena, and willemite was first described by Brown (10), who gave compelling evidence for the origin of these minerals by supergene process rather than by a second stage of primary mineralization. Cited as proof of deep oxidation by meteoric waters were the presence of solution cavities and of massive earthy hematite to considerable depths but in diminishing abundance as the structures were followed downward. It must be emphasized that the oxidation is by no means complete throughout the Hanging Wall ore zone at any elevation, but it is certainly more intense near the surface and gradually decreases with depth. Oxidation effects at 2000 feet below surface are very slight and localized, showing virtual disappearance of supergene activity at this level, although one occurrence of hematite and magnetite, with residual sphalerite, has been noted in drill core 2300 feet below surface (1650 feet below sea level).

The secondary sphalerite occurs as masses of minute white to yellowish grains replacing pyrite, primary sphalerite, and gangue minerals. Secondary galena and chalcopyrite are of minor importance and usually can be detected only in polished sections, although there are rare showings of massive chalcopyrite with subordinate bornite, evidently of supergene origin because of their occurrence within intensely oxidized areas. Magnetite appears only in the most intense areas of alteration, usually intimately intermixed with secondary sphalerite in an extremely fine-grained texture. Hematite is very abundant as bands and masses of earthy red variety in the wall-rock and as both earthy and hard varieties in partial replacement of residual pyrite. Chlorite may be in part a product of the original metamorphism, but, since it is largely restricted to the Hanging Wall ore zone and gradually diminishes with depth, chloritization is believed to be essentially a supergene effect as is hematitization. Chlorite pervades both the tremolite and diopside wall-rocks, as well as gangue minerals within the ore bodies. Willemite, occurring in chloritic tremolite beneath thoroughly oxidized ore bodies, has not definitely been identified below the 700-foot level.

The typical oxidized ore has a "birdseye" texture in which centripetal replacement of pyrite grains is common. It is very common to see concentric bands of hematite, magnetite-sphalerite, and white sphalerite surrounding the residual pyrite. Usually primary sphalerite without any alteration may be found very close to even the most intensely oxidized ores. Doe (20, p. 27) noted dark-brown sphalerite with white sphalerite rims adjacent to chlorite and magnetite.

There is no convincing proof of actual supergene enrichment in zinc content either throughout the plunge of the ore zone or at any particular elevation. Although the zinc grade of the Hanging Wall is at least 50 per cent higher than the average grade for the No. 2 mine, it is believed that the higher grade is a result of favorable lithologic and structural controls in localization of the primary ore.

Brown (10, p. 351) cites the colloform structures of hematite, centripetal replacement of pyrite, and the fine-grained nature of the secondary minerals as evidence for supergene replacement at low temperatures, even by cold

waters. In regard to the mineralogy, it is significant that the secondary products are all anhydrous instead of the usual assemblage of such minerals as limonite and hydrozincite. It must be concluded that the secondary minerals were formed below the water table or at least in an environment deficient in free oxygen. Conditions must have been ideal in this highly crumpled, brecciated, and sheared portion of the structure for the slow, very deep circulation of oxygen-deficient ground water. The ore provided abundant pyrite for reaction and was undoubtedly more permeable than other ore bodies of the district. The dating of the formation of secondary ores is obscure and cannot be resolved with the available data.

## ORE GENESIS

Newland (3, p. 644) in 1916, suggested that the Edwards ore bodies were introduced as replacements of the Grenville marbles by agencies related to the granite intrusion at great depth, following the silication and recrystallization of the marbles.

Smyth (5) made a largely microscopic study of the Edwards ore bodies in 1918 at a very early stage in the development of the mine. His conclusions were influenced greatly by the opinion (5, p. 10) that the ores had the form of veins or irregular masses that cut across the stratification and were the result of later emplacement long subsequent to the sedimentary series formed under marine conditions. Smyth (5, p. 39) concluded that the Edwards ores were developed as replacements of the Grenville marbles by hot gasses and solutions evolved by intruding or related magmas.

In 1936, Brown (9, p. 244) noted that the ore bodies, commonly designated "veins" in mining parlance, are actually replacements along true bedding or modified flow banding. He was in agreement (9, p. 257) with Smyth and Newland that the ore solutions were derived from the underlying igneous masses. Later Brown (11, p. 514) observed that structurally the ore bodies are frequently associated with areas or bands of unusually complex folding or crumpling and that remnants of sharply flexed and folded bands are preserved at many places in the ore in striking attitudes. He concluded (11, p. 535–536) that field and microscopic data made it reasonably clear that the primary ore deposits were localized in definite channels of microbrecciation in impure silicated marbles, rather than chemically purer facies, for the reason that such heterogeneous mixtures were more susceptible to microbrecciation than pure marbles. It is suggested that the impure silicated marbles possessed a higher porosity under the particular temperature-pressure conditions that prevailed during ore deposition. Brown (11, p. 539) believed that an ore medium was required that was more concentrated in metallic ingredients than a hydrothermal vehicle in the conventional sense.

Engel and Engel (18, p. 1385) confirmed Brown's (9, p. 254) earlier studies in observing that the more iron-rich sphalerites were introduced along with some of the pyrite and pyrrhotite during the anthophyllite state of retrograde metamorphism in the marble. The formation of less iron-rich sphalerites and of iron sulfides were thought to have continued throughout the period in which serpentine and early talc formed.

Doe (20, p. 52), on the basis of a comparison of trace elements in pyrite and information on the isotopic composition of lead between the sulfide ores and the surrounding units of the Grenville series, suggests, but does not conclusively prove, that the sulfide deposits at Balmat were not derived from the surrounding rocks.

Brown (26, p. 45, 30, p. 66) has revised his earlier concepts and now believes that the Balmat-Edwards ores were not only present before, but took a complete part in the Grenville metamorphism, thus explaining their present high-temperature characteristics. Solomon (27, p. 154, 155) proposed the formation, in a rather shallow basin where carbonates were being precipitated, of periodic accumulations of fine-grained sulfides; later these sulfide concentrations were compacted into the sedimentary sequence. Subsequent folding and metamorphism caused the sulfides to be recrystallized, isotopically homogenized, and locally concentrated to give their present distribution.

Detailed studies of the ore bodies as outlined in this paper would, in the opinion of the authors, tend to support Brown and Solomon in their concept of pre-metamorphic sulfides.

## CONCLUSIONS

An attempt has been made by the authors to give the reader some idea of the concentrated geological studies that have continued for a period of some 50 years in the Balmat-Edwards district. During this time, contributions have been made by many individuals in regional, district, and mine geology, complemented by laboratory work on many facets of the problems presented in the course of this work. After these many years of study, the evolution of geological ideas continues, with many questions still to be answered. Much fertile ground remains for continued geological research.

## REFERENCES CITED

1. Emmons, E., 1838, Report of the second geological district of the State of New York: N.Y. Geol. Surv. Ann. Rept., v. 2, p. 185–252.
2. Cushing, H. P., et al., 1910, Geology of the Thousand Islands regions: N.Y. State Mus. Bull. no. 145, 194p.
3. Newland, D. H., 1916, The new zinc mining district near Edwards, New York: Econ. Geol., v. 11, p. 623–644.
4. Uglow, W. L., 1916, Ore genesis and contact metamorphism at the Long Lake zinc mine, Ontario: Econ. Geol., v. 11, p. 231–245.
5. Smyth, C. H., 1918, Genesis of the zinc ores of the Edwards district, St. Lawrence County, New York: N.Y. State Mus. Bull. no. 201, p. 7–39.
6. Fairchild, H. L., 1919, Pleistocene marine submergence of the Hudson, Champlain, and St. Lawrence valleys: N.Y. State Mus. Bull. nos. 209, 210, 76p.
7. Cushing, H. P., and Newland, D. H., 1925, Geology of the Gouverneur quadrangle: N.Y. State Mus. Bull. no. 259, 122p.
8. Buddington, A. F., 1934, Geology and mineral resources of the Hammond, Antwerp, and Lowville quardrangles: N.Y. State Mus. Bull. no. 296, 251p.
9. Brown, J. S., 1936, Structure and primary mineralization of the zinc mine at Balmat, New York: Econ. Geol., v. 31, p. 233–258.
10. ——— 1936, Supergene sphalerite, galena, and willemite at Balmat, New York, Econ. Geol., v. 31, p. 331–354.
11. ——— 1947, Porosity and ore deposition at Edwards and Balmat, New York: Geol. Soc. Amer. Bull., v. 58, p. 505–546.
12. Engel, A. E. J. and Engel, C. G., 1953, Grenville Series in the northwest Adirondack Mountains, New York: Part I, General features of the Grenville Series: Geol. Soc. Amer. Bull., v. 64, p. 1013–1047.
13. Engel, A. E. J. and Engel, C. G., 1953, Grenville Series in the northwest Adirondack Mountains, New York: Part II, Origin and metamorphism of the major paragneiss: Geol. Soc. Amer. Bull., v. 64, p. 1049–1097.
14. Prucha, J., 1953, Pyrite deposits of Jefferson and St. Lawrence counties, New York: N.Y. State Sci. Serv., R. I. no. 8, 66p.
15. Brown, J. S. and Engel, A. E. J., 1956, Revision of Grenville stratigraphy and structure in the Balmat-Edwards district, northwest Adirondacks, New York: Geol. Soc. Amer. Bull., v. 67, p. 1599–1622.

16. Aldrich, L. T., *et al.*, 1957, Ages of rocks in the Canadian Shield: *in Ann. Rept. Dir.* Geophys. Lab., Carnegie Inst. Washington Year Book 56, p. 105–108.

17. Knight, C. L., 1957, Ore genesis and the source bed concept: Econ. Geol., v. 52, p. 808–817.

18. Engel, A. E. J. and Engel, C. G., 1958, Progressive metamorphism and granitization of the major paragneiss, northwest Adirondack Mountains, New York: Geol. Soc. Amer. Bull., v. 69, p. 1369–1414.

19. Brown, J. S. and Kulp, J. L., 1959, Lead isotopes from Balmat area, New York: Econ. Geol., v. 54, p. 137–139.

20. Doe, B. 1960, The distribution and composition of sulfide minerals at Balmat, New York: Unpublished Ph. D. Thesis, Calif. Inst. Tech., 151p.

21. Buddington, A. F. and Leonard, B. F., 1962, Regional geology of the St. Lawrence County magnetite district, Northwest Adirondacks, New York: U.S. Geol. Surv. Prof. Paper 376, 145p.

22. Doe, B., 1962, Distribution and composition of sulfide minerals at Balmat, New York: Geol. Soc. Amer. Bull., v. 73, p. 833–854.

23. ——— 1962, Relationships of lead isotopes among granites, pegmatites, and sulfide ores near Balmat, New York: Jour. Geophys. Res., v. 67, p. 2895–2906.

24. Engel, A. E. J., 1962, The Precambrian geology and talc deposits of the Balmat-Edwards district, northwest Adirondack Mountains, New York: U. S. Geol. Surv. Open File Rept., 357p.

25. Wynne-Edwards, H. R., 1962, Descriptive Notes, Map 27, Gananoque map-area Ontario, Geol. Surv. Canada, 1:63,360.

26. Brown, J. S., 1963, Personal communication to P. J. Solomon: *in Unpublished Ph. D. Thesis,* Harvard Univ., 162p.

27. Solomon, P. J., 1963, Sulfur isotopic and textural studies of the ores at Balmat, New York, and Mount Isa, Queensland: Unpublished Ph. D. Thesis, Harvard Univ., 162p.

28. Wynne-Edwards, J. R., 1963, Descriptive Notes, Map 7, Brockville-Mallorytown map-area, Ontario, Geol. Surv. Canada, 1:63,360.

29. Leonard, B. F. and Buddington A. F., 1964, Ore deposits of St. Lawrence County magnetite district, northwest Adirondacks, New York: U.S. Geol. Surv. Prof. Paper 377, 259p.

30. Brown, J. S., 1965, Oceanic lead isotopes and ore genesis: Econ. Geol., v. 60, p. 47–68.

31. Bloomer, R. O., 1965, Precambrian Grenville or Paleozoic quartzite in the DeKalb area in northern New York: Geol. Soc. Amer. Bull., v. 76, p. 1015–1026.

# 3. The Benson Mines Iron Ore Deposit, Saint Lawrence County, New York

ROBERT M. CRUMP,* EDWARD L. BEUTNER*

## Contents

## Illustrations

## Tables

* Jones & Laughlin Steel Corporation, Pittsburgh, Pennsylvania.

## ABSTRACT

Benson Mines low-grade iron ore reserve is a replacement deposit within the Grenville gneisses of the Adirondacks. The average grade of the crude ore is about 23 per cent iron. The iron minerals are principally magnetite and hematite disseminated throughout the gneisses as discrete grains about 1 mm in size. Both the ores that are beneficiated by magnetic means and the non-magnetic ores that are treated by a gravity method of concentration are found in stratigraphically different positions but are mined from one open pit, which is about 2.5 miles long and 800 feet wide and are treated separately. The gneisses are described and named according to the distinctive mineralogy of each. Recognition of the gneissic types permits the interpretation of the structure as a northward-plunging syncline overturned to the west. A large subsidiary anticline-syncline complex occurs on the east limb of the major syncline. Iron minerals in sufficient quantity to be considered ore are confined to areas where the footwall rock has been overturned to become the hanging wall.

## INTRODUCTION

The Benson Mines open pit iron ore mine and concentrating plant of Jones & Laughlin Steel Corporation are located in St. Lawrence County, New York, between the villages of Star Lake and Newton Falls. New York State Highway 3 traverses the southern part of the property, and a branch line of the New York Central Railroad serves the area. The present report draws heavily on data obtained from the extensive drilling program that preceded development of the property and was continued for 12 years after mining started.

## ACKNOWLEDGMENTS

The authors are indebted to the Benson Mines personnel for their assistance in assembling the data presented here and to Jones & Laughlin Steel Corporation for permission to publish. Many Jones & Laughlin geologists have contributed to an understanding of the Benson deposit. Worthy of particular mention are L. P. Barrett, retired (Chief Geologist of Jones & Laughlin at the time of the acquisition of the property); S. A. Tyler, deceased (on leave from the University of Wisconsin at the time of his work at Benson); and W. M. Fiedler (presently General Manager of Raw Materials for Jones & Laughlin). These three did much of the early exploration work on the property and directed the later work.

J. E. Tusa of the Jones & Laughlin Ore Research Laboratory contributed the photomicrographs, and Miss Betty Rollings prepared the illustrations. Miss Mary Ammon prepared the manuscript and was invaluable in her constructive criticism. We are grateful for their help.

## HISTORICAL INFORMATION

### Mining Operations in the Area

The history of the Benson deposit goes back to about 1810, when engineers were surveying the site of a military road to connect Albany with Ogdensburg on Lake Ontario. The road is supposed to have passed directly over the ore body, and the effect of the magnetic ore on the engineers' instruments was sufficiently strong to direct their attention to the deposit. Thirty years later, the state geologist of New York mentioned the Benson (then called "Chaumont") deposit in his annual report on mineral resources of the state (1, p. 318).

It was not until 1889, however, when lumbering brought a railroad into the vicinity of the deposit, that any attempt was made to operate the property. At that time Magnetic Iron Company explored the deposit and moved its equipment from its mine at Jayville to the site of the present Benson pit. The mine produced a small tonnage annually for four years, but, in 1893, the depression and the competition of the low-cost Lake Superior ores led to the abandonment of the property (Table I).

In 1900, the property was reopened but only for a short time. It was not until 1907 that the Benson Mines Company was organized and

*TABLE I.   Product Shipments from Benson*

| Years | Gross Tons Magnetite Product Shipped | R/C[1] | Grade | | Gross Tons Non-Magnetic Product Shipped | R/C[1] | Grade | |
|---|---|---|---|---|---|---|---|---|
| | | | Fe | SiO$_2$ | | | Fe | SiO$_2$ |
| 1895–1896 | 25,169 | NA[2] | NA | NA | None | | | |
| 1899–1902 | 87,042 | NA | NA | NA | None | | | |
| 1907–1908 | 8,521 | NA | NA | NA | None | | | |
| 1917–1918 | 61,069 | NA | NA | NA | None | | | |
| Sub-total | 181,801 | | | | | | | |
| 1943–1951 | 5,438,844 | 3.01 | 62.58 | 7.17 | None | | | |
| 1952–1960 | 8,273,685 | 2.94 | 63.24 | 6.32 | 3,767 813 | 3.24 | 61.06 | 6.08 |
| 1961–1965 | 4,888,284 | 2.78 | 64.39 | 5.08 | 1,937,891 | 3.18 | 61.89 | 4.91 |
| Total 1943–1965 | 18,600,813 | 2.93 | 63.34 | 6.24 | 5,705,704 | 3.23 | 61.34 | 5.68 |

[1] R/C equals gross tons of crude ore required to produce one gross dry ton of concentrates.

[2] NA = Not Available.

started open-pit operations. A small production was maintained until 1914, when the impetus of the First World War led to reorganization and expansion of the plant. Production continued until the close of the war, but in 1919 the mine again was abandoned.

Jones and Laughlin Ore Company leased the Benson mineral properties in 1941, and construction of the present plant facilities was started by the Defense Plant Corporation in 1942. Ore shipments to the Pittsburgh furnaces began in February 1944. The holdings of the Defense Plant Corporation were purchased by Jones & Laughlin Steel Corporation in December 1946, and the property was operated as the Jones and Laughlin Ore Company until it merged with the parent corporation in 1952. Since then it has been known as the New York Ore Division of Jones & Laughlin Steel Corporation.

At the time the property was leased, only the magnetic ore was known and estimated as reserves. Subsequent diamond drilling revealed the presence of non-magnetic ore, and, in 1952, a gravity concentrator using Humphrey spiral separators was built to recover the non-magnetic ore. Today, about 40 per cent of the open pit reserves are classified as non-magnetic.

## PHYSIOGRAPHIC HISTORY AND PRESENT TOPOGRAPHY

The Benson Mines property is located in the Adirondack portion of the Grenville Subprovince in the Canadian Shield. The physiographic and topographic features of this area

in relationship to the broader features of the Adirondacks have been well described by Buddington and Leonard (4, p. 7–8). The area lies northwest of the main and highest topographic axis of the Adirondack Mountains in the so-called "Childwold Rock Terrace" as defined by Buddington and Leonard. Elevations in the vicinity of the deposit range between 1400 and 1700 feet. Little River, a tributary of the Oswegatchie River, runs through the property and skirts the ore deposit to the south.

The area was affected by glacial erosion and deposition. Glacial striations indicate that at least the last advance came from about N20°W. The more friable members were eroded in some places to a depth of 60 feet below the more competent gneisses. Ground moraine covers the area, the scattered outcrops accounting for less than 1 per cent of the surface. West of the crusher, surficial oxidation—limonite and hematite stain and martitization—extends to a depth of 300 feet, and the gneisses are quite friable. North of the crusher, oxidation is confined to areas of brecciation, fault zones, and open joints. Such local oxidation has been observed in drill core to a depth of about 300 feet also. The oxidized, friable material west of the crusher apparently escaped glacial erosion because it was protected by the footwall hill that rises to an elevation of about 300 feet above the ore gneiss level. In the area now north of the crusher, the advancing glacier had an unobstructed path and gouged out the friable non-magnetic ore in a trough about 1300 feet long, 200 to 300 feet wide, and 40 to 60 feet deep.

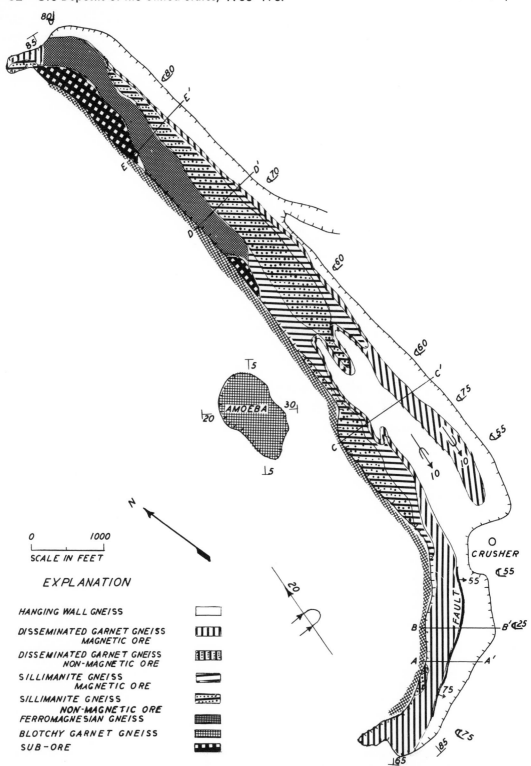

FIG. 1. *Benson Mines Pit—Plan Diagram.*

The oxidation observed in the Benson rocks is interpreted as being preglacial but has not been dated more closely. The amount of oxidation which took place in post-glacial time may have extended to some depth below the surface following fractures and open joints, but it did not affect the solid rocks as did the pre-glacial oxidation. The gneisses north of the crusher, on freshly stripped surfaces, were unaltered by weathering save for about ⅛ inch of limonite encrusting some of the larger areas of magnetite. Outcrops are iron stained to the depth of from 1 to 2 inches, and some have spalled from freezing and thawing conditions.

## GEOLOGIC HISTORY

No attempt will be made here to interpret the complex geologic history of the rocks of the Benson Mines deposit. The gneisses are considered to be a part of the Grenville series, and the mineralized gneisses are interpreted as metasediments. The rocks of the hanging wall are metasediments with some granitic material injected generally conformable with the gneissic banding. The only recognizable igneous rocks within the mineralized zone are pegmatitic; they strike across the banding and are steeply dipping to vertical. These dikes are certainly later than the folding and appear to be later than the mineralization of the Grenville gneisses. Ore emplacement followed igneous injection, metamorphism, and folding.

The age of the Grenville orogenic belt is postulated by Wilson (2, p. 24) to be approximately 1000 m.y. Buddington and Leonard (5, p. 24) cite age determinations on a number of Adirondack crystals that give results ranging from 1094 to 1200 m.y.

## ECONOMIC GEOLOGY

### Structure of the Ore Body

The structure of the ore-bearing horizon of Benson Mines is well known from extensive core drilling, magnetic work, and outcrop mapping in the vicinity. The major structure is a syncline overturned to the west and plunging about 20° to the north in the vicinity of the mine (Figures 1, 2). The west limb of the syncline can be traced by magnetics 1 mile beyond the outcrop of the trough of the syncline. The east limb can be traced magnetically and by outcrop mapping for a distance of about 5 miles beyond the Benson Mines open pit area. The total known strike length of rocks

of similar lithology is, thus, about 8 miles. The Brown's Falls section and the Skate Creek section of the magnetic anomaly are on the east limb that, in these areas, is not overturned (5, Plate I). Diamond drilling for ore on Brown's Falls by Jones & Laughlin and on Skate Creek extensions by the United States Bureau of Mines has confirmed the lithology and structure but did not find sufficient ore to be of interest to Jones & Laughlin.

A subsidiary roll forms an anticline and syncline on the east limb of the major synclinal structure. Both folds plunge south or in a reverse manner to the major structure. These subsidiary structures can be traced for about 4000 feet and are of considerable economic importance because they more than double the volume of rock in this area favorable to ore emplacement. A given unit in the trough of the syncline is now overturned 180°, and the east limb of the subsidiary syncline, where vertical now, is 270° from its original position.

The structure of the Amoeba pit (Figure 1) is believed to be another expression of the roll on the east limb. The Amoeba is inter-

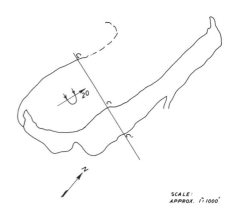

PLAN OF MAJOR STRUCTURE

SCALE:
APPROX. 1" 1000'

SECTION OF MAJOR STRUCTURE

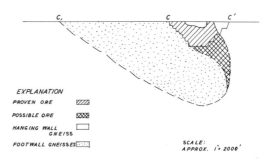

EXPLANATION
PROVEN ORE
POSSIBLE ORE
HANGING WALL GNEISS
FOOTWALL GNEISSES

SCALE:
APPROX. 1" 2000'

FIG. 2. *Plan of Major Structure.*

preted to be the bottom of one of the synclinal rolls that occur in the area. Deep drilling confirms the presence of footwall rock below the mineralized unit, and all of the structure mapping confirms this interpretation. However, this is the only place in the vicinity where this gneiss type is mineralized to a sufficient extent to be ore.

## Explanation of Sections

The numbers above or to the left of the indicated drill holes are the acid soluble iron assays. The numbers under or to the right of the indicated drill holes are the magnetic iron assays. The magnetic iron numbers are that portion of the soluble iron which can be concentrated by the magnetic separators. A portion of this iron may be hematite attached to magnetite, and it is therefore not a true measure of the magnetite present.

The five-foot sample analyses are grouped into zones of similar grade. Grade zones persist for hundreds of feet along the strike and some for the depth of the pit down dip. However, variation in thickness and in averages of the zones is evident from the sections.

The lines of elevation indicate the scale of the sections.

Sections AA′ (Figure 3) and BB′ (Figure 4) illustrate the lack of ore where the hanging wall gneiss is not overturned (Holes 464, 466, and 677). Section AA′ (Figure 3) also contains a section of non-magnetic ore in disseminated garnet gneiss (Hole 466). Section CC′ (Figure 5) illustrates the anticline-syncline complex and the distribution and variation in thickness of the lithologic and ore units. Sections DD′ (Figure 6) and EE′ (Figure 7) illustrate the distribution of the lithologic and ore units and the wedging out of the ferromagnesian gneiss down dip. Magnetic ore is present under the hanging wall on Section EE′ (Figure 7) but disseminated garnet gneiss is absent in Hole 1018 (Figure 7).

Figure 8 presents the data accumulated for each drill hole. Acid soluble iron was run on each 5-foot sample. A Davis tube concentrate was weighed for one out of each three samples. These concentrates were accumulated and an acid soluble iron was run on the pulverized material. All the remaining numbers were derived by calculation.

Details of the structure may be studied on the included maps and sections (Figures 1 through 7). The major features are all due to folding. Faults are present but have quite small displacement. Only strike faults dipping

45° to 70°E are known. One can be traced for a strike distance of about 2000 feet. This is recognizable as a black mineralized gouge between the ore and the hanging wall.

Structural control is inferred from the present position of the ore; the mechanics of the control are not understood. Ore is found in the Benson area everywhere that the stratigraphic footwall has been overturned to become the structural hanging wall. Ore cuts off on the north end of the property where the hanging-wall rock strikes west and dips south and is therefore no longer structural hanging wall. At the south or trough end of the structure, the ore also stops where the hanging-wall rock ceases to be hanging wall. Ore can be traced in depth to the synclinal trough as long as the hanging-wall rock has an overturned attitude. There is not a recognizable change in lithology that could explain why the ore stops where the hanging-wall rock type ceases to be overturned.

The location of the ore cut-off in depth for the northern half of the property is less well known because the synclinal trough plunges far below open-pit depths, and only limited drilling has been done below the proposed open pit. The deeper drilling indicates that the grade of the mineralized rock decreases, but nowhere in the northern area has the bottom of mineralized gneiss been determined.

This pattern of hanging-wall structural control for the ore would be more convincing if it could also be said that mineralized rock occurs only where the hanging wall rock is overturned. This, however, is not true. The magnetic anomaly continues about 5 miles beyond the north end of the Benson open pit, and drilling north of the pit and at Brown's Falls proved no overturning but indicated about 35 feet of ore-grade material on top of the hanging wall gneiss rock. The development of ore minerals here is similar to that in the open pit but is not extensive enough to be considered ore in the economic sense.

An additional structural observation for which no explanation has been developed, but which would appear to have some pertinent significance if understood, is the presence of pink-feldspar pegmatite between ore and hanging-wall rock wherever penetrated by drilling and observed in the pit. This coarse-grained pegmatite ranges in thickness from about 6 inches to more than 3 feet and is present everywhere that there is ore beneath the hanging wall. The age of the pegmatite is best interpreted as essentially contemporaneous with that of the ore and thereby may be genetically

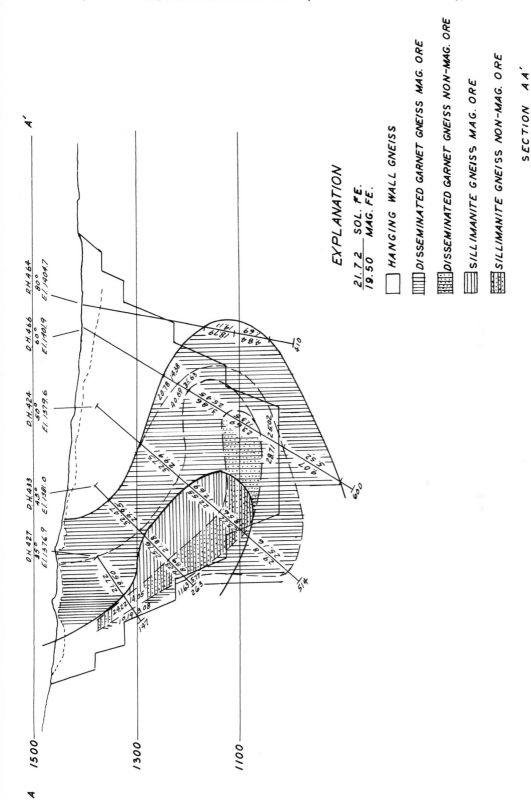

FIG. 3. *Section AA'*.

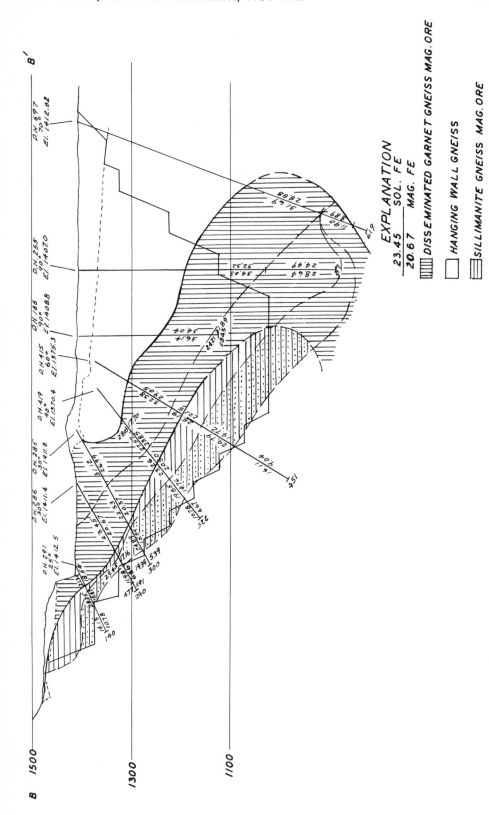

Fig. 4. *Section BB'*.

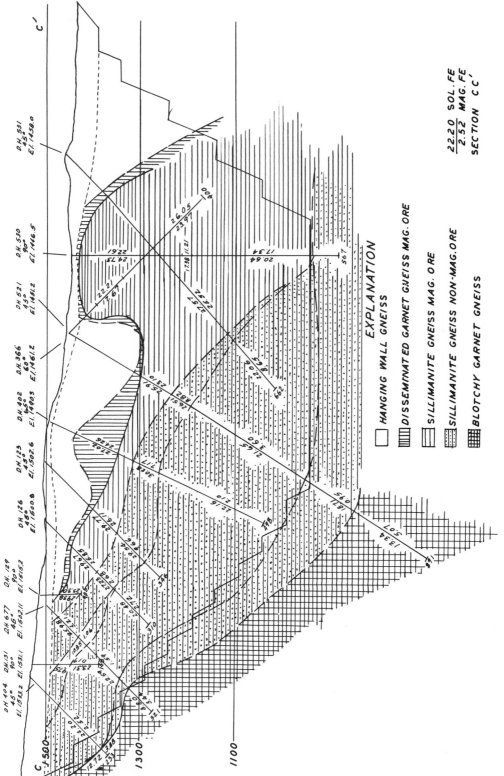

FIG. 5. *Section CC'.*

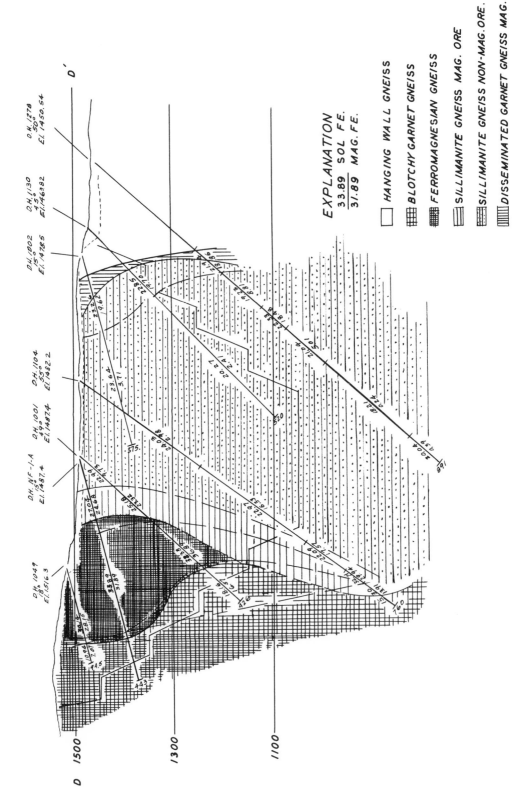

EXPLANATION

$\dfrac{33.89}{31.89}$ SOL FE.
         MAG. FE.

HANGING WALL GNEISS

BLOTCHY GARNET GNEISS

FERROMAGNESIAN GNEISS

SILLIMANITE GNEISS MAG. ORE

SILLIMANITE GNEISS NON-MAG. ORE.

DISSEMINATED GARNET GNEISS MAG.

SECTION D D'

Fig. 6.  Section DD'.

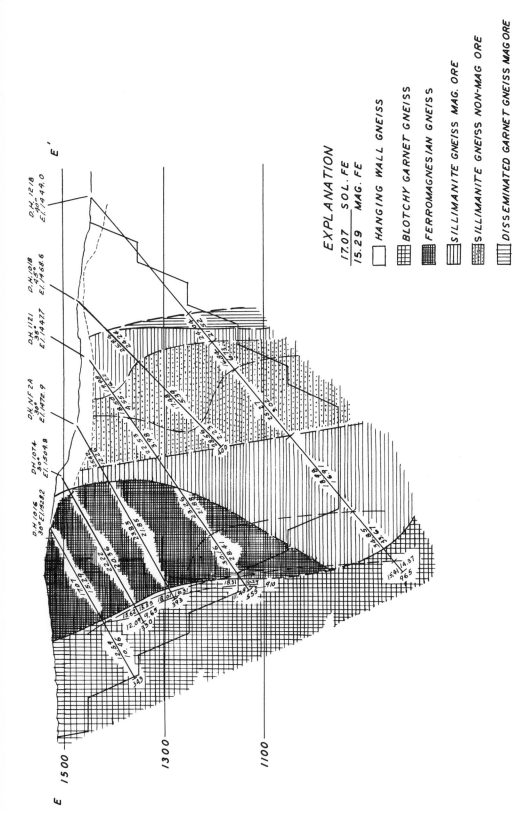

FIG. 7. Section EE'.

## JONES AND LAUGHLIN ORE CO.

### DIVISION OF DIAMOND DRILL SAMPLES OF EASTERN MAGNETITE INTO MAGNETIC & NON-MAGNETIC ORES

MINE _NEWTON FALLS_    STARTED _AUG. 5, 1942_    ANGLE _45°W_    SIZE OF CORE _1⅛"_    HOLE NUMBER _1018_

SEC. NO. _26_ (E E')    FINISHED _AUG. 17, 1942_    ELEVATION _1468.6'_    DRILLED BY _SPRAGUE & HENWOOD_ LOCATION _14,200.5N - 6093.4 E_

| GEOLOGIC DESCRIPTION OF CORE | SOL | SOLUBLE IRON CRUDE | FOOTAGE | MAGNETIC IRON CRUDE | NON-MAG. IRON CRUDE | PERCENT MAGNETIC IRON | TUBE CONC. (WT) | TUBE TAILS (WT) | TUBE CONC. (SOL) | TUBE TAILS (SOL) |
|---|---|---|---|---|---|---|---|---|---|---|
| SAND AND BOULDERS | | | 0-15 | | | | | | | |
| PINK FELDSPAR, BIOTITE GNEISS, WITH | | | 15-80 | | | | | | | |
| AMPHIBOLE AND PYROXENE. | | | | | | | | | | |
| BRECCIATED FELDSPAR PEGMATITE-78-80 | | | | | | | ASSUMED | | | |
| MOTTLED MAGNETITE GNEISS, | 35 | 26.27 | 80-85 | 24.75 | 1.52 | 94.21 | 35.61 | 69.50 | | |
| NO GARNET | 30 | 22.28 | 85-90 | 21.39 | .89 | 96.01 | | | | |
| " | 25 | 32.22 | 90-95 | 31.51 | .71 | 97.81 | | | | |
| " | | 31.24 | 95-100 | 31.12 | .12 | 99.61 | 44.77 | | | |
| MOTTLED MAGNETITE, SILLIMANITE GN | | 30.85 | 100-105 | 30.38 | .47 | 98.48 | | | | |
| MOTTLED MAGNETITE GNEISS | 20 | 27.56 | 105-110 | 26.83 | .73 | 97.35 | | | | |
| " | 15 | 25.63 | 110-115 | 24.66 | .97 | 96.22 | 35.48 | | | |
| MOTTLED MAGNETITE, SILLIMANITE GN | 15 | 27.79 | 115-120 | 26.35 | 1.44 | 94.82 | | | | |
| " BORNITE-124 | 30 | 25.25 | 120-125 | 23.59 | 1.66 | 93.43 | | | | |
| " | 25 | 24.87 | 125-130 | 22.89 | 1.98 | 92.04 | 32.93 | | | |
| " | 35 | 28.86 | 130-135 | 26.38 | 2.48 | 91.41 | | | | |
| " | 25 | 23.13 | 135-140 | 21.00 | 2.13 | 90.79 | | | | |
| MOTTLED MAGNETITE GNEISS | 10 | 26.64 | 140-145 | 24.02 | 2.62 | 90.17 | 34.56 | | | |
| " | 20 | 24.80 | 145-150 | 22.35 | 2.45 | 90.13 | | | | |
| MOTTLED MAGNETITE, SILLIMANITE GNEISS, BORNITE-155 | 20 | 16.12 | 150-155 | 14.52 | 1.60 | 90.09 | | | | |
| MOTTLED MAGNETITE GNEISS. | 35 | 26.33 | 155-160 | 23.71 | 2.62 | 90.05 | 34.12 | | | |
| " | 25 | 19.36 | 160-165 | 15.64 | 3.72 | 80.81 | | | | |
| MOTTLED MAGNETITE, SILLIMANITE GNEISS | 25 | 17.66 | 165-170 | 12.64 | 5.02 | 71.57 | | | | |
| " | 25 | 11.52 | 170-175 | 7.18 | 4.34 | 62.33 | 10.33 | | | |
| " | 15 | 22.70 | 175-180 | 14.85 | 7.85 | 65.41 | | | | |
| " | 30 | 21.49 | 180-185 | 14.72 | 6.77 | 68.49 | | | | |
| " | 35 | 21.73 | 185-190 | 15.55 | 6.18 | 71.56 | 22.38 | | | |
| " | 30 | 18.12 | 190-195 | 13.22 | 4.90 | 72.95 | | | | |
| MAGNETITE GNEISS | 30 | 11.52 | 195-200 | 8.56 | 2.96 | 74.34 | | | | |
| MOTTLED MAGNETITE GNEISS | 25 | 23.33 | 200-205 | 17.67 | 5.66 | 75.74 | 25.42 | | | |
| SOME HEMATITE | 30 | 25.04 | 205-210 | 17.61 | 7.43 | 70.31 | | | | |
| " | 40 | 11.90 | 210-215 | 7.72 | 4.18 | 64.89 | | | | |
| " | 55 | 18.75 | 215-220 | 11.15 | 7.60 | 59.47 | 16.05 | | | |
| " | 45 | 17.52 | 220-225 | 8.20 | 9.32 | 46.80 | | | | |
| MAGNETITE, HEMATITE, BIOTITE, | 45 | 13.51 | 225-230 | 4.61 | 8.90 | 34.12 | | | | |
| SILLIMANITE GNEISS | 30 | 16.51 | 230-235 | 3.54 | 12.97 | 21.44 | 5.09 | | | |
| " PEGM 238-239 | 45 | 17.48 | 235-240 | 4.34 | 13.14 | 24.81 | | | | |
| " PEGM 241-242 | 50 | 20.19 | 240-245 | 5.69 | 14.50 | 28.18 | | | | |
| " | 30 | 20.50 | 245-250 | 6.47 | 14.03 | 31.56 | 9.31 | | | |
| " | 45 | 15.76 | 250-255 | 3.59 | 12.17 | 22.79 | | | | |
| " | 45 | 18.35 | 255-260 | 2.58 | 15.77 | 14.03 | | | | |
| " | 45 | 16.14 | 260-265 | .85 | 15.29 | 5.27 | 1.23 | | | |
| " | 30 | 13.28 | 265-270 | .60 | 12.68 | 4.51 | | | | |
| " | 35 | 18.58 | 270-275 | .70 | 17.88 | 3.76 | | | | |
| " | 40 | 22.26 | 275-280 | .67 | 21.59 | 3.01 | .96 | | | |
| " | 45 | 16.35 | 280-285 | .47 | 15.88 | 2.90 | | | | |
| " | 45 | 18.66 | 285-290 | .52 | 18.14 | 2.79 | | | | |
| " | 25 | 14.91 | 290-295 | .40 | 14.51 | 2.68 | .57 | | | |
| " | 45 | 19.04 | 295-300 | .48 | 18.56 | 2.49 | | | | |
| " | 65 | 13.74 | 300-305 | .32 | 13.42 | 2.31 | | | | |
| " | 40 | 12.67 | 305-310 | .27 | 12.40 | 2.13 | .39 | | | |
| " | 60 | 17.50 | 310-315 | .34 | 17.16 | 1.96 | | | | |
| " | 60 | 13.90 | 315-320 | .25 | 13.65 | 1.78 | | | | |
| " | 30 | 18.12 | 320-325 | .29 | 17.83 | 1.60 | .42 | | | |
| " | 35 | 11.82 | 325-330 | .35 | 11.47 | 2.96 | | | | |
| " | 55 | 17.33 | 330-335 | .75 | 16.58 | 4.32 | | | | |
| MAGNETITE, HEMATITE, GNEISS | 50 | 23.78 | 335-340 | 1.35 | 22.43 | 5.68 | 1.94 | | | |
| " | 55 | 34.00 | 340-345 | 2.02 | 31.98 | 5.92 | | | | |
| " | 30 | 31.33 | 345-350 | 1.93 | 29.40 | 6.16 | | | | |
| " | 40 | 29.09 | 350-355 | 1.86 | 27.23 | 6.39 | 2.68 | | | |
| " | 55 | 27.56 | 355-360 | 1.54 | 26.02 | 5.60 | | | | |
| " | 55 | 22.85 | 360-365 | 1.09 | 21.76 | 4.80 | | | | |
| MAGNETITE, MUSCOVITE, | 45 | 28.77 | 365-370 | 1.15 | 27.62 | 4.00 | 1.65 | | | |
| HEMATITE GNEISS | 35 | 30.87 | 370-375 | 2.36 | 28.51 | 7.63 | | | | |
| " | 35 | 27.94 | 375-380 | 3.15 | 24.79 | 11.26 | | | | |
| " PEGM 383-384 | 40 | 37.07 | 380-385 | 5.52 | 31.55 | 14.89 | 7.94 | | | |
| " | 35 | 26.56 | 385-390 | 3.04 | 23.52 | 11.42 | | | | |
| " | 40 | 28.25 | 390-395 | 2.25 | 26.00 | 7.96 | | | | |
| " | 30 | 18.44 | 395-400 | .83 | 17.61 | 4.50 | 1.20 | | | |
| COMPOSITE | | 21.75 | 80-400 | 9.89 | 11.86 | 45.47 | 14.26 | | | |

GRAPHIC PERCENTAGE — MAG. IRON / NON-MAG. IRON: 0 10 20 30 40 50 60 70 80 90 100

PERCENT MAG. FE

PERCENT NON-MAG. FE.

| | SOL. IRON | FOOTAGE | MAG. IRON | PHOS. | SILICA. | MANG. | ALUM. | TI. | SULPH. |
|---|---|---|---|---|---|---|---|---|---|
| COMPLETE ANALYSIS | | | | | | | | | |

Fig. 8. _Drill Hole Data._

related to it. In places, pegmatitic pink feldspar, associated in space with the pegmatite, is found both in the ore gneiss and in the hanging wall gneiss. Pegmatitic stringers in the ore would make it appear that at least some of the pegmatite is younger than the ore. In other places, the pegmatite contains enough magnetite, in grains of size comparable to the pegmatite minerals, to be of ore grade and fairly convincing evidence that the pegmatite is older than the period of ore emplacement unless magnetite also was a component of the pegmatite melt. If, in fact, pegmatite is both older and younger than the ore as interpreted from these observations, a contemporaneity of age is indicated, and the pegmatite came from the same source-magma as the mineralizing solutions and followed a structural break between present ore gneiss and hanging wall gneiss. This speculation has not been pursued but might be amenable to a trace-element study. The principal minerals of the pegmatite are microcline, perthite, and quartz.

Again, the mechanics of the process are difficult to visualize. All the ore in the mine area is under the pegmatite, but a genetic relationship seems to be indicated.

## Lithology

GENERAL MINERALOGY The Benson gneisses contain a wide variety of minerals. Feldspar, sillimanite, garnet, quartz, hornblende, pyroxene, biotite, magnetite, and hematite are the principal rock-forming minerals. Apatite, zircon, titanite, spinel, and monazite are the most abundant accessory minerals. The first three minerals mentioned above are of special interest because they have been useful in differentiating lithologic units and establishing certain relationships of the units to the ore.

Some of the magnetic ore areas contain abundant pyrite, and there is one zone, rich in pyrrhotite, that cuts across the banding of the gneisses. Copper occurs very sparsely and sporadically as native copper, chalcopyrite, chalcocite, bornite, covellite, and the carbonates, azurite and malachite. The copper minerals do not appear to be related to the pegmatite dikes. Molybdenite occurs both sparsely disseminated in some of the ore gneiss and as large flakes and crystals in some of the pegmatites. Fluorite, calcite, muscovite, biotite, and octohedral pyrite are the most common minerals in the pegmatites except for quartz and feldspar. An unidentified amber-colored hydrocarbon has been observed in drill core and in the broken ore. Limonite and martite are common in some portions of the mine. The penninite variety of chlorite has been identified (6) from an ore sample particularly high in $TiO_2$. Rutile and ilmenite occur sparsely as ex-solution laths in magnetite and hematite.

Many of the hematite grains contain an unctuous film on the cleavage surfaces. Refractive indices and X-ray analyses indicate the film is a mixture of illite and kaolinite (6). Zoisite, scapolite, calcite, and sericite have been recognized microscopically.

No detailed paragenetic sequence has been worked out for the minerals. The magnetite and hematite are believed to have replaced

gangue minerals, but there is no clear evidence that this is the case. The sulfides, in general, are in zones cutting across the gneissic banding and, in polished sections, appear to be later than the iron oxides.

FELDSPARS A limited amount of optical data has been obtained on the feldspars using the five-axis universal stage and the X-ray precession method (3). X-ray analysis indicates that the untwinned feldspar of the magnetic ore corresponds to an orthoclase that is slightly triclinic geometrically. The twinned feldspar of the non-magnetic ores is an intermediate microcline twinned by the albite and pericline laws according to X-ray data.

Microscopic determinations of the optical properties of the feldspars from non-magnetic ore were less precise than desirable because the strain shadows and undulatory extinction interfered with accurate orientation. Typical analyses are as follows:

|  | Feldspar | |
| --- | --- | --- |
|  | From Non-Magnetic Ore | From Magnetic Ore |
| $\alpha$ | 1.5188 | 1.5195 |
| $\beta$ | 1.5228 | 1.5233 |
| $\gamma$ | 1.5241 | 1.5245 |
| 2 V Calculated | $-60°$ | $-59°$ |
| F-C | 0.0073 | 0.0064 |

The minor element content has not been investigated.

The feldspars of the Adirondack iron ore deposits are of great academic interest and are a fertile field for scientific research. A. F. Buddington and B. F. Leonard (5, p. 45) noted a phase change in the potassic feldspar: "Where the quantity of iron oxides reaches about 10 to 12 per cent microcline disappears. . . . Instead we find a clear, untwinned potassic feldspar that very rarely shows a 'shadowy' or undulatory extinction and never shows grid structure."

At Benson, there may also be a phase change, but the relationships are quite different from those described by Buddington and Leonard (5, p. 45). The magnetic ore contains a green or colorless untwinned potassic feldspar. The non-magnetic ore contains pink or colorless microcline with shadowy and undulatory extinction.

Much of the twinning of the microcline and strain shadows in the feldspar and quartz appear to have been initiated at the contact with hematite grains (Figure 9A). Because these phenomena are absent in the magnetic ore, it is reasoned that there may be a genetic relationship between the strain in the feldspar and quartz and the presence of hematite.

The microcline of the hanging wall gneiss and the adjacent pegmatite has good grid structure, and undulatory extinction has not been observed in the microcline of these units.

The Benson feldspars also served a practical role in the exploration and development drilling and mapping. It was recognized early that the ore gneiss contained predominantly potassic feldspar and the hanging-wall gneiss had plagioclase as the principal feldspathic constituent. With a ten-power lens, it was possible to pick the probable contact beween the two rock-types even though ore minerals were lacking and the appearance of the two rocks was otherwise similar. This was helpful when drilling at depth for the hanging-wall rock where it was not overturned and for the west limb margins near the southern limits of the deposit. It also was used for outcrop mapping along the sparsely mineralized west limb.

SILLIMANITE   Sillimanite occurs in all units of the gneisses studied—from hanging wall through blotchy garnet gneiss. It ranges from very tiny crystals to sizes 0.25 inches in cross section by 3 to 4 inches in length. Much of it is altered to a mixture of sericite and chlorite. It almost always has well-developed prisms and appears needle-like. Some has planar orientation, some cuts across the banding (random orientation), and some is in veinlets across the gneissic banding. Some sillimanite is found as poorly developed crystals rimming the iron-oxide minerals. Magnetic ore and zones near contacts with non-magnetic ore contain the largest concentrations of sillimanite. Fine, disseminated sillimanite, generally oriented parallel to the foliation of the gneiss, is common in the non-magnetic ore. One zone in the footwall area consists of knots of fine-grained sillimanite, generally with a pink-stained feldspar.

GARNET   By composition, the garnet is a mixture of almandite and spessartite. It occurs in the hanging wall and, locally, is common to abundant in the high-quartz phase next under the hanging wall; it is a common but erratic constituent of the magnetic gneiss adjacent to, or perhaps a part of, the ferromagnesian gneiss. It is common in the ferromagnesian gneiss. Large grains or knots are common con-

stituents in the blotchy garnet gneiss. In some places, it rims magnetite.

SUBDIVISIONS OF THE LAYERED ROCK UNITS   Six layered rock units, each distinguished by characteristic minerals, occur within the mine area (Figure 1 through 7). As indicated in the earlier discussion on structure, the Benson Mines ore body is localized in the east limb of a syncline that is overturned to the west. The upper or hanging-wall unit in the mine would thus be the lower unit in the west-limb structure. In this paper, designation of sequence is structural. From upper to lower, the names applied to the units are:

(1) Hanging-wall gneiss.
(2) Disseminated-garnet gneiss—zero to 350 feet—average about 150 feet.
(3) Sillimanite gneiss—zero to 650 feet—average about 350 feet.
(4) Ferromagnesian gneiss—zero to 280 feet—average about 200 feet.
(5) Blotchy garnet gneiss—maximum 300 feet—poorly known.
(6) Biotite gneiss.

The minerals from which the compound names are derived are not considered sufficiently unique to prove a continuity of original composition because additions and subtractions of material during metamorphism could have altered the final mineral compositions. The units as defined serve the practical purpose of providing a continuity of descriptions through a succession of geologists and have also emphasized the desirability of attempting to explain the distribution of magnetic and non-magnetite ore within units of differing mineralogy.

Of these six units, numbers (2), (3), and (4) contain important portions of the ore body; some ore occurs in number (5) in the Amoeba body.

Differences within units are great and complicate the identification of the units found as isolated outcrops above glacial drift. Sequence is relied upon heavily for identification of the units within the mine.

*Hanging-Wall Gneiss*   This is a heterogeneous rock some of which is igneous, some metasedimentary, and most of it not certainly recognizable as either. Quartzite, marble, diopside rock, and hornblendite are common as bands 5 to 50 feet thick. It is rarely that such units can be traced laterally between drilled sections. The bulk of the rock is a well-banded hornblende and/or pyroxene gneiss either pink or gray and having granitic texture. The hornblende gneiss can be traced for the length of

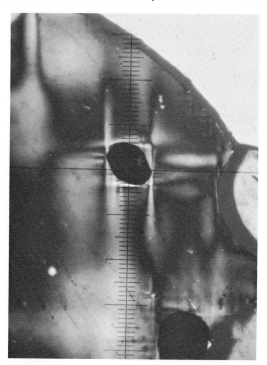

*A-HEMATITE IN MICROCLINE*

*120 μ*

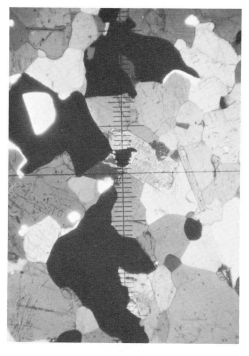

*B- MAGNETITE ORE*          *C- HEMATITE ORE*

Fig. 9A. *Hematite in Microline.* 9B. *Magnetite Ore.* 9C. *Hematite Ore.*

the anomaly with its banding parallel to the ore and hanging-wall-gneiss contact. The immediate hanging wall is invariably a pink microcline pegmatite. The thickness of this unit differs, but it is present in every drill hole cutting the ore-hanging wall contact.

The quartzite in this unit generally is glossy gray in color but takes on brownish and greenish tints where varying amounts of biotite and diopside are included with the quartz. The auxiliary minerals give the rock an appearance of bedding that is parallel to the gneissic banding of adjacent units. Plagioclase (albite or oligoclase) generally is present and may amount to more than 50 per cent of some interbedded layers. The bedded character and the various mineralogies of the rock are believed to be related to original composition. However, if it is truly a quartzite, it is so thoroughly recrystallized that all traces of detrital characteristics of the grains have been eradicated.

Marble is less common than the quartzite and nowhere is found as a traceable unit. It consists of coarse-grained, white and pink calcite with granular diopside and biotite common in some of it.

Hornblendite and a rock composed almost wholly of diopside are known only from isolated drill holes. They are associated with, and appear to grade into, the marble in some places.

The pink and gray granitic gneisses make up the bulk of the hanging-wall gneiss encountered by drilling in the pit area. Quartz, orthoclase, microcline, perthite, and oligoclase are common to almost all of the different phases of the gneiss. Biotite, muscovite, hornblende, pyroxene, and chlorite are present in various amounts and emphasize the gneissic appearance of the rock. Garnet and sillimanite are present locally near the contact with ore. In general, the gneiss is of granitic texture and composition, but some phases are so low in quartz as to be sysenitic. No plagioclase more calcic than oligoclase has been found. The common accessory minerals are apatite, zircon, leucoxene, magnetite, and hematite.

The pink microcline pegmatite at the ore contact ranges widely in thickness and in its content of minerals other than the microcline and quartz. Books of muscovite and biotite are common. Magnetite, similar in size to the other minerals, is present in some areas in irregular patches.

The contact between the microcline pegmatite and the disseminated-garnet gneiss is quite sharp from the mining standpoint. There is, however, a contact zone 1 to 2 feet wide within

which grade determines which portions are mined and which are left in place. Toward the ore from this contact, there is a decreasing amount of pegmatitic pink microcline mixed with the disseminated-garnet gneiss. Beyond this contact toward the hanging-wall gneiss, patches of magnetite may be found in the pegmatite.

*Disseminated-Garnet Gneiss* The disseminated-garnet gneiss is absent locally, attains a thickness of 350 feet in some places, and averages about 150 feet thick. Near the hanging wall contact, it is uniformly high in quartz and, in places, is a quartzite with much magnetite and a minor amount of disseminated garnet and biotite. The common feldspar is orthoclase, but, near the hanging-wall contact, some oligoclase and microcline may be present. In general, it is a potash feldspar-quartz gneiss with grains of red garnet disseminated throughout. Pyroxene, hornblende, chlorite, muscovite, sillimanite, and biotite are present locally. Apatite, zircon, pyrite, and titanite are common accessories. This unit carried the highest-grade ore in the southern half of the ore body, and its ore is predominantly magnetic. Some non-magnetic ore (due to presence of martite, hematite, and limonite) occurs within this zone.

The disseminated garnet defines the limits of this unit in our present classification. For much of the strike length, magnetic ore extends from the hanging wall, through the disseminated garnet gneiss, and for many feet into the next unit (sillimanite gneiss).

*Sillimanite Gneiss* The name "sillimanite gneiss," is a good field term because sillimanite is almost everywhere present in the outcrops of this unit in the vicinity of the mine. Within the pit, diamond drill core and cut faces, however, demonstrate that this gneiss has thick sections without sillimanite. Also, nearly every phase of the Grenville gneisses in the mine area may carry some sillimanite. The sillimanite gneiss is distinguished from other units not so much by its sillimanite content as by its lack of ferromagnesian minerals.

The average thickness of the sillimanite gneiss is about 350 feet; it attains a maximum thickness of 650 feet and, toward the north end, is absent locally. It is characteristically a high-orthoclase or high-microcline, low-quartz rock, with magnetite and hematite. Biotite, chlorite, and muscovite are present locally and some areas have a little garnet near the contact with the next higher unit. The sillimanite gneiss carries the bulk of the non-magnetic ore, and, within this unit, some of the ore contains less than 1 per cent magnetite.

Apatite, zircon, titanite, and leucoxene are common accessories.

The mineralogical and physical characteristics of the sillimanite gneiss differ with the percentage of non-magnetic iron present. These variations will be considered under the discussion of the ores.

The texture of the sillimanite gneiss ranges from aplitic to pegmatitic with most of the rock having a fairly uniform granitic texture. The pegmatitic areas are elongated parallel to the gneissic banding but have irregular boundaries against adjacent units. They are interpreted as replacement pegmatites. Fine-grained phases are fairly uniform in thickness and may be traced for long distances along the strike.

*Blotchy Garnet Gneiss*  This rock forms the footwall of, and occurs as a continuous unit along, the strike length of the Benson-Mines ore body. In the northern half, it succeeds the ferromagnesian gneiss, and, in the southern half where ferromagnesian gneiss is absent, it follows the sillimanite gneiss. In the main pit area, concentration of ore minerals nowhere extends more than a few feet beyond the contact with blotchy garnet gneiss; however, it is the host rock for ore in the Amoeba area. Blotchy garnet gneiss is known from outcrops and from several long, flat diamond drill holes that explored the footwall. Maximum thickness of the known section is about 300 feet.

Garnet is characteristic of this unit, which derives its name from the habit of the garnet to occur in knots or blotches. The rock is more coarse grained than any of the ore-bearing gneisses. It should be emphasized that the descriptions apply to the rock units and not to single specimens. No specific characteristic has been discovered that makes it possible to classify certainly a specimen or even an outcrop as disseminated garnet gneiss or blotchy garnet gneiss. The mineralogy of one is very similar to that of the other, and the variations in each are great enough so that characteristics of the two overlap.

In general, the blotchy garnet gneiss is a light gray, coarse-grained rock with orthoclase, quartz, and biotite as the chief constituents. Microcline and labradorite have been recognized microscopically but are very rare. Pyrite, pyrrhotite, and sillimanite are common, and magnetite is more abundant than in the hanging-wall gneiss. Apatite and zircon are the common accessories with a little titanite and/or leucoxene.

*Ferromagnesian Gneiss*  The ferromagnesian gneiss is a lens-shaped unit present for only a limited strike length of about 3300 feet with an average width of about 200 feet. Diamond drilling has established the limits by proving it wedges out in depth. It is confined to the northern half of the ore property.

The high-grade magnetic ore of the northern part of the ore body occurs in this unit. It is characterized by an abundance of hornblende and pyroxene where magnetite is sparse. The rich (30 to 40 per cent soluble iron) sections carry correspondingly less of the ferromagnesian minerals. Orthoclase and quartz commonly make up about 25 per cent of the rock. Scapolite is recognized in some thin sections, and labradorite is found rarely. Garnet occurs sporadically, being plentiful in some areas where it generally is coarser than in the disseminated garnet gneiss. It rarely forms the knots or blotches so common in the blotchy garnet gneiss. Zircon, spinel, and apatite are the common accessory minerals. Pyrite and pyrrhotite are abundant locally and generally are present in small amounts. Sillimanite occurs in some places.

A magnetic zone, without ferromagnesian minerals other than biotite but containing garnets locally, generally is found between the ferromagnesian gneiss and the sillimanite gneiss. It is mapped with the sillimanite gneiss because sillimanite is commonly present and because hornblende and pyroxene are absent. The designation is arbitrary and open to criticism. It has been pointed out that the contact between disseminated-garnet gneiss and sillimanite gneiss is drawn at the horizon where garnet is absent. The upper boundary of the sillimanite unit is drawn below the point where hornblende or pyroxene is first recognized although garnet may be present.

*Biotite Gneiss*  This unit is poorly known because it is not mineralized, and there has been little reason for drilling it. It has been studied in outcrop and in two long holes drilled to release an area for dumping waste rock. Biotite is found throughout the unit but is never conspicuous or abundant. The rock is medium to fine grained, composed mostly of feldspar and quartz. Disseminated garnet, sillimanite, chlorite, and hornblende are found locally. Plagioclase, oligoclase to andesine, is present throughout but much of the feldspar is orthoclase or microcline. Phases of the hanging-wall gneiss are indistinguishable from this rock, and classification depends to a large extent upon position.

The units described above are, in general, parallel to the hanging wall-ore contact and parallel to the general banding. They serve as lithological units, but the extent to which present lithology reflects composition of sediments prior to metamorphism is not known.

Original composition of the sediments must have been altered by additions and leaching due to the igneous activity, and metamorphism need not have been equally intense throughout. The extreme thickening and thinning of the disseminated garnet gneiss seems better explained by metamorphism than by original composition. Perhaps much of the unit marked "Sillimanite Gneiss" here would be better mapped as disseminated-garnet gneiss. Biotite in the sillimanite gneiss could have been metamorphosed to garnet. It is concluded that the ore gneiss, here divided into four lithologic units (counting "blotchy garnet gneiss" as ore gneiss also) was originally a thick unit of clay sediments that differed appreciably in composition, particularly in the amount of bases present. The ferromagnesian gneiss must have been quite high in calcium and magnesium. Original composition was sufficiently uniform and metamorphism was varied enough so that it is now very difficult to map the original units accurately.

The boundaries between defined units are gradational everywhere. The disseminated garnet-hanging wall gneiss contact is the sharpest, but even here it is not possible to draw an irrefutable line at the contact. The lithology is described and designated without rigid classification, and no exact boundaries can be drawn. The classification of the ore gneisses is of importance in indicating the great length over which units of similar description may be traced and is of economic interest because the magnetic and non-magnetic ore-types are confined, in general, to specific lithologic units.

### The Ore Minerals and Ore Classification

GENERAL  Magnetite and hematite are the essential ore minerals of the Benson-Mines ore deposit. Their distribution is such that two types of ore can be mined and milled separately. This distribution, based on recovery efficiency of the magnetic and gravity separations and upon the profit margins for the two products, is outlined on the included maps and sections (Figures 1 through 7). If, upon analysis, a block of ore is found to be 80 per cent or more magnetic, the ore is concentrated by magnetic separation. If less than 80 per cent of the iron is magnetic, the ore is concentrated by gravity, with auxiliary magnetic separation to collect the fine magnetite or flotation for fine magnetite and hematite. In present practice, 80 per cent is an average figure and changes in both directions depending upon the variation of the soluble iron analysis from the

average. Characteristics of the two types of ore are described separately although it will be realized that the types are defined on an economic rather than on a scientific basis. There is a complete gradation from plus 99 per cent magnetic iron to less than 1 per cent magnetic iron in the ore.

The distribution of the two types of ore appears to be related to lithologic units. Magnetic ore occurs in the disseminated-garnet gneiss, in the sillimanite gneiss, in the ferromagnesian gneiss, and in the blotchy garnet gneiss. Non-magnetic ore occurs principally in the sillimanite gneiss. The distribution of the ores is a very important consideration economically because of bulk mining requirements, but the many exceptions to these generalizations may be even more important from the standpoint of interpretations of the geology. Almost as much of the sillimanite gneiss is classified as magnetic ore as is classified as non-magnetic. Non-magnetic ore is found in places in the disseminated-garnet gneiss.

The average grade of the ore reserve is a function of economics rather than occurrence. The non-magnetic ore averages about 23 per cent acid-soluble iron and the magnetic ore about 22 per cent. However, marginal to the reserve—in depth and on the footwall—the ore minerals continue gradationally beyond the assay cutoff to values below 5 per cent iron. The bulk of this material is classified as non-magnetic and, if all mineralization in the Benson area were included, the average for the non-magnetic material would be considerably below 23 per cent iron. This is true to a lesser extent for the magnetic material because the cutoff on the magnetic footwall and in depth is sharper—less gradational—and the bulk of the lean magnetic material is therefore more limited. Except for selected hand specimens, none of the material could be considered high enough grade for profitable present-day use without benefication.

Microscopic studies indicate that, of all the host-rock minerals present, only the sillimanite was not replaced somewhere. The sillimanite has a distinctive needle-like shape with the length many times the thickness. Grains of magnetite and hematite have not been observed with such a shape. All other host rock minerals are without recognizable external form and were subject to replacement with possibly a preference for the ferromagnesian minerals. Thin sections through rich ore specimens seem to retain the various gangue minerals in approximately the same proportions as in specimens that are much lower in iron or are barren. Rich specimens do not consist of magnetite and/or hematite with the predominant gangue mineral garnet or quartz or feldspar.

MAGNETIC ORE    The distribution of the magnetic ore may be seen on the plan map and sections (Figures 1 through 7). One continuous strip of magnetic ore lies adjacent to the hanging wall for the entire strike length of the mine. The second major zone corresponds closely to the ferromagnesian gneiss. On the north end of the ore zone, where the gneissic banding changes strike from north to west, all of the ore is magnetic. Non-magnetic ore is found again on the footwall side for the last 700 feet. The Amoeba, in blotchy garnet gneiss, is all magnetic ore. The magnetic ore differs in mineralogy and other characteristics depending upon the lithologic unit within which it occurs. Each will be considered separately.

The magnetite ore band adjacent to the hanging wall includes all of the disseminated garnet gneiss, except locally, and for most of the strike length includes some of the sillimanite gneiss unit. Magnetite is disseminated throughout the rock, but the ore has recognizable banding and some bands are much richer in magnetite than others. Rich bands grade into leaner bands rather than occurring with sharp contacts or mica partings such as are common in the Clifton deposit. Although some bands as wide as 2 feet may persist for strike lengths of more than 100 feet with a grade above 50 per cent iron, these are not common and the richer bands are not selectively mined. All information on ore grades is obtained from analyses of diamond drill core samples split into 5-foot lengths. Details of the analyses can be seen on the sample drill sheet (Figure 8). The variations would be greater for smaller samples. A natural grouping of the analyses proves a zone of plus 25 per cent recoverable iron immediately adjacent to the hanging wall and a zone of minus 25 per cent recoverable iron succeeding it. This is true from the south end of the pit to nearly the north end of the anticline. This natural division is much less marked north of the anticline where the hanging-wall magnetite zone is much thinner and steeper.

Grain size in this zone differs considerably; none is extremely fine, however, or very coarse. Magnetite grains are similar in size to the gangue minerals (Figures 9B and 9C). The minus 20-mesh grind releases essentially all of the magnetite. This ore characteristically breaks between grains (within the area mined to date) and concentrates easily. Only the tiny accessory grains of magnetite are euhedral, but the coarser material may have well developed octahedral parting.

Magnetic ore in the sillimanite gneiss zone is adjacent to and continuous with the disseminated-garnet gneiss zone. Ferromagnesian minerals characteristically are sparse or absent, and the gangue tends to be high in potash feldspar and low in quartz. The contact with non-magnetic ore, to the west, is gradational and is not marked by any recognizable lithologic change other than by the increased percentage of hematite, the increase in friability, and the change in the optical properties of the feldspars. The transition between magnetic and non-magnetic ore in the vicinity of the ferromagnesian gneiss is generally within sillimanite gneiss but in places is at or near the contact of the two rock types.

The ferromagnesian magnetite unit occurs only in the north half of the open pit area. This has rich zones and lean zones, but it is without the continuity of the zoning in the disseminated-garnet ore. There is, however, a fairly well-defined rich ore shoot that plunges to the north and generally is adjacent to the footwall blotchy garnet horizon.

This unit provides ample opportunity for comparing the unmineralized gneiss with the ore. The ferromagnesian gneiss on the upper cuts has so little magnetite that it must be stripped and dumped as waste. If the same gneissic band is followed downward, it changes to high-grade ore. Mineralogically, the ore at depth has more magnetite and less hornblende and pyroxene. The percentage of hornblende and pyroxene differs within the unmineralized gneiss to such an extent that it is difficult to generalize on the relation of the magnetite to the feldspar and quartz. The richest ore has a gangue of feldspar and quartz only; as the amount of magnetite decreases more and more of the gangue is composed of ferromagnesian minerals. Green spinel, recognized only in thin sections and there as small grains associated with magnetite, is a common accessory mineral. It is less abundant in other units.

The ferromagnesian zone material does not have the characteristic of breaking between grains, but a good concentrate can be made without grinding finer than minus 20 mesh.

Amoeba magnetite ore is located in a nearly elliptical area of about 700 feet by 1400 feet in the footwall zone. The name "Amoeba" was applied to this isolated block of ore because of the appearance of the contoured dip needle anomaly, which consists of a scattering of highs and lows so that the contours spread out very irregularly. The irregularity of the anomaly is believed due to the nearly horizontal attitude of the magnetic ore. Lithologically, the ore appears to be of the blotchy garnet type and has the coarsest grain size of any of the magnetic ore. It is, however, a very tough horizon and breaks across the grains so that finer grinding is required for a grade of concentrate comparable to that from the other two magnetic horizons.

NON-MAGNETIC ORE   The non-magnetic ore is defined by economics on the basis of recovery in the two mills and by the profit margin for the two products. In general, ore which contains less than 80 per cent magnetic iron is beneficiated by gravity methods and produces a concentrate with about 2 per cent $TiO_2$ and 0.4 per cent phosphorus.

The non-magnetic ore for the northern half of the property is confined between the hanging wall magnetic ore and the footwall or ferromagnesian magnetic ore. The southern portion lacks magnetic ore on the footwall, and the non-magnetic ore extends to the assay footwall. The ores are designated by the analyses of the split drill core. All contacts with magnetic ores are gradational, and, because of structure, blocks of ore near the contacts are necessarily mixtures of the two.

Zones of richer ores and leaner ores may be traced along strike and down dip, just as in the case of the magnetic ore. The magnetite and hematite grains are the same size and shape as the gangue grains and sufficient liberation is developed at a 14-mesh grind to yield about a 62 per cent iron concentrate.

The bulk of the non-magnetic ore is within the sillimanite gneiss but a portion is within the disseminated-garnet gneiss. None of the ferromagnesian gneiss is classified as non-magnetic ore. Metallurgically, the general absence of ferromagnesian minerals in the non-magnetic ore gneiss allows a gravity separation. Garnet has a sufficiently high gravity to be concentrated with the metallics when the disseminated-garnet gneiss non-magnetic ore is being milled.

A distinctive feature of the non-magnetic ore is its friability. No quantitative data are available, but there appears to be a direct correlation between the relative proportions of hematite and magnetite and the degree of friability. The ore that is less than 5 per cent magnetic is as friable as an uncemented sandstone. As the proportion of magnetite increases, the ore becomes more massive and tougher and increasingly difficult to crush.

The following table compares and contrasts the salient features of the two ore types. However, these features overlap somewhat, perhaps because the sum of the magnetic plus non-magnetic iron has a considerable range, and this also has an influence on the characteristics of the rocks.

The hematite of the non-magnetic ore megascopically and microscopically is typical specularite. It has good basal parting with well developed polysynthetic twinning in most grains. Observed needles of ilmenite and rutile in general parallel the (0001) direction. The minor martite that has been observed—interpreted as martite because of well-developed octohedral parting—occurs in areas high in limonite and almost surely is of surficial oxidation origin and much later than the bulk of the hematite.

If the hematite is a martitization product of magnetite, it has undergone a recrystalliza-

TABLE II.   *Summary of Ore Characteristics*

| Magnetic | Non-Magnetic |
| --- | --- |
| (1) Generally difficult to crush. | Friable. |
| (2) Distribution generally confined to lithologic horizons. | Same. |
| (3) Occurs in all the lithologic horizons. | Not found in ferromagnesian gneiss and only to minor extent in disseminated garnet gneiss. |
| (4) Generally high quartz gangue. | Generally low quartz gangue. |
| (5) Sulfides present. | Sulfides absent but has many limonite-filled vugs of a size corresponding to sulfides. |
| (6) Titanite present. | No titanite but some leucoxene. |
| (7) Contains green or colorless untwinned potash feldspar. | Contains pink or colorless twinned and strained potash feldspar. |
| (8) Rare labradorite grains in ferromagnesian gneiss ore. | No plagioclase. |
| (9) Has fairly uniform gneissic structure and texture. | Ptygmatic folding and pegmatitic lenses are quite common. |
| (10) Average grain size > 1 mm | Same. |

tion from its cryptocrystalline state to its present appearance of specularite. It would be futile to speculate about the occurrence of such an event and the mechanics of such a process unless it were known that the hematite did in fact have an oxidation origin.

Certain characteristics of the non-magnetic ore suggest that the hematite may have been derived from oxidation of magnetite. If the magnetite were altered to hematite by the addition of oxygen there could be an increase in volume of the iron oxide mineral exceeding 4 per cent. This might explain the correlation of increasing friability with increasing proportions of hematite in the ore. It might also explain the strain shadows in quartz which is adjacent to or includes hematite and the shadowy extinction and curved twinned planes in the "microcline." These features are absent from the magnetic ore, which carries little hematite.

An oxidation origin for the hematite is also compatible with the pink or colorless feldspar in the non-magnetic ore and the green or colorless feldspar with the magnetic ore. It could also explain the lack of sulfides in the non-magnetic ore.

For lack of criteria with which to prove that the hematite is really martite, the above observations are of limited significance and certainly are insufficient to prove an oxidation origin for the hematite. As stated before, the hematite is equally as convincing as specularite as the magnetite is as magnetite. They are both without regular external crystal form.

TITANIUM IN ORE Benson is considered a non-titaniferous iron ore deposit. However, the gravity separation by Humphrey spirals concentrates the $TiO_2$ to about 2 per cent by weight of the concentrate and at this percentage is undesirable in the blast furnace and steel making facilities. Because it creates a metallurgical problem, there has been much emphasis placed on the occurrence of $TiO_2$, and it deserves a special section.

The crude ores classified as magnetic average about 0.75 per cent $TiO_2$, from available analyses, and range from 0.37 to 1.28 per cent $TiO_2$. In the nonmagnetic crude ores the extremes are less wide and average 0.85 per cent $TiO_2$. It has been established that most of the $TiO_2$ is associated with the hematite and that the lower the percentage of hematite in an ore sample, the higher will be its content of $TiO_2$. The non-magnetic iron concentrated by laboratory techniques from an ore that is 98 per cent magnetic will be in the range of from 10 to 14 per cent $TiO_2$. The comparable non-magnetic concentrate from an ore that is only 2 per cent magnetic will be about 2 per cent $TiO_2$. The gravity concentrates from ores low in iron (from 12 to 15 per cent) tend to be higher in $TiO_2$ than concentrates from rich ores because additional crude ore is required for a ton of concentrate and the $TiO_2$ is concentrated from these additional tons. Also, about 30 per cent of the $TiO_2$ contained in a magnetite concentrate at the normal 14 mesh grind can be eliminated by grinding through 325 mesh and reconcentrating by magnetic means. More of the hematite is liberated by fine grinding, and some of the $TiO_2$ is eliminated with the hematite in the magnetic separation.

The $TiO_2$-bearing minerals—titanite, ilmenite, anatase, rutile, and leucoxene—are present in the ore but not to the extent of accounting for all or most of the $TiO_2$. A large portion, and perhaps most, of the $TiO_2$ appears to be in solid solution in the hematite and much less in magnetite. Titanite and leucoxene occur as minor accessory minerals. The anatase, rutile, and ilmenite occur as exsolution laths, predominantly in hematite and to a lesser extent in magnetite. The exsolution laths are generally parallel to crystal faces (0001) in hematite and (010) in magnetite and are rarely more than 5 microns thick (7).

Because the unmineralized Grenville gneiss normally contains about the same amount of $TiO_2$ as the mineralized gneiss and because there is no determinable relationship between the amount of iron in the crude ore and the percentage of $TiO_2$, it is believed that most or all of the $TiO_2$ was present in the host rock at the time of mineralization; that is, it did not come in with the iron-depositing solutions but rather occurred as detrital minerals in the original sediment. The significance of this observation is not clear. It may mean that hematite has a greater affinity for $TiO_2$ than has magnetite. It may indicate that the hematite-depositing solutions preceded those depositing magnetite. If $TiO_2$-bearing magnetite is more readily oxidized than is pure magnetite, it might also suggest an oxidation origin for the hematite.

## ORE GENESIS

The Benson deposit is believed to be of replacement origin and the replacement to be structurally controlled. This interpretation of the local geology was accepted and advanced by the eight Jones & Laughlin geologists intimately acquainted with the property from the start of J&L's exploration to the present. However, other interpretations have been suggested, and not infrequently visiting geologists have expressed dissatisfaction with a replacement origin.

The indications for replacement are not numerous nor are they unequivocal. However, the magnetite and hematite grains are the same general size and shape as the gangue minerals and, except for some of the smallest grains, are all anhedral (Figures 9B and 9C). Unmineralized gneiss with similar gangue minerals is recognized both down dip and along the strike. Ore minerals are not sufficiently abundant to be open-pit ore except where the east limb of the syncline has been overturned to become hanging wall. This is true along the strike and has proved to be true wherever drilling has penetrated to the trough of the major syncline. This apparent structural control suggests replacement but would also conform to an interpretation of iron movement from a sedimentary source during or after folding and metamorphism. A further indication for a general replacement origin and proof of some replacement are specimens of books of biotite from pegmatites, within the ore, that have been replaced partially by magnetite. The magnetite is micaceous with cleavage conformable to the biotite, and the books are sufficiently euhedral to allow little doubt that the biotite was replaced. Partial replacement within the ore itself is more difficult to detect. The minerals are

anhedral and the grains of iron oxide and gangue are discrete. Limited microscopic examination has not revealed the iron oxides replacing grains along structural features such as cleavage and parting in the feldspars or ferromagnesian minerals.

Table III shows analyses representative of the crude ore and concentrates. They are very different from analyses of Lake Superior type iron formation in their high $Al_2O_3$ and alkalies. If the deposit originated as an iron formation, it must be concluded that it was very different from other iron formations or that extensive chemical changes took place during metamorphism.

The minor accessory minerals such as apatite, zircon, and monazite have rounded outlines characteristic of detrital minerals. The host rock is interpreted as a series of metamorphosed clay and silt sediments. This does not exclude the possibility of an original sediment sufficiently high in iron which, during metamorphism, was distributed as found today. There is, however, no criterion known that would recommend this mode of origin over replacement by iron-rich solutions from an igneous source.

## CONCLUSIONS

The Benson Mines deposit is not unique, but it is the largest known deposit of its kind. About 23 per cent acid-soluble iron is disseminated as grains of magnetite and hematite throughout a 400 to 600 foot thickness of Grenville gneiss for a strike length of about 2.5 miles. The gangue minerals are potash feldspar, quartz, and a variety of ferromagnesian minerals including biotite, hornblende, pyroxene, and chlorite. The hanging wall, composed of syenitic, granitic, and metasedimentary materials, is the stratigraphic footwall. Everywhere along the strike length, the ore-hanging wall contact is sharp and is marked by a reddish microcline pegmatite. The footwall rocks are metasediments, and the ore contact is drawn at the limit of economic ore as determined by analyses.

The ore deposit presently is interpreted as a replacement in Grenville gneisses. Iron-rich solutions from an igneous source selectively replaced minerals of the Grenville gneiss during a period of intrusion following metamorphism and folding of a series of clay sediments of pre-Grenville age. A structural control is suggested because ore occurs only where the stratigraphic footwall rocks have been overturned to become hanging wall. The ore

TABLE III. *Representative Chemical Analyses*

|  | Magnetite | | Non-Magnetic | |
|---|---|---|---|---|
|  | Crude | Concentrate | Crude | Concentrate |
| Fe | 25.80 | 63.62 | 20.80 | 63.04 |
| FeO | 12.35 |  | 1.74 |  |
| $SiO_2$ | 42.28 | 5.44 | 46.28 | 4.35 |
| MnO | 0.45 | 0.338 | 0.10 |  |
| $P_2O_5$ | 0.506 | 0.067 | 0.55 | 1.008 |
| CaO | 1.07 | 0.210 | 0.83 |  |
| $Al_2O_3$ | 9.56 | 3.53 | 12.78 | 3.51 |
| MgO | 0.79 | 0.314 | 0.44 | 0.45 |
| NiO | 0.02 | 0.01 | 0.009 |  |
| $C_2O_3$ | Trace | Trace | 0.012 |  |
| $V_2O_5$ | 0.018 |  | 0.027 |  |
| $TiO_2$ | 0.61 | 0.725 | 0.55 | 1.88 |
| S | 0.70 | 0.262 | 0.028 | 0.078 |
| CuO | 0.066 |  | 0.016 |  |
| Alkali | 5.45 |  | 6.97 |  |

mineral grains are disseminated throughout the host rock, and only in small hand specimens is replacement nearly complete.

There are perhaps more questions than answers for the Benson Mines deposit.

What was the nature of the structural control indicated by the distribution of ore in relation to the hanging wall?

Why do magnetite and hematite each occur as the predominant mineral with a general stratigraphic distribution but are gradational between types and with each mineral being present throughout the deposit?

Why is the friability of the non-magnetic ore related to the amount of total iron and to the proportion of hematite to magnetite?

Why were grains selectively replaced?

Why was replacement nowhere more complete?

Why is plagioclase the predominant feldspar in the hanging-wall gneiss and rare in the ore gneiss, which last has a predominance of potash feldspar?

Why are the characteristics of the feldspars apparently related to the ore type regardless of the host rock type?

What is the significance of the pegmatite at the hanging wall?

Why are the subsidiary folds not parallel to the major structural features?

Why is $TiO_2$ associated more abundantly with the hematite although it does occur with the magnetite also?

There are other questions of a more specific nature that are too detailed for listing here. Probably, these questions will be answered some day.

## REFERENCES CITED

1. Emmons, E., 1839, Report of the Second Geological District: N.Y. Geol. Surv. 2d Ann. Rept., p. 185–252.
2. Wilson, M. E., 1958, Precambrian classification and correlation in the Canadian Shield: Geol. Soc. Amer. Bull., v. 69, p. 757–773.
3. Bailey, S. W., 1959, Personal communication.
4. Buddington, A. F. and Leonard, B. F., 1962, Regional geology of the St. Lawrence County magnetite district, northwest Adirondacks, New York: U.S. Geol. Surv. Prof. Paper 376, 145 p.
5. ———— 1964, Ore deposits of the St. Lawrence County magnetite district, northwest Adirondacks: U.S. Geol. Surv. Prof. Paper 377, 259 p.
6. Hagni, R. D., 1966, Personal communication.
7. ———— 1966, Personal communication.

# 4. Triassic Magnetite and Diabase at Cornwall, Pennsylvania

DAVIS M. LAPHAM*

## Contents

* Pennsylvania Geological Survey, Harrisburg, Pennsylvania.

## Illustrations

## Tables

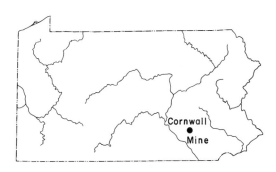

## ABSTRACT

Ore bodies at Cornwall, Pennsylvania, have been mined since 1742 principally for iron from magnetite, but also for copper (in chalcopyrite), silver (in chalcopyrite), gold (in chalcopyrite), cobalt (in pyrite), sulfur, and crushed limestone aggregate.

Metallization occurs as a limestone replacement above the essentially conformable, south dipping portion of a discordant saucer-shaped diabase sheet. Pre-Triassic and Triassic faulting have emplaced host Cambrian limestone and two Ordovician(?) hornfels units, now for the most part resting above diabase.

Thermal metamorphism of carbonate-bearing country rocks largely to calc-silicates, is ubiquitous. Metasomatism, particularly of potassium, is unique to ore zones. Associated magnetite-actinolite mineralization replaces, in a preferential sequence, all earlier metamorphic and metasomatic minerals and rarely diabase chilled margin. Mineralogical (diopside, actinolite, mica, sulfides, oxides) and chemical (Fe, Cu, S, Ni) zoning of the eastern ore body is focused down dip, near its western edge, along a major vertical fault through diabase. This zoning is spatially related to magnetite veins in diabase, to an ore body within diabase, and to a shear zone that probably extends through the sheet.

Ore localization controls include: (1) faulting that provided an ore solution channel through diabase to limestone, (2) replaceable laminated limestone, (3) textures, distribution, and replaceability in previously mineralized host and relatively impermeable hornfels units, (4) the composition, temperature, oxidation state, and migration rate of ore fluids, and (5) the early fractionation and later differentiation of iron, alkali-rich volatiles, and silica from primary diabase magma, partly determined from volume, composition, and distribution of granophyre.

Ore is believed to originate by a high Eh-controlled partition and fractionation of iron and alkalies within a compositionally normal diabase magma before intrusion of the diabase sheet. Magnetite mineralization followed diabase intrusion. Triassic magnetite ore bodies are hypothesized to have originated in two ways: by early diabase magma fractionation and separate iron introduction (Cornwall type), and by expulsion from late-phase diabase differentiates (Dillsburg type).

## INTRODUCTION

The magnetite deposit at Cornwall is located in Lebanon County, in east-central Pennsylvania. The mine is near the north border of the Triassic Lowlands Province, just south of lower Paleozoic Great Valley limestone formations, 23 miles east of Harrisburg and 65 miles west of Philadelphia.

Magnetite occurs as a limestone replacement immediately above the south-dipping limb of a saucer-shaped diabase sheet. It is one of several such Triassic occurrences throughout the Pennsylvania Triassic belt. Each major sheet is associated with at least one important magnetite deposit overlying diabase and replacing limestone. All of these deposits, comprising a single metallogenic province, were grouped first by Spencer (3, p. 10) as "deposits of the Cornwall type." The Cornwall ore body was considered by Lindgren (10, p. 696) to typify the pyrometasomatic class of ore deposits; that is, an ore deposit resulting from "emanations" which issued from a crystallizing igneous pluton. In this historical framework, the Cornwall deposit has acquired considerable geologic significance, and its understanding is of significance in both the theoretical and practical aspects of geology. The data and genetic theories presented here represent a summary of a more detailed report in preparation (47).

The author is especially grateful to the mining and geological staff of Bethlehem Steel Corporation for their close and valued cooperation throughout all phases of this work. The structural geologic and stratigraphic mapping are largely the work of Carlyle Gray and Alan Geyer. This project was initiated by the Pennyslvania Geological Survey under the direction of Carlyle Gray and was continued under the direction of Arthur Socolow. Grateful acknowledgement is made to them and other Survey staff members for their assistance and encouragement.

## HISTORY

The Cornwall, Pennsylvania, magnetite mine is the oldest, continuously operated mine in North America. The land was originally a 5000 acre tract purchased from the Province of Pennsylvania and later assigned to William Allen. Magnetic iron, i.e., magnetite, was discovered by Peter Grubb in 1732. Outcrops of magnetite were found on three hills: Grassy Hill to the west, Middle Hill, and Big Hill to the east. These are the present site of the now abandoned open pit. Between 1734 and 1737 Peter Grubb purchased land from William Allen and named the site Cornwall after his English home county.

Actual mining and processing of the magnetite ore began in 1737 with the construction of the first ore furnace and has continued ever since. Cannon, shot, and stoves for the Continental Congress and for use during the American Revolution were among its first products.

Peter Grubb died in 1754, and the property was divided into a large number of tracts with independent furnaces at both Cornwall and nearby Lebanon. In 1853, a new furnace was built on the site of the earlier one. From 1853 to 1883 it produced cold-blast charcoal iron. Currently it is operated as an historical site by the Pennsylvania Historical and Museum Commission. Consolidation of the separate mining operations occurred in 1864 when the Cornwall Ore Banks Company purchased 95 of the 96 holdings. Between 1916 and 1921, Bethlehem Steel acquired ownership of the Cornwall Ore Banks Company, but not until 1926 did they acquire the last of the separate enterprises, the Robesonia Iron Company. From this time to the present, Bethlehem Steel Corporation has been the sole owner.

All early mining was by open pit methods. In 1919, an eastern ore body, forecast by Spencer in 1908 (3), was discovered by dip needle survey. Shortly thereafter, in 1921, underground mining of the western ore body beneath the open pit began by sublevel and shrinkage stoping. In 1927 and 1928, the first shafts were sunk for underground mining of the eastern ore body. At this time the mining method used was panel caving, a modification of block caving, in which the ore was undercut by bells. More recently, fan-hole undercutting has been adopted. Between 1940 and 1953, open pit mining of the western ore body was resumed; it in turn was succeeded by the present underground mining. Because of dangerous slumping, the open pit now is closed completely.

In 1960, the Elizabeth Mine, a small open pit operation, was begun near the eastern ore body. This small deposit now has been mined out. When this operation was initiated, the main concentrator was moved from Lebanon to the Cornwall mines where the ore is pelletized. Recently, the mining of a small open pit ore body was begun at the eastern end of the old open pit. This operation completes the mining of known surface magnetite bodies.

Several small magnetite deposits in this area have been operated in the past. The Doner mine, northeast of Cornwall, differed from the Cornwall type only in that the ore lies below the diabase sheet. The Carper mine, eight miles west of Cornwall, yielded about 1500 tons of ore. Further west, and about two miles southeast of Hummelstown, there was a small occurrence of magnetite and hematite. This is the only known ore occurrence wholly within Triassic sediments and not immediately adjacent to diabase.

An historical survey of geological theories concerning the origin of the magnetite reflects a change in geological thinking from an early emphasis on sedimentary iron to a reliance on a magmatic source. All, however, have noted and utilized the spatial significance of magnetite adjacent to diabase. Early theories (1,2) proposed a remobilization of sedimentary limonite caused by the heat of diabase intrusion. Later, Spencer (3) proposed, and Hickok (9) concurred, that iron-bearing solutions originated within the diabase sheet. Spencer (3) referred to such occurrences as "deposits of the Cornwall type." Hotz (16,18) in his study of the Dillsburg deposits, has concurred essentially, attributing the ore to expulsion during late pegmatitic-granophyric diabase differentiation (18, p. 703). More recently Gray and Lapham (29, p. 15) and Lapham (32, p. 37–38) have suggested that both the iron and the diabase at Cornwall came from a common magmatic source rather than requiring an in situ derivation of iron from the crystallizing diabase.

## REGIONAL GEOLOGY

### Introduction

A Triassic belt of red, continental shale, siltstone, and conglomerate extends continuously from south-central to southeastern Pennsylvania. For the most part it follows the trend of curvature of the Appalachian salient (Figure 9). At least four major diabase sheets, and numerous nearly vertical diabase dikes, are intrusive into these sediments. Except in the Morgantown-French Creek area (Figure 9) these sheets lie near the partially faulted north border of the Triassic basin. In the Cornwall area of central Pennsylvania, recumbently folded, northeast-striking Cambrian carbonate and Ordovician carbonate or pelitic sediments form the north Triassic contact. Similar units lie to the south and are presumably stratigraphically and structurally continuous, beneath the Triassic basin, with the lower Paleozoic sediments to the north.

The record of structural deformation and metamorphism previous to Triassic time is exceedingly complex. Multiple deformations from Taconic, Acadian, and Appalachian tectonism have resulted in large-scale nappe structures, multiple cleavage, and considerable thrust faulting. The major structures are probably at least as old as the Acadian, and may, in part, be Taconic in age. By comparison, Triassic, or post Triassic, deformation was slight. Graben and normal faulting, northward basin tilting, and largely permissive diabase intrusion (with consequent contact metamorphism) comprise the major deformational elements. Recently, structural and magnetic data on the vertical diabase dikes have been interpreted to indicate post-Triassic intrusion (42, p. 13), or at least an age significantly younger than that of the sheets (35, p. 521). The sheets and associated magnetite have been dated radiogenically within the Triassic at about 190 m.y. (31, p. 741).

### Stratigraphy

The stratigraphic units in the Cornwall area are illustrated in Figure 1. The middle and late Cambrian carbonates at the northern Triassic border are largely limestones with dolomitic or shaly to sandy interbeds, and impure dolomite (29, 24). Chert, oolites, and stromatolites are present at some horizons. With respect to the Cornwall ore deposit, their description is relevant chiefly for identification of the host limestone.

Although metamorphically recrystallized and metasomatically replaced, the host limestone retains some features of its pre-Triassic character. Tight isoclinal folds are outlined by laminae of shaly limestone and thin dolomitic or shaly interbeds. Locally, the Buffalo Springs, Snitz Creek, and Millbach are the only formations with all of these characteristics. A general lack of extensive dolomite and sandy dolomite interbeds in the host limestone prob-

FIG. 1. *Stratigraphic Column for the Vicinity of Cornwall, Pennsylvania.*

ably eliminates the Snitz Creek Formation from consideration. However, the Millbach Formation can be eliminated only on structural grounds. More extensive thrust faulting than is believed present would be required to bring the more distant Millbach Formation into contact with diabase. Thus, the structural position of the Buffalo Springs Formation immediately below diabase (Figure 3) favors it as the host limestone.

Two other units, the "Blue Conglomerate" and the "Mill Hill Slate", are of questionable correlation and origin. The "Mill Hill Slate" lies both above and below diabase, whereas the more restricted "Blue Conglomerate" is found only above diabase. Diamond drill hole data from the hanging wall of the eastern ore body show that up dip "Blue Conglomerate" passes into "Mill Hill Slate", but the transition is irregular and marked by repetitious "slices" or "fingers" of the two units with each other and with limestone. Similarly, to the west near Quentin, an outcrop of "Mill Hill Slate" contains interbeds of "Blue Conglomerate" lithology. Thus the two lithologies were apparently once facies of a single formation. They are now essentially hornfels units.

The present "Mill Hill Slate" is a dense, dark brown to gray-black, laminated hornfels, frequently rich in carbonate or calc-silicate minerals. Alternating bands are silica-poor and silica-rich or calcium-poor and calcium-rich. These thin, remarkably persistent laminae are the relics of an originally diverse bedding composition. Primary cross-bedding structures have been preserved.

Because of a probable thrust contact between "Mill Hill Slate" and the Buffalo Springs

Formation (Figures 2,3), and because the Cambrian carbonate sequence is overturned to the south, correlation of the "Mill Hill Slate" cannot be deduced structurally, but must be based on lithologic correlation alone. The original formation must have been a laminated, in portions conglomeratic, limy shale. There is no unit in the immediate vicinity which exactly fits this description, but certain shaly sections of the basal Martinsburg Formation most closely approach it. Other units similar to the "Mill Hill Slate" are the shale at Schaefferstown to the east and the Cocalico Shale to the south. Both of these units are probably correlative with shale of the Martinsburg Formation (47).

The "Blue Conglomerate" is a breccia, at least in part of tectonic origin. It contains subangular to sub-rounded quartzitic and pelitic fragments in a hornfels matrix of mica, chlorite, actinolite, and feldspar. The matrix contains little or no quartz.

The Triassic red sandstones and shales belong to the Hammer Creek Formation in the upper part of the Newark Group. In this area, the sediments are coarser-grained and bedding planes less continuous than elsewhere in the Triassic basin. Near the Triassic diabase sheets, they have been bleached. Several authors have suggested that the intrusion of diabase sheets occurred before major basin tilting, while the Triassic sediments were still approximately horizontal. This proposal, first advanced by Stose (7, p. 54), has been supported more recently by Sanders (35) and Beck (39). However, although pole rotation of site-mean directions of magnetization about a pole whose axis parallels the Triassic strike was found for the Gettysburg and Morgantown plutons, it was not evident in the Yorkhaven-Cornwall pluton (39, p. 2853). Lack of any linear distribution at Cornwall might result from a local tectonism during or after intrusion or from a lack of linearity of the pluton such as might accrue by compound sheet injection. Stratigraphic, structural, petrographic, and chemical evidence all indicate that the Cornwall area is something of a special case so that there appears to be no simple sequential relationship between angle of sedimentary tilt and attitude of the present pluton (47). The younger diabase dikes are chemically and mineralogically distinct from the sheets.

### Structure

Structural relationships in the Cornwall area are complex. A detailed discussion will be published by Lapham and Gray (47). Although of considerable import to the deformation and origin of the Triassic province, they bear only marginally on the mechanisms of ore introduction and localization at Cornwall.

The areal geology and an interpretative cross section are shown in Figures 2 and 3, respectively. At Cornwall, a south dipping diabase sheet about 1200 feet thick is enveloped by "Mill Hill Slate" with minor "Blue Conglomerate" and host limestone above the diabase. "Blue Conglomerate" is inferred to exist below the sheet (Figure 3). The hornfels units dip conformably, or nearly so, with the diabase. In places, axial planes of the host limestone folds have been rotated to a northerly dip by faulting and/or diabase intrusion. Cambrian sediments to the north are folded intensely. They are overturned and dip south somewhat less steeply than the hornfels units. The Triassic clastics generally dip 40° to 60°N but flatten to about 22° in the vicinity of the mine. They are paralleled by the southern lip of the saucer-shaped diabase sheet (Figure 3). Thus the diabase contacts here are nearly always conformable: to the south with Triassic sediments

and to the north with host limestone and hornfels. These attitudes have probably exerted some control on the form of the diabase intrusive.

Correlation of the "Mill Hill Slate" with the Martinsburg Formation necessitates a thrust fault to bring the "Mill Hill Slate" in contact with the much older overturned sequence of Cambrian carbonates (fault 1, Figure 3). The initial thrusting at this horizon may have occurred well before Triassic time and is presumably either Taconic or Acadian in age (43, 45). The tectonic nature of the "Blue Conglomerate" phase of the "Mill Hill Slate" may have originated with this deformation, although later Triassic thrusting is believed to have been significant in its development.

At least local border faulting (fault 2, Figure 3) occurred in Triassic time after deposition of the New Oxford Formation and before the completion of Hammer Creek sedimentation. Although there is no direct evidence of a border fault, it is inferred from the sedimentation patterns (34, 47). Erosion at this time along the north border resulted in a fault scarp

GEOLOGIC MAP OF CORNWALL, PENNSYLVANIA

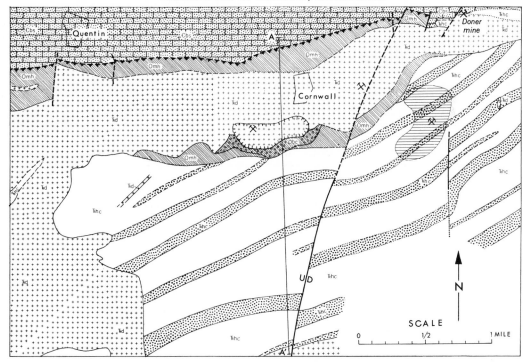

Fig. 2. *Geologic Map of the Cornwall Area (Revised after Geyer and others (24)). In the area of the open pit, the map has been restored to the surface pattern that existed before the pit was opened.*

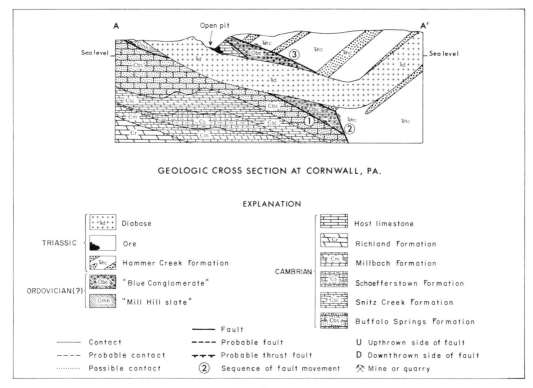

GEOLOGIC CROSS SECTION AT CORNWALL, PA.

EXPLANATION

TRIASSIC
- Diabase
- Ore
- Hammer Creek Formation

ORDOVICIAN(?)
- "Blue Conglomerate"
- "Mill Hill slate"

CAMBRIAN
- Host limestone
- Richland Formation
- Millbach Formation
- Schaefferstown Formation
- Snitz Creek Formation
- Buffalo Springs Formation

——— Fault
——— Contact
– – – Probable contact
············ Possible contact
– – – Probable fault
▼▼▼ Probable thrust fault
② Sequence of fault movement

U Upthrown side of fault
D Downthrown side of fault
⚒ Mine or quarry

FIG. 3. *Geologic Cross-Section of the Cornwall Area. Inferred and questionable faults are shown* (cf. *Figure 2*).

against which compression from the south was later thrust, resulting in tilting, and renewed thrusting (fault 3, Figure 3) as was suggested originally by A. C. Spencer (personal communication to Carlyle Gray). The diabase sheet was intruded either before this compressional tilting or as an integral part of the deformation. It is interesting to note that, if the major portion of the sheet were intruded essentially along a horizontal plane before northward tilting, the present south dipping portion of the sheet at Cornwall might have been considerably steeper than it now is. If true, movement of the "Mill Hill Slate" hornfels complex at its contact with the Buffalo Springs Formation (fault 1, Figure 3) would be required during Triassic Time. Lack of gravitational settling and approximately equivalent widths of contact metamorphism below and above diabase may be a consequence of such an originally steeper dip.

Later, normal faults transect the diabase. A major fault between the western and eastern ore bodies (Figure 2) is related to mineral and chemical zonation and to an ore-mineralized fault system at depth within the diabase.

The two major ore bodies are separated by a hump in the upper surface of the sheet. Diabase is in contact with "Mill Hill Slate" hornfels on the eastern termination of the ore and with limestone at the western termination. The Elizabeth open pit ore body was situated in a small depression in the top of the sheet.

No feeder dike has been found for the Cornwall sheet. However, to the southwest there is a strong magnetic anomaly over diabase (rather than over the contact) which might indicate the presence of such a feeder (30, p. 15). On the other hand, as Hotz (17, p. 387) has pointed out, a feeder may not be required for permissive, lateral spread of diabase magma.

### Plutonism and Metamorphism

The next sections present descriptions of the diabase, the metamorphism, and the relationships between them and the ore mineralization. The mineralogical and chemical composition of the Cornwall diabase is grossly similar to other occurrences of diabase such as the Dillsburg sheet (16,18), the Palisade sheet (12), and the Karroo dolerites (15). Essentially, it is a quartz bearing, sub-ophitic diabase with chilled margins, pegmatitic and granophyric facies near the top, and some vertical mineralogical variation. Both thermal metamorphism and alkali metasomatism have reconstituted the adjacent sediments above the metamorphosed and metasomatized units after the solidification of an apparently coherent chilled margin.

## ECONOMIC GEOLOGY

### Production and Ore Grade

Except for the recovery of supergene-enriched copper in the early days of mining, iron was the only metal recovered until 1920. At that time, magnetic separation and froth flotation were initiated to remove sulfides from the iron ore. Chalcopyrite is separated from pyrite and cobalt is recovered from the latter mineral. Sulfur, for sulfuric acid, also is recovered. Some gold (1700 oz. in 1953) and silver are recovered from the chalcopyrite during refining. Limestone overburden removed from the open pit was crushed and sold for aggregate until the cessation of open pit mining in 1953.

The record of mining in natural net tons of iron ore from 1740 to 1964 is shown in Table I. The present production averages over one million tons per year. Nearly 100 million tons have been removed to date (1966). A typical recovery analysis is shown in Table II. The ore grade averages 40 to 42 percent iron and pelletized magnetic concentrates 62 to 64 per cent. Cobalt in pyrite concentrate is not constant but averages about 1.2 to 1.4 per cent Co. Phosphorus and titanium are not present in any significant amounts. Electron

TABLE I.   Ore Production (Natural Net Tons)

| Year | Open Pit | Western Mine | Eastern Mine | Totals |
|---|---|---|---|---|
| 1740–1880 | 6,124,921 | | | 6,124,921 |
| 1881–1900 | 12,196,540 | | | 12,196,540 |
| 1901–1927 | 17,644,908 | 292,648 | | 17,937,556 |
| 1928–1939 | 10,898,867 | 1,802,179 | 628,295 | 13,329,341 |
| 1940–1950 | 12,171,092 | 345,689 | 9,072,641 | 21,589,422 |
| 1951–1960 | 1,776,928 | 3,551,131 | 10,860,095 | 16,188,154 |
| 1961–1964 | | 1,881,294 | 3,923,451 | 5,804,745 |
| | 60,813,256 | 7,872,941 | 24,484,482 | 93,170,679 |

microprobe studies (40) and unit cell measurements show a small variation in Fe, Mg, and Al in magnetite, dependent upon its geologic history, but they are insignificant in terms of any affect upon ore grade.

### Description of Ore Bodies

The Cornwall mine area comprises two major ore bodies, an eastern and a western mine, two minor ore pockets, the Elizabeth mine and a small open pit at the eastern end of the main open pit (western ore body), and an undeveloped concentration of ore within diabase (the "footwall" ore body) between the two major ore bodies.

The western or body, the site of the original open pit operations and now largely mined out, lies directly on diabase except for small, intervening blocks of limestone. It originally cropped out up dip and hence its upper limit is the present erosion surface. The strike length of this ore body before mining was nearly 4000 feet. At its western limit it had a dip length of about 1600 feet, decreasing eastward. Maximum ore thickness here exceeded 100

TABLE II.   Product Analyses, 1964 (in %)

| | Fe | S | Cu | Ni | Co | Mn | P |
|---|---|---|---|---|---|---|---|
| Ore (Mill Feed) | 39.4 | 1.29 | 0.29 | | | | |
| Fe concentrate | 64.8 | 0.06 | 0.05 | | | | |
| Final Tailing | 7.7 | 0.61 | 0.04 | | | | |
| Pellets | 63.6 | 0.02 | 0.05 | | | 0.06 | 0.006 |
| Cu concentrate | 28.7 | 33.1 | 26.68 | 0.046 | | | |
| Pyrite concentrate | 41.5 | 49.4 | 0.09 | 0.15 | 1.37 | | |

Recovery: Magnetic Iron—99%
Copper —94%
Pyrite —73%

feet. At its eastern end, the ore was divided by a roll, or ridge, in the diabase sheet. If the two bifurcations were once continuous, their juncture has since been eroded. A rise in the diabase at the western end terminates both limestone and ore. Near the eastern end of the open pit, ore fingered out into limestone. At the extreme eastern end and down dip, "Blue Conglomerate" cuts out host limestone and forms a somewhat irregular ore boundary (Figures 2 and 3).

The eastern ore body is lenticular with a lima bean shape (Figure 6). It extends 3000 feet along strike and more than 2400 feet down dip. Ore thicknesses average about 100 feet but attain a maximum of 240 feet. Along the extreme western edge, hanging wall ore tongues out into limestone. Elsewhere either "Mill Hill Slate" (up dip) or "Blue Conglomerate" (down dip) (Figure 3) forms the ore hanging wall.

The two smaller ore concentrations replaced limestone where there was a small trough in the upper surface of the diabase sheet. For the most part, the hanging wall has been eroded. At the Elizabeth open pit, ore rested on diabase that was frequently of pegmatitic texture. Granophyre was not observed here.

A large mass of lean ore ("footwall" ore body), not mined at present, lies beneath the eastern end of the western ore body, almost wholly enclosed by diabase. This ore body is ovoid and intersects the main ore above through a narrow ore bridge. The ore mass strikes about N35°W and plunges about 40°NE. At its intersection with the main ore body, banded and brecciated limestone replacement textures are common. Within the ore, frequent blocks of barren diabase have been intersected by diamond drill cores. No other textural information is available. Shearing and brecciation in diabase are present down dip from this footwall ore. Strike projections of ore veins in diabase observed in the open pit also are on strike with the footwall ore body.

## Ore Textures

Megascopic ore textures at Cornwall reflect very well the textural and compositional variation of the units that they replace. However, with a few minor exceptions, there is very little regularity or distributional significance to them. Mapping of textures has not yielded discrete, large scale, textural zonation. It does, however, yield much information on the conditions governing ease of replacement of one rock type relative to another.

Of the many ore textures, the most striking is banded, green and black ore, found throughout the eastern and western ore bodies. Magnetite-rich laminae alternate with green, gangue-rich laminae; or, in gangue, more siliceous and Ca-poor laminae alternate with those that are less siliceous and more Ca-rich. They commonly vary in width from tenths of inches to several inches, are discontinuous beyond a few feet (commonly, a few inches) are offset by small microfaults, and exhibit numerous small scale folds. In every respect, they resemble the laminae in unreplaced limestone. Bedding and axial plane attitudes, however, are oriented with remarkable irregularity,—a consequence of brecciation and small scale faulting largely restricted to the ore zone. Particularly in the lower parts of the eastern ore body, bands of essentially gangue-free magnetite are separated by planar vugs into which dodecahedral or octahedral terminations project, touching the crystal projections of an adjacent band.

In the eastern ore body, textures are present that are characteristic of replacement of "Blue Conglomerate", or more rarely of "Mill Hill Slate". "Blue Conglomerate" replacement has resulted in a spotted ore that frequently resembles a phenocrystic texture. Especially where the "Blue Conglomerate" has pinched out limestone and approaches diabase, this ore may be a mottled pink resulting from the presence of ferriferous hematitic orthoclase (47). This, usually lean, ore has acquired the misnomer of aplitic ore. Only certain gangue minerals, usually actinolite or mica, have been replaced by magnetite; the feldspar or quartz breccia fragments remain unreplaced. Replaced laminae of the "Mill Hill Slate" are difficult to differentiate from replaced limestone laminae. Usually, however, there is much less magnetite, and the laminae are less contorted than in limestone.

Another characteristic texture is that of platy magnetite. It is most common at the footwall and hanging wall of the ore where it has replaced pre-existing specular hematite. Minor compositional variation and optical zoning in this magnetite are common and have been discussed by Davidson and Wyllie (40). In part, the zoning may result from the replacement of calcite by hematite, and finally by magnetite. Interrupted magnetite crystallization is indicated by the zonal growth of bluish cores capped by brownish, more impure magnetite (40).

Vein magnetite, massive or vuggy, occurs in diabase and more rarely transects banded gangue-ore textured rock. Such veins are usually down dip along the footwall. Magnetite here tends to be more euhedral than up dip or toward the hanging wall. "V"-shaped contraction cracks in the top of the diabase commonly are filled with magnetite and diopside; only rarely is the chilled margin actually replaced (i.e., along the southwestern footwall of the eastern ore body). The diabase-ore contact is visually sharp except where "Blue Conglomerate" or "Mill Hill Slate" hornfels in contact with diabase is nearly indistinguishable from chilled diabase. In the south-central part of the eastern ore body, a tongue of diabase extending into ore has been bleached and chloritized by ore zone mineralization. Small magnetite-actinolite fractures extend into the diabase tongue.

Although mylonitic, brecciated, and sheared ore textures are found throughout the mine, they are especially prominent in the southwest and south-central parts of the eastern ore body. On a microscopic scale, shear flowage postdates the crystallization of metamorphic mica, tremolite-actinolite, and at least some of the magnetite. Sericitization of plagioclase at the top of the diabase sheet, present throughout the chilled margin, is especially noticeable in this area.

In general, both metallic and gangue minerals are finer-grained toward the hanging wall and mixed coarse- and fine-grained at the footwall. This is particularly true for diopside and actinolite. Grain size also crudely follows the laminated texture of the replaced units, alternating coarser- with finer-grained laminae. This textural influence is less pronounced for magnetite replacement than for gangue silicates.

## Mineralogy and Chemistry
## of the Ore Zone

The composition of the ore zone is the result of several crudely consecutive processes: thermal metamorphic reconstitution, metasomatic permeation, ore mineralization, and secondary alteration.

The major metallic minerals present, listed in approximate order of decreasing abundance, are magnetite, pyrite, chalcopyrite, and hematite. Minor metallic minerals include bornite, chalcocite, covellite, native copper, galena, marcasite, millerite, pyrrhotite, sphalerite, and wurtzite. The major gangue minerals associated with the main stage of ore deposition are actinolite and chlorite.

Cornwall magnetite varies in morphology from anhedral-granular to euhedral. Most euhedral crystals are of dodecahedral habit, but many show predominantly octahedral faces. Platy magnetite pseudomorphs after hematite are common at the footwall and hanging wall. Both brownish and bluish magnetite can be differentiated in reflected light, and most crystals are zoned either continuously or discretely (4,6,9,40). Cores with a bluish cast tend to be richer in Fe and poorer in Mg, Al, Ti, V, Ca, and Si than more brownish-yellow rims (40, p. 770). Minute spinel inclusions tend to be richer in Mg-Al or Mg than the matrix magnetite (40, p. 768–769). Spectrographic analyses show the presence of Be, B, Cr (rarely), Co, Cu, Ga, Mg, Mn, Ni, Na, Ti, V, and Zn (rarely). There is no known consistent spatial variation of these types with the exception of the zoned platy magnetite. Unit cell dimensions of magnetite, measured from $a_0 - 2\theta$ (hkl) and extrapolated $a_0 - \cos^2\theta$ plots, show a range from 8.391 Å to 8.411 Å and average about 8.400 Å. At least in part, this variation probably results from different proportions of the observed optical and compositional zones.

The most common ore mineral associations are magnetite-chalcopyrite-pyrite(-sphalerite), magnetite-pyrite-actinolite(-phlogopite), magnetite-chlorite(-pyrite-zeolite), magnetite-hematite-calcite(-zeolite), and magnetite alone. Pyrite may be present in any of the above associations but is frequently of more than one generation. Sulfide minerals appear to be more abundant here than at the similar Morgantown occurrence.

Pyrite is ubiquitously present in simple and exceedingly complex crystals. Simple cubes, frequently as limonite pseudomorphs, have been found in the area. The "Mill Hill Slate" west of Cornwall is commonly pyritized. In the ore zone, pyrite crystallization has spanned the period of ore mineralization through that of zeolite crystallization (colloidal pyrite "dust" on stilbite). Crystals of pyrite are notable for their cobalt content, which, based on semi-quantitative X-ray fluorescence analyses, may exceed two per cent Co. The possibility of spatial or morphological control on the cobalt content of pyrite is not known. Other spectrographically determined trace elements include As, Bi, Cu, Pb, Ag, Zn, and Ni.

Pyrrhotite in crystals of hexagonal outline have been found in serpentinized and steatitized portions of the eastern ore body, but its distribution is not known. Two X-ray patterns indicate that its symmetry is now monoclinic.

Chalcopyrite occurs as massive pods or veins replacing pyrite along preferred crystallographic planes (as does magnetite) and in veins transecting the laminae of magnetite and gangue silicates. The other metallics are of minor importance and occur as fracture fillings or interstitially among the major metallic and gangue components. Millerite, marcasite, cuprite, malachite, azurite, some pyrite, and probably bornite and covellite are secondary minerals. Native copper was found at the surface on Big Hill at the eastern end of the open pit, but has not been reported from underground operations.

Actinolite, and to a lesser extent chlorite and phlogopite, is almost universally associated with magnetite ore. Together, they impart the green color so characteristic of non-magnetite bands throughout the ore body. The actinolite associated with magnetite occurs in slender prisms or needles with nearly parallel extinction and is the youngest of two types of actinolitic amphibole. Byssolite is characteristic of vug ore associated with zeolites. Chlorite associated with magnetite has anomalous blue interference colors and is iron-rich. Earlier chlorite associated with alteration of early actinolite and late chlorite associated with zeolitization are less iron-rich. Chlorite associated with serpentinization and steatitization is magnesium-rich. Cross-cutting magnetite veins are usually zoned with actinolite walls and chlorite-magnetite cores. This relationship is especially characteristic of ore veins in diabase.

Chemical changes in the ore zone reflect the mineralogical alterations consequent upon both metamorphism and ore mineralization (Table III). Comparison of: (1) the Buffalo Springs Formation (assuming it to be the host limestone), (2) the "Blue Conglomerate," (the eastern ore hanging wall, which is in contact with diabase down dip) and (3) gangue from the ore footwall and hanging wall illustrate only gross chemical changes in the original carbonate unit. Most such changes occurred before ore-stage mineralization.

There is very little chemical evidence of the composition of the original host limestone. Laminae preferentially replaced by gangue, rather than ore, were richer in silica and magnesia; perhaps also in alumina. It is possible that silica and alumina have migrated up dip by ore-fluid leaching of the "Blue Conglomerate" hornfels where it is in contact with diabase. Lack of ground mass quartz in this hornfels could be explained in this manner. Calcium and carbon dioxide have been removed from the system (Table III). If leaching of the "Blue

TABLE III. *Chemical Changes in the Ore Zone of the Eastern Ore Body\* (in Per Cent)*

|               | (1)    | (2)    | (3)    |
|---------------|--------|--------|--------|
| $SiO_2$       | 12.61  | 52.11  | 42.76  |
| $TiO_2$       | N.A.   | N.A.   | 0.64   |
| $Al_2O_3$     | 2.24   | 23.75  | 10.78  |
| $Fe_2O_3$     | 0.95   | 3.66   | 5.82   |
| FeO           | N.A.   | 5.90   | 4.02   |
| MnO           | N.A.   | N.A.   | 0.133  |
| MgO           | 12.88  | 2.72   | 20.13  |
| CaO           | 32.24  | 2.54   | 6.83   |
| $Na_2O$       | N.A.   | 3.98   | 0.27   |
| $K_2O$        | N.A.   | 2.41   | 5.43   |
| $PO_4$        | N.A.   | N.A.   | 0.190  |
| $CO_2$        | 39.09  | 0.66   | 0.37   |
| F             | N.A.   | N.A.   | 0.163  |
| S             | N.A.   | N.A.   | 0.013  |
| $H_2O$ comb.  | N.A.   | 2.27   | 2.45   |
|               | 100.00 | 100.00 | 99.999 |

\* N.A. = Not Analyzed.

(1) Avg. of 8 samples of Buffalo Springs Formation (37, p. 29).

(2) "Blue Conglomerate" (47).

(3) Avg. of 2 gangue samples (footwall and hanging wall) (47).

Conglomerate" occurred, sodium also has left the system. There is no necessity for migration of magnesium if ore replaced lime-rich portions of the Buffalo Springs Formation and gangue replaced magnesium-rich portions. The $Fe^3/Fe^2$ ratio in the ore zone is approximately that of magnetite (Table III). Comparative chemical analyses of the hornfels and limestone units show that in the ore area iron has been added, rather than subtracted, from them (47).

## Mineral Paragenesis

A generalized paragenetic diagram for the eastern ore body (Figure 4) is similar in broad outline to that derived by Hickok (9, p. 232) for the western ore body. Maxima are indicated crudely by line thickness. Mineral associations have been divided, somewhat arbitrarily, into four groups: 1) an outer halo chiefly representing contact thermal metamorphic minerals; 2) the development of metamorphic and metasomatic silicates in the ore zone before the onset of metallic mineralization; 3) the main ore stage with continued silicate mineral crystallization; and 4) final low temperature mineralization and weathering products. Only the major minerals are considered; many more trace minerals, especially of secondary origin, are present (41,47). However, such a generalized diagram obscures significant details. Notably erroneous is the impression of a consistent time sequence throughout the ore zone, whereas in actuality mineralization proceeded by the migration of chemical and thermal fronts such that the development of minerals near the hanging wall postdates the formation of the same minerals nearer the footwall.

Only a few of the more significant details can be discussed here. Predominant among them is the consistent replacement of contact metamorphic minerals by metasomatic silicates and their replacement, in turn, by ore minerals. Near the footwall early diopside and garnet are mantled by later diopside (fine-grained), phlogopite, and actinolite (coarse-grained). These, in turn, are replaced or mantled by pyrite, magnetite, fine-grained actinolite, and chlorite. This sequence, consistent at any one level above the footwall and across the dip, has been for years the only really obvious reason for the establishment of a noticeable time lapse between the cooling of the emplaced diabase sheet and ore deposition.

The two periods of diopside formation within the ore zone (Figure 4) are largely a spatial phenomenon. Coarse-grained diopside predominates along the footwall and finer-grained diopside attains a maximum about midway between footwall and hanging wall (Figure 5). Actinolite, on the other hand, is present along the footwall in both coarse blades (extinction angle up to 28°) and fine felted needles (extinction angle usually 0° to 10°) associated with magnetite. The latter pair replace coarse actinolite along the principal "C" axis cleavage resulting in ragged, feathered terminations. Away from the footwall, only fine-grained actinolite is present.

Magnetite and chalcopyrite replace pyrite along the principal cleavage directions and also embay it, resulting in highly corroded pyrite anhedra. Chalcopyrite and a later generation of pyrite fill fractures or interstices between magnetite grains or, more rarely, replace the pre-existing magnetite, yielding etched magnetite crystal faces. Magnetite replaces coarse-grained footwall amphibole where it has been protected from chloritization and fine-grained amphibole replacement. Magnetite only rarely replaces diopside and then only the coarser-grained crystals. More commonly, isolated diopside grains have acted as a nucleation growth center for magnetite crystallization. Where pink, hematitic or ferriferous orthoclase is present, it frequently is mantled or cut by

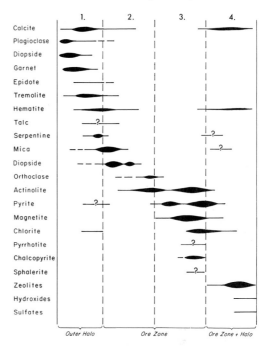

FIG. 4. *Generalized Paragenetic Diagram.*

magnetite along pre-ore fractures. Where calcite is present with specular hematite, magnetite has preferentially replaced hematite. Where calcite is present with amphibole, magnetite has selectively replaced calcite, or, if replacing both, is euhedral toward calcite and anhedral toward amphibole.

These observations lead to the following generalized selective series, the minerals of which are successively more difficult of replacement by magnetite: hematite—calcite—pyrite—mica—amphibole—feldspar—pyroxene—quartz.

Secondary, or at least very late, magnetite is present. Occasionally vugs of zeolite crystals, particularly analcime, are lightly dusted with magnetite. Such vugs are usually rich in byssolite.

## Mineralogical and Chemical Zonation

Zonation is not a conspicuous feature of the Cornwall ore body. However, subtle variations do exist and are genetically as well as spatially significant.

The only previous studies of zonation have been concerned with diopside-actinolite and metallic-silicate mineral distribution from the western ore body. Hickok (9, p. 230–231) noted a concentration of diopside near the diabase gradually giving way to a zone of ac-

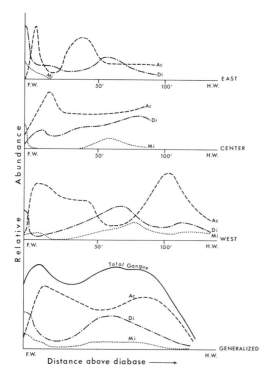

FIG. 5. *Mineralogical Zoning from Footwall to Hanging Wall of the Eastern Ore Body (given in feet above diabase). Relative abundance was calculated from thin section observation. The generalized distribution graph presents an idealized summary of the three major gangue minerals. Early and late paragenetic types of diopside and actinolite are not differentiated.*

tinolite near the hanging wall. He also notes platy magnetite along the hanging wall and garnet along the footwall. Earlier, Callahan and Newhouse (6) had noted a similar distribution and also stated "towards the upper part of the ore body the pyrite and chalcopyrite increase in amount. . . ." (6, p. 409), and again that "the silicate minerals are most concentrated toward the diabase" and "magnetite, pyrite, and chalcopyrite are all most abundant in that part of the orebody which is farthest from the diabase" (6, p. 409).

Data for the following discussion were all taken from diamond drill cores and hand-specimen sampling of the eastern ore body. The mineral distributions are not strictly in accordance with those cited above for the western ore body. Because of the presence of "Mill Hill Slate" and "Blue Conglomerate" above this eastern ore body, composition of host rocks has exerted a greater control than to

the west where both ore zone and hanging wall are largely limestone.

Grain size and textural distributions already have been noted. In general, contact metamorphic minerals form a tabular aureole approximately paralleling the upper surface of the diabase sheet. Less tabular is later silicate and metallic mineralization that extends farther above the sheet but is less extensive up dip. Other controls on grain size and textural distribution include mylonitization and brecciation, composition of the host rocks, texture of the host rocks, and the time migration of thermal and chemical fronts.

The distribution of the three major gangue minerals, diopside, actinolite and mica, with respect to the footwall (diabase) are illustrated in Figure 5. Three sections from diamond drill cores illustrate variation from footwall to hanging wall for eastern, central, and western parts of the eastern ore body. A generalized distribution diagram is also included. Diopside (and to a lesser degree mica) shows two maxima: one, at the footwall, of coarser-grain size, and a second, near the center of the ore zone, of finer-grain size. Actinolite shows a similar bimodal distribution but displaced toward the hanging wall, beyond diopside maxima. Distribution of diopside and actinolite is not spatially inverse but is inverse when together they are compared with metallic mineral concentration. This is a consequence of the appreciable difficulty of replacement of these earlier minerals by magnetite, rather than a function of ore solution distribution. An exception to this replacement rule may be the concentration of magnetite at the footwall along the western and southwestern portions of the ore body. Here the silicates, particularly actinolite, have been noticeably replaced by ore. From this southwestern portion of the ore body, there also appears to be a crude outward concentric zoning first of late diopside, then of late actinolite, and finally of chlorite from the footwall up dip and toward the hanging wall.

Zoning of metallics is less pronounced. Chalcopyrite appears to have penetrated the least distance outward, magnetite next, and pyrite is commonly outside the ore zone. However, pyrite is ubiquitous and chalcopyrite is much less abundant than the other two, so that this distribution would seem to be more apparent than real. It is supported, however, by the distribution of total Fe and Cu (Figure 6).

One of the most interesting types of zoning is shown by the paragenetic relationship between the finer-grained actinolite of the ore stage and magnetite. At the footwall in the central to southwestern part of the ore body, magnetite replaces actinolite (which in turn has replaced coarse-grained actinolite and/or diopside). Elsewhere, up dip along the footwall and also toward the hanging wall, magnetite preceded the finer-grained actinolite. Thus, magnetite-depositing ore solutions moved more rapidly and eventually overtook the migrating front of actinolite crystallization.

Other types of zoning, the significance of which are not so clear, also are arranged spatially outward from the southwestern and southcentral quadrants of the eastern ore body. Unit cell dimensions of magnetite tend to decrease laterally and vertically from this area (32, p. 39; 47). This decrease may be a function of the relative proportions of optically blue and brown magnetite in zoned crystals which in turn is, at least in part, chemically controlled (40). Vug ore and magnetite veins in diabase are concentrated in this area.

Variations in total Cu, Fe, S, and Ni follow the mineralogical zoning pattern and are illustrated by contours of relative concentrations in Figure 6. The major concentration of total Fe has proceeded from a down-dip maximum at the footwall through an intermediate minimum to an up-dip maximum at the hanging wall. Total Cu distribution is similar except that vertical migration dominated over lateral migration. Two representative east-west cross sections of total Ni distribution, one up dip and the other down-dip, illustrate the same movement up and out from the footwall.

These zoning data are considered indicative of an ore-solution movement from a discrete source area along the footwall and along the southwestern edge of the eastern ore body. The chemical data are also permissive of a once extant connection between the eastern and western ore bodies through the hanging wall along the northwestern edge of the eastern ore body.

## Ore Localization Controls

As the previous discussions illustrate, many factors have combined to control the locus of ore deposition at Cornwall. The two most obvious, because they are uniformly present throughout this metallogenic province, are the presence of a diabase sheet (as contrasted to later, barren dikes) and of replaceable carbonate beds above the diabase. Others, of variable importance, are briefly summarized below.

(1) Compositional stratigraphy: Limestone, or dolomitic limestone was replaced more easily than the "Mill Hill Slate" or "Blue

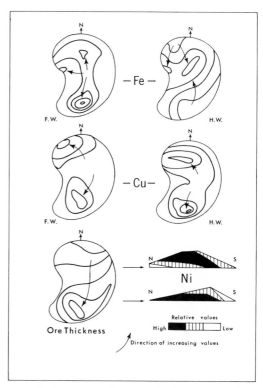

FIG. 6. *Lateral and Vertical Chemical Zoning from Footwall to Hanging Wall of the Eastern Ore Body. Generalized contours delimit maxima and are based on quantitative analyses.*

Conglomerate" hornfels. Compositional laminae in the limestone have yielded laminae of ore alternating with Ca-Mg silicate gangue.

(2) Structure: Faulting has emplaced the replaced host Cambrian and Ordovician (?) units. The form of the sheets and of the ore laminae are, in part, the result of the dips and small scale folds, respectively, of the associated sedimentary and hornfels rocks. Faulting also has cut off limestone and resulted in a damming effect on the ore by hornfels on the east. A major fault offset of the diabase between the eastern and western ore bodies apparently is related genetically to a mineralized shear zone 660 feet below the top of the diabase, to the plunge of an ore body within diabase, and to chemical and mineralogical zonation along the western edge of the eastern ore body.

(3) Mineralogy: Magnetite selectively has replaced earlier minerals and their quantitative distribution has controlled ore concentration. Diopside and actinolite bear an inverse relationship with magnetite abundance. At least

in part, the distribution of these earlier minerals is temperature dependent with respect to distance from the cooling diabase sheet. Magnetite also has replaced preferentially medium grain sizes rather than the very coarse-grained or the very fine-grained metasilicates.

(4) Geochemical environment of the ore fluid: Changes in Eh are reflected in the distribution of hematite at formational contacts; a decrease in this Eh is reflected by the replacement of sulfides and magnetite. Presumably, a change in pH by solution of the $CO_3^=$ radical has played a role both in the migration and precipitation of the invading ore solutions. For the magnetite-pyrite-pyrrhotite association, alkaline ore fluids may be required (26). Somewhat earlier, the crystalization of halogen-bearing minerals (datolite, fluorite, idocrase) may indicate a supercritical or gaseous phase. They might also be used to suggest that halogen complexing was important in the fluids that accomplished the metasomatism previous to the ore stage.

(5) Finally, the composition and differentiation history of the diabasic magma played a primary role. Certain limiting conditions are believed to have prescribed the particular chemical fractionation required to yield both diabase and ore as discrete entities.

## DIABASE AND ASSOCIATED ALTERATION

### Diabase Petrography

Through use of the definitions of basalt types, the Cornwall sheet belongs to the tholeiitic magma type (33, p. 353–354) even though hypersthene is present in the ground mass and in normative calculations. There is no noticeable gravitative layering, although some vertical mineralogical zonation (cryptic layering) is present. Textures, for the most part, are uniformly holocrystalline and sub-ophitic throughout. Exceptions are the chilled margins and the coarser pegmatitic and granophyric differentiates. The Cornwall diabase is essentially normal in that both its mineralogy and chemistry are similar to that of other diabase sheets (47). Variational details, however, can be related to its associated ore mineralization.

The major minerals present are pigeonite, augite, hypersthene, and plagioclase. Minor and trace minerals include olivine (largely in the chilled margin and there amphibolitized), biotite, sericite, chlorite, Fe-Ti oxides, pyrite, enstatite, clinohypersthene (12, p. 103; 47),

hornblende, serpentine, microcline, and quartz. The latter two are present as an intercumulus, micropegmatitic intergrowth. Free quartz is rarely present as discrete grains, although it occurs in the norm and may compose up to 5 per cent of the diabase near the top of the sheet. Numerous exsolution features in pigeonite and hypersthene are characteristic of variable growth rates, decreasing temperature, and progressive compositional change.

Plagioclase occasionally shows both oscillatory and progressive zoning. Excluding the most sodic plagioclase of the granophyre, the average composition is $An_{51.5}$. Composition varies with depth, becoming more sodic 50 to 100 feet below the top, averaging about $An_{55}$ through the major portion of the sheet, and again becoming somewhat more sodic near the base. The overall plagioclase composition is less calcic than for Palisades (12, p. 1070) and Dillsburg (18, p. 685) sheets, but is about the $An_{50}$ composition of typical tholeiites (33, p. 353).

Mapping of the upper diabase surface in the open pit shows that here the chilled margin is unbroken. Although it has been suggested that small granophyric veins may have penetrated into country rock (9, p. 206), there is no real evidence to support this claim. Alteration of this zone is intense, and includes sericitization of plagioclase phenocrysts, amphibolitization (rarely serpentinization) of olivine, dusting of magnetite on the aphanitic ground mass, and, rarely, magnetite replacement immediately below the ore zone.

About 100 feet above the base there is evidence of a discontinuity with characteristics of a second injection of diabase (47). However, more data are required before multiple injection can be firmly substantiated.

The chemical composition of the Cornwall sheet (Table 4) is very similar to that of the Dillsburg sheet (18, p. 690), the Palisade sill (12, p. 1080), the Karroo dolerite (15, p. 649), and average tholeiite (20, p. 134). The oxides MgO and CaO (the "excess" CaO is in pyroxene rather than in plagioclase) are slightly more abundant, and $Al_2O_3$, $TiO_2$, and total Fe slightly less abundant at Cornwall. With respect to variations within the rock type as a whole, this divergence is quite minor. The Dillsburg average is probably not accurate because only samples from near the top of the sheet were analyzed. The Cornwall average was obtained from six samples from the vertical core drilled through the sheet.

Mineralogical data on pyroxene exsolution and inversion indicate that the minimal tem-

*TABLE IV.  Comparative Chemical Compositions of Diabase (Anhydrous) (in Per Cent)*

|                    | (1)    | (2)    | (3)    | (4)* | (5)* |
|--------------------|--------|--------|--------|------|------|
| $SiO_2$            | 51.93  | 53.73  | 51.82  | 52.5 | 51.0 |
| $TiO_2$            | 0.83   | 1.46   | 1.16   | 1.0  | 1.4  |
| $Al_2O_3$          | 13.42  | 16.04  | 14.28  | 15.4 | 15.6 |
| $Fe_2O_3$          | 8.69   | 1.74   | 1.82   | 1.2  | 1.1  |
| $FeO$              | 1.76   | 9.34   | 9.31   | 9.3  | 9.8  |
| $MnO$              | 0.19   | 0.18   | 0.09   | 0.2  | 0.2  |
| $MgO$              | 9.77   | 4.40   | 9.06   | 7.1  | 7.0  |
| $CaO$              | 11.11  | 9.93   | 9.51   | 10.3 | 10.5 |
| $Na_2O$            | 1.79   | 2.53   | 2.17   | 2.1  | 2.2  |
| $K_2O$             | 0.51   | 0.65   | 0.78   | 0.8  | 1.0  |
|                    | 100.00 | 100.00 | 100.00 | 99.9 | 99.8 |
| $Fe_{tot}$         | 7.46   | 8.57   | 8.60   | 8.07 | 8.38 |

* Excludes $P_2O_5$ of 0.1 to 0.2%

(1) Avg. of 6 normal diabase analyses, Cornwall, Pa. (47).

(2) Normal diabase, Dillsburg, Pennsylvania (18, p. 690).

(3) Avg. of 12 diabase analyses, Palisades, N. J. (12, p. 1080).

(4) Avg. Karroo dolerite (15, p. 649).

(5) Avg. tholeiite (20, p. 134).

perature at intrusion was about 1050°C (47). A maximum contact thermal metamorphic temperature of about 600°C, deduced from mineralogical phases and reactions, is in accord with this intrusion temperature for a sheet 1200 feet thick (22).

## Diabase Differentiates and Geochemistry

Differentiation of the Cornwall sheet, while not a conspicuous feature, has produced diabasic pegmatite and granophyre in addition to late-crystallizing interstitial micropegmatite. Because of late stage iron and alkali enrichment in these phases, they merit brief discussion for their pertinence to ore deposition.

With the exception of some diabase pegmatite near the base of the sheet, all the differentiates are within 400 feet of the top. Granophyre more commonly is nearer the top than pegmatite. Both occur as pods, or pockets within normal diabase. The transition between diabase and pegmatite is gradual; between pegmatite and granophyre only slightly transitional, and between diabase and granophyre quite sharp. Pegmatitic diabase also occurs as schlieren and granophyre as small veins extending toward the top but apparently terminating at the chilled margin. Bodies of granophyre have not been found outside the diabase. These two differentiates probably comprise

considerably less than 10 per cent of the total intrusive.

Mineralogy of the diabase pegmatite is less complex than for normal diabase. Augite and hypersthene are the major pyroxenes, with minor pigeonite (usually partly inverted), ferrohypersthene, and hedenbergite. The normal plagioclase of this phase averages about $An_{56}$ and the range of substitution is narrower than in normal diabase. Biotite, micrographic quartz, and microcline, chlorite, and hornblende also may be present. There is a noticeable increase in the amount of Fe-Ti oxide minerals. The amount of Fe and Ca/Mg in the pyroxenes also increases over that in normal diabase. In general, diabase pegmatite represents only a slight chemical and mineralogical differentiation from normal diabase.

Granophyre, on the other hand, evidences a sharp mineralogical and chemical separation. Pyroxene and amphibole are characteristically absent. Plagioclase composition usually ranges between $An_{10}$ and $An_{20}$ but may be considerably more sodic. Myrmekite, graphic granite, and rarely discrete quartz are present. Perthite is absent, Fe-Ti oxide minerals are much less abundant than in diabase pegmatite. Ferriferous and hematitic orthoclase, frequently with optical properties similar to sanidine (47), is abundant and characteristic. It imparts a distinctive pink color to the rock. Fluorite has been noted. Contacts between granophyre and diabase or pegmatitic diabase evidence recrystallization and alteration of earlier feldspars and pyroxenes. Textures are moderately coarse and holocrystalline.

The average chemical composition of the Cornwall granophyre is compared with other diabase granophyres in Table V. All are silica-rich and contain more $Al_2O_3$ than other diabase phases. Alkalies exhibit enrichment. Total iron and $TiO_2$ are variable. Abundances are dependent upon the fractionation, differentiation, and assimilation history of each particular magma, especially with respect to alkali and iron content.

The differentiation trend for the Cornwall sheet compared with that for the Skaergaard, Dillsburg, and Cascades plutons is illustrated on a typical differentiation diagram (Figure 7). The trend follows the tholeiitic trend of the Skaergaard, rather than the calc-alkali trend of the Cascades pluton. A maximum per cent FeO (FeO and $Fe_2O_3$ expressed as total FeO) occurs during the pegmatitic phase at Cornwall. This maximum is considerably less pronounced than at Dillsburg where iron enrichment occurs in rock transitional between diabase pegmatite

TABLE V.  *Comparative Chemical Compositions of Granophyre (Anhydrous) (in Per Cent)*

|  | (1) | (2) | (3) | (4) | (5) |
|---|---|---|---|---|---|
| $SiO_2$ | 66.01 | 64.49 | 73.56 | 64.96 | 58.69 |
| $TiO_2$ | 0.36 | 1.26 | 0.43 | 1.04 | 1.26 |
| $Al_2O_3$ | 16.15 | 12.78 | 12.01 | 13.97 | 11.99 |
| $Fe_2O_3$ | 1.80 | 2.76 | 1.25 | 1.65 | 5.76 |
| $FeO$ | 0.53 | 7.51 | 0.88 | 5.91 | 9.36 |
| $MnO$ | 0.04 | 0.13 | 0.06 | 0.10 | 0.21 |
| $MgO$ | 1.26 | 0.67 | 1.50 | 2.40 | 0.72 |
| $CaO$ | 3.17 | 3.38 | 2.58 | 4.11 | 5.02 |
| $Na_2O$ | 2.19 | 5.22 | 7.22 | 3.13 | 3.90 |
| $K_2O$ | 8.49 | 1.43 | 0.36 | 2.58 | 2.38 |
| $P_2O_5$ | — | 0.36 | 0.14 | 0.15 | 0.71 |
|  | 100.00 | 99.99 | 99.99 | 100.00 | 100.00 |
| $Fe_{tot.}$ | 1.67 | 7.84 | 1.56 | 5.80 | 11.40 |

(1) Avg. of 2, Cornwall, Pa. (47).

(2) Avg. of 2, Dillsburg, Pa. (18, p. 690).

(3) "Diabase Aplite," Palisades, N.J. (12, p. 1080).

(4) Avg. of 7 "metasomatic granophyres," Karroo (15, Table 15).

(5) Hedenbergite granophyre, Skaergaard (11, p. 210).

and granophyre (18, p. 690). That part of the Cornwall trend sub-parallel to the MgO-FeO join is the same as the average tholeiite trend (33, p. 424; 47).

Two points concerning Cornwall diabase differentiation are of special interest. Their significance will be amplified in the discussion of ore genesis. First, although alkalies show a normal total enrichment, this enrichment is in $K_2O$ rather than the more usual $Na_2O$ (Table V) and is present as orthoclase rather than sodic plagioclase. Secondly, the FeO/

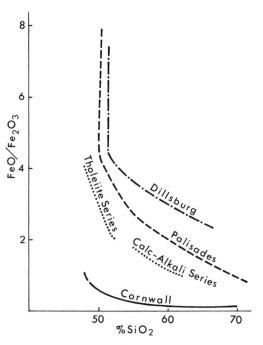

FIG. 8. *Variation of Iron Oxidation State with Silica (compared for diabase sheets and related igneous rock series).*

$Fe_2O_3$, ratio for the Cornwall sheet (Figure 8) is significantly lower than for other diabase. Normally, FeO exceeds $Fe_2O_3$, but at Cornwall (except for the later dikes, broken line, Figure 8), the reverse is true. This diagram emphasizes that with respect to the oxidation of iron during silica enrichment the Cornwall and Dillsburg sheets represent extreme situations. Such an anomalously high Eh, especially early in diabase crystallization, could not help but have had a restrictive influence on the incorporation of iron in the early pyroxenes and olivine and thus also on the iron content of the magma which became the present diabase sheet. Thus, Eh has acted as a partition function with respect to iron incorporation. In this regard, it should be noted again that the total iron content of the Cornwall sheet is rather low, even though the normal differentiation trend of iron enrichment, followed by alkali enrichment, has occurred. Similarly normal (33, p. 429–430) is the slope of the depletion curve (Figure 8) whereby the remaining magma during crystallization becomes enriched in $Fe^3$ relative to $Fe^2$ as differentiation proceeds. The slope suggests an increasing $Po_2$ that approximates a closed system with respect to H and O as proposed by Osborne (25) and modified by Yoder and Tilley (33, Figure 20, p. 429).

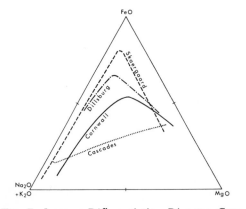

FIG. 7. *Igneous Differentiation Diagram Compared for Diabase Sheets and Related Igneous Rock Series.*

## Contact Metamorphism

Throughout the Triassic belt, mineralogical and textural changes in host rock accompanied the intrusion of diabase sheets. They have been obscured in the vicinity of the ore bodies by metasomatism and hydrothermal alteration superimposed on strictly thermal effects. Furthermore, there is not always a clear time separation among these processes. Although contact metamorphism began earlier, it undoubtedly continued well into the period of metasomatic crystallization. A more detailed account is given by Lapham and Gray (47).

The most abundant contact metamorphic minerals are the ubiquitous diopside, tremolite, and phlogopite, present in all three hornfels rocks: the "Mill Hill Slate", the "Blue Conglomerate", and host limestone. Of the three, diopside is usually the earliest and the closest to diabase contacts. In host limestone, there are, in addition, calcite and garnet. The "Mill Hill Slate" and "Blue Conglomerate" also contain plagioclase, hematite, and epidote. The red Triassic clastics are bleached gray from 300 feet to nearly 1000 feet from diabase contacts. Disseminated red hematite in the ground mass has been converted to black specular-hematite flakes.

Throughout most of the area, the "Mill Hill Slate", rather than the "Blue Conglomerate", is in direct contact with diabase. As such, it represents a somewhat higher temperature facies. Here feldspar exhibits lower obliquity, there is greater solid solution of the sodium molecule in potash feldspar, its grain size is finer, and it contains a mixture of equilibrium assemblages (such as hornblende hornfels, and albite-epidote hornfels) beyond the number that might be predicted by the mineralogical phase rule (47).

Contact metamorphism may have resulted in the loss of volatiles, Si, and K from the hornfels, if it is assumed that these units were originally similar to shale of the Martinsburg Formation (47). However, this assumed loss is dependent upon the unknown original lime content for this facies of the Martinsburg Formation. Total iron has not changed from a composite shale analysis of the Martinsburg Formation to average hornfels, although there has been a considerable Eh reduction (47).

The development of calcic plagioclase was dependent in part upon the original calcium content previous to hornfelsing. Thus the "Mill Hill Slate" contains more plagioclase than the "Blue Conglomerate". There was also competition for Ca or K dependent upon original Si/Al ratios, which in part determined the plagioclase/alkali feldspar ratio (47).

The maximum temperature of contact metamorphism is estimated at about 600°C from the following lines of evidence:

(1) presence at the diabase contact of the hornblende hornfels facies in the temperature range of 550 to 700°C (28, p. 520);

(2) mineral reactions such as that between quartz and muscovite to yield potash feldspar (23) in the neighborhood of 600 to 700°C;

(3) the temperature of the microcline-sanidine transition for ferriferous and normal varieties (36) which places a maximum of about 700°C on the assemblage; and

(4) a probable temperature of about 600°C at the contact of a diabase sheet intruded at about 1100°C according to the calculation method of Jaeger (22).

## Metasomatism

More important to an understanding of the genesis of the Cornwall deposit than thermal metamorphism is the introduction of new constituents into the hornfels zone, particularly into the ore zone. This is especially true from a time viewpoint, since metasomatism bridges the gap between thermal metamorphism and ore deposition. A detailed discussion is given by Lapham and Gray (47). This summary deals largely with potash metasomatism.

Diabase granophyre is enriched in potash. Ferriferous pink orthoclase is characteristic of this facies. Pink, sometimes ferriferous, orthoclase is also present in certain sections of the ore zone, particularly down dip in the eastern ore body where "Blue Conglomerate" lies near diabase. It is thus necessary to examine potash metasomatism for any genetic connection between diabase granophyre and ore.

The so-called Fe-orthoclase in these units is actually intermediate between Fe-sanidine and Fe-microcline. In granophyre and along the ore footwall, its optical properties more closely match those of sanidine; farther up in the ore zone they approach the properties of Fe-microcline (47), or merely of microcline with hematite inclusions. It is not found outside the ore zone. It has not been reported from the ore zone of the Dillsburg area (18), although it occurs sparsely within the granophyre, nor from Morgantown, either within or outside of the ore zone (46). Paragenetic relationships show that it replaces, or transects in veinlets, the diopside-phlogopite-tremolite metamorphic assemblage. In turn, it probably is replaced by magnetite ore, although the relationships often are not definitive.

There are three possible sources of potash metasomatism: (1) release and migration during thermal metamorphism of shale; (2) volatile escape from diabase during granophyre crystallization; and (3) hydrothermal deposition as an early part of the ore-fluid stage.

The chemical changes from Martinsburg shale (assuming it to be the precursor of the present "Mill Hill" and "Blue Conglomerate" hornfels units) to contact metamorphic and metasomatic units are compared in Table VI. In general terms, oxides which increased in weight per cent from shale to metamorphic hornfels, decreased again during metasomatism. This effect is most noticeable for $K_2O$. Thus loss during contact metamorphism could quantitatively explain a complementary gain during metasomatism, even on a one-to-one weight per cent basis:

$$\text{Metamorphic Loss} \cong \text{Metasomatic Gain}$$
$$(5.00) - (3.04) \cong (7.04) - (5.00)$$
$$1.96\% \cong 2.04\%,$$

where 3.04 per cent $K_2O$ is the metamorphic hornfels average and 7.04 per cent $K_2O$ is the metasomatic hornfels average.

However attractive this closed system balance might seem, it is too simple an explanation to be completely satisfactory. The hornfels units would have had to loose potassium throughout a considerable area, yet retain it as potash feldspar only within the ore zone. This seems unlikely. Considerable enrichment in potash feldspar would be expected wherever temperature and the Si-Al content of the primary shales were suitable. This condition should have been met easily outside the limits of the ore zone (47). Such a potassium migration also would require not only that potassium move parallel to, but also up a thermal gradient provided by the cooling diabase. This too seems unlikely. Finally, this process might have produced microcline (and perhaps did, to a small extent), but there is no reason why the rather unusual Fe-microcline and Fe-sanidine should result. For these, and other reasons (47), it is believed that this mechanism was largely ineffective with respect to potash metasomatism.

Assuming for the moment that ore deposition is unrelated to granophyre crystallization (see "Diabase Differentiates and Geochemistry" and "ORE GENESIS"), a choice between them as a source of potassium is difficult to make. Normal diabase is slightly $K_2O$ deficient (Table IV) and granophyre is not only $K_2O$-rich but also contains the Fe-orthoclase found in the ore zone. On the other hand, granophyre at Cornwall is of quite small extent, is usually well below the top of the sheet, and does not transect the chilled margin as far as is known. In addition, Fe-orthoclase-bearing granophyre appears sparsely elsewhere in Triassic diabase, but Fe-orthoclase beyond the confines of diabase has not been reported. The restriction of its occurrence to this ore zone, and, furthermore, its concentration in that ore zone area where chemical and mineralogical zonation are focused, is taken to be indicative of a genetic association between potash metasomatism and ore rather than between granophyre and ore. Certainly an alkali-rich (and alkaline) ore fluid is compatible with the meagre knowledge of magnetite ore fluid transport. If this mechanism is correct, the slight diabase deficiency in $K_2O$ would represent a rather early separation of potassium (with iron) from the diabase magma.

TABLE VI.    *Metasomatic Chemical Variation\* (in Per Cent)*

|  | Martinsburg Shale Avg.\*\* | "Blue Conglomerate" | | "Mill Hill Slate" | |
|---|---|---|---|---|---|
|  |  | (1) | (2) | (1) | (2) |
| $SiO_2$ | 56.50 | 52.11 | 55.08 | 58.91 | 59.69 |
| $TiO_2$ | 0.99 | N.A. | N.A. | 0.91 | 0.88 |
| $Al_2O_3$ | 19.52 | 23.75 | 23.90 | 14.76 | 16.45 |
| $Fe_2O_3$ | 7.53 | 3.66 | 3.58 | 3.49 | 2.45 |
| FeO | 1.48 | 5.90 | 3.29 | 3.47 | 1.46 |
| MnO | N.A. | N.A. | N.A. | 0.20 | 0.15 |
| MgO | 1.93 | 2.72 | 2.06 | 1.86 | 0.90 |
| CaO | 0.40 | 2.54 | 0.48 | 5.84 | 4.98 |
| $Na_2O$ | 0.71 | 3.98 | 3.72 | 3.66 | 2.89 |
| $K_2O$ | 5.00 | 2.41 | 6.09 | 3.67 | 7.99 |
| $CO_2$ | 0.59 | 0.66 | 0.29 | 0.15 | 0.74 |
| $H_2O$ | 4.84 | 2.27 | 1.51 | 3.07 | 1.42 |

\* N.A. = Not Analyzed.

\*\* Average of 8 Lebanon County samples (38, p. 234–249)

(1) Contact thermal metamorphic hornfels samples (47)

(2) Metasomatic hornfels samples (47)

## ORE GENESIS

### The Cornwall Deposit

The close spatial association of magnetite ore and diabase at Cornwall, and elsewhere in the Triassic Province, always has been considered conclusive evidence of a genetic relationship. Of the several theories proposed (1,2,3,9,29), two major hypotheses will be discussed here: 1) that both a spatial and chronological genetic association exist, and 2) that the spatial association is the result of a time and chemico-physical process separation.

The only compelling reason for proposing that iron-bearing ore fluids issued directly from the crystallizing diabase is the constancy of their association. Additional lines of evidence lend largely superficial support: the presence of ore generally above diabase, crude mineralogical zoning above the top of the sheet, and the presence of late-stage differentiates in diabase which exhibit iron enrichment.

The data previously cited vitiate the diabase explusion theory for the Cornwall deposit and are summarized as follows:

(1) The paragenetic sequence indicates a time lapse (of unknown extent) between solidification of the chilled margin and ore deposition. This lapse first included contact thermal metamorphic minerals and, somewhat later, metasomatic minerals. The chilled margin is apparently continuous (except possibly for minor granophyric escape) and contains contraction cracks filled by contact-metamorphic diopside that, in turn, is replaced by magnetite.

(2) Although a pegmatitic differentiate is enriched in iron, there is no evidence that it penetrated the upper chilled margin. The much less iron-rich granophyre may penetrate this margin but only to a very small extent. Neither type is confined to the ore zone. Furthermore, both are present at localities elsewhere in the diabase whether or not there is ore.

(3) There is no textural or feeder evidence from within the diabase for the concentration or expulsion of material. The composition of the sheet (except possibly near the base) appears to be uniform. An originally high concentration of iron in one part of the sheet seems unlikely.

(4) The ore is spatially related to a metasomatic aureole, principally of chlorite and actinolite, rather than to the distribution of earlier minerals.

(5) Diabase expulsion of iron fails adequately to explain the chemical and mineralogical zoning, especially with respect to a major fault through diabase and the spatially related ore veins, shears, and footwall ore within diabase located approxiately between the two major ore bodies.

For the above reason it is concluded that the source of iron was not the immediately adjacent diabase but was a fractionated magma

TABLE VII.  *Averages of Total Iron in Diabase and Basalt*

| Location | %$Fe_{tot.}$ |
| --- | --- |
| Tasmania (14) | 6.7 |
| Cornwall, Pa. (47) | 7.5 |
| Downes Mountain (13) | 7.8 |
| 5 North American Triassic dikes (47) | 7.9 |
| 9 North American Triassic sheets (47) | 8.0 |
| Dillsburg (18) | 8.0 |
| Karroo (15) | 8.1 |
| Palisades (12) | 8.3 |
| Avg. Tholeiite (20) | 8.4 |
| Avg. Basalt (8) | 8.8 |
| Avg. Tholeiite (19) | 9.1 |
| Whin Sill (5) | 9.3 |

that yielded, separately, diabase and ore solutions. The Cornwall portion of the diabase sheet appears to be somewhat depleted in total iron, on the order of 0.5 per cent to 0.7 per cent Fe (Table VII) assuming a normal iron content of about eight per cent Fe for diabase magma. Volume calculations show that the volume of the sheet in the vicinity of Cornwall is sufficient to have yielded the Cornwall ore body based on this magnitude of iron deficiency (47).

From this theory it follows that the spatial association of diabase and ore is primarily dependent upon the presence of a replaceable horizon, upon a channel along the base of the sheet, and upon faulting through the diabase as access to the replaceable horizon. The nature of this faulting is probably not fortuitous but is a requirement of the structural geometry at the depressed center of the saucer-shaped sheet where Triassic and pre-Triassic sediments are in thrust contact.

Although the objections to an expulsion theory are satisfied by this proposal, a plausible mechanism for magma fractionation is required. Such a mechanism is discussed in greater detail by Lapham and Gray (47). In brief, this mechanism is dependent upon Eh conditions in the initial magma. From the high $Fe^3/Fe^2$ ratio of Cornwall diabase, note has been made of the difficulty of incorporation of ferric iron in the early silicates. Pockets of high Eh in otherwise normal magma have recently been noted in Hawaiian tholeiite (44). The magma source site of the Cornwall diabase may well have contained such pockets. Iron accumulation could have been promoted by gravity or convection gradients and/or the accumulation of alkali-rich volatile differentiates.

The fundamentally important role of oxidation with respect to the course of iron crystallization, such as the Bowen (calc-alkali) and Fenner (tholeiitic) trends, has been discussed recently by Muan and Osborn (21), Osborn (25), Yoder and Tilley (33) and many others. Normally, a tholeiitic magma could be expected to trend toward iron enrichment without noticeable total silica enrichment, a trend exemplified by the Skaergaard pluton (11) and the Palisade diabase (12). The Cornwall diabase follows this differentiation trend (Figure 7) but with two subtle distinctions: there has been a small volume of considerable silica enrichment in granophyre and the late stage pyroxenes are less iron-rich than, for example, in the Dillsburg (16, p. 13; 18, p. 685–687) or Karroo (15) sheets. Thus, a uniquely high Eh has obviated somewhat the iron enrichment trend. It is probably also responsible for the low-olivine content (33, p. 375, 425–426) and perhaps for the formation of hypersthene (33, p. 425–426) in the Cornwall sheet.

Investigation of one drill hole, which penetrated the diabase, indicates that the basal 100 feet may represent a later injection of diabase, partially chilled against the main, overlying pluton. A genetic relationship could exist between such a younger diabase and the ore process, but there are insufficient available data for substantiation. Some late diabasic magma could be expected if separation by fractionation of the main mass was not complete at the time of initial intrusion. The late diabase dikes are probably additional evidence of an incomplete physicochemical separation within the chambers of the source magma.

## Magnetite of the Triassic Metallogenic Province

Somewhat unusual conditions have been hypothesized during the formation of the Cornwall diabase magma. If such conditions led to the formation of magnetite ore as an entity separate from diabase crystallization, then the supposedly similar magnetite deposits throughout the Triassic Province require re-examination. Although such a discussion lies outside the scope of this report, a few summary points should be mentioned.

The magnetite deposits of the Triassic basin are illustrated in Figure 9. Any comparison of the magnetite districts must recognize that there are four distinct plutons composed of smaller sheets. Each pluton has at least some concentration of magnetite associated with it. The Dillsburg sheet contains considerable grano-

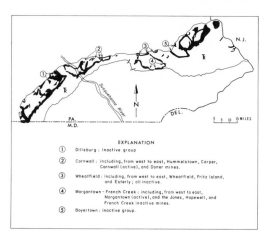

EXPLANATION

① Dillsburg : inactive group

② Cornwall : including, from west to east, Hummelstown, Corper, Cornwall (active), and Doner mines.

③ Wheatfield : including, from west to east, Wheatfield, Fritz Island, and Esterly ; all inactive.

④ Morgantown - French Creek ; including, from west to east, Morgantown (active), and the Jones, Hopewell, and French Creek inactive mines.

⑤ Boyertown : inactive group.

FIG. 9. *Province Map of Triassic Magnetite Districts.*

phyre and scattered, quite small volume, magnetite deposits. Late differentiates evidence considerable iron enrichment and the diabase as a whole is not noticeably or only slightly, iron deficient (Table VII). The Cornwall sheet contains much less granophyre, less late iron enrichment, exhibits notable iron deficiency, and is associated with a considerably larger magnetite concentration. The Morgantown-French Creek district apparently contains little granophyre, if any, and the concentration of magnetite is even more extensive than at Cornwall. Ore abundance thus appears to bear an inverse relationship to quantity of diabase granophyre and presumably a direct relationship to alkalic volatiles withheld for the medium of later ore transport. Such variation from southwest to northeast is more likely to be causal than coincidental. However, there is no reason to impose the same mechanism of ore emplacement upon each district. It is entirely compatible for magnetite to be directly related to late diabase differentiates at Dillsburg, while being related to a fractionated source at Cornwall. Retention of volatiles, partition of iron, and initial Eh conditions, therefore, could have played the primary role in the mode of Triassic magnetite formation.

To refer to all of these ore bodies as deposits of the Cornwall type is probably something of a misnomer. Ore resulting directly from late stage diabase differentiates should be termed Triassic magnetite of the Dillsburg type; ore resulting primarily from source-magma fractionation is believed to be exemplified by Triassic magnetite of the Cornwall (and, presumably, Morgantown) type.

## REFERENCES CITED

1. Rogers, H. D., 1858, The geology of Pennsylvania, a government survey: Lippincott, Phila. [(First) Geol. Surv. Pa.], v. 2, 1046 p., particularly p. 718–720.

2. Lesley, J. P. and d'Invilliers, E. V., 1886, Report on the Cornwall iron ore mines, Lebanon County: (Second) Geol. Surv. Pa., Ann. Rept. for 1885, p. 491–570.

3. Spencer, A. C., 1908, Magnetite deposits of the Cornwall type in Pennsylvania: U.S. Geol. Surv. Bull. 359, 102 p.

4. Newhouse, W. H. and Callahan, W. H., 1927, Two kinds of magnetite: Econ. Geol., v. 22, p. 629–632.

5. Holmes, A. and Harwood, H. F., 1928, The age and composition of the Whin sill and the related dikes of North of England: Min. Mag., v. 21, p. 493–542.

6. Callahan, W. H. and Newhouse, W. H., 1929, A study of the magnetite ore body at Cornwall, Pennsylvania: Econ. Geol., v. 24, p. 403–411.

7. Stose, G. W., 1932, Geology and mineral resources of Adams County, Pennsylvania: Pa. Geol. Surv., 4th ser., Bull. Cl, 153 p.

8. Daly, R. A., 1933, Igneous rocks and the depth of the earth: McGraw-Hill, N.Y., 598 p.

9. Hickok, W. O., 4th, 1933, The iron ore deposits at Cornwall, Pennsylvania: Econ. Geol., v. 28, p. 193–255.

10. Lindgren, W., 1933, Mineral Deposits: 4th ed., McGraw-Hill, N.Y., 930 p.

11. Wager, L. R. and Deer, W. A., 1939, Geological investigations in East Greenland, Part III, The petrology of the Skaergaard intrusion, Kangerdluzssuaq, East Greenland: Medd. om Grønland, v. 105, 335 p.

12. Walker, F., 1940, Differentiation of the Palisade diabase, New Jersey: Geol, Soc. Amer. Bull., v. 51, p. 1059–1106.

13. Walker, F. and Poldervaart, A., 1940, The petrology of the dolerite sill of Downes Mountain, Calvinia: Geol. Soc. Africa Tr., v. 43, p. 159–173.

14. Edwards, A. B., 1942, Differentiation of the dolerites of Tasmania, Part I: Jour. Geol., v. 50, p. 451–480.

15. Walker, F. and Poldervaart, A., 1949, Karroo dolerites of the Union of South Africa: Geol. Soc. Amer. Bull., v. 60, p. 591–706.

16. Hotz, P. E., 1950, Diamond-drill exploration of the Dillsburg magnetite deposits, York County, Pennsylvania: U.S. Geol. Surv. Bull. 969-A, p. 1–25.

17. ———— 1952, Form of diabase sheets in southeastern Pennsylvania: Amer. Jour. Sci., v. 251, p. 375–388.

18. ———— 1953, Petrology of granophyre in diabase near Dillsburg, Pennsylvania: Geol. Soc. Amer. Bull., v. 64, p. 675–704.

19. Nockolds, S. R., 1954, Average chemical compositions of some igneous rocks: Geol. Soc. Amer. Bull., v. 65, p. 1007–1032.

20. Poldervaart, A., 1955, Chemistry of the earth's crust: Geol. Soc. Amer. Spec. Paper 62, p. 119–141.

21. Muan, A. and Osborne, E. F., 1956, Phase equilibria at liquidous temperatures in the system $MgO-FeO-Fe_2O_3-SiO_2$: Amer. Ceram. Soc. Jour., v. 39, p. 121–140.

22. Jaeger, J. C., 1957, The temperature in the neighborhood of a cooling intrusive sheet: Amer. Jour. Sci., v. 255, p. 306–318.

23. Winkler, H. G. F., 1957, Experimentelle Gesteinsmetamorphose, I: Geochim. et Cosmochim. Acta, v. 13, p. 42–69.

24. Geyer, A. R., *et al.*, 1958, Geology of the Lebanon quadrangle: Penna. Geol. Surv., 4th ser., Geol. Atlas Pa., Atlas 167C, 1:2000.

25. Osborne, E. F., 1959, Role of oxygen pressure in the crystallization and differentiation of basaltic magma: Amer. Jour. Sci., v. 257, p. 609–647.

26. Barnes, H. L. and Kullerud, G., 1960, Equilibrium between pyrite, pyrrhotite, magnetite, and aqueous solutions: Ann. Rept. Dir. Geophys. Lab., Carnegie Inst. Wash. Year Book 59, p. 135–137.

27. Lapham, D. M., 1960, Photomicrography of the Cornwall ore body, Cornwall, Pennsylvania: Geol. Soc. Amer. Bull., v. 71, pt. 2, p. 1913.

28. Turner, F. J. and Verhoogan, J., 1960, Igneous and metamorphic petrology: McGraw-Hill, N.Y., 694 p.

29. Gray, C. and Lapham, D. M., 1961, Guide to the geology of Cornwall, Pennsylvania: Pa. Geol. Surv., 4th ser., Bull. G35, 18 p.

30. Socolow, A. A., 1961, Geologic interpretation of certain aeromagnetic maps of Lancaster, Berks and Lebanon Counties: Pa. Geol. Surv., 4th ser., I. C. 41, 19 p.

31. Fanale, F. P. and Kulp, J. L., 1962, The helium method and the age of the Cornwall, Pennsylvania magnetite ore: Econ. Geol., v. 57, p. 735–746.

32. Lapham, D. M., 1962, Magnetite mine, Cornwall, Pennsylvania: Int. Mineral. Assoc., Knoxville Meeting, Northern Field Excursion Handbook, 3rd Congress, p. 36–40.

33. Yoder, H. S. and Tilley, C. E., 1962, Origin of basalt magmas: an experimental study of natural and synthetic rock systems: Jour. Petrol., v. 3, p. 342–532.

34. Glaeser, J. E., 1963, Lithostratigraphic nomenclature of the Triassic Newark-Gettysburg Basin: Pa. Acad. Sci. Pr., v. 37, p. 179–188.

35. Sanders, J. E., 1963, Late Triassic tectonic history of northeastern United States: Amer. Jour. Sci., v. 261, p. 501–524.

36. Wones, D. R. and Appleman, D. E., 1963, Properties of synthetic triclinic $KFeSi_3O_8$, iron microcline, with some observations on

the iron microcline ⇌ iron-sanidine transition: Jour. Petrol., v. 4, p. 131–137.

37. O'Neill, B. J., 1964, Limestones and dolomites of Pennsylvania, Part I: Pa. Geol. Surv., 4th ser., Bull. M50, 40 p.

38. O'Neill, B. J., *et al.*, 1964, Properties and uses of Pennsylvania shales and clays: Pa. Geol. Surv., 4th ser., Bull. M51, p. 234–249.

39. Beck, M. E., Jr., 1965, Paleomagnetic and geological implications of measured magnetic properties of the Triassic diabase of southeastern Pennsylvania: Jour. Geophys. Res., v. 70, p. 2845–2856.

40. Davidson, A. and Wyllie, P. J., 1965, Zoned magnetite and platy magnetite in Cornwall type ore deposits: Econ. Geol., v. 60, p. 766–771.

41. Lapham, D. M. and Geyer, A. R., 1969, Mineral collecting in Pennsylvania: Pa. Geol. Surv., 4th ser., Bull. G33, 3d ed., p. 97–102.

42. de Boer, J., 1966, Paleomagnetic-tectonic study of the Mesozoic dikes in the Appalachians (abs.): Geol. Soc. Amer. NE Sect. Program, p. 13.

43. Pierce, K. L. and Armstrong, R. L., 1966, Tuscarora fault, an Acadian (?) bedding plane fault in central Appalachian valley and ridge province: Amer. Assoc. Petrol. Geols. Bull., v. 50, p. 385–390.

44. Wright, T. L. and Sato, M., 1966, Oxygen fugacities of magmatic gases from crystallizing Hawaiian tholeiite (abs.): Amer. Geophys. Union Tr., v. 47, p. 209.

45. MacLachlan, D. B., 1967 (in press), Structure and stratigraphy of the limestones and dolomites of Dauphin County, Pennsylvania: Pa. Geol. Surv., 4th ser., Bull. G44.

46. Sims, S. J., in press, The Grace mine magnetite deposit, Berks County, Pennsylvania: Graton-Sales Volume, A.I.M.E., N.Y.

47. Lapham, D. M. and Gray, C., in preparation, Geology and origin of the Triassic magnetite deposits and diabase at Cornwall, Pennsylvania: Pa. Geol. Surv., 4th ser., Bull. M56.

# 5. Geology of the Friedensville Zinc Mine, Lehigh County, Pennsylvania

WILLIAM H. CALLAHAN*

## Contents

## Illustrations

## Tables

* The New Jersey Zinc Company, Franklin, New Jersey.

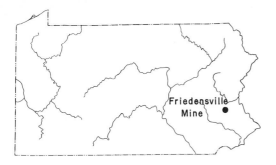

## ABSTRACT

The Friedensville zinc mine of The New Jersey Zinc Company is located about four miles south of Bethlehem, Pennsylvania in the Saucon Valley, an infolded and down faulted block of Cambro-Ordovician carbonate sediments surrounded by hills of pre-Cambrian gneiss and Cambrian quartzite except where breached by Saucon Creek at its entrance into the Lehigh River.

The Beekmantown formation of lower Ordovician age is the host for zinc and limonite ores, the former occurring in the lower dolomitic portion thereof and the latter in the upper interbedded dolomite and limestone portion of the section. The Beekmantown is overlain unconformably by the Jacksonburg of middle Ordovician age and according to Miller (2) underlain conformably by about 2500 feet of Cambrian dolomites and 200 feet of the basal Cambrian Hardyston quartzite. Igneous rocks, younger than the Cambro-Ordovician carbonate rocks, are absent in the vicinity of the ore deposits.

The major structure is a N60°E-trending saddle-shaped doubly plunging anticline overturned to the north. A syncline lies to the north of the anticline. A thrust fault of substantial displacement repeats the south limb of the anticline.

The ore consists of sphalerite and pyrite, along with gangue dolomite and quartz, filling and replacing the sedimentary type matrix of a strata-bound solution collapse breccia in the dolomitic portion of the lower Beekmantown. This breccia was produced in pre-Jacksonburg time as a consequence of uplift, erosion, and development of a karst topography and subsurface drainage system in the Beekmantown.

Deposits of both zinc and limonite are present on the vertical north limb, on the crest and on the gently dipping (25°) south limb of the Friedensville anticline; consequently, there is no preferred structural setting for mineralization. Rather this distribution supports the view that the folding is post-ore. Features confirming this view are the development of flow cleavage in the breccia matrix, the deformation of some of the sphalerite, and the fracturing of the pyrite mineralization therein. There is nothing to indicate that the age of mineralization is other than Ordovician and older than Taconic deformation. In the absence of intrusive igneous activity younger than the ore host, the writer suggests that submarine volcanic exhalations associated with the volcanic activity recorded in the eastern eugeosynclinal facies of the Ordovician may be the source of the metal ions.

## HISTORICAL INFORMATION

Early in the 19th Century an unusual mineral was observed on the farm of Jacob Ueberroth. Test thereof in an iron furnace in 1830 yielded no metal as the zinc volatilized and escaped. W. T. Roepper in 1845 identified the mineral as calamine, zinc silicate. Wetherill in 1853 erected furnaces at South Bethlehem to produce zinc oxide from the ore. F. Osgood in 1859 opened the Correll Mine and built a spelter furnace to produce metal. During the period 1860 to 1893, operations were conducted intermittently from the Ueberroth, Old Hartman, New Hartman, Correll and Triangle mines; in 1893, mining was abandoned. Competition from the high grade New Jersey ores and the increasingly substantial inflows of water encountered as the mines were deepened contributed to their demise. For a period during the 1870's, a Cornish pump, reputedly the largest built, operated at the Ueberroth Mine pumping 12,000 to 16,000 gallons per minute from a depth of the order of 250 feet. The serious hydrological problem disclosed by this pumping inhibited exploration and development of the area for many years. The New Jersey Zinc Company acquired the Ueberroth, Old Hartman, New Hartman, and Triangle mines in 1899 through merger with the previous owner, The Lehigh Zinc Company. Additional property was purchased intermittently thereafter.

Several calyx (shot) drill holes were used to explore for extensions of ore in 1899 and again in 1914. During 1916 and 1917, water was pumped from the New Hartman to permit sampling. Drilling in 1924 to explore for an extension of the New Hartman ore body demonstrated, in fairly widely spaced holes, a tonnage and grade that, in a more normal hydrologic environment, would be considered economic to mine. However, exploitation was deferred. During the period 1937 to 1942 drilling was done to prove up the tonnage and grade by more closely spaced holes within the previously delineated ore outline, and hydrologic studies and pumping tests were conducted, all to reduce the uncertainties involved in again undertaking exploitation of the deposits. The writer was in charge of field activities during this period.

Preparation for shaft sinking started in 1945 with the drilling of holes to test the proposed site. Pressure grouting of these holes started in 1947, and, following completion of this work, shaft sinking, placing of concrete lining, and grouting were done in stages. During this operation, level stations were turned off, and pump stations, crusher station, and ore storage pocket were excavated. The shaft was bottomed at 1261 feet in August, 1952. More than the usual tribulations were encountered in this operation.

Excavation of sumps, which involved some pressure grouting, and installation of permanent pumping plants were completed in 1955, and development was started. This involved drifting on the 400 level and the sinking of an incline in drained ground to provide another opening for an escapeway, ventilation, and access for mobile equipment. Flows encountered in drifting reached 15,000 gallons per minute with occasional total pump burdens up to 24,000 gallons per minute. Development work and drawdown of the water table were sufficiently advanced early in 1958 to permit mining to begin. Pump capacity is now rated at 26,000 gallons per minute. Pumpage is presently stabilized at 19,000 to 20,000 gallons per minute.

Miller (2, p. 74) reports production from Friedensville during its early history as estimated at 50,000 tons of spelter and 90,000 tons of zinc oxide valued at approximately $20,000,000. Production by The New Jersey Zinc Company started in 1958. 2,859,000 tons of ore averaging 6.5 per cent zinc were milled up to the end of 1964. The operation is designed to mine and mill 2500 tons of ore per day.

Through the years many geologists have contributed information regarding the deposits or their environment. H. S. Drinker, F. A. Genth, F. L. Clerc, J. P. Lesley, J. F. Kemp, B. L. Miller, Robert D. Butler, Donald M. Fraser, and Geo. W. Stose—Anna I. Jonas, have published their observations. J. F. Kemp, Frank L. Nason, W. O. Hotchkiss, and A. C. Spencer in their capacity as consultants contributed private reports as have our employees,

A. W. Pinger, R. B. Hoy, R. W. Bridgman, R. L. Schumacher, Johnson Crawford, W. T. Forsyth, Noel Moebs, A. H. Willman, and R. W. Metsger. Josiah Bridge (1940) identified fossils from the area and during 1946 to 1947, W. O. Robinson and F. J. Herman of the U.S. Department of Agriculture studied the geobotanical-geochemical aspects of the soils and of the high-zinc slime ponds resulting from earlier operations. F. W. Scheidenhelm consulted on hydrology.

## GEOGRAPHY AND TOPOGRAPHY

Friedensville is located in Upper Saucon Township, Lehigh County, Pennsylvania, in the southcentral part of the Allentown quadrangle of the U.S. Geological Survey. It is situated in the Saucon Valley, a small re-entrant of the Great Valley Province into the Reading prong of the New England upland which bounds it to the southeast as shown on Figure 1. This re-entrant is a Paleozoic lowland almost entirely rimmed by ridges of Precambrian rocks and Cambrian quartzite except where it is connected with the Great Valley through a breach in the rim east of Bethlehem.

The enclosing ridges rise to a maximum of about 500 feet above the valley floor and, as a part of the Reading prong, are continuous with the Highlands of New Jersey. The Saucon Valley, immediately underlain by lower Paleozoic carbonate rocks, is an area of low relief, plus or minus 100 feet, at an average elevation of 375 feet above sea level. The entire valley, a watershed of approximately 58 square miles, is drained by Saucon Creek, a northeasterly flowing tributary of the Lehigh River which it enters through the breach referred to above. The topography of the valley is of the karst type, characterized by rounded hills and sink holes. The presence of many sink holes, the fractured, cavernous, and weathered character of the upper 500 feet penetrated in the exploration drilling, and the severe hydrologic conditions encountered in mining all indicate that at least locally a substantial volume of water is transmitted underground by percolation and permeation.

## GEOLOGY

In reading what follows regarding the features associated with ore and its environment, it is to be remembered that much of the data are biased, having been obtained in the immediate vicinity of the mineralization. This situa-

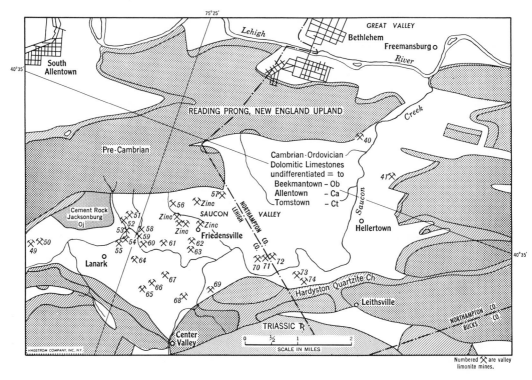

FIG. 1.  *Portion of Miller, et al. Geologic Map, Allentown Quadrangle Pennsylvania.*

tion is not uncommon in mineralized areas where effort is directed to avoid barren ground. To what extent some of these features are common in the same geologic environment remote from mineralization is indeterminate because of lack of outcrops, drill holes, and mine openings. This being so, there is a tendency to try to interpret all features of the mineralized environment in terms of their relation to the ore-forming process whereas it is probable that very few of these features operated to cause or control that process. The remaining features either were completely neutral in the relationship to ore formation or perhaps even are distractions to the observer. Much of the controversy regarding ore genesis stems from this difficulty of distinguishing, in a mineralized environment, the unique from the ubiquitous and the temporal aspects of features now spatially associated.

## Stratigraphy

The zinc deposits at Friedensville, and the limonite deposits concentric thereto, occur in the Beekmantown formation, a carbonate unit of Lower Ordovician age. Above the Beekmantown, and separated therefrom by an erosional unconformity, is the Jacksonburg formation of Middle Ordovician age. Conformably below the Beekmantown in succession downward are the Cambrian formations designated locally as Allentown and as Tomstown, both dolomitic, and the Hardyston quartzite. The latter is unconformable with the underlying Precambrian.

Figure 2 shows the regional stratigraphic column as originally determined by Miller (2) and modified as to Beekmantown units by the writer, Hoy, and Spencer in private reports to The New Jersey Zinc Company. Because the Jacksonburg, Allentown, Tomstown, and Hardyston formations are not ore-significant they will not be described here. Attention will be focused on the Beekmantown formation, the ore host, about which considerable information is available from a quite small area.

The Beekmantown section at Friedensville is pieced together from information obtained in mine workings, surface, and underground and from about 170 drill holes, none of which penetrated the entire section. Mine workings have contributed the least information inasmuch as they are largely confined to ore-bearing strata. No complete section of the Beekmantown is exposed in outcrop. The thickness is indicated to be of the order of 1000 to 1400 feet.

The Beekmantown is a limy dolomite with

| Thickness | System | FORMATION | MEMBER NAME | MEMBER FEET | LITHOLOGY |
|---|---|---|---|---|---|
| | | MARTINSBURG | | | Shales and slates |
| 700' | ORDOVICIAN | JACKSONBURG Oj | Cement Rock | 300' | Thin-bedded, highly argillaceous limestone. |
| | | | Cement Limestone | 400' | Thin-bedded, argillaceous limestone. |
| 1225' | ORDOVICIAN | BEEKMANTOWN Ob | | 180' | Limestone and Limy Dolomite. |
| | | | U.Sg | 200' | Limestone and Dolomite, local sandgrains. |
| | | | | 300' | Limestone and Dolomite. |
| | | | C.M. | 125' | Dolomite, local sandgrains. |
| | | | | 100' | Light Gray Dolomite |
| | | | Thc. | 200' | Dolomite, local breccia, Ore horizon. |
| | | | E.M. | 120' | Shaley Dolomite or limestone. |
| 1600' | CAMBRIAN | ALLENTOWN Ca | | 130' | Dolomite, local shale partings. |
| | | | | 10' | Dolomite, local sandgrains. |
| | | | | 360' | Dolomite, Some intervals resemble the local layers Sandstone, Oolites and Cryptozoa. |
| | | | | 1100' | Interbedded cryptozoan reefs, oolite beds, shaley beds, and sandstone layers. Result of "cyclic" deposition. |
| 900' | CAMBRIAN | TOMSTOWN Ct | | | Massive dolomites interbedded with calcareous shales. Occasional sandstone beds. Oolites near top. |
| 200' | | HARDYSTON | | | White to purple grits becoming locally arkosic to jaspery. |
| | PRE-C | PRE-CAMBRIAN PЄ | | | Gneiss, schist, and marble. |

FIG. 2. *Generalized Stratigraphic Column Lehigh and Saucon Valleys Pennsylvania.*

local zones of limestone and dolomite, ranging in color from creamy through shades of gray to black. In general, the lower portion is dark-colored and dolomitic (the Rickenbach unit of Drake, 11, p. L4) and the upper portion light-colored and limy (the Epler unit of Drake, 11, p. L5). The section contains a considerable variety of beds, characterized as ribbony, thin, cherty, sandy, conglomeratic, and brecciated, that are interstratified with the more massive ones. A few of them, specifically the light-colored beds containing scattered silica sand grains, such as the Callahan Marker (CM) and the Upper Sand Grain Maker (USgM), have been useful as key beds. One dark colored sandy bed, the Trihartco Marker (Thc), is in the ore zone and, therefore, is quite useful even though thin, somewhat discontinuous, and difficult to identify in such a mineralized environment. Because of the paucity of distinctive features in the section to serve as key beds, the knowledge of sedimentation and structure in the area is quite general. Insoluble residue studies failed to find additional key beds.

The bottom and top boundaries of the Beekmantown are placed at easily recognized lithologic changes. The bottom is placed at the base of the Evans Marker (E.M.), a medium-to dark-gray thinly bedded dolomite that is locally a limestone, characterized by thin anastomosing shale partings. This is to be contrasted with the underlying beds designated Allentown which are interbedded light, medium and dark gray dense saccharoidal, generally massive dolomites. The upper Beekmantown is characterized by light-colored interbeds of limestone, dolomitic limestone and limy dolomite, all of which contrast sharply in appearance with the overlying dark-gray to black shaly limestone of the Jacksonburg.

One fossil found in drill core about 6 feet stratigraphically below the base of the Evans Marker was identified by Josiah Bridge (private communication) as Ophileta levata and another as probably referable to Gasconadia. He commented that "Both forms are characteristic of the Stonehenge and the equivalent of the Tribes Hill limestone of New York. We consider these forms to mark the horizon at or very close to the base of the Ordovician." Whether or not the Evans Marker is at the exact base of the Ordovician, it is a useful geologic key bed of distinctive character and wide extent, having been found where expected in drilling in the Saucon Valley and in regional mapping in the Lehigh Valley. The top of the Beekmantown is not characterized by any particular bed because the upper portion was eroded prior to deposition of the Jacksonburg. However, there can be little doubt as to the nature of this formational boundary or as to the identification of the overlying formation as Jacksonburg because the features of the Jacksonburg have been studied intensively in connection with its use as a raw material for the Lehigh Valley cement industry.

Bridge also identified Lecanospira of Beekmantown age among fossils collected by Lawrence Whitcomb of Lehigh University from a quarry identified by the writer as having a stratigraphic position between the Callahan and the Upper Sand Grain Markers. Fossils collected by the writer from drill core which penetrated this interval were reported on by Bridge as follows:

"I see sections which may be those of small cephalopods and gastropods. They resemble forms which occur in the Stonehenge and lower Nittany divisions of the Beekmantown, and that is about all that can be said at present."

Consequently, all the evidence available confirms and none denies that the ore host is the Beekmantown formation of Lower Ordovician age as reported by Miller (2).

From the standpoint of mineralization, the most important unit in this formation is that designated "Breccia Sedimentary" which brackets the Trihartco Marker. The term "Breccia Sedimentary" is used to emphasize that the breccia did not result from deformation but rather is composed of fragments derived from solution-collapse cemented by finer detritus resulting from the solution process. The breccia was formed as a consequence of the development of a karst topography and subsurface drainage system during uplift and erosion in late Beekmantown time (9, p. 229). The mineralization generally is restricted to the sedimentary matrix of the breccia, and fractures in the fragments are post-mineralization. Incidentally, it is worth noting that stalactites and stalagmites are not present in this breccia whereas they are the common filling of modern caverns in consolidated carbonate rocks. This difference in the two cases suggests that the solution collapse breccias developed as solution progressed in poorly consolidated beds subject to failure when unsupported.*

---

* Editor's Note: May not the solution caverns have developed below the water table, thus explaining the absence of stalactites (Bretz, J. H., 1950, G. S. A. Bull., v. 61, p. 789–833)?

Author's Comment: Yes, but such caverns commonly are not filled with breccia fragments. Perhaps those that are so filled or partly filled remain to be discovered.

## Structure

A cursory examination of the distribution of the Precambrian rocks relative to the Paleozoics and the pattern of the trace of the contact between them as shown on Figure 1 suggests that the Saucon Valley is a breached fenster, resulting from the partial erosion of a sheet of Precambrian rocks which had been thrust north over the Paleozoics. This was the writer's tentative first interpretation, and one he later discovered was postulated by Stose and Jonas (5). Miller (6) did not concur. Because the overthrust concept implied continuity of the Paleozoics of the Saucon and Lehigh Valleys under the Precambrian of South Mountain, with possible short circuiting of the Lehigh River into Saucon Valley, thus accounting for the heavy inflows of water into the mines, the situation north of Friedensville was investigated with drill holes. The block diagram (Figure 3) shows the results, confirming Miller's (6) view that the Paleozoics on the north side of Saucon Valley have been displaced downward relative to the Precambrian on a normal fault which dips to the south. The drilling results were presented to Miller (7) and subsequently published by him.

Knowledge of the structure of the Saucon Valley is quite generalized because of the paucity of rock exposure in outcrop or quarry. The structure in the vicinity of the mine was determined from exposures in mine open pits and underground workings together with drill hole information proximate thereto. It is shown in progressively greater detail on Figures 4 and 5. In the western portion of the Saucon Valley, drilling in the vicinity of the Schneider limonite mines (Nos. 49 and 50, in Figure 1) confirms the presence of Beekmantown in the area and supports the view that the gross structure between the Colesville normal fault to the north and the Saucon reverse fault to the south is a double plunging saddle shaped anticline overturned to the north. Such structure accounts for the localization of the Jacksonburg formation in the vicinity of Lanark, athwart the trend or gain of the structure.

Figure 3 shows the structure in the vicinity of the mines to be a southwesterly plunging (18°) overturned anticline, with vertical north limb at the Ueberroth mine, and a moderately inclined south limb (25°) at the Triangle, Correll, and New Hartman mines. The Old Hartman mine is on the crest of the anticline. To

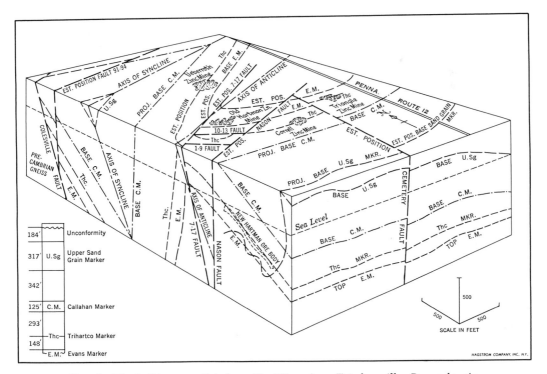

FIG. 3. *Block Diagram, Friedensville Mine Area Friedensville, Pennsylvania.*

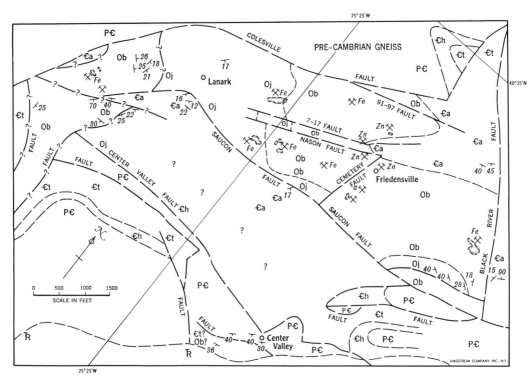

FIG. 4. *Geologic Map—West End of Saucon Valley Lehigh County, Pennsylvania.*

the north of the Ueberroth mine is a syncline, the features of which have not been investigated. The existence of these structures is documented by the facing of beds as disclosed by the stratigraphy and by the relation of axial plane fracture and flow cleavage to bedding on all elements of the structure. To the west of the area depicted in the diagram, the dip of the limbs and the plunge of the axis appear to be reduced as the anticline flattens into a saddle structure.

The major faults (the Colesville and the Saucon) are of the strike type; their existence has been proven by drilling. Stratigraphic displacement on the Colesville fault cannot be determined because of the absence of identifiable key beds on opposite sides, but it cannot be less than 2700 feet, the thickness of the Cambrian formations which have been eroded from the Precambrian rocks north of the fault. Displacement on the Saucon fault, which dips south at an undetermined angle, juxtaposes Allentown formation in the hanging wall against upper Beekmantown in the footwall. The stratigraphic displacement thereon could be equivalent to the thickness of the Beekman-

town; that is, 1200 to 1400 feet, in the absence of information as to which portion of the Allentown, lower, middle, or upper, abuts the fault.

Strike faults of lesser displacement from north to south are the 91–94, the 7–17, and the Nason about which little is known. They are of the postulated or "necessary" type and are invoked because of indicated stratigraphic displacements as shown on plan and section. Only the Nason fault has been intersected by mine workings, and its expression is not impressive.

The 91–94 fault is indicated to be of the normal type as is its neighbor the Colesville fault. The block between the 7–17 and the Nason faults is indicated to be a graben.

The only cross fault indicated is the Cemetery, the existence and position of which are based on the relative position of the Trihartco Marker in the Triangle and Correll mines. The attitude of this fault has not been established. It appears to be post-mineral and responsible for the two outcrops of the New Hartman ore body in the Correll and the Triangle mines.

The diagonal faults designated 1–9 and

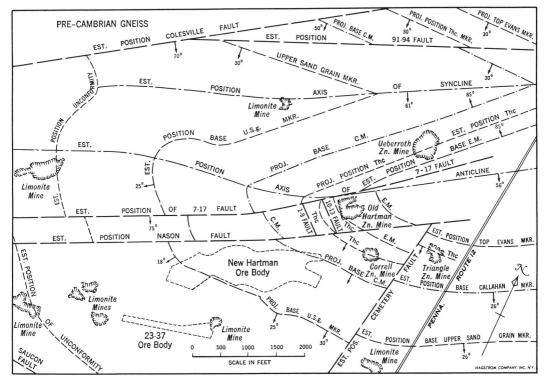

FIG. 5. *Geologic Map—Friedensville Mine Area Friedensville, Pennsylvania.*

10–13 are of the "necessary" variety, the attitude of which is uncertain.

Fracturing is prominent in the open pits and in the mines. The most prominent attitude is N40° to 60°E with dips ranging from 75°NW thru vertical to SE. Some of the fractures are along ore boundaries. Within the range of mining, most fractures are weathered and consequently quite obvious. Some small scale faulting is also present in the ore body, but whether it is tectonic or an expression of solution collapse is debatable.

Fracture cleavage in the beds and flow cleavage in the mineralized interstices of the breccia fragments, while present locally, is less prominent in fresh faces underground than in weathered exposures on surface.

It is the writer's opinion that structure per se did not control the localization of ore inasmuch as ore deposits are present on all elements of the fold and are not related to faulting. At most, tectonics merely deformed horizontally disposed ore bodies and determined their subsequent pattern and attitude of occurrence. Because the Taconic revolution is the earliest post-Beekmantown tectonic event in the area and the sulfide minerals themselves show evidence of deformation as well as indifference to structural setting, it is suggested that the mineralization is pre-Taconic; that is, Ordovician in age.

## MINERAL DEPOSITS

### Zinc

The zinc ore bodies are strata bound and all occur in a stratigraphic range of about 100 feet. The New Hartman body is the only one sufficiently delineated to provide information regarding form. It is a rather thin tabular body of varying thickness, the planar extent of which has not been completely delimited. The azimuth of its major axis and the plunge thereof appear to conform to the attitude of axis of the anticline.

The minerals identified to date are dolomite, calcite, chert, rare microcline feldspar, sphalerite, pyrite, rare chalcopyrite, calamine, smithsonite, rare greenockite, sauconite (a questionable zinciferous clay), and carbonaceous material and sericite in streaks. The dolomite is present as fragments in the breccia, medium to dark gray in color, and as white veinlets. Calcite occurs as white veinlets as does the quartz. Chert occurs as ellipsoids, generally flattened. The frequency of occurrence and distribution of the microcline has not been determined as it is not readily recognized in mine mapping. To date it has been identified only in thin section. There is uncertainty whether it is authigenic or detrital.

Sphalerite is generally dark gray in color and on a wet face in the mine is difficult to distinguish from the dark gray host. While in general the grain size is medium to coarse, locally it is flint-like and is characterized by conchoidal fracture. Butler (4) and Fraser (3) have described the textures observed in polished and thin sections, respectively. Butler reports that the flint-like or mylonitized sphalerite is twinned. Both agree that such sphalerite has suffered crushing and deformation.

Pyrite is widely distributed in the ore as masses and in subhedral to euhedral crystals. It commonly occurs banded with the sphalerite and locally is found veining it or veined by it suggesting at least two periods of pyrite deposition. Crushing and fracturing of the pyrite has been observed in polished sections, another example of the effect of post-mineral deformation.

Qualitative spectrographic analyses of two specimens of Friedensville pyrite carefully separated from the rock by specific gravity methods are as shown in Table II.

Calamine and less commonly smithsonite occur in greatest abundance in the shallow portions of the ore body. However, to the 800-foot depth currently reached by mining, oxidized zinc persists. It has been observed in drill core at depths of 900 to 1000 feet below surface, or about 600 feet below sea level and 850 to 950 feet below the prepumping elevation of the water table. Weathering and limonite have been encountered in one

TABLE I.   *Analytical Data on Carefully Hand-Picked Sphalerite**

| Analysis | Equivalent to |
|---|---|
| 66.2% Zn | 98.67 ZnS |
| 0.40 Fe | 0.63 FeS |
| 0.05 Cd | 0.06 CdS |
| 0.36 SiO$_2$ | 0.36 SiO$_2$ |

*Spectrographic analysis disclosed the following elements to be present in the approximate order of magnitude listed:

| | |
|---|---|
| Mg, Ca, Al, Ga | 0.01 to 0.001% |
| Ge | 0.001% |
| Cu, Pb, Mn | 0.0001% |

TABLE II.   *Spectrographic Analyses of Friedensville Pyrite*

| Element | Sample No. 1 | Sample No. 2 |
|---|---|---|
| Fe | >10% | >10% |
| Si | 1–10% | 1–10% |
| Zn | 1% | 1% |
| Mg | 0.1% | 0.1% |
| Al | 0.01–0.1% | 0.01% |
| Pb | 0.01–0.1% | 0.01–0.1% |
| As | 0.01–0.1% | 0.01–0.1% |
| Ca | 0.01% | 0.01% |
| Mo | 0.01% | 0.01–0.1% |
| Mn | 0.001–0.01% | 0.01% |
| Cu | 0.001% | 0.001% |
| Ti | 0.001% | Not Detected* |
| Ni | 0.001% | 0.0001% |
| Co | Not Detected† | Not Detected† |

\* Titanium not detected. If present it is less than 0.001%.
† Cobalt not detected. If present it is less than 0.001%.

drill hole to depth of 1400 feet. Greenockite and sauconite were reported to be present in the surface workings.

The boundaries of the ore are irregular but clean cut as might be expected where mineralization is localized in a breccia filled pre-existing cavern. The fragments in the body are barren, of a variety of sizes, and randomly oriented as evidenced by the attitude of bedding therein. This breccia may be designated as rubble type. The fragments are commonly fractured and healed by white gangue dolomite and quartz *without* sulphides. The fractures do not extend into the matrix, suggesting that the matrix was incompetent relative to the fragments during deformation. The sulphide mineralization is confined to the matrix of the breccia.

In the hanging wall, and locally in the footwall, is a mosaic breccia in which the fragments are only slightly disturbed, suggesting in the case of the hanging wall insufficient weakening to permit collapse of the blocks and in the footwall insufficient solution to free them for movement. The best-grade ore is in the matrix of the rubble breccia whereas the mineralization filling the openings in the mosaic breccia is low grade in comparison with that in the rubble.

The mineralization involves filling of open space as well as some replacement thus denying the classification of such deposits as simple limestone replacements. The sulfides commonly coat breccia fragments, occur as bands parallel to bedding, and in the matrix of the breccia, and also as disseminations in the matrix. Except as noted above regarding the distribution of higher- and lower-grade ores, there is no evident zoning or segregation of the sulfides. The wall rock of the body and the fragments therein are unaltered on a megascopic scale. Locally, the long dimensions of the rubble fragments are oriented parallel to the dip of the axial plane of the anticline and the matrix mineralization therein is streaked in the same pattern. It is this type of evidence, together with that cited by Butler (4) and Fraser (3), and the absence of mineralization in the fractured breccia fragments which supports the conclusion that the ore has suffered post-mineral deformation.

Leaching-bleaching is a feature of the hanging-wall environment of the ore at Friedensville, Pennsylvania as is also the case at Austinville, Virginia, and Jefferson City and Treadway, Tennessee. A similar feature has also been observed in the southeast Missouri lead district. It occurs in beds that ranged in original color from dark gray to light gray. Whatever the process involved, the usual result is a rock of lighter color than normal. Original textures are commonly preserved. Where the bleaching has been studied intensively the bleached rock appears to have undergone no significant change in chemical composition nor in grain size.

The significance of bleaching as a guide to prospecting is indeterminate because it may be associated with ore or may occur where ore is unknown. It has been interpreted as alteration related to the ore-forming process, as a mere weathering phenomenon either of recent or ancient age, or possibly as a product of original sedimentation or diagenesis. And finally, it is not known how common such a feature may be in unmineralized areas.

## Iron

The iron industry in the Lehigh Valley was based originally on limonite ores that are roughly strata bound. They were categorized as either Mountain or Valley type on the basis of their topographic-stratigraphic position. The Mountain ores occur at the boundary between the Great Valley and the mountains at the contact between the base of the Cambrian carbonate section and the top of the Hardyston quartzite. The Valley ores occur in the Great Valley in Cambrian carbonate formations and in the upper part of the Beekmantown. The deposits are reported to have pyritic roots in some instances, although such have not been observed by the writer because the mines are now caved or filled with water.

Several Valley limonite deposits in the upper

Beekmantown occur in the Saucon Valley, of which eight are stratigraphically above and concentric relative to the Friedensville zinc deposits as shown on Figures 3 and 4. These limonites contained sufficient zinc to plague the early operators of iron blast furnaces with zinc oxide build up in their stoves. What then is the significance of this stratigraphic-spatial relation of the limonite to the zinc mineralization?

Miller (2, p. 82) comments—

"In forming a theory to account for the formation of the sphalerite, pyrite and marcasite the connection between the Friedensville zinc deposits and the limonite deposits that occur elsewhere in the Saucon Valley should be recognized. Lesley (1) suggested such a connection but did not enter into detail.

"In the iron mines that lie 1½ miles west of the zinc mines considerable pyrite was found in the lower depths worked and more would undoubtedly have been found if operations had continued. . . . Small amounts of zinc were also present in the iron ores of these mines."

Concepts of genesis for the limonites range from that of direct precipitation of iron oxide as recent bog deposits to that involving the oxidation of syngenetic pyrite. The later view has been advanced by the writer; to-wit: that the upper Beekmantown limonites are the oxidized derivative of pyritic deposits localized in solution collapse breccias formed during Ordovician time. This concept postulates the existence of two levels of solution collapse breccias in the Beekmantown, the lower one being the host for zinc mineralization with relatively minor pyrite content whereas the upper one is the host for pyrite mineralization with trace zinc content. This situation is analagous to that in southwestern Wisconsin where the upper level mineralization is primarily galena whereas that at the lower level is sphalerite, galena, and pyrite.*

The stratigraphic-spatial association of limo-

* Editors Note: Does not the presence of Maquoketa shale (at the time of formation) imply that this Wisconsin situation differed materially from that postulated as existing at Friedensville?

Author's Comment: No. In reference (9), I point out that the disconformity between the Galena and the overlying Maquoketa shale is evidence that the Galena beds were uplifted and subjected to some erosion. During the erosion period, a subsurface drainage system developed in the O, P, and Q beds, expressed by the thinning thereof and the collapse of the overlying beds as evidenced by the pitches and flats. The pitches and flats may be considered as the first stage in the development of a solution collapse breccia.

nite and zinc mineralization in the Saucon Valley, as observed by early students of the area, was judged to be sufficiently significant on an empirical basis to warrant the writer premising a prospecting program thereon in 1941. This involved sampling of a large area and the determination of the zinc content of upper Beekmantown limonites by conventional chemical analytical procedures; all of this prior to the development of the highly sensitive geochemical methods for determination of the heavy metals and before White's (8) discovery of the significant role of limonite in accumulating zinc in soils of the Appalachians. Drilling was done in the vicinity of two limonite mines in areas of no outcrop. In one case the beds were right side up but in the other, to our surprise and dismay, they were upside down. While the prospecting was nonproductive and also noncommittal regarding the genetic significance of the spatial relation of the zinciferous limonite to the zinc deposits at Friedensville, nevertheless the fact that these two limonite deposits occurred in such contrasting structural situation suggests that the pyritic sources thereof suffered post-mineral deformation as did the zinc deposits.

## GEOLOGIC HISTORY

(1) Precambrian events.

(2) Successive deposition of Hardyston quartzite, Tomstown, Allentown, and Beekmantown dolomites.

(3) Uplift and erosion of pre-Jacksonburg formations with consequent development of karst topography thereon and of a subsurface drainage system and solution collapse breccias in the Breekmantown and possibly lower formations.

(4) Mineral deposition prior to or shortly after submergence of pre-Jacksonburg terrain and the beginning of Jacksonburg deposition.

(5) Taconic Revolution. The ore and its environment were deformed by folding and perhaps by faulting.

(6) Silurian, Devonian, and Carboniferous formations are not present in the area.

(7) Appalachian Revolution. No attempt has been made to distinguish between the effects of the Taconic and of the Appalachian deformations.

(8) Post-Paleozoic sedimentation and volcanism are not recorded in Saucon Valley.

(9) Uplift, erosion, development of present topography, exposure of the ore deposits, and oxidation of them.

(10) Local deposition of glacial outwash.

## GENESIS

Study of the Friedensville zinc deposits has contributed nothing regarding the source of the metal ions. It has, however, provided evidence that the deposits are localized in solution collapse breccias and were formed prior to deformation of the environment. The absence of igneous intrusives younger than the ore host or volcanics contemporary therewith in the vicinitiy of the deposits is negative evidence that such may not be invoked as a proximate source of the metals. This leaves open to speculation the location of the source and nature of the plumbing system from source to site of deposition.

Confronted with this unresolved remoteness of source, and the degree of freedom thus granted, one may choose a subvertical source in an igneous intrusive at indeterminate depth and of age younger than the ore host, an ambient source in the pre-ore environment as did Miller (2, p. 82), or a lateral source in extrusives of age about contemporaneous with the ore host. Currently I choose the latter and postulate submarine volcanic exhalations as the source of the metal ions, such being derived from the volcanic members of the eugeosynclinal facies of the Ordovician to the east. This facies contains base metal mineral deposits in New Brunswick, Maine, New Hampshire and Virginia, the first of which is tentatively considered by Kalliokoski (10, p. 499) to be of Ordovician age.

Though remote laterally such source warrants consideration because: (1) Metal ions can travel more easily through a sea than they can along a tortuous path through a great thickness of rock and (2) It is probable that a sea of solution is required to form an ore deposit, metal ions being quite insoluble.

With respect to plumbing systems, it is self-evident that prior to lithification the inherent permeability of the host rocks are most available and effective for the drainage or explusion of connate water during uplift and erosion and for the subsequent infiltration of the fresh water required to produce a subsurface drainage system and solution collapse breccias. The major permeability anomalies thus produced are then available to control the pattern of infiltration of sea water following resubmergence of the area.

Consequently it seems sensible to seek a source of metal ions in the Ordovician sea rather than to postulate, passively, such source in an intrusive of post-Ordovician age with its accessory plumbing system from source to site of deposition, phantoms respectively as to presence, position, and mineral potential and as to persistence and pattern of permeability.

## CONCLUSION

The mineralization and environment of the zinc deposits at Friedensville, Pennsylvania are Mississippi Valley type. The ore is localized in the Beekmantown formation of Lower Ordovician age in solution collapse breccias developed during the erosion interval marking the unconformity between the Beekmantown and the overlying Jacksonburg formation of Middle Ordovician age. Mineralization occurred during the Ordovician while the beds were essentially horizontal. Subsequent deformation is responsible for the present distribution and attitude of the deposits and the current classification of them as of the Appalachian type.

## REFERENCES CITED

1. Lesley, J. P., 1892, The Saucon zinc mines of Lehigh County: Penna. Second Geol. Surv. Sum. Final Rept., v. 1, p. 436–439.
2. Miller, B. L., 1925, Mineral resources of the Allentown quadrangle, Pennsylvania: Penna. Geol. Surv., 4th ser., Topo. and Geol. Atlas no. 206, 195 p.
3. Fraser, D. M., 1935, Microscopic investigation of Friedensville, Pennsylvania zinc ore: Amer. Mineral., v. 20, p. 451–461.
4. Butler, R. D., 1935, Mylonitic sphalerite from Friedensville, Pennsylvania: Econ. Geol., v. 30, p. 890–904.
5. Stose, G. W. and Jonas, A. I., 1935, Highlands near Reading, Pennsylvania, an erosion remnant of a great overthrust sheet: Geol. Soc. Amer. Bull., v. 46, p. 757–779; authors reply to disc., p. 2038–2040.
6. Miller, B. L. and Fraser, D. M.: 1935, Comments on "Highlands near Reading, Pennsylvania, an erosion remnant of a great overthrust sheet" by George W. Stose and Anna I. Jones: Geol. Soc. Amer. Bull., v. 46, p. 2031–2038.
7. Miller, B. L., 1944, Specific data on the so-called "Reading overthrust": Geol. Soc. Amer. Bull., v. 55, p. 211–254.
8. White, M. L., 1957, Occurrence of zinc in soil: Econ. Geol., v. 52, p. 645–651.
9. Callahan, W. H., 1964, Paleophysiographic premises for prospecting for strata bound base metal mineral deposits in carbonate rocks: *CENTO Symposium on Mining Geology and Base Metals,* Ankara, Turkey, p. 191–248.
10. Kalliokoski, J., 1965, Metamorphic features in North American massive sulfide deposits: Econ. Geol., v. 60, p. 485–509.
11. Drake, A. A., 1965, Carbonate rocks of Cambrian and Ordovician age, Northampton and Bucks counties, eastern Pennsylvania, and Warren and Hunterdon counties, western New Jersey; U.S. Geol. Sur. Bull. 1194-L, p. L1-L7.

# 6. The Grace Mine Magnetite Deposit, Berks County, Pennsylvania

SAMUEL J. SIMS*

## Contents

## Illustrations

* Bethlehem Steel Corporation, Bethlehem, Pennsylvania

## Tables

## ABSTRACT

The Grace mine magnetite deposit, located 2 miles north of Morgantown in Berks County, Pennsylvania, was discovered in 1948 by an aerial magnetometer survey. It is situated on the southern border of the Triassic lowlands section of the Piedmont province. Cambrian sandstones, shales, and limestones overlie Precambrian crystalline rock and were folded and thrust-faulted during the late Paleozoic Appalachian revolution. Late Triassic continental sediments were deposited on the deformed basement rocks and were intruded by Late Triassic diabase dikes and sills before regional tilting and normal faulting occurred.

The Grace mine magnetite ore is characteristic of the Cornwall type. The ore has replaced a Cambrian limestone lens isolated between Triassic diabase footwall and Triassic sedimentary rock hanging wall. The ore body is roughly tabular in shape, dips 20° to 30°NE and plunges about 20°N80°E. The ore is granular, fine- to medium-grained, typically consisting of unevenly distributed magnetite in a matrix of light green to white serpentine gangue. Magnetite is the major ore mineral, pyrite and chalcopyrite are common accessory minerals, and pyrrhotite occurs locally. Sphalerite, marcasite, galena, hematite, digenite, and goethite are uncommon. Serpentine, talc, and chlorite are the major gangue minerals; calcite and dolomite are common, and phlogopite, tremolite, and biotite are present locally. Apatite is a common accessory; sphene is rare.

Calcium-magnesium silicates were formed first by contact metamorphism of impure limestone by the intruding diabase. Later, the silicates were altered hydrothermally to the serpentine-talc-chlorite assemblage now present. Magnetite, the first ore mineral, formed by replacing serpentine and was followed by pyrite-pyrrhotite and then by chalcopyrite-galena-sphalerite. Marcasite, secondary pyrite, goethite, and hematite are secondary minerals resulting from alteration of pyrrhotite and magnetite.

It is estimated that confining pressure was about 1500 bars and the temperature range was 500° to 675°C during ore formation.

## INTRODUCTION

The Grace mine magnetite deposit, one of the most recently discovered major iron ore bodies in the United States, is owned and operated by Bethlehem Steel Corporation and is located in a region of historically important iron ore mines in southeastern Pennsylvania, 2 miles north of Morgantown, in Berks County, as shown in Figure 1.

The deposit, discovered in 1948, was the first such discovery to result from an airborne magnetometer survey. Diamond-drilling was initiated in 1949, and encountered ore in the first hole on December 19, at a depth of 1524 feet. By January, 1951, 17 drill holes had delineated the ore body. Ore was first produced in 1958. The mine is named after the late Eugene C. Grace, Chairman of Bethlehem Steel at the time of the discovery.

Thanks are due to D. M. Fraser, G. L. Hole, and G. K. Biemesderfer of Bethlehem Steel Corporation for their comments and criticism during the preparation of this paper, to C. F. Eben and the Engineering staff at the Grace Mine for their assistance at the mine site, and to E. S. Erickson of Bethlehem's Homer Research Laboratory for X-ray analyses and polished section preparation.

## HISTORICAL INFORMATION

Iron mining began in southeastern Pennsylvania before the Revolutionary War, reached

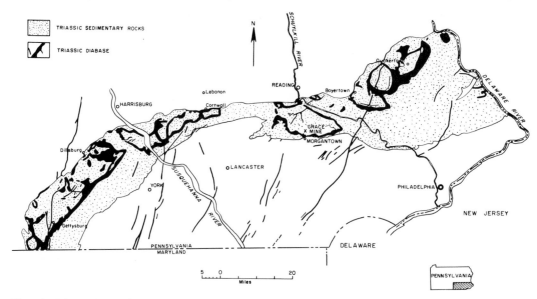

FIG. 1. *Map of Southeastern Pennsylvania, showing Triassic sedimentary rocks and diabase and location of the Grace Mine. Modified from Geologic Map of Pennsylvania, Pennsylvania Geologic Survey, 1960.*

a peak of activity during the 1880's, and then diminished (before the discovery of the Grace mine) until only the Cornwall Mine was operating. Similarities in occurrence and mineralogy led Spencer (3) to classify the ores of southeastern Pennsylvania as iron ores of the "Cornwall type" after the Cornwall deposit in Lebanon County.

This group of deposits is in the Triassic basin of Pennsylvania and extends from Boyertown southwestward to Dillsburg (Figure 1). It consists of the once important mines of Boyertown; the Wheatfield and Fritz Island deposits south and southwest of Reading; the Jones, Warwick, Hopewell, and French Creek mines east of Morgantown, and the Dillsburg deposits southwest of Harrisburg. The zone including these deposits is about 75 miles long and from 5 to 20 miles wide.

The Cornwall type of iron ore normally contains 40 to 50 per cent iron, chiefly in magnetite, and about 2 per cent sulphur, mainly in pyrite and chalcopyrite. Gangue minerals include tremolite, diopside, phlogopite, chlorite, serpentine, calcite, and dolomite. The deposits generally occur at, or near, the borders of the Triassic basin where diabase is in contact with carbonate rocks.

The association of the deposits with contacts between diabase and carbonate rocks suggested to Spencer (3, p. 13–16), that they had formed by the replacement of carbonate rock by iron ore minerals introduced from the diabase. Prior to Spencer's work, it was believed that the ores had originated by activating iron already present in the country rock and concentrating it near the diabase (1,2).

The host rocks for the Cornwall type of iron ore consist of Precambrian marble at French Creek (7), Cambrian limestone at Cornwall (3,6,8,19), and Triassic limestone conglomerate at Dillsburg (3,4,17). These rocks, therefore, exhibit a common chemical composition rather than similar rock type or age. The Grace mine deposit has the characteristics of Cornwall-type iron ore, is mineralogically similar, and occurs at the contact between diabase and Paleozoic limestone along the southern border of the Triassic basin.

Almost all the iron ore produced from this region has come from the Cornwall deposit, which, since mining began in 1742, has yielded 93,000,000 tons of ore, compared with 3 to 4 million tons from all the other deposits, excluding Grace. Production of iron ore from the Grace mine is shown in Table I.

TABLE I.   Production of Crude Ore from Grace Mine, Short Tons

| 1958 | 1959 | 1960 | 1961 | 1962 | 1963 | 1964 |
|---|---|---|---|---|---|---|
| 125,136 | 437,297 | 920,011 | 1,664,002 | 2,911,059 | 2,483,701 | 2,827,360 |

## PHYSIOGRAPHIC HISTORY AND TOPOGRAPHY

The Grace mine magnetite deposit is in the Piedmont physiographic province and is located on the southern border of the Triassic lowlands and northern border of the Conestoga Valley. The topography of the area is best described as mature, characterized by rounded hills separated by irregular valleys. Prominent topographic features of the area are Conestoga and Little Conestoga creeks and Welsh Mountain, a linear ridge trending east-northeast which reaches an altitude of 1000 feet above sea level and passes 2 miles south of Morgantown. About one mile north of Welsh Mountain, Conestoga Creek flows southwestward, parallel to the ridge, towards the Susquehanna River, but changes course from southeast to west-southwest at Morgantown. Little Conestoga Creek flows parallel to Conestoga Creek about one mile to the north. North of the Conestoga Valley, the open and hilly terrain is characteristic of the Traissic lowlands. The elevation of the mine site is about 700 feet above sea level and that of Conestoga Creek at Morgantown is about 520 feet.

The topography of the area reflects the relative resistance to erosion of the underlying formations. Triassic sedimentary areas have open hilly topography, and the higher areas are underlain by more resistant rock units. The Paleozoic beds of the Conestoga Valley show linear ridges and open valleys. Welsh Mountain is underlain by a resistant quartzite, Conestoga Valley by less resistant limestones.

The physiographic history of the area is one of repeated cycles of uplift and erosion. A series of erosion surfaces was developed after deposition of Triassic sedimentary rocks, each of which was, in turn, uplifted and dissected. Bascom and Stose (9, p. 103–105) cite the following successive erosion cycles: (1) development of the Schooley peneplain (Cretaceous), (2) partial development of the Harrisburg peneplain (Miocene), and (3) development of the Bryn Mawr and Brandywine berms (Late Tertiary).

Granted the existence of these peneplains, the general land surface at the Grace mine magnetite deposit and northward probably corresponds with the Harrisburg surface, while the older Schooley peneplain may be represented by the top of Welsh Mountain. The younger berms are present as terraces along Conestoga Creek.

## GEOLOGIC HISTORY

The rocks surrounding and including the Grace mine consist of Precambrian crystal-

TABLE II.   Stratigraphic Column for Grace Mine and Vicinity. Modified from Bascom and Stose (9).

| SYSTEM | SERIES | | FORMATION | SECTION | THICKNESS | DESCRIPTION |
|---|---|---|---|---|---|---|
| TRIASSIC | UPPER TRIASSIC | Newark Group | Brunswick Formation | Ⓡbc | top not exposed | Poorly sorted interfingering sandstone, shale and conglomerate; indurated adjacent intrusive Triassic diabase. |
| | | | Stockton Formation | Ⓡs Ⓡd Ⓡs | 1000' | Micaceous and arkosic conglomerate and sandstone with some red shale; indurated adjacent intrusive Triassic diabase. |
| | | | — Unconformity — | | | |
| CAMBRIAN | UPPER | | Conococheague Limestone | Ɛcg | top not exposed | Dark and light limestone and interlayered lenticular dolomite; with beds of sandy limestone and Cryptozoon reefs |
| | MIDDLE | | Elbrook Limestone | Ɛe | 800' | Thin-bedded, fine-grained, earthy limestone with layers of marble and dolomite; hard, sandy limestone bed at base. |
| | LOWER CAMBRIAN | | Ledger Dolomite | Ɛl | 1000' | Pure granular gray dolomite with pure white lower beds. |
| | | | Vintage Dolomite | Ɛv | 600' | Massive gray dolomite with thin-bedded limestone at top. |
| | | | Antietam Quartzite | Ɛa | 450' | Gray fine-grained quartzite, grades downward to phyllite, contains fossil casts and moulds. |
| | | | Harpers Phyllite | Ɛhp | 800' | Gray sandy phyllite and mica schist with thin quartzite beds. |
| | | | Chickies Quartzite | Ɛc | 700' | Thick-bedded vitreous quartzite with thin-bedded sericitic quartzite and quartz schist in upper part. |
| | | | Hellam Conglomerate | Ɛh | 300' | Coarse-grained quartzite and quartz conglomerate with interbedded black slate |
| PRE Ɛ | | | — Unconformity — grd | | | Medium-grained gneissic granodiorite. |

lines, lower Paleozoic mildly metamorphosed sedimentary rocks, Triassic sediments, and Triassic intrusive diabase. Table II is a stratigraphic section and Figure 2 is a geologic map of the Grace mine area which was mapped originally by Bascom and Stose (9) at a scale of 1:62,500 as part of their quadrangle mapping. During the development of the mine, the more detailed map shown in Figure 2 was compiled.

The Precambrian basement rock is a medium-grained gneissic granodiorite showing a moderate cataclastic texture (9, p. 35) and is exposed on the south flank of Welsh Mountain; unconformably overlying it are Cambrian sandstone, shales, and limestones as shown in Table II, the Elbrook limestone being the uppermost Cambrian unit exposed (Figure 2). Bascom and Stose (9) mapped a small patch of Conococheague limestone northwest of Morgantown at the lower diabase contact, but this was not verified in the more recent mapping, although it may well be present beneath the Triassic sedimentary rocks.

Before deposition of Triassic sediments, the Paleozoic rocks were subjected to compressive deformation to give folds that trend northeast and are overturned to the northwest, producing isoclinal dips to the southeast. Low angle thrust faults that strike approximately with the formations bring older units over younger ones. Compression, which took place both before and after thrusting, since some fault planes are folded (9, p. 81), produced a schistosity and partial recrystallization. The deformation (younger than Ordovician because Ordovician rocks are involved in the structures), is believed to be part of the Appalachian revolution that took place during the late Paleozoic (9, p. 96).

Triassic continental conglomerates, sandstones, and shales were deposited on the Paleozoic and Precambrian basement in response to the formation of the Triassic basin. Two formations of the Newark group (as shown in Table II) are assigned to the Upper Triassic (9, p. 66). Near the end of Triassic time, but before the end of Newark sedimentation and before regional tilting, diabase dikes and sills were intruded into the Triassic sedimentary sequence (9, p. 97; 25, p. 2854); some narrow diabase dikes also were injected into pre-Triassic rocks (Figure 1). In some places along the edge of the Triassic basin, diabase

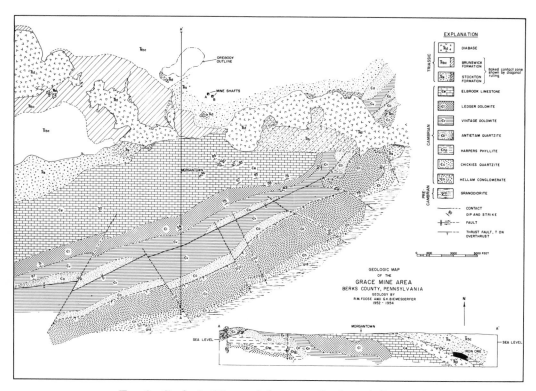

FIG. 2. *Geologic Map and Structure Section of Grace Mine Area.*

intersected Paleozoic limestone and lenses of it were thereby isolated between the diabase and overlying Triassic sedimentary rocks. It is in these blocks that Cornwall-type iron ore deposits were most likely to form.

Triassic sedimentary rocks were not folded but have gentle northward dips and are cut by many vertical normal faults, typically with small offsets. The southern portion of the Triassic basin was uplifted relative to its northern part. This movement resulted in a down-faulting of the northern edge of the basin along a line of eastward-trending normal faults (9, p. 97), the tilting being accompanied by fracturing and normal faulting of the Triassic rocks. Some of the normal faults extend into the older rocks south of the Triassic basin.

The close association of Cornwall-type magnetite deposits and late Triassic diabase indicates that the age of mineralization was likewise late Triassic. As will be shown in a later section, there is evidence to suggest a hiatus between diabase intrusion and ore deposition rather than that the ore was deposited by contact metasomatism of the intruding magma. The magnitude of this time lapse is unknown. Because the ore is cut by normal faults, it is believed that the ore deposit formed before regional tilting.

The Grace mine magnetite deposit occurs in a lens of Cambrian limestone that is overlain unconformably by Triassic sedimentary rocks and is separated from the main unit of limestone by the diabase sheet. Iron ore has formed by replacement of contact metamorphic minerals in the limestone lens.

## ECONOMIC GEOLOGY

### Form and Nature of Ore Body

The Grace mine magnetite deposit is a single ore body that has no outcrop; it is illustrated in Figure 3 by means of a generalized isometric block diagram and in Figure 4 by a surface plan.

It is roughly tabular in shape, strikes about N60°W, dips 20° to 30° northeast and plunges about 20°N80°E. The northeastern portion is not yet known in detail, so the shape shown in Figures 3 and 4 for that region is of necessity only approximate. The ore body is approximately 3500 feet long by 700 to 1500 feet wide (Figure 4), and it ranges from less than 50 feet to more than 400 feet in thickness. It lies between 600 feet and 2200 feet below sea level.

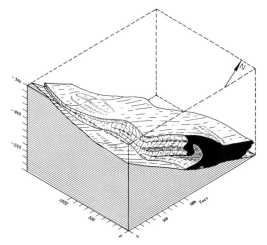

FIG. 3. *Generalized Isometric Block Diagram of the Grace Ore Body. Ore Body is shown resting on footwall diabase. Cambrian limestone and Triassic Hanging wall rock have been removed to show details of top and sides of ore body.*

In detail, the ore body swells and pinches and interfingers into the surrounding limestone host rock. Figure 3 illustrates some of these features as well as a roll-like structure on the southeast side of the body. Because of the mining method, the form of the ore body is interpreted from exploratory diamond drilling.

The ore body consists predominantly of magnetite and irregularly distributed zones of hydrous calcium-magnesium silicates, scattered lenses of limestone, and a few veins of milky quartz. The ore differs in its physical properties

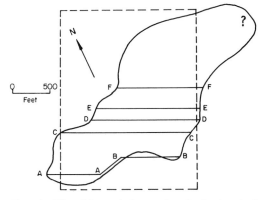

FIG. 4. *Plan Map of Grace Ore Body. Dashed lines show horizontal axes of block diagram (Figure 3). Section lines correspond to sections shown in figure 5.*

and typically is moderately friable and easily broken when struck with a hammer. It can, however, range from very crumbly to very hard. The specific weight of the ore ranges from 3.0 to 4.7 and averages 3.6. The ore has about 14 per cent porosity and, in places, is very vuggy.

The ore is granular, fine- to medium-grained, typically consisting of unevenly distributed magnetite grains in a matrix of light green to white gangue minerals. In places, magnetite is concentrated along parallel layers giving the ore a banded aspect, but elsewhere magnetite is concentrated along fractures that intersect the layers; or it may be evenly distributed in the gangue. On a small scale, the ore texture differs from place to place, but on the large scale of the ore body, it is quite uniform.

Post-ore fractures, undoubtedly associated with tilting and faulting of Triassic sedimentary rocks, seemingly are concentrated near the footwall contact of the ore in the quite brittle chilled margin of the diabase. Fractures in the ore body are inconspicuous, but under the microscope, are apparent and contribute to the friable nature of the ore.

## Mining Method

The mining method, which is dependent on the form and nature of the ore body, is described briefly because it limits access to the ore body proper. Panel caving (14), a modification of block caving, is used in combination

## Stratigraphic Relations

In general, the Grace mine magnetite body has a footwall of Late Triassic diabase and a hanging wall of Late Triassic Stockton formation, but locally the walls may be unreplaced Cambrian limestone. The ore body is surrounded laterally by Cambrian limestone.

Contacts between ore and surrounding rocks typically are sharp. The footwall contact, which is most frequently exposed, is very sharp and well defined between the chilled margin facies of diabase and ore. In places, hydrous calcium-magnesium silicate contact minerals separate ore from diabase. Contacts with limestone are relatively abrupt, commonly with a transitional zone of less than 10 feet. Contacts between ore and hanging wall, based on drill-hole data, appear to be sharp.

It is not certain which limestone unit has been trapped above the diabase and replaced by iron ore. The structure section of Figure 2 implies a replacement of Elbrook limestone, which some workers think resembles the limestone exposed underground. It is also possible that the limestone could be the Conococheague, which overlies the Elbrook. Because the limestone associated with the ore deposit is recrystallized, it is difficult to determine its original nature. Published analyses of Conococheague and Elbrook limestones (23, p. 25–28) in the vicinity of Grace mine are averaged and shown below compared with the average of two available analyses of limestone from Grace mine:

|               | CaO  | MgO  | $SiO_2$ | $Al_2O_3$ | $R_2O_3$ | CaO/MgO |
|---------------|------|------|---------|-----------|----------|---------|
| Conococheague | 37.0 | 12.9 | 7.0     | —         | 2.6      | 2.87    |
| Elbrook       | 26.9 | 14.9 | 13.8    | —         | 8.4      | 1.81    |
| Grace Mine    | 34.3 | 19.2 | 1.5     | 0.9       | —        | 1.79    |

with cutoff shrinkage stopes and open stopes to remove ore from fringe areas that do not have the desired thickness for caving. Undercutting of the ore body was initially begun with finger raises and bells. This method has recently been superseded by fan-hole undercutting. The ore is removed through a system of transfer, slushing, production, and haulage drifts and is hoisted to the surface in a vertical shaft. All development work is done in the diabase so that, except for exploration drifts, access to the ore body is limited to the footwall contact exposed in finger raises.

The CaO to MgO ratio may be significant because it is similar for Elbrook and Grace mine limestones, but much larger for Conococheague. Until further work shows otherwise, it is assumed that the Grace mine limestone is Elbrook formation.

## Petrology of Associated Rocks

Three rock types are intimately associated with the Grace mine ore deposit, i.e., diabase footwall, replaced limestone, and Triassic sedimentary hanging wall. These three units will be examined in more detail.

DIABASE   The diabase that forms the footwall of the Grace ore body is a uniform sheet about 1200 feet thick and is typical of the Late Triassic diabase exposed throughout southeastern Pennsylvania (Figure 1). It is a fine- to medium-grained tholeiitic rock showing characteristic subophitic diabase texture and is composed mainly of plagioclase and pyroxene with accessory apatite, magnetite, ilmenite, and pyrite. Amphibole and biotite in places form reaction rims on pyroxene, and micrographic intergrowths of quartz and potash feldspar are not uncommon in interstices between plagioclase and pyroxene. Besides the normal diabase, a very fine-grained chilled margin facies and a pegmatitic facies are present. Granophyric segregations as described at Dillsburg (12) and Cornwall (8,13,19) have not been observed underground in the Grace mine but may exist. A liberal use of gunite and concrete in the haulage ways underground limits observation.

Little or no internal deformation was observed in the diabase except for faults, shear zones, and joints, typically with a thin coating of dark green chlorite on their walls that is especially conspicuous on slicken-sides. Mineral grains, however, though altered, are not bent or broken. One xenolith was noted, a fist-sized piece of quartzite with a reaction rim of amphibole and associated magnetite, pyrite, and chalcopyrite, found in a drill core that came from pegmatitic facies diabase near the top of the igneous body.

The chilled margin facies of the diabase ranges in thickness from about 1 inch to 1 foot. It is characterized by a very fine-grained holocrystalline groundmass of pyroxene and plagioclase. The texture is subophitic with fine-grained (0.5 mm) phenocrysts of euhedral pyroxene and olivine, the latter commonly altered completely to serpentine, and laths of euhedral to subhedral plagioclase. Apatite, magnetite-ilmenite, and pyrite are accessory minerals. Plagioclase measurements are difficult because of the high degree of alteration in the chilled margin but, where possible, indicate that the feldspar is $An_{55-60}$. Pyroxene consists of augite and pigeonite in about equal amounts; olivine is the magnesium-rich variety. Plagioclase typically is altered to saussurite and pyroxene to chlorite and epidote. The chilled margin is cut by numerous veinlets about 0.5 mm thick of chlorite or serpentine, many with associated magnetite and biotite; one contained soda scapolite. It is noteworthy that the veinlets are thicker at the contact and pinch out within the diabase. In places, the chilled margin is completely altered to epidote. As the marginal phase grades to normal diabase, the groundmass grain size approaches that of the phenocrysts.

Normal diabase rock is fine- to medium-grained, grains ranging from 0.5 to 1.0 mm, and has a subophitic texture. Plagioclase ranges from $An_{50}$ to $An_{74}$, averages about $An_{60}$, and may be slightly altered to saussurite, particularly along twin planes. Zoning is common, and the cores are normally more calcic (one crystal showed a range from $An_{45}$ inward to $An_{62}$); some reverse zoning was noted, and one crystal showed a cyclical zoning of about ten cycles.

Pyroxene is predominantly augite and pigeonite, with some hypersthene. Biotite and, less commonly, amphibole form reaction rims on pyroxene in places. Biotite also is associated with grains of magnetite-ilmenite that range from euhedral to skeletal grains. Interstitial micrographic quartz and potash feldspar typically are present. Apatite and pyrite are minor accessories.

Normal diabase locally is cut by veinlets of serpentine or chlorite and alteration of plagioclase and pyroxene is conspicuous adjacent to these veinlets. Veins up to 1 inch thick of hydrous calcium-magnesium silicates are found near the upper part of the diabase. Talc, serpentine, actinolite, chlorite, and phlogopite were identified in these veins, with some pyrite and chalcopyrite. One vein predominantly of chalcopyrite and phlogopite was noted. Diabase adjacent to these veins is altered to epidote and actinolite.

Exposures of the pegmatitic facies are uncommon and volumetrically minor. The pegmatitic facies grades abruptly to normal diabase and seemingly is distributed irregularly. It is much coarser-grained than the normal diabase, having crystals over 2 mm long; it is highly altered and has a higher percentage of interstitial micrographic quartz and potash feldspar. Plagioclase is altered mainly to saussurite so that no determinations of composition were possible. Pyroxene likewise is highly altered, mainly to biotite, amphibole, and chlorite. Actinolite and a blue-green amphibole were observed replacing pyroxene. Magnetite-ilmenite, pyrite, apatite, sphene, epidote, and calcite are minor constituents.

LIMESTONE   The unreplaced limestone isolated above the diabase body is recrystallized and in places changed to a contact-meta-morphic mineral assemblage, ranging from white adjacent to the ore body to various shades of gray farther away. It is cut throughout by veins of calcite. The grain size ranges from 0.05 to 0.5 mm, the coarser grains being associated with white varieties and the finer with darker types.

Elsewhere, the limestone unit consists mainly of granoblastic calcite but may contain dolomite, serpentine, talc, diopside, and forsterite. Calcite is anhedral and may have inclusions of the other minerals. Dolomite typically is euhedral and finer grained than calcite. Serpentine occurs as layers and lenses in the limestone and along fractures. It also occurs as rhombic to rounded masses, suggesting that it is pseudomorphous after dolomite and diopside. Talc, where present, is intergranular to calcite. Diopside occurs as individual scattered grains, as concentrations along layers associated with serpentine, or as nodular aggregates of grains. Forsterite, as scattered grains, is uncommon and brucite was identified at one place. Serpentine and talc, with some magnetite and pyrite, are the common accessories near the ore body. Locally, where limestone is in contact with diabase, a thin zone of magnetite and sepentine is present.

The limestone unit shows a high degree of plastic deformation with flowage folds, pinching and swelling of layers, boudinage structures, and slip structures. Under the microscope, calcite grains tend to be slightly elongated parallel to layering, which seems to be defined by slip planes. Mineral grains have recrystallized after deformation since the twin lamellae show no sign of bending or breakage.

The upper contact of the limestone with Triassic sedimentary rock is marked by a zone several feet thick of limestone nodules in a limestone matrix. This is interpreted to be an intraformational conglomerate in the limestone.

TRIASSIC SEDIMENTARY ROCK   The Triassic sedimentary rock that forms the hanging wall of the ore deposit is a gray sandstone to conglomerate, baked and hardened by the diabase intrusion. The sandstone varies from arkose to graywacke depending upon the amount of micaceous material in the groundmass. Quartz, plagioclase, and microcline, the main constituents of the sandstone, are angular to subrounded, moderately well sorted, average 0.1 to 0.2 mm in diameter, and occur in a groundmass of much finer quartz, feldspar, and mica. Muscovite and biotite are also present as clastic components. Calcite and chlorite, which replaced the groundmass in places, act as cementing agents. Tourmaline, epidote, pyrite, and garnet, scattered throughout, are secondary metamorphic minerals. Hematite is common in the sandstone as plates along fractures. Near the limestone contact, the sandstone characteristically has a nodular aspect caused by masses of tremolite and serpentine. These masses probably are metamorphosed pebbles of limestone.

The conglomerate is similar to the sandstone but has pebbles of quartz in a sandstone matrix. At some places, magnetite forms pseudomorphs after calcareous pebbles.

One sample of arkose showed a cataclastic texture of quartz and feldspar grains in a groundmass of crushed quartz, feldspar, and mica, indicating brittle deformation due to normal faulting. Joints are widespread, and many are filled with natrolite, apophyllite, or gypsum.

## Mineralogy

ORE MINERALS   Magnetite is the major ore mineral in the Grace mine. Pyrite and chalcopyrite are common accessories, and pyrrhotite occurs locally. Sphalerite, marcasite, galena, hematite, digenite, and goethite were identified but are uncommon.

At the Cornwall deposit, two types of magnetite have been described (6, p. 629–632; 5; 8, p. 218–223; 26) designated "blue" and "brown" by Hickock (8). Tsusue (24, p. 5) states that he found only one type of magnetite from the Grace mine, which had a unit cell $a_0 = 8.396$ Å. However, the writer observed two types of magnetite resembling the blue and brown varieties of Cornwall. In oil immersion, the color difference is distinct—blue-gray versus light yellow-brown. The blue variety is by far the more common; the brown was observed in only a few sections taken near the footwall. The brown color seemingly is caused by many sub-microscopic inclusions. The blue, or normal, type is very faintly anisotropic. All euhedral magnetite crystals observed were either dodecahedra or octahedra, the former being predominant. Inclusions typically are concentrated in the core of magnetite crystals and commonly are serpentine.

Pyrite was observed as octahedra, cubes, and pyritohedra, some cubes showing penetration twins. A spectographic analysis of pyrite crystals from the Grace mine showed the presence of magnesium, silicon, aluminum, sodium, calcium, manganese, copper, silver, nickel, cobalt, and phosphorus and traces of lead, molybdenum, zinc, and titanium. Analysis specifically for cobalt showed an average of 0.51 per cent in pyrite. Pyrite is isotropic, shows no zonal structures, and is quite free of gangue inclusions.

Pyrrhotite is highly anisotropic and shows polysynthetic twinning. Preliminary X-ray data indicate that it has the monoclinic form, although platy crystals with hexagonal outlines indicate it was originally hexagonal and has inverted. It contains small amounts of nickel, cobalt, and copper. Chalcopyrite is moderately anisotropic and shows lamellar twinning, in places spear-shaped biconcave, and at right angles to one-another, indicating inversion from the cubic to tetragonal form (Kullerud, personal communication, 1965). Sphalerite and galena are primary minerals and occur both as individual grains and fracture filling in pyrrhotite and pyrite. Digenite was identified with chalcopyrite in filled fractures. Marcasite, goethite, and hematite are secondary minerals, formed by the alteration of pyrrhotite and magnetite. Secondary pyrite was also observed. Covellite tentatively was identified at an interface between marcasite and pyrrhotite.

GANGUE MINERALS  Serpentine, talc, and chlorite are the major gangue minerals. Calcite and dolomite are common; phlogopite, tremolite, and biotite are present locally. Apatite is a common accessory; sphene is rare.

Tsusue (24, p. 5) tentatively identified the Grace mine serpentine as the undifferentiable pair chrysotile-lizardite; however, petrographic observation strongly suggests antigorite. Chlorite consists of two varieties distinguished by optical properties: (1) penninite, with anomalous blue interference color and length-slow flakes, and (2) pyrochlorite, with slightly anomalous brown interference color and length-fast flakes.

Natrolite, apophyllite, fluorite, and gypsum fill fractures in diabase, limestone, and Triassic sedimentary rock but were not observed in the ore zone.

MINERAL ASSEMBLAGES  The most common primary ore-mineral assemblages are magnetite-pyrite plus chalcopyrite, sphalerite, or galena; magnetite-pyrrhotite plus chalcopyrite, sphalerite, or galena; and magnetite alone. The assemblages magnetite-pyrite-pyrrhotite and magnetite-chalcopyrite were noted in places. In general, the assemblages magnetite-pyrite-chalcopyrite and magnetite alone are the most widespread.

Gangue mineral assemblages are typically antigorite-talc-calcite-dolomite, antigorite-chlo-rite-calcite-dolomite, antigorite-talc-chlorite, antigorite-phlogopite-talc-chlorite, and tremolite-chlorite-biotite. In general, the gangue assemblage antigorite-talc-chlorite-calcite-dolomite is most common.

ZONING  No systematic zoning of either ore or gangue minerals was discovered although some zoning trends are suggested. Pyrrhotite seemingly occurs in the lower part of the ore body near the footwall contact. Sphalerite and galena seem to follow pyrrhotite. Preliminary studies by engineers at Grace mine on magnetic sulfides (pyrrhotite) indicate a concentration along the lower parts of the northwest and southeast extremities of the ore body. The "brown" variety of magnetite also was found only near the footwall. Diamond drilling sometimes shows zones of chlorite and tremolite directly beneath the hanging wall or directly above the footwall, but this zoning is not widespread. In general, however, mineral distribution throughout the ore body is irregular.

WALL-ROCK ALTERATION  Alteration of rocks surrounding the ore body can be divided into two categories: (1) contact metamorphism by the diabase intrusion and (2) subsequent hydrothermal alteration. Contact metamorphism has altered much of the Cambrian limestone to assemblages of calcium-magnesium silicates and has recrystallized the calcite and dolomite of the remainder. Triassic sedimentary rocks were partly converted to garnet and epidote. Hydrothermal alteration affects the diabase, limestone, and Triassic sedimentary rocks. The chilled-margin and pegmatitic facies of the diabase, especially, shows a high degree of hydrothermal alteration. Plagioclase is altered to saussurite; pyroxene, biotite, and amphibole to chlorite; and olivine to serpentine. Veinlets of chlorite and serpentine in diabase also indicate hydrothermal activity. In limestone, diopside is altered in places to serpentine. Veinlets of serpentine and recrystallized calcite also reflect hydrothermal activity as do tourmaline and chlorite in Triassic sandstone and conglomerate.

MINERAL TEXTURES  Magnetite is characteristically fine- to medium-grained (0.1 to 1.0 mm) poikiloblastic euhedral to anhedral equant crystals. Single crystals may be as coarse as 1 cm in diameter. The crystals occur individually or, more typically, in a granoblastic mosaic. Magnetite grain boundaries are sharp in most places. One polished section showed euhedral crystals with serrated borders.

Where magnetite is in contact with sulfides, the grain boundaries are so straight as to indicate that they are probably crystal faces. Crystal mosaics, of two or more to a great number of crystals, range in size, shape, and orientation from irregular ameboid masses to planar aggregates, and from parallel layers to arcuate lenses. In general, magnetite masses show little or no structure and are irregularly distributed, and inclusions are almost exclusively gangue minerals; rarely pyrite and pyrrhotite are found near the crystal boundary. The numerous inclusions impart a spongy aspect to the magnetite masses. Most magnetite grains are cut by fractures that, in places, are abundant, producing splintery fragments of magnetite.

Pyrite is euhedral to subhedral and in general is coarser-grained than magnetite, averaging 1 to 2 mm in diameter, with some crystals up to 10 mm. It usually contains inclusions of euhedral magnetite but is free of gangue. Pyrite tends to be intergranular to, and grows around, magnetite. In places, it has replaced magnetite but not extensively. Pyrite typically is associated with, and replaces, carbonate minerals. Almost everywhere, it is fractured to a higher degree than the associated magnetite. Locally, pyrite is shattered and consists of splinters arranged in a crude crystal shape and seemingly is preferentially fractured compared to other minerals.

Chalcopyrite occurs as irregularly shaped stringers growing around and between the other ore minerals. It also occurs as blebs in pyrrhotite and fracture fillings in pyrite. Replacement of pyrite by chalcopyrite in places, is controlled by fractures. Chalcopyrite also typically forms a partial rim on, and in some places replaces, pyrrhotite. Locally, chalcopyrite has inclusions of euhedral magnetite.

Pyrrhotite is commonly in tabular grains, these ranging in size from less than 1 mm to 15 mm long. Some, however, are equant to irregular in shape. Pyrrhotite is intergranular to, and also grows around and includes, euhedral magnetite. It is quite free of inclusions, except for magnetite and rare blebs of chalcopyrite, and is cut by fractures to a lesser degree than pyrite but more so than magnetite. Good planar cleavage, accentuated by secondary alteration is characteristic. Pyrrhotite masses intergranular to magnetite have strong optical continuity even in separate areas.

Sphalerite occurs chiefly as fracture fillings associated with chalcopyrite but also is present locally in anhedral grains, some with blebs of chalcopyrite. Galena occurs as individual anhedral grains and as rims on some pyrrhotite grains that are associated with chalcopyrite.

Digenite occurs only as fracture fillings associated with chalcopyrite.

Secondary sulfide minerals are uncommon in the Grace ore body and are associated mainly with pyrrhotite alteration. Marcasite most commonly forms on the rims of, and along cleavage traces in pyrrhotite. It also borders primary pyrite crystals and is itself rimmed with secondary pyrite. In places, secondary pyrite and marcasite are intermixed, especially where the two minerals have replaced pyrrhotite. Primary and secondary pyrite are distinguished by the nature of the polish they take—primary an even, smooth polish while secondary as does marcasite, polishes poorly and has many pits. In places, secondary pyrite forms a feathery overgrowth on primary crystals.

Goethite occurs as an oxidation product of pyrrhotite, generally along cleavage traces and associated with marcasite and secondary pyrite. Hematite is an oxidation product along fractures in magnetite, in places forming a matrix between splintery magnetite fragments. Hematite also forms needle-like aggregates on magnetite crystal faces associated with marcasite or secondary pyrite.

The common gangue minerals (serpentine, talc, tremolite, and chlorite), form matted or felted intergrowths as a matrix for ore minerals. Calcite and dolomite are intergranular to silicates. Gangue minerals range in size from less than .01 mm to more than 3 mm and average between 0.5 to 1.0 mm. In general, where the gangue is coarser-grained, the ore minerals are coarser. Gangue minerals, with the exception of typically euhedral apatite and dolomite, are anhedral, show no preferred orientation, and have straight to wavy extinctions. Individual species tend to be glomeritic and in places form layers but may be evenly distributed throughout the rock. Replacement textures were observed, with serpentine and chlorite replacing talc and, in one instance, talc replacing tremolite. Occasionally, there are aggregates of serpentine with crystal outlines of tremolite suggesting that serpentine has replaced tremolite. A marked mutual association suggests a preferential replacement of carbonate by pyrite. Likewise, magnetite and serpentine are closely associated.

MINERAL PARAGENESIS The early calcium-magnesium silicates, spatially related to limestone-diabase contacts later were altered hydrothermally to talc-serpentine-chlorite-tremolite assemblages which then were replaced by ore minerals. Mineral relics in limestone of diopside and forsterite as well as the shapes of serpentine aggregates and evidence of talc replacing tremolite suggest that the first mineral assemblage developed after diabase intrusion was calcite-dolomite-diopside-forsterite-tremolite-mica. Further hydrothermal, or retrograde alteration, converted diopside, forsterite and some tremolite and mica to serpentine, talc, and chlorite. This alteration was probably the beginning of the phase that resulted in the ore mineralization. The assemblage that was replaced by ore minerals was serpentine-talc-chlorite-tremolite-calcite-dolomite.

Magnetite was the earliest ore mineral to form as indicated by the great predominance of inclusions of gangue over those of sulfides and by the occurrence of sulfides inter-granular to magnetite. Pyrite and pyrrhotite then grew between and around magnetite (including euhedral crystals of magnetite) and locally replaced magnetite. Chalcopyrite follows the last-formed pyrite and pyrrhotite, replacing pyrrhotite margins and filling fractures in pyrite. Sphalerite, galena, and digenite were probably contemporaneous with chalcopyrite.

The secondary minerals were formed last. Pyrrhotite first was oxidized to goethite plus marcasite, and marcasite then inverted to pyrite. Where pyrrhotite is oxidized, an aureole of iron staining is produced in the gangue minerals for a distance of several centimeters around the pyrrhotite grain.

Overgrowths of marcasite and/or pyrite on primary pyrite grains and hematite on magnetite are interpreted as secondary mineralization. Hematite along fractures in magnetite is considered as secondary oxidation of magnetite.

## Grade of Ore in Valuable Metals

The chief product of value at the Grace mine is iron, as magnetite, which is concentrated magnetically; the final product is pellets. Sulfide concentrates, by-products of the milling of magnetite, yield valuable amounts of iron, copper, cobalt, and gold. Table III shows average chemical analyses of crude ore, magnetic concentrates, and sulfide concentrates.

It is interesting to compare the ratios of nickel, copper, and cobalt to sulfur in the magnetic and sulfide concentrates.

| Ratio | Magnetic | Sulfide |
|-------|----------|---------|
| Ni/S  | 0.13     | 0.006   |
| Cu/S  | 0.107    | 0.030   |
| Co/S  | 0.03     | 0.014   |

TABLE III.   Average Chemical Analyses of Grace Mine Crude Ore, Magnetic Concentrates, and Sulfide Concentrates; Weight Per Cent.

| Component | Crude Ore | Magnetic Concentrate | Sulfide Concentrate |
|-----------|-----------|----------------------|---------------------|
| Fe          | 43.7   | 68.3   | 39.9  |
| Magnetic Fe | 40.5   | 67.5   | —     |
| S           | 1.9    | 0.13   | 46.5  |
| P           | 0.027  | 0.005  | —     |
| Mn          | 0.14   | 0.06   | 0.01  |
| Cu          | 0.06   | 0.14   | 1.41  |
| Co          | 0.02   | 0.004  | 0.68  |
| $SiO_2$     | 16.0   | 2.4    | 5.9   |
| $Al_2O_3$   | 2.8    | 0.58   | 0.39  |
| MgO         | 12.3   | 1.92   | 3.70  |
| CaO         | 3.0    | 0.42   | 0.39  |
| $Na_2O$     | 0.20   | —      | —     |
| $K_2O$      | 0.41   | —      | —     |
| $TiO_2$     | 0.15   | —      | —     |
| Ig Loss     | 2.4    | —      | —     |
| Ni          | 0.009  | 0.017  | 0.26  |
| Zn          | 0.008  | —      | 0.22  |
| Pb          | 0.006  | —      | 0.20  |

The ratios are much higher in the magnetic concentrates, especially for nickel, showing the affinity of the three components for the magnetic sulfide, pyrrhotite. This assumes, of course, that they are not contained in magnetite. If pyrrhotite could be separated from magnetite, the final sulfide concentrate could be upgraded in nickel, cobalt, and copper.

To show lateral variations of components, plots of chemical analyses of composite samples from diamond drill stations are shown in Figure 5. These show that, at least from the available information, there is little variation in iron content throughout the ore body. However, these plots are averages of several holes drilled radially from a single station and therefore do not show maximum and minimum values. Analyses of 5-foot intervals from surface drill holes show variations from 3.1 to 64.7 per cent iron, 0.03 to 12.99 per cent sulfur, and nil to 0.647 per cent copper. However, the Grace ore body is considered so uniform that selective mining methods are not necessary.

Samples of ore from the first diamond drill hole, which penetrated the greatest thickness of ore, were studied. Figure 6 shows the vertical variation in chemical analyses as well as the different mineral assemblages plotted with depth. The vertical and lateral ranges of iron are similar. Mineralogically and genetically, it

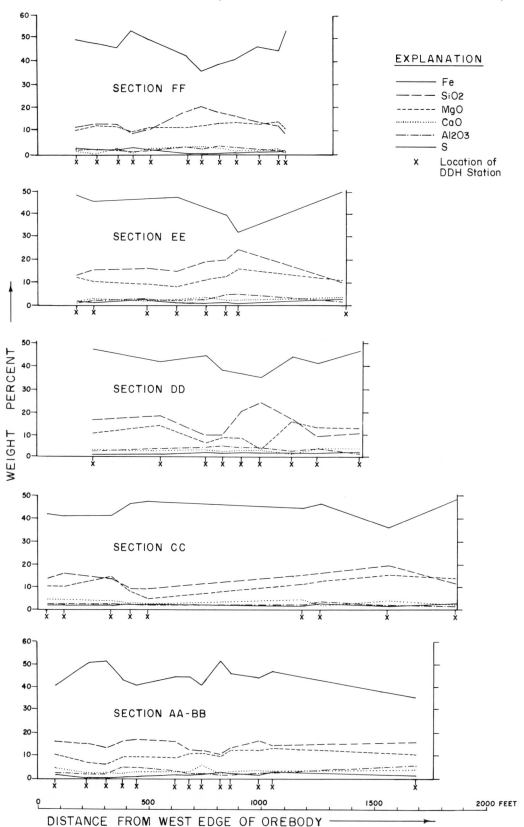

FIG. 5. *Lateral Variation Plots for Fe, SiO₂, MgO, CaO, Al₂O₃, and S. Analyses are composites from one or more holes drilled at a single station. Section locations are shown on figure 4.*

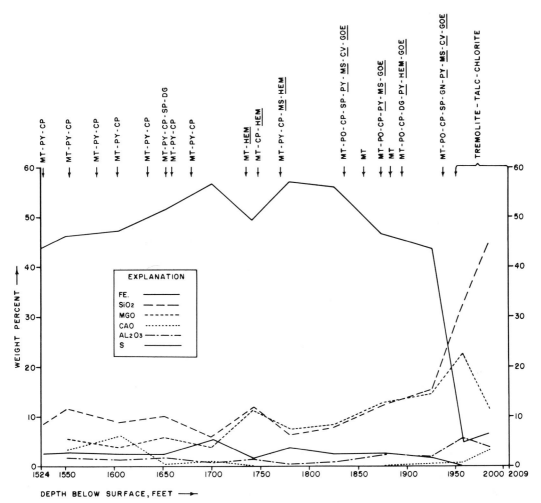

FIG. 6. *Vertical Variation of Ore Minerals and Chemical Analyses from Top (1524') to Bottom (2009') of the Grace Ore Body. Analyses are composites of up to 50 feet. Diamond drill hole No. 1. mt = magnetite, py = pyrite, po = pyrrhotite, cp = chalcopyrite, sp = sphalerite, gn = galena, dg = digenite, ms = marcasite, cv = covellite, hem = hematite. Secondary minerals are underlined.*

is noteworthy that pyrrhotite occurs near the bottom of the ore body.

## Factors Controlling Form and Location of Ore Body

The two most important factors controlling the form and localization of the Grace mine magnetite deposit are the position of the diabase intrusion and the presence of a carbonate unit above the diabase, the localization being adjacent to the diabase and the form that of the replaced carbonate. All important Cornwall-type magnetite deposits conform to this pattern.

The restriction of the ore to carbonate beds above diabase suggests some relationship between ore and the diabase intrusion, but this probably is a matter of channelways and favorable host rocks rather than of the ore fluids having been formed in the diabase and driven into the limestones. The diabase at Grace mine, as well as throughout southeastern Pennsylvania, is quite free of inclusions and is exceptionally uniform (12, p. 678). This uniformity over a wide area indicates the diabase was originally liquid and that intrusion was forceful rather than passive (11, p. 385–387; 12, p. 700). The diabase at Grace mine was introduced at a moderate depth because it in-

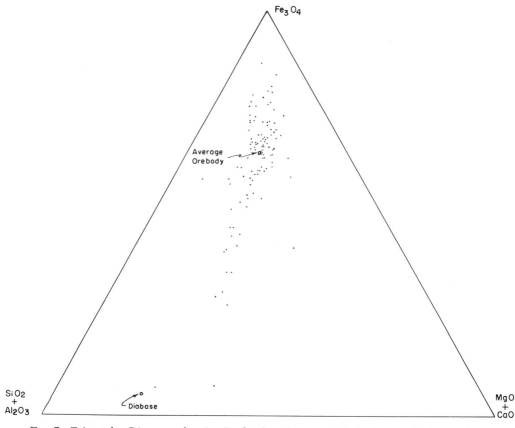

FIG. 7. *Triangular Diagram, showing $Fe_3O_4$ plotted versus $SiO_2 + Al_2O_3$ and $MgO + CaO$.*

truded along the bottom of a section of sedimentary rock estimated at about 20,000 feet in thickness (9, p. 67). The intrusion of such a horizontal sill-like body (25, p. 97; 25, p. 2854), would have uplifted and disturbed the overlying sedimentary rocks. It is suggested that the deformation of the carbonate unit overlying the diabase facilitated access of ore fluids to the actual site of ore deposition.

Petrographic study shows magnetite has replaced preferentially hydrous-calcium-magnesium silicates rather than carbonates. This selectivity also can be shown by plotting $Fe_3O_4$ versus $SiO_2 + Al_2O_3$ and $MgO + CaO$ on a triangular diagram (Figure 7). The range of points defines a linear trend and shows magnetite has replaced both $SiO_2 - Al_2O_3$ and $MgO - CaO$. If magnetite had replaced only carbonates, the linear trend should coincide with the $Fe_3O_4 - MgO + CaO$ leg of the triangle. Another factor, then, is the presence of hydrous calcium-magnesium silicates in the carbonate rock. Still another probably is the presence of an impermeable barrier above the carbonate unit acting to concentrate ore fluids. Baked and hardened Triassic sedimentary rock

hanging wall at the Grace mine deposit would have served this purpose.

The final factor in the formation of the ore body is the nature and source of mineralizing fluids. Because ore minerals replace silicate minerals and occur in fractures in the diabase, and because the ore is essentially free of titanium, it is concluded that mineralization took place after diabase intrusion and crystallization. However, the close association of diabase and ore indicates mineralization was related in some way with diabase magma. It has been suggested that at Dillsburg (Figure 1) ore fluids were released from the cooling diabase as the result of differentiation (12, p. 701) but that at Cornwall, on the other hand, ore fluids came in after diabase intrusion and cooling and were not derived directly from the immediate diabase (22, p. 37–38). No evidence was noted at the Grace deposit to suggest that iron-rich fluids were derived from the diabase during its cooling process. The lack of granophyric rock, the sparsity of pegmatitic facies, and the quite uniform texture of the Grace diabase all indicate that differentiation during cooling was minimal.

The source of ore fluids probably was from the primary magma source of the diabase. The fluids may represent late-stage differentiates or segregations from the magma chamber which escaped after crystallization of the diabase intrusion had created a path and internal pressure had built up high enough to force out the fluids. The diabase intrusive provided a favorable path along its upper surface, and ore deposits were formed where favorable rocks were encountered.

The nature of the fluids is unknown although certain clues suggest they were either gaseous or super-critical liquids. The presence of soda-scapolite in a vein in diabase, tourmaline in sandstone, and fluorite in limestone indicates introduction of Cl and F among other components so that iron may have been introduced as $FeCl_3$ or $FeF_3$ (18, p. 573). It is also possible that the iron was introduced as $Fe^{+2}$ and $Fe_2O_4^{-2}$ and magnetite was precipitated by neutralization of one ion by the other. The ore mineral assemblages show that iron, sulfur, copper, lead, and zinc were added and calcium, magnesium, silica, and $CO_2$ were removed during ore deposition. Water probably was a constant component. Recent preliminary experimental work by the Geophysical Laboratory indicates that the assemblage magnetite-pyrite-pyrrhotite requires an alkaline aqueous environment (16, p. 137).

## Temperature and Pressure Conditions of Ore Formation

It was suggested in an earlier section that the diabase was intruded at a depth of about 20,000 feet. This corresponds to a confining pressure of about 1500 bars, assuming an average specific weight of 2.5 for the overlying sediments.

Several lines of evidence suggest temperatures. The metamorphic mineral assemblages including diopside-forsterite-tremolite-talc-chlorite-calcite-dolomite are assigned to the hornblende-hornfels facies and corresponds to a temperature of 500° to 600°C (18, p. 516). This is in agreement with a preore temperature of 650°C determined by Tsusue (24, p. 8) using calcite-dolomite relations. The upper limit of serpentine in the presence of water was established as 500° ± 10°C (10). Because magnetite has inclusions of serpentine and has grown in a matrix mainly composed of serpentine, it would be logical to conclude that magnetite grew at a temperature of 500°C or less. However, chalcopyrite twinning caused by inversion from cubic to tetragonal form may indicate a temperature of at least 545°C (Kullerud, personal communication, 1965).

Since textural evidence strongly indicates chalcopyrite grew after magnetite, it follows that magnetite may have been formed at temperatures as high as 545°C. An attempt to use pyrite-pyrrhotite equilibrium relations to determine an ore temperature (20) failed because X-Ray diffraction patterns showed only monoclinic pyrrhotite, which cannot be used for this purpose (21). Experimental work by Kullerud (15) with the system Fe-S-O established a maximum temperature of 675°C for the stability of the assemblage pyrrhotite-pyrite-magnetite.

It is clear from the above statements that no single temperature of ore formation can be assigned reliably to the Grace mine deposit, but 545°C indicated by chalcopyrite twinning is considered to be the most valid (Kullerud, personal communication, 1965). A maximum of 675°C and a minimum of about 500°C is believed to cover the temperature range for primary ore formation.

## SUMMARY OF GEOLOGIC EVENTS REQUIRED FOR FORMATION OF THE ORE DEPOSIT

In summary, the following events resulted in the formation of the Grace mine magnetite deposit:

(1) Intrusion of late Triassic diabase along an unconformity between Cambrian limestone and Triassic sedimentary rock isolated and deformed a lens of limestone above the diabase.

(2) After the limestone lens had been deformed, it was altered by early, high-temperature fluids to a calcium-magnesium silicate assemblage stable at about 600°C.

(3) After the diabase had crystallized completely, hydrothermal solutions altered the silicates to hydrous varieties and considerably modified the diabase margin.

(4) Additional increments of hydrothermal fluids invaded the fractured host rock and formed the ore deposit in the temperature range 500°–675°C.

(5) Tectonic deformation caused further tilting of the Triassic sedimentary rocks and normal faulting.

## REFERENCES CITED

1. Rogers, H. D., 1858, The geology of Pennsylvania: (First) Pa. Geol. Surv., v. 2, p. 718.
2. Leslie, J. P. and D'Invillers, E. V., 1886, The

Cornwall iron mines: (Second) Pa. Geol. Surv. Ann. Rept., p. 491–570.

3. Spencer, A. C., 1908, Magnetite deposits of the Cornwall type in Pennsylvania: U.S. Geol. Surv. Bull. 359, 102 p.

4. Harder, E. C., 1910, Structure and origin of the magnetite deposits near Dillsburg, York County, Pennsylvania: Econ. Geol., v. 5, p. 602–612.

5. Newhouse, W. H. and Callahan, W. H., 1927, Two kinds of magnetite: Econ. Geol. v. 22, p. 629–632.

6. Callahan, W. H. and Newhouse, W. H., 1929, A study of the magnetite ore body at Cornwall, Pennsylvania: Econ. Geol. v. 24, p. 403–411.

7. Smith, L. L., 1931, Magnetite deposits of French Creek, Pennsylvania: Pa. Geol. Surv., 4th ser., Bull. M14, 52 p.

8. Hickok, W. O., IV, 1933, The iron ore deposits at Cornwall, Pennsylvania: Econ. Geol. v. 28, p. 193–225.

9. Bascom, F. and Stose, G. W., 1938, Geology and mineral resources of the Honeybrook and Phoenixville quadrangles, Pennsylvania: U.S. Geol. Surv. Bull. 891, 145 p.

10. Bowen, N. L. and Tuttle, O. F., 1949, The system MgO-SiO₂-H₂O: Geol. Soc. Amer. Bull. v. 60, p. 439–460.

11. Hotz, P. E., 1952, Form of diabase sheets in southeastern Pennsylvania: Amer. Jour. Sci., v. 250, p. 375–388.

12. ———— 1953, Petrology of granophyre in diabase near Dillsburg, Pennsylvania: Geol. Soc. Amer. Bull. v. 64, p. 675–704.

13. Gray, C., 1956, Diabase at Cornwall, Pennsylvania: Penna. Acad. Sci. Pr., v. 30, p. 182–185.

14. Bingham, J. P., 1957, Grace Mine: Min. Eng., v. 9, no. 1, p. 45–48.

15. Kullerud, G., 1957, Phase relations in the Fe-S-O system: Ann. Rept. Dir. Geophys. Lab., Carnegie Inst. Wash. Year Book 56, p. 198–200.

16. Barnes, H. L., and Kullerud, G., 1960, Equilibrium between pyrite, pyrrhotite, magnetite, and aqueous solutions: Ann. Rept. Dir. Geophys. Lab., Carnegie Inst. Wash. Year Book 59, p. 135–137.

17. Hotz, P. E., 1960, Diamond-Drill exploration of the Dillsburg magnetite deposits, York County, Pennsylvania: U.S. Geol. Surv. Bull. 969-A, 25 p.

18. Turner, F. J. and Verhuogen, J., 1960, Igneous and metamorphic petrology: McGraw-Hill, N.Y., 694 p.

19. Gray, C. and Lapham, D. M., 1961, Guide to the geology of Cornwall, Pennsylvania: Pa. Geol. Surv., 4th ser., Bull. G 35, 18 p.

20. Arnold, R. G., 1962, Equilibrium relations between pyrrhotite and pyrite from 325 to 743° C.: Econ. Geol., v. 57, p. 72–90.

21. Kullerud, G., et al., 1962, Heating experiments on monoclinic pyrrhotites: Ann. Rept. Dir. Geophys. Lab., Carnegie Inst. Wash., Year Book 61, p. 210–213.

22. Lapham, D. M., 1962, Magnetite mine, Cornwall, Pennsylvania: Internat. Mineral. Assoc., 3d Gen. Meeting, Excusion Guidebook, Natl. Capital Area, p. 36–40.

23. O'Niel, B. J., Jr., 1964, Limestones and dolomites: in Atlas of Pennsylvania's Mineral Resources, pt. 1, Pa. Geol. Surv. 4th ser., Bull. M50, 40 p.

24. Tsusue, A., 1964, Mineral aspects of the Grace Mine magnetite deposit: Pa. Geol. Surv., 4th ser., Bull. M49, 10 p.

25. Beck, M. E., Jr., 1965, Paleomagnetic and geological implications of magnetic properties of the Triassic diabase of southeastern Pennsylvania: Jour. Geophys. Res., v. 70, p. 2845–2856.

26. Davidson, A. and Wyllie, P. J., 1965, Zoned magnetite and platy magnetite in Cornwall type ore deposits: Econ. Geol. v. 60, p. 766–771.

# 7. Mineral Exploration and Development in Maine

ROBERT S. YOUNG*

## Contents

## Illustrations

## Table

## ABSTRACT

During the last quarter-century, exploration for metallic deposits in Maine has been sporadic with peaks generally coinciding with periods of high metal prices. Known cases of regional or semi-regional evaluations are few; most efforts have been local and usually intensive.

On the basis of broad geologic groups, the State has been divided into five fundamental environments: (1) coastal metamorphic belt; (2) southern volcanic belt; (3) Silurian slate belt; (4) northern volcanic belt; and (5) Devonian slate belt. Although not restricted to these areas, exploration in recent years has been concentrated in the northern and southern volcanic belts.

Notable examples of recent exploration activity, directed to sulfide minerals, include: (1) the Knox County nickel-copper-bearing ultramafic rocks; (2) the Harborside (Cape Rosier) copper-zinc deposit; (3) the Second Pond (Black Hawk) zinc-copper-lead-silver

* University of Virginia, Charlottesville, Virginia.

deposit; (4) the Big Hill-Barrett lead-zinc-silver-copper-gold zones; (5) the Parmachenee copper-zinc belt; (6) the Moxie Lake nickeliferous peridotite; and (7) the Attean quartz-monzonite molybdenum-copper occurrences. The Second Pond and Harborside prospects appear to be close to production.

Systematic investigations for non-metallic deposits appear to have been aimed principally toward limestone, asbestos, diatomaceous earth, and spodumene-bearing pegmatites.

Advances in prospecting methods, high metal prices, a favorable business outlook, increasing cooperation from large landholders, and some discoveries support the probability of increased exploration activity in Maine during the coming decade.

## INTRODUCTION*

As with so many of the eastern states, Maine has been the subject of frequent, although sporadic, periods of exploration and mine development. Also in keeping with exploration trends in the East, mineral investigations nearly always coincided with periods of high metal prices, hence were often short-lived. Most of the exploration efforts were designed to examine an individual mineral occurrence or a prospect with its immediate environs; known examples of regional evaluations are rare. However, at some time during the past 30 years, nearly every major United States and Canadian mining company has evinced at least a minor degree of interest in the mineral potential of the State; unfortunately, little is known of their efforts prior to the past 15 years or so. Many of the more recent prospecting efforts, especially those involving the smaller Canadian companies, have been publicized in the various trade journals.

Land ownership distribution has always presented a problem in planning exploration in Maine. In the southern area and along the entire coast, private ownership predominates and multiple ownership of a single, small prospect area is commonplace. The same is true, but to a lesser extent, in the central portion of the State. In the northern part of Maine, the vast majority of the land is held by large paper or pulp companies, some of which have displayed extreme reluctance to negotiate exploration and mining agreements. It must be pointed out that there are significant tracts of land (more properly, mineral rights) that

* The assistance of Robert G. Doyle, State Geologist of Maine, and A. M. Hussey II, Bowdoin College, is gratefully acknowledged.

are controlled by the State of Maine, through the Maine Mining Bureau (28). For the most part, these consist of: (1) "public lots" of different sizes in the so-called "unorganized townships" and (2) naturally-occurring water bodies of ten acres or more, known as "great ponds". Mineral rights under such tracts are available for staking and several, notably Second Pond and Goose Pond, Hancock County, and Crawford Pond, Knox County, recently have become very important pieces of real estate.

Although geologic simplifications are frequently misleading, in many general cases, such as this, there is no other choice. Doyle's (50, p. 139) presentation of the various basic geologic environments of the State may serve as an acceptable model (Figure 1). He depicts, from south to north, (1) a small coastal *metamorphic belt;* (2) a narrow but more extensive *southern volcanic belt* which extends from the longitude of Bucksport northeastward into New Brunswick; (3) a *Silurian slate belt,* which covers approximately one-half the State; (4) a narrow, discontinuous *northern volcanic belt;* and (5) a *Devonian slate belt* covering the northern one-third of the State. There is a close coincidence between the distribution of the more significant base-metal sulfide deposits (Figure 2) and the known limits of the "volcanic belts" (Figure 1). Additional coincidence exists between the spatial relations of sulfide deposits and large intrusives, both within and outside the volcanic belts.

There are a number of Federal and State publications directly oriented toward encouraging mineral exploration in Maine, too many to cite herein. Perhaps the most significant regional contribution is the geologic-aeromagnetic compilation for northern Maine by Boucot *et al.* (40). Covering much less area, but aimed more economically, was the aeromagnetic survey covering the Katahdin Iron Works pyrrhotite body and its environs (17). Aeromagnetic surveys, sponsored by both the U.S. Geological Survey and the Maine Geological Survey, are available for a number of quadrangles and smaller areas in Maine. Recent maps, showing metal content for a large number of stream sediment sites, taken in reconnaissance by U.S. Geological Survey exploration teams have proved highly valuable in guiding exploration (44). The Maine Geological Survey began an effort to attract attention to the State's mineral potential in 1958 with the publication of a compilation of metal mines and prospects (27). This was followed by a series of four "Special Economic Studies",

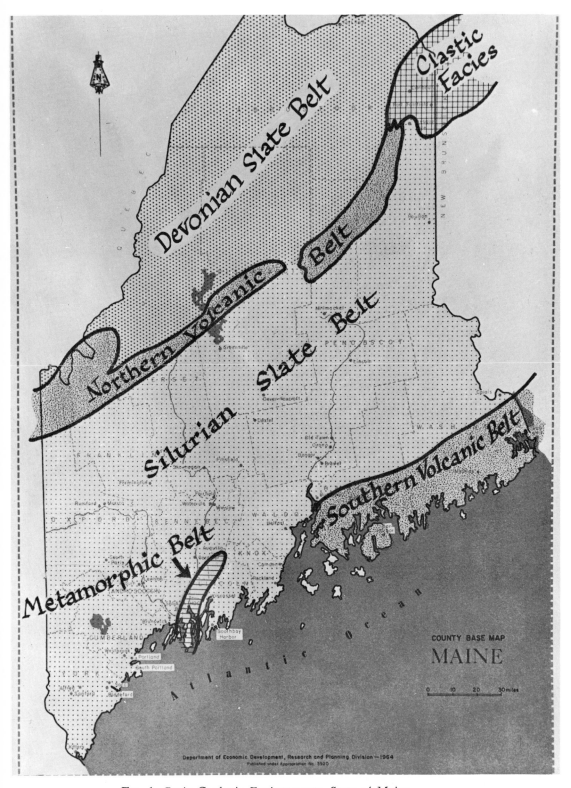

FIG. 1. *Basic Geologic Environments, State of Maine.*

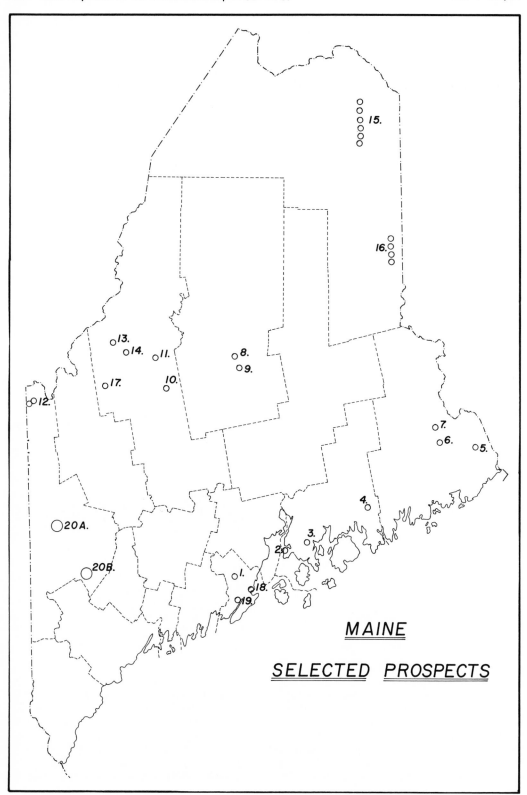

FIG. 2. *Index Map of Selected Properties.*

based on four years of State-sponsored exploration in Hancock, Washington, Penobscot, Piscataquis, and Somerset counties (30,35, 39,46). These latter publications appear to have been directly responsible for the 1962–65 wave of exploration among the small, privately-owned prospects in the southern volcanic belt.

Although hundreds of the base-metal-bearing sulfide occurrences are known in Maine and most of these have been evaluated in some manner, exploration in recent years has been directed to four principal areas, each representing a different geological environment and each showing considerable promise.

## METALS

### Metamorphic Belt

One or more belts in Knox County, southern Maine, have been explored extensively for sulfide-bearing ultramafic bodies, principally copper-nickel in "peridotite" masses. Beers *et al.* (32) reported on the characteristics of two such intrusives, near Union, within which the mineralized portions appear to approach ore grade and dimensions. The nickeliferous sulfides of this prospect occur in lherzolite masses intruded into biotite and hornblende schists of the Penobscot Formation (Cambrian?). As now understood, the mafic bodies conform to

the general regional structure but locally show both rolling and plunging. Metamorphism of the lherzolite is low-grade and of various types; in many cases, it is best described as "alteration", probably related to intrusion by granitic pegmatites. The sulfide suite consists principally of pentlandite and chalcopyrite in a pyrrhotite base; niccolite and cobaltite are present in very minor amounts. These deposits, apparently unique for the State, are unusual in at least three aspects: (1) the high sulfide-to-mafic rock ratio; (2) the low pentlandite-to-pyrrhotite ratio in the unremobilized state; and (3) the presence of a graphitic alteration halo. The original prospect, as described by Houston (24, p. 58), and its trend for several miles were thoroughly explored geophysically with positive results. A detailed description of the geophysical methods and techniques used is presented by Beers *et al.* (32) with an evaluation of individual method applicability. The ultramafic belts of Knox County remain under intense study. It is quite typical that notice of these occurrences appeared in the literature more than a century before they were accorded thorough exploration (1, p. 135–136).

### Southern Volcanic Belt

In the past decade, the southern volcanic belt certainly has received more attention, largely from Canada-based companies, than any other area of equal size in the State. Sulfide prospects are abundant in the coastal strip between the Castine Peninsula and Eastport. The sulfide occurrences differ highly in geological relations, metals present, and tenor. Several former producers, notably the Harborside (Cape Rosier) mine and the Mammoth property, undoubtedly focused initial attention to this area, as did published prospect evaluations (35,39). The U.S. Bureau of Mines carried out limited drilling programs at a number of the larger showings during World War II and the immediately ensuing years (8,10,11). According to recently published notes, two properties are nearing the production stage, Harborside and Second Pond.

Although little is known of recent exploration on the Harborside property, outside the Bureau of Mines report (8), the mineralized zones are thought to be structurally-controlled replacement lenses of high-grade chalcopyrite and sphalerite in a sequence of various volcanic rocks, principally tuffs and agglomerates. The ore body, estimated to be a million tons or less, is shallow and present plans call for mining the deposit at least partially from an open pit. These plans necessitated legislation to allow destruction of a local "great pond" to facilitate development of the mine. The plant is scheduled for 1967 production.

Due to a considerable number of press releases (in the *Northern Miner,* for the most part), much information is available on: (1) the Second Pond prospect at Blue Hill, Hancock County, and (2) the Barrett-Big Hill prospects near Pembroke, Washington County. The Second Pond property, Canadian-owned, has been significantly developed underground, but the long anticipated production has not taken place. According to Doyle (52), the Second Pond (or Black Hawk) deposit occurs as a series of shallow-dipping (26°), parallel ore zones wrapping around the nose of a gently folded, southerly plunging anticline. The ore bands are in highly altered schists and amphibolites of the Ordovician(?) Ellsworth Formation. Ore concentrations are found: (1) at contacts of various members of the folded sequence; (2) as partial replacements of the contorted amphibolite member; and (3) as openspace fillings in a metaquartzite member. The ore zones dip, on the north side, into a binary granite, which forms an irregular contact with the Ellsworth Formation. The main mass of the granite lies slightly more than a mile from the shaft. The sulfide suite consists of pyrite, chalcopyrite, sphalerite, argentiferous galena, and minor pyrrhotite. As now known, the overall deposit appears to be distinctly zoned; from top to bottom, the sulfide zones are: (1) Mammoth zone: approximately 10 feet of zinc ore with minor copper; this zone occurs close to the surface in the old Mammoth shaft; (2) Second Pond zone: the main copper body in this zone is disseminated chalcopyrite in a massive quartzite; the ore is located preferentially in the base of the quartzite in thicknesses up to 30 feet; (3) Amphibolite zone: this is the principal ore zone of the deposit, 4 to 6 feet of partially replaced, highly deformed amphibolite that averages 14 per cent Zn (after 30 per cent dilution); (4) Douglas ("Lower") copper zone: an erratic zone of chalcopyrite pods and intervening stringers, not well explored underground; and (5) Lead zone: a massive pod of silver-rich galena lying on the east side of the main ore body (3. above).*

Unofficial reserve estimates run in the order

of 2.2 million tons of combined zinc-copper-lead ore (52). This figure is at some variance with that cited by the Bureau of Mines (41, p. 487), which states that "According to Denison Mines the indicated reserve, allowing 10 per cent for dilution, was 4.5 million tons comprising 1.2 million tons containing 16.75 per cent zinc, 0.55 per cent copper, and 0.37 ounce silver per ton; 2.8 million tons containing 1.97 per cent copper and 0.41 ounce silver; and 450,000 tons containing 6.75 per cent lead, 0.95 per cent copper, and 4.25 ounces silver."

The property, reportedly, has been developed sufficiently for immediate operation, but no commitment has been made for mill construction. A three-compartment shaft has been sunk to 1300 feet, and several thousand feet of development completed on the 380, 480 and 580 levels. (Figure 3)

An unusual continuity of the mineralized zone is suggested by narrow sulfide intersections at depths of 2135 and 2286 feet in a hole collared approximately 1 mile south of the Black Hawk shaft. The lower intersection ran 9.0 feet grading 2.25 per cent copper per ton and 0.34 ounce silver per ton (53).

Several other prospects in the Blue Hill-Castine area have also been tested in recent years; the *Northern Miner* has noted geophysical surveying and drilling at the Highland, North Castine-Emerson, and Hercules prospects. From published reports, it would appear that the Highland prospect emerged with a "possibility" rating, with one million tons of low-grade copper (0.6 per cent) indicated (41, p. 488). The Tapley copper prospect was drilled by the Bureau of Mines (10), and at least two extensive electromagnetic surveys have been run, covering the apparent northward trend.

In the area east of Blue Hill, the major activity appears to have been leasing and surficial examination of a large number of small, rather isolated prospects. Two silver-bearing zones were the subject of limited drilling programs during the period 1960–1965, the Gouldsboro prospect and the Cherryfield "Mine". Neither examination produced encouraging results.

As with much of the exploration performed in eastern Maine since 1961, the intense evaluation of the Barrett-Big Hill trend, near Pembroke in Washington County, was the direct result of publication of State-supported exploration data. Rhyolitic tuffs and flows, diabasic flows, and shales underlie this general area, but the sulfide deposits in the Pembroke-to-Ayers Junction trend appear to be restricted

---

* Engineering and Mining Journal (May, 1967, p. 156) states that the Second Pond project has been placed on an indefinite standby until "economic and other conditions are more favorable for the production of copper and zinc."

Fig. 3. *Surface Installations, Second Pond (Black Hawk) Deposit, near Blue Hill, Hancock County.*

to the igneous rock members of the sequence. Two prospects (Barrett and Big Hill) have been tested, in considerable detail, through diamond drilling during the past two years. Although these two outcropping mineralized areas are in close proximity, they appear to represent different geologic types. At the Barrett prospect, chalcopyrite and sphalerite were introduced into an andesite host as replacements and breccia fillings. Gold content appears unusually high in certain parts of the mineralized area. According to the *Northern Miner* (55), drilling indicates reserves at the Barrett zone of 385,000 tons that averaged 0.60 ounce silver per ton, 1.75 per cent copper per ton, and 0.23 ounce gold per ton.

Examination of the nearby Big Hill area suggests a large number of mineralized zones, some of which may be interconnected at depth (39, p. 58). Galena and sphalerite are the most abundant ore minerals; chalcopyrite, pyrite, and, probably, silver sulfides are present in minor amounts. The ore sulfides are intimately associated with quartz, generally in irregular, erratic veins, but they also occur as small disseminated blebs throughout the host rocks—rhyolite, dacite, and diabase flows. One small, isolated vein, apparently of galena and quartz, sampled in 1962, yielded an assay of 153.5 ounces of silver per ton (39, p. 60). Recent tonnage estimates (55), based on drill core data, are:

*Big Hill zone:* 650,000 tons averaging 5.08 ounces Ag per ton, 0.14 per cent Cu per ton, 1.17 per cent Zn per ton, 0.77 per cent Pb per ton; *"A" zone:* 180,000 tons averaging 3.0 ounces Ag per ton, 0.12 per cent Cu per ton, 5.20 per cent Zn per ton, 1.20 per cent Pb per ton.

According to published reports, underground development is planned for the Big Hill property.

Although not nearly as well publicized, two major mining companies carried out extensive airborne geophysical (magnetic and electromagnetic), ground geophysical, geological, and geochemical programs in the area immediately around, and to the south and east of, Alexander, Washington County. The areal extent of coverage is not known, but at least one of the programs culminated in the diamond drilling of several mineralized (nickel-copper-cobalt) ultramafic bodies. Unofficial reports indicate that a sulfide body of significant proportions is present at the Frost prospect (39, p. 32), but that the nickel-copper content is considerably below ore grade. There is no known exploration activity in this area at present.

Because of the present market popularity of molybdenum, mention must be made of the Catherine Hill prospect, Hancock County, and Cooper prospect, Washington County. Both of these molybdenum occurrences are well described in readily available literature, hence descriptions are omitted here (2,3,4,5, 7,14,21,39). It is obvious that both prospects have been examined frequently, but no work of significance has been done for 40 years. Doyle (50, p. 139) states that a grade and tonnage estimate is being prepared on the Cooper "Mine," but no other information is available.

## Silurian Slate Belt

Although numerous sulfide prospects are known in the area designated by Doyle (50) as the "Silurian slate belt", few have proved to be of significant proportions or grade. The size of the so-called "Katahdin Iron Works" pyrrhotite body is certainly anomalously large for this area. Apparently, no reserve figures have been published for this sulfide body, but the surface dimensions (6,24) and limited geophysical data (46, p. 57) indicate that it may be one of the largest in the eastern United States. The gossan over the "barren" pyrrhotite supported a local iron furnace for almost a half-century (24, p. 67). Although there is abundant evidence of previous drilling on the property, there is no available information as to recent exploration. The township containing the pyrrhotite mass is the property of a major mineral producing company, and presumably is being held as a sulfur reserve. The Katahdin Iron Works area was included in a Federally-sponsored airborne magnetometer survey (17), and, although several magnetic anomalies were located, no other such pyrrhotite-gabbro bodies have been identified, despite at least one major private effort in this direction.

## Northern Volcanic Belt

Much less information is available, in general, on mineral occurrences in the northern volcanic belt. This is principally because the landholders are large paper companies, and the exploration has been performed by large mining companies. Some of the paper companies, however, responded to requests for information pertaining to mineral discoveries, and much of the following discussion is based on factual-data reports presented to landowners by the exploration units. Because of the spectacular sulfide discoveries in New Brunswick, the northern volcanic belt has been the subject of intense exploration for more than a decade. Contrary to obvious hopes and expectations, such sulfide discoveries as have been made in the central and west-central parts of the State show few characteristics in common with the New Brunswick-type of mineralization.

The southwestern end of the northern volcanic belt was evaluated intensely by major exploration companies during two periods, 1953 to 1956 and 1960 to 1965. From available data, it appears that the exploration was well planned and executed, utilizing detailed geologic mapping, sophisticated geochemical methods, and up-to-date, thorough geophysical surveying. The subject area extends from Parmachenee Township, on the New Hampshire border, to Tim Pond Township, near Eustis (45). The belt is one of complex geology, including metamorphosed sediments, pyroclastics, and mafic flows, all intruded by numerous plugs of intermediate to silicic composition. Even after years of study, the stratigraphic relations are not entirely clear, but the entire layered sequence has been correlated tentatively with the Arnold River complex, of possible Precambrian age, and the Ordovician Ammonoosuc and Albee formations of New Hampshire. It is possible that some of the lower-grade metasediments of the Parmachenee and Rangeley areas are of Silurian-Devonian age. Although occurrences of sulfides appear to be abundant throughout the area, only two zones of limited size seem to have any potential interest. In the Rump Pond-Thrasher Peak area, base-metal sulfide bodies were found to be extensive along a belt of interbedded quartz-sericite schists and slates. Ten drill holes tested this favorable zone, and at least seven had sulfide intersections with traces to abundant amounts of chalcopyrite and sphalerite. Significant thicknesses of low-grade base-metal sulfides were penetrated on the slope of Thrasher Peak. Although detailed studies were made to identify the control(s) of the Thrasher Peak mineralization, the results were inconclusive. However, the major sulfide emplacements, as seen in drill cores and outcrop, are concentrated in stringers that preferentially parallel one set of cleavage. In some localities, the latest-stage folding appears to have exercised some control on the mineralization, with sulfide-bearing quartz veins restricted to, and conforming with, the crestal portions of folds. In the second area, southwestern Parmachenee Township, scattered outcrops of galena-arsenopyrite in quartz veins were discovered through geologic mapping in an area of geochemical anomalies. Type samples of this material assayed: (1) 0.555 ounce Au per ton; 1.9 ounces Ag per ton; 1.9 per cent Pb per ton; 0.20 per cent Zn per ton and (2) 0.070 ounce Au per ton; 1.7 ounces Ag per ton; 5.5 per cent Pb per ton; 0.20 per cent Cu per ton; 1.6 per cent Zn per ton. The country rock for the mineralized vein systems, which are up to 15 feet wide, is meta-arkose with fine-grained sericitic zones. The vein systems—sulfide-bearing quartz with lenses of sericite schist—generally parallel the local schistosity, and the entire mass appears to have undergone post-vein brecciation.

In view of the extent and nature of past exploration in this area, the need for additional work is problematic. However, the sulfide occurrences here are of economic interest, and re-evaluations of good work have, in the past, led to the discoveries of ore bodies. It is understood that Canadian interests plan a large scale airborne electromagnetic survey for this region, to be carried out in the Spring of 1967.

The most conspicuous structural feature of the northern volcanic belt is the so-called "Lanigan" or "Chase Stream-Churchill Stream" lineament in Somerset County. This 40-mile shear zone, with spectacular topographic expression, puts Ordovician(?) metamorphics in contact with Devonian sediments. Significant sulfide concentrations have been located in at least two places near the southwestern terminus of the "break." Outcroppings of chalcopyrite-bearing tuffs(?) at the Squirtgun Flowage dam (in the 10,000 Acre Tract) led to several years of multiple approach exploration in a rather limited area (34). Although no additional sulfide zones of importance were discovered during this initial exploration phase, the Squirtgun prospect remains an interesting one. Assay reports on several grab samples from the mineralized Squirtgun "ash" bed show copper contents that range between 4.05 and 8.57 per

cent copper per ton. Fourteen drill holes were put down locally around the mineralized outcrop and its trend, with the best intersection running 3.3 per cent copper per ton and 1.12 ounces silver per ton from 283 feet to 299 feet in drill hole SG–2. Several intersections contained significant amounts of zinc over mining widths. After the initial exploration program was terminated, a second, independent effort was made to extend and redefine the Squirtgun zone, based on geochemical and induced polarization surveying, apparently without success. In 1966, a Canada-based company secured mining rights on the township and initiated exploration, on trend, southwest of the Squirtgun occurrence. The *Northern Miner* (54) reports that a "belt of mineralization" some 200 feet wide and 500 to 600 feet long was discovered in virgin ground some 2 miles southwest of the Squirtgun Flowage outcrop. This discovery was apparently made on the basis of geochemistry. Core drilling is reported underway, with two holes having been completed; the best intersection is 6.0 feet averaging, per ton, 1.68 per cent Cu; 0.02 ounce Au; 2.7 ounces Ag (56). Perhaps most interesting is the statement that "the geological environment is described as quite similar to the ore occurrences in the Bathurst area of nearby New Brunswick."

Also in 1966, another Canadian company carried out a vertical-loop electromagnetic survey over several miles of the lineament in the Chase Stream Tract, immediately northeast of the known area of mineralization. Several conductors were located and drilled; in each case conductivity was related to low-grade graphitic zones.

Despite the fact that no other sulfide occurrences are known northeast of the Squirtgun Flowage, the geological setting of the remaining 30 miles of the lineament makes this a prime prospecting area.

ATTEAN QUARTZ-MONZONITE With the discovery of outcropping molybdenite and chalcopyrite in the Sally Mountain and Catheart Mountain areas of west-central Somerset County, in 1963, the broad expanse of outcropping Attean quartz-monzonite (40) became the principal prospecting area in Maine (57). Both discoveries were the result of geochemical surveys, based on stream sediment analysis. The Sally Mountain zone was first noted in a U.S. Geological Survey regional geochemical evaluation (44); the Catheart Mountain anomaly was located through industry-sponsored exploration. Since 1964, no less than three major mining companies have had exploration teams active in this area, resulting in the location of a number of interesting anomalies. Drilling of the Catheart Mountain anomalies has been underway since June 1966, and two other areas, Pyrite Creek and Fourmile Brook, have been tested, through limited shallow drilling.

Due to incomplete mapping, many of the details of the geology of both the region and individual anomalies remain to be determined. The Attean quartz-monzonite, exposed in an outcrop area of more than 200 square miles, has been assigned a tentative Ordovician age. If this age concept is valid, the batholith is intrusive into pre-Ordovician metasediments on the west and is lapped by Silurian and Devonian clastics on the east. As might be expected of an intrusive mass of these dimensions, the compositional and textural relations differ widely from place to place. However, in the best known mineralized areas, the process of mineralization appears to have been structurally controlled; breccia developments (Catheart Mountain) and shear zones (Sally Mountain) are locally prominent. The deformation that produced these structures is post-Devonian and may be expressed in the Devonian sediments to the east. The best example, and most obvious, of such an instance is the Bean Brook Mountain re-entrant syncline, the axis of which passes through Catheart Mountain (58).

The sulfide suites of the known mineralized areas are quite simple and give some indication of large-scale zonation, much as does a typical porphyry-type deposit. Molybdenite and chalcopyrite are the principal ore minerals and may occur in low concentrations over large areas. Molybdenite, with attending secondary ferrimolybdite, is found as high-grade masses in quartz veins and as disseminated grains throughout highly altered country rock. Pyrite accompanies both modes of occurrence. There are no known outcrops carrying chalcopyrite, or other copper minerals, in quantities sufficient to explain the copper content in stream sediments derived from either Sally or Catheart Mountains, but chalcopyrite is widespread in minor amounts. In a crude way, the copper-bearing areas may be peripheral to a molybdenite "core." As mentioned, host rock alteration is locally intense, is noted only in $MoS_2$-bearing zones, and consists of silicification and sericitization; low-grade saussuritization, probably deuteric, is widespread. In each of the two cases of molybdenum-copper mineralization, nearby minor showings of galena and sphalerite have been noted. At present, these are interpreted as products of the same stage of mineralization, lying at some distance from the "center." If this premise is true, both structural patterns and the spatial distribution of the various elements of the sulfide suites may prove to be valid guides to exploration.

On the basis of known prospects, it is safe to assume that exploration, principally geological and geochemical, will be intense in the area for some years to come. The region is especially well suited to evaluation through stream sediment analysis, having generally moderate relief, discrete, well-defined drainage basins, and a low contamination factor. Of the geophysical methods, only induced polarization or dipole-dipole resistivity appear to offer any potential promise as exploration tools.

## Miscellaneous Areas

There are a few prospect areas or trends that deserve mention but that may not be characteristic of any single one of the broad geologic belts.

MINERALIZED PERIDOTITES Foremost among these is a series of mineralized peridotites (copper-nickel-cobalt) that are found along the eastern border of the Moxie Lake-Moosehead Lake gabbro intrusive (sill?). At least three such zones are known: (1) Moore's Bog, near Caratunk; (2) Black Narrows, on Moxie Lake; and (3) Burnt Nubble, in Squaretown Township. Of these, that at Black Narrows is best exposed, and probably best known, having been rather thoroughly explored by two or three major mining companies.

The only known evaluation of the Moore's Bog "gossan" was carried out for the Maine Geological Survey (46, p. 66), and this revealed self-potential and magnetic anomalies with some associated high-value geochemical sites.

Although known to have been investigated through private efforts, the only available publications covering the Burnt Nubble mineralized gabbro-norite are those by Canney and Post (36) and Canney and Wing (49). In both of these latter publications, some aspect of geochemical prospecting is emphasized.

Houston (24, p. 80–81) briefly describes the petrography of the norite and peridotite phases of the mafic intrusive, as well as the metasedimentary host, at Black Narrows on Moxie Lake. Sulfide minerals present are identified as pyrrhotite, pentlandite, and chalcopyrite. Houston (24, p. 82) states that the deposit is of magmatic origin and thus origi-nated in the same manner as those near Union, Maine. The Black Narrows peridotite trend has been thoroughly covered through multiple geophysical methods and emerged as a Class "A" anomaly. A strong conductor, verified by both vertical-loop and horizontal-loop electro-magnetic surveying configurations, extends over several thousand feet, roughly centered over Black Narrows. The extent and width of the conductor were well confirmed in subse-quent self-potential, magnetic, and potential-drop-ratio surveys. Approximately a dozen core holes have been drilled into this local peridotite mass, but none produced ore-grade nickel-copper concentrations over mining widths. In fact, it would appear that not a single drill hole produced core that satisfac-torily explained the geophysical anomalies It seems certain, therefore, that further attempts will be made to determine the extent and char-acter of the Black Narrows mineralized zone.

CHAIRBACK MOUNTAIN-HAY BROOK REGION
Several geochemical anomalies or anomaly concentrations, discovered during the 1963 to 1964 regional reconnaissance performed for the Maine Geological Survey (46), in the Chairback Mountain-Hay Brook region of Pis-cataquis County suggest this as a priority pros-pecting area. Local landowners relate that the Chairback Mountain-Gulf Hagas area was geochemically evaluated in the early 1960's, but the results of this survey are not known. Canadian interests resampled the Hay Brook anomaly during 1966, and drilled two holes that failed to determine the heavy metals source. A subsequent induced polarization sur-vey, by the same interests, outlines a high-chargeability zone within the limits of the geo-chemical anomaly. Core drilling of the IP anomaly is scheduled for early 1967.

SHIN POND, RAINBOW LAKE, AND DEBOULLIE LAKE AREAS　Because of the apparent simi-larity of the rock type in the Shin Pond intru-sive, north of Patten (40,43,48), to that of the Attean quartz-monzonite, and the presence of molybdenum in stream sediments in a nearby area (42), a reconnaissance geochemi-cal program was completed in this area during the 1966 field season. Traces of molybdenite and moderate amounts of pyrite are known to occur in certain phases of the intrusive, but the potential of the area has not been determined.

Major mining interests are known to have supported independent geochemical surveys of semi-regional scale in, and around, the gran-ite-gabbro complex near Rainbow Lake, Pis-cataquis County.

Although reported here without the benefit of first hand knowledge, it has been verified that a copper-bearing syenite of large propor-tions was discovered in the Deboullie area, northern Aroostook County, during the 1950's. Outside the initial discovery evaluation, no other work is known to have taken place at Deboullie Lake, although an excellent study of the origin of the syenite has been published (33).

The presence of anomalously large amounts of molybdenum in stream sediments collected in the vicinity of a small intrusive, northern Aroostook County, was revealed in a U.S. Geological Survey release (29). Whether pri-vate industry has followed up this announce-ment is not known, although such seems highly likely.

MANGANESE DISTRICT, AROOSTOOK COUNTY
The problems attending the exploitation of the iron- and manganese-bearing sediments in Aroostook County are of long standing and as yet unresolved. Even after decades of re-search and exploration, the basic problem re-mains that emphasized by Trefethen (22, p. 6), "The question of manganese production from these rocks. . . . hinges upon solving the difficult metallurgical problems of eco-nomically extracting the manganese from its combinations with other elements." In the same reference, Trefethen points out that "the vast deposits of low grade manganese bearing rock in Aroostook County are widely known and publicized." Obviously, deposits bearing such a strategic metal were the subject of con-siderable testing, geologically and metallurgi-cally, during World War II. According to Hus-sey et al. (27), exploration work during the war period was carried on principally by the State of Maine and one mining company. A considerable amount of post-war work was done by the U.S. Bureau of Mines and Geolog-ical Survey.

Most of the known manganese prospects are in the Houlton and Caribou quadrangles and consist of belts of low-grade hematitic and manganiferous sediments, principally shales. Hussey et al. (27) mention numerous man-ganese prospects and very briefly indicate the basic geology and extent of exploration for each. Exploration methods appear mainly to have been geologic mapping, trenching and pitting, diamond drilling, and magnetic sur-veying.

The Dudley, Maple Mountain, and Pierce prospects are typical of those on which considerable effort was expended. The Dudley prospect, Castle Hill Township, was investigated by both State and Federal agencies, as well as private interests. The mineralized sediments are quite homogenous, with reserves of more than 21 million tons running 11.4 per cent manganese. Forty-eight diamond drill holes and three shallow shafts proved strike continuity of more than 4600 feet and depths of several hundred feet (25). In the period 1949 to 1951, the U.S. Bureau of Mines thoroughly evaluated the Maple and Hovey Mountains deposits in TDR2 and T9R3; trenching totalled nearly 6 linear miles and diamond drilling nearly 5 miles. Reserves of 256 million tons averaging 8.87 per cent manganese are estimated, with the possibility of mining 121 million tons of 11.16 per cent manganese "high-grade" (27, p. 27–28). Beneficiation tests were included in this study. Pavlides and Milton (37) reported on the geology and mineralogy of these deposits. The Pierce prospect, Hammond Plantation, also was tested by the Bureau of Mines but to a much lesser extent than the Maple and Hovey Mountains deposits (27, p. 31). Magnetic surveying was utilized to define the deposits, and six drill holes, aggregating 945 feet, were put down. Three manganiferous sedimentary beds were tested, with average grades between 9.5 and 12.9 per cent manganese. Individual mineralized beds may have lengths of 8400 feet.

Several reports on ore dressing and basic recovery procedures of the Aroostook-type manganese ores are available from the U.S. Bureau of Mines (15,18).

Although large tracts of mineral rights are known to be held by private interests on the manganese belts of Aroostook County, in anticipation of eventual production, no extensive exploration is known to have taken place in the region in recent years.

## NON-METALS

For reasons not completely apparent, much less information is available concerning exploration for non-metallics than for metallics. Yet, traditionally, the mineral production of Maine has always been made up predominantly of non-metallics. In 1955, Trefethen (22) cited production income from sand and gravel, crushed and dimension stone, cement rock, structural clays, slate, peat, feldspar, mica, beryl, and agstone. He mentions the possibility of exploiting lightweight aggregate,

spodumene, and sillimanite sources known to occur in the State.

The 1964 Minerals Yearbook (41, p. 484) lists production in Maine of clays, gem stones, peat, sand and gravel, stone, cement, and feldspar with a total declared value of $17,574,000 (the highest in at least 25 years). As might be expected, with the accelerated highway building program, sand and gravel led in value of production. Because of the abundance of these construction items in this glaciated country, little exploration, as such, is carried on for sand-gravel deposits. It is known that some contractors utilize geophysical methods, resistivity and seismograph, to determine till thickness.

Descriptive material on the pegmatite occurrences of the southern part of the State is abundant in State and Federal reports and technical journals. The distribution of feldspar-rich pegmatites is apparently now well known (26), and investigations other than purely scientific are limited. Gem stone production is restricted to the pegmatite region, centering in Oxford County, and apparently stems from small enterprises and individuals.

According to newspaper reports, several attempts have been made in the recent past to develop the northern portion of the Rockland "lime belt", including a proposal for a new railroad and pier to service the development. Cement rock production from the south end of the district, near Thomaston, has increased steadily in recent years; it is reasonable to assume that local exploration is preceding quarry development. Allen's investigations (12,14,19) of various limestone belts in the State include the Rockland and Rockport belts. Doyle (51, p. 18–22) recently described a high-calcium limestone occurrence at Owen Brook, Penobscot County. Rand (31, p. 103–104) published on the successful application of magnetic surveying in delineating contacts and structures in certain limestone environments. The geology of both the Rockland belt and the Owen Brook area indicates that Rand's method might well be applied to good advantage. The present status of exploration in the limestone areas of the State is not known.

In the mid-1950's, a great deal of attention was focused on the discovery of spodumene-bearing pegmatites that outcrop on the State Prison Farm, near Warren in Knox County. Being "public land", some of the area was staked by a group of local individuals, and the claims were transferred to a large mineral-producing company. Sundelius (38) described the pegmatite zone in considerable detail. The prospect was extensively drilled and mapped, apparently delineating some "ore". A reversal in the anticipated trend of lithium market requirements effectively halted development of the pegmatites, and no work, outside research efforts, has been evident for more than five years.

The slate industry of Maine, centered near Monson, Piscataquis County, has not been able to meet competition from substitute materials and has declined greatly in the past two decades. Some attempts have been made to develop a lightweight aggregate market, based on Monson slate, but with little apparent success to date.

The diatomaceous earth-bearing great ponds of Maine, scattered throughout much of the southern half of the State, have been examined frequently with respect to production possibilities. Allen and Pratt (20) and Beck (47) published surveys of many of the known deposits. Those in eastern Maine, principally Hancock and Washington counties, have been staked several times in the past ten years, and several of the larger deposits tested for both reserves and possible recovery methods. Exploration appears to be in a state of quiescence at the present time.

The existence of "asbestos" in the west-central part of Maine has been a matter of record for decades, but descriptive literature on the subject is quite scarce (9,13,23). The geology of the area containing serpentine belts, along the Franklin-Somerset county line, is not available in complete, detailed form from public agencies. However, aeromagnetic coverage, on a broad scale, is available and may assist in prospecting. Ground magnetic surveys show that the serpentine masses may present striking anomalies. One such belt, exposed along Spencer Stream, Somerset County, was drilled recently, but drilling results are not known. Geologic mapping and magnetic surveying have been in progress in the general serpentine-bearing region for the past several years. The close proximity of the Thetford asbestos district, Quebec, should continue to provide a strong incentive for exploration.

## SUMMARY

Because of the paucity of geologic map coverage in the State, mineral exploration has been somewhat retarded. This deficiency will be partially overcome with the publication of a new edition of the State geologic map in 1967. Despite the fact that various portions of the State have been examined in differing degrees of thoroughness by many mining companies, it remains a frontier of exploration in the eastern United States, and much remains to be done. The apparent successes of re-evaluation at Blue Hill and Harborside and primary exploration in west-central Somerset County are most encouraging and are sure to attract attention in this period of industrial expansion and high metal prices. However, the low population density and absence of industrial centers would indicate that mineral exploration will likely be focused on high-unit-value raw materials, generally metallic minerals, unless the lower-value products can be found adjacent to ports affording access to low-cost water transportation.

In summary, mineral exploration in Maine, although cyclic, has increased markedly during the past 20 years. With the advent of great advances in exploration methods, principally geophysical and geochemical, and the prospect for a continuing healthy market situation, exploration for, and development of, mineral deposits in Maine are certain to increase in the coming decade. Diversification aims of the large landholding organizations or their increasing cooperation with mining companies will facilitate such advances.

## REFERENCES CITED

1. Jackson, C. T., 1838, Second report on the geology of the State of Maine: Augusta, p. 135–136.
2. Smith, G. O., 1905, A molybdenite deposit in eastern Maine: U.S. Geol. Surv. Bull. 260, p. 197–199.
3. Hess, F. L., 1908, Some molybdenum deposits of Maine, Utah, and California: U.S. Geol. Surv. Bull. 340, p. 231–240.
4. Hills, B. W., 1909, The molybdenite deposits of Tunk Pond, Maine: Min. World, v. 31, p. 323–324.
5. Emmons, W. H., 1910, Some ore deposits in Maine and New Hampshire: U.S. Geol. Surv. Bull. 432, p. 42, 47.
6. Miller, R. L., 1945, Geology of the Katahdin pyrrhotite deposit and vicinity, Piscataquis Co., Maine: Maine Geol. Surv. Bull. 2, 21 p.
7. Trefethen, J. M. and Miller, R. N., 1947, Molybdenite occurrence, Township 10, Hancock County, Maine: Maine Dev. Comm., Rept. State Geol. p. 54–56.
8. Levin, S. B. and Sanford, R. S., 1948, Investigation of the Cape Rosier zinc-copper-lead mine, Hancock County, Maine: U.S. Bur. Mines, R.I. 4344, 8 p.
9. Wing, L. A., 1949, Preliminary report on asbestos and associated rocks of northwestern

Maine: Maine Dev. Comm., Rept. State Geol. p. 30–62.

10. Earl, K. M., 1950, Investigation of the Tapley copper deposit, Hancock County, Maine: U.S. Bur. Mines, R.I. 4691, 7 p.

11. ——— 1950, Investigation of the Douglas copper deposit, Hancock County, Maine: U.S. Bur. Mines, R.I. 4701, 17 p.

12. Allen, H. W., 1951, Report of the limestones of a portion of Knox County (Maine): Maine Dev. Comm., Rept. State Geol., p. 78–90.

13. Wing, L. A., 1951, Asbestos and serpentine rocks of Maine, Maine Dev. Comm., Rept. State Geol. p. 35–46.

14. Allen, H. W., 1953, Progress report of limestone survey, Knox County (Maine): Maine Dev. Comm., Rept. State Geol., p. 11–12.

15. Lamb, F. E., 1953, Ore dressing tests of Aroostook County, Maine, manganese ores: U.S. Bur. Mines R.I. 4951, 10 p.

16. Wing, L. A., 1953, Preliminary report on eastern Maine granites: Maine Dev. Comm., Rept. State Geol. p. 47–51.

17. Balsey, J. R., Jr. and Kaiser, E. P., 1954, Aeromagnetic survey and geologic reconnaissance of part of Piscataquis County, Maine: U.S. Geol. Surv. Geophys. Invest. Map, GP-116, 1:62,500.

18. MacMillan, R. T. and Turner, T. L., 1954, Recovery of manganese from ores of Aroostook County, Maine: U.S. Bur. Mines R.I. 5082, 41 p.

19. Allen, H. W., 1955, Limestone investigation, 1953–54: Maine Dev. Comm., Rept. State Geol., p. 11–29.

20. Allen, H. W. and Pratt, E. S., 1955, A survey of several reported diatomaceous earth deposits in Maine: Maine Dev. Comm., Rept. State Geol., p. 87–96.

21. Forsyth, W. T., 1955, Airborne magnetometer survey in eastern Maine: Maine Dev. Comm., Rept. State Geol., p. 31–45.

22. Trefethen, J. M., 1955, Review of Maine's mineral outlook: Maine Dev. Comm., Rept. State Geol., p. 5–10.

23. Ellingwood, S. G., 1956, Occurrence and origin of serpentine and chrysotile asbestos in northwestern Maine: Unpublished B. S. thesis, Bates College.

24. Houston, R. S., 1956, Genetic study of some pyrrhotite deposits of Maine and New Brunswick, Maine Geol. Surv: Bull. 7, 117 p.

25. Eilertsen, N. A. and Earl, K. M., 1957, Bulk sampling by diamond drilling, Dudley manganese deposit, northern district, Aroostook County, Maine: U.S. Bur. Mines R.I. 5303, 26 p.

26. Rand, J. R., 1957, Maine pegmatite mines and prospects and associated minerals: Maine Geol. Surv. Mineral Res. Index no. 3, 53 p.

27. Hussey, A. M. II, *et al.,* 1958, Maine metal mines and prospects: Maine Geol. Surv. Mineral Res. Index no. 3, 53 p.

28. Anon., 1959, The Maine mining law for State-owned lands: Public Laws of 1959, Chapter 135, Maine Mining Bur., revised Jan. 1, 1963, Legislative Document No. 348, 16 p.

29. Canney, F. C., *et al.,* 1961, Molybdenum content of glacial drift related to molybdenite-bearing bedrock, Aroostook County, Maine: U.S. Geol. Surv. Prof. Paper 424-B, p. 276–278.

30. Doyle, R. G., *et al.,* A detailed economic investigation of aeromagnetic anomalies in eastern Penobscot County, Maine: Maine Geol. Surv. Spec. Econ. Ser. no. 1, 69 p.

31. Rand, J. R., 1961, Geology for non-metallic producers—a case history: Pit and Quarry, v. 54, no. 6, p. 103–104.

32. Beers, R. F., *et al.,* 1962, Exploration of the Crawford Pond nickel deposit—a case history: A.I.M.E. Preprint 62L90, 11 p.

33. Boone, G. M., 1962, Potassic feldspar enrichment in magma: origin of syenite in Deboullie district, northern Maine: Geol. Soc. Amer. Bull., v. 73, p. 1451–1476.

34. Scott Paper Co., 1962, Wm. L. Philbrick Co., Factual data report.

35. Young, R. S., 1962, Prospect evaluations, Hancock County, Maine: Maine Geol. Surv. Spec. Econ. Ser. no. 2, 113 p.

36. Canney, F. C. and Post, E. V., 1963, Preliminary geochemical and geological map of part of Squaretown, Somerset County, Maine: U.S. Geol. Surv. Open-File Rept.

37. Pavlides, L. and Milton, C., 1963, Geology and manganese deposits of the Maple and Hovey Mountains area, Aroostook County, Maine: U.S. Geol. Surv. Prof. Paper 362, 116 p.

38. Sundelius, H. W., 1963, The Peg Claims spodumene pegmatites, Maine: Econ. Geol., v. 58, p. 84–106.

39. Young, R. S., 1963, Prospect evaluations, Washington County, Maine: Maine Geol. Surv. Spec. Econ. Ser. no. 3, 86 p.

40. Boucot, A. J., *et al.,* 1964, Geologic and aeromagnetic map of northern Maine: U.S. Geol. Surv. Geophys. Invest. Map, GP-312, 7 p.

41. Kusler, D. J., 1964, The mineral industry of Maine: U.S. Bur. Mines Minerals Yearbook, v. 3, p. 487–488.

42. Pavlides, L. and Canney, F. C., 1964, Geological and geochemical reconnaissance, southern part of the Smyrna Mills quadrangle, Aroostook County, Maine: U.S. Geol. Surv. Prof. Paper, 475-D, p. D96–D99.

43. Pavlides, L., *et al.,* 1964, Outline of the stratigraphic and tectonic features of northeastern Maine: U.S. Geol. Surv. Prof. Paper 501-C, p. C28–C38.

44. Post, E. V. and Hite, J. B., 1964, Heavy metals in stream sediment, west-central Maine: U.S. Geol. Surv. Mineral Investiga-

tions, Field Studies Map, MF-278, 1:250,000.

45. Andes Exploration Co. of Maine, 1965, Brown Company, Factual data report: 83 p.

46. Stickney, W. F., *et al.,* 1965, A detailed economic investigation of geochemical and aeromagnetic anomalies, north-central Maine: Maine Geol. Surv. Spec. Econ. Ser. no. 4, 81 p.

47. Beck, F. M., 1966, Diatomite in Maine: Maine Geol. Surv. Bull. 18, Contributions to the geology of Maine, p. 10–17.

48. Caldwell, D. W. (Editor), 1966, The Mount Katahdin region: New England Intercol. Geol. Conf. Guidebook, 61 p.

49. Canney, F. C. and Wing, L. A., 1966, Cobalt: useful but neglected in geochemical prospecting: Econ. Geol., v. 61, p. 198–203.

50. Doyle, R. G., 1966, Base metal environments and exploration history in Maine: Canadian Min. Jour., v. 87, no. 4, p. 138–140.

51. ——— 1966, The Owen Brook limestone prospect, Penobscot County Maine: Maine Geol. Surv. Bull. 18, Contributions to the geology of Maine, p. 18–22.

52. ——— 1966, Personal communication, Nov. 10.

53. 1966, The Northern Miner, Toronto, Jan. 27.

54. 1966, The Northern Miner, Toronto, Aug. 18.

55. 1966, The Northern Miner, Toronto, Nov. 17.

56. 1967, The Northern Miner, Toronto, Jan. 5.

57. Young, R. S., 1967, Sulfide mineralization in the Attean quartz-monzonite, Somerset County, Maine (abs.): Geol. Soc. Amer., Northeastern Section, March 1967, p. 68.

58. Delaney, J. R., 1967, Reconnaissance prospecting method for sulfide-bearing zones in the Attean quartz-monzonite, Somerset County, Maine, (abs.): Geol. Soc. Amer., Northeastern Section, March 1967, p. 22.

# 8. Titaniferous Ores of the Sanford Lake District, New York

STANFORD O. GROSS*

## Contents

## Illustrations

## Table

* National Lead Company, Tahawus, New York.

## ABSTRACT

The Sanford Lake district encompasses an area covering 24 square miles in the central Adirondack Mountains of northern New York State. Discovery of the titaniferous magnetite deposits dates back to 1826. Several attempts to exploit the ores for iron prior to 1942 proved uneconomical because of difficulties with the associated titanium, lack of transportation facilities, and the isolated location. Since 1942, the National Lead Company has produced ilmenite concentrates for the titanium pigment industry, and a magnetite by-product used in the steel and refractory industries.

Rocks of the area are Precambrian in age and of igneous origin. The area has a history of erosion, deformation, and metamorphism, subsidence, and finally uplift and glaciation. The resulting topography is extremely rough with drainage patterns controlled by a combination of fault patterns and glacial modifications.

The district is within the large anorthosite massif making up the central high peak area of the Adirondacks. All of the various low-silica rock types associated with the massif are found within the boundaries of the district. These consist of both the Marcy and Whiteface types of anorthosite and of gabbroic anorthosite, gabbro, and different grades of titaniferous magnetite ores. All of the rocks contain the same minerals and differ only in their percentages of these consituitents.

Gabbro, and ore associated with it, demonstrate flow structure as both have foliation and lineation. Anorthosite and its associated ores are massive and show very little primary structure. Minor pegmatites and diabase dikes provide evidence of remobilization in selected areas and of late intrusion as well.

Faulting is prevalent and follows a regional pattern of major faults trending northeast-southwest and of minor faults at nearly right angles to them.

There are four mineralized areas where an economic grade of ore has been found. Three of these have both gabbroic-type ore and anorthositic-type ore. The fourth has only gabbroic ore. Ores are graded according to their $TiO_2$ content, and these range from 9.5 per cent to over 30.0 per cent $TiO_2$. Ore bodies of both types are related to gabbro and conform to the configuration of the gabbro bodies within the anorthosite.

At least three distinct theories of ore genesis have been expounded by different investigators. Others propose some variations of one or more of these. Although complete agreement probably never will be reached, it is certain that no one process of ore emplacement can explain the complicated picture as seen today. Further understanding of the ores and their origin will require detailed mapping, sampling, and mineralogical studies of the entire district, and then it may never be understood without qualification.

## INTRODUCTION

The Sanford Lake district lies in western Essex County in northern New York, in the heart of the Adirondack Mountains at the headwaters of the Hudson River, of which Sanford Lake is a part. The lake is in the southeastern corner of the Santanoni quadrangle. The district is quite isolated, and it was necessary to build a 40-mile power line, a 30-mile railroad, 8.5 miles of highway, and housing facilities before economic operations were feasible.

Climate is distinctive of the northern United States or southern Canada. Temperatures drop to a minimum of minus 30° to 40°F in winter with averages of 0° to 10°F above during January and February. Summers are comfortable with temperatures rarely going above 90 degrees. Rainfall is 30 to 40 inches annually including an average snowfall of more than 100 inches.

The locations of the four important mineral bodies in the district are shown in Figure 1. These bear the following names with former designations shown in parentheses: Sanford Hill—South Extension (Sanford Hill), Cheney Pond, Mt. Adams (Iron Mountain or Ore Mountain), and Upper Works (Calamity—Mill Pond). The Sanford Hill—South Extension ore body was developed by the Titanium Division of National Lead Company, as the

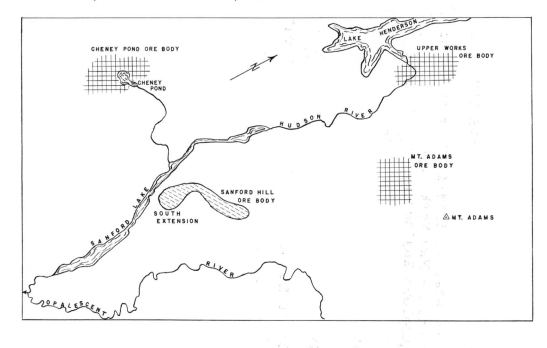

FIG. 1. *Sanford Lake District.*

MacIntyre Development, in 1941. Other than a few thousands of tons mined from the Upper Works ore body prior to 1900, all ore production has come from the Sanford Hill—South Extension ore body.

## HISTORICAL INFORMATION

### Mining Operations

The history of the area has been described by A. H. Masten (7) and W. C. Watson (2). The Sanford Lake deposits were discovered in 1826, when an Indian led a group of prospectors into the area. Ore was discovered first in the stream bed at the very headwaters of the Hudson River. Samples were shipped out for test work and proved to make excellent iron. Following this, operations increased, and, during a brief period about 1834, it is reported that 1500 to 2000 pounds of iron a week were produced from forges. In 1838, the first blast furnace was built, and the enterprise was incorporated as the Adirondack Iron and Steel Company. In 1839, the deposits were visited by Ebenezer Emmons (1), who predicted that this district would turn into a major steel pro-

ducing area of the United States. It is interesting to note that the Adirondack Steel Company made the first cast steel in America, utilizing iron from the Sanford Hill ore body. Steel from this plant was awarded a gold medal at the World's Fair in London in 1851.

In 1848, some 22 years after the discovery of these deposits, the presence of titanium in the ore was recognized. This element, which contributed largely to the lack of success of the early operators, is now the element that has caused the development of the deposits on a large scale.

In 1854, a new and larger blast furnace was put into service; the remnants are still standing about 2 miles upstream from the present operation. The period 1850 to 1858 saw the greatest activity during the early history of the deposits.

The enterprise was reorganized in 1894 as the MacIntyre Iron Company. In the early 1900's, extensive development work was undertaken in the district. Dip needle surveys over the ore bodies were followed by considerable diamond drilling. At this time, plans were made for a railroad and other necessary production facilities. As late as 1912 and 1913, a small crushing and concentrating plant was erected on the east shore of Sanford Lake, at the site of the Sanford Hill ore body, to produce concentrate ore for testing. Some 15,000 to 20,000 tons of ore were reported shipped and tested in modern furnaces at Mineville, New York under the supervision of F. E. Bachman (6), General Manager of the Northern Iron Company. As had proved true in the past, transportation difficulties were encountered, the region still being quite inaccessible. This, coupled with the ever present problems caused by titanium in the smelting process, ended any further development.

In 1892, A. J. Rossi (3), a French metallurgist, was hired to study and improve the methods of smelting titaniferous ores. Further work, done in 1908 by Rossi on the ores of the MacIntyre Iron Company, led to the discovery of the suitability of titanium oxide as a white paint pigment, the first of a series of discoveries of uses for titanium oxide.

National Lead Company acquired the property of the MacIntyre Iron Company in May 1941, and undertook the development of the Sanford Hill ore body. This work consisted of a dip needle survey over the Sanford Hill ore body, followed by an extensive diamond-drilling program, the opening of an open-pit mine, and the construction of a crushing and concentrating plant. In addition, a village, complete with water and sewage systems, was built to accommodate the employees of the company. A modern highway, 8.5 miles long, was built to connect the operation with the nearest state highway. Construction of a 40 mile power line was also necessary to provide power for the plant and the village. Shipment of ilmenite concentrates to the railhead at North Creek, 30 miles away, started in July 1942. Shortly thereafter, a railroad siding was completed to the plant thereby eliminating the 30-mile truck haul.

Since the beginning of ilmenite production, the operation has run on a continuous basis at a generally increasing production rate. A Greenawalt sintering plant was put into operation in 1945. Since that time, magnetite has been shipped both as a concentrate and a sintered product.

Continued exploration and development work has extended the ore body on Sanford Hill to the southwest where it was overlain by glacial till and Sanford Lake and consequently did not outcrop. A stripping program was started in 1961 to render this South Extension of the ore body available for future mining. The development program is continuing at the present time, and considerable ore production is coming from this new area.

## Statistics of Mine Production

No accurate records of ore production by early operators in the district now exist. Reference has been made to the few thousands of tons produced for blast furnace tests in the early 1900's. The amount of material mined prior to 1941 must have been limited to a small tonnage as evidenced by the minor excavations in existance at the time National Lead Company took over the properties.

Beginning with the development of the ore bodies for ilmenite production, and extending through the year 1964, a total of 80,900,000 tons of material have been removed from the Sanford Hill—South Extension area alone. This total is made up of 33,500,000 tons of ore to the crusher and 44,700,000 tons of waste.

## PHYSIOGRAPHIC HISTORY AND PRESENT TOPOGRAPHY

Very little evidence in the Sanford Lake district offers a clue to other than the most recent physiographic history of the area. To investigate further into the past, it is necessary to encompass the entire area of northern New York.

It is generally agreed that beginning in the Precambrian there was a long period of stability which left the land surface with a mature erosion profile. The Grenville orogeny uplifted the entire region with severe metamorphism and deformation of the rocks that now show on the surface. The stratagraphic thickness of Precambrian metasediments in the Adirondacks is indicated to be many thousands of feet thick. Therefore, a very long period of erosion must have taken place subsequent to the major deformation so that former basement rocks could be exposed at the surface.

Paleozoic time saw regional subsidence as is indicated by sediments of that age deposited on all sides of the central highlands of the Adirondacks. Evidence is inconclusive as to whether or not the high peak region was completely submerged. Movement in the later Taconic and Appalachian uplifts to the east undoubtedly affected the Adirondack region, resulting in further development of the fault and fracture patterns previously set.

A succession of uplifts in the area maintained a surface with a characteristically early erosion pattern developed along the general fault systems in the region. The result of this is a drainage system, particularly in the resistant anorthosites, wherein the major streams follow the major faults and tributary streams occupy secondary cross faults.

Little is known of the detailed history of the district during the glacial period. The entire region is well within the sphere of action of the Labrador ice field. The number of times this ice sheet advanced and retreated is unknown, the last advance having erased all or nearly all evidence of its predecessors. It is interesting to note that stripping operations in 1962 over the South Extension of the Sanford Hill ore body revealed a layer of several feet of compacted mud and clays sandwiched between 30 to 50 feet of glacial tills above and 20 to 30 feet of glacial tills below. This compacted material is of the same general composition as the recent lake bottom mud and clays removed in stripping a portion of the Sanford Lake basin. Wood particles from this buried layer have been identified as evergreen of the white pine species, and have been carbon dated at plus 40,000 years of age. It has not been determined whether this occurrence represents an inter-glacial period or simply a stage between minor fluctuations in the ice front. It was, however, of sufficient duration to allow recovery of evergreen vegetation from some region to the south.

The fact that the entire region was covered by the ice sheets is confirmed by the presence of glacial striae over widespread areas, including the highest peaks. Direction of these striae in the locality of Sanford Lake is almost due north-south. Final disappearance of the ice left a topography modified by glacial scouring and glacial deposits. General topographic features were little changed and quite similar to what we see today. Ridge slopes were smoothed, summits rounded, and numerous lakes formed either by deposition of glacial debris or excessive ice erosion. With continued uplift, stream erosion has been active in the post-glacial period, but it is doubtful if there has been more than slight modification in the Sanford Lake district because of the competency of the anorthosite rock.

The central topographic feature within the outline of the district is the Hudson River drainage. Although the drainage basin extends to the north of the district, the stream is first named the Hudson at the outlet of Henderson Lake. From this lake, it continues south across the district, Sanford Lake being a stillwater portion of the Hudson River with a maximum depth of 20 feet. The area immediately bordering the river and the lake has very little relief, but hills rise sharply both to the east and the west of this rather narrow valley to heights

of 400 to 600 feet above the lake, which is 1720 feet above sea level. Maximum relief within the district of approximately 1800 feet is afforded by Mt. Adams rising to the northeast of Sanford Lake. The district is within the high peak area of the Adirondacks, several of the highest elevations in New York State being visible within a few miles to the west, north, and east.

The ground surface is extremely uneven. The sides of the hills are most often a series of rock ledges left by differential erosion along fault or joint patterns. Swampy areas are common, being caused either by ice eroded depressions in bedrock or depressions in the impervious glacial till. This entire area is quite densely forested with hardwood and evergreen trees.

## GENERAL GEOLOGY

### Rock Types

Rocks of the Sanford Lake district generally are regarded as members of a genetically related anorthositic series, the whole being part of the large Adirondack anorthosite massif. In turn, the anorthosites of the Adirondacks are divided into two types. One of these, which forms the major portion of the massif, is the Marcy anorthosite. The second, which is found most abundantly around the border of the mas· sif, is the Whiteface anorthosite. Locally, anorthosite grades into gabbro by an increase in the content of mafic minerals. Buddington (10) has divided this sequence into four rock types in accordance with the amount of ferromagnesian minerals present as follows:

| Rock Type | Per Cent Mafic Minerals |
|---|---|
| Anorthosite | 0–10 |
| Gabbroic Anorthosite | 10–22.5 |
| Anorthositic Gabbro | 22.5–35 |
| Gabbro | over 35 |

For mapping purposes at the mining operation, the term anorthositic gabbro is not used. Rock types are mapped as anorthosite, gabbroic anorthosite, gabbro, and the various grades of ore. Many of the maps used in the mining operation, such as plan and section ore maps, are based strictly on sample assays. For such mapping, the following classification has evolved:

| Classification | Per Cent $TiO_2$ |
|---|---|
| Anorthosite | 0– 5.4 |
| Gabbro | 5.5– 9.4 |
| Low grade protore | 9.5–13.4 |
| Medium grade ore | 13.5–17.4 |
| High grade ore | 17.5 plus |

This relationship of $TiO_2$ content to rock type has proved quite reliable for mining purposes.

The anorthosite in the Sanford Lake district is principally of the Marcy type. This is a coarse-grained, somewhat porphyritic, dark, blue-gray to greenish rock composed mainly of labradorite feldspar. The dark color is attributed to an abundance of very fine inclusions of mafic minerals within the crystals of labradorite. The feldspar grains in the intergranular areas range from 0.5 to 3 cm in greatest dimension. The phenocrysts average about 5 cm in greatest dimension; however, isolated crystals up to 20 cm long have been measured. Albite twinning striations usually are visible to the unaided eye. In thin section, the larger plagioclase grains often show zoning. Again, these large grains will be bent and fragmented at their edges. The anorthosite may change character within a distance of a few feet. This change is a transitional one whereby the intergranular areas, consisting of fine- to medium-grained plagioclase, increase to as much as 30 per cent of the total rock. This finer-grained material becomes a lighter bluish-gray in color, undoubtedly a result of movement and fragmentation while the rock was in a near-solid state.

Stephenson (11) determined the plagioclase feldspar in the anorthosite to range from 37 to 64 per cent An. Avenius (12), on the other hand, found the plagioclase ranged from 50 to 53 per cent anorthite, the large grains being more calcic than the fine-grained material.

Pyroxene and hornblende are the major accessory minerals in the anorthosite. Biotite, grossularite, and sulfides also are observed. Ordinarily these accessory minerals are of the same grain size, or smaller than, the fine-grained plagioclase and tend to concentrate in intergranular areas of the finer-grained anorthosite. The dark opaque minerals, magnetite and ilmenite, occur in minor amounts. Hornblende is present as both a primary mafic mineral and as a secondary alteration product of the pyroxenes.

In some areas, particularly near faults, the anorthosite shows a gneissic structure due to mafic minerals concentrated in thin lenses oriented parallel to the structure. These gneissic zones often contain stringers of magnetite and ilmenite and range in thickness from a few inches up to 5 or more feet. Areas of anorthosite, ranging in size from small pieces the size of a hand specimen to extremely large blocks, occur within the ore mass. The ore and anorthosite nearly always are separated by a narrow band of garnet; in many places, the garnet is associated with hornblende and pyroxene. These reaction rims ordinarily are limited to narrow zones ranging in thickness from something discernible only with a microscope up to several inches. They occur around individual plagioclase crystals in massive ore, as well as between ore and wall rock.

The Whiteface facies of the anorthosites, which is generally considered a border phase, occurs within the Sanford Lake district in various amounts. It is of a medium-gray color and is considerably finer-grained than the Marcy anorthosite. The plagioclase feldspar in this rock is predominately of the andesine variety rather than labradorite. It is more equigranular and phenocrysts are less abundant. Pyroxene is the major accessory mineral with hornblende as the next most plentiful, followed by garnet and minor amounts of biotite. The Whiteface anorthosite shows a more distinct mineral lineation and has a slightly higher content of ferromagnesian minerals than does the Marcy type. Areas of greater shearing and fragmentation in the district conform more to the description of Whiteface anorthosite in type of feldspar, texture, and accessory minerals. Boundaries between the two types of anorthosite seldom are distinct. The rock can change from one type to the other, and back again, within short distances.

Anorthosite grades into gabbroic anorthosite with an increase of fine- to medium-grained intergranular plagioclase and mafic minerals. The anorthite content of plagioclase in the gabbroic anorthosite is in the range of 40 to 50 per cent. There may be a content as high as 50 per cent of plagioclase phenocrysts showing a preferred orientation paralleling that of the ore bodies. The phenocrysts commonly are dark green in color, contrasting with the light greenish-gray color of the finer-grained plagioclase matrix. Mafic minerals occur within the matrix and comprise from 10 to 35 per cent of the rock. The plagioclase occurs in broken fragments, the larger grains being bent. In most cases, the plagioclase grains are normally quite free from inclusions, compared to the plagioclase in the anorthosite. Pyroxenes still predominate as the major mafic mineral, their combined percentage commonly ranging from 10 to 15 per cent of the total rock. Garnet, ore minerals, and hornblende constitute the remainder of the mafics. As in the anorthosite, hornblende may predominate near faults as the major mafic mineral.

Gabbroic anorthosite is distributed in a band parallel to the strike of the ore body on the west side of Sanford Hill pit. It also occurs in minor amounts throughout other areas of the pit as elongate lenses at the borders of anorthosite included in the gabbroic rocks and locally in anorthosite near ore contacts. As an example of the difficulties of detailed rock classification, the same local occurrence has been classified by Gilson (13) as Whiteface anorthosite, by Heyburn (15) as gneissic anorthosite and by Kays (16) as gabbroic anorthosite showing lineation of mafic minerals.

Gabbro occurs as a fine- to medium-grained rock having a uniform texture. It consists essentially of plagioclase feldspar with 35 to 65 per cent mafic minerals. It may on occasion contain a few plagioclase phenocrysts. The color ranges from gray to brownish-gray to black, depending upon the amount and composition of the ferromagnesian and ore minerals present. The gabbro is, for the most part, finely banded due to preferred elongation of the ferromagnesian silicate minerals and a segregation between fine-grained plagioclase and mafics. Fine-grained garnet occurs throughout the gabbro and is more abundant as the fine-grained ore minerals increase. Hornblende increases as the percentage of ore minerals decreases. Locally, reddish-brown biotite is abundant near shear planes and along shear surfaces. The composition of plagioclase in the gabbro ranges from 30 to 40 per cent anorthite. Pyroxene commonly rims magnetite and ilmenite grains and is followed by rims of fine-grained garnet or hornblende, or both.

The gabbroic units parallel the ore lenses. Those units closely associated with ore are finer-grained and generally show sharp contact with anorthosite or gabbroic anorthosite, thus indicating an intrusive relationship. It is not uncommon to find elongate inclusions of anorthosite within the gabbro.

When the $TiO_2$ content of gabbro reaches 9.5 per cent by assay, it is considered locally as low-grade ore. Gabbro occurs on the hanging wall, or west side, of the Sanford Hill—South Extension ore body. It is found in a rather thin band on the hanging wall of the Mt. Adams prospect. There are occur-

rences of gabbro in anorthosite between Sanford Hill and Cheney Pond. These bodies are poorly exposed; however, magnetic results do not indicate ore mineralization in economic quantities. Diamond drill results at Cheney Pond show the ore body to be entirely within gabbroic rock with definite sharp contacts between gabbro and anorthosite at both upper and lower limits. Gabbro is found at the Upper Works prospect; however, lack of sufficient diamond drilling and a limited amount of detailed geologic mapping prevent a thorough knowledge of the extent of this rock type and its relationship to ore and anorthosite.

A few small pegmatites have been observed in the district, principally in the operating pits. Composition is mainly large anhedral plagioclase grains of a brownish-pink color. Large columnar crystals of hornblende are not uncommon. The pegmatites are generally irregular in shape and limited to less than one foot in width. The mineral composition and shape suggests a possible remelting of localized areas within the rock mass due to heat of deformation pressures.

Several diabase dikes of a greenish-gray to black color occur in the district. In the Sanford Hill—South Extension ore area, diabase has been encountered in diamond drill holes and mapped on the pit benches; intersection intercepts range from a few inches to 2 feet. Drill holes at Cheney Pond intersected diabase for several feet, but, since the orientation of this dike was not determined, the thickness is unknown. Diabase was encountered in short intersections in drill holes at Mt. Adams. Dikes also are observed in small drainages where they follow minor faults reflected in surface depressions. These dikes are very fine-grained and contain fractures with bleached margins that probably were formed as tension cracks during cooling. Stephenson (11) determined the mineral composition as labradorite in a ground mass of pyroxene, the latter mineral being altered to finely disseminated carbonate and chlorite.

## Structure

Primary geologic structure within the district is exemplified by foliation or by banding and linear flow structure. The Marcy anorthosite shows little or no foliation. The plagioclase phenocrysts commonly are tabular and locally may show evidence of planar flow. However, as stated by both Balk (9) and Stephenson (11), planar structure in the anorthosites ordinarily is very obscure. This is also true of the gabbroic anorthosite.

Gabbro shows very distinct foliation. Dark minerals tend to concentrate in bands or layers, alternating with bands higher in plagioclase.

Linear flow, or the orientation of individual mineral grains with their long axis parallel, is very difficult to see in the anorthosites. Stephenson (11) worked out a system whereby he recorded the strike and dip of the side pinacoid or (010) face of several apparently unoriented plagioclase phenocrysts in an outcrop and plotted them on a projection net. He found that the traces of these (010) faces would intersect in a common point, or nearly so. This suggests that the crystals are oriented around a common axis, with the side pinacoid faces parallel to this axis. This he interpreted as a primary flow structure.

Linear flow structure is well defined in most of the gabbro masses with the tabular and prismatic minerals in the bands oriented with their long axes parallel to the general strike of the ore body.

Faulting in the district is plentiful and conforms to fault patterns present throughout the region as a whole. Major breaks have a general northeast-southwest strike with steep dips to the northwest. There is a fault system of lesser magnitude at nearly right angles to the major system. Both of these systems are readily discernible in aerial photographs.

One of the major faults is nearly parallel with the Sanford Hill—South Extension ore body. It strikes 44°NE and dips 54°NW. The surface expression of this fault can be identified on aerial photographs a distance of a mile or more to the northeast and southwest of the ore body. In the Sanford Hill pit, there is a branch fault which curves from northeast to the east until it strikes approximately S84°E and dips 54°N where it intersects the southeast wall of the pit. This fault, in turn, has a branch which strikes approximately N70°E and dips 65°NW in the vicinity of the southeast wall of Sanford Hill pit. None of these faults has a single plane of fracture over any appreciable distance. The major fault in Sanford Hill pit ranges from a narrow break of 6 to 12 inches in width to a highly fractured zone up to 50 feet across. Branch faults are on curving planes and have generally narrower fracture zones than the large fault.

In the South Extension pit the major fault carries through from the Sanford Hill pit; however, the branch faults are more numerous and curve from southwest to the southeast.

Surface mapping, magnetic surveys, and diamond drilling have disclosed faulting to be prevalent in both the Mt. Adams and Cheney Pond ore bodies. Faulting cuts the ore bodies, with displacements of over 200 feet.

Extensive jointing has been observed in drill cores from all depths penetrated throughout the district. Although core recoveries are in the 90 per cent plus range, it is seldom that a solid piece of core over 3 to 5 feet long is recovered. Three major joint patterns are exposed in the pits, two of which are vertical and one is horizontal. Locally, the joint orientation is very complex, particularly as affected by the faulting. However, the joint patterns in the district generally are parallel to the regional structure.

## Age of Mineralization

The ore deposits of the Sanford Lake district are of Precambrian age. Although there is a definite difference of opinion among investigators as to the time of emplacement of both ore bodies and gabbro, relative to the anorthosites, there is general agreement as to placement in the geologic time table. Geologic history shows that the central Adirondacks have been an area of high relief since Cambrian time. Since the structure and texture of the rocks and ores show them to have been formed at considerable depths, this must have taken place prior to erosion of thousands of feet of overlying material. There has been no period since Precambrian time when differences in land elevations would have allowed for deposition of the necessary thickness of overlying sediments required for formation of the metaigneous rocks seen at the surface today.

## ECONOMIC GEOLOGY

### Forms of the Ore Bodies

The titaniferous magnetite ore bodies of the district are of two types. One type, referred to locally as "anorthositic ore," is associated with, and in contact with, anorthosite waste rock. These ore bodies occur as massive lenses with irregular dimensions. They show little or no flow structure, are coarse grained, and have definite, though irregular, contacts with the anorthosite. The second type ore body, "gabbroic ore," occurs as oxide-enriched bands within the gabbro. These bodies are fine- to medium-grained and have well-defined structures similar to those in the gabbro. The thickness of individual ore bands ranges from fractions of an inch to tens of feet. The percentage content of metallic oxides, the frequency of occurrence, and the thickness of these individual bands determine whether or not any particular block of ground is ore. Contacts range from distinct ore—gabbro interfaces to gradational changes from ore to waste material.

In the Sanford Hill-South Extension ore body both types of ore occur. The anorthositic type forms a footwall ore body and the gabbroic type forms a hanging-wall ore body. These are separated by various widths of anorthosite and/or gabbro rock. In some instances, the two types of ore are in direct contact.

The two ore lenses generally are parallel throughout. At the northerly, or Sanford Hill end, they strike approximately N30°E and dip 45°NW. What was formerly thought to be a pinching out of the ore lenses to the south proved to be the narrowing of the ore lenses at the center of a large fold structure. The strike of the South Extension is approximately north-south and the dip flattens to about 30° overall. Figures 2 and 3 show the configuration and relationship of the ore lenses.

This ore body is designated by two names because it has been developed in two pits. Mining started where ore was exposed in outcrops on the west side of Sanford Hill. Diamond drilling indicated the ore lenses as becoming very narrow where they extended underneath

the basin of Sanford Lake. Later magnetometer results, both aerial work in 1955 and ground follow-up in 1959, revealed a change in strike and width of magnetic anomalies. Drilling has proved this South Extension of the Sanford Hill ore body to be of commercial grade and tonnage.

The Mt. Adams ore body is a mineralized lens outcropping on the southerly slope of the mountain of that name and was known by early operators as Iron Mountain. This occurrence is quite small, averaging 70 feet in width over a length of 1750 feet. It strikes N58°W and dips 85°SW. The portion of the mineralized lens that is considered of ore grade is nearly all of the anorthositic, coarse-grained, high-iron ore type. There are minor widths of gabbro and gabbroic ore along the hanging wall. Both easterly and westerly along the strike, the rich ore gives way to low-grade mineralized gabbro.

At the Upper Works or Calamity-Mill Pond ore body, both rich massive ore in sharp contact with anorthosite and gabbro with ore-rich bands can be observed in outcrops. There has not been enough diamond drilling to determine the exact relationship of the two types of ore in this area. Outcrops alone are too few and too widely spaced to afford a detailed picture.

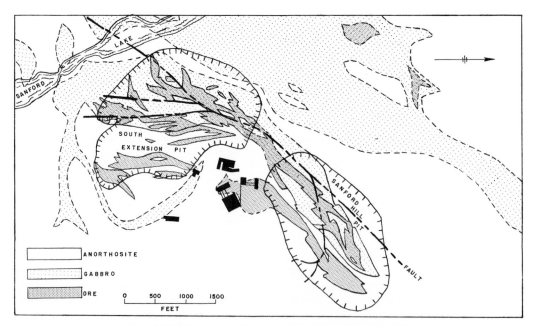

ANORTHOSITE

GABBRO

ORE

0      500     1000     1500

FEET

FIG. 2. *Geologic Map of the Sanford Hill—South Extension. Orebody.*

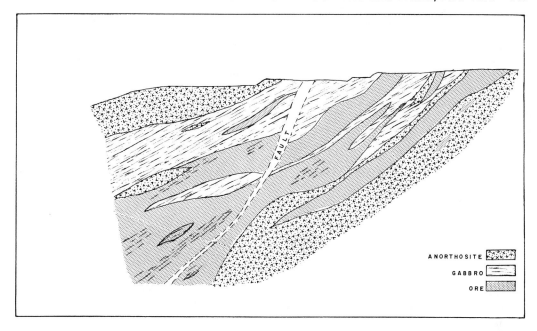

100    0    100   200   300   400   500
FEET

FIG. 3. *Typical Geologic Section Sanford Hill Pit.*

At Cheney Pond, a development drill program has revealed none of the massive anorthositic type of ore; instead, only oxide-rich bands up to 100 feet in thickness are found within a synclinal gabbro body. It is fine grained throughout. Stephenson (11) did an excellent job of mapping available outcrops and interpreting the structure of the parent gabbro and included ore. The ore referred to by him, however, is a narrow, low-grade band that outcrops on the hillside above the main ore bands and is of little economic importance. Drilling has shown the gabbro to be in sharp contact with underlying anorthosite.

## Mineralogy of the Deposits

The ores of the Sanford Lake district are titaniferous magnetite deposits in the sense of Osborne (8, p. 735). There are two ore types: (1) anorthositic and (2) gabbroic. Grain size and per cent content of ilmenite and magnetite differ between the two ore types and within each ore type. Generally speaking, the ratio of total Fe:TiO₂ in the anorthositic type of ore is 2:1 or greater, Gabbroic ore normally has a Fe:TiO₂ ratio of less than 2:1. There are areas, particularly in the South

Extension of the Sanford Hill ore body, where anorthositic ore has a low Fe:TiO₂ ratio.

Throughout the district, magnetite and ilmenite always occur together. Individual grains of magnetite seldom occur as the pure mineral. They normally contain some percentage of ilmenite in solid solution and up to 35 per cent of ilmenite as exsolution intergrowths. These intergrowths occur as tabular plates oriented parallel to the octahedral planes of the magnetite or along the boundary between magnetite grains.

A second highly magnetic mineral has been identified in the Sanford Hill-South Extension ores. This is ulvöspinel (Fe₂TiO₄). Ramdohr (14) identified a fine network within the magnetite of Sanford Hill ores as ulvöspinel. He used oil immersion techniques with reflected light and high magnification of polished specimens. Kays (16), using X-ray techniques with recorded chart tracings and chemical assay data, suggests a magnetite phase corresponding to an ulvöspinel content of 34 per cent. The ulvöspinel content is also expressed in the excess of FeO when comparing assay computations and values of total weight per cent magnetics in Davis tube measurements.

Vanadium in solid solution within the magnetite lattice structure is found in these deposits as in other titaniferous deposits throughout the world. No vanadium mineral has been identified.

Ilmenite occurs as a matrix around magnetite and is observed under the microscope to be corroded by magnetite. It is finer-grained than magnetite and can be distinguished megascopically in coarse-grained anorthositic ore by its high lustre and conchoidal fracture, compared to the dull lustre and parting planes in magnetite. In fine-grained ore, the ilmenite and magnetite are more equigranular and cannot be recognized individually in hand specimens.

The typical gangue minerals and their weight percentages in both ore types are given in Table I. In addition, minor amounts of apatite and spinel occur as primary minerals. A number of secondary minerals are found as the result of metamorphism, alteration due to faulting, and deposition in post-ore fault and joint openings. These include the chlorites and calcites along the numerous fault and joint planes. Among the minor minerals identified in the Sanford Hill-South Extension deposit are chalcopyrite, sphalerite, molybdenite, prehnite, barite, pyrite, leucoxene, scapolite, pyrrhotite, epidote, orthoclase, and quartz.

Green hornblende occurs predominately in the anorthosite and gabbroic anorthosite as both a primary magmatic mineral and as an alteration of pyroxene. Brown basaltic hornblende is most prevalent in the gabbro and occasionally in gabbroic anorthosite. It is associated with pyroxenes, apatite, and garnet, and, in some instances, occurs as a reaction skarn in the transition phase between ore minerals and plagioclase.

Clinopyroxene of various compositions oc-

curs to some extent in every rock type of the district. It is most prevalent in gabbro (as much as 30 per cent by volume), less abundant in gabbroic anorthosite, and least common in anorthosite. Although it is the most abundant femic mineral in anorthosite, it rarely exceeds 5 per cent by volume.

Orthopyroxene of various compositions is normally associated with, and frequently appears as inclusions in, clinopyroxene. It rarely attains 25 volume per cent in ore-bearing gabbro and ordinarily comprises less than 5 per cent in other rock types.

Garnet of mixed grossularite-andradite-almandite composition is present to some extent in all major rock and ore types. Although it occurs as fine-grained disseminations, garnet usually is associated with ore minerals, either as a rim of skarn of varied thickness around the ore or as scattered grains in plagioclase near the ore. The largest concentrations are found between massive anorthositic ore and anorthosite.

Biotite of the iron-rich variety occurs throughout the gabbro and gabbroic ore as an alteration of hornblende and as a primary mineral. It is most prevalent in shear zones in the gabbroic rocks but seldom exceeds a few volume per cent. The occurrence and abundance of biotite appears to be affected strongly by shearing.

Apatite is present in all rock types, usually as small anhedral grains interstitial to plagioclase. It has been found in quantities up to 5 and 10 per cent by volume in ore-rich gabbro but normally rarely exceeds 1 per cent.

All of the ore and gangue minerals occur throughout the ore bodies, but lateral compositional variations have been detected in some places. As previously noted, there is a variation of $Fe:TiO_2$ ratios apparent between ore associated with anorthosite and ore associated with gabbro. However, evaluation of the most recent drilling in the South Extension of Sanford Hill reveals a large ore lens associated with anorthosite which has a low $Fe:TiO_2$ ratio. No explanation for this is apparent. Likewise, there are areas in gabbroic ore where the $Fe:TiO_2$ ratio is high. This relationship of $Fe:TiO_2$ has been studied more extensively than other compositional variations due to its economic effect on the mining and milling operation.

The anorthositic ore lens in the Sanford Hill pit varies laterally in total $TiO_2$ content. At the southerly end the $Fe:TiO_2$ ratio is high due to the low proportion of ilmenite relative to magnetite. As the ore lens is followed to

TABLE I.  *Typical Mineral Analysis of Ore Types in Sanford Hill Ore Body*

| Gabbroic or Hanging Wall Ore | | Anorthositic or Footwall Ore | |
|---|---|---|---|
| Mineral | Weight % | Mineral | Weight % |
| Black Opaques | 61.5 | Black Opaques | 77.3 |
| Magnetics 25.7 | | Magnetics 40.8 | |
| Non Magnetics 35.8 | | Non Magnetics 36.5 | |
| Feldspar | 19.2 | Feldspar | 10.3 |
| Garnet | 8.1 | Garnet | 3.4 |
| Pyroxenes | 6.6 | Pyroxenes | 4.7 |
| Amphiboles | 2.3 | Amphiboles | 1.5 |
| Sulfides | 1.5 | Sulfides | 1.7 |
| Biotite | 0.8 | Biotite | 1.1 |

Source: Heyburn (15)

the north, the proportion of magnetite remains nearly constant while that of ilmenite increases. Near the northern end, within the high-iron ore, is a mass of extremely high ilmenite ore having a Fe:TiO$_2$ ratio ranging down to 1.2:1.0. The reason for this occurrence is not evident. The changes in mineral composition are very pronounced over extremely short distances within any one gabbroic ore lens. This observation of constant change applies throughout, regardless of location in respect to depth or lateral extent.

Mineral textures of the ore and gangue minerals are as changeable over short distances as is the mineral composition. The two characteristics are not necessarily related.

Gabbroic ore ranges from an aphanitic equigranular rock to a phanerite showing porphyritic or reticulated textures. Anorthositic ore always is phaneritic but ranges from equigranular to porphyritic to reticulated to possibly pegmatitic in texture. In this ore type, porphyritic refers to subhedral plagioclase phenocrysts in a matrix of finer-grained metallic oxides. Reticulated signifies coarse-grained subhedral to euhedral magnetite in a mesh of fine-grained anhedral ilmenite.

Stephenson (11) has done as thorough a study of mineral paragenesis as any investigator of the Sanford Lake district. His crystallization sequence shows the large plagiocase phenocrysts as having developed first and the finer-grained intergranular plagioclase having formed later. After the plagioclase, the other minerals in sequence are: apatite, hypersthene, augite, hornblende, garnet, and ore minerals. Biotite, appearing as a primary mineral, apparently was later than garnet since it occurs along fractures in some of the garnet in gabbro. The chlorites, carbonates, and scapolite formed as later alteration minerals as the result of deformation and minor hydrothermal activity.

The sulfides—pyrrhotite and pyrite—are most frequently associated with minerals of the reaction zone between ore and anorthosite. Pyrite occurs as very thin late veinlets that cut ore and gabbro; these veinlets seldom reach 0.5 inch in thickness. In drill cores at Cheney Pond, pyrite commonly is found over narrow widths at the lower contact between mineralized gabbro and anorthosite. It would appear to be the last of the minerals in the magmatic sequence to solidify.

Kays (16), in his detailed study of Sanford Hill pit, agrees that the sequence given by Stephenson generally is valid. However, he points out that no single paragenetic sequence

holds rigorously for all rock types of the Sanford Lake district.

Grade of ore, or degree of mineralization, commonly is determined by chemical assay for TiO$_2$ content and total Fe content. These are directly related to content of ilmenite and magnetite. Since ore must contain a sufficient amount of a desired mineral(s) that it can be mined at a profit, only the ilmenite and magnetite can be considered as ore minerals.

The early operators in the Sanford Lake district were interested only in iron. The present operator, National Lead Company, primarily is interested in ilmenite. Magnetite concentrate is a by-product produced in the milling process. To achieve the desired crusher grade, it is necessary to blend the many grades of available ore. In daily operations, this must be tempered with a knowledge of the type of ore used. As an example, a gabbroic ore represents a mineralized gabbro with disseminated ore and gangue minerals while an anorthositic ore of the same grade would be high-grade ore diluted with pure anorthosite. The two types of ore behave differently in magnetic cobbing with the low-grade anorthositic ore yielding a much cleaner concentrate than the low-grade gabbroic ore.

## Factors Controlling Form and Location of Ore Bodies

All bodies of titaniferous magnetite in the district are associated with gabbro masses. Gabbroic type ore is within the gabbro and therefore assumes the attitude of the flow structure in the lost rock. Anorthositic type ore is in anorthosite, but it is in contact with, or very close to, mineralized gabbro. These ore lenses also assume the strike and dip of the adjacent gabbro. Ore in gabbro does not favor any particular position in that rock. At Sanford Hill and Mt. Adams, the gabbroic ore appears near the footwall of the gabbro. At Cheney Pond, the ore-grade lenses within the gabbro occur approximately centered between footwall and hanging wall.

The one factor governing the location and form of the ore bodies is the presence and flow structure attitudes of the gabbro bodies. Genesis of the gabbro is a controversial subject and is discussed later.

## ORE GENESIS

As previously mentioned, the rocks are all considered to be of Precambrian age. There are, however, some differences in opinion as to the sequence of geologic events leading to the mineralogical and structural relationships as we see them today. The anorthositic series of the Sanford Lake district can be generally classified as anorthosite, gabbroic anorthosite, gabbro, and titaniferous magnetite. Since the basic minerals of all of these rock types are identical, differing only in proportion, they are all considered as originating from the primary magma which created the Adirondack anorthosites.

Earlier workers, including Kemp (4), Cushing (5), and Osborne (8), have all considered the anorthosites as having been intruded at great depth into sedimentary formations now exemplified by the metasediments in the larger Adirondack region. These men also related a large part of the metamorphism of these metasediments to the time of intrusion of the plutonic rocks. In recent work, Walton and DeWard (17) have proposed a newer theory, supported by detailed mapping, wherein evidence is given to show that a major part of the metasediments were laid down on a complex older basement which had formed in some pre-Grenville orogenic cycle and contained prototypes of the plutonic igneous rocks. All are in general agreement that with intense Grenville deformation and metamorphism certain elements of both igneous and sedimentary rocks, whether already metamorphosed or not, were remobilized in various degrees. The basic relationships between rock types and lithologic units that we see today were formed during this orogenic period.

Prior to 1947, all investigators of the Sanford Lake ores considered them to be the result of primary magmatic segregation. Concentration of oxides has been explained by various processes, such as: simple segregation, filter pressing of residual liquid with injection into the wall rock, and gravitational purification of the oxide melt by floating out of crystallized silicates with later injection. Some refer to mineralizers and replacement as aids to, or a part of, magmatic emplacement.

Gilson (13), though not questioning the origin of the original anorthosite, does relate the variations in the present anorthosites and the origin of both gabbro and ore facies to a series of pneumatolytic replacements. This involves andesinization of Marcy anorthosite to Whiteface anorthosite. Later solutions were the source of the ferromagnesian and ore minerals.

Heyburn (15) makes reference to two stages of gabbro emplacement. One of these gabbros is associated with and possibly contempora-

neous with the anorthosites. The second, a finer-grained gabbro, was determined to be in sharp contact with the anorthosite and closely associated with the ore. Considering sufficient volatiles to be combined with this iron- and titanium-rich gabbroic intrusive, there is then a basis for partial intrusive placement of gabbro and ore and partial replacement of anorthosite to form those parts of the ore body found within anorthosite country rock.

Kays (16), in his very comprehensive study of the petrography of the Sanford Hill pit, has suggested yet another means of formation for the ore bodies. In this case, the original anorthosite is considered to have been homogeneous Marcy type with all constituents necessary to form the ore body present but sparsely distributed. Strong regional stresses then caused local deformation and the development of sheared and granulated zones. As the rocks were ruptured, granulated, and sheared, large amounts of kinetic energy were released. This in turn aided the release of calcium and aluminum from the original labradorite plagioclase, effecting its transformation to the more sodic andesine plagioclase. The overall result is andesinization but by a different means than proposed by Gilson. The fracturing of the rock also set up low-pressure centers compared to the high pressure remaining in the unfractured areas. The resultant pressure gradients caused those elements in least thermo-chemical equilibrium with the mass constituents of anorthosite to migrate. Thus Fe, Mg, and Ti migrated from the high stress areas to the low stress areas. As Fe, Mg, and Ti were introduced, reactions took place with the Ca and Al released by andesinization to form ferromagnesian silicate minerals and garnet. The quantity of reaction minerals formed depended upon the degree of andesinization of the plagioclase and its subsequent release of Ca and Al. Utilization of all available Ca, Al and activated Si then resulted in wholesale crystallization of ore minerals. Therefore, two modes of ore emplacement are suggested; one is the partial replacement of pre-existing silicate rocks combined with the formation of ferromagnesian minerals as represented by the gabbro and gabbroic ore; a second is the wholesale replacement of pre-existing silicate rocks with little or no reaction as represented by the massive ore lenses in anorthosite. Kays' hypothesis is based upon a very detailed investigation of the distribution of basic elements throughout the ore lenses and varous rock types. This work did corroborate the studies by Avenius (12) in the Cheney Pond area, which indicated that the plagioclase feldspar in the gabbro of that area is andesine rather than labradorite and that hornblende is the most abundant ferromagnesian mineral.

This writer does not have any new or different theories regarding the ore genesis. However, after many observations, it is clearly evident that many theories can be proved if only a portion of the conditions are, or can be, taken into consideration. After a number of years working at the operation, the question of ore genesis becomes more and more complex. No one theory can explain satisfactorily all of the rock and mineral relationships now in evidence.

It is quite certain that all rocks of the district are genetically related as shown by their common mineral constituents. The primary mode of emplacement is partially or wholly obscured by subsequent metamorphic processes. It is likely that certain portions of the rocks have been remobilized one or more times by deformation pressures. Different degrees of plasticity are evident. Definite flow structure is seen in some of the gabbroic rock and gabbroic ore, indicating nearly liquid flow around xenoliths of anorthosite. Again small tight folds are seen in this rock, demonstrating deformation in a plastic state without rupture. In thin section, many of the feldspar crystals show bending and breaking.

There is evidence that volatiles were present at some point in the history of the deposits, probably in both early and late stages. The faults and joints now exposed show movement along many surfaces. Along the major breaks, there are breccia fragments of ore or gabbro in a matrix of carbonates. These are also the areas of recrystallized magnetite, large ilmenite crystal masses, and minor sulfide minerals. Near these faults are schlieren structures in the competent anorthosite showing remobilization and drag. In the gabbros, the jointing usually is more pronounced, and these joints show drag near major faults.

The present faults and associated hydrothermal minerals represent the last minor sequence in a very long and complicated series of geologic events.

## REFERENCES CITED

1. Emmons, E., 1842, Geology of New York, part II: Comprising the survey of the second geological district: Albany, N.Y., p. 247–263.

2. Watson, C. W., 1869, The military and civil history of the county of Essex, N.Y.: Albany, N.Y.

3. Rossie, A. J., 1893, Titaniferous ores in the blast furnace: A.I.M.E. Tr., v. 21, p. 832–867.

4. Kemp, J. F., 1899, Titaniferous iron ores of the Adirondacks: U.S. Geol. Surv. 19th Ann. Rept., pt. III, p. 409–416.

5. Cushing, H. P., 1907, Geology of Long Lake quadrangle, New York: N.Y. State Mus. Bull. no. 115, p. 451–531.

6. Bachman, F. E., 1914, The use of titaniferous ores in the blast furnace: Iron and Steel Industry Yearbook, p. 370–419.

7. Masten, A. H., 1923, The story of Adirondac: Princeton Press, Princeton, N.J., 199 p.

8. Osborne, F. F., 1928, Certain titaniferous iron ores and their origin: Econ. Geol., v. 23, p. 728–740.

9. Balk, R., 1931, Structural geology of the Adirondack anorthosite: Mineral. und Petrog. Mitt., Bd. 41, H. 3–6, p. 308–434.

10. Buddington, A. F., 1939, Adirondack igneous rocks and their metamorphism: Geol. Soc. Amer. Mem. 7, p. 19–48.

11. Stephenson, R. C., 1945, Titaniferous magnetite deposits of the Lake Sanford area, New York: N.Y. State Mus. Bull. no. 340, 95 p.

12. Avenius, R. G., 1948, Petrology of the Cheney Pond area: Unpublished MS Thesis, Syracuse University, 80 p.

13. Gilson, J. L., 1956, Genesis of titaniferous magnetites and associated rocks of the Lake Sanford district, New York: A.I.M.E. Tr., v. 205, p. 296–301 (*in* Min. Eng., v. 8, no. 3).

14. Ramdohr, P., 1956, Die Beziehungen von Fe-Ti Erzen aus Magmatischen Gesteinen: Comm. Géol. Finlande Bull. no. 173, p. 1–18.

15. Heyburn, M. M., 1960, Geological and geophysical investigation of the Sanford Hill ore body extension, Tahawus, New York: Unpublished M. S. Thesis, Syracuse University, 48 p.

16. Walton, M. S. and DeWaard, D., 1963, Orogenic evolution of the Precambrian in the Adirondack highlands, a new synthesis: Kon. Akademie van Wetenschappen, Sec. Sci. (Afdeeling Naturkunde) Pr., v. 66, p. 98–106.

17. Kays, M. A., 1965, Petrographic and modal relations, Sanford Hill titaniferous magnetite deposit: Econ. Geol., v. 60, p. 1261–1297.

# 9. Ore Deposits of the Southern Appalachians[*]

ROBERT A. LAURENCE[†]

## Contents

## Illustrations

## ABSTRACT

Ore deposits in the Southern Appalachians are (1) sedimentary or syngenetic, (2) epigenetic, and (3) residual. In general, deposits characteristic of high temperature and pressure are found in the Blue Ridge and Piedmont provinces and those of low and intermediate temperature and pressure, including the sedimentary ores, are found in the Valley and

[*] Publication authorized by the Director, U.S. Geological Survey.
[†] U.S. Geological Survey, Knoxville, Tennessee.

Ridge and Appalachian Plateaus provinces. Residual deposits are present throughout the area.

Recognition of several periods of intrusion and metamorphism in the Appalachians has changed former concepts which held that most mineralization was of late Paleozoic age in a zonal pattern including all four provinces. Many deposits probably are of early or mid-Paleozoic age.

Similarly, recent studies in geomorphology indicate that the formation of residual deposits is a continuous process and not related to specific cycles of peneplanation.

## GENERAL GEOLOGY OF THE REGION

The Southern Appalachian region, extending southwest from the Potomac River to the Coastal Plain in Alabama and Georgia, has long been an important contributor to the mineral economy of the United States. Both fuels and industrial minerals of the region outrank the metallic ores in tonnage and in value. However, the region does contain many valuable ore-producing areas, including the nation's currently leading zinc and pyrite districts. The region is also an important source of iron, copper, lead, and titanium. Historically, the Southern Appalachians were important as the site of early mining of gold, iron, manganese, and other ores; here, much of the basic knowledge, technology and skilled personnel was developed, and later applied to more extensive or higher grade deposits farther west.

The Southern Appalachians comprise four major geologic provinces; the boundaries of these are almost identical to those of the four physiographic provinces, which are the (1) Piedmont, (2) Blue Ridge, (3) Valley and Ridge (also called Appalachian Valley, and Folded Appalachians), and (4) Appalachian Plateaus. Because of this near-identity, the names of the physiographic provinces are applied to the corresponding geologic provinces (Figure 1).

### Piedmont Province

The Piedmont province includes a wide variety of Precambrian and Paleozoic metamorphic rocks, intrusive rocks of at least four different ages, and several basins of Triassic sedimentary rocks. The area has been deeply weathered, and saprolite conceals most of the unweathered bedrock. The Piedmont physiographic province is bounded on the east and south by the younger rocks of the Coastal Plain and on the west by the steep front of the Blue Ridge.

In Virginia, the rocks of the Piedmont province apparently merge with those of the Blue Ridge, but, in the Carolinas, Georgia, and Alabama, they are bounded on the west by the Brevard fault zone (46). In the Carolinas, this fault zone lies a few miles west of the physiographic boundary for about 100 miles of its length (35). It is a long straight zone, thought by many to be a major strike-slip fault zone (46), and by others to be an Alpine-type root zone (55).

The province broadens from about 50 miles wide in northern Virginia to 100 miles or more in the Carolinas, where it is divisible into four belts (26,51) which are, from east to west: (1) the Carolina slate belt, containing chiefly volcanic and sedimentary rocks of low metamorphic rank, cut by mafic and felsic intrusives, (2) the Charlotte belt of plutonic and metamorphic rocks, (3) the Kings Mountain belt, of metasedimentary and metavolcanic rocks of low to moderate metamorphic rank, and (4) the Inner Piedmont belt of high-rank metasedimentary and metavolcanic rocks, with minor felsic and mafic intrusives. Overstreet and Bell (51, p. 112–113) have suggested that the rocks of all four belts originated as three sequences of sedimentary and volcanic rocks, of Precambrian to Mississippian age which were separated by two major erosional unconformities, intruded by igneous rocks of at least four different ages, and subjected to at least four periods of regional deformation and metamorphism during the Paleozoic Era. This concept is a complete reversal of the earlier interpretation of the Piedmont as an area of Precambrian metamorphic rocks which were strongly deformed, further metamorphosed, and invaded by large granitic bodies in one massive post-Paleozoic Appalachian orogeny (5).

### Blue Ridge Province

The Blue Ridge is a rugged mountainous province consisting of a single narrow ridge in northern Virginia but widening to more than 30 miles in Tennessee and North Carolina. Its southern termination is in northern Georgia. The rocks of the province range from crystalline gneisses and schists of the amphibolite metamorphic facies in the eastern part, to unmetamorphosed clastic sedimentary rocks of late Precambrian and early Cambrian age in the western part. Igneous rocks of Paleozoic

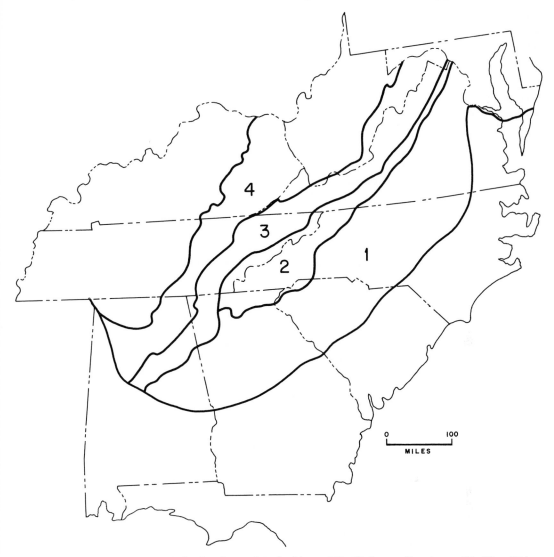

FIG. 1. *Physical Divisions of the Southern Appalachians.* (1) *Piedmont Province;* (2) *Blue Ridge Province;* (3) *Valley and Ridge Province;* (4) *Appalachian Plateau Province.*

age, from peridotite to granite, are widespread throughout the Blue Ridge.

The Blue Ridge province is an area of extensive westward overthrusting, as evidenced by its western boundary where it is separated from the Valley and Ridge by major thrusts; also, several windows in North Carolina and Tennessee contain Cambrian and Ordovician sedimentary rocks which are surrounded by older rocks of different metamorphic grades. The presence of sedimentary rocks of probable early Cambrian age, in a probable window surrounded by highly metamorphosed rocks of the Inner Piedmont belt, in Stokes County,

North Carolina, suggests that westward horizontal movement of the Blue Ridge may have been several tens of miles (33, p. D61–D63). However, in northern Virginia (20) and possibly at Cartersville, Georgia (19), there is no thrust fault at the western front of the Blue Ridge.

### Valley and Ridge Province

The Valley and Ridge province, or Folded Appalachians, is a long narrow belt, generally less than 50 miles wide, that contains more than 50,000 feet of folded and faulted sedimentary rocks (9, p. 502) which range in age from early Cambrian to Pennsylvanian. The Cambrian, Ordovician, and Mississippian Systems include thick and extensive deposits of carbonate rocks; clastic rocks predominate in the Lower Cambrian, Middle Ordovician (locally), Silurian, Devonian, Lower Mississippian and Pennsylvanian.

The province is characterized by folds in the northern part and thrust faults in the part southwest of Roanoke, Virginia. Some of these thrusts are continuous for more than 100 miles, and have a horizontal displacement of several miles. Most geologists now believe these thrusts are "thin-skinned" and involve only the sedimentary rocks, not the basement. Drill holes have not yet penetrated to the basement in the Valley and Ridge, but, in a few places, basement rocks are brought to the surface in the Blue Ridge along thrust faults that border the Valley and Ridge province on the east.

Small bodies of peridotite and alkalic rocks penetrate the Paleozoic strata in the northern part of the province in Virginia, West Virginia, and in Union County, Tennessee. Mafic lava flows occur in the Lower Cambrian Chilhowee Group in northeastern Tennessee and Virginia, and altered volcanic ash beds occur at several horizons in the Middle Ordovician rocks throughout the province.

## Appalachian Plateaus Province

The Appalachian Plateaus province, lying west of the Valley and Ridge province, consists principally of flat-lying sandstones, shales, and coal of Pennsylvanian age; underlying Mississippian shale and limestone are exposed along the bounding escarpments and in some of the deeper stream gorges. Although the rocks are generally undeformed, they are involved in the Pine Mountain thrust in southeastern Kentucky and adjacent parts of Virginia and Tennessee, the Cumberland thrust, the Sequatchie anticline, and lesser structures in Tennessee and Alabama. The only known igneous rocks are small mica-peridotite bodies that cut the Pennsylvanian beds in Elliott County, Kentucky.

## SEDIMENTARY DEPOSITS

### "Clinton" Iron Ores

Probably the most important sedimentary ore deposits of the region are the so-called "Clinton" hematite ores of middle and early Silurian age, which occur at many localities from New York to Alabama (2). The major development of these ores is in the Birmingham district, Alabama (Figure 2), where more than 300 million tons of iron ore have been produced from the various seams between 1840 and today. Remaining resources are estimated to exceed 2.1 billion long tons averaging 31 per cent iron, of which nearly a billion tons, averaging over 36 per cent iron, are in the major ore bed known as the Big Seam that has a maximum thickness of 30 feet (24, p. 13). This district is described in another chapter of this volume.

The "Clinton" ores generally are considered to be of sedimentary origin, and contain up to about 37 per cent iron as hematite that was deposited directly from sea water and also replaced limestone, including fossils, during or immediately after deposition. Removal of calcium carbonate by weathering has increased the iron content of the near-surface portions to as much as 57 per cent.

Recent work by R. P. Sheldon has demonstrated a close relationship between the character of the "Clinton" ores and the physical conditions of sedimentation (37, p. A-1). The more calcareous, highest-grade ores were deposited on the lagoon side of barrier islands; on the seaward side, the hematite grades through hematite-quartz sandstone to a marine nonferruginous sandstone.

Similar hematite ores occurring at various stratigraphic positions in Upper Ordovician to Middle Silurian rocks have been mined at many other localities, especially near Gadsden, Alabama, and Chattanooga and Rockwood, Tennessee. In general, the ore beds are thinner and the iron content slightly less at these places than at Birmingham.

### Alluvial Placers

Alluvial placers occur both in flood plains of the present streams and, to a much lesser extent, in higher terraces of former streams. The principal gold-bearing placers are in the Dahlonega district, Georgia, the South Mountain district, North Carolina, and the Coker Creek district, Tennessee, but alluvial placers have been worked to some extent near practically all the Piedmont and Blue Ridge gold districts (Figure 2).

Records of alluvial gold production of the region are not available, but it is doubtful if as much as half the total gold produced in the region was recovered from alluvial placers.

Alluvial placers in which monazite is the principal economic mineral have been mined and explored in part of the western Piedmont province, especially in south central North Carolina and South Carolina (29). The placers are in locally derived sediments that cover the floors of narrow valleys in an area of monazite-bearing igneous and metamorphic rocks that are cut by monazite-bearing pegmatites. The alluvial sediments are well bedded and similar in stratigraphy throughout the area. The highest tenors are in the smallest headwater deposits and in the coarsest grained basal sediments. Ilmenite, rutile, zircon, and other heavy minerals are present in most of the deposits. This area, especially near Shelby, North Carolina, was an important producer of monazite in the late 19th and early 20th centuries, but there has been little, if any, output since 1917.

A "fossil placer" in the Snowbird Group of late Precambrian age in Cocke County, Tennessee, has been reported recently (53). It is of potential economic significance because it contains 30 to 60 per cent heavy minerals, chiefly ilmenite and zircon.

## EPIGENETIC DEPOSITS

### Gold

Gold was sought by the Spaniards in the Southern Appalachians in the 16th century; discovery is said to date from 1799 (15, p. 27), but the principal period of mining and exploration was between 1810 and 1880. Not much interest has been shown in gold mining during the past quarter century but, in the light of new concept of Piedmont and Blue Ridge geologic history, it is a subject worthy of renewed attention and re-study. Present knowledge is well summarized in two classic papers (1,15).

Total production of gold in the region exceeds 1.6 million ounces. Since 1941, nearly all gold produced in the region has been as a by-product from the smelting of copper ores.

Most of the gold deposits are in the Piedmont province in a series of belts that generally are conformable to the regional structure, extending from Virginia to Alabama. The deposits occur as veins and "mineralized zones," usually along shear zones (15, p. 35) in schists and gneisses. A notable exception is the Kings Mountain ore body in North Carolina, which is a replacement deposit in limestone (15, p.

74). Gold is principally associated with pyrite, pyrrhotite and other sulfides; quartz, carbonates, sericite, and biotite are the principal gangue minerals. The mineral suite is characteristically hypogene.

At present, no gold deposits of economic importance are known in the Blue Ridge, though there are many minor mines, prospects, and occurrences.

For a long time, most geologists thought that the gold deposits were of late Paleozoic age, directly related to intrusion of granite batholiths accompanying the Appalachian orogeny (7, p. 110). The recognition of granites of earlier Paleozoic age requires a revision of this concept. Thus, Bell (30, p. 191) has suggested that gold in the Concord, North Carolina, area probably is related to small granite plutons that are older than Devonian.

### Tin

Tin deposits occur in three main localities or "belts" in the Southern Appalachians: (1) the Carolina tin-spodumene belt between Lincolnton North Carolina, and Gaffney, South Carolina, (2) Coosa County, Alabama and (3) Irish Creek, Rockbridge County, Virginia. Total production of recoverable tin from the Carolina belt, mostly from placers, is less than 300 tons (10, p. 251); the other areas have yielded even less (Figure 2).

The Carolina and Alabama deposits are in pegmatites and associated greisens (10,12); those in Virginia are in quartz veins with greisen borders (11). Cassiterite is the only tin mineral, but significant amounts of beryl are present (25, 40). Spodumene is the principal economic mineral of the Carolina tin-spodumene belt, where many of the large pegmatites contain 15 to 20 per cent spodumene (31, p. 525).

### Molybdenum

No molybdenum ore has been mined in the Southern Appalachians, but the Boy Scout-Jones and Moss-Richardson prospects, near the eastern edge of the Piedmont province in Halifax County, North Carolina, have been explored (50, p. 81–84). These deposits are molybdenite-bearing quartz veins in and near an ovoid granite body that is about 2 miles long by half a mile wide. Near the veins, molybdenite is disseminated in both granite and schist. Pyrite is common; chalcopyrite and fluorite also are present.

FIG. 2. (*1*) *Birmingham;* (*2*) *Ducktown;* (*3*) *Mascot-Jefferson City;* (*4*) *Copper Ridge;* (*5*) *Austinville-Ivanhoe;* (*6*) *Cartersville;* (*7*) *Sweetwater;* (*8*) *Tin-Spodumene Belt;* (*9*) *Roseland;* (*10*) *Hamme;* (*11*) *Dahlonega;* (*12*) *Other Piedmont Gold Districts;* (*13*) *Shenandoah Valley;* (*14*) *Ore Knob;* (*15*) *Gossan Lead;* (*16*) *Boy Scout-Jones;* (*17*) *Ga.-Ala. Bauxite;* (*18*) *Chattanooga Bauxite;* (*19*) *Crimora;* (*20*) *Northeast Tenn. Manganese;* (*21*) *Central Carolina Monazite Placers.*

## Tungsten

The Hamme tungsten district, in Vance County, North Carolina, and extending slightly into Mecklenburg County, Viriginia, is in the eastern part of the Carolina slate belt of the Piedmont province (Figure 2).

Tungsten minerals were reported in Vance County at the beginning of the 20th century, but mining in the district dates from 1942,

when Joseph and Richard Hamme discovered huebnerite-quartz veins near Townsville and developed the Hamme mine. This mine was operated by Haile Mines, Inc., and later by Tungsten Mining Corporation, until February 1963, when it was closed. It produced about a million short ton units of $WO_3$, more than 1000 tons of lead, and considerable amounts of gold, silver, and zinc. For several years, it was the nation's leading tungsten mine.

There has been minor tungsten production from a few other deposits in the district.

The district is underlain by a lenticular pluton of granodiorite that is about 6 miles wide and 20 miles long and trends north-north-east. The pluton is flanked by biotite and hornblende gneisses and phyllites, possibly derived from sedimentary rocks of probable early Paleozoic age (41, p. 46). The regional metamorphism and the granodiorite probably are of middle or late Paleozoic age. Slightly younger aplite dikes cut the granodiorite, and many pegmatites cut the older gneisses. Triassic diabase dikes occur throughout the district.

The ore deposits are in quartz veins in which huebnerite is the principal ore mineral (14), accompanied by some scheelite; other vein minerals include fluorite, rhodochrosite, sericite, pyrite, tetrahedrite, galena, sphalerite, and chalcopyrite.

Quartz veins are abundant throughout the district, especially in the granodiorite (41, p. 46–48), but they contain huebnerite only near contacts in the western part of the granodiorite pluton and in nearby phyllite. The principal tungsten-bearing veins are in a north-northeast-trending belt of shearing and sericitization that is about 8 miles long and 1.5 miles wide (18, p. 59–60). Most of the veins trend nearly parallel to this zone and several are almost at right angles to it. They are more abundant and richer in tungsten near the central part of the belt. The veins range from a few feet to more than 400 feet in length, and most dip steeply to the east. Some groups of closely spaced en echelon veins have a total length of more than 1700 feet. The deposits have been explored and mined to a depth of more than 1600 feet, with no apparent change in size, character, or grade of mineralization.

The order of deposition of the minerals, with considerable overlap according to Espenshade (14, p. 9), is: quartz, sericite, pyrite, huebnerite, rhodochrosite, fluorite, scheelite, galena, chalcopyrite, sphalerite. Quartz and sericite were formed throughout practically the entire period of vein deposition. The vein-forming solutions probably were closely related to the magma that formed the granodiorite. Parker (41, p. 32–38) believes that the granodiorite may be of metamorphic rather than magmatic origin, and derived partly from phyllite (41, p. 27), but that the veins probably are related to the same processes which formed the granodiorite (41, p. 48).

Tungsten minerals occur in many of the gold veins throughout the Piedmont gold belts, but not in economic quantities. Probably the most important occurrence is at the Furniss mine, in Cabarrus County, North Carolina, where scheelite occurs with gold and sulfides in a gangue of quartz, barite, and carbonates.

## Zinc and Lead

Although there are many zinc and lead prospects in the Valley and Ridge province, between West Virginia and Alabama, practically all the production has come from four areas: (1) the Mascot-Jefferson City district, Tennessee, (2) the Austinville district, Virginia, (3) the Copper Ridge district, Tennessee, and (4) the Shenandoah Valley district, Virginia, The first two are described in detail in other chapters of this volume. Zinc and lead ores have also been mined at Embreeville, in the Straight Creek and Powell River Districts, and at the Hardwick mine near Cleveland, all in Tennessee, and at several minor localities. Sphalerite is recovered as a by-product of copper mining at Ducktown, Tennessee (Figure 2).

The deposits occur mainly at two stratigraphic positions: (1) in the upper part of the Knox Group, within the lower part of the Kingsport Formation of early Ordovician age and (2) in the Shady Dolomite of early Cambrian age. Minor deposits occur in the Lower Cambrian Rome Formation, Upper Cambrian Maynardville Formation, and Copper Ridge Dolomite. No intrusive rocks are known to be associated with any of these deposits.

Although there are exceptions, deposits in Ordovician rocks generally contain light-yellow sphalerite, dolomite gangue, and very little pyrite and galena; those in Cambrian rocks contain darker sphalerite, significant amounts of galena and pyrite, and also dolomite gangue.

The ore and gangue minerals occur chiefly as breccia and vein fillings and, to a lesser extent, as replacements of the carbonate country rock. They have long been regarded as "telethermal" deposits, similar in origin to the well-known "Mississippi Valley type." As stated by Oder and Ricketts (34, p. 18), "ore genesis in the East Tennessee zinc deposits is scarcely better understood than it was in the day of Watson (1906)."

Evidence pointing to either a hydrothermal or a "lateral secretion" origin for the sphalerite is about equally inconclusive. However, the long-prevailing opinion that the host structures of the ore bodies were produced at the close of the Paleozoic by Appalachian tectonic forces, has largely been replaced by the theory that these structures were formed when the host rocks were essentially flat, probably by solution and collapse related to karst development on the post-Lower Ordovician unconformity (48). As stated by Callahan (42), the Appalachian orogeny simply disturbed and complicated the earlier structures. This is more difficult to apply to the deposits in Cambrian rocks, and it is probable that the deposits in the two stratigraphic positions may be unrelated genetically.

Deposits in the Copper Ridge district of east Tennessee are in the Lower Ordovician Kingsport Formation (48). These are sphalerite-bearing breccias that are similar in form and extent to those of the Mascot-Jefferson City district; however, the strata in the Copper Ridge district dip 30° to 45°, whereas in most of the Mascot-Jefferson City district, the strata dip 15° or less. George Ruskell (oral communication) has remarked that cross sections in the Copper Ridge district "look like southwest Wisconsin, tipped up 45 degrees." A small amount of galena and barite is present in one mine, and all the deposits probably contain more pyrite than those of the Mascot-Jefferson City district.

Robert S. Young (written communication) has provided this brief description of the Shenandoah Valley, Virginia, deposits:

"Significant occurrences of zinc and lead sulfides in Shenandoah Valley, Virginia, have been a matter of record for seventy-five years and three periods of limited mining are known. There are no current mining operations.

"The principal sulfide is brown to yellow sphalerite; associated epigenetic minerals are galena, pyrite, dolomite, fluorite, and rare barite. Smithsonite and greenockite are common secondary products. Sphalerite and galena are found in beds of Cambrian (Tomstown) and Ordovician (Chepultepec and Beekmantown) age as (1) replacement masses, (2) breccia cement, (3) vein fillings, and (4) disseminated crystals. Important concentrations of ore minerals are restricted to the Beekmantown Formation, usually the upper two-thirds, and without exception in a dolostone host.

"The Bowers-Campbell Mine, Tri-State Zinc, Inc., operated in a ZnS-bearing breccia from 1957 to 1962. The ore breccia body, apparently unique for the Appalachian District, is approximately 1000 feet long, 200 feet wide, and extends from the present surface to undetermined depth (more than 800 feet). The breccia body trends with enclosing strata; it stands perpendicular to bedding and portions have been offset on bedding thrusts as much as 100 feet. Breccia mass shape, nature of the walls and internal "stratigraphy" suggest a tectonic origin. Mineralogy, spatial distribution of sulfides, multiple ZnS stages, and geothermometric studies support a hydrothermal origin and a telethermal classification."

Little is known about the sulfide ores of Embreeville in northeastern Tennessee, as no deposits of economic size have been found. The ores consist of sphalerite and galena, with pyrite and sparry ("gangue") dolomite within the Lower Cambrian Shady Dolomite and are located in a complexly faulted syncline (16). Similarity to Austinville is suggested. The economic minerals are hemimorphite and cerussite, which have been derived from the sulfides during a long period of weathering. The residual clay in which these minerals occur is 300 feet thick in places.

The deposits of the Powell River area, which occur chiefly in the Upper Cambrian Maynardville Formation and Copper Ridge Dolomite, are small and generally closely related to steeply dipping strike-slip faults. At the largest of these, the New Prospect mine (from which less than 200,000 tons of crude ore and concentrates were produced), sphalerite, galena, dolomite and pyrite occur as replacement and fissure fillings in a brecciated and dolomitized limestone block of the Maynardville Formation that is between two vertical faults that are 65 feet apart (52).

## Copper

The principal copper deposits of the Southern Appalachians are massive sulfides (6). The most notable deposits are those of the Ducktown district, the Ore Knob mine in northwestern North Carolina, and the Gossan Lead in southwestern Virginia. These are copper-bearing pyrrhotite deposits in metamorphic rocks of the Blue Ridge, and generally are conformable with regional structure. Although gold-bearing, they are west of and entirely separate from the gold deposits of the Piedmont (Figure 2).

Copper was discovered at Ducktown in 1843 and was mined intermittently between 1847 and 1890. Since 1890, mining has been continuous. Production of copper, through 1963, is valued at $202,924,000 (54, p. 10); significant amounts of sulfuric acid, zinc, iron sinter, gold, and silver also have been recovered or produced from Ducktown ores.

The ore bodies at Ducktown (which are described in a separate chapter of this volume) are complex tabular lenses of massive sulfides (75 per cent pyrrhotite, 20 per cent pyrite, and 1 to 2 per cent each of magnetite, chalcopyrite, and sphalerite). The lenses average about 100 feet in width and generally are conformable with the enclosing graywacke and schist of the late Precambrian Great Smoky

Group (49, p. 25–30). Gangue minerals are chiefly quartz, carbonate, and actinolite-tremolite. The principal controlling factor in ore deposition probably is the replacement of calcareous beds (22, p. 70).

Arthur R. Kinkel, Jr. (written communication), has contributed the following description of the Ore Knob mine:

"The Ore Knob Copper mine in Ashe County, North Carolina, opened in 1855 has had a long history of intermittent operation. Prior to its most recent reopening in 1955 by Appalachian Sulphides, Inc., the major production period was from 1873 to 1883, when 200,000 tons of ore was mined to produce 12,500 tons of copper. Part of this ore was secondarily enriched. Total production to the end of 1961 has been about 35,000 tons of copper, 9,400 ounces of gold, and 145,000 ounces of silver. Average grade of the primary ore is 2.22 per cent copper. The mine was closed in 1962.

"The Ore Knob deposit is in mica gneiss that appears to have been a sedimentary rock which contained some interlayered mafic volcanic material. The wall rock is gray, foliated, locally crumpled gneiss and schist in which the principal variation is in the relative proportions of quartz, biotite, muscovite, and plagioclase (mainly oligoclase). Amphibole and garnet are present locally. Practically all biotite and muscovite crystals are oriented with the planar structure of the gneiss and schist. Linear structure plunging about 20 degrees southwest is marked by minor folds, shear cleavage that forms crinkles in foliation, mullion structure, and locally by alined amphibole.

"The ore body is along a narrow shear and breccia zone; it slightly transgresses the dip of gneissic banding in the upper part of the mine and locally cuts small folds. The ore shoot is at least 4,000 feet long and is alined along the plunge of lineation. The gneiss along the vein zone was altered to a maximum distance of five feet on either side of the sulfide-bearing ore zone. In the walls of the ore, and in fragments of gneiss in ore, the principal changes in the rock are recrystallization of biotite and muscovite (and hornblende where present) to coarse-grained, largely unoriented grains and crystals; coarsening and clumping of plagioclase; addition of minor calcite and tourmaline; the formation of actinolite, hornblende, epidote, and clinozoisite, and the addition of sulfides. The ore zone contains more calcite, garnet, biotite, and amphibole, and less muscovite, than the wall rock.

"The veinlike to lenticular ore body at Ore Knob is a mixture of massive sulfide, disseminated sulfides, and anastomosing veinlets, stringers, and pockets of sulfides in coarse, recrystallized, largely unoriented silicates which are the same as those in the gneiss. Massive sulfide makes up approximately 50 to 60 per cent of the ore and ranges from 100 per cent to less than 25 per cent. Pyrrhotite, pyrite, chalcopyrite, quartz, biotite, and amphiboles are the principal minerals in the vein and are found in widely different proportions. Copper, gold, and silver are recovered. Much of the ore is mainly pyrrhotite, but pyrite is always present and in a few areas is predominant. The ore is characterized by abundant coarse, unoriented or crudely foliated black biotite, concentrations of garnet and of plagioclase, and lesser amounts of coarse amphibole crystals in random orientation.

"At Ore Knob, the marked difference in texture between the silicates in gneiss and the same silicates in the ore zone indicates complete reorganization of rock minerals along the vein zone with only minor addition of material other than sulfides. Coarsening of grain and randomly oriented growth in the ore zone indicates reorganization of minerals in a non-stress environment, but mobility was limited to the immediate vein zone. Textures that show dilation of ore, and others that indicate mobility of sulfides and recrystallization of silicates after ore deposition, are common. Sharply bounded fragments and layers of gneiss in massive pyrrhotite-chalcopyrite ore were recrystallized to coarse-grained unoriented silicates, but during subsequent movement and rotation of fragments in the ore the minerals along the borders of the fragments were further recrystallized and oriented tangent to the fragments, probably during movement in the sulfide matrix. Reorganization and movement after ore formation is shown by local shattering of ore at Ore Knob and at many of the other Appalachian deposits by local fracturing of coarse silicates in ore; by flowage in ore to form pull-apart and boudinage structures along bands of gneiss in ore; by rotation of rock fragments in ore and accompanying development of pressure shadows into which chalcopyrite, pyrrhotite and quartz have moved; by cementation of fractured pyrite and silicates by chalcopyrite and pyrrhotite; by local replacement of unoriented silicates and carbonate by sulfides; by small apophyses of softer sulfides in wall rock; by growth of pyrite porphyroblasts and sheeting and rotation of these by later movement; by local recrystallization to plagioclase, calcite, and quartz to unstrained individuals or granoblastic mosaics; and by randomly oriented growth of biotite, clinozoisite, quartz, garnet, and amphibole in sulfides."

Kinkel (36) concludes that "the Ore Knob deposit appears to have been formed by replacement of crushed and sheared material along a fault that locally cuts the gneissic banding at a low angle. . . . A rising temperature gradient along the channel during the following initial sulfide deposition recrystallized minerals in the wall rocks, in smaller rock fragments in ore, and in the early sulfides, and formed recrystallized rims on larger rock fragments. . . . Present textures at most places thus give little information on primary textures."

According to Ross (6, p. 84), the Ore Knob deposit (as well as others of the Ducktown type deposits) was formed at a considerable depth, near the close of the late Paleozoic Appalachian orogeny. Vein material, derived from a much deeper intrusive body nowhere exposed in the region, was deposited in six different stages: (1) aplite and pegmatite, (2) quartz, (3) high-temperature ferromagnesian silicates, (4) carbonates, (5) late lime-rich silicates and (6) sulfides.

Brown (28) observed a decrease in the iron content of biotite and total amount of biotite in the wall rock as the ore body is approached, and that this decrease is greatest where the ore body is thickest. Noting no secondary iron minerals in the wall rock, he concluded that iron removed from wall rock during alteration may have been deposited in the sulfides of the ore body or carried out of the system entirely.

The Gossan Lead is a group of *en echelon* ore bodies occupying a zone about 20 miles long in Carroll and Grayson Counties, Virginia (27, p. 184). Secondary chalcocite ores were mined as early as the 1850's. After virtual exhaustion of the chalcocite ores, the gossan was mined as an iron ore, and pyrrhotite ore was mined later for making sulfuric acid. Copper and zinc were not recovered during this last period of mining, which ended in 1962.

The primary ores are mainly pyrrhotite, chalcopyrite, sphalerite, and other minor sulfide minerals. Gangue minerals include quartz, oligoclase, amphiboles, biotite, carbonates and garnet.

At least 16 other deposits, believed to be of the Ducktown type, have been mined or explored. They occur along a line from Floyd County, Virginia to Randolph County, Alabama (23,38) and are in the western part of the Blue Ridge and in the western Piedmont.

The Virgilina district extends for about 50 miles along the regional strike of the Carolina slate belt in the Piedmont of North Carolina and Virginia. Copper was mined as early as the 1850's, but principal activty was in the late 19th and 20th centuries.

The deposits are in quartz veins, usually 3 or 4 feet wide, which contain chalcocite, bornite, chalcopyrite and pyrite as the principal ore minerals. The district is best known from the classic work of Laney (3) in which he demonstrated that the chalcocite and bornite, occurring in graphic intergrowths, are primary hypogene minerals.

The host rocks of the Virgilina district are the volcanic and sedimentary rocks of the slate belt, generally considered to be of early Paleozoic age. Age of the veins is uncertain, as they may be related to either middle or late Paleozoic intrusives.

Other types of copper deposits in the Southern Appalachians include (1) veinlets and disseminations in the Hillabee Chlorite Schist in Alabama, said by Clarke (43) to be similar to the porphyry coppers, (2) native copper in basaltic rocks in the Blue Ridge of Virginia, (3) minor occurrences of chalcopyrite in carbonate rocks of the Valley and Ridge, usually associated with sphalerite and barite, and (4) very minor occurrences of disseminated chalcopyrite in sandstones of the Rome Formation.

## RESIDUAL DEPOSITS

### Bauxite

Bauxite was first recognized in North America in Floyd County, Georgia in 1887 (45, p. 3). All economic occurrences of bauxite in the Southern Appalachians are in the Valley and Ridge province, from Alabama to Virginia, but the major districts are near the discovery site, in northwestern Georgia and adjacent parts of Alabama and Tennessee. Most individual deposits are small but total production of the region from 1887 to 1965 probably exceeded 750,000 tons (Figure 2).

The Appalachian bauxite deposits occur in residuum which overlies carbonate rocks of the Knox Group (Cambrian and Ordovician) and the Shady Dolomite (Lower Cambrian), the former containing most of the important deposits. Exploration and mining of many of the bauxite districts during World War II showed that the deposits are approximately inverted cones, seldom more than 150 feet in maximum diameter or vertical depth (17, p. 189). The bauxite usually is enclosed by kaolin that is, in turn, surrounded by residual clay from the surrounding country rock. The deposits usually are found slightly below the so-called "Valley Floor peneplain," and range in altitude from 750 feet in Alabama to 2200 feet in Virginia.

The bauxite deposits were long thought to be residual concentrations of aluminous materials in the host dolomites and limestones, but the work of Adams (4), and later that of Bridge (17), demonstrated that the bauxite probably formed during the interval between the Paleocene and Eocene epochs by desilication of transported clays that filled ancient sink holes. Bauxite deposits in Tertiary sediments of the nearby Gulf Coastal Plain were formed at the same time.

Later work by Knechtel (39) supports the idea that the bauxite formed from weathering of residual, not transported, material (terra rossa) in the sink holes. Knechtel's work suggests formation at many different times rather than during a single period; in fact, bauxitization may be taking place now. This is supported by observations of H. M. Cofer in Georgia (oral communication, 1961) and the recent work of Hack (47), who demonstrated the formation of residuum in Virginia during continuous weathering rather than in cycles related to peneplain levels.

## Manganese

A wide variety of manganese oxide, silicate, and carbonate minerals has been reported in each of the geologic provinces of the Southern Appalachians. These manganese minerals occur in crystalline and sedimentary rocks and in their residual products. Only the oxide minerals in the residuum have been of economic importance, and most of these are associated with Paleozoic carbonate rocks in the Valley and Ridge province, especially the Shady Dolomite and Rome Formation (Lower Cambrian), the Knox Group (Cambrian and Ordovician), the Holston Marble (Middle Ordovician), the Oriskany and Helderberg Formations (Lower Devonian), and the Fort Payne Formation (Lower Mississippian). Deposits have been mined at many localities from northern Virginia to Alabama (21). These deposits have produced more than 850,000 tons of concentrate containing more than 35 per cent manganese, and 400,000 tons of ferruginous manganese concentrate. By far the largest part of the manganese has come from three main areas: (1) the Crimora mine and vicinity, Augusta County, Virginia, (2) the Cartersville district in Bartow County, Georgia, and (3) northeastern Tennessee, especially Bumpass Cove in Unicoi and Washington Counties (Figure 2).

In practically all these deposits, the manganese minerals are closely associated with oxides of iron, chiefly goethite ("brown ore"). Other mineral deposits of residual origin, such as barite, bauxite, zinc, and lead occur in some of the same areas, but these usually are not closely associated with the manganese deposits. The Crimora mine produced only manganese and ferruginous manganese ores; Bumpass Cove produced iron, lead, and zinc from secondary deposits, and the Cartersville district produced several minerals besides manganese ores.

The Cartersville district is mostly within the eastern part of the Valley and Ridge province in northwestern Georgia, but an adjacent section of the Piedmont province properly may be considered part of the district. Mining began about 1840. The district produced more than 5 million tons of brown iron ore, about 2 million tons of barite, more than half a million tons of manganese concentrates (including ferruginous manganese), a small amount of specular hematite, and significant quantities of ocher and umber for use as pigments (19). Practically all this output was produced from secondary deposits in the thick residuum which covers most of the bedrock of the district.

The Valley and Ridge part of the district is underlain by sedimentary rocks of early and middle Cambrian age, and the Piedmont part is underlain by amphibolite and gneisses, which were apparently derived from the Cambrian sedimentary rocks in Carboniferous time (19). Nearly all published geologic maps of this part of Georgia show a "Cartersville thrust fault" extending across the state, separating the crystalline rocks of the Piedmont from the sedimentary rocks of the Valley and Ridge province, but Kesler's detailed mapping clearly indicates that no such fault is recognizable within the Cartersville district (19, p. 30–33). As shown on the 1961 Tectonic map of the United States (35), it seems probable that the late Paleozoic granitization of the sedimentary rocks in this area has obscured evidence of the Cartersville fault within the district, though it is present nearby both to the north and southwest. It is thus reasonable to infer a genetic relationship between the metamorphism and the primary mineral deposits of the Cartersville district.

The economic minerals of the district occur almost entirely in the residual blanket which overlies the Lower Cambrian carbonate rocks. Whatever the ultimate source of the iron, barium, and manganese, the ore minerals were formed and concentrated by the action of ordinary ground waters which dissolved the soluble carbonates during a long period of weathering and erosion. Earlier workers thought that these deposits, as well as all others in the Valley and Ridge province, accumulated chiefly during earlier cycles of weathering and were therefore closely related to peneplains which were later uplifted and dissected. However, recent work by Hack (47) in the Shenandoah Valley of Virginia discounts this and substitutes the concept of dynamic equilibrium, whereby the landforms and residual blanket are products of long, continued weathering. Instead of occurring only at certain altitudes on old erosion surfaces, manganese deposits actually occur at any altitude at which the weathered surface of the manganese-bearing dolomite occurs. According to Hack (47, p. 76), deposition of manganese is still taking place.

Similarity of the residual manganese deposits of Cartersville to the many manganese deposits overlying lower Cambrian carbonate rocks, along the east side of the Valley and Ridge province from Pennsylvania to Alabama, strongly suggests a common mode of origin. Most recent workers consider that very minor amounts of syngenetic manganese carbonate in the original dolomite are the ultimate source of the manganese in the residual deposits (16,13).

Because of the deep weathering, it is difficult to learn much about the primary ore minerals. Barite, in veins and irregular bodies, is exposed in pinnacles of dolomite in the deeper open cut mines. Pyrite is abundant; galena, sphalerite, chalcopyrite, enargite and tennantite are present but not common. These sulfides, in places, are enclosed by barite indicating a common mode of origin (19, p. 46). Fluorite has not been reported. Specularite, in the Shady Dolomite, apparently has formed by recrystallization of iron minerals in the dolomite. Abundant quartz and calcite and strengite are the only other primary minerals. No primary manganese mineral has been observed or reported. Kesler considers the primary deposits to be of hydrothermal origin; most earlier workers thought that iron, barium, and manganese were original minor impurities in the carbonate bedrock. The probability that some of the residual minerals, especially barite, may be derived from epigenetic deposits and others, chiefly the iron and manganese oxides, from minor syngenetic concentrations, seems to have been overlooked.

## Brown Iron Ores

Brown iron ore deposits (chiefly goethite), which have been utilized since the earliest days of settlement in the Appalachians, occur in the residuum which overlies thick carbonate rocks, especially those of early Cambrian and early Ordovician age. The deposits occur mainly along the eastern side of the Valley and Ridge province from Pennsylvania to Alabama. Brown iron ores in the Oriskany Sandstone of Devonian age and residuum of the overlying Helderberg limestones were once mined extensively in western Virginia (Figure 2).

Deposits in residuum of the Lower Cambrian dolomites, which are commonly associated with manganese ores, seem to have formed by residual concentration during weathering, in the same manner as the formation of the manganese deposits (47, p. 76). However, deposits in residuum overlying the Lower Ordovician Kingsport Formation which contain very little manganese, may be derived from oxidation of epigenetic pyrite deposits the origin of which seems to be related to that of the much more extensive zinc deposits (44, p. 22–28).

## Barite

Residual deposits of barite in clays, overlying and derived from carbonate rocks that are chiefly of early Cambrian and early Ordovician age, have been mined at many localities, but mainly at Cartersville, Georgia, Sweetwater, Tennessee, Calhoun County, Alabama, and Tazewell and Russell counties, Virginia (Figure 2).

At all these places, the residual deposits are underlain by carbonate rocks that contain vein and breccia deposits of barite, and minor sphalerite, pyrite and galena (19,8,32). It is obvious that the commercial deposits accumulated during the long period of weathering and erosion that removed the soluble parts of the original host rocks.

## REFERENCES CITED

1. Becker, G. F., 1895, Reconnaissance of the gold fields of the Southern Appalachians: U.S. Geol. Surv. 16th Ann. Rept, pt. 3, p. 251–331.
2. Burchard, E. F., *et al.*, 1910, Iron ores, fuels, and fluxes of the Birmingham district, Alabama: U.S. Geol. Surv. Bull. 400, 204 p.
3. Laney, F. B., 1917. The geology and ore deposits of the Virgilina district of Virginia and North Carolina: Va. Geol. Surv. Bull. 14, 176 p.
4. Adams, G. I., 1927, Bauxite deposits of the southern States: Econ. Geol., v. 22, p. 615–620.
5. U.S. Geological Survey, 1932, Geologic map of the United States: scale 1:2,500,000 [reprinted 1960].
6. Ross, C. S., 1935, Origin of the copper deposits of the Ducktown type in the southern Appalachian region: U.S. Geol. Surv. Prof. Paper 179, 165 p.
7. Emmons, W. H., 1937, Gold deposits of the world: McGraw-Hill, N.Y., 562 p.

8. Edmundson, R. S., 1938, Barite deposits of Virginia: Va. Geol. Survey Bull. 53, 85 p.

9. Butts, C., 1940, Geology of the Appalachian Valley in Virginia, Part 1, Geologic text and illustrations: Va. Geol. Surv. Bull. 52, 568 p.

10. Kesler, T. L., 1942, The tin-spodumene belt of the Carolinas, a preliminary report: U.S. Geol. Surv. Bull. 936-J, p. 245–269.

11. Koschman, A. H., *et al.,* 1942, Tin deposits of Irish Creek, Virginia: U.S. Geol. Surv. Bull. 936-K, p. 271–296.

12. Hunter, F. R., 1944, Geology of the Alabama tin belt: Ala. Geol. Surv. Bull. 54, 61 p.

13. King, P. B., *et al.,* 1944, Geology and manganese deposits of northeastern Tennessee: Tenn. Div. Geol. Bull. 52, 283 p.

14. Espenshade, G. H., 1947, Tungsten deposits of Vance County, North Carolina and Mecklenburg County, Virginia: U.S. Geol. Surv. Bull. 948-A, 17 p.

15. Pardee, J. T. and Park, C. F., Jr., 1948, Gold deposits of the southern Piedmont: U.S. Geol. Surv. Prof. Paper 213, 156 p.

16. Rodgers, J., 1948, Geology and mineral deposits of Bumpass Cove, Unicoi and Washington Counties, Tennessee: Tenn. Div. Geol. Bull. 54, 82 p.

17. Bridge, J., 1950, Bauxite deposits of the southeastern states: *in* Snyder, F. G., *Editor, Symposium on mineral resources of the southeastern United States:* Univ. Tenn. Press, Knoxville, p. 170–201.

18. Espenshade, G. H., 1950, Occurrences of tungsten minerals in the southeastern states: *in* Snyder, F. G., *Editor, Symposium on mineral resources of the southeastern United States:* Univ. Tenn. Press, Knoxville, p. 56–66.

19. Kesler, T. L., 1950, Geology and mineral deposits of the Cartersville district, Georgia: U.S. Geol. Surv. Prof. Paper 224, 97 p.

20. King, P. B., 1950, Geology of the Elkton area, Virginia: U.S. Geol. Surv. Prof. Paper 230, 82 p.

21. Miser, H. D., 1950, Manganese deposits of the southeastern states: *in* Snyder, F. G., *Editor, Symposium on mineral resources of the southeastern United States:* Univ. Tenn. Press, Knoxville, p. 152–169.

22. Simmons, W. W., 1950, Recent geological investigations in the Ducktown mining district, Tennessee: *in* Snyder, F. G., *Editor, Symposium on mineral resources of the southeastern United States:* Univ. Tenn. Press, Knoxville, p. 67–71.

23. Kendall, H. F., 1953, Some copper-zinc bearing pyrrhotite ore bodies in Tennessee and North Carolina: *in* McGrain, P., *Editor, Proceedings of the Southeastern Mineral Symposium 1950:* Ky. Geol. Surv. Ser. 9, Spec. Pub. no. 1, p. 112–123.

24. Thoenen, J. R., *et al.,* 1953, The future of Birmingham red iron ore, Jefferson County, Alabama: U.S. Bur. Mines R.I. 4988, 69 p.

25. Griffitts, W. R., 1954, Beryllium resources of the tin-spodumene belt, North Carolina: U.S. Geol. Surv. Circ. 309, 12 p.

26. King, P. B., 1955, A geologic section across the southern Appalachians—an outline of the geology in the segment in Tennessee, North Carolina and South Carolina, *in* Russell, R. J., *Editor, Guides to Southeastern Geology:* Geol. Soc. Amer., p. 332–373.

27. Stose, A. J. and Stose, G. W., 1957, Geology and mineral resources of the Gossan Lead and adjacent areas in Virginia: Va. Div. Mineral Res. Bull. 72, 233 p.

28. Brown, H. S., 1959, Biotite alteration in the country rock at Ore Knob, North Carolina (abs.): Geol. Soc. Amer. Bull., v. 70, p. 1757.

29. Overstreet, W. C., *et al.* 1959, Thorium and uranium resources in monazite placers of the western Piedmont, North and South Carolina: Min. Eng., v. 11, p. 709–714.

30. Bell, H., 3d, 1960, A synthesis of geologic work in the Concord area, North Carolina: U.S. Geol. Surv. Prof. Paper 400-B, p. B189–B191.

31. Kesler, T. L., 1960, Lithium raw materials: *in Industrial minerals and rocks,* 3d ed., A.I.M.E., N.Y., p. 521–531.

32. Laurence, R. A., 1960, Geologic problems in the Sweetwater barite district, Tennessee: Amer. Jour. Sci., v. 258-A (Bradley Volume), p. 170–179.

33. Bryant, B. and Reed, J. C., Jr., 1961, The Stokes and Surry Counties quartzite area, North Carolina—a window?: U.S. Geol. Surv. Prof. Paper 424-D, p. D61–D63.

34. Oder, C. R. L. and Ricketts, J. E., 1961, Geology of the Mascot-Jefferson City district, Tennessee: Tenn. Dept. Conserv. and Commerce, Div. Geol., Rept. Inv. 12, 29 p.

35. United States Geological Survey; American Association of Petroleum Geologists, 1961, Tectonic map of the United States, exclusive of Alaska and Hawaii, 1:2,500,000.

36. Kinkel, A. R., Jr., 1962, The Ore Knob massive sulfide deposit, North Carolina: an example of recrystallized ore: Econ. Geol., v. 57, p. 1116–1121.

37. United States Geological Survey, 1962, Geological Survey Research, 1962: U.S. Geol. Surv. Prof. Paper 450-A, 257 p.

38. Espenshade, G. H., 1963, Geology of some copper deposits in North Carolina, Virginia and Alabama: U.S. Geol. Surv. Bull. 1142-I, 50 p.

39. Knechtel, M. M., 1963, Bauxitization of terra rossa in the southern Appalachian region: U.S. Geol. Surv. Prof. Paper 475-C, p. 151–155.

40. Lesure, F. G., *et al.,* 1963, Beryllium in the tin deposits of Irish Creek, Virginia: U.S. Geol. Surv. Prof. Paper 475-B, p. 12–15.

41. Parker, J. M. III, 1963, Geologic setting of the Hamme tungsten district, North Caro-

lina and Virginia: U.S. Geol. Surv. Bull. 1122-G, 66 p.

42. Callahan, W. H., 1964, Paleophysiographic premises for prospecting for strata bound base metal mineral deposits in carbonate rocks: *CENTO Symposium on Mining Geology and Base Metals,* Ankara, p. 191–248.

43. Clarke, O. M., Jr., 1964, Structural geology of the Hatchet Creek copper prospect, Clay County, Alabama (abs.): Geol. Soc. Amer. Spec. Paper 76, p. 240.

44. Maher, S. W., 1964, The brown iron ores of east Tennessee: Tenn. Div. Geol. R.I. 19, 63 p.

45. Overstreet, E. F., 1964, Bauxite deposits of the southeastern states: U.S. Geol. Surv. Bull. 1199-A, 19 p.

46. Reed, J. C., Jr. and Bryant, B., 1964, Evidence for strikeslip faulting along Brevard fault zone, North Carolina: Geol. Soc. Amer. Bull., v. 75, p. 1177–1195.

47. Hack, J. T., 1965, Geomorphology of the Shenandoah Valley, Virginia and West Virginia and origin of the residual ore deposits: U.S. Geol. Surv. Prof. Paper 484, 84 p.

48. Hoagland, A. D., *et al.,* 1965, Genesis of the Ordovician zinc deposits in east Tennessee: Econ. Geol., v. 60, p. 693–714.

49. Kingman, O., and Diffenbach, R. N., 1965, The Ducktown copper district: *in Guidebook for field trip,* Joint Meeting of the American Crystallographic Association and the Mineralogical Society of America, Gatlinburg, Tenn., p. 24–31.

50. Kirkemo, H., *et al.,* 1965, Investigations of molybdenum deposits in the conterminous United States 1942–60: U.S. Geol. Surv. Bull. 1182-E, 90 p.

51. Overstreet, W. C. and Bell, H. 3d., 1965, The crystalline rocks of South Carolina: U.S. Geol. Surv. Bull. 1183, 126 p.

52. Brokaw, A. L., *et al.,* 1966, Geology and mineral deposits of the Powell River area, Claiborne and Union Counties, Tennessee: U.S. Geol. Surv. Bull. 1222-C, 56 p.

53. Carpenter, R. H., *et al.,* 1966, Fossil placers in the Precambrian Ocoee Series, Cocke County, Tennessee: Geol. Soc. Amer., Southeastern Section, Program for 1966 Annual Meeting, p. 16–17.

54. Maher, S. W., 1966, The copper-sulfuric acid industry in Tennessee: Tenn. Div. Geol. I.C. 14, 28 p.

55. Burchfiel, B. C. and Livingston, J. L., 1967, Brevard zone compared to Alpine root zones: Amer. Jour. Sci., v. 265, p. 241–256.

# 10. Geology of the Austinville-Ivanhoe District, Virginia

W. HORATIO BROWN,* EDGAR L. WEINBERG†

## Contents

## Illustrations

* formerly of Wytheville, Virginia; Dr. Brown died on May 16, 1966.
† The New Jersey Zinc Company, Austinville, Virginia.

# Tables

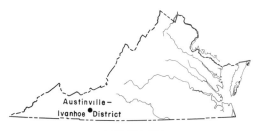

## ABSTRACT

The Austinville-Ivanhoe lead-zinc deposit occurs in the Lower Cambrian Shady dolomite. This deposit is located in southwestern Virginia in the faulted and folded Appalachian Valley and Ridge province. The ore bodies resemble the chimmey and manto deposits of northern Mexico and are formed by replacement and filling. They are confined to certain favorable zones of the Austinville and Ribbon members of the Shady. Structural and stratigraphic factors combine to control the location and form of the ore bodies. In certain places secondary oxide enrichment of the zinc minerals has contributed to the economic development of the area.

It is concluded that the ore is epigenetic and is deposited by hydrothermal solutions rising through the fault "plumbing." Classification is epithermal-telethermal.

## INTRODUCTION

Past discussions interpreting Austinville-Ivanhoe ore bodies in terms of genesis, localizing factors, or age of mineralization have engendered more heat among its many geologic workers than is likely to have been stored within the ore-forming fluids themselves. This is because of a somewhat unique relationship Austinville-Ivanhoe shows in its stratigraphic and structural setting compared with most other Mississippi Valley type deposits. It is the only major base metal deposit of Cambrian age in the eastern U.S. It encompasses a wider vertical range than most of its cousins in the Trenton-Knox age group. It probably shows most strongly the effects of orogeny upon its host rocks.

In this challenging environment, many workers within The New Jersey Zinc Company have contributed to the overall geologic picture. In the early 1900's Frank Nason, W. O. Borcherdt, A. W. Pinger, and C. E. Taylor were prominent in such labors. Systematic staff work continued with Geoffrey Gilbert, W. H. Callahan, W. H. Brown, H. Lary, S. S. Goodwin, Johnson Crawford, Howard Miller, A. D. Hoagland, J. D. Ridge, R. B. Fulton, F. H. Main, E. L. Weinberg, R. C. Gilbert, J. R. Foster, C. G. Van Ness, M. A. Chaffee and others.

Among the tasks these men accomplished were regional mapping of the entire Shady belt in southwestern Virginia, detailed mapping of the mine, pioneering in soil sampling techniques, and investigation of specific problems in sedimentation, stratigraphy, and structure as they applied to ore finding. The staff also has participated actively towards solving problems of hydrogeology.

It is with this background and through participation in these problems that the authors present the following paper.

## HISTORICAL INFORMATION

### Mining Operations in the Area

The only mining of consequence in the general area of Austinville-Ivanhoe has been for residual iron and for zinc and lead. Iron mining was begun shortly after 1800 and continued into the 1900's. It contributed substantially to the moderately high position Virginia maintained in pig iron production through the nineteenth century.

Mining for lead and later for zinc was begun shortly after the discovery of the Chiswell's Hole ore body by Colonel Chiswell about 1756. At least three different periods of mining can be recognized in the time that has elapsed since 1756.

The first, from pre-colonial through Civil War time, was almost entirely concerned with lead mining. Following Colonel Chiswell, Colonel Lynch of "Lynch Law" infamy and

Moses Austin, father of Texas' Stephen Austin, briefly managed the mines. Famous colonial visitors included Daniel Boone, General Andrew Lewis, and Thomas Jefferson. During the early period, the "Fincastle Resolutions," precursors of the Declaration of Independence, were written at the Lead Mines by the local Committee of Public Safety. Mining degenerated into a multiple, wasteful, handicraft type of operation from 1800 until 1848, when incorporation and consolidation prepared the way for later increased production during the Civil War. At that time, the mines became the chief domestic suppliers of lead to the Confederacy under the name of the Wythe Union Lead Mine Company.

The second period of mining began shortly after the Civil War when it was realized that zinc was present in large quantities with the lead. So-called "oxidized" or "soft" zinc (hemimorphite, mainly) was mined during a boom that lasted until 1902.

The third period began with the purchase of the mine in 1902 by The New Jersey Zinc Company. The property was bought mainly for its zinc oxide potential. However, reserves of sulfide ore were developed and, with the advent of the flotation process, production revived. A modern plant was built in 1927, and sulfide ores of zinc and lead have been produced regularly ever since. Ivanhoe, largely a prospect until recently, was put into production in the early 1950's. It is one of several new ore areas peripheral to the original mine that have been opened up to maintain reserve tonnages.

### Statistics of Mine Production

Information on the early production of the Austinville Mine is scant. The Kohler manuscript (1) gives the lead production from 1838 to 1868 as 20,938,746 pounds. It seems probable that in the previous 80 years at least as much or more had been produced.

The first zinc production started immediately after the Civil War and continued until 1905 when the mine appeared to be exhausted. It seems probable that in this period about 350 million pounds of zinc were produced.

Production between 1905 and 1927 was mostly development ore and mill-test concentrates. Full-scale production started in 1927. It is estimated that the mine has produced 868,000 tons of zinc and 185,000 tons of lead since its inception. Present mill feed is 3.7 per cent zinc, 0.7 per cent lead. Mill feed has varied widely from this in the past, mostly

on the higher side. Present mill feed is 2600 tons per day of crude ore.

## PHYSIOGRAPHY HISTORY AND PRESENT TOPOGRAPHY

The topography of the Ivanhoe-Austinville area has been developed on moderately folded sedimentary rocks of the Appalachian Valley and Ridge Province. The structure is further complicated by imbricate faulting. On this complex base three general erosion levels are developed. The highest ridge line, that of Poplar Camp Mountain and Lick Mountain (Figure 2), is described as the Upland Peneplain by Wright (4, pp. 9–21) and the Cretaceous Peneplain by Hayes and Campbell (2). Cooper (14) disputes any interpretation which correlates these ridge crests as remnants of prior erosion cycles. Rather, he states "the ridge crests are actually a series of knobs and sags obviously the product of present erosional activities."

The next lowest level, 500 to 1000 feet below the ridges, has variously been called the Valley or Valley Floor Peneplain. A considerable thickness of gravel has been deposited on this valley surface, much of it originating far upstream in the Blue Ridge Province.

From 100 to 200 feet below the Valley level, the New River is incised into a mature floodplain.

The development of these various levels is closely tied in with the historical development of the New River, its association with local, temporary baselevels, and the modifying effect of the periglacial climate on its regimen.

The association of glacial influence with the Valley surface is suggested not only by the extensive gravel deposits on that surface but also by the discovery of Pleistocene vertebrates in a cave just below that surface.

All of the oxidized lead and zinc deposits, as well as the residual iron deposits worked in the area, are associated with the Valley Floor Peneplain.

## GEOLOGIC HISTORY

### Stratigraphic Column

A composite stratigraphic column is shown in Figure 1 covering the detailed stratigraphy of the Shady Formation, the rock unit to which the base metal mining is confined.

Figure 2, Regional Geology-Austinville Area, indicates the position of the Shady in

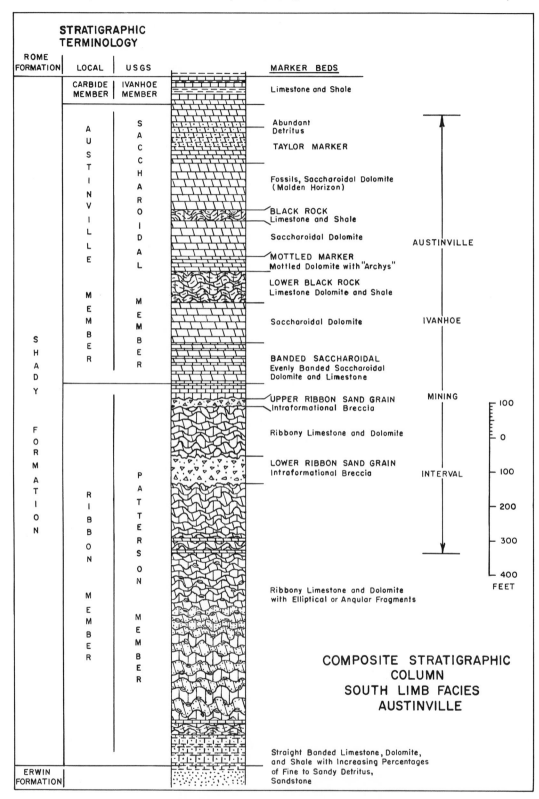

FIG. 1. *Composite Stratigraphic Column—Austinville.*

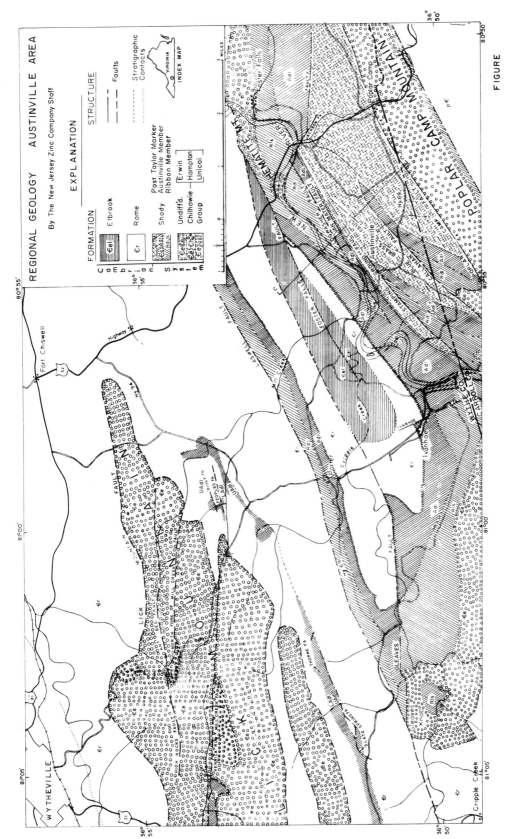

FIG. 2. *Regional Geology—Austinville Area. The U.S. Geological Survey uses the spelling "Chilhowee" in the name of the Chilhowee Formation; local usage at Austinville is "Chilhowie."*

the regional geologic column. The Elbrook approximates 2000 feet in thickness and consists of impure, thin-bedded dolomite, calcareous shale, and dolomitic limestone. Below it and superjacent to the Shady is the Rome Formation. In places, it has been estimated to be 2000 to 3000 feet thick, although few sections have been measured in the area. It consists of dolomite, limestone and variegated thick shales, dominantly red and black.

Both the Elbrook and the Rome suggest a shallow water, basin-shelf environment.

Except for local shaly horizons, the Shady is predominantly a carbonate. In the mine area its thickness approximates 2000 feet.

Two sets of roughly equivalent terminology exist for the three members of the Shady. Currier (6) proposed in 1935 the names labeled U.S.G.S. on Figure 1. The "local" names were in use on Company maps and reports at least five years earlier. Because of the precedence of time and usage, the authors propose to be stubborn and to use the local names in this report.

The upper member of the Shady, the Carbide, is rather restricted in occurrence, being mapped mainly in the immediate vicinity of Austinville and Ivanhoe. There it is approximately 100 feet thick and consists of calcilutite, locally intercalated with black shale. In places, a rich reef fauna is evident and has been described by Resser (8) as lower Cambrian. In other places the rock consists mostly of detrital calcarenite and calcirudite, much more than 100 feet in the aggregate. This is especially true in the area southeast of Austinville labeled "PTM" (Post Taylor Marker) on Figures 2 and 3. It is possible that a portion of this material may be correlative with basal Rome. Alternatively it may be roughly equivalent to Carbide but represent a much higher rate of deposition than is normal for the Carbide member.

The differences between the upper Shady zones areally are also evident in the middle and lower members. Facies variations are both gradual and abrupt across fault zones, along strike, and along dip. The major variations are along dip and thus across the structure. At least three major facies variations in the Austinville member are known in the vicinity of Austinville itself. These are the south-limb facies and the north-limb facies of the Austinville anticline and the Laswell facies. They are located, respectively, south of the Logwasher Fault, between the Logwasher and Laswell Faults, and northwest of the Laswell

Fault. The south-limb facies is shown on Figure 1. It is generally a thick (800 to 1000), saccharoidal dolomite with detrital, mottled, and shaly marker beds. The amount of shale and limestone in the section shows an increase to the southeast. Across the Logwasher Fault, the north-limb facies is also a thick saccharoidal dolomite and limestone. It has fewer distinctive marker beds but does show a marked increase in the amount of detritus over the south-limb facies. Deposition appears to have been quite irregular in this area as is evidenced by the difficulty in correlating individual markers for any distance.

Recent closely-spaced drilling across the anticlinal axis near Austinville has indicated that the transition between the south-limb and north-limb facies is abrupt although certain markers can be correlated across the axis. Callahan (16) emphasizes the facies change in this general area as a major one and as one that may well control the area of ore deposition. In further discussions (personal communication), he also suggests the facies change may influence the position of structural elements. The authors prefer to consider the facies change as one that may be significant among several localizing ore controls. Whichever emphasis may be more correct, it appears likely that the change is related to distance from shore line. In this case, it appears that the source must have been to the northwest of the north-limb facies.

Northwest of the Laswell Fault (and also southwest within the Speedwell Syncline) the Austinville member is locally thinned, ranging from 516 to 913 feet. For field purposes it can generally be divided into three units. The upper is a dark grey, thin-bedded dolomite, the middle a dolomite with abundant shale partings, and the lower a light-grey, massive dolomite. The middle unit weathers to a shaly appearing rock and has been designated the "parting shale."

The Ribbon member has many similar characteristics throughout the district. As is shown in Figure 1, it grades from thin partings in its upper portion to much thicker partings in the lower portion. Near the lower third of the formation, the ribbony texture gradually gives way to a straight banded texture, although, in places, ribbony beds reappear. The predominant carbonate lithology is diluted slightly, in places, by detrital material delineating certain marker beds. The Lower Ribbon Sand Grain marker is one of these beds; it has only been found on the south-limb facies.

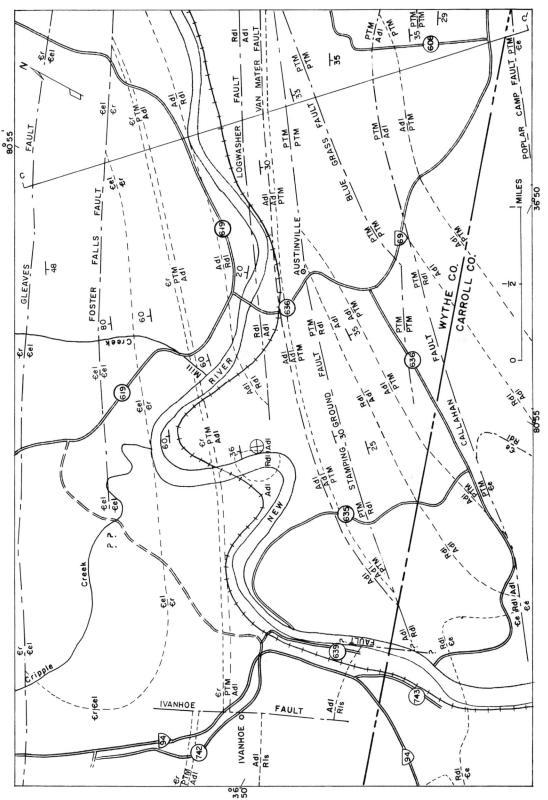

FIG. 3. *Detailed Geologic Map—Austinville-Ivanhoe.*

A somewhat equivalent group of marker beds, the TOMS, is predominantly known on the north-limb facies. Four measured sections of the Ribbon on the north-limb facies show thicknesses ranging from 821 to 849 feet. The composite stratigraphic column, representing the south-limb facies, shows a thickness of approximately 1100 feet.

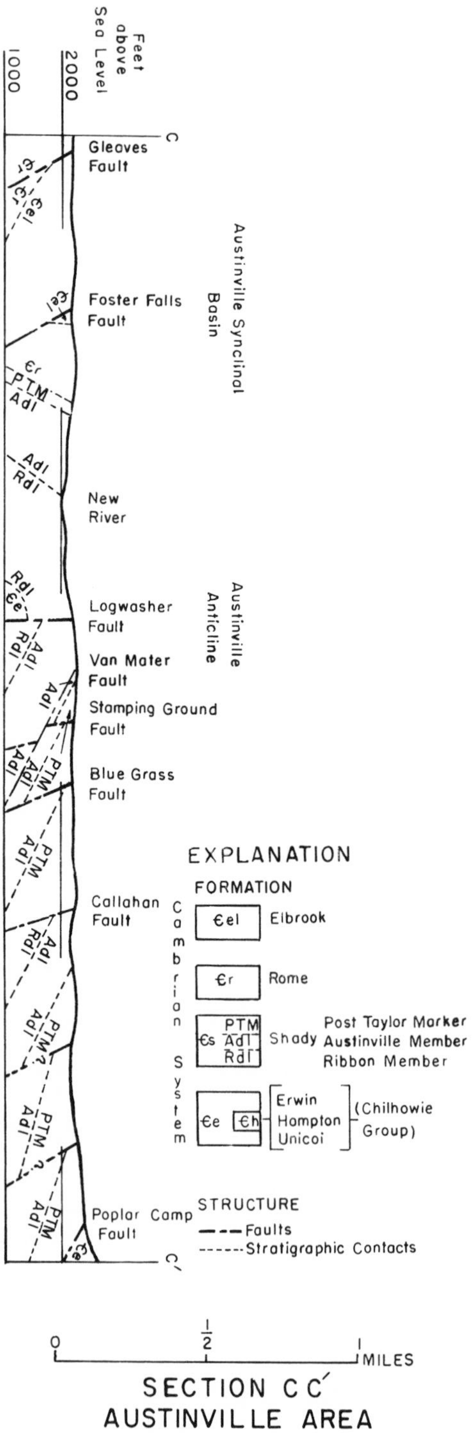

The contact between the Ribbon member of the Shady and the Erwin Formation is gradational within a few feet. In places considerable disseminated pyrite is found in the Ribbon within 50 to 150 feet of this contact. In fact, a local marker bed in this interval, showing contorted ribbony texture, in places contains up to 20 per cent pyrite. It has been designated the "Py" bed, in consequence. The lowermost Ribbon also contains locally abundant linguloid brachiopods.

The Erwin, Hampton, and Unicoi formations are grouped together as the Chilhowie group. Each is approximately 1500 to 2000 feet thick in the vicinity according to Cooper (14), and Stose and Stose (11), although the formations have been described as being much thicker elsewhere.

The Erwin is an impure shaly quartzite with local shale and clean sand members. The Hampton is an arkosic shale with intercalated quartzite beds. The Unicoi can be characterized as predominantly an arkosic sand with basalt, conglomerate, siltstone, and shale alterations.

## General Structure of the Area

The mine area lies near the southeast edge of the folded, sedimentary Appalachian Valley and Ridge province. The metamorphic Blue Ridge province lies just to the southeast and the flat sedimentary Appalachian Plateau at some distance to the northwest. In this portion of the Valley and Ridge province the folds are relatively open although they are broken by a complex series of imbricate faults. The major faults are reverse, dipping to the southeast. Some normal, diagonal, and cross faulting is also known, as indicated on Figure 3.

In the vicinity of Section C-C', as shown on Figure 4, the structures from southeast to northwest include the Poplar Camp Fault, multiple thrust (imbricate) faulting on the southeast limb of the Austinville anticline; the graben-like crest of the anticline; and the faulted Austinville synclinal basin. Similar folded and faulted sequences exist on strike. Here, as elsewhere in the southern Appalachians, the beds tend to be steeper on the northwest limbs of the folds. Pennsylvanian beds are the youngest involved in the folding in this portion of the Valley and Ridge province. By classical orogenic theory, then, one could say the age of folding and faulting is

FIG. 4. *Section CC'—Austinville Area.*

Pennsylvanian or younger. The evidence of lithologic changes and large unconformities scattered through the section indicate that vertical movement, at least, was more or less continuous. Recently, Cooper (15) has advocated the existence of structural highs and lows throughout the Paleozoic in some portions of this area.

Because of the general facies changes within the Austinville basin and anticline and because of some discordance with the fold axes, the authors are inclined to believe the major folding postdated the deposition of the Shady rather than accompanied it. If the folding and faulting are associated and essentially synchronous rather than sequential, then the age for the major orogenic changes appears to be Pennsylvanian or younger, possibly Appalachian revolution. If the folding and faulting are associated, but sequential, then large orogenic changes could be postulated several times after the deposition of the Beekmantown.

### Age of Mineralization

Until recently, structural and stratigraphic associations of the ore minerals were the only clues to the relative age of the mineralization relative to these features. Most of the ore lenses appear to be unmatcheable across the fault planes, although closely associated with faults. Very little mineralization has been noted to be faulted although some is known on slickensided surfaces. It is assumed therefore that the ore is post-faulting or synchronous with the faulting.

The ore textures are typical of replacement and filling. Both are post-depositional features. The principal question is how much later than deposition?

Three samples of galena from Austinville have recently been isotopically analyzed. The laboratory analyses given below were calibrated against NBS-200.

Cannon (14) suggests that age approximations may be made from such lead-isotope analyses providing certain assumptions are made. These include the composition and relative abundance of radioactive parent elements, the lead-isotope ratio at the earth's formation (primeval lead), the cessation of isotope generation at the particular time of extraction and deposition and the non-differentiation of the isotopes during deposition. However, expert advice inclines us not to hazard such approximations at this time because of the possible large error involved.

## ECONOMIC GEOLOGY—PRIMARY ORE

### Forms of the Ore Bodies

The ore bodies form lenses elongated along strike and restricted to certain favorable horizons (see Figure 5). In general, their footwalls are conformable with bedding whereas their hanging walls commonly cut across bedding contacts. In places, this crosscutting can be quite spectacular, with transgressions of 100 feet or more. The dip lengths of the lenses can be measured from 50 to 400 feet. Strike lengths have been known from less than 100 feet to over 2000 feet. Continuity of some of the lenses is interrupted, especially in the dip direction, by minor strike or diagonal faults. These can displace the favorable horizons as much as ten feet or more.

The size and type of the ore lenses are affected by the horizons in which they occur. For example, some of the larger, more massive lenses occur in the Malden horizon in the upper portion of the Austinville south-limb facies. The unit just above the Mottled Marker horizon commonly is host to irregular, anastomosing lenses, in keeping with its essentially biostromal character. The Banded Saccharoidal ore of the north-limb facies of the Austinville is generally in small, erratic lenses, in keeping with the general nature of that member.

Some of the ore bodies are confined to one side of a fault zone. Others enlarge greatly at intersections with certain cross faults. Their

| Sample | Atom Per Cent | | | | 206/204 | 207/204 | 208/204 |
|---|---|---|---|---|---|---|---|
| | $Pb^{204}$ | $Pb^{206}$ | $Pb^{207}$ | $Pb^{208}$ | | | |
| 1 | 1.329 | 25.55 | 21.06 | 52.06 | 19.22 | 15.85 | 39.17 |
| 2 | 1.333 | 25.45 | 21.09 | 52.12 | 19.09 | 15.82 | 39.10 |
| 3 | 1.326 | 25.54 | 21.09 | 52.04 | 19.26 | 15.90 | 39.25 |

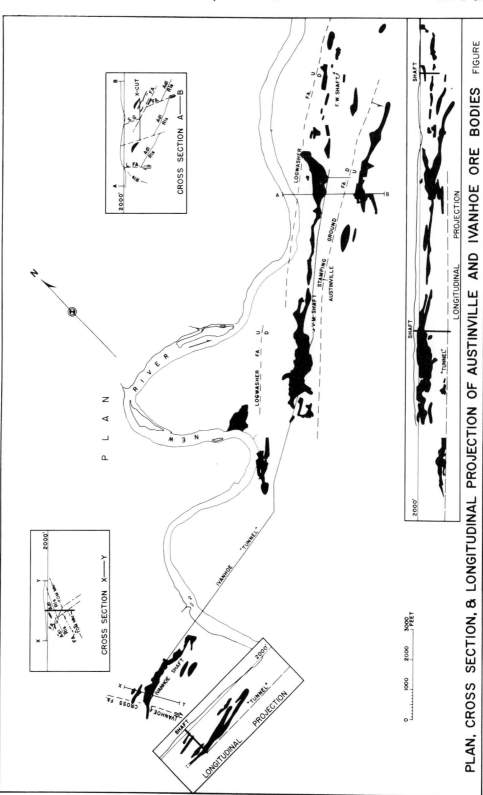

PLAN, CROSS SECTION, & LONGITUDINAL PROJECTION OF AUSTINVILLE AND IVANHOE ORE BODIES

FIG. 5. *Plan, Cross Section, and Longitudinal Projection of Austinville-Ivanhoe Ore Bodies.*

forms are sufficiently varied to reflect several different types of controls.

## Stratigraphic Relations of the Ore Bodies

Figure 1, the composite stratigraphic column, shows the Austinville-Ivanhoe mining interval in terms of stratigraphic position. Most of the favorable ore horizons at Austinville and Ivanhoe are contained therein.

Several years ago, soon after the detailed stratigraphy of the Austinville member was worked out at Austinville, a statistical survey was made to determine the stratigraphic distribution of the ore. The ore area on each of many, equally-spaced, vertical cross sections was determined for each stratigraphic interval. The results of that study are shown on Figure 6A. It shows clearly that the basal portion of the Malden zone is an outstandingly favorable ore horizon. Other horizons appear less favorable. The reason for this horizon being favorable is not entirely clear. However, a difference in competency between the Black Rock transition bed and the massive Malden bed could have contributed towards brecciation along their contact during folding and thus provided the ore solutions with through-going access to this zone.

The Ribbon member shows major ore lenses in association with brecciated marker beds. The ore itself is generally in ribbony textured rock but is almost always in juxtaposition to persistent breccia zones. Two such zones are the TOM markers and the Lower Ribbon Sand Grain marker. (Figure 1).

While the ore definitely is more abundant in certain beds than in others, it also transgresses stratigraphy in places. Faulting is almost always evident in such places.

## Mineralogy of the Deposit

MINERALS PRESENT   Ore minerals are chiefly sphalerite and galena. Secondary oxidized equivalents include hemimorphite, smithsonite, hydrozincite, cerussite, and anglesite. Pyrite is abundant locally and is disseminated ubiquitously through the ore. Coarse, white, gangue dolomite commonly accompanies the ore. It has familiarly been designated "met" since it supposedly originated as recrystallized or "metamorphosed" wall rock.

The iron content of the sphalerite ranges from 0.3 to 1.7 per cent. In places, both dark-brown and honey-colored transparent sphalerite occur together, although certain sphalerite color shadings are more or less characteristic of distinctive mineralized areas. No rational

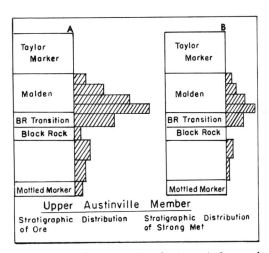

FIG. 6. *Stratigraphic Distribution of Ore and Met.*

scheme governing the changes in this coloration is evident, possibly because of the structural complications or perhaps because of the lack of systematic information. The very dark sphalerite, almost always, is in contact with pyrite. Only trace amounts of cadmium and silver occur, respectively, in the sphalerite and galena.

Some less common minerals occur with the ore but are mostly confined to the Ribbon member. These include chalcopyrite, gypsum, anhydrite, and fluorite. Barite and aragonite occur in both Ribbon and Austinville members. Clear, black or red chert occurs, to some extent, in both members.

WALL-ROCK ALTERATION   Dolomite is always the host rock. Three types of alteration modify this dolomite. (1) Metamorphism or recrystallization, (2) Bleaching, and (3) Leaching. These will be discussed in the above order.

The most important of these is metamorphism. It occurs near ore and in places, one can observe a fairly sharp transition from limestone, to barren dolomite, to dolomite with increasing metamorphism, to ore. This is more common in the Ribbon member. The Austinville member has relatively little limestone in the vicinity of ore, although it does show the other gradations laterally and vertically from ore lenses.

In the early 1930's, a quantitative study was made of the stratigraphic distribution of metamorphism. Results are plotted on Figure 6B.

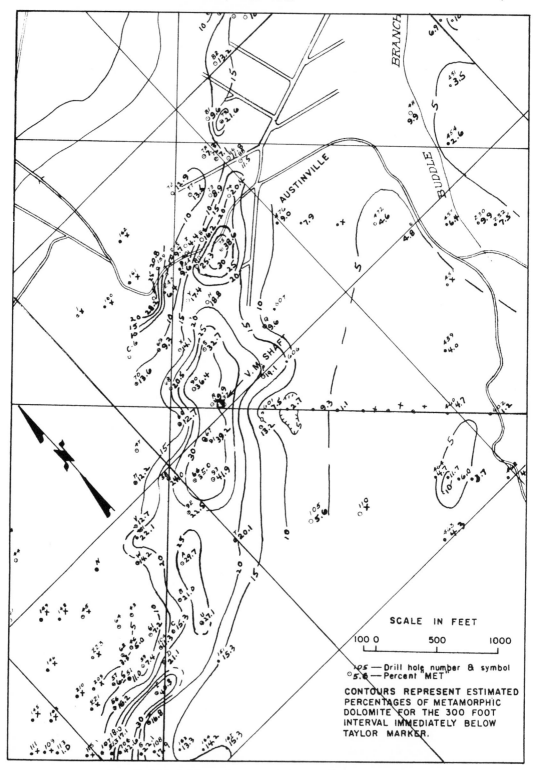

FIG. 7. *"Met" Map—Portion of Austinville Mine Surface.*

The configuration of 6A and 6B is so close that it is difficult to believe that the ore and metamorphism did not have a similar genesis. The similarity of distribution was checked further by plotting Figure 7 which shows a practically identical distribution in plan for a portion of the Austinville Mine. Analysis of the data was also made to determine the degree of metamorphism next to the ore on the hanging-wall, footwall, up-dip, down-dip, and on strike. Results are noted in Table I. This shows clearly that the strongly metamorphosed ground is concentrated everywhere around the ore except on the footwall. This concept has been valuable in prospecting the Austinville member. For some unknown reason, it does not apply to the Ribbon.

Bleaching involves the alteration of a dark grey or black dolomite to a light grey. Both light and dark grey dolomite are fresh-looking rocks. It is well developed at the northeast end of the main Austinville ore zone in an area known as the Flatwoods and above the Austinville dolomite ore body at Ivanhoe.

The bleaching does not follow the bedding but has a random distribution through the dark dolomites. Attempts to correlate light or dark dolomites between holes as closely spaced as 100 feet invariably leads to failure. The relationship of bleaching to ore is unknown.

Leaching is not clearly understood. It occurs in medium- to fine-grained, brown or buff dolomite. In one part of the mine area, it very definitely overlies ore and can be taken as a guide to ore.

The brown or buff color of the dolomite is clearly a result of oxidation. The dolomite is very porous, and under the microscope it can be seen that it originally contained abundant disseminated pyrite crystals. In places, many of these have oxidized, but enough pyrite cores of the crystals are present to indicate the origin proposed for the present texture.

MINERAL TEXTURES Mineral textures have been studied in detail only in a limited area of the Austinville member. Since the ore is replacement ore with subsidiary filling of open spaces, it tends to mimic pre-existing structures or sedimentary features. Thus, various types of breccia ore have been recognized. "Fracture ore" has also been described in which the sphalerite and galena occur along fine shear fractures. "Ribbon ore" is common in the Ribbon member. In this type of ore, the sphalerite and galena replace portions of ribbony beds, preserving a perfect replica of the original sedimentary texture. In places where isolated dolo-

TABLE I. *Statistical Study—Metamorphism Adjacent to Ore in Austinville Member—South-Limb Facies*

| Met | Hanging Wall[1] | Footwall[1] | Down Dip[1] | Up Dip[1] | On Strike[1] |
|---|---|---|---|---|---|
| None | 31% | 31% | 6% | 7% | 0% |
| Slight | 29% | 69% | 23% | 43% | 53% |
| Moderate | 8% | 0% | 12% | 20% | 10% |
| Strong | 32% | 0% | 59% | 30% | 37% |
| (Cases involved) | 38 | 42 | 34 | 30 | 19 |

[1] Results expressed as per cent of total cases measured.

mite grains have been replaced rather than the whole sedimentary structure, the ore is referred to as disseminated or "cornmeal" ore. Another variation resembling organic or crust-like structures is called "crustiform" ore. This commonly is in a matrix of strong met with little original dolomite remaining.

All of these designations are somewhat oversimplified since there are all gradations between the various types, and a clear cut example of one type is seldom found.

The only detailed study of ore texture distribution was made in 1931 in the south-limb facies of the Austinville member. In the upper levels of the Austinville mine, ore was classified at that time into four main textural types: rubble breccia, mosaic breccia, rosette ore, and disseminated ore. Of these types, all except the rosette represent types of ground preparation. Rosette ore may represent intensely mineralized ground of any type, especially rubble breccia.

The geology mine maps at a scale of 1 inch equals 30 feet were used in this study. In the area of investigation, the mine is developed by footwall strike drifts and by cross cuts spaced every 100 feet. Widths of each type of ore along the cross cuts were measured and classified as follows: footwall ore, hanging wall ore, ore in the central part of the orebody, ore near (i.e., against) faults, ore in metamorphic ground, ore in thin bedded horizons, and whole orebody ore (ore in which one textural type composed the entire thickness in the crosscut). From these measurements of the horizontal dimensions ore in the different structural positions given above was calculated as percentages of the whole. Results are shown in Table II.

Several striking features of distribution stand out on the tabular summary:

TABLE II. *Percentage Distribution of Ore*

| Occurrence | Textures | | | Disseminated Ore | Total |
|---|---|---|---|---|---|
| | Mosaic Breccia | Rosette Ore | Rubble Breccia | | |
| Foot Wall | 3.3 | 14.6 | | | 17.9 |
| Center | 2.0 | 1.6 | 12.8 | | 16.4 |
| Hanging Wall | 12.3 | 0.4 | | | 12.7 |
| Whole Thickness | 20.6 | 6.5 | | | 27.1 |
| Near Fault | 5.1 | 8.3 | 1.9 | | 15.3 |
| In Strong and Moderate Met | | | | 3.2 | 3.2 |
| Thin Bedded | | | | 7.4 | 7.4 |
| Totals | 43.3 | 31.4 | 14.7 | 10.6 | 100.0 |

(1) The footwall is predominantly rosette ore.

(2) The hanging wall is predominantly mosaic ore.

(3) Rubble ore occurs near faults and in the central part of ore bodies above a rosette footwall.

(4) Disseminated ore is minor in quantity.

(5) Mosaic ore is the most abundant type.

Since rosette ore does not represent a type of ground preparation but an intensity of mineralization, it should be possible to distribute the rosette among the other textures. About half the rosette ore lies on the footwall of ore bodies. In several cases this is beneath well developed rubble ore. Suggestions of an original rubble texture are common in this intensely mineralized rosette ground. The next largest portion of the rosette, one fourth, occurs near faults, also a location for rubble. It is quite probable that much of the rosette was derived from rubble and should be so assigned.

The picture derived from this study is that of a block of ground most intensely broken in the footwall portion and near faults. The breaking dies out into mere shattering upward from the footwall. The block rests on a relatively unbroken footwall.

Figure 6 showing the distribution of ore and metamorphism in the upper Austinville member indicates a picture parallel with that outlined above. The ore and metamorphism are at a maximum in the Malden footwall and decrease upward toward the Taylor marker.

PARAGENESIS The banded and rosette ore textures indicate a certain paragenesis for the mineral bands contained therein. A complete sequence from center outward is:

(1) Barren country rock (dolomite).

(2) Coarse, white, recrystalline dolomite (met).

(3) Pyrite.

(4) Sphalerite.

(5) Galena.

(6) Met.

The sphalerite in such banded specimens is generally dark or medium brown. A light yellow sphalerite is sometimes present as crystalline druses in cavities, along fine hair-like fractures, or as disseminated blebs in a coarse-grained matrix. All these occurrences, except the last point to a late stage of deposition.

The probable sequence of sulfide deposition, then, is from earliest to latest:

(1) Pyrite.

(2) Medium to dark brown sphalerite.

(3) Galena.

(4) Light yellow sphalerite.

Where chalcopyrite is present it appears to be somewhat later than the galena.

### Factors Controlling Form and Location of Ore Bodies

(1) Detailed structural patterns. Faults are associated with almost every ore lens in the area. The Austinville member at Austinville exhibits the most complex system of faults known in the district. This is illustrated, in general, by cross section C-C' on Figure 4 and, in particular, by Figure 8, the geologic map of a portion of the Austinville fifth level. The general structure of the portion of the mine southeast of the Logwasher Fault is that of a graben. This graben is on the south limb of the Austinville anticline between two major strike faults. Cutting the center of this graben is a persistent diagonal fault known as the

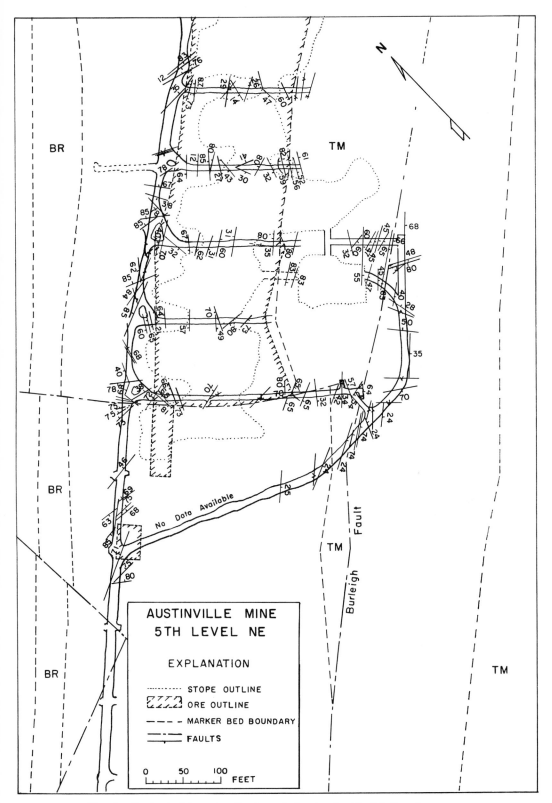

FIG. 8. *Geologic Map of a Portion of the Austinville 5th Level.*

Burleigh. Cutting this are at least three cross faults.

In places, the strike ore pencils follow the Burleigh Fault and appear to be clearly controlled by the fault. Elsewhere, the strike pencil ore bodies follow monoclinal strike terraces. This folding can be seen on the Taylor Marker in the hanging wall of the ore bodies.

At the intersection of the strike faults and monoclinal terraces with the cross faults, there are locally great enlargements of the ore bodies (see Figure 8). At these enlargements in some places the ore bodies are cross cutting through a large extent of the Austinville member.

Steep strike faults tend to be less prominent in the Ribbon member than in the Austinville member. This can probably be ascribed to the less competent nature of the Ribbon. Not only are the faults somewhat different in character and spacing in the Ribbon, but they appear to be accompanied by more drag folding and less brecciation than the faults in the Austinville. One of the bedding faults in the Ribbon ore is mineralized in and directly above the fault plane but not below it. The mineralization extends outward from, and parallel to, the fault plane, cutting across a prominent drag fold.

Thus, both in the Austinville and Ribbon members, the strike of the ore, the host rocks and many of the prominent faults are parallel. Where there is transgression, there appears to be fault rather than stratigraphic control.

It has been suggested that the ore bodies from Austinville to Ivanhoe lie in an en echelon pattern. The pattern would extend across the axis of the anticline and across at least three major fault blocks. If this regional pattern is indeed real, its now obscure cause would be another structural control for the ore.

(2) Favorable beds. As mentioned in the section on stratigraphic relations of the orebodies, there are definitely favorable beds in both the Austinville and Ribbon members. In the south-limb facies, they range from Taylor Marker in the Austinville member to below the Lower Ribbon Sand Grain Marker in the Ribbon member. Lithologically all are now dolomites. With the exception of the Ribbon units, most of these beds are massive in their unfaulted state. In all cases, they show a distinct textural variation with the subjacent beds. In the Ribbon, more massive, autobrecciated beds constitute the subjacent units. Whether the effective control is a contrast of competent-incompetent beds or chemical or permeability differences is hard to say.

(3) Paths of ore and gangue material in solution. Among the several methods known for determining the direction of flow of mineralizing solutions (see Gross, 10), the distribution of metal ratios or mineral zoning is the most applicable to Austinville.

Brown (7) studied the change in the $Zn/Pb$ ratio upward and outward from several zones of intense fracturing and ore deposition. Statistical analyses of the results indicated that the ratio gradually dropped within the individual zones as one proceeded upward from the most intense centers of mineralization and faulting, outward along strike and thence upward once again.

Specifically, a portion of the mine area was studied near the intersection of the Burleigh Fault with two other steeply dipping faults, the Shot Shaft and Big. D. In those locations a crosscutting chimney of ore is formed in the breccia. At the Malden zone the ore extends outward along strike. This coincides with severe fracturing along the base of the Malden, now shown mostly as a high grade rosette ore. Formerly, this was presumably a rubble breccia. Higher in the Malden a mosaic breccia is mineralized. This presumably developed through mineral stoping upward from the fracture zone. It is theorized that solutions moved upward through the intersecting fault zone to the base of the Malden. Here they moved through the fractured zone, gradually opening a mosaic breccia as far up as the Taylor Marker by collapse of the undermined ground. This path of travel appears most likely since it follows a gradual dropping of the $Zn/Pb$ ratio from the deepest ore in the chimney to the Malden footwall and thence to the Malden hanging wall.

A comparison of the orebodies at Austinville to those described by Prescott (5) in northern Mexico leads the senior author to conclude that Austinville can be classified as a similar series of chimneys and mantos, obeying nearly all of Prescott's "Basic Laws". There are a series of seven of these, the most important of which states, "The ore bodies are essentially chimneys or pipes . . . . The ore deposited farthest from the source has passed thru and inside the wall of the orebody nearer the source."

## ECONOMIC GEOLOGY—SECONDARY ORE

### Oxide Enrichment

Since supergene sulfide enrichment does not exist to any extent in a deposit of sphalerite and galena such as this, it is oxide enrichment which has significantly influenced the pattern of mining in the district in those places where it has enriched the surface ore.

The known secondary zinc minerals include hemimorphite, smithsonite and hydrozincite. Of these hemimorphite is by far the most abundant.

Weathering along the Shady-clay interface generally proceeds most rapidly along the joint planes. They apparently provide handy places for the run-off waters to collect. These waters are probably weakly acidic because of dissolved organic materials and carbon dioxide content and thus accelerate solution of the carbonate matrix. At any rate, the typical subsurface rock topography is made up of pinnacles and troughs showing a maximum relief of sixty feet. The hemimorphite and smithsonite occur encrusted on the surface of the dolomite pinnacles. The crust of "soft ore" may vary from a few inches to a few feet thick. It is thickest on top of and in the troughs between pinnacles and thinnest on the sides of the pinnacles. The crust is banded parallel with the clay-dolomite interface. Besides being banded it also appears brecciated.

Several explanations are available to account for the texture and location of the "soft ores," predominantly hemimorphite. One may interpret the crustified banding as a solution-reaction-reprecipitation phenomenon. The zinc sulfide in the dolomite, according to this explanation, goes into solution in the clay. The zinc ions then recombine with silica and precipitate on the dolomite causing the crustified banding. Brecciation is caused by further solution of the dolomite and slumping of the crusts. Another explanation relates the brecciation to the original breccia form of the sphalerite mineralization. Solution of the sphalerite would leave a subterranean talus slope along the pinnacle surface as the base for the oxidized mineralization. A third possibility, which is suggested here for the first time, is that variations of the pyrite content of the primary ore may cause local variations of pH along the pinnacle-clay interface. This could contribute to highly irregular deposition of hemimorphite and lead to the appearance of a brecciated surface. Takahashi (12) has shown, that with other factors constant, solubility of the secondary zinc minerals is highly sensitive to pH. At atmospheric conditions, hemimorphite is least soluble below a pH of 6.2, smithsonite is least soluble between 6.2 and 8.1 and above 8.1, hydrozincite holds the honors. The soils of the Austinville area have a pH between 2.5 and 5.0 as noted by Fulton (9, p. 661),

although locally the pH may rise as high as 8.0 in areas of ground water saturation. This soil pH could account for the predominance of hemimorphite and the original pyrite distribution could account for the irregular pattern of deposition. The sulfide ores in bedrock average from three to 20 per cent zinc. The high grade "soft ores" average about 40 per cent zinc.

Oxidized lead minerals include anglesite, cerussite, and very scarce massicot and minium. Because of the limited mobility of the lead ion in the carbonate environment most of these minerals are found in pockets in the clay or in contact with the galena itself where oxidizing waters have moved along channelways in the rock. Oxidation and solution have extended in places to the seven hundred level (700 feet below the Valley surface) although most of the oxidized area is confined to the upper 4 or 5 mine levels.

## ORE GENESIS

Lindgren (5) aptly included Austinville with Mississippi Valley type zinc-lead deposits. The relative simplicity of its mineral suite, its occurrence in an early Paleozoic carbonate sediment, and its adherence to the stratabound habit all tend to confirm Lindgren's estimate.

As he noted then, there were two main schools of thought on ore genesis for this type of deposit, the "descensionists" and the "ascensionists." The former favored deposition by percolating surface waters while the latter favored deposition by ascending thermal waters.

Today there is little argument that the ores are epigenetic. The textures alone demand this. However, the modern day descensionists apparently favor preparation by ground water solution and karst development while the rocks were still flat lying. One vocal branch of this school prefers to discount greatly the localizing effect of structural deformation and instead finds evidence that now-deformed ores were relatively flat lying when deposited. Oceanic emanations or cold water lateral secretions would have to serve as ore fluids in this interpretation.

Modern day ascensionists need both original sedimentary variants and later, imposed structures to form loci for hydrothermal fluid to move and deposit along. Needless to say, the authors are among this group.

There is such an obvious structural control at Austinville (and Ivanhoe) that it would be well-nigh impossible to postulate any deposition while the beds were underformed. The majority of ore lenses are vertically stacked near fault zones. Many of the ore bodies are in tectonic breccias. A general halo of coarse, recrystalline dolomite surrounds the ore lenses in a manner difficult to ascribe to percolating meteoric waters or diagenetic processes. Metal ratios indicate a rational gradient from depth upward along fault intersections. The age of the mineralization is apparently much later than the age of sedimentation and also apparently somewhat later than the deformation. It seems clear, from the stratigraphic-ore studies, that sedimentary processes played a definite part in ore localization. It is also apparent, though, that without the through-going "plumbing" represented by the ubiquitous fault system, little or no ore would be in its present position.

The mineral assemblage of sphalerite, galena, pyrite, fluorite and local, trace chalcopyrite suggests the low pressures and temperatures of Lindgren's epithermal or Graton's telethermal zone.

## CONCLUSIONS

To summarize, it is believed that these deposits are essentially a low temperature form of manto formed by solutions traveling through the ore bodies from their source to their termination. This is evidenced by the distribution of metal ratios through paths of maximum breaking. This theory of origin gives hope for deposits of greater length than those limited by downward percolating waters. The localization of the mineralizing solutions in this portion of the Shady was probably influenced, of course, by original sedimentary variations as well as structural features.

## REFERENCES CITED

1. Kohler, E., 1870, History of Austinville, Va.: unpublished manuscript in The New Jersey Zinc Company files.

2. Hayes, C. W. and Campbell, M. R., 1894, Geomorphology of the southern Appalachians: Nat. Geog. Mag., v. 6, p. 63–126.
3. Wright, F. J., 1925, The physiography of the upper James River basin in Virginia: Va. Geol. Surv. Bull. no. 11, 67 p.
4. Prescott, B., 1926, Underlying principles of limestone replacement deposits of the Mexican province: Eng. and Min. Jour., v. 122, no. 7, p. 246–252; no. 8, p. 289–296.
5. Lindgren, W., 1933, Mineral Deposits: 4th ed.: McGraw Hill, N.Y., 930 p.
6. Currier, L. W., 1935, Zinc and lead region of southwestern Virginia: Va. Geol. Surv. Bull. no. 43, 122 p.
7. Brown, W. H., 1935, Quantitative study of the zoning of ores at Austinville Mine, Wythe County, Virginia: Econ. Geol., v. 30, p. 425–433.
8. Resser, C. E., 1938, Cambrian System (restricted) of the southern Appalachians: Geol. Soc. Amer. Spec. Paper 15, 140 p.
9. Fulton, R. B., 1950, Prospecting for zinc using semiquantitative chemical analyses of soils: Econ. Geol., v. 45, p. 645–670.
10. Gross, W. H., 1956, The direction of flow of mineralizing solutions, Blyklippen Mine, Greenland: Econ. Geol., v. 51, p. 415–426.
11. Stose, A. I. J. and Stose, G. W., 1957, Geology and mineral resources of the Gossan Lead district and adjacent areas in Virginia: Va. Div. Mineral Res. Bull. 72, 233 p.
12. Takahashi, T., 1960, Supergene alteration of zinc and lead deposits in limestone: Econ. Geol., v. 55, p. 1083–1115.
13. Cannon, R. S., Jr., et al., 1960, The data of lead isotope geology related to problems of ore genesis: Econ. Geol., v. 56, p. 1–38.
14. Cooper, B. N., et al., 1960, Grand Appalachian field excursion: V.P.I., Eng. Ext. Ser., Geol. Guidebook No. 1, 187 p.
15. Lowry, W. D., Editor, 1964, Tectonics of the southern Appalachians: V.P.I., Dept. Geol. Scis. Mem. 1, 114 p.
16. Callahan, W. H., 1964, Paleophysiographic premises for prospecting for strata-bound base metal mineral deposits in carbonate rocks, CENTO Symposium on Mining Geology and Base Metals, Ankara, p. 230–235.

# 11. The Birmingham Red-Ore District, Alabama*

THOMAS A. SIMPSON,† TUNSTALL R. GRAY‡

## Contents

## Illustrations

* Publication authorized by the State Geologist of Alabama and by the Woodward Iron Company.
† Geological Survey of Alabama, Tuscaloosa, Alabama.
‡ Woodward Iron Company, Woodward, Alabama.

## Table

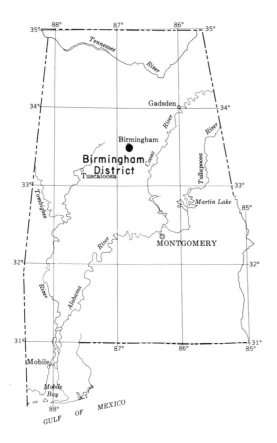

to an annual rate of about 1.5 million gross tons. Reserves calculated for the district are sufficient to last many years. Mining in the district is now confined to Woodward Iron Company's Vance and Pyne mines.

The rocks exposed consist of about 15,000 feet of consolidated Paleozoic sediments with a thin veneer of unconsolidated Cretaceous sediments in the Vance area. The district is characterized by a distinctively parallel and subparallel, slightly arcuate series of faults and joint systems.

In view of available information, it seems reasonable to conclude that the Birmingham red ores are of sedimentary origin, modified to a small extent by diagenetic replacement after original deposition.

## INTRODUCTION

The Birmingham district first produced steel from Alabama hematite ores in 1899. Prior to that time, ore from the district was first smelted in 1864 near Irondale and the first coke pig iron was produced in 1876. In the early 1900's about 60 mines, both underground and surface pits, were producing red ore. To 1967, the mines in the district have produced about 300 million tons of ore.

In 1962, the Birmingham mines were producing more than 6.0 million gross tons of ore a year, and the district was the third largest producer of hematite iron ore in the United States. However, since the influx of high-grade foreign ores, only two mines remain in operation and production has decreased to about 1.5 million tons annually, and the district now ranks fifth in the nation in the production of iron ore.

## ABSTRACT

The Birmingham district first produced steel from Alabama hematite ores in 1899. Since then, the district generally produced more than 6.0 million gross tons of ore a year to the late 1950's. Production has declined since then

"Red paint" rock was first discovered on Red Mountain by frontiersmen Friley and Jones from Tennessee in the early 1800's. Later discoveries of coal and limestone within a radius of 5 to 15 miles of Birmingham provided the resources for the district eventually to become recognized as the "Pittsburgh of the South." The importance of this iron and steel industry to the south and the nation is evidenced by the world-wide distribution of products from its blast furnaces and steel mills.

The first geologic studies of the iron ores were begun by the U.S. Geological Survey in 1906. Since then, many reports concerning the geology, mining methods, milling practices, and iron and steel production have been prepared and published about the district.

This paper does not attempt to review the extensive literature that has been written on the district, but extensive bibliographies are listed in the references cited at the end of this report.

## Acknowledgments

Acknowledgment is made to many officials of the mining firms, who over the years have freely discussed the geology and practices of the district. The authors also are grateful and wish to acknowledge the assistance obtained from authors of the many publications about the district.

## Location and Extent

The Birmingham red-ore district is in the southern part of Jefferson County and the northeastern part of Tuscaloosa County in north-central, Alabama (Figure 1). Early prospecting and initial mining activity were scattered along exposures of the iron ore seams in the Red Mountain Formation from south of Bessemer to Dekalb County, Alabama. As iron and steel technology improved in the late 1800's, charcoal pig iron gave way to coke pig iron, and mining activity became centered in the Birmingham-Bessemer areas near abundant supplies of coking coal and "self-fluxing" red ore.

In 1967, the Pyne mine of Woodward Iron Company, an underground operation, and a strip mining operation near Vance, leased and operated by Woodward, are the only active red-ore mines. The operation at Vance is not in the Birmingham district as described in this paper, but the ore is mined from the Red Mountain Formation, the same formation that is mined at Pyne mine.

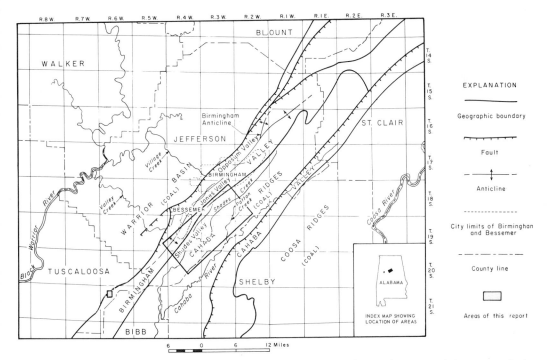

FIG. 1. *Index Map of Alabama, showing generalized geographic and structural features of the Birmingham red-ore district.*

## Physiography and Topography

The causes of the physiographic and topographic characteristics of the district lie in the succession of geologic events recorded in the stratigraphic column and the subsequent interpretation of that column (Table I).

The Birmingham district lies in the southern part of the Appalachian system and falls within two of the natural subdivisions of that system—the Appalachian plateau to the northwest and the Ridge and Valley province to the southeast (12, p. 274). Sand Mountain is located on the northwestern margin of Birming-

ham Valley and marks the boundary between the Ridge and Valley province to the southeast and the Appalachian Plateau to the northwest (Figure 2). Each subdivision is characterized by distinctive geomorphologic features controlled by the underlying rock types and by the geologic structure.

The plateau region is underlain by rocks that are essentially flat lying, although they have been disturbed somewhat by folding and faulting. Generally, the folding in the plateau region is open and shallow, and the structures encompass large broad areas. Faulting generally is confined to normal faults that are

TABLE I.   *Generalized Summary of Geologic Formations*

| Era | System | | Formation | Thickness (feet) | Description |
|---|---|---|---|---|---|
| PALEOZOIC | Carboniferous | Pennsylvanian | Pottsville Formation | 750 to 5,500 | Sandstone, shale, conglomerate and coal beds; thin- to thick-bedded. |
| | | Mississippian | Parkwood Formation | 900 to 1,300 | Shale, massive, with interbedded hard sandstone layers. |
| | | | Floyd Shale | 750 to 1,200 | Shale, black, soft, fissile, with interbedded sandstone and limestone beds |
| | | | Bangor Limestone | 0 to 300 | Limestone, bluish-gray, coarsely crystalline, thick-bedded. |
| | | | Hartselle Sandstone | 0 to 120 | Sandstone, white to tan, medium-grained, quartzose, locally friable. |
| | | | Gasper Formation | 0 to 125 | Shale and sandstone, dark gray, thin-bedded. |
| | | | Limestone unit of Warsaw Age | 80 to 150 | Limestone, blue-gray, coarsely crystalline, fossiliferous, cavernous. |
| | | | Fort Payne Chert | 90 to 140 | Chert and limestone, massive, iron and manganese stained bands, cavernous. |
| | | | Maury Formation | 1 to 3 | Shale, green to red, glauconitic, with phosphate nodules. |
| | Devonian | | Frog Mountain Sandstone | 6 to 22 | Sandstone, yellow to gray, fine- to medium-grained ferruginous, quartzitic. |
| | Silurian | | Red Mountain Formation | 300 to 500 | Shale, sandstone, limestone, thin- to thick-bedded, generally red to dark brown, with iron ore seams. |
| | Ordovician | | Chickamauga Limestone | 110 to 340 | Limestone, dark to light gray, fine-grained to coarsely crystalline, extremely fossiliferous. |
| | Cambrian | | Copper Ridge Dolomite | 800 to 2,000 | Dolomite, fine-grained, massive; chert, tough and compact. |
| | | | Ketona Dolomite | 150 to 900 | Dolomite, white to tan, crystalline, massive. |
| | | | Conasauga Formation | + 2,000 | Limestone, dark gray, thick-bedded, locally shaly. |

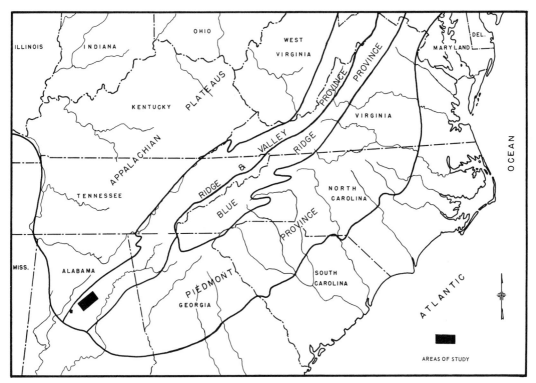

FIG. 2. *Physiographic Divisions of the Southern Appalachian System.*

arranged in an en echelon pattern and to thrust faults that are shallow and of slight throw. The rocks consist chiefly of shales and sandstones interspersed with conglomerates and coal beds. Extensive stream erosion has cut deeply into the rocks, forming large, broad, flat-topped ridges of various sizes and orientations. The ridges are separated by deep gorges and crooked valleys, and dendritic drainage patterns have been developed. Topographic relief from the base of the flat-topped ridges to their crests is as much as 300 feet (6, p. 12).

The ridge and valley region consists of distinctively parallel and gently arcuate land forms of northeasterly trend that reflect the underlying rock types and subsurface structures. Folds are close and generally limited in areal extent, and they too are long, narrow, northeasterly trending forms. Thrust faults are common and of large magnitude; at some places displacement is as much as several thousand feet. Normal faults and strike-slip faults generally are of limited extent with throws that are from a few to as much as several hundred feet (17). The ridges are typical hogbacks with steep northwest escarpments and gentle southeast dip slopes. The flat, gently

rolling valley floors are underlain by beds of limestone and chert. Topographic relief from the valley floors to the crests of the ridges ranges from 100 to 200 feet to as much as 1000 feet. The drainage system has adjusted itself to the structure, and the resulting trellis-type drainage pattern is typical of an area of strongly folded and faulted rocks.

The plateau region is drained by the Locust and Mulberry forks of the Black Warrior River and their tributaries. The ridge and valley region is drained by the Cahaba and Coosa River systems. The major river systems of both regions flow in a southerly direction and eventually merge south of the district and empty into the Gulf of Mexico.

## MINING

Mining activity in the past was confined generally to "slope mines" opened by driving slopes that followed the iron-ore seam down dip from the outcrop. There were, however, two shaft mines, the Shannon and Pyne. The Shannon mine was abandoned about 1930, but it had an inclined shaft that penetrated the iron-ore seam at a depth of 2100 feet beneath the surface. The Pyne mine, the only red-ore

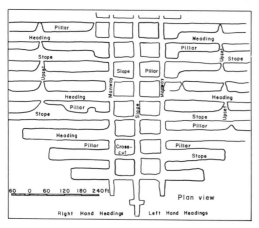

FIG. 3. *Typical Plan, showing development for a red-ore mine along the upper slope.*

roof support, have given way to roof bolting systems.

Normal development work is termed "first mining," and generally 40 to 60 per cent of the ore is left to support the roof. The extraction of ore from these pillars is the last stage of mining in a given area and is termed "second mining." Pillar extraction is started at a point farthest from the main haulageways and slopes and retreats toward the shaft. About 70 to 80 per cent of the original ore has been removed after "second mining" (Figure 5).

The Pyne mine has been completely mecha-

mine at present operating underground, has a vertical shaft and penetrates the ore seam at about 1200 feet below the shaft collar.

As slope mining proceeded down the dip, developments on the upper slope were opened to the right and left of the main haulageways and upsets were driven upward and a stope opened. Pillars were left as support as required by the roof and wall conditions (Figure 3).

As mining methods and machinery improved, ore washing and concentration also improved, and the development lent itself to panel systems on the "middle" slopes and "bottom" slopes (Figure 4).

Wooden cribs and stulls, although occasionally still employed for local supplementary

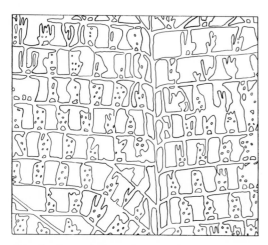

FIG. 4. *Typical Plan, showing development for for a red ore-mine along the middle and bottom slopes.*

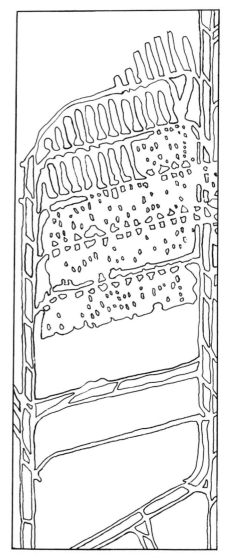

FIG. 5. *Pyne Mine Plan of Development, showing pillar layout after "first" and "second" mining.*

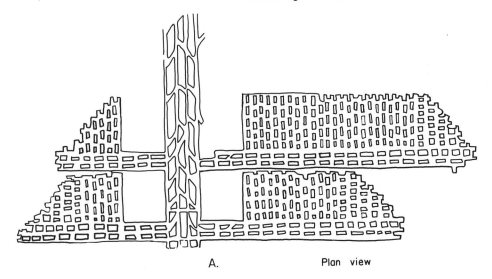

A.               Plan view

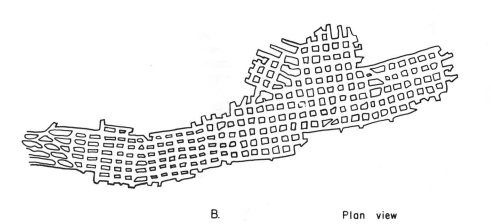

B.              Plan view

FIG. 6. *Plan View of Pyne Mine, showing: (1) development on foot wall side of the Shannon fault and (2) development on hanging wall at the Shannon fault.*

nized to accommodate electrically operated mechanical loaders and shuttle cars. A typical layout of mine development in two parts of the Pyne mine is shown in Figure 6.

## Vance Strip Mine

A description of the Woodward Iron Company Vance operation is presented at this point because the location is considerably southwest of the major Birmingham district, and this surface operation is smaller than the underground system developed at the Pyne mine property. Geologically and mineralogically, the rock formations and ore are the same at the two prop-

erties except for the thin veneer of upper Cretaceous rocks overlying the ore-bearing Red Mountain Formation in the Vance area.

The Vance strip mine is located 35 miles southwest of Birmingham (Figure 1). The Red Mountain Formation strikes generally northeasterly, and the main structural features, where observable, consist chiefly of faulted synclines. To the southwest of the open pit, the Coker Formation of the Tuscaloosa Group of Cretaceous age unconformably overlies the Silurian Red Mountain Formation (Figure 7). The Tuscaloosa Group consists of sands, clays, and gravels.

The ore mined from the Red Mountain For-

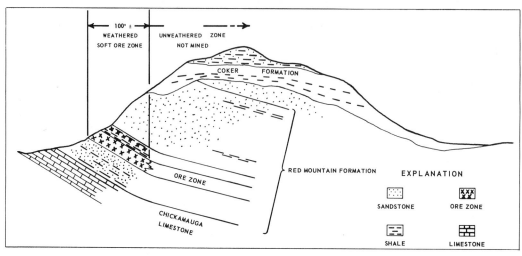

FIG. 7. *Generalized Section near Vance, Alabama.*

mation occurs in the weathered outcrop and averages about 10 feet in thickness and about 100 feet in width. It consists of a poorly cemented clayey-hematitic sand. Down dip from the weathered outcrop, the ore becomes harder and well cemented with the phosphorous and lime contents increasing as it does so.

The Big Sandy Iron and Steel Company sank four slopes and mined the hard ore in this area between 1900 and 1909. Several thousand tons of ore from these mines were reported to have been shipped to furnaces in the Birmingham area. The Southeastern Coal and Iron Company began strip mining the ore in 1958, built a washer, and shipped the ore to furnaces in the Birmingham areas. In 1962, the Woodward Iron Company acquired the property and processes the ore for use in its furnaces at Woodward, Alabama.

Sands and clays are removed from the weathered outcrop area with a seven-yard drag line and pans. Sandstone overlying the ore seam is drilled, and the nine-inch holes are loaded with ammonium nitrate and fuel oil and detonated. The fractured sandstone is removed with the seven-yard drag line.

The soft ore in the outcrop is loaded into trucks by shovel or loaded by pans for hauling to the raw-ore stockpile at the washer. Ore from the stockpile is pushed by a bulldozer to the washing plant. The raw ore is passed through a double roll crusher to a scrubber and onto a vibrating screen. The fine part passing the screen is routed to a 42-inch classifier, and the coarser ore passes to an attrition machine and then to the 42-inch classifier. The

ore then passes through a 30-inch classifier to the washed ore storage bin.

From the storage bin, the ore is trucked to a railroad siding where it is loaded for shipment.

*Typical Analysis of Vance Ore*
*(Per Cent)*

|  | Iron | Calcium oxide | Insolubles | Phosphorous |
|---|---|---|---|---|
| Raw | 36.14 | 1.07 | 36.70 | 0.30 |
| Washed | 49.97 | — | 23.40 | 0.084 |

## GENERAL GEOLOGY

The rocks exposed in the district consist of about 15,000 feet of consolidated Paleozoic sediments ranging in age from Cambrian to Pennsylvanian and of a thin veneer of unconsolidated sand, gravel, and clay of Cretaceous age. The distribution of rocks in the Vance and Birmingham areas is generalized in Figures 8 and 9, respectively. The oldest rocks exposed are carbonate shelf-type sediments, and the youngest are chiefly unconsolidated sediments of alluvial origin. All of the rock units of Paleozoic age have been complexly folded and faulted into typical Appalachian structures.

### Summary of the Geologic History

The Birmingham red iron ore district is a small but integral part of the Appalachian sys-

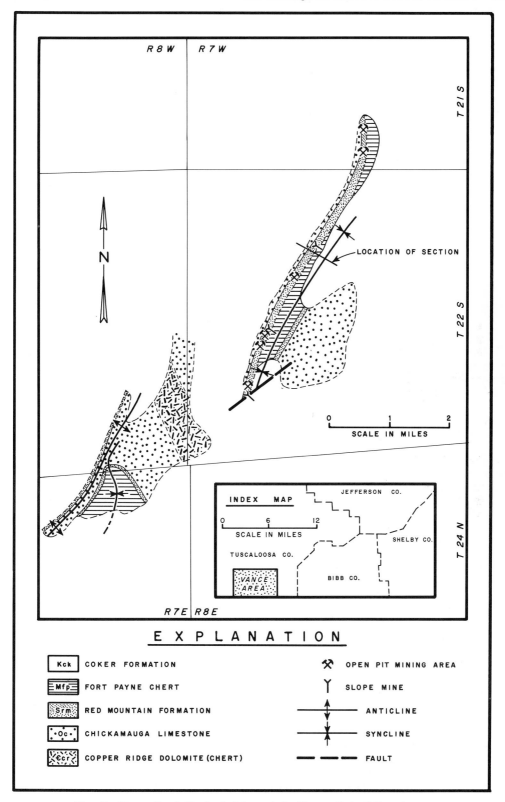

FIG. 8. *Generalized Geologic Map of the Vance Strip Mine Area.*

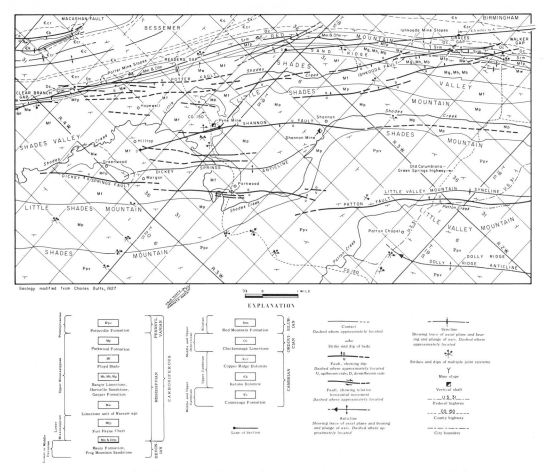

FIG. 9. *Generalized Geologic Map of the Birmingham Red-Ore District, Alabama.*

tem that extends from Alabama to New England. The southern Appalachian system is divided into four physiographic provinces as shown in Figure 2.

A generalized summary of the geologic formations exposed in the area is given in Table I, and the following brief discussion of the geologic history of the area is taken largely from Butts (8) and Adams, *et al.* (7, p. 41–230) and from the writers observations.

A geosyncline is believed to have existed about 500 million years ago in the area of the present Appalachian Plateaus, Ridge and Valley, Blue Ridge, and part of the Piedmont physiographic provinces. Its length is not known, but it is thought to have extended from Newfoundland to Alabama for 3000 miles and was about 100 to 500 miles in width. Bordering the geosyncline on the west was the stable interior of the North American continent and to the east lay a borderland of series of island

arcs, a main source area for the 30,000 feet of sediments that accumulated in the basin. During the filling of the basin, different periods of emergence and submergence occurred which resulted in unconformities and stratigraphic gaps.

Toward the close of the Paleozoic era, compressive forces developed in the crust that caused the marginal borders of the source area to shear and push westward up and over the strata that had accumulated to a great thickness in the geosyncline. As a result, the western and eastern borders of the Ridge and Valley province are marked by distinctive tectonic lines of thrust faults.

The sediments underlying the Appalachian plateaus are the unfolded or weakly folded parts of the geosyncline, whereas the sediments underlying the eastern borderlands are now belts of highly folded and faulted rocks.

Following the Paleozoic era, a long period

of time elapsed during which the continent underwent denundation by erosion. This erosional interval lasted through the Mesozoic era, rocks up through those of lower Cretaceous age being unrecorded in the district. Toward the latter part of the Mesozoic era, the southern part of the State began to sink, and the ocean gradually encroached upon the land. The Birmingham district was adjacent to, and immediately north of, the Cretaceous sea, and deltaic and continental sediments of Cretaceous age were deposited in the district. The eroded remnants of these sediments still are present in the area.

Deposits of Recent age consist of stream alluvium formed in, and adjacent to, present stream channels.

## Structural Geology

The structural features in the area are typical of those found in the Appalachian system. During the Paleozoic era and subsequent waning periods of structural development, the essentially flat-lying strata were extensively folded and faulted. The degree of folding and faulting was dependent upon local stratigraphic and initial fracture control, and the results are vividly portrayed in prominent surface features and well-defined structural trends.

Burchard, et al. (6, p. 21–25), and Butts (8) recognized and described many of the geologic structures in the Birmingham district. Subsequent investigations have produced few modifications of the original work except for additions to, and more detailed delineations of, some of the structural features. The essential flat-lying sedimentary strata that had accumulated in the Appalachian geosyncline were folded into narrow parallel folds. The northwestern and southeastern margins of the Ridge and Valley province are marked by distinctive tectonic lines, the northwestern margin defining the outer limit of Appalachian overthrusting (Figure 1). This Ridge and Valley tectonic zone of overthrusting and associated structural elements is suggestive of superficial folding (decollement) above a basal shearing plane. Rich (11) suggested this means of deformation for the Cumberland thrust block in Tennessee. The shearing plane may have been in the Precambrian basement rocks or along a plane in one of the incompetent shaly beds of the Conasauga Formation.

FOLDS   There are two folds well developed in the district, the Dolly Ridge anticline and the Dickey Springs anticline (Figure 9).

The Dolly Ridge anticline, or Aldrich dome (9, p. 115), is an asymmetrical structure in the eastern part of the area; the beds on the southeast flank dip more steeply than those on the northwest flank. The anticline has a northeast strike and plunges sharply to the southwest. A high-angle reverse fault, the Patton fault, shears the formation along the northwestern flank of the anticline and runs parallel to the anticlinal axis.

The Dickey Springs anticline in the vicinity of Greenwood and Morgan (central part of Figure 9) is not visible on the surface but can be delineated from diamond drill-hole logs and mine openings in the Pyne mine. The southeastern end of the anticline has been sheared by a high-angle fault. The scarp formed by the fault stands out in bold relief against the surrounding terrain.

Butts (8, economic geology map) indicated that the axis of the Dolly Ridge anticline extended much farther to the southwest, with folds dying out near the Morgan-Greenwood area. Butts also showed the axial trace of a syncline farther to the northwest parallel to the axial trace of the Dolly Ridge anticline, which he termed the Little Valley Mountain syncline. Subsurface data show that the Dolly Ridge anticline merges into a sharp monoclinal structure not visible on the surface. The southwest extension, which Butts labeled the Dolly Ridge anticline, is in fact the Dickey Springs anticline, which plunges to the northeast and attenuates in the vicinity of Shannon. This fold is easily traced in the openings in the Pyne mine.

FAULTS   The district is characterized by three distinctive parallel and subparallel slightly arcuate series of faults. These series have been termed the Ishkooda-Potter, the Shannon, and the Dickey Springs-Patton fault systems, respectively.

*Ishkooda-Potter Fault System*   The Ishkooda-Potter fault system strikes generally N40° –50°E and is made up of high-angle gravity faults (Figure 9). The Ishkooda mine slope on Red Mountain in the northeastern part of the area mapped and the Potter mine slope in the northwestern part of the area mapped both penetrated the fault system. A major feature of this fault system is a well-developed graben formed between some of its major fault components. There are many smaller faults in the rocks, which overlie the underlying mine openings, that are not visible on the surface or in the mine workings but have been inferred from diamond drill-hole cores and logs.

The faults of this system cut obliquely across Red Mountain in the northeastern part of the area and form a slightly arcuate pattern, convex to the southeast. They then cut obliquely back across Red Mountain in the southwestern part of the area. The faults are parallel and range in throw from several feet to as much as 400 feet, and the dips of the fault planes change greatly from place to place.

*Shannon Fault System*   The Shannon fault system, in the central part of the district, consists of one large normal fault and several smaller subsidiary faults (Figure 9). The Shannon fault strikes about N50°E, and its trace is marked by a sinuous line across a basal sandstone unit of the Parkwood Formation. The Shannon fault has been penetrated and crossed by headings in the Pyne and Shannon mines. Large amounts of water under high pressure were discovered in the upthrown and downthrown blocks of both mines as development work penetrated areas contiguous to the fault (Figure 9). The dip of the fault plane exposed in the mine averages about 60°SW. The throw ranges from about 100 feet in the Pyne mine to about 400 feet in the Shannon mine. The probable extension of the Shannon

fault to the northeast is buried beneath the talus deposits of the Shades Mountain escarpment.

Slip or shear planes within the upthrown block of the Shannon fault occur about 150 feet northwest of the main fault. Within the downthrown block, shear planes occur about 50 feet beyond the main fault. The Shannon fault zone is filled with breccia and gouge produced by drag (Figure 10).

Other faults with throws of 20 to 80 feet have been penetrated and exposed in the Pyne mine. These small faults parallel the trend of the Shannon fault and occur east of the main fault. The displacements die out upward and are taken up by folds in the overlying rocks before the faults reach the surface. Recent exploratory drilling in the Pyne mine area indicates the existence of at least one major fault zone.

*Dickey Springs-Patton Fault System*   The Dickey Springs-Patton fault system is exposed in the southeastern and southwestern parts of the area and is composed of a large normal fault and two reverse faults with associated smaller faults.

The system derives its name from the high-

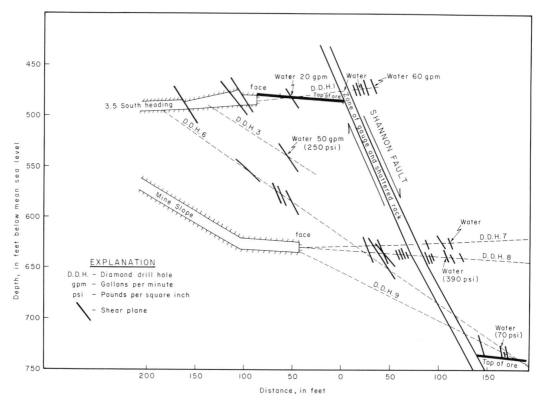

FIG. 10. *Cross Section, looking northeasterly, perpendicular to the Shannon fault in Pyne mine.*

angle reverse Patton fault exposed in the north bank of a cut along the Old Columbiana-Green Springs highway near Patton Chapel in the NE1/4 NW1/4, Sec. 1, T19S, R3W and from the high-angle normal Dickey Springs fault identified by Butts (8) in the vicinity of Morgan and Greenwood.

The Patton fault has a general strike of N50°W, and a dip of 58°NW. The throw of the fault attenuates to the southwest of its surface trace and dies out in the NE1/4 NE1/4, Sec. 21, T19S, R3W. A maximum throw of about 400 feet occurs in the NE1/4, Sec. 1, T19S, R3W. In the SW1/4 NW1/4, Sec. 11, T19S, R3W, the Pottsville Formation has been extensively shattered and fractured as the result of movement on this fault.

JOINTS   All formations in the district are cut by conspicuously parallel and subparallel joint sets in diverse orientations. Some of the sets are much weaker in expression than others. At most localities, the joints appear to be part of a conjugate system of two or more sets.

The lineal extent of most of the joints ranges from a few feet to as much as several hundred feet. The sets are arranged in narrow zones, and, at the terminus of each trace, another joint is slightly offset and runs in a parallel or subparallel direction to the previous one.

The angles of dip of the joints commonly range from 70° to 90°, but a few northwest dips are as low as 30°, and some southeast dips are as low as 10°. However, the low-angle dips are very rare.

Joints in plan for any set commonly will intersect at various angles without any obvious consistency. Joints in section of the same, as well as different sets also will intersect without any obvious consistency.

The joints in thin-bedded rocks generally do not extend vertically across layers of adjacent strata but are confined to the one bed. This is not always true for thick-bedded rocks in which some of the joints extend vertically across boundaries of such beds into adjacent strata for indeterminate distances.

Spacings between the joint traces in beds less than 3 feet thick range from 3 to 14 inches and the spacings of joints in beds greater than 3 feet thick range from 1 to 30 feet. Most of the joint faces are straight and well-defined, but, occasionally, the weaker sets display curved surfaces.

Joints observed at underground mine openings in the "Big Seam" of the Red Mountain formation generally are uniform and evenly spaced and serve as guides for mine develop-ment work. When the ore is blasted, a straight vertical face is left at the juncture of the joint trace and block of unmined ore. Further observations underground disclose that the frequency and width of openings between the joint surfaces increase near folds and faults.

Because of the distinctive regional parallelism and continuity, there is little doubt but that these joints were produced by horizontal compressive forces acting in the earth's crust from a generally southeast direction.

ORIGIN OF THE FAULT SYSTEMS   Continued application of the compressive forces that caused the regional jointing overturned the Birmingham anticline to the northwest with the eventual shearing of the formations on the northwest margin along a low-angle thrust plane. Thus, the origin of the three series of fault systems in the area can be attributed to an elastic-release mechanism initiated when the major thrust plane developed along the northwest flank of the Birmingham anticline. When the thrust plane developed, the regional compressive stress ceased to operate and locally induced stress fields were dissipated by the development of the three series of fault systems. The slightly arcuate forms of the fault systems could have been the result of an overextended frontal lobe on the upper plate of the overriding thrust block. After the compressional release stress readjustment occurred, the lines of faulting so developed took on a slightly curving pattern concave to the northwest.

The consistency of angular relation and parallelism of the joint traces in plan at each locality of joint measurement indicate that the joint sets probably were part of a conjugate shear system. Whether or not there is more than one system present cannot be determined from the field study. It seems likely that at least one conjugate shear system would develop concurrently with the folding. It is not possible to tell from field studies, at least with any degree of confidence, whether the joints were caused by tensional or shearing forces. Some joints could have been opened by a component of shear oblique to the plane of fracture, as well as by a component normal to the plane of fracture.

The weaker sets and increased number of joint sets contiguous to large folds and major faults can be attributed to elastically induced local stress fields. This accounts for the diversification of joint trends in some parts of the area.

It may be concluded that the major structural elements were developed by regional, essentially horizontal compressive forces from a southeast direction. It appears also that the joint system was capable of exercising some degree of control on the direction of fault systems, as shown by coincidence of joint and fault alignment. This initial or primary stress field has been termed the "F₁" force on Figure 11.

The possibility of a secondary and later stress field is also postulated. This stress field has been termed the "F₂" force on Figure 11. The joints formed by the two different stress fields are overlapping, owing to the diversification of the stray joint sets at each locality of measurement.

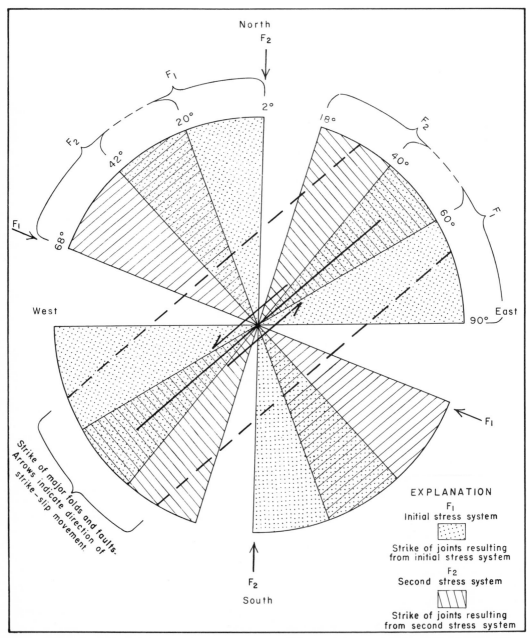

Fɪɢ. 11. *Stress Model Diagram, showing strike of joints, folds, and faults in the Birmingham red-ore district.*

The south compression postulated would cause strike-slip movement along pre-existing fractures as the result of a shear couple. However, the amount of slip could not be determined.

## Ore-Bearing Red Mountain Formation

The name Red Mountain Formation was restricted by Smith to the Silurian sequence in Alabama. This sequence contains strata of Niagara and Medina ages, possibly correlating with the upper and lower Llandovery (1). Butts (6, p. 14) referred to the Silurian rocks and described the Clinton (Rockwood) Formation as being approximately the equivalent of the Clinton Formation of New York. In this connotation, the term has been used by geologists for many years and the "Clinton ores" of Alabama are frequently referred to by that nomenclature.

In the vicinity of Vulcan Park in Birmingham, the Red Mountain Formation is about 260 feet thick and consists of crossbedded lenticular sandstone and shale beds of various thicknesses. There are three seams of hematite iron ore: the Ida or "Hickory Nut" Seam, the Big Seam, and the Irondale. The middle or Big Seam, the thickness of which ranges from 3 to 30 feet, is economically the most important.

In the Vulcan area, the Irondale Seam consists of about 5 feet of ferruginous sandstone and is the uppermost stratum of early Silurian age in the district. The Irondale is successively overlain by about 4 feet of "flat" pebble conglomerate and by the Big Seam of middle Silurian age, which is about 18 feet thick. The Ida Seam is not mined in the district at present (1967).

Southwest of the Pyne mine (Figure 9), the flat pebble conglomerate lenses out, and the Big Seam is in contact with the Irondale Seam. In the Pyne mine area, the conglomerate is approximately 7 inches thick. Northeast of the Pyne mine, near Vulcan Park, the conglomerate is approximately 4 feet thick; and, to the northeast between Irondale and Springville, 20 feet or more of shale separate the Big and Irondale seams and indicate a lithologic facies change.

The imbricate flat pebbles average about 1 inch in thickness. The length and width of pebbles differs locally. The maximum size is in the Pyne mine area where some clasts have a long dimension of 20 inches. Generally, the pebbles become gradually smaller to the north and northeast.

The lithology of the pebbles in most areas is the same as that of the thin ferruginous sandstone lenses underlying the Irondale Seam. The cementing material of the pebbles differs locally; most commonly it is ferruginous shale and fine-grained sandstone. Near Bessemer and Vulcan Park, the conglomerate is composed of flat limestone pebbles that possibly were derived from the underlying thin-bedded limestone of Ordovician or Cambrian age.

Apparently, there was a local uplift southeast of Bessemer that resulted in the erosion of lower Silurian shale and thin-bedded ferruginous sandstone, producing the rounded flat cobbles and pebbles of this conglomerate. Figure 12 shows a generalized stratigraphic section in the Birmingham-Bessemer area. The locations of the drill holes are shown in Figure 9.

## Mineralogy of the Ore

The Clinton iron ore beds, referred to as red ore, occur in the Red Mountain Formation of Silurian age. The two main textural and compositional varieties of the ore are known as fossil and oolitic. The mineralogy is amorphous hematite mixed with minor amounts of limestone, dolomitic limestone, quartz, clay materials, and other minerals in lesser quantities.

The ore occurs in lenticular beds, showing extreme textural variations and cross-bedding, and generally is intermixed with sandstone, shale, and limestone and is intercalated with these rocks.

The fossil ore consists chiefly of aggregates of fossil remains in such forms as bryozoa, crinoids, corals, and brachiopods. In some instances, the fossils themselves have been replaced by iron oxide, and, in many cases, there is no calcium carbonate. Much of the fossil material has been broken into fragments, but, in several areas, the fossils are well preserved and easily distinguishable.

The oölitic ore consists chiefly of aggregates of flat grains with rounded edges somewhat the size and shape of flax seed (6, p. 26). The nucleus of each oölite consists of quartz upon which successive layers of iron oxides and very thin layers of silica and some aluminous material have been deposited. In a part of the district, the fossil ore predominates, but, in other parts, the oölitic ore is more abundant In some instances, they are mixed together.

Lindgren (10) lists four well-defined types of primary ore as occurring in the Birmingham district: (1) fine-grained, pebbley conglomerate or sandstone, each pebble or grain coated

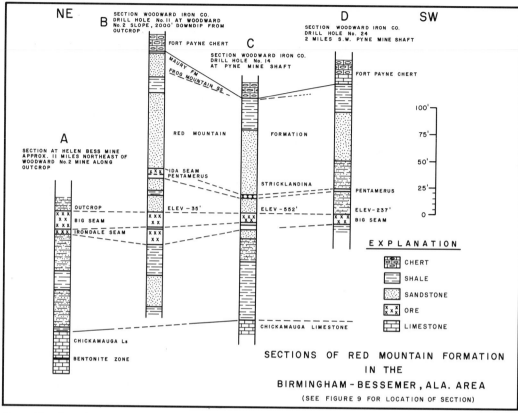

FIG. 12. *Correlation of Columnar Sections along the Strike, Birmingham Red-Ore District, Alabama.*

with hematite and the rock cemented with calcite and hematite; (2) fossil fragments, mainly of bryozoa, shells, and trilobites, partly coated and replaced by hematite, with abundant hematite oölites (usually each with a sand grain center) and hematite cement; (3) all-hematite oölites in a calcite matrix—the oölites average 1 to 2-mm in diameter; and (4) small, flattened concretions of hematite with fragments of fossils replaced by hematite and with hematite cement—known as flax-seed ore and very common in the district. The various ores contain only a little siderite, but fragments of quartz and other minerals are common. The beds differ markedly along strike and dip in calcium-carbonate and silica content.

In the earlier days of mining, the terms "soft" and "hard" ores commonly were used to denote low or high calcium carbonate content, respectively. The "soft" ore had been leached of most of its lime by weathering, whereas the "hard" ore contained lime in amounts sufficient to be termed "self fluxing." At first, it appeared that the iron content de-

creased about one per cent for each 100 feet down the dip, however, this is valid at only certain points on the slopes. After an iron content of about 35 per cent is reached at depth, the ore maintains a rather constant composition throughout the district.

## Chemical Composition

In general, the major constituents of the iron ore are as follows:

|  | Per Cent |
|---|---|
| Metallic iron | 32 to 45 |
| Calcium oxide | 5 to 20 |
| Silica | 2 to 25 |
| Alumina | 2 to 5 |
| Magnesium oxide | 1 to 3 |
| Phosphorous | 0.25 to 1.5 |
| Sulfur | Trace to 0.5 |
| Water | 0.5 to 3 |
| Manganese | Less than 0.25 |

Specific gravity of the ore ranges from 2.93 to 3.56, depending upon composition and textural characteristics. Factors used in calculating tonnage have ranged from 183 pounds per cubic foot with a volume of 12.25 cubic feet per long to 225 pounds per cubic foot with 10 cubic feet per long ton.

## Origin

Three opposing theories have been advocated to explain the origin of these Birmingham red ores:

(1) *Original deposition (sedimentary):* The ores were formed essentially contemporaneously with, and in the same rock sequence as, the associated limestone, sandstone, and shale, the ore, therefore, being one of the rock types deposited in the basin.

(2) *Residual enrichment:* The ore is thought to have been developed by groundwater attack on ferruginous limestone, the lime having been leached out above the water table, leaving the insoluble (iron-rich) part of limestone in a concentrated body thinner than the bed from which it was derived.

(3) *Replacement:* The ores are of much later origin than the rocks containing them, having been formed by the replacement of lithified beds of limestone by iron brought in by percolating waters.

Since the early development of these theories, much information bearing on them has been gained by deeper mining in the district.

Exploration drilling to determine the lateral extent and depth to the ore horizon farther to the southeast has shown that the iron content does not decrease with depth, as some early workers thought but, rather, remains fairly consistent except for isolated pockets of lower-grade, fine-grained ores. The content of metallic iron in the primary ore ranges from about 32 to 36 per cent down to depths as great as 2800 feet below the land surface; such depths are reached from 3 to 5 miles down from the outcrop along Red Mountain.

The residual theory apparently was the result of an original error of observation and one of interpretation as has been pointed out by Eckel (6, p. 28–31). Porter (2, p. 189) made a statement which Eckel (6, p. 28–31) quotes, that "lime [meaning calcium carbonate] increases in the red ore as it is mined downward . . ." What Porter's table (that follows the quotation) shows is that there is an abrupt change in CaCO$_3$ content 250 feet below the surface from a trace to between 21 and 37 per cent. This change, of course, marks the boundary between soft ore (above) and hard ore (below). Russell, however, put a meaning on Porter's statement that he (Porter) could not have intended and says (3, p. 22–23) that "the outcrops of the beds are soft, porous, highly fossiliferous ore . . . [that] is easily worked and easily smelted. The ore . . . [at] a depth of about 250 feet, measured down the slope, changes to a hard, compact ferruginuous limestone, rich in fossils. The marked difference in the character of the ore [as supposed by Russell] in the upper portions of the mine as compared with that from the lower . . . is due entirely to weathering." Russell goes on to give an analysis of the hard "ore" as containing 7.75 per cent Fe, 46.75 per cent CaO, and 34.90 per cent CO$_2$. Where Russell got a specimen of hard ore that contained less than 8 per cent iron is incomprehensible since the hard ore averages well over 30 per cent iron.

It is easy to see, however, how Russell (accepting this 8 per cent as the iron content of the hard "ore") would have concluded that, below the water table, the hard ore was actually only a ferruginous limestone and that the reserves of ore below the level of drainage (the water table) were negligible. This mistake, in itself, would have done no great harm had it not been taken uncritically from Russell and used as the basis of such statements as Kemp's (4, p. 114) that "the [Birmingham] ore in many places is really a highly ferruginous limestone, and below the water table in the unaltered portion if often passes into limestone, while along the outcrop it is quite rich." Thus, Russell's errors of observation (identifying 7.75 per cent iron bearing limestone as hard ore) and of interpretation (producing the soft ore from the weathering of this ferruginous limestone) were adopted by an eminent authority and sanctified in print. This error was repeated by Ries in 1905 and, only by Eckel's work (6) was it conclusively demonstrated to be false.

Russell concluded that an increase in thickness from 2.5 feet in soft ore to 3.0 feet in hard ore meant that the soft ore, after removal of calcium carbonate in solution, would have been porous enough to make its compression to a 2.5-foot thickness possible while the 3.0-foot hard ore remained unaffected by the earth forces involved (whatever they may have been). Eckel (6, p. 32) has pointed out that 20 feet of ferruginous limestone, containing 5 per cent iron before leaching, would have been needed to produce 2.5 feet of soft ore and that the thinning would have caused structural effects that never have been observed. Certainly, had such large-scale leaching occurred, the thick limestones overlying the ore seams would, through the development of solution-enlarged openings, have shown slump and collapse structures not evident in the district.

To support the sedimentary origin of the ore, it can be pointed out, for example, that the abundant oölitic ore appears to have been deposited in a near-shore environment in which the elliptical semi-rounded quartz grains were only slightly agitated by current action. Such conditions permitted deposition around the quartz grains of concentric rings of iron oxide derived from the large amounts of iron oxides and hydroxides (possibly in the colloidal state) present in the Clinton sea. In time, the particles accumulated to form fairly thick layers of iron-bearing material; these layers later were compacted and cemented by additional iron oxide.

The fossil ore, on the other hand, shows a preponderance of calcareous fossil fragments that are replaced only in part by iron oxide. If the ore had been formed entirely by diagenetic replacement of the carbonate sediment, secondary replacement of fossil remains should have been as complete as that of the inorganically precipitated calcium carbonate. Instead, it appears probable that much of the hematite was a primary deposit from sea water even though some diagenetic replacement, particularly of fossil fragments, did occur. Had the origin of the red ore been entirely by diagenetic replacement, however, the iron content of the ore seams should have been more regular than it is.

The hematite in the Red Mountain formation appears, therefore, to have been: (1) precipitated directly from sea water or (2) deposited by replacement of calcium carbonate during the diagenetic stage of the sedimentary process. The presence of dwarfed forms of well-known fossils in the iron ore of the Red Mountain formation strongly suggests that the iron was present in the sea water in sufficient amount to affect the growth rate of organisms that secreted calcium carbonate shells; but the iron probably was not added to these shells until the organisms producing them had died and the shells had fallen to the sea floor. Similarly, the hematitic oölites may have been normal calcareous oölites with a silica nucleus when they reached the sea bottom and only later, through replacement reactions, did iron oxide replace to the calcium-carbonate outer portions of the oölitic structures. Thus, some of the hematite in the deposits may have been

introduced as replacements of chemically, rather than organically, precipitated calcium carbonate. It is possible, however, that some, or even all, of the oölites may have been formed with a hematite outer shell and were not affected by replacement processes at all. It seems certain, however, that at least the hematite cement found in Lindgren's ore types (1), (2), (4) was a direct precipitation from sea water and not a replacement of an original calcium-carbonate cement. The all-hematite oölites of his ore type (2) also probably were formed by direct precipitation. It is not definitely known whether small amounts of siderite in the ore were produced by replacement of calcium carbonate or by direct precipitation. In short, the Clinton seas, at certain times, carried a larger content of iron than the seas normally did after the Precambrian, containing enough iron both to effect the direct precipitation of hematite and to drive forward the replacement reactions already described. It is likely that the iron reached the sea in the ferrous form but that, probably prior to its deposition from oxygen-rich, agitated, near-shore ocean water, it was almost entirely oxidized to the ferric state. Castaño and Garrels (13) have demonstrated that, under proper conditions of iron activity and pH and Eh, ferrous ions will be converted to ferric and precipitated as ferric oxide both directly as solid hematite and by replacement of solid calcium carbonate already present on the sea floor. As the hematite in the Clinton ores appears to have been formed both by direct precipitation and by replacement of solid calcium carbonate, it would seem probable that the activity of ferric ion and the Eh and pH of the Clinton seas on occasion approximated the experimental conditions set up by Castaño and Garrels.

It seems most reasonable, therefore, to conclude, in view of these arguments, that the Birmingham red ore is of direct sedimentary origin, modified only to a moderate extent by diagenetic replacement after deposition. The depositional environment of the ores was one in which shallow gulfs and reefs predominated. During the development stages, marine organisms and their detritus were worked and reworked by waves and currents in the presence of chemically precipitated calcareous and ferruginous material. The degree of saturation of the sea water in calcium carbonate and iron oxides necessary for the formation of the precipitates would have required a large-scale removal in solution of these constituents from the surrounding land mass, a result that would have been made more certain by a warm tropical or subtropical climate. These conditions of sedimentation and climate, together with transgressive and regressive movements of the shallow epeiric seas and the concomitant downward and upward movements of the adjacent land masses, would have provided an environment favorable for the deposition of the Birmingham red ores.

Sheldon's work (16, p. Al) indicates that the character of the ores was controlled in large measure by the spatial relations of their

were indicated by geologic conditions observed up to that time.

The U.S. Bureau of Mines published an extensive and exhaustive report on the future of the Birmingham district in 1953 (15). The estimates of reserves were made according to the following classifications: (1) virgin deposits, (2) idle mines, and (3) operating mines. These classes were further divided into three categories of recoverable commercial reserves, as follows:

|  | Indicated | Inferred<br>(Gross tons) | Total | Per Cent |
|---|---|---|---|---|
| Virgin deposits | 74,210,000 | 163,610,000 | 237,820,000 | 43.4 |
| Idle Mines | 18,538,000 | 6,860,000 | 25,398,000 | 4.6 |
| Operating mines | 187,613,000 | 97,054,000 | 284,667,000 | 52.0 |
| Total | 280,361,000 | 267,524,000 | 547,885,000 | 100.0 |

loci of deposition to the topography of the near-shore sea floor. Sedimentary structures in the ores, plus their lithology and stratigraphic and facies relationships indicate to him that a line of barrier islands, located immediately southeast of the site of ore deposition, was the major controlling factor in the ore-forming process. On the lagoonal side of these barrier islands, built by southwesterly longshore currents, the calcite-hematite facies was deposited. This ore facies graded seaward (farther to the southeast) into less oxidized siderite-chamosite and dark claystone facies. Toward the northeastern (landward) side of the barrier, the quartz-hematite sandstone of the islands graded, in turn, into marine non-ferruginous sandstone.

## RESERVES

Burchard (6, p. 133) estimated the red-ore reserves of the Birmingham district to be 358,470,700 long tons available under commercial conditions prevailing at that time. In addition, he estimated that an additional 438,426,100 long tons of potential ore were present. This made a total of 796,896,000 long tons of red ore as compared to a previous preliminary estimate by Eckel (5) of about a billion long tons for Alabama. These estimates were made when the geology and structural conditions in the subsurface were imperfectly known, and assumptions had to be made concerning the persistence of grade in depth and essential constancy of structure as these

Considerable interest was directed at this time (1953) toward the potential development of ferruginous sandstone lenses within, and overlying, the three major iron ore seams being mined in the district.

During the life of the district up to 1953, mining had been carried out only in the three main seams: the uppermost seam, the Ida; the Big Seam; and the lowermost seam, the Irondale; these had been mined in the district in different combinations depending upon the grade of ore contained in them in various locations. In 1953, however, neither the Ida nor the Irondale were being mined. The Big Seam is divided in the district into an upper and lower bench; the marker bed that forms the division ranges from a thin shaly pebbly parting to a shaly pebble conglomerate bed about 2 feet thick. In general, in mining northeast of the midpoint of the district (Figure 9), both the upper and lower benches were worked; whereas to the southwest, only the upper bench was worked.

Tentative overall reserve estimates made by Clemmons, *et al.,* (15), adding the marginal ore and ferruginous sandstone ore to the commercial ore, indicated a combined total of red ore reserves of about 2,160,469,000 long tons.

Mining life for the district was based on 7 million gross tons of production annually and was shown to range from 41 to 235 years, utilizing all classes and divisions at that rate of production.

Chapman (14), in a paper on the iron and steel industry in the South, showed recoverable ore and potential ore for the Birmingham district to be 1,855,000,000 and 2,680,000,000 long tons, respectively.

Considering these previous calculations and the decrease in production in the late 1950's and early 1960's, there appears to be an extensive reserve of iron ore available in the district.

## RETROSPECT AND OUTLOOK

The fortuitous circumstance of the occurrence of substantial deposits of iron ore, coal of coking quality, and fluxing stone within close proximity of one another continue to make the city of Birmingham and its surrounding environs one of the largest mining camps in the world. For more than 167 years, red ore, coal, and limestone have been produced from the mines of the district.

The growth of the iron and steel industry in the area has been marked progressively through the years by successively improved technological developments in geological exploration, mining methods, concentration and milling procedures, and blast furnace practices. For years, the district produced a low-cost pig iron that placed its product on a competitive basis with its northern counterparts. Little did the industry realize, when coke-pig iron was first produced in 1876, that, nearly a century later, blast furnaces of the distirct would be producing over 4 million net tons of basic and foundry pig, valued at more than 165 million dollars (U.S. Bureau of Mines, Minerals Yearbook, 1950). By the same token, little could they realize that by 1965, foreign ore would have created such inroads on the local red-ore mine production that 65 per cent of the ore consumed in the district would be foreign imports.

It would appear that technological advances and economics have, on a national and international scale, removed the competitive advantage of local ore. Although production of pig iron contributes over 235 million dollars to the Birmingham regional market at present, red-ore production has dropped from about 7 million gross tons, to about 1.5 million gross tons annually (U.S. Bureau of Mines, Minerals Yearbook, 1965).

It seems unlikely, as well as unreasonable, to believe that the red-ore reserves in the district will remain untouched. It has been forecast that, by the year 2000, the United States' consumption of iron ore will be at least 240 million tons, as compared to the 120 million tons at present. The known reserves, which constitute proven reserves, are, therefore, still

a major local reserve of iron for the mills of the district.

The Birmingham district red-ore reserves are available for exploitation but will remain in the ground until technological advances and economic conditions change.

## REFERENCES CITED

1. Smith, E. A., 1876, Report of progress for 1876: Ala. Geol. Surv., 100 p.
2. Porter, J. B., 1887, The iron-ores and coals of Alabama, Georgia, and Tennessee: A.I.M.E. Tr., v. 15, p. 170–218.
3. Russell, I. C., 1889, Subaerial decay of rocks and the origin of the red color of certain formations: U.S. Geol. Surv. Bull. 52, 63 p.
4. Kemp, J. F., 1903, The ore deposits of the United States and Canada: McGraw-Hill, N.Y. 481 p.
5. Eckel, E. C., 1906, A review of conditions in the American mining industry: Eng. Mag., v. 30, 527 p.
6. Burchard, E. F., et al., 1910, The iron ores, fuels and fluxes of the Birmingham district, Alabama: U.S. Geol. Surv. Bull. 400, 198 p.
7. Adams, G. I., et al., 1926, Geology of Alabama: Ala. Geol. Surv. Spec. Rept. 14, 312 p.
8. Butts, C., 1927, Description of the Bessemer-Vandiver quadrangles (Alabama): U.S. Geol. Surv. Geol. Atlas, Folio 221, 22 p.
9. Semmes, D. R., 1929, Oil and gas in Alabama: Ala. Geol. Surv. Spec. Rept. 15, 408 p.
10. Lindgren, W., 1933, The oölitic hematite ores: Mineral deposits, 4th ed., McGraw-Hill, N.Y., p. 273.
11. Rich, J. L., 1934, Mechanics of low-angle overthrust faulting as illustrated by Cumberland thrust block, Virginia, Kentucky, and Tennessee: Amer. Assoc. Petrol. Geols. Bull., v. 18, p. 1584–1596.
12. Fenneman, N. M., 1938, Physiography of eastern United States: McGraw-Hill, N.Y., 714 p.
13. Castaño, J. R. and Garrels, R. M., 1950, Experiments on the deposition of iron with special reference to Clinton iron ore deposits: Econ. Geol., v. 45, p. 755–770.
14. Chapman, H. H., 1953, The iron and steel industry in the South: Univ. Ala. Press, 927 p.
15. Clemmons, B. H., et al., 1953, The future of Birmingham red iron ore, Jefferson County, Alabama: U.S. Bur. Mines R.I. 4988, 19 p.
16. U.S. Geological Survey, 1962, Iron deposits in the Birmingham district, Alabama: U.S. Geol. Surv. Prof. Paper 450-A, p. 1.
17. Simpson, T. A., 1963, Structural geology of the Birmingham red-iron ore district, Alabama: Ala. Geol. Surv., Circ. 21, 17 p.

# 12. Geology and Ore Deposits of the Ducktown District, Tennessee

MAURICE MAGEE*

## Contents

* Tennessee Copper Company, Ducktown, Tennessee.

# Illustrations

# Tables

Ducktown
District

## ABSTRACT

The Ducktown ore deposits have been known, explored, and mined for 120 years. Eight massive sulfide ore bodies occur in highly folded and metamorphosed graywacke, graywacke conglomerate, mica schist, chlorite-garnet schist, and staurolite schist of Precambrian age. The ore deposits are tabular bodies that have been extensively folded, generally conform to the enclosing rocks, but show local complex differences.

The ore deposits, ranging in size from 250,000 tons to over 20 million tons, are composed principally of the minerals pyrrhotite, pyrite, chalcopyrite, sphalerite, and magnetite. Gangue minerals are quartz, calcite, actinolite, tremolite, hornblende, garnet, and masses of schistose wall rock. Three fault systems are present that have influenced ore control and ore body configuration. Hydrothermal effects are evident in wall-rock alteration. Evidence of retrograde metamorphism is extensive in the wall and ore associated rock. Preliminary isotopic age dates have indicated four possible metamorphic events during the Paleozoic Era.

Ore genesis is considered to be hydrothermal replacement of receptive beds, predominantly highly calcareous zones, quartzitic zones, and brecciated shear zones. Ore deposition is thought to have occurred during the Devonian Period, with later mobilization and recrystallization of the ore and gangue during Middle and Late Paleozoic times.

## INTRODUCTION

The Ducktown district is located in the extreme southeast corner of Tennessee. The zone of economic mineralization is entirely in Polk County, Tennessee, but minor mineralization extends southwest into Fannin County, Georgia. Locally, the Ducktown district is referred to today as the Copper Basin (the area is a physiographic basin), but, in mining and geological circles, it is still referred to as the Ducktown mining district, or Ducktown Basin.

The "Geology and Ore Deposits of the Ducktown district, Tennessee" of Emmons and Laney (4), published in 1926, is the most comprehensive record of the district. Concepts of structure, age, and ore genesis have evolved in the forty years since this publication. Also, considerable knowledge of several of the ore bodies has been developed since 1926. This report presents new information and ideas on the district, but the reader is referred to the above publication for other details.

## Acknowledgments

This report is a compilation from the studies of a number of geologists, many of whom are listed in the references. Thanks are due particularly to R. N. Diffenbach, Edward Swanson, and John Hill—members of the Tennessee Copper Company geology staff—for assistance in map preparation, a critical reading of the paper, and many helpful suggestions. Also thanks are due to H. F. Kendall, General Manager of the Tennessee Copper Company, and Owen Kingman, Chief Exploration Geologist for Cities Service Minerals Corporation for a critical review of the paper, to R. G. Clay, Superintendent of Mines of the Tennessee Copper Company, for his cooperation and assistance, and to the Tennessee Copper Company for permission to use Company resources in preparing this paper.

## HISTORY OF THE DISTRICT

The discovery of copper in the Ducktown district was first recorded (1,4) as 1843, when a prospector found native copper in a stream draining from the southwest end of the Burra Burra ore body. The first ore shipment (4) left the district in 1847, when 90 casks of ore weighing 31,000 pounds were hauled by mule to the railroad at Dalton, Georgia. Average grade of the first shipment was about 25 per cent copper. That same year, an iron furnace, built in an attempt to produce iron from the gossan (iron oxide) ore failed because of the high copper content.

The first mining companies were incorporated in 1852, and by 1855, more than thirty corporations were chartered (7). Most of these corporations did little more than prospect for ore, but several contributed significantly to the early development of the district. Records of actual production during this period are not complete. A boom in the development of the district followed construction of a wagon road

to Cleveland, Tennessee in 1853, and the building of smelters near the mines in 1854 and 1855.

In 1858, the Union Consolidated Company acquired several properties, and began mining the "black copper" ores from the supergene enriched ore zones. They also recovered copper from the mine waters and built two smelters. A copper refinery was built at Cleveland, Tennessee at about the same time.

Federal troops destroyed the refinery and rolling mill at Cleveland in 1863, and mining at Ducktown was suspended. The Union Consolidated Company resumed production in 1866, under the management of Julius E. Raht; other companies also resumed operations. By 1877, the high-grade "black copper" ores were exhausted; 24 million pounds of copper had been produced (4). Unsuccessful attempts were made to recover copper from the primary sulfides. Lower copper prices, depletion of high grade ores, depletion of the nearby wood for fuel, litigation concerning the several mining operations, and a lack of understanding of sulfide smelting technology all contributed to the closing of the mines. The smelters were shut down a year later.

The advancements in the art of metallurgy and the subsequent building of a railroad into the area, making available low-cost fuel and supplies, enabled the mines to return to production. In 1889, the Ducktown Sulfur, Copper, and Iron Company acquired the holdings of the Union Consolidated Company (4) and started mining and smelting operations of the primary sulfide ores for copper. Between 1890 and 1907, an estimated 1.5 million tons of iron ore were mined from the gossan outcrops of several ore bodies.

The Tennessee Copper Company was organized in 1899. It acquired the holdings of the Pittsburgh and Tennessee Copper Company, the Burra Burra Company, and the Polk County mine properties.

In 1904, open roasting of ore was replaced by the pyritic process of smelting. Sulfuric acid manufacturing started in 1907, which ended the loss of sulfur dioxide to the air. Milling and flotation processes started in the early 1920's, to treat separately the fines and the lumps of direct smelting ore. The flotation and separation of iron sulfide concentrates led to the manufacture of iron oxide sinter in 1925. The recovery of zinc sulfide concentrates started in 1927.

The Ducktown Sulfur, Copper, and Iron Company was reorganized as the Ducktown Chemical and Iron Company in 1925. Their operations somewhat paralleled those of the Tennessee Copper Company until 1936, when the final consolidation of properties took place—the Tennessee Copper Company acquired the holdings of the Ducktown Chemical and Iron Company.

## Present Operations

Today the Tennessee Copper Company, a division of the Tennessee Corporation, a subsidiary of the Cities Service Company, is the operator of all mines and production facilities in the Ducktown district. The Tennessee Copper Company produces annually more than 1,500,000 tons of sulfide ore from five mines. The ore is milled, and concentrates of copper, iron, and zinc sulfides and iron oxide occurring as magnetite are recovered. Small quantities of lead sulfide have been recovered during milling experiments and also a small amount in gold and silver is recovered from smelted copper. Products from the complex ores include metallic copper, iron oxide sinter, sulfuric acid, zinc concentrates, sulfur dioxide, copper sulfate, and a host of related chemical products.

## Tabulation of Production

Because of the 120-year history of the district with several operating companies and incomplete records, it is impossible to present a complete production table. Tabulated below is an estimate compiled from various sources. Much of the early ore produced was shipped directly to Wales and to smelters in the northeastern United States. Early production records were kept in pounds of copper produced, rather than tons of ore mined. Not until the early 1900's were comprehensive ore tonnage records kept. Also, records of the iron, sulfur, and zinc content of the ore were not tabulated until the 1920's.

## PHYSIOGRAPHY

The Ducktown district is located in a topographic basin with an area of more than 100 square miles. Elevations within the basin range from 1500 to 1800 feet above sea level; surrounding mountains rise to 4200 feet. The district is drained by the Ocoee River, a tributary of the Hiwassee River. The open roasting of ores and the cutting of wood for fuel before the turn of the century resulted in the loss of most of the vegetation within a 36 square mile area. A result of this devegetation is an area of almost completely denuded and severely eroded hills with deeply incised gullies. Parts of the area are being reclaimed by the planting of pine trees by the Tennessee Copper Company and grasses for pasture. Natural grasses and trees are returning in response to control of smelter gases.

The mountains around the district are heavily wooded, and the slopes and relief are typical of the southern Appalachian mountains.

Outcrops of the ore deposits tend to occur along resistant ridges that strike generally northeastward. These ridges are dissected in several places by minor streams. Drainage patterns are influenced somewhat by fault zones. Certain superposed stream courses evidently have been modified on encountering faults.

## REGIONAL GEOLOGIC SETTING

The map of the regional geology shown in Figure 1 is a compilation from several sources and the work of many geologists who have investigated the area. The Athens Shale and the Knox Group of dolomitic limestones of Ordovician age are 15 miles northwest of the mining district. These are separated from the underlying Chilhowee Group of early Cambrian age by the Great Smoky fault zone. The Chilhowee Group consists of sandstones, shales, and conglomerates and occurs as slivers in the Great Smoky fault zone. The Chilhowee

*TABLE I.   Chronology of Ore and Copper Production*

| Period | Product | References |
|---|---|---|
| 1847–1853 | 14,291 tons of ore with grade better than 25 per cent copper | 2 |
| | A small quantity of gossan iron ore | 1 |
| 1853–1862 | No reliable records known. Production essentially copper sulfides and oxides from the supergene zone | |
| 1862–1865 | Mines idle | |
| 1866–1879 | 24,000,000 pounds of copper | 4 |
| 1877–1891 | Mines idle | |
| 1896–1919 | Mary mine production 3,000,000 tons of 2.25 per cent copper | 28 |
| | An estimated 2,000,000 tons of gossan iron ore from Isabella-Eureka mine | 28 |
| 1831–1963 | 547,095 short tons copper recovered—total Tenn. (Bureau of Mines statistics) | 24 |

is underlain unconformably by the older Precambrian Sandsuck Formation of the Walden Creek Group. The Walden Creek consists of quartzites, phyllites, a narrow impure limestone, and calcareous shales.

The older Great Smoky Group is conformable to the overlying Walden Creek Group. The Great Smoky Group consists of interbedded metamorphosed graywackes, conglomerates, mica schists, quartzites, chlorite-garnet schist, and chlorite-garnet-staurolite schist. In Figure 1, the Dean and Hughes Gap Formations of Hurst (16) and Units 1 and 2 of Hernon (21), northwest of Ducktown, are shown in proper juxtaposition, but a correlation has not been proposed. The slaty unit shown in Figure 1 is a dark graphitic phyllite and crinkled mica schist and is probably younger than the Copperhill Formation. Hernon divided the Copperhill into Units A and B, Unit B referring to the upper (younger) section of more fine-grained metasediments.

An increase in metamorphic intensity and a greater structural complexity is noted in traversing the Great Smoky Group from west to east. Transitions through the chlorite, biotite, and garnet zones are noted northwest of Ducktown. The staurolite zone is first noted

in the hanging wall of the Burra Burra ore body, and kyanite has been found in the center of the Burra anticline about 2000 feet southeast of the first staurolite occurrence. It is interesting to note that the first occurrences of massive sulfides on the northwest of the mining district are approximately coincident with the staurolite isograd. A higher degree of metamorphism is suggested by the presence of kyanite in the wall-rock near some of the ore deposits. Carpenter (26) noted an abnormal curve in the kyanite isograd that isolates the Ducktown district from the surrounding area.

The mining district is located within the Copperhill Formation of Late Precambrian metasediments, consisting of interbedded graywackes, graywacke conglomerates, and mica schists, with sericite-biotite schists, chlorite-garnet schists, chlorite-staurolite-garnet schists, and biotitic quartzites conspicuous near the ore deposits.

There is an amphibolite sill 1 mile southeast of the district. This sill, considered an altered gabbro by Emmons (4), is generally conformable to the bedding and extends, with several breaks, for 12 miles southwest into Georgia. Several smaller amphibolite occurrences are near the Boyd and Eureka ore bodies.

TABLE II.   *Stratigraphic Sequence Across the Ducktown Region of the Great Smoky Mountains*

| Age | Series | Group | Northwest formations | Southeast formations |
|-----|--------|-------|---------------------|---------------------|
| Ordovician | | Chickamauga<br>Knox | Athens Shale<br>Undifferentiated | |
| Cambriam | | Chilhowee | Nebo Sandstone<br>Nichols Shale | |
| | | | | Nottely Quartzite<br>Andrews Schist<br>Murphy Marble — Murphy Belt<br>Brasstown<br>Tusquitee Quartzite<br>Nantahala Slate |
| | | | —————Disconformity————— | —————Conformable————— |
| | | Walden Creek | Sandsuck | |
| Precambrian | | | (16)        (21)<br>Dean       Unit 2<br>            Unit 1 | Dean |
| | Ocoee | Great Smoky | Hughes Gap   Slaty Unit | Hughes Gap<br>Hot House |
| | | | Copperhill   Unit B<br>            Unit A | Copperhill |

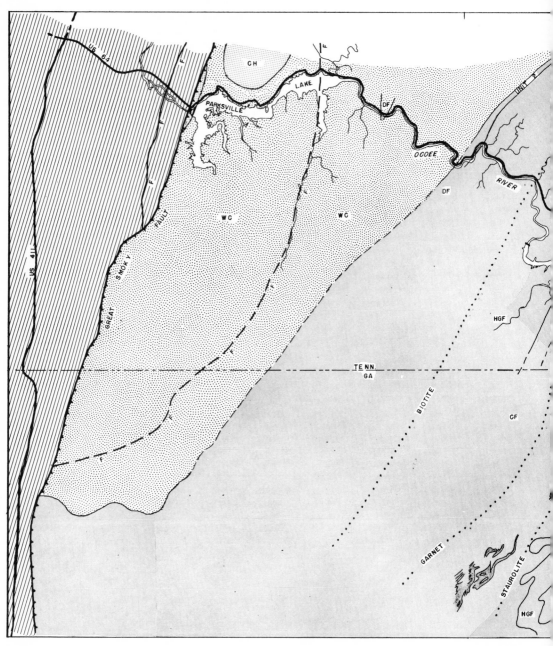

Fig. 1. *Regional Geologic Map of the Ducktown Area, compiled from the Geologic Map of
Hernon,*

These are the only evidences of igneous activity in the district.

Southeast of the mining district are the Hughes Gap, Hot House, and Dean Formations of Hurst (11) that overlie the Copperhill Formation. The formations are correlated by Hurst with chemically similar, but mineralogically (metamorphically) different, rock forma-

tions to the northwest of the mining district in a major anticlinorium 15 miles across. The Hughes Gap, Hot House, and Dean Formations are mineralogically similar to the Copperhill Formation. Boundaries of these formations are transitional and, in part, interpretive. To the southeast, and overlying the Great Smoky Group, are the Nantahala Slate and the rocks

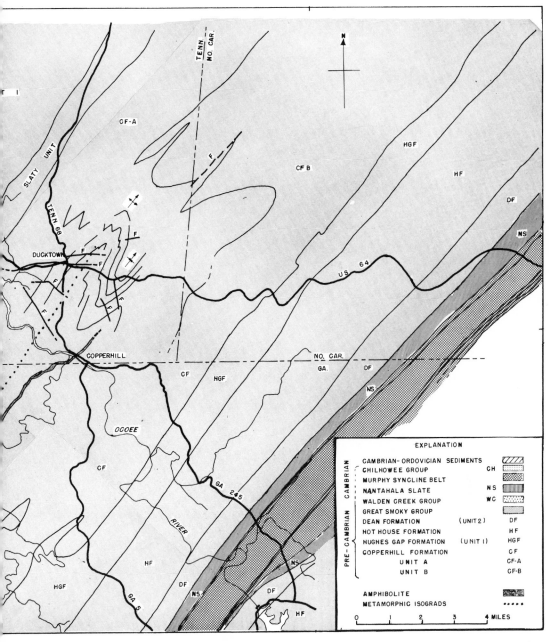

*Tennessee, 1966 (25); Tennessee Copper Company map, 1947; Hurst, V. J. (11, 12, 16); R. N. (21).*

of the Murphy syncline, which consist of quartzite, schist, slate, and the Murphy Marble. The base of the Nantahala is tentatively recognized as the top of the Ocoee Series (11,13). The equivalent relationships and correlations northwest of the Ducktown district are not established. The stratigraphic sequence of the above rock units is shown in Table II.

## DISTRICT GEOLOGY

The geology of the mining district has been mapped in detail by the geologic staff of the Tennessee Copper Company. Figure 2 illustrates the principal features. Structural section A-B (Figure 3) shows the relative positions of the ore deposits.

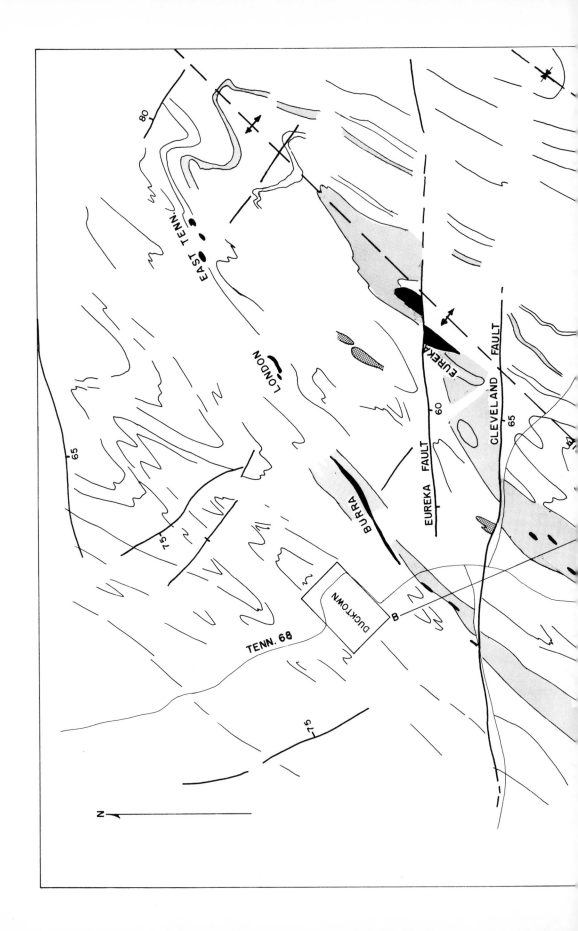

Fig. 2. *Generalized Geologic Map of the Ducktown District from the Tennessee Copper Company Map, 1947. Gossan widths are exaggerated to show their configuration relative to each other.*

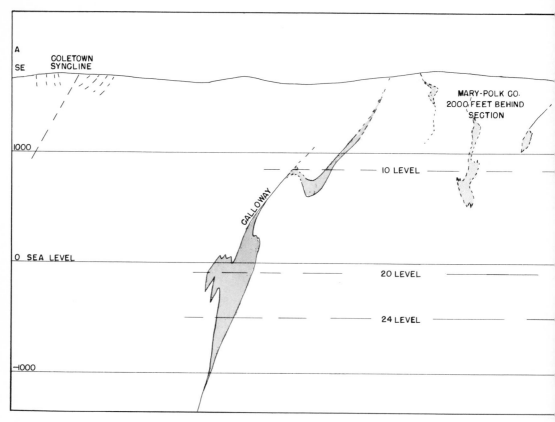

FIG. 3. *Structural Section A-B, looking southwest across the Ducktown district, showing the posi-hanging wall is overturned; along the*

There are no stratigraphic marker beds in the district, but there are characteristic schist beds, usually graphitic or staurolitic (Figure 2, bedding traces), that can be followed for distances up to 3000 feet along strike. The stratigraphic units (Figure 2, shaded) are the results of a detailed heavy mineral study by Snyder (30,31). These zones are composed of groups of interbedded graywackes and schists the ilmenite content of which is higher than the adjacent rocks.

### Structure

The area has undergone intense deformation and recrystallization. Structural complexities of folding and faulting are well exposed by the lack of vegetation in the area. However, the absence of stratigraphic markers renders a rigorous interpretation impossible.

The district is characterized by two major northeast plunging folds—the Burra anticline and the Coletown syncline. Smaller folds are superimposed on the major fold pattern. The general dip of schistosity and bedding is about 45° to 80°SE, and the general strike, N35°E. The Coletown synclinal axis is well defined, but the Burra anticlinal axis is not. Many of the smaller folds in the central and southwest portions of the anticline plunge to the southwest. Small domes and basins 100 to 1000 feet in length exist in the central anticlinal area. Such structures may have had some controlling influence in ore localization and configuration.

The internal structure of the Burra anticline may be a series of isoclinal folds, which cause repetitions of the same stratigraphic sequence. Figure 4 presents this interpretation. It is drawn to indicate that the ore deposits could occur at one stratigraphic horizon. Emmons (4) believed that all the ore deposits of the district occur in the same geologic horizon with repetition from faulting and folding. More recent work has indicated two, or possibly three, stratigraphic zones containing the ore deposits. The Burra Burra, London, and

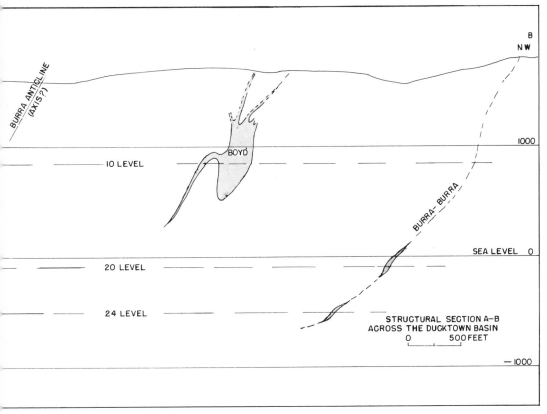

*tions of four of the ore bodies relative to each other. Graded bedding along the Burra Burra Boyd hanging wall, bedding is normal.*

East Tennessee ore bodies occur along the northwest limb of the Burra anticline; the Eureka, Boyd, and Cherokee deposits are near the center and Calloway and Mary-Polk County occur on the southeast limb of the Burra anticline. The Calloway and Mary-Polk County ore bodies possibly are stratigraphically equivalent to one of the above two groups, but structural complexity and absence of satisfactory stratigraphic markers have prevented definite correlation.

The schists are strongly foliated and crinkled. The angle of schistosity to bedding is generally 0° to 15° but may exceed 15° in the noses of folds; however, the foliation of schist beds sometimes wraps around folds. Two sets of minor folds or crinkles in the schist have been observed, with occasional third- and fourth-order folding superimposed. Hurst (11) presents a detailed discussion of these features. Grant (32) noted a general parallelism of one set of minor fold axial planes to some of the larger fold axial planes and a parallelism of strike and plunge of minor fold axes in schists

associated with ore folds in some of the deposits.

## Faulting

Three fault systems are evident (8). They are northeast, northwest, and east-west systems. The northeast, or strike, faults are thought to be pre-ore and possibly, in part, contemporaneous with ore deposition and were influential in localization of the ore. Shearing in this strike-fault system is evident in the mines but not easily recognized on the weathered surface. Fault breccia is observed in all of the massive sulfide deposits and along the walls of the ore bodies.

The northwest and east-west fault systems are considered post-ore. These cross faults have displaced several of the veins and segments of veins and influenced the final position of the ore deposits, but they evidently were not influential in ore deposition. The Eureka fault (east-west) has displaced the Eureka ore body horizontally by 350 feet. The Cherokee

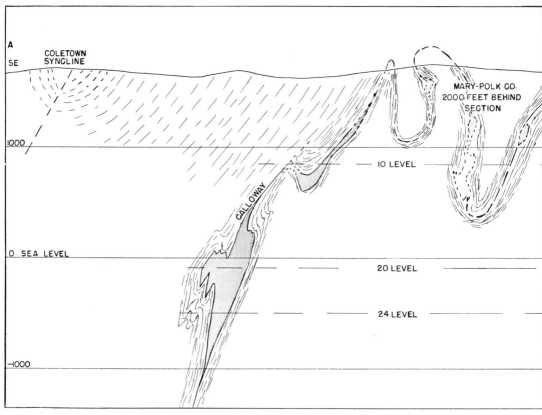

Fig. 4. *An Interpretation of Structural Section*

ore body has been offset horizontally about 250 feet by a northwest fault. The amount of vertical displacement along these faults is not known. The east-west faults consistently offset the north side to the east.

### Rock Types

The rocks in the district consist of metamorphosed Precambrian sediments composed of about 60 per cent metagraywacke, 30 per cent mica schist, and 10 per cent metaconglomerate, metaarkose, quartzite, chlorite-garnet schist, and chlorite-garnet-staurolite schist. These are described briefly here; more comprehensive descriptions have been presented in Emmons (4) and Hurst (11).

METAGRAYWACKE The metagraywacke is a light gray, usually fine-grained rock composed of 50 to 60 per cent angular-to-subrounded quartz, with feldspar, biotite, and muscovite, and minor calcite, garnet, ilmenite, zircon,

sulfides, and other accessory minerals. Thickness of beds ranges from a few feet to more than 500 feet. Graded bedding is common, with gradations from metaconglomerate, with up to 0.37 inch-quartz pebbles, to fine-grained graywacke, graywacke schist and biotite-muscovite schist. Top-bottom criteria are frequently observed. The metaarkose and quartzite beds are essentially compositional variants of the metagraywacke.

Bands of quartz, up to several feet wide and several hundred feet long, occur in the metagraywacke. These quartz bands sometimes parallel the bedding and follow fold structures, but, more commonly, they parallel the schistosity. They are possibly a result of metamorphism of the graywacke.

Nodular masses called pseudodiorite are found in the metagraywacke. This rock usually occurs in ellipsoidal masses a few inches to 3 feet across but is occasionally found in lenses or dike-like bodies up to 1 foot wide and 50 feet long. The nodular pseudodiorite is frequently found aligned parallel to the bedding along certain horizons. Pseudodiorite is com-

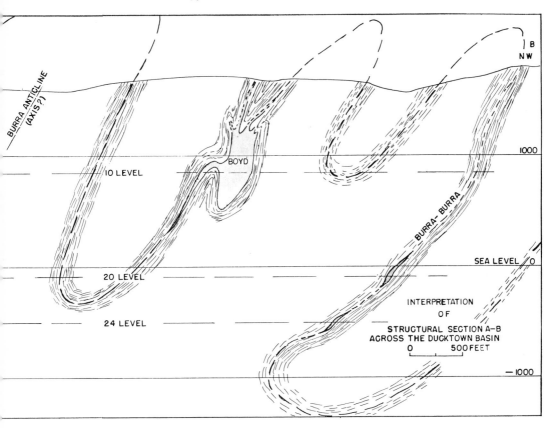

*A-B, illustrating the one horizon concept.*

posed of quartz, plagioclase, hornblende, garnet, and zircon. Concentric banding or zoning of the hornblende and garnet is common. Pseudodiorite nodules are thought to be metamorphosed calcareous concretions. They occur in graywackes elsewhere in the Great Smoky Group and are not unique to the Ducktown district.

MICA SCHIST   The mica schist beds are generally a few inches to 20 feet in thickness but occasionally exceed 200 feet in width. They are composed of biotite and muscovite in various amounts and frequently grade into an intermediate rock called a graywacke-schist, containing up to 40 per cent quartz with some feldspar and accessory minerals similar to those in graywacke. Garnet, staurolite, and graphite may be present in various amounts in some mica schist beds. These units are stratigraphic markers, but, as is true of all the beds in the district, they lack continuity and unique identity. These beds also usually lens out or change in mineralogical composition in a few hundred feet. There is a general proximity of

the staurolite schist to the ore horizons, but the association is not persistent.

AMPHIBOLITE   An amphibolite sill (Figures 1, 2) occurs southeast of the Coletown syncline. This body ranges from a few inches to several hundred feet thick and extends southwest for 12 miles. There is no apparent association of this sill to the ore deposits. Similar amphibolite masses have been found near the Boyd and Eureka ore bodies but not in contact with the ore. The age and any possible genetic relationship to the ore are not known.

The amphibolite sill generally is conformable to the bedding; some masses, however, cut sharply across the bedding. It is fine- to coarse-grained and has an average composition of 70 per cent hornblende, 20 per cent andesine, and 10 per cent accessory clinozoisite, titanite, sulfides, and quartz. Rock masses of a more coarsely crystalline hornblende with a similar mineral assemblage are found as gangue in the ore of the Calloway and Mary-Polk County deposits.

## Wall-Rock Types

The rock types described above, except the amphibolite, frequently occur along the walls of the ore deposits and as fragments in the ore. However, there are also rock units that are uniquely associated with the ore as wall rock.

SERICITE SCHIST  Sericite schist is composed of a very fine-grained white muscovite; it presents a satin-like appearance. Sulfides (pyrite and pyrrhotite) in amounts up to 20 per cent, but usually less than 2 per cent, may be disseminated through it. Biotite and sometimes chlorite are intermixed and occur as porphyroblasts in the sericite schist. This rock usually is in beds up to 30 feet thick. It is equally abundant on hanging wall and footwall, but is also more common along the walls of the upper parts of the ore deposits. Sericite schist also occurs as fragments within the ore lenses.

BIOTITE SCHIST  A dark brown to black, almost pure, biotite schist, with alteration to phlogopite, occurs similarly to sericite schist, but is less abundant. Porphyroblasts of biotite and euhedral garnet up to about 0.5 inches across occur in the biotite schist. This rock is found more frequently as infolds and waste inclusions in the ore than as wall rock.

CHLORITE SCHIST  Chlorite schist is a common wall rock; it is frequently separated from the ore by a few feet of biotite-sericite schist, but it is in contact with the ore over much of the hanging wall of the Burra Burra ore body. Chlorite is more abundant on the hanging wall but also occurs on the footwall and below the ore deposits. It is most frequently found from 500 to 1000 feet below the surface to the bottom of the ore bodies and attains thicknesses of 500 feet. The associated mineral assemblage is quite varied—biotite, sericite, and quartz are frequent admixtures; also porphyroblasts of euhedral garnet to 0.5 inches and staurolite to several inches in length occur and occasionally constitute up to 50 per cent of the rock. Magnetite and kyanite porphyroblasts also are found in some areas in the chlorite schist.

A sandy chlorite schist is extensive in the hanging wall of the Burra Burra ore body; chlorite-garnet schists several hundred feet wide occur on the hanging wall and footwall in the lower Boyd ore body; chlorite-staurolite-garnet schist is extensive as wall rock along the lower Calloway ore body, with widths up to 500 feet. Chlorite schist is extensive over much of the hanging wall at the Cherokee ore deposit. The iron content of the chlorite changes with its proximity to ore; this will be discussed under wall-rock alteration.

QUARTZITE  A fine-grained, friable, biotitic quartzite with well-rounded quartz grains less than 1 mm in size, is found widely associated with most of the ore deposits. The biotite content usually is less than 5 per cent and frequently exhibits lineation parallel to the bedding. Sulfides are present up to 30 per cent but usually are less than 10 per cent. Some quartzite in Calloway mine contains sufficient chalcopyrite to be classed as ore. This type of quartzite has been observed only adjacent to or in the ore; it rarely is found elsewhere in the district. A recrystallized form of quartzite, in which the quartz grains are poorly defined or obliterated, occurs in the ore.

## Metamorphism

Primary mineral assemblages of the rock in the Ducktown district place them in the staurolite-almandine and kyanite-almandine-muscovite subfacies of the almandine-amphibolite facies (27).

Alterations of garnet to chlorite, biotite to chlorite, staurolite to sericite, and feldspar to sericite are observed in the wall rock. These evidences of retrogressive metamorphism are not limited to the wall rock; Hurst (11) observed similar occurrences in the overlying Hughes Gap and Dean Formation 6 miles southeast of the mineralized district. He considered this alteration phenomenon indicative only of the pervasion of the rocks by water and not necessarily a retrogressive process.

Rocks in the district have undergone at least three periods of deformation and recrystallization, including the retrogressive metamorphism. The time span of these events is not known. Unaltered staurolite porphyroblasts coexist with partially and completely sericitized pseudomorphs of staurolite. Kyanite prophyroblasts coexist with staurolite. The retrogressive metamorphic phenomena may have been produced contemporaneously with the ore-forming process.

## Isotopic Age Determinations

Nine age determinations, using the potassium-argon radiometric method, have been made on minerals from the Ducktown area.

TABLE III.    *Potassium/Argon Isotopic Ages from Rocks in the Ducktown District*

| Rock | Location | Mineral | Age m.y. |
|------|----------|---------|----------|
| Schist | Two miles southeast of Calloway ore body | Muscovite | 323 |
|  |  | Biotite | 347 |
| Phyllite-mica schist | 4.5 miles west of Ducktown | Biotite | 434 |
| Biotite-muscovite schist | 250 feet in footwall of Calloway ore body | Muscovite | 327 |
| Biotite schist | 15 feet in footwall of Calloway ore body | Biotite | 1199 |
| Biotite schist | Inclusion in Calloway ore body | Biotite | 374 |
| Hornblende-biotite gneiss | Inclusion in Calloway ore body | Hornblende | 1045 |
| Hornblende-biotite-actinolite schist | Inclusion in Boyd ore body | Hornblende | 387 |
| Hornblende in sulfides | Inclusion in Eureka ore body | Hornblende | 478 |

Minerals were taken from schist inclusions in the ore, associated wall rock, and a mica schist, 4.5 miles from the district but still in the Great Smoky Group. These data in Table III were obtained from J. Laurence Kulp, F. Donald Eckelmann, and Arthur R. Kinkel Jr. (written communications, 1959–1965). Parts of the data have since been published ( 14,15,23).

The isotopic compositions of three galena specimens from the ores of the Calloway, Mary-Polk County, and Eureka deposits have been determined. These indicated an isotopic composition similar to galena specimens having a known origin related to the Appalachian orogeny.

Interpretations of these data suggest four possible metamorphic events affecting the rocks in the district after original deposition. The two ages over 1 b.y. associated with the Calloway ore deposit may reflect an original rock-forming age but may also be anomalous due to entrapment of radiogenic argon during formation. However, it is interesting to note a similar age was indicated (23) associated with the Ore Knob sulfide deposit 190 miles northeast of Ducktown.

## GEOLOGY OF THE ORE DEPOSITS

The eight ore bodies in the district range widely in size and shape and, to some degree, in the percentages of the sulfide and gangue minerals present as well as in the wall-rock characteristics. The general similarities will be presented here and the peculiarities discussed under the description of each deposit.

### Structure

The deposits are generally tabular in shape and conform to the regional trend of the strike and dip of the bedding. Folding has produced thickening of the lenses; ore configurations tend to conform to the dragfold structures. Conformity of the deposits to the bedding, while true in general, is not found everywhere. Up to 15° divergence between the bedding and the ore deposits frequently is observed, but rarely is it more than 15°, except in cross faults and minor fracture fillings.

The ore bodies generally plunge southwestward, while the major anticlinal structure appears to plunge to the northeast. However, parts of the ore bodies do not follow this generalization. Minor folds in schists associated with the ore generally plunge southwest. It appears that the major Burra anticlinal fold envelopes the ore deposits in the district and that the internal smaller folding influences the configurations of the individual deposits.

### Faulting

The three faults systems shown on the district map (Figure 2), and discussed under District Geology, are evident in the ore deposits. The strike faulting was a most influential factor in ore control, but the magnitude of its importance in providing loci of ore deposition is not known. Evidence of fault breccia, shearing, and slickensides is abundant in the ore and along the ore-wall contacts. Recemented fracture zones, fault gouge, and small fault openings occasionally are observed.

Fracturing and small faults are evident in the ore. Silicate minerals frequently are broken and recemented with sulfides. Minor displacements of a fraction of an inch are seen in these minerals. Bands of quartz in the ore frequently are seen with displacements of several feet.

The tabular and less deformed ore lenses frequently have a sharply-defined footwall commonly of graywacke, with the contact occasionally a slickenside. Almost no alteration is observed in the graywacke footwall. The hanging wall usually is schistose, and identification of strike faulting in the incompetent schist is rare, since movement was, for the most part, along the planes of schistosity.

It is probable that strike faulting aided localization of some of the ore lenses, and the deposits resulted from the pervasion of hydrothermal solutions along northeast-striking fault zones into favorable depositional areas.

Cross faulting (east-west and northwest faults) is evident in all of the ore deposits. Displacement of the ore along these faults is generally from 1 to 50 feet and occasionally is up to 350 feet. This faulting is usually tight, with very little mineralization penetrating the fault zones. There is a tendency for most of the ore deposits to terminate or lens out on one extremity, near a southerly-dipping cross fault. Burra Burra and Eureka ore deposits lens out on the southwest a few hundred feet from the Cleveland fault (Figure 2); the Boyd ore body terminates on the northeast at the Cleveland fault. The Calloway ore body lenses out of the southwest near an east-west fault; the Mary-Polk County ore body lenses out on the northeast at a fault zone; Cherokee lenses out on the northeast near a northwest cross fault.

## Mineralogy of the Primary Ore

The ore deposits are approximately 65 per cent massive sulfides and 35 per cent gangue minerals. The ore minerals, in approximate abundance, are 60 per cent pyrrhotite, 30 per cent pyrite, 4 per cent chalcopyrite, 4 per cent sphalerite, 2 per cent magnetite, traces of silver and gold, and other metallic minerals. Wide differences of these percentages are found in the several ore deposits and in areas of each deposit. For example, there are areas of massive, almost pure, pyrrhotite, zones of nearly pure granular pyrite, and areas containing up to 20 per cent magnetite or 15 per cent sphalerite. There are areas with large euhedral pyrite cubes up to 12 inches across engulfed in pyrrhotite. Similarly, well-rounded pyrite masses, from 0.5 to 2 inches in diameter, occur in massive pyrrhotite.

Chalcopyrite and sphalerite occur, disseminated in the massive pyrrhotite, and as small veinlets in the ore and in associated gangue minerals. Magnetite is widely disseminated but occurs most frequently as fine-grained masses in the pyritic areas of the ore.

Associated with the ore are various gangue minerals. These occur as scattered discrete crystals or aggregates in lenses or masses from a few inches to 100 feet wide and several hundred feet long. The sulfide content in the gangue ranges from ore grade to almost none. Quartz is abundant in all of the deposits, disseminated in the ore and occurring as veins and large masses of milky-white quartz up to 50 feet wide. Other minerals are predominately the lime silicates—actinolite, tremolite, and hornblende—and occasionally well-developed but highly-fractured crystals of diopside, garnet, and zoisite. Actinolite and tremolite occur in radiating and long bladed crystalline masses. Calcite is present as fine-grained crystals disseminated in the sulfides and also as large coarsely-crystalline masses, up to 30 feet across. Some calcite is fluorescent. Dolomite is less abundant but occurs similarly to, and mixed with, calcite. Other gangue minerals, occurring in minor amounts, include feldspar, talc, asbestos, epidote, clinozoisite, gypsum, tourmaline, graphite, rhodonite, and apatite.

Much of the gangue is made up, in some areas, of lenses and large folds of schistose wall rock. Biotite-sericite schist, with well-developed euhedral garnet crystals, is quite common in the ore. Blocks of graywacke several feet across occur in the massive sulfide ore. Biotitic quartzite is abundant in some sections of the deposits. A banded quartz-hornblende-garnet gneiss occurs in the Calloway ore body as large, angular to partially-rounded blocks, several feet wide. Distribution of the gangue minerals and rock inclusions in the ore is widely diverse, and no zoning pattern has been established.

## The Primary Ore Minerals

PYRRHOTITE Pyrrhotite, the most abundant ore mineral, usually is fine-grained and massive but does occur in crystal growth up to 0.25 inches in size. The crystal form is hexagonal. Its composition is about $Fe_{11}S_{12}$, with an average 47.5 atomic per cent iron. It is only very slightly magnetic. The color is generally a metallic bronze, and the mineral frequently exhibits an internal "schistosity." This may be a relict texture of replacement of a schist or the result of post-ore deformational stress.

Another form of pyrrhotite, called "slaggy" pyrrhotite, is dull gray to black and very fine grained. It is probably a low temperature redeposition. "Slaggy" pyrrhotite frequently occurs near cross faults and in brecciated zones in the ore deposits.

Pyrrhotite frequently contains inclusions of

rounded-to-euhedral pyrite, fine-grained inclusions and veinlets of chalcopyrite and sphalerite, and small inclusions of the various gangue minerals. Gangue minerals in pyrrhotite frequently show a lineation, while the pyrite distribution in pyrrhotite appears random.

PYRITE   Pyrite, the second most abundant ore mineral, usually occurs in fine-to-coarsely-crystalline aggregates of 0.125 to 0.5 inches but it occasionally is found in larger crystals several inches across. Masses of nearly pure pyrite up to 50 feet wide are found. Pyritic zones generally are low in chalcopyrite and sphalerite, but magnetite frequently is associated with the pyrite.

Pyrite has been found only in the cubic form, occasionally with small octahedral modifications. The rounded crystals and well-developed euhedral forms may represent two generations of pyrite.

CHALCOPYRITE   Chalcopyrite is associated most frequently with pyrrhotite but also is abundant in quartz, calcite, and the lime silicate minerals. It occurs finely intergrown with, and as veinlets in, pyrrhotite as well as veinlets in the gangue minerals. It is found penetrating along fractures in quartz, calcite, and the lime silicates. Veinlets of chalcopyrite occur in biotite-schist inclusions in the ore, usually along the planes of schistosity. It occurs similarly in biotitic quartzite, paralleling the lineation of biotite.

Chalcopyrite is the only significant copper mineral in the primary ore, although bornite and cubanite have been reported (4).

SPHALERITE   Sphalerite is the only zinc mineral in the ore. It occurs as does chalcopyrite and frequently is associated with it. Masses of fine-grained sphalerite often are found in quartz and calcite in the ore.

Sphalerite in the district is the variety marmatite, which has a high iron content, 7 to 20 molecular per cent FeS, and is dark brown to black in color.

MAGNETITE   Magnetite, the only oxide ore mineral, occurs both as very fine-grained magnetite disseminated in the sulfide and as massive magnetite in pockets in the ore. Well-developed octahedral crystals up to 0.25 inches in size are found in the ore; crystals of magnetite up to 1 inch in size have been found in chlorite-schist wall rock. Magnetite is most abundant in the pyritic ore and less common in pyrrhotite. It is found to be replaced by

sulfides, and it probably is earlier than most of the sulfides in the ore. Magnetite commonly is associated with the lime silicate minerals but is rare in quartz, calcite, and biotitic quartzite.

GOLD AND SILVER   Minute traces of gold and silver occur, primarily associated with chalcopyrite and galena. Gold and silver have not been identified megascopically or in polished surface examination.

OTHER SULFIDES   Galena occurs in small amounts as well-developed crystals associated with chalcopyrite and sphalerite. It is found occasionally as small veinlets in quartz, silicates, and schistose rock inclusions in the ore. Native bismuth has been found in galena. Bornite, cubanite, molybdenite, arsenopyrite, and stannite have been reported in the ore. Nickel and cobalt occur in trace amounts, but no nickel or cobalt minerals have been identified. None of the minerals in this group is recovered at present.

## Mineral Zoning in the Primary Ore

The data presented in this section are based, in part, on actual ore assays, visual observations, and detailed zoning studies in the more recently-developed Boyd and Calloway ore bodies. Many incongruities within these generalizations can be found. The emphasis on copper and zinc zoning is, naturally, an economic one. The geologic meaning of these small percentage changes may not be as significant as the economic importance.

In the district, the three ore deposits—Burra Burra, London, and East Tennessee on the northwest flank, as well as the Calloway and Mary-Polk County deposits on the southeast flank of the Burra anticline—are referred to as the high-copper ores, generally containing about 1.6 per cent copper. The three central deposits—Eureka, Boyd, and Cherokee—contain an average of 0.7 per cent copper. Zinc content follows a similar pattern, with about 1.2 per cent zinc in the outer deposits, and 0.5 per cent zinc in the central deposits.

The ore is more massive (less gangue) and contains more pyrite in the central deposits. Ore minerals in the central deposits constitute about 70 per cent of the ore, while in the outer deposits, the ore mineral content is about 55 per cent. Magnetite content of the central deposits is about 3 per cent; the outer deposits contain less than 1 per cent magnetite.

MAGNETITE DISTRIBUTION  A study of the distribution of magnetite in the Boyd ore body revealed a central area with 10 per cent magnetite, diminishing along strike and dip to less than 1 per cent. No zoning from hanging wall to footwall was noted. An isopleth map revealed a gradual southwest plunge of the contours of the magnetite concentration coincident with the plunge of the ore body. More magnetite is found in the thicker, more massive porions of the ore.

A similar study of the Eureka deposit revealed a concentration of over 20 per cent magnetite near the southwest end of the ore body. The magnetite diminishes along strike and dip to less than 1 per cent. A similar concentration of over 15 per cent magnetite is evident in the southwest end of the Cherokee deposit. In the Calloway ore body, magnetite content is very low. The upper part contains less than 0.2 per cent; the lower section has a core of 5 per cent magnetite that diminishes to zero along strike. Magnetite rarely is found in the Mary-Polk County ore body.

Ratios of copper and zinc to magnetite indicate no correlation, except the general district pattern of higher magnetite concentrations in the central (low-copper) deposits.

COPPER DISTRIBUTION  Copper zoning in the Calloway deposit reveals a general parallelism of zones of high copper with the strike and dip of the ore body. Zones of up to 4 per cent copper are found in the lower part of the ore body along the footwall, but there also are high-copper zones in the center and along the hanging wall of the lower deposit. Vertical zoning as well is evident in the Calloway deposit. The upper part, above 14 level, contains an average of 2 per cent copper and the lower zone averages about 1.6 per cent copper.

In the Boyd ore body, a general parallelism of copper zones with the strike of the ore body is evident. In addition, there is an increase in the copper content in the northeast and southwest extremities to over 1 per cent, while the thicker, more massive central zone contains about 0.6 per cent copper. Concentrations of copper were noted in the upper parts of the East Tennessee and Mary-Polk County ore bodies.

ZINC DISTRIBUTION  Zoning of zinc usually is parallel to the zoning pattern of copper, but zinc is not necessarily coincident with the copper. Zones containing over 6 per cent zinc are noted in the Calloway deposit parallel to

and near the footwall. There also are internal zones of high-zinc content. Zinc is more abundant in the upper Calloway vein, with over 3 per cent zinc, while in the lower vein, the zinc decreases vertically, with depth, to less than 1.8 per cent below the 20 level. By contrast, the Burra Burra ore body, a high-copper deposit similar to Calloway, contained much less zinc, with no zones over 1.5 per cent.

## Secondary Ore Development

The massive sulfide deposits have been subjected to weathering and chemical decomposition by percolating groundwater and a fluctuating water table. This process, discussed more fully by Gilbert (3) and Emmons (4) produced a cap of limonite gossan (Figure 5), extending downward to the water table, a depth in some places of 100 feet below surface. Below the gossan, a zone of rich secondary copper minerals was deposited. This supergene zone, enriched in copper, was the target of the first mining and a factor in the early exploitation of the district.

Composed essentially of chalcocite, the supergene ore was called "black copper" ore and had a sooty, porous texture. This zone ranged from a few inches to about 8 feet thick and carried a wide range of copper content, from 10 to 50 per cent, averaging about 25 per cent copper (4). These differences resulted from changes in the copper content of the original sulfide ore, the depth of weathering, and the localized conditions of chemical precipitation. A wide variety of other secondary copper minerals was reported (4) in this zone, including malachite, azurite, chrysocolla, cuprite, and native copper. Remnants of resistant pyrrhotite and pyrite were also present in the supergene zone.

The supergene zone was tabular, with well-defined boundaries between the overlying gossan and massive sulfide below and between the impervious hanging wall and footwall. The zone followed the water table, which conformed somewhat to the topography. The primary sulfide below the supergene zone reportedly showed some local enrichment in copper for a few feet downward. A notable exception is the East Tennessee deposit, in which secondary chalcocite associated with primary sulfides was noted 220 feet below the surface.

The secondary copper ore was mined out most thoroughly before 1900, and very little mineralogic evidence of this zone can now be found.

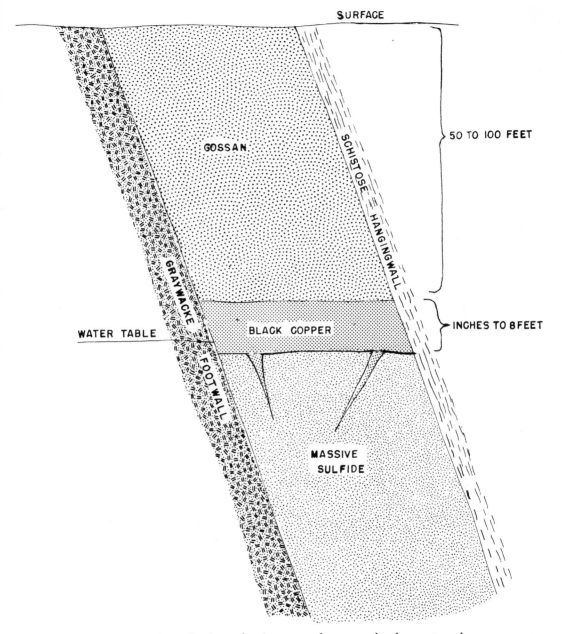

FIG. 5. *Diagrammatic Cross Section, showing secondary ore development and supergene-enriched copper zone.*

GOSSAN  The overlying gossan is composed predominately of limonite, with iron content of about 45 per cent and less than 0.08 per cent phosphorous. An estimated 2,000,000 tons of this ore was mined from several of the ore bodies, notably Eureka and Cherokee, before 1907. No significant amount of gossan remains to justify mining.

The gossan is highly porous, containing a pore space of about 40 per cent. It exhibits a typical boxwork texture characteristic of decomposed massive sulfides. Residual quartz and silicate minerals are abundant in the gossan. A copper content of 0.1 to 0.8 per cent is present as secondary copper minerals and is somewhat proportional to the original copper content of the massive sulfide ore body from which it was derived. In general, however, the gossans from the various deposits are all quite similar and cannot be identified as belonging to a particular vein.

## Temperature of Ore Formation

Coexistence of pyrrhotite, pyrite, and sphalerite in equilibrium in the ore deposits makes the ore suitable for comparison to laboratory-simulated pressure and temperature conditions of formation. Metamorphism and recrystallization of the ore after original deposition probably occurred within the temperatures observed.

Investigations by R. G. Arnold (written communication, 1958) of a reaction rim of hexagonal pyrrhotite around a pyrite crystal from the Burra Burra ore body indicated a minimum temperature of formation of 470°C, uncorrected for pressure. The iron content was 46.8 atomic per cent Fe. An investigation of eight ore samples from the Boyd and Calloway ore deposits revealed no reaction rim of pyrrhotite around pyrite; the iron in pyrrhotite ranged from 47.4 to 48.0 atomic per cent, and a minimum temperature of 280°C was indicated. This temperature does not reflect the actual temperature of formation, since the pyrrhotite was not necessarily formed in a sulfur-rich environment. The pyrite and pyrrhotite in these samples apparently crystallized at different times.

Moh and Kullerud (18) investigated 130 ore specimens from the Calloway ore deposit and found a similar low temperature range for pyrrhotite, which they consider was a readjustment to retrogressive metamorphic conditions. Investigation of the pyrrhotite-sphalerite assemblage indicated a temperature range from 200°C to 600°C for sphalerite, uncorrected for pressure. In the Calloway ore body, a temperature pattern was found that indicated temperatures up to 600°C on the northeast end and temperatures as low as 200°C on the southwest end in the lower levels of the deposit, 1400 to 1800 feet below surface. Existence of kyanite in the adjacent wall rock at Calloway indicates a higher pressure-temperature condition that must have pre-dated the emplacement of the ore in its present form.

Investigations by Carpenter (20) of sphalerite from Burra Burra and Calloway confirm the range of temperature values found by Moh and Kullerud. Carpenter concluded, however, that the temperature variations of 150°C indicated in adjacent sphalerite samples were caused by differences in the FeS content in the sphalerite of up to 9 molecular per cent.

This would make the results only generally indicative of the temperature of formation. Determinations of iron in 31 samples of pyrrhotite from several of the deposits indicate a range from 47.37 to 47.68 atomic per cent iron, which represents a minimum temperature of 315°C. Chlorite and epidote, both present in the ore, have an upper stability limit of 490°C and 460°C, respectively.

## Wall-Rock Changes and Alteration

Various changes have been noted in the character of the wall rocks associated with the ore deposits. These changes include metamorphic, mineralogic, major chemical, and trace-element variations. These features appear to be uniquely associated with ore deposits and some may be related to the ore-forming process. It is probable that some of the changes observed represent original mineralogic changes caused by metamorphic processes, as well as changes produced from hydrothermal solutions that were not ore-associated.

Snyder (31) observed differences in the ilmenite content of the metagraywacke and schist beds in the district. The ilmenite is detrital and the differences probably were due to original compositional differences. Stratigraphic zones with a high ilmentite content were recognized by Snyder as the ore-producing zones. These zones are composed of several litholigic units of schist and graywacke. Snyder concluded that the Burra Burra, London, and East Tennessee deposits were in the same producer zone but not in the same position in the stratigraphic horizon.

Changes in the zircon content of the wall rock were noted by Gibson (10) at the Calloway ore body. By counting the zircon grains observed under ultraviolet light, he determined that detrital zircon is abundant in metagraywacke, less abundant in the mica schists, and almost non-existent in the ore and gangue. The zircon content of the hanging wall and footwall schists was very low. The low zircon-bearing beds in proximity to the ore deposits may reflect a changing depositional environment in which deposition of more calcareous minerals took place.

Investigations by Brown (27,33) revealed several mineralogic changes in the wall rock at the Boyd and Calloway ore bodies. In petrographic examination of the rocks from the country rock, through an alteration zone up to 60 feet wide, to the ore, Brown noted changes in the chlorite, from ripidolite to amesite, and in the biotite, from biotite to phlogopite. These changes were identified by noting the lowering of the index of refraction (Ny) in both mineral groups. These changes also represent a reduction in iron content in the altered zone. Brown also noted, in the altered zone, an increase in twinning and zoning in plagioclase, alteration of biotite to chlorite and sericite, recrystallization resulting in the formation of porphyroblasts of chlorite, muscovite, and phlogopite, and an increase in epidote and clinozoisite. The lowering of the index of refraction (Ny) of chlorite and biotite in the wall rocks adjacent to the ore body was also noted (34) at the Eureka and Cherokee deposits.

Mineralogic changes in wall-rock composition are observed with increasing depth of the ore deposits. Biotite and sericite are more abundant near the upper parts of the Boyd and Calloway deposits; chlorite increases with depth and is the most abundant mineral in the walls of the lower parts of the Boyd, Eureka, and Calloway deposits. Garnet increases, with depth, in the walls at Boyd; while staurolite is more abundant at depth at the Calloway deposit.

Geochemical analyses by Kingman (9) indicated an increase in the trace quantities of copper and zinc in barren channelways that represent possibly ore-bearing areas. Spectrographic analyses by Moore (19) of mineral fractions of footwall rock collected from 10 to 200 feet away from the Boyd ore body, indicated an increase in tin in rocks adjacent to the ore (stannite has been identified in the ore) relative to those away from it. A similar increase was noted in nickel, cobalt, and titanium in the metallic fraction of the rocks away from the ore body.

## Origin of the Primary Ore

Emmons (4, p. 61), 40 years ago, commented "Few districts of equal area in this country have been studied by so many investigators of high authority, and there are few concerning which opinions differ so widely." Investigations of the past 40 years support this view.

Emmons' theory of hydrothermal replacement of lenses of limestone along one horizon is a widely accepted view. He considered the lime silicates and calcite gangue in the ore to be recrystallized remnants of the original host rock.

Ross (6) concurs with hydrothermal replacement but believes that deposition was primarily along shear zones. He concluded that the gangue minerals—quartz, ferromagnesian minerals, lime silicates, and calcite—were introduced hydrothermally and deposited with the ore in genetic stages.

Mineralogic studies by Carpenter (20) support the view of Ross. Carpenter observed three stages of deformation that affected the ore minerals and were superimposed on the original replacement features. He also suggests the possibility of a genetic relationship between the amphibolite sills near the ore deposits and the ore-forming process.

Recrystallization and remobilization features are present in the ore and effectively mask the original ore-forming features. Such recrystallization characteristics may be misinterpreted as original crystallization features. Kinkel (17) observed recrystallization and remobilization features in the ore at Ore Knob and similar features in the ore from Ducktown (written communication, 1963.) Kalliokoski (22) observes a "metamorphic overprint" in the Ducktown ore and elsewhere that may have resulted from regional metamorphism after emplacement.

Henry S. Brown (written communication, 1967) concluded from rock alteration studies (27,33) that hydrothermal solutions moving upward through shear zones in advance of ore deposition produced much of the rock alteration features. Later, sulfurous vapors penetrated the shear zones, removing iron from biotite and chlorite schists along the channelways and replacing the gangue minerals with iron sulfides. Brown states that the copper and zinc may have been derived from the rock along the channelways.

Some geologists, particularly European and Australian, visiting the district in the past decade, support a syngenetic origin for the deposits.

This writer believes that hydrothermal replacement of favorable zones is evident. These zones consist of both folded lime-bearing schistose beds and shear zones, occurring in three separate stratigraphic horizons. The shear zones were channelways to depositional areas as well as favorable zones for deposition. The favorable beds were probably impure lime-bearing muds that were metamorphosed to calcite and lime-silicate minerals, biotitic quartzite lenses, and biotite-sericite schist lenses before their replacement by sulfides. The abundance of partially-replaced breccia fragments in the ore, evidence of shear faulting along the walls of the deposits, and the low-angle crosscutting of the ore across the original beds, indicate the fault-zone control of ore deposition and the probable replacement along shear zones.

Hydrothermal solutions probably carried most of the components of the ore minerals from depth in several successive stages. Some of the iron may have been derived from rocks along the channelways. The ore-bearing solutions probably changed in composition as deposition and replacement took place; a decrease in the temperature of these solutions was probable. Consequently, the degree of replacement and the rock alteration effects were partially a function of changing temperature and chemical composition of these solutions.

## Geologic History

The geologic history outlined here is offered as the likely sequence of geologic events. Obviously, other interpretations are possible and improvements will evolve as future studies resolve some of the many questions. Interpretations of the sequence and time of metamorphic crystallization, recrystallization, hydrothermal effects, and retrograde metamorphism are subject to debate. Isotopic age dates are inserted for general correlation. They do not reflect a time span of the periods of deformation.

(1) Late Precambrian deposition of alternating coarse and fine clastic sediments and muds from older Precambrian land mass (1199 m.y.). These were mixed with a few calcareous layers on a continental shelf environment. Slumping and turbidity currents resulted in intermixing and graded bedding.

(2) Overlying deposition of similar Cambrian sediments occurred to a combined Precambrian-Cambrian thickness exceeding 30,000 feet.

(3) Early Paleozoic uplift and orogeny (478, 434 m.y.) took place, accompanied by regional and dynamic metamorphism of the rocks from chlorite to kyanite grade. Folding of lime-silicate assemblages, calcite, magnetite, biotitic quartzites, chlorite schists, and other host rocks into approximately the present configuration followed. Major thrust faulting produced brecciated zones and channelways for ore-forming solutions.

(4) Devonian deformation (387, 374, 347, m.y.)—hydrothermal solutions followed channelways and replaced favorable zones with sulfides. Hydrothermal solutions altered the wall rocks and produced local retrogressive metamorphic effects. Intrusions of amphibolite sills and dikes probably occurred during the Devonian period.

(5) Late Paleozoic deformation (323,

327 m.y.), produced east-west and northwest faulting that offset and rotated the ore deposits; mobilization and recrystallization of the ore and gangue minerals took place in response to metamorphosing conditions. Further hydrothermal wall-rock alteration phenomena probably were developed at this time. Final mobilization of sulfides occurred along some faults and fractures.

(6) Permian to present—the land mass was uplifted; erosional processes formed the present topography. Weathering during the Quaternary produced decomposition of the sulfides and formation of the gossan and supergene-enriched copper zones.

## DESCRIPTION OF THE ORE BODIES

### The Northwest Deposits

The Burra Burra, London, and East Tennessee ore deposits (Figure 2) occur along strike, probably in one shear zone and possibly within the same stratigraphic horizon, on the northwest limb of the Burra anticline. There is a general similarity in these deposits, but a vast difference in size, as can be seen in Figures 6 and 7. Mining has been completed in these deposits.

BURRA BURRA  The largest production of massive sulfide, to date, has been from the Burra Burra ore body. The deposit contained 19,000,000 tons of ore, of which 15,636,000 tons were mined. The remaining ore is in support pillars that cannot be economically recovered. Mining was completed in 1958. The vein is over 3400 feet long and extends 2400 feet vertically below the surface and almost 3600 feet down the dip of the vein. However, not all of this area was composed of ore.

The deposit is generally a thin, warped, tabular lens, but local folding, as shown in Figure 7 at section 12N, produced sulfide accumulations over 200 feet thick. The plunge of the deposit is almost parallel to the 70°SE dip; minor fold axes plunge 45° to 65° northeast. The vein cuts across the bedding of the footwall rocks, in many places, at an acute angle of 15° or less, but the deposit generally is conformable to the hanging wall. Snyder's (31) stratigraphic studies tend to confirm this disconformity. The vein flattens to almost 30° near the bottom of the deposit before lensing out. If the vein is conformable to bedding at depth in the overturned anticlinal limb, it should turn back to the northwest as shown

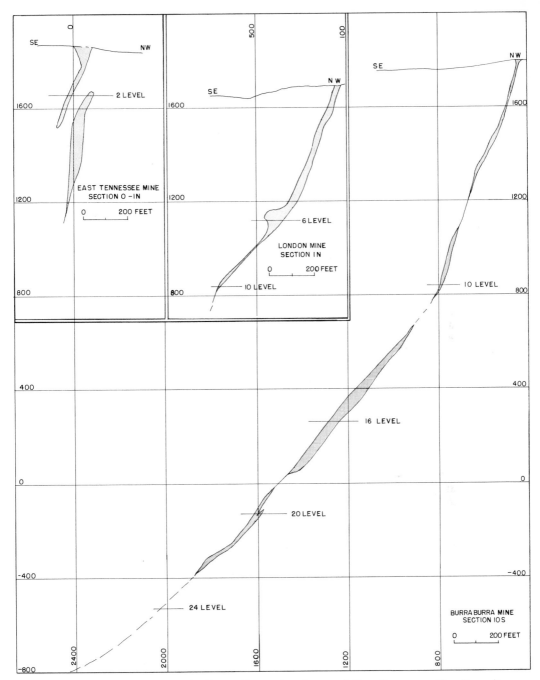

FIG. 6. *Cross Sections of the Burra Burra, London, and East Tennessee Ore Deposits.*

in Figure 4. This has not been established.

The footwall rocks are graywacke and mica schist with some zones of "bleached" sericite schist. The hanging wall is composed of a sandy, quartz-chlorite schist with some garnet and a few staurolites. This rock is up to 200 feet thick in places.

The ore and gangue mineral assemblage in the Burra Burra ore body is typical of the district; chalcopyrite is slightly more abundant than the average of the district. The average copper content produced was 1.6 per cent and the total sulfide mineral content was about 57 per cent. Quartz, calcite, and the lime-sili-

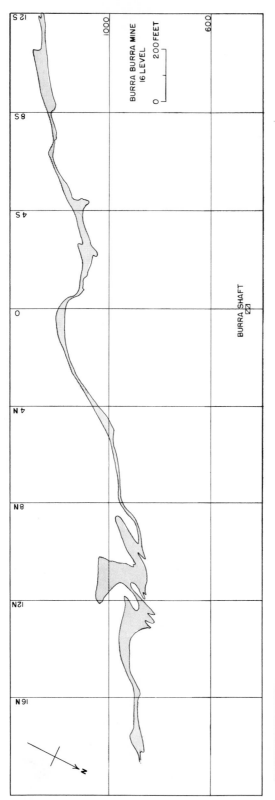

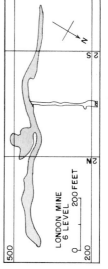

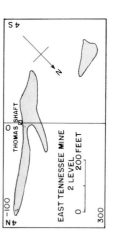

Fig. 7. *Horizontal Plan Views of the Burra Burra, London, and East Tennessee Ore Deposits.*

cate minerals are the most abundant gangue. Breccia fragments of wall rock in the ore are common; slickensides along the walls are less common. The ore minerals probably were deposited along a shear zone and as a replacement in fold structures.

LONDON   The London ore body (Figures 6 and 7) is 2300 feet northeast of the Burra Burra deposit. Production, completed in 1926, totaled 1,600,000 tons of ore containing 1.8 per cent copper. This ore body was about ⅒th the size of the Burra Burra deposit. The vein is 1000 feet long and extends 1000 feet down a dip of 60°. The mineral assemblage was typical of the district; however, the total sulfide content produced comprised less than 50 per cent of the ore. Quartz and calcite were the most abundant gangue minerals. The structure and wall rocks were similar to the Burra Burra ore body. Exploration between the Burra Burra and London indicated the same wall rocks present but no sulfide mineralization or associated gangue minerals.

EAST TENNESSEE   The East Tennessee ore body (Figures 6 and 7) is 2000 feet northeast of the London ore body. Ore production from this deposit was about 250,000 tons, the smallest deposit in the district. The East Tennessee deposit is unique in several ways. The mine was developed to a depth of 500 feet, well into the primary ore, by 1877. This was done at a time when production elsewhere in the district was predominately from the secondarily enriched zone. The supergene zone at East Tennessee extends more than 200 feet below surface, and enrichment of the primary ore apparently went much deeper. Henrich (2) stated that early production averaged about 7.5 per cent copper; however, this ore was hand sorted, which consequently upgraded the product assayed. The gangue was primarily slaty or schistose rock. Emmons (4) observed an abundance of calcite and lime-silicate minerals, particularly tremolite and diopside, as gangue. This is apparent in the existing dump material.

East Tennessee was more a disseminated than a massive sulfide deposit and probably was composed of about 30 to 40 per cent sulfides disseminated in a matrix of silicate minerals and schist. The mineral assemblage was typical of the district but with notably more chalcopyrite, calcite, and lime silicates. The wall rock was graywacke and mica schist. The sandy chlorite schist found in the hanging

wall of Burra Burra and London was absent at East Tennessee.

## The Central Deposits

The Eureka, Boyd, and Cherokee ore bodies comprise the central group of deposits (Figure 2). They are similar to each other in mineralogy and wall-rock types but differ in size and structure. They are thought to be along the same stratigraphic horizon striking N30° to 40°E. These deposits are lower in copper than the deposits on the northwest and southeast limbs of the Burra anticline. Total sulfide content, however, is higher than that found in the outer deposits, and pyrite is more abundant, although pyrrhotite is still the predominant sulfide. Sulfur content of the ore is about 25 per cent and iron content about 35 per cent. Sphalerite is less abundant than in the outer deposits.

The low copper content of these deposits prevented their early development. Sulfide production from Eureka has been essentially continuous since 1925; Boyd was first developed in 1941 and is still in production. The Cherokee deposit is only currently reaching significant production.

EUREKA   The Eureka ore body consisted of two mines, the Isabella and Eureka, operated by two companies until 1936. Both these names have been used in describing the ore body, which is actually one vein. The name Eureka prevails today. The ore body (Figures 8, 9) is an elongated irregular lens transected and offset horizontally 350 feet by the Eureka fault, a post-ore, east-west fault. Slickensides are found in the ore along this fault zone. The ore deposit strikes N30°E and dips generally 45°SE. The northeast fold of the deposit is shaped like a prow of a ship that curves downward to the southwest and flattens as a keel to near horizontal at the 8 level (Figure 8). Much of the original deposit has been eroded away, exposing a gossan up to 250 feet wide and 1500 feet long. Perhaps as much ore was eroded away as was found in the deposit.

The wall rock is predominantly sericite schist on the hanging wall and graywacke and graywacke schist on the footwall in the upper parts of the ore body. Chlorite-garnet schists abound on both walls near the bottom and below the deposit. Staurolite and sericitized pseudomorphs of staurolite are abundant on surface and down to 8 level and also are found

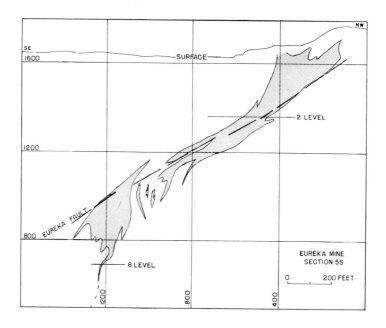

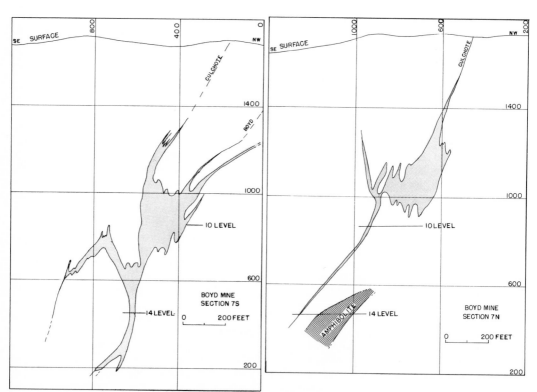

FIG. 8. *Cross Sections of the Boyd and Eureka Ore Deposits. The drag fold pattern of the Boyd ore body is evident.*

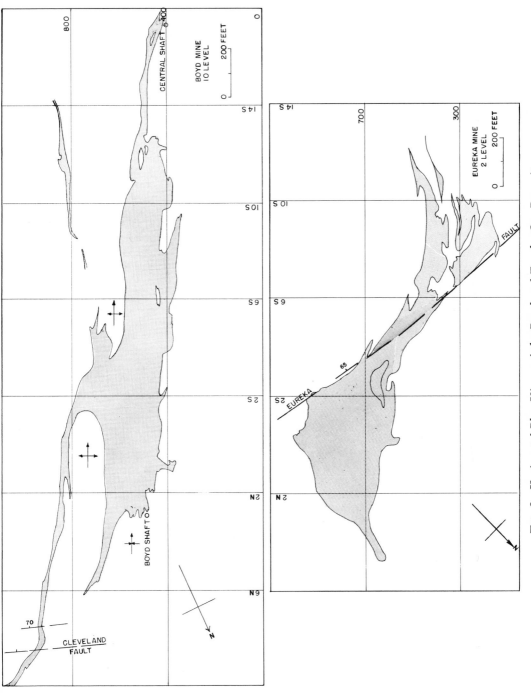

FIG. 9. *Horizontal Plan Views of the Boyd and Eureka Ore Deposits.*

below the ore body. Kyanite has been found in the wall rock as well.

Exploratory drilling has found a massive amphibolite sill more than 300 feet wide under the ore body, as well as off the northeast end. This occurs within 600 feet of the ore. The genetic relationship to the ore deposit, if any, is not known.

The ore body is conformable to the wall rock with generally sharp contacts and very little sulfide disseminated in the schistose walls. The deposit has a higher ratio of sulfide ore to gangue than any of the other deposits in the district. Pyrite is quite abundant and exceeds pyrrhotite in parts of the deposit. Magnetite is abundant locally near the south central part of the vein; it occurs as fine-grained crystalline masses up to 2 feet across.

The gangue assemblage is predominantly actinolite and tremolite with lesser amounts of calcite and quartz. Masses of schistose wall rock are infolded in the ore. Unique to Eureka are well-developed diopside and hedenbergite crystals that are up to 6 inches in length. These crystals have been fractured and broken and recemented by sulfides. Masses of this gangue are found widely scattered in the ore.

The ore is probably a replacement of a favorable zone in a large fold structure.

BOYD   In the early years of the district, two veins, the Boyd and Culchote, were explored and originally considered to be two separate deposits, 400 feet apart. These veins were 10 to 20 feet wide and contained about 0.7 per cent primary copper; they were too narrow and too low in copper for economic development. Exploration in 1940 revealed (Figure 8) a massive sulfide deposit 200 feet wide occurring 600 feet below the surface. Later exploration revealed that the two veins exposed at surface were part of the larger deposit below.

The Boyd ore body is an elongate tabular body, 2400 feet long, that has been complicated and thickened by isoclinal folding. The major fold axes associated with the vein plunge gradually about 20°SW. The northeast nose of the ore body is similar to Eureka, with a prow and keel that flattens to almost horizontal below the 14 level. Cross faulting is frequently seen but has had little effect on the deposit. The Cleveland fault (Figures 2, 9) on the northeast end of the ore body marks the apparent terminus of the deposit, although small occurrences of sulfides are found north of the fault.

The wall rock is composed mostly of sericite

and biotite schists above the 10 level and predominantly a chlorite-garnet schist below 10 level. The lower portion of the ore deposit has an almost complete envelope of chlorite-garnet schist, which is known to continue 600 feet below the ore body and is over 400 feet thick in some places. Staurolite and sericitized pseudomorphs of staurolite are observed occasionally, but are not abundant. The ore body generally is conformable to the enclosing rock. An amphibolite sill (Figure 8) occurs near the northeast end and near the footwall of the ore body.

The ore is the typical massive sulfide; however, there are large areas of biotitic quartzite and biotite-sericite schist containing from 10 to 50 per cent of disseminated sulfides. These latter rocks are most abundant on the hanging wall side and in the folded limb of ore on the hanging wall. Well-developed pyrite cubes, measuring up to 12 inches across, are found enclosed in pyrrhotite. Magnetite is abundant in the central part of the deposit with some areas containing up to 15 per cent magnetite. The gangue minerals are similar to those of the Eureka deposit.

The Boyd ore body is thought to replace a favorable horizon in a fold structure. Very little fault breccia has been noted in this deposit, except in the extreme north end near the Cleveland fault zone.

CHEROKEE   The northeast part of the Cherokee ore body has been referred to in the past as the Old Tennessee mine and the School Property mine, and the southwest extension as the Cherokee mine. These divisions are political; the ore body is one deposit.

The Cherokee ore body (Figures 10, 11) is the longest deposit in the district; it is over 6000 feet in strike length and is known to extend to a depth of 2400 feet below surface. Much of the area, however, consists of subeconomic mineralization or vein widths too narrow to mine. Mining development, to date, extends 1000 feet below the surface with some exploration drilling to 2400 feet below the surface; the entire size and shape of the deposit has not been fully outlined.

The vein is a long, thin, warped, tabular lens. Folding has produced thickened ore widths up to 150 feet, but much of the vein is less than 30 feet wide. The Cherokee deposit has much less structural complexity than the other deposits in the district. A northwest cross fault bisects the deposit near the center and offsets the deposit horizontally about 250 feet. Rotation of the ore body about this fault may

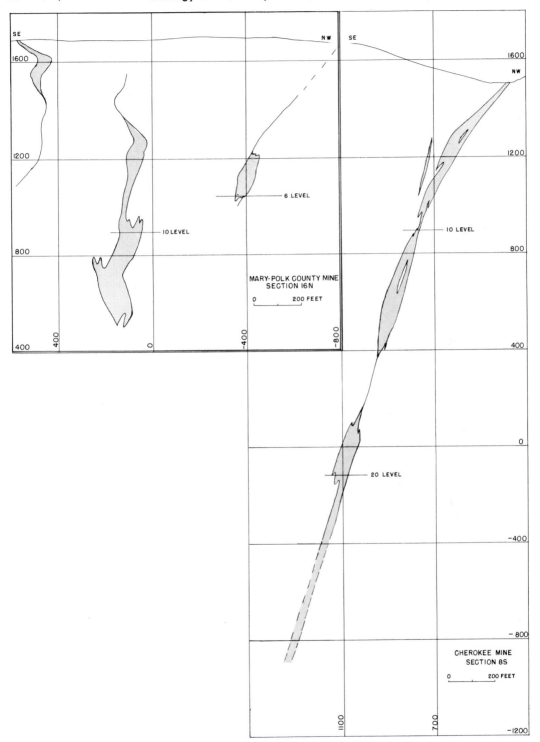

FIG. 10. *Cross Sections of the Cherokee and Mary-Polk County Ore Deposits. The three segments of the Mary-Polk County deposit are part of one vein system. The Cherokee cross section is interpretive below the 10 level.*

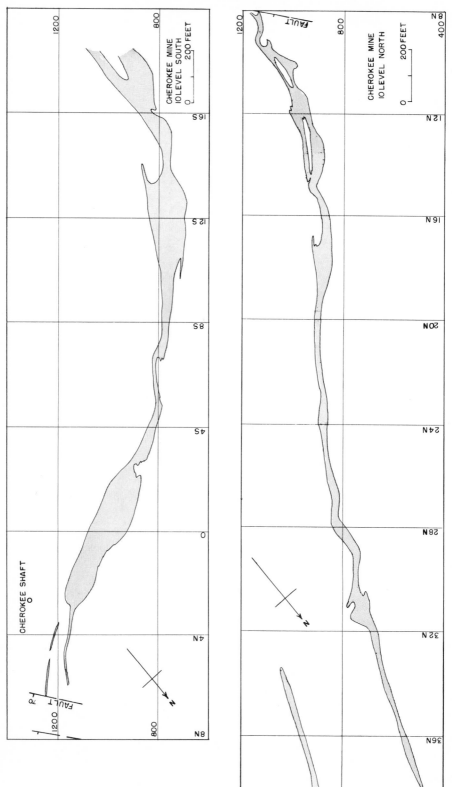

FIG. 11. *Horizontal Plan View of the Cherokee Ore Deposit. The north segment of the 10 level is a continuation of the south segment, across a fault offset of 250 feet.*

have taken place as is evidenced by the change of dips across the fault. The ore deposit north of the fault generally dips about 45°SE, while the southern part dips 65°SE.

The wall rocks are essentially graywacke and graywacke schist on the footwall and biotite-sericite schist on the hanging wall; however, chlorite-garnet schist increases with depth on the hanging wall and is up to 200 feet wide in some lower parts of the deposit. Sericitized staurolite pseudomorphs are occasionally found in the chlorite schist. The ore and gangue minerals are similar to those of the Boyd and Eureka deposits.

## The Southeast Deposits

The Calloway and Mary-Polk County ore bodies comprise the deposits on the southeast limb of the Burra anticline. They generally are similar in mineralogic composition to the deposits on the northwest limb of the anticline. They contain about 1.7 per cent copper and considerably more gangue minerals than the central deposits. Actinolite, tremolite, quartz, and calcite are the predominant gangue minerals. The structures of these two deposits are quite different, but they are believed to occur in the same drag fold, a structure that has been offset by faulting.

CALLOWAY   The Calloway ore body (Figures 12, 13) is 1500 feet long and extends downward more than 3000 feet below surface. There are two parts to the deposit; the structures above and below 14 level (about 700 feet elevation) are quite different. Above 14 level, the deposit is controlled by a doubly-plunging isoclinal anticline and syncline. The plunge ranges from steeply northeast to near vertical. Within this structure, part of the ore occurs in a synclinal fold plunging 45°E. Below 14 level, the deposit becomes less complex. It develops into an arcuate, elongate, tabular shape that steepens to a dip of 70°SE; the plunge reverses to about 70°SW.

The ore generally is conformable to the wall rock. However, the structure has not followed the Coletown synclinal structure (Figure 3) as was earlier anticipated. Folding in the lower segment of the deposit has produced thickening of the ore zones with massive sulfide up to 400 feet wide being found on the 20 level.

The wall rocks of the upper body are essentially biotite-sericite schists on both walls of the three upper limbs (Figure 12 section 0) of the deposit. The walls of the lower vein, below 14 level, are similar to the upper vein, with 1 to 20 feet of biotite-sericite schist that is generally conformable to the ore body. Adjacent to the biotite-sericite schist is an unusual rock composed of chlorite, quartz, garnet, staurolite, and biotite. It is generally schistose, and the schistosity and bedding parallel the strike and dip of the ore body. Staurolite development in this rock is extensive, with porphyroblasts up to 8 inches long and 3 inches across; many of these have been partially or totally sericitized. Kyanite porphyroblasts locally are abundant. Small 0.125 to 0.25 inch garnets occur in bands that cut across the staurolite and kyanite porphyroblasts. These bands of garnet parallel the schistosity and bedding; the garnet is locally chloritized. This rock extends up to 300 feet in width on both hanging wall and footwall of the lower Calloway deposit.

The mineral assemblage of the Calloway ore is similar to that of the district, but it is notably higher in chalcopyrite and sphalerite. Copper and zinc both average about 2 per cent in the upper vein and about 1.6 per cent in the lower vein. Pyrrhotite and pyrite are the predominant minerals. Magnetite is less than 1 per cent. Gangue minerals are the same as elsewhere in the district; however, there are large areas of biotitic quartzite with disseminated sulfides, occurring particularly on the strike extremities of the lower vein. In the lower Calloway vein, large breccia fragments of banded quartz-hornblende gneiss have been found. Calcite masses several feet wide occur in the ore. Unique to Calloway and somewhat unusual in Precambrian metasediments are small pockets of methane detected in the wall rock and in the ore of the lower vein. Graphite is found locally in the ore and wall rock, particularly in fractures and along slickensides. The Calloway deposit probably was a replacement of a favorable highly-folded zone and an intensely brecciated shear zone.

MARY-POLK COUNTY   This deposit (Figures 10, 13) has been treated previously as two separate ore deposits, based in part on a political division and in part on the presence of several veins. Exploration has indicated that the separate mineable ore lenses probably are part of one vein system in a complex series of isoclinal folds. The main part of the Mary deposit is continuous for a strike length of 3000 feet, but it is composed of several ore lenses and narrow non-economic veins. The plunge of the ore body is gradual to the southwest.

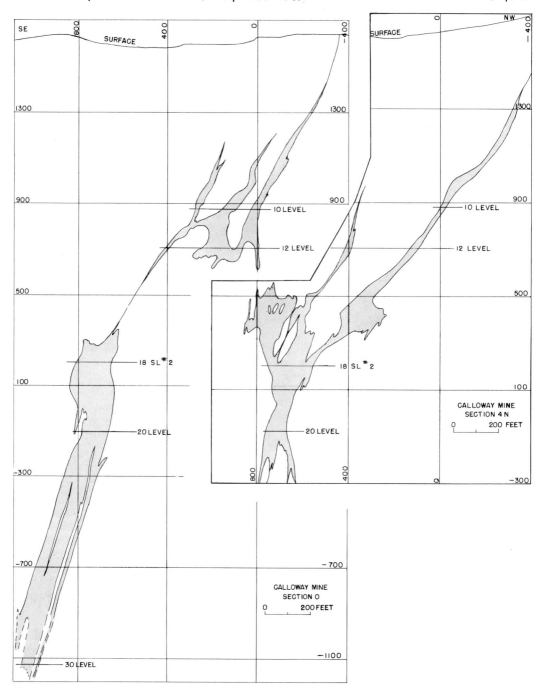

FIG. 12. *Cross Sections of the Calloway Ore Deposit. Section 0 illustrates the structural contrast of the upper and lower veins.*

Graywacke, biotite schist, and sericite schist occur as the wall rock along much of the deposit. Staurolite schist is abundant on the southwest limb of the Polk County vein. Chlorite-staurolite schist, similar to the lower Calloway wall rock, is abundant several hundred feet below the north end of the Mary vein.

The secondary copper ore was very rich at both the Mary and Polk County lenses, and the primary sulfides, for several hundred

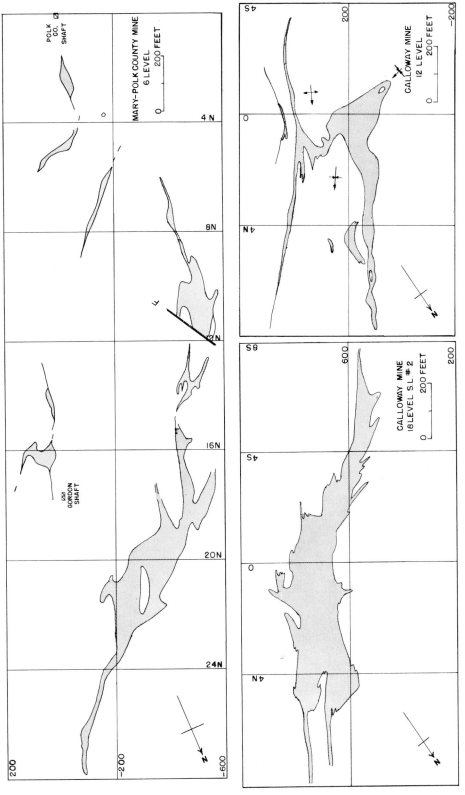

FIG. 13. *Horizontal Plan Views of the Calloway and Mary-Polk County Ore Deposits. The two plan views of the Calloway deposit illustrates the structure of the upper and lower veins.*

feet down, contained about 2.5 per cent copper. The lower parts average about 1.5 per cent copper. Pyrrhotite is the predominant sulfide in the primary ore with very little pyrite or magnetite. Sphalerite is abundant locally with some areas containing over 2 per cent zinc. Traces of galena and molybdenite have been found in some parts of the deposit and are evidently more abundant in the Mary-Polk County than in the other deposits in the district. The only molybdenite identified has come from the Mary-Polk County deposit.

The gangue is composed of quartz, actinolite, tremolite, calcite, and large masses of dark green-to-black hornblende that are somewhat similar to the amphibolite sills in the district. Large well-developed garnet crystals up to 1 inch in size are observed in schistose fragments in the ore. Breccia fragments of graywacke, graywacke schist, and quartz-hornblende gneiss are found locally in the ore. The deposit is essentially a replacement of favorable lenses in a folded schistose structure.

## Other Sulfide Occurrences

Some 4000 feet southwest of the Calloway deposit are the Meek and U.S. prospects (Figure 2). These are small occurrences of gossan with actinolite; no massive sulfide was found at either prospect.

Southwest of the mining district, in an area extending for 5 miles southwest into Georgia, are several small occurrences of sulfides. Two of these, the Number 20 and Mobile mines, produced a small quantity of secondary copper and some sulfide ore. Both veins were too narrow and too low in grade to support continued economic mining. The ore-mineral assemblage of these occurrences is similar to the ore in the mining district. Exploration of other prospects further southwest has indicated less massive sulfide and a more disseminated type of sub-economic mineralization in schists.

## SUMMARY

This paper has presented only some of the highlights of the geology of the Ducktown district. Words cannot fully describe the complexity of structure, the diversity of mineralogic details, and the massive sulfide ore deposits as they are observed. The geologic history, the structure, the paragenetic sequence of mineralogy, and the origin of the sulfides never may be understood fully, but this complex district always will be a geologic challenge.

Currently in progress are a variety of geological and geochemical investigations that will add to the knowledge of the district. These include chemical partitioning in the sulfide and silicate minerals, whole rock radiometric analyses, sulfur-isotope stulies, internal stress analysis in the sulfide minerals, metal ratios in the ore, structural analysis of first- and second-order folds, and petrologic studies of the wall rocks. Geologic mapping of new mine development will add to the knowledge of the individual deposits.

The complete geologic story remains to be written.

## REFERENCES CITED

1. Safford, J. M., 1869, Geology of Tennessee; Nashville, 550 p.
2. Henrich, C., 1895, The Ducktown ore deposits and the treatment of Ducktown copper ores: A.I.M.E. Tr. v. 25, p. 173–245.
3. Gilbert, G., 1924, Oxidation and enrichment at Ducktown, Tennessee: A.I.M.E. Tr. v. 70, p. 998–1023.
4. Emmons, W. H. and Laney, F. B., 1926, Geology and ore deposits of the Ducktown mining district, Tennessee: U.S. Geol. Surv. Prof. Paper 139, 114 p.
5. Schmedeman, O. C., 1931, Geology of the Isabella ore body: Unpublished M.S. thesis, Univ. Wis. 49 p.
6. Ross, C. S., 1935, Origin of the copper deposits of the Ducktown type in the southern Appalachian region: U.S. Geol. Surv. Prof. Paper 179, 165 p.
7. Barclay, R. E., 1946, Ducktown back in Raht's time: Univ. of North Carolina Press, Chapel Hill, 283 p.
8. Simmons, W. W., 1949, Recent geologic investigations in the Ducktown mining district, Tennessee: p. 67–71 in Snyder, F. G., Editor, Symposium on mineral resources of the southeastern United States, Univ. Tenn. Press, Knoxville, 236 p.
9. Kingman, O., 1951, Geochemical techniques and results at Ducktown [Tenn.]: Min. Cong. Jour., v. 37, no. 10, p. 62–65.
10. Gibson, O., 1953, Heavy accessory mineral study in the Ducktown basin: Ga. Geol. Surv. Bull. no. 60, p. 278–288.
11. Hurst, V. J., 1955, Stratigraphy, structure, and mineral resources of the Mineral Bluff quadrangle, Georgia. Ga. Geol. Surv. Bull. no. 63, 137 p.
12. Hurst, V. J., 1955, Geologic map of the Epworth, Georgia quadrangle: Ga. Dept. of Mines, Mining, and Geol., Unpublished.
13. King, P. B., et al., 1958, Stratigraphy of Ocoee series, Great Smoky Mountains, Tennessee and North Carolina: Geol. Soc. Amer. Bull., v. 69, p. 947–966.
14. Long, L. E., et al., 1959, Chronology of major

metamorphic events in the southeastern United States: Amer. Jour. Sci., v. 257, p. 585–603.

15. Kulp, J. L. and Eckelmann, F. D., 1961, Potassium-argon isotopic ages on micas from the southern Appalachians: N.Y. Acad. Sci. Ann., v. 91, p. 408–419.

16. Hurst, V. J. and Schlee, J. S., 1962, Ocoee metasediments, northcentral Georgia, southeast Tennessee: Guidebook No. 3, Ga. Dept. of Mines, Mining, and Geol. 28 p.

17. Kinkel, A. R. Jr., 1962, The Ore Knob massive sulfide copper deposit, North Carolina, an example of recrystallized ore: Econ. Geol., v. 57, p. 1116–1121.

18. Moh, G. H., *et al.,* 1964, Studies of Ducktown, Tennessee, ores and country rocks: *in. Ann. Rept. Dir. Geophys. Lab.,* Carnegie Inst., Washington, Year Book 63, p. 211–213.

19. Moore, D. P., 1964, Distribution of trace elements in selected samples of wall rock from the Boyd ore body: Unpublished M.S. thesis, Va. Poly. Inst., 100 p.

20. Carpenter, R. H., 1965, A study of the ore minerals in cupriferous pyrrhotite deposits in the southern Appalachians: Unpublished PhD dissertation Univ. Wisc., 70 p.

21. Hernon, R. N., 1965, Geologic maps of the Ducktown, Tenn., Isabella, Tenn., and Persimmon Creek, N.C., quadrangles: U.S. Geol. Surv., unpublished.

22. Kalliokoski, J., 1965, Metamorphic features in North American massive sulfide deposits: Econ. Geol., v. 60, p. 485–505.

23. Kinkel, A. R., Jr., *et al.,* 1965, Age and metamorphism of some massive sulfide deposits in Virginia, North Carolina, and Tennessee: Geochim. et Cosmochim. Acta, v. 29, p. 717–724.

24. Maher, S. W., 1966, The copper-sulfuric acid industry in Tennessee: Tenn. Div. of Geol., I.C. 14, 28 p.

25. Hardeman, W. D., *et al.,* 1966, Geologic Map of Tennessee: Tenn. Div. of Geol., 1:250,000.

26. Carpenter, R. H., 1967, Regional metamorphism in the western Blue Ridge of Tennessee and North Carolina: Tenn. Val. Auth., Knoxville.

27. Brown, H. S., in preparation, Some petrographic features of rock alteration related to ore bodies at Ducktown, Tennessee.

*28. Westervelt, W. Y., 1919, Proposed consolidation of Ducktown properties.

*29. Gebhardt R. C., 1946, Surface prospecting in Ducktown.

*30. Snyder, F. G., 1948, Petrographic studies of the Great Smoky Formation of the Ducktown basin.

*31. Snyder, F. G., 1949, Correlation of beds of the Great Smoky Formation by heavy mineral methods.

*32. Grant, W. H., 1956, Minor folds in the Ducktown district, Tennessee.

*33. Brown, H. S., 1960, Wall-rock alteration at the Boyd mine, Ducktown, Tennessee.
—— 1961, Wall-rock alteration at the Calloway mine, Ducktown, Tennessee.

*34. Magee, M., 1961, Wall-rock alteration of the Cherokee vein.

* This citation is to a private report to the Tennessee Copper Co.

# 13. The Mascot-Jefferson City Zinc District, Tennessee

JOHNSON CRAWFORD,* ALAN D. HOAGLAND†

## Contents

## Illustrations

* The New Jersey Zinc Company, Jefferson City, Tennessee.
† The New Jersey Zinc Company, Sarasota, Florida.

## ABSTRACT

Zinc mining at Jefferson City began in 1854 with small scale production of oxidized ore from open pits. Significant production began in 1913 with the development of the Mascot Mine by the American Zinc Company. In 1965, four flotation mills were treating crude ore at the rate of about 12,000 tons per day and the Mascot-Jefferson City district currently is the largest zinc producing area in the United States. The producing mines, Mascot, Young, North Friends Station, New Market Zinc, Grasselli, Davis-Bible, Coy, and the Jefferson City, are 15 to 28 miles northeast of Knoxville in the Valley and Ridge physiographic province, an area of gentle relief lying between the more rugged Great Smoky Mountains and the Cumberland escarpment.

The ore is strata bound, occurring essentially within a stratigraphic range of 200 feet in the Lower Kingsport and Upper Longview formations of Lower Ordovician age. In the lower strata of the mineralized section, the ore generally is found in coarsely crystalline clastic dolomite breccia containing silica and chert debris within a sequence of aphanitic limestones. The upper strata of the mineralized section are fine-grained primary dolomites that, in the ore zones, have been fragmented in mosaic patterns by solution collapse with the interfragmental space filled by white gangue dolomite and sphalerite.

The mineralogy is unusually simple, sphalerite being the only important sulfide. Pyrite is very sparse. Dolomite gangue is abundant, silica was deposited in quite small amounts, calcite is minor, and fluorite and barite are very rare.

The principal ore controlling structures are rubble breccia zones that were formed during the post-Lower Ordovician-pre-Middle Ordovician erosion interval by supergene solution as a phase of the development of a regional karst system of underground drainage that extended to depths of at least 800 feet. The major Appalachian orogenic structures are post-ore and have been superimposed on the ore bodies. There is substantial evidence to support the current thinking of the writers that the ore deposits were formed at a depth not exceeding 800 feet.

## HISTORICAL

### Mining Operations

The first zinc mining in the Mascot-Jefferson City district was in 1854 at Jefferson City (then known as Mossy Creek). At this time open pit mining for zinc silicate and carbonate ores began at the site of the old Mossy Creek mine. A zinc smelter for the manufacture of zinc oxide was erected at Jefferson City in 1867, and operations were conducted intermittently on a small scale until 1894 by the Edes, Mixter and Heald Zinc Company (7).

The first substantial mining operation in the district was at Mascot. In 1890, open pit mining of oxidized ores was undertaken by the Edes, Mixter and Heald Zinc Company. In 1900, the Roseberry Zinc Company undertook shaft mining of zinc sulphide ores at the site of the present No. 3 shaft of the American Zinc Company. In 1911, the American Zinc, Lead & Smelting Company acquired the Mascot properties and in May, 1913, began operations on the basis of 1000 tons per day mill capacity (1).

With the successful operation of the Mascot mine, the American Zinc Company of Tennessee undertook explorations in the area from Jefferson City southwest through Friends Station to Mascot. The efforts resulted in the development of substantial tonnages of zinc ore and enabled the Company gradually to expand the Mascot milling operation to its present capacity of 5500 tons per day.

The Mascot mill has been supplied not only by ores mined at Mascot but also by ore tonnages from the Jarnigan and Athletic mines (now depleted), and from the Coy, Grasselli, North Friends Station, and Young mines. The crude ore production from these mining operations has been shipped by rail and truck to the Mascot mill.

Intermittent mining of oxidized ores by shallow shafts and open pits on properties east of New Market began in 1892 and was carried on until 1913. In 1907, the properties were acquired by the Grasselli Chemical Company, which company conducted exploratory drilling for zinc sulfide ores and began shaft sinking and mine development in 1925. In 1937, the American Zinc Company of Tennessee began production of zinc sulfide ores at the Grasselli mine under contract with E. I. du Pont de Nemours, which had acquired the property. Ore produced was shipped by rail to the Mascot mill. In 1947, American Zinc, Lead & Smelting Company purchased the property from du Pont.

From 1901 to 1922, exploration for zinc sulfide ores was conducted by a number of companies including American Metal Company, Osgood Exploration Company, American Smelting & Refining Company, and The New Jersey Zinc Company. These efforts did not result in the development of any new mines.

In 1926, Universal Exploration Company, a subsidiary of United States Steel Corporation, entered the district and conducted exploration on a larger scale than any heretofore undertaken. This resulted in extensive discoveries immediately south of Jefferson City. In May, 1930, a 900 ton-per-day mill was completed, and operations began at the newly developed Davis-Bible mine. The property has operated continuously from that date and is currently producing at the rate of 1500 tons per day.

The Davis-Bible mine has been developed extensively over a strike length of several miles and includes workings within the city limits of Jefferson City. Mining under residential lots within the city limits was accomplished under a "Community Lease Agreement" wherein each property owner shared proportionally in royalties on ore produced from the leased area.

In 1928, the American Zinc Company developed the Jarnigan mine east of Jefferson City, and mining operations were conducted intermittently for approximately 20 years. During World War II, the American Zinc Company undertook the development of the Athletic mine on property of Carson-Newman College. This deposit, which had been discovered by the Universal Exploration Company, sustained a mining operation for approximately 5 years.

At the close of World War II, exploration for zinc sulfide deposits was undertaken on a much larger scale than any heretofore by The New Jersey Zinc Company, the American Zinc Company and the United States Steel Corporation. These efforts resulted in significant ore discoveries and proved the extensions of ore deposits under the low angle Koppick Knob or Bays Mountain overthrust fault that bounds a large portion of the district.

Substantial discoveries were made by The New Jersey Zinc Company at Jefferson City and Beaver Creek (Hodges). In 1953, the Company started development of the Jefferson City mine and the construction of a 1000 ton-per-day mill. In 1956, production operations commenced, and, in 1959, the mill capacity was increased to 2000 tons per day.

Discoveries made by the American Zinc Company, principally subsequent to World War II, resulted in the development of the North Friends Station, Young, Immel, and Coy mines. The New Market mine was developed under lease agreement with the American Zinc Company by the Tri-State Zinc Company, a subsidiary of Consolidated Gold Fields, Ltd. Construction of a 3000 ton-per-day mill at the New Market mine was undertaken in 1961. The geographic location of the district is shown in Figure 1 as are those of the principal mines.

In 1954, the United States Steel Corporation explored the Mitchell Bend area near Richland and proved the presence of a substantial zinc deposit.

## Statistics of Mine Production

Four zinc mills are currently in operation in the Mascot-Jefferson City district:

| | | | |
|---|---|---|---|
| (1) | Mascot Mill | 5500 | Tpd Capacity—American Zinc Company |
| (2) | New Market Mill | 3000 | Tpd Capacity—Tri-State Zinc Company |
| (3) | Davis-Bible Mill | 1500 | Tpd Capacity—U.S. Steel Corporation |
| (4) | Jefferson City Mill | 2000 | Tpd Capacity—The New Jersey Zinc Co. |
| | Total | 12,000 | Tons-per-Day Capacity |

The U.S. Bureau of Mines reports the total zinc production from Tennessee during the period 1850 through 1962 at 1,597,014 tons of recoverable zinc. The total value at the time of production is recorded at $322,381,000 through 1962. In 1963, Tennessee produced 95,847 tons of recoverable zinc valued at $22,045,000. In 1964, the U.S. Bureau of Mines records the Tennessee production at 115,943 tons with an estimated value of $31,536,000. The total Tennessee zinc production from 1850 through 1964 is 1,788,661 tons of recoverable zinc valued at $374,426,000.

Of the total Tennessee zinc production, approximately 85 per cent was from the Mascot-Jefferson City district. The balance of the production came mainly from four areas: the Ducktown area in Polk County, the Copper Ridge belt in Hancock County, the Powell River area in Claiborne and Union counties, and the Embreeville area in Washington and Unicoi counties.

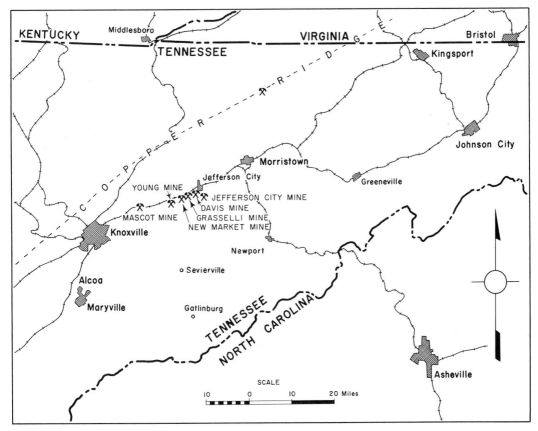

FIG. 1. *Map of Location of Mascot-Jefferson City Zinc District, Tennessee.*

## PHYSIOGRAPHIC HISTORY AND PRESENT TOPOGRAPHY

The Mascot-Jefferson City district lies in the Appalachian Valley, which is bounded on the southeast by the Great Smoky Mountains with elevations ranging from 5000 feet to more than 6000 feet and on the northwest by the Cumberland Plateau with a general surface elevation of from 2000 to 2500 feet. The topography of the mining district is a subdued expression of the valley and ridge type, greatly modified locally by highly developed karst features. Maximum relief in the district is between 1050 feet to 1500 feet above sea level, but generally slopes are gentle. The Paleozoic sedimentary lithologies and Appalachian orogenic structures provide the matrix from which weathering has sculptured the land surface.

Present topography is the result of essentially continuous erosion since the Appalachian orogeny, and remnants of the oldest recognizable surfaces were formed at least as early as the Cretaceous. The residual soils, which are commonly 50 to 100 feet thick over the Kingsport and Mascot formations, appear to have been produced by the dissolution of solubles throughout the full span of Cenozoic time. These dominantly dolomite formations, host rocks of the productive zinc deposits in East Tennessee, form belts of quite flat terrain with rich soils developed on a thick residuum of red clay. Thin alluvial material related to terrace deposits is widely, but rather sparsely, distributed in the mining district. The finest farms and the most desirable terrain for urban use are found on the Kingsport and Mascot formations. The superimposition of moderate folding and faulting of the rocks has resulted in a somewhat intricate pattern of soil types and land forms. In the Mascot-Jefferson City district, the geologic factors favoring mineral wealth coincide with other favorable agricultural and economic factors.

The Chickamauga series of Middle Ordovician age, which overlies the ore beds, is composed predominantly of alternating limestone, marble, and shaly limestone. These sediments generally are characterized by very thin soils, extensive outcrop or rocky terrain, rough topography of low relief, and locally extreme sink hole development and sub-surface drainage. In general, these are forest lands poorly suited for agriculture.

The rocks underlying the Kingsport-Mascot formations, the lower Knox dolomites and the shales and limestones of the Conasauga group, are, likewise, characterized by thin, rocky, poor soils and by rough terrain or moderate relief. Farms tend to be small and poor in this geologic environment that is largely devoid of urban development.

The influence of structure is pronounced, and, where the dip of the beds is steep or the folding is compressed, the unfavorable aspects of the lithologies are accentuated.

The weathering of the zinc ore bodies is most active within a few tens of feet of the interface between rock and clay residuum. Ground water percolation is controlled largely by joints and fractures and lessens with depth. Moderate to locally strong circulation extends to depths in excess of 1200 feet below the surface (14), but oxidation of ore is quite insignificant below a depth of about 300 feet. Local bodies of oxidized zinc ore have been formed at the rock-clay interface. With further weathering, a considerable fraction of the zinc becomes, after some dispersal, fixed in the clay in association with the hydrous iron oxides and in the crystal lattice of the clay minerals. Minor zinc as base-exchangeable ions associated with the clay minerals remains moderately mobile (12, 6). With time and the development of a mature clay residuum, extensive zinc enriched halos were developed from the weathering of the mineralized rock. The resulting distribution of the anomalous zinc in the soil is a useful guide in the search for zinc ore bodies that may be covered by a residual mantle of clay and soil.

## GEOLOGIC HISTORY

### Stratigraphy*

The zinc ore mined to date in the Mascot-Jefferson City district occurs in the Longview and lower Kingsport members of the Knox dolomite. The upper part of the Knox dolo-

* Editor's Note: For a detailed discussion of the stratigraphy of the district, reference should be made to: Bridge, J., 1956, Stratigraphy of the Mascot-Jefferson zinc district: U.S. Geol. Surv. Prof. Paper 277, 76 p., particularly p. 7–60.

mite in this district is composed of the Longview, Kingsport, and Mascot formations of early Ordovician age. The upper Kingsport formation is composed dominantly of fine-grained dolomite beds that, because of their regional extent and uniformity, are regarded to be of primary or diagenetic origin. The lower part of the Kingsport and the uppermost part of the Longview are dense limestones interbedded with fine-grained dolomites, the R,S,T and U beds (3, Figure 1; 4, Table 2; 15, Table 1) (Figure 2).

The ore-grade deposits in the northeast portion of the district (Jefferson City) have a somewhat lower stratigraphic position than those in the southwest portion (Friends Station, Hodges, and Mascot).

In the mining operations near Jefferson City, the "U" bed of the upper Longview is an important producer, while, in the area between New Market and Mascot, only a very minor ore tonnage is known to occur in this bed. The ore in the "U" bed averages 10 to 12 feet in thickness.

The "S" bed is a prominent producer throughout the district and is one of the major sources of ore. In the northeast portion of the district, the thickness of ore in the "S" bed averages about 15 feet. In the southwest portion of the district, the thickness is somewhat greater, probably averaging 25 to 30 feet.

The "R" and post-"R" beds are more extensively mineralized in the southwest portion of the district (Mascot and Young mines), than in the northeast portion.

In the "breakthrough" areas, the stratigraphic range of maximum ore development is locally in excess of 200 feet, extending from slightly below the base of the "U" bed to approximately 100 feet above the top of the "R" bed. The "breakthrough" areas are considered to occur along zones of major solution and collapse. In general, they have somewhat of an irregularly circular configuration with appreciably greater height than width. The linear extent, however, is substantial and in excess of 500 feet. Figure 6 shows a generalized cross section of a portion of a "breakthrough" ore body.

### Structure*

The ore deposits of the Mascot-Jefferson City district occur in the Kingsport formation

* Editor's Note: The various structural belts of which the district is composed are treated in Bridge, J., 1956, Stratigraphy of the Mascot-Jefferson City zinc district: U.S. Geol. Surv. Prof. Paper 277, 76 p., particularly p. 61–69.

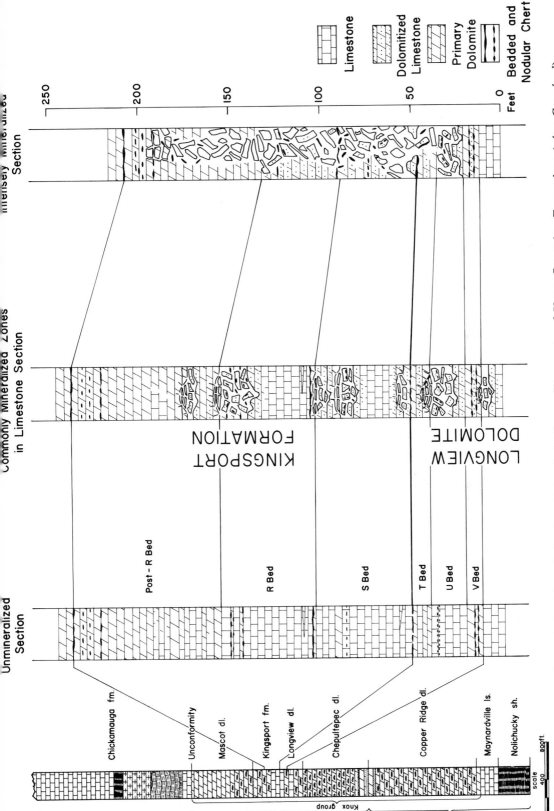

FIG. 2. *Detailed Stratigraphic Relationships of the Ore Beds of the Lower Kingsport and Upper Longview Formations (after Crawford).*

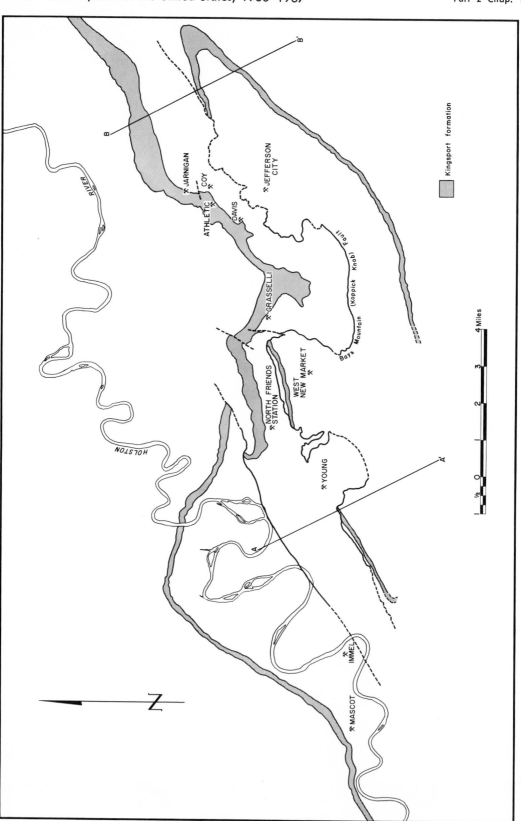

FIG. 3. Map Showing Location of Mines in the Mascot-Jefferson City Zinc District and Relationship to Kingsport Formation Outcrop.

in the footwall block of a major low angle overthrust fault. (Figure 3). This overthrust fault, originally known as the Koppick Knob fault, has also been called the Rocky Valley fault and the Bays Mountain fault. It extends along strike for about 20 miles, dying out at each extremity. At the Jefferson City mine, the horizontal displacement is about 4 miles, an approximate measure of the maximum movement along this structure.

For years, it has been traditional to associate the localization of the ore deposits with structures formed by the Appalachian orogeny. Recent geologic evidence has thrown a great deal of doubt on this interpretation and indicates the ore-controlling structures to be due to solution and collapse and to be pre-Appalachian orogeny in age.

This evidence, which is presented in a separate section, provides a basis for logically dividing the structural development of the area into two separate periods:

(1) Pre-Appalachian structures which may have been significant in ore localization and

(2) Appalachian structures of much greater intensity which were superimposed over, and greatly confused and largely obliterated the early structural situation.

The pre-Appalachian structures of early Ordovician age were initially formed by processes of diagenesis involving dehydration and lithification of the sediments. This was followed by gentle folding and doming with resultant jointing. Possibly major tear faults developed with horizontal slip and essentially no vertical displacement of the bedding. Minor flexures and folds developed, and areas of solution and collapse were localized along these early structures. These pre-Appalachian structures were well developed prior to the deposition of the middle Ordovician Chickamauga limestone.

The dominant structural features of the district are those that were formed during the Appalachian orogenic period and are post-Mississippian in age. These are extensive gentle folds, with axes slightly overturned and dipping to the southeast, and extensive low angle thrust faults also with a gentle southeasterly dip. These structures are modified by local transverse doming and cross tear faults. The Mascot-Jefferson City district occurs on a prominent cross fold which constitutes a type of structural anomaly. The detail of the relationships of the folds and thrust faults are illustrated in Figure 4. The structural interpretations have been modified by F. T. Fischer.

Based on available stratigraphic evidence, the Bays Mountain overthrust fault must be later than the middle Ordovician Chickamauga limestone. However, parallel and similar faults both northwest and southeast cut formations as young as Mississippian. By analogy, therefore, the Bays Mountain thrust is commonly considered to have occurred during the Appalachian orogenic period.

## Age of Mineralization

Until Kendall's observations (9) showed that zinc mineralization in the Jefferson City mine preceded the main structural disturbances of the Appalachian orogeny, it was generally held that the ore minerals were introduced late in the Appalachian orogeny.

Kendall noted the presence of detrital material in certain of the ore structures, and the term sand has been applied to it. The sand is generally light grey to almost white and is made up mainly of particles of dolomite with some detrital sphalerite. Locally, however, the sphalerite content is high enough to impart a yellowish cast to the sand, which commonly exhibits graded bedding. The sand is clearly post-ore in age and has not been found brecciated nor cut by later fracturing. This sand is a minor constituent of total volume of ore, probably less than 1 per cent.

The most significant observation with respect to this sand is that the laminations are parallel to the strike and dip of the country rock. This parallelism of the bedding of the formation and the sand laminations indicates that the formations were essentially horizontal when the sand was deposited. Hence the ore must have been deposited before any tilting of the strata occurred. It follows, therefore, that the ore deposition must have taken place before the period of the Appalachian orogeny. Subsequently this evidence of pre-folding age has been abundantly substantiated by being observed in other mines in the district. As a result of this very significant discovery by Kendall, ore textures and structures and regional relationships were viewed in a new light. Present interpretations suggest the age of mineralization to be as early as the pre-Chazyan unconformity (15). This interpretation is supported by the fact that no zinc occurrences of magnitude are known above this unconformity.

Since galena is essentially absent from the district, samples of galena from another East Tennessee mine (Flat Gap) were treated by lead isotope analysis to determine if the isotopic relationships would reveal significant information respecting the age of the mineralization in these districts. These measurements, made by Isotope, Inc., of Westwood, New Jersey, are tabulated below:

| Sample Number | Atom Per Cent | | | |
|---|---|---|---|---|
| | 204 | 206 | 207 | 208 |
| Flat Gap #1[1] | 1.325 | 25.38 | 21.00 | 52.29 |
| Flat Gap #2[1] | 1.327 | 25.40 | 21.01 | 52.26 |
| Flat Gap #3[1] | 1.327 | 25.35 | 21.04 | 52.28 |

[1] Mineralization at the Flat Gap mine in the Copper Ridge district of East Tennessee was essentially contemporaneous with the mineralization in the Mascot-Jefferson City district.

Apparently these lead isotope relationships do not provide a basis for age determination.

The age and nature of the ore breccias is one of the keys to the problem of age determination. Ulrich (2) regarded these to be cave or sink hole breccias and emphatically rejected the theory that they were fault breccias. Nevertheless, there was at that time an almost unanimity of opinion supporting the orogenic origin of these great structures. But as Oder and Ricketts point out (11), Oder (8), Laurence (10), Wedow (16), and others have in the past several years suggested that a significant relationship exists between the post-Knox, pre-middle Ordovician unconformity and the formation of the reservoirs for the zinc deposits of East Tennessee. This idea was further developed and documented by Hoagland, *et al.*, (15) who conclude that the age of zinc deposition is approximately contemporaneous in time with the formation of the post-Mascot pre-middle Ordovician breccias. The breccias are considered to be dissolution collapse structures that resulted from the pre-Chazyan karst system and were modified by pre-orogenic hydrothermal mineralization phenomena.

Kendall (9) and others have suggested that a significant part of the sphalerite may be syngenetic. The writers agree that there are widespread traces of syngenetic sphalerite, especially that associated with unaltered chert. Probable syngenetic sphalerite is rarely, if ever, identified in the ore zones but is commonly found in minor amounts in weakly mineralized areas. Whether it may have been sufficiently abundant on a regional basis to have been a source for the local concentration into ore bodies is an interesting but speculative possibility.

Although nearly all observers have agreed that the sphalerite in the ore is epigenetic, all but the earliest of the breccias contain brecciated fragments of ore. It is apparent, therefore, that, as Odell suggested (5, p. 78), much of the ore breccia was formed by dissolution during the mineralizing process. The better-grade ore bodies are commonly authigenic ore breccias in which ore fragments of various sizes are veined by ore and gangue minerals, indicating that the mineralizing process by dissolution and collapse maintained a permeable environment concurrently with the deposition of the ore and gangue minerals.

## ECONOMIC GEOLOGY

### Forms of the Ore Bodies

Ore bodies of the Mascot-Jefferson City district are characteristically irregular in shape. They are, however, stratiform in that their major dimensions are planar and are controlled by stratigraphy. The "U" bed manto ore bodies in the eastern (Jefferson City) part of the district are striking examples of this. These ore bodies are largely confined to a limestone bed ("U" bed), about 20 feet thick, between two layers of fine-grained primary dolomite. The mantos extend for many hundreds, or even thousands, of feet in a network of runs and crossings (Figure 5). In cross-section, the "U" bed ore zones are quite irregular, often being mushroom shaped.

Ore bodies in the R, S and upper Kingsport beds are also manto-like, but they are much more irregular in cross section and are generally of greater vertical extent than those of the "U" bed (Figure 6).

### Mineralogy of the Deposits

The Mascot-Jefferson City Zinc district is unusual in the simplicity of the mineralogy. Sphalerite is the only primary metallic mineral of economic significance, and it is of unusual purity, generally containing less than 0.5 per cent iron. Concentrates from the mines of the district commonly contain better than 64 per cent Zn. These concentrates contain on the average of .025 per cent cadmium. Pyrite, the next most abundant sulfide, is sparse and, in the eastern part of the district, is exceedingly so. On the average, pyrite makes up probably less than 1 per cent of the total sulfide. Chalcopyrite and galena have been found, but these minerals are extremely rare.

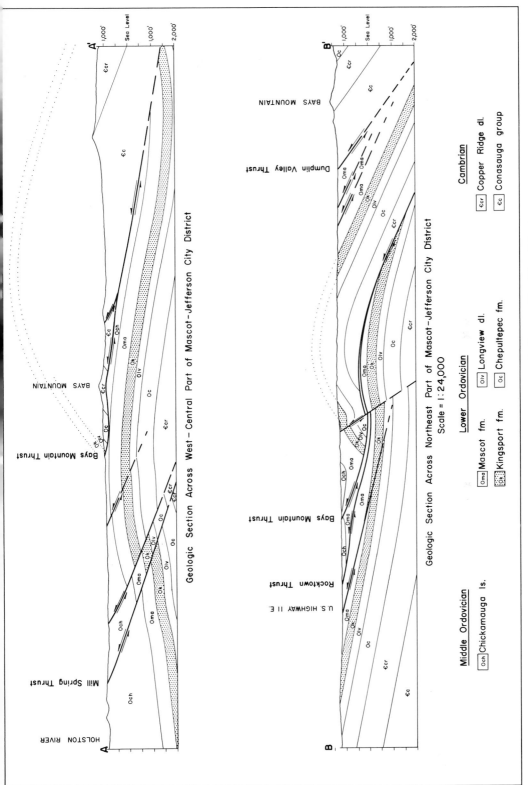

FIG. 4. *Geologic Sections across Mascot-Jefferson City Zinc District.*

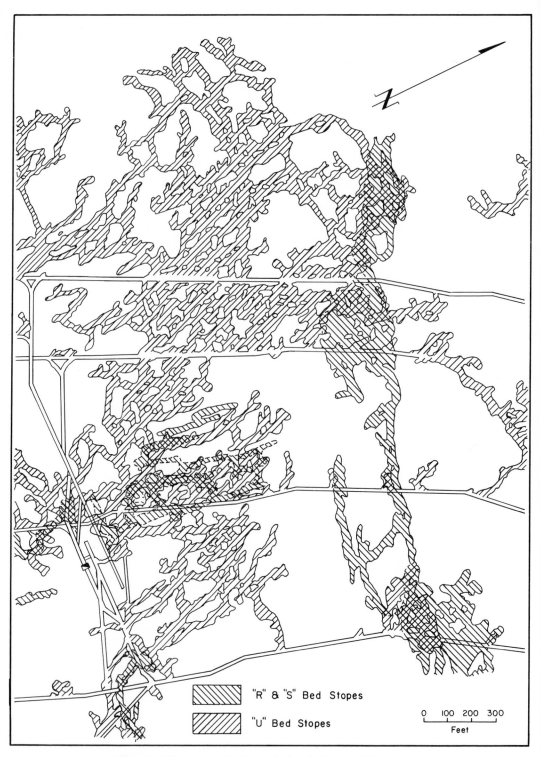

"R" & "S" Bed Stopes

"U" Bed Stopes

0   100  200  300
Feet

FIG. 5. *Plan of a Portion of the Jefferson City Mine.*

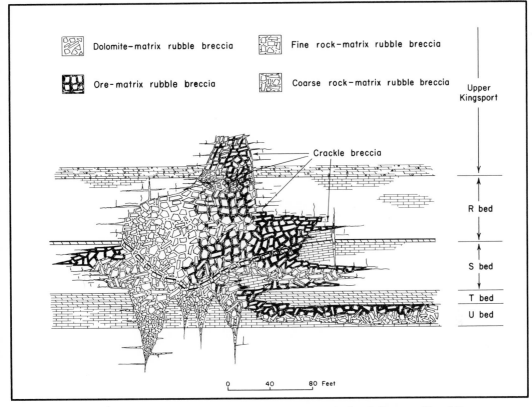

FIG. 6. *Generalized Section through a Portion of the Jefferson City Mine.*

Dolomite is the most abundant gangue mineral as would be expected in a dolomite rock host. There appears to be a gradation from re-crystallized sedimentary dolomite or dolomitized limestone to the coarsely crystallized white dolomite which fills, along with the sphalerite, many of the open spaces of the crackle and rubble breccias. Mineral dolomite formation began at the inception of, or before, sulfide deposition and continued until after the major sulfide deposition was completed. Minor calcite was introduced subsequent to dolomite deposition, and the lesser amounts of fluorite and barite were brought in after the dolomite and before the calcite. Post-dating the late fluorite, traces of sphalerite were deposited as minute crystals on the surfaces of the earlier minerals in vugs as filling in minute cracks within the ore zones. The zinc in this late deposition is quantitatively insignificant.

Silicification of the limestone walls accompanied the deposition of sphalerite. Silica is the second most abundant gangue mineral, but it is very subordinate in quantity compared with dolomite.

Overlying the ore zones the normally grey, bedded massive dolomite is frequently bleached to a light-gray of chalky white color. Although the cause of the bleaching is not understood, it has a striking association with ore zones in the "R" and "S" beds and must be related in some way to the mineralizing process.

There is a suggestion of a regional zoning of minerals, but, because of the complex regional structure, the fragmentary data are not sufficient to justify a firm conclusion that zoning exists. Pyrite is somewhat more abundant in the western part of the district. Northeast and southwest of the district, the Fall Branch and the Sweetwater districts have dominant barite, accessory fluorite, and minor sphalerite. To the north, in the Copper Ridge Belt, and also at the same stratigraphic horizon as the Mascot-Jefferson City ores, the Flat Gap mine is characterized by moderate amounts of associated pyrite and appreciable, though minor, galena. These are believed to be elements of a regional mineral zoning pattern on a lateral base. Depth zoning, however, is not apparent over a vertical range of about 3000 feet. This is not surprising as deposition is believed to have been accomplished at a quite uniform depth of from 500 to 800 feet below the then-existing surface and variations of depth today result from the post-mineral Appalachian orogeny.

## Factors Controlling Form and Location of Ore Bodies

The significant commercial ore of the Mascot-Jefferson City district is concentrated in a stratigraphic range of about 220 feet which, for practical purposes, may be thought of as the Kingsport formation. Although scattered ore minerals are widely dispersed some distance above and below this approximately 220-foot zone (that is, below the "U" bed and above the 197 bed at the top of the 220-foot zone), such mineral development rarely is of mineable tonnage and grade. The question is, why has the Kingsport been so favored? The supposedly favorable lithologies of the Kingsport are not unique to that formation and quite similar beds both above and below were bypassed by the mineralization. The important effect of the pre-Chazyan erosion on the Mascot and Kingsport formations has been suggested by a number of geologists such as Oder, Laurence, Ricketts, Wedow, and Maher. Oder and Ricketts (11), and Hoagland, et al. (15) have presented evidence that establishes the significance of this relationship. Callahan (13), summarizing a wide suite of stratiform mineral deposits, suggests that there is a great degree of dependence on surfaces of unconformity in the localization of this class of deposit.

The significant basis for the stratigraphic control of the Mascot-Jefferson City district ore bodies is considered to be keyed to two principal factors as follows:

(1) The development in post-Mascot time of a mature karst system involving Mascot, Kingsport, and locally upper Longview carbonate sediments to depths of 800 feet or more below this surface. The extensive circulation of meteoric water that produced the karst system resulted in unusual permeability and complete sub-surface drainage along major systems or networks, partially filled with debris. A crude analogy may be drawn between this pre-middle Ordovician surface and sub-surface in East Tennessee and the present day conditions of the Florida and the Yucatan peninsulas. The minor joints and shears resulting from the uplift, dehydration, and lithification doubtless controlled the major trends of the channels of circulation.

(2) Mineralizing water responsible for dolomitization and zinc deposition found and made use of the permeable system developed by the meteoric waters. It is suggested that the vertical limits of mineralization were controlled by the static head and by the temperature drop resulting from the intermingling of mineralizing waters with the surface waters.

## Stratigraphic Relations of the Ore Bodies

The lower Kingsport is predominantly a limestone in its unaltered state, whereas the upper Kingsport is composed of bedded fine-grained primary dolomite. Essentially all of the ore (except the U-bed ore) is produced from the Kingsport and in general may be divided into two types. In the lower Kingsport, the ore bodies are made up of sand and breccia derived from rock dolomite, bleached chert, and fragments of ore cemented by veins and veinlets of gangue dolomite and sphalerite. This ore is generally sharply bounded laterally by limestone (commonly silicified) walls. Disseminated sphalerite grains in the coarse dolomite matrix are in part clastic debris and in part chemical precipitates filling interstitial spaces. The amount of replacement is quite minor.

The upper Kingsport ore is predominantly a crackle breccia of the brittle fine-grained primary dolomites. The interfragmental space is filled with white dolomite gangue and sphalerite. The crackle-breccia filling is later in age than the ore breccias described in the preceding paragraph.

The strong stratigraphic control in the Kingsport ore bodies of East Tennessee is related to the Kingsport lithologies in part, but the relationship of the ore bodies to the post-Mascot pre-Chickamauga unconformity appears to be the dominating factor. Two things were particularly important. These were:

(1) The development of extensive zones of sub-surface permeability and circulation in the Kingsport formation by meteoric waters of Chazyan age to depths of approximately 800 feet below the Mascot surface.

(2) The temperature-pressure relationship at depths of from 600 to 800 feet below the surface and the mixing with meteoric waters are postulated to have been favorable for the

deposition of sphalerite from the mineralizing waters.

No speculation is presented here as to the source of these mineralizing waters nor as to the source of the zinc that was deposited by them. It seems clear, however, that the mineralizers were not the meteoric waters that caused the early development of the paleokarst system in the Kingsport formation.

## Detailed Structural Patterns

A serious effort has been made over a period of many years to discover a structural key to the localization of the ore bodies in the expectation that such knowledge would greatly assist development planning and exploration. Because the structures formed by the Appalachian orogeny have all but obliterated the more subdued earlier structural features and because it was generally held that the ore is post-orogeny in age, the search for this key has been directed towards the possible relationship of ore to faults, lineaments, and various types of folds. This search has not disclosed any unique or significant association of ore with structures of this class. Ore bodies are found trending across structural axes along flanks of folds and in domal and synclinal areas as well. Nor is there a definitive relationship of ore trends to faults, most of which are cross-cutting and offset the ore zones. Thus, the relationship of ore to the prominent and well-defined structures is random or accidental. This is, of course, the pattern that would develop with the structural elements super-imposed on pre-existing mineralized areas.

The significant ore controls appear to be the large rock-matrix rubble breccia areas that have been described previously as the products of weathering during post-Mascot, pre-Chickamauga time. Undoubtedly pre-ore structures controlled the localization of the great breccia zones. These subtle structures, coordinate systems of joints for the most part, were largely obliterated by the immense breccia bodies that were formed. The breccias were the permeable zones when mineralization occurred. Some were mineralized, most were not.

Rock-matrix rubble breccias are of complex shape. In their upper parts, above the stratigraphic horizons of the ore zones, they are sometimes circular or oval in plan and pipe-like in section. In the Kingsport they commonly are rudely tabular linear bodies that extend horizontally for thousands of feet. Adjacent to these, with many irregularities, pipes, offshoots, and spurs, are the structures along which ore zones of great size and length have been localized.

## Favorable Beds

The limestone horizons of the lower Kingsport are favorite loci for ore. The ore, however, invariably is in coarse crystalline dolomite, coarse dolomite breccia, or hybrid breccia types. Only in the highly silicified ore-contact zones will the limestone be mineralized for a foot or two at most. It appears that the coarse dolomite breccias and the much less abundant coarse dolomite sands are, for the most part at least, epigenetic. They are clastic, and they are unconformable with the enclosing sediment. These channels of dolomite and siliceous debris are the favorable host rocks.

Overlying these coarse dolomite breccia channels, or associated with areas in which the limestone has been removed or greatly thinned, are bodies of mosaic or crackle breccia. These are collapse structures and are invariably associated in one way or another with the channels of coarse dolomite debris. Important ore zones were formed in the crackle breccias with sphalerite and white gangue dolomite filling the spaces between the rock dolomite fragments. The ore here, as in the coarse dolomite environment of the limestone horizons, filled open spaces. Although these deposits are persistently, though erroneously, referred to as replacement deposits, replacement is quantitatively a very minor phenomenon in these ores.

While the continued studies of data revealed by mining and exploration have contributed much to the geological understanding of the district, there still is no convincing evidence as to the source of the mineralization. However, with persistence in the exploration effort, the Mascot-Jefferson City district has continued to grow and is now the leading zinc producing area of the United States.

## REFERENCES CITED

1. Secrist, M. H., 1924, Zinc deposits of east Tennessee: Tenn. Div. Geol. Bull. 31, 165 p.
2. Ulrich, E. O., 1931, Origin and stratigraphic horizon of the zinc ores of the Mascot district of east Tennessee: Wash. Acad. Sci. Jour., v. 21, p. 30–31.
3. Crawford, J., 1945, Structural and stratigraphic control of zinc deposits in east Tennessee: Econ. Geol., v. 40, p. 408–415.
4. Oder, C. R. L. and Miller, H. W., 1948, Strati-

graphy of the Mascot-Jefferson City zinc district: A.I.M.E. Tr., v. 178, p. 223–231.

5. Oder, C. R. L. and Hook, J. W, 1950, Zinc deposits in the southeastern states: p. 72–87, particularly p. 73–82, *in* Snyder, F. G., *Editor, Symposium on mineral resources of the southeastern United States,* Univ. Tenn. Press, Knoxville, 236 p.

6. White, M. L., 1956, The occurrence of mineralization in soil: Unpublished Rept. to the N. J. Zinc Co.

7. Maher, S. W., 1958, The zinc industry of Tennessee: Tenn. Div. Geol. I. C. no. 6, 28 p.

8. Oder, C. R. L., 1958, How American Zinc's Tennessee DMEA project proved 35,000,000 tons ore: Min. World, v. 20, no. 7, p. 50–53.

9. Kendall, D. L., 1960, Ore deposits and sedimentary features—Jefferson City Mine, Tennessee: Econ. Geol., v. 55, p. 985–1003; 1961, v. 56, p. 1137–1138; disc., v. 56, p. 444–446.

10. Laurence, R. A., 1960, Geologic problems in the Sweetwater barite district, Tennessee: Amer. Jour. Sci., Bradley Volume, v. 258-A, p. 170–179.

11. Oder, C. R. L. and Ricketts, J. E., 1961, Geology of the Mascot-Jefferson City zinc district, Tennessee: Tenn. Div. Geol. R. I. no. 12, 29 p.

12. Hoagland, A. D., 1962, Distribution of zinc in soils overlying the Flat Gap Mine: Min. Eng., v. 14, no. 1, p. 56–58.

13. Callahan, W. H., 1964, Paleophysiographic premises for prospecting for strata bound base metal mineral deposits in carbonate rocks: *CENTO Symposium on Mining Geology and Base Metals,* Ankara, p. 191–248.

14. Miller, H. W. and Jolley, D. H., 1964, Flooding and recovery of the Jefferson City Mine: Min. Cong. Jour., no. 1, p. 16–21.

15. Hoagland, A. D., *et al.*, 1965, Genesis of the Ordovician zinc deposits in east Tennessee: Econ. Geol., v. 60, p. 693–714.

16. Wedow, H., Jr. and Marie, J. R., 1965, Correlation of zinc abundance with stratigraphic thickness variations in the Kingsport formation, west New Market area, Mascot-Jefferson City mining district, Tennessee: U.S. Geol. Surv. Prof. Paper 525-B, p. B17–B22.

# 14. Geology and Mineral Deposits, Midcontinent United States

FRANK G. SNYDER*

## Contents

* 603 S. Gables Blvd., Wheaton, Illinois.

## Illustrations

## Tables

## ABSTRACT

The Precambrian of Midcontinent United States includes a metamorphic belt of probable Middle Precambrian age, a belt of Keweenawan volcanics and sediments, and widespread igneous activity that extended from Iowa to Texas.

Most of the present structural configuration of the Midcontinent was developed by epeirogenic movements during the Paleozoic. The dominant elements are basins and arches.

Three major metallogenic provinces occur in the Midcontinent. They include the Missouri Precambrian iron ores, the Lake Superior copper ores, and the Mississippi Valley lead-zinc-barite-fluorite deposits.

Several large iron deposits, occurring in rhyolite and andesite host rocks, are known. Both discordant massive bodies and replacement-type bodies occur. The iron deposits are believed to have been formed during the period of igneous activity. They are essentially the same age as the volcanics and older than the granites that intrude the volcanics.

The Lake Superior copper ores include two established districts, the Keweenawan district and White Pine, and the copper-nickel deposits in the Duluth gabbro.

The lead-zinc-barite-fluorite deposits include four major districts, Southeast Missouri, Upper Mississippi Valley, Tri-State, and Illinois-Kentucky, and at least eight minor districts. Ore may occur in stratiform deposits and in veins. The major districts all contain stratiform deposits, but several have produced appreciable quantities of vein ore. Only vein deposits are known in some of the smaller districts.

The major features of all districts can be catalogued in terms of host rocks, structural patterns of districts, types of ore-bearing structures, and mineralogical characteristics. From these features, a Mississippi Valley type ore body, in its type area, can be defined.

## INTRODUCTION

The Midcontinent embraces the area from the folded Appalachians on the east to the Rocky Mountain front on the west, from Lake Superior on the north to the Gulf Coast embayment on the south. In area, it comprises about one-third of continental United States.

Geologically, the region is largely one of

Paleozoic outcrop; of "layer cake" structure, in which formations are flat-lying or only gently folded; of sedimentary rocks uncomplicated by widespread metamorphism and igneous activity. Topographically, the area is one in which plains predominate. Upland areas include the Cumberland and Alleghany Plateaus, dissected plateaus like the Ozarks, and a few monadnocks on the Precambrian surface. The Ouachita province of Oklahoma and Arkansas, an extension of the Appalachian structural belt, forms the only folded mountains.

Mineral deposits of three distinctive types are important. These are the Missouri magnetite bodies in Precambrian volcanic rocks, copper deposits of the Lake Superior area, and the Mississippi Valley type deposits of lead, zinc, barite, and fluorite. These three types of deposits account for most of the metal production of the region and represent an important segment of the United States mineral industry.

## GEOLOGIC SETTING

### Precambrian

The exposed Precambrian in the northern Midcontinent ranges in age from granite gneisses of the Minnesota River valley dated at 3550 m.y., the oldest dated rocks in North America (34), to late Keweenawan sediments. Recent studies based on age dating of basement rocks intersected by drillholes (44) permit definition of major events in the buried Precambrian (Table I).

Rocks of early Precambrian age extend southwestward from the outcrop area across the Dakotas, through the Black Hills, and into Wyoming. Age dating (45) suggests that early Precambrian may be present in parts of Nebraska and Kansas.

Rocks of middle Precambrian age appear to form an arcuate belt across Missouri, Kansas, and Nebraska and into the Rocky Mountain area (55). Within this belt, steeply-dipping Huronian-type metasediments are cut by intrusive diorite dated as 1460 m.y. age (49). The metamorphic belt is not exposed, and no mineral deposits within it are known.

Several events of late Precambrian time can be delineated (Figure 1). The abundant granites and silicic volcanics, encountered in many drillholes and exposed in a few areas, are part of a belt of igneous activity, in which the rocks produced all become progressively

TABLE I.   *Major Midcontinent Precambrian Events*

| Era | Approx. age | Event |
|---|---|---|
| Late | 600 | Uplift and erosion |
| | | Faulting |
| | 1100 | Grenville orogeny |
| | | Development of Keweenawan basin extending from Lake Superior, across Minnesota, Iowa, Kansas. Extrusion and intrusion of mafic igneous rocks; deposition of clastic sediments. |
| | 1350 to 1150 | Widespread igneous activity from Iowa to Texas. Includes St. Francois, Nemaha, Spavinaw, and Panhandle rhyolites and granites. |
| | 1500 | Intrusion of intermediate to mafic magmas in Missouri |
| Middle | 1700 | Penokean orogeny |
| | | Folding, metamorphism, and intrusion |
| | | Deposition of clastic sediments ments across central Missouri Kansas, Nebraska. |
| Early | 2500 | Algoman orogeny in northern states |
| | | Possible belt of metasediment in Nebraska and Kansas. |

EXPLANATION
Metamorphic belt of Middle Precambrian age
Keweenawan belt.
Major fault lineament.

Fig. 1. *Precambrian Structural Features.*

younger southward (49). The iron deposits of Missouri were formed at this time. Numerous mafic intrusives, ranging in age from 1460 m.y. to 1100 m.y. and in composition from diorite to gabbro, including norite, are known.

Rocks of Keweenawan age, exposed in the Lake Superior area, are projected, below the surface, on geophysical data across Iowa and into Kansas. In subsurface, they form the Midcontinent gravity high, the largest positive gravity anomaly in North America. The narrow gravity high is flanked by gravity lows, and is interpreted as caused by mafic volcanics; the lows as clastic Keweenawan sediments; a pattern that conforms to the exposed Keweenawan. Gravity data also suggest that the Keweenawan basin extended southeastward across the Michigan basin. The Keweenawan igneous activity, closely related in time to the Grenville orogeny, consisted largely of mafic lavas and intrusives. White (26) regards this as one of the world's great periods of plateau basalt flows.

Basement rocks of the eastern Midcontinent fall within the 800 m.y. to 1,000 m.y. age range (48), representative of the Grenville orogeny. Those of the western Midcontinent fall within the 1100–1500 m.y. range. The boundary between the two ages of basement appears to be a fault extending from northeast Arkansas into the St. Lawrence Valley. Woolard (20) considered this one of the great fault lineaments of North America. Faulting was initiated prior to deposition of Upper Cambrian sediments; present day activity along the zone is reflected by a belt of earthquake epicenters.

Remnants of a belt of volcanics extend from western Ohio and eastern Indiana where they are encountered as subsurface rocks, across southeast Missouri as the St. Francois and Eminence outcrops of rhyolites and andesites, through northeast Oklahoma, and into western Texas. The belt includes early to middle Cambrian volcanics of southern Oklahoma as well as those of Precambrian age (55). Local relief of over 2000 feet at the beginning of late Cambrian time is indicated (30). This belt appeared to form a "continental divide" during late Precambrian and early and middle Cambrian time, from which sediments were carried to the Keweenawan basin to the northwest and the Appalachian basin to the southeast. This volcanic belt, in which the higher peaks form the igneous outcrop belt of the Midcontinent, has been termed the "Ancestral Ozarks" (55).

## Paleozoic and Younger Rocks

At the beginning of late Cambrian time, the Precambrian surface of the Midcontinent consisted of the narrow highland belt described above and broad undulating plains sloping gently toward the Keweenawan and Appalachian basins. Most of the present relief on the Precambrian surface is a result of local epeirogenic movements during Paleozoic time.

During the early Paleozoic, until early Ordovician time, the Midcontinent subsided as a broad unit that received thin blankets of sediments; following Jefferson City time, individual arches and basins developed. The basins received thick accumulations of sediments; the arches received thinner accumulations and periodically were uplifted and truncated.

The major elements of the stratigraphic succession for the central and western Midcontinent are shown in Table II. Other areas may use different names for correlative formations, but no attempt is made here to define and correlate units. Of importance to mineral deposits is the general character of the ore-bearing formations. The Upper Cambrian units, except for the basal Lamotte-Reagan Sandstone and Davis Shale, are carbonates, predominantly dolomites. Lower Ordovician formations are chiefly cherty dolomites; Middle and Upper Ordovician units are mainly limestones with minor sandstone and shale. Silurian units, except for evaporites in subsurface, were thinned or removed by erosion over much of the Midcontinent; the Devonian, where preserved, consists largely of limestone.

The Chattonooga shale (and its correlatives) was deposited over most of the Midcontinent. Widespread submergence in Mississippian time led to deposition of a thick section of fossiliferous, clastic, and cherty limestones. Many of these units maintain a consistent character over much of the Midcontinent. The Pennsylvanian consists largely of shales, clays, and coal beds with minor limestones and sandstones. Much of the subsidence in the basins, shown in Figure 2, occurred during Pennsylvanian time and Pennsylvanian formations comprise a large part of the sedimentary section.

Permian and younger formations are quite unimportant over the eastern part of the Midcontinent. In the western states and in the Gulf Coast embayment, Permian and younger formations are present; in other areas, if deposited, they have been largely removed by erosion.

The thick, and structurally complex, section

*TABLE II.*   *Generalized Stratigraphic Chart*
[Adapted from (29)]

| Era | System | Series | Group | Formations |
|---|---|---|---|---|
| | Quaternary | Recent | | Alluvium |
| | | Pleistocene | | Glacial deposits |
| | Tertiary | Pliocene | | Ogallala |
| Mesozoic | Cretaceous | | Montana | Pierre |
| | | | Colorado | Niobrara |
| | | | Dakota | |
| | Jurassic | | | Morrison Sundance |
| | Triassic | | | Dockum |
| Paleozoic | Permian | Guadalupe | | Red Beds |
| | | Leonard | | Flower Pot |
| | | Wolfcamp | | |
| | Pennsylvanian | Virgil | | |
| | | Missouri | | |
| | | Desmoines | Marmaton Cherokee | |
| | | Atoka | | Atoka |
| | | Morrow | | |
| | Mississippian | Chester | Jack Fork (Ouachitas) | Pitkin Fayetteville Batesville Hindsville |
| | | Meramac | Stanley (Ouachitas) | St. Genevieve St. Louis Salem Warsaw |
| | | Osage | Boone | Keokuk Reeds Spring St. Joe |
| | | Kinderhook | | Sedalia Compton Hannibal Sylamore Chattanooga |

TABLE II.   *Generalized Stratigraphic Chart (Continued)*

| Era | System | Series | Group | Formations |
|-----|--------|--------|-------|-----------|
| | | Upper | | |
| | Devonian | Middle | | |
| | | Lower | | |
| | Silurian | Niagara Alexandria | | |
| | | | Cincinnati | Orchard Creek Thebes Maquoketa Fernvale |
| | Ordovician | | Champlain | Kimmswick Decorah Plattin Rock Levee Joachim St. Peter Everton |
| | | | Canadian | Smithville Powell Cotter Jefferson City Roubidoux Gasconade |
| | Cambrian | Upper | | Eminence Potosi Derby-Doe Run Davis Bonneterre Lamotte-Reagan |
| Precambrian | | | | |

in the Ouachita area of Arkansas and Oklahoma, except for Mississippian and Pennsylvanian groups, is not shown in Table II. This section, embracing some 20,000 feet of sandstones, siltstones, and shales with interbedded volcanic ash units was deposited in a rapidly subsiding eugeosyncline. The depositional environment and orogenic folding at the close of the Paleozoic are more related to the Appalachians than to the Midcontinent epeirogenic environment.

The major structural features of the area are shown in Figure 2. The arches and basins began to develop early in the Paleozoic era. Minor ones were active for short periods, particularly in the late Paleozoic. The arches are areas of less subsidence than the basins; they received a thinner section of sediments than the more rapidly subsiding basins. Intermittently, the arches were positive features undergoing erosion.

The amount of vertical movement in the Midcontinent is reflected in thicknesses of sediments in the basins. These include approximately 14,000 feet in the Michigan basin, 12,000 feet in the Illinois basin, 4000 feet in the Forest City basin, and 3500 feet in the Cherokee basin.

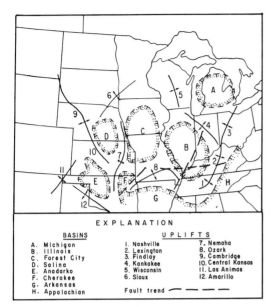

FIG. 2. *Prominent Structural Features.*

## Cryptoexplosion Structures

Cryptoexplosion structures are the circular, uplifted, often intensely brecciated areas, from one to several miles in diameter, that are anomalous to their setting in flat-lying, undisturbed sediments. Some 15 to 20 of them are known in the Midcontinent (55), the number depending on definition. Their origin is debatable; they are interpreted by different schools of thought as being formed by forces from below, in which case they are termed cryptovolcanic structures (7), and extraterrestrial forces, in which case they are regarded as meteorite impact scars (21).

Viewed individually either mode of origin appears plausible. Some of them, however, are aligned along pre-existing fault zones (42); some fall on the major positive structures shown in Figure 2; some structures show more than one age of brecciation. It appears likely that both internal and external forces can result in somewhat similar structural features and that more definitive criteria are needed to correctly interpret their origin.

## Post-Cambrian Igneous Activity

The lower to middle Cambrian volcanics of southern Oklahoma (38) belong to the pre-Upper Cambrian history of the area; they antedate recognized Paleozoic sedimentation. Younger igneous activity of the Midcontinent includes the well-known syenites and exotic

rock types exposed in Arkansas and the numerous peridotite dike or pipe intrusives known in other areas. Several of the intrusives occur along fault zones and apparently are associated with cryptoexplosion structures.

The alkali syenites occurring near Little Rock, Arkansas, at Magnet Cove, and in adjacent areas, are adequately described in the literature. The intrusives are of Cretaceous age.

Extrusion of basaltic lava and intrusion of peridotite dikes occurred intermittently throughout the Paleozoic. Three localities indicating late Cambrian igneous activity have been described (42, Snyder and Gerdemann, this volume). The Avon diatremes have been dated as of Devonian age; the southern Illinois peridotites as Permian in age, and the Silver Hill, Kansas, peridotites as Paleocene (43).

## Metallogenic Provinces

The Midcontinent embraces three major and several minor metallogenic provinces. The entire area from the Appalachians to western Kansas is a lead-zinc province. It is the classic example for the lead-zinc deposits in carbonate rocks that have been designated the Mississippi Valley type. Several major districts and numer-

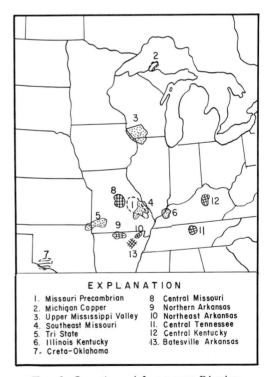

FIG. 3. *Location of Important Districts.*

ous minor ones occur in the area. Almost every formation from basal Upper Cambrian sandstone to Pennsylvania shales contains lead and zinc sulfides, and production from many formations is recorded.

The copper deposits of the Lake Superior area form another major metallogenic province. This includes both the deposits occurring in mafic volcanics and those found in Late Precambrian sediments.

The Precambrian iron deposits of Missouri form a third major metal province. Other provinces, economically important but more restricted in size and production, include the Red Beds copper deposits, the exotic minerals associated with nepheline syenite intrusives in the Ouachitas, the ubiquitous brown iron deposits, and the Batesville, Arkansas, manganese district (Figure 3).

## PRECAMBRIAN IRON DEPOSITS OF MISSOURI

Occurrences of hematite and magnetite in Precambrian igneous rocks of Missouri have long been known. They include the Iron Mountain mine described in this volume, Pilot Knob which produced a small amount of hematite ore, Iron Knob, and the Greasy mine. All outcrop in the exposed Precambrian of the St. Francois Mountains.

During the past 12 years, exploration drilling of large magnetic anomalies revealed by airborne surveys has led to the discovery of a number of magnetite bodies in buried Precambrian. Deposits that have been announced include Pea Ridge, Pilot Knob subsurface, Bourbon (2 deposits), Kratz Springs, Boss, and Camels Hump (Figure 4). In addition, at least three other magnetite deposits are known.

The Precambrian succession in southeast Missouri, as defined by surface mapping (25), consists of two series of felsites, separated by a tuff bed, and two stages of granitic intrusion. The small hematite deposit, known as the Greasy Mine, is in granite. All other known deposits occur within the volcanic sequence.

### Greasy Mine

The Greasy Mine is located in sec. 8, T34N, R5E, St. Francois County, Missouri. The ore body consisted of veins of micaceous hematite in granite (12). A small amount of ore was mined through a vertical shaft 80 feet deep and from drifts along the veins.

### Iron Knob

The Iron Knob deposit, located in the northwestern part of Wayne County, Missouri, on the surface, consists of scattered hematite boulders in an area of rhyolite outcrop. The deposit is too small to be of economic interest. It is important only in indicating the southern limits of known iron mineralization.

### Pilot Knob (Outcrop)

Pilot Knob is a conical peak located 1 mile east of a village by that name. The ore beds, outcropping at the peak of the knob, represent replacements of tuff and breccia. Approximately 1.5 million tons of ore have been mined from open pit and from adits driven downdip from the outcrop.

The ore occurs in two beds. The upper ore bed ranges from 10 to 20 feet in thickness and is separated from the lower ore bed by 2 to 3 feet of sericitized rhyolite. The lower ore bed ranges from 5 to 30 feet in thickness. Underlying the lower ore bed is a felsite, called the Pilot Knob felsite, that is referred to the older volcanic group (28). Dip of the felsite is from 15° to 30°SW.

Specular hematite is the ore mineral. The hematite replaces tuff and breccia matrix and veins breccia fragments. Where the host rock is largely tuff, grade of ore is 45 to 50 per cent iron. Breccia units become thicker and more numerous upward in the section and grade of iron decreases as the amount of breccia increases.

Phosphorus and sulphur content are very low. Steidtman (4) reported that shipments of ore over a six month period averaged 58.11 per cent Fe; 17.02 per cent $SiO_2$; 0.013 per cent P, and 0.077 per cent S.

### Iron Mountain

Production at Iron Mountain mine, more fully described by John Murphy, this volume, began in 1845 and was more or less continuous until 1966 when the mine was closed. The ore occurred in an andesite, intrusive into rhyolite, dacite, and pyroclastics. The ore body outcrops, and conglomerate ore on the Precambrian surface was the first ore mined.

The primary ore occurred in two distinct ore bodies. One, known as the Main ore body, was shaped like an inverted cup with a flat top and steep sides. The other, called the

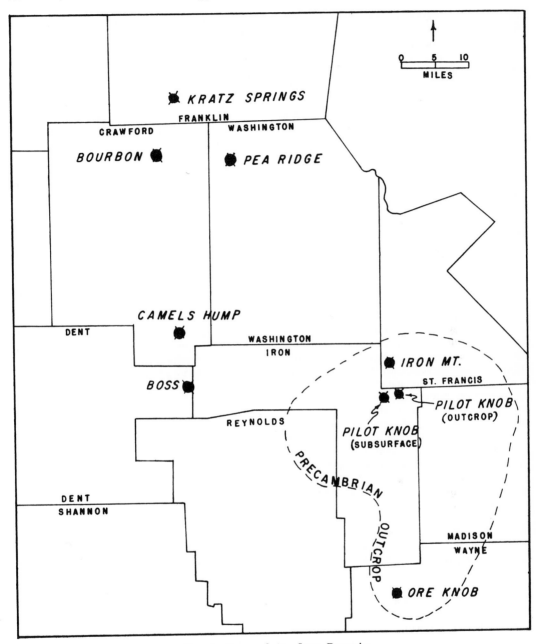

FIG. 4. *Precambrian Iron Deposits.*

Northwest ore body, consisted of a thick vertical segment connected with a nearly horizontal lens. The main ore minerals are hematite and magnetite. Part of the ore consisted of felsite breccia with an iron oxide matrix. Gangue minerals include pyroxene, actinolite, garnet, quartz, and calcite. Minor amounts of sulfides occur as galena, bornite, and pyrite.

## Pilot Knob (Subsurface)

The subsurface Pilot Knob ore body, now being developed, lies about a mile west of the Pilot Knob outcrop. The ore body has an arcuate form, striking nearly north-south in the southern part and curving westward in the northern part. In horizontal plan, it has a half moon shape, opening to the southwest. The dip is about 30°W.

The main ore mineral is magnetite; hematite is subordinate. The ore is low in phosphorus. Fluorite is abundant. Small amounts of chalcopyrite occur sporadically in the magnetite. Silicates reportedly are rare.

Much of the ore is massive iron oxide, but breccia ore, in which magnetite cements breccia fragments, forms a part of the deposit.

The hanging wall and footwall are classed as andesites but are different volcanic units. The hanging wall andesite shows no appreciable alteration but the red footwall andesite is altered in places to a grey rock. The footwall andesite appears to be strongly fractured; the hanging wall andesite is compact and unbroken. A mafic sill occurs below the ore body.

## Bourbon Ore Bodies

The Bourbon anomaly in the northern part of Crawford County was outlined by Missouri Geological Survey ground-magnetic work in the 1930's. In 1943, the U.S. Bureau of Mines drilled 3 holes to depths of 2400 feet on the southern part of the anomaly (11). Low grade magnetite ore, averaging 30 to 40 per cent iron, was cut by two of the drillholes. The ore is a fine-grained, siliceous magnetite.

The northern part of the anomaly has been extensively drilled as a joint venture by the American Zinc Company and the Granite City Steel Company. The American Zinc Company very kindly provided the following description of the ore body.

The Precambrian is at depths of 1300 to 1600 feet, beneath Lower Paleozoic sediments. Drilling has outlined a large, nearly circular body of magnetite. It has the form of an inverted cone, the apex of which outcrops at the Precambrian surface. Vertical extent of the ore is 2000 feet, and a down-dip continuation of the ore to the southwest is indicated. The ore body is estimated to contain approximately 200 million tons of ore grading 40 per cent iron.

Hematite is present at the Precambrian subcrop, but the ore is predominantly magnetite. Four distinct felsite units are recognized in drillcores, and the ore replaces parts of three of the volcanic units. A north-south trending fracture, now occupied by a mafic dike, cuts across the ore body. West of the fracture the ore is uniform in grade; east of it the magnetite content gradually decreases (Figure 5).

Throughout most of the deposit, the iron oxides replace the volcanic rock; no breccia ore is present. A tabular mass, conformable with flow contacts, extends eastward from the main ore body.

## Pea Ridge

This ore body is described by Emery elsewhere in this volume. It is cited here only to indicate its place in the district pattern.

The Pea Ridge ore body is rudely tabular in form and nearly vertical in attitude. It strikes approximately east-west crosscutting the rhyolite host beds which dip 70°N. (Figure 5). A small part of the ore body outcrops at the Precambrian surface which is overlain by 1200 to 1400 feet of Paleozoic sediments. Magnetite is the main ore mineral, but hematite makes up approximately 20 per cent of the ore in the upper part of the body. Much of the ore is massive magnetite in knife-edge contact with the rhyolite host, but ore on the hanging wall side (south) is composed of fresh unaltered fragments of rhyolite cemented by magnetite. Replacement textures are rare.

The phosphorus content of the ore body is in large apatite crystals and is high; chalcopyrite is a minor constituent. A quartz-amphibole zone is present on the hanging wall and a quartz-sericite zone on the footwall.

## Kratz Springs

This ore body, located some 15 miles north west of Pea Ridge, has a gentle dip and appears to be conformable to flow structures in the rhyolite host. The ore is dominantly magnetite. Numerous mafic dikes cut the ore body.

No breccia ore is known. The ore body could be a replacement type, but, on the limited information now available, it appears to be a sill-like injection.

## Camels Hump

The Camels Hump deposit, located a few miles west of the Viburnum Lead Belt, is a massive magnetite body overlain by some 1200 feet of Paleozoic sediments. The ore body is at the Precambrian surface, and the upper part is oxidized to hematite.

The ore body strikes nearly north-south and has a steep dip. The host rock is a rhyolite, but the relations of ore to host are now unknown. In form and general character, the

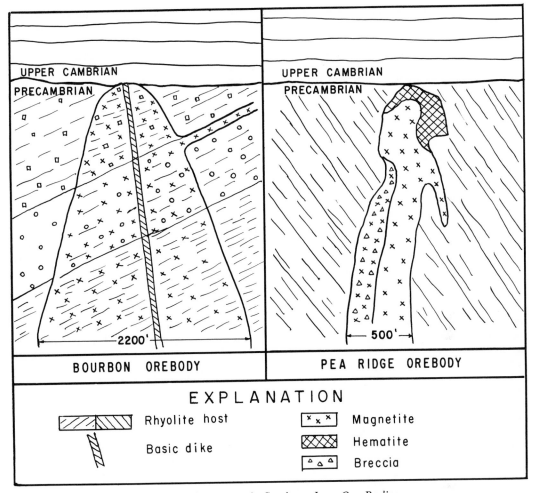

UPPER CAMBRIAN

PRECAMBRIAN

UPPER CAMBRIAN

PRECAMBRIAN

2200'

500'

BOURBON   OREBODY

PEA   RIDGE   OREBODY

EXPLANATION

Rhyolite host

Basic dike

Magnetite

Hematite

Breccia

FIG. 5. *Diagrammatic Sections, Iron Ore Bodies.*

ore body appears to resemble that at Pea Ridge.

### Boss

The Boss deposit in eastern Dent County consists of low-grade magnetite with appreciable amounts of copper. It is regarded as an iron-copper deposit. The copper minerals are chalcopyrite and bornite.

The host rocks are a series of rhyolite flows that dip 50° to 60°S. The flows are cut by a syenite intrusive. The ore apparently occurs in a series of steeply dipping zones from 25 to 200 feet thick that are parallel to the syenite-rhyolite contact. Mineralization occurs along the contact and as replacement masses within the intrusive. Ore extends downward

from the Precambrian surface for more than 2000 feet.

### Origin of the Iron Deposits

The major features of the deposits occurring in rhyolite host rocks may be summarized as follows:

(1) The occurrence of hematite is closely related to the Precambrian surface. It is the predominant ore mineral only on the outcrop, in the shallow deposits, or in the upper part of the deeper deposits.

(2) Magnetite is the main ore mineral, and often the only iron oxide, in the deeper deposits.

(3) Minor constituents in all deposits are copper, fluorine, phosphorus, and iron as sulfide.

(4) All deposits are virtually devoid of titanium.

(5) Ore bodies may be disconformable with the host rock structures, in which case they contain blocks of brecciated host rock cemented by iron oxide, or they may be conformable with the host rock and show replacement textures. In the disconformable ore bodies, replacement is absent or minor in amount.

Recent work on age dating of Missouri Precambrian rocks (49) contributes significantly to knowledge of time of origin of the iron deposits and their relationship to igneous activity. Three determinations on rhyolites give ages of 1260, 1290, and 1330 m.y.; the 1290 m.y. age is from host rock rhyolite at Pea Ridge. Four determinations on granites from southeast Missouri give ages of 1120, 1190, 1210, and 1220 m.y. A post-ore aplite dike at Pea Ridge mine gave an age date of 1310 m.y., essentially the same, within the limits of error, as the host rock rhyolite (50). Emplacement of the Pea Ridge ore body must have followed closely after formation of the volcanic rocks, and the iron may have been derived from the same magmatic source. The age of the post-ore dike at Pea Ridge indicates the deposit was formed prior to crystallization of the granites.

Field studies (27) indicate that extensive breaking accompanied extrusion of the volcanics; rhyolite flows were brecciated and the blocks of rhyolite breccia were cemented by rhyolite. Chemical analyses of the igneous rocks (22) reveal that the granites consistently contain less iron than the felsites; other constituents are essentially the same. The granites are intrusive into the felsites and are regarded as products of the same period of igneous activity (30).

It is suggested that the iron is a differentiate from the magma that formed the rhyolites and granites. Differentiation was taking place during extrusion of the volcanics, leaving the remaining magma deficient in iron. Block faulting of the volcanics, induced by magmatic pressures, permitted escape of the iron-rich fluid that filled fracture zones and cemented rhyolite breccia fragments. When no open brecciated zone was available, the ore solutions followed flow structures and contacts in the volcanics giving rise to conformable replacement-type bodies.

Iron forming the small hematite body at Greasy Mine may have been derived from the same source but was emplaced in the granite host later than time of formation of the Pea Ridge deposit and probably later than other ore bodies in felsite host rocks.

Recent studies (53) indicate that the Boss ore body has a different trace element suite than the other magnetite deposits, suggesting a different mode, and possibly time, of origin.

## LAKE SUPERIOR COPPER DEPOSITS

Both of the major copper districts, the Keweenawan Peninsula and the White Pine deposits, are the subjects of detailed papers in this Volume, and only a general statement of major relationships is merited here.

The Keweenawan sediments and volcanics were deposited in an arcuate trough that also had been the site of late Huronian deposition. The Huronian sequence was folded and metamorphosed prior to Keweenawan time.

The threefold Keweenawan sequence is remarkable for its great thickness and for its great abundance of mafic volcanics. The lower part of the sequence is thin and consists of sandstones, shales, and marls. It outcrops as the Barron Quartzite in northwestern Wisconsin (46).

The Middle Keweenawan portion embraces a great thickness of lava flows with thin interbedded clastic sediments. The flows outcrop on the south limb of the geosyncline from Keweenaw Point to, and beyond, the western end of Lake Superior and along the north limb from Duluth to Thunder Bay, Ontario. The Duluth gabbro intrusive complex was intruded at or near the base of the volcanic sequence and has been dated at 1000 to 1200 m.y. (45).

The volcanics of the South Shore, because of the copper mineralization, have been closely studied. The sequence, termed the Portage Lake lava series, is estimated to be 30,000 feet thick. Only the upper 15,000 feet, dipping toward Lake Superior, is exposed. The lavas consist entirely of basalts and andesites; rhyolites are lacking except at the top of the series (51). The flows average about 43 feet in thickness (51) and are continuous for many miles. Dips of flows toward the center of the basin decrease upward in the section, indicating the basin was subsiding during extrusion of the lavas.

The Upper Keweenawan consists of approximately 20,000 feet of clastic sediments, divided into the basal Copper Harbor Conglomerate, the Nonesuch Shale, and the Freda Sandstone. The Freda Sandstone is overlain with slight discordance by the Upper Cambrian Mt. Simon Sandstone.

In the Keweenawan district, copper occurs in conglomerate lenses interbedded with the volcanics and in the upper permeable parts of flows classed as amygdaloids. A single conglomerate unit, the Calumet and Hecla Conglomerate, has been the largest single productive unit in the district. Several amygdaloidal zones, each representing a flow top, have been important producers.

The deposits have been ascribed to rising hydrothermal solutions derived from underlying mafic intrusives (2). White (this volume) postulates migration of the metals from the center of the syncline to their present sites.

The stratiform deposits at White Pine, Michigan, in the Upper Keweenawan sedimentary sequence represent a different type of mineralization. The ore occurs in siltstone in the basal beds of the Nonesuch Shale. The ore zone, ranging in thickness from 4 to 25 feet, extends over many square miles. The host rock formation is flat-lying and unmetamorphosed, and no igneous activity younger than the host formation is known in the area.

Pyrite, probably of syngenetic or diagenetic origin, is common in the Nonesuch shale. The ore sulfides include chalcocite (the main ore mineral), bornite, chalcopyrite, and native copper. The deposit has been interpreted as of syngenetic origin. The authors of the detailed study that follows present evidence suggesting an epigenetic origin for the deposit.

Current exploration activity in northern Minnesota indicates the Duluth gabbro may become a new large mining district. Nickel and copper sulfides are present in the lower part of the gabbro in an occurrence similar to the Sudbury ores. Mineralization is as high as 1.5 per cent combined copper and nickel in the proportion of 3:1.

Detailed studies have shown that the Duluth gabbro, once believed to be a differentiated lopolith (a simple layered intrusion), is a series of multiple intrusions of mafic magma.

The main rock types (52) are anorthositic gabbro and layered gabbros intrusive into the anorthositic phase. The intrusive gabbroic rocks include troctolite, and the sulfides occur in this phase, usually just above the base of the complex. Granophyres are common in the upper part of the body. According to Sims (52, p. 110):

"Chalcopyrite and pyrrhotite are the main sulphide minerals with pentlandite and cubanite intergrowths.

"Bornite and pyrite are local sulphides. Violarite is a common alteration product of pentlandite. The oxides magnetite and ilmenite are ubiquitous. Biotite and locally hornblende are commonly associated with the primary silicate minerals—calcic plagioclase, olivine, and pyroxene—in the mineralized zones. In general, sulphide mineral grain size varies with the size of the silicate mineral. Tiny irregular and discontinuous veinlets of sulphide mineral and silicate gangue can be seen locally, particularly in hornfels inclusions. Rarely found are small pod-form bodies of massive sulphides. The textures of the sulphide minerals and their paragenetic relations to the silicate minerals of the gabbro indicate that the two essentially formed contemporaneously."

## CRETA, OKLAHOMA COPPER DEPOSIT

The occurrence of copper minerals in sandstones and shales in Oklahoma, Texas, and states to the west has long been known. The copper minerals generally are contained in gray to light brown members within red sandstones and shales of Permian, Triassic, and Jurassic age. Because of their occurrence with red-colored formations, the deposits are known as red-beds copper. Numerous attempts to work the deposits have been unsuccessful because of their low grade and small size.

In 1964, the Oklahoma Geological Survey announced the discovery of a copper deposit near Creta, Oklahoma (39). The deposit lies in the southwestern part of Jackson County in the extreme southwestern part of the state. Nearly flat-lying rocks of Permian age form the surface. The copper occurs in 2 thin beds in the Flowerpot Shale of Permian age.

The Flowerpot Shale is approximately 200 feet thick in the ore area. The formation is of marine origin and consists of shales interbedded with gypsum and dolomite. The gypsum units, the Chaney bed and the Kiser bed, form important marker horizons in the formation. The Chaney bed, about 3 feet thick, occurs 45 to 50 feet below the top of the Flowerpot. The Kiser bed, normally an impure gypsum, is a light gray dolomite in the Creta area. It is about 6 inches thick and is 25 to 30 feet below the top of the Flowerpot. The two copper beds lie between the Chaney and Kiser units.

The upper copper bed is 4 inches thick and occurs 5 feet below the Kiser bed. The lower copper bed is about 6 inches thick and lies 10 feet below the Kiser. Copper content of the lower copper bed ranges from 2.65 to 4.45 per cent and averages 3.8 per cent for nearly 3 miles along outcrop (39). The upper bed ranges from 0.38 to 1.27 per cent copper and averages 0.79 per cent (39).

Malachite is the copper mineral seen in outcrop and was the basis for discovery of the deposit and field mapping of its distribution. The main ore mineral is chalcocite, which is preserved under very shallow cover. On the outcrop below the lower copper bed are numerous nodules and encrustations composed of malachite, azurite, brochantite, and small opaque grains tentatively identified as chalcocite and cuprite (39). The nodules and encrustations occur only on the outcrop; they are not present within the unweathered copper-bearing shale.

Ham and Johnson investigated copper and boron content of other shales in the Flowerpot Formation. They reported (39, p. 13) that the lower copper bed contains 2500 to 4000 times as much copper and twice as much boron as other shales in the formation. The boron content of marine shales, contained in illite, is believed to indicate the salinity of the sea water (23). Because of the widespread distribution of the thin copper-bearing beds, their uniformity in character, and the copper-boron relationships of these beds to non-copper bearing shales, Ham and Johnson (39) postulated a syngenetic origin in a hypersaline environment for the deposit.

Over a considerable area, the deposit is covered by less than 40 feet of overburden. Ham and Johnson (39) estimated approximately 70,000 tons of copper under less than 40 feet of overburden in the lower copper bed. To the southwest the overburden becomes thicker because of the regional dip, but the copper-bearing shales continue for an unknown distance. The property is now in operation on a 1000 ton-per-day basis.

Aside from economic considerations, the Creta copper deposit is of interest because of its similarity to the Kupferschiefer of Germany, also Permian in age. Unlike the Kupferschiefer, which contains appreciable amounts of zinc and lead, the Flowerpot contains only copper. Semiquantitative analyses show maxima of 0.02 per cent lead, 0.05 per cent barium, and 0.02 per cent vanadium (39). Zinc was not reported.

## LEAD-ZINC-BARITE-FLUORITE DEPOSITS

Deposits of lead, zinc, barite, and fluorite, occurring singly or in various combinations, are found throughout most of the Midcontinent in rocks ranging from Upper Cambrian to Pennsylvanian in age. They occur both as vein fillings and as stratiform deposits. Some districts are major ones with a production record of many millions of tons. Some are minor ones with only limited production. A few are unimportant economically but are significant in terms of genesis of the deposits.

No simple sub-division by metal composition, by form of ore bodies, by geographic distribution, or by age of host rock is logical. The deposits will be described individually, in approximate order of age of host rock in which the deposits occur, beginning with the oldest host rocks first. Only brief attention will be given to those districts that are the subjects of individual papers in this volume.

### Southeast Missouri Lead District

The Southeast Missouri Lead District is arbitrarily defined (Snyder and Gerdemann, this volume) as consisting of the stratiform deposits occurring in the Bonneterre Formation and the upper part of the Lamotte Sandstone. Such deposits are known in Madison, St. Francois, Washington, Crawford, Iron, Reynolds, and Shannon counties, Missouri. A few minor deposits of this type are known in adjacent counties. The deposits occur on the east, north, and west sides of the outcropping Precambrian high known as the Francois Mts.

Lead was discovered at Mine Lamotte, Madison County in 1720, and the district has been the largest and most productive lead district in the United States. New deposits, discovered in the past decade, insure the place of the district as a major producer for many decades.

The ore occurs in the Bonneterre Formation near or within a few miles of Precambrian topographic highs that cut out the underlying Lamotte Sandstone. Close to the igneous knobs ore may occur in the upper 100 feet of the Lamotte Sandstone. The Bonneterre Formation is dolomitic near and over the basement highs, grades into limestone in local basins and away from knobs, and is shaley in deeper parts of the Illinois and Forest City basins. The dolomite bears features indicating shallow water deposition and contains numerous vertical and lateral facies variations. Major unconformities occur at the base of the Lamotte Sandstone and at the top of the Jefferson City Formation. A minor disconformity is found at the top of the Eminence Formation and numerous diastems are present in the stratigraphic sequence, including several in the Bonneterre formation.

The area has been strongly faulted. There is evidence of repeated fault activity, at least as early as Bonneterre time (42); probably occurring again as the main fault episode in post-Jefferson City time; and reflected in recent earthquakes that represent adjustment along old fault lines.

Ore bodies, ranging in size from a few tons up to tens of millions of tons, are scattered throughout the Bonneterre host over an area of several hundred square miles. In any given area, certain stratigraphic zones are preferred to others, but ore may occur in any part of the formation. Individual ore-bearing structures are largely sedimentary traps consisting of pinchout zones with distinctive facies differences from adjacent beds, usually associated with a basement high or an underlying sand or carbonate high, thereby simulating an anticlinal structure; algal reef structures and associated facies; and breccia or pseudobreccia masses.

Mineralization is believed to have occurred during the time of the post-Jefferson City faulting. Some ore appears to be pre-faulting; some is in ground prepared by faulting; some is present as undistorted crystals in fault openings. Isotopic studies possibly suggest mineralization continued over a long period of time.

The mineralogy of the ore is simple. Galena is the dominant sulfide; sphalerite and chalcopyrite locally are abundant but are restricted to certain parts of the district; cobalt and nickel in the sulfide, siegenite, are minor in amount; silver and cadmium are present as minute constituents. Iron sulfides are common but not abundant.

Isotopic studies indicate a wide range in radiogenic character. Samples from the Bonneterre formation range in Pb 206/204 composition from 19.56 to 22.08 and average 20.99.

## Southeast Missouri Barite-Lead Deposits

The barite-lead deposits occurring in Washington and Jefferson counties, Missouri, represent one of the largest barite districts in the United States. A few minor deposits occur in adjacent counties.

Most of the ore occurs within a 100 foot thick zone that embraces the upper part of the Potosi Formation and the lower part of the Eminence Formation. The deposits lie within the geographic area of the Southeast Missouri lead district, but they are distinctly different. The barite-lead district contains barite as the dominant and often the only economic mineral; no barite has been found in the lead district, although it has been mapped in great detail and hundreds of thin and polished sections have been studied.

Both the Potosi and Eminence Formations are dolomite throughout southern Missouri. Rock types include algal reef, and sand, silt, and clay size carbonate, now entirely dolomite. Preference for a particular lithology has not been recognized.

During the early history of the district, veins containing mainly galena were mined, and several sub-districts were recognized (1). Nearly all of the present barite production is from scattered chunks and masses in weathered residuum. Productive pits are restricted to certain parts of Washington and southwestern Jefferson counties; usually several pits are clustered within a small area.

Galena occurs in minor amounts with the barite. Iron sulfide, partially oxidized to limonite, is abundant in some pits.

Barite, in the form of thin veins and fracture fillings, occurs in unweathered dolomite in the floors of pits. Chunks of barite occurring in the residuum often are many times larger than any vein thickness seen in outcrop. The Potosi Dolomite is a vuggy cherty unit. Vugs are lined with quartz druse, often in masses more than a foot in diameter. The author has found druse-lined vugs in which the quartz druse was coated with barite. The quartz druse and chalcedony, presumably, are groundwater phenomena. The large size of barite chunks and masses, and barite lining druse cavities, suggest that the barite is mobilized and redeposited in the zone of residuum.

Isotopically, galena from the Potosi and Eminence Formations overlaps the range of Lead Belt galena but generally is more radiogenic. Seven samples from these formations within the barite district, range in Pb 206/204 composition from 21.61 to 22.41 and average 21.80.

## Central Texas Deposits

The lead deposits occurring in central Texas around the Llano uplift are currently of no economic importance. The galena occurs in minor amounts in the Cap Mountain Limestone where the underlying Hickory Sandstone pinches out against Precambrian topographic highs, an occurrence reminiscent of the Southeast Missouri district. Some mineralization occurs in younger Upper Cambrian formations.

Galena is the most abundant sulfide, but

minor amounts of sphalerite are present. Barnes (16) reported that the sulfides occur in dolomite, dolomitic limestone, limestone, calcareous sandstone, and glauconitic sandstone. The sulfides occur as disseminations and as crystalline aggregates. Replacement is minor in amount. The most strongly mineralized feature is a collapse structure (the Silver Creek prospect), but most of the sufides occur in pinchout zones adjacent to Precambrian knobs.

## Sphalerite in Western Kansas

Sphalerite has been recognized in oil wells in the Upper Arbuckle Dolomite over a large area in western Kansas at depths of 4900 to 6440 feet (32). The mineralization is in the Jefferson City-Cotter member of the Arbuckle, which is described as a cherty, gray to white, medium crystalline dolomite. The only sulfide known is sphalerite. It occurs sparsely over some 150 square miles and, in one well, reaches a grade of 7 to 8 per cent zinc.

The areas of mineralization are associated with broad anticlinal highs which are regarded by Evans (32) as a major line of folding.

Evans (32) distinguished amber, canary-yellow, and black sphalerite. Chert ranges in color from gray-white through various shades of gray to a dark jasperoid type. In one well, he recognized (32, p. 560) a sequence consisting of blue-grey chert, followed by sphalerite, glassy quartz, and white opaque chert with black jasperoid below.

The deposit is too deep to be of economic interest, but it indicates widespread mineralization of Mississippi Valley type at depth.

## Central Missouri District

Throughout a large part of central Missouri, many small deposits of lead, zinc, and barite occur in Lower Ordovician formations. They are found in several formations, but the Jefferson City Dolomite is the most widely mineralized.

Many of the deposits are circular in form and from 70 to 150 feet in diameter. They are interpreted as sinks in which the ore occurred in brecciated rock around the outer margin of the central subsided zone. Seams of galena adhere to the unbroken outer dolomite walls, fill openings between breccia blocks, and vein blocks of dolomite in the core.

Other deposits are fracture fillings, forming vertical veins up to several feet in thickness,

in the Jefferson City Dolomite. Ore occurring along bedding planes is minor in amount.

Barite is the most abundant economic mineral, followed by galena, then sphalerite. Clear crystalline masses of calcite form the main gangue mineral.

Most of the deposits were worked during the middle and late 1800's. Little attention has been given to them in recent years, although a similar deposit, the Alice mine, in southern Missouri was drilled by the U.S. Bureau of Mines during World War II.

None of the deposits known to the writer is very large, but in aggregate they contained a large quantity of metal.

## Upper Mississippi Valley District

This district has long been one of the important zinc producing regions in the country. Like the Tri-State district, it has been intensely studied, and a voluminous literature on the district is available.

The zinc-lead ores occur mainly in the Galena, Decorah, and Platteville Formations of Middle Ordovician age. Small deposits also occur in formations as old as Upper Cambrian and as young as Silurian, the youngest present in the area. The ore-bearing formations were deposited in a shallow marine environment on the southwest flank of the Wisconsin arch.

Heyl *et al.* (15) recognized folds of three orders of magnitude and several types of faulting in the district.

The first deposits mined in the district were gash veins containing galena. These are discontinuous veins filling joints in the Galena Dolomite, often stratigraphically above the more productive sphalerite-galena deposits. Other fissure veins are found in the northern part of the district in formations stratigraphically below the main ore host rocks.

The most productive deposits are the pitch and flat type, in which the pitch is a steeply inclined fracture, the flat is a bedding plane. Ore bodies of this type occur along reverse and bedding plane faults of small displacement (15). Solution thinning of the Platteville and Decorah is regarded as an important factor in the development of the ore-bearing structures and breccias. The pitch and flat ore bodies often overlie breccia and bedded replacement deposits.

The major ore minerals are sphalerite and galena, but chalcopyrite, enargite, and barite are found. Iron sulfides are abundant. Chert and jasperoid are present, as are crystalline dolomite and calcite.

Russell and Farquahar (24) report 12 lead isotope determinations made by the Columbia laboratory. The minimum Pb 206/204 ratio is 21.20, the maximum is 22.73, and the average is 22.12. The Toronto laboratory (24) reports a single determination for a sample near Dodgeville, Wisconsin, which shows the Pb 206/204 ratio is 23.96, well outside the range for 12 samples reported by the Columbia laboratory.

Heyl, *et al.* (47) in a study of a number of galenas from Mississippi Valley type deposits report 19 determinations from the Wisconsin-Illinois district ranging from a minimum Pb 206/204 ratio of 20.83 and a maximum of 24.44. These, combined with analyses reported by Russell and Farquahar (24), total 32 determinations with an average Pb 206/204 content of 22.43.

Heyl (this volume) states that the 206/204 ratio increases across the district to the northeast.

## Northeast Arkansas District

Mineral deposits in this district occur in the Lower Ordovician Smithville Formation in Sharp and Lawrence counties. The earliest zinc mining was done in the 1850's; small scale operations or prospecting have been carried out sporadically since that time.

The Smithville Formation is a fine- to medium-grained cherty dolomite with interbedded sandstone and limestone. Although some ore is associated with fractures or minor faults, no extensive fault system has been recognized in the area.

Sphalerite is the main ore mineral, although much of the early, near-surface production was on calamine ore. Galena is minor in amount. Traces of chalcopyrite are present. Pink crystalline dolomite is a common accessory. Some of the deposits are filled fissures, but most are of the replacement type and are restricted to certain stratigraphic zones in the flat-lying host.

The most extensively explored deposit was the Campbell Zinc Company's operation is Lawrence County. Three shafts, to depths of 277, 156, and 40 feet were sunk. Mineralization was indicated in numerous drillholes. Two levels of ore were reported (6) at one shaft. The lower ore zone, up to several feet thick, was conformable to bedding. The host rock is a dark, brecciated, or conglomeratic dolomite. Limestone underlies the ore. The upper ore zone ranged from 2 to 58 feet in thickness, with the upper surface of ore transgressing

bedding (6). At the second shaft, a mile distant, two ore levels also were present, with ore again in brecciated dolomite.

Numerous other prospects are known in the district. Most were opened on surface showings of zinc carbonate and silicate, which, like the Campbell mine, usually were in brecciated dolomite.

A lead isotope analysis (24) gave a Pb 206/204 ratio of 21.89.

## Northern Arkansas District

The Northern Arkansas district lies on the southern flank of the Ozark uplift. Ore deposits occur in Boone, Marion, Newton, Searcy, and Sharp counties, Arkansas, and extend for a short distance into southern Missouri.

Formations in the area include a nearly complete Lower Ordovician section overlain by Mississippian and Pennsylvanian units. During the pre-Mississippian erosional period, all of the Devonian and most of the Silurian formations were removed. McKnight (6) stated that there are 12 well-established unconformities within the section, indicating repeated periods of uplift and erosion during the Paleozoic.

Scattered minor mineral deposits occur in the Cotter and Powell Formations. The Everton is the most important host rock and has produced about 70 per cent of the ore mined in the district. The Everton has a maximum thickness of 400 feet but is eroded to a feather edge on the flank of the Ozark uplift. The Everton is composed of gray, usually fine-grained limestone, gray, fine- to coarse-grained dolomite, and thin lenticular sandstones. Several sandstone beds are extensive enough to be recognized as members.

The medium- to coarse-grained facies of the dolomite are the chief ore hosts. The fine-grained dolomites and the limestones of the Everton are mineralized less often.

Most of the remaining production from the district came from the "Boone" formation, a term that embraces the Warsaw, Keokuk, Reeds Spring, and the St. Joe units. In the Northern Arkansas district, the Mississippian section includes the thin, basal Sylamore Sandstone, the St. Joe Limestone, and the Keokuk. Most, if not all, of the Warsaw has been removed by erosion. No mineralization is known in the Sylamore. The St. Joe Limestone consists of red and gray thin-bedded limestones containing abundant crinoid remains. The formation contains a number of prospects, but only one mine that yielded an appreciable amount of ore. The Keokuk consists of light gray, medium- to coarse-grained limestones with abundant chert.

Extensive faulting in the district is embraced in two systems, one striking northeast, the other northwest. Displacements of several hundred feet are recognized. The major faults include the St. Joe with a strike length of 15 miles, the Rush Creek with a strike length of 8 miles, and the Tomahawk fault. The faults are normal, have steep dips, and form narrow grabens with adjacent faults. The faulting is dated as post-Atokan and probably is a part of the late Paleozoic deformation.

The district has been divided into several sub-districts by McKnight (6), in which the sub-districts are related to the major fault blocks or to stream valleys. Some of the ore deposits are aligned along major faults in shattered and brecciated rock adjacent to the fault; other deposits occur as "runs" and consist of ore confined to a particular stratigraphic zone in an unfaulted area.

The Rush Creek district, in the southeastern part of Marion County, extends for several miles along the Rush Creek and related faults. The deposits are ore runs adjacent to the faults, not in the faults. McKnight (6) reported a production of approximately 26,000 tons of zinc concentrates from this subdistrict. Renewed mining operations in 1960–62 resulted in some additional production of ore.

The ore occurs in the Everton Formation in dolomite, sandy dolomite, and limestone. Ore production was obtained from about 15 mines and from several different stratigraphic zones embracing a section of over 150 feet in the Everton. Both lateral and vertical lithologic variations are common. The medium-grained dolomite may grade into fine-grained ore-bearing dolomite or into limestone. Chert is abundant in the formation.

Individual ore zones range from 2 to 20 feet in thickness, up to 200 feet in width, although usually less than 100, and up to 600 feet in length. Within many ore zones, the ore is rich in small pockets that are separated by low-grade disseminated ore. McKnight (6) reported that many ore zones appear to be developed along obscure fractures.

The Rush Creek sub-district is typical of several that are closely related to faulting and in which the trend of the district and of individual ore bodies appear to be fault-controlled. Deposits in the Zinc sub-district represent a different type in that they occur as "blanket-veins" (stratiform deposits) unrelated to faults.

The Zinc sub-district is in eastern Boone County. Ore occurs in the Everton Formation in several different stratigraphic zones and also at one mine in the overlying St. Joe Limestone. Silicified limestone is host rock for some ore deposits in the Everton; the coarser grained dolomite is the host at others.

McKnight (6, p. 240) described the ore occurrence at the Madison mine in the Zinc sub-district as follows:

"The ore-bearing rock in the lower tunnel is a medium to coarse grained dolomite that grades in places into limestone and that may carry unaltered blocks of limestone within it. The bed averages about 5 feet in thickness and is capped by 15 inches of sandstone, slightly mineralized in spots. The ore in the portal of the upper tunnel, extending for 15 feet above the floor, is in quartzite and dolomite. Part of the dolomite shows a greenish cast, owing to the development of a greenish clay mineral between the grains of dolomite. Away from the mine the dolomite may grade into limestone. The rock between the two mine levels is comparatively barren, but where it is exposed it does not differ greatly in character or in type of mineralization from the rocks that carry the ore."

This mine, like many others in this sub-district, shows a stratigraphic and lithologic control of ore. Tectonic structures are lacking in most ore bodies. The predominant control appears to have been a subtle primary structure that was destroyed in mining or unrecognized. Many of the mines are no longer accessible, but, at some, facies changes in the ore-bearing bed can be seen at the mine portal. Cryptozoan structures are preserved in the Everton and were recorded by McKnight at the Madison mine.

The Coker Hollow mine in the Zinc sub-district produced ore from both the St. Joe and Everton Formations. Three levels of ore were developed, the upper in the St. Joe, the lower two in the Everton. The ore-bearing structure in the St. Joe Formation was a circular breccia mass regarded by McKnight (6, p. 249) as a pipe-slump. Breccia fragments included Boone chert as well as fragments of St. Joe limestone, indicating appreciable vertical displacement of blocks. The structure is ascribed to solution of underlying beds.

Primary ore minerals in the district include sphalerite, galena, chalcopyrite, and pyrite. A few crystals of enargite have been reported. The chief gangue minerals include jasperoid, pink crystalline dolomite, quartz, and calcite. Much of the ore produced in the district was in the form of the oxidized zinc minerals, calamine and smithsonite, from shallow pits and short drifts into hillsides.

An isotopic analysis of lead from the Rush Creek district gave a Pb 206/204 ratio of 22.73 (24).

## Central Kentucky District

The Central Kentucky district embraces an area of 3600 square miles that approximately coincides with the Blue Grass lowland. Small amounts of barite, lead, zinc, calcite, and fluorite have been mined. McFarlan (10) reported a maximum single year's production of 11,068 short tons of barite in 1914.

The ore minerals occur principally as vein deposits in Middle Ordovician limestones. Most of the known deposits are in the High-bridge and Lexington Groups of early middle Ordovician age, the oldest exposed formations on the dome. However, vein deposits occur in younger Middle Ordovician and in Upper Ordovician formations, and fluorite, barite, galena, and sphalerite are found in vugs and geodes in rocks as young as Mississippian.

Disconformably underlying the High Bridge group is the St. Peter-Knox sequence. The High Bridge is composed of thick-bedded limestones; the Lexington group of thin-bedded limestones. Overlying the Lexington group is a dominantly shale sequence, also of middle Ordovician age.

The main structural feature of the area is the Lexington dome. The Rough Creek-Kentucky River fault zone (Figure 2) transects the uplift. This, and related faults, form a complex system that underwent several periods of movement. Most of the mineral deposits occur as fissure veins in and near the major faults.

The mineral veins consist of simple fissure fillings and breccia replacement veins. Both types can occur in the same vein system. The veins are nearly vertical and contacts with vein walls usually are sharp. The Hayden vein, south of Danville, Kentucky, ranges from 8 to 14 feet in width. Many veins have a width of 2 to 6 feet. The Gratz vein system in the northwestern part of the district has been traced for a distance of approximately 10 miles and represents the largest and best known concentration. The fissure veins pinch and swell along strike. Many deposits consist of a series of mineralized en echelon fractures along a fault zone (40).

Jolly and Heyl (40) define a lateral zoning pattern for the district ranging from fluorite at the center to barite-galena-sphalerite on the periphery. They recognize (p. 616) three zones, differing in mineralogy and in ore textures.

Zone 1—Fluorite-calcite veins with minor amounts of barite. Black sphalerite may be present, but galena is rare. The ore shows coarse comb-crystal textures.

Zone 2—Fluorite-barite-calcite veins with yellow and black sphalerite. Also present are fluorite-barite veins with some galena, calcite, and yellow and black sphalerite, as well as barite veins with sphalerite and calcite. Both comb-crystal and colloform textures are present.

Zone 3—Barite-galena-sphalerite veins without fluorite. Sphalerite includes yellow, black, red, and orange varieties. Some barite-galena-sphalerite veins contain traces of fluorite. Colloform textures are common.

Sphalerite in the district is low in iron but rich in cadmium, germanium, and gallium. The black sphalerite appears to be early and, in places, is replaced by yellow and red-orange sphalerite. Chalcopyrite is present in minute amounts. The limestone walls show only very minor alteration, represented by quartz, cherty jasperoid, and ferroan dolomite (40).

Isotopic measurements on three galena samples from this district are reported by Russell and Farquhar (24). Pb 206/204 ratios for the three ranges from 19.91 to 20.22 and averages 20.04.

## Central Tennessee District

Deposits of this district are very similar in geologic setting, in types of veins, and in mineralogy to those of Central Kentucky. The deposits occur northeast, east, and southeast of the city of Nashville in Middle Ordovician rocks of the Central Basin. Several formations, correlative with the formations in Central Kentucky, are the host rocks.

The major domal structure of the two areas is similar, but a major fault system, comparable to the Kentucky River system, is absent in the Tennessee district. Numerous minor faults forming a northeast-trending system and a northwest-trending system, both with small throw, are present. Most of the mineral veins are associated with the northeast faults, and veins are displaced by the northwest faults (13).

Veins range in thickness from a few inches to 7 feet. Most of them strike about N40° to 45°E and are nearly vertical. Up to 200 feet vertical extent has been indicated by drilling in some veins. As are those in Kentucky, the veins are both filled-fissure and breccia-replacement type. Most veins contain open cavities.

The vein minerals are barite, fluorite, and calcite with small amounts of galena and sphalerite, and traces of chalcopyrite. The chalcopyrite is known only from the northern part of the district. Quartz and chert are minor in amount.

The most productive vein was the Hoover vein in Cannon County. The vein is in the cherty Carters Limestone to a depth of 82 feet, is up to 7 feet wide, and was mined for 900 feet strike length. Below the Carters Limestone, the softer, non-cherty Lebanon Limestone carried no ore, but a two inch thick vein of calcite marked the deeper extension of the vein. The ore consisted of light red sphalerite, some dark sphalerite, a small amount of barite, and traces of galena, but no fluorite.

No zoning study, comparable to that done for the Kentucky district, has been reported. Jolly and Heyl (40), however, report that their Zone 3 can be traced into the Central Tennessee district.

A single isotopic analysis (24) reported a Pb 206/204 ratio of 20.04, the same as the average for the Central Kentucky district.

## Tri-State District

The Tri-State district, long recognized as one of the great mining districts of the world, is described in detail in a paper by Brockie, *et al.*, that follows.

The ore, roughly in the proportion of Zn:Pb = 5:1, occurs in Mississippian formations on the southwest flank of the Ozark uplift. The Mississippian, unconformably lying on the Jefferson City-Cotter erosional surface, consists of the St. Joe Limestone, the Reeds Spring, Keokuk, and Warsaw beds, and limestones, sandstones, and shales of the Chester Group. Members of the productive formations are designated by a letter system in which the Warsaw embraces B through J beds, the Keokuk includes K through Q beds, and the Reeds Spring is designated as the R bed.

The M bed has been the most important producing horizon. The main ore-bearing structures in the M bed are flat-lying breccia zones. Ore-bearing areas in the M and the overlying J bed show a zontal pattern that, from the center outwards, consists of a dolomite core, the ore zone in brecciated jasperoid, and limestone. Both the dolomite and limestone contain abundant chert. The ore zone contains abundant jasperoid as well as chert. The ore-bearing body in the contact zone between the limestone and dolomite is strongly fractured and brecciated. Reference to Figures 8 to 10, (Brockie *et al.*, this volume) shows the pattern of changes from one rock type to another and the structural setting in which the facies changes occur.

The O bed contains blanket type ore bodies, referred to in the district as "sheet ground." These deposits were extensive but generally of lower grade than the breccia ore bodies.

The chief ore minerals are sphalerite and galena. Chalcopyrite and a minor amount of enargite are present. Cadmium, gallium, germanium, and indium are contained in the sphalerite. One of the striking features of the district is the very large amount of jasperoid and chert occurring with the ores.

Galenas from this district were among the first studied by Nier (8) who found that they contained unusually high amounts of the radiogenic isotopes of lead. He termed this type of lead J-type, after the Joplin occurrence, and the name has been generally applied to all leads with similar high radiogenic character.

Samples of ore from the district have been analyzed for isotopic character by many different laboratories. Unfortunately, many of the samples were taken from museum or private collections, so specific locations in the district and the geologic setting of the samples are unknown; thus, use of the data in interpreting mineralization history is limited.

Russell and Farquahar (24) report (for Toronto, Columbia, and Minnesota laboratories) a total of 34 determinations for the district. The Columbia analyses show the greatest extremes with Pb 206/204 ratios of 20.07 representing the minimum and 23.10 representing the maximum. Average for the 34 analyses is 22.01 Pb 206/204.

## Illinois-Kentucky Fluorspar District

The Illinois-Kentucky district (Grogan and Bradbury, this volume) differs from the ones described above in that fluorspar is the dominant mineral in this district but is rare to absent in the others. In addition, veins, not stratiform deposits, have been the most productive ore structures.

The deposits, both vein and bedded, occur in the Renault and Ste. Genevieve Formations of Mississippian age. Bastin (3), however, stated fluorspar deposits are found in all formations in the area from Devonian to Pennsylvanian.

The district has an area of about 1000 square miles and is located on a faulted anticlinal structure. Hicks Dome, one of an east-west trending line of cryptoexplosive structures (42) is located on the apex of the anticline, immediately north and northwest of the district. The Rough Creek-Shawneetown fault zone, a short distance north of the district, is part of one of the major fault lineaments in the Midcontinent.

A prominent system of faults extends from the Gulf Coast overlap northeastward. Most of the vein deposits are fissure fillings in this fault system or in a related east-west system. Some of these deposits show very little replacement of wallrocks; others have replacement bodies on one or both sides of the vein. Fluorite is the dominant mineral in most of the veins, but, in two veins, the Hutson and the Old Jim, both in Kentucky, sphalerite is the chief ore mineral.

The bedded deposits occur in the Cave-In-Rock area. The deposits are elongate, trough-shaped ore bodies up to 2 miles in length. The ore-bearing structures are of solution origin (14). Brecke (31) recognized three main member contacts in which ore occurs in a limestone beneath what he terms a blanket formation, usually a refractory member. Phenomena regarded by Brecke (31) as important phases of ore solution activity include decalcification, dolomitization, and silicification.

Fluorite is the dominant ore mineral in the district, and five stages of fluorite are recognized by Hall and Friedman (35). Four are regarded as earlier than chalcopyrite, sphalerite, and galena; the fifth as later than the sulfides. Bitumen is a common constituent of fluid inclusions and apparently was present throughout the mineralization process.

Because fluid inclusions in the fluorite are easier to observe than in most ore minerals, more data are available for this district on composition of inclusions and temperature of formation than for most districts. Hall and Friedman (35) state that the fluid inclusions in early ore minerals are sodium-calcium-chloride brines. The yellow fluorite, the earliest to form, contains brine similar in composition to connate water in the same formation in the Illinois basin. Later fluorite shows changes in composition of fluid inclusions thought to be due to mixing with magmatic waters. The last minerals to form show the effect of mixing of ore fluid with meteoric waters.

Two Pb isotope analyses (24) gave 206/204 ratios of 19.97 and 20.20. Eight analyses reported by Heyl *et al.* (47) gave values ranging from 19.91 to 21.01 Pb 206/204. The average for the 10 analyses is 20.30.

## Characteristics of Deposits of Mississippi Valley Type

INTRODUCTORY STATEMENT    The lead-zinc deposits of the Midcontinent are the classic examples that give rise to the term "Mississippi Valley type." Because they have long provoked argument and because they are of such great importance to the mineral economy of the area, the United States, and the World, a summary of the major characteristics of the deposits is in order.

The productive deposits are restricted to certain formations and usually are found within particular facies zones of the host. Features of individual ore bodies so often are conformable with host rock structures that a syngenetic, diagenetic, or epigenetic origin, according to the preference of the observer, can be postulated. Indeed, in some districts, visitors are facetiously offered their choice of the "syngenetic" trip or the "epigenetic" trip for the single standard mine tour.

The debate over origin of the deposits stems largely from the stratiform character of the ore bodies, the restriction of ore to host rock lithologies formed in certain sedimentary environments, the simulation by ore minerals of sedimentary features and textures, the simple mineralogies represented, the low temperatures of ore deposition, and the absence of an obvious source—igneous or otherwise. Those features of the deposits that are subject to dual interpretation can be called *permissive evidence* (54); they fit several possible modes of origin and do not support one in preference to another. In all of the major districts and in many minor ones, sufficient evidence is available to define for each district the features that point to a particular mode of origin or preclude a particular origin. These features can be called the *diagnostic evidence* (54). Analysis of the evidence for the major Midcontinent districts that have been mapped and studied in detail indicates these districts are of epigenetic origin.

Individual districts extend over areas of as much as 1000 square miles with mineral deposits occurring in several to many concentrations or sub-districts. A sub-district may contain hundreds of individual ore bodies, often with no apparent connection between them.

The major features of the districts, briefly described above and more fully in papers that follow, can be catalogued under the subjects of host rocks, structural patterns, types of ore traps, and mineralogical characteristics.

THE HOST ROCKS   *Shallow Water Carbonates:* The lead-zinc-barite-fluorite deposits of the Midcontinent occur in shallow water transgressive carbonate formations and occasionally in "dirty" sandstones associated with the carbonate host. The carbonate lithologies in which the ore occurs are those formed in very shallow marine environments. The sediments include abundant sand-size clastic carbonate composed of intraclasts, oolites, and fossil fragments; organic reef and reef detritus; and lagoonal deposits. Black argillaceous shales, frequently containing abundant diagenetic iron sulfide, are associated with the lagoonal and reef sediments. In all of the districts containing stratiform deposits, numerous lateral and vertical facies variations occur. The lateral changes often coincide with the boundaries of individual ore bodies; the vertical variations are present where ore occurs as blanket-type deposits on successive bedding planes.

*Basin Position*   The lithologic types define the sedimentary environments and, therefore, the basin position, in which the host rock sediments were deposited. Invariably, the ore deposits are in the shelf environment on the basin rim (the often-cited relationship to ancient shorelines) or over a topographic high within the basin in the shallow water sediments deposited over and around that high. None of the Midcontinent districts occurs in basin-type sediments.

*Unconformities*   Repeated diastems, often approaching the rank of disconformities, are recorded in shallow-water carbonates. Several major unconformities are present in the Midcontinent. In addition to the pronounced unconformity on the Precambrian, prominent erosional breaks include the post-Arbuckle (Jefferson City-Cotter-Powell surface), post-Devonian, and post-Mississippian.

Unconformities serve three roles that affect mineralization. One is the effect on formations below the unconformity; one on the character of sediments deposited above the unconformity; and one in providing a pathway for solution movement. During a long erosional interval, karst topography developed on the carbonate surface resulted in enlargement of fractures by groundwater action and formation of sinkholes and collapse breccias. In a carbonate host below a major unconformity, filled sinks and collapse breccias are important ore traps (36).

Above the unconformity, irregularities on the erosional surface provided a variety of depositional environments for younger sediments. The facies pattern and the sedimentary traps in formations above the unconformity result from these local variations in environment. The Southeast Missouri facies pattern of calcarenite bars and algal reefs developed over the maturely dissected Precambrian surface is a prime example. Equally apparent, although less well-known, is the role of the post-Jefferson City surface on deposition of Mississippian formations in the Tri-State district.

The unconformity itself usually is the most pronounced break in the sedimentary section; and probably is the path for lateral movement of ore solutions.

*Limestone-Dolomite Relationships*   The limestone-dolomite interface appears to be an important factor in localizing mineralization in stratiform deposits. The host formation is deposited as limestone, parts of which, in structurally favorable positions, are dolomitized. Dolomitization probably is diagenetic.

The favored locus for ore occurrence is at or near the dolomite-limestone interface. Mineralization in undolomitized limestone in the major Midcontinent districts is rare. Ore is almost as rare well within the dolomitized area. As shown by Brockie, *et al.* (this volume), ore in the Tri-State district occurs in the fractured jasperoid zone at the limestone-dolomite contact. In the Southeast Missouri district, the relationship is not so sharp, but haulage drifts and many drillholes cut limestone. The orebodies are in dolomite, but limestone can occur above or below ore or as islands surrounded by large ore bodies. In the Wisconsin-Illinois district, the Galena formation bearing ore is dolomite; away from ore bodies it is limestone.

*Stratigraphic Distribution of Ore*   An important feature of Midcontinent mineralization is its great stratigraphic distribution. In southern Missouri, the stratigraphic succession includes all formations from the Upper Cambrian Lamotte Sandstone to and including the Lower Ordovician Jefferson City. Younger Ordovician and Silurian and Devonian formations were removed by erosion and Mississippian and Pennsylvanian formations were deposited on the Jefferson City surface. Mineralization occurs in all formations present from the Lamotte Sandstone to the Cherokee Shale of Pennsylvanian age. Within any given district a single formation or group contained most of the ore mined (or known), but formations above and below the host contain numerous sulfide occurrences.

The Bonneterre formation in Southeast Missouri; the Galena-Platteville and Decorah Formations in Wisconsin-Illinois, the Osage-Meramec formations in the Tri-State and the Meromec-Chester formations in the Ill.-Ky. district have been the major producers; but sulfide occurrences are known and small amounts of ore have been produced from many different units. In the Northern Arkansas district, ore has been produced from the Ordovician Jefferson City, Cotter, Powell, and Everton Formations and from the Mississippian St. Joe and "Boone." Mineralization theory must recognize that a wave of ore solution dumped large amounts of metal in a single or a few formations and left its imprint on many others, or that mineralization occurred repeatedly throughout Paleozoic and probably Mesozoic time, affecting different formations at different times.

*Depth of Ore Occurrence* The Mississippi Valley type ore bodies long have been regarded as shallow deposits. Because they do not fit the threefold Lindgren classification, Graton (5) proposed the term *telethermal* for the group. Telethermal deposits are defined (17, p. 293) as "the ore deposits produced at or near the surface from ascending hydrothermal solutions and representing the terminal phase of its activity."

The depth connotation which the term carries may have appeared valid when the known deposits were restricted to the arbitrarily shallow depths represented by economic mining limits; but it is now recognized that galena-sphalerite mineralization of similar type can occur at great depths. Evans study (32) of sphalerite mineralization in lower Paleozoic sediments in western Kansas at depths of 4900 to 6440 feet was reviewed earlier. Similar widespread sphalerite mineralization is present in the Leduc Formation of Devonian age in Alberta, Canada, at depths in excess of 4500 feet. I have seen both galena and sphalerite of similar type in drillcores of carbonate rocks of Tertiary age at depths of 9000 feet. A recent oil test, Pan-American No. 1 Benevides, Webb County, Texas, encountered galena, sphalerite, and cerussite in dolomite in the Sligo Formation of Lower Cretaceous age at a depth of 14,860 to 15,140 feet (R. E. Rohn, personal communication).

Obviously, there is no evidence to indicate the thickness of the sedimentary cover at the time the deposits were formed. It is apparent, however, that Mississippi Valley type mineralization *can occur* at any depth; the only depth limitation is thickness of the sedimentary section.

THE STRUCTURAL PATTERN  Each major district that has been mapped in detail reveals its own structural pattern. The pattern consists of the tectonic features peculiar to the district superimposed upon, and to some extent controlled by, the district sedimentary facies relationships. The location of the district is facies controlled (the basin position). The trend of the district or of sub-districts often is fault controlled. The distribution of ore within the district to a large degree is controlled by sedimentary features and structures or by lithologic differences that affect the behavior of rocks under stress.

*The Major Fault Zones* Major faults crossing the mineral district, such as the Federal-Schultz system in the Southeast Missouri, the Miami trough in the Tri-State, or the strong northwest trending faults in the Northern Arkansas district, rarely are mineralized. Usually, these faults are high-angle normal faults; they are tight, and dense gouge and breccia mark the fault zone. The district or sub-district may be aligned along the fault trend, but the fault does not carry mineralization.

*The Minor Fracturing and Ground Preparation* The minor faults and numerous fractures subsidiary to the major faults play an important role in localizing ore within the stratiform deposits. Fractures may be filled with ore minerals to form veins ranging in thickness from a paper-thin zone to several inches. The chief function of the minor faults and fractures, however, is *ground preparation*, the development of numerous closely spaced openings through which solutions can move easily.

Several types of ground preparation are described in the chapter on the Southeast Missouri district. These include zones along en echelon fault systems where movement is translated from one fault segment to another through development of numerous tear fractures; broad zones with horsetail and braided fracture patterns developed where a major fault changes direction; and zones of closely spaced high-angle fractures where bedding plane movement is interrupted by massive, structureless rock. In any of these situations, a zone is created where the host rock is intensely broken. The "prepared" rock is more extensively mineralized and carries higher-grade ore than the same host rock outside the fractured area.

Other forms of ground preparation related to faulting include the development of extensive fracturing along a major fault, an important control in the Rush Creek sub-district of Northern Arkansas, and the formation of crackle breccias that carry ore in several districts. Bedding plane separation can be developed on the downthrown side of a normal fault. Where faulting is pre-ore, mineral seams can be present on successively higher beds near the fault, resulting in an increasing thickness of the ore zone toward the fault.

*Ground Preparation Through Subsidence* Solution collapse breccias, the important ore traps in several districts, in part, are fracture controlled. They are formed in some areas by differential solution controlled by lithologic differences, in other areas along fractures enlarged by groundwater action. In the latter case, the trend of ore bodies may reflect the district fracture pattern.

Subsidence of beds, regardless of the cause, results in the development of bedding-plane openings as well as brecciation. I have had the opportunity of studying caved mine areas in several districts. Caving over inadequately supported mine openings may take the form of a sudden plunger-type drop of a large thick block; it may be in the form of continued brecciation to a height of many times the original opening; or it may be a gradual subsidence of beds that dies out upward. Arching to the point of stability does not appear to be common across wide openings in flat-lying, bedded rocks.

In one caved area observed, a roughly circular block 200 feet in diameter dropped suddenly approximately 20 feet into a mine opening. The area caved to surface and the pattern of fracturing was clearly revealed. On three sides of the block the separation is a sharp, vertical break with the 20 feet displacement occurring on one fault, although successive discontinuous ring fractures appeared up to 30 feet outward from the dropped block. On the fourth side of the block, beds were folded downward from their original position; no fault break occurred. Undoubtedly, separation along bedding planes would occur in the folded segment in this type of adjustment.

Another caved area encompassed mine workings over 1000 feet across in a mine opening ranging in height from 10 feet to 40 feet and averaging about 15 feet. The opening was approximately 600 feet below surface and did not cave to surface. After a history of some 10 years of spalling and caving, and continued expansion of the caved area, several cored holes were drilled over the cave. The drilling

revealed that separation along bedding planes, with repeated openings of 6 to 10 inches, are present up to 150 feet above the mine level.

A third example involved a mine opening 300 feet across and 15 feet in height. The mine was filled with brecciated blocks of all sizes that extended upward beyond the limits of observation. Drilling over the caved area showed the brecciated rock extended over 100 feet above the mine opening. From the sides of the stope, caving extended upward in a step-like, not arch, fashion, due to separation on bedding planes. At any given point, the base of the undisturbed rock is a bedding plane, separated by an opening from the collapsed, brecciated caved rock. Eventually, the cave reached the point of arching where the "steps" from opposite sides of the opening converged.

The observed behavior of fractures, brecciation, and separation along bedding planes above mine openings suggests that similar features may develop above natural openings. Openings developed along bedding planes by groundwater action or by mineralization stoping need not be large in the vertical dimension to prepare the rock for a considerable distance above the actual opening. Mine studies indicate that effects on the rocks above the mine opening extend to a distance up to 10 times the height of the opening. The known vertical dimensions of collapse breccia-type ore structures would necessitate the removal only of a thin segment of the host formation. Theoretically, an ore solution, moving along a bedding plane and dissolving some carbonate rock, could prepare its own repository above the level along which solution occurred.

*Associated Vein Systems* Within the geographic boundaries of several of the major districts containing stratiform ore bodies, filled-fissure vein systems are present. Often the vein system provided the first ore mined in the district. A substantial tonnage of ore was produced from the veins in several districts, but, in most instances, the volume of vein ore was much less than was produced from the nearby stratiform deposits.

Notable examples of such vein systems are the Rosiclare fluorite-galena veins relative to the Cave-In-Rock fluorite-sphalerite deposits in the Illinois-Kentucky district (the only vein system of the group that has produced more ore); the Washington County, Missouri barite-galena deposits relative to the Southeast Missouri district; and the galena-bearing fissure veins and lodes on the northern fringes of the Wisconsin-Illinois sphalerite-galena district. Although veins may occur directly above the bedded deposits, they more commonly are miles distant laterally from the stratiform ore bodies, and often are in different host rocks. In the Wisconsin-Illinois district, gash-vein lead deposits occur above the typical pitch and flat sphalerite bodies; but veins and lodes are found in fractures of stratigraphically older rocks 10 to 30 miles north of the main productive area. The Washington County, Missouri, deposits are 10 to 30 miles from the Lead Belt and in stratigraphically younger rocks. The vein deposits in the Illinois-Kentucky district are approximately 10 miles distant from the stratiform ore bodies and in rocks of the same age.

The details of vein occurrence, the geographic position of veins relative to bedded deposits, and the stratigraphic position of the two types of deposits differ from one district to another. These features appear to be controlled by tectonic and mineralization history in the particular district. The geographic proximity of the vein and bedded deposits, however, suggests a genetic relationship. In most cases, the mineralogy of the two types of deposits is different, indicating a mineralogical zonation that is not yet clearly defined.

TYPES OF ORE-BEARING STRUCTURES    Each district has its own complement of ore-bearing structures. In each case, the combination of features making up the ore repositories is distinctive for a given district. One familiar with several districts can readily recognize each, from map patterns alone, without identification of ore minerals or age of host rock.

The ore-bearing structures in stratiform deposits, however, can be grouped into three basic types. They represent primary sedimentary features modified to various degrees by later events. The basic types of structures are stratigraphic traps, organic reefs and associated features, and breccias of diverse origins. The primary features that form ore traps are normal ones developed in shallow water carbonates. They differ to some degree, just as the host formations outside the ore districts differ, but all reflect the environment of deposition and the combination of lithologies making up the rock.

Most of the distinctive features of a district are due to events that have modified the original rock. This is a matter of local control, of type and degree of fracturing, and of extent of solution activity imposed upon the particular lithology. In the three major lead-zinc districts, Southeast Missouri, Tri-State, and Wisconsin-Illinois, the differences in modifying effects are great. The Southeast Missouri district shows the least modification. Stratigraphic

traps, facies controlled, are modified to only a minor degree by fracturing, and alteration effects are slight. In the Tri-State district, the great abundance of jasperoid and chert and the degree of brecciation obscure the primary features, but the position and character of the siliceous breccia zone is controlled by primary differences. In the Wisconsin-Illinois district, both solution activity and fracturing have taken a different form than in the Tri-State, resulting in the differences in ore structures.

*Stratigraphic Traps*    Traps of this type are best developed over an irregular depositional surface. They include pinchout of beds against the flank of an underlying high, thinning of beds over a high, and lateral changes in composition of a carbonate bed. An abundance of mud, either calcareous or argillaceous, results in compaction effects that greatly alter the attitude of the original sediments. The ore-bearing structures described and illustrated in the Southeast Missouri paper that follows indicate the variety of traps due to facies differences.

*Reefs and Associated Structures*    Reefs provide a rigid framework around which less competent clastic carbonate sediments are deposited. They may be either more or less receptive to mineralization than the adjacent clastic sediments. In southeast Missouri, parts of the reef are mineralized; in some districts, reefs form the unmineralized areas between ore bodies.

In addition to being a possible ore host, the reefs, often dolomitized, react differently to tectonic stresses than adjacent bedded sediments. The most extensive fracturing often is at the reef-bedded rock contact.

*Breccias*    Four types of breccia can be recognized in Midcontinent districts. Each has a different origin. The types differ somewhat in geometry of the breccia body, but the resulting breccias in hand specimen or limited exposure are often similar in appearance.

One type of breccia is that formed by submarine slides during deposition of the host formation (19). The breccias occur along the flank of a local depositional high; contours on the base of the breccia body coincide with the position of the high and the basin outline. The breccia bodies are elongate, arcuate, and sometimes lobed. Ore bodies in breccias of this type form long narrow zones, often with the height greater than the width. Breccia fragments and matrix are the host formation. Vugs and cavities are rare.

A second type of breccia is the collapse breccia developed by solution action. The solution effects take place most readily along a fracture zone or at a contact of two different lithologies. Where the carbonate host has been exposed to erosion and parts have been removed, breccias of this type form major ore structures. Forms of the breccia bodies differ widely. They may be narrow zones, fracture controlled, and have the trend of the district fracture pattern. They may, if lithologically controlled, show the district facies pattern.

A third type of breccia should more properly be called a pseudobreccia. It has the appearance of a breccia but much, probably most, of the material has not been moved from its original position. Slight discordances in bedding are often apparent, indicating some rotation of fragments, but material foreign to the formation is lacking.

This type of breccia appears to be formed in place by differential dissolution of carbonate. Many carbonate rocks contain material of two distinctly different grain sizes giving a mottled, irregular texture to the rock. Material of one grain size is removed more easily than the other resulting in a vuggy rock. When the process is carried to an extreme, most of one constituent is removed; the undissolved fragments lend an appearance of a vuggy breccia. The vugs may be open, may contain some insoluble residue, or may be filled with white crystalline carbonate.

Breccias of this type can occur at any depth and are not related to an erosional surface. They show no oxidation effects. Usually the base of the breccia body is flat; the sides are irregular. No material foreign to the breccia-bearing formation is present because general subsidence has not occurred. The breccia forms in a particular facies of the host, often parallel to some sedimentary feature that controlled the facies pattern. Many so-called collapse breccias that show no disturbance of overlying beds and no thinning of the breccia-bearing member probably are of dissolution origin.

A fourth type of breccia is that developed by fracturing. This may take several forms, as described earlier, and result in various degrees of brecciation. Where fracturing is intense, small-fragment crackle breccias are developed.

*Fissures*  Fissures due to jointing or minor faulting represent an ore structure in most of the major districts, an important ore structure in the Illinois-Kentucky district, and the chief ore structure in several of the small districts. The types of fissures, their size, and their abundance differ from one district to another in accord with the local fracture pattern.

*Sizes of Ore Bodies*  The size of the entrapping structure is the chief control of size of ore bodies. The Southeast Missouri and Tri-State districts contain numerous large structures in the Bonneterre and Keokuk-Warsaw Formations, respectively, and large and closely grouped ore bodies are the rule. The Wisconsin-Illinois district, in the Platteville-Decorah-Galena Formation, contain numerous smaller, scattered structures. Ore bodies are small and scattered over a large area. The Northern Arkansas district has a great number of individual ore bodies but the entrapping structures are small, greatly limiting the potential of the district.

The potential size of the structures expected and their degree of concentration or grouping into sub-districts are basic factors in evaluating an area for exploration.

MINERALOGICAL CHARACTERISTICS  Much attention has been given in the literature to mineralogy of the several districts and to comparisons of certain features of all districts. No attempt will be made here to present a complete discussion of all features. The major items of concern will be reviewed briefly to indicate some of the similarities that can be documented in the papers that follow and in previous literature.

*Elements Present*  The ore minerals are present in simple form as sulfides, sulfates, and halides. The metals that appear to be universally present are zinc, lead, copper, and iron. Zinc predominates over lead in all except the Southeast Missouri (and probably Central Texas) district. Both barite and fluorite often are present as minor constituents but are the chief ore mineral only in a few districts. Cobalt and nickel are found in a few districts. Cadmium and germanium are contained in most of the sphalerites. Silver usually is present but is a very minor constituent. Arsenic has been reported from the Tri-State, Northern Arkansas, and Wisconsin-Illinois but not from the other districts.

*Isotopic Character of Lead*  All of the districts for which lead isotope data have been reported show that the lead is the radiogenic J-type. Where any appreciable number of analyses is reported, a wide range in isotopic character is indicated. Where sufficient isotopic information is available, the data show an orderly distribution pattern throughout the district.

The isotopic data are open to differences in interpretation. They may suggest, however, by the wide range in isotopic character, that mineralization occurred over a long period of time. They also suggest, by their high radiogenic character and the lack of homogeneity, that the mineralizing solutions did not arise by simple derivation from a magma. The data indicate that the simple pattern of lead isotopes demanded by a syngenetic hypothesis does not exist for the districts studied, and the deposits cannot be syngenetic in origin.

*Isotopic Character of Sulfur* Less attention has been given to sulfur determinations for mineral districts than for lead. The data available suggest that the sulfur is of biogenic origin (Snyder and Gerdemann, this volume).

*Paragenesis* Studies on the order of crystallization of the ore minerals in most of the districts have been made. In several districts results of studies reported by different authors give different interpretations. Bastin and Ridge stated the problem succinctly (9, p. 109), "Concerning the relative order of deposition of the several sulfides, the published reports indicate the utmost diversity of opinion."

The differences in interpretation appear to arise from the long period of mineralization involved and the reversals, interruptions, and surges of new ore solutions, that modify or replace earlier minerals. Hagni and Grawe (37) recognized 8 periods of chalcopyrite mineralization in Tri-State ores, 4 periods of galena, and 5 of sphalerite, although the bulk of a given metal was deposited during one stage. More detailed study might reveal additional stages.

It seems apparent that, during the long period of mineralization, changes in metal content of solutions occurred as well as changes in conditions of deposition. A paragenetic study of a given specimen defines the situation at that point in space and time. In a different part of the district and at a different time, considerably different mineral relationships may exist and a different paragenetic order would result. I have defined elsewhere (54) the time zoning reflected by lead, zinc, and copper in the Southeast Missouri district, as suggested by isotopic data. If time is a major factor in Mississippi Valley type mineralization, paragenetic studies need to be interpreted with caution. Mineral relationships at any particular point in the district do not necessarily indicate the position in a changing sequence over a long period of time.

*Composition of Fluid Inclusions* A rapidly growing body of data on composition of fluids contained in ore minerals indicates that saline waters, probably of connate origin, played an important role in mineralization. Hall and Friedman, after a study of the Wisconsin-Illinois and Illinois-Kentucky districts, state (35, p. 908), "Our data indicate that the fluorite and lead-zinc ores were deposited from a concentrated Na-Ca-Cl-type water that is about the composition of oil-field brines (connate water) present in many basins underlain by pre-Tertiary marine sedimentary rocks." Similar conclusions have been reached in studies of fluid inclusions from other districts.

*Temperatures of Deposition* Temperature data, developed also by study of fluid inclusions, indicate most deposits were formed in the temperature range from 75 to 150°C (35). The later-formed minerals reveal the lower temperatures of formation.

Temperature measurements made in sedimentary basins during oil-well drilling indicate that brine temperatures have a range comparable to the temperatures of crystallization of the ore minerals. Buckley *et al.* (18) show for the East Texas oil field temperatures of 85°C at 1000 feet depth increasing regularly to 220°C at 10,000 feet depth. At 5000 feet, the approximate depth of several Midcontinent basins, temperatures of 170° to 180°C are recorded. Similar measurements are recorded for the upper Texas Gulf Coast and Mississippi-Alabama areas (18).

*Silicification* The mineral districts show all degrees of silicification from an almost negligible amount in the Southeast Missouri district to extreme silicification in the Tri-State district. Silicification may take the form of quartz cement around sand grains, a quartz mosaic pervading carbonates, and chert and jasperoid.

The amount of silica occurring with the ore appears to be related to the proximity and abundance of silica in the host formation. The chert-free Bonneterre formation shows no silicification accompanying ore throughout most of the district. Only where the ore horizon is near the Lamotte Sandstone or basement knobs of granite or rhyolite porphyry does a minor amount of silicification occur (Snyder and Gerdemann, this volume). In those districts that are in a moderately cherty formation, the ore is accompanied by appreciable amounts of jasperoid, and sometimes quartz crystals. Examples include the Wisconsin-Illinois district and the Northern Arkansas district. Where the host rock is very cherty, as is the case with the Mississippian formations in the Tri-State, the ore is highly siliceous.

Other examples outside the Midcontinent could be cited. No quantitative data are available, but the amount of silica with the ore appears to be closely related to the amount of silica contained in the host rock, suggesting that much of the silica occurring with the ore is derived locally.

## Definition of Mississippi Valley type deposits

Mississippi Valley type deposits in Midcontinent United States, the type locality, may be defined as follows:

The deposits are stratiform in character and concentrated within a single or a few formations, but with minor amounts of similar sulfides occurring in all formations above and below the major host. Fissure veins may be associated with the stratiform bodies in the same formation or in older or younger formations.

TABLE III.   *Characteristics of Mississippi Valley (Midcontinent) Lead-zinc Deposits*

(1) Ore occurrence in host rock
  A. In shallow water carbonates.
  B. On rim of basin or over high in basin.
  C. Unconformities above, below, or within the host.
  D. At or near the limestone-dolomite interface.
  E. At any depth.
(2) Structural pattern of districts
  A. Ore does not occur on major faults.
  B. Ground prepared by minor faults and fractures.
  C. Ground prepared by subsidence and dissolution.
  D. Filled fissure vein systems usually associated with stratiform deposits.
(3) Types of ore-bearing structures
  A. Stratigraphic traps.
  B. Reefs and associated structures.
  C. Breccias.
    1. Submarine slide.
    2. Solution collapse.
    3. Pseudobreccia formed by dissolution.
    4. Fault breccias.
  D. Fissures.
  E. Size of ore bodies directly related to size of entrapping structures in host.
(4) Mineralogical characteristics
  A. Elements present
    1. Zn, Pb, and Cu always present.
    2. Ba, Co, Ni, Cd, Ge, Ag usually present. Fluorite may be present.
  B. Lead is J-type with considerable range in Pb 206/204 ratio.
  C. Sulfur probably of biogenic origin.
  D. Paragenetic studies are difficult to interpret because of change of solutions with time.
  E. Fluid inclusions are mainly connate water.
  F. Temperatures of formation range from 70° to 150°C.
  G. Silica occurring with the ore is roughly proportional to the abundance of silica in the host.

The host rocks are shallow water marine carbonates, and the ore deposits occur on the rim of the sedimentary basin or over highs within the basin at or near the limestone-dolomite interface and at any depth.

The orebodies are contained in stratigraphic traps, in organic reefs and associated sediments, and in breccias of submarine slide, solution collapse, dissolution, or fault origin; all of the structures "prepared" to some degree by post-lithification events.

The ore minerals are simple sulfides of Zn, Pb, and Cu, with or without Ba, Fl, Co, Ni, Cd, Ge, and Ag. The lead is J-type in radiogenic character. The ore metals appear to have been brought in by saline solutions at temperatures ranging from 70° to 150°C. Silica occurring with the ore was probably derived from the host formation and re-deposited. The characteristics of the Mississippi (Midcontinent) lead-zinc deposits are summarized in Table III.

Earlier a definition for telethermal ore bodies was cited. The term implies a shallow depth for deposits formed by solutions of hydrothermal (igneous) origin. Because Mississippi Valley type deposits can occur at any depth and there is serious doubt that the ore solutions were derived from below the sedimentary cover, the term *telethermal* should be dropped for deposits of this type.

## REFERENCES CITED

1. Winslow, A., 1894, Lead and zinc deposits: Mo. Geol. Surv., v. 7, 763 p.
2. Butler, B. S., *et al.*, 1929, The copper deposits of Michigan: U.S. Geol. Surv. Prof. Paper 144, 238 p.
3. Bastin, E. S., 1932, The fluorspar deposits of southern Illinois: 16th Int. Geol. Cong. Guidebook 2, p. 32–44.
4. Steidtmann, E., 1932, The iron deposits of Pilot Knob, Missouri: 16th Int. Geol. Cong. Guidebook 2, p. 68–73.
5. Graton, L. C., 1933, The hydrothermal depth zones: *in Ore deposits of the western states* (Lindgren Volume), A.I.M.E., N.Y., p. 181–197.
6. McKnight, E. T., 1935, Zinc and lead deposits of northern Arkansas: U.S. Geol. Surv. Bull. 853, 311 p.
7. Bucher, W. H., 1936, Cryptovolcanic structures in the United States: 16th Int. Geol. Cong. 1933, Rept., v. 2, p. 1055–1084.
8. Nier, A. O., 1939, Variations in the relative abundances of the isotopes of common lead from various sources: Amer. Chem. Soc. Jour. v. 60, p. 1571.
9. Bastin, E. S. and Ridge, J. D., 1939, Paragenesis of the Tri-State zinc district: p. 105–110, *in* Bastin, E. S., *Editor, Lead and zinc de-*

posits of the Mississippi Valley region, Geol. Soc. Amer. Spec. Paper no. 24, 156 p.

10. McFarlan, A. C., 1943, Geology of Kentucky: Univ. Kentucky, Lexington, 531 p.

11. Cullison, J. S. and Ellison, S. P., Jr., 1944, Diamond drill core from Bourbon high, Crawford County, Missouri: Jour. Geol., v. 28, p. 1386–1396.

12. Hayes, W. C., Jr., 1947, Precambrian iron deposits of Missouri: Unpublished Ph. D. dissertation, Univ. Iowa.

13. Jewell, W. B., 1947, Barite, fluorite, galena, sphalerite veins of middle Tennessee: Tenn. Div. Geol., Bull. 51, 114 p.

14. Grogan, R. M., 1949, Structure due to volume shrinkage in the bedding replacement fluorspar deposits of southern Illinois: Econ. Geol., v. 44, p. 606–616.

15. Heyl, A. V., et al., 1955, Zinc-lead-copper resources and general geology of the Upper Mississippi Valley district: U.S. Geol. Surv. Bull. 1015-G, p. 225–243.

16. Barnes, V. E., 1956, Lead deposits in the upper Cambrian of central Texas: Texas Bur. Econ. Geol., R.I. 26, 68 p.

17. Howell, J. V., Coord. Chairman, 1959, Glossary of Geology, Amer. Geol. Inst., Washington, 325 p.

18. Buckley, S. E., et al., 1958, Distribution of dissolved hydrocarbons in subsurface waters: in Weeks, L. G., Editor, Habitat of Oil, A.A.P.G. Symposium, Tulsa, 1384 p.

19. Snyder, F. G. and Odell, J. W., 1958, Sedimentary breccias in the southeast Missouri lead district: Geol. Soc. Amer. Bull., v. 69, p. 899–925.

20. Woolard, G. P., 1958, Areas of tectonic activity in the United States as indicated by earthquake epicenters: Amer. Geophys. Union Tr., v. 39, p. 1135–1150.

21. Dietz, R. S., 1959, Shatter cones in cryptoexplosion structures (meteorite impact?): Jour. Geol., v. 67, p. 496–505.

22. Hayes, W. C., Jr., 1959, Compilation of chemical analyses, Precambrian rocks of Missouri: Mo. Geol. Surv. and Water Res., Misc. Pub. (multilithed).

23. Frederickson, A. F. and Reynolds, R. C., Jr., 1960, Sedimentary deposits of copper, vanadium-uranium and silver in southwestern United States: Econ. Geol., v. 32, p. 906–951.

24. Russell, R. D. and Farquahar, R. M., 1960, Lead isotopes in geology: Interscience Publishers, N.Y., 243 p.

25. Tolman, C. and Robertson, F., 1960, Precambrian geologic map of Missouri (preliminary copy): Mo. Geol. Surv. and Water Res.

26. White, W. S., 1960, The Keweenawan lavas of Lake Superior, an example of flood basalts: Amer. Jour. Sci., 258-A (Bradley Volume), p. 367–374.

27. Anderson, R. E. and Scharon, H. L., 1961,

Notes on the geology of the Tom Sauk area: in Hayes, W. C., Jr., Editor, Geology of the St. Francois Mountain area, Mo. Geol. Surv. and Water Res. R.I. 26, p. 119–121.

28. Hayes, W. C., Jr., 1961, Precambrian rock units in Missouri: in Hayes, W. C., Jr., Editor, Geology of the St. Francois Mountain area, Mo. Geol. Surv. and Water Res. R.I. 26, p. 81–82.

29. Koenig, J. W., 1961, Editor, The stratigraphic section in Missouri: Mo. Geol. Surv. and Water Res. [Rept.], v. 40, 2d ser., 185 p.

30. Snyder, F. G. and Wagner, R. E., 1961, Precambrian of southeast Missouri, status and problems: in Hayes, W. G., Jr., Editor, Geology of the St. Francois Mountain area, Mo. Geol. Surv. and Water Res. R.I., 26, p. 84–94.

31. Brecke, E. A., 1962, Ore genesis of the Cave-In-Rock fluorspar district, Hardin County, Illinois: Econ. Geol., v. 57, p. 499–535.

32. Evans, D. L., 1962, Sphalerite mineralization in deep-lying dolomites of Upper Arbuckle age, west central Kansas: Econ. Geol., v. 57, p. 548–564.

33. Bucher, W. H., 1963, Cryptoexplosion structures caused from without or from within the Earth? ("astroblemes" or "geoblemes?"): Amer. Jour. Sci., v. 261, p. 597–649.

34. Catanzaro, E. J., 1963, Zircon ages in southwestern Minnesota: Jour. Geophys. Res., v. 68, p. 2045–2048.

35. Hall, W. E. and Friedman, I., 1963, Composition of fluid inclusions. Cave-In-Rock fluorite district, Illinois and Upper Mississippi Valley zinc-lead district: Econ. Geol., v. 58, p. 886–911.

36. Callahan, W. H., 1964, Paleophysiographic premises for prospecting for strata-bound base metal mineral deposits in carbonate rocks: CENTO Symposium on Mining Geology and Base Metals, Ankara, p. 191–248.

37. Hagni, R. D. and Grawe, O. R., 1964, Mineral paragenesis in the Tri-State district Missouri, Kansas, Oklahoma: Econ. Geol., v. 59, p. 449–457.

38. Ham, W. E., et al., 1964, Basement rocks and structural evolution of southern Oklahoma: Okla. Geol. Surv. Bull. 95, 302 p.

39. Ham, W. E. and Johnson, K. S., 1964, Copper deposits in Permian shale, Creta area: Okla. Geol. Surv. Circ. 64, 32 p.

40. Jolly, J. L. and Heyl, A. V., 1964, Mineral paragenesis and zoning in the Central Kentucky mineral district: Econ. Geol., v. 59, p. 596–624.

41. Heyl, A. V., et al., 1965, Regional structure of the southeast Missouri and Illinois-Kentucky mineral districts: U.S. Geol. Surv. Bull. 1202-B, 20 p.

42. Snyder, F. G. and Gerdemann, P. E., 1965, Explosive igneous activity along an Illinois-

Missouri-Kansas axis: Amer. Jour. Sci., v. 263, p. 465–493.

43. Zartman, R. E., *et al.*, 1965, K-Ar and Rb-Sr ages of some alkalic intrusive rocks from central and eastern United States (abs.): Geol. Soc. Amer. Program, p. 187–188.

44. Goldich, S. S., *et al.*, 1966, Geochronology of the Midcontinent region, United States, Part I: Jour. Geophys. Res., v. 71, p. 5375–5388.

45. Goldich, S. S., *et al.*, 1966, Geochronology of the Midcontinent region, United States, Part 2: Jour. Geophys. Res., v. 71, p. 5389–5408.

46. Halls, H. C., 1966, A review of the Keweenawan geology of the Lake Superior region: p. 3–27, *in* Steinhart, J. S. and Smith, T. J., *Editors, The earth beneath the continents,* Amer. Geophys. Union, Mono. 10, 663 p.

47. Heyl, A. V., *et al.*, 1966, Isotopic study of galenas from the Upper Mississippi Valley, the Illinois-Kentucky, and some Appalachian Valley mineral districts: Econ. Geol., v. 61, p. 933–961.

48. Lidiak, E. G., *et al.*, 1966, Geochronology of the Midcontinent region, United States, Part 4: Jour. Geophys. Res., v. 71, p. 5427–5438.

49. Muehlberger, W. R., *et al.*, 1966, Geochronology of the Midcontinent region, United States, Part 3: Jour. Geophys. Res. v. 71, p. 5409–5426.

50. Snyder, F. G., 1966, Precambrian iron deposits in Missouri (abs.): Econ. Geol., v. 61, 799 p.

51. White, W. S., 1966, Geologic evidence for coastal structure in the Western Lake Superior basin: p. 28–41 *in* Steinhart, J. S. and Smith, T. J., *Editors, The earth beneath the continents,* Amer. Geophy. Union, Mono. 10, 663 p.

52. Anon., 1967, Eleven firms win state leases in Minnesota Cu-Ni area: Eng. and Min. Jour. v. 168, no. 3, p. 108, 110.

53. Kisvarsanyi, G. and Proctor, P. D., 1967, Trace element content of magnetites and hematites, southeast Missouri iron metallogenic province, U.S.A.: Econ. Geol., v. 62, p. 449–471.

54. Snyder, F. G., 1967, Criteria for the origin of stratiform ore bodies with application to southeast Missouri: Econ. Geol., Mono. 3, p. 1–13.

55. Snyder, F. G., in press, Tectonic history of Midcontinent United States: Univ. Mo. at Rolla.

# 15. The Iron Mountain Mine, Iron Mountain, Missouri

JOHN E. MURPHY,* ERNEST L. OHLE†

## Contents

## Illustrations

* The Hanna Mining Company, Ironton, Missouri.
† The Hanna Mining Company, Cleveland, Ohio.

## ABSTRACT

Hematite-magnetite ore bodies at Iron Mountain, Missouri, have produced nearly 9 million tons of iron ore concentrates since 1844. The ore minerals occur principally as open-space filling in fractured and brecciated andesite porphyry with very minor replacement of the wall rocks. The enclosing rocks are Precambrian in age and consist of a series of dacite, rhyolite, and andesite flows aggregating about 2000 feet in thickness. All of the ore is in andesite and the mineralization is Precambrian in age.

The shape of the ore bodies, the breccia character of the ore, and the gangue mineralogy are believed to be unique. One of the two main ore bodies is crudely dome-shaped with a tendency for development of concentric mineralized shells; the other is an elongated lens in plan but hook-shaped in cross section. Both have strong linear elements striking northeast, east-west, or northwest. Various modes of origin for the structures and for the tension breccias have been suggested, but none is completely satisfactory.

Hematite is the main ore mineral; magnetite is subordinate. Important gangue minerals are andradite, quartz, calcite, actinolite, apatite, epidote, and chlorite. Although some evidence favors injection of an iron-rich melt, other evidence suggests ore formation by a high temperature, water-rich solution.

## INTRODUCTION

The Iron Mountain Mine of The Hanna Mining Company exhibits a number of unusual geologic features which make its description and inclusion in this volume appropriate even though it has not been of great significance economically. The property was closed in September, 1966, upon exhaustion of its commercial ore, after 122 years of intermittent production. The total output over this period was nine million tons of iron concentrates, approximately half of which was produced during the last period of operation, 1940 to 1966. Near surface portions of the ore were mined by open pit methods but the majority of the tonnage was extracted from underground, using a room and pillar system. The average crude iron grade in recent years has varied from 30 to 38 per cent.

The unusual geologic features of the deposit are its shape, structural setting, and gangue mineralogy, which are believed to be unique among iron ore bodies. So far no entirely satisfactory explanation has been found for the dome shape and the hook shape of the two principal ore bodies in cross-section, or the pronounced linear elements which are conspicuous on plan maps. Any explanation of these gross structural patterns must also account for the striking appearance of the ore itself which consists of massive veins of solid hematite or magnetite with sharp walls or of irregular bodies of host rock breccia with a predominantly iron oxide matrix. Proponents of several schools of geological thought may be found in the literature—offering explanations for both the structure of the ore bodies and the occurrence of the ore minerals. Rarely has such a small deposit attracted so much scholarly investigation.

## LOCATION AND GEOLOGIC SETTING

The Iron Mountain Mine is located 80 miles south of St. Louis, Missouri, on the north side of the St. Francois Mountains. This location at the east end of the Ozark Dome is in gently rolling country with a total relief of 300 to 500 feet. Most of the hills are composed of Precambrian igneous rocks while the valleys are filled with Paleozoic sediments. These sediments were deposited on a mature erosion surface that was developed during the Lipalian interval, and original dips as high as 20° in all directions away from the buried, igneous hills are not uncommon. At one time, sediments covered all of the Precambrian. Subsequent uplift and erosion have removed the younger formations so that today, in the vicinity of the iron deposits, only the basal Lamotte Sandstone and the overlying Bonneterre Dolomite remain.

## DETAILED GEOLOGY

Figure 1 shows the major geologic features in the vicinity of Iron Mountain Mine and the locations of the two principal ore bodies which are called "Main" and "Northwest." Smaller ore bodies were mined in Big Cut, Hayes Cut, and Newman Cut during the early history of the operation, but the details of their geology are not known.

### Rock Types

All of the Precambrian rocks are igneous. They consist of a series of flows and pyroclas-tics at least 2000 feet thick that were fractured, selectively mineralized, and finally intruded by andesite and felsite dikes. There are three principal igneous rock units in the mineralized area: a lower pyroclastic, a middle (?) dacite, and an upper andesite. The latter contains all of the ore. A thousand feet north of the ore bodies, there is a body of rhyolite which may be older than any of the above-named units. Two very deep holes drilled to explore beneath the ore body, bottomed in a felsite which appears to be cross-cutting the pyroclastics and andesite. It may be a dike. Since most drill holes have been stopped without penetrating more than a few tens of feet below the ore,

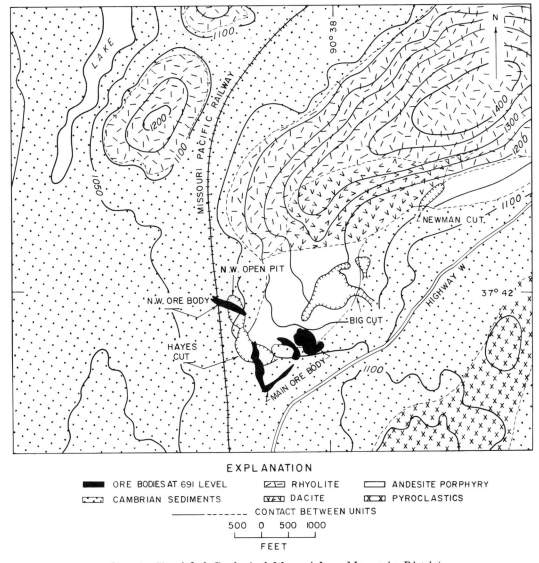

EXPLANATION

| | | |
|---|---|---|
| ■■■ ORE BODIES AT 691 LEVEL | ▱▱ RHYOLITE | ▭ ANDESITE PORPHYRY |
| ▱▱ CAMBRIAN SEDIMENTS | ▱▱ DACITE | ▱▱ PYROCLASTICS |

———— - - - - - - - CONTACT BETWEEN UNITS

500   0    500   1000

FEET

FIG. 1. *Simplified Geological Map of Iron Mountain District.*

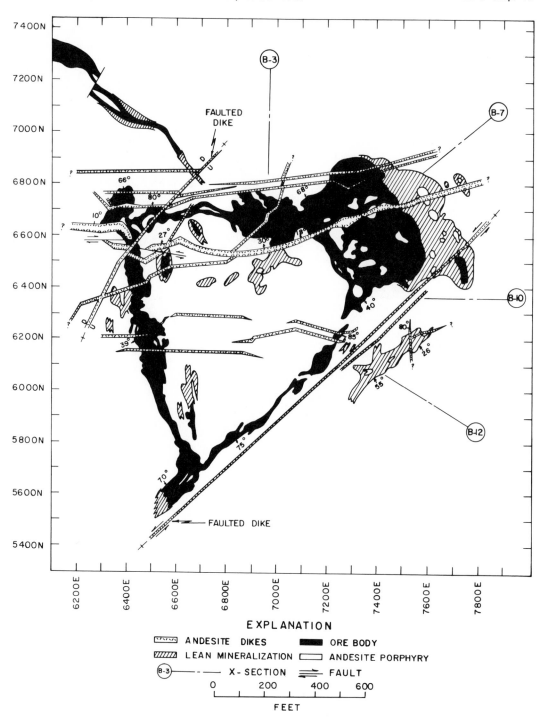

FIG. 2. *Plan Map 654 Level Main Ore Body.*

there is only a limited amount of information available on the distribution and relations of the rock units underlying andesite.

The pyroclastic zone consists of red, brown, and black, well-bedded, lithic and lapilli tuffs and volcanic breccias. It is at least 500 feet thick, and individual beds range from 4 to 40 feet in thickness. The rocks strike north-south and dip 20° to 30° to the west.

The dacite unit is as much as 300 feet thick.

It overlies the pyroclastic zone in the northern part of the mine area but pinches out beneath the Main Ore Body so that, to the south, the ore-bearing andesite rests directly on the pyroclastic beds. The dacite has a general east-west strike and a 20 degree dip to the south. Most of it is porphyritic.

The host rock for all of the mineralization is an andesite porphyry flow that is at least 1100 feet thick. Various authors have called this andesite intrusive but recent drilling has shown it to contain at least one zone of lithic tuff and occasional amygdules. The unit thickens to the west, but its total original thickness is unknown because the top contact with a younger formation has not been found.

The rock is gray, purple, or brownish red in color, and fine to medium grained. Most of the plagioclase phenocysts are randomly ori-

ented, but in places they exhibit a preferred orientation. Not uncommonly, the rock has a mottled appearance caused by irregular clots and grains of chlorite and epidote alteration.

Intruding all of the above rock units is a series of post-ore andesite dikes which range from a fraction of an inch to 10 feet in thickness. They may be gray or green, and all of them are very fine grained. Most of the dikes strike nearly east-west and dip steeply northward, but one particularly prominent dike dips at the relatively flat angle of 5° to 45°. It is younger than the steeper intrusives.

## Faults and Joints

Many near-vertical strike-slip faults intersect the ore body, but most have displacements of less than a foot. Faults are especially com-

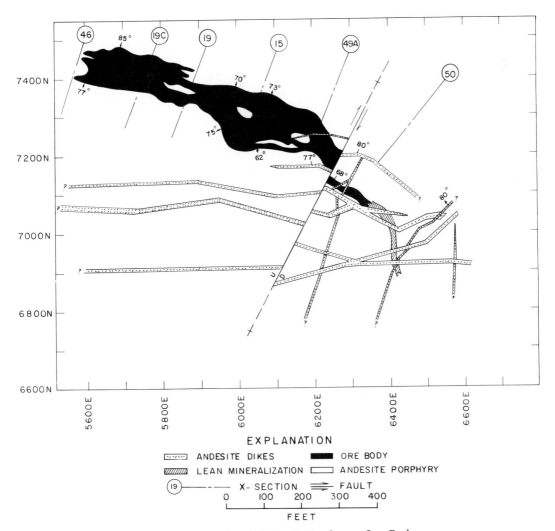

FIG. 3. *Plan Map 693 Level Northwest Ore Body.*

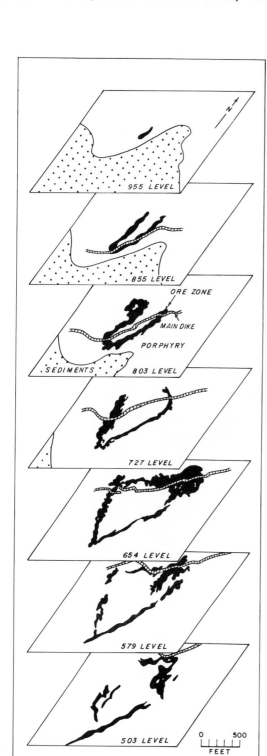

FIG. 4. *Main Ore Body.*

mon along the andesite dikes with movement both preceding and following the intrusion. Two good examples are found on the northwest and southeast sides of the Main Ore Body (Figure 2). In the case of the former, the northwest side dropped 30 feet; in the latter, there has been 10 feet of displacement. All of the faults are Precambrian.

The largest and most prominent fault zone in the mine displaces the Northwest Ore Body about 40 feet horizontally (Figure 3). The fact that the ore in the 693 level is narrower southeast of the fault indicates that there has been vertical movement also. This zone contains up to three feet of gouge mixed with calcite, dolomite, pyrite, marcasite, and minor amounts of chalcopyrite.

Joints are very common, with two sets that strike northeast and northwest being most abundant. These dip 82°NW and 87°NE, respectively. In many areas, the joints are closely spaced; some are tight, and some are filled with calcite or dolomite.

## Age of the Mineralization

There is absolutely no question but that the mineralization is Precambrian. This is indicated by the occurrence of abundant hematite pebbles incorporated in the basal Paleozoic conglomerate. More specifically the ore is later than the andesite host rock but earlier than the andesite dikes.

## ECONOMIC GEOLOGY

### Forms of the Ore Bodies

The shapes of the Iron Mountain ore bodies and the associated structures that influenced these shapes constitute the most interesting aspects of the ore occurrence. Both the Main Ore Body and the Northwest Ore Body are characterized by unusual curvilinear forms that are strikingly displayed in plans and cross-sections and have called forth a variety of explanations from the geologists who have studied the mine.

The Main Ore Body has been compared by various authors to (1) a dome-shaped shell, (2) a three-sided inverted cup, or (3) a baseball hat. As shown in the series of plan maps at different elevations (Figure 4) and the cross sections (Figures 5, 6, 7, 8), there is some semblance to all three.

The top of the structure is a small, elliptical dome with ore zones 30 to 50 feet thick that roll outward and downward. One hundred feet below the top, the plan view has the shape of a horseshoe with its open end pointing north. This shape is retained for another 100 feet of depth, at which point the sides become straight and the body assumes the shape of a crude triangle partially open to the north with the apex pointing south. At 300 feet below the top, or, at 654 elevation above sea level (Figure 4), the north limb has developed to close the structure and the ore body reaches its maximum dimensions. Here each leg of the triangle is about 1200 feet in length. Below the 654 level, the west limb becomes steeper and turns under with reversed dip so that the limbs become disjointed and split into small

segments. Also they tend to become so narrow that they are not economical to mine.

Both inside and outside the main ore-bearing dome there are, in certain areas, segments of low-grade mineralization that are partial developments of other roughly concentric ore structures. Thus, there is a strong suggestion that the minable ore body is part of an overall dome-in-dome pattern. None of these weaker structures has yielded important tonnages from an economic standpoint, but their presence must be considered when attempting to explain the origin of the ore body. The iron mineralization in both the richer and the leaner triangle shrinks in size. At the same time, the shells occur principally as massive veins and

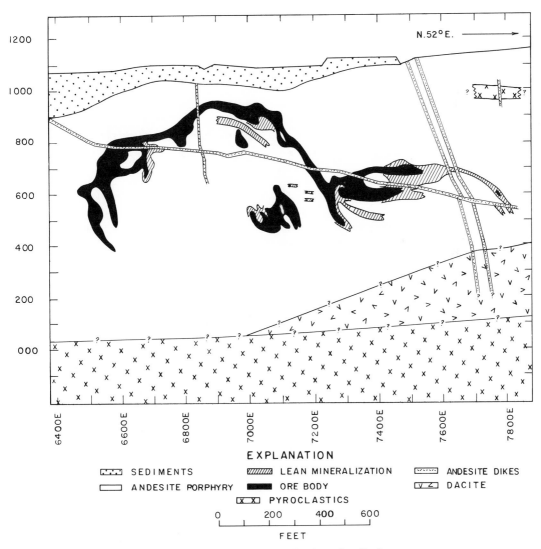

EXPLANATION

| | | |
|---|---|---|
| SEDIMENTS | LEAN MINERALIZATION | ANDESITE DIKES |
| ANDESITE PORPHYRY | ORE BODY | DACITE |
| | PYROCLASTICS | |

0    200    400    600

FEET

FIG. 5. *Cross Section B-7 Main Ore Body.*

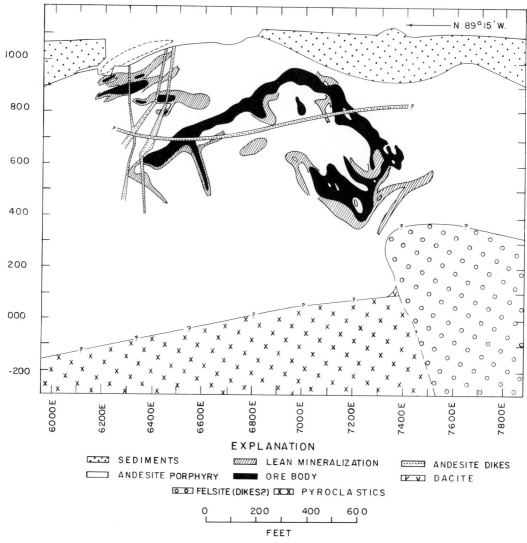

N. 89° 15' W.

EXPLANATION

| | | | |
|---|---|---|---|
| SEDIMENTS | LEAN MINERALIZATION | ANDESITE DIKES | |
| ANDESITE PORPHYRY | ORE BODY | DACITE | |
| FELSITE (DIKES?) | PYROCLASTICS | | |

0    200    400    600

FEET

FIG. 6. *Cross Section B-10 Main Ore Body.*

breccia matrix with only a minor amount of disseminated replacement.

The pattern of the Northwest Ore Body is different but equally interesting (Figures 3, 9, 10). The ore body in plan is a rather simple northwest-trending lens, but in cross section it resembles a hook, a sickle, or the profile of a pelican.

The top of the ore body is located 20 feet below the present erosion surface, immediately below the residual clay. The curving structure rolls downward to the north and south to form two legs which strike N70°W. The nearly horizontal top part is made up of several minor veinlets of massive hematite, each less than 5 feet thick, and two larger veins that range from 10 to 20 feet thick.

The south limb pinches out 200 feet below the surface (Figure 10). The north limb also narrows and is only 10 feet thick just above the 805 level; at greater depth, however, it widens dramatically to as much as 150 feet before it, too, shrinks to less than minable thickness below the 616 level. As the width narrows, the length increases, to as much as 1180 feet on the 616 level (Figure 11), and the overall shape in plan is that of an elongated teardrop. As the length increases, the ore body extends farther to the southeast and almost connects with the Main Ore Body. The deepest drill hole intersection shows that the mineralization persists to the 500 level or about 600 feet below the surface.

In the upper part of the Northwest Ore

Body, the ore is largely massive hematite with only a few blocks of included andesite porphyry host rock. Below the 700 level, the andesite blocks become more numerous and the massive ore grades into breccia.

Another type of ore, which was important during the early history of the deposit, consisted of accumulations of hematite boulders in the basal Paleozoic conglomerate. These were eroded from the bed rock ore outcrop during the Lipalean interval and incorporated in the Lamotte Sandstone and Bonneterre Dolomite in upper Cambrian time. Bodies containing as much as 600,000 crude tons were mined over the Main Ore Body and east of Big Cut. They are thin, blanket ore bodies that range from less than 1 foot up to 40

feet thick and spread over several acres. All of the mineralized boulders are hematite.

## Mineralogy of the Deposit

The Iron Mountain deposit has a fairly typical high-temperature mineral assemblage. The main minerals found and the sequence of their deposition is shown in Figure 12. Hematite and, to a lesser extent, magnetite are the ore minerals recovered. The gangue minerals in the order of decreasing abundance are: andradite, quartz, calcite, actinolite, apatite, dolomite, fluorite, pyrite, chalcopyrite, and barite. Chlorite and epidote are prominent as alteration products of the wall rocks and included breccia fragments.

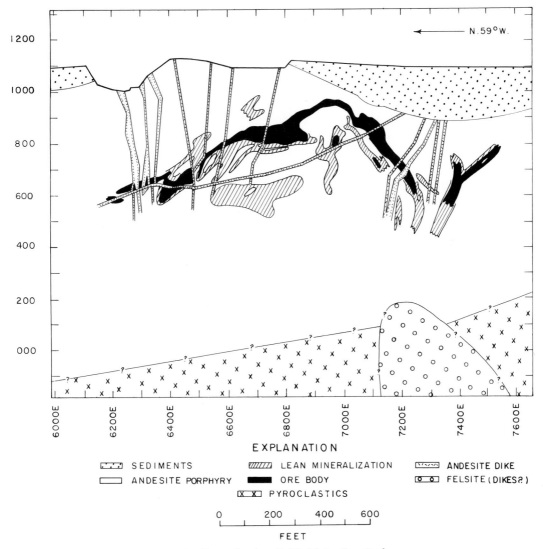

FIG. 7. *Cross Section B-12 Main Ore Body.*

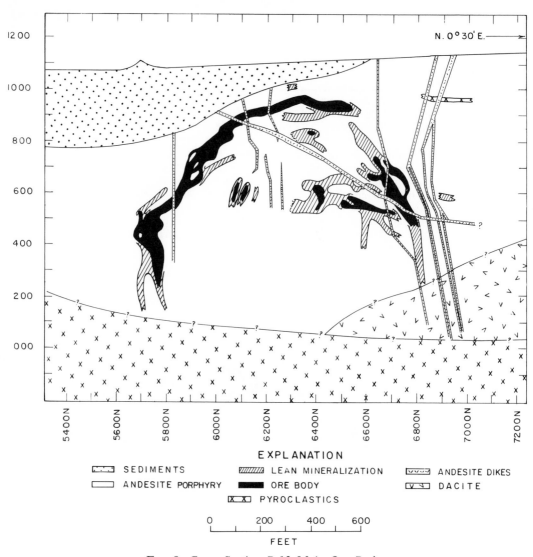

FIG. 8. *Cross Section B-13 Main Ore Body.*

ORE MINERALS  Hematite makes up 80 per cent of the iron oxide in the Main Ore Body and 95 per cent in the Northwest Ore Body. It occurs primarily as veins, ranging from hairline thickness up to 40 feet, but it also occurs as the cementing material in brecciated andesite, and as disseminated particles in andesite, wall rock, or breccia blocks. Disseminated hematite grains and clots range in size from 0.1 to 10 mm. They are dark steel gray and have a metallic luster. Three types of hematite can be distinguished megascopically and in polished sections. These are: (1) massive hematite; (2) large platy hematite crystals attached to the massive hematite; and (3) flakes of hematite that often are thin enough to appear

red in transmitted light. Small traces of magnetite can be seen in polished sections of many specimens of the massive hematite, and it may amount to 1 or 2 per cent of the total iron present. A few suggestions of octahedral shapes are seen in polished sections of hematite that may have resulted from oxidation of magnetite, but detailed mapping has not otherwise identified martite anywhere in the ore body. The coarse grain size and lack of conspicuous martite suggests that most if not all of the hematite is primary.

Magnetite makes up the remaining part of the iron oxide. It also is found as veins, as the cementing matrix in breccia, and as disseminated particles. It is black and very fine

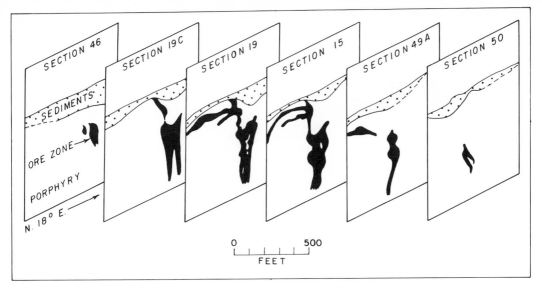

FIG. 9. *Northwest Ore Body.*

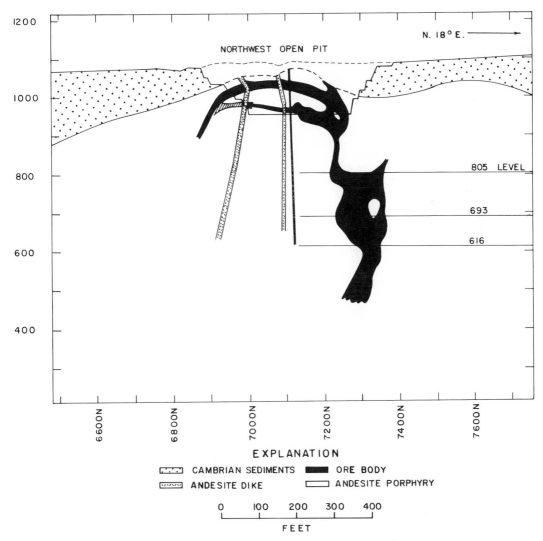

FIG. 10. *Cross Section 15 Northwest Ore Body.*

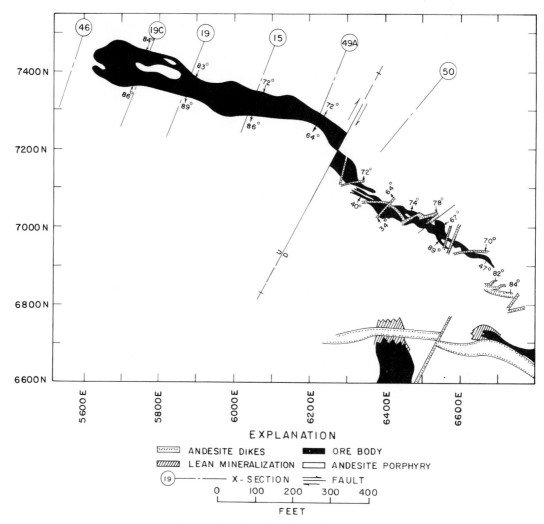

FIG. 11. *Plan Map 616 Level Northwest Ore Body.*

to fine grained. Most magnetite is massive, but rare octahedral crystals occur in vugs. Magnetite occurs on all levels, but it is abundant only on the eastern side of the Main Ore Body. Near the top of the dome structure, it is concentrated in a zone 40 feet long. This zone gradually enlarges in depth to a maximum length of 700 feet at the 503 level. Rare isolated magnetite pods less than 1 foot in their longest dimension have been noted within hematite at the northern end of the west limb on the 616 and 579 levels.

The contacts between magnetite-rich and hematite-rich zones are gradational. On approaching the magnetite-rich zone on the east flank of the Main Ore Body, the first occurrences of magnetite encountered are as cores in the center of hematite veinlets. These are believed to indicate late introduction of magnetite, a conclusion that is substantiated by the rather common occurrence of magnetite veins crossing hematite veins and breccia bodies.

GANGUE MINERALS  Gangue minerals and rock fragments make up 30 to 50 per cent of the veins in the Main Ore Body and about 20 per cent in the Northwest Ore Body where the ore is more massive.

Andradite garnet is the most abundant gangue mineral in hematite ore but ranks third behind quartz and calcite in magnetite ore. It is conspicuously yellow in color. An area of quite massive garnet concentration 200 feet long and 40 feet wide occurs in a vein on the north limb of the Main Ore Body. It is located in Figure 8 between the top of the dome and the point where economic mineralization starts on the 803 level. Massive actinolite accompanies the garnet and, along with some hematite, constitutes the matrix of ore breccia.

In many places, garnet occurs as veins cutting through hematite, as the cement in hematite breccia, or in the center of hematite veinlets as partitions between hematite and quartz or hematite and calcite. Thus there are many indications that it is later than hematite. Garnet pseudomorphic after amphibole occurs in hematite and magnetite. In many instances, these pseudomorphic crystals have garnet rims with calcite centers.

Near-euhedral to subhedral crystals of andradite garnet have been noted in the quartz and calcite that fill the centers of hematite veinlets. Both quartz and calcite are believed to be younger than the garnet and to have reached their present seemingly anomalous location through recurrence of brecciation during the mineralizing period. Many actinolite crystals have a cap or rim of garnet. Ridge (6, p. 3) suggests that this is a reaction rim between amphibole and molten hematite, but the finding of pseudomorphs of garnet after amphibole, as noted above, indicates that this is more likely a true replacement relationship. This conclusion is supported by Allen and Fahey (5, p. 740), who found "zoned dodecahedra of garnet that cut across actinolite and contained fibrous inclusions of actinolite."

Where andradite garnet is in contact with apatite, the garnet appears younger, as the apatite has definite crystal shape, but no positive replacement by garnet has been noted.

Allen and Fahey state (5, p. 741) that "garnet is cut by and replaced by hematite." However, no other occurrences of this have been noted, and the reverse situation certainly is more prevalent.

It is common to find an increase in garnet content at the ends of ore limbs. This is particularly true at the northern end of the west limb and the western termination of the northern limb. As the percentage increases, the ore gradually becomes uneconomic.

Quartz is the second most abundant gangue mineral in the hematite ore and the most abundant in magnetite ores. It is massive to crystalline and has a color ranging from milky white to clear. It is found as linings in vugs, as crystals pseudomorphic after amphibole or apatite, and locally as cement in late breccia containing fragments of hematite and magnetite. In the vugs, quartz is, in places, the last mineral de-

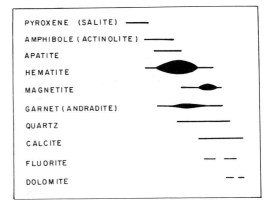

FIG. 12. *Major Vein Mineral Paragenesis at Iron Mountain, Missouri.*

posited, and it is often seen in the form of low-temperature quartz crystals. Locally, apatite crystals are partially replaced by quartz, and garnet commonly is veined by quartz.

Calcite is the third most abundant gangue mineral in hematite ore and the second in magnetite ore. It is usually massive except in vugs, where it forms scalenohedral and rhombohedral crystals. It occurs as irregular masses surrounded by garnet, as linings in vugs, and as the centers of crystals pseudomorphic after apatite or amphibole in which the outer rim is garnet. Allen and Fahey (5, p. 741) found that calcite replaced garnet crystals centripetally by entering through channels penetrating the crystal rims.

Actinolite was first noted by Allen and Fahey (5, p. 737). They state that actinolite was formed from salite $[Ca(MgFe)(SiO_3)_2]$ as the temperature of iron-bearing solutions decreased and that in all of the commercial ore the change to actinolite is complete.

Actinolite occurs as fibrous, gray, feathery phenocysts in both hematite and magnetite ore at scattered intervals in the ore bodies. Along the contact of porphyry and the iron oxides, the crystals extend lengthwise into the veins and grow larger away from the rock walls. Some of the actinolite found within magnetite ore appears fresh and has a dark green to black color that resembles hornblende. In hematite ore, the actinolite is chloritized or altered to clay. As noted above, a thin garnet rim commonly separates hematite and actinolite; in some cases the garnet has pseudomorphically replaced all the amphibole. Actinolite also is found as irregular shaped masses composed of non-oriented and sometimes interlocking crystals in a matrix of hematite and magnetite. These range from a few inches across in hematite up to several feet across in magnetite.

In a few drill core specimens, calcite has replaced actinolite and is partially pseudomorphic after it. Some of these "actinolite" crystals have calcite in the center, remnant chloritized amphibole at one end, and garnet at the other end.

At the northern end of the Main Ore Body, actinolite occurs as massive veins up to 10 feet thick with crystals over a foot long extending inward from the andesite porphyry vein contacts. In plan view, the eastern limb of the ore body going northward grades from quite massive magnetite to quite massive amphibole and then into the garnet zone discussed previously.

A late stage of amphibole has been noted by Singewald and Milton. It occurs as minute pale green prisms in quartz.

Pink apatite crystals occur in hematite and magnetite at erratically distributed points within the ore body. The most prominent location is in the easternmost vein shown in Figure 7. Here, apatite crystals up to several inches in length are scattered randomly throughout the magnetite vein. One massive apatite vein a foot thick has been exposed for 200 feet.

Where it has been noted in contact with amphibole, apatite fills open spaces between amphibole crystals. This suggests that the apatite was later than the amphibole. As noted previously, calcite and quartz have replaced apatite. One occurrence has been noted in which calcite pseudomorphic after apatite was rimmed by garnet crystals.

Chlorite occurs mainly as an alteration product of actinolite and to a lesser degree of garnet. It also has attacked blocks of andesite porphyry which are incorporated in the ore veins.

Epidote likewise occurs as a replacement of andesite fragments in the ore zones. It is also associated with disseminated hematite in low-grade areas. Near the ends of some breccia ore lenses with hematite as the matrix, epidote becomes the main cementing mineral.

Several other gangue minerals occur in minor amounts. Dolomite occurs as a pink to white filling in fractures that cross the ore veins, especially in the upper portion of the dome structure. Minor amounts of purple fluorite occur in magnetite ore, although fluorite does not occur actually in contact with magnetite. It is usually mixed with quartz, filling and lining vugs. A single mass of barite about eight inches in diameter has been found in a vug and rare occurrences of galena, pyrite,

and bornite have been noted. Bornite and pyrie are found only near the hanging-wall contact with the andesite porphyry, but galena is scattered more widely. Some crystals are enclosed in quartz and thus are judged to be late in the mineral sequence.

## Wall-Rock Alteration

Wall-rock alteration is not a conspicuous feature at Iron Mountain. Much of the wall rock and many of the andesite blocks included in the breccia are fresh. In some areas, however, epidote and chlorite have attacked the available host rock to a considerable degree, and, in a few places, sericitization and silicification have been noted, along with bleaching and staining that are thought to be related to the ore-forming process. Epidote occurs only with hematite, never with magnetite, but the other minerals are found with either iron oxide.

Epidote occurs in andesite, mainly as disseminated clots that replace the ground mass and feldspar phenocrysts, and as fracture filling veinlets. Where abundant, the clots give the rock a light shade of green.

Alteration of the host rock is erratic. Locally epidotization has been intense enough to destroy completely all evidence of the character of the original rock. There is no consistent zoning pattern, but locally a halo of more or less intense epidotization surrounds hematite concentrations, and in some areas disseminated epidote extends for more than 200 feet into the hanging wall. Many hematite veins up to 6 inches thick are bounded by epidote-rich zones about a half-inch wide. There are also many examples of hematite veinlets crossing epidote alteration halos.

Alteration by epidote on the footwall side of the Main Ore Body within the porphyry core, appears to be slightly stronger than that on the hanging wall. An increase in concentration has been noted near the northwest corner of the Main Ore Body on the 579 level. The rock in this area has been much replaced by disseminated epidote, hematite, and quartz, which occur in separate bands in an area over 100 feet in length and 50 feet in width. This banded rock resembles bedded tuff, as the texture and general appearance of the andesite porphyry have been completely destroyed. The banding in most places parallels the associated veins.

Chlorite is the second most abundant alteration mineral generally, and, in magnetite areas, it is the most abundant because epidote is absent. Most commonly it occurs as disseminated clots in the porphyry with the clots replacing the ground mass and parts or all of the feldspar phenocrysts. This gives the rock a dark brown color. The mineral most commonly replaced has been identified by Steidtman (1, p. 90) as hypersthene. Chlorite occurs in weak to moderately strong concentrations for at least l00 feet away from magnetite ore and for over 200 feet from hematite ore.

Bleaching is not a characteristic feature in the Iron Mountain Mine, but, in one area on the north side of the Main Ore Body near the 750 elevation, a portion of the footwall 20 feet wide and 100 feet long has been intensely bleached to a gray-white color. It is accompanied by a highly developed banding of disseminated hematite. The banding parallels the overlying brecciated vein.

Staining to a red or reddish-brown color is typical along the margins of most breccia zones and in places near the massive veins. Both the feldspar phenocrysts and the ground mass are affected. There is no definite pattern and no systematic width to the colored halos; some are less than an inch wide while others, adjacent to veins of the same size, are several feet in width. The red alteration is also found rimming epidote disseminations or veins.

## Vein Structures and Textures

The ore minerals at Iron Mountain fill open spaces and, to a lesser extent, replace the host andesite porphyry. In most areas, the contacts with the wall rock are sharp. However, on the inner margin of the north and northwest parts of the Main Ore Body and in the flat vein area of the Northwest Ore Body, they are gradational so that massive hematite passes into disseminated ore and finally into barren porphyry.

Even the most massive concentrations of hematite and magnetite usually have some "included" fragments of andesite porphyry. These range in size from tiny particles to blocks 20 feet across. Most of the blocks are at least slightly altered to chlorite or epidote, and some are replaced almost completely by soft, black chlorite. Even the soft areas usually have sharp boundaries. In addition to the rock inclusions, the thick hematite and magnetite veins and pods commonly contain crystals of garnet, actinolite, and apatite.

In a zone of particularly intense brecciation, extending nearly horizontally northeastward from the Main Ore Body (Figure 5), there are many excellent examples of banded veins.

Most of them are less than a foot wide. Characteristically, both the iron oxide and gangue mineral crystals have their long axes normal to the walls and layers of hematite, garnet, quartz, and calcite occur in sequence from the walls inward. This same area contains many vugs up to 4 feet in their greatest dimension; commonly, they are lined with comb quartz, and some also contain excellent crystals of hematite, magnetite, and calcite.

The inclusions of magnetite, hematite, and wall rock fragments in veins of late minerals indicates that brecciation continued as the ore fluid circulated. The disagreements as to mineral sequence among the various authors who have described the deposit result from the fact that overlaps and, locally, even reversals in the order of deposition are common. This may be evidence that the entire ore-forming epoch was rather short.

## Factors Controlling the Form and Location of the Ore Bodies

The shape and size of the Iron Mountain ore bodies were determined by the shape and size of the open spaces available at the time the ore solution was circulating. Hence, the problem consists of explaining the pattern of these openings. Although the unusual configuration of the two principal ore bodies has excited considerable discussion and speculation, no general agreement has resulted.

As indicated by the maps, there is a definite development of tabular features trending N70°W, N10°W, N50°E, and east-west. Several simple stress systems could be invoked to explain these orientations. The cross sections, however, indicate that a more complicated system of mechanics is required to explain the dome shape of the Main Ore Body and the hook shape of the Northwest Ore Body. It is significant that the walls in many parts of these ore bodies would fit together almost perfectly if the ore were removed, and it seems almost undeniable that the bodies of breccia formed by collapse or spalling into open spaces that quite probably were filled with ore solution at the time. There is no simple and obvious explanation of all the relationships.

Lake (4, p. 65) attributes the openings to dissolution of porphyry by the mineralizing solution and draws a comparison to breccia pipes in western United States and Mexico. Amstutz (8, p. 904), on the other hand, compares the ore bodies to ring dikes and calls on convection currents in the deep crust as the cause of the subsidence. While the ring dike comparison seems plausible to explain the gross features of the Main Ore Body, it faces difficulties in accounting for the linear segments mentioned above, for the near-horizontal breccia sheet on the northeast side of the Main Ore Body, and for the hook-shaped cross section of the Northwest Ore Body.

Relaxation of an upthrusting force (intrusive ?) could tend to form the arcuate pattern of tensional openings now filled with breccia and cemented by ore and gangue minerals. It is true that no intrusive column has been identified below the Main Ore Body, but it is also true that there has not been a great deal of exploration at depth, and such a plug conceivably could be present. It is also possible that the upward thrusting and relaxation was caused by pulsating injection of the ore fluid itself. There is some similarity between the arcuate shape in cross section of the Iron Mountain ore bodies and the shape of the Climax molybdenum ore body, and theories that have been invoked for Climax may be applicable to the Missouri ore structures.

In any event, no matter what the ultimate conclusion as to the cause of the brecciation is, it is a fact that ore occurs only where the fractured and brecciated bodies of host rock were available as a repository, and it probably is true that the brecciation and mineralization were very close in time, if not actually simultaneous. The fact that all of the ore occurs in the andesite porphyry could be the result of a greater tendency for this rock to fracture; certainly the small amount of replacement that occurred does not point to its great chemical potency as a precipitant. However, little is known of the properties of the various rocks relative to each other in the area, and it is possible that the localization of ore in the thick sheet of andesite porphyry simply may have resulted from that rock having been in the critical location when the brecciating forces were applied.

### Nature of the Ore Solutions

Various opinions have been expressed as to the nature of the ore solutions which deposited the Iron Mountain ore. Ridge (6, p. 4), Amstutz (7, p. 905), and Geijer (3, p. 29) have advocated injection of a water-poor melt while Singewald and Milton 2, p. 7) and Allen and Fahey (5, p. 740) have favored a high temperature water-rich (hypothermal) or solution.

Ridge believes that all the mineral relationships can be explained as reaction products within a melt containing a minor amount of mineralizers, and that the amount of wall-rock alteration is too limited for the area to have been exposed to large volumes of a thin, reactive ore fluid. On the other hand several arguments have been put forward to support the later proposal. These are:

(1) The alteration to chlorite indicates that water was an active and abundant component of the ore fluid.

(2) The presence of disseminated replacement hematite and magnetite in a few areas indicates that significant diffusion took place, and this is much more likely to have occurred in the case of a tenuous hydrothermal solution than of a melt.

(3) The banded veins with occasional vugs showing incomplete fillings are difficult to explain as the result of injection of a melt.

## ACKNOWLEDGMENTS

The authors are indebted to The Hanna Mining Company and F. M. Chace for permission to publish this description of the Iron Mountain deposit. Also to F. M. Chace, J. T. Cumberlidge, and J. A. Noble for critically reviewing the manuscript and to J. Mayes who drafted the maps and sections. None of the reviewers is necessarily in agreement with the ideas expressed.

## REFERENCES CITED

1. Steidtman, E., 1928, Geologic studies of the Iron Mountain district: Unpublished report to The Hanna Mining Company, p. 1–130.
2. Singewald, J. T. and Milton, C., 1929, Origin of iron ores of Iron Mountain and Pilot Knob, Missouri: A.I.M.E. Tr., v. 85 (1929 Yearbook), p. 330–340.
3. Geijer, P., 1930, The iron ores of the Kiruna type: Sveriges Geol. Undersök, Ser. C, No. 367, Arosbok 24, no. 4, p. 1–39.
4. Lake, M. C., 1933, The iron ore deposits of Iron Mountain, Missouri: *in Mining districts of the eastern states*, 16th Int. Geol. Cong., Guidebook 2, p. 56–67.
5. Allen, J. T. and Fahey, J. J., 1952, New occurrences of minerals at Iron Mountain, Missouri: Amer. Mineral., v. 37, p. 736–743.
6. Ridge, J. D., 1957, The iron ores of Iron Mountain, Missouri: Mineral Industries, v. 26, no. 9, p. 1–8.
7. Amstutz, G. C., 1960, Polygonal and ring tectonic patterns in the Precambrian and Paleozoic of Missouri, U.S.A.: Eclogae Geol. Helv, v. 52, no. 2, p. 904–913.

# 16. The Native-Copper Deposits of Northern Michigan*

WALTER S. WHITE†

## Contents

* Publication authorized by the Director, U.S. Geological Survey.
† U.S. Geological Survey, Beltsville, Maryland.

## Illustrations

## Table

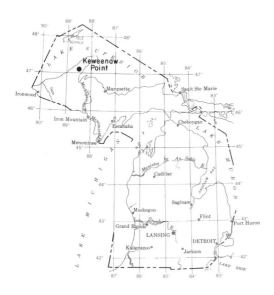

## ABSTRACT

The Michigan native-copper district has produced about 5,400,000 tons of copper since mining began in 1845. The copper occurs primarily as open-space fillings and replacements in amygdaloidal flow tops and conglomerate beds of the Portage Lake Lava Series, of middle Keweenawan age. Secondary minerals with which the copper is associated include microcline, chlorite, epidote, pumpellyite, prehnite, quartz, and calcite; zeolites, other than laumontite, are not abundant. A zonal pattern of the secondary minerals transects bedding and suggests regional metamorphism mainly to the grade of the prehnite-pumpellyite graywacke facies of Coombs. Most of the major copper deposits are near the upper limit of abundant quartz in this zonal pattern.

The distribution of copper in both amygdaloid and conglomerate deposits is controlled partly by primary permeability of the rocks and partly by later tectonic fracturing that was concentrated in the mechanically weak flow tops and sedimentary beds. Ore shoots and deposits are therefore controlled partly by primary structural features, such as areas of thick and thin fragmental amygdaloid or the boundaries of conglomerate lenses, and partly by regional structure: a number of ore deposits are in synclinal downwarps.

The copper-bearing solutions that formed the deposits were ascending. They may have come from a large concealed intrusive, or may have been driven out of the porous tops of lava flows when the rocks were compacted and metamorphosed at great depth in the central part of the Lake Superior syncline. The second hypothesis seems more consistent with the widespread distribution of the copper, its relation to mineral zoning that represents truly regional metamorphism, its native state, and the likelihood that copper deposition was later than most or all of Keweenawan volcanism.

## INTRODUCTION

The native-copper deposits of northern Michigan (Figure 1), which were mined in prehistoric time and constituted the principal

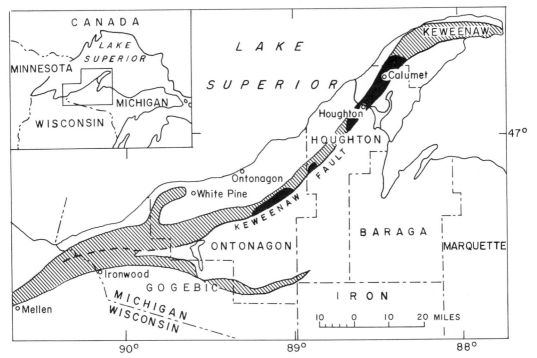

Fig. 1. *Location of the Michigan Native-Copper District. Ruled pattern represents area in which mafic lavas of Keweenawan age crop out. All the known major native-copper deposits are in the northeastern area indicated in solid black; the two smaller black areas contain minor deposits.*

United States source of copper for 40 years, have been the subject of intensive geologic study for more than a century. Despite the many detailed descriptions of these unique deposits and their host rocks, the ultimate source of the copper and some of the conditions that have led to its concentration are still debated. The most comprehensive report on this district (11) was published 38 years ago. This review will emphasize more recent work that has shed new light on the environment and origin of the deposits.

The copper-shale deposit in the base of the Nonesuch Shale at White Pine, Mich., is described separately in chapter 21 of this volume.

## Previous Work

Since the first significant geologic account of the native-copper deposits by Douglas Houghton, published in 1841 (1), there have been six comprehensive reports on the district and its geologic setting (2,3,4,7,8,11), each of which made substantial additions to what had been known before. The latest of these, by Butler and Burbank (11), was published in 1929 as an outgrowth of the privately sup-

ported Calumet and Hecla Special Survey, begun in 1920 under the direction of L. C. Graton. That monographic report presents a detailed picture of the copper deposits and their geologic setting that should remain authoritative for many years to come.

The literature on the Keweenawan rocks and their copper deposits is enormous. Butler and Burbank (11, p. 3–14) list nearly 500 books and papers published prior to 1929. Articles published since 1929 total more than 250. Many of these are general accounts or reviews, without original data, and many others are of limited or specialized scope. Since 1929 Broderick (10,13,16,17,22), Cornwall (23,29), Stoiber and Davidson (34), and White (31,39) have described facets of the geology of the copper district that have direct bearing on the origin and setting of the native-copper deposits. Many others have made contributions to Keweenawan geology that have indirect bearing; those of Hamblin (43,44) are particularly significant in this category.

The U.S. Geological Survey has published a series of 13 geologic quadrangle maps at a scale of 1:24,000 covering the most productive part of the native-copper district (24,30).

Geologic and magnetic maps by Lamey (19) and Spiroff (21) are products of an extensive mapping program carried out by the Michigan Geological Survey in the 1930's.

## Sources of Information and Acknowledgments

This review draws mainly upon the published literature, particularly the comprehensive report by Butler and Burbank (11), but does include some unpublished observations by members of the U.S. Geological Survey, including H. R. Cornwall, E. S. Davidson, J. J. Runner, R. E. Stoiber, A. A. Stromquist, R. W. Swanson, J. C. Wright, and myself. I am also indebted to M. A. Lanphere of the U.S. Geological Survey for a K/Ar age determination.

Officials of Calumet and Hecla, Inc., and the Copper Range Co. have generously permitted unrestricted access to mine workings, drill core, and extensive collections of drill logs and mine maps and have provided data on copper production. The U.S. Geological Survey group owes a special debt of gratitude to successive members of the Geological Department of Calumet and Hecla, Inc., for countless courtesies and much stimulating discussion of the native-copper deposits; these individuals include C. D. Hohl, J. P. Pollock, R. G. Weege, and particularly the late T. M. Broderick. Mr. Hohl's meticulous logs of many hundreds of Calumet and Hecla drill holes deserve special mention as an invaluable source of basic data.

I am grateful to J. C. Wright, H. T. Morris, and D. E. White for many thoughtful and constructive suggestions concerning the manuscript and to P. B. Barton, Jr., for helpful discussion and some of the phrasing of the paragraph on hematite-native copper relationships.

## HISTORICAL INFORMATION

### Mining Operations in the Area

Indians were mining copper in the Lake Superior region as early as 3000 B.C. (42, particularly p. 127–129), and their extensive pits later led prospectors to most of the known deposits. Although 17th-century accounts of Jesuit missionaries contain references to copper, serious mining was not attempted until 1844, three years after Douglas Houghton's principal report (1) was published. The first successful mine, the Cliff Mine, was opened in 1845.

The copper at the Cliff Mine occurred along a steeply dipping vein, or "fissure." For the first 6 years of mining, deposits of this type were the sole source of copper and produced more than half the total district copper each year until 1863. These deposits rapidly diminished in importance thereafter, relative to those of other types, and have only yielded about 2 per cent of all the copper mined to the present.

Native copper disseminated in the amygdaloidal tops of lava flows accounts for about 58 per cent of the total district production. The principal productive flow tops, and the dates of first and most recent significant mine production from each, are as follows: Isle Royale, 1852–1948; Pewabic (mostly Quincy Mine), 1856–1945; Ashbed (Atlantic Mine), 1872–1906; Kearsarge, 1882–present; Osceola, 1891–present; Baltic, 1898–present.

Exploitation of the famous Calumet and Hecla Conglomerate began in 1865, and large-scale production ceased in 1939. This ore body, by far the largest in the district, has produced 39.1 per cent of all the copper. Exploration along the northern continuation of the bed, beginning in 1957, has been encouraging.

All the principal ore deposits were thus known before 1900, and each was discovered in exposures at or close to grass roots. Exploration since 1900, primarily by diamond drilling, has led to additional discoveries, some of which have been commercial successes. Although these later discoveries have not rivaled the major deposits listed above, they offer encouragement for the future of the district by showing that all ore deposits do not necessarily crop out, as a result, for example, of superior resistance to erosion.

## Production

The production of the Michigan native-copper district is summarized in Table I. The grand total includes 1,175,390 tons of copper mined and 282,948 tons reclaimed from tailings since 1925, the cutoff date for statistics given by Butler and Burbank (11, p. 64–98). The Iroquois Amygdaloid and Houghton Conglomerate mines are in new ore bodies discovered after 1929.

There is a marked tendency for the average grade of large deposits to be appreciably larger than that of small deposits (11, p. 147). This tendency is suggested by the figures in the

TABLE I.   *Production of Copper, Michigan Native-Copper District, 1845 through 1964*

| Stratigraphic unit(s) | Rock Treated (Thousands of Short Tons) (Partly Based on Estimates for Early Years) | Copper Produced, Including Copper from Tailings (Short Tons) | Grade (Per Cent Copper) |
|---|---|---|---|
| Calumet and Hecla Conglomerate | 79,838 | 2,105,472 | 2.64 |
| Kearsarge Amygdaloid | 98,173 | 1,033,062 | 1.05 |
| Baltic and adjacent amygdaloids | 61,359 | 922,356 | 1.50* |
| Pewabic and adjacent amygdaloids (mostly Quincy Mine) | 42,878 | 538,473 | 1.26 |
| Osceola Amygdaloid | 30,241 | 280,377 | 0.93 |
| Isle Royale Amygdaloid | 18,879† | 170,457 | 0.90 |
| Ashbed Amygdaloid (mostly Atlantic Mine) | 10,904† | 71,42ϳ | 0.66 |
| Evergreen "series" (Ontonagon County) | 5,673† | 42,525 | 0.75 |
| Allouez Conglomerate | 5,783† | 36,262 | 0.63 |
| Iroquois Amygdaloid | 3,208 | 23,927 | 0.75 |
| Houghton Conglomerate | 1,696 | 18,400 | 1.08 |
| Forest Amygdaloid (Victoria Mine) | 1,740 | 10,012 | 0.58 |
| Winona Amygdaloid | 1,294 | 8,842 | 0.68 |
| Fissure mines | — ‡ | 106,700 | |
| Other mines and explorations | | | |
|    Houghton and Keweenaw Counties | 1,290† | 8,923 | 0.69 |
|    Eastern Ontonagon County | 938† | 7,577 | 0.81 |
|    West of Ontonagon River (excluding Nonesuch Shale) | 98§ | 570 | 0.58 |
|    Isle Royale (island) | 86§ | 489 | 0.57 |
| Total production | 364,078 | 5,385,849 | 1.48 |

\* For many years, ore from Baltic Amygdaloid was upgraded by hand sorting underground (14).

† Total based partly on estimates.

‡ A figure for "rock treated" would not be meaningful because so much of the copper occurred in small to large masses and was smelted directly.

§ Total based wholly on estimates.

table and is even clearer when individual deposits are compared. The relationship of size to grade may be roughly expressed as follows for deposits in amygdaloidal rock:

$$\text{Grade (in per cent)} = 0.36 + 0.101 \log \text{total copper (in tons)}$$

The standard deviation of actual grades of 20 individual deposits from grades calculated by means of the equation, given the tonnage of copper, is 0.14 per cent copper. Some of the estimates used in preparing Table I were based on this equation where no better basis exists.

## PHYSIOGRAPHIC HISTORY AND PRESENT TOPOGRAPHY

The Michigan Copper district is on the south side of the Lake Superior topographic basin, in the southern part of the Canadian Shield. It is an area of rather low relief, in which

very few hills rise as much as 1000 feet above the level of Lake Superior. The so-called Copper Range, a narrow upland belt upheld by mafic igneous rocks of middle Keweenawan age, stands several hundred feet above lowlands on either side that are underlain by younger sandstone. For nearly 200 miles southwest from its eastern end at Keweenaw Point, 37 miles east-northeast of Calumet, (Figure 1), this upland is a more or less continuous topographic ridge 2 to 5 miles wide.

The region has probably been subject to erosion through much of post-Ordovician time, and the dominant topographic forms primarily reflect differences in resistance of the underlying bedrock. Glacial and post-glacial landforms, however, make conspicuous minor topographic features. At most places, bedrock is mantled by a layer of glacial till and fluvioglacial deposits ranging in thickness from a few inches to at least 500 feet. Late-glacial and postglacial lakes that occupied the Lake Superior basin have left conspicuous shorelines, particularly along the northern flank of the Copper Range, and a veneer of lacustrine sand, silt, and clay is common at lower elevations. The Pleistocene and Recent history of the region has been described by Leverett (12), Hack (51), and Hughes (48).

## GEOLOGIC HISTORY

### Stratigraphic Column

The rocks of the Michigan Copper district belong to the Keweenawan Series of late Precambrian age. Figure 2 is a generalized columnar section showing the formations of the Keweenawan Series exposed in the vicinity of the district.

The host rocks of the native-copper deposits are certain amygdaloidal flow tops and conglomerate beds of the Portage Lake Lava Series. This lava series, which is at least 15,000 feet thick in the Michigan Copper district and probably exceeds 30,000 feet in thickness in parts of the Lake Superior region, is made up of several hundred flood-basalt flows (39). Beds of rhyolite conglomerate, from less than an inch to tens of feet thick, separate some of the flows. Figure 3 is a longitudinal stratigraphic section of the Portage Lake Lava Series between Victoria and the end of the Keweenaw Peninsula. It shows the relative position of certain flows and conglomerate beds that are useful stratigraphic horizon markers. Many horizons can be identified for distances up to 90 miles.

Figure 4 is an enlargement of the area enclosed by a dotted line in Figure 3; the larger scale makes it possible to show all the flow tops identified in drill holes within this block of ground. In this particular part of the geologic column, the flows are distinctly more regular than average, however, and the spacing of drill holes here is many times closer than for the region as a whole.

Half of the volume of lava in the Portage Lake Lava Series occurs in flows more than 74 feet thick. The thickness of individual flows or flow units ranges from a foot or so to as much as 1500 feet, and many of the thicker flows were formed by individual outpourings of tens or even hundreds of cubic miles of basalt (39).

The lava flows are predominantly basalt or basaltic andesite, containing essential calcic plagioclase (generally labradorite), augite, and

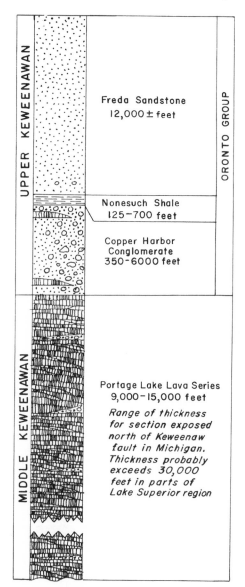

FIG. 2. *Columnar Section of the Keweenawan Rocks. The Jacobsville Sandstone, separated from the Portage Lake Lava Series by the Keweenaw fault, is not shown. It is probably younger than the Freda Sandstone.*

minor olivine. About 40 per cent of the mafic lava has ophitic texture, coarsest in the thickest flows. The remaining flows are quite fine grained but show textural differences that depend on the degree to which feldspar grain sizes depart from uniformity (11, p. 23–25; 23, p. 160–161).

The uppermost 5 to 20 per cent of most individual lava flows is conspicuously amygda-

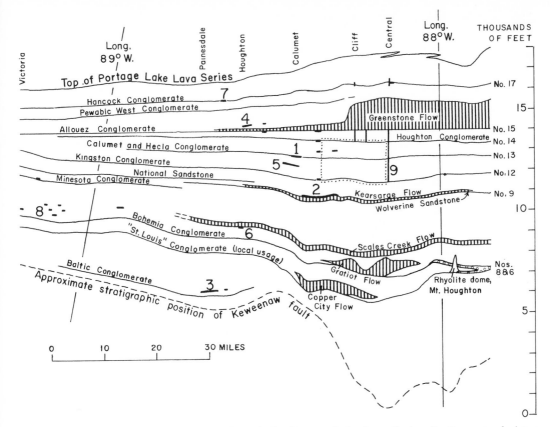

FIG. 3. *Longitudinal Stratigraphic Section of the Portage Lake Lava Series. Section extends from Victoria (13 miles south-southeast of Ontonagon—see Figure 1) to eastern end of Keweenaw Peninsula (37 miles east-northeast of Calumet). Conglomerate beds are numbered according to local usage (column at right-hand edge of diagram). Heavy black lines represent copper deposits, the more important of which are identified by the large numerals as follows: 1, Calumet and Hecla Conglomerate; 2, Kearsarge Amygdaloid; 3, Baltic Amygdaloid; 4, Quincy Mine; 5, Osceola Amygdaloid; 6, Isle Royale Mine; 7, Atlantic Mine; 8, mines of Evergreen and associated amygdaloids; 9, Central Mine (fissure deposit). Dots outline area of Figure 4.*

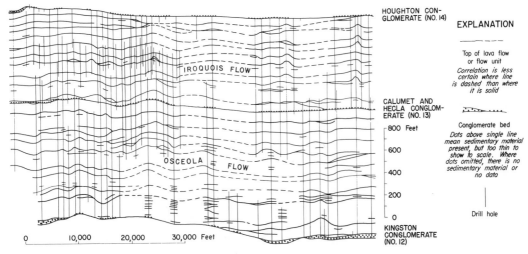

FIG. 4. *Longitudinal Stratigraphic Section of a Group of Keweenawan Lava Flows. For location, see Figure 3. (From White (39, Figure 1)).*

loidal and contains 5 to 50 per cent vesicles filled with secondary minerals. The abundance of amygdules decreases downward toward the amygdule-free massive basalt of the middle and lower part of the flow. In 21 per cent of the flow tops, the vesicular crust was brecciated, presumably as the lava was poured out, and now consists of rubbly or fragmental material in which both amygdules within fragments and interstices between fragments are filled with secondary minerals.

Many of the thicker flows are differentiated. The principal differentiation mechanisms appear to have been crystal settling (complicated by convection in some of the very thick flows) and volatile transfer (16; 17, p. 532–534; 23, p. 190–193). Volatile transfer had several results that bear on the genesis of the native-copper deposits. (1) Flow tops are enriched in iron and oxygen and are commonly red. (2) The massive basalt in the interiors of certain thick flows that have been studied in detail contains only a quarter to half as much indigenous copper as the basal chill zone, suggesting that copper has been lost from the interior along with other volatiles during differentiation (16, p. 309; 23, p. 183, 188). The same conclusion is supported by Cornwall's (23, p. 198–199) observation that the trace-copper content of thin, more quickly cooled flows is two to three times greater than that of thicker, more differentiated flows. Broderick and Hohl (16) and Cornwall (23, p. 189–190) both note, also, the tendency of copper to be concentrated in pegmatitic facies of these thicker flows. It is likely that a similar primary concentration of copper occurred in the flow tops, where the vesicles and the concentration of volatile elements indicate similar activity of magmatic gases during cooling of the flows. The basal chill zones of three flows studied in most detail by Broderick and Hohl (16, p. 308) and Cornwall (23, p. 185, 188) contain 140, 220, and 480 ppm copper, so it is conservative to infer that, as a result of enrichment by volatile transfer, the tops of thicker flows contained 300 to 400 ppm copper on cooling. Such concentration is difficult to prove, however, because some of the trace copper now present in flow tops may have been brought in by postcooling epigenetic processes. (3) The sulfur content of the lava flows is appreciably less than that of comagmatic intrusives, suggesting loss of sulfur as gases during extrusion (15, p. 418, 420–421; 23, p. 197).

Rhyolite flows make up about 10 per cent of the middle Keweenawan lava series in Minnesota (20, p. 806), but are less common in Michigan. In the general area of the native-copper district, at least three small extrusive rhyolite domes are exposed in cross section (25,28). More extensive rhyolite flows are present in the upper part, and particularly at the top, of the Portage Lake Lava Series in and south of the Porcupine Mountains, which are west of the area of significant native-copper deposits in the lava series (40; 55, Figure 2). J. C. Wright (unpublished data) has shown that some, but not all, of these flows are welded tuffs.

Beds of conglomerate or, less commonly, sandstone locally lie between mafic lava flows of the Portage Lake Lava Series. A few such beds contain ore deposits, one of them the largest in the district. The conglomerate of these beds is thoroughly lithified and normally consists of subangular to rounded pebbles, cobbles, or boulders of rhyolite and subordinate basalt in a sand matrix. Individual beds of conglomerate or sandstone range from less than an inch to more than 100 feet in thickness and tend to be highly lenticular. In places where the bed itself is missing, its horizon generally can be recognized in drill core by the presence of detrital material in the underlying amygdaloidal flow top, and, as a result, these conglomerate beds provide excellent horizon markers within the Portage Lake Lava Series (Figure 3). The interval between conglomerate beds or their horizons ranges from about 500 to 3000 feet and averages about 1000 feet; the intervals between the three conglomerate horizons of Figure 4 are about 800 and 1200 feet.

The gross shoestring form, coarse texture, crossbeds, and imbricated pebbles suggest that the conglomerates are stream deposits, laid down on top of essentially flat-lying lava flows somewhat in the manner of pediment gravels (39, p. 369–371). Within the area of the copper district, these streams flowed generally northward from the margins toward the center of the Lake Superior basin (31,40,43,44).

### Intrusive Rocks

Intrusive rocks are rather rare in the Michigan Copper district, though common in the Portage Lake Lava Series in other parts of the Lake Superior basin. Geologic maps of the district (24,30) show a few small intrusive bodies of rhyolite. An unusual body of syenodiorite and granophyre on the south side of the Copper Range, 23 miles east-northeast of Calumet (6,24), is of interest because it is the only large body of mafic intrusive rock exposed in the vicinity of the native-copper district and because there are exceptional concentrations of copper sulfides in veins and flow tops in its vicinity (6, p. 392–394; 36, p. 22).

The large Duluth Gabbro Complex of Minnesota is well known, as is a gabbro-granophyre complex in the vicinity of Mellon, Wis. (9,27), on the south side of Lake Superior. Pebble counts in the Copper Harbor Conglomerate provide indirect evidence for another large mafic complex south of the native-copper district. In an area extending 15 to 20 miles northeast and southwest from Calumet, 30 to 70 per cent of the pebbles in the Copper Harbor Conglomerate are granophyre; the proportion is much lower farther away from Calumet. The streams that deposited most of the Copper Harbor Conglomerate flowed northward (40), so it is reasonable to infer that a gabbro-granophyre complex once existed and still may be partly preserved unconformably beneath the younger Jacobsville Sandstone south of the native-copper district. Granophyre pebbles are also fairly abundant in conglomerate beds in the upper part of the Portage Lake Lava Series, suggesting that part or all of this gabbro-granophyre complex is at least as old as the lavas.

Dikes, mafic or felsic, are very rare in the Keweenawan rocks of Michigan, except in the region south of the Porcupine Mountains. It seems remarkable, in view of the enormous outpourings of mafic lava flows, that the extensive mine workings and drill holes in the Michigan Copper district have not revealed numerous mafic dikes. Near the syenodiorite mentioned above, a dike several tens of feet thick, subconcordant with the flows it intersects, has been traced for several miles (J. P. Pollock, oral communication), but it is almost unique. The Michigan Copper district clearly is many miles distant from the vents from which its lava flows poured.

Volcanism was most active in middle Keweenawan time but persisted into the beginning of late Keweenawan time. Lava flows and volcanic ash (40) are found near the top of the Copper Harbor Conglomerate. A rhyolite plug intruded into the Freda Sandstone 7 miles north of Hancock (Figure 5) is the only igneous rock that is unequivocally younger than the Copper Harbor Conglomerate, and the unique occurrence of abundant biotite in this rock could mean that it is not comagmatic with typical Keweenawan rhyolites.

## Ages of Rocks

Radiometric ages should ultimately give the time and duration of Keweenawan volcanic activity, but the results to date are equivocal. Goldich and others (41, p. 95–96; 54, p. 5404–5405) have summarized some of the un-

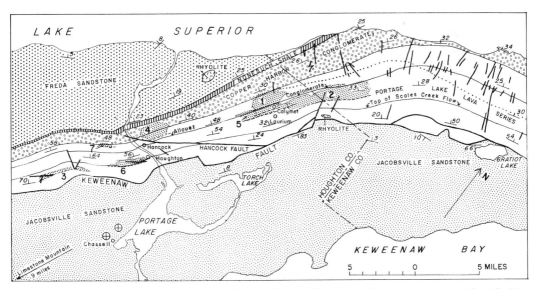

FIG. 5. *Geologic Map of Area of Major Native-Copper Deposits. Major deposits are identified by large numerals as follows, in order of production: 1, Calumet and Hecla Conglomerate; 2, Kearsarge Amygdaloid; 3, Baltic Amygdaloid; 4, Quincy Mine; 5, Osceola Amygdaloid; 6, Isle Royale Mine; 7, Atlantic Mine. Heavy dotted line is northwestern limit of abundant quartz in amygdaloids.*

certainties. U/Pb ages of zircons reported by Silver and Green (46) suggest that the igneous episode represented by the Duluth Gabbro Complex, the Endion sill, and the greater part of the middle Keweenawan lava flows of Minnesota occupied a geologically short period of time approximately 1115 ± 15 m.y. ago. Biotite from the rhyolite body intruded into the Freda Sandstone north of Hancock (Figure 5) gave a K/Ar age of 1054 ± 34 million years (M. A. Lanphere, written commun., April 1966), which is compatible with the U/Pb age above. Whole-rock Rb/Sr ages, on the other hand, are consistently younger. The following may be mentioned: Duluth Gabbro Complex, 1080 ± 25 m.y. (47); Nonesuch Shale (maximum age), 1046 ± 46 m.y. (49); Endion sill, Minnesota, 925 ± 40 m.y. (47); quartz porphyry intruded into Portage Lake Lava Series south of White Pine (Figure 1), 978 ± 40 m.y. (57) This quartz porphyry is lithologically identical with pebbles in conglomerate beds of the Portage Lake Lava Series.

The Rb/Sr ages on the Endion sill and the quartz porphyry are the only evidence that volcanism involving typical Keweenawan types of igneous rock may have persisted after the end of Copper Harbor Conglomerate deposition. Until the cause of discrepancies between methods has been determined, it is, as Goldich and others (54) say, "clearly premature to attempt to set a time limit on the duration of the middle Keweenawan activity," to which may be added that of the late Keweenawan.

### Structure

The native-copper district is underlain by the Portage Lake Lava Series on the south flank of the Lake Superior syncline. Figure 5 is a geologic map of the area that contains the major mines, and Figure 6 shows the traces of axial planes of folds in the same part of the copper district.

The principal structural features of the district are as follows:

(1) The Keweenaw fault is a low- to high-angle reverse fault, more or less parallel to the strike of bedding on its northwestern side, along which rocks of the Portage Lake Lava Series have been thrust over the younger Jacobsville Sandstone.

(2) Dips in the Portage Lake Lava Series and overlying formations generally increase in steepness toward the Keweenaw fault. This steepening is due in part to actual curvature of individual beds, but much of the steepening reflects the fact that stratigraphic units are wedges that thicken northward, as described below.

(3) Dips in individual horizons of the Portage Lake Lava Series tend to steepen 20° to 30° along strike from the northeastern to the southwestern end of the area of principal mines (Figure 5). Each horizon is, thus, a gently twisted surface. It may be significant that most of the copper production has come from this area of twisting.

(4) Minor folds are superposed on the broad curvatures described above. These folds have wave lengths in the range 5 to 10 miles and very gentle curvature; as Figure 6 shows, their axial planes have a variety of orientations.

(5) Reverse faults strike more nearly east-west than bedding and dip a little more steeply. Faults of this type are typified by the Hancock fault and are most prominent southwest of Calumet.

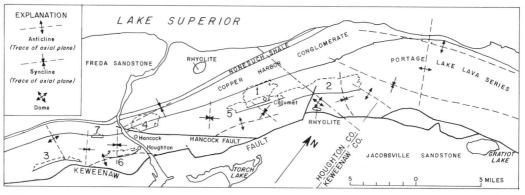

FIG. 6. *Folds in the Area of Major Native-Copper Deposits. For explanation of numerals identifying major copper deposits, see Figure 5.*

(6) High-angle faults strike north to northwest. Faults of this type are most abundant in the eastern half of the area of Figure 5 and farther east, a region characterized by a gross anticlinal structure; they are also common on minor anticlines with north- to northwest-striking axial planes. Many faults of this type, particularly, are shown in Figure 5, but the number shown would be much larger if exposures were better or if the area were explored more completely by drill holes and underground workings. Most of the native-copper deposits in fissures follow faults or fractures of this type.

(7) The Jacobsville Sandstone is nearly horizontal, except near the Keweenaw fault, where bedding is steep to overturned because of drag on the fault.

## Age of Deformation

A major problem, as yet not fully resolved, is the time or times of deformation. Resolution of the problem is important because the copper and associated minerals occupy faults and fractures and were clearly introduced during or after much of the deformation.

Some deformation, mainly tilting, took place during deposition of the lava series. Northward thickening of stratigraphic units, measurable in mines and drill holes, indicates that the basin was being downwarped as the rocks were laid down (31,39). The basement beneath the Portage Lake Lava Series was gradually tilted northward 7° or 8° in the time interval between deposition of the conglomerate beds numbered No. 6 and No. 17 (Figure 3). The interval between these two units represents 55 to 60 per cent of the thickness of the exposed part of the Portage Lake Lava Series, so angular divergence between the top and bottom of the whole lava series may well amount to 12° to 15°.

In the Copper Harbor Conglomerate, measurable northward divergence of the top and bottom of the formation amounts to 8° above the Quincy Mine. Elsewhere the effects of folding and divergence cannot be reliably separated, but the difference in dip between the top and bottom of the Copper Harbor Conglomerate due to tilt before or during its deposition is probably in the range 5° to 10° at most places.

Angular divergence between the lowest exposed flows of the Portage Lake Lava Series and the top of the Copper Harbor Conglomerate thus amounts to about 20° and implies at least this much tilting prior to deposition of the Nonesuch Shale. The Nonesuch Shale and lowermost Freda Sandstone now dip 20° to 45°NW, so the amount of tilting during or after deposition of the Freda Sandstone was equal to, or greater than, the amount of tilting that took place before deposition of the Nonesuch Shale.

Although much of the prevailing homoclinal dip thus originated by progressive tilting during deposition of the Portage Lake Lava Series, many of the structural features are very late Keweenawan or Paleozoic. The Jacobsville Sandstone, which is almost surely younger than the Freda Sandstone (43, p. 6–8), is strongly deformed along the Keeweenaw fault, and at Limestone Mountain (southwest corner of Figure 5), lower Paleozoic rocks are sharply folded.

Most of the structural features of the copper district, except for tilting prior to deposition of the Nonesuch Shale, do seem to be related, directly or indirectly, to the deformation that produced the Keweenaw fault. In and near the copper district, dips exceeding 30°NW are found only within 4 or 5 miles of the Keweenaw fault, and 10 miles from the fault, dips are mostly less than 5°. The most acute folds are those close to the Keweenaw fault, and anticlines with northwest-striking axial planes are sharpest close to the fault, becoming subdued or dying out away from it. Faults and fractures are much more abundant in the mines close to the fault than in those farther away. Whether the Keweenaw fault and these seemingly related structures formed wholly in post-Jacobsville time or began to form earlier is not known (see 11, p. 50–53).

To sum up, 5° to 10° of the total present dip of the top of the Portage Lake Lava Series and about 20° of the dip of the bottom antedates the deposition of the Nonesuch Shale. Most of the tilting, folding, and faulting probably occurred during or after deposition of the lowermost Freda Sandstone. Some faulting and folding followed deposition of the Jacobsville Sandstone, but how much of the post-Nonesuch deformation is post-Jacobsville is not known.

## Age of Mineralization

Mineralization of the Portage Lake Lava Series was contemporaneous with or later than much or most of the deformation, as attested by several kinds of evidence.

(1) Native copper occurs in veins accompanied by the same gangue minerals that occur with it in conglomerate beds and amygdaloidal

basalt. Some of these mineralized fissures cut thousands of feet of lava flows. Other fissures, as Broderick (13) has pointed out, represent paths by which copper-bearing solutions leaked out of conglomerate beds. In many veins, the earlier minerals have been fractured and healed by later minerals (11, p. 53–54). Slickensided native copper is present locally but is sufficiently rare to suggest that post-copper movement on fissures was a minor part of the whole movement.

(2) At many places, individual ore lenses terminate abruptly at small joints or faults with little visible displacement, strongly suggesting that jointing antedated the ore.

(3) The pattern of mineralization in some of the amygdaloidal deposits appears to be related to the geometry of folds in their host rocks, as discussed further below.

(4) Preliminary fluid-inclusion studies of quartz (34, p. 1277) from the Kearsarge Amygdaloid suggest that the dip of the lava flows was already fairly steep when the quartz was deposited. The observed gradient measured along the dip of the lava flows is about 1.9°C/100 feet, uncorrected for pressure. If the rocks had their present dip of about 40° at the time of quartz deposition, the vertical gradient was about 2.8°C/100 feet, or more

than four times the present gradient. If the quartz was deposited at the end of Copper Harbor time, when the rocks dipped 10° to 15°, the observed gradient represents a most improbable vertical gradient of more than 9°/100 feet. This line of reasoning is promising but needs further work to be definitive.

On the basis of the argument in the preceding section on the age of the deformation, therefore, most of the mineralization probably occurred after deposition of much or all of the Freda Sandstone and was later than most, if not all, of the Keweenawan volcanism. Whether or not it was also post-Jacobsville has not been determined.

## ECONOMIC GEOLOGY

### Forms of the Ore Bodies

The principal ore deposits are stratiform because their host rock is either a conglomerate bed or the amygdaloidal basalt at the top of a lava flow. Figure 7 is a diagrammatic cross section of one lava flow (Flow B) and parts of the next higher and lower flows, illustrating some of the common features of host- and wall-rock relationships. The following general-

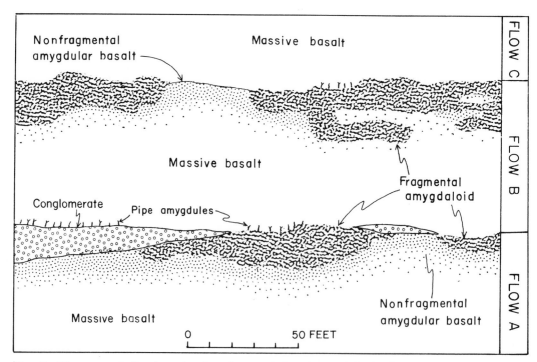

FIG. 7. *Diagrammatic Section, showing characteristics of amygdaloidal flow tops and conglomerate beds.*

izations apply to copper deposits in amygdaloidal basalt:

(1) Both the footwall and hanging wall of an amygdaloidal ore deposit are composed of massive basalt. The basalt below belongs to the same flow as the amygdaloidal layer and the basalt above to the next succeeding flow.

(2) Fragmental amygdaloid, the most common host rock, grades downward and laterally into nonfragmental amygdaloid; intervening areas of nonfragmental amygdaloid tend to be rather lean or barren.

(3) The upper boundary or hanging wall of the amygdaloidal layer is the base of the next overlying flow and is fairly sharp, except where tongues of the overlying lava have penetrated irregularly down into fragmental amygdaloid. This sharp boundary may be marked by pipe amygdules.

(4) Figure 7 shows how fragmental layers pinch and swell. The amount and abruptness of this pinching and swelling is less pronounced in some flows, such as the Kearsarge Flow, than in others, such as the Osceola Flow. In many of the flows with very thick fragmental tops (Osceola, Iroquois, and Isle Royale Flows), tongues and layers of massive or slightly amygdular basalt commonly break the continuity of the fragmental layer, as shown in the top of Flow B at the right-hand side of Figure 7.

(5) Copper tends to be more abundant near the top and bottom of the fragmental layer than in the middle. In exceptionally rich shoots, the entire thickness of fragmental amygdaloid is copper bearing. Where tongues of fragmental anygdaloid lie below layers of more massive basalt, as in the right-hand part of Flow B, exploration for the rich shoots that some of these tongues contain creates a special problem for the mining geologist (45).

(6) Because the distribution of fragmental amygdaloid is irregular and because copper is far from uniformly distributed in fragmental amygdaloid, Figure 7 shows the difficulty of assigning a meaningful average thickness to many ore deposits. In general, stope heights (measured perpendicular to bedding) of 10 to 15 feet are common; the average height is less in some mines and greater in others.

Copper deposits in conglomerate beds occupy lenticular bodies of conglomerate such as the one shown between Flows A and B in Figure 7. The footwall of a conglomerate bed is fragmental or nonfragmental amygdaloid, the interstices of which commonly are filled with fine sandstone or siltstone. The hanging wall of the conglomerate bed is the massive basalt of the next succeeding flow. The distribution of copper may or may not be coextensive with the conglomerate lens, laterally or stratigraphically. The margin of the Calumet and Hecla Conglomerate ore body is, at least in part, a line of pinchout, like the one shown in Figure 7; the lenticular character of this conglomerate bed may be one reason for its exceptionally rich concentration of copper (11, p. 115–116). In thick persistent conglomerate beds, such as the Allouez and Kingston, ore deposits are leaner and terminate laterally at vague assay boundaries.

Stratigraphically, copper tends to be concentrated along certain layers of a conglomerate bed. These more or less concordant ore lenses are separated from one another by lean or barren conglomerate or sandstone. Where the entire conglomerate bed is only 5 or 10 feet thick, most of it may be copper bearing. Where the conglomerate bed is thicker, copper is ordinarily confined to one or more ore lenses ranging from less than a foot to as much as 15 feet in thickness; in the new Kingston Conglomerate mine (shown in Figure 3 by the short heavy line just above the large "2") lenses may occur in the lower, middle, or upper part of the 40-foot bed and range from a few feet to hundreds of feet in strike length (R. W. Weege, written communication, Dec. 1966). Ore lenses are not well correlated with any conspicuous textural or compositional features of the conglomerate beds; in some places, sandy layers in a conglomerate bed are richer in copper than adjoining conglomerate and, in other places, poorer.

Within the plane of an amygdaloidal or conglomeratic layer, ore deposits and ore shoots display a variety of sizes and shapes. The strike length of major deposits ranges from about a mile for the Atlantic Mine to about 7 miles for the main part of the deposit in the amygdalodial top of the Kearsarge Flow. The workings of the Quincy Mine have a dip length of 8800 feet, and the bottom level is about 6000 feet below the surface. The workings of the Calumet and Hecla Mine have a dip length of 9200 feet, and the bottom level is 5500 feet below the surface. The Kearsarge Amygdaloid has been mined for as much as 7500 feet down the dip, and the Baltic Amygdaloid for as much as 5200 feet. The two deepest mines and the Atlantic Mine, which are stratigraphically quite high, have dip dimensions comparable to their strike lengths. In contrast, the dip dimensions of ore deposits lower in the stratigraphic column are only a fraction of their strike lengths. This difference may be related to the pattern of mineral zoning, discussed below.

The ore shoots that make up the different deposits in flow tops display a variety of shapes. Some deposits contain many highly elongate shoots with a strong preferred orientation. Such shoots range in width from 100 to more than 500 feet and in length from a few hundred to a few thousand feet. In other deposits, the areas of rich copper have very irregular shapes with no apparent preferred orientation. As described above, areas with abundant copper correlate very roughly with areas of thick fragmental amygdaloid and are commonly separated by areas in which the flow top is thin and possibly also nonfragmental. The differences apparently stem from primary irregularities in the degree of fragmentation of the flow top at the time the lava was poured out. Figure 8 compares two parts of the same mine in which the shoots differ in their regularlty.

Ore deposits in conglomerate beds tend to be nearly coextensive with the conglomerate itself where the bed is thin and lenticular, as has been noted above. The axes of both the Calumet and Hecla Conglomerate and the Houghton Conglomerate ore deposits follow the axes of conglomerate lenses that strike a little west of north. These lenses are interpreted as elongate alluvial-fan deposits, laid down by streams flowing toward the center of the Lake Superior basin (39, p. 370–371). The Houghton Conglomerate ore body consists of subparallel shoots separated by areas in which the conglomerate bed is thin or absent; these areas may represent interfluves in the ancient drainage system.

Ore shoots are less clearly defined in conglomerate beds that are thick and widespread. Their shapes and controlling factors may become better understood as the new mine on the Kingston Conglomerate (Figure 3) is developed.

## Stratigraphic Relations of the Ore Bodies

Figure 3 shows the stratigraphic position of major and minor native-copper deposits, including several that have produced less than 10,000 tons of copper. The larger deposits are identified by name; names of the lesser deposits are shown on a similar diagram by Stoiber and Davidson (34, Figure 7). Figure 3 shows only a few key horizons within the Portage Lake Lava Series, as can be seen by a comparison with Figure 4, an enlargement of the area

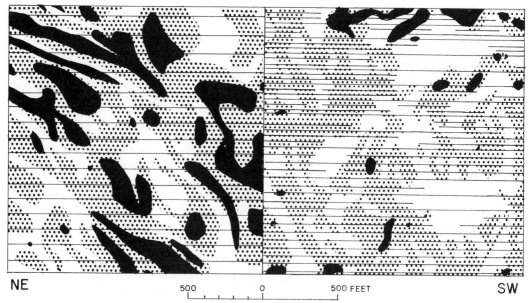

NE                                              SW
500          0          500 FEET

FIG. 8. *Distribution of Copper in an amygdaloidal Flow Top. Sections in the plane of the flow top, Baltic Amygdaloid, southern part of Champion Mine. Left side, showing more regular shoots, goes from 5th to 20th level. Right side, showing less regular shoots, goes from 30th to 46th level. Horizontal lines are drifts. Solid black indicates rich ore; stippled pattern indicates generally minable ground; lean to barren areas left blank.*

enclosed by a dotted line in Figure 3; Figure 4 shows all the individual flow tops that have been identified within this enclosure.

Figure 3 indicates that: (1) ore deposits are found both high and low in the Portage Lake Lava Series; (2) a rather small proportion of all the lava flows and conglomerate beds of the Portage Lake Lava Series contain economic concentrations of native copper; (3) individual ore deposits are confined to a quite small part of the strike length of the flow tops or conglomerate beds in which they occur; (4) certain stratigraphic units contain concentrations of copper at more than one place, but most of the units that contain a major ore deposit do not appear to be particularly favorable at a distance from that ore deposit; and (5) there is a conspicuous tendency for certain units to contain copper in places that are directly above or below ore deposits in other units.

Most of the ore deposits are confined to a single stratigraphic unit, and the next higher or lower flow top or conglomerate bed tends to be more or less barren. In the mines on the Baltic Amygdaloid, however, the next higher amygdaloid and the two amygdaloids next below have been stoped over considerable areas. In different parts of the Quincy Mine, seven or more different amygdaloids have been extensively stoped. More than one flow top has been stoped in many of the small mines of the Evergreen and associated amygdaloids in Ontonagon County.

The heavy vertical lines in Figure 3 show stratigraphic units intersected by some of the larger fissure deposits. Not shown are important copper-bearing fissures intersecting bedding at a low angle in certain mines of the Evergreen and associated amygdaloids.

## Ore Minerals

Native copper is, for practical purposes, the only ore mineral of the district. Native silver everywhere accompanies the copper but in minute quantities that probably do not exceed 0.1 per cent of the copper at most places. Chalcocite is fairly common: (1) in small veins cutting the Isle Royale and Baltic Amygdaloids; (2) with calcite coating joints in conglomerate beds; and (3) in veins near the mafic intrusive mentioned earlier. It is negligible as a source of copper. Copper arsenides are common only in certain late veins that cut the Kearsarge Amygdaloid (11, p. 56–58; 34, p. 1275–1277).

The native copper occurs primarily as small to large grains disseminated in amygdaloid or conglomerate. Some occurs as small to large masses, ranging in weight from less than a pound to many tons; the largest masses occur in well-defined veins and were a major source of copper in the fissure mines. Some of the copper fills amygdules and interfragmental openings and shows crystal faces. Most of the copper, however, either fills irregular discontinuous fractures in, or replaces host rock and earlier formed secondary minerals (see below).

Much of the native copper is slightly arsenical (10 p. 149–162). The amount of arsenic ranges up to about 0.5 per cent of the copper and differs not only from mine to mine and grain to grain but even in the different growth zones of individual crystals (26).

## Associated Minerals

The minerals associated with the native copper include the primary rock-forming minerals of the basalts and conglomerates and a large suite of secondary minerals that, as the copper, occur primarily in amygdules and interfragmental spaces. The massive basalt of the interiors of the lava flows is remarkably fresh for rocks a billion years old—olivine for example is still preserved in many—and the metamorphic alteration described below is conspicuous only in the amygdaloidal tops.

The modes of occurrence of the secondary minerals in the flow tops have been described in greatest detail by Butler and Burbank (11) p. 55–62) and Stoiber and Davidson (34, particularly p. 1256–1270). Figure 9 lists the most prominent of these minerals and shows their sequence of formation as determined primarily from the order of superposition in amygdules, interfragmental spaces and veins (11, p. 53–62, Figure 11). These secondary minerals, particularly chlorite, epidote, pumpellyite, and prehnite, also replace their amygdaloidal host rock; the extent of such replacement is erratic both within and between flow tops, but in general the amount of replacement increases with stratigraphic distance downward in the section. At many places, the original glassy basalt of the flow tops has been replaced almost entirely by mixtures of pumpellyite, epidote, quartz, and chlorite in various proportions. Epidotization, is the most common form of replacement in rhyolite conglomerate, but locally certain pebbles have been selectively chloritized. The extent of replacement and the size and abundance of the areas involved tend to increase with stratigraphic depth.

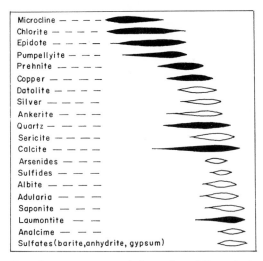

Microcline — — — —
Chlorite — — — —
Epidote — — — —
Pumpellyite — — —
Prehnite — — — —
Copper — — — —
Datolite — — — —
Silver — — — —
Ankerite — — —
Quartz — — — —
Sericite — — — —
Calcite — — — —
Arsenides — — —
Sulfides — — —
Albite — — — —
Adularia — — —
Saponite — — —
Laumontite — —
Analcime — — —
Sulfates (barite, anhydrite, gypsum)

FIG. 9. *Paragenesis of Secondary Minerals in Flow Tops and Veins. Solid black symbols indicate the more abundant minerals, open symbols the less abundant. Modified from Butler and Burbank (11, Figure 11).*

Although much native copper occurs in rock that appears generally unaltered, except for secondary mineral fillings, the rock enclosing or immediately adjacent to copper is very commonly bleached or otherwise altered. As previous authors have shown (11, p. 131–136), the principal chemical changes involved are reduction and partial removal of iron.

Calcite, chlorite, and, throughout most of the district, quartz, are the most abundant secondary minerals. Epidote, pumpellyite, and prehnite are also common and suggest temperatures of formation in the vicinity of 300°C (32, p. 83–92). Native copper is typically later than pumpellyite and epidote, and at many places is intimately intergrown with prehnite and datolite. Although the native-copper deposits are commonly called "zeolitic," zeolites, except for laumontite, are not abundant (10, p. 325). They occur principally in veins and as late-filling minerals in amygdules and vugs and are a low-temperature mineral assemblage superposed on the earlier formed higher-temperature assemblage (11, p. 60–61). There is nothing, however, to suggest an important break in a rather continuous sequence of mineralization governed by falling temperature.

The secondary mineral assemblages are stratified roughly within individual flow tops. The dominance of chlorite amygdule fillings in the lowermost part of flow tops has long been recognized. Stoiber and Davidson (34, p. 1270–1273) made detailed studies of the distribution of minerals at various places in mines on the Kearsarge Amygdaloid and found

that the other common secondary minerals also tend to be concentrated in certain parts of the flow top. Microcline, for example, commonly is found near the base of the amygdaloidal zone and also at the top, including the basal part of the overlying flow. Between these upper and lower microcline-bearing bands are a calcite-epidote band, above, and a quartz-epidote band below. The bands containing these various assemblages are discontinuous and their boundaries irregular (34, Figure 9). Stoiber and Davidson (34, p. 1273) attribute the banding to sequential deposition from hydrothermal solutions along the irregular channel provided by the flow top.

Neither the distribution nor the abundance of copper shows any discernible spatial relation to the mineral banding (34, p. 1273). In many places, patches of abundant copper are elongate at a high angle to the flow top and may cross two or more bands. Copper does, however, tend to be more abundant in amygdaloidal layers that have a large assemblage of secondary minerals than in those of simple mineralogy (11, p. 149). A diverse assemblage, especially one containing copper, probably signifies a more open channelway through which large quantities of hydrothermal solution of changing composition passed. Simple mineralogy and little or no copper suggests low permeablity and an almost closed system.

## Mineral Zoning

The most important result of recent mineralogical studies of the district has been the recognition and delineation of a regional pattern of mineral zoning. Broderick (13) was the first to emphasize the existence of regional zoning of the silicate minerals, which he correlated with the average arsenic content of the ore from various mines. More recently, Stoiber and Davidson (34) charted the distribution of minerals that are keys to the zonal pattern and investigated the chemical changes involved. They were able to plot, on maps and longitudinal stratigraphic sections, the northwestern (stratigraphically highest) limit of abundant epidote and of abundant quartz, and the southeastern (stratigraphically lowest) limit of abundant prehnite. The northwestern limit of quartz, much generalized from their data, is shown as a dotted line on Figure 5. The northwestern limit of abundant epidote lies a short distance north or south of this same line. The southwestern limit of prehnite lies just southeast of the Scales Creek Flow throughout most of the area of Figure 5. The zones defined by these limits resemble zones of regional metamorphism. They are exactly comparable to regional amygdule-mineral zoning described by Walker from Ireland (38) and Iceland (37), although the minerals involved are different; the mineral assemblages have much in common with those of the low-grade metamorphic sequence described by Coombs (32, p. 59–63) from the Triassic Hokonui Facies of the Taringatura area, New Zealand. At a stratigraphic depth of about 17,000 to 23,000 feet in the Hokonui Facies, heulandite and analcime give way to laumontite and albite as the principal metamorphic alteration products of volcanic glass and plagioclase; pumpellyite, prehnite, and epidote appear as prominent accessories.

Even though the assemblages described by Coombs probably represent neomineralization under near-equilibrium conditions, and those of Michigan, by contrast, are minerals deposited in an essentially open system, the actual minerals present throughout much of the native-copper district are remarkably like those of the lower part of the Hokonui Facies. Zeolites other than laumontite are prominent only high in the section toward the east (34, p. 1262–1263), that is, above the dotted line of Figure 5, suggesting an approach to the conditions above 23,000 feet in New Zealand. The upper limit of abundant quartz (dotted line of Figure 5) is comparable to the upper limit of abundant quartz veins that lies near the zeolite-greenschist facies transition in New Zealand (32, p. 83). The lower limit of prehnite in Michigan would be analogous to a point somewhere below the base of the Hokonui Facies at 30,000 feet, at the base of the quartz-prehnite zone (subfacies) of the prehnite-pumpellyite metagrawacke facies proposed by Coombs (35, p. 340–342). The top and bottom of the zone with prehnite and pumpellyite in Michigan are at least 10,000 feet apart stratigraphically, and their stratigraphic separation in New Zealand exceeds 7000 feet by an unknown amount. (In both localities this stratigraphic separation is measured at the surface and may appreciably exaggerate the actual thickness of the prehnite-pumpellyite zone if zone boundaries dip much more gently than bedding.)

The distribution of copper is related to this regional zoning pattern. Stoiber and Davidson (34, p. 1263–1266) show that five of the seven major deposits and many of the smaller deposits lie close to the upper limit of abundant quartz. Only the mines on the Baltic and Isle Royale Amygdaloids appear to lie far below it. All the stratiform ore deposits lie below the upper limit of abundant epidote.

Areas of significant copper, defined as areas in which more than 60 per cent of the amygdaloids and conglomerate beds intersected by drill holes contain at least traces of megascopically visible copper, tend to be stratigraphically higher than the level of the five major ore deposits just mentioned (34, Figure 7 and p. 1263–1266). The zone containing most of the large areas of significant copper has its lower boundary close to the limit of abundant quartz and its upper boundary 3000 to 5000 feet stratigraphically higher.

The distribution of copper, therefore, can be correlated in part with the distribution of other minerals in an overall pattern that is broadly analogous to a pattern of regional metamorphic zones. To various degrees, this metamorphism has affected all Keweenawan flow tops throughout the Lake Superior region and is not a local phenomenon. Because of the existence of many rather open channelways, however, zone boundaries and the distribution of certain constituents such as copper are far more irregular than they would be if the system had been less permeable and more nearly closed.

## Factors Controlling Form and Location of Ore Bodies

Host-rock permeability was the principal factor controlling form and location of ore bodies. Temperature and possibly depth appear to have controlled the upper and lower limits of a zone most favorable for copper deposition on a regional scale, as suggested by the relation to mineral zoning just described. The oxidation state of host rocks has exerted some control on the specific sites of copper deposition within ore deposits (areas of well-oxidized amygdaloid are generally more favorable than areas of weakly oxidized amygdaloid), but not, so far as can be ascertained, on the size, shape, or location of ore deposits; other compositional differences of the host rocks seem to have had negligible influence, as copper is found in all the different types of lava and sedimentary rock recognized in the district.

Broderick (10,22), in particular, has stressed the probable complexity of the pathways along which copper-bearing solutions have moved and in which they have deposited their copper. The "plumbing system" of a given deposit may involve passage of solutions along the permeable fragmental tops of lava flows, along conglomerate beds, or through fissures that cut across the layered sequence. He recognized combinations of channelways, exemplified by a situation in which solutions apparently had leaked from a conglomerate bed by way of some vertical fractures and formed small mantolike ore bodies in amygdaloidal layers several tens of feet stratigraphically above the conglomerate bed. Where several adjacent amygdaloidal layers are copper bearing, all in the same limited area and not outside it (as at the Quincy Mine, the mines on the Baltic Amygdaloid, and some of the smaller deposits), it is reasonable to conclude that the ore bodies in the different layers represent alternate routes of copper-bearing solutions from a common source. The remarkable concentration of major and minor ore deposits in the vicinity of Calumet (Figure 3) suggests the same sort of dispersion from a common source on a grand scale; it is surely more than coincidence that ore deposits occur in four different stratigraphic units directly above the central part of the large deposit on the Kearsarge Amygdaloid (Figure 3), in an area containing many crosscutting faults and fractures.

The importance of primary permeability in amygdaloidal and sedimentary layers relative to permeability of tectonic origin has not been studied quantitatively. Butler and Burbank (11, p. 115–116), presumably influenced by Graton's interest in the topic (18), lay considerable stress on the primary permeability. Several facts, however, suggest that permeability due to fracturing as the lava series was tilted may have been equally or more important: (1) Copper was definitely deposited after the primary permeability had been drastically reduced by deposition of gangue minerals, principally microcline, pumpellyite, chlorite, epidote, quartz, and some calcite and prehnite (Figure 9). (2) Although an appreciable amount of copper occurs in vesicles and interfragmental spaces, more occurs in irregular discontinuous fracture fillings and replaces the host rock. Individual small particles of copper freed from the enclosing rock commonly have shapes that indicate formation in cracks. Polished sections and slabs show much copper in tiny irregular veinlets, and the pattern of copper stains in many drift walls similarly indicates that zones of crackling, parallel or at an angle to bedding, partly have localized copper deposition. (3) Some copper fills tiny cracks in secondary minerals, thus proving that some crackling took place during the mineralization period. (4) Similarly, tiny copper-filled veinlets cutting across two or more pebbles in conglomerates demonstrate tectonically created permeability.

One would expect amygdaloidal layers and sedimentary beds to be loci of tectonic fracturing because they are mechanically weak compared with the layers of massive basalt that lie above and below them. During tilting of the rocks, displacements analogous to bedding slip in folded sedimentary rocks would be concentrated in these already porous and fragmental layers. Both primary and tectonic permeability thus tend to be greatest in the same rocks, and the two types of openings are, in fact, often difficult to distinguish in outcrops or underground exposures of fragmental amygdaloid.

Inasmuch as some, perhaps a large part, of the permeability of amygdaloidal and conglomeratic layers at the time copper was introduced was of tectonic origin, structural control of ore deposits is a possibility. Some, but not all, of the ore deposits do indeed appear to be related to structural features, as described in the following paragraphs.

Several major ore deposits are in areas of gentle synclinal warping. The two deposits with the greatest strike lengths are those on the Kearsarge and Baltic Amygdaloids. Both cross anticlines (Figure 6), but, in both, the deposit is lean in the axial region of the anticline and richest in adjacent areas that are gently synclinal. In the Quincy Mine, the dip changes from 55° near the surface to 35° in the lower levels, defining a sharply synclinal flexure with an axis that plunges very gently northeastward (this flexure probably resulted from a local steepening of dips adjacent to the Hancock fault). The Isle Royale Mine occupies the axial region of a prominent syncline that plunges almost directly down-dip. The other major ore deposits—those on the Calumet and Hecla Conglomerate, on the Osceola Amygdaloid, and at the Atlantic Mine—are in areas where the strikes or dips of opposite limbs of synclines differ by only a few degrees, and bending may not be significant. Areas of anticlinal bending may be unfavorable because the many fractures accompanying them permitted copper-bearing solutions to be drained off and dissipated into the overlying strata.

The hypothesis of structural control has additional support in an apparent relationship of certain ore deposits to thick flows. A number of copper-bearing layers are at the top of, or just below, an exceptionally thick flow (11, p. 118). In the areas where their tops contain ore deposits, the Osceola Flow is 140 feet thick, the Kearsarge Flow 180 feet thick, and the Baltic Flow 200 feet thick; the Scales Creek Flow, just above the Isle Royale Amygdaloid, is 230 feet thick. Very thick flows overlie the Allouez and Houghton Conglomerate beds, which locally are copper rich. Presumably, during folding, the presence of thick flows leads to a concentration of bedding slip in the mechanically weak amygdaloidal or sedimentary layers immediately adjoining. Where no thick flows are present, these weaker layers are more closely spaced, and the amount of bedding slip on each is much smaller; the amount of fracturing is not enough to lead to formation of an ore deposit except where the amount of folding is extreme, as at the Quincy Mine.

Sites most favorable for ore deposits on the basis of structure alone may be defined as follows: layers of fragmental amygdaloid or conglomerate 5 to 30 feet thick (where the layers are thicker, fracturing is too dissipated) that are continuous and fairly regular for distances measured in miles and bordered above or below by a very thick flow. If such layers are closely spaced stratigraphically, the favorableness of each is diminished. Areas of strong synclinal curvature are especially favorable. The axial regions of anticlines are quite unfavorable, but synclinal areas adjacent to them are favorable.

A structural feature of greater magnitude than those just described may be related, in part, to the location of the whole district. As noted earlier, the area of principal mines (Figure 5) coincides with an area in which the dip of individual horizons steepens 20° to 30° along strike from northeast to southwest. Because dips also tend to steepen toward the Keweenaw fault, lines of equal dip trend more nearly east-west than the strike of bedding. The line connecting points along which the dip is 30°NW has the same trend and lies very close to the northwestern limit of abundant quartz (Figure 5) as defined by Stoiber and Davidson (34, p. 1259–1266). Whether this coincidence is significant or not remains to be determined.

## ORE GENESIS

### Time of Copper Deposition

The native copper is clearly epigenetic, introduced into the lavas and conglomerate beds after they were buried; some of the deformation definitely occurred during the period of copper deposition. For reasons given earlier, it is likely, but difficult to prove conclusively, that the rocks were tilted to almost their present attitude at the time of copper deposition.

### Source of Copper

The source of copper was down-dip, as advocated by Smyth in 1896 (5), in opposition to then-current theories involving meteoric circulation. The patterns of leakage from large deposits into small (22, p. 696–697) and the funneling effects of impermeable barriers (11, p. 115–116) are consistent only with up-dip movement of copper-bearing solutions. The most likely sources are either concealed intrusives (11, p. 124–126, 141) or mafic lavas in the central part of the basin that lost copper on metamorphism at depth.

Butler and Burbank (11), presenting evidence developed by Graton and his colleagues, advocated a direct magmatic source for the copper, and Broderick supported this view in a series of noteworthy papers (10,16,17,22). This magmatic source is assumed to be a large mafic intrusive, akin to the Duluth Gabbro Complex, or its felsic differentiates. The strongest direct arguments are based on the abundance of copper sulfide veins associated with Keweenawan intrusive rocks (11, p. 124–125, 130). A second argument is based on the presumed magmatic origin of arsenic, nickel, and cobalt, in particular, in veins that cut copper-bearing amygdaloid (22, p. 710–719); as Stoiber and Davidson have shown (34, Figure 9 and p. 1274–1277), the veins bearing arsenide minerals, nickel, and cobalt formed late in the history of mineralization, long after the copper was deposited (Figure 9), and are very local and exceptional features that may well be only distantly related or unrelated, genetically, to the native-copper deposits. The other arguments (11, p. 118–127; 22, p. 710–719) are based primarily on the inadequacy of theories involving meteoric circulation or syngenetic deposition but are not incompatible with any theory involving ascending solutions.

The principal arguments against a magmatic origin are not conclusive. Chief of these is the likelihood that much or all of the copper was introduced during or after deposition of the Freda Sandstone, long after the peak of Keweenawan volcanism, and possibly even long after its termination. Abundant pebbles of granophyre in conglomerate beds of the Portage Lake Lava Series and Copper Harbor Conglomerate indicate intrusion and unroofing of a gabbro-granophyre complex prior to deposition of the Nonesuch Shale.

Secondly, although economic concentrations of copper are confined to about 1 per cent of the total area of Keweenawan lava flows exposed in the Lake Superior region, native copper in lesser amounts occurs in amygda-loids throughout the Lake Superior region. In the absence of quantitative data on the abundance of copper in these occurrence, it is not unreasonable to conclude that grades of 0.1 to 0.2 per cent copper are common. The source of this low-grade copper, which differs only in abundance from that in economic deposits, is therefore of regional extent; if the source is an intrusion, it must be almost coextensive with the Lake Superior basin. A gabbro-granophyre complex underlying the lavas throughout the Lake Superior region, and exposed in the Duluth Gabbro Complex, has been suggested (11, p. 124–126, 141), but its existence seems doubtful (53, p. E19).

Finally, if the copper-bearing solutions were of magmatic origin, one would expect the copper to occur as sulfides, as it does throughout the world in deposits associated with intrusive mafic rocks (11, p. 129–130). The abundance of iron sulfides in nature in general, and in ore deposits in particular, normally leads to a copper mineralogy that is dominated by chalcopyrite and to a lesser extent by bornite. In Michigan, the wall-rock alterations most commonly associated with the native copper involved reduction and partial removal of iron in hematite-bearing rocks (11, p. 133–136). These considerations led Butler and Burbank and their associates to infer that the hematite acted as an oxidizing agent to remove sulfur by converting it to $SO_4^{-2}$ ion. Although the stability field of hematite does lie entirely within the field of predominance of the sulfate (or bisulfate) ion, in "normal" magmatogenic ores, the coexistence of hematite with chalcopyrite, bornite, or chalcocite and its presumed stability with covellite and digenite (50, p. 166–170; 33, p. 992–993) shows that the hematite itself would not remove the sulfur from these copper minerals and, therefore, could not be solely responsible for altering an otherwise "normal" hydrothermal solution to one that deposited native copper. It may, therefore, be more reasonable to infer, following Lindgren (15, p. 418) and Cornwall (23, p. 197; 29, p. 619–621) that the solutions did not represent a "normal" magmatic hydrothermal fluid, but had a much lower sulfur content, for a reason suggested below.

An alternative to a magmatic source is a source in the lavas themselves. This is an essential feature of most of the earlier theories discussed by Butler and Burbank in an excellent review of previous work (11, p. 120–130). In these theories, lateral secretion of copper was commonly linked with downward circula-tion of meteoric water and could therefore, be refuted by the evidence for ascending solutions. More recently, my colleagues and I (29, p. 621; 31, p. 11–15; 34, p. 1457–1459) have suggested that copper-bearing solutions may represent water driven from the voids and interstices of flow tops as the lava flows were crushed and metamorphosed in the axial region of the Lake Superior syncline, down-dip from the present copper district. Under this hypothesis, water buried with the flows dissolved a small part of the trace copper in the flow tops as the latter were recrystallized during metamorphism.

This hypothesis is favored for the following reasons:

(1) The geometry of the Lake Superior basin indicates that lava flows now extend down-dip to depths in the range 30,000 to 50,000 feet (31, p. 8–11; 53, Figure 3). The amount of subsequent erosion is not known, and the depth at the time of deformation may well have been appreciably greater. At these depths, temperatures in the range of 300° to 500°C and pressures in the range 2.5 and 4 kilobars are probable; crushing and metamorphism of the flow tops and driving out of interstitial water seem inevitable.

(2) This hypothesis is compatible with evidence that the ore deposits are related to deformation near the end of or after the deposition of the Freda Sandstone and are later than most or all Keweenawan volcanism.

(3) The hypothesis does not require a hypothetical intrusive of enormous dimensions for which there is no supporting evidence.

(4) It satisfies completely the requirement that the source of the copper and the ore-forming process be regional in scope.

(5) The abnormally low sulfur content of the solutions that deposited the native copper is readily explained (29, p. 619–621): lavas should contain less sulfur than intrusions of similar composition because of escape of sulfur at the time the lava congeals; solutions generated by leaching of lava, therefore, should have less sulfur than solutions leaving an intrusion.

The greatest potential weakness of this hypothesis lies in its vulnerability to quantitative tests. The lavas down-dip are a more than adequate source of copper, but the volume of water buried in the original voids of lava flows and conglomerate beds is of more doubtful adequacy. The following order-of-magnitude data and computations are pertinent:

(1) Assuming an average thickness of 4 miles of lava flows and 2000 square miles for

the area that includes the copper district and the deepest part of the Lake Superior syncline (31, Figure 4), the total volume of rock involved is of the order of 8000 cubic miles, or $10^{14}$ tons.

(2) Assuming that 16 per cent of the lava column is amygdaloid, and that this amygdaloid contained 300 ppm copper, the total amount of copper available for leaching from amygdaloid was $5 \times 10^9$ tons or about 1000 times the total district production. The amygdaloids thus appear to be an adequate source of copper.

(3) Assuming, as above, that 16 per cent of the lava column is amygdaloid, and that 16 per cent of average amygdaloid was primary voids, then 2.6 per cent of the lava column, by volume, was primary voids, and 0.87 per cent of the column by weight was water filling these voids. The total volume of water in the inferred source area was therefore a little less than $10^{12}$ tons.

The adequacy of $10^{12}$ tons of water depends on the quantity of copper it contained. If it contained only 10 ppm Cu, as some of the brine from thermal wells in the Salton Sea area (52, Table I), the total copper in solution was only about twice the total district production. This ratio is clearly inadequate, because only a small part of the copper present in all deposits, high and low grade, discovered and undiscovered, is represented, and because the efficiency of the concentrating process can be assumed to be low. If concentrations of the order of 1000 ppm Cu are possible, the hypothesis is on firmer ground. Other variables are less critical; the amount of available water is only doubled or trebled, for example, by assuming that the conglomerate beds grade down-dip into rather extensive bodies of playa-type sedimentary rocks.

The problem of an adequate supply of water may apply with equal cogency to a potential magmatic source, but there is no sound basis for estimating the possible volume of magma involved. It would take 8000 cubic miles of magma to yield $10^{12}$ tons of water (the amount deduced above for the yield from flow tops) if 1 per cent water is driven off.

In summary, the theory that the ore-bearing fluid was water of metamorphic origin satisfies the evidence for ascending hydrothermal fluids and avoids the main shortcomings of a magmatic hydrothermal theory. But the quantitative adequacy of the metamorphic-water theory remains to be tested, and for the present it, as the magmatic-water theory, is simply a useful working hypothesis.

## Localization of Ore Deposits and the District

The factors that seem to govern the location of individual ore deposits have been discussed above. Ore deposits occur in amygdaloidal or conglomeratic layers that had high initial permeability, although much of the permeability remaining at the time of copper deposition may have been due to tectonic crackling, localized in these layers because they were physically inhomogeneous and weak. Several authors (11, p. 115–116; 22, p. 688–690) have described how various kinds of impermeable barriers within these layers have channelled copper-bearing solutions into certain rich shoots. There are strong suggestions that areas of synclinal folding are particularly favorable loci for ore deposits. And finally, the ore deposits show a relation to regional zoning of the secondary minerals in amygdaloids: a majority of the large ore deposits lie within, and close to, the margin of what Stoiber and Davidson (34, p. 1260) have called the "quartz zone." This favorable mineral zone is probably one in which the temperature and pressure were ideal for precipitation of copper at the time of mineralization.

There are many layers and areas that satisfy the above criteria but do not contain ore deposits—many more, in fact, than those that do. Whether the copper-bearing solutions are magmatic or formed by metamorphism of rocks containing interstitial water, these solutions have probably traveled for distances measurable in miles (15, p. 525). The structural features that have channeled copper-bearing solutions away from some beds and into others, therefore, may lie far down the dip, beyond hope of recognition or detection.

This last consideration obscures the ultimate reason for localization of individual deposits, but need not prevent the discovery of why groups of deposits (i.e., districts) have a certain place in the framework of the Lake Superior basin as a whole. Geophysical data will be indispensible in this effort. If the source of the copper is, indeed, a large body of mafic rock in depth, the only hope of locating it and studying its spatial relation to known ore deposits lies in geophysical measurements. If the source is in the lavas themselves, the location of districts is presumably controlled primarily by factors related to the geometry of the basin, and this can be determined most clearly by geophysical techniques.

The Michigan native-copper district lies more or less up-dip from a particularly wide and deep part of the Lake Superior basin (31, p. 11–13). The shortest path to the surface from this deep area along any given bedding plane, however, would reach the surface in or southwest of the area of principal mines. Direct up-dip movement of solutions, therefore, will not explain the precise distribution of known major ore deposits—there should be more mines southwest of the Baltic Amygdaloid mines (large numeral "3" in Figure 5) and fewer to the northeast. It is suggested (31, p. 12) that because of a predominant nearly east-west orientation of primary irregularities in flow tops, the flow tops were more permeable in an east-west direction than north-south. Solutions moving up-dip would, therefore, be continually diverted off to the east, and the major deposits would thus lie at the upper end of the easiest (rather than the shortest) passage from the deep part of the Lake Superior syncline to the surface. Recent geophysical work has tended to confirm the general basin configuration on which this hypothesis depends (c.f. 31, Figure 4, and 56, Figure 1) but is not yet adequate to define the shape of the basin in detail.

## REFERENCES CITED

1. Houghton, D., 1841, [Fourth] annual report of the State Geologist: Mich., House of Representatives [Doc.] no. 27, 184 p.
2. Foster, J. W. and Whitney, J. D., 1850, Report on the geology and topography of a portion of the Lake Superior land district in the State of Michigan; Part 1, Copper lands: U.S. 31st Cong., 1st Sess., House Exec. Doc. 69, 224 p.
3. Pumpelly, R., 1873, Copper district [Upper Peninsula]: Mich. Geol. Surv., v. 1, pt. 2, 143 p.
4. Irving, R. D., 1883, The copper-bearing rocks of Lake Superior: U.S. Geol. Surv. Mon. 5, 464 p.
5. Smyth, H. L., 1896, On the origin of the copper deposits of Keweenaw Point [abs.]: Amer. Jour. Sci., new ser., v. 3, p. 251–252.
6. Wright, F. E., 1909, The intrusive rocks of Mt. Bohemia, Michigan: Mich. Geol. and Biol. Surv. Rept. for 1908, p. 355–402.
7. Lane, A. C., 1911, The Keweenaw series of Michigan: Mich. Geol. and Biol. Surv. Pub. 6 (Geol. ser. 4), 2 v., 983 p.
8. Van Hise, C. R. and Leith, C. K., 1911, The geology of the Lake Superior region: U.S. Geol. Surv. Mono. 52, 641 p. (see particularly Chap. 15, The Keweenawan Series, p. 366–426).
9. Aldrich, H. R., 1929, The geology of the Gogebic Iron Range of Wisconsin: Wisc. Geol. and Nat. Hist. Surv. Bull. 71, 279 p.
10. Broderick, T. M., 1929, Zoning in the Michigan copper deposits and its significance: Econ. Geol., v. 24, p. 149–162, 311–326.
11. Butler, B. S. and Burbank, W. S., 1929, The copper deposits of Michigan: U.S. Geol. Surv. Prof. Paper 144, 238 p.
12. Leverett, F., 1929, Moraines and shore lines of the Lake Superior region: U.S. Geol. Surv. Prof. Paper 154-A, 72 p.
13. Broderick, T. M., 1931, Fissure vein and lode relations in Michigan copper deposits: Econ. Geol., v. 26, p. 840–856.
14. Mendelsohn, A., 1931, Mining methods and costs at the Champion copper mine, Painesdale, Michigan: U.S. Bur. Mines I.C. 6515, 16 p.
15. Lindgren, W., 1933, Mineral deposits: 3d ed., McGraw-Hill, N.Y., 930 p.
16. Broderick, T. M. and Hohl, C. D., 1935, Differentiation in traps and ore deposition: Econ. Geol., v. 30, p. 301–312.
17. Broderick, T. M., 1935, Differentiation in lavas of the Michigan Keweenawan: Geol. Soc. Amer. Bull., v. 46, p. 503–558.
18. Graton, L. C. and Fraser, H. J., 1935, Systematic packing of spheres, with particular relation to porosity and permeability: Amer. Assoc. Petrol. Geols. Bull., v. 43, p. 785–909.
19. Lamey, C. A., 1938, A dip-needle survey of the Toivola-Challenge mine area, Michigan: Econ. Geol., v. 33, p. 635–646.
20. Sandberg, A. E., 1938, Section across Keweenawan lavas at Duluth, Minnesota: Geol. Soc. Amer. Bull., v. 49, p. 795–830.
21. Spiroff, K., 1941, Dip-needle survey of Wyandotte-Winona area, Houghton County, and Cherokee area, Ontonagon County: Mich. Geol. Surv. Prog. Rept. no. 7, 17 p.
22. Broderick, T. M., et al., 1946, Recent contributions to the geology of the Michigan copper district: Econ. Geol., v. 41, p. 675–725.
23. Cornwall, H. R., 1951, Differentiation in lavas of the Keweenawan Series and the origin of the copper deposits of Michigan: Geol. Soc. Amer. Bull., v. 62, p. 159–201.
24. Cornwall, H. R., 1954, Bedrock geology of the Delaware quadrangele, Michigan: U.S. Geol. Surv. Geol. Quard. Map GQ-51, 1:24,000. (This is one of nine colored geologic maps covering the Keweenaw Peninsula northeast from Calumet. The others are numbered as follows: GQ-27, 34, 35, 36, 52, 54, 73, 74.)
25. Cornwall, H. R., 1954, Bedrock geology of the Lake Medora quadrangle, Michigan: U.S. Geol. Surv. Geol. Quad. Map GQ-52, 1:24,000.
26. Dreier, R. W., 1954, Arsenic and native copper: Econ. Geol., v. 49, p. 908–911.
27. Leighton, M. W., 1954, Petrogenesis of a gabbro-granophyre complex in northern Wisconsin: Geol. Soc. Amer. Bull., v. 65 p. 401–442.

28. Cornwall, H. R., 1955, Bedrock geology of the Fort Wilkins quadrangle, Michigan: U.S. Geol. Surv. Geol. Quad. Map GQ-74, 1:24,000.

29. Cornwall, H. R., 1956, A summary of ideas on the origin of native copper deposits: Econ. Geol., v. 51, p. 615–631.

30. White, W. S. and Wright, J. C., 1956, Geologic map of the South Range quadrangle, Michigan: U.S. Geol. Surv. Min. Inv. Field Studies Map MF-48, 1:24,000. (This is one of four black-and-white geologic maps covering the area of major copper deposits southwest from Calumet. The others are numbered as follows: MF-43, 46, 47.)

31. White, W. S., 1957, Regional structural setting of the Michigan native copper district, *in* Snelgrove, A. K., Editor, *Geological exploration—Institute on Lake Superior geology:* Mich. Coll. Mining and Technol. Press, Houghton, p. 3–16. Reprinted, 1961, in Mich. Basin Geol. Soc. Ann. field excursion, Guidebook, 1961, p. 3–16.

32. Coombs, D. S., *et al.*, 1959, The zeolite facies, with comments on the interpretation of hydrothermal synthesis: Geochim. et Cosmochim. Acta, v. 17, p. 53–107.

33. McKinstry, H. E., 1959, Mineral assemblages in sulfide ores: the system Cu-Fe-S-O: Econ. Geol., v. 54, p. 975–1001.

34. Stoiber, R. E. and Davidson, E. S., 1959, Amygdule mineral zoning in the Portage Lake Lava Series, Michigan copper district: Econ. Geol., v. 54, p. 1250–1277, 1444–1460.

35. Coombs, D. S., 1960, Lower grade mineral facies in New Zealand: 21st Int. Geol. Cong. Rept., pt. 13, p. 339–351.

36. Pollock, J. P., *et al.*, 1960, A geochemical anomaly associated with a glacially transported boulder train, Mount Bohemia, Keweenaw County, Michigan: 21st Int. Geol. Cong. Rept., pt. 2, p. 20–27.

37. Walker, G. P. L., 1960, Zeolite zones and dike distribution in relation to the structure of the basalts of eastern Iceland: Jour. Geol., v. 68, p. 515–528.

38. —— 1960, The amygdale minerals in the Tertiary lavas of Ireland. III. Regional distribution: Mineral. Mag., v. 32, p. 503–527.

39. White, W. S., 1960, The Keweenawan lavas of Lake Superior, an example of flood basalts: Amer. Jour. Sci., v. 258A (Bradley Volume), p. 367–374.

40. White, W. S. and Wright, J. C., 1960, Lithofacies of the Copper Harbor conglomerate, northern Michigan: U.S. Geol. Surv. Prof. Paper 400-B, p. B5–B8.

41. Goldich, S. S., *et al.*, 1961, The Precambrian geology and geochronology of Minnesota: Minn. Geol. Surv. Bull. 41, 193 p.

42. Griffin, J. B., *Editor,* 1961, Lake Superior copper and the Indians; miscellaneous studies of Great Lakes prehistory: Univ. Mich., Mus. Anthropology, Anthropological Papers no. 17, 189 p.

43. Hamblin, W. K., 1961, Paleogeographic evolution of the Lake Superior region from Late Keweenawan to Late Cambrian time: Geol. Soc. Amer. Bull., v. 72, p. 1–18.

44. Hamblin, W. K. and Horner, W. J., 1961, Sources of the Keweenawan conglomerates of northern Michigan: Jour. Geol., v. 69, p. 204–211.

45. Weege, R. J. and Schillinger, A. W., 1962, Footwall mineralization in Osceola Amygdaloid, Michigan native copper district: A.I.M.E., Tr., v. 223, p. 344–350.

46. Silver, L. T. and Green J. C., 1963, Zircon ages for middle Keweenawan rocks of the Lake Superior region [abs.]: Amer. Geophys. Union Tr., v. 44, p. 107.

47. Faure, G., 1964, The age of the Duluth gabbro complex and the Endion sill by the whole-rock Rb-Sr method: Mass. Inst. Technol., Dept. Geol. and Geophysics, 12th Ann. Prog. Rept. for 1964, p. 255–257.

48. Hughes, J. D., 1964, Physiography of a six quadrangle area in the Keweenaw Peninsula north of Portage Lake (abs.): Dissert. Abs., v. 25, p. 406.

49. Faure, G. and Chaudhuri, S., 1964, Whole-rock Rb-Sr age of the cupriferous parting shale member of the Nonesuch Formation, Michigan: Geol. Soc. Amer., Program, 1964 Ann. Meeting, p. 54–55.

50. Garrels, R. M. and Christ, C. L., 1965, Solutions, minerals, and equilibria: Harper and Row, New York, 450 p.

51. Hack, J. T., 1965, Postglacial drainage evolution and stream geometry in the Ontonagon area, Michigan: U.S. Geol. Surv. Prof. Paper 504-B, p. B1–B40.

52. White, D. E., 1965, Saline waters of sedimentary rocks, *in* Young, A. and Galley. J. E., *Editors, Fluids in subsurface environments:* Amer. Assoc. Petroleum Geols. Mem. 4, p. 342–366.

53. White, W. S., 1966, Tectonics of the Keweenawan basin, western Lake Superior region: U.S. Geol. Surv. Prof. Paper 524, 23 p.

54. Goldich, S. S., *et al.*, 1966, Geochronology of the midcontinent region, United States, [pt.] 2, northern area: Jour. Geophys. Res., v. 71, p. 5389–5408.

55. White, W. S. and Wright, J. C., 1966, Sulfide-mineral zoning in the basal Nonesuch Shale, northern Michigan: Econ. Geol., v. 61, p. 1171–1190.

56. White, W. S., 1967, Geologic evidence for crustal structure in the western Lake Superior basin: *in* Steinhart, J. S., and Smith, T. J., Editors, *The earth beneath the continents,* Amer. Geophys. Union Mono. 10, p. 28–41.

57. Choudhuri, S. and Faure, G., 1967, Geochronology of the Keweenawan rocks, White Pine, Michigan: Econ. Geol., v. 62, p. 1011–1033.

# 17. Geology of the Southeast Missouri Lead District

FRANK G. SNYDER,* PAUL E. GERDEMANN†

## Contents

## Illustrations

\* 603 S. Gables Blvd., Wheaton, Illinois.
† St. Joseph Lead Company, Bonne Terre, Missouri.

## Tables

## ABSTRACT

The Southeast Missouri lead district, located about 70 miles south of St. Louis, embraces four important sub-districts and several minor ones. The important sub-districts, in order of discovery, are Mine La Motte, the Old Lead Belt, Indian Creek, and the Viburnum Lead Belt. Descriptions that follow are of geology and ore deposits of the Old Lead Belt.

The ore deposits are stratiform in character and are localized in a narrow carbonate bar and algal reef environment on the flanks of exposed Precambrian of the St. Francois Mountains. Ore structures include a variety of primary depositional features such as pinch-out zones, disconformities, ridge structures, reefs, and submarine gravity slides. Faulting, prior to and during ore deposition, "prepared" the host rock and strongly fractured areas are mineralized much more intensively and ex-

tensively than unfractured parts of the same sedimentary structures.

The deposits are epigenetic in character. Mineralization is believed to have occurred at the time that post-lower Ordovician faulting took place. The mineralizing solutions are believed to be concentrated brines from adjacent basins that moved out of the basins during faulting, uplift, and erosion of the St. Francois positive area.

## INTRODUCTION

The Southeast Missouri Lead Belt is one of the world's largest lead mining districts, having produced more than 9,000,000 tons of pig lead. At least 90 per cent of this production has been from the area which might properly be called the Old Lead Belt, since it has operated for more than 100 years and is now near its economic death. It is located about 70 miles south of St. Louis in St. Francois County, centering on the town of Flat River.

About 50 miles farther west, a new district, the Viburnum Lead Belt, is springing to life and promises to be as productive as the older area. This district forms a narrow north-south band extending from a few miles north of the town of Viburnum to about 12 miles northwest of the town of Ellington.

The Southeast Missouri ore deposits are stratiform in the upper Cambrian Bonneterre dolomite and represent a classical example of the "Mississippi Valley type." Extensive geological study has been carried on since 1947 and has contributed a wealth of information concerning the characteristics and controls of these deposits.

The authors gratefully acknowledge the contributions of the large number of geologists who have participated in the underground mapping program and studies of ore controls during the past 18 years. We also are grateful to the St. Joseph Lead Company for permission to publish this paper and for its continued support of research in both applied and theoretical geology.

## HISTORICAL INFORMATION

### Mining Operations in the Area

Philip Francis Renault led a party of French miners and slaves up the Mississippi River from New Orleans in the year 1720 and established himself near Kaskaskia, Illinois. From that point, parties were sent out in search of precious metals. One of these led by M. La Motte discovered the deposits that bear his name in Madison County, Missouri, near Fredericktown. Gradually other mines were opened in Washington and St. Francois Counties, and mining has been continuous nearly to the present time. This early mining was entirely for chunk galena found in the residuum, which required no milling.

The history of modern mining, which led to the opening of the Lead Belt in St. Francois County, began on a 946-acre tract of land, located at the present site of Bonne Terre. It was known as La Grave Mines. On May 2, 1864, the St. Joseph Lead Company purchased this tract of land, and mining began in open pits. These gradually sloped downward until the operation developed into an underground mine; shafts were later sunk for better access to the ore.

In 1869, the first diamond core drill was brought to Bonne Terre. Through its use, large bodies of ore were discovered that assured the development of the area as a great mining district. Since that time, more than 100,000 diamond drill holes have been completed; from these were obtained 50,000,000 feet of core.

As many as 15 companies were in production during the early years from the 1860's to the early 1900's. The St. Joseph Lead Company gradually acquired the holdings of these companies and became the only operator in the district in 1933.

The mine at Bonne Terre, which began production in 1864, was closed July 3, 1961, after 97 years of continuous mining.

At present, the mining complex in the Flat River area consists of eight mine units hauling by underground railroad to one central mill.

Production averages about 12,000 tons of ore per day, and lead and copper are recovered. The locations of active mines, mills, mines in development, and inactive mines are shown in Figure 1.

### Statistics of Mine Production

Mine output from St. Francois County reached a peak in 1942 when 197,430 tons of lead metal were produced. A strike during 1962 and 1963 forced production to its lowest tonnage since 1915. With only 36,301 tons produced in 1962, Missouri lost its leadership for lead production for the first time since 1907. Table I lists total mine production for St. Francois County by company and years of operation.

## PHYSIOGRAPHIC HISTORY AND PRESENT TOPOGRAPHY

The Lead Belt is located on the northeastern edge of the Precambrian igneous core of the St. Francois Mountains. Topography is some-

TABLE I.   Total Tons of Pig Lead Produced by Fifteen Mining Companies in St. Francois County—1865 through 1965[1]

| Company | Period of Operation (Incl.) | Total Tons of Pig Lead Produced |
|---|---|---|
| St. Joseph Lead Co. | 1865–1965 | 5,889,529 |
| Desloge Lead Co. | 1877–1886 | 28,916 |
| Doe Run Lead Co. | 1887–1914 | 337,601 |
| Leadington Lead Co. | 1893–1894 | 329 |
| Desloge Consolidated Lead Co. | 1893–1928 | 400,812 |
| Central Lead Co. | 1894–1904 | 51,404 |
| Taylor Place or Flat River Lead Co. | 1894 | 499 |
| St. Louis Smelting and Refining Co. | 1897–1933 | 716,278 |
| Union Lead Co. | 1900 | 22 |
| Columbia Lead Co. | 1900–1904 | 11,713 |
| Derby Lead Co. | 1902 | 336 |
| Irondale Lead Co. | 1902 | 498 |
| Federal Lead Co. | 1902–1923 | 721,303 |
| Baker Lead Co. | 1913–1918 | 36,921 |
| Boston-Elvins Lead Co. | 1924 | 2,000 |
| Total | | 8,198,161 |

[1] Does not include production from Mine La Motte, Indian Creek, Viburnum Lead Belt, or other outlying areas. Production figures from St. Joseph Lead Company records.

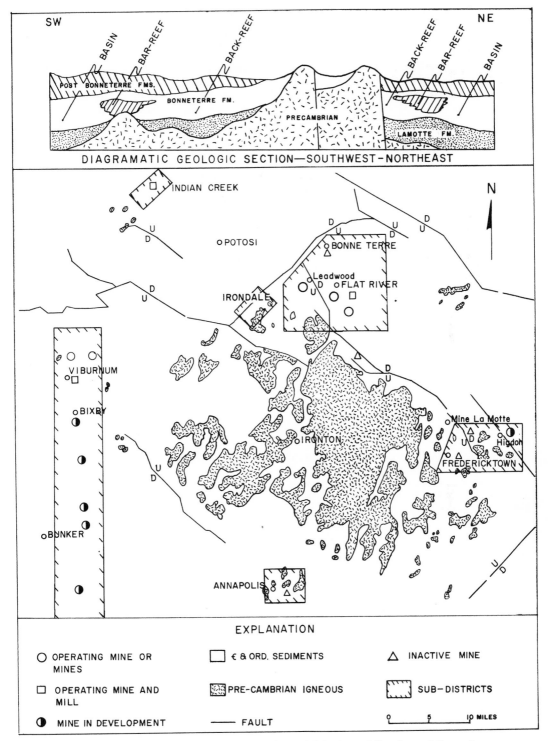

FIG. 1. *Major Geologic Features and Lead Districts of Southeast Missouri.*

what dependent on the rock formations that are at surface in any given area, but, in general, it is classed as mature.

The igneous rocks were eroded into mature topography of considerable relief prior to deposition of the upper Cambrian sediments. Later erosion stripped the sedimentary cover from many of the higher knobs and modified their surface expression. They appear as large, rounded, "sugarloaf" hills and represent the highest elevations in the area.

The non-cherty Bonneterre, Davis, and Derby-Doe Run Formations erode more easily than the younger cherty formations and have the gently rolling appearance of late maturity. The cherty formations develop a thick residuum on narrow sharp ridges with underdeveloped flood plains along streams.

Big River and its tributaries drain most of the district except for the extreme southwestern part, which is drained by the St. Francis river system. Both streams flow in an entrenched meander system indicating uplift and rejuvenation.

## GEOLOGIC HISTORY

### Stratigraphy

The lead-producing area of Missouri is located on the flank of a somewhat circular-shaped, positive structure known as the St. Francois Mountains, the northeastern part of the Ozark dome. The core of this structure is formed by Precambrian igneous rocks, mainly rhyolitic porphyries intruded by granites and diabasic rocks; they are the oldest rocks in the mining region. Many of the erosion-formed igneous knobs are exposed at surface, and others remain buried by Cambrian and Ordovician sedimentary formations.

The sedimentary formations in order of deposition are: Lamotte Sandstone, Bonneterre Formation, Davis Shale, Derby-Doe Run Dolomite, Potosi Dolomite, Eminence Dolomite, Gasconade Dolomite, Roubidoux Formation, and Jefferson City Dolomite. The Cambrian-Ordovician division is considered to be at the top of the Eminence Dolomite.

The Lamotte Sandstone is mainly a pure orthoquartzite with occasional siltstone sections or dolomitic beds in the upper part. It is generally fine-grained, friable, crossbedded, and often porous and permeable. Near the base or on the flanks of igneous knobs it may be conglomeratic or arkosic. The maximum thickness is about 450 feet.

The Bonneterre Formation is composed almost entirely of dolomite in the mining region although it is predominantly limestone beyond it. Dolomite, thought to be of diagenetic origin, forms a halo of uneven size and shape around Precambrian igneous knobs that are high enough to extend into or through the Bonneterre. The paleotopographic highs also affected sedimentation, giving rise to a different Bonneterre facies on or near the structure as compared with that farther into the basin. Sea currents, influenced by the larger igneous masses, built ridges or spits of clastic carbonate material.

During an early period of geologic study J. E. Jewell and R. E. Wagner subdivided the Bonneterre Formation into zones numbered from 1 through 19, from the top down. The major units are shown in Figure 2. Detailed descriptions of the units are given by Ohle and Brown (11) and Snyder and Odell (18). Snyder and Odell (18, p. 899) describe the zones as follows:

"All zones are generalized lithologic types. The zones are predominantly tan crystalline calcarenite* alternating with gray or brown shaly carbonate of varied texture. The lower units, 10 through 19, differ considerably in thickness or may be absent. Where a unit thins, the adjacent one thickens, so that the overall thickness of the formation is about the same throughout the central productive area.

The 19 bed is a sandy dolomite regarded as the transitional phase between the Lamotte quartz sand and the Bonneterre carbonates. The transitional unit is a member of the Bonneterre formation and, in many areas, is separated by a sharp hiatus from the underlying sandstone. Quartz sand of the transition unit represents reworked Lamotte. Quartz-bearing beds of the 19 unit intertongue with quartz-free Bonneterre units on the flanks of depositional ridges.

The 15, 10, and 5 units are composed predominantly of tan, crystalline calcarenite; in many places the 10 and 5 are oolitic. The 15 and 10 units are markedly lenticular, strongly crossbedded, and variable in thickness. The 12 and 7 beds are gray to brown shaly dolomite of varied lithologies. The 7 bed represents the main period of algal reef formation (11). Where algal formation is restricted or

---

* All of the Bonneterre dolomite was deposited as limestone. The terms *calcarenite* for lime sand and *calcilutite* for lime mud are used to indicate character at deposition although these rocks are now dolomite.

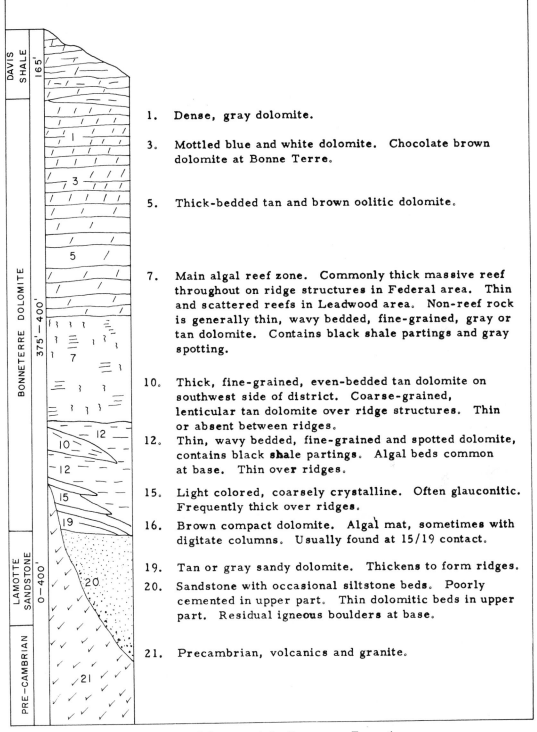

1. Dense, gray dolomite.

3. Mottled blue and white dolomite. Chocolate brown dolomite at Bonne Terre.

5. Thick-bedded tan and brown oolitic dolomite.

7. Main algal reef zone. Commonly thick massive reef throughout on ridge structures in Federal area. Thin and scattered reefs in Leadwood area. Non-reef rock is generally thin, wavy bedded, fine-grained, gray or tan dolomite. Contains black shale partings and gray spotting.

10. Thick, fine-grained, even-bedded tan dolomite on southwest side of district. Coarse-grained, lenticular tan dolomite over ridge structures. Thin or absent between ridges.

12. Thin, wavy bedded, fine-grained and spotted dolomite, contains black shale partings. Algal beds common at base. Thin over ridges.

15. Light colored, coarsely crystalline. Often glauconitic. Frequently thick over ridges.

16. Brown compact dolomite. Algal mat, sometimes with digitate columns. Usually found at 15/19 contact.

19. Tan or gray sandy dolomite. Thickens to form ridges.

20. Sandstone with occasional siltstone beds. Poorly cemented in upper part. Thin dolomitic beds in upper part. Residual igneous boulders at base.

21. Precambrian, volcanics and granite.

FIG. 2. *Subdivision of the Bonneterre Formation.*

lacking, the unit is composed of gray shaly dolomite like the 12 unit.

Calcitic limestone, designated as the 11 bed, rarely is cut by mine workings. Away from mine workings and off the ore-bearing structures it is penetrated by many drill holes, most commonly within the upper part of the 12 and the lower part of the 7, although in places the entire section from the top of the 19 to the base of the 5 is limestone. Lithologic character of the limestone has had little detailed study, but most of the textural variations occurring in dolomite also are found in limestone.

One unusual rock type, classed as the 9 unit, is a sedimentary breccia of limited areal and stratigraphic extent. It represents a primary or penecontemporaneous product of slump or intrastratal flow in shaly 12 or 7 beds. It is most common in the 12 but may occur at any position in the lower 200 feet of the formation, except in the tan crystalline units."

The lithologies of the Bonneterre dolomite have been described in terms of four end member components (18). These are algal material, sand size grains (carbonate, fossil fragments, quartz), calcareous mud, and argillaceous mud. A given bed may be composed of any one of the components or of any combination, depending upon environment of deposition. In addition, finely divided, diagenetic iron sulfide is a common constituent of the fine-grained carbonates and argillaceous sediments (9). It imparts a gray color to the carbonates and a black color to the shales; as contrasted with the tan and brown colors due to iron oxide in the coarse-grained carbonates.

The Davis Shale conformably overlies the Bonneterre Formation. It is composed of interbedded shales, carbonates, fine-grained glauconitic sandstones, and glauconitic siltstones. Flat-pebble and edgewise conglomerates are characteristic of the Davis. The formation averages 170 feet in thickness.

The Derby-Doe Run Dolomite consists of two major units. The lower is composed of thin-bedded, argillaceous, dolomite and the upper of massive oolitic dolomite or algal reef dolomite. This formation is transitional into the Potosi Formation above. The Potosi is a brown, massively bedded, siliceous dolomite consisting to a large degree of algal reef and recrystallized oolitic dolomite. The silica is present in the form of banded chalcedonic and quartz druses. The Eminence and Gasconade Formations contain clean, light gray, medium to coarsely crystalline, cherty dolomites. Widespread masses of silicified cryptozoon beds are present. The Roubidoux consists of sandstone, dolomitic sandstone, and cherty dolomite. The Jefferson City Formation is composed principally of tan and gray, medium to finely crystalline, cherty and argillaceous dolomite. A pronounced erosional unconformity is present at the top of the Jefferson City group.

## Structure, General for the Area

The main structural feature of southern Missouri is the Ozark dome. As stated above, the core of this structure is composed of Precambrian granites, volcanics, and minor mafic intrusives, all of which have remained as a positive area since the Cambrian and Ordovician sediments were deposited upon them. The diameter of this core is about 40 miles and is somewhat elongated in a northwest-southeast direction. A subsidiary structure, also elongated in a northwest-southeast direction, is located to the southwest near the town of Eminence. Sedimentary units dip outward circumferentially from the dome except where modified by later faulting. These dips are greater on the northeast side.

Thicknesses of the Cambrian units indicate that tilting to the southeast was taking place during deposition. The Bonneterre Formation is about 400 feet thick in the Lead Belt, in excess of 500 feet at Mine La Motte, and Grohskoph (12) reports from 1270 to 1580 feet of Bonneterre in the extreme southeast part of Missouri.

A system of faults roughly peripheral to the dome are present. Surface geologic mapping has not indicated this faulting on all sides of the dome, but exploration drilling proves its presence. The Old Lead Belt is bounded by the Simms Mountain fault on the southwest side, the Big River fault on the northwest, and the Farmington anticline on the northeast (7). This triangular-shaped area contains many faults with vertical displacements of 100 feet or less. The larger of these strike northwest and many smaller ones trend northeast. Even though subsurface geology is better known within the sub-district than without, there is no doubt that many more faults are present in the mineralized area. The faults usually have steep dips and are characterized by near-horizontal or low-angle slickensides.

The Simms Mountain fault, which strikes northwest along the face of the igneous mountains, drops the north side down about 600 feet and reverses the original northeast dip of the sediments. These beds now dip about 2°SW.

Paleozoic explosive igneous activity and intrusive diabasic dikes and plugs are known at several locations in southeast Missouri. The well-known Avon dikes and diatremes are located on the south end of the Farmington anticline in western Ste. Genevieve and eastern St. Francois counties, about 20 miles east of Flat River. These structures were classed into three types of Kidwell (6): (1) basic igneous dikes of non-explosive origin; (2) breccia bodies of fragmented pre-existing rocks with included igneous material; and (3) breccia bodies of explosive origin consisting of fragmented pre-existing rocks without primary igneous material. Time of dike formation was dated as post-Devonian by igneous coated Devonian fossils (3). Recent isotopic age determinations on dike rocks give a Devonian age (Allen Heyl, personal communication).

The Furnace Creek structure, a buried, ejecta-filled, volcanic crater is located in Washington County about 15 miles west of the Lead Belt (28). The structure consists of a funnel-shaped crater in the Lamotte Sandstone, about 1.5 miles in diameter, surrounded by a thin layer of volcanic "ash" in the Bonneterre Formation. Most of the "ash" bed consists of sharply angular, very fine fragments of quartz, undoubtedly shattered Lamotte sand grains. The event occurred in lower Bonneterre time and is buried by Bonneterre and younger formations. This structure and the Avon diatremes are among eight unusual structures aligned in a 400-mile east-west trending structural zone across southern Illinois, Missouri, and eastern Kansas. Five of the eight structures contain intrusive or extrusive igneous material, ranging in age from upper Cambrian to Cretaceous (or Paleocene).

Two additional structures have been discovered in recent years. One, located in the south part of the Viburnum Lead Belt, gives indications of being almost identical to Furnace Creek. The other, located about 5 miles south-southwest of Furnace Creek, is known only from surface outcrops of interbedded sediment and volcanic material and extensive brecciation.

## ECONOMIC GEOLOGY

### Sub-Districts

In southeast Missouri, lead-zinc-barite mineralization occurs in every formation from upper Cambrian Lamotte Sandstone to lower Ordovician Jefferson City, the youngest present in the area. The nine stratigraphic units in this interval total some 2400 feet in thickness.

Mineral production is recorded from seven of the nine formations.

Because mineralization is so widespread, both geographically and stratigraphically, the Southeast Missouri lead district is arbitrarily defined as the area in which stratiform deposits of lead-zinc-copper occur in Bonneterre Formation (and to a minor extent in upper Lamotte Sandstone) around the exposed Precambrian of the St. Francois Mountains (Figure 1). These deposits contain no barite.

Mineral deposits in the upper formations contain barite as a major constituent and usually are in veins. The Potosi Formation is the chief source of Missouri barite production, with most of the ore coming from residual deposits within the geographic limits of the Southeast Missouri district. Appreciable amounts of galena occur with the barite, and, in the 19th century, numerous small vein deposits were worked for galena. However, these deposits by their mineralogy, their form, and their stratigraphic position represent a separate problem from the main Southeast Missouri lead ores and will receive no further attention herein.

The Southeast Missouri district is composed of four important sub-districts and several minor ones (Figure 1). The major sub-districts are Mine La Motte, the Old Lead Belt, and the Viburnum Lead Belt. Each of these embraces an area of many square miles and contains hundreds of individual ore bodies. Indian Creek, the fourth important sub-district, is of lesser size but has substantial production. Minor sub-districts include Annapolis, Irondale, and other small mines, all now inactive. In all of these, mineralization occurs in the Bonneterre Formation. The Shirley-Palmer area, also a minor one, represents an exception; mineralization occurs in the Bonneterre Formation, but veins in the Potosi above provided substantial production in the early history of the district. Other areas, notably Valle Mines, also produced galena from deposits in the Potosi.

At Mine La Motte, only the lower 50 feet of the Bonneterre is mineralized. A shaly member low in the formation marks the upper limit of ore, although minor amounts of mineralization occasionally are found above the shaly unit. Location of ore bodies is strongly controlled by Precambrian knobs that cut out the Lamotte sand, with ore lying immediately above, or near, the Lamotte pinchout. Many ore bodies have a narrow linear form, arcuate in plan, following the Lamotte pinchout around a knob. So characteristic of this sub-district are the long narrow ore bodies, low in the host formation, that the term "Mine Lamotte type" is in general use to describe size, shape, and stratigraphic position of similar ore bodies elsewhere in the district.

Galena is the major ore mineral, but minor amounts of sphalerite, chalcopyrite, and siegenite occur closely related to the galena. Occurrences of the less abundant metals are sporadic throughout the sub-district, and no zoning pattern has been established.

The Indian Creek sub-district mineralization is in and adjacent to a fringing reef. The algal reef of 7-zone type was developed along and over a northeast-trending Precambrian high. Mineralization occurs along the northwest edge of the reef, over the top of the reef, and in adjacent fore-reef sediments consisting of tan oolitic dolomite and reef debris. The main mineralized zone is several thousand feet long, up to 300 feet wide, and has a stratigraphic range of approximately 150 feet. Mineralization of ore grade extends in the reef for a few hundred feet along internal reef structures.

The Viburnum Lead Belt, comparable in size and extent of mineralization to the Old, has most of the ore in the upper part of the Bonneterre. West of the St. Francois Precambrian high, a shale-silt bed from 2 to 10 feet thick occurs in the Bonneterre approximately 50 feet below the Davis/Bonneterre contact. A large part of the newly-discovered ore lies just below this Davis-type bed, representing a minable zone averaging 20 feet thick. In parts of the new district, ore also lies above this shale-silt unit, extending through the upper 50 feet of Bonneterre and into the lower few feet of Davis Shale. In the northern and central part of the Viburnum Lead Belt, lower Bonneterre ore is very minor in amount; in the southern part it may be a large part of the ore.

The Irondale and Annapolis sub-districts are small concentrations of sulfides around igneous knobs. Mineralization at Irondale is similar to lower Bonneterre ore in parts of the Old Lead Belt. The Annapolis sub-district is unique in that mineralization occurs in a vuggy, porous, "white-rock" facies of the Bonneterre.

As noted above, each of the sub-districts bears ore within a restricted part of the Bonneterre Formation. Only the Old Lead Belt is mineralized through the full stratigraphic range. In this sub-district, ore occurs in Lamotte Sandstone, 100 feet below the Bonneterre/Lamotte contact, through the upper Lamotte, and throughout the 400 foot thickness of the Bonneterre. This sub-district is the area that has been mapped and studied in

greatest detail; the descriptions that follow represent studies made in that sub-district with only infrequent reference to other areas. For brevity in the descriptions that follow, the term, district, will refer specifically to the Old Lead Belt sub-district.

## Stratigraphic and Facies Relationships of Ore Bodies

THE BONNETERRE FACIES PATTERN The forms of ore bodies, basically, are the forms of sedimentary depositional structures in the Bonneterre host. Almost every conceivable primary structural feature that could be developed during shallow water carbonate deposition is represented. In places in the district, postlithification faulting has disrupted primary structures; mineralization on opposite sides of the fault may not be uniformly related to sedimentary features. In places, faulting has resulted in mineralization of beds not normally mineralized. In places, also, solution and oxidation of carbonate host, some of it probably pre-ore, disturbs the intimate relationship of ore and primary structures. However, recognition of the Bonneterre facies pattern is essential to understanding lateral and vertical variations in form, characteristics, and stratigraphic position of ore bodies.

The Old Lead Belt is contained in a barrier reef and related carbonate facies developed along the north edge of the St. Francois Mountains that formed an island complex in late Cambrian time (Figure 1). The Precambrian surface of the Missouri platform was one of low relief with scattered areas that stood several hundred to over a thousand feet above the general erosional level (25). Over this irregular surface, Lamotte Sandstone was deposited. Irregularities on the Precambrian are reflected by subdued but definite irregularities on the top of the Lamotte.

The Bonneterre Formation in the Old Lead Belt consists of: (1) a dolomite back reef facies several miles in width, composed almost entirely of medium- to coarse-grained calcarenite and oolite; (2) a dolomite reef facies, 3 to 4 miles in width in which the lower 100 feet of the Bonneterre shows extensive lateral facies changes from calcarenite bars to local basin sediments, with the bar facies overlain by the 100 foot thick algal reef zone; and (3) a narrow dolomite fore-reef zone that interfingers into limestone.

The back-reef area is unmineralized, except along its northeastern fringe where depositional structures have a northwest-trend parallel to the old shoreline. Mineralization in this northwest-trending belt is at the 7/10 contact and locally extends up to 50 feet into the 7 zone reef.

The lower 100 feet of Bonneterre in the reef facies zone contains a wide variety of depositional structures; disconformities, overlap features, and pinchout zones, and great lateral variations in lithology. A series of northeast-trending calcarenite bars was developed, over which the algal reefs grew, so the lower 200 feet represents a complex of stratigraphic and reef structures. The bar-reef complexes, sometimes called structural centers (15), were major sites for ore deposition.

Outward from the bar-reef complex the dolomite passes into limestone, in part oolitic but dominantly shaly and fine-grained. There is no mineralization in the limestone, although some limestone "islands" are present in the district and are cut by haulage drifts.

At the close of the deposition of the lower 200 feet of Bonneterre, the control exerted by the irregular basement suface had been largely eliminated. The upper Bonneterre over many hundred square miles is largely tan oolitic dolomite, with only minor variations in lithology and some narrow, irregularly-trending bars.

TYPES AND FORMS OF ORE BODIES　Within the district, ore bodies take many different forms, usually controlled by some sedimentary feature or structure. Many areas show a close fit of ore and sedimentary structure; so that, in mining, it is essentially the sedimentary structure that is mined.

To demonstrate the range in form and character of the different types of ore-containing structures, several will be described. These, with many variations and combinations, contain most of the mineralization in the district. Early descriptions of the district were given by Winslow (1) and Buckley (2). Many of Buckley's sketches of mine faces accurately show details of ore occurrence.

*Pinchout-type*　The pinchout-type of ore body is common where a Precambrian knob stands high enough to cut out the Lamotte and lower Bonneterre. A simple type, illustrated by a few Lead Belt and many Mine La Motte ore bodies, is shown in Figure 3. Mineralization is in the lower Bonneterre in 19, 15, and 12 beds above the Lamotte pinchout line. The ore body partially encircles the buried knob; igneous rock forms one wall of the stope. The ore body is narrow, from 50 to 150 feet wide. Mineralization, in the form of disseminated

and bedding plane ore, may be only a few feet thick away from the knob and up to 40 feet thick close to the knob.

*Variations on Pinchout Type*　One of the important variations of this type occurs where the knob is high enough to cut out 100 feet or more of the Bonneterre. In this case, a variety of lithologies, including reef, are present, and several levels of ore may occur.

In the southwest part of the district a knob cuts out 150 feet of lower Bonneterre. On the north and northwest sides of the knob, mineralization occurs at the sediment/porphyry contact (Bonneterre or Lamotte) in every one of several hundred holes drilled. On other sides of the knob, no mineralization is present at the sediment/porphyry contact in a similar number of drill holes. Mineralization of economic grade occurs on the northwest side of the knob at the Bonneterre/Lamotte contact, the 15/19 contact, in the 15 zone, the 12/15 contact, in the 12 zone, and in the 7 zone where algal reef grew against the knob. There is no mineralization at these levels except on the northwest side. Over the peak of the knob, some 40 feet above the highest porphyry, ore occurs in small slide breccias in the 5 zone.

Mineralization is continuous along the sediment-knob interface; ore bodies represent areas where ore minerals spread along bedding planes outward from the knob. Individual ore bodies parallel the knob, and igneous rock forms one wall of the stope.

A second variation of the pinchout type is revealed at Hayden Creek mine in the western part of the district. Here, the difference is not in position but in type of host rock (8). The ore occurs in granite boulder beds on the flank of a granite knob in lower Bonneterre and Lamotte beds. (Figure 4). Two adjacent knobs are present, each with minera'ized boulder beds on its west side. Descriptions that follow are of relations shown by the eastern knob.

Granite boulders occur as debris from the nearby knob. A thin boulder bed, resting on granite, has sand matrix with Lamotte sand above it. Above the Lamotte is the main boulder bed, up to 100 feet thick, with a matrix of Bonneterre dolomite. Most of the boulders are fresh red granite, although, in some areas of the mine, these show extensive alteration to gray and white kaolinized types. The boulder bed developed during normal sedimentation as accretions of boulders off the knob, rudely stratified, and containing interbedded, undisturbed sediment. It is not a talus pile or gravity slide.

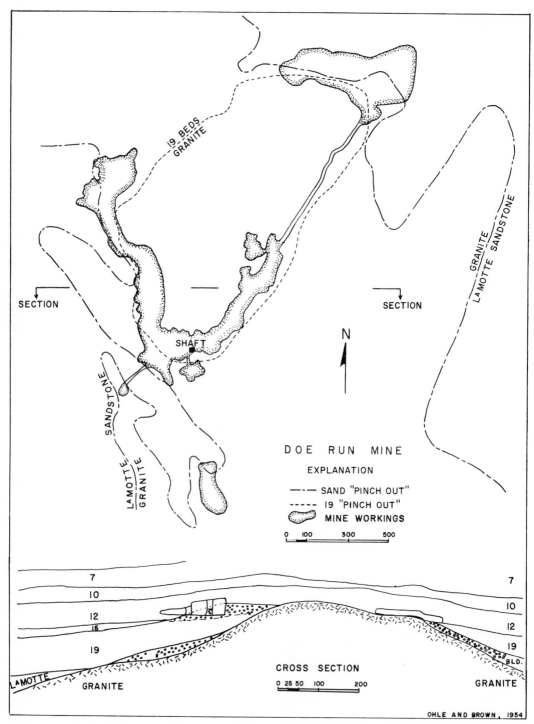

FIG. 3. *Plan and Section of Pinchout-Type Ore Body, Doe Run Mine.*

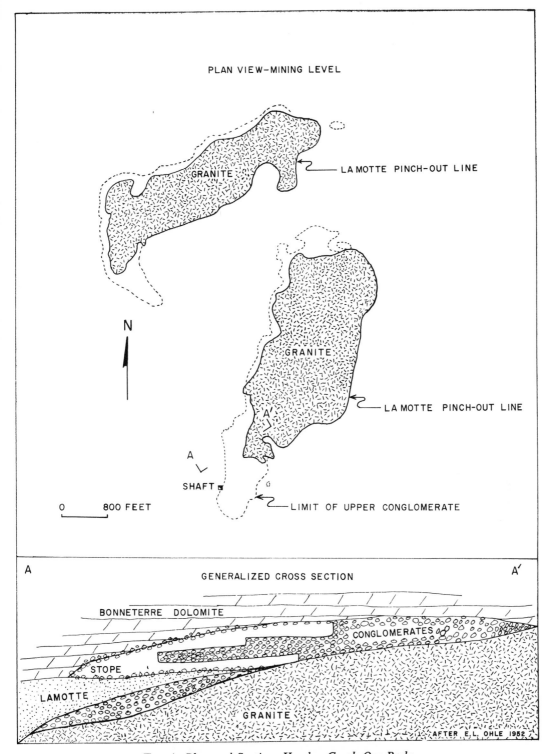

PLAN VIEW—MINING LEVEL

GRANITE

LA MOTTE PINCH-OUT LINE

N

GRANITE

LA MOTTE PINCH-OUT LINE

A'

A

SHAFT

0     800 FEET

LIMIT OF UPPER CONGLOMERATE

A         GENERALIZED CROSS SECTION        A'

BONNETERRE DOLOMITE

CONGLOMERATES

STOPE

LAMOTTE

GRANITE

AFTER E.L. OHLE 1952

FIG. 4. *Plan and Section, Hayden Creek Ore Body.*

Mineralization occurs as replacement of the dolomite matrix and in fractures in boulders. Ore occurs in the lower boulder bed in limited amounts; throughout the entire mineralized area in the lower part of the upper boulder bed; throughout the entire boulder bed thickness at its outer margin where boulders are in contact with 12 zone type Bonneterre; and in boulders at the top of the boulder bed below normal Bonneterre sediments.

An important and revealing feature of the deposit, not commonly seen in the district, is the extent of silicification. Sandstones against the granite knob and sand matrix in the lower boulder bed close to the knob are silicified to quartzite. In the upper boulder bed, the granite boulders in parts, but not all, of the mine are rimmed with black jasperoid.

The Hayden Creek area is the main area where granite boulder beds were mineralized, although some ore occurs in thin granite boulder beds on the flanks of Doe Run and Bonne Terre knobs. Rhyolite porphyry knobs are about as abundant as granite knobs, but instances of extensive mineralization of porphyry boulder areas are unknown.

*The Ridge Structure* A large part of the ore mined from lower Bonneterre in the western part of the district was associated with ridge structures at the 15/19 and 12/15 contacts. The ridge is a bar of coarse-grained sediment that has an anticlinal form. Throughout the district most of the ridge structures have a northeast trend. They may be huge bars with a relief of several tens of feet, a strike length of several thousand feet, and visible dips on the flanks; or they may be small structures a few hundred feet in length that are detected only by isopach maps of units with good elevation control.

Ore may occur on the flank of a ridge, related to pinchouts of certain units and transgressive overlap of succeeding beds; or it may occur over the crest of the ridge, where black shale over the crest grades into thick-bedded, shaly, gray carbonates on the flanks. In some areas, ridges mineralized at the 15/19 or the 12/15 contacts contain the only mineralization present; in other areas, ridges may be a component of the large bar-reef complexes with mineralization at several levels.

Ore bodies, where the only control appears to be the ridge structure and related facies differences, can occur almost anywhere in the section. As stated, they are common at the 15/19 and 12/15 contacts. They also occur in the 12 zone; the 10 zone is a ridge or bar facies overlain by reef, and much of the Lead

Belt 5-zone production is from narrow, sharp ridge structures.

Figure 5 illustrates a compound ridge structure. The lower level of mineralization is at the 15/19 contact on the gently-dipping northwest flank of a 19 ridge. Ore is associated with black shale that locally grades into thin lenticular algal mats. The galena commonly is above and in the upper part of the black shale. Upper level ore is associated with a 12/15 contact. The crest of the 15 ridge lies to the northwest of the upper mineralized area, which itself is on the northwest side of the 19 ridge. Mineralization of the upper structure is restricted to its southeast flank (30). Disseminated mineralization, occasionally of ore grade, is present in the 15 bed; but most of the ore in both structures is along contacts. In the upper level structure, the main ore seam follows the cross-cutting lithologic contact.

A similar structure, with mineralization largely restricted to one level, is Stope 4587, a leading example cited by Tarr (4). This ore body was so high grade that, after mining, the bedding-plane floor was swept with brooms to recover the coarse-grained galena not picked up by shovels.

Mineralization is associated with black shale on the northwest flank of a 15/19 ridge. Down the flank of the ridge the black shale grades into algal mat. The algal mat represents a small fringing reef; the black shale is a lagoonal deposit in a local euxinic environment between the fringing reef and the ridge crest (30).

The stratigraphic section in this stope is a transgressive 12 bed, a thin 15 zone, black shale with galena, and the 19 bed. In the central part of the ore body, the 19 bed bears disseminated galena throughout a thickness of 15 feet. The main ore seam is up to 18 inches thick. In places it is a single massive seam; in places it splits to two and sometimes three seams, but it is continuous over an area some 2500 feet long and up to 600 feet wide. Some features suggest that mineralization stoping and open space filling along bedding planes were important in development of the thick ore seam. Displaced blocks of host rock occur within galena (Figure 6a). The undisturbed counterparts of the bed from which the blocks came may be untouched by mineralization or may be partially replaced by galena but are still in their original position. Calcite-lined cavities within the galena seam also indicate that openings existed along bedding planes.

The transgressive fore-set beds of the 12 zone, above the ore seam, are sharply and abnormally truncated (Figure 6b). Tarr (4)

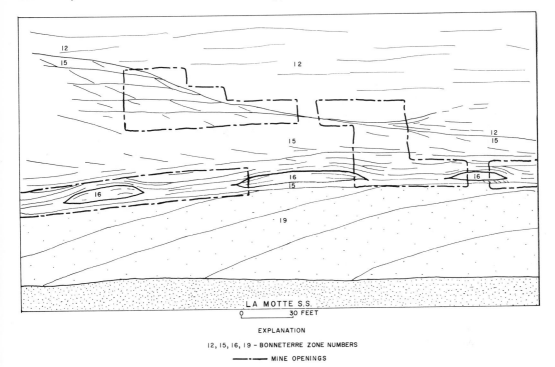

EXPLANATION

12, 15, 16, 19 – BONNETERRE ZONE NUMBERS

—·—— MINE OPENINGS

FIG. 5. *Section Through Compound Ridge Structure, No. 8 Mine.*

interpreted this as a solution effect and regarded the black shale as a stylolitic residue.
The shale is now recognized as a primary sedimentary facies; but solution undoubtedly was
important in developing the abnormally sharp
and ragged dolomite/shale contact.

A widespread pattern of fracturing, in which
small fractures persistently strike northeast
parallel to the stope trend, and dip 30°NW,
indicates pre-ore deformation. The fractures
frequently are filled with galena; the galena
veins extend from the main ore seam upward
for 12 to 15 feet, with ore spreading as thin
seams on higher bedding planes.

*Bar-Reef Complex* The northeast-trending
bar-reef complexes (15) are gigantic compound ridges and represent the main ore structures of the district. The structures range in
size up to 15,000 feet long, 1000 feet wide,
and include the lower 200 feet of the Bonneterre. These features exhibit arch structures
at one or more horizons due to local abundance of carbonate sand that grades into, and
intertongues with, gray, shaly carbonates. Epigenetic minerals may occur at any of the major
bedding plane contacts or disseminated
throughout any of the Bonneterre units except
the tan crystalline dolomite of the 10 zone.

Virtually the only introduced mineral in this
unit is galena filling thin fractures.

Two major types of bar-reef mineralization
are recognized. In one type the lower Bonneterre consists of alternating gray and tan
dolomite units; mineralization occurs on successive contacts of gray over tan carbonate
and disseminated in the gray rock but not at
tan over gray carbonate contacts. Reef mineralization is directly above the bedding plane
and disseminated ore of the lower units. In
such occurrences, the entire 200 feet, the lower
half of the formation, may be mined. Axes
of structures at different stratigraphic positions
may show some divergence or the crest of
the structure may be shifted southeastward at
successively higher levels (11, p. 218); but
essentially the sedimentary features that are
mineralized are vertically stacked to give a
great thickness of almost continuous mineralization.

A second type of bar-reef complex has a
continuous section of unmineralized tan crystalline dolomite below the reef. Lower Bonneterre mineralization occurs at gray over tan
dolomite contacts along the flank of the calcarenite bar (Figure 7). Numerous such contacts close to the bar, where tan units inter-

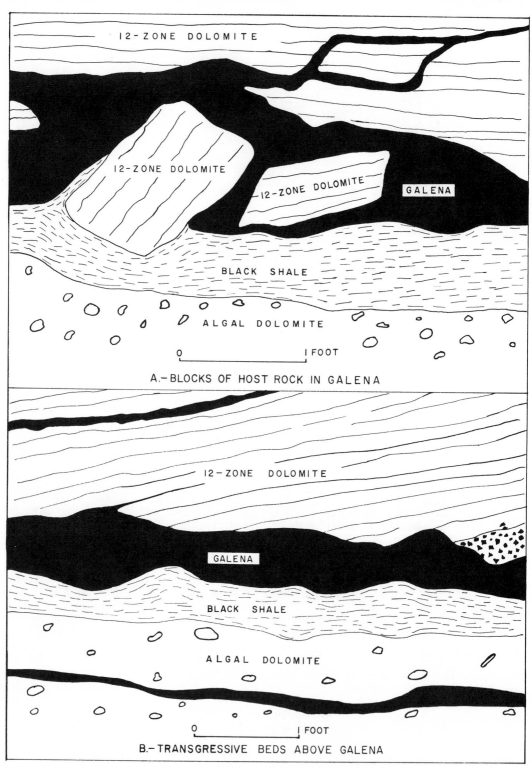

Fig. 6. *Details of Mineralization, 4587 Stope.*

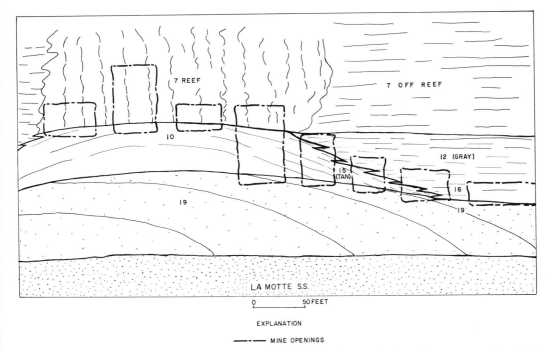

FIG. 7. *Mineralization in Bar-Reef Complex, in Lower Beds on Flank of Structure, and in Reef Over Crest.*

tongue with gray, result in mineralization on closely spaced bedding planes and a zone that was mined from the 19 bed up to the 7/10 contact. Away from the bar, the tan units feather out until only mineralization at the 15/19 unit is present.

Over the crest of the bar the lower part of the reef is mineralized. Internal reef features exerted a strong control over mineralization, so that parts of the reef are mineralized only in the lower 20 feet; parts are mineralized up to 50 feet into the reef. Ore within the reef is variable in grade due to the influence of internal reef features.

*The Algal Reef* As used in the Lead Belt, the term *reef* means a major mass showing organic structures that contain entrapped and interbedded sediment. Where best developed, the reef mass is roughly comb-shaped in plan with a nearly vertical northwest wall and a trailing southeast edge (Figure 8). Over much of the district, calcarenite bars provided the reef foundation; in a few places igneous knobs not buried by 7-bed time provided the foundation, and the reef grew directly on the igneous knob.

The major ore-controlling structure within the reef is the contact between the superimposed colonial algal structures and clastic carbonates. The narrow, convex algal zone, usu-

ally trending at right angles to the direction of reef elongation, is termed a roll. The contact zone of organic and clastic sediments, containing abundant black shale, invariably is more strongly mineralized than either the organic rock or the clastic carbonate away from the contact. Where rolls are narrow and the better grade zones are closely spaced, the entire reef may be mined; where roll and inter-roll zones are wide, roll edges may be mined as individual ore bodies transverse to the main reef trend.

The ore minerals occur along bedding-plane and growth-line contacts, in fracture zones, and disseminated in the organic carbonates. Vast amounts of reef rock are unmineralized. In the western part of the Old Lead Belt, unmineralized reef structures occur above bedded ore. In the eastern part of the district, mineralized reefs are known with no bedded ore below them.

To the southwest, the several reefs that are a part of the bar-reef complexes merge into a broad reef meadow that, in turn, grades into the back-reef facies. Within the reef meadow, mineralization is largely of the disconformity type.

*Disconformities* Minor disconformities within the Bonneterre are the loci for blanket-type mineralization and, apparently, provided avenues for widespread lateral migration of

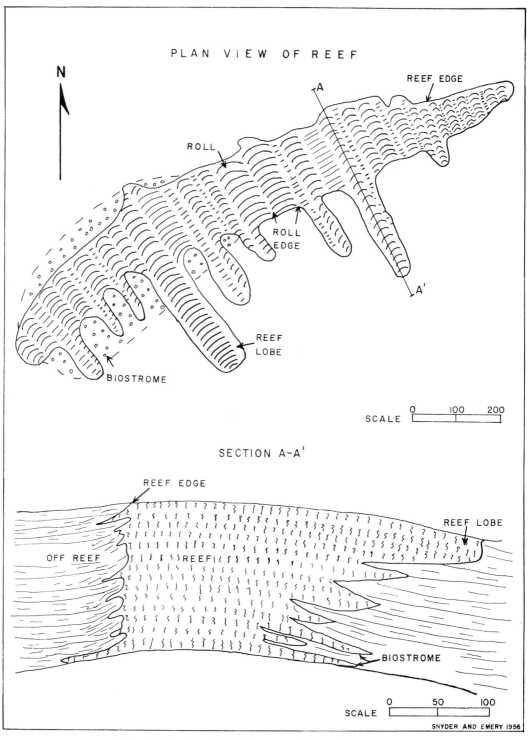

FIG. 8. *Details of Reef Structure, Plan and Section.*

ore solutions. On the southwest side of the district, along the edge of the back-reef zone, persistent traces of ore minerals occur at the 7/10 contact. Mineralization at this position becomes ore grade where slight relief on the disconformity simulates a ridge or domal structure. The shale bed present over the topographic rise is lacking or less pronounced in adjacent lower areas. Beds above and below the disconformity usually show some discordance, although the over-all thickness of the Bonneterre is constant.

*Submarine Slides* Breccia bodies formed by submarine gravity sliding along the flanks of calcarenite bars are quite common throughout southeast Missouri and are important ore structures in the district (18).

Gravity sliding occurred repeatedly, with a stratigraphically higher slide frequently intersecting and truncating a lower one. Continuous ore zones in compound slides have been mined that are up to 6000 feet long and 160 feet in height. The ore bodies are long and narrow and parallel a calcarenite bar and often are continuous around the nose of the bar.

The slides are mineralized only where the breccia mass has intersected one of the major contacts. Mineralization is restricted to the contact and the piled up mass at the basin margin. The sloping up-dip part of the slide rarely is mineralized. The glide plane, along which the mass moved, may be seen passing into the back of the stope, but mineralization continues below the glide plane as contact ore only (Figure 9).

## Fracture Control of Mineralization

DISTRICT FAULT PATTERN Earlier publications on the district have given a preponderance of attention to the control exerted by primary sedimentary features. As described above, the ore occurs in sedimentary structures, and mining essentially follows the sedimentary structure. However, faulting played a very important role and is probably the major factor controlling differences in grade of ore and interruptions in the pattern of mineralization within the sedimentary structures.

A major fault, known as the Federal-Schultz fault system, transgresses the district. (Figure 10). This is one of a series of related faults that ring the Precambrian outcrop. Displacements along the main segments of the Federal-Schultz system range from 40 to 100 feet, downthrown to the northeast. Most of the major faults show considerable left lateral movement, marked by the offset of ridge axes

and the low angles of slickensides. On one minor fault, a lateral to vertical ratio of 8:1 was measured by correlation of specific features in mine workings on opposite sides of the fault; probably on all faults within the district, the lateral component is greater than the vertical.

Where the strike of the major fault zone is northwest, the system consists of an overlapping series of en echelon faults, with the main displacement passing from one to another of successive segments. Where the major fault zone swings to an east-west trend, as it does in the central part of the district, intense horse-tailing was developed.

A strong northeast-trending fault system, the Ashbank system, is present in the eastern part of the district. Displacement is down on the northwest side. The graben between the Federal and Ashbank systems is one of the main areas of copper mineralization.

Jointing, commonly with N50°W trend, is widespread. Some joints are healed and appear to be early; some are filled with galena, calcite, and marcasite; some with calcite and marcasite only; some are open, having been enlarged by post-ore groundwater action. Northeast-trending, galena-filled fractures are common in mineralized areas. In calcarenite bars below mineralized reefs, they often provide low-grade ore that can be mined with the higher-grade contact and reef ore above.

Some of the fracturing and faulting clearly are post-ore. Individual ore bodies may be displaced by faults with no perceptible difference in mineralization on opposite sides of the fault. Galena with curved cleavage surfaces, representing slightly deformed crystals, is common.

However, much, probably most, of the fracturing occurred before the cessation of mineralization. Some faults are filled with seams of galena, up to three inches thick, that also may be deformed, indicating both pre-ore and post-ore movement.

In any given part of the district, the age of faulting relative to mineralization can be established. Evidence will be presented later indicating that mineralization took place over a long period of time. We believe that faulting was initiated in the early stages of mineralization; continued—probably intermittently—during mineralization; and continued, at least to a minor extent, after the main period of galena deposition.

GROUND PREPARATION The most important role played by faulting is *ground preparation,* intense fracturing that permitted easy move-

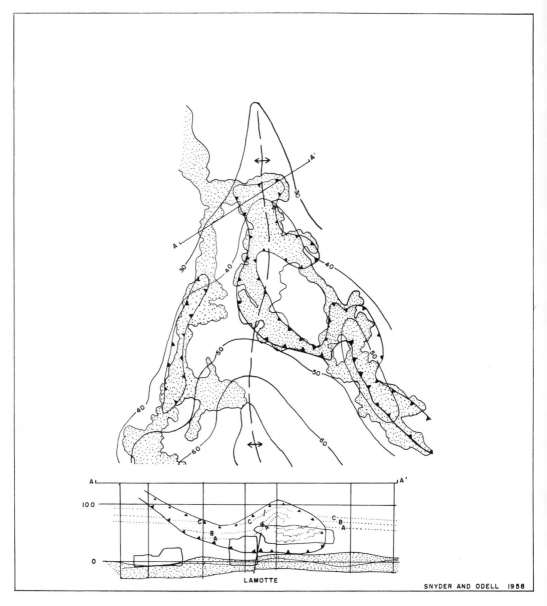

SNYDER AND ODELL 1958

EXPLANATION

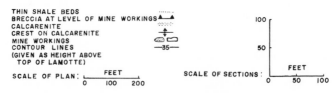

THIN SHALE BEDS
BRECCIA AT LEVEL OF MINE WORKINGS
CALCARENITE
CREST ON CALCARENITE
MINE WORKINGS
CONTOUR LINES
(GIVEN AS HEIGHT ABOVE
 TOP OF LAMOTTE)

SCALE OF PLAN:   FEET
                0   100   200

SCALE OF SECTIONS:   FEET
                     0   50   100

FIG. 9. *Submarine Slide Ore Body, Leadwood.*

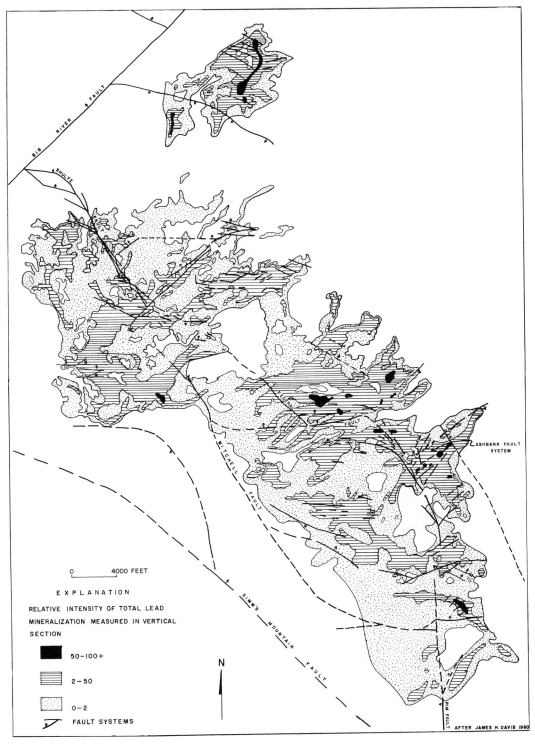

FIG. 10. *Fault System and Intensity of Mineralization.*

ment by solutions through the host rock structures and provided access to lithologic types not normally mineralized.

One type of ground preparation is shown by a ridge that is intersected by the northwest segment of the Federal fault, at a point where displacement on the fault was transformed to the adjacent en echelon segment. The two fault segments are about 100 feet apart; vertical displacement is 40 feet. The area between the segments and for some distance beyond the fault is intensely broken by east-west tear fractures. Within this intensely broken area the height of mining is approximately 40 feet; the width of mining is approximately 300 feet; the grade of ore, calculated from drillhole assays and pillar exposures, is 8 to 10 times minimum mine grade. Away from the faulted area (to the southwest), along the same sedimentary structure, height-of-mining and width-of-stope gradually diminish. At a distance of 1200 feet from the fault, height-of-mining is 10 feet; width-of-mining is 150 feet; the grade of ore is minimum mine grade. The ore body in the "prepared" zone had a cross section area of 12,000 square feet; in the normal unfaulted sedimentary structure, 1500 square feet. The vast differences in cross sectional area and in ore grade indicate the role

of faulting in mineralization of the sedimentary structures.

A second type of ground preparation is that in the central part of the Federal area where the Federal fault trends east-west. The fault horsetails into a braided pattern. This strongly fractured zone, approximately 1000 feet in width, extends across three parallel, northeast-trending bar-reef complexes that normally are separated by 600 to 800 feet of unmineralized "basin" sediments (Figure 11). Outside of the fault zone, each of these sedimentary complexes has its normal types of mineralized sedimentary structures. Within the faulted area, mineralization extends as a broad and rich WNW-ESE belt that is a combination of sedimentary trends and fault trends. Mineralization spreads across the bar-reef complexes and the intervening areas; again, it is of far better grade than the district average.

A third type of ground preparation is attributed to bedding-plane movement. Within the district, there are three situations in which normal bedded carbonates, often thin-bedded and shaly, pass into massive structureless rocks. These are the reef-bedded rock contact, the slide breccia-bedded rock contact, and the boulder-bedded rock contact. In each case, the massive unit contains high-angle fractures dip-

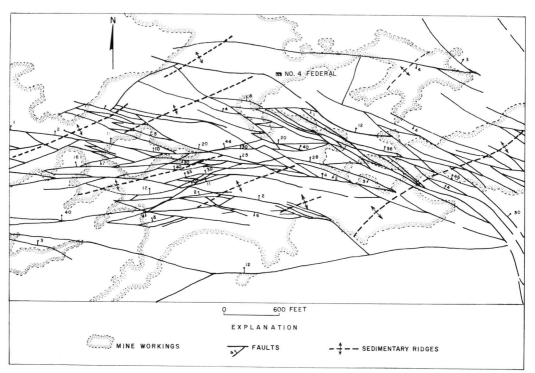

FIG. 11. *Braided Fault Pattern and Mined Areas.*

ping toward the bedded rock contact. Within the convex reef structures, fractures start on an algal growth surface but continue without change in attitude when the growth surface curves. In the slide breccias, the outer (basin) margin of the mass shows repeated fractures not present in adjacent bedded rock. In the boulder type of ore body, the boulders are abundantly fractured close to the outer contact, and the outer edge of the boulder mass is the only area that is mineralized throughout the full boulder height (Figure 4).

In these instances, it is postulated that slipping along innumerable bedding planes, probably common in slightly deformed rocks but difficult to detect, was translated into high-angle fracturing wherever the bedding plane movement was interrupted. In reef, slide breccia, the boulder bed, the fractures cut across primary features, providing easy access for mineralizing solutions. Again, the result is a greater than usual height of mineralization and a higher-grade ore.

## Discordance of Mineralization on Opposite Sides of Fault

Contouring intensity of mineralization, in terms of top of mineralization and base of mineralization, and isopaching the vertical interval of mineralization are useful tools in deciphering mineralization-fault history. Where ore is pre-faulting, mineralization patterns on opposite sides of the fault are virtually identical. The same structures are mineralized; ore is at the same stratigraphic position on both sides of the fault; intensity of mineralization is the same.

Where faulting is pre-ore, the base of mineralization may be at different stratigraphic positions on opposite sides of the fault; the top of mineralization very commonly is at different stratigraphic positions; the intensity of mineralization may be profoundly different. Maps showing intensity of mineralization show numerous instances of interruptions in the normal pattern due to control exerted by pre-ore faulting.

## Mineralogy of the Deposits

PRIMARY SULFIDES  The primary sulfide minerals include galena, sphalerite, chalcopyrite, siegenite, bravoite, pyrite, marcasite, and possibly bornite and millerite.

Galena occurs as bedded or sheet deposits along disconformities and bedding planes, in part as replacements but locally as open space fillings; as disseminated crystals and crystalline aggregates in several types of dolomite and black shales; and as fracture fillings. Massive bedding seams of solid galena, up to 18 inches thick, are a mosaic of very fine- to coarse-grained crystals, frequently with a layered appearance due to alternately fine and coarse sulfides. Curved cleavage planes are common. In a few areas, galena along bedding planes is thoroughly crushed and smeared. However, uncrushed and undeformed galena also is abundant.

Well-formed euhedral crystals, more than 1 inch in diameter are rare in the Old Lead Belt, although some occur in small solution cavities and in fractures. These usually are simple cubes or cubes modified by octahedral faces. Galena occurring in Lamotte Sandstone or in lower Bonneterre sandy dolomite may occur as "sand-lead" crystals in which galena filled the pores between quartz grains or replaced the dolomite matrix in the sandy Bonneterre.

Sphalerite, although a minor constituent for the district as a whole, may be abundant locally. It is usually dark in color, fine-grained, and occurs as disseminations in the dolomite host. Davis (23) reports that many sphalerite crystals have alternating wedges that appear yellow and purple in transmitted light, presumably, a twinning effect. Euhedral crystals are rare and, when found coating fractures, are seldom more than ⅛ inch in diameter.

The sphalerite is restricted to certain parts of the district (Figure 12), and in these areas commonly is associated with gray shaly dolomite and black shale. In this respect, it appears to be much more closely tied to a particular host rock facies than is galena. The sphalerite usually occurs with galena but invariably is with or above the stratigraphically highest galena mineralization.

Chalcopyrite is abundant in the eastern part of the district and is rare in the western part (Figure 12). The copper mineral occurs as disseminations in nearly all Bonneterre lithologies and as thin seams along bedding planes. Where it occurs, chalcopyrite usually is with or below the stratigraphically lowest galena; its most common occurrence is in lower Bonneterre (14). Near the junction of the Federal and Ashbank fault systems, copper mineralization occurs with galena throughout a stratigraphic interval of 200 feet. This is the only area known in the Old Lead Belt in which copper mineralization penetrates more than a few 10's of feet above the Lamotte Sandstone.

The cobalt-nickel sulfide, siegenite [(Co,-

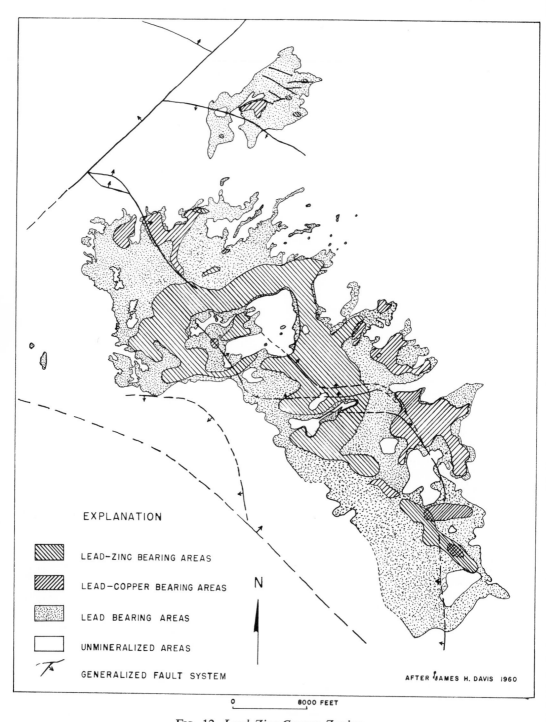

EXPLANATION

LEAD–ZINC BEARING AREAS

LEAD–COPPER BEARING AREAS

LEAD BEARING AREAS

UNMINERALIZED AREAS

GENERALIZED FAULT SYSTEM

N

AFTER JAMES H. DAVIS 1960

0       8000 FEET

FIG. 12. *Lead Zinc-Copper Zoning.*

Ni)$_3$S$_4$] is present in minor amounts, usually as small disseminated crystals. In a few areas in the eastern and southern part of the district, it occurs as seams up to 3 inches thick along bedding planes. The siegenite is present in many copper-bearing specimens, and its district occurrence is closely related to that of copper. In some instances, siegenite is associated with pyrite rather than chalcopyrite, but its occurrence is restricted to the copper-bearing area.

Bravoite [(Fe,Ni)S$_2$] is present in very small amounts. Bravoite was discovered in Missouri ore by Rasor (5), who stated that this was the first reported occurrence of this mineral in North America. The mineral occurs as minute crystals in, and thin coatings on, pyrite and also replaces siegenite (see excellent photographs in 5, pp. 403–405).

Pyrite is associated with all ore minerals. Polished sections of iron sulfides frequently show the material to be an intimate mixture of pyrite and marcasite. Pyrite from nickel-bearing areas commonly contains a small amount of nickel. This may be due to the presence of bravoite in minute grains or may be nickel present in solid solution.

Marcasite is abundant. It occurs intimately mixed with all ore minerals and also as massive aggregates without associated ore minerals.

Gray shaly carbonate beds of the Bonneterre owe their color to finely divided iron sulfide; fossil relicts often have a thin coating of iron sulfide; black shales occurring as thin beds in the lower Bonneterre or along algal features in reef contain abundant iron sulfide. The iron sulfides in these situations are widespread throughout southern Missouri; they appear to be of diagenetic origin and unrelated to the lead-zinc-copper mineralization.

The only minor elements of economic importance are cadmium and silver. Examination of many polished sections has not revealed any visible cadmium or silver minerals. The elements are completely contained in solid solution in other sulfides, cadmium in sphalerite and silver in both galena and sphalerite. Sphalerite, on the average, contains about 1 per cent cadmium. However, the actual amount present differs from one mine area to another. Davis (23) reported that zinc-cadmium ratios ranged from 38:1 to 110:1 and found that the highest cadmium values occurred near the main central fault system and decreased gradually outward from the fault.

Silver is contained in both sphalerite and galena but is far more abundant in the sphalerite. Davis (23) reported the silver-zinc ratio

TABLE II.    *Estimated Metal Ratios, Old Lead Belt*

| | |
|---|---|
| Lead | 25,000 |
| Zinc | 1,500 |
| Copper | 500 |
| Nickel | 40 |
| Cobalt | 20 |
| Cadmium | 10 |
| Silver | 1 |

in sphalerite is approximately 20 times the silver-lead ratio in galena. Total silver content in the ore differs within the Lead Belt and apparently follows a pattern similar to that for cadmium in its relationship to the central fault system.

Very minute amounts of antimony, arsenic, and bismuth are reported in concentrates, but they are so minor in amount that the ore is essentially free from these elements.

METAL RATIOS In a few individual ore bodies, ratios of zinc or copper to lead may be 1:1 or greater, but the district is predominantly a lead district. The district map showing mineral zoning (Figure 12) reveals that large areas contain no visible zinc or copper.

Estimated metal ratios, given in terms of multiples of the least abundant metal, silver, are shown in Table II. Kilburn (24), in a study using mill concentrates sampled during one week's production, had approximately similar results for zinc, copper, nickel, and cobalt, but his lead figures were unrealistically low.

INTENSITY OF MINERALIZATION To correlate areas of intense mineralization with geologic features, drill hole data and mine estimates of ore grade were compiled for the Old Lead Belt (Figure 10).

The intensity map is an isopach map showing all mineralization in a vertical section regardless of the stratigraphic interval embraced. Where several mineralized zones with intervening barren zones are present, the mineralized zones are combined to give a single figure. The map was developed using feet-per cent mineralization; for example 20 feet per cent could consist of 20 feet of 1.0 per cent lead, 10 feet of 2.0 per cent, 2 feet of 10.0 per cent, or any possible combination of thickness and grade to arrive at the total.

The general correlation of intensity of mineralization with the bar-reef complexes is apparent. Equally apparent is an inconsistency of the mineralization pattern with primary sedimentary features. Some basement knobs show intense mineralization in surrounding sediments; others do not. Several areas show that mineralization is most intense in strongly faulted zones. Total intensity along a given sedimentary structure may be distinctly different on opposite sides of a fault.

THE GANGUE MINERALS The only gangue material of volumetric importance is the host dolomite. Introduced (or recrystallized) gangue minerals, minor in amount, include calcite, dolomite, dickite, quartz, and jasperoid.

Crystal-lined vugs and cavities are rare in the Old Lead Belt, compared to other major districts. Well-formed crystals of calcite are quite common in the occasional vugs; white to yellow-brown crystalline calcite occurs in aggregates and masses often a foot or more in diameter. Several stages of calcite growth, with intervening marcasite, can be distinguished, but no detailed study of the calcite zoning has been made.

White, gangue dolomite is rare. Over large parts of the district none can be seen. In a few cases, as at Hayden Creek mine, it is rather abundant.

The Lead Belt is unique among mid-continent base metal districts in its apparent lack of silicification. The Bonneterre Formation is chert-free, and, in most drill core and most mine workings, no silicification can be seen. However, silicification on a minor scale occurs in a few geologic situations. Some ore bodies in the sandy transition zone of the Bonneterre show thin beds that have been silicified. Fault zones cutting lower Bonneterre and Lamotte beds may be silicified, and, in places, the normally friable Lamotte Sandstone becomes a quartzite in and near the fault.

The most striking example of silicification is the development of jasperoid in the Hayden Creek ore body. This ore body, in granite boulder beds of early Bonneterre age on the flank of a Precambrian knob, contains extensive areas in which granite boulders are rimmed with black jasperoid up to one inch thick. The jasperoid contains abundant fine pyrite. It has a columnar texture radiating from the boulder-jasperoid contact.

The lack of silicification in most of the host rock and its occurrence in sandy beds or near igneous knobs suggest that the silica was derived from adjacent siliceous rocks and transported only a very short distance before being redeposited.

WALL-ROCK ALTERATION Alteration consists of the minor silicification described above and

dissolution and oxidation of the carbonate host. Dolomitization is regarded as diagenetic; there is no evidence to suggest that post-lithification dolomitization occurred or that limestone-dolomite relationships are genetically related to sulfide mineralization.

The history of dissolution and oxidation is complex and appears to have occurred in several stages. Specific features that are evidence of solution activity include:

(1) Vugs, from a fraction of an inch up to several feet in diameter, usually aligned along some primary features such as internal reef structures, are present. The dolomite surrounding the vug is slightly oxidized and often "sanded".* Vugs in mineralized areas contain, but are rarely lined with, galena and calcite crystals.

(2) Certain features suggest that some thick seams of galena may have formed as open space filling. The seams often contain vugs in the middle of the seam that are partially filled with calcite.

Blocks of dolomite at various attitudes are suspended in galena in thick ore seams as shown in Figure 6A. The open space was developed prior to or during mineralization; the presence of host rock blocks suspended in ore suggests that the process was concurrent with mineralization.

(3) Solution collapse breccias are present in two areas in the district, a large one in the Leadwood area, a smaller one in the extreme eastern part of the Flat River area. Both are in mineralized areas and are almost completely surrounded by ore, but the collapse breccias are unmineralized.

The breccias contain abundant calcite and marcasite but no ore minerals. It seems unlikely that the breccias would be barren if they had been present at the time of mineralization. They appear to have formed during the final stages of calcite-marcasite deposition, although they could be later than, and unrelated to, the main sulfide mineralization process.

(4) Fractures crossing ore bodies and unmineralized areas alike are enlarged by solution activity and have oxidized walls. When first cut by mine workings, they are filled with water and mud, often under high pressure. In mineralized areas, the fractures contain cerussite, anglesite, aragonite, manganese oxides, and strongly oxidized host rock dolomite.

---

* The sanding process is generally recognized as partial dissolution in which some carbonate is removed, the carbonate grains remaining form an incoherent or loosely-bonded carbonate "sand".

Deeper ore bodies, particularly those covered by several formations younger than the Bonneterre, do not show this type of oxidation and enlargement of fractures. This type of solution and oxidation is a result of groundwater action related to the present erosion surface.

(5) Within ore bodies, an oxidation of gray carbonate beds is observed that provides one of the most puzzling and debatable features of the district. The oxidized areas are almost barren of mineralization; the adjacent unoxidized beds of the same unit may be richly mineralized. The senior author has mapped (No. 12 mine) a 10-inch black shale band containing 20 per cent lead passing within a foot into oxidized (green) shale with no lead. The shale, and similar oxidized dolomite zones, show no porosity differences from unoxidized rock that would indicate removal of galena. Galena adjacent to the oxidized rock shows no oxidation effects similar to those caused by post-ore groundwater action described above.

It seems possible that the Bonneterre Formation was subjected to a period of groundwater oxidation, related to some earlier erosion surface, prior to mineralization. The extent of oxidation is minor; but, apparently, some areas of gray, shaly, iron sulfide-rich Bonneterre were oxidized; the oxidized Bonneterre constituted an environment unfavorable for mineralization, and the rock is barren.

The events in the solution process may be listed as follows:

(1) A possible pre-ore period of groundwater solution and oxidation, either related to the post-Eminence or a later erosion surface.

(2) Solution and stoping, with the development of openings (but without oxidation) during the process of mineralization.

(3) Solution and development of collapse breccias without oxidation. This could have occurred in a late stage of mineralization history and have been a continuation of the mineralization stoping process.

(4) Solution and oxidation by groundwater related to the present erosion surface.

SECONDARY MINERALS   Secondary minerals are quite unimportant in the district and are of no economic importance, except in minor milling losses of oxidized lead. Galena in oxidized parts of ore bodies, or in the residuum over shallow ore bodies, commonly has a film of cerussite coating the sulfide. Minor amounts of anglesite have been recognized.

Chalcopyrite in oxidized areas may be altered to bornite, chalcocite, covellite, and malachite. Native copper has been observed in minor amounts in the Lamotte Sandstone and is regarded as an alteration product.

No oxidized zinc ores have been recorded. Aragonite, probably secondary, is present in some oxidized fracture zones.

ISOTOPIC DATA AND MINERAL ZONING   In 1955, J. L. Kulp initiated a comprehensive study of lead and sulfur isotopes of the district. The research was supported by a Geological Society of America grant, the Lamont Geological Observatory, and the St. Joseph Lead Company. J. S. Brown and F. G. Snyder planned the sampling program and provided geologic data; F. D. Eckelmann, assisted by St. Joseph Lead Company staff geologists, collected the samples. Eckelmann supervised the lead-isotope determinations and their interpretation and is senior author of an unpublished report of the work (17); W. U. Ault and D. S. Miller performed the sulfur-isotope study (13,17,19, 20,21,22).

Approximately 250 samples were taken from all geologic situations in the Southeast Missouri district, most of the samples from the Lead Belt, Mine La Motte, and Indian Creek mines but with representation from vein deposits in Precambrian igneous rocks and from vein deposits and barite-galena deposits in the Potosi and Eminence Formations. The isotopic determinations and the geologic data necessary for their interpretation are too voluminous to include in this report. The data shed light on some problems but, in many respects, are open to conflicting interpretations. No attempt will be made to review the data completely; only those features that appear to shed light on the history of mineralization will be cited.

Ordinary lead consists of four stable isotopes; lead 204, 206, 207, and 208. Lead 204 is considered primordial lead; lead 206 is a disintegration product of uranium 238; lead 207 of uranium 235; and lead 208 of thorium 232. Lead composition is changed with time by additions of the radioactive decay products; the radiogenic character of lead usually is expressed in terms of the ratio of lead 206 to 204. The Pb 206/204 composition of modern ocean lead, given by Patterson (10), is 18.45. Lead occurring in the Southeast Missouri district, like that in other mid-continent districts, is more radiogenic than modern lead and is termed J type after the Joplin (Tri-State) occurrence.

Southeast Missouri lead, occurring in sediments from Bonneterre to Eminence age,

TABLE III.   *Comparison of Galenas Relative to Faulting*
(Data from Brown and Snyder, 21)

| Galena | Pb 206/204 | | S 32/34 |
|---|---|---|---|
| Older, 5 samples | Minimum | 20.72 | 21.81 |
| (deformed) | Maximum | 21.23 | 22.11 |
| | Average | 20.95 | 21.94 |
| | | | |
| Younger, 7 samples | Minimum | 21.21 | 22.03 |
| (undeformed) | Maximum | 21.82 | 22.65 |
| | Average | 21.48 | 22.19 |
| | | | |
| Neutral, 1 pair | Bedded | 21.07 | 22.61 |
| | Fracture | 21.02 | 22.69 |

ranges from 19.56 to 22.41 Pb 206/204 with an average value for 172 samples of 20.99. Samples from the Bonneterre Formation show a range from 19.56 to 22.08 Pb 206/204. Lead in the younger formations is more radiogenic than lead in the Bonneterre, but its range overlaps the range of Lead Belt ores.

The range in isotopic character is believed to represent a change in composition of the metal with time. Actual time span represented by the isotopic range cannot be estimated, but ranges in composition relative to certain geologic events can be determined.

The only geologic situation that offers a specific dating point in history of mineralization is the occurrence of "galena pairs", consisting of crushed and deformed bedded galena near a fault and undeformed galena crystals in the fault. The deformed bedded ore is geologically older than the last movement on the fault; the undeformed crystals in the fault are younger. Radiogenic comparison of geologically older and younger galenas is shown in Table III.

The average value for the deformed galenas, 20.95, is virtually the same as the average for all galenas, 20.99, and indicates the deformed bedded ore is representative for the district.

The neutral pair, with similar composition for bedded and fracture galena, is only slightly more radiogenic and, therefore, younger than the deformed galena. The fact that bedded ore and galena in a nearby fracture are virtually identical in composition suggests that mineralization and fracturing occurred at the same time in that area.

The district-wide pattern of isotopic variation is complex. Generally, the lower 206/204 values occur along the fringes of the mineralized areas. Samples with 206/204 ratios of less than 20.2 occur on the eastern edge of the Flat River area, the northwest edge of the Leadwood area, and the southwest edge of the Bonneterre area. Slightly more radiogenic lead occurs in several parts of the district but is most abundant in the eastern part and coincides closely with copper occurrence. Lead comparable in isotopic character to the deformed galenas cited in Table III occurs over large parts of the district; it roughly parallels the central fault zone; and it coincides closely with the pattern of zinc distribution. The most radiogenic leads occur along the southwest side of the district and in outlying areas.

Inferences drawn from the isotopic data are: (1) the early mineralization occurred on the fringes and eastern part of the district and was prior to the main period of faulting; (2) a large part of the mineralization took place during the main period of faulting, with much of it occurring prior to the last fault movement; and (3) some mineralization was subsequent to the main period of faulting.

Further data on the trend of radiogenic character with time is given by several determinations made on a single galena crystal from Indian Creek mine. The crystal grew in a solution cavity; had overgrown other smaller crystals in the cavity; and obviously is younger than the overgrown crystals. Sulfur ratios from this crystal were reported by Ault (19), who stated, p. 252–253: "The wall contacts and small cubes have low sulfur ratios and are considered to have been formed early. Extremities of growth (last forming free edges and surfaces) gave higher ratios, and a profile diagonally across the cube showed a consistent trend or compositional variation of 0.75 per cent." Data reported by Ault (19) and data for the same crystal but different determinations, by Eckelmann (17) are given in Table IV.

For the Indian Creek ore body, 13 lead isotope determinations were made, with Pb 206/204 ranging from 20.43 to 21.91 and averaging 21.26. The range within the single crystal from 20.71 to 21.91 encompasses most of the range shown for the entire large ore body and indicates the crystal was growing throughout most of the mineralization period.

The isotopic data provide a simple and logical answer to the problem of lead-zinc-copper zoning (Figure 12). As stated earlier, copper usually occurs with or below the stratigraphically lowest lead; zinc with or above the stratigraphically highest lead. Copper is largely restricted to the eastern part of the district; zinc

TABLE IV.   *Isotopic Character of Indian Creek Galena Crystal*

| Location of Sample | Ault (19) | Eckelmann (17) | | Location of Sample |
|---|---|---|---|---|
| | S 32/34 | S 32/34 | Pb 206/204 | |
| Wall contacts, 3 samples | Min.  21.98 | 21.98 | 20.71 | Small overgrown crystals, 2 samples |
| | Max.  22.06 | 21.99 | 21.44 | |
| | Av.   22.01 | 21.98 | 21.08 | |
| Extremities, 6 samples | Min.  22.20 | 22.10 | 21.25 | Large crystal over small ones, 6 samples |
| | Max.  22.24 | 22.23 | 21.91 | |
| | Av.   22.22 | 22.18 | 21.58 | |

to the central and western part. In the few places where zinc and copper overlap in plan view, they are separated by a considerable vertical interval. Distribution of the metals relative to radiogenic character of associated lead is shown in Table V.

The problem of zoning in Mississippi Valley type districts has been a puzzling one in terms of temperature-pressure-depth controls. If, as seems apparent in this district, the range in radiogenic character of lead is a function of time, the metal zoning is a time zoning, unrelated to temperature, pressure, or sedimentary environment.

Determinations on sulfur isotopes were made for 124 samples, part of the same samples for which lead determinations were made. S 32:34 ratios vary from 21.60 to 22.69 and average 21.94. It is difficult to find any clear cut pattern for sulfur, although in a general way heavy sulfur appears to be early and lighter sulfur later. The pattern is not consistent either laterally along bedding planes or vertically in a given section within any mine; although a study of a single ore body gave consistent results, with five galenas showing an S 32:34 range from 21.78 to 21.89 and seven associated marcasites showing a range from 21.75 to 21.80 (22).

The sulfur data do not indicate whether the sulfur was introduced, that is brought in with the lead, or whether it was present in the sediments when the metals were introduced. The wide S 32:34 range in the Southeast Missouri samples is close to that given by Ault (19) for native sulfur of biogenic origin, 21.73 to 22.57. It also is encompassed within the reported range for sulfides in sedimentary rocks in U.S. and Canada, 21.58 to 23.20 (19). The range is considerably greater and the average heavier than for a number of other mining districts (22).

PARAGENESIS Paragenetic studies of the Southeast Missouri ores have been made by Tarr (4) and Davis (16,23).

Davis (23), using veining relationships of one sulfide in another, reported the sequence, beginning with the first formed, as: pyrite and marcasite, sphalerite, siegenite, galena, and chalcopyrite, with a galena stage later than chalcopyrite and some later marcasite. Tarr (4) reported sphalerite and galena as later than chalcopyrite.

The dominant factor, we believe, in paragenetic history is the wide range shown in radiogenic character of total lead and the restricted range in radiogenic character of lead with either sphalerite or chalcopyrite. If the radiogenic character of lead has a time significance in this district, lead was deposited throughout a long period of time, although possibly intermittently. Both chalcopyrite and sphalerite were deposited within a shorter span of time, with chalcopyrite being earlier than sphalerite. Both chalcopyrite and sphalerite could vein or replace galena or be veined or replaced by galena.

The paragenetic sequence as we interpret it is given in Figure 13. Minute crystals of chalcopyrite occasionally are seen coating all other minerals, including marcasite. The exact relationship of this late stage chalcopyrite to the main sulfide mineralization has not been determined.

TABLE V.   *Radiogenic Character, Lead with Copper and Zinc*

| | Minimum | Maximum | Average |
|---|---|---|---|
| Pb 206/204 with Cu | 19.77 | 20.91 | 20.54 |
| Pb 206/204 with Zn | 20.53 | 21.33 | 20.98 |

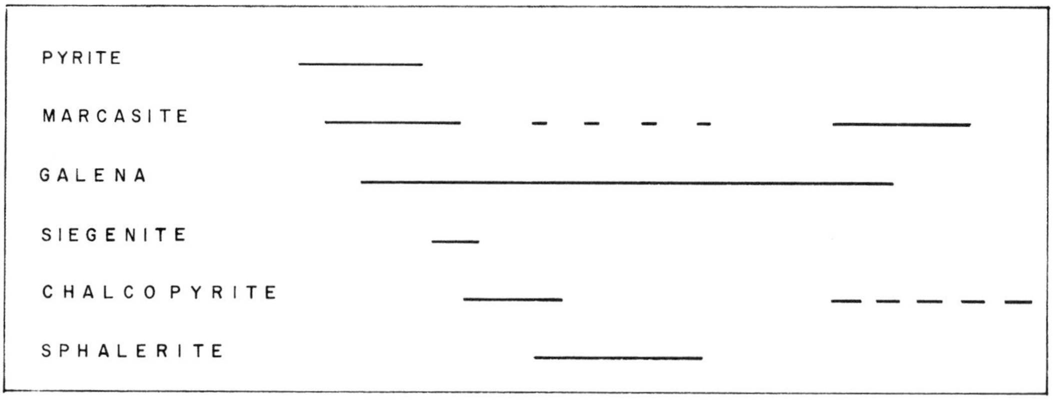

| | | | |
|---|---|---|---|
| PYRITE | ———— | | |
| MARCASITE | ———— | — — — — | ———— |
| GALENA | ———————————————— | | |
| SIEGENITE | — | | |
| CHALCOPYRITE | ———— | | — — — — — |
| SPHALERITE | ———— | | |

FIG. 13. *Paragenetic Sequence.*

## ORE GENESIS

The lead ore bodies of the Southeast Missouri district belong to the group of low temperature, sometimes conformable, deposits, lacking in any obvious igneous source, the origin of which long has been a subject of controversy.

Theories that have been proposed to explain the origin of these deposits include deriving the metals: from a deep-seated basic igneous source; from an unknown granitic magma; by "sweating out" from a heated, but unmobilized, basement rock; by lateral secretion; by artesian circulation; and by connate waters from nearby sedimentary basins. These theories call for the metals to be introduced after deposition of the host formation, and the deposits would be epigenetic. Theories that call for the metals to be deposited with the host rock sediments, and therefore syngenetic, include deriving the metals from sea water or by volcanic exhalations on the sea floor.

### Epigenetic Character of the Southeast Missouri Deposits

Criteria for definition of a stratiform ore body as syngenetic, diagenetic, or epigenetic in origin have been defined by Snyder (29). To be classed as epigenetic in origin, a deposit must exhibit several of the following features:

(1) Mineralization of post-lithification structures.

(2) Marked changes in height, width, and tenor of ore that can not be related to sedimentary or diagenetic features.

(3) Extensive open space filling; vein, breccia, or bedding.

(4) Tectonic control over distribution of mineralization.

(5) District-wide lack of close control of mineralization by specific sedimentary environment.

(6) Presence of J-Type lead.

In the preceding pages, evidence has been presented showing that post-lithification faulting occurred prior to and during mineralization. Vast differences in height, width, and grade of ore in many individual ore bodies are related to faulting, not to sedimentary features. Mineralization departs from sedimentary structures in intensely faulted areas; it is controlled by the fault pattern, not by specific rock types or depositional environment.

Numerous examples of conformability—a close fit—between ore and host rock features can be seen on individual mine faces and in thin sections. If *only* these features are considered, a syngenetic origin would appear plausible. However, numerous ore bodies crosscut sedimentary features, including the major stratigraphic-member contacts. Almost every lithology in the upper Lamotte Sandstone and throughout the Bonneterre Dolomite is mineralized in some areas. Sediments deposited in both reducing environments and oxidizing environments may carry ore in some areas and be barren in others.

The lead is J-type in radiogenic character. Zoning of lead with copper and zinc bears no relation to the host rock sedimentation pattern. The deposits are epigenetic in origin.

### Summary of Features That Must Be Considered in Mineralization Theory

The following demonstrable features must be considered in any theory to explain the source, transport, and deposition of metals in the Southeast Missouri deposits; the theory offered must be compatible with these observations.

AREAL LIMITS OF MINERALIZATION   Mineralization is restricted to a small number of concentrations within a 5000 square-mile area. In any area of concentration, there is a sharp cutoff from ore of economic grade to barren rock as illustrated in Figure 10. An area of concentration, like the Old Lead Belt, contained millions of tons of metal; the same formation, outside the area of concentration, contains no visible sulfides, other than the ubiquitous iron sulfide.

VERTICAL EXTENT OF MINERALIZATION   Lead-zinc mineralization occurs in every sedimentary formation present in the area. The formations span virtually all of late Cambrian and early Ordovician time.

STRATIGRAPHIC CONTROL   Mineralization, like that in many districts, is within a stratigraphic sequence that includes sandstone below (Lamotte), carbonate host (Bonneterre), and shale above (Davis). Within the Bonneterre, still closer stratigraphic control can be demonstrated; such as, restriction of mineralization to the lower 50 feet of the Bonneterre below a shaly member at Mine La Motte, to the lower 100 feet of the host below a fine-grained limestone at Leadwood, and to the upper part of the Bonneterre immediately below the "false Davis" shale-silt unit in parts of the Viburnum Lead Belt.

RELATIONSHIP TO REEF FACIES   Within the Old Lead Belt, mineralization is restricted, almost exclusively, to the reef, the underlying bank facies, and similar structures above the reef. Only minor mineralization occurs within the back-reef and fore-reef sediments. The reef facies area is the one where abundant facies changes occur and where numerous stratigraphic and primary structural traps are present. Only a small fraction, less than 10 per cent, of known reef is mineralized. Within reefs, as within other features, the primary structure continues without apparent change after the end of mineralization is reached.

THE ROLE OF DIAGENETIC IRON SULFIDE   The association of ore with a gray/tan dolomite interface, or with black shales, regardless of stratigraphic position or attitude, is so common that the difference in the state of iron in the host rocks must be considered a possible factor in mineral deposition. This may have been a result of sulfur available in the sulfide-rich beds for reaction with metal-bearing solutions or it may have been solely a difference in chemical environment of the rock being traversed.

MINERALIZATION AT SEDIMENT-BASEMENT INTERFACE   Diamond drilling through Lamotte sandstone near basement knobs indicates that mineralization is present at the sediment-knob interface in numerous areas. Where it is present, it is restricted to certain parts of the basement knob, suggesting a directional component to solution movement.

The ore body described above, where mineralization is restricted to one side of the knob, becoming ore grade at several levels in the Bonneterre, is a prime example. In this instance, every drillhole on the northwest side of the knob bears mineralization at the sediment-basement contact; none on the opposite side does. The presence of continuous mineralization at the sediment-porphyry contact on the northwest side and the lack of it on other sides, the regular pattern of ore on the northwest side, and ore above the knob peak suggest that mineralizing solutions from the northwest moved along the sediment-basement interface and were channeled upward by basement topography. Along several bedding plane contacts or in Bonneterre lithologies favorable for solution movement, mineralization spread laterally outward from the knob for distances up to 150 feet.

It is almost inconceivable that any pattern of solution movement in this area, other than an upward movement from the northwest, could account for the ore distribution. Downward-moving solutions inevitably would have resulted in at least trace mineralization on the opposite side of the knob, even though structures to provide ore of economic grade might not have been present. Any general circulation in the Lamotte would have resulted in widespread trace mineralization on other parts of the knob. The concept of ore solution movement at the sediment-basement interface, guided by basement topography, is inescapable in this instance.

DISTRIBUTION OF MINERALIZATION IN UPPER LAMOTTE   A study of some 60,000 drillholes in the Flat River and Leadwood areas revealed that no mineralization is present in the upper Lamotte or at the Bonneterre-Lamotte contact over large parts of the district, even though higher beds are well-mineralized. In limited areas, continuous trace mineralization is present in the upper Lamotte. In other areas, mineralization at this position is in narrow channels, up to 3 miles in length and from 150 to 400 feet in width, with parts being of ore grade. It is apparent that no general circulation of ore fluids in the Lamotte existed; the solutions were confined to limited areas, probably controlled by permeability differences and basement configuration.

Where mine workings cut the Lamotte, differences in permeability control present-day groundwater movement. Some sand beds are tight and make little water; some are tight with water movement restricted to bedding planes. Some beds are porous and continuously "bleed" water into mine openings. Undoubtedly, dilute aqueous solutions carrying metals would have followed the same permeability control.

AGE OF FAULTING   Faulting, that has been shown to be a factor in controlling mineralization, occurred after Potosi, and probably later than Jefferson City time. Near Mine La Motte, faults cut the Potosi, the youngest formation present in that area. Near the Viburnum Lead Belt, faulting cuts the Roubidoux Formation; north of the Old Lead Belt, the Jefferson City Formation is faulted. Within the boundaries of the Old Lead Belt, the Potosi Formation, the youngest now present, is faulted. As stated earlier, some ore is pre-faulting; ore bodies are displaced by faults and galena is deformed. Some mineralization appears to be concurrent with faulting. Some ore bodies are post-faulting and mineralization is controlled by fault-prepared ground. In a general way, the eastern-most of the Old Lead Belt ore bodies are pre-faulting; those in the central area closely coincided with faulting (prepared ground, mineralization that has undergone deformation in faults and fractures); outlying ore bodies, particularly to the south and west, are post-faulting.

TIME DURATION OF MINERALIZATION PERIOD The difference in radiogenic character of lead in the district indicates a time span for the mineralization period. For a single galena crystal, the difference between lead composition of the point of attachment of the crystal and its outer edges in indicative of a change in the solutions during the time required for the crystal to form. The difference in radiogenic character between deformed bedded ore and undeformed galena crystals in a nearby fault indicates a lapse of time sufficient for faulting to take place.

Although no specific time interval can be correlated with the range in isotopic composition of lead, the period of mineralization appears to have been of long duration. The period of lead deposition must have been long enough to include a period of copper deposition and a period of zinc deposition with only slight overlap, and long enough to span most of a period of extensive faulting and fracturing.

VOLCANIC ACTIVITY   The volcanic events described earlier represent possible, although unlikely, sources for the metals. The Avon diatremes, only a few miles from Mine La Motte, are Devonian in age. The Furnace Creek and south Viburnum volcanic events are early Bonneterre in age. They are short-lived events, not repeated in later formations, and could not account for the widespread mineralization in younger formations over much of southern Missouri. Some of the main areas of Bonneterre mineralization are on the opposite sides of Precambrian knobs from the volcanic centers and separated by miles of unmineralized Bonneterre Formation.

MINERALIZATION IN BASEMENT ROCKS   Mineralization in trace amounts is present in basement knobs where they protrude into Bonneterre Formation. At the Pea Ridge iron mine, 40 miles northwest of Bonne Terre, a massive magnetite body occurs in Precambrian rhyolites, below 1200 feet of Paleozoic sedimentary rocks. When the mine was opened, water in the igenous rocks was saline, probably connate water from the upper Cambrian seas. Saline water has been intersected at depths of 1600 feet below the top of the Precambrian. The depth of penetration by saline waters suggests that trace amounts of metals in the upper part of the Precambrian, below galena ore bodies in Bonneterre Formation, could have been introduced by these waters and need not have come from depths within the basement. The presence of the trace metals could be similar to the presence of hydrocarbons in fractured basement rocks adjacent to oilfields and bears no genetic connotation.

TYPES OF HOST ROCK STRUCTURES   Every type of feature present in the carbonate host that represents an interruption to blanket-type units is an ore trap at some point in the district. The structures may be primary, diagenetic, or post-lithification. This suggests that solutions entering or moving through the host rock could be trapped by any structural feature interrupting their path. Structures in which sulfides follow and mimic the original depositional features are common because of the great influence of the sedimentary fabric on solution movement and host-rock replacement.

The size of individual ore bodies is almost entirely a function of size of individual structures or of structures modified by faulting; the form of ore bodies is controlled by the type of structure.

ALTERATIONS AS A REFLECTION OF CHARACTER OF ORE SOLUTION　The simplicity of alteration accompanying, or associated with, mineralization indicates the solutions were nearly in equilibrium with the carbonate host. The solutions moving through the dolomite had no effect except to remove sufficient carbonate to make room for ore and gangue minerals, with a slight excess of solution over ore deposition. The major part of the solution accompanied sulfide mineralization; a minor part continued after sulfide mineralization with cavities along bedding planes, in fractures, and along reef features being filled or partially filled with calcite.

The ore solutions were not oxidizing in terms of altering diagenetic iron sulfide to iron oxide. The only extensive oxidation is that related to the present erosion surface.

## Origin of the Deposits

The evidence cited indicates the metals were brought into the Bonneterre formation not earlier than post-Potosi time and, probably, not earlier than post-Jefferson City time. Mineralization appears to have been a long slow process, but time of termination of the sulfide-deposition stage is indeterminable.

The solutions were simple in character, carrying only lead, zinc, copper, nickel, cobalt, cadmium, silver, and iron to form metallic sulfides. In addition, they may have carried magnesium, potassium, calcium, and sodium. The sulfur appears to be of biogenic origin and probably was derived from the host.

We believe the preponderance of evidence indicates that the solutions originated in deep, adjacent sedimentary basins, possibly from both the Forest City basin and the Illinois basin. The abundance of cobalt and nickel in Mine La Motte and eastern Lead Belt ores and their scarcity in western Lead Belt and Viburnum ores, suggest an initial difference in ore-bearing fluids. There is no basis for relating the metals to the brief volcanic episode in southeastern Missouri, and the distribu-

tion of ore relative to volcanic centers argues against a genetic connection.

The metals carried by the solutions are thought to have been deposited with the basin sediments that contain abundant shales. In addition to the well-known granites and rhyolites of the mid-continent basement, mafic intrusives, including norite, are known in the subsurface rocks and could have been the source provenance for the nickel.

It is unlikely that water of compaction, as postulated by Noble (27), was the ore fluid. Water ejected during compaction would have antedated the time of ore deposition as dated by faulting. In addition, the brines occurring in sulfides in other mid-continent ore deposits are too concentrated to fit a water of compaction hypothesis. This greater concentration of dissolved substances may have been acquired during the time the connate water remained in the basin before migration.

The brines, believed to have carried the metals, were driven out of the basins after compaction of the basin sediments. Brine movement was probably initiated by uplift and erosion of the Ozark dome, accompanying circumferential faulting, that upset hydrodynamic stability. Much of the brine movement was along the sediment-basement interface with only limited migration through sedimentary aquifers. Guided by basement topography, the metal-bearing brine entered the Bonneterre Formation where the Lamotte sand feathered out and Bonneterre rested directly on the Precambrian rocks. There is no basis for postulating either juvenile or groundwater mixing with the basin brines, either as additional sources of metals or factors in transportation or deposition of metals. Likewise, there is no evidence to indicate that juvenile waters may not have contributed to the ore fluid.

The metals were precipitated as sulfides when they encountered iron-sulfide and probably hydrogen sulfide rich zones in the Bonneterre Formation, in any type of trap available.

## REFERENCES CITED

1. Winslow, A., 1894, Lead and zinc deposits: Mo. Geol. Surv., v. 7, 763 p.
2. Buckley, E. R., 1909, Geology of the disseminated lead deposits of St. Francois and Washington Counties: Mo. Bur. Geol. and Mines, v. 9, pt. 1, 259 p.
3. Tarr, W. A. and Keller, W. D., 1933, A post-Devonian igneous intrusion in southeastern Missouri: Jour. Geol., v. 41, p. 805–823.
4. Tarr, W. A., 1936, Origin of the southeastern

Missouri lead deposits: Econ. Geol., v. 31, p. 712–754, 832–866.

5. Rasor, C. A., 1943, Bravoite from a new locality: Econ. Geol., v. 38, p. 399–407.

6. Kidwell, A. L., 1947, Post-Devonian igneous activity in southeastern Missouri: Mo. Geol. Surv. and Water Res., R. I. no. 4, 83 p.

7. Wagner, R. E., 1947, "Lead Belt" geology: Min. and Met., v. 28, no. 488, p. 366–368.

8. Ohle, E. L., 1952, Geology of the Hayden Creek lead mine, southeast Missouri: A.I.M.E. Tr., v. 193, p. 477–483 (*in* Min. Eng., v. 4, no. 5).

9. Beaumont, D. F., 1953, Significance of clays in southeast Missouri lead ores: unpublished Ph. D. thesis, Columbia Univ., 48 p.

10. Patterson, C. C., 1953, The isotopic composition of meteoric, basaltic, and oceanic leads, and the age of the earth: *in* Inghram, M. G. Wasserburg, G. J., and Aldrich, L. T., *Editors, Conf. on nuclear processes in geologic settings* (Sept. 1953), Pr., Williams Bay, Wis., 82 p.

11. Ohle, E. L. and Brown, J. S., 1954, Geologic problems in the southeast Missouri lead district: Geol. Soc. Amer. Bull., v. 65, p. 201–221, 935–936.

12. Grohskoph, J. G., 1955, Subsurface geology of the Mississippi embayment of southeast Missouri: Mo. Geol. Surv. and Water Res. Repts., 2d ser., v. 37, 133 p.

13. Eckelmann, F. D., *et al.*, 1956, Lead isotopes and the pattern of mineralization in southeast Missouri: Geol. Soc. Amer. Bull., v. 67, p. 1688–1690.

14. Emery, J. A. and Seiberling, T. O., 1956, Copper Occurrence in the Federal district, southeast Missouri Lead Belt: Unpublished Staff Rept., St. Joseph Lead Co., 25 p.

15. Snyder, F. G. and Emery, J. A., 1956, Geology in development and mining, southeast Missouri lead belt: A.I.M.E. Tr., v. 208, p. 1216–1224 (*in* Min. Eng., v. 8, no. 12).

16. Davis, J. H., 1958, Distribution of copper, zinc, and minor metals in the southeast Missouri lead district: Econ. Geol., v. 53, p. 917–918.

17. Eckelmann, F. D., *et al.*, 1958, Lead and sulfur isotopes and the history of mineralization in southeast Missouri: Prelim. Rept. Lamont Geol. Observatory, 84 p.

18. Snyder, F. G. and Odell, J. W., 1958, Sedimentary breccias in the southeast Missouri lead district: Geol. Soc. Amer. Bull., v. 69, p. 899–926.

19. Ault, W. U., 1959, Isotopic fractionation of sulfur in geochemical processes: p. 241–259 *in* Abelson, P. H., *Editor, Researches in Geochemistry,* Wiley, N.Y., 511 p.

20. Kulp, J. L. and Eckelmann, F. D., 1959, Lead isotopes and ore deposition in the S. E. Missouri lead district: Econ. Geol., v. 54, p. 1344–1345.

21. Brown, J. S. and Snyder, F. G., 1959, Discussion of lead isotope data for southeast Missouri: Econ. Geol., v. 54, p. 1345.

22. Ault, W. U. and Kulp, J. L., 1960, Sulfur isotopes and ore deposits: Econ. Geol., v. 55, p. 73–100.

23. Davis, J. H., 1960, Mineralization in the southeast Missouri lead district: Unpublished Ph. D. thesis, Univ. Wisc., 68 p.

24. Kilburn, L. C., 1960, Nickel, cobalt, copper, zinc, lead, and sulfur contents of some North American base metal sulfide ores: Econ. Geol., v. 55, p. 115–137.

25. Snyder, F. G. and Wagner, R. E., 1961, Precambrian of southeast Missouri: status and problems: p. 84–94 *in* Hayes, W. C., *Geology of the St. Francois Mountain area,* Mo. Geol. Surv. and Water Res., R. I. no. 26, 137 p.

26. Evans, D. L., Sphalerite mineralization in deep lying dolomites of upper Arbuckle age, west central Kansas: Econ. Geol., v. 58, p. 548–564.

27. Noble, E. A., Formation of ore deposits by water of compaction: Econ. Geol., v. 58, p. 1145–1156.

28. Snyder, F. G. and Gerdemann, P. E., 1965, Explosive igneous activity along an Illinois-Missouri-Kansas axis: Amer. Jour. Sci., v. 263, p. 465–493.

29. Snyder, F. G., 1966, The origin of stratiform ore bodies: what constitutes evidence: *in Symposium on stratiform ore deposits of lead-zinc-barite-fluorite,* New York, 21 p.

30. Myers, H. E., 1966, Guidebook, S. E. G. Mine Tour: St. Joseph Lead Company for *Symposium on stratiform ore deposits of lead-zinc-barite fluorite,* 11 p.

# 18. Geology of the Pea Ridge Iron Ore Body

## Contents

## Illustrations

* Meramec Mining Company, Sullivan, Missouri.

## ABSTRACT

The Pea Ridge iron ore deposit near Sullivan, Missouri, is a dike-like mass of magnetite enclosed in Precambrian porphyries. The ore body tops at the Precambrian surface at a depth of 1300 feet below the present surface and extends to an unknown depth. It dips nearly vertically, intersecting the dip and strike of the enclosing porphyries at an extremely sharp angle. Faulting, both pre- and post-ore, appears to be of minor consequence, and present knowledge indicates that no preferred structural conditions existed prior to emplacement.

The ore body is composed of magnetite with specular hematite, quartz, apatite, and pyrite occurring as accessory primary minerals. Specular hematite in appreciable quantities occurs on parts of the cap and the footwall of the ore body. Fresh porphyry breccia is enclosed in the magnetite along the hanging wall and in the western portion of the ore body.

Hanging-wall quartz-amphibole and footwall quartz-hematite occur as replacements of certain porphyries. Further alteration has caused sericitization of parts of these zones.

The ore deposit is interpreted as having been formed essentially by a single injection of a magmatic differentiate with hydrothermal end phases forming the quartz-amphibole and quartz-hematite portions of the deposit.

## INTRODUCTION

The Pea Ridge ore body, a major iron ore deposit, is located on the north-east flank of the Ozark uplift in Washington County near Sullivan, Missouri. It is about 60 miles southwest of St. Louis. The deposit was developed and is operated by the Meramec Mining Company, a joint venture of the St. Joseph Lead Company and the Bethlehem Steel Corporation.

Development of the property was started in 1957 with the sinking of two shafts and the construction of milling and pelletizing facilities capable of producing 2,000,000 tons of pellets per year. The operation went into production in February, 1964, and has been brought into full production.

The discovery of the deposit dates back 36 or more years with many private, state, and governmental organizations participating. In 1929, the Missouri Bureau of Geology and Mines instituted a program of using geophysical tools to delineate ore deposits and geological structures within the state. In 1930, they started magnetic surveys along the highways of the state and, with the aid of WPA funds, completed the surveys in 1932. These initial surveys disclosed two major magnetic anomalies in the area of the Pea Ridge deposit, designated as the Bourbon and the Kratz Spring anomalies (12). The Pea Ridge anomaly was not discovered at this time due to the inaccessibility of the area. Magnetic maps of the state were published in 1943.

During 1943 and 1944, the U.S. Bureau of Mines drilled four holes to a depth of from 1800 feet to 2300 feet in the Bourbon anom-

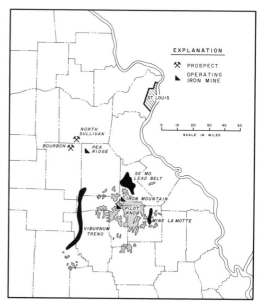

Fig. 1. *Location of Pea Ridge and Other Precambrian Iron Ore Deposits in Southeastern Missouri.*

aly. This drilling showed a large mass of magnetite of middling grade intimately mixed with quartz. From that time to the discovery of the Pea Ridge deposit, it failed to interest mining companies.

An airborne magnetometer survey of the Sullivan-Bourbon area was flown in 1950. This project was a cooperative venture among the Missouri Bureau of Geology and Mines, the U.S. Geological Survey, and private companies of which the St. Joseph Lead Company was the principal contributor. This survey disclosed a new and larger anomaly in the vicinity of Pea Ridge (14). This survey was later augmented by ground surveys of the three anomalies, the two discovered earlier and the one at Pea Ridge.

The St. Joseph Lead Company, in an expanding program of exploration in southeast Missouri, started drilling the Pea Ridge anomaly in 1953 and, in the following 5 years, completed a total of 19 holes having depths from 1800 feet to 3200 feet (14). The drilling located a major deposit of magnetite with lesser quantities of specular hematite, extending from the Precambrian surface to an unknown depth. In 1957, a corporation was formed for the purpose of developing, mining, and processing the ore deposit; this corporation, named the Meramec Mining Company, is owned equally by St. Joseph Lead Company and Bethlehem Steel Corporation (Figure 1).

## GEOLOGIC SETTING

The Pea Ridge iron ore deposit of Precambrian age is a dike-like structure, faintly crescentic in plan, and has a distinct "hump" on the northern side. It strikes approximately N60°E and dips nearly vertically. The deposit is approximately 0.5 miles in length and has a maximum width of 600 feet.

Except for a portion of the top of the deposit, it is enclosed in Precambrian rhyolite porphyries. The porphyries strike N80°W and dip 75°NE and, as a result, make an extremely sharp angle with the deposit.

The rhyolites generally show well-developed flow lines, mainly from concentrations of magnetite but also with quartz oriented parallel to the flow directions. As the porphyries are markedly different, one from the next, they formed distinct mappable units. Contacts between the porphyries are knife-edge sharp with no evidence of alteration or erosion. It is assumed that these igneous rocks are intrusive rather than extrusive.

Each of the four porphyries has been designated by the level on which it was first exposed, and, to date, no attempt has been made to compare these with the exposed porphyries in the St. Francois Mountains just to the south (16).

As more than one porphyry has been observed to fill minor fractures in another along the zone of their contact, their relative ages have been ascertained. The sequence of porphyries is such that successively younger rocks lie farther from the ore body toward the north.

The oldest mapped unit is the 1975 porphyry, a gray-brown porphyry with small brownish-pink orthoclase phenocrysts; locally, the color may range to a reddish-brown. The groundmass is extremely fine, and magnetite concentrations have produced distinct flow lines in this porphyry. It is found in contact with the deposit on and below the 1975-foot level on the footwall side and composes the hanging wall of the deposit. Its contact with the ore body is very sharp.

Making sharp contact with the 1975 porphyry is the 2275 porphyry. This rock is a coarse to very coarse, light, bright-red rhyolite with prominent quartz and orthoclase. Where observed, it contained only very small quantities of magnetite. It makes very sharp contacts with the ore body, on all but the uppermost level, and with the younger 1825 porphyry.

The 1825 porphyry is an extremely fine, almost aphanitic, deep chocolate brown-gray porphyry. Extensive fracturing has formed permeable channels in the porphyry, allowing reddening and bleaching of much of this igneous mass. This porphyry is not in contact with the ore body.

An assumedly still younger porphyry, the 1675 porphyry, is a coarse red quartz monzonite with very prominent feldspar. Minor but ubiquitous hornblende is found wherever the porphyry is exposed. Quartz is found in minor quantity. Much epidote is found as fracture fillings. Due to faulting, the exact relationship of this rock type to other porphyries is not known.

Extensive erosion prior to the deposition of the earliest Paleozoic rocks developed a relief on the Precambrian surface that exceeded 300 feet. The upper portion of the eastern end of the deposit was exposed along an escarpment 250 feet high. Minor arkosic beds were developed within the Lamotte sandstone, containing pebbles and boulders of specular hematite. The western portion of the deposit is covered by as much as 200 feet of porphyry. The Precambrian is covered by a thickness of from 1100 feet to 1400 feet of Cambro-Ordovician limestones, dolomites, sandstones, and shales. The oldest of these flat lying sediments, the basal Lamotte sandstone, erased virtually all of the irregularities in the Precambrian surface. The overlying Cambrian Bonneterre formation has a relief of less than 60 feet across the deposit; thus there is no surface indication of the ore deposit that lies beneath.

## THE ORE BODY

The deposit is composed of five distinct zones; although only three are of economic interest, all play a part in explaining the mineralization process. These are the magnetite zone, the specular-hematite zone, the porphyry-breccia zone, the quartz-hematite zone, and the hanging-wall quartz-amphibole zone. Contacts and the relationships between these zones range from sharp to gradational, depending upon the zone and their location within the deposit (Figures 2A,2B,3A,3B,3C).

### The Magnetite Zone

The major and dominant portion of the ore-body is the magnetite zone. It occupies the center of the deposit and occasionally makes up the entire ore-body from footwall to hanging-wall. The major mineral species is magnetite with lesser quantities of specular hematite, quartz, apatite, pyrite, fluorite, barite, calcite, and sericite, as well as other less prominent minerals (8,11). The zone as such contains iron in excess of 60 per cent with the magnetic iron content being approximately 55 per cent.

The magnetite in general is fine-grained, dense, and massive. Occasional zones will contain magnetite so fine that it has a porcelanic texture. Only rarely is it found as coarse discrete crystals. A characteristic of this zone is the small- to medium-sized vugs that are found throughout its extent; these are lined with a variety of minerals, mostly quartz, specular hematite, and apatite. The magnetite content in relationship to the other minerals increases with depth. No grain or linearity has been found within this zone.

Specular hematite occurs in the magnetite as discrete crystals (sometimes as large as 3 inches in diameter), as fine lacey incipient veinlets, as fine to medium sized irregular inclusions, and as crystals in vugs. It would be difficult to consider this hematite as being anything but essentially contemporaneous with the magnetite. Occasional tabular masses of specular hematite are found following fracture sur-

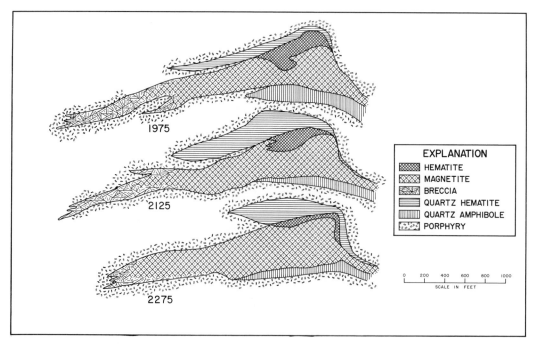

EXPLANATION

HEMATITE
MAGNETITE
BRECCIA
QUARTZ HEMATITE
QUARTZ AMPHIBOLE
PORPHYRY

0 200 400 600 800 1000
SCALE IN FEET

Fig. 2. *Geologic Plan of Pea Ridge Iron Ore Body at Various Depths Below Surface.*

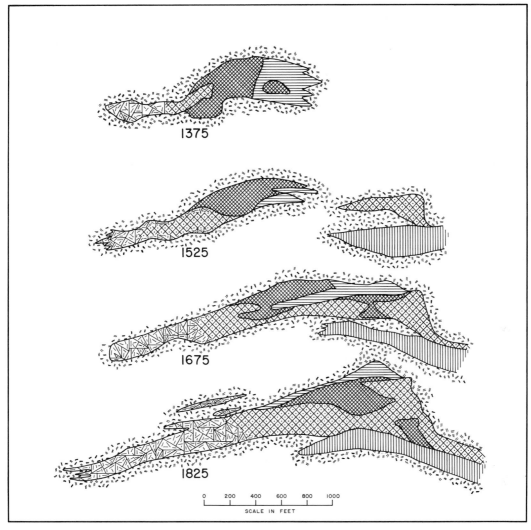

FIG. 2. *Continued.*

faces and making somewhat indistinct contacts with the magnetite. Such hematite appears to have been formed by secondary or later oxidation of the magnetite (15). Specular hematite within the magnetite is far more common in the eastern and upper zones of the deposit than in the lower and western portions, with some zones being completely lacking in hematite.

Quartz occurs as a fine mixture with the magnetite, as fine to medium-sized irregular inclusions, as small veins, and as crystals in small vugs. It is considered to be contemporaneous with the magnetite. Apatite occurs as inclusions in the magnetite, in veins, and

as crystals in vugs. Invariably it is coarser grained than the magnetite and occasionally forms large crystals in incipient vein-like structures. Apatite appears to be consistent in quantity throughout the deposit.

Pyrite is found as discrete crystals in the magnetite, as incipient veins, and in vugs. It is far more important in the western portion of the ore body than the eastern. The pyrite also is considered to be contemporaneous with the magnetite.

Fluorite, barite, and calcite are found in association with quartz, calcite, and hematite in the ubiquitous vugs found throughout the deposit. The occurrence of these minerals

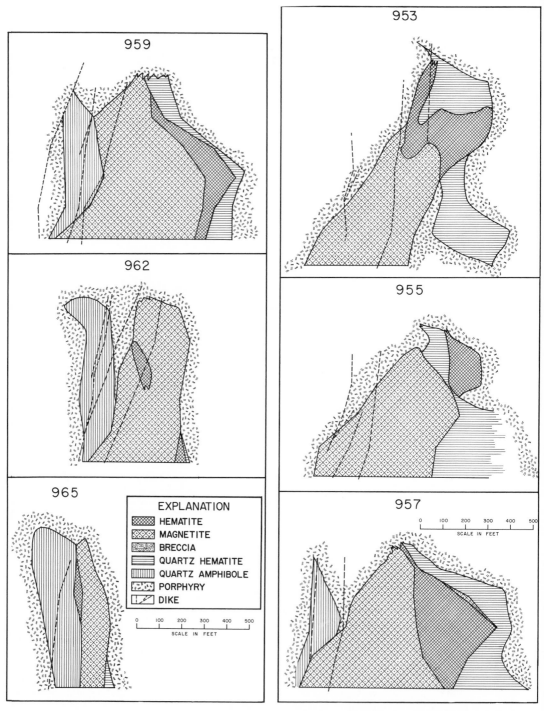

Fɪɢ. 3. *Geologic Cross Sections of Pea Ridge Iron Ore Body. Sections in N-S direction looking west, extend from east (965) to west (947).*

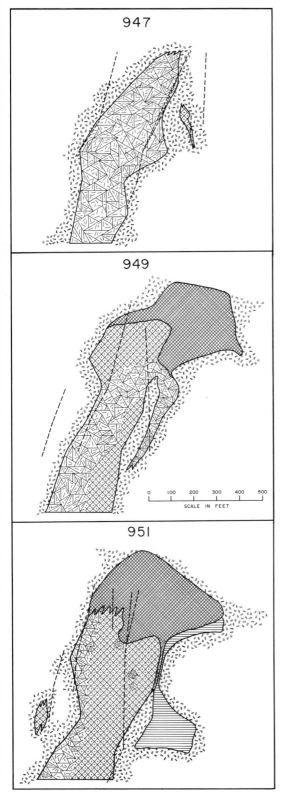

947

949

0 100 200 300 400 500
SCALE IN FEET

951

FIG. 3. (*Continued*)

appears to be consistent throughout the magnetite zone. These minerals also are considered to be essentially contemporaneous with the magnetite.

Sericite, when found, occupies the position normally occupied by quartz. It is considered to be an alteration product.

The magnetite zone is in contact with the porphyries both to the east and to the west of the specular-hematite zone; locally, near a contact, a small increase in quartz content of the magnetite is found. Nevertheless, contact between porphyry and magnetite is strikingly sharp with no attendant alteration of the porphyries and is completely healed with no fracturing or fault gouge. The contact of the magnetite zone with that of the specular-hematite zone is gradational and differs from one place to another. Between the magnetite zone and the hematite zone, the contact is mapped strictly only where economic factors require the ore to be classed as magnetic or non-magnetic. The magnetite zone is not in contact with the quartz-hematite zone.

Contact with the porphyry-breccia zone also is gradational and, in many areas, is merely a matter of definition. Random porphyry inclusions are found throughout the ore body, but porphyry breccia becomes a mappable zone only where the porphyry content is appreciable. For economic reasons, the magnetite-porphyry-breccia contact is mapped where the magnetic iron content drops below 45 per cent. On the hanging-wall side, the transition may occur over a distance of from 5 feet to 10 feet; but, in the western portion of the deposit, porphyry inclusions may be found for several hundred feet before the porphyry content becomes appreciable.

The magnetite zone is found in contact with the hanging-wall quartz-amphibole zone in a very few localities. In general, a thin porphyry breccia zone separates the two.

## The Specular-Hematite Zone

The specular-hematite zone is designated as that containing over 45 per cent total iron and less than 30 per cent magnetic iron. It occupies portions of the footwall of the deposit, being centered in the "hump" and along the cap where it is exposed on the Precambrian erosion surface. It pinches down sharply with depth, being a very minor portion of the deposit below the 2125-foot level. Minor tabular sections extend along fracture zones into the main magnetite zone. Minor occurrences of hematite are not designated as such unless they have an effect on the mining operation.

The specular hematite is fine to coarsely crystalline, compact, and massive. The crystals in general have parallel orientation, giving a platy appearance to the rock. With the hematite are martite, quartz, sericite, apatite, and the lesser amounts of accessory minerals, these being in occurrence and quantity, essentially the same as in the magnetite. Pyrite, however, is virtually absent.

The hematite zone does not make contact with the porphyries but is blanketed on the footwall side by the quartz-hematite zone; its contact with the quartz-hematite zone is sharp. At the Precambrian surface, it is in direct contact with the Lamotte formation. Some thin arkosic beds containing hematite are found in the Lamotte.

## The Porphyry-Breccia Zone

The porphyry-breccia zone is situated on the hanging-wall side of the massive magnetite in the eastern portion of the deposit. In the western areas, it is found on the footwall, on the hanging-wall, and above the magnetite zone. It is designated as that part of the deposit that contains porphyry fragments and iron, the latter ranging between 30 and 45 per cent. The porphyry fragments generally are sharply angular, of various sizes, and fresh. The quantity of such fragments may range from a few random pieces within a mass of magnetite to a shattered but not widely dislocated porphyry in which the interstices have been filled with magnetite. In the western portion of the deposit, mafic fragments are found that have undergone various degrees of alteration. In some places, the mafic material will compose the major quantity of porphyry present. In none of the mine workings nor in any of the deep or prospect drilling has any of this mafic rock type been found in place.

Starting westward from approximately the midpoint of the deposit, random porphyry fragments are found in the massive magnetite zone. The porphyry content gradually increases to the point at which it is a dominant portion of the rock. Magnetite content decreases until it becomes merely a fracture-filling material. This gradual increase in porphyry content takes place over a distance of 800 feet or more. The actual end of iron mineralization has not been found and the furthest drilling still shows porphyry with occasional fracture fillings of magnetite.

Contact with the hanging-wall quartz amphibole is sharp and is marked by the appearance of both amphibole and quartz. In the upper reaches of the deposit, porphyry breccia is found with the interstices filled by specular hematite.

## The Quartz-Amphibole Zone

The hanging-wall quartz-amphibole zone is composed of coarse quartz and actinolite with heavy accumulations of apatite and minor magnetite (10). It is an extremely varied zone in content, its composition ranging from almost massive quartz to a fine green meshwork of amphibole crystals. Along the eastern end of the deposit, the zone is approximately 120 feet wide, with the outer dominant portion being a massive white quartz. Its widest portion is found where the magnetite pinches out. To the west, it gradually thins to nothing and becomes merely amphibole fracture fillings in the hanging-wall porphyries, around which fractures there is minor silicification of the porphyry. The zone appears to widen with increased depth. Contact with the porphyry on the eastern end, however, is sharp, the rock passing over a short distance from quartz to felsite.

## The Footwall Quartz-Hematite Zone

The footwall quartz-hematite zone is also an extremely varied division of the deposit. It forms a narrow border between the specular hematite and the porphyry, along the "hump" of the deposit, gradually expanding and separating from the deposit to the west. Along the "hump" it forms a 10 feet to 30 feet wide zone of fine-grained quartz and specular hematite, the outer portion of which is extensively altered and sericitized. Fluorite, barite, molybdenite, and tourmaline are found in this sericitized zone. It grades into the porphyry along a narrow alteration front. To the west, the hematite decreases to the point of virtual nonexistence, leaving a thick zone of white quartz. From the hematite-magnetite contact along the footwall, it separates from the main deposit and strikes due west. The zone appears to be thickening with depth. The full extent of the quartz-hematite zone after it separates from the main deposit is not known.

## DIKES

The Pea Ridge deposit, in common with all other Precambrian iron deposits in the Ozark Province, is transected by both mafic and aplitic dikes. The dikes also cut the enclosing porphyries.

The mafic dikes are found in two systems, one striking N60°E and the other N85°E. Deviations from these strikes occur where the dikes split to form parallel dikes. The N60°E dikes are confined to the hanging-wall side of the deposit. The N85°E dikes transect the deposits, being on the hanging-wall side in the eastern portion of the deposit and passing into the footwall to the west of the deposit. Both types of mafic dikes tend to follow for short distances along the hanging-wall contact of the massive magnetite, deviating locally from the general strike trend. Each system has a flatter dip than the deposit (60°-80°SE), and each tends to pass out into the hanging-wall. The various dips indicate that these dikes are part of a larger system, not as yet located. Thickness of the dikes ranges from a fraction of an inch to more than 11 feet, with the thicker dikes being more coarsely crystalline. Both systems have been subjected to various degrees of alteration.

The aplitic dikes also transect the ore deposit and the enclosing porphyries. They form a system striking N25°W and dip deeply to flatly to the northeast. They tend to be more regular in thickness, ranging from 4 inches to 2 feet. They are composed of equigranular quartz and orthoclase, although locally dikes are composed almost entirely of quartz.

The age relationship between the mafic and aplitic dikes has been established, since aplitic dikes have been found transecting the mafic ones.

## FAULTING

Only one fault of major displacement is found in the area of the deposit. This fault has been named the "Rock Burst" fault because of frequent spontaneous rupturing of the rock in its hanging-wall side. It strikes N80°W and dips 45°NE. The fault zone ranges from 2 feet to 10 feet wide and has a well developed fault gouge; the amount of throw is not yet known. The fault has not yet been exposed above the 1675-foot level, and its projection does not show it to approach closer than 300 feet north of the deposit.

Parallel to this fault surface and separated from it by approximately 300 feet is an escarpment of the Precambrian surface extending into the Lamotte. Part of the deposit is exposed on this escarpment. No actual evidence of faulting along this surface has been found. If faulting has occurred, the tonnage of ore displaced by this possible fault must be minor as the magnetics do not indicate any such displacement.

Other faulting within the deposit is of a minor nature, displacement being less than 6 feet on such faults.

No minor offshoots of magnetite have been found that might have followed fault surfaces. Contacts of the zones of mineralization show no displacement either in exposed mine openings or from diamond drill information.

## FRACTURING

Fracturing of both the deposit and the porphyries has been exceedingly intense, the porphyries being somewhat more intensively fractured. The porphyries show two dominant trends of high-angle fractures, one striking N60°E and the other east-west. Sheeting strikes N45°W and dips 10°NE. Many other fractures are seen to change from vertical to horizontal in dip and markedly in strike in very short distances.

The ore deposit proper exhibits two dominant trends of fracturing that strike N45°W and N60°E, and each has widely divergent dips. Often heavily fractured areas will not exhibit a single fracture extending the width of the drift. Because of this, fractured areas often are mapped showing only the direction of the fracture trend.

## GENESIS

The geologic data accumulated to date at the Pea Ridge mine, coupled with experimental data and theoretical considerations, indicate that the genesis of the deposit was the forceful injection of an iron-rich differentiate from a silicic or intermediate magma. This interpretation is comparable to the interpretation made by other geologists in their study of somewhat similar deposits, such as Kiruna and Grängesberg in Sweden and Iron Mountain in Missouri (1,2,3). The deposit differs from most other iron ore deposits described as being magmatic segregations, in that it exhibits end-phase hydrothermal activity as well. This feature has been the subject of much theoretical discussion on magmatic deposits but appears seldom to have been found in nature (5,13), although the Rektor deposit near Kiruna may be a case in point.

The breccia sections and the inclusions of porphyry within the deposit and, more particularly, the presence of exotic particles, the source of which is yet unknown, are strongly indicative of magmatic injection. Also, the extremely sharp unaltered contact between the ore and the enclosing porphyries as well as the breccia particles indicate that little alteration or replacement occurred during the emplacement of the deposit. The entry of the ore fluid in an apparently unprepared porphyry, following neither a fault surface or the contact between porphyries indicates the forceful nature of the depositional process. The lack of flow lines within the ore mass indicates a rather short duration of injection. The presence of amygdules and vesicles in the magnetite, however, indicates the presence of volatile elements in the ore fluid. The presence of apatite and fluorite in the vugs is an indication of an appreciable content of volatiles and suggests low temperature and low viscosity. The experimental data of Bowen and Fischer, as well as the theoretical considerations of Niggli, Spurr, Park, Lindgren, and others, indicate that the separation of an iron-oxide differentiate from a magma is possible and that additional fluidizers such as apatite and fluorite would accompany this separation, allowing fluidity at lower temperatures than in a drier magma. Desborough's concept (17), on the other hand, would supply the iron of the deposits from the alteration of titaniferous magnetite.

At this time, or essentially contemporaneous with the solidification of the melt, minor quantities of specular hematite, quartz, apatite, flourite, barite, pyrite, and other minerals were precipitated. As the mass started to crystallize, the more volatile elements passed off as a hydrothermal end-phase forming the quartz-amphibole of the hanging-wall and the quartz-hematite of the footwall. Both of these features can be considered as alteration and replacement of the enclosing porphyries.

The large mass of massive specular hematite seems to offer the possibility of development by one or more than one process. Evidence suggests genesis of either the hematite epigenetically by meteoric water or by hydrothermal fluids.

Within the massive hematite zone are anomalous masses of partially magnetic ore. The contacts between the magnetite zone and the hematite zone are gradational. Masses of martite are found. The overall iron content of the hematite zone is equal to that of the magnetite zone and the phosphate content of the two zones is comparable. The sulphur content of the hematite zone, however, is almost zero. This suggests that the massive specular hema-

tite zone was originally emplaced as magnetite. Spatially the massive specular hematite is found in the cap of the ore deposit where it is exposed on a 250 foot escarpment on the Precambrian erosion surface. It pinches rapidly with depth forming a thin deposit along the footwall. Where the deposit stops within the porphyry, it remains as a magnetite. This evidence points to an epigenetic genesis, possibly formed by meteoric waters. On the other hand, the hematite zone is found only in association with the quartz-hematite zone, unquestionably of a hydrothermal origin.

It appears that late-stage hydrothermal activity with excess available oxygen could also explain the genesis of the massive hematite, possibly both environments played a part in the formation of this zone.

As no evidence of metamorphism is found either in the folding of the deposit or in the development of metamorphic minerals and as no consequent faulting is found, it is considered that the deposit is in essentially the same position as that in which it was originally emplaced.

## REFERENCES CITED

1. Geijer, P., 1910, Igneous rocks and iron ores of Kiirunavarra, Luossavaara and Tuolluvaara: Econ. Geol., v. 5, p. 699–718.
2. Crane, G. W., 1912, The iron ores of Missouri: Mo. Bur. Geol. and Mines, 2d ser., v. 10, p. 1–13, 107–145.
3. Geijer, P., 1915, Some problems in iron ore geology in Sweden and in America: Econ. Geol., v. 10, p. 299–329.
4. Spurr, J. E., 1923, The ore magmas, McGraw-Hill. N.Y., 2 vols. 915 p.
5. Niggli, P. (Boydell, H. C., Trans.), 1929, Ore deposits of magmatic origin: Murby, London, 93 p.
6. Bowen, N. L., 1929, The evolution of the igneous rocks: Princeton Univ. Press, 334 p. (reprinted by Dover Pubs., 1956).
7. Lindgren, W., 1933, Mineral deposits: 4th ed., McGraw-Hill, N.Y., 930 p.
8. Frank, A. and Monihan, C. S., 1947, Occurrence of barite at Pilot Knob, Missouri: Amer. Mineral., v. 32, p. 681–683.
9. Fischer, R., 1950, Entmischungen in Schmelzen aus Schwermetalloxyden, Silikaten and Phosphaten; ihre geochemische and lagerstättenkundliche Bedeutung: Neues Jb. f. Mineral. Abh., Abt. A., Bd. 81, S. 315–364.
10. Allen, V. T. and Fahey, J. J., 1952, Salite and actinolite at Iron Mountain, Missouri: Amer. Mineral., v. 37, p. 283.
11. ——— 1952, New occurrences of minerals at Iron Mountain, Missouri: Amer. Mineral., v. 37, p. 756.

12. Searight, T. K., *et al.*, 1954, The structure and magnetic survey of the Sullivan-Bourbon area, Missouri: Mo. Div. Geol. Surv. and Water Res. R.I., no. 16, 14 p.

13. Mutch, A. D., 1956, A critical evaluation of the classification of ore deposits of magmatic affiliations: Econ. Geol., v. 51, p. 665–685.

14. Brown, J. S., 1958, Iron in Missouri: Unpublished lecture to joint meeting of St. Louis Section A.I.M.E. and Engineers Club.

15. Basta, E. Z., 1959, Some mineralogic relationships in the system $Fe_2O_3$-$Fe_3O_4$ and the composition of titanomaghemite: Econ. Geol., v. 54, p. 698–720.

16. Hayes, W. C., *et al.,* 1961, Guidebook to the geology of the St. Francois Mountain area: Mo. Div. Geol. Surv. Water Res. R.I. no. 26, 137 p.

17. Desborough, G. A., 1963, Mobilization of iron by alteration of magnetite-ülvospinel in basic rocks in Missouri: Econ. Geol., v. 56, p. 332–347.

18. Park, C. F. and MacDiarmid, R. A., 1964, Ore deposits: W. H. Freeman, San Francisco, 475 p.

# 19. Fluorite-Zinc-Lead Deposits of the Illinois-Kentucky Mining District

ROBERT M. GROGAN,* JAMES C. BRADBURY†

## Contents

\* E. I. duPont de Nemours & Company, Wilmington, Delaware.
† Illinois State Geological Survey, Urbana, Illinois.

# Illustrations

# Tables

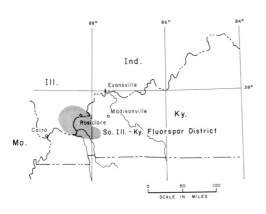

## ABSTRACT

The Illinois-Kentucky mining district has, since 1880, accounted for 80 per cent of all U.S. production of fluorspar.

Exposed strata range from Devonian through Pennsylvanian in age. Igneous intrusive rocks occur as mafic dikes and sills and intrusive breccia dikes and plugs. Structurally, the district is situated on a generally northwest-trending arch that has been sliced into a series of long, narrow blocks by normal faults that trend northeastward. The structural apex of the district is formed by Hicks Dome, a sharply-defined oval-shaped uplift that contains many outcrops of intrusive breccia, some of which are mineralized. Hicks Dome is presumed to have been formed by explosive forces of volcanic origin. The mineralization has been dated as Permian, or younger, possibly early late Cretaceous, by isotopic analysis and structural-stratigraphic relations.

The ore deposits are of two types: vein deposits formed by fissure fillings along faults and bedding-replacement deposits formed by the replacement of strata at certain favored stratigraphic positions along minor faults and fractures. Fluorite is the chief valuable mineral in the deposits. Sphalerite and galena commonly are present in small amounts but, in some ore bodies, may form a substantial part of the valuable constituents of the ore. Calcite is the chief gangue mineral; quartz and barite are present in various amounts.

The localization of the ore deposits by faults and fractures, the association of fluorite mineralization with intrusive breccias, and various lines of petrographic and geochemical evidence indicate that the fluorite-zinc-lead deposits were epigenetic and deposited by solutions emanating from a deep alkalic magma.

## HISTORY OF MINING

### Mining Operations in the Area

The Illinois-Kentucky mining district encompasses an area of about 1000 square miles (Figure 1). It is the most important producer of fluorspar in the United States and is also the source of important amounts of lead and zinc. The occurrence of fluorspar and lead was known to the early settlers from about 1812 (3, p. 115), although, long before that time, prehistoric peoples had found and carved fluorspar into ornaments and figures. Substantially all the early exploration and mining was for lead, though the Columbia and Royal mines in Kentucky were first worked for silver (3, p. 115). Fluorspar was regarded as an essentially valueless gangue mineral.

Lead mining started in Kentucky in 1835 near the site of the Columbia mine in Critten-den County (3, p. 115) and in Illinois in 1842

on the Pell farm half a mile northwest of Rosiclare (1, p. 366–372). President Andrew Jackson headed the Kentucky company. Eventually, a number of small mines were being operated in both states, and many deposits, opened up for lead, later became important producers of fluorspar. The period in which lead was the dominant mineral mined continued until the 1870's, stimulated first by Civil War demands and depressed later by declining prices and costly transportation.

Shipments of fluorspar apparently were commenced in the 1870's (2, p. 13; 4, p. 45), and went, just as now, to manufacturers of hydrofluoric acid, glasses, and enamels and into use as a metallurgical flux. The amounts shipped were very small, however, until about 1889 when a rising trend began. By 1910, there were concentrating plants capable of treating 500 tons of crude ore daily by jigs and tables. Froth flotation for the production of acid-grade fluorspar was introduced in 1929.

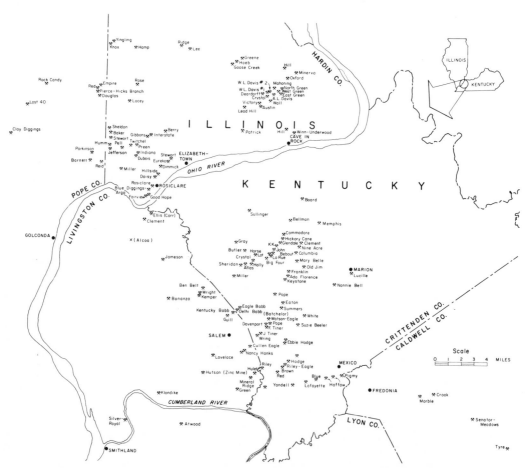

FIG. 1. *Illinois-Kentucky Mining District, showing locations of mines.*

TABLE I.   Fluorspar Shipped from Mines

| Year | Illinois Kentucky | | Total U.S. | |
|------|------------|---------|------------|---------|
| | Short Tons | Value | Short Tons | Value |
| | (Thousands) | (Millions) | (Thousands) | (Millions) |
| 1880–1936 | 3544 | $ 51.2 | 3824 | $ 54.6 |
| 1937–1948 | 2754 | 78.0 | 3442 | 94.8 |
| 1949 | 184 | 6.6 | 237 | 8.3 |
| 1950 | 235 | 8.7 | 302 | 10.6 |
| 1951 | 273 | 11.6 | 374 | 14.4 |
| 1952 | 236 | 11.3 | 331 | 15.4 |
| 1953 | 211 | 10.7 | 318 | 15.7 |
| 1954 | 144 | 7.5 | 246 | 12.3 |
| 1955 | 175 | 8.1 | 280 | 12.6 |
| 1956 | 193 | 9.1 | 330 | 14.3 |
| 1957 | 191 | 9.8 | 329 | 15.8 |
| 1958 | 178 | 9.1 | 320 | 15.1 |
| 1959 | 131 | 6.8 | 185 | 8.7 |
| 1960 | 160 | 8.1 | 230 | 10.4 |
| 1961 | 156 | 7.7 | 205 | 9.3 |
| 1962 | 167 | 7.9 | 206 | 9.2 |
| 1963 | 167 | 8.1 | 200 | 9.0 |
| 1964 | 166 | 8.1 | 217 | 9.7 |
| 1965 | 194* | 9.4* | 241 | 10.9 |
| Total | 9459* | 277.8* | 11,817 | 351.1 |

Sources: Hatmaker and Davis (12), Weller, et al. (19), U.S. Bureau of Mines Minerals Yearbook (17).

* Estimated.

The development of the district was promoted first by the growth of basic open-hearth steel production, in which fluorspar serves as the fluxing agent, then by the steel industry's enlarged requirements during World War I, and later by the growth of the aluminum and fluorine chemical industries. World War II brought a period of intense exploration and mine expansion with the result that production of fluorspar rose from 165,000 tons in 1939 to its all time high of nearly 309,000 tons in 1943.

For years, a group of large mines at Rosiclare, Illinois, were the district's principal producers. In addition to having persistent veins, these mines enjoyed low transportation costs because of their proximity to the Ohio River.

Discovery in the late 1930's of large and numerous replacement deposits near Cave in Rock, Illinois, not only added many years life to the district but turned it into a substantial lead and zinc producer. Discovery, since World War II, of several large vein deposits in Kentucky also has helped considerably to offset the exhaustion of the mines at Rosiclare.

## Statistics of Mine Production

In Table I, shipments of fluorspar from mines in Illinois and Kentucky are compared with the United States' total in terms of tons and of dollar value. Since 1880, this district has yielded a little less than 9.5 million tons of finished fluorspar concentrate of all grades, accounting for 80 per cent of all U.S. production. Although for many years it provided 30 to 40 per cent of world production, its relative importance has declined since about 1950

TABLE II.   Recoverable Lead and Zinc Produced

| Year | Lead | | Zinc | |
|------|------|--------|------|--------|
| | Short Tons | Nominal Value* | Short Tons | Nominal Value* |
| | (Thousands) | (Thousands) | (Thousands) | (Thousands) |
| 1939 | .4 | $ 40 | 1.2 | $ 127 |
| 1940 | 1.9 | 193 | 6.1 | 772 |
| 1941 | 2.5 | 294 | 7.9 | 1183 |
| 1942 | 2.5 | 330 | 8.1 | 1336 |
| 1943 | 2.2 | 286 | 5.6 | 929 |
| 1944 | 2.0 | 266 | 5.9 | 975 |
| 1945 | 2.6 | 344 | 4.7 | 781 |
| 1946 | 3.7 | 598 | 5.0 | 881 |
| 1947 | 1.9 | 554 | 5.7 | 1203 |
| 1948 | 3.0 | 1070 | 7.4 | 2016 |
| 1949 | 2.8 | 867 | 6.5 | 1589 |
| 1950 | 1.5 | 406 | 6.6 | 1844 |
| 1951 | 2.5 | 880 | 9.6 | 3448 |
| 1952 | 2.8 | 919 | 8.0 | 2583 |
| 1953 | 1.8 | 498 | 5.6 | 1214 |
| 1954 | 1.3 | 379 | 5.0 | 1064 |
| 1955 | 2.7 | 812 | 8.6 | 2119 |
| 1956 | 2.3 | 748 | 9.8 | 2647 |
| 1957 | 1.6 | 466 | 7.1 | 1620 |
| 1958 | 1.2 | 280 | 7.7 | 1579 |
| 1959 | 1.6 | 393 | 7.9 | 1804 |
| 1960 | 2.3 | 523 | 10.1 | 2614 |
| 1961 | 2.9 | 626 | 9.5 | 2198 |
| 1962 | 3.5 | 677 | 11.5 | 2679 |
| 1963 | 3.3 | 727 | 11.0 | 2644 |
| 1964 | 3.0 | 820 | 11.2 | 3052 |
| Total | 59.8 | $13,996 | 193.3 | $44,901 |

Source: U.S. Bureau of Mines Minerals Yearbook (17).

* Based on average yearly quoted price of lead metal at New York and zinc metal at St. Louis.

when Mexico and Europe began to expand production, and it now turns out only about 6 or 7 per cent of annual world output.

Individual veins or replacement deposits of major size may have contributed a total of 100,000 to 800,000 tons of finished fluorspar concentrate to the total. Most of the 6.4 million tons produced in Illinois has come from about 20 mines, an average of 320,000 tons per mine.

Although the production of lead and zinc (Table II) is by no means so important as that of fluorspar when judged by U.S. and world standards, it is important in a local sense, having in recent years contributed 35 to 45 per cent of the total sales value of the three minerals combined.

## PHYSIOGRAPHIC HISTORY AND PRESENT TOPOGRAPHY

The Illinois-Kentucky mining district is bisected by the Ohio River, the 1- to 2-mile-wide valley and flood plain of which have an elevation of 300 to 340 feet above sea level. The land surface on both sides of this valley is characterized by a pleasantly varied mature topography, the highest points of which reach elevations in excess of 900 feet above sea level.

The sedimentary rocks of the district differ greatly in their resistance to erosion. Cyclic alternation of resistant sandstones with less resistant limestones and shales of upper Mississippian age has resulted in the development of a series of ridges and valleys where the sequence has not been disturbed by major faulting. In the major fault grabens, highly resistant Pennsylvanian sandstone and conglomerate stand up in long, bold, landscape-dominating ridges. Broad expanses underlain by the St. Louis and Ste. Genevieve Formations of Mississippian age are gently rolling and have abundant sinkholes. At Hicks Dome, a central higher area of Devonian limestone is concentrically surrounded by a broad valley on the shale of the New Albany Group and by a ridge of chert and siltstone of the Fort Payne Formation.

Four erosional levels above the Ohio flood plain have been recognized (6, p. 47–52) at elevations of 400 to 420 feet, 500 to 540 feet, 600 to 640 feet, and 860 to 900 feet above sea level. These presumably are remnants of once extensive plain-like flat areas. Three constructional terraces along the Ohio River and its tributary valleys were built by glacial meltwater that flooded the valley in late Wisconsinan time (40, p. 3). These have elevations of 350 to 355, about 360, and 380 to 390 feet, respectively.

## GENERAL GEOLOGY

### Stratigraphy

The bedrock formations exposed in the Illinois-Kentucky mining district range in age from Devonian through Pennsylvanian. Terrace deposits of silt, sand, and gravel and a general covering of loess on the uplands are Pleistocene in age. Figure 2 summarizes graphically the characteristics of the bedrock units.

Intrusive rocks, represented in the district by dikes and sills of mica-peridotite and lamprophyre and bodies of intrusive breccia, have been described by Clegg and Bradbury (29) and by Bradbury (38). Another type of igneous rock, associated with a recently found exposure of intrusive breccia in the northwestern part of the district (54) has been identified by Bradbury as a fine-grained nepheline-feldspar rock.

The mica-peridotites are dark gray to dark greenish gray, fine- to medium-grained, porphyritic rocks, characterized by olivine (mostly serpentinized), phlogopite, and pyroxene. Alteration of the olivine and pyroxene to serpentine and of mica to a green chlorite ranges from slight to intense. Apatite is a common accessory.

The lamprophyres are medium to dark gray or medium to dark greenish gray, very fine grained and porphyritic to nonporphyritic. They differ from the mica-peridotites chiefly in the absence of olivine or of recognizable serpentine pseudomorphs after olivine. Replacement of the silicates by carbonate was extensive in the lamprophyres and may have obliterated evidence of primary olivine. However, other bits of evidence, such as the presence of nearly 3 per cent $Na_2O$ in one of the lamprophyres as compared to a maximum of 0.46 per cent in the mica-peridotites, (38, Table II) suggests that the lamprophyres, or at least some of them, do indeed represent something other than mica-peridotite.

The intrusive breccias are fragmental rocks occurring in bodies of dike-like or presumed plug shapes. Many are composed solely of sedimentary rock fragments in a matrix of ground-up sedimentary rock; some contain scattered igneous rock fragments in addition to the predominantly sedimentary ones in a matrix of ground-up sedimentary and igneous

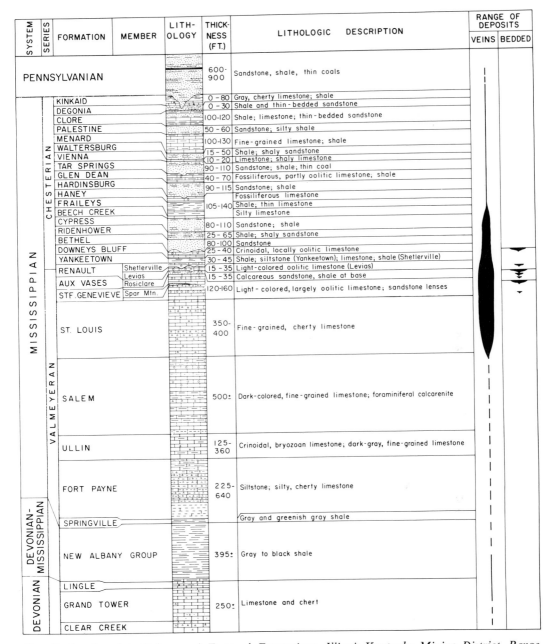

| SYSTEM | SERIES | FORMATION | MEMBER | LITH-OLOGY | THICK-NESS (FT.) | LITHOLOGIC DESCRIPTION | RANGE OF DEPOSITS VEINS | RANGE OF DEPOSITS BEDDED |
|---|---|---|---|---|---|---|---|---|
| PENNSYLVANIAN | | | | | 600-900 | Sandstone, shale, thin coals | | |
| MISSISSIPPIAN | CHESTERIAN | KINKAID | | | 0 - 80 | Gray, cherty limestone; shale | | |
| | | DEGONIA | | | 0 - 30 | Shale and thin-bedded sandstone | | |
| | | CLORE | | | 100-120 | Shale; limestone; thin-bedded sandstone | | |
| | | PALESTINE | | | 50 - 60 | Sandstone; silty shale | | |
| | | MENARD | | | 100-130 | Fine-grained limestone; shale | | |
| | | WALTERSBURG | | | 15 - 50 | Shale; shaly sandstone | | |
| | | VIENNA | | | 10 - 20 | Limestone; shaly limestone | | |
| | | TAR SPRINGS | | | 90 - 110 | Sandstone; shale; thin coal | | |
| | | GLEN DEAN | | | 40 - 70 | Fossiliferous, partly oolitic limestone; shale | | |
| | | HARDINSBURG | | | 90 - 115 | Sandstone; shale | | |
| | | HANEY | | | | Fossiliferous limestone | | |
| | | FRAILEYS | | | 105-140 | Shale; thin limestone | | |
| | | BEECH CREEK | | | | Silty limestone | | |
| | | CYPRESS | | | 80 - 110 | Sandstone; shale | | |
| | | RIDENHOWER | | | 25 - 65 | Shale; shaly sandstone | | |
| | | BETHEL | | | 80 - 100 | Sandstone | | |
| | | DOWNEYS BLUFF | | | 25 - 40 | Crinoidal, locally oolitic limestone | | |
| | | YANKEETOWN | | | 30 - 45 | Shale; siltstone (Yankeetown); limestone; shale (Shetlerville) | | |
| | | RENAULT | Shetlerville | | 15 - 35 | Light-colored oolitic limestone (Levias) | | |
| | | | Levias | | | | | |
| | | AUX VASES | Rosiclare | | 15 - 35 | Calcareous sandstone, shale at base | | |
| | | STE. GENEVIEVE | Spar Mtn. | | 120-160 | Light-colored, largely oolitic limestone; sandstone lenses | | |
| | VALMEYERAN | ST. LOUIS | | | 350-400 | Fine-grained, cherty limestone | | |
| | | SALEM | | | 500± | Dark-colored, fine-grained limestone; foraminiferal calcarenite | | |
| | | ULLIN | | | 125-360 | Crinoidal, bryozoan limestone; dark-gray, fine-grained limestone | | |
| DEVONIAN-MISSISSIPPIAN | | FORT PAYNE | | | 225-640 | Siltstone; silty, cherty limestone | | |
| | | SPRINGVILLE | | | | Gray and greenish gray shale | | |
| | | NEW ALBANY GROUP | | | 395± | Gray to black shale | | |
| DEVONIAN | | LINGLE | | | | | | |
| | | GRAND TOWER | | | 250± | Limestone and chert | | |
| | | CLEAR CREEK | | | | | | |

Fig. 2. *Stratigraphic Column of Exposed Formations, Illinois-Kentucky Mining District. Range of Deposits columns are scaled to correlate with Lithology column, e.g., the three main bedded horizons are the top part of the Downeys Bluff, top part of the Ste. Genevieve, and the Spar Mountain Member. (Adapted from Baxter, et al., 54, Plate 1).*

rock, and one (Grants Intrusive) contains, in addition to metamorphosed sedimentary rock fragments, a large amount of material of igneous derivation, including crystals of hornblende and biotite as much as 0.5 inches long and fragments of an aegerine-bearing syenite up to 1.5 inches long. The matrix in the Grants Intrusive is carbonate (identified as dolomite by X-ray diffraction), raising the possibility that the intrusive may not be entirely a fragmental rock, because the carbonate could be either primary or a replacement of a crystalline silicate matrix such as characterizes many of the lamprophyres.

The nepheline-feldspar rock is dark green, very fine grained, and contains many white and light-green inclusions and streaks, which represent metamorphosed and partly digested rock fragments. Grains of augite, aegerine-augite, and hornblende appear as dark-colored inclusions a few tenths to a few millimeters in diameter. Although mineral identifications in the very fine-grained groundmass are tentative, it appears to consist of feldspar and nepheline, with a profusion of tiny needles of a sodic pyroxene that show a strong tendency to cluster around the borders of inclusions. A chemical analysis by the Illinois State Geological Survey, showing over 9 per cent $Na_2O$, tends to support the mineral identifications. Flow lines are conspicuous both in the hand specimen (shown by light and dark streaks) and thin section (indicated by alignment of long axes of inclusions and by streaks of very tiny dark-colored inclusions). Fluorite is found in many of the rock inclusions but never by itself in the groundmass. It is most abundant in those inclusions composed primarily of tremolite and calcite, which are probably metamorphosed limestone fragments, but it also is present in partly altered feldspathic inclusions. Many of the feldspathic inclusions appear to be made up of the same primary minerals as the much finer-grained groundmass and presumably represent partial crystallization of the magma at depth prior to its intrusion into the Paleozoic strata. The alteration of the fluorite-bearing feldspathic inclusions, in contrast to the unaltered condition of the groundmass, and the presence of fluorite in metamorphosed limestone inclusions suggest that fluoritization accompanied intrusion of the nephelinitic magma.

The age of igneous intrusion in the mining district can be dated as later than the middle part of the Pennsylvanian (29, p. 15) from stratigraphic relations. K-Ar dating on phlogopite, biotite, and hornblende from several of the peridotite dikes and an intrusive breccia (Grants) from the Illinois-Kentucky district showed an age of $265 \pm 15$ m.y. (early Permian) (51). However, as the dikes were presumably intruded as a crystal mush (27, p. 17), it follows that the main components crystallized at some time prior to actual intrusion, and therefore the early Permian age must be regarded as a maximum. The same reasoning applies to the Grants Intrusive, in which the large biotite and hornblende inclusions surely crystallized before the body of material was intruded to its present position. A lead-alpha age determination on a monazite from a breccia on Hicks Dome (33) gave an early late Cretaceous age of 90 to 100 m.y. (35). Perhaps not too much significance should be attached to a single age determination, but this Cretaceous age conceivably could represent: (1) the age of all intrusive igneous activity in the district; (2) the age of a period of explosive igneous activity and breccia intrusion younger than mafic dike intrusion; or (3) the date of mineralization genetically related to, but younger than, igneous intrusion.

## Structure

The Illinois-Kentucky fluorspar district is located in the most complexly faulted area of the central United States (Figure 3). It is situated on a generally northwest-trending structural high that is flanked on the north and northeast by the Illinois Basin and on the southwest by the Mississippi River Embayment. The Shawneetown-Rough Creek Fault Zone, chiefly a high-angle thrust in the vicinity of the fluorspar district, enters Illinois from the east a few miles north of the mining district, bends to the southwest to form the west edge of the district, and finally merges with the northeast-trending New Madrid Fault Zone (shown as a zone of bending and faulting in Figure 3). Within the district, northeast-trending normal faults slice the area into a series of long, narrow blocks (Figure 4). These northeast-trending faults have been traced southwestward into the embayment sediments (45), where they presumably become a part of the New Madrid Fault Zone.

Hicks Dome, an oval structure with about 4000 feet of uplift at its highest point, forms the structural apex of the district (Figure 4). The oldest rocks in the district, limestone and chert of Devonian age, crop out in a square mile area at the apex of the dome. Ringed by the less resistant Mississippian-Devonian New Albany Shale Group, the central limestone area occupies a topographic as well as a structural high. Outside the ring of New Albany, the Ft. Payne Formation (siltstone, limestone, and chert) forms an encircling cuesta that rises to approximately the same elevation as the central limestone area. Many occurrences of intrusive breccia are known within the central area and around its periphery and are particularly well-exposed as dikes around the inner face of the upper part of the cuesta. Radial and ring faults have been mapped that extend as much as 4 miles from

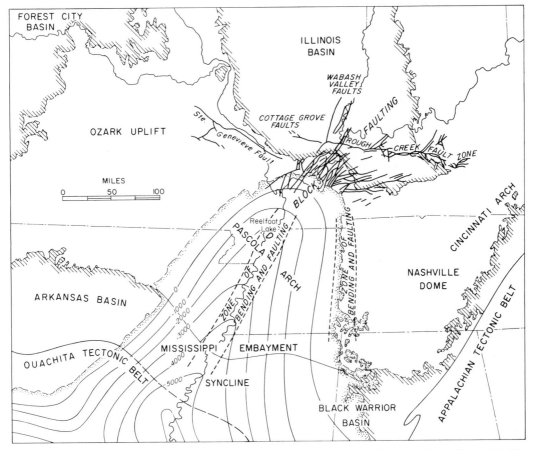

FIG. 3. *Regional Structural Setting of the Illinois-Kentucky Mining District (from Ross, 45, Figure 2). Diagonal shading outlines areas of Pennsylvanian strata; dotted shading outlines area of Cretaceous-Tertiary strata. Contours are drawn on top of Cretaceous.*

the apex of Hicks Dome (49,54). The presence of the intrusive breccias and radial and ring faults suggest that Hicks Dome is of explosive origin (22, p. 897; 50, p. 11). An aeromagnetic survey of the Illinois portion of the mining district showed no magnetic anomaly under Hicks Dome, thus ruling out the likelihood of a laccolithic intrusion (48).

The mafic dikes occupy fractures trending N10° to 45°W, with the single exception of one with a northeast strike near the apex of Hicks Dome (Figure 4). Where relations between dikes and faults are observable, the dikes are cut by the faults and, therefore, pre-date the block faulting (18, p. 321; 19, p. 72).

Intrusive breccia exposures are known only from the Illinois part of the district, all lying within 6 miles of the apex of Hicks Dome (Figure 4). Within the area of Devonian strata at the apex of Hicks Dome and around

its immediate periphery, the intrusive breccia occurs chiefly in dike-like bodies with northeast and northwest strikes (not shown in detail on Figure 4). The occurrences away from the central area of the dome are bodies of undetermined shape but are presumed, from their roughly circular or oval outcrop areas, to be plugs.

The structure of the district has been interpreted as being a northwest-trending domal anticline pushed up by igneous intrusion and broken by gravity faults on shrinkage of the cooling magma (6, p. 56–57; 50, p. 11–12). Inasmuch as the northwest-trending mafic dikes indicate tension in a northeast-southwest direction, due, presumably, to igneous intrusion, and the northeast-trending block faulting indicates tension aproximately normal to this, the relation between igneous intrusion and block faulting is not simply one of an upward

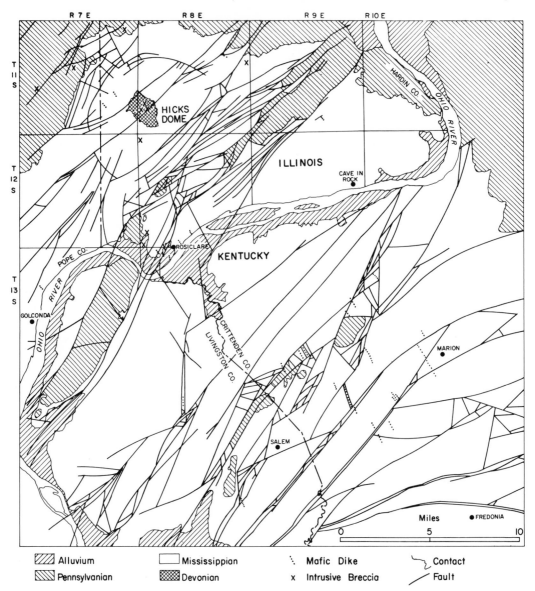

FIG. 4. *Geologic map of the Illinois-Kentucky Mining district (compiled from Weller, et al., 6; Weller, et al., 19; Baxter, et al., 42; Baxter and Desborough, 49; Heyl, et al., 50 and Baxter, et al., 54).*

push followed by collapse. Heyl, *et al.* (50, p. 12) suggest that faulting on the northeast-trending New Madrid System intervened between intrusion and cooling of the magma and set the pattern for the subsequent block faulting. It also is possible that initial faulting and intrusion were contemporaneous; if the early faulting was compressional, with major movement occurring along the Shawneetown Fault Zone, the dikes could have been intruded at

the same time into tension fractures normal to the compression and the block-faulting initiated along release fractures on relaxation of the compresion. Whatever the time relations between initial faulting and intrusion, however, it is apparent from the geologic map (Figure 4) that the subsequent block faulting post-dated arching. As field relations indicate that it also was later than igneous activity, the block faulting probably was the last major tectonic event. In fact, movement has recurred on faults of the New Madrid System to the present day, as evidenced by earthquakes (44).

## Age of Mineralization

The fluorite-sulfide mineralization of the Illinois-Kentucky mining district is later than the faulting that cuts Pennsylvanian strata. That the mineralization also is later than at least most of the igneous activity is shown by the veining of mafic dikes by fluorite and sulfides (18; 19, p. 72) and by the occurrence of fluorite crystals in the open spaces of intrusive breccias (22, p. 897; 50, p. 12). As the age of the igneous activity has been established by isotopic dating to be at least as young as Permian, it follows that the mineralization is Permian or younger.

Other lines of evidence suggest further possible refinements to the dating of the mineralization but are not unequivocal. For example, evidence from the embayment area indicates that most of the block faulting took place before Cretaceous sediments were deposited, as Pennsylvanian and Mississippian beds are displaced much more than the overlying Upper Cretaceous strata (45, p. 21). This suggests that the mineralization took place before Upper Cretaceous sedimentation*, with the minor displacements of Upper Cretaceous strata related to the post-mineral movement on the faults in the fluorspar district. Furthermore, many geologists familiar with the southern Illinois region feel that the Cretaceous sediments were not laid down on as flat a surface, as postulated by Ross (45, p. 18), but on a faulted surface of considerable relief, suggesting that much of the block faulting took place just prior to Upper Cretaceous sedimentation. Another bit of information, which appears to fortify a Cretaceous age for the mineralization, is the early late Cretaceous age derived from a single determination on monazite from a Hicks Dome breccia.

---

* It is not known which part of the Upper Cretaceous is represented by these beds. The base gravel or conglomerate is correlated with the Tuscaloosa Formation (40, p. 18), deposition of which probably began fairly early in the Upper Cretaceous, perhaps at about 90 m.y., based on Kulp's estimate of 100 m.y. as the beginning of late Cretaceous time (36, Figure 1). However, at the head of the embayment the conglomerate is a transgressive deposit at the base of the McNairy Formation, and its time of deposition would be closer to the middle part of the Upper Cretaceous.

From the evidence at hand, it may be concluded that the period of mineralization that led to the fluorite-sulfide deposits was of Permian age or younger, possibly early late Cretaceous.

## FORMS OF THE ORE BODIES

Fluorspar in the Illinois-Kentucky district occurs in fissure veins, in bedding-replacement deposits, and in residual deposits resulting from the weathering of near-surface portions of the primary forms.

### Vein Deposits

The veins follow steeply inclined faults or fault zones, predominantly of the normal type. The vertical displacement of faults occupied by workable veins is mostly in the range of 50 to 500 feet. The actual direction of fault movement, as indicated by slickensides, usually has a strong vertical component but is in places nearly horizontal. In some places, the wall rocks of the fault have been only slightly brecciated, and the vein extends massively from one smooth, well-defined wall to the other. Elsewhere, there has been much brecciation along a general fault zone, and, in such areas, veins may occupy only a part of the zone or may split around large slabs of wall rock and send small branch veinlets out 50 to 100 feet into the walls. The various degrees of brecciation may succeed each other along the course of a single ore shoot. The dip and strike of the vein also may change rapidly from place to place.

Most veins consist of calcite locally replaced to a greater or lesser degree by fluorspar, which forms ore shoots of various dimensions and characteristics. In other veins, calcite is much less prevalent, and the ore occurs in pockets, where it may have resulted from replacement of the wall rock by fluorspar (28, p. 22). In still others, fluorspar cements brecciated wall rock, which is usually a sandstone or siliceous limestone (3, p. 136–137). This last type is minor in occurrence.

The veins range in width from a fraction of an inch to at least 45 feet. The fluorspar ore shoots alone, which may occupy the whole or only a part of the vein, are in a few places as much as 25 to 35 feet wide but more commonly range from 1 to 15 feet. Some of the ore shoots revealed by mining were remarkably long and continuous, whereas others pinched and swelled in an erratic manner. The major ore shoot in the Blue Diggings vein in the

Fairview Mine at Rosiclare was mined continuously for 5700 feet. It was 600 to 700 feet high, averaged 3.5 feet wide, but in places reached 30 feet. The Rosiclare vein ore body in the Rosiclare Mine was mined continuously for 4850 feet, from the surface to 720 feet in depth, and averaged 5 to 6 feet wide throughout, with a maximum width of 14 feet. The vein in the Daisy Mine, also at Rosiclare, had an ore shoot 2000 feet long and 250 feet high with an average width of 3 feet and a maximum of 9. These, however, are exceptional cases. In the better fluorspar mines of the Illinois-Kentucky district, more typical dimensions for average ore shoots are 250 to 400 feet long, 100 to 200 feet high, and 5 feet wide.

Ore shoots terminate laterally and with depth by lensing out into barren calcite, although sometimes they are cut off by cross-faults or approach and join another ore shoot on another fault or another branch of the same fault. Major ore shoots, separated by 200 feet or more of barren vein material, succeed one another along many faults (the Blue Diggings near Rosiclare is a good example) but no instance is known to the authors where this has been the case in the vertical direction. Perhaps this is only because of the rather limited depths to which the mines penetrate. The deepest workings in the entire district are on the 900-foot level of the Blue Diggings vein in the

Fairview Mine, where the vein is mostly calcite with small amounts of fluorspar. Actually, only a little ore was mined below the 800-foot level in this mine. There have been many exploratory holes drilled below the bottoms of ore shoots in the major mines in the Rosiclare area, but nothing of promise has been found. A breccia zone of calcite and limestone was found below the veins on the Alcoa property (37, p. 6), for example, and barren continuations of productive veins were encountered in drilling at intervals down to 500 or 700 feet below the overlying ore in the Daisy and Hillside mines (14, p. 11–15). These latter holes reached depths of 1000 to 1360 feet below the surface.

### Bedding-Replacement Deposits

In contrast to vein deposits, which occur in all parts of the district, bedding-replacement deposits occur only in two areas, one north and northwest of Cave in Rock, Illinois, and the other south and east of Carrsville, Kentucky, across the Ohio River from Rosiclare, Illinois. As little mining has been done in the Carrsville deposits, the knowledge of the geology of the replacement deposits comes largely from studies in the Cave in Rock district (19,39).

The bedding-replacement deposits typically are linear in plan and crescentic or wedge-

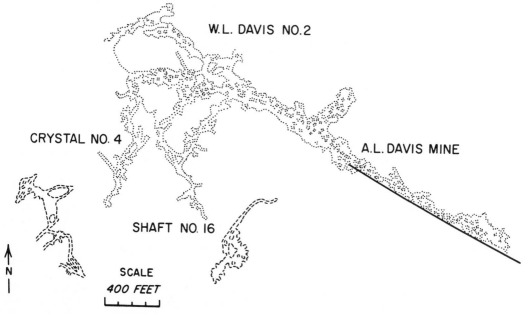

Fig. 5. *Underground Workings in a Portion of the Cave in Rock Area, illustrating typical ore-body shapes (modified from Brecke, 39, Figure 5).*

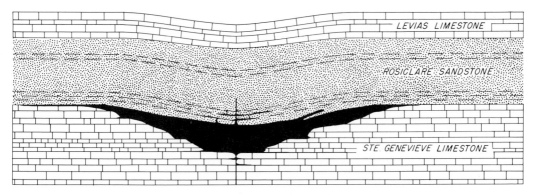

ORE BODY SYMMETRICAL ABOUT CENTRAL FISSURE

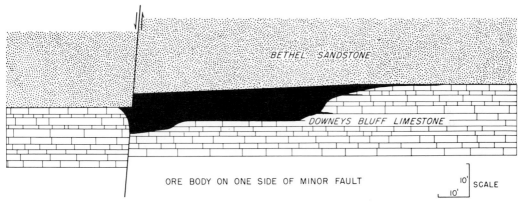

ORE BODY ON ONE SIDE OF MINOR FAULT

FIG. 6. *Schematic Cross Sections, showing the two general types of bedding-replacement deposits in the Cave in Rock area (from Grogan, 15, Figure 2).*

shaped in cross section (Figures 5 and 6). Ore bodies are commonly 50 to 200 feet wide but may reach as much as 500 feet. They commonly are from 5 to 20 feet thick, although, in restricted areas, the ore may reach 30 feet. The deposits wedge out laterally at their margins, and mining is terminated when the ore becomes too thin to mine profitably. Most of the ore bodies range from 200 to 1500 feet long, but several more than 3000 feet long have been found, and one, consisting of the Oxford Mine of Ozark-Mahoning Company in its southwest portion and the No. 1 Mine of Minerva Company in its northeast portion, has a length of over 2 miles.

As indicated by the cross sections (Figure 6), the ore minerals replace specific limestone beds outward from a central (or marginal) joint or fault of small displacement. The ore is thickest along the main fracture and wedges out at the margins; it is widest at the top of the ore body. In the larger deposits, the main fracture may be a single break or several parallel breaks in a zone as much as 10 feet wide.

Many small deposits, such as those southwest of the W. L. Davis No. 2 Mine (Figure 5), consist of interconnecting ore bodies along individual intersecting fractures.

## Residual Deposits

Weathering of the limestone host rock and calcite gangue of vein and bedding-replacement deposits has resulted in the concentration of the more insoluble fluorite into residual deposits, consisting of pieces of fluorite mixed with residual clay and boulders of host rock. Because of this concentrating action on the fluorite, a vein too narrow to be mined profitably may be overlain by a workable residual deposit. The overburden above a major vein also may contain a residual deposit as much as 60 feet wide.

Most of the residual deposits have been associated with veins rather than with bedding-replacement deposits, probably largely because of the much greater number of deposits of the vein type in the fluorspar district and their much greater vertical dimension. However, the bedding-replacement deposits also probably are less subject to disintegration by weathering than the vein deposits because of the less fractured nature of their host rock.

The tonnage produced from residual deposits throughout the history of the district is not known, but it probably has been small compared to that from the primary deposits. Residual ore may have formed a significant part of the production from the Kentucky portion of the district (23,24,25,26,31,41), but only a minor tonnage has come from residual ore in Illinois. This difference probably is related to the apparently greater degree of fracturing and brecciation along the faults in the Kentucky part of the district.

## STRATIGRAPHIC RELATIONS OF THE ORE

Although fluorspar mineralization in the Illinois-Kentucky mining district spans a considerable part of the stratigraphic column (Devonian to Pennsylvanian), most of the commercial deposits occur chiefly in strata of Mississippian age, from the St. Louis Limestone upward to the middle part of the Chesterian Series (Figure 2).

### Vein Deposits

Vein deposits show the widest stratigraphic range. The very productive veins of the Rosiclare district were mined from the lower part of the St. Louis to the middle part of the Chesterian Series. Minor commercial production also has come from outlying deposits as low stratigraphically as Middle Devonian (Rose Mine just southeast of the apex of Hicks Dome) and as high as the lower part of the Pennsylvanian (mines along the Lusk Creek Fault Zone at the northwest margin of the mining district). However, the most productive parts of the vein deposits have been those in which one or both walls were Ste. Genevieve or St. Louis Limestone, presumably because of the superior competency of these formations.

### Bedding-Replacement Deposits

The bedding-replacement deposits are more restricted stratigraphically than the veins and tend to occur at three principal stratigraphic levels. Although mineralization has been found in the lower part of the Ste. Genevieve as much as 170 feet below the top of that forma-

tion in a probable solution-collapse breccia (39, p. 513), the lowest important ore horizon is at the level of the Spar Mountain Sandstone Member, approximately 60 feet below the top of the Ste. Genevieve. The two other principal ore horizons are at the top of the Ste. Genevieve Limestone immediately below the Rosiclare Sandstone and at the top of the Downeys Bluff Limestone immediately below the Bethel Sandstone. Other horizons may be ore-bearing locally in the upper part of the Ste. Genevieve and in the Levias Member of the Renault Limestone or in limestones in the lower part of the Shelterville Member immediately above the Levias. To the authors' knowledge, no bedding-replacement ore has been discovered above the base of the Bethel.

## MINERALOGY

### General

Although the same minerals generally are present in both the vein and bedding-replacement deposits, the different modes of emplacement of the minerals of the two types of deposits has led to certain differences in the proportions of the various minerals present, their spatial relations to one another, and their textures. The various aspects of the mineralogy of the deposits are discussed separately in the following paragraphs for each type of deposit.

### Minerals Present

VEIN DEPOSITS   The most important economic mineral in the vein deposits is fluorite. Sphalerite and galena are present in various, but generally small, amounts. The principal gangue mineral is calcite. Other minerals present include quartz, pyrite, chalcopyrite, and barite.

*Fluorite* is chiefly coarse grained and massive, though occasional cavities in the veins yield groups of cubic crystals. Most of the fluorite is white or light gray, but shades of purple or blue are common, and green may be found in places. The fluorite generally occurs within the calcite as tabular or lenticular masses, more or less parallel to the fault walls, and may be said to form veins within the calcite veins. In the richer parts of the fluorite-calcite veins, the fluorite may occupy most of the space between the fault walls; as the amount of fluorite decreases, it forms only narrow veins or veinlets within the calcite. In parts of some veins, calcite may be virtually absent, and, as mentioned in the section on "Forms of the Ore Bodies," wall rock may have been replaced by fluorite in these places.

*Sphalerite* and *galena* occur chiefly as small masses, disseminated grains, and veinlets in the fluorite or in the wall rock bordering the veins. The sphalerite generally is reddish brown and the galena a typical steel gray. The distribution of the two sulfides is erratic even within an individual fluorspar vein, ranging from none to several per cent, but ore from most vein mines will average only 1 to 2 per cent combined sphalerite and galena.

Two exceptional concentrations of zinc minerals are found in the Kentucky part of the district. At the active Hutson mine (18), a considerable tonnage of high-grade sphalerite ore has been produced, and, at the abandoned Old Jim mine (7, p. 26–28), a presumably minor tonnage of zinc carbonate was mined by open-cut methods. Both deposits occur adjacent to, and partly traverse, mafic dikes and are characterized by the absence of fluorite and galena. Other important concentrations of zinc minerals in vein deposits appear to be zinc-rich portions of fluorspar veins, commonly with accompanying galena. Some of these zinc concentrations are associated with mafic dikes, others are not.

The distribution of galena in the vein deposits is similar to that of sphalerite, the two normally occurring together, with the two principal exceptions noted above. Abnormal concentrations of galena are found in places but generally are not large enough to support profitable operations for galena alone. It is believed, however, that several of the productive fluorspar veins originally were mined for their lead ore. This particularly may have been true of the Rosiclare vein system, if an 1872 prospectus that spoke glowingly of the rich lead deposits at "Rose Clare" is to be believed.

Unlike galena from other Mississippi Valley region deposits, that from the fluorspar district carries some silver. Published figures indicate amounts as high as 14 ounces per ton of galena concentrates with an average of 5 to 7 ounces per ton for the Illinois portion of the district and up to 3 ounces for the Kentucky portion (7, p. 45). Heyl, *et al.* (52, p. 950) reported amounts of 12 ounces per ton from Hicks Dome and Rosiclare galenas, 4 to 6 ounces per ton from deposits near Salem and Marion, Kentucky, and 2 to 3 ounces near Princeton, Kentucky, at the southeastern edge of the district.

*Calcite* is the most abundant mineral in the vein deposits. It generally is white to light gray, coarse grained, and massive. As is true of fluorite, it also may be found lining cavities, where it may form scalenohedral crystals as large as 3 or 4 inches in diameter, though diameters of 1 or 2 inches are more common. The proportion of calcite to fluorite increases in the deeper parts of the veins, and the fluorspar deposits generally are said to "bottom in calcite."

*Quartz* is present in minor amounts. Within the veins, it most commonly occurs in vugs, where small crystals may coat calcite or fluorite. It also is present in the wall rock at many places, where it is found as euhedral crystals replacing limestone next to the veins.

*Pyrite* is a widespread but minor constituent of the fluorspar veins. It may occur as crystals on fluorite cubes or as veinlets in fractures in fluorite or in the wall rock. Crusts of radiating pyrite have been found lining cavities in a few places (11, p. 29).

*Marcasite* is abundant in the sphalerite veins of the Hutson mine (18, p. 325). Currier (7, p. 27) reported abundant pyrite and masses of limonite in the waste piles at the Old Jim mine but had no way of knowing whether it was associated with intrusion of the dike or with the zinc mineralization.

*Chalcopyrite* is fairly common as small crystals in fluorite or lining in vugs but is quantitatively unimportant. Bastin (11, p. 30) reported the occurrence of a 4-pound mass of chalcopyrite in fluorite in the Hillside Mine but stated that such an occurrence was exceptional.

*Barite* is a rather common mineral in the vein deposits, but, because of its sporadic occurrence and its generally minor quantity in any one deposit, little attempt has been made to recover it commercially (30; 7, p. 43). It is most commonly found as compact crystalline masses within the veins or as bladed crystals coating fluorite. It apparently is much more common in the upper parts of veins than in the deeper portions. This fact, together with the occurrence of veins in which barite predominates in the marginal portions of the district, has led Brecke (46) to postulate barite zoning.

Supergene minerals are present but, on the whole, are economically unimportant. Local concentrations of *smithsonite* and *cerussite* have been mentioned in the discussions of sphalerite and galena. *Cuprite* and *malachite* can be found as the oxidation products of the small amounts of chalcopyrite in the ores. *Gypsum,* reported from the upper levels of the Hillside mine (11, p. 31), also is probably a secondary mineral, resulting from reaction of the vein calcite with sulfate ion derived from the oxidation of sulfides.

BEDDING-REPLACEMENT DEPOSITS   As in the vein deposits, the principal economic mineral in bedding-replacement deposits is fluorite. Sphalerite and galena are abundant in some deposits but virtually lacking in others. Other primary minerals include calcite, quartz, barite, chalcopyrite, pyrite, marcasite, witherite, and strontianite. The more common secondary minerals are smithsonite, cerussite, and malachite. A little pyromorphite has been found at the Patrick Lead mine.

*Fluorite* occurs in coarse- and fine-grained layers in banded replacement ore, as disseminated grains in partly replaced rock, as large or small cubic crystals lining cavities, as massive bodies of various shapes within the ore bodies, and as veinlets in the ore and the adjacent strata above and below the ore. Some fluorite is colorless, white, or gray but much is tinted various shades of purple, yellow, or rose, with blue and green varieties much less common. Zonation of color parallel to the cube faces is conspicuous in many crystals, the commonest arrangement being a purple or blue exterior around clear or yellow cores. Individual crystals of considerable size occasionally are found and highly prized by collectors. One such crystal in a group from the Crystal mine measures 12.5 inches on an edge.

*Sphalerite,* practically absent in some deposits, is abundant enough in others to constitute a valuable part of the ore, as, for example, in the Davis-Deardorff, West Green, and Minerva #1 mines. The mineral occurs principally as a replacement of limestone in the fine-grained layers of banded ore and, as such, generally is fine grained. In open spaces in the ore, larger crystals are found but seldom exceed 0.25 inches in diameter. The sphalerite generally is dark brown to black, suggesting a high iron content, but reddish brown and yellowish brown crystals are not uncommon as disseminated grains in low-grade fluorspar (incompletely replacing limestone). Cadmium is present in the sphalerite to various degrees; Bailie (21) reported that some Ozark-Mahoning ore bodies contained considerable quantities of cadmium. Bailie stated that the high-cadmium zinc concentrates ran 0.8 to 1.08 per cent cadmium, but he gave no figure for the low-cadmium zinc concentrates. Sphalerite from Minerva #1 mine has averaged about 1 per cent cadmium since the mine opened in 1943 (19, p. 119; 20, p. 76). Germanium also is present in recoverable amounts, running 0.025 per cent in zinc concentrates from

Minerva #1 (20, p. 76). No figures have been published for germanium in Ozark-Mahoning ores, but the amount of germanium probably is about the same as in Minerva sphalerite. Gallium is reported to be present in at least some of the sphalerite, but little is known of its amount or distribution (19, p. 119).

*Galena* is more widespread in its occurrence than sphalerite but, like the latter, is abundant enough to pay for its systematic recovery in only a few deposits. Among the major deposits, for example, the Davis-Deardorff contained substantial amounts of galena, while the nearby West Green and the Minerva #1 ore bodies contained virtually none. The mineral occurs for the most part as well crystallized masses from 0.25 to 3 inches in diameter in coarse-grained layers of banded ore and in crystal-lined cavities. It contains some silver, the general amount of which is indicated by an assay obtained by the Ozark-Mahoning Company. It showed 7.38 ounces of silver per ton from 24 carloads of galena concentrates, principally from the Davis-Deardorff mine.

*Quartz* differs greatly in abundance in various deposits. It occurs in two principal forms: (1) as tiny, doubly terminated crystals in the fine-grained layers of banded ore, in disseminated ore, and in the limestone adjacent to the ore in some deposits and (2) as small to large crystals incrusting other minerals in cavities. Quartz is very abundant in some deposits and practically absent from others. Some of the banded ore from the Davis-Deardorff and the Lead Hill mines consists of alternating bands of fluorspar and quartz, some of the latter being coarsely crystalline in some specimens and fine-grained in others (19, p. 122; 39, p. 530). Because of its abundance in certain deposits and its close association with the ore, there is little doubt that quartz was introduced during the mineralizing process, but it is also possible that some of it was recrystallized from sand grains and silt particles originally present in the limestone.

*Barite,* like quartz, is sporadic in its occurrence, being exceedingly abundant in some deposits, or parts of deposits, and essentially lacking in others. It occurs principally as fine grained aggregates replacing limestone, calcite, and fluorite, to a lesser extent as bunches of small-bladed crystals in cavities, and more rarely in stalactitic forms in cavities. At several places, it has completely replaced the coarsely crystalline fluorspar in banded ore, leaving the fine-grained layers almost untouched. In general, the barite seems to have replaced massive or coarsely crystalline fluorspar preferentially over fine-grained fluorspar. Barite appears to have formed late in the period of mineral deposition and to have been deposited principally along the top, bottom, and margins of the deposits.

*Chalcopyrite, pyrite,* and *marcasite* are present only in minor amounts, generally incrusting the surfaces of fluorite cubes. Chalcopyrite crystals with a nail-like form have been noted within fluorite cubes. The "nails" commonly are oriented at right angles to the cube faces of the fluorite crystals with their points toward and their heads away from the center. It is common for the heads of a number of the "nails" to lie in a common plane that represents an earlier cube face at one stage in the growth of the fluorite crystal.

*Witherite* has been found in the Minerva and West Green mines and *strontianite* in the Minerva mine. The witherite occurs in grayish pseudohexagonal twinned mosaic crystals as much as 3 inches long and 1.5 inches wide in cavities in the ore. In some places, it appears to be an alteration product of barite. The strontianite occurs as slightly pinkish bladed grains and brownish aggregates, both with radial structure. No well formed crystals of strontianite have been found.

*Smithsonite, cerussite* and *anglesite,* and *malachite* are found in small quantities in the oxidized portions of near-surface deposits. They result from ground-water alteration of sphalerite, galena, and chalcopyrite, respectively.

*Pyromorphite* occurs as tiny yellowish to bluish green crystals together with galena, cerussite, and anglesite in oxidized ore in the Patrick open-pit lead mine. Although difficult to find in mine exposures, it showed up conspicuously on concentrating tables in the mill as a broad fringe of green material at the upper margin of the lead concentrate. The mineral is of secondary origin.

*Bitumen* occurs in the deposits in several different forms. "Live" oil oozes from the host rock in some mine openings and is found as globular inclusions in fluorite crystals. "Dead" oil may be found as black films in fractures or pore spaces in the host rock and on crystals of fluorite and calcite in the ore. Asphalt of various degrees of viscosity occurs in vugs in ore and wall rock. Solid, brittle asphaltite is reported by Brecke (39) as being the most common form of bitumen in the ores. A rather large mass of asphaltite was discovered partially filling a cavity in the limestone host rock in the Hill mine. According to B. L. Perry, Ozark-Mahoning geologist, the cavity was 10 by 3 feet in plan, 1 to 1.5 feet high, and approximately half filled with asphaltite from the bottom up. The top surface of the asphaltite mass was essentially level. The bituminous material was mostly vitreous, but a layer of dull frothy material approximately 1 inch thick lay at the top. Samples donated by Perry and F. H. Hansen were examined in the laboratories of the Illinois State Geological Survey and identified as the asphaltite, grahamite.

In hand specimen, the 1-inch layer of frothy material makes a sharp but gently undulating contact with the vitreous material. The vitreous material itself exhibits two well-defined textures: (1) irregularly shaped fragments with a smooth, glass-like lustre in (2) a matrix that appears finely granulated to the naked eye. Under the binocular microscope, the matrix material is seen to be somewhat vesicular. The granular aspect is due to a minutely hummocky or "lumpy" surface, such as one would find in gelatin that had been well stirred after having become partly set. Small local displacements of less than 0.5 mm also are common in the matrix along a myriad of subparallel fractures a millimeter or less apart and provide discontinuities that help give it the finely granular appearance.

The only mineralization noted in the samples of grahamite were some cleavage pieces of fluorite in one specimen and crystals of quartz up to 1 mm in length, both singly and in small groups, in frothy material on another specimen. Some of the quartz crystals contained inclusions of apparently solid bitumen and may be presumed to have grown in place.

The brecciated nature of the grahamite and the "lumpy" character of its matrix suggest that the bitumen had become at least partly hard when it was disrupted by additional degassing, either as part of a more or less continuous process or because of an influx of heat supplied, perhaps, by warmer mineralizing solutions. Hall and Friedman (43, p. 905) postulated an influx of magmatic water following the major part of fluorite deposition and preceding or coinciding with quartz and sulfide deposition. Although such an influx does not necessarily imply an appreciable heating up of the mineralizing solutions, the age relations between the grahamite and included fluorite and quartz suggest that the grahamite formed after the fluorite and before or during quartz deposition.

## Wall-Rock Alteration

VEIN DEPOSITS Alteration of wall rock in the vein deposits consists chiefly of *silicification,* which appears to have affected both limestone and sandstone.

Hardin (28, p. 24) reported replacement of limestone along veins of the Babb Fault (Kentucky) by small doubly terminated crystals of quartz and small masses of chalcedony, estimating that "5–20% of silica has been added to most limestones within 20 feet of the veins." Trace (41, p. 17) found that limestone wall rocks in the Keystone Mine (Kentucky) were "silicified to varying degrees 10–30 feet from the vein." Harrison (32) reported an increase of silica in the limestone of the hanging wall within 18 inches of the veins in Alcoa's Rosiclare workings. Sandstones are commonly altered to quartzite, resulting in the surface expression of some faults as quartzite reefs, particularly in the Kentucky portion of the district.

Bastin (11, p. 15) noted *ferruginous calcite* in the limestone adjacent to the vein at two places in the Hillside mine.

BEDDING-REPLACEMENT DEPOSITS Wall rock alteration, in the sense of mineralogical changes in the host rock, is quite minor in the replacement deposits. Some solution of calcite in excess of its partial replacement by fluorite in the more impure limestone beds has resulted in a softening of these strata. Brecke (39, p. 525–530), who refers to this solutional removal of calcite as *decalcification,* points out the presence of chalky-appearing, silty, dolomitic beds at the edge of the Oxford ore body that apparently are the residue from impure limestone beds that have suffered decalcification.

*Dolomitization* appears to have been of little importance during the mineralization period. Brecke (39, p. 504–505) mentions certain strata of the host limestone within the ore that are chiefly dolomite but feels that these occurrences are the result of residual concentration of dolomite following solutional removal of calcite from dolomitic limestone beds. Occasional dolomite veinlets, some iron-bearing, in the ore suggest mobilization and redeposition of dolomite present in the host limestone. Brown-weathering layers in limestone host rock have been noted within a few yards or tens of yards of fluorite deposits and appear to be due to the presence of iron-bearing dolomite. Although these brown layers may have been formed by the introduction of iron-bearing carbonates, they may have been caused merely by increased susceptibility to weathering brought about by decalcification and residual concentration.

*Quartz* is quite common as tiny doubly terminated crystals in limestone that contains disseminated fluorite and in the fine-grained layers between coarse, quite-pure fluorite layers in banded ore. Quartz of the same habit also may be found in the limestone at the edges of some ore bodies. Silicification is intense, however, in only a few places. In the Davis-Deardorff mine and in the Lead Hill deposits on the southwest extension of the Davis-Deardorff fracture zone, quartz has completely replaced layers of limestone, producing a banded texture. The intervening layers of limestone may or may not be replaced by fluorite. The Hill mine, in the northeastern part of the Cave in Rock district, showed replacement of limestone by brown microcrystalline silica at the horizon of the Spar Mountain Sandstone Member (39, p. 531). As the Davis-Deardorff and Lead Hill deposits also are at the level of the Spar Mountain (the deepest of the major replacement ore zones), the distribution of quartz suggests a vertical zoning in the Cave in Rock district.

## Mineral Structures and Textures

VEIN DEPOSITS Fluorite and calcite, the most abundant, and in places the only, vein minerals, occur chiefly in massive form, with lenticular masses of fluorite occupying the central portions of the veins. Where there is only a small amount of fluorite, it fills narrow veins and veinlets in the calcite.

Fluorite may show various kinds of contacts with massive calcite. In a single hand specimen may be found both smooth and ragged calcite crystal faces projecting into fluorite, fluorite crystal faces projecting into calcite, calcite replaced along its cleavages by fluorite, and smoothly curving contacts between calcite and fluorite. Such a mixture of contact relations suggests that the solutions that brought about replacement of calcite by fluorite also redeposited calcite, at least locally. It seems reasonable to suppose that the solutions could at times have become saturated, or even supersaturated, with the carbonate ions that were being removed from the vein calcite, thus temporarily becoming calcite-depositing instead of calcite removing solutions.

The narrow veinlets of fluorite in calcite generally have walls that do not match, a condition that is commonly ascribed to replacement. However, the fact that many of the veinlets show crystal faces of calcite projecting into the fluorite suggests that the nonparallelism of the walls may have been caused in part by additional deposition of calcite on the veinlet walls subsequent to fracturing but prior to deposition of fluorite.

Fluorite cut by veinlets of late calcite also has been found, but this type of relation is quantitatively unimportant.

Another textural feature common in many veins is a banding parallel to the walls. The bands may be narrow or wide, or both, and may be displayed in massive fluorite or as alternating bands of fluorite and calcite or fluorite and wall rock. In massive fluorite, the bands generally show up by color differences, for example, as alternating bands of purple and light-colored fluorite. In places, alternate bands of fluorite may differ in sulfide mineral content. Many veins exhibit pronounced sheeted structure in addition to the banding, and there may be slickensides on some of the fracture surfaces, indicating post-ore movement on the fault. Even where the veins are not conspicuously sheeted, a more or less centrally located parting seam of gouge 1 to 6 inches wide is quite common.

Galena commonly occurs as disseminated grains and replacement veinlets in fluorite. The veinlets are generally less than an inch wide and are bounded by galena cube faces. Currier (5, Plate II-D) showed photographs of galena along cleavage planes in calcite.

Sphalerite is found as individual crystals, small masses, and veinlets in fluorite. In places, sphalerite may show replacement of fluorite along fluorite cleavage directions (5, Plate II-E). In some vugs, late fluorite has been deposited on sphalerite. Sphalerite also may be disseminated in the wall rock next to the veins. Its relations to galena are not entirely clear. Because of their commonly scattered distribution, the two sulfides usually do not show contact relations with each other, even though they both occur in the same local areas. In one specimen from the Jefferson mine, in which galena veined fluorite that contained sphalerite in individual crystals and small masses, the galena also contained sphalerite crystals as islands, suggesting that the sphalerite in the galena was residual from replacement of the host fluorite by galena.

Chalcopyrite generally occurs as small grains, both in the fluorite and in the other sulfides but is in such small amounts that no unequivocal contact relations could be established.

Pyrite, minor in amount, occurs as crystals or coatings on the faces of fluorite or other sulfide crystals. It also may be found as small veinlets in fractures in the vein or wall rock.

Barite, in scattered crystalline masses within the fluorite veins, bears replacement relations to the fluorite. Barite crystals also have been found in cavities in the veins.

BEDDING-REPLACEMENT DEPOSITS The ore structures in the bedding-replacement deposits may best be described as banded, imperfectly banded, and disseminated.

Banded ore (Figure 7A) consists of layers of coarsely crystalline, prevailingly light-colored, practically pure fluorite, alternating with finer-grained, darker layers containing fluorite, other minerals, and unreplaced limestone in various proportions. The layering gives to faces of ore a conspicuously banded appearance, leading to the local name of "coontail ore." The layers range from a fraction of an inch to several inches thick, most being 0.5 to 1 inch thick. Some layers are continuous for many feet, whereas others terminate abruptly or split into two parts within short distances. In most places, the layering is prevailingly parallel to the major bedding planes or it curves and is inclined much like the ordinary cross-bedding in sediments. In some exposures, however, the banding is contorted, bent, or broken in a manner difficult to associate with any possible original sedimentary structures.

The fluorspar in the coarse layers consists of a pair of crusts, the crystal-studded surfaces of which face inward toward a center seam. In some places, the crystals have intergrown so that they interlock, and the crusts meet along an irregular, approximately centrally located plane; elsewhere they project into an open center seam or cavity. This comblike structure is believed to indicate that the crystals grew into an open space.

The fine-grained darker layers have no comb structure. They contain fluorite, various impurities, such as grains of calcite, ferroan dolomite, and quartz, and in some deposits also contain sphalerite. Of the impurities, the calcite and round-grained quartz are unreplaced remnants of the original limestone, whereas the quartz in tiny needle-like crystals is believed to have been introduced during the mineralization. The status of the ferroan dolomite as either a primary or introduced mineral has not been established.

In some of the banded ore from the Lead Hill mines or from the Cave in Rock mine, the coarse layers are composed of comb quartz or comb quartz on fluorite, and the fine-grained bands consist of fine-grained quartz that completely silicified limestone. In other deposits, the fluorite in the coarse layers is more or less completely replaced by barite.

FIG. 7. *Photographs of Banded and Imperfectly Banded Ore in the Victory Mine, Cave in Rock Area. (A) Banded ore, Carlos ore body; (B) imperfectly banded ore, Addison ore body.*

Although some deposits consist chiefly of banded ore, others may be composed partly or almost wholly of imperfectly banded ore (Figure 7B), a textural type characterized by layers of various colors, grain sizes, and purity but without the regularity or sharp definition of banded ore. In general, imperfectly banded ore contains less fluorite than banded ore, but this, as is the textural distinction between the two, is largely a matter of degree. It is common to find beds of imperfectly banded ore interlayered with beds of banded ore.

Occasional ore beds, which appear essentially nonlaminated, contain more or less abundant fluorite, and sometimes sphalerite, in disseminated form. The beds occur within sets of imperfectly banded beds but most commonly are found at the outer margins of deposits, and generally show weak or incipient mineralization of limestone that ultimately might have been converted to imperfectly banded ore.

Massive fluorite is common as horizontal veins and lenticular masses, primarily at the tops and margins of deposits, and as irregular masses having the general shape of vertical veins of various widths. In some ore bodies, particularly those in the Ste. Genevieve Limestone, massive fluorite may be found filling

"V-structures," features occurring at the tops of ore bodies and defined by banded ore or limestone strata that dip inward towards a central fracture (15, figures 9 and 10). Much of the massive fluorite appears to have been deposited in cavities, but some may have formed by direct replacement of limestone.

Structures relict from limestone that are still recognizable in the deposits definitely indicate a replacement origin for the ore, as has been generally recognized (2,5,8,10,11,13). Those structures on which there has been general agreement include fossils, stylolites, oolites, outlines of original calcite and quartz crystals, and evidences of the original large-scale stratification preserved by shale and clay partings.

Banded ore appears to have been formed by replacement of laminated, and in places cross-bedded, limestone of quite high purity. Although not apparent in fresh exposures, the laminated character of the limestone is well displayed in weathered outcrops at the margins of ore bodies in the area of the Victory and Austin mines. The coarser-grained, purer fluorite bands appear to have formed from alternate layers in the limestone that were coarser grained and, perhaps, of higher purity. Imperfectly banded ore, on the other hand, was derived from limestone less perfectly laminated, not so conspicuously oolitic, generally finer grained, and apparently containing somewhat greater amounts of clayey and siliceous impurities. The Ste. Genevieve Limestone, in which most deposits of banded ore are found, differs notably in lithologic character from bed to bed and also within short distances laterally, making it largely impossible to correlate beds on the basis of lithology for more than a few hundred yards. These rapid lateral changes are reflected in the ore of different deposits and parts of deposits in the same stratigraphic position. It is demonstrated by the almost exclusively banded nature of the ore in the main Carlos ore body of the Victory mine and the contrastingly high proportion of imperfectly banded ore in the adjoining Addison ore body in the same mine, both of which occupy the topmost beds of the Ste. Genevieve Limestone. The vertical interlaying of banded and imperfectly banded ore beds in many deposits also shows rapid lateral change.

Brecke (39, p. 515–525) presents some strong arguments for the origin of the banding by diffusion and rhythmic replacement, first proposed by Bastin (11, p. 49–55), and refers to certain of the ores in the Leadville, Colorado, district (9, p. 202–205) in support of his theory. That diffusion banding did take place locally in the Cave in Rock ore bodies is demonstrated by the uncommon occurrence of limestone inclusions in the ore that show concentrically banded fluorite around their margins. However, it is the authors' opinion that by far the greatest part of the banded Illinois ores have important structural differences from the Leadville ores, and that, on the whole, the banding in the Cave in Rock

deposits was caused by sedimentary structures originally present in the oolitic limestones.

## Mineral Paragenesis

Although there are some differences in the relative proportions of the various minerals present in the *vein* and *bedded-replacement* deposits, the sequence of deposition of the minerals appears to be the same in the two types of deposits. In the vein deposits, calcite was the first mineral to be deposited and remains the most abundant mineral. In the bedding-replacement deposits, some calcite was deposited early (43, p. 890–891) but is not conspicuous in the ore bodies. Fluorite followed calcite and replaced it to form the fluorite-bearing veins. In the bedding deposits also, fluorite replaced early calcite, but most of the ore formed by replacement of the limestone host rock. Some overlap of fluorite and calcite deposition occurred in both the vein and bedding deposits. Hall and Friedman (43, p. 890–892) worked out a detailed paragenesis for the Cave in Rock deposits and distinguished five stages of fluorite based on successive color bands (Figure 8).

Quartz and the sulfides, sphalerite and galena, appear to have been closely related in time and to have overlapped fluorite deposition to a considerable extent, as well having overlapping relations to each other. Chalcopyrite and pyrite, present only in very minor amounts, occur both within and on fluorite crystals and appear to bear the age relations indicated in Figure 8.

Late scalenohedral calcite is found in both veins and bedding deposits, perched on fluorite

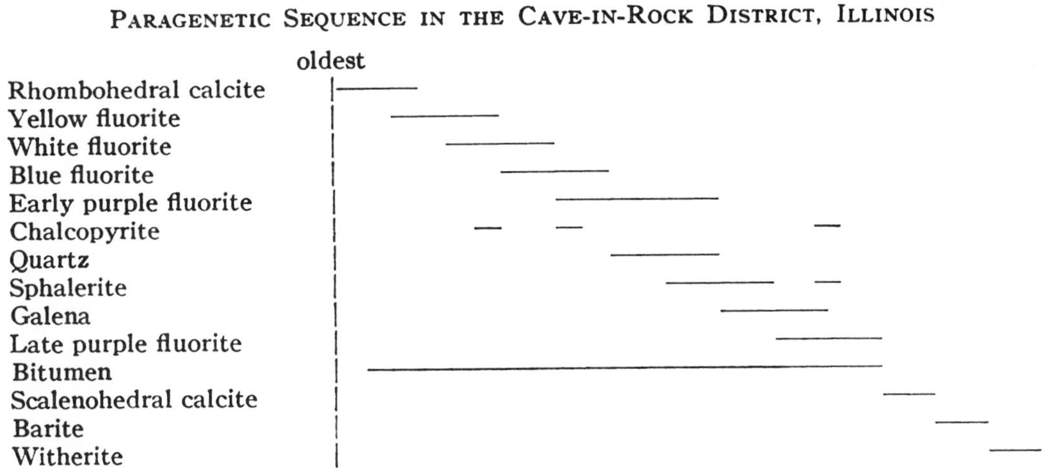

PARAGENETIC SEQUENCE IN THE CAVE-IN-ROCK DISTRICT, ILLINOIS

Fig. 8. *Paragenetic Sequence in Cave in Rock Area (From Hall and Friedman, 43, Table 1).*

quartz, and sulfides, but is abundant in neither type of deposit. Cavities lined with "dogtooth" calcite have been found associated with both vein and bedding deposits.

Barite occurs in cavities and replaces limestone, calcite, and fluorite. Witherite and strontianite (found only in the Minerva No. 1 mine) also occur in cavities in the ore. Witherite in some places appears to be an alteration product of barite.

Bitumen is found as inclusions in fluorite throughout the mineralizing sequence. It also has been found coating late fluorite.

## Grades of Ore

For several reasons, such as the low price of fluorite as a commodity ($CaF_2$) compared to that of most metals, the variety of grades in which it is marketed, and its mode of occurrence (massive) in vein deposits, the grade of ore, stated as its percentage of $CaF_2$, has not had the traditional significance in fluorspar mining that it has had in metal mining. In the vein mines, grade of ore is generally expressed in width of $CaF_2$ in the vein. For example, a mine map may show such notations as "$CaF_2$ 1 ft," "$CaF_2$ 5 ft." To the experienced observer, this means that, at a specific place, the fluorite in the vein adds up to the stated amount, whether it is 15 or 100 per cent of the vein filling at that place, the balance of the vein width being generally calcite and slices of rock. It is probable that each company has its own cut-off widths, dependent on mining costs, but a rule of thumb in the vein mines of the district for many years has been that a vein should average 2 feet of fluorspar to be considered minable. This, of course, is a very general figure and can differ in actual practice, depending on mining conditions, kind and distribution of impurities (particularly silica), and other factors. Also, the presence or absence of sphalerite and, to some extent, galena may have some effect on the minability of a portion of a vein. However, the sulfides are so erratically distributed and generally of such low average concentration that, with only a few exceptions, they have little bearing on the economics of vein mining.

In the bedding-replacement deposits, however, the percentages of $CaF_2$, sphalerite, and galena have come to have more significance. There are no available figures on cut-off grades, but it is generally conceded that an ore body should average at least 30 per cent $CaF_2$ to be minable if no sulfides are present. In a few instances, however, bodies of ore

that averaged only a little over 20 per cent $CaF_2$ have been mined where they lay close to other underground workings. If sulfides are present in the amount of at least 4 per cent, the percentage of fluorite becomes of secondary importance in determining mineability.

## FACTORS CONTROLLING FORM AND LOCATION OF ORE BODIES

### Vein Deposits

STRUCTURAL CONTROLS    The chief controlling factor for the vein deposits, in regard to both location and form, was faulting. As the vein deposits are primarily fissure fillings, the faults simultaneously provided the channelways for solution movement and the open spaces for mineral deposition.

Figure 9 shows the fault pattern in the Rosiclare area, the most productive vein system in the Illinois-Kentucky district. Figure 10 shows the detail of the workings of the Aluminum Company of America in the southern part of the Rosiclare vein system. The veins occupy a complex system of branching and intersecting normal faults. The Rosiclare Fault (marked by shafts 11, 12, 13, and 14) and the Blue Diggings Fault (marked by shafts 8, 9, and 10) are roughly parallel, trending N20° to 30°E, whereas the Hillside Fault (shaft 5) trends nearly north-south and cuts across both of them. The Argo (shaft 7) and Daisy (shaft 6) Faults appear to be branches of the Blue Diggings Fault. Northeast of the Hillside Fault, mineralization continues on a more nearly northeast course along what appears to be a system of intersecting faults. The Hawkins Fault (shaft 4) has been tentatively mapped as an extension of the Argo Fault and has the largest displacement of any fault in the Rosiclare district. A small amount of ore was taken from the shaft on the Hawkins Fault.

The faults generally dip 70° to 80°W, with the exception of the Blue Diggings Fault, which has an eastward dip that rarely exceeds 65° and locally flattens to 45°. Because of this opposing dip, the Blue Diggings Fault cuts off the Daisy Fault at a depth of about 800 feet.

Vertical displacement along the faults of the Rosiclare vein system ranges from a few feet to 650 feet. Between 25 and 500 feet, amount of displacement has apparently had little effect on vein width. Along faults of 25 feet or less displacement, however, veins generally are nar-

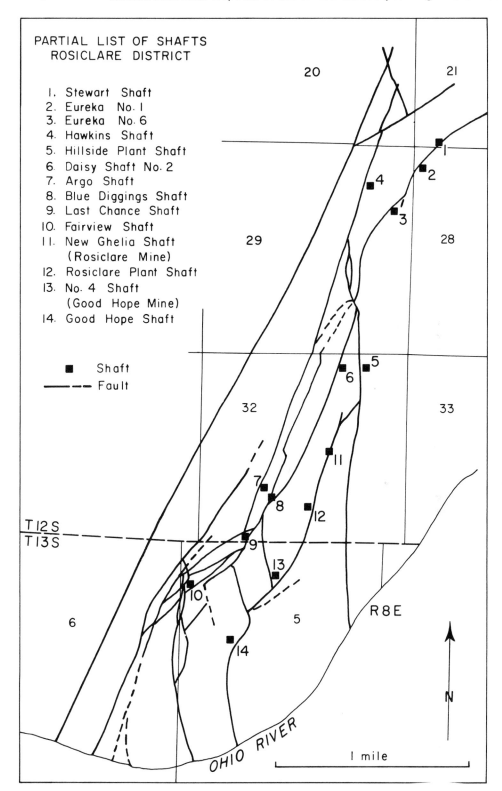

FIG. 9. *Faults and Mine Locations, Rosiclare Area (modified from Baxter and Desborough, 49, Plate 1).*

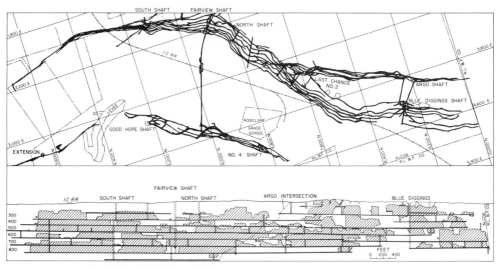

FIG. 10. *Plan and Longitudinal Sections, showing Aluminum Company of America workings on Rosiclare vein system (modified from Schaefer, et al., 37, Figure 6 on basis of data furnished by Aluminum Company of America, 1965).*

row, presumably because there has not been enough differential movement for irregularities along the fault plane to force the walls any considerable distance apart. Where differential movement has exceeded 500 feet, excessive breaking and grinding of the displaced rocks has tended to fill the faults with broken rock and gouge.

Post-mineral faulting, both as renewed movement along the veins and as cross faulting, is common in mines in the Resiclare area but appears to be of little importance economically. Faulting within, and parallel to, the veins has sheared the vein filling in places. Cross faults may offset the veins slightly but rarely enough to cause mining problems. Some cross faults may be mineralized, as is the one that cuts the Blue Diggings vein and contains sphalerite cementing a breccia of wall rock, but none of economic importance is known.

STRATIGRAPHIC CONTROLS  Although "favorable beds," in the physico-chemical sense, had some local effect on ore formation in the vein deposits, they were not an important ore control in these deposits as a whole. Lithology of the wall rocks, nevertheless, appears to have exerted control in a general way on the size and persistence of the ore shoots and, concomitantly, on the stratigraphic position of the minable deposits. The largest ore shoots, in regard to length, vertical extent, and average width, generally are found where the fault walls are the competent Ste. Genevieve and

St. Louis Limestones. The generally massive Bethel and Cypress Sandstones, commonly quartzitized along faults, also are associated with the more productive parts of veins. The shales and shaly limestones and sandstones, on the other hand, make poor fault walls, and generally are the sites of the narrower parts of veins. Consequently, the more productive parts of faults normally are those portions below the top of the Cypress Sandstone and more particularly below the top of the Ste. Genevieve Limestone (Figure 2). This lithologic effect, however, is not a stringent control, and many good ore shoots have been found where one fault wall was shale. Also, a few commercial deposits have been found in the upper Chesterian strata with little or no mineralization at deeper levels in the older, generally more favorable strata.

### Bedding-Replacement Deposits

STRUCTURAL CONTROLS  The bedding-replacement deposits follow the course of groups of *joint-like fractures* and *minor pre-mineral faults* trending N45° to 60°E and N30° to 85°W (Figure 11). The northeast-trending fracture zones clearly are the most persistent and support the largest ore bodies; for example, the ore body along the most southeastern of the pair of fracture zones in the northeast part of Figure 11 is more than 2 miles long.

Although most of the ore formed by direct replacement of limestone along these fracture zones, certain features associated with the ore bodies indicate that there also was solutional removal of limestone. Some of these features (for example, thinning of limestone strata within the ore bodies and accompanying formation of sag synclines in beds directly overlying ore (Figure 6), crystal-lined cavities in the ore, and minor occurrence of fluorite-cemented breccia at the tops of ore bodies) are more aptly termed ore accompaniments than ore controls. One type of solutional feature, however, the slump structure or breccia pipe, appears to have exerted at least local control over ore emplacement.

Brecke (39, p. 511–514) described two breccia pipes, one in the North Green and the other in the Hill mine. In the North Green mine, the Bethel Sandstone was found 100 feet below its normal position and the Rosiclare Sandstone 60 feet below, with mineralized breccia persisting to the limit of core drilling 110 feet below the Spar Mountain Sandstone Member. The Hill mine breccia pipe showed Rosiclare Sandstone as much as 75 feet below its normal position in the mine. Additional slump structures have been discovered in the northeastern part of Minerva No. 1 mine (D. B. Saxby, personal communication, 1967). Figure 12 shows a portion of the Minerva mine, locations of the slumps,

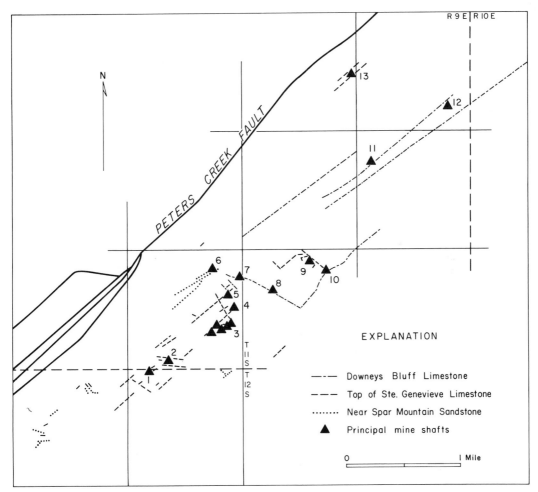

FIG. 11. *Principal Deposits and Mineralized trends, Cave in Rock Area (modified from Grogan, 15, Figure 1). 1. Addison shaft, Victory mine. 2. Carlos shaft, Victory mine. 3. Crystal mine shafts, nos. 1, 2, 5, 6, 7. 4. Crystal shaft no. 3. 5. Crystal shaft no. 4. 6. Davis-Deardorff mine. 7. W. L. Davis no. 2 mine. 8. A. L. Davis mine. 9. West Green mine. 10. East Green mine. 11. Oxford mine. 12. Minerva No. 1 mine. 13. Hill-Ledford mine.*

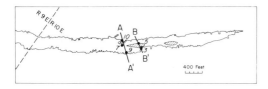

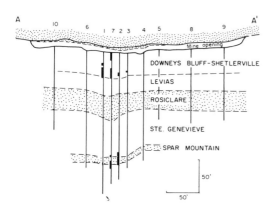

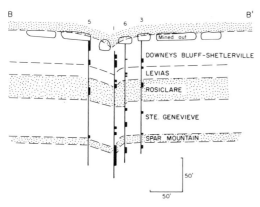

FIG. 12. *Location of Slump Features in Minerva No. 1 Mine and Cross Sections, showing details of structure. Numbered vertical lines in sections indicate drill holes. Holes with duplicate numbers in the two sections are from different series of drilling. Hole No. 1 in BB' is projected into line of section from behind a pillar (from drawings furnished by D. B. Saxby).*

and two cross sections illustrating the relation of slump 1 to normal structure. In section A-A', outside the slump area, the base of the Bethel Sandstone dips gently inward, describing the typical synclinal structure of the bedded replacement deposits. Section B-B', across one of the slumps, shows the base of

the Bethel plunging into the slump. As seen underground, the basal Bethel and the underlying Downeys Bluff Limestone dip as steeply as 45° near the center of the slump. Both the Bethel and the Downeys Bluff are broken and cemented with ore minerals, but, in general, there is only minor displacement of blocks of Bethel and Downeys Bluff in relation to one another. Evidently, the slumping was a gradual process rather than a sudden collapse.

Increased activity by the mineralizing solutions in the vicinity of the slump structures is expressed by ore-grade mineralization (fluorite and sphalerite) in strata below the main ore body. According to Saxby (personal communication, 1967), this stratigraphically lower ore extends much farther to the southwest, opposite the regional dip, from the slump structure than it does to the northeast, in the direction of the regional dip. This same spatial relation of mineralization to slump structures in the Hill and North Green-East Green mines was pointed out by Brecke (39, p. 511–514). However, it seems unlikely that the slump structures in the Minerva ore body, occurring as they do at the northeast end, were the main feeding conduits for the 2-mile-long ore body, as suggested by Brecke for the Hill and North Green-East Green ore bodies. It seems more probable that, in general, the mineralizing solutions rose along the main Minerva fracture zone with more intense local activity in the vicinity of the slump structures.

The possible role of sedimentation in the formation of "sag synclines" has been suggested by Saxby (personal communication, 1967), who pointed out the thickening of basal beds in the Bethel Sandstone towards the center of the syncline in the Minerva mine. As he has seen this gradual thickening of sandstone beds toward the central portion of the ore body in several places in the Minerva mine and at the Rosiclare-Ste. Genevieve contact in the Crystal mine, Saxby feels that the "sag synclines" are in reality shallow channel fillings by the sand on the top of the underlying limestones. The coincidence of the ore-bearing fracture zones with the channels suggests that the fracture zones were in existence during Mississippian time and guided the erosion of shallow channels on top of the limestone prior to deposition of the sandstone. As the Bethel-Downeys Bluff contact is recognized as an unconformity (54, p. 12), the existence of such erosion channels is not unlikely. The suspected presence of an unconformity at the Rosiclare Sandstone-Ste. Genevieve Limestone contact

(54, p. 10–11) suggests that features seen in the Crystal mine may also be sandstone-filled channels.

Even though the inward dip of the sandstone roof rocks conceivably could be at least partly depositional, the presence of collapse breccias in the ore bodies provides abundant evidence of solution of the limestone host rocks during ore deposition.

STRATIGRAPHIC CONTROL    The constant association of bedded replacement deposits with certain limited stratigraphic intervals demonstrates clearly the existence of beds that were particularly favorable sites for replacement ore. As mentioned in previous sections, the three major mineralized horizons were, from the lowest to the highest: (1) within the Ste. Genevieve at the level of the Spar Mountain Sandstone Member; (2) the top of the Ste. Genevieve at the base of the Rosiclare Sandstone; and (3) the top of the Downeys Bluff Limestone at the base of the Bethel Sandstone. The factor or factors that made these three limited intervals uniquely suited for mineralization, however, are not obvious. The only evident common denominator among the three is that they are limestone, but occurring as they do in a succession that is largely limestone, this hardly is a unique property. Lithologically, the lowest two generally are light colored and oolitic and of quite high purity; the highest is gray to brownish gray, fine to coarse grained, fossiliferous, locally oolitic, and generally contains more or less argillaceous impurities. With regard to stratigraphic relations, the upper two are similarly situated, i.e., overlain by sandstone that is shaly at the base and conceivably could have acted as a dam to the mineralizing solutions. The lower one, although in places overlain by the calcareous Spar Mountain Sandstone Member, actually is most strongly mineralized where the sandstone is missing and the overlying beds are dark-colored limestone (39, p. 501). This apparent difference in stratigraphic relations, however, suggests a second possible common denominator. It may be that each favored interval was exposed to erosion and/or reworking during a depositional hiatus, as proposed by Brecke (39, p. 501). The Downeys Bluff-Bethel contact generally is agreed to be an unconformity, and there is evidence that the top of the Ste. Genevieve may have been, at least in part, exposed to erosion. The Spar Mountain Sandstone Member horizon, however, shows no evidence of erosion, both its upper and lower contacts being gradational

(54, p. 10). Nevertheless, the fact that the Spar Mountain is a lens, and, consequently, not deposited continuously over the area, suggests that the underlying oolite may have been subjected to reworking in areas where no sand was being deposited. Unfortunately, strict textural comparisons between oolitic beds overlain by the Spar Mountain and those at the same horizon where the Spar Mountain is missing cannot be made because of the scarcity of outcrops. As a matter of fact, the only exposures of this interval with the sandstone missing are in mines, where the oolitic beds have been mineralized.

Admittedly, the presence of a depositional hiatus at the top of the three most favored intervals, even if it could be supported by field evidence, does not provide a clearly demonstrable reason for the favorability of these intervals. It might be mentioned that it is not even unique. Another unconformity in the ore-bearing part of the section exists between the Levias Limestone Member and the overlying Shetlerville Member, but, as the Levias carries commercial amounts of ore in places, this is not a serious objection. In the final analysis, the most powerful argument for a depositional hiatus as an influencing factor is that it appears to be the only property that may be possessed uniquely by all three favored horizons.

### Paths of Solution Movement

Because of the occurrence of the vein deposits in faults that have great persistence both along strike and in depth, the faults may be assumed quite logically to have been the channels along which the mineralizing solutions moved. In regard to the Cave in Rock area of replacement deposits, it generally is agreed (19, p. 127; 55, p. 380) that the major Peters Creek Fault, which lies parallel to, and half a mile from, the mineralized area, provided the main conduit for the mineralizing solutions. It is likely that solutions rising along this fault, which has displacements of over 1000 feet, encountered places filled with gouge and were forced into the wall-rock system of minor fractures that conducted them farther outward and upward and ultimately brought them into contact with favorable host rocks.

### SUMMARY OF SEQUENCE OF GEOLOGIC EVENTS

Although there is doubt about the time relations of some of the events connected with fluorite mineralization, it is possible to outline a logical sequence of events that does not conflict, at least, with the several lines of evidence bearing on the problem.

1. Compression from the southeast, causing thrusting and probably some lateral movement along the Shawneetown Fault Zone.

2. Intrusion of an alkalic magma contemporaneously with the compressive force, along a northwest-trending line of weakness in the basement, causing an arching of the overlying sedimentary strata and penetrating these strata as mafic dikes along tension fractures.

3. Relaxation of compressive forces and cooling and shrinking of the magma, causing collapse of the overlying strata and block-faulting along the northeast-trending fractures that were initiated during the previous period of compression.

4. Explosive igneous activity causing the formation of Hicks Dome and the intrusive breccias. This event probably was nearly contemporaneous with the block-faulting.

5. Mineralization following, but closely related in time to, the igneous activity.

## SECONDARY ORE

Processes of weathering, particularly solution of carbonate rocks, have produced residual concentrations of fluorite that have been mined in many parts of the district. Although none of these deposits has been a large producer, and most have been quite small, the residual deposits are attractive prospects because the fluorite, embedded in clay, can be easily won with equipment no more sophisticated than a power shovel or dragline and a log-washer. Although there are no accurate figures on the amount of fluorite that has been produced from residual deposits in the Illinois-Kentucky district, it probably is small compared to total production.

Oxidized sulfide deposits have played a very minor role in the history of the Illinois-Kentucky district. One operation, the Patrick mine of the Alco Lead Corporation in the Cave in Rock district of Illinois, attempted to mine a galena-cerussite deposit but had difficulty making a salable product from the cerussite (19, p. 130). According to Currier (7, p. 48), smithsonite occurred in large enough amounts at several places to support mining operations in Kentucky. Apparently, the most notable of these were the Hutson mine, where sphalerite is now being recovered below the zone of oxi-

dation, and the Old Jim mine, which has been inactive for many years.

## ORE GENESIS

Most students of the Illinois-Kentucky fluorite district for many years have postulated that the deposits were formed from fluids that emanated from an igneous magma (5,11,19). The presence of igneous intrusives in the district and the fact that fluorine was considered as typically volcanic in origin were held to be sufficient evidence to indicate magmatic affiliations. Although Brecke (47) presented some interesting arguments for a lateral secretion type of origin, recent publications on the mineralogy and chemical analyses of the igneous intrusives, composition of the fluid inclusions in the ore and gangue minerals, and isotopic and trace-element analyses of galenas from the mining district appear to strengthen the case for a magmatic-hydrothermal origin.

The recent find of a nepheline-feldspar intrusive and the recognition of fragments of aegerine-bearing syenite in a carbonatized intrusive indicate the presence of a body of syenitic magma at depth. The presence of abnormal concentrations of fluorite, thorium, rare earths, beryllium, zirconium, niobium, and titanium in breccias near the center of Hicks Dome suggest mineralization affiliated with alkalic rocks (50, p. 11). Thus, a likely source of the fluorine-bearing solutions, a body of alkalic magma, is firmly established.

Studies of temperature of deposition, carried out by Freas (34) and Grogan and Shrode (16) on fluid inclusions in the various minerals of the fluorite deposits, indicated a formation temperature of 90° to 140°C, including a +13° correction for the lithostatic pressure of the 3000 feet of rock believed to have been present over the forming deposits. This is roughly two to three times the rock temperature that would be expected on the basis of a normal geothermal gradient. There are indications that the temperature fluctuated as much as 15°C during the growth period of individual crystals with the highest temperatures generally prevailing in early stages of crystal growth. Sphalerite and quartz were deposited at slightly lower temperatures than fluorite. Freas (34) concluded from his temperature data and his interpretation of the paragenetic sequence that there were two pulses of hydrothermal solutions of similar composition and temperature from a single source during the ore-forming period.

The composition of the ore-forming fluids was investigated intensively by Hall and Friedman (43) and on a reconaissance basis by Grogan and Shrode (16). In general, the fluid was a concentrated Na-Ca-Cl brine, but the total solids content and composition fluctuated from time to time during the successive formation of the various minerals.

In inclusions of the early fluorite, the fluid closely resembles oil field brines, from which it is concluded that the ore fluid probably was connate water to which fluorine had been added from a magmatic source. The fluid in the quartz, however, is markedly different from that in early fluorite, containing less total solids and deuterium than the fluorite fluid. This quartz fluid does not resemble either present day connate water or meteoric waters of the region, and it could not have been produced by mixing the two. Hall and Friedman (43) concluded it resulted from the addition of fluorine-poor magmatic water to the system. This conclusion was based in large part on their knowledge from previous work that water of magmatic origin from various volcanic glasses and from hydrous minerals in granites and volcanic rocks has a low deuterium content. The characteristics of the liquids from sphalerite and galena are transitional between those of quartz water and fluorite water, suggesting an increase of the proportion of connate water in the system. The composition of the fluid in inclusions in late fluorite resembles that of early fluorite. This has been interpreted as indicating the completion of the purging of magmatic water from the system and its replacement by connate water charged with fluorine. The late sulfate gangue minerals contain brine that is much less concentrated and whose Cl⁻ ion content is quite low. The deuterium concentration is as low as it was in the quartz fluid. Perhaps this composition reflects the addition of a mixture of magmatic and meteoric waters to the system.

A recent study of the lead isotopes in eight galenas from the Illinois-Kentucky district by Heyl *et al.* (52) included three samples from the Cave in Rock deposits and one from a mineralized brecia on Hicks Dome. All of the galenas were of the J type, being quite rich in radiogenic isotopes, relative to ordinary leads. The galena from Hicks Dome, however, was less radiogenic than the others. When plotted geographically, the $Pb^{206}/Pb^{204}$ and $Pb^{207}/Pb^{204}$ ratios show a general rise in values toward the southeast away from Hicks Dome. Although there is no observable zonation

in the abundance of fluorspar, sphalerite, and galena throughout the Illinois-Kentucky district, it has been reported (52, p. 950) that the silver content of the galena decreases in a southeastward direction from the Hicks Dome-Rosiclare area, and that barite is more common on the fringes of the district (46). Heyl further reported (personal comunication, 1967) that additional galena analyses show similar decreases of silver to the south and west away from Hicks Dome. This apparent silver and barite zonation, together with the increase of radiogenic lead southeastward from Hicks Dome, supports the concept that the ore fluids rose in its vicinity and migrated radially outward. In the case of lead, the magmatic lead became progressively more contaminated with the more radiogenic indigenous lead as it moved away from the source.

Most recently, a syngenetic origin has been proposed for the bedding replacement deposits (53), based on a petrographic examination of the ores. We feel, however, that the ore textures attributed by Amstutz and Park to diagenetic processes are just as reasonably explained by preservation of original rock textures during replacement by the introduced minerals with modifications caused by solution and redeposition of materials during mineralization.

The fluorite-zinc-lead deposits of the Illinois-Kentucky mining district, then, have a combined connate-magmatic-hydrothermal origin. Hypotheses involving lateral secretion or a syngenetic origin can be supported by certain lines of evidence, but all the features of the deposits, including localization by faults and fractures, association of fluorite mineralization with igneous activity of Permian or later age, and zoning with respect to a center of explosive igneous activity, can best be explained by the theory outlined in this paper.

## CONCLUSIONS

1. The fluorite-zinc-lead deposits of the Illinois-Kentucky mining district occur along faults and fractures, chiefly in Middle and Upper Mississippian formations. Known igneous rocks are mafic dikes and intrusive breccias.

2. The mining district occupies an area of complex normal faulting, bordered on the north and northwest by the Shawneetown-Rough Creek Fault Zone and apparently merging to the southwest with the northeast-trending New Madrid Fault Zone.

3. A gentle northwest-trending arch under-lies the district and has as its apex Hicks Dome, a well-defined domal structure characterized by exposures of intrusive breccia. It is postulated that the arch was pushed up by intrusion of an alkalic magma at depth with the emplacement of mafic dikes in northwest-trending tension fractures; during cooling of the magma, accumulation of gases led to the formation of Hicks Dome, and their explosive release created dikes and plugs of intrusive breccia. Block faulting was initiated partly by collapse of the arched strata on shrinkage of the cooling magma and partly by relaxation of compressive stresses that formed the Shawneetown-Rough Creek Fault Zone.

4. The ore deposits consist of two primary types: vein deposits and bedding-replacement deposits. The vein deposits are fissure fillings along vertical or near-vertical faults and the bedded deposits are replacements of certain limestone strata outward from fractures or minor faults. Residual deposits develop from the weathering of both primary types.

5. Age of mineralization has been established as Permian or younger by isotopic dating of minerals in the igneous intrusives. Other considerations suggest that the mineralization may be early late Cretaceous.

6. The localization of the deposits along faults and fractures and the association of fluorite with abnormal concentrations of rare earths, beryllium, niobium, and titanium in Hicks Dome intrusive breccias indicate that the deposits are epigenetic. The source of heat and of fluorite and metals is presumed to be a body of alkalic magma at depth.

## REFERENCES CITED

1. Norwood, J. G., 1866, Report on the Rosiclare lead mines, *in* Worthen, A. H., Geology: Geol. Surv. of Ill., v. 1, p. 366–372.
2. Bain, H. F., 1905, The fluorspar deposits of southern Illinois: U.S. Geol. Surv. Bull. 225, 75 p.
3. Ulrich, E. O. and Smith, W. S. T., 1905, The lead, zinc, and fluorspar deposits of western Kentucky: U.S. Geol. Surv. Prof. Paper 36, 218 p.
4. Fohs, F. J., 1906, Fluorspar: Eng. and Min. Jour., v. 81, no. 1, p. 45–46.
5. Currier, L. W., 1920, Igneous rocks; fluorspar, lead, and zinc: *in* Weller, S., *et al., The geology of Hardin County and the adjoining part of Pope County,* Ill. Geol. Surv. Bull. 41, p. 235–244, 247–304.
6. Weller, S., *et al.,* 1920, Geology of Hardin County: Ill. Geol. Surv. Bull. 41, 416 p.
7. Currier, L. W., 1923, Fluorspar deposits of Kentucky: Ky. Geol. Surv., ser. 6, v. 13, 198 p.
8. Spurr, J. E., 1926, The Illinois-Kentucky ore magmatic district: Eng. and Min. Jour., v. 122, nos. 18, 19, p. 695–699, 731–738.
9. Emmons, S. F., *et al.,* 1927, Geology and ore deposits of Leadville mining district: U.S. Geol. Surv. Prof. Paper 148, 368 p.
10. Schwerin, M., 1928, An unusual fluorspar deposit: Eng. and Min. Jour., v. 126, no. 9, p. 335–339.
11. Bastin, E. S., 1931, The fluorspar deposits of Hardin and Pope Counties, Illinois: Ill. Geol. Surv. Bull. 58, 116 p.
12. Hatmaker, P. and Davis, H. W., 1938, The fluorspar industry of the United States, with special reference to the Illinois-Kentucky district: Ill. Geol. Surv. Bull. 59, 128 p.
13. Currier, L. W. and Wagner, O. E., Jr., 1944, Geology of the Cave in Rock district: U.S. Geol. Surv. Bull. 942, pt. 1, p. 1–72.
14. Muir, N. M., 1947, Daisy fluorite mine, Rosiclare Lead and Fluorspar Mining Co., Hardin County, Illinois: U.S. Bur. Mines R. I. 4075, 16 p.
15. Grogan, R. M., 1949, Structures due to volume shrinkage in the bedding-replacement fluorspar deposits of southern Illinois: Econ. Geol., v. 44, p. 606–616.
16. Grogan, R. M. and Shrode, R. S., 1949, Formation temperatures of southern Illinois bedded fluorite as determined from fluid inclusions: Amer. Mineral., v. 37, p. 555–566.
17. U.S. Bureau of Mines Minerals Yearbooks 1949–1965: U.S. Bur. Mines, Dept. Interior, Washington, D.C.
18. Oesterling, W. A., 1952, Geologic and economic significance of the Hutson zinc mine, Salem, Kentucky: Econ. Geol., v. 47, p. 316–338.
19. Weller, J. M., *et al.,* 1952, Geology of the fluorspar deposits of Illinois: Ill. Geol. Surv. Bull. 76, 147 p.
20. Anderson, O. E., 1953, How Minerva Oil Company produces high-grade ceramic fluorspar: Eng. and Min. Jour., v. 154, no. 3, p. 72–78.
21. Bailie, H. E., 1954, Concentrating operations of the Ozark-Mahoning Company at Rosiclare, Illinois, Cowdrey and Jamestown, Colorado: A.I.M.E. Reprint, 18 p.
22. Brown, J. S., *et al.,* 1954, Explosion pipe in test well on Hicks Dome, Hardin County, Illinois: Econ. Geol., v. 49, p. 891–902.
23. Klepser, H. J., 1954, Fluorspar deposits in western Kentucky, Part 3, Senator-Schwenk area, Tabb fault system, Caldwell County: U.S. Geol. Surv. Bull. 1012-F, p. 115–130.
24. Thurston, W. R. and Hardin, G. C., Jr., 1954, Fluorspar deposits in western Kentucky, Part 3, Moore Hill fault system, Crittenden and Livingston counties: U.S. Geol. Surv. Bull. 1012-E, p. 81–113.

25. Trace, R. D., 1954, Fluorspar deposits in western Kentucky, Part 2, Central part of the Commodore fault system, Crittenden County: U.S. Geol. Surv. Bull. 1012-C, p. 39–57.

26. Trace, R. C., 1954, Fluorspar deposits in western Kentucky, Part 2, Mineral Ridge area, Livingston and Crittenden counties: U.S. Geol. Surv. Bull. 1012-D, p. 59–79.

27. Clegg, K. E., 1955, Metamorphism of coal by peridotite dikes in southern Illinois: Ill. Geol. Surv. R. I. 178, 18 p.

28. Hardin, G. C., Jr., 1955, Babb fault system, Crittenden and Livingston Counties: *in Fluorspar deposits in western Kentucky:* U.S. Geol. Surv. Bull. 1012-B, pt. 1, p. 7–37.

29. Clegg, K. E. and Bradbury, J. C., 1956, Igneous intrusive rocks in Illinois and their economic significance: Ill. Geol. Surv. R. I. 197, 19 p.

30. Bradbury, J. C., 1959, Barite in the southern Illinois fluorspar district: Ill. Geol. Surv. Circ. 265, 14 p.

31. Hardin, G. C., Jr. and Trace, R. D., 1959, Geology and fluorspar deposits, Big Four fault system, Crittenden County, Kentucky: U.S. Geol. Surv. Bull. 1042-S, p. 699–724.

32. Harrison, W. H., 1960, Slusher operations on narrow veins solves stoping problem: Min. Cong. Jour., v. 46, no. 10, p. 32–34.

33. Trace, R. D., 1960, Significance of unusual mineral occurrence at Hicks Dome, Hardin County, Illinois: U.S. Geol. Surv. Prof. Paper 400-B, p. 63–64.

34. Freas, D. H., 1961, Temperatures of mineralization by liquid inclusions, Cave in Rock fluorspar district, Illinois: Econ. Geol., v. 56, p. 542–556.

35. Heyl, A. V. and Brock, M. R., 1961, Structural framework of the Illinois-Kentucky mining district and its relation to mineral deposits: U.S. Geol. Surv. Prof. Paper 424-D, p. 3–6.

36. Kulp, J. L., 1961, Geologic time scale: Science, v. 133, no. 3459, p. 1105–1114.

37. Schaefer, R. W., *et al.,* 1961, Mining, milling, and water-control methods, Rosiclare Fluorspar Works, Aluminum Company of America: U.S. Bur. Mines I. C. 8012, 52 p.

38. Bradbury, J. C., 1962, Trace elements, rare earths, and chemical composition of southern Illinois igneous rocks: Ill. Geol. Surv. Circ. 330, 12 p.

39. Brecke, E. A., 1962, Ore genesis of the Cave in Rock fluorspar district, Hardin County, Illinois: Econ. Geol., v. 57, p. 499–535.

40. Pryor, W. A. and Ross, C. A., 1962, Geology of the Illinois parts of the Cairo, LaCenter, and Thebes quadrangles: Ill. Geol. Surv. Circ. 332, 39 p.

41. Trace, R. D., 1962, Geology and fluorspar deposits of the Levias-Keystone and Dike-Eaton areas, Crittenden County, Kentucky: U.S. Geol. Surv. Bull. 1122-E, 26 p.

42. Baxter, J. W., *et al.,* 1963, Areal geology of the Illinois fluorspar district, Part 1—Saline Mines, Cave in Rock, Dekoven, and Repton quadrangles: Ill. Geol. Surv. Circ. 342, 43 p.

43. Hall, W. E. and Friedman, Irving, 1963, Composition of fluid inclusions, Cave in Rock fluorite district, Illinois, and Upper Mississippi Valley zinc-lead district: Econ. Geol., v. 58, p. 886–911.

44. McGinnis, L. D., 1963, Earthquakes and crustal movement as related to water load in the Mississippi Valley region: Ill. Geol. Surv. Circ. 344, 20 p.

45. Ross, C. A., 1963, Structural framework of southernmost Illinois: Ill. Geol. Surv. Circ. 351, 27 p.

46. Brecke, E. A., 1964, Barite zoning in the Illinois-Kentucky fluorspar district: Econ. Geol., v. 59, p. 299–302.

47. Brecke, E. A., 1964, A possible source of solutions of the Illinois-Kentucky fluorspar district: Econ. Geol., v. 59, p. 1293–1297.

48. McGinnis, L. D. and Bradbury, J. C., 1964, Aeromagnetic study of the Hardin County area, Illinois: Ill. Geol. Surv. Circ. 363, 12 p.

49. Baxter, J. W. and Desborough, G. A., 1965, Areal Geology of the Illinois fluorspar district; Part 2—Karbers Ridge and Rosiclare quadrangles: Ill. Geol. Surv. Circ. 385, 40 p.

50. Heyl, A. V., *et al.,* 1965, Regional structure of the southeast Missouri and Illinois-Kentucky mineral districts: U.S. Geol. Surv. Bull. 1202-B, 20 p.

51. Zartman, R. E., *et al.,* 1967, K-Ar and Rb-Sr ages of some alkalic intrusive rocks from central and eastern U.S.: Amer. Jour. Sci., v. 265, p. 848–870.

52. Heyl, A. V., *et al.,* 1966, Isotopic study of galenas from the Upper Mississippi Valley, the Illinois-Kentucky, and some Appalachian districts: Econ. Geol., v. 61, p. 933–961.

53. Amstutz, G. C. and Park, W. C., 1967, Stylolites of diagenetic age and their role in the interpretation of the southern Illinois fluorspar deposits: Mineralium Deposita, v. 2, p. 44–53.

54. Baxter, J. W., *et al.,* 1967, Areal geology of the Illinois fluorspar district, Part 3—Herod and Shetlerville quadrangles: Ill. Geol. Surv. Circ. 413, 41 p.

55. Brecke, E. A., 1967, Sulfide and sulfur occurrences of the Illinois-Kentucky fluorspar district: Econ. Geol., v. 62, p. 376–389.

# 20. The Geology and Ore Deposits of the Tri-State District of Missouri, Kansas, and Oklahoma

DOUGLAS C. BROCKIE,* EDWARD H. HARE, JR.,* PAUL R. DINGESS*

## Contents

## Illustrations

* Eagle-Picher Industries, Inc., Miami, Oklahoma.

## Tables

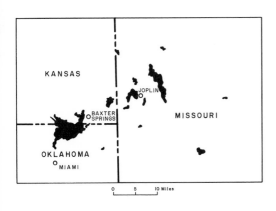

## ABSTRACT

Mining in the Tri-State district of Missouri, Kansas, and Oklahoma has been nearly continuous from about 1848 until the present day, although the major activity was from about 1880 to 1955. The district, which has produced over $2 billion in zinc and lead concentrates, ranks as one of the greatest mining districts in the world. Unlike other Mississippi Valley-type deposits in the United States, which occur in Cambrian and Ordovician limestones and dolomites, the Tri-State district ores occur in Mississippian limestones containing abundant chert. Zinc is 5 to 6 times more abundant than lead. In the Picher Field, the principal sub-district which has accounted for 61 percent of the total district production, most of the ore has been mined from the single horizon, M bed. Most ore bodies, other than those of the "Sheet Ground" or blanket-type, are in large, essentially flat-lying breccia zones and have a definite mineral zonal pattern. An irregular, generally elongated central dolomitic core is surrounded progressively outward by the main ore run, the jasperoid zone, the muddy or shaly and bouldery zone, the sparry calcite limestone zone, and the fossiliferous, dominantly crinoidal, limestone zone. Some of the zones may be absent or nearly so and may overlap to some extent with an adjacent zone. Although the district has been studied by many authors, much controversy still exists about such items as genesis, paragenesis, and time of emplacement. These subjects are discussed in the paper.

## INTRODUCTION

The Tri-State Mining District of Missouri, Kansas, and Oklahoma extends from east of Springfield, Missouri, to slightly west of the Picher Field of Kansas and Oklahoma, an east-west distance of approximately 100 miles. The north-south dimension is less than 30 miles.

Since most of the production has come from the western half of the district, this paper describes the geology of the four-county area consisting of Ottawa County, Oklahoma; Cherokee County, Kansas; and Newton and Jasper Counties, Missouri, with only minor comment on the eastern part of the district lying outside these four counties. Figure 1 shows the major

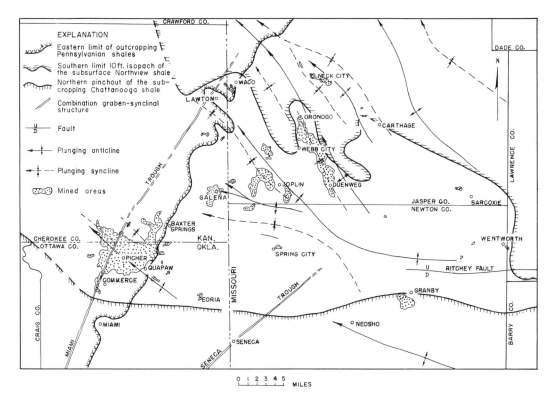

FIG. 1. *Map of the Tri-State District, showing the major structural and geological features.*

geological features and mining areas in the western half of the district. However, the most detailed data presented pertain to the Picher Field of Oklahoma and Kansas, since most of the rest of the district has been idle from about the middle forties.

Mining began in the extreme eastern part of the district and involved smaller mining areas compared to the larger ones that were mined in later years. As new areas were discovered, especially under the Pennsylvanian shales, the mining activity moved westward until today almost all of the mining is in the Picher Field. The major mining areas in their approximate order of importance are the Picher Field, Oronogo-Duenweg, Joplin, Galena, Granby, and Lawton-Waco.

As of 1964, the ore production from this large district is estimated to be in excess of 500,000,000 rock tons, from which in turn were produced 22,639,000 and 3,732,000 tons of zinc and lead concentrates respectively, valued at $1,477,192,000 and which, when smelted, had a value of $2,051,719,000.

Since this district was discovered, many noted geologists have studied the various areas, and, consequently, different theories and opin-ions have resulted as to the origin of the ore-bearing solutions, paragenesis, chertification, the importance of fracturing and shearing, the origin of the gangue dolomite, and the origin of the jasperoid.

At the present time, a United States Geological Survey manuscript on the district by E. T. McKnight is under review prior to publication.

## HISTORY AND PRODUCTION

### Mining Operations

Winslow (6, p. 274) mentions that H. M. Breckenridge, Jr. wrote in 1810 that hunters informed him of the presence (in Missouri) of a great abundance of lead ore on the Osage River and in the country drained by the White River and that H. R. Schoolcraft found lead ore in Stone County, Missouri in 1818 and smelted it for bullets. Schoolcraft, in 1819, also mentions lead ore discoveries on Strawberry River in Lawrence County, Missouri, as well as along other rivers. Some lead was mined from the Pierson Creek area southeast of Springfield, Missouri as early as 1844. Lead was first mined at Leadville, near Joplin, Missouri in 1848. In 1850, lead ore was found outcropping near Granby, but the first zinc ore was not sold until 1869. Prior to 1869, all zinc ore was thrown aside. The Galena, Kansas, area was discovered in 1876 and the Aurora, Missouri, area in 1886. Ore was first mined in Oklahoma at Peoria in 1891. The Badger-Peacock area in Kansas was developed in 1889, and, by 1900, the Lawton area to the north was prospected although it did not attain prominence until about 10 years later. Meanwhile in Oklahoma, mining and prospecting had spread from the Peoria area to near Quapaw by 1904.

Ore was discovered by drilling near Commerce, Oklahoma, in 1905, and mining and milling commenced by 1907. Mining spread as far north as the Kansas state line by 1913, and, by 1916, drilling was in progress in Kansas near Blue Mound and Baxter Springs. Also by 1916, large-scale operations existed in the Picher Field. By 1917, the Kansas part of the Waco area became productive, but it was completely overshadowed by the opening of the Picher Field. The decline or cessation of mining in practically all the other major areas coincides very closely to the meteoric rise of the Picher Field. Many mills in Missouri were closed down and moved to mines in Oklahoma (Picher Field). The major period of overlap covers approximately three years.

Many factors have affected, and will continue to affect, the mining operations in the district and especially the Picher Field. The six most important factors are, or have been: (1) milling practices, (2) mechanization of the mines, (3) ore prices and government subsidies, (4) commingling of ores, (5) reduction in royalties, and (6) detailed geological mapping and underground long-hole drilling.

(1) Milling practices—During early days of mining the Joplin-type mill was used, which was effective enough to separate about 90 per cent of the galena but produced a very poor sphalerite recovery. C. O. Anderson (13, p. 112, 113) reported that, in 1921 and 1922, he found a number of mills that extracted less than 50 per cent of the zinc mineral. He stated that, "Johnson and Heinz in their handbook, Milling Details and Milling Practices, issued in 1919, set forth that as a result of much testing, the average local mill was only 63 per cent efficient". Although there was some experimentation with the flotation process in 1915, the early spread of flotation originally was hindered somewhat by high royalty rates

($4.80/ton of flot zinc concentrate), but a 50 per cent reduction, plus subsequent reductions, made flotation methods much more attractive. However, it did not become an important part of the recovery process in the district until 1924 and 1925 when the first filter, the American disc type, for dewatering flotation concentrates was installed.

One of the biggest milling events, which prolonged the life of the Picher Field, was the erection of a large, 15,000 ton-per-day Central Mill by The Eagle-Picher Company to treat much of the ore produced in the Picher Field. Formerly, a mill had to be constructed on each landowner's property—usually every 40 acres or less. Obviously, had such a situation continued to prevail, many small but rich, or large low-grade ore bodies or portions of ore bodies on adjacent lands never would have been mined.

(2) Mechanization of the mines—Mining has evolved from single jacking, hand shoveling, and hand tramming to multi-drill ex·ension jumbos, some reaching to 70 feet in height, rubber-tired front-end loaders and large diesel haulage units. Between these two extremes of transportation have been mule, rope, battery truck, and locomotive haulage.

When a high roof required trimming, the famous Tri-State District "roof trimmer" came into action, doing his job from a high ladder manipulated into position by fellow workmen using rope lines. Can hoisting is a curiosity that has stood the test of time, and is still in use today.

(3) Ore prices and government subsidies— The ore bodies in the District are characterized by assay-type walls, since the grade of the ore gradually diminishes outward until barren ground is reached. Although the gradation is more abrupt—in some cases only a few feet— in M and higher beds, the "Sheet Ground" ore bodies commonly are very gradational. Gradational-type ore bodies similar to those in the "Sheet Ground" are abandoned during times of low metal prices but are reworked when metal prices are higher and government subsidies are granted; World War II subsidies especially permitted the mining of much low-grade ore.

(4) Commingling of ores from two or more properties often has permitted a mine to remain open that otherwise would not have been profitable. In some cases commingling was necessary in order to mine small ore bodies on adjacent properties that did not justify separate development.

(5) A reduction in royalty rates likewise

has allowed mining companies to re-open a mine or to open a new mine—especially a low-grade "Sheet Ground" type ore body where the mineralization lessens outward very gradually.

(6) Detailed geological mapping and underground long hole drilling—In the early fifties, detailed geological mapping, coupled with underground long-hole drilling, developed many small hidden ore bodies in formations above or adjacent to mined areas at a small cost (40, p. 27–29). Considerable attention was given to the dolomite-jasperoid contacts during long-hole prospecting programs.

## Statistics of Mine Production

The Tri-State Mining District ranks as one of the major mining districts of the world, having produced over $2 billion of recoverable metal from 1850 to 1964, inclusive. Converting all the metal values to more recent prices of zinc (14.5¢ lb.) and lead (16¢ lb.), the recoverable metal value would be $4.28 billion. During 1925 and 1926, average weekly pro-

duction of concentrates from the district was valued at over $1,000,000. Approximately 400 million short tons of crude ore were treated from 1907 to 1964. Based on the amount of zinc and lead concentrates produced as compared to later production figures, a conservative estimate of an additional 100 million tons to perhaps over 150 million tons of equivalent grade ore may have been treated between 1850–1906. Hence, the total crude ore produced from 1850–1964 probably is in excess of 0.5 billion short tons.

Present production in the district is almost exclusively from the Picher Field and will be slightly less than 1,000,000 tons of ore for the year 1965. No production has been reported from the Missouri part of the Tri-State District since 1957.

Table I briefly summarizes the production as compiled by the United States Bureau of Mines, in the Tri-State district that, as originally defined, extended from Springfield, Missouri, westward to include the Picher Field. Production from the Missouri mining areas located outside the four-county area of this

*TABLE I. Tri-State District Production—Missouri, Kansas, and Oklahoma, 1850–1964 (incl.).*

| Years (Inclusive) | Crude Ore Treated (4) | Metal Recovered from Ore Heads % Lead | Metal Recovered from Ore Heads % Zinc | Old Tailings Treated (4) | Mill Concentrates Lead (5) | Mill Concentrates Zinc (5) | Mill Concentrates Total Value $(7) | Recoverable Metal at Smelters Lead (6) | Recoverable Metal at Smelters Zinc (6) | Recoverable Metal at Smelters Total Value $(7) |
|---|---|---|---|---|---|---|---|---|---|---|
| 1850–1906 | (3) | (3) | (3) | None | 831 | 4,355 | 157,241 | 599 | 2,042 | 255,852 |
| 1907–1910 | 34,192 | .40 | 1.64 | (1) | 171 | 1,130 | 52,067 | 132 | 559 | 72,254 |
| 1911–1920 | 109,179 | .47 | 1.81 | (1) | 652 | 3,811 | 249,168 | 505 | 1,978 | 405,001 |
| 1921–1930 | 107,311 | .75 | 3.00 | (2) 16,119 | 1,030 | 6,137 | 339,770 | 799 | 3,195 | 524,015 |
| 1931–1940 | 45,772 | .72 | 3.23 | 57,853 | 436 | 3,389 | 131,551 | 332 | 1,808 | 210,848 |
| 1941–1950 | 75,656 | .39 | 1.75 | 76,678 | 394 | 2,832 | 304,058 | 298 | 1,519 | 390,048 |
| 1951–1960 | 29,935 | .48 | 1.56 | 1,985 | 198 | 894 | 133,676 | 145 | 476 | 177,647 |
| 1961–1964 | 1,959 | .73 | 2.71 | None | 19 | 90 | 9,661 | 14 | 53 | 16,054 |
| — | — | — | — | — | — | — | — | — | — | — |
| 1907–1964 | 399,004 | | | 152,635 | 2,901 | 18,284 | 1,319,951 | 2,226 | 9,588 | 1,795,867 |
| — | — | — | — | — | — | — | — | — | — | — |
| 1850–1964 | ? | ? | ? | 152,635 | 3,732 | 22,639 | 1,477,192 | 2,825 | 11,631 | 2,051,719 |

(1) Some old tailings were treated in the Tri-State.

(2) Some old tailings were treated in the Tri-State. 1921–1925, incl., but are not included here. The tailings figures for 1926 pertain only to Oklahoma.

(3) Not available.

(4) Thousands of short tons of ore or tailings.

(5) Thousands of short tons of zinc and lead concentrates.

(6) Thousands of short tons of zinc and lead metal.

(7) Thousands of dollars.

paper, such as Ash Grove-Everton, Aurora, Springfield, Stark City, and Stotts City, plus minor production from Christian, Green, Ozark, Hickory, Howard, Taney, and Wright counties, is less than $14 million in concentrate values out of a total district production of $1,477,192,000. Over $9 million of the $14 million came from the Aurora camp.

The Picher Field has contributed 61 per cent of the total district production of zinc-lead concentrates. The other major camps have produced as follows: Oronogo-Webb City-Duenweg—14 per cent; Joplin—9.5 per cent;

Granby—5 per cent; Galena—5 per cent; Waco-Lawton—2 per cent; Aurora 1.5 per cent; Neck City—1 per cent.

Only one per cent of the total district concentrate production has come from areas other than those mentioned above.

The extensively mined area of the Picher Field is illustrated in Figure 2. Table III shows the recoverable grade of ores and yearly tons of concentrates produced in the Picher Field related to the prevailing prices of zinc and lead metal.

A graphic chart by Brown (35, p. 789)

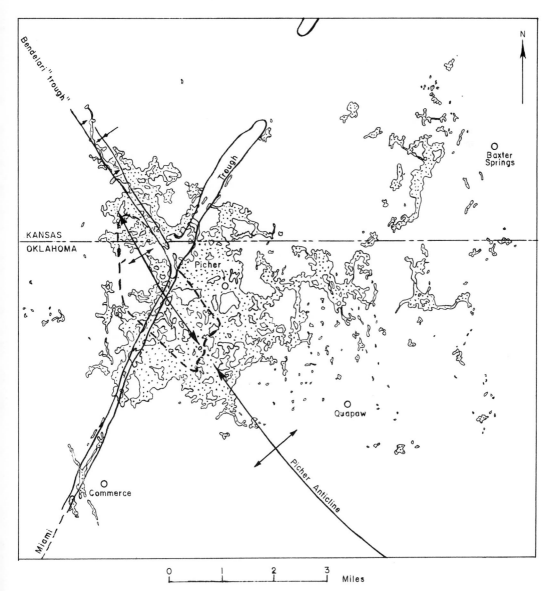

FIG. 2. *Map of Picher Field, showing underground mine workings and axes and outlines of the principal structural features. A large broad dome is shown by the dashed lines.*

shows yearly production and grades of ore related to the average yearly price of zinc metal from 1870 to 1950 for nine of the major camps in the District.

## PHYSIOGRAPHY AND TOPOGRAPHY

The Tri-State district is situated in two physiographic provinces. The Missouri portion and the eastern part of the Oklahoma and Kansas portion lie within the Ozark Plateau province, whereas the western part of the Oklahoma and Kansas portions of the district are within the Central Lowland province. The demarcation line between these provinces is approximately along Spring River, which is nearly coincident with the eastern outcrop line of the Pennsylvanian shales, as shown in Figure 1. Elevations in the district vary from a little over 1200 feet in the east to around 750 feet along Spring River, the principal drainage. Relief in the Central Lowland portion, which is underlain by Cherokee shale of Pennsylvanian Age, is less than 100 feet; however, in the Ozark Plateau to the east, relief varies from 100 to 350 feet. The outcropping rocks in these more hilly and higher elevated portions are principally limestones and cherts of Mississippian age.

## GEOLOGIC HISTORY

### Stratigraphy

A generalized stratigraphic column of the Tri-State district is shown in Figure 3. The rocks range in age from Precambrian to Pennsylvanian. Some Quaternary and Tertiary (?) gravels are present. All rocks lying below Mississippian sediments are identified only from drill-hole information.

#### Precambrian

Recent drilling into the Precambrian has revealed the presence of altered andesite porphyry, rhyolite porphyry, and microgranite porphyry. The surface of the Precambrian is very irregular, with depths below the surface ranging from 290 feet to nearly 2000 feet and with an average depth of 1700 to 1800 feet.

~~~~~ Unconformity ~~~~~
#### Cambrian and Ordovician

Since no commercial ore bodies have been found in the Cambrian and Ordovician sediments of the district to date, only a general

discussion of them is included here. Detailed descriptions of the Cambro-Ordovician Formations have been presented adequately in published literature. (32,36,39,41,43,50,51,52,53, 54).

The formations consist principally of dolomites with sandstone and sandy and shaly dolomite members. It was originally believed that the Potosi, Derby-Doe Run, and Davis Formations were not present in the district; however, recent deep drilling in the Picher Field has revealed their presence. The Roubidoux Formation contains several sandstone layers that are the principal aquifers in the District.

Special mention should be made of the nature of the Bonneterre in the Tri-State district as this formation is the principal host for the large lead deposits of southeast Missouri. The sediments that have been tentatively identified as Bonneterre appear to represent a transition zone between the Lamotte Sandstone and the Davis Formation. They consist of fine-grained silty dolomites, dolomitic siltstones, and dolomitic sandstones.

~~~~~ Unconformity ~~~~~
#### Unassigned Devonian-Mississippian

CHATTANOOGA SHALE This unit consists of a black to dark-brown, fissile, carbonaceous and slightly arenaceous shale. The Chattanooga is largely absent in the district; however, a few small, thin, isolated patches may be present. The northern subcrop of the shale occurs immediately south of the district (Figure 1).

~~~~~ Unconformity ~~~~~
#### Mississippian

In some previous literature, the Boone formation has been designated as a collective term involving some or all of the Warsaw, Keokuk, and Reeds Spring Formations and the St. Joe Group. (2, 3, 7, 21, 22). The sediments involved are described under the Kinderhook, Osage, and Meramec Series. On the basis of lithologic differences, Fowler and Lyden (15, p. 218) subdivided the Meramecian Warsaw and the Osagean Keokuk and Reeds Spring Formations into several letter-designated units or beds as illustrated in Figure 4. The present erosional cycle has removed some of these beds in the eastern part of the district. Practically all zinc-lead ore production in the district is from the Boone formation.

#### Kinderhook Series

CHOUTEAU GROUP The Chouteau Group consists of the Compton Limestone and the overlying Northview Formation. The term St.

| | SERIES | FORMATION | REMARKS | |
|---|---|---|---|---|
| PENNSYLVANIAN | Des Moines | Cherokee Group | Cherokee Group is the surface formation in the Picher Field | |
| MISSISSIPPIAN | Chester | Fayetteville Shale<br>Batesville Sandstone<br>Hindsville Limestone | The Chester Series is referred to as the Mayes Formation in older literature. The **Carterville Formation in the Joplin area** is faunally related to the Fayetteville and the Batesville Formations. | |
| MISSISSIPPIAN | Meramec | Warsaw | B-J bed= Warsaw Formation J bed is equivalent to Cowley Formation of southeastern Kan. | All Meramec and Osage sediments are referred to in some literature as the Boone Formation |
| MISSISSIPPIAN | Osage | Keokuk | K-Q beds (N,O,P&Q= Grand Falls Chert) | |
| MISSISSIPPIAN | Osage | Reeds Spring | R bed | |
| MISSISSIPPIAN | Osage | Fern Glen (St. Joe Group) | Corresponds to Pierson Formation in recent literature | |
| MISSISSIPPIAN | Kinderhook | Northview Shale (St. Joe Group / Chouteau) | Thin to absent in Picher Field | |
| MISSISSIPPIAN | Kinderhook | Compton Limestone (St. Joe Group / Chouteau) | | |
| MISSISSIPPIAN DEVONIAN | | Chattanooga Shale | Absent except as local patches in Picher Field Subcrop edge is located a few miles south of district | |
| ORDOVICIAN | Beekmantown | Cotter Dolomite<br>Jefferson City Dolomite<br>Roubidoux<br>Gasconade Dolomite<br>Gunter Member | | |
| CAMBRIAN | Ozark | Eminence Dolomite | | |
| CAMBRIAN | Ozark | Potosi | | |
| CAMBRIAN | Ozark | Derby – Doe Run Davis (Elvins Group) | | |
| CAMBRIAN | Ozark | Bonneterre (?)<br>Lamotte Sandstone | | |
| | | Precambrian Complex | | |

FIG. 3. *Generalized Stratigraphic Section, Tri-State District.*

Joe Group is still in use today to designate the Compton, Northview and Fern Glen Formations; however, due to an overlap in Series, the Fern Glen is described as a part of the Osage Series. Some mineral production has come from the Chouteau Group in the Springfield area mines.

COMPTON FORMATION   The Compton Formation is principally a limestone, gray to dove

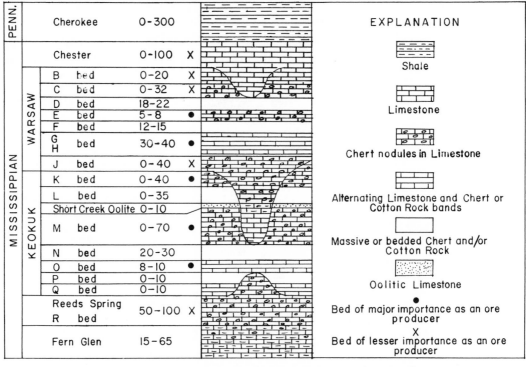

| PENN. | Cherokee | | | 0-300 | |
|---|---|---|---|---|---|
| | Chester | | | 0-100 | X |
| MISSISSIPPIAN | WARSAW | | B | bed | 0-20 | X |
| | | | C | bed | 0-32 | X |
| | | | D | bed | 18-22 | |
| | | | E | bed | 5-8 | ● |
| | | | F | bed | 12-15 | |
| | | | G H | bed | 30-40 | ● |
| | | | J | bed | 0-40 | X |
| | KEOKUK | | K | bed | 0-40 | ● |
| | | | L | bed | 0-35 | |
| | | | Short Creek Oolite | 0-10 | |
| | | | M | bed | 0-70 | ● |
| | | | N | bed | 20-30 | |
| | | | O | bed | 8-10 | ● |
| | | | P | bed | 0-10 | |
| | | | Q | bed | 0-10 | |
| | Reeds Spring R bed | | | 50-100 | X |
| | Fern Glen | | | 15-65 | |

EXPLANATION

Shale

Limestone

Chert nodules in Limestone

Alternating Limestone and Chert or Cotton Rock bands

Massive or bedded Chert and/or Cotton Rock

Oolitic Limestone

● Bed of major importance as an ore producer

X Bed of lesser importance as an ore producer

Fig. 4. *Geologic Section of Mississippian Formations, Picher Field. (Modified after Lyden, 32)*.

in color, containing calcareous shale partings. Thicknesses range from 5 to 20 feet.

NORTHVIEW FORMATION The Northview Formation is a green to dark gray-green and blue-green, silty, partly calcareous, shale. It is thin to absent in the District, but thickens to the north.

### Osage Series

FERN GLEN FORMATION The St. Joe Limestone, which corresponds to the lower part of the Fern Glen of eastern Missouri, is correlated with the Pierson Limestone as identified in the Springfield, Missouri, area. Moore, Fowler and Lyden (26, p. 7) tentatively identified a somewhat cherty fossiliferous limestone near Joplin, Missouri, as the Pierson Formation. Sub-surface information in the Picher Field indicates that the Fern Glen (Pierson or St. Joe) consists of a few blue and gray chert nodules in a very fine-grained, flaky, soft, silty and shaley limestone. A greenish hue is common to many occurrences. An average thickness of about 25 feet prevails in the main part of the Picher Field.

REEDS SPRING FORMATION This is the R bed of Fowler and Lyden (15, p. 218; 20, p. 109-110). It typically is composed of blue,

gray to dark gray, and black, bedded, irregular chert nodules in a dark gray and brown fine-grained limestone. Considerable brown fine-grained dolomite is present in some areas. C. E. Robertson of the Missouri Geological Survey (personal communication) believes that the O, P, and Q horizons are actually Upper Reeds Spring. Average thicknesses are from 50 to 100 feet.

〜〜〜 Unconformity 〜〜〜

KEOKUK FORMATION The Keokuk Formation consists of K (youngest) through Q (oldest) beds. The upper part consists primarily of gray and brown, medium- to coarse-crystalline limestone, containing considerable amounts of light-gray to white dense chert to cottonrock (a soft white calcareous chert or siliceous limestone) in the form of layers and nodules. The lower part, which involves the "Grand Falls" (N, O, P, Q beds), consists of white to light-brown and gray bedded chert and brown to gray fine-grained dense limestone. Away from mineralized areas, the "Grand Falls" becomes more limy and the cherts become more cotton-rocky. Several tens of miles south of the district and higher on the flanks of the Ozark uplift, Huffman (41, p. 44) states, the "Keokuk is a massive, white to buff and gray mottled chert locally interbedded with irregular stringers and masses of blue gray, dense fine-grained limestone." He also points out that large crinoidal reefs or bioherms are found in places. In Missouri, portions of the Keokuk are used as a source for road metal and for dimension stone.

*K bed* is a gray crinoidal limestone containing abundant chert nodules. The nodules in the upper portion are more rounded, whereas the lower portion contains larger elongated nodules. Normal thicknesses vary from 0 to 12 feet but in places reach 40 feet. It is absent in the western part of the Picher Field.

*L bed* is a white cottonrock and dense light-gray chert, but, in unmineralized areas, it is composed partly of limestone. Its thickness ranges from 0 to 35 feet but is absent in the western part of the Picher Field. Where present, its thickness is uniformly about 30 feet.

*M bed,* the principal ore horizon, consists of white and light-gray chert nodules in gray and pale-brown crystalline and fossiliferous limestone. The Short Creek Oolite, a persistent oolitic limestone zone found at the top of this unit, is recognizable mostly outside the areas of mineralization. The oolite zone varies from 2 to 10 feet in thickness. M bed varies in thickness from 0 to 90 feet in the Picher Field. Pre-Warsaw erosion has removed K, L, and parts of M beds leaving J bed resting directly on M bed in some areas, especially in the western part of the Picher Field. In a few areas, J rests on N bed. Thicknesses of unmineralized M bed limestone capped by the Short Creek Oolite, as determined from drill hole data, differ greatly, ranging from 17 feet in an area west of the Picher Field, to over 100 feet in the Joplin area; the thickening to the east is quite irregular.

GRAND FALLS CHERT    The Grand Falls Chert is described by Smith and Siebental (10, p. 4) as a member of the Boone Formation. They indicated that the chert member could not be traced persistently from place to place where it would be expected to occur stratigraphically. Fowler and Lyden (15, p. 218) correlated the Grand Falls with O, P, and Q beds. Current usage of the Grand Falls by the writers and others includes N, O, P, and Q beds.

*N bed,* the upper part of the Grand Falls Chert, is typically a bedded gray to white cottonrock and glassy gray to light-gray and light-brown chert. In mineralized areas and in areas where deformation is intense, a gnarly and knotted structure is predominant with a loss of the bedded character. Away from the deformed and mineralized areas, it is interbedded brown to gray fine-grained limestone and white cottonrock to light-gray chert. It is usually from 20 to 30 feet thick.

*O bed* is characterized by thin gray cherts beds alternating with thin brown commonly fine-grained limestone beds. A thickness of 8 to 10 feet is typical. It is the unit containing most of the "Sheet Ground" ore bodies of the district.

*P bed* is characterized by light gray cottonrock chert and glassy chert as beds and large flat nodules. Thicknesses range from 0 to 10 feet.

*Q bed,* the lowest member of Grand Falls Chert, is similar to O bed. It is thin-bedded, consisting of gray chert beds alternating with brown limestone beds. It is from 0 to 10 feet thick.

~~~~~ Unconformity ~~~~~
Meramec Series

WARSAW FORMATION    There has been considerable controversy in the past whether the Warsaw Formation should be assigned to the Meramec or Osage Series. The Warsaw, according to Moore, Fowler and Lyden (26, p. 2, 10–11) and Spreng (44, p. 66) is assigned to the Meramec Series, and is so used in this paper. The Warsaw is primarily a light gray and light brown to brown abundantly cherty, fine- to some coarse-grained limestone. Some beds consist almost entirely of chert and/or cottonrock. Its thicknesses range from 86 to 195 feet. The Warsaw is the source of "Cathage Marble" in Jasper County, Missouri, and locally is quarried for use as an agricultural limestone and road metal. Most of the tripoli mined in Ottawa County, Oklahoma, comes from the Warsaw and Upper Keokuk Formations. The Warsaw is the second most important ore-producing formation in the district. The Warsaw Formation is composed of Fowler and Lyden's designated lithologic units B bed (youngest) through J bed (oldest).

*B bed* contains light-gray and brown limestone with some white and blue chert. Depending upon the character of the post-Meramec pre-Chester unconformity, its thickness lies between 0 and 20 feet.

*C bed* is characterized by white and blue chert nodules in gray and light-brown limestone. There is generally more dark blue chert in the lower 10 feet. Because of the above mentioned differences in the unconformity, the thickness ranges from 0 to 32 feet.

*D bed,* in mineralized areas, consists primarily of white cottonrock and dense white and light-gray chert. Greater amounts of calcareous material seem to characterize the bed in unmineralized areas. This bed ranges in thickness from 18 to 22 feet.

*E bed* is a gray to light-brown limestone containing gray and brown chert nodules. Locally a brown dolomite is present. It ranges in thickness from 5 to 8 feet.

*F bed,* near areas of mineralization, consists primarily of light-gray cottonrock and dense white and light-gray chert. Elsewhere it is composed partly of a light-brown limestone. F bed has a thickness range of 12 to 15 feet.

*GH bed* was originally described as two separate beds; however, due to their lithologic similarity, they have been grouped together as one unit. They are thin-bedded gray, brown, and blue chert beds alternating with brown fine-grained limestone beds. The thickness ranges from 30 to 40 feet. In parts of the Picher Field, small amounts of dark-brown, fine'y speckled chert occur near the GH-J bed contact.

*J bed* unconformably overlies the beds of the Keokuk Formation. It consists of gray and dark-gray glauconite-speckled chert (in many places containing abundant sponge spicules), in a gray and dark-gray to dark-brown glauconite-speckled, nodular phosphatic, fine-grained muddy to coarse-crystalline limestone. Where J bed occupies depressions in the underlying Keokuk, it commonly is a dark, shaly limestone, especially near its base. This bed furnishes a good marker throughout most of the Picher Field. Thicknesses range from 0 to 40 feet depending upon erosional relief on the underlying Keokuk. However, the usual thickness for the area lies between a few inches to 5 feet. J bed thickens westward from the Picher Field and is correlated with the Cowley Formation of Kansas. J bed commonly is mined in conjunction with GH and/or K bed and is partly mined with M bed where K and L beds are absent.

Unconformity
Chester Series

The Chester, formerly called the Mayes, lies unconformably on the Warsaw. It is present over a fairly continuous belt through the Oklahoma and Kansas portions of the district; however, in the Joplin area, it is present only as outliers that are believed to be sink-fill deposits in the older formations. These sink-fill deposits are represented by the Carterville Formation as described by Smith and Siebenthal (10, p. 5–6). The Carterville, which is composed of inhomogeneous varieties of limestones, sandstones, and shales, is closely related to the

Fayetteville and Batesville Formations. It is of limited areal extent, occurring in patches filling depressions in the underlying Boone formation. Local thicknesses of up to 200 feet have been reported in the Joplin area. The Chester comprises the Hindsville, Batesville, and Fayetteville Formations. Several usually small, but rich, zinc and lead ore bodies have been mined in the Chester.

HINDSVILLE FORMATION   The Hindsville consists of alternating limestones and shales in the Oklahoma-Kansas portions of the district. The limestone is partly siliceous, oolitic, and platy, but it is mostly chert-free and fossiliferous with small amounts of glauconite. The shaly phases are calcareous and sandy. There is generally a thin chert-pebble conglomerate at its base. Weidman (17, p. 19) states that the Mayes Limestone has been "silicified and dolomitized in the vicinity of the ore deposits in the mining district." Part of the limestone in the Carterville of the Joplin area probably corresponds to the Hindsville of the Oklahoma-Kansas areas.

BATESVILLE FORMATION   The Batesville is a yellow brown fine-grained, friable, calcareous sandstone in its outcrops. In drill holes, a light-brown to light-gray fine-grained quartzitic sandstone is characteristic of it. Weidman (17, p. 19) described the sandy phase of the Mayes as being silicified in the vicinity of the ore deposits and hence developing into a quartzitic sand and quartzite. The formation has a maximum thickness of 10 feet in the Oklahoma area.

FAYETTEVILLE SHALE   The Fayetteville Shale has not been mapped in the Tri-State district, but there is some speculation that it exists in the mines in the Miami Trough structure. Weidman (17, p. 21) stated that the Fayetteville consists of shale and interbedded limestone and sandstone, some of the shaly beds containing abundant bryozoa similar to Archimedes He gave a thickness in Ottawa County, Oklahoma as between 40 and 50 feet.

Unconformity
Pennsylvanian
Des Moines Series

CHEROKEE GROUP   The Cherokee Group, often referred to as the Cherokee shale, consists primarily of shales with some interbedded sandstones and several economically important coal beds. These rocks cover much of the Oklahoma and Kansas portions and a part of the Missouri portion of the district. In the eastern portion of the District, it is represented by outliers of various sizes, occupying depressions in the underlying "Boone" Sediments. The thickness of the Cherokee ranges from 0 to over 300 feet in the western part of the District.

### Unconformity
### Tertiary (?) or Quaternary (?)

LAFAYETTE GRAVELS   Rounded and concretionary chert nodules are exposed in limited areas in the District.

### Quaternary

ALLUVIUM AND TERRACE DEPOSITS   Stream deposited sand, gravel, silt, and clay occupy the flood plains and bottom land near the principle drainages.

## General Structure

The Tri-State district is located along the edge of the northwestern flank of the Ozark uplift, which is a broad geanticline. The strata in the district are located far from the axis of the Ozark Dome and dip, in general, to the northwest at low angles of only 15 to 20 feet per mile. Folding and normal faulting are the principal structures that cause irregularities in the regional dip and are illustrated in Figure 1.

FAULTING   Three major faults—the Ritchey, Seneca, and Miami Faults—occur within the district as defined in this paper. The Ritchey Fault extends from Monett, and perhaps as far east as Verona in Lawrence County, almost due west to at least 4 miles west of Ritchey in Newton County, Missouri for a total distance of about 25 miles. This normal fault, downthrown on the south side, has a maximum displacement of at least 150 feet (38). There is no apparent relationship between ore mineralization and the Ritchey Fault.

The Seneca Fault is practically a continuous structure that extends from near Tipton Ford in Newton County, Missouri, approximately S40°W in a linear direction for about 70 miles to a point about 3 miles east of Pryor in Mayes County, Oklahoma. It most commonly is a graben, occasionally a syncline, varying in width from less than 200 feet to over 0.5 miles. The vertical displacement of the graben ranges from several feet as a faulted syncline (11, p. 198) or from being indistinguishable in the residual "Boone" cherts in the Oklahoma area to less than 300 feet, bringing the Fayette-

ville and younger beds in contact with the Boone. Beds having dips of 10° to 15° or even 25° are common close to the faulting (41, p. 90). Small ore bodies have been mined from a few miles southwest of Seneca, Missouri, northeastward along the Seneca Fault.

The Miami Trough is a linear structure that becomes evident as an elongated syncline near the junction of Ottawa, Craig, and Delaware counties in Oklahoma and extends N15°E through the intensely mineralized Picher Field where it is, in different places, a graben, a hinged fault, or a syncline. Figure 2 shows the axis of the trough extending through the Picher Field. At one point, the Pennsylvanian shales are displaced nearly 350 feet in the graben. A vertical cross-section through the trough is shown in Figure 5. Drill holes that have penetrated the Precambrian igneous complex at the northern edge of the Picher Field show that the graben structure extends into the Precambrian rocks. The Miami Trough gradually assumes an ill-defined course N35°E to N40°E just north of the Picher Field to as far northeast as the Waco-Lawton mining area near the Missouri-Kansas state line.

FOLDING   A parallel series of more or less evenly spaced, narrow linear structural folds that generally have a northwesterly trend and a gentle pitch to the northwest are present in the Missouri part of the Tri-State district.

The Neck City-Purcell-Alba, Oronogo-Webb City-Duenweg, and part of the Joplin mineralized trends are confined largely to the syncline portions of the folds. Structural contour maps on deeper formations indicate that the Oronogo-Webb City-Duenweg mineralized area may be faulted at depth, becoming a graben-like structure.

The most pronounced structure in the Missouri area is the Joplin anticline, which was mapped by Siebenthal (10, p. 9). It first becomes evident on the north side of the Ritchey Fault in Newton County, Missouri, assumes a northwesterly course from a point about 2 miles northeast of Tipton Ford in Newton County, and extends to the vicinity of Lawton, Kansas. Siebenthal (10, p. 9) states that, "It is, in general, much steeper on the western limb than on the eastern, and possibly along its course between Joplin and Smithfield may break down into a fault." Structural contours on deeper stratigraphic horizons as determined from drill hole information confirm Siebenthal's suggestion of faulting.

In Kansas, a northwest projection of the strike of the sharply downwarped side of the

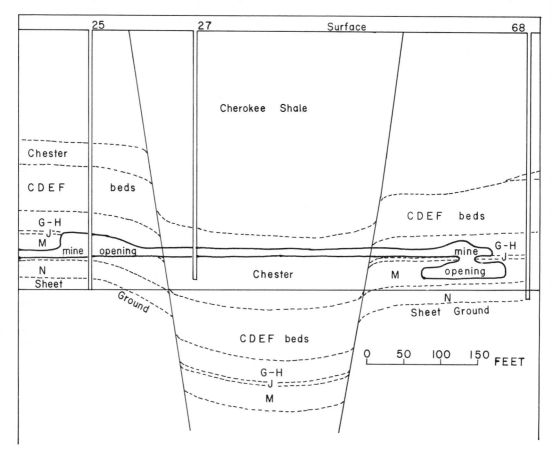

Fig. 5. *Vertical Section through Miami Trough in Blue Goose Mine.* (*After Fowler and Lyden,* 15).

Joplin anticline coincides with the Lawton syncline. The Cow Creek anticline (10, p. 9), slightly over 2 miles long, is a sharp fold in the Cherokee Group and is located about 2 miles southwest of, but parallel to, the Lawton Trough.

In the Picher Field, the Bendelari "trough" and the Picher anticline are the principal folds. The Bendelari "trough", a monoclinal fold with a maximum amplitude of 80 feet, trends N30°W from its intersection with the Miami Trough in the heart of the Picher Field. The downfolded side is on the northeast. Recent deep core drilling indicates a fault in the Precambrian just southwest of the "trough" that, if projected up dip, would coincide with the sharply folded part of the monocline as observed in the mines. It appears as if this "trough" is merely a result of tectonic adjustment along this line of faulting. The Picher Field has a northwest mineralized extension along the Bendelari "trough."

The Picher anticline, first named in this report, is a broad gently plunging fold that first becomes evident in the eastern part of Ottawa County, Oklahoma, extends in a northwesterly direction through the Picher Field, and finally loses its identity just west of the Picher Field. Structure maps on the top of N bed (28, 42) show an interruption in this fold in the southeast portion of the Field. It also is interrupted by the Miami Trough. Structure maps on the top of the Short Creek oolite and on the Cotter Dolomite also reflect the presence of the Picher anticline.

ALIGNMENTS   A study of Figure 2 showing mined-out areas of the Picher Field reveals several striking lineations. The two more obvious ones bound the northeastern edge and the western edge of the Field. In addition to these lineations, others criss-cross the Field and are more easily discerned on maps of greater detail. McKnight (57) believes that these alignments "represent breaks in the Pre-cambrian basement . . . along which ore solutions arose."

## Age of Mineralization and Lead-Isotope Studies

Zinc and lead mineralization is known to occur in the Cherokee shale of Pennsylvanian Age, which contains the youngest strata occurring in the Tri-State district. Hence, at least part of the mineralization is post-Cherokee in age.

Age dating of the district ores by lead isotopes gives anomalous results. The values so obtained have been identified characteristically in the literature as anomalous Joplin, or J-type leads.

In 1953, G. L. Cumming (37,) of the University of Toronto in Canada tested some (Picher Field) galena samples and obtained an age of 1000 m.y. by statistical analysis. Even with a large error assigned to the result, he states that the seemingly indicated age is much too old. Cumming goes on to say:

"It is, of course, feasible to say that the ores are derived from reworked deposits of a much older age and in the end this may be the only possible solution. There are some other possibilities which should be considered however.

"It seems highly doubtful that the anomalous nature of the leads can be explained by any process as outlined above which took place in the sediments themselves. If the time of contact were very short, the value of 1000 million years is of course much too old since the ores are at least as young as the Mississippian. On the other hand, if the time of contact of the lead and uranium were long, then the sediments would have to have been in place for a similar long period of time and this is again impossible on geological grounds. This would seem to indicate that the anomalous nature of the ores can not be accounted for by a syngenetic process."

Cannon, et al., (47) sampled individual growth zones within a large single crystal of galena from the Picher Field. From the data obtained, he suggests that the crystal may have grown over a very long span of time, perhaps one hundred million years or more.

Cannon, et al. (48, p. 576) state ". . . . these isotopic variations seem to imply progressive leaching of lead, probably during a protracted period of time, from rocks with appropriate content of lead, uranium, and thorium. The data seem to imply source rocks younger than Precambrian, presumably some of the Paleozoic sedimentary rocks within which the Tri-State ores were deposited. Growth of the zoned crystal would seem to have occurred intermittently, during part or all of a period of some 300 to 100 m.y. ago, equivalent to late Paleozoic and Mesozoic time on the Holmes time-scale."

Brown (55, p. 58) also believes that the processes involved in concentrating the ore fluids of Mississippi Valley deposits in general occupied a long period of time, as much as 100 to 200 million years but that the "final episode of deposition as ores probably is much shorter, perhaps only tens of millions of years".

Recent studies by Heyl, et al. (60) of lead isotopic data from the central and eastern interior of the United States support previous studies that the galena is anomalously radiogenic (J-type). Further it is stated that "Slopes from Pb206/Pb204 versus Pb207/Pb204 plot are compatible with a contribution of lead by the 1300 ± 300 m.y. basement rocks that underlie the deposits. . . . Galena occurring in shales in the Mississippi Valley region outside mineralized districts contain ordinary lead. Theories that call upon such shales to be the major source of lead in the deposits of the Mississippi Valley type must explain this distinct difference in isotopic composition."

It is generally agreed that deposition of the zinc-lead ores in the District involved a long span of geologic time. The anomalous nature of the lead isotopes complicates the origin of the ores. However, they were emplaced in the Mississippian sediments as late as (or even much later than) post-Cherokee (Des Moines) time.

## ECONOMIC GEOLOGY

### Stratigraphic Relations of Ore Bodies

Ore bodies of the Tri-State district are restricted almost entirely to the Mississippian sediments, specifically the Keokuk and Warsaw Formations. The Mississippian limestones are overlain unconformably by the Pennsylvanian Cherokee shale. This shale "cap" generally has been regarded to have acted as an impermeable barrier to rising mineralized solutions and to have aided in the localization of the ore deposits. Cherokee shale outliers associated with some of the sink-type deposits of Missouri were sufficiently broken to permit mineable grade ore to be deposited in conjunction with the ore in the underlying Mississippian formations.

Siebenthal (12, p. 193–199) was the principal exponent in pointing out the relationship between ore in the Mississippian sediments and the north subcrop edge of the Chattanooga Shale that lies just to the south of the mineralized areas in the district. Figure 1 shows that the shale is absent or only locally present under the principal mineralized areas of the district. South of the district, a few small areas that have been mineralized are known to occur immediately beneath or several tens of feet below the shale. In areas where there is no underlying Chattanooga Shale, the Mississippian formations commonly contain some indications of mineralization; however, where the Chattanooga is present, these same formations are nearly devoid of mineralization.

A similar relationship also appears to exist with the 10 foot isopach of the Northview Shale that is located north of the district. The more intensely mineralized areas are where there is less than 10 feet of underlying Northview Shale, as depicted in Figure 1. In fact, the Oronogo-Duenweg mineralized trend follows a definite pattern of thinner Northview Shale trending north-northwest.

Significant, but non-commercial, mineralization has been observed in the Cotter, Jefferson City, Roubidoux, and Gasconade Formations of Ordovician age.

Practically no mineralization has been found in the Eminence through Bonneterre and Lamotte Formations of Cambrian age, nor in the Precambrian basement complex.

Figure 4 shows the alphabetical breakdown of the Mississippian age Warsaw and Keokuk Formations and also depicts the relative importance of the mineralized beds. These beds, where unmineralized, consist principally of limestone with some cottonrock or chert as nodules or beds.

Beds consisting primarily of massive chert are mined only where intensely fractured and where they are subjacent or superjacent to an ore body in one of the more favorable beds. Considerable chert is present where ore bodies are found in the favorable horizons. In fact, "chertification" maps have been used in the past as an aid to exploration.

A noteworthy observation is that the principal ore horizon, M bed, is near an unconformity at the base of the Warsaw. Also, ore minerals in the Reeds Spring Formation, the upper contact of which is defined by an unconformity, are confined to its upper 20 feet.

The areal extent of mined ore bodies within individual lettered beds in the Picher Field is shown in Figure 6. M bed is by far the most important mineralized horizon. GH and K beds are next in importance. Very minor M bed ore bodies have been mined beneath the two large mined-out areas in GH bed. Most of the mined areas shown on the Chester and E bed map are in E bed. The Sheet Ground workings lie at the fringes of the Picher Field, and generally the overlying beds are unmineralized. A few small ore bodies have been mined in the Reeds Spring Formation at scattered places in the field.

On the northeast, or downfolded, side of the Bendelari "trough," GH bed was the main ore horizon, but, on the southwest side, ore bodies were dominant in M bed. The K bed ore body relationships are complicated by post-K bed erosion.

## Mineralogy of the Deposits

PRIMARY DEPOSITS   Only two minerals, sphalerite and galena, are commercially important in the Tri-State district; however, many other minerals have been identified, as shown in Table II. Some of the minerals in this table have been observed at only one mine. The following discussion, which is limited to the major minerals, was taken largely from Hagni (45) as his study is the most recent and most comprehensive concerning minerals and paragenesis.

*Sphalerite* occurs as yellowish-brown modified tetrahedral crystals in vugs, as light-yellow crystals disseminated through jasperoid and dolomite, as small reddish-brown crystals lining vugs and fractures and finely disseminated in jasperoid, and rarely as stalactites. The iron content is usually less than 0.5 per cent. Minor amounts of germanium, gallium, and cadmium are associated elements and are recovered from the concentrates during smelting. Sphalerite is by far the most important mineral in the district having at least a 5.26:1 recovery ratio with galena for the period 1907–1964.

*Galena* is present as disseminated crystals in jasperoid and gray gangue dolomite, as well formed crystals deposited in vugs, and rarely as anhedral grains deposited contemporaneously with zinc sulphides in stalactites.

*Chalcopyrite* is found most commonly as well-formed crystals deposited in vugs on pink dolomite and on sphalerite. Although a commonly occurring mineral, it is not present in sufficient quantity to be recovered commercially.

*Pyrite* occurs as small crystals, pellets, and

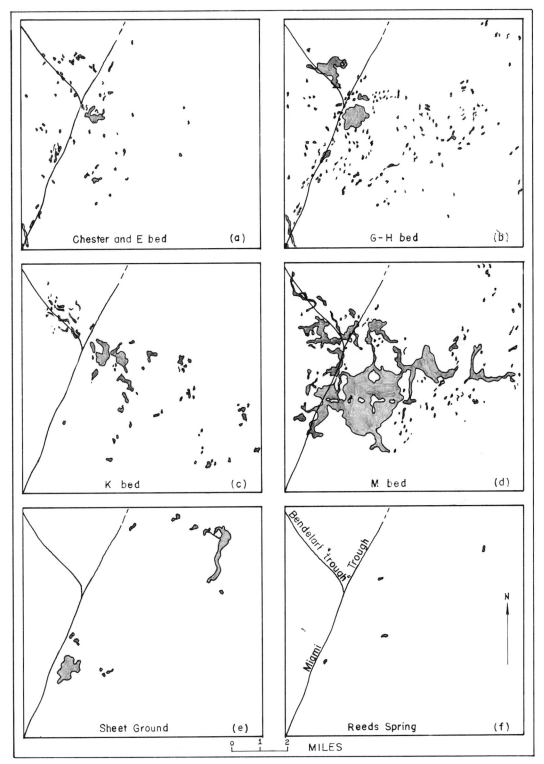

FIG. 6. (*a through f*) *Series of Maps of the Picher Field, showing the areal distribution of mined-out ore bodies in their respective beds.*

TABLE II.   *Minerals in the Tri-State District Ores*

## MAJOR MINERALS

| Sulfides: | Carbonates: |
|---|---|
| Galena | Calcite |
| Sphalerite | Dolomite |
| Chalcopyrite | |
| Pyrite | Silicates: |
| Marcasite | Quartz |

## MINOR MINERALS

| Native Elements: | Oxides: |
|---|---|
| Sulfur | Cuprite |
| | Hematite |
| Sulfides: | Pyrolusite |
| Bornite | Limonite |
| Wurzite | |
| Greenockite | Carbonates: |
| Millerite | Smithsonite |
| Covellite | Aragonite |
| | Cerussite |
| Sulfosalts: | Hydrozincite |
| Enargite | Aurichalcite |
| | Malachite |
| Sulfates: | Azurite |
| Barite | Leadhillite |
| Anglesite | |
| Gypsum | Arsenates: |
| Starkeyite | Picropharmacolite |
| Chalcanthite | Mimetite |
| Melanterite | |
| Epsomite | Phosphates: |
| Goslarite | Vivianite |
| Linarite | Pyromorphite |
| Jarosite | Wavellite |
| Plumbojarosite | |
| Aluminite | Silicates: |
| Copiapite | Hemimorphite |
| Caledonite | Allophane |
| | Chrysocolla |
| | Kaolinite |
| | Glauconite |

From: R. D. Hagni (45, p. 33)

anhedral grains disseminated in jasperiod, chert, gray dolomite, and limestone, as small crystals intergrown with marcasite and chalcopyrite, as small crystals in vugs on pink dolomite, sphalerite, galena, chalcopyrite, and marcasite, and as small crystals included in calcite. It rarely occurs in large masses.

*Marcasite* formed as crystals in vugs on pink dolomite, sphalerite, galena, and chalcopyrite, on and in late calcite, as small grains disseminated in jasperoid, gray dolomite, and chert, and as intergrowths with stalactitic sphalerite.

*Calcite* generally is transparent, nearly colorless to amber and is in crystals that are as much as 2 feet long. This mineral is abundant, especially along the fringes of the Picher Field as well as near the fringes of an ore body from which ore runs extend into limy areas.

*Dolomite* is present as: (1) uncommon, fine-grained gray sedimentary dolomite in beds; (2) coarse-grained gray gangue dolomite, generally in massive beds that commonly contain abundant brecciated chert fragments and nodules and also as disseminated crystals in jasperoid; (3) pink dolomite, usually as crystals in vugs and as fracture-fillings that cut both gray gangue dolomite and jasperoid. Individual crystals in the coarse-grained gray dolomite are white; however the large amounts of jasperoid and jasperoidal mud interstitial to the crystals gives it an overall gray color. The fine-grained gray "sedimentary" dolomite has no apparent relationship to ore mineralization; however the coarser gray gangue dolomite and the later, more coarsely crystalline pink gangue dolomite are associated very closely with zinc-lead mineralization.

*Quartz* is the most abundant gangue mineral of the ore deposits and occurs as chert, jasperoid, quartz druses, and well-formed quartz crystals.

Chert consists mainly of microcrystalline quartz, occurring principally as: (1) massive beds as exemplified by L and N beds; (2) beds 2 to 6 inches thick intercalated with equal or slightly thicker limestone beds as found in GH, O, Q, and R beds; (3) nodules and lenses in limestone parallel to the bedding such as those in K and M beds; and (4) irregular replacement blebs in limestone. The abundance of chert can not be overstated as it is everywhere present in tremendous quantities in the breccia zones containing the ore deposits.

Beds overlying an orebody in M bed, for example, commonly will be silicious, or "cherty", with no apparent thinning of the beds as compared to the thickness in an unmineralized area. A close examination of the drill cuttings from the beds above an ore body occasionally reveals that some of the chert has a texture very similar to that of jasperoid. This jasperoidal-textured chert grades into a fine-grained limestone that is practically identical in color and grain size. Because the color of this silicified limestone is identical to that of the original limestone (i.e., light brown, light gray), and is similar in color to the "original" chert, the cuttings generally have been described as chert. The writers believe that this material may be jasperoid, but it does not have the interstitial organic matter or bitumen that is associated with the jasperoid as so designated in the next paragraph and described by others in previous literature.

Jasperoid consists mostly of microcrystalline quartz crystals. Most jasperoid is to essentially brown to black in color, the coloring agent being a brown opaque matter reported by Smith and Siebenthal (10) to be bitumen. Jasperoid, which is closely associated with sulfide minerals, is massive but is characteristically very thinly banded and occurs interbedded with chert and fills the spaces between broken chert fragments in the mineralized breccia zones. Chert fragments commonly appear to be "floating" in jasperoid. Soft muddy phases of the jasperoid are common and close examination reveals them to consist mostly of quartz crystals with minor sulfides, glauconite, and organic material.

Fine quartz druses coat many of the vugs but are evident only upon close examination.

Late quartz crystals which may be as large as a 1 or 2 cm were deposited on sulfides and other minerals in some vugs. They are not common and are not present everywhere but are more abundant in the Reeds Spring and Sheet Ground ore bodies.

SECONDARY DEPOSITS    In the eastern part of the Tri-State district where some oxidation of the primary ores has occurred and especially where the Pennsylvanian shales were eroded and the deposits are now close to the surface, secondary zinc, lead, iron, and copper minerals were formed. Small economic ore deposits consisting of some of the secondary minerals such as smithsonite, hemimorphite, and cerussite were mined in the early days. This was especially true of the Granby camp.

## Mineral Zoning and Wall Rock Alteration

Mineral zoning in the Picher Field for the most part has a definite areal pattern and is found in all of the main ore horizons with the exception of the Sheet Ground. It is developed only rarely in the Chester Series.

In the main part of the field, the intense mineralization has resulted in overlap of adjacent zoning patterns so that limestone rarely is observed in the mineralized horizons. However, in the fringe areas where mineralization and deformation were less intense, the complete zonal relationship of an individual ore body can be observed.

The typical zonal pattern in K and M beds is a central dolomitic core surrounded progres-

sively outward by the main ore run, then by the jasperoidal zone, and finally by the unmineralized limestone. Typical mineral zoning patterns are illustrated in Figures 7 and 8 and should be studied with Figure 9, which last shows the relationship of these ore bodies to structure.

The dolomitic core may be oval, round, or elongated but commonly has an irregular outer boundary and may range from less than 10 feet to several hundred feet in width and up to several thousand feet in length. It consists of fragments and nodules of chert in a matrix of gray gangue dolomite and minor jasperoid. Limestone occasionally is found within the dolomitic core. In places, the dolomite is massive or bedded. It also occurs as large broken blocks in a matrix of jasperoid near the so-called "contact" of the dolomite zone with the jasperoid zone.

The "contact" area is the principal ore zone, generally associated with a series of strong attendant vertical to near vertical fractures. Pink gangue dolomite commonly fills the fractures on the dolomitic side of the "contact" and may overlap a few feet into the jasperoid zone. Farther from the "contact," but within the jasperoid zone, some of these fractures commonly are filled with later calcite.

The jasperoid zone is characterized by the absence of dolomite and by abundant broken angular chert fragments, from 1 inch to 2 feet in diameter, in a matrix of, and cemented by, jasperoid. In the intercalated chert and limestone beds, such as is typical of GH bed, very little brecciation is evident, and the jasperoid replaces the thin limestone beds.

At the outer edge of the jasperoid zone, the jasperoid commonly becomes muddy, calcite is abundant, and the ground generally is bouldery and open. This transition zone overlaps the limestone that first becomes evident near the base of the workings.

It has been suggested that the shaly or muddy phase between the jasperoid and limestone zones is confined mainly to areas where K and L beds are absent so that J bed, which generally is a shaly limestone, rests directly on M bed and is a probable source for the shale-mud. In the western part of the Picher Field, J bed does rest on M bed, and the zonal relationship of dolomite to ore to jasperoid to shale-mud to limestone is very prevalent. When an ore run "muds-out," the zonal relationship becomes dolomite to shale-mud to limestone and both the jasperoid and the ore are absent. At the eastern edge of the Picher Field where K and L beds are present, the

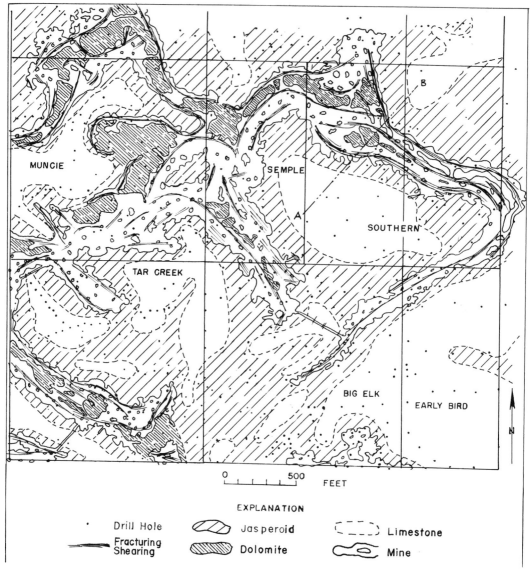

FIG. 7. *Map of GH, J and K Beds of the Bilharz Mine showing mineralization and fracturing or shearing. (After Lyden, 33).*

smaller M bed ore runs in the St. Louis No. 4 mine have a zonal relationship of dolomite to ore to jasperoid to limestone. In this mine, the jasperoid zone is very narrow, but, where the ore run diminishes, the zonal relationship is dolomite to limestone. Such a relationship, however, is not restricted to all areas where the overlying L bed is present. Other mines also contain L bed, but a shale-mud or muddy jasperoid phase commonly exists between the jasperoid and limestone, even though the zone may be quite narrow. Hence, the suggestion of a restrictive relationship of the shale-mud

to the overlying shaly J bed limestone appears invalid. In addition, both stratigraphically higher beds (GH, etc.) as well as lower beds (Reeds Spring) also have muddy or shaly phases. The writers believe that some of the mud and shale could have been an original depositional feature.

Hagni and Saadallah studied the M bed limestone adjacent to the mineralized areas and concluded that a zonal relationship exists in the limestone. Their studies showed that "Distally from ore bodies, the M bed limestone consists predominantly of allochems, mainly

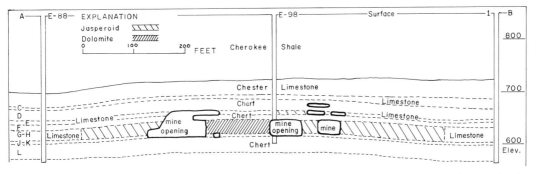

FIG. 8. *Vertical Section A-B, showing relationship of jasperoid and dolomite to mined areas in the Bilharz Mine. (After Lyden, 33).*

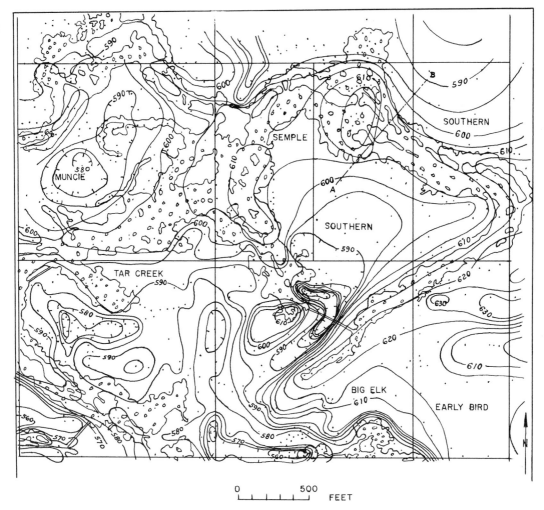

FIG. 9. *Map of the Bilharz Mine, showing the relationship of mine workings in GH, J and K beds to structural contours drawn on the top of underlying L bed. Small dots represent drill hole control points. (After Lyden, 33).*

crinoids, while microcrystalline calcite, sparry calcite, and microspar are much less abundant constituents" (58, p. 1607). Further, they found that the limestone nearer the ore bodies has a lesser allochem content and an increased amount of sparry calcite and that this gradual change commonly begins at a distance of from 40 to 80 feet from the ore bodies.

Caves are found throughout all of the zones described above, and it is from these caves that the fine mineral specimens of the district have been obtained. Mined areas on the fringes of the Picher Field often exhibit calcite-lined caves along the outer edges of the ore bodies.

Other changes in the mineral composition of certain horizons in the Tri-State district are shown by the many, generally unmineralized massive chert horizons found in the Picher Field that consist of limestone with limy cottonrock and chert in areas far removed from mineralization. The presence of much greater amounts of silica (chert and jasperoid) in the vicinity of ore bodies does represent a zonal relationship, and as stated earlier, "chertification" maps have been used in the past as an aid to exploration.

## Mineral Paragenesis

Many paragenetic studies have been made of the minerals associated with ore occurrences in the Tri-State district but the most recent, as well as the most detailed, microscopic and macroscopic work is that of Hagni (45). Figure 10 is a paragenetic diagram by Hagni showing his seven repetitive periods of mineralization. The paragenetic sequence given below is very generalized and concerns the main period(s) of deposition only.

The origin of the chert has been controversial. Weidman (17, p. 84–85), Bastin (18), Giles (21), and Fowler, Lyden, Gregory and Agar (20) believe that all or most of the chert is epigenetic, replacing limestone during the early stages of ore deposition from hydrothermal solutions. Hovey (5, p. 733), Haworth (8, p. 88), Smith (23), Tarr (19), Hagni (45, p. 146) claim most or all of it is syngenetic. Buckley and Buehler (9, p. 40) hold that the cherts are syngenetic and epigenetic, the latter source being silica carried in ground waters. Dake (14, p. 242) believed that most of the cherts formed as a product of weathering. Sev-

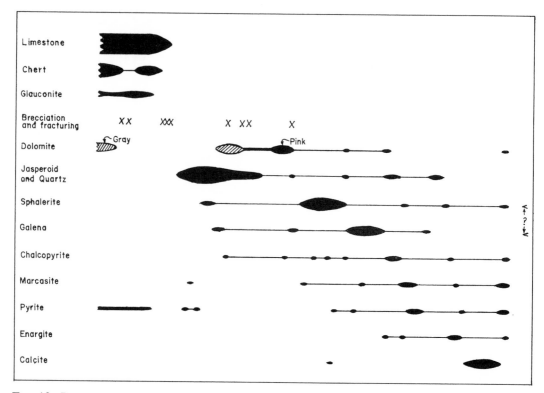

FIG. 10. *Paragenetic Diagram for the Tri-State District Minerals.* (*Taken from Hagni and Grawe, 48*).

eral papers go into considerable detail and contain elaborate discussion of this controversial subject. The writers believe that most of the chert is primarily syngenetic and diagenetic in origin. Suffice it to say that there is general agreement that most, if not all of the Mississippian chert was already present before deposition of the Mississippian Chester limestone. Ridge (24), in his paragenetic diagram on page 305, shows the chert as being questionably in part syngenetic and certainly in large part later than the dolomite but before the brecciation, i.e., diagenetic.

The exact occurrence of dolomite in the paragenetic sequence is another controversial issue. It has been considered as pre-jasperoid by some (1, 4, 6, 15, 24, 31, 34) and as contemporaneous with the jasperoid by others (10, 33, 45). The writers believe that the gray gangue dolomite is much earlier than the jasperoid and ore minerals. Angular to subrounded blocks of gray gangue dolomite are found in the jasperoid zone near the dolomite "contact," suggesting that jasperoid solidification occurred after dolomitization. Also, an ore run along a dolomite-jasperoid contact may be persistent for a great distance, but when the jasperoid "muds-out," the grade of ore quickly diminishes, even though the dolomitic core persists. Dolomitic cores and their associated ore bodies within the stratigraphically-lettered beds are not always superimposed nor do they necessarily have the same general strike as subjacent or superjacent ore bodies. This fact repudiates somewhat a hydrothermal origin for the dolomite and suggests that original sedimentary features rather than vertical fractures were the main control of dolomitization.

The jasperoid appears to have deposited contemporaneously with most of the ore, although some ore continued to be deposited afterwards. The jasperoid has replaced limestone, but the writers believe it also has replaced a considerable amount of mud and shale.

The pink gangue dolomite is found primarily as: (1) fracture fillings in gray dolomite and jasperoid, (2) in open cavities, (3) rimming patches and fragments of gray dolomite, and (4) grading into gray dolomite. Spectrographic analyses of the two varieties of dolomite show that the gray gangue dolomite has slightly greater trace amounts of certain elements than are found in pink dolomite. This may indicate that the pink variety is merely a remobilized and recrystallized product of the gray variety.

The writers believe that the sphalerite and galena formed more or less contemporaneously with the jasperoid and continued to be formed after the deposition of both the jasperoid and the later pink dolomite.

Most of the chalcopyrite was deposited as crystals after crystallization of pink dolomite, sphalerite, and galena but preferentially deposited on pink dolomite and on sphalerite rather than on galena. Hagni suggests that the "sphalerite has acted as a seed crystal for the chalcopyrite or that the galena is younger than the chalcopyrite" (45, p. 163). However, he goes on to say that most of the chalcopyrite is later than most of the sphalerite and galena.

Pyrite and marcasite are present in minor quantities in the limestone, chert, and shale. In mineralized areas, they occur mainly as crystals on sphalerite, galena, dolomite, and chalcopyrite, and also as crystal growths in the bouldery jasperoid zones.

The main period of calcite deposition was after most of the sphalerite, galena, and chalcopyrite and after much of the pyrite and marcasite. Some calcite was deposited very late in the paragenetic sequence because brecciated blocks containing jasperoid, dolomite, galena, and sphalerite commonly sit in a matrix of coarsely crystallized calcite.

The quartz, omitting that present as jasperoid and druses, is in fairly coarse crystals from about 2 mm. up to 1 or 2 cm. in size and was late in the paragenetic sequence but before the main calcite deposition. These coarse quartz crystals typically occur in the Sheet Ground and Reeds Spring Formation but are rare in M bed and stratigraphically higher ore horizons.

No discussion is given here on the other less common minerals that are present. A comprehensive discussion of them can be found in Hagni (45).

## Forms of Ore Bodies

Ore bodies in the Tri-State district assume three basic shapes: (1) irregular, relatively narrow, long "runs" of varying heights; (2) circular "runs"; and (3) flat-lying more or less tabular bodies of considerable areal extent.

The elongated "run"-type ore bodies are by far the most important in the district. Ore bodies of this type are found in all the important ore horizons with the exception of the "Sheet Ground" (primarily O bed). These "runs" follow curved fracture patterns in juxtaposition with the dolomitic core zones previously described. "Runs" commonly are present on either side, and occasionally extend up into overlying beds and over the top, of the dolomitic core. Occasionally mineralization is sufficiently rich to permit the dolomitic core to be mined with the "runs." Certain "runs," or groups of "runs," if followed far enough, tend to occupy a large semi-circular pattern, the most conspicuous one being located southwest of Baxter Springs, Kansas, as seen in Figure 2. These patterns are more noticeable along the outer edges of the Picher Field; however, others exist but have been masked by the extensive mining. The open ends of the arcuate mined areas are oriented away from the center of the Picher Field.

The "run" type ore bodies range from 10 feet to as much as 500 feet in width where the dolomitic core is sufficiently mineralized to be mined, from 5 feet to over 100 feet in height, and from a few hundred feet to several thousand feet in length. In a number of places, areas where ore minerals were emplaced from the base of M bed through all the beds to the base of the Chester Series, mining has been carried out through a vertical height of as much as 120 feet.

Along the edges of ore bodies, it is not unusual to find that ore minerals extend farther into the walls at the top of the favorable horizons than at the bottom of them. This is true even if the limestone contact has not been reached. Especially near the top of an ore run in the GH horizon, a galena-rich band of mineralization often extends outward beyond the edge of the ore body.

Circular "runs" are present throughout much of the district, the most famous one being the Oronogo Circle, north of Joplin, Missouri. Smith and Siebenthal (10, p. 16) describe the circular deposits as grading into country rock on the outside, with dolomite forming a ring inside the ore zone itself and as more or less completely filling the central mass or core of the circle. The ore zone itself dips quaquaversally from the center of the core. In plan, the ore bodies are circular, but, in section, they are truncated cones. These circular ore bodies also are found throughout the Picher Field.

Ore bodies in the O bed horizon are flat-lying, tabular deposits, thus the term "Sheet Ground." These deposits have a uniform thickness of from 12 to 15 feet and an areal extent of from 40 to 200 acres. The ore minerals and associated jasperoid are found in horizontal layers, 0.5 to 4 inches thick, intercalated with chert beds 6 to 12 inches thick, and are partially in broken ground resulting from slight folding and fracturing of the chert. Sheet Ground ore bodies in the Picher Field generally have a striking absence of intense fracturing and little to no brecciation. In certain areas, the GH, K, and M bed horizons contain flat-lying tabular-type ore bodies in which ore minerals are found over a large horizontal area.

## Factors Controlling Forms of the Ore Bodies

BRECCIATION AND "THINNING" Considerable brecciation is evident throughout the mineralized areas of the district and is not confined to a single episode but has occurred several times. Hagni (45, p. 180) states, "The main period of brecciation took place after lithification of the limestone and chert, and preceded the introduction of the jasperoid and gray dolomite. Brecciation recurred after some of the jasperoid and gray dolomite had been introduced. Breccia fragments of jasperoid and gray dolomite are cemented by these same minerals. Brecciation followed the formation of pink dolomite, for pink dolomite forms phenoclasts in the Grace B Mine. Most recent fracturing accompanied by brecciation has produced fragments of all the above minerals and those fragments are now cemented locally by calcite."

Smith & Siebenthal (10, p. 9) described three types of breccias: (1) basal breccias, (2) sheet breccias, and (3) zonal breccias. They associate brecciation in the Joplin district to minor faulting, solution readjustment, warping, and horizontal thrust. Intraformational breccias are also recognized, but their importance is unknown.

Basal breccias are known in the western part of the Picher Field along the J bed-M bed erosional contact. So-called sheet breccias are present in the Sheet Ground horizon, but the beds are only slightly disturbed, with little brecciation. What the writers term zonal breccias in the Picher Field are the M and K bed breccias, which are the most important structures in the Tri-State district. The greatest dimension of this type breccia is its length that can be up to several thousand feet. The width ranges from less than 50 feet to over 500 feet and the height from 5 to 30 feet.

Some breccias are mineralized, others are not. Breccia zones are found between unmineralized limestones, and the outer edges of the breccia often extend up over the limestone. The dolomitic cores, jasperoidal areas, and attendant ore mineralization, all of which are described fully under Mineral Zoning, are an integral part of these breccia zones. Breccias typically are light-colored nodules of chert and chert fragments in a matrix of gray gangue dolomite, jasperoid, muddy jasperoid, or jasperoidal mud. Commonly, rock fragments of gray dolomite are found in the jasperoid matrix near the dolomite-jasperoid "contact." Sizes of the chert fragments range from less than 1 inch in diameter to more than 2 feet. In many cases, the chert fragments form a mosaic and were so little disturbed that the original shape of the nodule from which they were derived can be determined.

Where L bed is present in the Picher Field, the underlying M bed, where mineralized, is from 20 to 30 feet thick. Away from the ore bodies, the unmineralized M bed ranges from 50 to 70 feet thick. The reductions in thickness are referred to as "thinning." K bed exhibits a similar relationship.

Callahan (56, p. 202) and Hagni and Desai (59) believed that erosion and ground water solution-collapse breccias are the main types of breccias associated with the ore bodies and also would explain the "thinning" of the beds in this manner. Recent preliminary studies by the writers suggest that some of the "thinning" is due to original variations in depositional thickness of M bed on higher portions of an undulating N bed surface. The base of L bed shows only local minor depressions over the M bed ore bodies that have been studied thus far. Although solution-collapse of M bed is a factor involved in the thinning of M bed, the writers believe that it may be a less important feature than original thickness variations in deposition of the limestone.

FRACTURING AND/OR SHEARING  Generally closely spaced vertical and near-vertical fracturing or shearing, with little to no displacement, is prominent near and along the contacts between dolomite and jasperoid zones where the zinc-lead mineralization is most intense. The typical relationship between fracturing and mineralized zones is depicted in Figure 8. All fracturing and/or shearing in this paper will be referred to simply as fracturing. The fracturing may be discontinuous or weak to nearly absent in places, but it always is most prominent near the "contact." Other fracturing that strikes in different directions is of no value as an ore guide.

Changes in the strikes of fractures are usual; nevertheless the strikes still approximately parallel the edge of the dolomitic zone. It is not uncommon to observe curved fracturing to extend discontinuously around, and to the opposite side of, a dolomitic core.

Fractures dip steeply away from the dolomitic zones. Generally speaking, most fractures strike northeast and northwest. Fractures commonly appear to be devoid of epigenetic minerals but often contain minor amounts of sphalerite and galena, gangue pink dolomite, and calcite. Those containing ore minerals or pink dolomite offer excellent visual guides for underground mining and prospecting. Fracturing is most important in M bed and higher stratigraphic horizons. It does not appear to be very important in the localization of Sheet Ground ore bodies.

In the western portion of the Picher Field where J bed rests on M bed, the writers have observed horizontal striations in the shaly J bed roof that strike normal to the trend of the ore body. In most cases where intense mineralization and brecciation occur, such striations, if ever present, have not been observed and were probably obliterated.

FAULTING  Faulting has been discussed previously under the general structure of the district. Outside of the immediate area of the Miami Trough and the Bendelari "trough," faulting is not so prominent. Age of faulting is pre-ore as well as post-ore. Mineralizations are not particularly concentrated within these two troughs but are primarily distributed in the breccia zones adjacent to, as well as some distance away, from them.

Information obtained from deep exploratory holes drilled by Eagle-Picher Industries, Inc., proves that the Miami Trough, where present as a graben, extends into the Precambrian basement complex. Core drilling data suggest to the writers that the Bendelari "trough" is a reflection of faulting, originating in the Precambrian rocks. Other strong synclinal and/or anticlinal structures on horizons in Mississippian rocks probably reflect faulting in the deeper sedimentary horizons or in the Precambrian rocks. Such a relationship appears especially to be true of the Oronogo-Webb City-Duenweg mineralized trend, where structure maps show a progressively greater downward displacement of older formations with depth. Although the mineralized trend is reflected as but a weak syncline in Mississippian formations, the writers believe that it becomes faulted at depth and changes to a graben structure.

FOLDING  Structural contour maps have been made on selected beds in the Mississippian rocks, particularly N bed and L bed. Epigenetic minerals commonly appear to be associated with anticlinal structures, as depicted in Figure 9, even though they often are subtle and have less than 10 feet of closure. Although mineralization can be shown to be associated with anticlines and synclines, many ore bodies cross these structures.

Fowler (25, p. 58–59) notes that in beds that have undergone little folding, the ore is usually found in nearly vertical shear zones. In flexed strata, the ore may occur in any portion of the folds. Lyden states that in areas where both synclines and anticlines are mineralized, "it is common to find M bed mineralized only in the synclines and the upper beds GH and K mineralized only in the anticlines" (33, p. 1254). However, the opposite relationship exists along most of the anticlinal and synclinal limbs of the Bendelari "trough," which is a major structural feature (Figure 6).

A large, broad dome present in the central part of the Picher Field represents a portion of the Picher anticline as shown in Figure 2. It may have acted as a focal point for deposition from the ore-bearing solutions.

SLUMPING   Local slumps are present in several areas in the Picher Field and may involve several different horizons. In the northwestern part of the field, where much of the area has

J bed resting on M bed, some local areas have J bed slumping into M bed. A slumped J bed commonly will cut off an ore body only to have it recur on the opposite side of the slump. This type of slump should not be confused with an erosional depression.

Slumping has occurred before, during, and after various periods of brecciation and mineralization.

PIPE-SLUMPS   Several so-called pipe-slumps are known in the Tri-State district. Figure 11 illustrates a vertical section through two such pipe-slumps in the Webber Mine, Picher field. These pipe-slumps are cylindrically shaped bodies of strata 100 to 120 feet in diameter that have separated from the surrounding strata and subsequently dropped en masse through a vertical distance of over 30 feet. Many pipe-slumps in the Picher field are pre-Pennsylvanian in age, as shown in Figure 11. Pipe-slumps are surrounded by circular vertical shearing and the walls of the pipes themselves may show slickensided surfaces. Ore runs in M bed may be present up to the slump and the M bed which has been displaced downward in the pipe may or may not be mineralized. Lyden (33, p. 1255) has described post-Pennsylvanian pipe-slumps in which solution of M

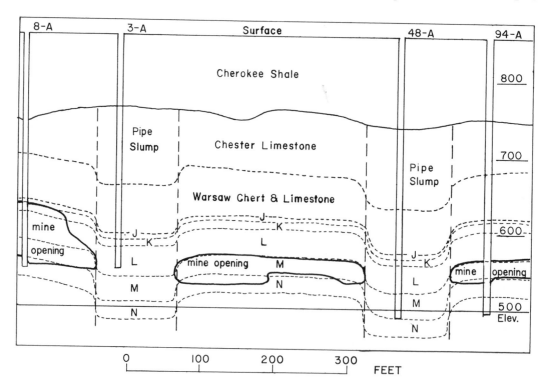

FIG. 11. *Vertical Section, Webber Mine, showing pipe slumps.* (*After Lyden, 33*).

bed has allowed the overlying horizons to slump. Recent drilling into a newly discovered pipe-slump on the Webber land disclosed its point of origin as occurring in the Reeds Spring-Fern Glen Formations. No displacement exists in the underlying Ordovician-age Cotter Dolomite. No mineralization was found in this particular pipe-slump.

## Fluid Inclusion Studies and Paths of Ore and Gangue Material

Fluid inclusion studies on sphalerite crystals from the Tri-State district by Schmidt (46) indicate that the temperatures at which the ore minerals were formed range from 83° C to 120° C. The composition of fluid inclusions was found to consist essentially of brine fluids of sodium and calcium chlorides (16, p. 429). An inference is drawn by us, therefore, that the ore deposits were precipitated from saline, hot water solutions.

Stoiber (30, p. 800), using the criteria of Newhouse (27) concluded, from the position of minerals and from symmetry of growth, that solution movements were effected by the Miami trough. Directions recorded west of the trough were mostly northeasterly or northwesterly whereas east of the Trough the directions were mostly southeasterly or southwesterly.

Certain areas that have vertical fractures extending through several of the lettered stratigraphic units are sufficiently mineralized in all the units involved to constitute one ore body. This indicates that vertical movement of ore solutions took place.

Considerable lateral movement may be inferred since ore minerals, not always in economic quantities, are found throughout an area of several hundred square miles. In the Picher Field, the fracture and breccia zones would allow for widespread lateral movement. In certain areas, one horizon may be mineralized, with the overlying and underlying horizons devoid of even trace amounts of mineralization (Figure 6). In certain cases, there is a suggestion that the fractured massive chert beds also allowed lateral solution movement since they may contain small amounts of mineralization, whereas the adjacent beds, which are ordinarily the principal ore horizons, may be completely barren. In Sheet Ground and the Reeds Spring Formation, a horizontal, 3- to 5-foot mineralized zone may be seen that is separated from limestone both above and below by a thin chert bed only 6 to 12 inches thick.

The richest areas of mineralization (Picher Field and Waco-Lawton area) are located along the Miami Trough and are situated more or less at the western edge of the district. Figure 1 shows that most of the major mined areas in the district are located along an axis having a northeasterly trend; however, many of the individual areas have a northwesterly trend corresponding to the regional structure.

The writers believe that the mineralizing solutions first passed through major structural zones and that lateral movement through the Keokuk and Warsaw Formations became a dominant factor. Where these solutions encountered breccia zones and other favorable structures, the zinc and lead and other minerals were precipitated.

## Grade of Ores in Valuable Metals

Table I shows the average grade of ore in terms of metallic lead and zinc for the period 1907 to 1964. The recoverable grade of ores from 1907 to 1920, inclusive, averaged 0.45 per cent Pb and 1.77 per cent Zn. From 1921 to 1940, the recoverable grade of ore averaged 0.74 per cent Pb and 3.07 per cent Zn, with most of this production coming from the Picher Field and some from the Waco-Lawton area.

During 1941 to 1960 the recoverable ore grade was 0.42 per cent Pb and 1.70 per cent Zn. The lower ore grade was due to several factors such as higher zinc and lead prices, subsidies during World War II, mine mechanization, and a natural decline in production from the Picher Field. The ratio of recoverable metallic zinc to lead from 1907 to 1964 was 4.06 to 1. In terms of sulfides, the recoverable ratio was 5.26 to 1. As mentioned previously, the recovery of zinc in the early day mills was only about 50 per cent. The overall grade of ores from the Picher Field and the Waco-Lawton area was higher than the other camps in the district. These two areas are the westernmost camps of the district, and both lie along the Miami Trough.

The yearly grades of recoverable metals as compared to metal prices and concentrate production in the Picher Field is shown in Table III. The graphs show a distinct inverse relationship of ore grade to prices and production. During the years 1921, 1932 to 1933, and 1959 to 1961, the average ore grade was high, but the metal prices and concentrate tonnages were lower than preceding years. These years conform very closely to national depression-recession periods. Lower grade ores, but greater concentrate production, coincide with higher

TABLE III.  *Metal Prices, Ore Grade, and Tons of Concentrates Produced-Picher Field, 1904–1964.  Date compiled from U.S. Bureau of Mines and Engineering and Mining Journal.*

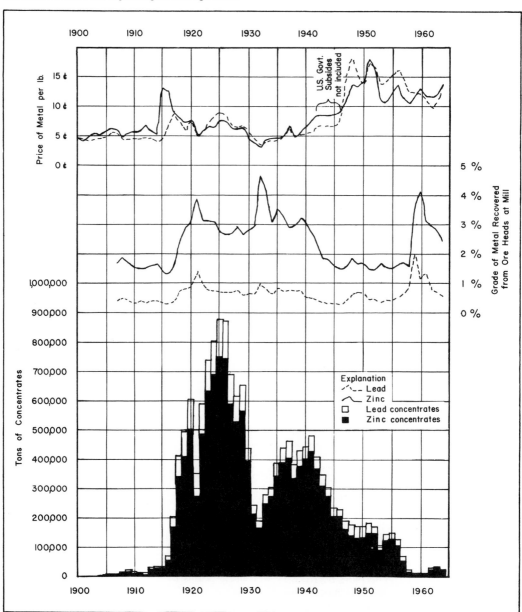

metal prices. This was especially true during World War I (mainly 1915 and 1916) and also during World War II (late 1941 to mid-1946) when United States Government subsidies were in effect. Increased demand, the Korean War, and general inflation produced continued high prices beyond World War II but because the richer ore bodies largely had been mined, the grade of ore and concentrates produced remained low.

Weidman (17, p. 49, Table XV) has compiled production, grade, and price statistics of concentrates produced in the Miami-Picher district from 1908 through 1930. In 1910, nearly 0.25 million tons of ore containing 4.3 per cent Zn and 2.4 per cent Pb were produced; these came from the Commerce area at the southwestern extension of the Picher Field. In 1921 and 1922, nearly 3 million tons of 4.54 per cent Zn and 1.43 per cent Pb, and over 6 million tons of 3.78 per cent Zn and 1.04 per cent Pb, respectively, were produced from the Picher Field.

It was not uncommon to mine ore runs and smaller pockets containing over 15 per cent combined zinc and lead sulfide.

Weidman (17, p. 76), in discussing the Picher Field, shows that "two significant facts appear to be brought out by a study of the relative distribution of lead and zinc ores: (1) There is a pronounced lack of uniformity in the composition of ore bodies as indicated by the marked variation in the ratio of zinc ore to lead ore in individual mines of the district; (2) In addition to the recognized tendency for the galena to form above sphalerite as observed in cavities and in large mined out ore bodies indicating a vertical zonal distribution, there is also a distinct lateral zonal distribution of the galena and the sphalerite, the ore bodies with highest proportion of galena being developed in the Miami syncline." Weidman shows in his Table XIX (17, p. 74) the production figures of some important mines. The tonnage of concentrates produced per mined acre ranges from 880 tons at the Lincoln Mine in Joplin to 9966 tons at the Woodchuck Mine in the Picher Field. His Table XIXA (17, p. 75), which gives the ratio of zinc ore to lead ore in 34 mines of the Picher Field, shows differences in ratios from 1 to 1.6 (Scammon Hill Mine) to 24 to 1 (Royal Mine).

## SUMMARY AND CONCLUSIONS

The Cambrian and Ordovician dolomites and sandstones were laid down unconformably on a highly irregular surface of Precambrian igneous and possibly metamorphic rocks. Then, a major uplift occurred just north of the Tri-State district so that some of the Ordovician sediments were truncated by erosion. This uplift, the Chautauqua arch, was the axis of the Ozark uplift during post-Ordovician, pre-Mississippian time. The Chattanooga Shale of Devonian-Mississippian age is absent over a similar but smaller area of uplift. However, by Mississippian time, carbonate sediments were being deposited over the entire area. Several minor periods of uplift and subsequent erosion occurred during this time. Many of the northwest trending folds were formed in late Mississippian time. The Miami Trough, which began to develop during this period, continued to form beyond post-Cherokee (Pennsylvanian) time. After deposition of Pennsylvania sediments, uplift took place that

shifted the axis of the Ozark uplift to its present position south of the district. Subsequent erosion has reduced the land surface to its present form.

Most of the Mississippian limestones are fine-grained and slightly silty. However, M and K beds and the Hindsville Formation generally are coarsely crystalline and consist of much crinoidal debris (calcirudites).

The origin of the chert has been controversial. It has been considered by various people as epigenetic, syngenetic, and a combination of the two processes. The writers favor the last theory but also believe that diagenesis was a very important factor in the formation of chert. Regardless of its origin, more chert is present in the vicinity of the ore bodies then away from them.

Although there were several periods of uplift and erosion, the most important ones were pre-Warsaw, pre-Chester, pre-Pennsylvanian, and post-Pennsylvanian. Many of the breccias are thought by some to have formed during these erosional intervals by residual chert accumulations and by solution-collapse. However, others believe most of the breccias were formed by the ore solutions dissolving the limestone, causing collapse and brecciation of the chert. Still others believe the chert breccias were formed as a result of tectonic forces, such as bedding slippage and movement along shear zones.

The gray gangue dolomite is considered by many who have studied the district to be epigenetic and hydrothermal and more or less contemporaneous with the jasperoid and ore minerals. Others and the writers consider the dolomite to have been deposited earlier. Although microscopically the dolomite may appear to be contemporaneous with the jasperoid, the macroscopic relationships suggest that the gray gangue dolomite is much earlier than the jasperoid. The writers believe that the dolomite even may be diagenetic, replacing some previously existing sedimentary structure. Such an hypothesis for the dolomite would permit a much simpler explanation of several related features; these are: (1) Insufficient structure exists in many places where ore bodies are found to permit segregation of dolomite to one side of a fracture zone if it were introduced contemporaneously with the jasperoid. (2) Very few fractures are evident within the central part of the dolomitic core. The fracture zones generally parallel the highly irregular dolomite-jasperoid contact and can be much easier explained by fracturing developed between two different rock types. (3) The width, length, and general shape of the dolomitic cores do not conform to a general zonal arrangement one would expect if the dolomite was introduced with the jasperoid and ore-bearing solutions. Gray gangue dolomite is found in a few areas (away from the Picher Field) that have no attendant jasperoid or zinc-lead mineralization. (4) The dolomitic cores in successive stratigraphic beds are not always superimposed, nor do they necessarily have the same general strike in relation to dolomitic areas in underlying or overlying beds. (5) Angular to subrounded blocks of gray gangue dolomite are found in the jasperoid zone near the dolomite "contact". (6) The "mudding-out" of the jasperoid along the dolomite-jasperoid contact within the Picher Field is further evidence that the dolomite is earlier than the jasperoid. (7) Where jasperoid is present and adjacent to dolomite zinc-lead mineralization is found, but where the jasperoid is absent, or where mud or muddy shale are adjacent to the dolomite, zinc-lead mineralization is conspicuously weak. The relationship just described also shows a distinct coincidence in space and time relationship between the silica forming the jasperoid and the zinc-lead ore minerals.

The coarsely crystalline pink gangue dolomite is believed by the writers to be remobilized and recrystallized gray gangue dolomite.

The origin of the jasperoid is also controversial. Practically all jasperoid is dark in color. It has replaced limestones and shales, and, in many places, contains related bituminous material. Several authors believe the shale to be transported Cherokee shale that worked its way downward and laterally into the crevices and openings in the Mississippian strata. Some consider the mud as residual from solution of the limestone. Still others believe the mud is weathered jasperoid. From examination of rock cuttings obtained from numerous cable-tool drilled holes, and from observations in the mine workings as well as in pull-drifts leading away from mined areas, it is the writers' opinion that some of the shale and much of the mud may have been deposited in Mississippian time.

After deposition of the dolomite, the jasperoid and ore-bearing solutions were introduced. The origin of these solutions has been ascribed to: (1) downward percolating ground waters, (2) artesian circulating ground waters, (3) hydrothermal solutions from depth, and, most recently, (4) migration of connate waters outward from basins due to compaction of sediments. If the anomalous J-type lead isotope results are interpreted correctly as to possible age of the galenas, then mode of origin (1) is untenable. However, (2), (3) and (4) or a combination of them, could account for the anomalous lead-isotope values, as the ore solutions would have been derived from older Ordovician and Cambrian rocks, and perhaps to some extent from the Precambrian rocks. Fluid inclusion studies suggest that the modes of origin that appear most tenable are (3) and (4).

That considerable lateral movement of ore solutions has occurred can not be doubted, as the formations above and below an ore horizon often are unmineralized limestones.

There still are problems of genesis and paragenesis of minerals associated with ore deposits in the Tri-State district that remain to be solved. Some of the minerals, such as chert and dolomite and the mud or shale, may be original sedimentary features or reflections of them. Some suggested studies which may shed light on these problems are: (1) a study of fossils found in cherts in the dolomitic zone compared to those in the jasperoid zone and in the unmineralized limestone, (2) a study of the jasperoid to mud to limestone relationships, with special emphasis on the mud or shale, and (3) a study of limestone remnants in dolomitic core areas compared to limestone away from ore bodies.

Many data have been presented in various parts of the paper that must be considered in any theory of the origin of the ore-bearing solutions, especially in the Picher Field: (1) the apparent control by the Chattanooga and Northview Shales of mineralization in Mississippian and Cambro-Ordovician formations, (2) the relationships of mineralization to major structural features intersecting a window in the Chattanooga Shale and a thin Northview Shale, (3 & 4) lead-isotope and fluid-inclusion studies, (5) Stoiber's data (30) suggesting that the flow of ore-bearing solutions in the Picher Field spread outward from the Miami Trough into Mississippian strata, (6) the vertical and lateral relationships of the ore bodies to the host rock as observed in the mines and from innumerable drill-hole cuttings. From the above information, the writers conclude that the warm, saline ore-bearing solutions evidently were derived from some distant source, migrating through the Cambro-Ordovician sediments until a zone of structural weakness such as the Miami Trough, combined with a window in the Chattanooga Shale (and a certain thinness of the Northview Shale), permitted concentration and access of the ascending solutions into the Mississippian formations. The solutions migrated laterally outward from the Trough and deposited in ground already prepared by dolomitization and brecciation.

Only recently a Symposium was held in New York on the "Genesis of Stratiform Lead-Zinc-Barite-Fluorite Deposits" that served to emphasize that considerable controversy still exists as to whether many ore deposits classified as Mississippi Valley-type are syngenetic or epigenetic.

## ACKNOWLEDGMENTS

The authors are indebted to Eagle-Picher Industries, Inc. for kind permission to publish this paper.

## REFERENCES CITED

1. Schmidt, A. and Leonhard, A., 1874, The lead and zinc regions of southwest Missouri: Mo. Geol. Surv. Rept. for 1874, p. 380–734.

2. Branner, J. C. and Simonds, F. W., 1891, Ark. Geol. Surv. Ann. Rept. 1888, v. 4, p. xiii, 27-37.

3. Penrose, R. A. F., Jr., 1891, Manganese, its uses, ores and deposits: Ark. Geol. Surv. Rept. for 1890, v. 1,642 p.

4. Jenny, W. P., 1893, The lead and zinc deposits of the Mississippi Valley: A.I.M.E. Tr., v. 22, p. 171–225.

5. Hovey, E. O., 1894, A study of the cherts of Missouri: *in Appendix A. Lead and Zinc. Deposits,* Winslow A. and Robertson, J. D.: Mo. Geol. Surv. Rept., v. 6, 7, p. 727–739.

6. Winslow, A. and Robertson, J. D., 1894, Lead and zinc deposits: Mo. Geol. Surv. Rept., v. 6, 7, 763 p.

7. Adams, G. I. and Ulrich, E. O., 1904, Zinc and lead deposits of northern Arkansas: U.S. Geol. Surv. Prof. Paper 24, 118 p.

8. Haworth, E., *et. al.,* 1904, History, geography, geology, and metallurgy of Galena-Joplin lead and zinc: Kans. Univ. Geol. Surv., v. 8, p. 1–126.

9. Buckley, E. R. and Buehler, H. A., 1905, The geology of the Granby area: Mo. Bur. Geol. and Min. Repts., 2d ser., v. 4, 120 p.

10. Smith, W. S. T. and Siebenthal, C. E., 1907, Description of the Joplin district: U.S. Geol. Surv. Geol. Atlas, Folio 148, 20 p.

11. Siebenthal, C. E., 1908, Mineral resources of northeastern Oklahoma: U.S. Geol. Surv. Bull. 340, p. 187-230.

12. ――――― 1916, Origin of the zinc and lead deposits of the Joplin region, Missouri, Kansas, and Oklahoma: U.S. Geol. Surv. Bull. 606, 283 p.

13. Anderson, C. O., 1932, Milling in the Tri-

State district: p. 110–129 *in* Weidman, S., *The Miami-Picher zinc-lead district, Oklahoma,* Okla. Geol. Surv. Bull. 56, 177 p.

14. Dake, C. L., 1932, The ore deposits of the Tri-State district (Missouri-Kansas-Okla.-homa): (disc.) A.I.M.E. Tr., v. 102, p. 241–242.

15 Fowler, G. M. and Lyden, J. P., 1932, The ore deposits of the Tri-State district (Missouri-Kansas-Oklahoma): A.I.M.E. Tr., v. 102, p. 206–251.

16. Newhouse, W. H., 1932, The composition of vein solutions as shown by liquid inclusions in minerals: Econ. Geol., v. 27, p. 419–436.

17. Weidman, S., 1932, The Miami-Picher zinc-lead district, Oklahoma: Okla. Geol. Surv. Bull. 56, 177 p.

18. Bastin, E. S., 1933, Relations of cherts to stylolites at Carthage, Missouri: Jour. Geol. v. 41, p. 371–381.

19. Tarr, W. A., 1933, The Miami-Picher zinc-lead district: Econ. Geol., v. 28, p. 463–479.

20. Fowler, G. M., *et. al.,* 1934, Chertification in the Tri-State (Oklahoma-Kansas-Missouri) mining district: A.I.M.E. Tr., v. 115, p. 106–163.

21. Giles, A. W., 1935, Boone chert: Geol. Soc. Amer. Bull., v. 46, p. 1815–1878.

22. McKnight, E. T., 1935, Zinc and lead deposits of northern Arkansas: U.S. Geol. Surv. Bull. 853, 311 p.

23. Smith, W. S. T., 1935, Chertification in the Tri-State (Oklahoma-Kansas-Missouri) mining district: (disc.) A.I.M.E. Tr., v. 115, p. 156–159.

24. Ridge, J.,1936, The genesis of the Tri-State zinc and lead ores: Econ. Geol., v. 31, p. 298–313.

25. Fowler, G. M., 1939, Structural control of ore deposits in the Tri-State lead and zinc district: *in* Bastin, E. S., *Editor, Contributions to a Knowledge of the Lead and Zinc Deposits of the Mississippi Valley Region,* Geol. Soc. Amer. Spec. Paper 24, 156 p.

26. Moore, R. C., *et. al.,* 1939, Stratigraphic setting-Tri-State District of Missouri, Kansas and Oklahoma: *in* Bastin, E. S., *Editor, Contributions to a Knowledge of the Lead and Zinc Deposits of the Mississippi Valley Region,* Geol. Soc. Amer. Spec. Paper 24, 156 p.

27. Newhouse, W. H., 1941, The direction of flow of mineralizing solutions: Econ. Geol. v. 36, p. 612–629.

28. McKnight, E. T., *et. al.,* 1944, Maps showing structural geology and dolomitized areas in part of the Picher zinc-lead Field, Okla. and Kan., U.S. Geol. Surv. Tri-State zinc-lead investigations, Prelim. Maps 1–6.

29. Martin, A. J., 1946, Summarized statistics of production of lead and zinc in the Tri-State (Missouri-Kansas-Oklahoma) mining district: U.S. Bur. Mines I.C. 7383, p. 1–67.

30. Stoiber, R. E., 1946, Movement of mineralizing solutions in the Picher Field, Okla-

homa-Kansas: Econ. Geol., v. 41, p. 800–812.

31. Behre, C. H., *et al.,* 1950, Zinc and lead deposits of the Mississippi Valley: 18th Int. Geol. Cong. Rept., pt. 7, p. 48–61.

32. Kercher, R. P., and Kirby, J. J., 1948, Upper Cambrian and lower Ordovician rocks in Kansas: Kans. Geol. Surv. Bull. 72, 140 p.

33. Lyden, J. P., 1950, Aspects of structure and mineralization used as guides in the development of the Picher Field, A.I.M.E. Tr., v. 187, p. 1251–1259.

34. Bastin, E. S., 1951, Paragenesis of the Tri-State jasperoid: Econ. Geol. v. 46, p. 652–657.

35. Brown, J. S., 1951, A graphic statistical history of the Joplin or Tri-State lead-zinc district: Mng. Eng., v. 3, no. 9, p. 785–790.

36. Moore, R. C., *et. al.,* 1951, The Kansas rock column: Kans. Geol. Surv. Bull. 89, 132 p.

37. Cumming, G. L., 1953, Written personal communication (10-28-53) to J. P. Lyden.

38. Bieber, C. L., 1955, Structural geology of southwestern Missouri: Unpubl. manuscript, Mo. Geol. Surv.

39. Reed, E. W., *et. al.,* 1955, Ground water resources of Ottawa County, Oklahoma: Okla. Geol. Surv. Bull. 72, 203 p.

40. Clarke, S. S. and Brockie, D. C., 1956, Jackleg drilling in the Tri-State district: longhole prospecting and production. Min. Eng. v. 8, no. 1, p. 27–30.

41. Huffman, G. G., *et. al.,* 1958, Geology of the flanks of the Ozark Uplift: Okla. Geol. Surv. Bull. 77, 281 p.

42. Fowler, G. M., 1960, Structural deformation and ore deposits: Eng. and Min. Jour., v. 161, no. 6, p. 183–188.

43. Howe, W. B. and Koenig, J. W., 1961, The stratigraphic succession of Missouri: Mo. Div. Geol. Surv. and Water Res. Repts., v. 40, 2d ser., 185 p.

44. Spreng, A. C., 1961, Mississippian System: p. 49–78, Mo. Div. Geol. and Water Res. Repts., v. 40, 2d ser., 185 p.

45. Hagni, R. D., 1962, Mineral paragenesis and trace element distribution in the Tri-State zinc-lead District, Missouri, Kansas and Oklahoma: unpublished Ph. D. Thesis, Univ. Mo. at Rolla, 252 p.

46. Schmidt, R. A., 1962, Temperatures of mineral formation in the Miami-Picher-district as indicated by liquid inclusions: Econ. Geol. v. 57, p. 1–20.

47. Cannon, R. S., Jr., *et. al.,* 1963, U.S. Geol. Surv. Prof. Paper 450-E, p. E73–E77.

48. ———— 1963, Lead isotope variation with crystal zoning in a galena crystal: Science, v. 142, no. 3592, p. 574–576.

49. Hagni, R. D. and Grawe, O. R., 1964, Mineral paragenesis in the Tri-State district, Missouri, Kansas, Oklahoma: Econ. Geol., v. 59, p. 449–457.

50. McHugh, J. W. and Broughton, M. N., 1964, Symposium on the Arbuckle: Tulsa Geol. Soc. Digest, v. 32, 186 p.

51. Burgess, W. J., 1964, Stratigraphic dolomitization in Arbuckle rocks in Oklahoma: p. 45–48 *in Symposium on the Arbuckle,* Tulsa Geol. Soc. Digest, v. 32, 186 p.

52. Clark, J. M., 1964, The Arbuckle of Northwest Arkansas: p. 76–90, *in Symposium on the Arbuckle,* Tulsa Geol. Soc. Digest, v. 32, 186 p.

53. Harlton, B. H., 1964, Surface and subsurface subdivisions in Cambro-Ordovician carbonates of Oklahoma: p. 38–42, *in Symposium on the Arbuckle,* Tulsa Geol. Soc. Digest, v. 32, 186 p.

54. McCracken, M. H., 1964, The Cambro-Ordovician rocks of northeastern Oklahoma and adjacent areas: p. 49–75, *in Symposium on the Arbuckle,* Tulsa Geol. Soc. Digest, v. 32, 186 p.

55. Brown, J. S., 1965, Oceanic lead isotopes and ore genesis: Econ. Geol. v. 60, p. 47–68.

56. Callahan, W. H., 1965, Paleophysiographic premises for prospecting for strata bound base metal mineral deposits in carbonate rocks: *CENTO, Symposium on Mining Geology and Base Metals,* Ankara, Turkey, p. 191–248.

57. McKnight, E. T., 1965, Personal communication to D. C. Brockie, Nov. 16.

58. Hagni, R. D. and Saadallah, A. A., 1965, Alteration of host rock limestone adjacent to zinc-lead ore deposits in the Tri-State District, Missouri, Kansas, Oklahoma: Econ. Geol. v. 60, p. 1607–1619.

59. Hagni, R. D. and Desai, A. A., 1966, Solution thinning of the M bed host rock limestone in the Tri-State District, Missouri, Kansas and Oklahoma: Econ. Geol., v. 61, p. 1436–1442.

60. Heyl, A. V., *et. al.,* 1966, Isotopic study of galenas from the Upper Mississippi Valley, the Illinois-Kentucky and some Appalachian Valley mineral districts: Econ. Geol. v. 61, p. 933–961.

# 21. The Upper Mississippi Valley Base-Metal District[*]

ALLEN V. HEYL[†]

## Contents

## Illustrations

* Publication authorized by the Director, U.S. Geological Survey.
† U.S. Geological Survey, Beltsville, Maryland.

## Tables

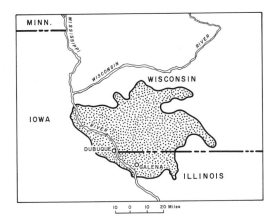

## ABSTRACT

This old district is a major zinc and lead source and minor copper and barite source. Ores are chiefly in the Galena Dolomite and in limestones and dolomites of the Decorah and Platteville Formations, all of Middle Ordovician age. Locally, deposits have been mined from Lower Ordovician dolomites and Upper Cambrian sandstones. Small deposits of lead, zinc, and iron sulfides have been found in Upper Ordovician Maquoketa Shale and in Lower and Middle Silurian Edgewood, Kankakee, and Hopkinton Formations. No post-Precambrian igneous rocks are known in the region, and the granitic and metasedimentary Precambrian basement rocks unconformably underlie the district at depths of 1500 to 2000 feet.

The strata are gently flexed and faulted, probably largely the result of gentle compressive and rotational adjustments in the underlying crystalline basement, especially along lineaments between basement blocks. Folds of three orders of magnitude are recognized, and many related reverse, strike-slip, and normal faults of small to moderate displacements are present.

Zinc-lead deposits are commonly linear, arcuate, or elliptical in plan. Most ore occurs as vein fillings in fractures and as breccia fillings and lining vugs, but some is in bedded replacements and impregnations of wall rocks. The ore is in shears, small reverse and bedding-plant faults, joints, and solution-slump structures related to intermediate and small folds. The main primary minerals in their general sequence of deposition are quartz, illite, dolomite, pyrite, marcasite, cobaltite(?), sphalerite, galena, chalcopyrite, millerite, barite, and calcite.

Wall-rock alterations include solution of the carbonate rocks, silicification, dolomitization, clay alteration, and sanded dolomite. Oxygen-isotope, trace-element, lead-isotope, and sphalerite-stratigraphy studies are in progress. Fluid inclusions in sulfide and gangue minerals are filled with concentrated brines at 120° to 40°C.

Sulfides are directly replaced by supergene minerals above the water table at 50 to 100 feet in depth. In this relatively cool climate and limestone environment, smithsonite is in pseudomorphs after sphalerite; solution transport and redeposition of supergene minerals are rare.

Ore genesis is postulated by rising hot-water solutions—a mixture of heated connate, magmatic, and meteoric waters.

## INTRODUCTION

The district is a major source of zinc in the United States. For the past 20 years, the district has been among the 10 largest sources of zinc nationally. It also is an important source of lead and was the most important national source during much of the 19th century. Copper, barite, and iron pyrites have also been produced.

The district is north of the Illinois basin and includes an area of about 4000 square miles in the southwest corner of Wisconsin, the northwest corner of Illinois, and the northeast fringe of Iowa. (Figure 1) It lies about 100 miles south of the main outcrop of the Precambrian shield in Wisconsin. The zinc, lead, and other mineral deposits are in the northern fringe of gently flexed Paleozoic sedimentary rocks that overlap the shield on the flanks of the broad Wisconsin Dome. Most of the known mineral deposits are restricted to dolomites, limestones, and shales of Middle Ordovician age, but deposits are found in all other Paleozoic rocks in the district.

## HISTORY AND PRODUCTION

The lead deposits were discovered by the French in the latter part of the 17th century. The French taught the Indians primitive mining methods, and for 150 years small-scale mining operations were carried on by the Indians and French (7). The first American settlers arrived in 1821. The district quickly boomed, and by 1840 most of the mineralized areas had been discovered. During the 1840's, this was the principal lead-producing district in the United States, but production afterwards declined, and now lead is recovered only as a byproduct of zinc mining. Most of the lead

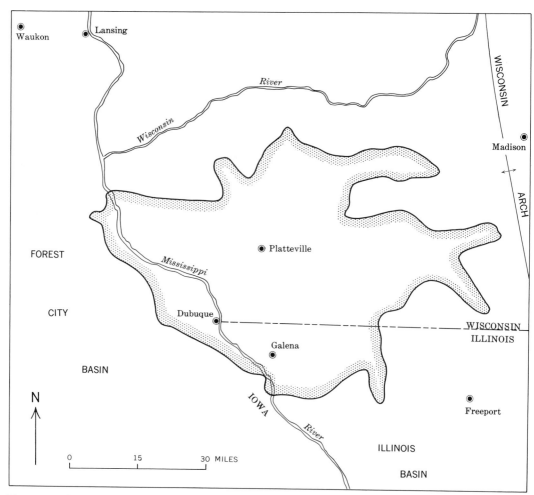

FIG. 1. *Index Map, showing the known boundaries of the Upper Mississippi Valley Zinc-Lead District and a few major regional structures.*

TABLE I. *Tonnage and Value of Recoverable Zinc and Lead Metals Produced in Upper Mississippi Valley District (Wisconsin, Northern Illinois, and Iowa) 1925–1964*

| Years | Zinc | | Lead | |
|---|---|---|---|---|
| | Short Tons | Value | Short Tons | Value |
| 1925–29 | 120,417 | $ 15,869,400 | 9,539 | $ 1,362,008 |
| 1930–34 | 47,784 | 3,923,002 | 4,173 | 336,064 |
| 1935–39 | 31,964 | 3,312,788 | 2,989 | 300,618 |
| 1940–44 | 57,639 | 11,278,212 | 5,218 | 752,182 |
| 1945–59 | 86,528 | 21,188,480 | 9,796 | 2,610,048 |
| 1950–54 | 145,047 | 41,907,902 | 14,173 | 4,343,234 |
| 1955–59 | 196,421 | 40,173,100 | 15,690 | 4,487,034 |
| 1960–64 | 160,775 | 39,502,286 | 10,600 | 2,359,918 |
| 1925–1964 Total | 846,575 | $177,155,170 | 72,178 | $16,551,106 |

Data: U.S. Bureau of Mines.

ore was produced from the gash-vein* deposits, and much of the lead-mining area shown in Figure 1 consists of groups of such deposits.

Zinc production began in 1860 and reached a peak during World War I. A revival in zinc mining by small companies occurred about 1938 and continued until shortly after World War II (Table I). Since then, a few large companies have developed ample reserves and are mining on a major scale. By 1952, the

* The term "gash veins" for joint-controlled lead deposits that are restricted to one formation was proposed by Whitney (2, p. 48) and was widely used in this sense before it was redefined as a type of tension fracture by structural geologists.

district had again taken one of the leading positions in United States production. The known reserves of ore increased from about 1,000,000 short tons in 1942 to nearly 15,000,000 short tons in 1958 but have declined since then.

Zinc production from 1860 to 1951 was about 1,200,000 short tons of metal from about 43,000,000 tons of ore. The value of the metal on the basis of 15 cents per pound metallic zinc is $363,000,000. Nearly 85 per cent of the zinc was mined in Wisconsin, 15 per cent in Illinois, and a small amount in Iowa. In recent years, however, Illinois and Wisconsin have produced nearly equal quantities of zinc.

The total recorded lead production from 1795 to 1951 was 821,00 short tons of metallic lead valued at $248,000,000 on the basis of 15 cents per pound. Slightly more than 50 per cent of the lead was mined in Wisconsin; the rest was produced in northern Illinois and Iowa.

Copper was mined in Wisconsin in the 19th century and as much as 10,000 short tons of 15 to 20 per cent copper ore may have been produced. Barite was mined in Wisconsin between 1919 and 1930, but data on the total quantity produced and its value are not available.

The lead and the copper ores and some of the zinc ores were smelted locally in the 19th century. In recent years, however, most of the zinc and lead ores have been concentrated in gravity and flotation mills at the mines or custom mills in the district, and the concentrates have been shipped to central and southern Illinois and to Missouri for smelting.

## PHYSIOGRAPHY AND TOPOGRAPHY

Most of the district lies within the Driftless Area (10, p. 2–3; 11, p. 1–2, 7), but glaciers of pre-Wisconsin age have crossed the eastern and western boundaries and part of the southern boundary. The nearest glacial deposits of Wisconsin age are found in the immediate vicinity of Madison, Wisconsin. The glacial deposits of pre-Wisconsin age thin gradually toward the boundary of the Driftless Area and are represented only by small remnants of drift that are restricted to the southwesternmost corner of Wisconsin near Dubuque, Iowa, and the northwesternmost corner of Illinois along the Mississippi River near Galena (Figure 1; 28, p. 20). Pleistocene bench and terrace gravels, sands, and clays are common in the valleys of the principal rivers that flow across

the Driftless Area from the surrounding country (29, p. 8–9). Deposits of interglacial loess of pre-Wisconsin age that are as much as 50 feet thick cap the bluffs along both sides of the Mississippi River and extend several miles away from the river.

The topography is more rugged than in most of the region surrounding the Driftless Area. The dominant topographic feature is an upland that is about 900 feet in altitude in the southern part of the district and that rises gently toward the north to 1250 feet. The local relief normally ranges between 100 and 300 feet, but the maximum relief is 1100 feet, in part resulting from a number of low escarpments and isolated hills or buttelike mounds that rise about 200 feet above the general level of the gently rolling upland. The highest of the mounds, West Blue Mound, has an altitude of 1710 feet. A generally dendritic network of streams, all tributary to the Mississippi River, drain the district and occupy valleys that are shallow in the central part of the district but that are much more deeply incised near the master river. The Mississippi flows in a cliff-walled valley incised 200 to 300 feet into the adjoining upland. The major tributary of the Mississippi River from the east is the Wisconsin River, which essentially bounds the district on the north; all its tributaries near the district are deeply incised. The east and west fringes of the district were glaciated and, therefore, show topography typical of areas formerly covered by continental glaciers and their outwash plains.

## GEOLOGIC HISTORY

### Stratigraphy

The rocks exposed in the district are Cambrian, Ordovician, and Silurian in age (Figure 2). Rocks of the Precambrian basement have been found only in wells at depths of 1500 to 2000 feet. The oldest rocks exposed—the Franconia and Trempealeau Formations of Late Cambrian age—crop out along the Wisconsin River and near Madison, Wisconsin. The Prairie du Chien Group of Early Ordovician age is exposed in some of the deep valleys and along the north and east margins of the district (10). Most of the outcrops in the district are Galena Dolomite, but exposures of the Platteville Formation and St. Peter Sandstone are common, particularly in the valleys. Maquoketa Shale of Late Ordovician age (10, 11) covers some of the high hills and un-

| System | Series | Group or formation | Description | Approximate thickness, in feet | |
|---|---|---|---|---|---|
| SILURIAN | Middle | Hopkinton Dolomite | Dolomite, buff, cherty; *Pentamerus oblongus* common | 190+ | 300+ |
| SILURIAN | Lower | Kankakee Formation | Dolomite, buff, cherty | 45–50 | |
| SILURIAN | Lower | Edgewood Dolomite | Dolomite, gray, argillaceous | 9–116 | |
| | | DISCONFORMITY | DISCONFORMITY | | |
| ORDOVICIAN | Upper | Maquoketa Shale | Shale, blue, dolomitic; phosphatic depauperate fauna at base | 108–240 | |
| ORDOVICIAN | Middle | Galena Dolomite | Dolomite, yellowish-buff, thin-bedded, shaly | 40 | 225 |
| | | | Dolomite, yellowish-buff, thick-bedded; *Receptaculites* in middle | 80 | |
| | | | Dolomite, drab to buff, cherty; *Receptaculites* near base | 105 | |
| | | Decorah Formation | Dolomite, limestone, and shale, green and brown; phosphatic nodules and bentonite near base | 35–40 | |
| | | Platteville Formation | Limestone and dolomite, brown and grayish; green sandy shale and phosphatic nodules at base | 55–75 | |
| | | St. Peter Sandstone | Sandstone, quartz, coarse, rounded; local anomalous variations in thickness | 28–340 | 280–340 |
| | | DISCONFORMITY | DISCONFORMITY | | |
| | Lower | Prairie du Chien Group (undivided) | Dolomite, light-buff, cherty; sandy near base and in upper part; shaly in upper part | 0–240 | |
| CAMBRIAN | Upper | Trempealeau Formation | Sandstone, siltstone, and dolomite | 120–150 | |
| | | Franconia Sandstone | Sandstone and siltstone, glauconitic | 110–140 | |
| | | Dresbach Group: Galesville Sandstone | Sandstone | 60–140 | 700–1050 |
| | | Dresbach Group: Eau Claire Sandstone | Siltstone and sandstone | 70–330 | |
| | | Dresbach Group: Mount Simon Sandstone | Sandstone | 440–780 | |

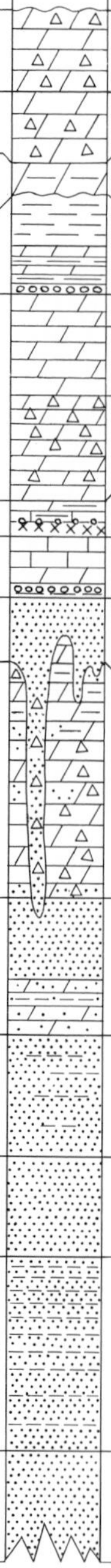

FIG. 2. *Stratigraphic Section of the Upper Mississippi Valley District.*

derlies the long slopes that rise to higher hills capped with Lower Silurian dolomite. Most outcrops of the youngest strata—dolomite of early Silurian age—occur along a partly incised escarpment along the southern and western boundaries of the district; these strata also form a few isolated erosional outliers within the district.

Some of the details of the stratigraphy are illustrated by the generalized stratigraphic section (Figure 2). The rocks of the Prairie du Chien Group, the Platteville, Decorah, and Galena Formations, and Lower Silurian rocks in the district are composed mostly of dolomite, although they include some limestone, shale, and a few beds of sandstone. Between these carbonate rocks are the predominantly clastic St. Peter Sandstone and Maquoketa Shale. The Franconia, Trempealeau, and St. Peter Formations are permeable aquifers that are widespread.

Most of the zinc deposits are in the lower part of the Galena Dolomite, the Decorah Formation, and the upper two thirds of the Platteville Formation. The principal lead deposits are restricted to Galena Dolomite. Small lead, zinc, and copper deposits are found locally in the northern part of the district in the Prairie du Chien Group (24, p. 292–295) and the Trempealeau and Franconia Formations. In places, the St. Peter Sandstone is pyritized; it is mostly heavily pyritized where it directly underlies zinc deposits.

The Platteville, Decorah, and Galena Formations represent an unusually widespread and uniform marine environment of a relatively shallow-water platform that had remarkable stability during deposition (24, p. 251). Throughout the 4000 square miles of the district, the variations laterally and vertically of the mappable units within the formations are remarkably small. Even where dolomitized locally or regionally, the same units can be distinguished within the dolomite with very small variations in thicknesses and other lithologic features. Reefs, bioherms, bars, channels, and other unconformities are notably absent, although the Lower Ordovician beneath and the Upper Ordovician and Silurian above have most of these stratigraphic variations.

All members of the Platteville Formation, except the Glenwood, decrease a total of about 20 feet in thickness toward the west, whereas all members of the Decorah Formation increase a total of about 5 feet toward the west. The Galena Dolomite remains remarkably uniform in thickness throughout the district but changes from a coarsely crystalline massive

vuggy dolomite to a somewhat argillaceous fossiliferous limestone westward and northwestward. Agnew (22) describes these changes throughout Iowa.

In the central and western parts of the district, the upper members of the Platteville and all the members of the Decorah Formation are in part limestone. These limestone members are the first limestones of any consequence above the Precambrian and thus provided a favorable host for later leaching and ore deposition. Two of the limestone units, the Quimbys Mill and Guttenberg Members, contain interbedded carbonaceous shales, a further probably favorable factor in ore deposition. One of the superimposed features of this mixed dolomite-limestone-shale lithology of the Middle Ordovician strata is the notable sparsity of caves and sinkholes within the district, even though a typical karst topography is widespread in these same units where they are mainly limestone, in southeastern Minnesota and northeastern Iowa. The Prairie du Chien below and Silurian dolomites above have many sinkholes and caves.

Brown and Whitlow have worked out in detail the stratigraphy of the well-exposed upper Ordovician Maquoketa Shale and Silurian System in the Iowa part of the district (29, p. 22–43). They have well established the apparently conformable relationship of the top of the Galena Dolomite to the overlying Maquoketa Shale within the district. Brown and Whitlow state (29, p. 1–2):

"The Maquoketa shale is divided into three lithologic units. They are, in ascending order: the brown shaly unit, the Brainard member, and the Neda member. The Neda member occurs in the Dubuque South quadrangle at the western part of the Niagaran escarpment where the Neda is as much as 5 feet thick. It is composed of grayish-red shale and beds and lenses of limonite oolites.

"After the deposition of the Neda member the seas regressed, and erosion of the Maquoketa shale began. The resultant erosional surface has at least 135 feet of relief in the Dubuque South quadrangle. The Edgewood dolomite of Early Silurian age disconformably overlies the Maquoketa shale and fills the irregularities of the erosion surface. Consequently, the aggregate thickness of the two units is nearly uniform.

"The Edgewood dolomite comprises two members, the Tete des Morts and the Mosalem. The Mosalem member is wavy-bedded olive-gray to tan argillaceous dolomite. Locally at its base is a thin interformational conglomerate. The Tete des Morts member is massive grayish-yellow argillaceous glauconitic dolomite that forms many cliffs.

"The Kankakee formation, also of Early Silurian age, overlies the Edgewood dolomite and consists of grayish-yellow dolomite and as much as 50 per cent white bedded chert. Hopkinton dolomite of Middle Silurian age overlies the Kankakee and consists of nearly pure coralline dolomite and cherty dolomite that locally includes as much as 60 per cent chert."

Limestones in the Middle Ordovician formations are progressively dolomitized from the central part of the district eastward towards the Wisconsin arch. The facies change to dolomite is not uniform in each stratigraphic unit, but, in general, all the units east of Mineral Point, Wisconsin (Figure 1), are mainly dolomite, but those to the west contain more limestone than dolomite. Allingham (36, p. 182–183) describes this change in the Platteville Formation, and Taylor illustrates it well in the Quimbys Mill Member of the formation (45, Figure 44, p. 292–295). These regional dolomite facies closely resemble physically the much more local dolomite alteration aureoles

that surrounded ore bodies, but they are probably not related.

No post-Precambrian igenous rocks are known in the district, and the nearest certain occurrences are dikes, sills, and diatremes in southern Illinois. Several cryptoexplosion structures are known to the west, south, southeast, and northeast of the district, as is shown on the revised tectonic map of the United States (32). The origin of these structures is unknown and very controversial, and theories range from meteoric impact from above to deep-seated gaseous explosion from beneath. At most, the bearing of these structures on the genesis of the district ore deposits is very questionable.

## Structure

The mining district lies within about 100 miles of the northern edge of Paleozoic sedimentary rocks that overlap the North American Precambrian shield. The Wisconsin arch, a broad northward-trending anticlinal arch, lies east of the area. The Illinois basin lies south of the district, and the Forest City basin lies west and southwest of the district, in Iowa.

The regional strike of the sedimentary for-

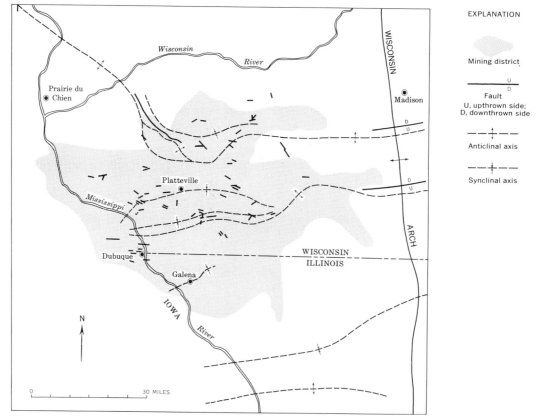

Fig. 3. *Map of the Upper Mississippi Valley District, showing larger folds and faults.*

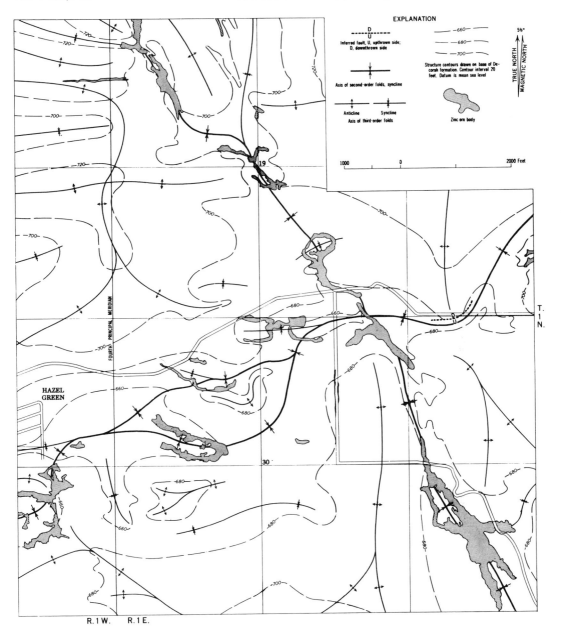

FIG. 4. *Structure Map of the Area East of Hazel Green, Wisconsin. Second- and third-order folds with related zinc ore bodies illustrate intricate crossfold patterns and interrelationships of ore bodies and folds. Contour interval 20 feet.*

mations is N85°W throughout most of the district, but it progressively changes to N45°W in the western part. The regional dip is about 17 feet to the mile toward the south-southwest. The rocks of the district are folded into low broad undulations that trend northeastward, eastward, or northwestward.

The larger broad folds range from 20 to 30 miles in length, 3 to 6 miles in width, and 100 to 200 feet in amplitude (Figure 3).

Rarely do the folds have dips greater than 15° on their limbs, and commonly the dips are much less than 15°. Generally the north limbs dip more steeply than the south limbs. The smaller folds trend either eastward to northeastward or northwestward. Folds with these two general trends occur throughout the area and form an unusual rhombic pattern (Figure 4).

Most faults in the district are small reverse,

bedding-plane, normal, and shear faults. Displacements commonly range from 1 to 10 feet, but small thrusts on the north limbs of folds have displacements of 25 to 50 feet, and quite a few strike-slip shear faults have displacements of 25 to possibly 1000 feet (28, p. 37–38; Figure 6).

Many of the ore deposits in the district are localized along small faults—chiefly bedding-plane and reverse faults on the limbs of folds. Bedding-plane faults are present in the incompetent uppermost part of the Platteville and lower part of the Decorah strata. Reverse faults curve upward from these bedding-plane faults, at first at low angles but steepening to about 45° in the overlying competent beds. In general, these two types of faults are confined to the flanks of the folds and tend to follow the outline of the folds to form arcuate or linear patterns. Most commonly the reverse faults dip toward the bordering anticlinal areas (Figure 7).

All the rock formations in the district contain well-developed vertical and inclined joints. The vertical joints are traceable for as much as 2 miles horizontally and for as much as 300 feet vertically. Quite a few of the long joints are actually strike-slip faults that have virtually no vertical component of displacement. Joints are especially well developed in the Galena Dolomite. Most of the vertical joints may be grouped into three average trends N77°W, N13°W, and N25°E (Figure 5). The joints of the N77°W group are generally more open than those of the other two

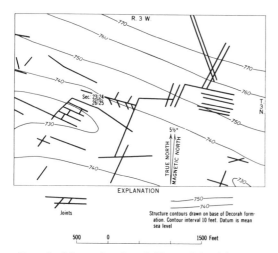

EXPLANATION

Joints

Structure contours drawn on base of Decorah formation. Contour interval 10 feet. Datum is mean sea level

500   0        1500 Feet

FIG. 5. *Map, showing folds and lead-bearing joints in parts of Sections 23, 24, 25, 26, T. 3 N., R. 3 W., northeast of Potosi, Wisconsin, where the joints fall into three groups.*

groups. The inclined joints are commonly tight fractures of local extent, and many of them are probably incipient reverse faults.

Most of the larger folds, faults, and joints observed in the district are probably a result of a general period of deformation by lateral compression and differential vertical uplift, accompanied by some rotational forces - all transmitted through the basement by adjustment of basement blocks. This deformation was preceded by earlier minor regional deformations. After the main period of deformation, some uplift and tilting of the rocks took place.

Many of the smaller folds, faults, and joints may be of similar tectonic origin, as described in detail by Heyl and others (28, p. 31–66), and reiterated more recently by Agnew (35, p. 258–260), but, in addition, probably many of the mineralized structures have been markedly accentuated by later solution and slump of the limestone beds just prior to and during mineralization. Some geologists, such as Carlson (31, p. 117–118), Mullens (44, p. 474, 489, 513–518), and Paul Herbert (oral communications 1945–1960) consider that some or all of the smaller structures have formed by solution and slump without tectonic control other than joints. Other geologists, such as Reynolds (26) and Klemic and West (43, p. 393, 405–407), suggest that the smaller structures may be controlled by initial tectonic fracture systems but that the structures were developed in their present form by solution and slump.

The widespread faults and joints are now known to provide a connected fracture system between permeable beds that could have provided the needed access for solutions. At least in a few places in the district, faults are of such large displacement that they very likely continue into the Precambrian basement. The exposed sedimentary rocks of all ages are transected by a complex fracture system of minor faults. The individual fractures of this system do not have any great vertical extent, but they connect with others and also with bedding-plane faults of considerable lateral extent, as well as with aquifers such as the St. Peter Sandstone and the sandstones of Cambrian age. Water movements through the system are thus essentially unimpeded both vertically and laterally. Similar evidence was presented by Percival (3, p. 68), but many geologists later misinterpreted his statements to mean that the smaller fractures continued without break to the basement.

Linear dislocations in the magnetic patterns

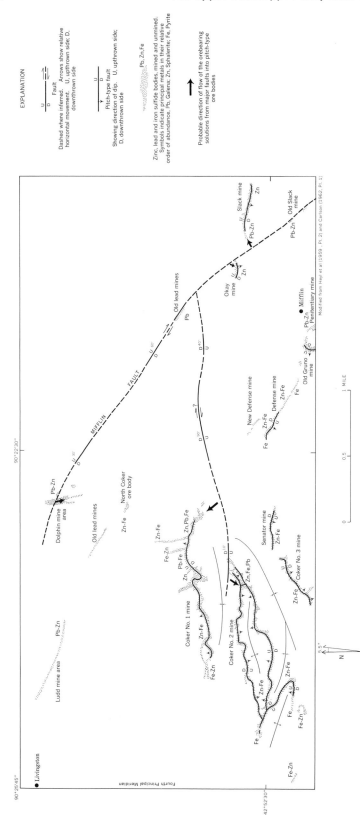

Fig. 6. *Coker Mines and nearby Ore Bodies in Iowa County, Wisconsin, showing probable manto-like feeding systems extending out from major faults to the pitch-and-flat ore bodies.*

of the basement (47) suggest that northwest and northeast trends are present beneath the Paleozoic rocks in the Precambrian basement. The coincidence of some of these lines with known major folds (for example, the main northeast-trending fold shown on Figure 4) and faults, such as the Mifflin fault (Figure 6), that deform the exposed Paleozoic strata, shows that tectonic movement was renewed along some of these linear dislocations in post-Precambrian time. Gentle upward, downward, and lateral adjustments took place in the basement crustal blocks intervening between the linear dislocations. These movements probably provided the necessary basement control of most of the larger faults, gentle folds, and flexures that are found in the overlying Paleozoic strata.

A regional tectonic deformation apparently began sometime before the deposition of the ores; it continued with diminishing intensity during the deposition of the ores and ceased at the end of the ore deposition, or probably shortly thereafter. This sequence of deformation produced the joints, the major and some of the minor folds, bedding-plane and reverse faults, shear faults, and some of the brecciation of the ores which occurred several times during their deposition, most probably in the order just given.

## Age of Mineralization

The age of the deposits cannot be definitely determined from any of the known relations in the Mississippi Valley. The rocks of the district were all gently warped during the main regional tectonic deformation, which was at least post-middle Silurian in age, and, to judge from nearby regional relations, probably Pennsylvanian or later. The known intrusions elsewhere in the Mississippi Valley are quite remote from the district, and most of their age relations are not especially clear. In the Illinois-Kentucky flurospar district, mafic intrusions cut the Pennsylvania rocks but not those of Late Cretaceous age (15, p. 131); they have been isotopically dated as Early Permian by Zartman and others (47). As the deformation that produced the La Salle anticline extension of the Wisconsin arch continued at least to the close of the Pennsylvanian, the best probable inference is that the ore deposition occurred sometime between late Paleozoic time and the end of the Mesozoic. Lead ore has been found in the glacial gravels of Clinton County, Iowa, so the deposits were in existence before the Pleistocene epoch.

## ECONOMIC GEOLOGY—PRIMARY ORE

The ore deposits of the Upper Mississippi Valley district are distributed over a wide area in Wisconsin, Illinois, and Iowa. Lead and zinc ores have been the principal metallic ores mined, but small tonnages of high-grade copper ores and barite have been produced. Iron sulfides have been mined for the manufacture of sulfuric acid both separately and as a by-product of zinc mining. Limonite and hematite have been mined for iron ore in the Iowa and Wisconsin parts of the district. The waste rock, both from the mines and the jig mills, has been widely used for road metal and railroad ballast.

### Stratigraphic Relations of the Ore Bodies

Deposits of lead, zinc, copper, and iron sulfides have been found in all the strata of Cambrian, Ordovician, and Silurian age that are exposed in the district. However, all the deposits of lead, zinc, and copper of commercial size occur in the Prairie du Chien Group and in the Platteville, Decorah, and Galena Formations of Ordovician age. Considerable limonite and hematite, formed by the oxidation of iron sulfides, and a little lead have been produced from deposits in the sandstones of Cambrian age (6, p. 49–56; 5, p. 518–520).

Most of the known deposits in rocks of Cambrian and Early Ordovician age are found in the main outcrop areas of these rocks in the northern fringe of the district. Recent exploratory drilling by the U.S. Geological Survey within the main district has located lean deposits of zinc, lead, and iron sulfides in these rocks (20,21).

The St. Peter Sandstone of Middle Ordovician age is locally impregnated with pyrite and traces of galena and sphalerite. None of the deposits is of commercial value. Many known deposits in the St. Peter Sandstone are directly below large sulfide deposits in the overlying formations.

Nearly all mineral deposits in the main part of the district are in the Galena, Decorah, and Platteville Formations of Middle Ordovician age. The ores are deposited over the same stratigraphic range in these formations both in the northern part of the district, where the formations are largely eroded, and in the southern part, where the beds are buried beneath the Maquoketa Shale and Silurian dolomite.

The Maquoketa Shale is only locally mineralized. Small quantities of galena, sphalerite, and barite were found at the Glanville prospect (23) near Scales Mound, Illinois.

The few localities where the Silurian rocks are known to contain small deposits of lead or iron sulfides are in Iowa, far to the southwest of the main part of the district. Only a very few tons of ore were produced from any one of these deposits.

## Forms and Controls of Ore Bodies

The mineral deposits can be classified into several types: (1) pitch-and-flat deposits, which are veins and replacements in and near reverse and bedding-plane fault zones (pitches and flats respectively) of small displacement controlled by gentle folds and accompanied by some sagging from solution thinning; (2) gash-vein deposits, which are discontinuous veins controlled by selective solution of beds along joints; and, less commonly, (3) stockworks; (4) bedded replacements; (5) solution-collapse breccias; (6) fissure veins and lodes; and (7) rare ore-lined giant vugs or small sulfide-encrusted caves.

The zinc and lead deposits of the pitch-and-flat type are found in systems of reverse and bedding-plant faults that show more advanced development in size and displacement in the southern part of the district (28, p. 109). The ore bodies of arcuate, linear (Figure 8), and elliptical form in plan are found on the flanks and around the ends of folds (Figure 6, Figure 7). More commonly the bodies are around the ends of synclines than around the ends of anticlines. Many bodies are between a quarter of a mile and a mile long and contain from 50,000 to 3,000,000 tons of ore. In places, the pitch-and-flat ore bodies are underlain by broad blankets of breccia and bedded

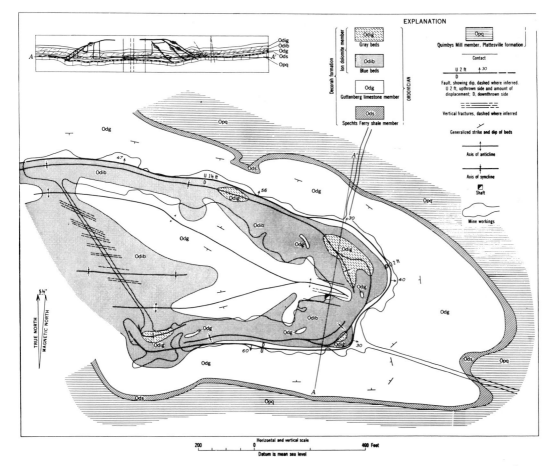

Fig. 7. *Geologic Mine Map and Section of the Hoskins Mine, New Diggings, Wisconsin, showing a typical arcuate ore body in a syncline. The geologic data and interpretation are shown as plotted on a horizontal section near the mine floor.*

replacement and bedded vein ores, in part controlled by bedding-plane faults, and by the basal parts of the reverse fault systems. These blanket ore bodies are in many ways similar to the "sheet ground" deposits that lie beneath fracture-controlled "runs" of the Tri-State district (12). The ore in the pitches and flats and in the underlying bedded blankets is deposited mainly as: (1) veins along the fractures and bedding planes; (2) cavity fillings in tectonic and solution breccias; and (3) disseminations by replacement and impregnation in favorable beds, particularly shaly strata.

The origin of the pitch-and-flat ore bodies has long been a subject of major controversy. The following ideas have been offered to explain their origin: (1) simple compressional or rotational tectonics (14, p. 788–802); (2) a combination of tectonics followed by subsiding solution and slump (28); (3) initial inclined compressional (26, 31) or tensional (5) tectonic fractures controlled by vertical joints followed by major solution and slump; (4) formation by simple slump with only initial joint control as seen in many a brick wall (9, 43, 50). Further study, especially very detailed geologic drill-hole and mine mapping

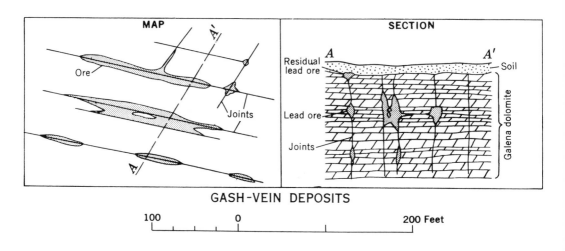

**GASH-VEIN DEPOSITS**

100          0                              200 Feet

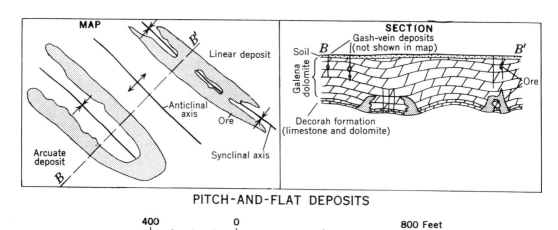

**PITCH-AND-FLAT DEPOSITS**

400          0                              800 Feet

FIG. 8. *Diagrammatic Plans and Sections, showing typical patterns of gash-vein lead deposits and underlying pitch-and-flat deposits of the arcuate and linear types and their stratigraphic positions relative to one another.*

may resolve some of these differences in the future. Simple collapse pitches formed by weathering along joints are common on a small scale in some joint-controlled deposits. They were formed in Recent geologic time above the water table. A few collapse pitches and tumbled breccias formed by simple solution sag by the ore solutions along vertical joints were observed by the writer.

Less common than the pitch-and-flat types of ore body are broad blankets of disseminations of sulfides that formed by replacement and impregnation in favored stratigraphic units. The blankets are controlled by vertical joints or by solution sagging near intersecting networks of major joints. In places, the master joints contain thin feeder veins of ore extending down through barren limestone or dolomite beds beneath the ore bodies (T. E. Mullens, oral communication, 1959).

Stockworks of ore have been described in a few places by Chamberlin (5, p. 467–468). Simple solution breccia ore bodies without marked fracture control are described by Mullens (44, p. 523–524) and Heyl and others (28, p. 133–134). In a few places these developed into solution collapse breccias.

The gash-vein lead deposits are in joints, which for the most part are in beds of the Galena Dolomite stratigraphically above (but not necessarily directly above) the beds containing pitch-and-flat deposits (Figure 8). The ore is in veins of limited extent filling vertical joints and in podlike deposits in breccias, replacements, and solution breccias of sanded dolomite in favorable beds along these joints. The podlike deposits, locally known as openings, may lie one above the other in different beds along the same joint connected by thin gash veins of galena in the joint. Single mineralized joints are traceable for more than a mile; they commonly have nearly constant strikes. Gash-vein deposits commonly contain between 100 and 10,000 tons of lead ore (less commonly zinc, copper, or barite). Structurally they are very similar to deposits described by Eckel (18, p. 78, Figure 6) in the southern San Juan Mountains of Colorado. Bradbury (27) has recently described very well and in detail several gash-vein or crevice lead-zinc deposits. Other fine descriptions are by Brown and Whitlow (29, especially Plates 5, 6, and 7) and by Chamberlin (5, especially p. 451–457). Mullens (44, p. 501, Plate 26) has described a less common type of ore body from the Hancock mine and Kittoe ore body near Shullsburg, Wisconsin, where the ore bodies which are mostly in the Platteville For-

mation are controlled mainly by vertical joints and associated solution alteration zones. Ore bodies that are similarly controlled are described by Heyl and others (28) in the New Mulcahy, Old Mulcahy, Rodham, and Fox mines, and in parts of the James and Helena-Roachdale mines, all in Wisconsin.

Most of the known fissure veins and lodes are in the north fringes of the district where they occur in faults cutting dolomites and sandstones of Early Ordovician and Late Cambrian age. Good examples of veins or lodes of lead and zinc in simple fissure veins are the Demby-Weist mines northeast of Dodgeville, Wisconsin, and the Lansing lead mine northwest of Lansing, Iowa. Descriptions of these veins in the Prairie du Chien and uppermost Cambrian sandstones are given in Heyl *et al.* (28, p. 280–282, 295, Figure 99).

## Wall-Rock Alterations

The ore-bearing solutions have formed alteration halos in the host limestones and, to a lesser extent, the dolomites in the immediate vicinity of the deposits. Alteration is of five types: solution of carbonate rocks, particularly the limestones; silicification; dolomitization; clay-mineral alteration; and formation of sanded dolomites. Solution of carbonate rocks has considerably thinned some of the favorable beds in many ore bodies (Figure 9). This

FIG. 9. *Guttenberg Limestone Member of Decorah Formation and Interbedded Carbonaceous Shale Altered by the Ore Solutions to Shaly Residues. Note the progressive change from unaltered sedimentary rock at right of knife to thin residue shale at Graham-Ginte Mine, Galena, Illinois. Photograph courtesy of H. B. Willman, Illinois State Geological Survey.*

thinning has caused fractures in the rock to open, thus providing spaces for deposition of the ore. In places, the rock has been so altered by solution that the overlying rock has collapsed to form tumbled breccia. Dolomitic rocks within and near the ore bodies have locally been "sanded" (17, p. 27), that is, changed to a friable or incoherent mass of dolomite crystals by the ore-bearing solutions. The cementing bond between the large dolomite crystals was weakened by intergranular solution, and many of the fine dolomite grains were dissolved by the ore-bearing solutions. The jasperoids of the silicified zones, the fine-grained dolomite halos, and the sanded dolo-

mites all closely resemble the same kinds of alterations in the Tintic district, Utah, except that they are not nearly as widespread or abundant in Wisconsin. Locally desilicated chert nodules have been noted; deep-green celadonite as well as thin quartz veins are present in deposits of the Prairie du Chien Group, and in one place crystallized flakes of sericite are in small wall-rock vugs (41).

X-ray diffraction of a series of samples from a thin carbonaceous basal shale bed of the Quimbys Mill Member of the Platteville Formation shows well-defined alteration aureoles around the two ore bodies of the Thompson-Temperly mine (Figure 10) (40). Clay min-

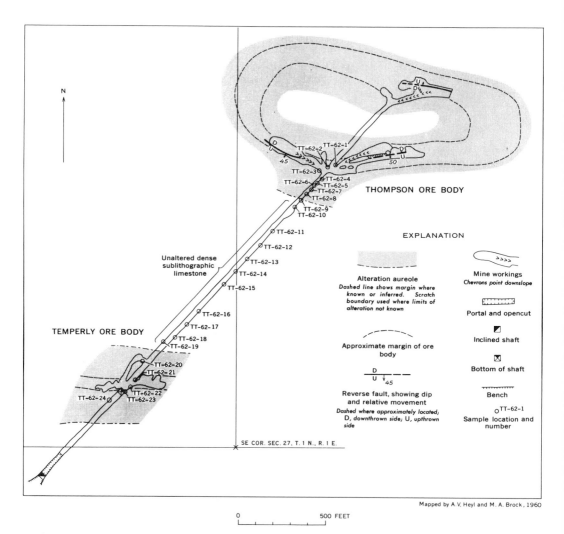

Fig. 10. *Sketch of the Thompson-Temperly Mine, showing approximate location of shale samples. Map is a horizontal section of the ore bodies at an elevation where they lie within the Quimbys Mill Member of the Platteville Formation of Middle Ordovician Age.*

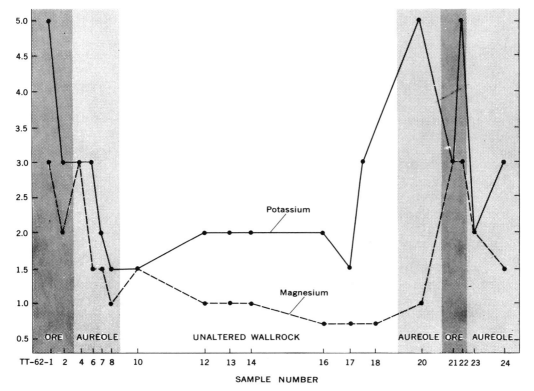

FIG. 11. *Results of Semiquantitative Spectrographic Analyses of Potassium and Magnesium in Samples From the Thompson-Temperly Mine. Results are reported in per cent to the nearest number in the series, 1, 0.7, 0.5, 0.3, 0.2, 0.15, and 0.1, and so on, which represent approximate midpoints of group data on a geometric scale. The assigned group for semiquantitative results will include the quantitative value about 30 per cent of the time.*

erals change progressively from the Md polymorph of illite in the unaltered rock to 1M and 2M illite in the aureoles, and the 2M polymorph increases markedly within the ore deposits themselves. In the same samples, calcite decreases, and dolomite and microcline increase toward the ore zone.

Stable isotope studies show little fractionation of $O^{18}/O^{16}$ and $C^{13}/C^{12}$ of coexisting limestone and dolomite samples from the same mine and nearby mines as compared with that from barren outcrops (41). However, the cacite in ore bodies contains markedly lower $O^{18}/O^{16}$ ratios than that in the barren limestone wall rocks of the Quimbys Mill Member beyond the alteration aureoles. The calcite and dolomite of the altered wall rock of the ore body and alteration aureole contain slightly lower $O^{18}/O^{16}$ ratios than the barren limestone of the Quimbys Mill Member. Further studies are in progress on possible fractionation of the oxygen isotopes between carbonates and silica.

Qualitative X-ray fluorescence and spectrographic analyses of the samples (Figure 10) show slight decreases in potassium, aluminum, magnesium, silica, titanium, iron, and manganese in the outer part of the alteration aureoles and marked increases in the ore zone (Figures 11, 12, and 13) (42). For example, aluminum shows a marked decrease at the outer edge of the altered zone from the amount of unaltered wall rock. At the inner edge, it increases slightly, and in the ore zone, markedly. In contrast, calcium increases slightly in the altered zone and decreases markedly in the ore zone. These facts indicate leaching in the outer fringe of the aureole and migration of the elements toward the ore zone. However, the proportionate volumes and contents of the leached zone as compared with those of the inner-alteration aureole and ore zone suggest that much of the aluminum, iron, silica, titanium, and manganese were added by the ore-bearing solutions from sources other than the narrow leached outer part.

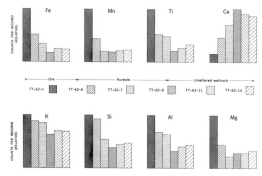

FIG. 12. *Graphs, showing relative amounts of 8 elements in 6 samples of shale from the Thompson-Temperly Mine as determined by X-ray emission analysis.*

## Ore Textures

The ore shows several varieties of texture dependent upon the types of deposition, structural controls, and wall rock alterations: (1) veins with comb structures or colloform bands with symmetrical or asymmetrical banding; (2) solid veins of only one mineral; (3) individual well-formed crystals impregnating host rock without notable replacement of the rock grains, (4) individual euhedral crystals that replaced the host rock; (5) coarse crystals lining open cavities; (6) reniform, colloform, nodular, and stalactitic masses; (7) replacement of fossils; and (8) pseudomorphic replacement of earlier formed minerals.

## Mineral Paragenesis

The principal primary minerals of the deposits, in general paragenetic sequence of deposition are quartz (chert, jasperoid, and other forms of cryptocrystalline silica), illite, dolomite, pyrite, marcasite, sphalerite, galena, barite, chalcopyrite, millerite, enargite, and calcite (Figure 14). A cobalt-nickel-arsenic mineral, not yet identified, but possibly having the composition and atomic structure of nickeliferous cobaltite, is known as fine intergrowths with marcasite. Recent mineralogic studies by Whitlow (49) of a late, blue metallic copper sulfide he found associated with chalcopyrite and enargite indicate that the blue mineral is digenite rather than chalcocite. E. H. Roseboom (U.S. Geological Survey, oral communication, 1963) examined the X-ray diffraction traces of this mineral and confirmed this identification. A similar blue chalcocite-like late mineral collected by Roseboom in 1962 has been identified by him as djurleite (oral com-

munication, 1965). It is closely associated with chalcocite. Silver occurs in the galena in quantities as much as 2 ounces per ton, and in sphalerite, and a very little gold has also been noted. Elements in small quantities in the ores are cadmium, mangenese, germanium, molybdenum, and zirconium.

Several studies have been made of minorelement distribution in the minerals of the district by Heyl and others (28, p. 94–96), Marshall and Joensuu (33), and Bradbury (27, 30). Bradbury selected his sulfide samples for relation to paragenesis and district geology and showed by trace-element analysis that there is little variation in the trace elements between the sulfide in the pitch-and-flat deposits and the sulfides in the overlying gash-vein deposits. He demonstrated that silver occurs in a few of the sphalerites in notable quantities and that some galenas contain more antimony than galenas from some other Mississippi Valley districts.

The paragenetic sequence (Figure 14) was

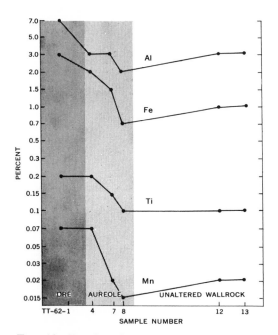

FIG. 13. *Results of Semiquantitative Spectrographic Analyses of Aluminum, Iron, Titanium, and Manganese in Samples From the Thompson Ore Body. Results are reported in per cent to the nearest number in the series, 1, 0.7, 0.5, 0.3, 0.2, 0.15, and 0.1, and so on, which represent approximate midpoints of group data on a geometric scale. The assigned group for semiquantitative results will include the quantitative value about 30 per cent of the time.*

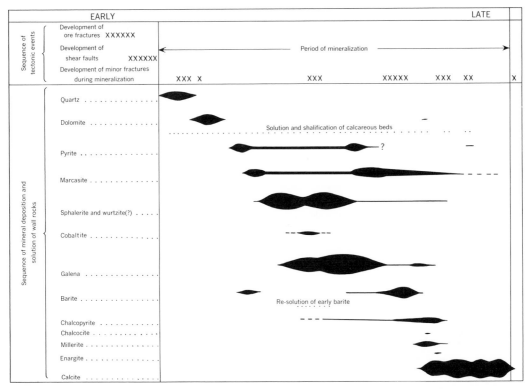

Fig. 14. *Paragenic Sequence of Primary Minerals and Sequence of Tectonic Events and Leaching of Wall Rocks.*

worked out by detailed geologic mine mapping in the district from 1943 through 1950 (28, p. 98–101). The underground studies of thousands of samples and vugs were followed by an examination of all open mines and of every important mine dump throughout the 3,000 square miles of the district. In the later stages of the program, many of the deposits on the fringe of the district were examined, and the knowledge obtained from all of these deposits was compiled and synthesized. Since 1950, and especially since 1959, the newly opened mines were visited, examined, and the paragenesis again studied in detail by the authors. Some further modifications were made to provide the diagram presented in Figure 14.

The sequence in the Upper Mississippi Valley district was developed through studying the superposition of the several minerals in unsheared vugs and veins. Most of the minerals in the Upper Mississippi Valley district were deposited in open spaces. Replacement of wall rock by sulfides is a quite minor type of mineral deposition. Deposition by impregnation is common; in this process, crystals or crystal clusters grow from nuclei in the soft

residum formed by solution and leaching during ore deposition (28, p. 103–105, 135). In such ores, the mineral crystals do not replace the rock but push the surrounding shales aside during growth.

In the Upper Mississippi Valley district, the sequence was investigated as follows: (1) the first step was to study those veins and vugs where the original walls were armored by pyrite, in such situations, the first sulfide was deposited, and later the rest of the minerals were crystallized in a simple succession of crystals and crusts from the walls to the vein center; (2) it was recognized that sphalerite had a very long period of deposition relative to other minerals, and, although multibanded, the early-deposited sphalerite is generally darker than the later deposited sphalerite; (3) three successive crystal habits of galena were noted: early-deposited galena is in cubes, later deposited galena is in cubo-octahedrons, and the latest galena is in octahedrons; (4) district-wide calcite stages were identified: calcite was deposited in four successive stages marked by distinctive crystal habits and it forms overgrown and zoned crystals in the later periods of ore deposition. The earlier calcite stages overlap with the sulfides, and all four stages are separated from each other by periods of leaching and etching which become progressively weaker. The crystal habits of the four stages are distinguished by a change from rhombohedrons or stubby scalenohedrons truncated by basal pinacoids, through two scalenohedral habits, to a final rhombohedral habit. Within each stage, many microcolor bands are detectable; some of the early stages can be further divided into at least two habit substages.

It should be emphasized that the paragenetic sequence is generalized. Deposition throughout the general period took place under conditions of rhythmic oscillations in both composition and temperature to form the hundreds of minute color bands characteristic of all the main minerals. There is abundant evidence, in addition, that deposition took place under conditions that pulsated without equilibrium (P. M. Bethke, oral communication, 1965) between precipitation and leaching of any one mineral. The banding is particularly well preserved in the translucent minerals such as sphalerite, barite, and calcite, but it can be seen upon etching in colloform pyrite and even in galena. Certain bands and stages were deposited in one vug, one vein, or one deposit and not in others; hence, many minor variations can be found by comparing the minerals of one locality with another, but the overall sequence remains the same and in the same superposition wherever studied. Detailed work started by P. M. Bethke and P. B. Barton (oral communication, 1963) on the banded sphalerites in the district has suggested that many of the minute bands can be traced and correlated in the sphalerites within one deposit and that a reasonable possibility exists that many of the bands can be correlated over large areas of the district. The U.S. Geological Survey is now engaged in further studies of sphalerite paragenesis to test these possibilities.

## Lead-Isotope, Deuterium, and Fluid-Inclusion Studies

Regional variations in the abundance of lead isotopes from galena in mineral deposits throughout the district are being investigated by A. V. Heyl, Maryse Delevaux, R. M. Zartman, and M. R. Brock (48). Preliminary lead-isotope results of a district-wide spread of galenas indicate a very large range for the isotopes of lead, all of which are radiogenic J-type leads. The data indicate a systematic pattern of wide variation throughout the district. The results suggest a regular progression in isotopic composition from the lowest $Pb^{206}/Pb^{204}$, $Pb^{208}/Pb^{204}$, and $Pb^{206}/Pb^{208}$ ratios in the west near the Forest City basin, and in the south near the Illinois basin, toward notably more radiogenic lead in the northeast near the crest of the Wisconsin arch. It appears from the pronounced regional isotopic variations that local variations would be of only secondary magnitude. No vertical variations in isotopic composition were observed which are comparable to those occurring laterally despite a rather complete sampling of various mineralized horizons. The lead isotope pattern crosses the district's copper-centered zonation without any distinctive local variations that would mark a focus of deep-seated magmatism.

Hall and Friedman (38) investigated the changes in chemical composition and deuterium concentration relative to hydrogen of primary fluid inclusions in ore and gangue minerals. The variations in chemical composition and deuterium concentrations relative to hydrogen were described and studied with respect to the sequence of deposition of the minerals of the deposits and to changes in composition of the ore solutions during formation of the deposits and with respect to the composition of typical concentrated oil-field brines in the Illinois basin.

In the Upper Mississippi Valley district, Hall and Friedman (38, p. 900–907) found that the inclusion fluid in the early ore minerals is a very concentrated Na-Ca chloride brine approximately of the same composition as that in the sulfide minerals in the Cave-In-Rock mineral district in southern Illinois. Inclusions could not be recovered from some of the early minerals, such as dolomite, pyrite, and early barite, in the Upper Mississippi Valley district, and hence only the inclusions of the later part of the paragenesis were analysed. Sphalerite and galena contain inclusions that are similar in most respects to oil-field brines, such as total concentration D/H, Mg/Na, and Cl/Na ratios, but differ in respect to high Ca/Na and K/Na ratios, and by high-calcium concentrations which are several times those found in most oil-field brines.

The inclusion fluids in the later calcite gangue minerals they found to be much less concentrated and to contain a deuterium concentration, relative to hydrogen about that of meteoric water. Hall and Friedman (38) conclude that the early ore minerals were deposited from a predominantly Na-Ca-Cl type of solution of approximately the same composition as an oil-field brine and that this solution was gradually flushed by meteoric water during the final stages of mineralization. They conclude that "the addition of magmatic water in the system in the [district] can neither be postulated nor refuted from the chemical evidence."

An independent extraction of fluid inclusions and analysis by Roedder, Ingram, and Hall (39, p. 364–365) from the core and rim of a galena crystal from Platteville, Wisconsin, provided main ratios of the principal constituents very similar to those obtained by Hall and Friedman.

Bailey and Cameron (19) measured the temperature of filling of fluid inclusions in the Upper Mississippi Valley district and found temperatures (uncorrected for pressure) of 121° to 75°C for several stages of sphalerite and 78° to 50°C for stage 2 and stage 3 calcite. They point out that the temperature decline from sphalerite to calcite is consistent with standard assumptions that temperature decreases during mineralization, and thus their study confirms the evidence of dilution by meteoric water noted during the calcite stage by Hall and Friedman (38).

Erickson (46) has made a detailed study of fluid-inclusion temperatures in specimens of calcite from four mines in the Upper Mississippi Valley district using the four stages of district-wide calcite deposition worked out by Heyl *et al.* (28, p. 97–102). Erickson's results strongly support and confirm those of Bailey and Cameron (20), for the general range of calcite temperatures is almost identical with theirs (74°C to 46.2°C uncorrected for pressure), and he was able to include measurements from the final stage 4 calcite. He states (46, p. 506):

"Results indicate a decline in the temperature of calcite deposition. The temperature ranges of deposition (uncorrected for pressure) are: calcite II, 74° to 51°C; calcite III, 62.5° to 46.5°C; and calcite IV 56° to 46.2°C."

Correction for pressure would raise the above ranges from 5° to 10°C.

Erickson's data (46) support the hypothesis that during later sulfide and post-sulfide calcite stages of deposition the fluids were being diluted and cooled by mixing with meteoric waters.

Barnes (37) has suggested that prior to ore deposition the ore-transporting fluid ranged in pH between weakly acid and moderately alkaline. He bases this conclusion on alteration mineralogy (in particular, the potassium feldspar and mica), carbonate stability, fluid inclusions, and sulfur and sulfide stabilities. He states that fluid inclusions in minerals deposited during ore deposition indicate that the ore fluid is a brine. Ore mineral stabilities indicate highly reducing conditions during transport and deposition. He concludes that experimental evidence suggests that the ore fluids are neutral to alkaline bisulfide solutions. Precipitation, he suggests, is by contact of such a solution with meteoric water which would cause precipitation simultaneously by oxidation (with resulting decrease in pH), by decrease in temperature, and, to a minor extent, by loss of pressure and dilution. Barnes (37) suggests that ore deposits would be localized close to the ore solution-meteoric water interface. Field relations (28) support the possibility of such a solution interface, and the wallrock alteration to potassium feldspar and high-order illite (40, 42) is also favorable evidence for Barnes' hypothesis.

## Zonation of Ore Deposits

The ore deposits show both an areal and vertical zoning. The areal zoning is illustrated in the east-central part of the district where copper, barium, nickel, and arsenic are quite abundant in the ores, and this unusual metal zone extends northwesterly through the east-central part of the district, irrespective of structure. The vertical zoning is shown by the greater concentrations of lead near the surface, whereas zinc, iron sulfides, nickel, silica, and dolomite are in greater abundance in the deeper deposits. Some of the vertical zoning may be the result of the completely different composition of carbonaceous limestone versus dolomite or sandstones in different stratigraphic units and in the way in which the ore-bearing solutions reacted with these chemically different lithologies.

## Grades of Ore in Valuable Metals

The pitch-and-flat zinc ore bodies commonly range between 3 and 10 per cent zinc before mining, but some rich pockets contain as much as 20 or 25 per cent zinc. The average zinc content is probably about 5 per cent and the average lead content about 0.5 per cent. However, the lead content of certain lead-rich ore bodies may be as much as 5 or 6 per cent. Owing to mass production mining methods used in recent years, the average grade of ore in the large mines is about 3 per cent zinc. The grade of hand-sorted copper ore when mined many years ago was 15 to 20 per cent copper. The grade of lead ore as mined from the gash-vein lead deposits is estimated (27) to have been 12 to 22 per cent lead before hand sorting, a quite reasonable figure from the data available.

## ECONOMIC GEOLOGY— SECONDARY ORE

The primary lead, zinc, iron, and copper sulfide deposits of the district are leached and oxidized where they are within 30 to 100 feet of the present land surface. The oxidation of the ore minerals is complete in many places, except for the galena, which remains mostly unoxidized and only is coated with lead carbonate. Incised streams have lowered the water table from 30 to 100 feet in many places, and thus have exposed primary ore deposits between valleys to supergene changes by downward-moving vadose waters. In places where stream valleys overlie primary deposits, the ores are not altered by supergene changes owing to protection afforded the underlying sulfide deposits by maintanance of the water table at the surface and by the neutral waters in the carbonate rocks beneath. The oxidized zone generally continues for a few feet below water table and, in a few places, extends along open fractures to a depth of 100 feet below the surface of the water table. Secondary sulfides are common only in the few copper deposits and are found at or just below water table. No secondary lead and zinc sulfides are known. Residual lead deposits and placer deposits of galena, barite, and smithsonite are commonly small and, at present, are mainly useful for locating ore deposits in the bedrock.

In this area, as in many other districts, supergene minerals occur in three zones, in descending order: the zone of leaching and oxidation, the zone of enrichment containing secondary oxidized ores, and the zone of secondary sulfide ore enrichment. Table II shows the main primary minerals of the ores, the elements they contain, and the supergene minerals produced from them.

*TABLE II.   Minerals Formed by Oxidation and Redeposition*

| Primary Minerals | Principal Elements and Minor Constituents in Minerals | Resulting Secondary Minerals | | | | | | |
|---|---|---|---|---|---|---|---|---|
| | | Elements | Sulfides | Oxides | Arsenates Phosphates | Silicates | Carbonates | Sulfates |
| Galena | Pb, S, also Ag | Sulfur | | | Pyromorphite | | Cerussite | Anglesite(?) |
| Sphalerite Wurtzite | Zn, S, also Fe, Cd, Mn, Ge | | Greenockite | Wad Psilomelane Pyrolusite | | Hemimorphite Sauconite Zincian montmorillonite | Smithsonite Hydrozincite Aurichalcite | Goslarite Gypsum |
| Pyrite Marcasite Cobaltite(?) | Fe, S, also Co and As | Sulfur(?) | | Limonite Hematite | Erythrite Vivianite | | | Melanterite Copiapite Gypsum |
| Chalcopyrite Chalcocite[1] Enargite | Cu, Fe, S, and Ag, As | Copper | Bornite Chalcocite[1] Covellite | Tenorite Cuprite Limonite | | | Malachite Azurite Aurichalcite | |
| Millerite | Ni, S | | Bravoite Violarite | | | | Honessite | |
| Dolomite | Mg, Ca, C, O | | | | | | | Epsomite |
| Calcite | Ca, C, O, also Mn | | | Wad Psilomelane Pyrolusite | | | Calcite (travertine) Aragonite | Gypsum |
| Quartz | Si, O, also Au | | | | | Hemimorphite Sauconite Zincian montmorillonite | | |

[1] Primary chalcocite is blue and metallic; supergene chalcocite is sooty. Primary chalcocite includes two other similar but rare minerals, digenite and djurelite, both recently found in the district.

## Oxidation Enrichment and Residual Concentration

Beds that contain the primary sulfide deposits in the district have been eroded in many places, and some ore bodies have been almost completely removed by erosion. Such erosion, accompanied by oxidation and leaching, is particularly common in gash-vein deposits and also in many pitch-and-flat deposits, especially in the northern part of the district where the regional rise of the beds exposes Galena, Decorah, and Platteville strata at the surface over much of the area. Sphalerite and marcasite deposits are leached and oxidized. The zinc sulfide has been oxidized to zinc carbonate (smithsonite), and less commonly to hydrozincite, hemimorphite, sauconite, and goslarite. Because oxygenated water acts much more slowly on sphalerite than on associated iron sulfides, oxidation is most probably accomplished by sulfuric acid developed from the decomposition of marcasite and pyrite in vadose water. For example, in the district the pH of ground water below water table ranges from 7 to 7.5 whether the water is collected from rich deposits of marcasite, pyrite, galena, and sphalerite or from barren rock (25, 34). Surface and vadose water in the zone of oxidation are more variable, having pH values that range from 7 to 8.4, including water that contains metals derived from oxidizing ore bodies. The water within oxidizing bodies is locally very acidic, but it is quickly neutralized by the dolomite and limestone wallrock and calcite gangue, so that nearly all the oxidized zinc minerals replace sphalerite, and little is lost or transported more than a few inches or feet.

Some zinc is carried away and precipitated as zinc carbonate elsewhere or, in a few places, forms other zinc minerals. Likewise, most of the calcium sulfate formed has apparently been removed in solution, but a little is precipitated as gypsum in nearby fractures. In most places, a direct replacement of sphalerite by smithsonite results in pseudomorphs that retain the original euhedral form of the sphalerite crystals. Smithsonite also is present as porous, cellular masses, locally called "drybone," that line and fill fractures and replace calcite or dolomite. Where it occurs as drybone, some zinc has been carried a short distance before deposition as the carbonate. Smithsonite so resists further alteration that, in places, it has been carried away by erosion and concentrated in placer deposits. However, much smithsonite that entered the zone of extreme leaching or

gossan because of the erosion of the overlying rocks has undoubtedly been broken up by weathering, and some of the zinc has been redissolved and removed by surface and descending vadose waters.

Lead, in contrast to zinc and iron, shows very slight mobility in the oxidized zone, especially in limestones and dolomites. The galena remains unaltered except for coatings of gray or white lead carbonate (cerussite) and lead sulfate (anglesite). The only other known lead mineral is pyromorphite which is in minute quantities at a few localities. All salts of lead are difficultly soluble, particularly in the slightly alkaline or nearly neutral meteoric waters of carbonate rocks. Lead carbonate is least soluble, the sulfate somewhat more, and the chloride the most soluble. Galena is attacked by dilute $H_2SO_4$ to a small extent, especially if $H_2SO_4$ is combined with ferric sulfate.

Very little galena has altered except that in small grains or around the borders of the large crystals. Large quantities of lead carbonate were seldom found even in the shallowest lead mines. A little lead has been dissolved and carried away in solution, as is indicated by small hexagonal crystals of grayish-white chlorophosphate of lead (pyromorphite) that infrequently coat other minerals away from the original galena. Phosphorus was probably obtained from small phosphatic nodules in certain beds of the Decorah Formation. Reduction with galena has released free sulfur which has been deposited in a few places as clear yellow crystals in cavities within the galena.

Residual masses of galena with a carbonate coating are found in large quantities in gossans and in oxidized veins. Also, considerable amounts have been removed by streams and transported into the valleys to be concentrated as placer deposits in natural riffles in the stream beds.

Chalcopyrite shows oxidation effects similar to sphalerite and marcasite. Supergene copper minerals are deposited in and below the zone of leaching, mainly as a variety of oxides and carbonates. Dolomite and calcite are the principal minerals of the wallrocks. Reaction of the wallrocks with the copper sulfate produced in primary oxidation caused copper carbonates to form before migration occurred to a great extent. This deposition of carbonates probably prevented much supergene sulfide enrichment below the water table even where copper was quite abundant. In a few places, zinc and copper in solution reacted with limestones to form the double zinc-copper carbonate aurichalcite.

Nickel sulfide (millerite) has altered to nickel iron sulfides (bravoite and violarite) by direct replacement; the iron probably came from ferric sulfate in solution formed from the decomposition of marcasite. Nickel sulfate (honessite) was formed by oxidation from violarite and bravoite and replaced them or was deposited as crusts and stains on the adjacent calcite.

Barite is completely unattacked by oxidation and appears just as unaltered at the surface as at depth. It is chiefly in primary deposits, but, similarly to galena, it is also found locally in residual and placer deposits.

Most calcite gangue is dissolved and carried away, but some is redeposited as travertine above the water table. In some places, however, calcite is replaced by limonite and smithsonite which form pseudomorphs. Locally, the calcium carbonate is redeposited in openings as aragonite; it also reacts with many of the minerals described above to form their carbonates.

Near the fluctuating water table, some minerals formed by oxidation, such as smithsonite, increase notably in quantity. Smithsonite in this lower zone commonly is white and is as well-formed crystals or crystallized incrustations. Where copper is present in the ores, quantities of copper oxides and carbonates are deposited with this smithsonite (for example, at the Eberle mine near Highland, Wisconsin). Some sphalerite and marcasite in all stages of oxidation generally remain. Many of these mixed ores, though rich, have remained unmined owing to the difficulties in separating the sphalerite from the smithsonite.

Several minerals that are generally not seen elsewhere are noted in this zone. Further oxidation probably removes them, and they are not present after the original sulfides have been oxidized. Some of these minerals are formed from impurities that have been dissolved from the primary sulfides. Wad, psilomelane, and pyrolusite are derived from sphalerite and calcite; erythrite (hydrous cobalt arsenate) is derived from an arsenic-cobalt mineral in marcasite; aurichalcite (copper-zinc carbonate) is formed from chalcopyrite and sphalerite. In places, associated with these minerals, are cavities containing brittle, waxy, brown tallow clays of which there are two types, sauconite, containing about 22 per cent zinc oxide, and zincian-montmorillonite carrying about 4 per cent zinc oxide. Similar tallow clays have been seen above in the leached zone and away from ore bodies, but it is not known if they also contain zinc.

## Supergene Sulfide Enrichment

Secondary sulfide enrichment is an unimportant feature of the ore bodies. Copper deposits and copper-bearing zinc deposits contain supergene copper sulfides. Chalcocite, bornite, and covellite are the principal secondary enrichment minerals. The areas of enrichment are local and lie near the zone of the fluctuating water table that changes slightly owing to varying seasonal precipitation. The secondary sulfides are commonly intermixed with partly oxidized primary copper, iron, and zinc sulfides, and in places with copper and zinc carbonates and oxides. The secondary copper sulfides replace primary chalcopyrite and sphalerite, and, in places, they fill cavities and form ceinlets. Large quantities of supergene chalcocite and covellite were noted by the writers in drill-hole samples at the McIlhon zinc deposit near Mineral Point, Wisconsin (28, p. 166; 37, p. 233). The chalcocite and covellite replaced chalcopyrite and sphalerite; they are most abundant near the top and sides of the ore body and are accompanied by abundant pyrite and marcasite.

Despite a careful search for lead or zinc sulfide deposits showing evidences of secondary sulfide enrichment, few signs of such deposits were noted in any of the many ore bodies studied, except for stains of greenockite and, very locally, violarite and bravoite. Nowhere in the abundant cavities and open veins were deposits of lead and zinc sulfides noted on surfaces of the last-deposited primary calcite crystals. Secondary galena and sphalerite were not observed in mines reopened after many years of being flooded, although the writer very carefully searched the walls where vadose waters descended, pools where ground water collected, and numerous iron and copper implements left in the mines when they were closed. Iron hydroxides and several sulfate minerals were deposited in many places where meteoric waters passed. Fragments of sulfides were found cemented to iron tools, but they were cleavage fragments or fragments of crystals cemented to the iron by an iron hydroxide coating.

The sulfide stalactitic forms (28) are deposits of primary minerals, even though superficially they resemble cave travertine deposits. They are everywhere direct outgrowths of regularly banded primary veins and show, without change, the same paragenetic sequences and crystal habits as the primary viens. Also most stalactitic forms found in place point upward, and all have central tubes; thus their formation by falling drops in open cavities is improbable.

## ORE GENESIS

The origin of the ores of the Upper Mississippi Valley zinc-lead district has been a subject of controversy for over a hundred years. Theories for the deposition of the ores by meteoric and artesian water were developed, and these later were adapted to the Tri-State and other similar districts. Thus, the district is the type area in the study of ore deposits of this nature in the United States. Many variations in theories and explanations of origin that fit into this general idea of cold-water deposition have been advanced, and the suggested sources of metal are numerous; similarly, numerous variations of these cold-water deposition theories have recently been developed and applied to the equally controversial bedded-uranium deposits of the Colorado Plateau and the Black Hills regions. Previous to the development of the meteoric water theories, the geologists who first studied the district, David Dale Owen (1) and James Gates Percival (3, 4), considered the ore deposits to have formed by "emanations" from somewhere in the basement rocks beneath the comparatively thin Paleozoic cover. Within the last decades a number of geologists have returned to this early view. Many geologists in Europe have recently accepted hypotheses of meteoric origin, or of secondary hydrothermal solutions as proposed by Schneiderhöhn (16) for somewhat similar deposits in Europe, Asia, and Africa.

The writer is among those whose studies in recent years have led him to return to support a hypothesis of origin by rising warm aqueous solutions accompanied by widespread distribution through the available sandstone aquifers that underlie the ore bodies. Unequivocally, all the lead, zinc, and copper deposits, and most of the iron sulfides, except perhaps the widespread pyrite disseminations in the Glenwood Shale Member of the Plattenville Formation and basal Maquoketa Shale are epigenetic and not syngenetic. Although the studies by the writer were undertaken with the meteoric and artesian hypotheses in mind, nearly all the evidence obtained that bears on the origin of the ores favored the genesis of the ores by deposition from rising thermal waters from the underlying rocks. Deposits of epigenetic sulfides extend down into the sandstones of Cambrian age. The ultimate source of the solutions therefore must be in these underlying beds, or the deposits must be derived either from widespread solutions of connate origin in these Cambrian sandstone aquifers, that moved into the district by artesian circulation as hypothesized by Van Hise and Bain (8), or from rising thermal waters whose ultimate heat source was magmatic bodies that intruded the rocks of Precambrian age, and perhaps heated the connate waters in the sandstones above and started them moving updip in the aquifers. The absence of any major deformation or metamorphism of the rocks of Paleozoic age excludes an origin by reheating of the rocks and associated ground waters to redissolve the metals from the Middle Ordovician wallrocks and deposit the ore minerals in their present form.

Probably the ore solutions were a mixture of connate and meteoric waters blended, heated, and mobilized by heated solutions from some deeper source that provided most of the base metals of the deposits. The ultimate source of the heated solutions is not known, but it could have been a magmatic body that intruded the Precambrian rocks but did not extend up into the sandstones of Cambrian age. The juvenile fraction of the solutions, which may have been small, may have risen from faults and lineaments in the Precambrian basement (either underlying the district as suggested by Heyl and King (48) or in the nearby basement of the basins as suggested by Pb-isotope studies). After entering into the thick sandstone aquifers of early Paleozoic age, the solutions may have heated and mobilized the connate brines in them and spread laterally and widely through the aquifers beneath the district. Then the mingled and heated solutions may have risen above the aquifers through the many available joints, faults, and permeable zones into the carbonate rocks of Middle Ordovician age. The solutions deposited their metal content in the first limestones and dolomites above the rocks of Precambrian basement that contained shale partings rich in hydrocarbons, and that lay beneath a thick impermeable shale cap. The fracture systems in which the ores were deposited had previously been formed by widespread gentle tectonic disturbances of the craton, possibly in late or post-Paleozoic time. During the later stages, the meteoric waters within the middle Ordovician carbonate rocks may have diluted and cooled the solutions, and eventually meteoric waters flushed the concentrated brines out of the district.

A summary follows of some of the more

pertinent points that support the hypothesis that the ore bodies were deposited from a mixture of connate, meteoric, and warm hydrothermal solutions, the exact source of which is not known.

(1) The district has definite limits beyond which the same limestone and dolomite beds are barren of ore, even though they are cut by similar but unmineralized fracture systems favorable for the deposition of ore, contain minute quantities of lead and zinc equal to those in the district's host rocks, and have undergone equivalent minor tectonic deformations. This regional center of ore concentration on the flanks of a structural dome is most easily explained by a hydrothermal source, possibly magmatic, beneath the mineralized area.

(2) Faults and fractures are abundantly present, and the district is underlain by major sandstone aquifers that today form the channels for the notably abundant ground waters of the district. These fractures and known aquifers provide ample channels to permit the upward and lateral flow of thermal solutions to the present locations of the ore deposits. Mineralized roots or feeders in underlying rocks are present in several places in the district, such as the Demby-Weist fissure veins previously described, or the system of intersecting major faults about which ore bodies of the Mifflin subdistrict (Figure 6) are closely clustered (28). Other lines of ore bodies and their controlling structures suggest that they overlie structural lines of weakness, possibly major faults and shears, in the underlying basement.

(3) Linear dislocations, probably representing faults, in the Precambrian basement have been located by aeromagnetic surveys (49). Marked concentrations of known lead and zinc deposits occur in lines near or above some of the lineaments. Leakages of rising and probably heated solutions along some of the basement fault lines and along the margins of the more magnetic basement rock bodies possibly have provided at least a component of the solutions that deposited ores in the district. The solutions may have entered the overlying Cambrian sandstone aquifer system and spread out laterally in it while mixing and heating the connate water already in the aquifer system. The mixed heated solutions and the metal fraction that they now contained probably spread laterally and up through faults, joints, and more permeable beds into the overlying carbonate rocks. Thus, they could have moved into structurally prepared areas, such as joints

and pitch-and-flat zones, dissolving the carbonate wallrocks to make their own open space while they were depositing the sulfides and gangue minerals. This hypothesis of feeding lineaments, lateral and upward transport, and then deposition is supported by the marked concentration of the known lead and zinc ore bodies near and above buried magnetic highs (probably monadnocks) and in lines near or above probable basement fault lineaments.

(4) The complex system of faults, fractures, and aquifers, plus the nearby impermeable cap of the Maquoketa Shale, would all tend to cause a great deal of lateral flow of the solutions in addition to their vertical rise; this lateral flow would tend to distribute the solutions and their resulting ore deposits widely to form the large mineralized district.

(5) The deposits of the district are not restricted to a single favorable stratigraphic unit overlying a Spechts Ferry Shale "ponding" layer, as they were formerly thought to be. Instead they are distributed, in some quantity, in all the known rocks within the district from the upper Cambrian to the Silurian, irrepective of depth.

(6) The ores were emplaced in the latter part of a period of tectonic deformation that continued during ore deposition; both processes ceased at approximately the same time. These age relations are very similar to those of many hydrothermal districts.

(7) The paragenetic relations in the district show a single, geologically short cycle of mineral deposition, unrepeated, that terminated in the past. The sequence is notably similar to the type generally present in low-temperature hydrothermal deposits.

(8) The mineral deposits are notably similar, except for the absence of fluorite, to the fluorite-zinc bedded replacement deposits of the Illinois part of the Kentucky-Illinois-fluorspar district, and also to many telethermal lead-zinc deposits on the fringes of many magmatic base-metal districts, such as Leadville, Colorado; Goodsprings, Nevada; and Tintic, Utah, in the western United States. The dolomites and jasperoids of the wallrocks are notably similar to those at Tintic, Utah, one of the best established zoned hydrothermal districts.

(9) The mineralogy of the ores is more complex than formerly recognized, as the ore solutions contained not only lead, zinc, iron and sulfur ions, but also barium, copper, silver, nickel, cobalt, arsenic, cadmium, silicon, magnesium, potassium, sodium, and chlorine.

(10) The well-defined lateral and vertical

mineral zonings of the ore bodies are evidence of hydrothermal deposition.

(11) The temperature of deposition of the sphalerite (75° to 121°C) and calcites, as determined by Newhouse (13, p. 748), Bailey and Cameron (19), and Erickson (46) from crystal inclusions (50° to 78°C), is about that of telethermal deposits; and this temperature, though low, is far too high for any conceivable meteoric deposition, even assuming that the now-eroded overlying beds were 1,000 feet thicker than their present maximum thickness.

(12) Saline connate waters of a composition and concentration similar to those in sphalerite at one time filled the thick Cambrian and lower Ordovician aquifers beneath the district. Remnants of these waters still remain at Prairie du Chien, Wisconsin. Such connate waters, heated by, and mixed and blended with, a fraction of magmatic hypogene solutions and later meteoric waters, supply a reasonable fluid of deposition as supported by Hall and Friedman's (38) fluid-inclusion data.

(13) Chemical studies (15, p. 139–140) show that the precipitation of lead sulfide from rising solutions is possible nearer the surface, and farther from the thermal source, than the greatest concentrations of zinc sulfide.

(14) Although large volumes of limestone and dolomite wallrock in the ore bodies were dissolved during ore deposition, the aggregate of the lead and zinc they contained, though undoubtedly a contribution to the total, are still much too small to supply the quantities of these metals concentrated in the ore deposits. Undoubtedly, much of the jasperoid, calcite, and dolomite deposited with the ore bodies had their source in the dissolved wallrocks and underlying sandstones.

## CONCLUSIONS

Localization of the ore deposits in crosscutting structures which postdate deposition of the middle Ordovician and probably the Silurian sedimentary rocks, makes it unequivocal that the deposits were emplaced, in their present form, after sedimentation of the host rocks, and are, therefore, epigenetic. Many questions remain unresolved, however, concerning the ultimate source of the metals and the nature of the mineralizing fluid.

The author believes that the balance of evidence is on the side of the hypothesis which postulates the deposition of the ores by hydrothermal (hot water) solutions, probably a mixture of heated connate and magmatic waters, the latter being probably the smaller fraction.

Later in the deposition of the ores, meteoric waters were drawn into the system and the other two fractions diluted progressively. The notable similarity between ore bodies of this epigenetic district and those of proven hydrothermal districts is apparent; however, the question cannot be considered closed, and the writer is well aware of the limitations of the evidence presented to support this hypothesis. A great deal more knowledge of the mineralogy, temperatures, and other physical and chemical factors controlling ore deposition, of the geology of the hitherto unexplored rocks beneath the known part of the geologic formations, and of the basement, will be necessary before the origin of these deposits can be proved beyond any doubt.

The district is very large and has many small- to moderate-sized deposits of good grade. A very large part of it is unexplored by modern physical and chemical exploration methods. Without question, the potential resources of both zinc and lead are large, and the district should be an important source of these metals for many years to come. Copper, barite, and pyrite should be future byproducts in places. Future prospecting will very probably include searches for major extension of the district, particularly to the south and east in Illinois and towards the southwest in Iowa. Many old areas, little prospected in recent years, such as many parts of the central and northern parts of the district, will again become productive, as will little-known outlying areas near Potosi and Beetown, Wisconsin, to the west and near Yellowstone and Wiota, Wisconsin, to the east.

If new smelting methods now being developed make possible the mining of oxidized or mixed oxidized-sulfide zinc-lead ores, then large resources of such ores remain in many parts of the district, mostly at shallow depths.

## REFERENCES CITED

1. Owen, D. D., 1840, Report of a geological exploration of a part of Iowa, Wisconsin, and Illinois . . . in 1839: U.S. 28th Cong., 1st Sess., Senate Ex. Doc. 407, 191 p.
2. Whitney, J. D., 1854, Metallic wealth of the United States: Philadelphia, Lippincott, Grambo and Co., 510 p.
3. Percival, J. G., 1855, Annual report on the geological survey of the State of Wisconsin: Wisconsin Geol. Surv., Madison, 101 p.
4. ———1856, (Second) Annual report on the geological survey of the State of Wisconsin: Wisconsin Geol. Surv., Madison, 111 p.
5. Chamberlin, T. C., 1882, The ore deposits

of southwestern Wisconsin: *in Geology of Wisconsin,* Geol. Surv. Wisc., v. 4, p. 365–571.

6. Strong, M., 1882, Geology of the Mississippi region north of the Wisconsin River: *in Geology of Wisconsin,* Geol. Surv. Wisc., v. 4, p. 1–98.

7. Thwaites, R. G., 1895, Notes on early lead mining on the Fever (or Galena) River region: Wisc. Hist. Colln., v. 13, p. 271–292.

8. Van Hise, C. R. and Bain, H. F., 1902, Lead and zinc deposits of the Mississippi Valley, U.S.A.: Inst. Min. Eng. Tr., v. 23, p. 376–434.

9. Grant, U. S, 1906, Report on the zinc and lead deposits of Wisconsin: Wisc. Geol. and Nat. History Surv., Bull. 14, Econ. Ser. 9, 100 p.

10. Grant, U. S. and Burchard, E. F., 1907, Lancaster and Mineral Point folio Wis.-Iowa-Ill.: U.S. Geol. Surv. Geol. Atlas, Folio 145, 14 p.

11. Shaw, E. W. and Trowbridge, A. C., 1916, Galena and Elizabeth folio Ill.-Iowa: U.S. Geol. Surv. Geol. Atlas, Folio 200, 13 p.

12. Fowler, G. M. and Lyden, J. P., 1932, The ore deposits of the Tri-State district (Missouri-Kansas-Oklahoma): A.I.M.E. Tr., v. 102, p. 206–251.

13. Newhouse, W. H., 1933, The temperature of formation of the Mississippi Valley lead-zinc deposits: Econ. Geol., v. 28, p. 744–750.

14. Behre, C. H., Jr., *et al.,* 1937, The Wisconsin lead-zinc district—preliminary paper: Econ. Geol., v. 32, p. 783–809.

15. Bastin, E. S., *et al.,* 1939, Contributions to a knowledge of the lead and zinc deposits of the Mississippi Valley region: Geol. Soc. Amer. Spec. Paper 24, 156 p.

16. Schneiderhöhn, H., 1941, Lehrbuch der Erzlagerstättenkunde, [Bd.1]: G. Fischer, Jena, 858 p.

17. Lovering, T. S., *et al.,* 1949, Rock alteration as a guide to ore—East Tintic district, Utah: Econ. Geol. Mono. 1, 65 p.

18. Eckel, E. B., 1949, Geology and ore deposits of the La Plata district, Colorado: U.S. Geol. Surv. Prof. Paper 219, 179 p.

19. Bailey, S. W. and Cameron, E. N., 1951, Temperatures of mineral formation in bottom-run lead-zinc deposits of the Upper Mississippi Valley as indicated by liquid inclusions: Econ. Geol., v. 46, p. 626–649.

20. Heyl, A. V., *et al.,* 1951, Exploratory drilling in the Prairie du Chien group of the Wisconsin zinc-lead district by the U.S. Geological Survey in 1949–50: U.S. Geol. Surv. Circ. 131, 35 p.

21. Agnew, A. F., *et al.,* 1953, Exploratory drilling program of the U.S. Geological Survey for evidences of zinc-lead mineralization in Iowa and Wisconsin, 1950–51: U.S. Geol. Surv. Circ. 231, 37 p.

22. Agnew, A. F., 1955, Facies of middle and upper Ordovician strata in Iowa: Amer. Assoc. Petrol. Geols. Bull, v. 39, p. 1703–1752.

23. Heyl, A. V., *et al.,* 1955, Zinc-lead-copper resources and general geology of the Upper Mississippi Valley district: U.S. Geol. Surv. Bull. 1015-G, p. 227–245.

24. Agnew, A. F., *et al.,* 1956, Stratigraphy of middle Ordovician rocks in the zinc-lead district of Wisconsin, Illinois, and Iowa: U.S. Geol. Surv. Prof. Paper 274-K, p. 251–312.

25. Kennedy, V. D., 1956, Geochemical studies in the southwestern Wisconsin zinc-lead area: U.S. Geol. Surv. Bull. 1000-E, p. 187–223.

26. Reynolds, R. R., 1958, Factors controlling the localization of ore deposits in the Shullsburg area, Wisconsin-Illinois zinc-lead district: Econ. Geol., v. 53, p. 141–163.

27. Bradbury, J. C., 1959, Crevice lead-zinc deposits of northwestern Illinois: Ill. State Geol. Surv. R. I. 210, 49 p.

28. Heyl, A. V., *et al.,* 1959, The geology of the zinc and lead deposits of the Upper Mississippi Valley district: U.S. Geol. Surv. Prof. Paper 309, 310 p.

29. Brown, C. E. and Whitlow, J. W., 1960, Geology of the Dubuque South quadrangle, Iowa-Illinois: U.S. Geol. Surv. Bull. 1123-A, 93 p.

30. Bradbury, J. C., 1961, Mineralogy and the question of zoning, northwestern Illinois zinc-lead district: Econ. Geol., v. 56, p. 132–148.

31. Carlson, J. E., 1961, Geology of the Montfort and Linden quadrangles, Wisconsin: U.S. Geol. Surv. Bull. 1123-B, p. 93–138.

32. Cohee, G. V., *et al.,* 1961, Tectonic map of the United States: U.S. Geol. Surv. and Amer. Assoc. Petrol. Geol., 1:2,500,000.

33. Marshall, R. R. and Joensuu, O. 1961, Crystal habit and trace element content of some galenas: Econ. Geol., v. 56, p. 758–771.

34. Heyl, A. V. and Bozion, C. N., 1962, Oxidized zinc deposits of the United States, Part I, General geology: U.S Geol. Surv. Bull. 1135-A, p. A1–A49.

35. Agnew, A. F., 1963, Geology of the Platteville quadrangle, Wisconsin: U.S. Geol. Surv. Bull. 1123-E, p. 245–277.

36. Allingham, J. W., 1963, Geology of the Dodgeville and Mineral Point quadrangles, Wisconsin: U.S. Geol. Surv. Bull. 1123-D, p. 169–244.

37. Barnes, H. L., 1963, Environmental limitations to mechanisms of ore transport: *in* Kutina, J., *Editor, Symposium-problems of postmagmatic ore deposition,* v. 2, Geol. Surv. Czechoslovakia, Prague, 1965, p. 316–326.

38. Hall, W. E. and Friedman, I. 1963, Composition of fluid inclusions, Cave-In-Rock fluorite district, Illinois, and the Upper Missis-

sippi Valley zinc-lead district: Econ. Geol., v. 58, p. 886–911.

39. Roedder, E., *et al.,* 1963, Studies of fluid inclusions III: Extraction and quantitative analysis of inclusions in the milligram range: Econ. Geol., v. 58, p. 353–374.

40. Heyl, A. V., *et al.,* 1964, Clay-mineral alteration in the Upper Mississippi Valley zinc-lead district, *in* Bradley, W. F., *Editor, Clays and clay minerals*—12th Nat. Conf. on Clays and Clay Minerals . . . Atlanta, Georgia, September 30–October 2, 1963: Macmillan, New York, p. 445–453.

41. ——— 1964, Clay-mineral alterations and other coordinated geochemical studies in the Upper Mississippi Valley zinc-lead district (abs.): 10th Annual Inst. on Lake Superior Geology, Pr.

42. Hosterman, J. W., *et al.,* 1964, Qualitative X-ray emission analysis studies of enrichment of common elements in wall-rock alteration in the Upper Mississippi Valley zinc-lead district: U.S. Geol. Surv. Prof. Paper 501-D, p. D54–D60.

43. Klemic, H., and West, W. S., 1964, Geology of the Belmont and Calamine quadrangles, Wisconsin: U.S. Geol. Surv. Bull. 1123-G, p. 361–435.

44. Mullens, T. E., 1964, Geology of the Cuba City, New Diggings and Shullsburg quadrangles Wisconsin and Illinois: U.S. Geol. Surv. Bull. 1123-H, p. 437–531.

45. Taylor, A. R., 1964, Geology of the Rewey and Mifflin quadrangles, Wisconsin: U.S. Geol. Surv. Bull. 1123-F, p. 279–360.

46. Erickson, A. J., Jr., 1965, Temperatures of calcite deposition in the Upper Mississippi Valley lead-zinc districts: Econ. Geol., v. 60, p. 506–528.

47. Zartman, R. E., *et al.,* 1965, K-Ar and Rb-Sr ages of some alkalic intrusive rocks from central and eastern United States (abs): Geol. Soc. Amer. Spec. Paper 87, p. 190–191.

48. Heyl, A. V., *et al.,* 1966, Isotopic study of galenas from the Upper Mississippi Valley, the Illinois-Kentucky, and some Appalachian Valley mineral districts: Econ. Geol. v. 61, p. 933–961.

49. Heyl, A. V. and King, E. R., 1966, Aeromagnetic and tectonic interpretation of the Upper Mississippi Valley zinc-lead district: U.S. Geol. Surv. Bull. 1242A, p. A1-A16.

50. Whitlow, J. W. and West, W. S., 1967, Geology of the Potosi quadrangle, Wisconsin-Iowa: U.S. Geol. Surv. Bull 1123-I, p. 533–571.

# 22. Copper Deposits in the Nonesuch Shale, White Pine, Michigan

C. O. ENSIGN, JR.,*   W. S. WHITE,†   J. C. WRIGHT,†
J. L. PATRICK,*   R. J. LEONE,*   D. J. HATHAWAY,‡
J. W. TRAMMELL,§   J. J. FRITTS,‡   T. L. WRIGHT,*

## Contents

\* Copper Range Company, New York, N.Y.
† U.S. Geological Survey, Beltsville, Maryland.
‡ White Pine Copper Company, White Pine, Michigan.
§ Copper Range Exploration Company, Inc., Seattle, Washington.

## Illustrations

## Table

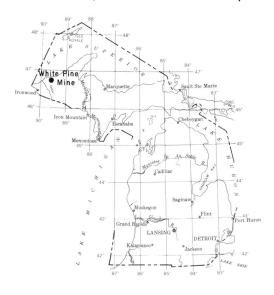

feet of the Nonesuch Shale are rich enough and sufficiently close enough together to constitute a mineable thickness where they lie within the cupriferous zone.

Pyrite is the characteristic sulfide of the Nonesuch Shale as a whole. The principal sulfide mineral of the cupriferous zone at the base of the formation is chalcocite. The succession of minerals outward and upward from the center of the cupriferous zone is native copper-chalcocite-bornite-chalcopyrite-pyrite. The bornite and chalcopyrite are confined to a narrow fringe at the margin of the cupriferous zone. This pattern of mineral zones is believed to represent reaction between syngenetic pyrite and introduced copper.

The main copper mineralization is independent of structure and apparently antedates deformation. Details of the geometry of the cupriferous zone indicate the possibility that copper was introduced from the underlying sandstone prior to lithification. The source of the mineralizing solutions and their copper is not known.

## ABSTRACT

The copper deposit at White Pine, Michigan, from which a little more than 5 per cent of United States primary copper currently is produced, is a large stratiform orebody, 4 to 25 feet thick and several miles across. The present ore column, containing about 1.2 per cent copper as chalcocite and native copper, is confined to the basal beds of the Nonesuch Shale. This upper Keweenawan (Upper Precambrian) formation is a distinctive gray siltstone unit, about 600 feet thick, overlying and overlain by thick red-bed sequences. All these rocks contain abundant volcanic debris, but the latest known igneous activity within 50 miles antedated the Nonesuch Shale. The rocks are unmetamorphosed and only moderately deformed, mainly by tilting and faults.

Pyrite, presumably syngenetic, is the only prominent sulfide throughout most of the Nonesuch Shale. The basal beds of the formation, however, contain disseminated copper minerals instead of pyrite and constitute the cupriferous zone. On a regional scale, the top of the cupriferous zone cuts across bedding, and ranges from an inch or so above the base of the formation in some areas to more than 50 feet in others. However, within the cupriferous zone the distribution of copper shows remarkable stratigraphic control. Certain beds are copper-rich compared to the beds that lie between them and the differences persist throughout the White Pine region. Several dark-gray thinly laminated beds in the lower 10 to 25

## INTRODUCTION

The copper deposit in the Nonesuch Shale of late Precambrian age at White Pine, Michigan (46°46′N, 89°34′W) (Figure 1) is the principal known copper-shale deposit in the United States. The orebody, mined by the White Pine Copper Company, a wholly owned subsidiary of Copper Range Company, is in Ontonagon County, in Michigan's Upper Peninsula. The village of White Pine, situated at the mine, is located on Michigan Highway 64, 6 miles south of Lake Superior, and southeast of the topographically prominent Porcupine Mountains.

White Pine lies within a belt of copper mineralization that extends from Mellen, Wisconsin, to the tip of the Keweenaw Peninsula, 160 miles to the northeast. The famous Michigan native copper deposits are concentrated in the Houghton-Calumet area, 70 miles to the northeast, although scattered native copper occurrences have been worked all along the belt.

The White Pine deposit is similar to the Kupferschiefer deposits of Germany and Poland and to the deposits of central Africa in the Roan Series in its stratigraphically controlled blanket form and many other characteristics, but the White Pine deposit is unique in other ways, such as the predominance of the single sulfide, chalcocite. Special features recognized in the White Pine deposit may con-

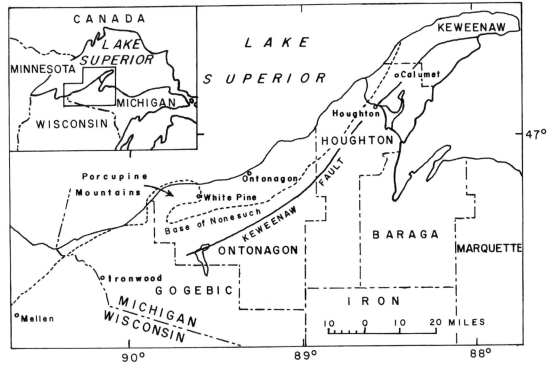

FIG. 1. *Location of the White Pine District.*

tribute to a broader understanding of the whole group of copper-shale deposits.

## Previous Work

The geologic account by White and Wright (8), though partly out of date owing to the large amount of later exploration and study, gives some detailed information not in the present summary account and lists most of the pertinent literature published prior to 1954. Many valuable accounts of special facets of the geology and discussions of the origin of the deposit published since 1954 are cited in the text below.

## Sources of Information and Acknowledgments

The data on which this paper is based represent observations or interpretations of one or more of the authors, except where work by others is cited.

So many competent geologists have been concerned with the deposits in the Nonesuch Shale that the original discoverer of some facts or relationships has often been forgotten. Many of us have joined profitably in study

and discussion with the following individuals, in particular, whose contribution to what we know about the deposits is gratefully acknowledged even if not identified: John R. Rand, Ernest L. Ohle, Ira B. Joralemon, and Peter C. Forbes. We also wish to express appreciation for the significant contributions that have been made to our understanding of the deposits by the studies of graduate students (10,16,21,38) and by the discussions of others (14,15,26,28,29). We are particularly indebted to American Metal Climax, Inc. for the permission to use data from its explorations in the Porcupine Mountains region. Finally, we wish to thank officials of the Copper Range Company, especially James Boyd, President, not only for encouraging publication of this account but also for their support of the exploration and investigations that made it possible.

It should be emphasized that among the authors of this paper there exists a broader spectrum of opinion concerning the origin of the deposit than is indicated by the presentation. Some of the authors believe that the obvious stratigraphic and lithologic control of copper distribution is strong evidence for a syngenetic origin of some, at least, of the copper.

## MINING AND EXPLORATION IN THE DISTRICT

The history of development of deposits in the Nonesuch Shale illustrates the impact on mining ventures of improving metallurgical technology and changing geological interpretation. Copper mineralization in the Nonesuch Shale was known in the White Pine district as early as the 1850's. Copper at the site of the Nonesuch mine 2 miles west of White Pine was discovered in 1865.

The first prospectors focused their attention on native copper* in sandstone beds at or near the base of the Nonesuch Shale. The sandstone, although barren throughout most of the region, locally constitutes rich ore, particularly where the rock is intensely faulted. This ore was the target at the Nonesuch mine. The less spectacular, but in the long run vastly more rewarding, chalcocite in shale was not appreciated until much later.

The town of Silver City, 6 miles north of White Pine, was the scene of a short-lived promotional boom from 1872 to 1876. Reports of high-grade silver in the Nonesuch "vein" prompted the formation of several dozen companies, the opening of a half-dozen small exploratory mines, and even the construction of a concentrator in 1874. It is reported that 22 tons of ore were processed, yielding 523 ounces of silver valued at $732.

The first economically successful operation, from 1915 to 1921, was by the old White Pine Copper Company. The ore was, again, native copper in sandstone, this time from a small faulted area immediately adjoining the White Pine fault. Operations ceased when exploratory drifts and drill holes suggested that the limits of the area of richly impregnated sandstone had been reached.

The success of the first White Pine mine inspired diamond drilling in a number of explorations north, west, and south of White Pine from 1916 to 1919, but only one, the White Pine Extension, had results encouraging enough to warrant sinking an exploratory shaft. In these investigations, the near absence of copper in sandstone focused attention on its presence in the shale and led to better appreciation of the intimate stratigraphic control of copper mineralization. The low grade

* "Native copper" here refers to the mineral, copper, in its uncombined state. "Copper" without the prefix "native" will be used to refer to the element without regard to its mineralogical or chemical state.

and fine dissemination of the copper minerals prevented exploitation at that time.

Recognition that the shale-copper constituted a major low-grade ore deposit because of its persistence over a large area apparently came slowly. The Copper Range Company acquired the White Pine property and began to explore it by drilling in 1937. The first drill holes tested a narrow zone close to the White Pine fault. By 1942, however, the potential value of widespread mineralization in the shale beds far from the White Pine fault had been recognized. Blocking out a shale-ore deposit averaging a little over 1 per cent copper became the objective of drilling. Concurrent research in application of sophisticated milling techniques and chemical flotation made it possible to economically recover the metal from this ore. Experimental mining at the Schacht shaft was begun in 1945. The exploration and milling research program led, by 1953, to the large-scale mining operations of the present White Pine Copper Company, a subsidiary of the Copper Range Company.

### Statistics of Mine Production

The grade of ore mined since 1961 has been slightly over 1.2 per cent, and mill recovery of about 85 per cent has yielded just over 1.0 per cent copper from the ore. A program of ore-dilution control, described by Ensign (32), has improved and maintained the grade

*TABLE I. Production from Nonesuch Shale and Immediately Underlying Rocks*

| | Rock Treated (short tons, dry) | Copper Produced (short tons) |
|---|---|---|
| Nonesuch Mine, 1868–1885 | — | 195 |
| Old White Pine Mine, 1915–1921 | 887,654 | 9,117 |
| Present White Pine Mine, 1955 | 2,971,900 | 31,427 |
| Do. 1956 | 3,892,760 | 36,639 |
| Do. 1957 | 3,892,056 | 35,855 |
| Do. 1958 | 4,229,611 | 43,330 |
| Do. 1959 | 3,967,751 | 36,987 |
| Do. 1960 | 3,868,010 | 39,792 |
| Do. 1961 | 5,370,575 | 56,441 |
| Do. 1962 | 5,610,563 | 58,742 |
| Do. 1963 | 5,435,076 | 57,547 |
| Do. 1964 | 5,428,071 | 57,017 |
| Total | 45,700,776 | 463,089 |

of rock mined. The production record of the larger mines in the Nonesuch ore is tabulated in Table I. Data for the Nonesuch and old White Pine mines are from Butler and Burbank (2, p. 91, 97). The estimated production for 1965 is slightly over 6,000,000 tons. The figures for 1962 and 1963 include small amounts of ore and copper produced from a development shaft on the southwest, down-thrown side of the White Pine fault.

The current rate of underground mining, more than 17,000 tons of ore per day, ranks White Pine as one of the largest underground copper mines in North America. It furnishes slightly more than 5 per cent of the United States primary copper. Reserves, which underlie more than 20 square miles of ground, are sufficient for more than 100 years at the current mining rate.

## PHYSIOGRAPHY HISTORY AND PRESENT TOPOGRAPHY

The Nonesuch Shale underlies a lowland sloping gently toward Lake Superior and away from higher land to the south underlain by middle Keweenawan lava flows. The lowland is due to the relative softness of the gently dipping upper Keweenawan sandstones and shales that underlie it. Surface features of the lowland, however, are due to glacial deposits, late glacial lake deposits, and their sculpture by streams and wave action. East and west of the Porcupine Mountains, the lowland is a plain into which local streams have incised themselves a few tens of feet, as described by Hack (37). Larger rivers with drainage basins heading up south of the lowland have eroded ravines and canyons almost 300 feet deep in clay and bedrock since the disappearance of the high-level late glacial lakes of the Lake Superior basin. Ancient shorelines marked by low wave-cut cliffs, beach deposits, spits, and related features are prominent in many places; these beach and related deposits are the principal source of sand and gravel in the area. The Porcupine Mountains, a glacially sculptured prominent bedrock dome 10 miles long, rise about 1000 feet above the lowland northwest of the mine.

The over-all slope of the lowland area underlain by the Nonesuch Shale is generally only a few tens of feet per mile. The mantle of glacial drift and lake deposits averages 65 feet thick in about 400 drill holes. As this is greater than the depth to which most streams are incised, bedrock exposures of the shale are very sparse indeed. Along the outcrop of

the Nonesuch Shale for 90 miles southwest from Calumet, we are aware of only 18 places where the approximate base of the formation can be seen in natural exposures.

## GENERAL GEOLOGY

### Stratigraphy

The Portage Lake lava series, of middle Keweenawan age (Figure 2) is a thick sequence of flood basalt flows that accumulated in the subsiding Lake Superior syncline (18). The thickness of the lava series is known to exceed 15,000 feet on Keweenaw Point (7), and structural inference supported by gravity data suggests that it may be of the order of 30,000 feet in places. Rhyolite flows, both extrusive domes and welded tuffs, occur within the sequence, but, in Michigan, are common only in the area south and west of White Pine, which was apparently a volcanic center in middle and possibly earliest late Keweenawan time.

Beds of conglomerate and sandstone from less than an inch to a few tens of feet thick are found between some of the lava flows. The pebbles and cobbles are predominantly Keweenawan rhyolite, and mafic debris is gen-

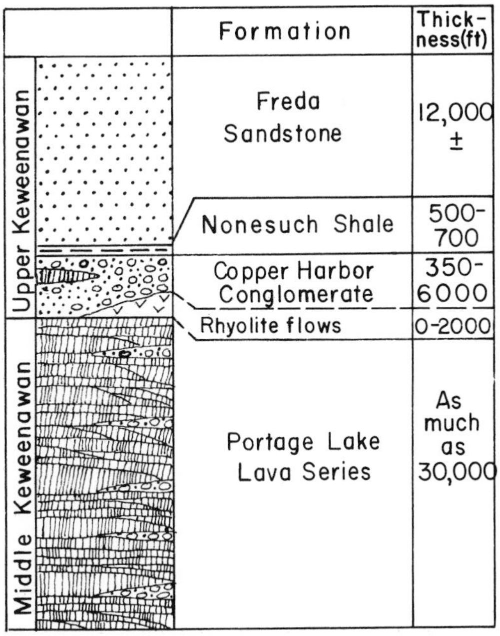

| Formation | | Thickness(ft) |
|---|---|---|
| Upper Keweenawan | Freda Sandstone | 12,000 ± |
| | Nonesuch Shale | 500-700 |
| | Copper Harbor Conglomerate | 350-6000 |
| | Rhyolite flows | 0-2000 |
| Middle Keweenawan | Portage Lake Lava Series | As much as 30,000 |

FIG. 2. *Columnar Section of the Keweenawan Series in the White Pine District.*

erally subordinate, despite the preponderance of mafic lava flows. Large clasts of pre-Keweenawan rocks, mostly quartzite and iron-formation, are rare at most places, but the abundance of quartz in the sand-size fraction suggests that the streams that deposited the conglomerates had headwaters in areas of pre-Keweenawan rocks outside the Keweenawan basin. Foreset beds and imbrication of pebbles show that these streams flowed generally north-ward toward the axis of the basin (25, 6). The famous native copper deposits of the Lake Superior region occur in some of the conglomerate beds and amygdaloidal lava flow tops of this series.

Throughout most of the Keweenawan province, the Copper Harbor Formation rests directly on mafic lava flows of the Portage Lake Lava Series. In the Porcupine Mountain area, on the other hand, the formation rests on an intervening sequence of rhyolite extrusives about 2000 to 3000 feet thick. In this same general area, the conglomerate overlying the rhyolite is locally less than 350 feet thick compared to its more common thickness of 3000 to 6000 feet. The observed relationships are most simply explained if the rhyolite flows formed a topographic high around a volcanic center on the surface underlying the conglomerate. The piles of rhyolite flows appears to be 10 to 15 miles across and possibly 3000 to 4000 feet high, centered a few miles west of White Pine; the area of the pile includes most of the present area of the Porcupine Mountains and the hills 5 to 8 miles south of White Pine. This eminence was completely buried by the beginning of Nonesuch time and was not an element in Nonesuch palegeography. Regardless of its origin, the eminence may be regarded as a buried hill surrounded and covered by Copper Harbor Formation (conglomerate).

The Upper Keweenawan of the White Pine area consists of three formations. They are, in ascending order, the Copper Harbor Formation, the Nonesuch Shale, and' the Freda Sandstone (Figure 2). The Copper Harbor conglomerate, which lies on the Portage Lake Lava Series, consists primarily of reddish-brown to grayish-brown fine-to-coarse-grained sandstone, conglomerate and minor reddish-brown siltstone. The formation locally contains a few scores to hundreds of feet of mafic lava flows (20). The main detrital constituents are quartz, volcanic rock fragments, both acidic and mafic, and feldspar; clasts of chert, jasper, and other crystalline rocks are present but rare. Crossbedding is very common. Studies by Hamilton (38) show that these sediments came from a south to southwesterly direction from a Precambrian igneous and metamorphic source area and were deposited in shallow water where current activity was strong. White and Wright (20) interpret the formation as a piedmont-fan deposit grading northward into finer grained, but thicker, flood-plain or lacustrine deposits. The copper Harbor conglomerate ranges from 350 feet or less south of White Pine to as much as 6000 feet near the shore of Lake Superior.

The Nonesuch Shale consists of a sequence of gray to brownish-gray siltstone and subordinate shale and sandstone. The principal rock type of the formation is a rippled and thickly to thinly laminated siltstone with reddish-gray partings. Gray to dark-gray siltstone predominates in the lower 100 feet of the formation. Similar rocks are found at intervals throughout the column, and make up about 10 per cent of the rock in the upper half of the formation. Thick beds of massive light-gray siltstone to very fine-grained sandstone are abundant in the lower-middle and also the uppermost parts of the formation.

The Nonesuch Shale is about 600 feet thick along a line that runs westward from the vicinity of White Pine to the northern part of the Presque Isle area. It thickens northward at a rate of about 25 feet per mile, partly by facies change and partly by tonguing with the formations above and below but probably mostly by thickening of individual stratigraphic units within the formation. The formation can be traced northeastward from White Pine for 70 miles to a point due north of Calumet, Michigan, where the base of the formation crosses the shoreline of Lake Superior. Its thickness and lithology are similar throughout this distance, insofar as they have been established by a few scattered drill holes and surface exposures, except that the basal 150 feet contains an increasing proportion of medium- to coarse-grained sandstone beds toward the northeast. West of the Presque Isle area scattered exposures in streams indicate that the outcrop trace of the formation forms a continuous belt to a point somewhere beyond Mellen, Wisconsin, 60 miles southwest of White Pine. The formation is about 380 feet thick where it crosses the Michigan-Wisconsin State line and 250 feet near Mellen. The ore-bearing beds of the White Pine mine are in the basal 60 feet of the Nonesuch Shale, which are described in greater detail in the section below tilted, "Stratigraphy of the Copper-bearing Rocks."

The Freda Sandstone consists of alternating layers of reddish fine-grained arkosic sandstone, siltstone, and micaceous silty shale. Hamblin (23,24) considers that the rocks were deposited on wide flats or flood plains by streams flowing northward. The minimum breadth of outcrop at the Michigan-Wisconsin state line, where the formation is almost vertical, is about 12,000 feet; the top is nowhere exposed, so this 12,000 feet is commonly taken as the minimum thickness.

Younger, more mature sandstones of late Keweenawan age crop out in Wisconsin (3, p. 1480–1482). In Michigan, the Jacobsville Sandstone, found only south of the Keweenaw fault (Figure 1), is probably contemporaneous with some of these younger formations.

## Intrusive Rocks

Some of the rhyolite in the rhyolite complex south of White Pine and in the Porcupine Mountains west of White Pine is surely intrusive, but insufficient geologic mapping has been done in these areas to establish the abundance or geometry of the intrusive bodies. Dike or sill-like masses have been observed in drill holes (2, Pl. 32) and scattered exposures. A distinctive quartz porphyry exposed in road cuts 9 miles south of White Pine seems to form a single large semiconcordant mass, perhaps 5,000 feet thick and 10 miles long, intruded into the Portage Lake Lava Series. Whole-rock Rb/Sr analyses suggest an age of 905 ± 53 million years for this rhyolite (G. Faure, written communication, November 4, 1965). The great bulk of the rhyolite exposed south and west of White Pine, however, is almost certainly extrusive in origin and, as noted above, lies stratigraphically between normal mafic lavas of the Portage Lake Lava Series and the Copper Harbor conglomerate. Irving (1, p. 91–107) and Thaden (5) describe these rocks in detail. Whole-rock Rb/Sr analyses of rhyolite from just beneath Copper Harbor conglomerate south of White Pine and from the main body of rhyolite in the Porcupine Mountains lie on the same isochron, which suggests an age of 1100 ± 30 m.y. (34). This is a little older than the apparent isochron age of 1046 ± 46 m.y. previously obtained by Chaudhuri and Faure (29) using the same method for the Nonesuch Shale. The domal shape of the mass of rhyolite in the Porcupine Mountains is due to tectonic deformation in post-Freda time, and not to intrusion of a "laccolith" of rhyolite, as has been suggested (2).

With two possible exceptions, all evidence points to a peak of igneous activity in Portage Lake time, followed by declining activity (increased rate of sedimentary to volcanic rocks) that ended at the close of Copper Harbor time (19). The first exception is a rhyolite plug that cuts the Freda Sandstone near Hancock, Michigan; this rhyolite is petrographically unlike the typical Keweenawan rhyolites and may be unrelated to them. The second possible exception is the quartz porphyry exposed south of White Pine. If the whole-rock Rb/Sr ages of this porphyry and the Nonesuch Shale, quoted above, are correct, the porphyry is younger than the basal Nonesuch Shale.

## Diagenesis

Postdepositional changes of upper Keweenawan sedimentary rocks in the White Pine area are better described as diagenetic than metamorphic. Sedimentary structures and fine clastic textures are unusually well preserved. The rocks have no metamorphic cleavage, except in a few local deformed areas where incipient fracture cleavage exists. X-ray diffraction studies of the fine siltstone show that well-ordered illite and chlorite (diffraction peaks 0.2° to 0.3° in width at half amplitude) are major components. These have probably been partly recrystallized and homogenized from a presumably heterogeneous detrital clay fraction of the original sediment. Microscopic examination shows that these minerals are extremely fine grained and have only a very weak and irregular preferred orientation. Sandstone is thoroughly cemented and indurated by calcite, chlorite, quartz (primarily as overgrowth), and locally laumontite. The clastic grains were mechanically crushed at many contact pressure points and the pore space was greatly reduced by intensive compaction, but the original sand grains were not conspicuously altered or homogenized, and new minerals were not formed except as cements.

An indication that the temperature of the Nonesuch Shale in the White Pine mine has never been very high is given by the presence of porphyrins in the organic fraction (31,34). Abelson (12, p. 85) presents a time-temperature curve suggesting the degree of stability of some porphyrins, calculated from the activation energy for experimental degradation of porphyrins in tar. This curve, though merely indicative, suggests that these porphyrins could not have survived a temperature as high as 200°C for 1000 years, 250°C for 100 years, or 300°C for 1 year.

## STRUCTURE

The Keweenawan Series of northern Michigan and Wisconsin lies on the south flank of the Lake Superior basin and generally dips north to northwest toward the center of the basin (Figure 1). A large high-angle reverse fault, the Keweenaw fault, thrusts the middle and upper Keweenawan rocks southward over still younger upper Keweenawan strata; it follows the south margin of a long prominent ridge of hills extending the entire length of the Keweenaw Peninsula. Southwest of White Pine, in the Wakefield Ironwood area, the Portage Lake Lava Series is in direct contact with older crystalline Precambrian metasedimentary rocks. Northwest of White Pine, the Porcupine Mountains mark an oval upwarp of the Keweenawan Series. Resistant felsic flows crop out in most of the mountain mass.

The regional dip of the upper Keweenawan strata, generally toward the axis of Lake Superior, is due partly to tilting contemporaneous with sedimentation, as indicated by northward thickening of stratigraphic units. Large dips due primarily to post-Freda deformation are, for the most part, confined to a narrow zone that only locally extends more than 2 to 3

miles down the dip from the outcrop of the Nonesuch Shale. The Keweenawan basin is a broad shallow syncline with sharply turned-up edges, more like an elongate pie plate than a bowl.

The Porcupine Mountain anticline (Figure 3) is a second-order fold located well inside the main upturned edge of the basin but very likely folded at the same time that the southern edge was thrust up. It is sharply asymmetric, with strata dipping 20° to 30° on the northern flank and very steeply or even overturned on the southern. At its eastern end, the anticline plunges off gently to the east so that its influence on surface attitudes, at least, disappears about 10 miles east of White Pine. To the west, the axis curves southward and joins the margin of the Keweenawan basin almost at right angles in an area of extraordinarily poor outcrop. South of the Porcupine Mountain anticline, the Nonesuch Shale underlies an elongate structural basin, the Iron River syncline. The Presque Isle syncline is a westward-plunging fold on the west side of the Porcupine Mountains. The base of the Nonesuch Shale is about 4500 feet deep along the axis close to the northern edge of the Iron River syncline and is about 1050 feet below the level of Lake

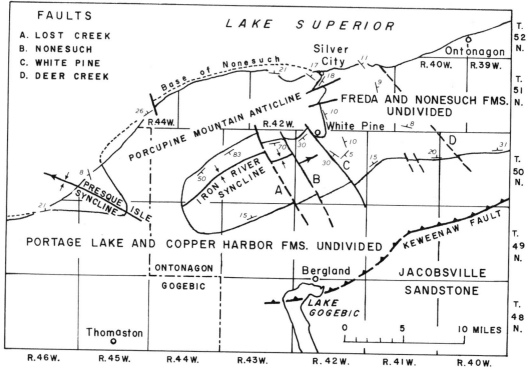

FIG. 3. *Geologic Sketch Map of the White Pine-Porcupine Mountain District.*

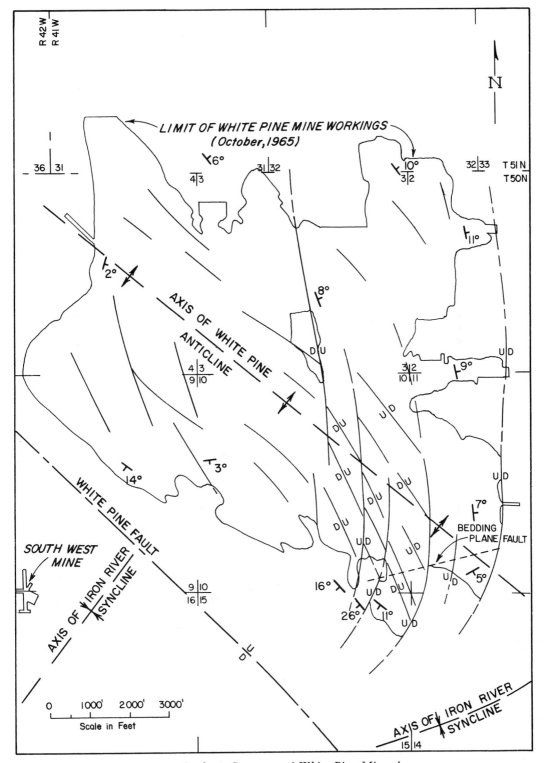

FIG. 4. *Geologic Structure of White Pine Mine Area.*

Superior where the axis of the Presque Isle syncline crosses the shore of Lake Superior.

Superposed on these second-order folds are folds of a lower order that are the result of drag along the major faults of the region. The most conspicuous and best documented example adjoins the White Pine fault (Figure 4), a reverse fault zone trending northwest and dipping steeply northeast. Gently dipping rocks northeast of the fault zone are generally 1500 to 2000 feet higher than equivalent strata to the southwest. The horizontal component of movement on the White Pine fault zone involves right-lateral displacement of the order of 3000 feet as estimated from the offset of a number of isopachs and lithofacies boundaries along the fault.

On the northeast side of the White Pine fault, drag on the bedding adjacent to it has produced the very asymmetric White Pine anticline. The axis of the White Pine anticline, nearly parallel to the fault, plunges 9° towards the southeast. Strata on the southwest limb are dragged to steep attitudes (in excess of 20°) near the fault, but beds on the northeast limb have gentle dips of 10° or less that gradually merge with regional attitudes.

The White Pine fault is accompanied by countless minor faults, which have been mapped in great detail in the mine workings because of their effect on operations. The larger of these faults are shown in Figure 4. Measured displacements range from a fraction of an inch to many tens of feet. Displacement may be primarily dip slip, strike slip, or diagonal. Northeast of the White Pine fault, normal fault movement is far more common than is reverse; most of the faults dip steeply and generally parallel each other. The prevailing strike well away from the White Pine fault is about N10°W, but as these faults are followed southward, within 1000 feet or so of the White Pine fault they curve rapidly to the right as though to merge with the White Pine fault. This sense of curvature is consistent with the right-lateral displacement on the White Pine fault.

A set of faults with vertical displacements generally less than 5 feet is oriented northwest nearly parallel to the anticlinal axis. These faults dip steeply and most have normal offsets. Some of them in the southeast part of the mine workings (Figure 4) appear to have been rotated from a N45°W strike to a N30°W strike in blocks distorted by the N10°W strike set.

Underground exploration southwest of the White Pine fault suggests that reverse faults of small displacement at a low angle to bedding are common. This may be an expectable consequence of compression in incompetent beds overlying competent beds in the core of a syncline. Meager data also suggest that more steeply dipping faults on the downthrown side of the White Pine fault are parallel to that fault.

Several major faults subparallel with the White Pine fault are shown on Figure 3. Their position is inferred from surface exposures, from aligned linear gaps through the bedrock ridges to the south and through the Porcupine Mountains, and from widely spaced drill-hole control. Among these faults are the Nonesuch, Lost Creek, and Deer Creek (Figure 3). The Lost Creek fault is probably comparable to the White Pine fault in sense and magnitude of motion; other indicated faults of the same group also may have large displacements. These faults strike about perpendicular to the south limb of the Lake Superior syncline and clearly belong to the set of steeply dipping fractures that cut the south limb of that syncline throughout its length (2, p. 49–50). Both the synclinal folds and the cross faults deform the Freda sandstone and so postdate its deposition. It seems probable that the cross faulting was concomitant with the main deformation of the region.

Two cross faults striking about N70°E on the east flank of the Porcupine Mountains (Figure 3) are defined by both stream exposures and by fairly closely spaced drill holes. On the northern fault, a large component of horizontal slip, probably right-lateral, is inferred from significant differences in the strata on opposite sides of the fault.

## STRATIGRAPHY OF THE COPPER-BEARING ROCKS

Copper-bearing rocks of the Copper Harbor Formation and the Nonesuch Shale comprise four informally named units of local usage (Figure 5): the Lower Sandstone, the Parting Shale, the Upper Sandstone, and the Upper Shale. These units, which aggregate about 40 to 80 feet thick, are recognized throughout the White Pine-Porcupine Mountain area. At the White Pine mine, individual beds within these units are identified by an unambiguous sequential numbering system (31, Figure 5). The first digit identifies the informally named unit of local usage ("1" and "3" for Lower and Upper Sandstone, respectively; "2" and "4" for Parting and Upper

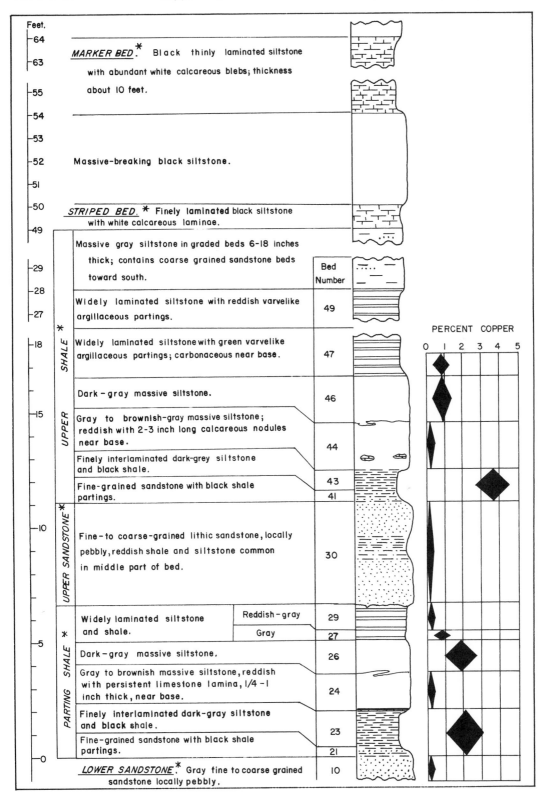

FIG. 5. *Columnar section of Copper-Bearing Rocks.*
\* Denotes names of local usage.

Shale). The second digit denotes the particular bed. Even-numbered beds are massive; odd-numbered beds, laminated. Not all possible numbers are used; for example, bed No. 21 is characteristically succeeded by bed No. 23 (both laminated, thus odd-numbered) without an intervening massive even-numbered bed.

Uniform copper tenor is a striking characteristic of each bed (Figure 5). The distribution of copper shown on Figure 5 represents data from nearly 300 drill holes at White Pine on both sides of the White Pine fault. The mean copper content is indicated by the position of the upper and lower corners of the diamond. One standard deviation on either side of the mean is included between the left and right corners; the spread of values indicated by the diamond includes about two-thirds of all the assays for each bed. It is a notably small range; for example, two-thirds of the samples from the laminated beds at the base of the Parting Shale (beds Nos. 21 and 23) range from 1 to 3 per cent; very few samples have extremely high or low copper content.

## Petrography

It is convenient to describe petrography common to both sandstone units and to both siltstone units before describing their individual stratigraphic features. The sandstone is immature volcanic-rich lithic arenite. Although volcanic detritus is abundant, detrital quartz equals or exceeds it in bulk. Clastic opaque iron-oxide grains, which are common as accessories in the main mass of the Copper Harbor conglomerate, are rare or absent in sandstone of the copper-bearing beds. The sandstone ranges from sparsely pebbly, poorly sorted, medium- and coarse-grained to moderately sorted, fine- and medium-grained. Chlorite, as radial blades rimming clastic grains and as cement in pore space, imparts a greenish-gray color. Calcite and quartz are also common cements. Locally black solid carbonaceous matter, with or without associated native copper, is an abundant cement. The sandstone of the copper-bearing rocks appears to be derived largely by reworking from the main mass of the Copper Harbor just beneath.

Most of the siltstone consists of minute angular detrital fragments of quartz, feldspar, and subordinate volcanic rock fragments set in an abundant matrix of chlorite, sericite, and microcrystalline quartz. Wiese (21) estimated the bulk mineralogy from X-ray diffraction analyses and from normative analyses computed from chemical analyses:

| | Approximate Per Cent |
|---|---|
| Quartz | 20–30 |
| Chlorite | 20–30 |
| Plagioclase | 15 |
| Muscovite | 10–15 |
| Orthoclase | 5 |
| Epidote | 4 |
| Hematite | 0–5 |
| Copper minerals | 0–8 |
| Carbonaceous matter | 0.5 |
| Other minerals | 7 |

This modal analysis shows the predominance of quartz, chlorite, and silicate minerals derived from volcanic detritus. Chemical analyses of the Nonesuch closely resemble analyses of a "typical" black shale such as Cretaceous Pierre Shale (17, p. B450) in silica and alumina content. Magnesia, soda, titania, and total iron-oxide content of the Nonesuch, however, are all more than double their abundance in the Pierre, undoubtedly because these elements are abundant in the volcanic rock detritus. Trace metals, except for abundant copper and silver, are remarkably like those of representative black shales without metallic mineralization (33, Table 1). The Kupferschiefer black shale, in contrast, contains more than a hundred times as much copper, lead, zinc, silver, and arsenic as representative black shales and about five times as much vanadium, cobalt, and molybdenum (33, Tables 1, 5, and 6). The organic carbon of the Nonesuch Shale, as in the Pierre Shale, amounts to only about 0.5 per cent.

## Lower Sandstone (of local usage)

Bed No. 10, the greenish-gray sandstone capping the Copper Harbor Formation, is the lowest unit that commonly contains copper at White Pine. Its basal contact with the underlying red sandstone typical of the Copper Harbor Formation is gradational. It ranges from 5 to 40 feet thick; it is commonly 15 to 20 feet thick. Large trough sets of crossbeds and scour channels are common; local accumulations of carbonaceous cement and native copper are characteristic. These sediments of the Lower Sandstone were probably reworked from the upper Copper Harbor in shallow subaqueous beaches, bars, shoals, and deltaic distributaries.

In the area of the Southwest mine (Figure 4), Hamilton (37) showed that the uppermost

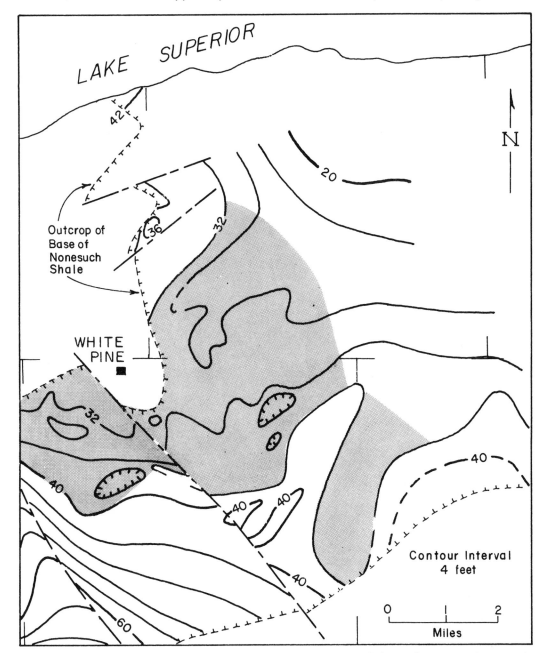

FIG. 6. *Inferred Paleotopography at the Base of the Nonesuch Shale. Embayment referred to in text is shaded.*

1 to 3 feet of the Lower Sandstone are fine grained, moderately to well sorted, fairly quartzose, chloritic, partly flat bedded, and partly crossbedded. This finer-grained sandstone contains very little copper.

The irregular topographic surface of the Lower Sandstone critically controlled the distribution of the overlying Parting Shale, par-

ticularly the thickness and lithology of the No. 23 bed. Figure 6 is a map of the inferred topography; the arbitrary datum is a surface 80 feet stratigraphically below the top of the Striped bed of local usage (Figure 5). Interpreting this as a paleotopographic map involves two assumptions that appear empirically valid: that the top of the Striped bed was a nearly level surface, and that compaction—particularly differential compaction of sandstone and siltstone—does not introduce serious distortion.

The old shoreline at the end of Lower Sandstone deposition, located from slope changes and other data, was probably near the 40-

and 44-foot contours. A large shallow embayment (shaded on Figure 6) straddles and is offset by the White Pine fault. It is partly enclosed on the southeast by a high interpreted as an old deltaic distributary or sandbar, on the south and southwest by the toe of what may be an ancient alluvial fan, and on the north

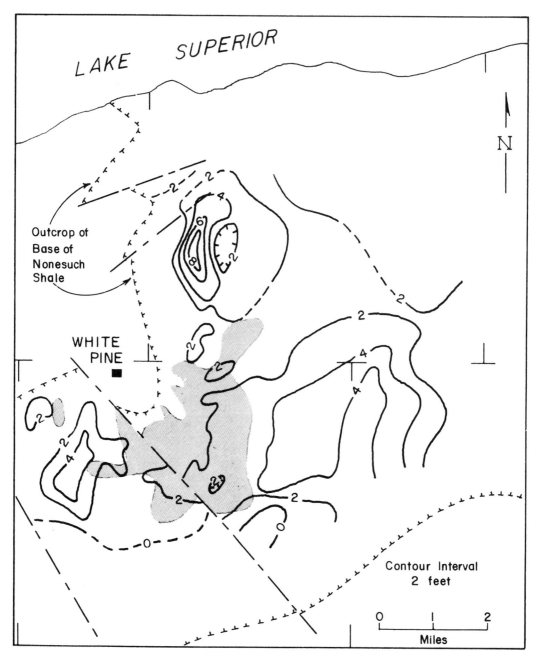

FIG. 7. *Thickness and Character of Combined Nos. 21 and 23 Beds. Silty No. 23 bed is thicker than sandy No. 21 bed within shaded area.*

by a high interpreted as an ancient shoal, sand-bar, or spit. It is open to the northeast where it appears to merge with a larger basin. The preserved basin is about 3.5 miles broad (northwest to southeast), 5.5 miles long, and less than 10 feet deep. The basin was the locus of deposition at White Pine of the thickest, finest-grained, most carbonaceous, and ulti-mately most cupriferous sediments.

## Parting Shale (of local usage)

This dark-gray evenly laminated silstone is commonly about 8 feet thick in the White Pine area and ranges from 0 to 30 feet thick in the western Presque Isle syncline. It pinches out southward fairly near the 40-foot contour of Figure 6. Near this pinchout, the Parting Shale is reddish, rather coarse grained, and its individual beds are not distinguishable.

The basal beds of the Parting Shale are laminated, prominently contrasting sand-silt and carbonaceous shale. The lower laminated bed (No. 21) is a relatively coarse sandstone, transitional from the Lower Sandstone, that contains conspicuous widely to closely spaced discontinuous carbonaceous siltstone partings. The No. 23 bed comprises gray siltstone ir-regularly interlaminated with dark carbona-ceous clayey siltstone. The combined beds are predominantly nonsandy and 2 to 4 feet thick in the central part of the basin (shaded on Figure 7). They become thick and sandy near the sandstone highs of the Lower Sandstone (Figure 6).

The middle of the Parting Shale comprises two massive-breaking siltstone beds. The faintly reddish lower bed, No. 24, rests on No. 23 bed with a sharp abrupt contact. Near the base is a distinctive thin calcareous bed, only 0.1 to 0.5 inch thick, which is present throughout the White Pine-Porcupine Moun-tain area and beyond. The upper No. 26 bed contains very fine and even carbonaceous lami-nae but is nonfissile and massive on outcrop. Each of the massive beds is typically 1.5 to 2.0 feet thick, but No. 26 bed thins erratically, in part due to post depositional slumping, near the highs on the Lower Sandstone surface.

The two top beds of the Parting Shale are siltstone, prominently, evenly, and widely inter-laminated with clayey siltstone. The beds are about 1.5 feet thick near White Pine, thicken-ing to more than 5.0 feet about 6 miles north-east of White Pine. In the lower bed (No. 27) the clayey siltstone is greenish gray; in the upper (No. 29), reddish brown. The silt-stone laminae are generally about 0.5 to 2.0

inches thick; the clayey siltstone laminae, less than an inch thick. The top of the upper bed is locally mudcracked.

The Parting Shale probably was deposited in a single epoch of transgression and infilling. The basal laminated beds are thought to have been deposited during the transgression in very shallow, poorly aerated water intermittently agitated by tides, storms, and deltaic currents. The middle massive beds probably record the initial stage of infilling as transgression reached its maximum; a stagnant reducing environment that permitted preservation of organic matter may have been enhanced by existence of off-shore bars or by a layering of the water. The final infilling by weak prograding deltaic cur-rents in shallow but protected quiet water is thought to have yielded the upper evenly lami-nated beds; mudcracks at the top suggest that sedimentary infill had reduced water depth to nearly zero at the end of Parting Shale deposi-tion.

## Upper Sandstone (of local usage)

No. 30 bed is a pebbly medium-grained chloritic, volcanic-rich lithic arenite, very simi-lar to the Lower Sandstone. The mine workings at White Pine magnificently expose north-northeast-trending channels several hundred feet broad and 0.5 to 2.0 feet deep (locally deeper) that have been described by Trammell and Ensign (39). Crossbeds, erosion channels, scour-and-fill structures, toroids, and groove casts document the strong currents that eroded and filled the channels. Subdued channel slope and cross-sectional profiles and the abundance of cuspate interference ripple marks suggest that the Upper Sandstone in the mine area represents an interplay of fluvial and shallow water processes.

The Upper Sandstone ranges erratically in thickness and grain size. But regionally it thins northeastward from about 4 to 8 feet thick southwest of the White Pine fault to 2 to 5 feet thick in the area of the present workings northeast of the fault. It is locally thicker in channel cuts; farther northeast it grades into widely laminated siltstone with reddish shale partings. Some very fine grained sandstone in the northeasternmost area was probably de-posited as offshore bars.

## Upper Shale (of local usage)

The Upper Shale and the sequence of beds within it are closely comparable to the Parting Shale and its component beds. The two units apparently record similar cycles of transgression and infilling. The subtle differences of Upper Shale from Parting Shale consist chiefly of a smoother more regular base, more regularly laminated beds, and predominance of finer-grained, more argillaceous, more carbonaceous sediments of the Upper Shale. The

top surface of the Upper Sandstone was aggraded or beveled to a very smooth surface when renewed transgression ended sandstone deposition. The basal laminated beds (Nos. 41 and 43) of the Upper Shale are thinner (Figure 8), areally more uniform in thickness, and in detail more regularly laminated than

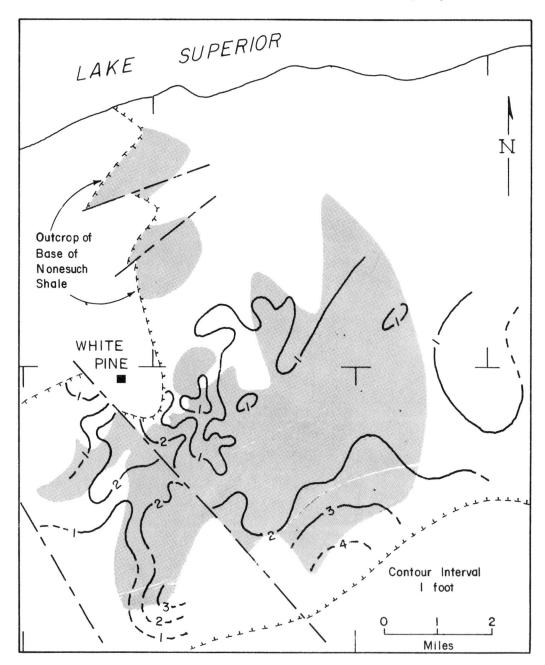

FIG. 8. *Thickness and Character of Combined Nos. 41 and 43 Beds. Silty No. 43 bed is thicker than sandy No. 41 within shaded area.*

the corresponding beds in the Parting Shale. The massive siltstone beds (Nos. 44 and 46) are also slightly thinner and areally more uniform than their Parting Shale counterparts. Near the base of the No. 44 bed is a zone of brown oblate calcareous nodules averaging 2 to 3 inches in their longest dimension, perhaps a counterpart of the limestone layer at the base of No. 24 bed. The widely laminated beds of the Upper Shale (Nos. 47 and 49) have prominent very agrillaceous layers, fairly abundant carbonaceous matter, and a distinct, very even, almost varvelike lamination. No. 47 bed is greenish-gray and commonly 8 feet thick; No. 49 bed is reddish-gray and about 3 to 5 feet thick. These are succeeded by a sequence about 15 feet thick of massive, graded siltstone beds, each about 6 to 18 inches thick; no numbers have been assigned to this unit. Fine- to coarse-grained sandstone is prominent in the upper part of this unit toward the south.

Very calcareous and carbonaceous silty shale succeeds the Upper Shale. The basal 1-foot-thick Striped bed of local usage and the slightly higher 10-foot-thick blebby Marker bed of local usage are easily recognized units used for control of exploratory drilling. We have used the top of the Striped bed as a datum plane in our stratigraphic studies because it was probably the smoothest, most nearly horizontal sharply defined horizon stratigraphically close to the ore zone.

## FORM OF THE ORE BODIES

The lowermost beds of the Nonesuch Shale contain significant amounts of copper over many tens of square miles. An ore body exists where one or more of these basal beds, taken together, constitute an ore column thick and rich enough to mine. Ore bodies, therefore, have the form of thin blankets, square miles in extent and generally 4 to 25 feet in thickness, conforming to the structure defined by the base of the gently folded Nonesuch Shale. Differences in the thickness and grade of an ore body represent the net effect of differences in the thickness and grade of individual beds making up the ore body.

### Copper Content of Individual Beds

Figure 5, as described above, shows the characteristic range in grade for each of the beds of the lowermost part of the Nonesuch Shale in the area of the White Pine mine. The occurrence of richest copper in the dark-colored carbonaceous beds is remarkably consistent throughout the district. Beds Nos. 21, 23, 26, 41, 43, and 46 are notably more copper rich than the intervening beds (32), not only in the White Pine area but also along the southern flank of the Porcupine Mountains and in the Presque Isle area. Even within beds, some differences from top to bottom are characteristic. In the No. 23 bed, for example, black shaly laminae near the top and bottom of the bed are typically richer than the lighter-colored, coarser laminae in the center of the stratum. Copper-rich laminae in the No. 26 bed maintain their richness through convolutions due to prelithification slumping.

On a district-wide basis, changes in grade of individual beds tend to be gradual and can generally be correlated with changes of sedimentary facies. Figure 9 shows the areal distribution of copper content in the combined Nos. 21 and 23 beds, and Figure 10 is a similar map for the Nos. 41 and 43 beds. Areas wi.h less than 20 pounds copper per ton in Figure 9 coincide, for the most part, with places where the bed is thin, reddish, and sandy and overlies ancient topographic highs in the underlying sandstone surface (Figure 6). The Nos. 41 and 43 beds (Figure 10) tend to have lower-than-average grades where they are notably sandy.

Grades in the other two rich beds, Nos. 26 and 46, are less evenly distributed. Abrupt local changes in the content of copper minerals can be observed underground in the No. 26 bed.

### Cupriferous Zone

The lowermost bed or beds of the Nonesuch Shale contain copper sulfides at most places, whereas pyrite is the characteristic sulfide of the hundreds of feet of Nonesuch Shale above these copper-bearing beds. It is convenient, therefore, to recognize a "cupriferous zone" and a "pyrite zone," terms that are used here in a nonstratigraphic sense to distinguish geologic bodies with notably different copper content and sulfide mineralogy. The boundary between these two zones is independent of lateral facies changes and transgresses bedding on a regional scale.

Within the cupriferous zone, the range in copper content from bed to bed is very large, as indicated in Figure 5. Some dark shale beds characteristically contain 2 to 4 per cent copper, whereas others, particularly those with a reddish cast, may have average copper contents as low as 0.05 per cent. The average

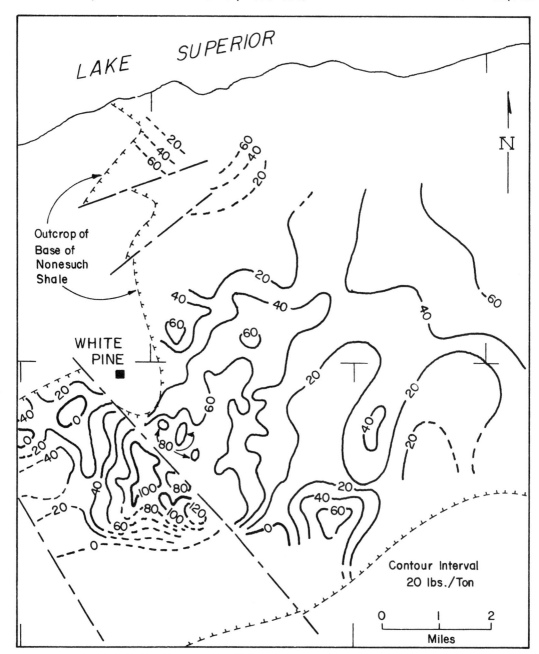

FIG. 9. *Grade of Copper in Combined Nos. 21 and 23 Beds.*

copper content of all beds in the pyrite zone, on the other hand, is between 0.01 to 0.02 per cent copper, and assays exceeding 0.04 per cent copper are rare except within a few feet of the boundary of the cupriferous zone.

These contrasts apply to individual beds, as well as to the zones taken as a whole. The top of the cupriferous zone, that is, the boundary between the cupriferous and pyrite zones,

is a distinct surface that cuts across bedding on a regional scale (Figure 11). The same bed, therefore, may lie within the cupriferous zone in one place and outside it in another. Where a particular bed lies within the cupriferons zone, it commonly has a copper content in the range indicated for that bed by the values in Figure 5. Where it lies outside the cupriferous zone, on the other hand, its average copper content generally is less than 0.02 per cent, no matter what the grade of the same bed within the cupriferous zone. Beds richest in copper (for example, No. 43 bed) commonly contain 100 to 400 times as much copper inside the cupriferous zone as they contain outside. Even the leanest beds, such as the red siltstone just above No. 43 bed, contain at least four times as much copper inside the cupriferous zone as they contain outside. The contrast between the copper content of a given bed in the two zones is, therefore, greatest for the dark-gray shale beds, but it is readily apparent even in those beds that are leanest within the cupriferous zone.

Detailed comparison of rocks inside and outside the cupriferous zone requires a precise definition of the top of the zone. Where assay

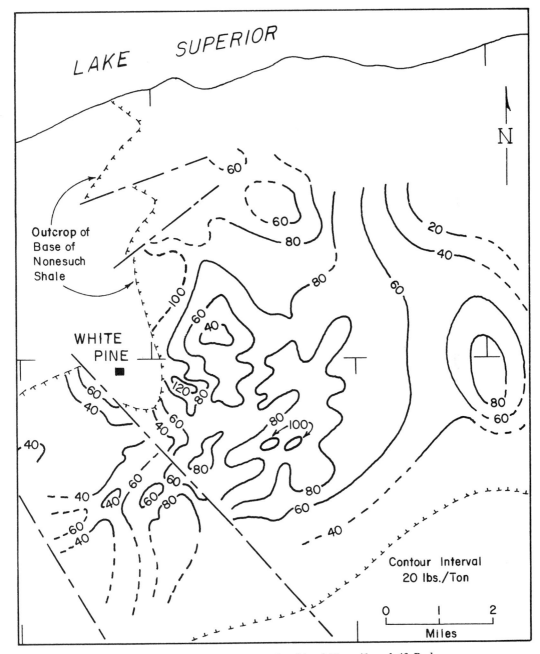

FIG. 10. *Grade of Copper in Combined Nos. 41 and 43 Beds.*

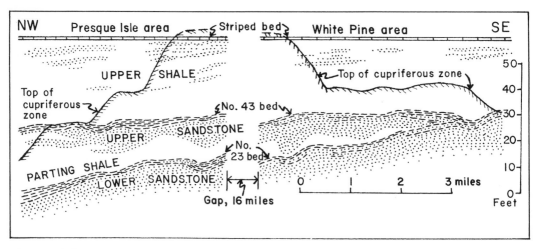

Fig. 11. *Schematic Cross-Section showing Form and Regionally Transgressive Top of Cupriferous Zone. The left part of the section follows the axis of the Presque Isle syncline; the right part crosses the White Pine ore body southwest of the White Pine fault.*

data are available for the rocks above the cupriferous zone, the top is arbitrarily taken at the top of the stratigraphically highest sample assaying 0.2 per cent copper or more. If assay data are not available, the top is placed just above the stratigraphically highest bornite, where present, or just below the lowest greenockite or pyrite where no bornite is observed under a 27-power binocular microscope. The position of the top defined by one of these methods commonly lies within a foot or so of the position defined by the other method, and a difference of 5 feet between the two positions is exceptional.

## MINERALOGY OF THE DEPOSITS

### Native Copper

Native copper is locally abundant, particularly in the Lower Sandstone and in the lowermost beds of the Nonesuch Shale. It probably constitutes from 7 to 9 per cent of the copper in current mill feed. Native copper is the only copper mineral in some laminae, but it is more commonly associated with chalcocite, generally as separate grains. Boundaries between rock with low and high ratios of native copper to chalcocite cut across bedding in many places. The average size of native-copper grains is notably larger than that of the sulfide minerals described below and ranges from 1.3 mm in sandstone to about 0.2 mm in finely laminated massive siltstone (27, Figure 12). Grains of native copper, highly elongate in the plane of the bedding, are common. Thin sheets of the metal up to 2 to 3 feet in length are fairly common in bedding planes in the No. 23 bed; locally these sheets cut across bedding.

### Native Silver

The only silver mineral that has been identified is the native metal. Detailed polished-section studies of the mineralogy of the top of the cupriferous zone by A. C. Brown (40) have revealed no silver-bearing sulfides, and none has been found in veins. The most conspicuous native silver occurs with native copper. Sheets and grains of native copper are commonly rimmed with silver—the opposite relationship has not been observed. Silver also occurs as discrete grains, with the same general habit as native copper, both in rocks with native copper and in rocks with only chalcocite. Silver tends to be rich in beds that are also rich in copper, particularly native copper, but its distribution is far more erratic than that of copper.

### Chalcocite

Chalcocite, finely disseminated in siltstone and shale, is the principal source of copper in the Nonesuch Shale. Blue and white varieties coexist, both as separate particles and together in the same grain. In sandstone and siltstone, chalcocite grains have irregular shapes and are primarily interstitial to the coarser detrital grains, but the larger chalcocite grains, in particular, have made some space for themselves by replacing the rock minerals. Wiese (21) concluded that much of the chalcocite replaces fine-grained matrix chlorite.

The size of most of the chalcocite grains in shale and siltstone ranges from less than 1 micron to about 200 microns, with an average size commonly somewhere in the range of 2 to 20 microns. The total number of grains less than a micron in diameter is very large, even though this very fine-grained chalcocite accounts for a small fraction of the chalcocite by weight. In sandstones, grains 1 mm or more in diameter are common.

Chalcocite also occurs in oblate ellipsoidal knots or lumps 2 to 50 mm across, generally elongate parallel to bedding. These knots, though quantitatively insignificant, are of interest because they are commonly surrounded by a poorly defined zone with earthy hematite and may enclose tiny euhedral plates of hematite. Some replacement of individual grains of the host rock is apparent. Layering in the enclosing rock is commonly deflected around these knots, suggesting that they were formed before the rock was fully compacted.

Figure 5 shows the strong tendency for the chalcocite to be concentrated in certain beds. Within these individual beds, there is a tendency for copper to be more abundant on certain laminae than on others. These differences do not appear to persist for long distances, as they do in the Kupferschiefer (9a, p. 20); an individual fine lamina may contain more chalcocite than its neighbors for a few inches, less commonly for a foot, beyond which point some of its neighbors are richer.

## Other Copper Sulfides

The principal copper minerals at the top of the cupriferous zone are bornite (both brownish and purplish varieties) and chalcopyrite. Covellite is present but not abundant. A. C. Brown (40) has also identified digenite and djurleite in these rocks.

Bornite and chalcopyrite have a mode of occurrence and textural relationships similar to those of the chalcocite. Individual grains are a little coarser, on the average, than those of chalcocite but, like the grains of chalcocite, tend to have irregular shapes.

## Pyrite

Pyrite is an ubiquitous constituent of dark shaly beds throughout the Nonesuch Shale

above the cupriferous zone. Most of the pyrite occurs as tiny euhedral grains, ranging from about 1 mm across to less than a micron, disseminated sparsely to abundantly through the rock. Grain sizes in the range 2 to 5 microns are particularly common. In particularly dark-gray to black shale, pyrite constitutes 2 to 3 per cent of the rock; in most of the dark-gray shale and siltstone pyrite probably lies in the range 0.5 to 1 per cent. Some layers contain grains that are loosely to closely grouped in more or less spherical aggregates 20 to 50 microns across, some of which suggest framboidal clusters.

Wiese (21) has suggested that pyrite is a syngenetic constituent of the shales and that the other sulfides were formed by reaction between introduced metals and the sulfur of the pyrite. The similar sulfur content of dark shale beds (1 ± 0.5 per cent), whether the sulfide is pyrite or chalcocite, supports Wiese's hypothesis. Quite apart from reasoning based on the worldwide association of pyrite with black shale, the presence of pyrite in dark shaly layers through hundreds of feet of Nonesuch Shale above the cupriferous zone, throughout an area of several hundred square miles, is a strong argument for a syngenetic origin of the pyrite. Sulfur isotope data are available for chalcocite from White Pine (27, p. 660); determinations by Jensen show the broad range in $S_{32}/S_{34}$ ratios that Jensen (13, p. 389) regards as characteristic of sedimentary sulfides.

Wiese's hypothesis is fully supported also by the upward sequence—chalcocite-bornite-chalcopyrite-pyrite—observed at the top of the cupriferous zone. Pairs of minerals adjacent to one another in this sequence are found together in individual layers, and specimens in which pairs of noncontiguous minerals (that is, chalcocite-chalcopyrite, bornite-pyrite) occur are uncommon. Pseudomorphs of chalcocite after pyrite, clusters of chalcocite grains similar in configuration to observed pyrite clusters, and pyrite crystals rimmed by bornite, though quantitatively unimportant, have all been observed in specimens from the upper part of the cupriferous zone.

The great bulk of the chalcocite in the cupriferous zone, however, shows no textural evidence whatever to suggest that it formed by direct replacement of pyrite. The grain size of the chalcocite shows a much wider range, on the whole, than that of pyrite, and very few grains, indeed, have the regular polygonal shapes that are so common among the pyrite grains. If Wiese's hypothesis is correct, therefore, it is necessary to assume either that when copper sulfide grains grew at the expense of pyrite they did not do so by replacement, or that copper sulfide replaced a primitive form of iron sulfide.

## Greenockite

Greenockite is a common but not abundant constituent of rocks in a zone a few inches to a few feet thick, at or close above the top of the cupriferous zone, particularly in rocks that are quite pyritic (19, p. 408). It tends to form clots ranging from a fraction of a millimeter to 20 mm long. Within these clots, the greenockite is either a cement or replaces primary matrix; it makes up only 10 to 50 per cent of the bulk of the clot. Like the copper minerals, greenockite is more abundant in some beds than others but does not appear to be persistent in individual laminae.

## Sphalerite

White to cream-colored sphalerite has been observed in layers that also contain greenockite, and it has the same general habit. It is far less common than greenockite. According to X-ray diffraction studies by W. C. Kelly (written communication, 1964), some of this sphalerite shows faint wurtzite lines, but we have not found this to be characteristic of all the sphalerite.

## Galena

Galena is a rare constituent of the shale. In drill core, it has been observed as thin film-like grains coating bedding planes.

## Rare-Earth Carbonates

Unlike all the foregoing minerals, which have been observed both disseminated in the rocks and as constituents of veinlets, the rare-earth carbonates, synchisite and bastnaesite, have been identified only from veins.

## Mineral Zoning

The existence and significance of mineral zoning in the cupriferous zone have been described by White (19, p. 407–408). The upward sequence, native copper-chalcocite-bornite-chalcopyrite-pyrite defines a pattern of mineral zones that are essentially parallel to the top of the cupriferous zone. Figure 12 shows the general spatial sequence for the individual minerals described above. Sphalerite, galena,

and the rare-earth carbonates occur in such small amounts that their precise range, particularly their upper limits, is not known.

The vertical dimensions of the various zones differ from place to place. A native-copper zone is present in the lower part of the cupriferous zone, especially in the Lower sandstone, in the White Pine area. In the western part of the ore body, close to an area where the top of the cupriferous zone is in or above the Striped bed, native copper occurs in the No. 47 bed. The upper limit of native copper, as is true of the top of the cupriferous zone, is found at progressively lower stratigraphic horizons toward the east and southeast, so that over most of the area of the White Pine mine, no native copper occurs in the Upper Shale. In areas where the cupriferous zone is only a few feet thick, there generally is no native copper zone.

Chalcocite is the dominant copper mineral almost to the top of the cupriferous zone, and, where the top is in a dark shale bed, the entire upward sequence chalcocite-bornite-chalcopyrite-pyrite may be encompassed in a few inches stratigraphically. Where the top is in rather massive light-gray siltstone, such as is commonly found in the 20 feet or so immediately below the Striped bed, the zone of transition from chalcocite to pyrite may be several feet thick and poorly defined.

The upper limit of chalcopyrite and the non-cupriferous minerals is not known. Greenockite commonly occurs in highly pyritic beds like No. 43 bed and the Striped bed, even where these lie 10 to 30 feet above the top of the cupriferous zone, but is most abundant in these beds where they lie at or just above the top of the cupriferous zone. Galena and chalcopyrite, however, are present in trace amounts hundreds of feet above the cupriferous zone and may be almost ubiquitously present throughout the section in these amounts.

## Native Copper in Sandstone

Native copper is locally abundant in sandstone, particularly the Lower Sandstone. Whether this copper is part of the mineral-zoning sequence is open to question. Where sandstone copper is abundant, the total copper content of the whole cupriferous zone is notably higher than normal; sandstone copper, therefore, does not simply represent copper that has been redistributed from the immediately overlying or underlying shale. These concentrations of copper in sandstone lie in a narrow belt on both sides of the White Pine fault, suggesting a relationship to the fault (8, p. 706–708), but also occur at a distance from known major faults, as at the old Nonesuch mine. Hamilton (38) concludes from a detailed study of the Lower sandstone on both sides of the White Pine fault that the occurrence of native copper is controlled by the occurrence of interstitial carbonaceous matter and that this, in turn, is closely related to the lithology and, hence, to the environment of deposition of the sandstone. Carpenter's (27, p. 663–664) study of alteration halos immediately adjacent to cross-cutting veins, discussed below, suggests that the sandstone mineralization antedates the veins, which are related to movement on the White Pine fault. It is possible, therefore, that the proximity of areas of sandstone copper to the White Pine fault is fortuitous.

## VEINS

Veins containing copper minerals cut the White Pine ore body in many places. Though they have no economic importance, they are of interest for their bearing on the time of ore mineralization. The veins represent material introduced along joints and small faults. They range from less than an inch to several feet in width and may persist for several thousand feet along the strike.

R. H. Carpenter (27) has studied these veins in some detail in one area of the White Pine mine close to the White Pine fault where they are particularly abundant. The following description and interpretation are based partly on Carpenter's work and partly on our own observations.

The veins are primarily calcite veins in which metallic minerals and other gangue minerals are generally subordinate. Chalcocite is the principal metallic mineral in the veins where they cut cupriferous strata. Its crystal form is orthorhombic where observed, indicating formation below 105°C. Other metallic minerals in veins cutting cupriferous strata include native copper and silver, bornite, chalcopyrite, covellite, and pyrite. The paragenetic sequence tends in the direction of decreasing copper-iron ratio. In pyritiferous strata above the top of the cupriferous zone, vein minerals include bornite, chalcopyrite , greenockite, sphalerite, galena, and rare-earth carbonates. In addition to dominant calcite, observed gangue minerals include quartz, dolomite, barite, potassium feldspar, chlorite, illite, and montmorillonite.

Where chalcocite veins cut beds containing native copper, they are commonly bordered by halos a fraction of an inch to 20 inches wide that contain no native copper. Within the halos, sulfur content decreases and copper content increases away from the vein wall.

Carpenter (27) concluded that the copper in the veins was obtained by lateral secretion from contiguous strata. Evidence for this conclusion is the scarcity of ore minerals in veins where they cross barren beds, halos of reduced copper content where veins cross copper-rich beds, and rock fragments apparently leached of copper minerals in fault breccias.

Whether the veins accompanied or followed the formation of the strata-bound ore, it seems evident that they were not significant "feeders." Where veins cut barren beds they tend to be barren, and where they cut ore-bearing beds the copper in the veins is largely balanced by the copper deficiency of adjacent halos, so the net effect of veins on total copper content of a given block of ground is negligible. Outside the halos, the grade of strata-bound ore does not appear to be related to distance from veins and faults (8, p. 708–712). Only in a narrow zone a few feet thick just above the cupriferous zone is there evidence that copper has been introduced along veinlets. Here bornite veinlets cut pyritiferous strata, but copper in the form of chalcopyrite extends laterally away from the veinlets for 1 or 2 cm. Greenockite is found in veinlets only where the contiguous strata contain disseminated greenockite.

## FACTORS CONTROLLING FORM AND LOCATION OF ORE BODIES

Distribution of copper in the lowermost part of the Nonesuch Shale is clearly influenced by the presence or absence of beds of favorable lithology, but the fact that a favorable bed of virtually constant lithology can contain abundant copper where the bed is within the cupriferous zone and only pyrite outside that zone is equally clear evidence that lithology is not the sole control. Factors controlling the form and location of ore bodies, therefore, are of two kinds: presence or absence of favorable lithology and the configuration of the cupriferous zone.

### Favorable Lithology

The most cupriferous beds are those that are fairly well-laminated and dark-gray to black in color. Reddish beds are lean. There

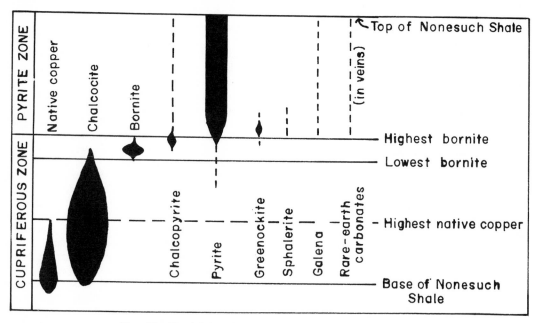

FIG. 12. *Spatial Sequence of Certain Metallic Minerals.*

appears to be a good correlation between metal content and organic matter, but it is not clear whether the organic matter has brought about the precipitation of metals directly by sustaining a reducing environment, or whether it is merely present as an inert relic of the milieu in which sulfate-reducing bacteria were active during sedimentation. The presence of appreciable native copper in the ore suggests that

precipitation of metals is not due to the presence of sulfide alone. It is reasonable to infer, therefore, that the organic matter has had a direct role in making certain beds favorable, even if Wiese's (21) hypothesis that the copper sulfides represent reaction between introduced copper and syngenetic iron sulfide is correct.

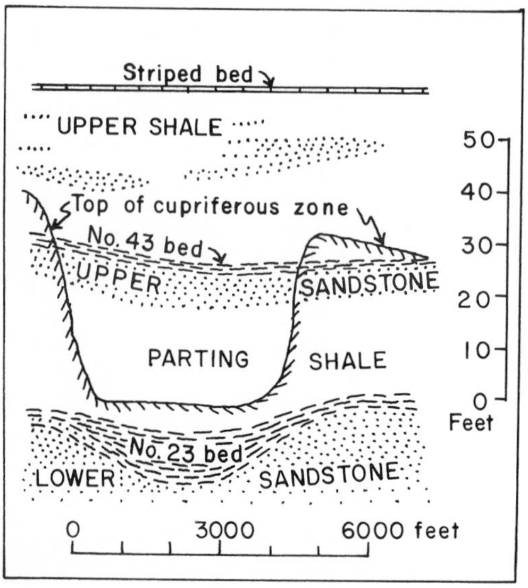

## Configuration of the Cupriferous Zone

Four features of the cupriferous zone deserve mention for their bearing on ore genesis.

(1) The slope of the top of the cupriferous zone is related to lithology. It is most nearly parallel to bedding (Figure 12) where the top of the zone is in a favorable horizon, and it is steep where it crosses relatively unfavorable groups of beds.

(2) Though the boundary is locally steep, it does not have conspicuous "overhangs," that is, places where rocks within the cupriferous zone are stratigraphically above rocks outside the zone. The existence of "overhangs" cannot be ruled out categorically, because cross-sections like Figure 11 are based on drill holes 1000 feet or more apart. But in all the individual drill holes for which adequate data are available, there are no unambiguous examples of an "overhang;" this indicates that the phenomenon is at most an exceptional rather than normal characteristic of the cupriferous zone.

(3) The configuration of the cupriferous

FIG. 13. *Cross-Section, showing cupriferous zone in a part of the Presque Isle area.*

zone is not related in any simple way to paleo-geography or lithofacies of the lowermost part of the Nonesuch Shale. Figure 11 is a generalized composite longitudinal stratigraphic section representing almost the extreme range of shoreward (right-hand end) to deeper water (left-hand end) facies observed in the Porcupine Mountain region. Had this section been drawn in another place, the stratigraphy could be similar and the position of the cupriferous zone different, or vice versa.

(4) In part of the Presque Isle area, a unique deep depression in the top of the cupriferous zone coincides areally with an exceptionally thick part of the No. 23 bed (Figure 13). Over the greater part of a square mile, the No. 43 bed, the Upper Sandstone, and the upper half to two-thirds of the Parting Shale are barren directly over the area in which the No. 23 bed is exceptionally thick and has a copper content normal for that bed (Figure 5). The barren area in the No. 43 bed and Upper Sandstone does not, on the other hand, correlate with any recognized paleogeographic features that existed during and following deposition of the Upper Sandstone.

## TIME OF MINERALIZATION

Most, if not all, of the copper in the Nonesuch Shale appears to have been present before the rocks were deformed. The principal pieces of evidence supporting this conclusion are as follows:

(1) There is a lack of correlation between shale-ore distribution and the abundance of faults and fissures (8, p. 708–712).

(2) Chemical and mineralogical relationships of copper in and adjacent to veins formed during or following deformation indicate that the veins postdate the copper in shale and, indeed, strongly suggest that the copper in veins was derived from that in the shale, rather than the reverse (27, p. 653–660).

(3) The stratigraphic position of the top of the cupriferous zone is generally independent of structure. The high shown in Figure 11 crosses the highly tilted and fractured ground adjacent to the White Pine fault almost at right angles but does not change in height or trend as it crosses. In addition, the high appears to be offset along the fault by an amount similar to the known right-lateral displacement on the fault.

(4) There are no known mantos on favorable beds above the cupriferous zone near faults such as might be expected if faults were

important channels for syntectonic or post-tectonic movement of ore-bearing solutions.

Neutral or equivocal evidence bearing on the age of mineralization includes the following points:

(1) Proximity of copper concentrations in sandstone to the White Pine fault led White and Wright to postulate a post-tectonic origin for the sandstone copper (8, p. 706–708). The sandstone copper appears to antedate the mineralized veins described by Carpenter (27, p. 663–664), however, and Hamilton (38) has correlated the abundance of copper in the Lower Sandstone with channels and lithologic features of the sandstone. Sandstone copper at the Nonesuch mine is not close to known faults comparable to the White Pine fault.

(2) The two parts of the White Pine ore deposit on opposite sides of the White Pine fault, as defined by contours of equal grade or total copper, appear to be simply the offset parts of a single pretectonic ore body. Though this relationship is necessary if the ore is pretectonic, it does not rule out the possibility that post-tectonic mineralization has mimicked something pretectonic to produce the same effect.

The only evidence suggesting that any significant amount of the mineralization is post-tectonic is the fact that in general the richest concentration of copper in the Lower Sandstone and in the Nos. 21 and 23 beds (Figure 9) seems to be more or less centered on the vicinity of the White Pine fault. This evidence is actually inconclusive because the same relationship would result if a fault cut across the middle of a preexisting ore body. The copper content of higher beds (see, e.g., Figure 11) does not show a similar relationship nor does the position of the top of the cupriferous zone.

In summary, the weight of evidence indicates that the copper antedates the main deformation of the area. Evidence that the copper is later than the deposition of the rocks in which it occurs is reviewed in the next section, on origin.

## ORIGIN OF THE COPPER DEPOSITS

The White Pine copper deposit has been variously classed as syngenetic (8,11), as epigenetic, but formed early in the history of the rocks (19,20), and as magmatic hydrothermal, formed later than the structure (2, p. 141, 169–172; 14, 15). The last-named hypothesis seems ruled out by the apparent pretectonic age of the mineralization and by the probability that significant igneous activity in the region had ceased long before the rocks were deformed.

Several features of the mineralization argue against a syngenetic origin:

(1) It is difficult to reconcile the persistence of syngenetic copper deposition through as much as 50 feet of beds when these same beds record major environmental changes. Models of syngenetic deposition such as those proposed by Richter (4, p. 29–50, 57–59) and Garlick (22, p. 146–165) require that the zone most favorable for copper deposition shift back and forth as sea level and the position of the shoreline change. In the areas represented by the center of Figure 11 each bed from the base of the Nonesuch Shale to the Striped bed contains 4 to 100 times as much copper as the same bed outside the cupriferous zone, so according to a syngenetic hypothesis, this area was persistently favorable. Yet the beds in this section record major changes in bottom configuration and depth of water, including one emergence at the end of Parting Shale deposition.

(2) The top of the cupriferous zone does not characteristically show the large "overhangs" that should result from alternating transgression and regression. Though the presence of overhangs, such as have been observed in the Rhodesian deposits (22, Figs. 47, 50) is not incompatible with an epigenetic origin, overhangs are virtually demanded by syngenetic theory where the sediments record major shifts in water depth and shoreline position.

(3) The absence of any recognized copper mineralization in the upper parts of the Nonesuch Shale, despite the presence of favorable rocks and depositional environments, argues against a syngenetic origin.

(4) The barrenness of the No. 43 bed in precisely the same area where the No. 23 bed is thick, as illustrated in Figure 13, is an extraordinary coincidence if the mineralization is syngenetic.

The characteristics of the copper deposits at the base of the Nonesuch Shale, therefore, seem more compatible with an epigenetic origin than with a syngenetic. The pretectonic age of the mineralization indicates that the copper was introduced while the sediments were still relatively undisturbed. Compaction around knots of chalcocite suggests that the mineralization probably occurred prior to or during their cementation and diagenesis. Copper occurs in more than trace amounts at the base of the Nonesuch Shale over an area that certainly exceeds 300 square miles (800 square kilometers). This extent implies great lateral spread of mineralizing solutions if the deposits are, indeed, epigenetic, and the logical aquifer for such spread is the underlying sandstone of the Copper Harbor Formation. Waters rising or migrating laterally in the Copper Harbor conglomerate were in contact with the basal part of the Nonesuch Shale over a wide area, and the sulfur-rich reducing environment in these shales would act as a sink for any metals carried by the water.

Movement of copper within the Nonesuch Shale, however, was primarily upward, as suggested by the lack of "overhangs" in the top of the cupriferous zone and by the relations in Figure 13. Under this hypothesis, the upper limit of the copper mineralization, with its succession of mineral zones, represents a diffusion front.

The direction from which the solutions came is not readily ascertained. The principal concentrations of copper in the basal part of the Nonesuch Shale seem to fall in a belt several miles wide that extends from the Presque Isle basin east through the area of the White Pine mine—the submarginal concentrations that led to early explorations along the south flank of the Porcupine Mountains may belong to the southern edge of this belt. The only broad geologic feature with which this belt appears to correlate is a zone of rapid southward thinning of the Copper Harbor conglomerate where it laps up on the volcanic pile described earlier. It is tempting to infer, but difficult to prove, that this wedge concentrated a flow of water moving southward within the Copper Harbor conglomerate toward the ancient margin of the Keweenawan basin (20).

The ultimate origin of the solutions and the source of their copper are unknown. The water circulating in the Copper Harbor Formation may have contained increments of magmatic water. If the circulation was relatively early in the history of the rocks, as suggested, it need not be far removed in time from known volcanic activity in the area. Similarly, existing evidence does not rule out the possibility that the solutions that mineralized the Nonesuch Shale were the same as those that formed the native copper deposits in the Portage Lake Lava Series, whatever their origin.

A reasonable alternative to a magmatic source for the mineralizing solutions is ground water buried with and introduced into the sedimentary rocks of the Keweenawan basin, as suggested earlier by two of us (19,20). Davidson (36) has recently called attention to the worldwide association of strata-bound copper deposits and evaporites; he reasons that the strong brines generated by solution of evaporites in ground water have leached metals from underlying rocks and thus become ore-forming fluids. It is likely that evaporites did, and perhaps still do, exist in the central parts of the Keweenawan basin under Lake Superior. The finely divided volcanic detritus of the Copper Harbor conglomerate would provide a more convenient source for the copper than the basement rocks suggested by Davidson.

More sophisticated geochemical investigations than have been undertaken hitherto probably offer the best hope for solving the problem of the ultimate source of the copper.

## REFERENCES CITED

1. Irving, R. D., 1883, The copper-bearing rocks of Lake Superior: U.S. Geol. Surv. Mono. 5, 464 p.

2. Butler, B. S. and Burbank, W. S., 1929, The copper deposits of Michigan: U.S. Geol. Surv. Prof. Paper 144, 238 p.

3. Tyler, S. A., *et al.*, 1940, Studies of the Lake Superior precambrian by accessory-mineral methods: Geol. Soc. Amer. Bull., v. 51, p. 1429–1538.

4. Richter, G., 1941, Geologische Gesetzmässigkeiten in der Metallführung des Kupferschiefers: Archiv f. Lagerstättenforschung, H. 73, 61S.

5. Thaden, R. E., 1950, The Porcupine Mountain red rock: unpublished M.S. thesis, Mich. State Univ.

6. White, W. S., 1952, Imbrication and initial dip in a Keweenawan conglomerate bed: Jour. Sed. Pet., v. 22, p. 189–199.

7. Cornwall, H. R., 1954, Bedrock geology of the Delaware quadrangle, Michigan: U.S. Geol. Surv. Geol. Quad. Map GQ-51, 1:24,000.

8. White, W. S. and Wright, J. C., 1954, The White Pine copper deposit, Ontonagon County, Michigan: Econ. Geol., v. 49, p. 675–716.

9. Messer, E., 1955, Kupferschiefer, Sanderz und Kobaltrücken im Richelsdorfer Gebirge, Hessen: Hessisches Landesamt für Bodenforschung, Hessisches Lagerstättenarchiv, H. 3, 125 S.

10. Doan, V. L., 1956, A petrographic study of the Nonesuch shale sectioned by the White Pine water intake tunnel, Ontonagon County, Michigan: unpublished M. S. thesis, Mich. Tech. Univ.

11. Rand, J. R., 1956, Copper mineralization at the White Pine mine, Ontonagon County, Michigan (abs.): p. 17 *in* Snelgrove, A. K., *Editor, Geological exploration,* Inst. on Lake Superior Geol., Houghton, Mich., 109 p.

12. Abelson, P. H., 1959, Geochemistry of organic substances: *in* Abelson, P. H., *Editor, Researches in Geochemistry,* John Wiley and Sons, N.Y., p. 79–103.

13. Jensen, M. L., 1959, Sulfur isotopes and hydrothermal mineral deposits: Econ. Geol., v. 54, p. 374–394.

14. Joralemon, I. B., 1959, The White Pine copper deposit: Econ. Geol., v. 54, p. 1127; disc. of (8).

15. Sales, R. H., 1959, The White Pine copper deposit: Econ. Geol., v. 54, p. 947–951; disc. of (8).

16. Brady, J. M., 1960, Ore and sedimentation of the Lower Sandstone at the White Pine mine, Michigan: unpublished M. S. thesis, Mich. Tech. Univ.

17. Tourtelot, H. A., *et al.*, 1960, Stratigraphic variations in mineralogy and chemical composition of the Pierre shale in South Dakota and adjacent parts of North Dakota, Nebraska, Wyoming, and Montana: U.S. Geol. Surv. Prof. Paper 400-B, p. B447–B452.

18. White, W. S., 1960, The Keweenawan lavas of Lake Superior, an example of flood basalts: Amer. Jour. Sci., v. 258-A (Bradley Volume), p. 367–374).

19. ——— 1960, The White Pine Copper deposit: Econ. Geol., v. 55, p. 402–409; reply to (13) and (14).

20. White, W. S. and Wright, J. C., 1960, Lithofacies of the Copper Harbor conglomerate, northern Michigan: U.S. Geol. Surv. Prof. Paper 400-B, p. B5–B8.

21. Wiese, R. G., 1960, Petrology and geochemistry of a copper-bearing Precambrian shale, White Pine, Michigan: unpublished Ph. D. dissertation, Harvard Univ.

22. Garlick, W. G., 1961, The syngenetic theory: p. 146–165: *in* Mendelsohn, F., *Editor, The geology of the Northern Rhodesian Copper belt,* MacDonald, London, 523 p.

23. Hamblin, W. K., 1961, Micro-cross-lamination in upper Keweenawan sediments of northern Michigan: Jour. Sed. Pet., v. 31, p. 390–401.

24. ——— 1961, Paleogeographic evolution of the Lake Superior region from late Keweenawan to Late Cambrian time: Geol. Soc. Amer. Bull., v. 72, no. 1, p. 1–18.

25. Hamblin, W. K. and Horner, W. J., 1961, Sources of the Keweenawan conglomerates of northern Michigan: Jour. Geol., v. 69, p. 204–211.

26. Ohle, E. L., 1962, Thoughts on epigenetic vs. syngenetic origin for certain copper deposits: Econ. Geol., v. 57, p. 831–836.

27. Carpenter, R. H., 1963, Some vein-wall rock relationships in the White Pine Mine, Ontonagon Co., Michigan: Econ. Geol., v. 58, p. 643–666.

28. Joralemon, I. B., 1963, Vein-wall rock relationships, White Pine Mine: Econ. Geol., v. 59, p. 160–161; disc. of (26).

29. Rand, J. R., 1963, Vein-wall rock relationships, White Pine Mine: Econ. Geol., v. 59, p. 160–161; disc. of (26).

30. Chaudhuri, S. and Faure, G., 1964, The whole-rock Rb-Sr age of the Precambrian Nonesuch Shale in Michigan: Mass. Inst. Technol. Dept. Geol. and Geophysics Ann. Prog. Rept. 1381-12, p. 221–223 (Rept. prepared for U.S. Atomic Energy Comm.).

31. Eglinton, G., *et al.*, 1964, Hydrocarbons of biological origin from a one-billion-year-old sediment: Science, v. 145, no. 3629, p. 263–264.

32. Ensign, C. O., Jr., 1964, Ore dilution control increases earnings at White Pine: A.I.M.E. Tr., v. 229, p. 184–191.

33. Wedepohl, K. H., 1964, Untersuchungen am Kupferschiefer in Nordwestdeutschland; ein Beitrag zur Deutung der Genese bituminoser Sedimente: Geochimica et Cosmochimica Acta, v. 28, p. 305–364 (with English abs.).

34. Barghoorn, E. S., *et al.*, 1965, Paleobiology of a Precambrian shale: Science, v. 148, no. 3669, p. 461–472.

35. Chaudhuri, S. and Faure, G., 1965, The whole-rock Rb-Sr age of the Precambrian Dept. Geol. and Geophysics Ann. Progress felsites of the Porcupine Mountain district of northern Michigan: Mass. Inst. Technol., Rept. 138-13 (Rept. in preparation for U.S. Atomic Energy Comm.).

36. Davidson, C. F., 1965, A possible mode of origin of strata-bound copper ores: Econ. Geol., v. 60, p. 942–954.

37. Hack, J. T., 1965, Postglacial drainage evolution and stream geometry in the Ontonagon area, Michigan: U.S. Geol. Surv. Prof. Paper 504-B, p. B1–B4.

38. Hamilton, S. K., 1965, Copper mineralization in the upper part of the Copper Harbor conglomerate at White Pine, Michigan: unpublished M. S. thesis, Univ. Wisc.

39. Trammell, J. W., and Ensign, C. O., Jr., 1965, Ancient stream channels and their effect on mine planning and grade control at the White Pine mine, Mich.: A.I.M.E. Tr., v. 231, p. 401–408.

40. Brown, Alexander C., 1965, Minerlogy at the top of the cupriferous zone, White Pine mine, Ontonagon Co., Michigan: unpublished M.S. thesis Univ. Mich.

# 23. Geology of the Iron Ores of the Lake Superior Region in the United States

RALPH W. MARSDEN*

## Contents

## Illustrations

* University of Minnesota, Duluth, Minnesota.

# Tables

## ABSTRACT

The natural iron ores of the Lake Superior Region in the United States are being replaced by iron-ore concentrates produced from magnetite- or hematite-rich horizons in the Precambrian cherty iron formations. The production of natural ores will soon be less than one-half of the total ore shipped annually. The tonnage of natural ore produced will continue to decline until it is a minor percentage of the ore shipped.

The natural ores include (1) soft ore, (2) hard ore, (3) conglomerate ore and (4) siliceous ore. The soft ores are the important ore type and are sometimes termed "Lake Superior type ore". These ores are in situ deposits of hematite and limonite formed by the leaching of silica and other gangue materials in the iron formations. The leaching is believed to have been done by downward moving surface waters. In deep ore deposits, generally over 1000 feet below the present surface, circulation and leaching action may have been assisted by hydrothermal activity.

Hard ore deposits occur on the Vermilion and Marquette ranges where hematite or magnetite has replaced the silica (and possibly some silicates) in cherty iron formation. These ores commonly occur in structural situations favorable for the movement of hydrothermal waters. The ores are believed to be the result of hydrothermal action with iron, possibly from the iron formation, transported by hydrothermal waters to areas favorable for the replacement of silica by iron oxides.

Certain stratigraphic horizons of the Lake Superior Precambrian cherty iron formations contain sufficient iron as magnetite or hematite, with a grain size and texture that will give adequate mineral liberation, to yield an ore-quality product after grinding and concentration. The magnetite-rich deposits are termed magnetite taconites. These ores occur on the Mesabi, Gogebic, and Marquette ranges. Magnetite taconite ores are unenriched cherty iron formation and are the result of the sedimentary deposition of the iron formation and the metamorphic history of the rock. The commercial quality hematite-rich iron formations in Michigan are termed jasper ores. The jasper ores are being mined on the Marquette and Menominee ranges. These ores are composed of specular or granular hematite associated with quartz and iron silicates. The jasper ores appear to be recrystallized iron formations in which iron, prior to recrystallization, occurred as hematite.

## INTRODUCTION

The Lake Superior Region is noted for its important iron ore deposits of Precambrian age that have supplied the iron and steel plants of the United States for over 100 years. The iron ores occur in cherty iron formations of early Precambrian age on the Vermilion Range in Minnesota, and in the middle Precambrian age on the Mesabi and Cuyuna Ranges in Minnesota, the Gogebic Range in Wisconsin and Michigan, and the Marquette and Menominee Ranges in Michigan (Figure 1). Lake Superior Iron Ranges were sources only of natural ores and natural ore concentrates until the mid 1950's, when the commercial production of iron ore concentrates from the magnetite taconites in Minnesota and hematite-rich jaspers in Michigan began. In 1965, pelletized iron-ore concentrates constituted about 40 per cent of the iron ore produced in the Lake Superior Region and, in 1968, when concentrating and pelletizing plants being built are in operation, production of iron-ore concentrates from taconites and jaspers should increase to 60 per cent of ore shipments. This change from natural ores to a concentrated iron ore will continue to increase, and the natural ores will become a minor source.

The change from natural ore has introduced a new concept of what constitutes an iron ore. Much of the ore mined today was considered

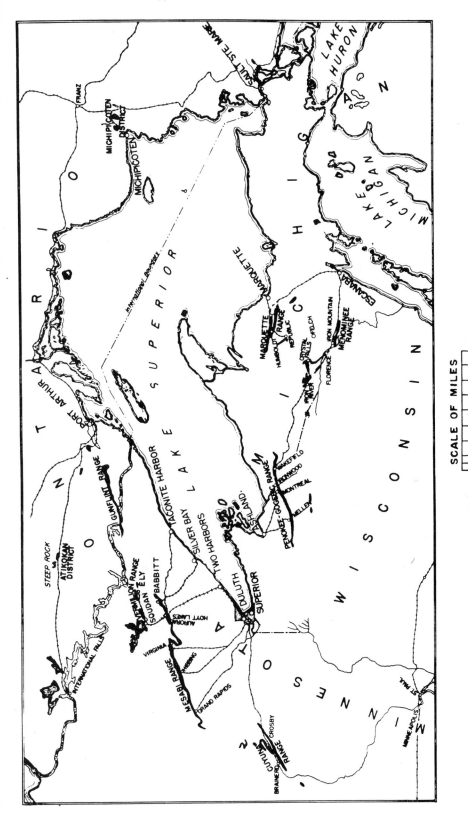

FIG. 1. *Map of the Lake Superior Region Showing the Location of the Iron Ranges.*

only a potential ore, or even waste rock, a few years ago. The important geological literature of the past, concerned with the Lake Superior iron ores, described the occurrence and origin of the natural ores. Today, a modern treatise on Lake Superior Region iron ores must emphasize the magnetite taconites of Minnesota, Wisconsin, and Michigan and the hematite jaspers of Michigan. Some iron formation materials may never be classed as ore, because the iron does not occur in an easily recoverable form or occurs in deposits too small for economic production. There are, however, large tonnages of potentially concentratable iron formation which technological advances in mineral dressing may change to ore in the future.

This paper and other related papers in the Graton-Sales Volume are primarily concerned with the currently usable iron ore materials. The chief contribution of these papers to the knowledge of Lake Superior iron ores will be in the description of the concentrating ores that now are being mined or are in deposits being brought into production. Descriptions of the natural ores are available in many earlier publications of which Van Hise and Leith (2), Gruner (8), Hotchkiss (4), Wolff (3), White (12), were important contributors.

## History

The mining history of the Lake Superior Region dates from an ancient people who mined copper on the Keweenaw Peninsula, Michigan, in pre-historic times. Knowledge of copper in the Lake Superior Region was passed by Indians to the early explorers and was reported by LaGarde in 1636. The discovery of iron ore came much later when William A. Burt, a government surveyor, found ore near Negaunee, Michigan in 1844. The first iron ore was mined by the Jackson Association in 1848 and was smelted locally. In 1850 and 1852, iron ore shipped to Pennsylvania began the movement of iron ore to lower Great Lakes furnaces. Regular shipments of iron ore started from the Negaunee area, Michigan in 1856, after the ship canal was opened at Sault Ste. Marie.

Exploration for iron ore spread from the Marquette Range, and it is likely that iron-bearing rocks were known on the Menominee, Gogebic, Mesabi, and Vermilion Ranges in the 1850's. Iron ore was discovered on the Menominee Range in 1867, on the Gogebic Range in 1884, the Vermilion Range in 1885, the Mesabi Range in 1890, and the Cuyuna

Range in 1903. The wide distribution of test pits and old shafts, in virtually all known areas of iron-bearing rocks in the Lake Superior Region, is proof of the thoroughness and the vigor of the early iron-ore prospectors. The dip needle, for magnetic surveys, and the pick and shovel were used effectively by the pioneer explorers. Diamond drilling was used extensively after the late 1880's.

The first mining was by small open pits, but soon the bulk of the iron ore came from underground mines, which have been dominant on the Marquette, Menominee, Gogebic, and Vermilion Ranges and were important in the early years on the Mesabi Range. Open-pit mines have produced the bulk of the iron ore shipped from the Mesabi and Cuyuna Ranges. Because of the relatively high cost of underground mining and the change in ore-quality requirements, most underground mines are closed even though substantial reserves remained in several and the ore was partly developed.

Interest in the possibility of producing high-grade iron-ore concentrates from the magnetite-rich iron formations—i.e. magnetite taconites—of Minnesota, Michigan, and Wisconsin and from the hematite-rich jaspers of Michigan, dates from the early 1900's. Pioneer efforts made to concentrate Mesabi Range taconite in the period from 1915 to 1924 were commercially unsuccessful but indicated a future potential. Interest in the concentration of iron-formation materials was renewed in 1945. Commerical production of concentrates from Michigan jasper ore started in 1954, at the Humboldt Mine on the Marquette Range. In 1965, about 70 per cent of the Marquette Range iron ore shipments were in the form of iron ore pellets produced from Michigan jaspers and magnetite taconites. Large-scale commercial production of magnetite taconite ore on the Mesabi Range started in 1956 at the Peter Mitchell Mine near Babbitt, Minnesota. In 1965, about 40 per cent of the iron ore shipped from the Mesabi Range was as iron-ore pellets produced from magnetite taconite. The production of natural iron ores is declining rapidly, and it is being replaced by iron-ore pellets produced from iron-ore concentrates.

## Production

Mines of the Lake Superior Region, in the United States, have yielded about 3,666,-000,000 gross tons of iron ore from 1848 through 1965. Table I summarizes the pro-

TABLE I.    *Iron Ore Shipments by Ranges in Gross Tons*

|  | Mesabi | Vermilion | Cuyuna | Gogebic | Marquette | Menominee | Total (U.S.) Lake Superior |
|---|---|---|---|---|---|---|---|
| 1854–1860 |  |  |  |  | 243,000 |  | 243,000 |
| 1861–1870 |  |  |  |  | 3,597,000 |  | 3,597,000 |
| 1871–1880 |  |  |  |  | 10,345,000 | 901,000 | 11,246,000 |
| 1881–1890 |  | 3,223,000 |  | 8,470,000 | 18,749,000 | 11,936,000 | 42,378,000 |
| 1891–1900 | 31,390,000 | 11,968,000 |  | 22,731,000 | 26,825,000 | 21,196,000 | 114,110,000 |
| 1901–1910 | 193,496,000 | 15,138,000 |  | 34,021,000 | 36,744,000 | 41,418,000 | 320,817,000 |
| 1911–1920 | 332,928,000 | 13,860,000 | 13,850,000 | 60,587,000 | 39,843,000 | 51,601,000 | 512,669,000 |
| 1921–1930 | 331,953,000 | 14,339,000 | 17,877,000 | 60,516,000 | 37,125,000 | 44,881,000 | 506,691,000 |
| 1931–1940 | 230,882,000 | 10,152,000 | 9,757,000 | 35,231,000 | 33,393,000 | 17,314,000 | 336,729,000 |
| 1941–1950 | 598,117,000 | 15,807,000 | 28,414,000 | 53,351,000 | 50,692,000 | 41,128,000 | 787,509,000 |
| 1951–1960 | 567,552,000 | 13,712,000 | 25,712,000 | 40,479,000 | 49,929,000 | 39,482,000 | 736,866,000 |
| 1961 | 42,196,000 | 865,000 | 1,320,000 | 2,190,000 | 4,152,000 | 3,886,000 | 54,609,000 |
| 1962 | 42,162,000 | 1,093,000 | 946,000 | 2,318,000 | 4,500,000 | 3,462,000 | 54,481,000 |
| 1963 | 43,637,000 | 766,000 | 839,000 | 1,314,000 | 5,850,000 | 4,304,000 | 56,710,000 |
| 1964 | 47,662,000 | 1,024,000 | 698,000 | 1,602,000 | 7,945,000 | 4,624,000 | 63,555,000 |
| 1965 | 49,148,000 | 803,000 | 609,000 | 330,000 | 8,925,000 | 4,361,000 | 64,176,000 |
| Total tons | 2,511,123,000 | 102,750,000 | 100,022,000 | 323,140,000 | 338,857,000 | 290,494,000 | 3,666,386,000 |

Sources: Reed, Robert C.—Annual Shipments by Ranges and Total Shipments of Individual Mines, Michigan Department of Conservation, Geological Survey Div., University of Minnesota Bulletin—Mining Directory Issue, U.S. Bureau of Mines—Minerals Yearbook (Metals & Minerals).

duction in gross tons for the principal iron ranges. The Mesabi Range has produced about 70 per cent of the total Lake Superior iron ore and is expected to maintain its dominant position.

Production of natural ores, including natural ore concentrates, from the Mesabi Range declined from 65.5 million gross tons in 1955 to 30.3 million gross tons in 1965. During this period taconite concentrate production increased from 1.5 million gross tons in 1955 to 18.9 million gross tons in 1965. The projected 1968 annual capacity of plants processing magnetite taconite in the Mesabi Range is 31.5 million gross tons. Additional plants, or plant expansions, are being considered to add further to this capacity. In Michigan, natural ore production declined from 14,000,000 gross tons in 1955 to 7,000,000 gross tons in 1965. During this period, jasper and magnetite-taconite concentrates increased from 174,000 gross tons in 1955 to 7,000,000 gross tons in 1965. It is estimated that the 1968 capacity of processing plants will be 10.3 million gross tons. This estimate includes a plant that pelletizes 1.2 million tons of natural ore fines.

Gruner (10) estimated reserves of magnetite taconite ore on the Mesabi Range, available

to open-pit mining, to be approximately 5 to 6 billion gross tons that should yield about 2,000,000,000 gross tons of iron-ore concentrates. Carr and Dutton (18) estimate the total magnetite taconite reserves of the Mesabi Range, to a depth of 500 feet below the surface, to be about 15 billion gross tons that should yield about 6 billion tons of concentrates. Published reserve estimates do not include a tonnage of commercial-quality, low-grade iron ore, recoverable by open-pit mining on the other Lake Superior iron ranges. Carr and Dutton estimate the following total tonnages for low grade (25 to 45 per cent Fe) iron-bearing materials within 500 feet of the surface in the other Lake Superior Ranges: Cuyuna Range 4,400,000,000; Vermilion Range—"Large"; Gogebic Range—7,750,000,000; Marquette Range—17,500,000,000; Menominee Range—4,320,000,000. The Carr Dutton reserve estimate does not include a tonnage for oxidized taconite on the Mesabi Range nor does it include the tonnage potential for magnetite taconites available to underground mining. The Carr and Dutton estimate of low-grade iron ore and iron-bearing materials in the Lake Superior Region totals 48,970,000,000 gross tons. Recognizing the tonnages of Mesabi Range oxidized taconite and the additional potential of underground magnetite taconite reserves, not included in the estimate, a total reserve in excess of 50 billion tons of iron-formation materials containing from 25 to 45 per cent Fe is indicated for the Lake Superior Region. Only a part of this large tonnage of iron formation is available to open-pit mining and is currently economic.

## Grades of Ore

The grades of Lake-Superior iron ores are determined by chemical composition and by physical structure. During the early days of mining, each mine had a small annual production, and the ore of each had a particular chemical composition and physical structure. The ore quality was designated by a name, usually that of the mine. As the tonnage and the number of individual mines increased, the variety of ores increased, and so it became necessary to mix ores to a standard grade of a specific composition and structure. The various mining companies now produce a number of ore grades that are designated by a grade name. In 1900, 160 grades of iron ore were shipped from the Lake Superior Region, the number of grades increased to 294 in 1920 and then decreased to 70 grades shipped in 1965. A further decrease in the number of ore grades shipped annually is expected because the chemical composition and physical structure of pellets from each processing plant are fixed within quite narrow limits. Table II gives a representative list of ore grades shipped during 1965 and Range averages for several classes of ore.

TABLE II.   *Representative Grades of Ore Shipped from the Lake Superior Region in 1965*

| | Grade name | Fe | $SiO_2$ | Phos. | Mn. | $Al_2O_3$ | CaO | MgO | S | Loss by Moisture ignition |
|---|---|---|---|---|---|---|---|---|---|---|
| **Mesabi Range** | | | | | | | | | | |
| Average Bessemer Ore | | 57.44 | 8.55 | .036 | .58 | .43 | | | | 5.85 |
| Average Non-Bessemer Ore | | 53.95 | 8.15 | .055 | .51 | 1.08 | | | | 8.95 |
| Average Pellets | | 61.17 | 8.53 | .021 | .29 | .37 | | | | 2.82 |
| Taconite Pellets | Reserve | 60.32 | 8.64 | .025 | .30 | .44 | .52 | .60 | | 3.41 |
| Taconite Pellets | Aurora | 62.24 | 7.87 | .016 | .26 | .25 | | | | 2.09 |
| Taconite Agglomerate | Oliver 31 T | 65.16 | 5.21 | .014 | .16 | .10 | 2.30 | .54 | .004 | .33 |
| Bessemer Ore | Beacon | 54.90 | 8.92 | .045 | .30 | .48 | .07 | .07 | .015 | 4.08 | 7.38 |
| Bessemer Ore | Oliver 16 | 54.60 | 8.31 | .045 | .21 | .90 | .05 | .11 | .013 | 3.53 | 8.62 |
| Non-Bessemer Ore | Oliver 13 | 51.00 | 6.61 | .056 | .47 | 1.77 | .04 | .10 | .008 | 4.78 | 12.89 |
| Non-Bessemer Ore | Shell | 53.44 | 5.53 | .054 | .61 | 1.00 | .05 | .06 | .005 | 6.51 | 9.59 |
| **Marquette Range** | | | | | | | | | | |
| Average Non-Bessemer Ore | | 52.40 | 4.47 | .532 | .27 | 2.56 | | | | 7.68 |
| Average Pellets | | 63.04 | 7.02 | .038 | .09 | .86 | | | | 1.20 |
| Average Siliceous Ore | | 36.92 | 43.71 | .028 | .07 | .58 | | | | 2.29 |
| Non-Bessemer | Tracy Blend | 52.68 | 6.48 | .079 | .33 | 2.44 | | | .324 | 11.45 |
| Lump Ore | Cliffs Shaft | 61.77 | 6.46 | .110 | .28 | 1.85 | | | .005 | .50 |
| Pellets | Empire | 63.31 | 7.45 | .012 | .07 | .38 | | | .003 | 1.31 |
| Pellets | Marquette | 64.52 | 5.34 | .038 | .07 | .54 | | | .003 | 1.31 |
| Siliceous Ore | Tilden Silica | 38.30 | 39.92 | .039 | .07 | .65 | | | .010 | 3.10 |
| **Menominee Range** | | | | | | | | | | |
| Average Non-Bessemer Ore | | 52.40 | 4.47 | .532 | .27 | 2.56 | | | | 7.68 |
| Average Manganiferous Ore | | 45.43 | 7.38 | .473 | 2.95 | 4.03 | | | | 7.69 |
| Non-Bessemer Ore | Wauseca | 51.52 | 3.66 | .594 | .36 | 2.15 | 1.90 | .82 | .190 | 7.38 | 7.80 |
| Pellets | Groveland | 60.51 | 8.95 | .027 | .73 | .52 | 1.25 | 1.02 | .003 | .04 | .68 |
| **Cuyuna Range** | | | | | | | | | | |
| Average Manganiferous Ore | | 32.71 | 18.19 | .136 | 14.83 | 1.74 | | | | 7.12 |
| Manganiferous Ore | Alstead | 46.67 | 8.68 | .108 | 5.56 | 1.63 | | | | 8.36 |
| Manganiferous Ore | Jill | 33.66 | 14.67 | .173 | 12.48 | 1.93 | .35 | .17 | .006 | 8.61 | 8.29 |
| **Vermilion Range** | | | | | | | | | | |
| Lump Ore | Oliver 41 | 61.15 | 5.20 | .054 | .10 | 3.02 | .06 | .23 | .042 | .84 | 2.90 |

Source: Iron Ore 1965—American Iron Ore Association.

Iron ores are classed as Bessemer ore, non-Bessemer ore, manganiferous ore, and siliceous ore on a basis of their chemical composition. Each class of ore has a defined composition. The Bessemer and non-Bessemer ores are determined by the phosphorous content. Since 1925, Bessemer ores have a maximum content of .045 per cent phosphorus (dry basis). The classification of iron ores as Bessemer or non-Bessemer may have little significance in future years as Bessemer steel practice is becoming extinct. Further, the abundance of taconite pellets with a phosphorus content of less than .045 per cent may eliminate the premium quality designation now associated with bessemer-grade ore. Manganiferous ores contain over 2.0 per cent manganese and commonly range from 2.0 to 10.0 per cent manganese. In general, ores containing over 18 per cent silica are classed as siliceous ores, although most iron ores that are mined specifically to yield a siliceous-grade ore contain from 38 to 44 per cent silica and from 35 to 40 per cent iron.

In recent years most of the natural ores are crushed to a top size of about 2 inches and screened into a coarse size, 2 inches to 0.25 inches, and a fine ore, minus about 0.25 inches in size, before shipping. The coarse ores are used directly in the blast furnace, whereas the fine ores commonly are sintered.

Certain compact hematite or magnetite ores of the Lake Superior Region are classed as lump-grade ores for use in open-hearth practice. Lump-ore production is declining rapidly, and, after 1967, little if any lump-ore will be produced. The lump-grade Lake-Superior ores are being replaced by agglomerated iron-ore concentrates and by imported high-grade lump ores.

## PHYSIOGRAPHY

The Lake Superior Region, in the United States, is a general upland area of moderate to low relief on the south and west sides of Lake Superior. The lake has a mean elevation of 602 feet above sea level. A small lowland area adjacent to the lake is bounded by rather steep slopes rising to the upland area that ranges from 600 to 1200 feet above Lake Superior. Most of the upland is characterized by low hills and ridges separated by flat, often swampy areas. The higher hills rise about 200 to 600 feet above the upland level.

The entire region is glaciated, so many of the surface features are glacial in origin or have been modified by glacial action. Some areas, such as northern Minnesota and local areas in Michigan and Wisconsin, have only a thin ground-moraine cover; other areas are heavily blanketed by glacial deposits. Rock exposures are common along ridges in areas of thin drift but are rare in areas of thick glacial accumulation.

The Lake Superior basin drains to the south and east through the lower lakes and the St. Lawrence River. A major part of the region is within this watershed. The area generally north of the Mesabi Range, Minnesota, drains northward into Hudsons Bay. The western end of the Mesabi Range and parts of northern Wisconsin are within the Mississippi River drainage. Thus, much of the upland area is in the general headwaters areas of three major drainage basins. The streams generally are small.

## GEOLOGIC HISTORY

The Lake Superior Region, in the United States, is an extension of the Canadian Precambrian Shield and has the same general rock sequences and lithologies. A time span of over 2 billion years is represented by rocks ranging from gneisses and schists of the Lower Keewatin of early Precambrian age to sandstones of the Upper Keweenawan of late Precambrian age. The geochronology, geologic column, and general correlation of the important iron ranges of the Lake Superior Region, are shown on Table III. Keweenawan rocks occur in and around the rim of the Lake Superior basin with Huronian rocks farther to the south and west. An extensive area of older rocks occurs still farther from the basin to the north and west (Figure 2).

The predominantly sedimentary rock sequence of middle Precambrian age is termed Huronian in this report following long established usage. James (17) in 1958, proposed, and the United States Geological Survey adopted, the term Animikie series for these rocks. Arguments for changing the terminology are presented by James. Geologists retaining the Huronian classification raise objections to the term Animikie, since the Animikie group, in its original area of definition, represents only a part of the middle Precambrian sequence found in Michigan.

## STRATIGRAPHY

The general rock sequence of the Lake Superior Region is known, but important details of the lower Precambrian sequence are yet to be determined. Extensive work has been

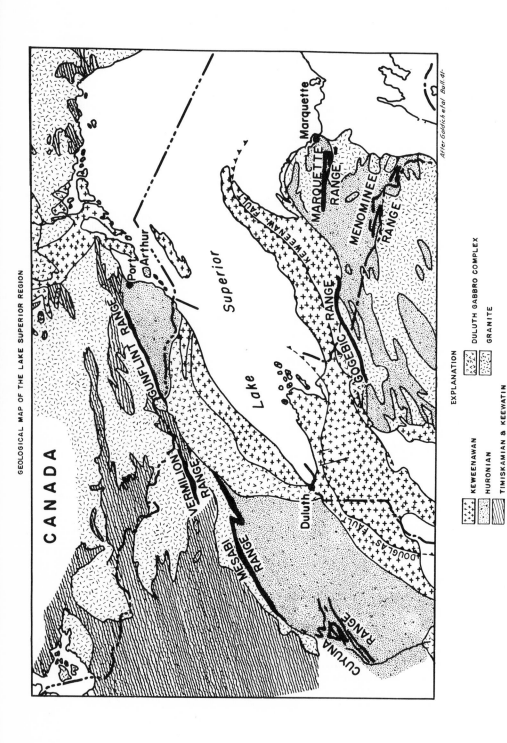

Fig. 2. *Geological Map of the Lake Superior Region in the United States and adjacent Canada.*

TABLE III. The Geochronology,[1] Geologic Column, and General Correlation of the Important Iron Ranges, Lake Superior Region in the United States

| Era | Period / System | Orogeny | Vermilion Range | Mesabi Range | Cuyuna Range | Gogebic Range | Marquette Range | Menominee Range | Michigan stratigraphic classification (after James [17]) | Precambrian division |
|---|---|---|---|---|---|---|---|---|---|---|
| Late Precambrian | Keweenawan (Upper) — Grenville | Grenville | Duluth Gabbro | Duluth Gabbro | Basic Intrusives | Sandstone; Gabbro and Granite | Sandstone | Rare Diabase Dikes | Keweenawan | Upper Precambrian |
| | Keweenawan (Middle) | | North Shore Volcanics | North Shore Volcanics; Puckwunge Sandstone | | Sandstone, Shale and Conglomerate; Volcanics; Sandstone | | | | |
| | Keweenawan (Lower) — 1.1 b.y. | Penokean | | | Granite | Granite | Granite and Basic Intrusives | Granite and Gabbro | Granite | Middle Precambrian |
| Middle Precambrian | 1.7 b.y. — Huronian (Animikie Group, Minnesota) | | | Virginia Argillite | Rabbit Lake Argillite | Tyler Slate | Michigamme Slate with some Iron Formation; Clarksburg Volcanics; Greenwood Iron Formation; Goodrich Quartzite | Fortune Lakes Slate; Stambaugh Formation; Hiawatha Graywacke; Riverton Iron Formation; Dunn Creek Slate; Badwater Greenstone; Michigamme Slate | Paint River Group; Baraga Group | ANIMIKIE SERIES (Michigan) |
| | | | | Biwabik Iron Formation; Pokegama Quartzite | Trommald Iron Formation; Mahnomen Slates and Quartzite | Ironwood Iron Formation; Palms Quartzite | — Unconformity — Negaunee Iron Formation; Siamo Slate; Ajibik Quartzite | — Unconformity — Vulcan Iron Formation; Felch Formation | Menominee Group | |
| | | | Granite; Knife Lake Slates | Granite; Knife Lake Slates | ? Dolomite | — Unconformity — Bad River Dolomite; Sunday Quartzite | Wewe Slate; Kona Dolomite; Mesnard Quartzite | — Unconformity — Randville Dolomite; Sturgeon Quartzite | Chocolay Group | |
| Early Precambrian | 2.5 b.y. | Algoman | Granite; Knife Lake Slates | Granite; Knife Lake Slates | | Granite | Gneissic Granite | Gneissic Granite | Granite | Lower Precambrian |
| | Timiskamian | | | | | | Six Mile Lake Amphibolite; Solberg Schist; East Branch Arkose | | Dickinson Group | |
| | ? b.y. | Laurentian | Saganaga Granite | | | Granite | Granite Gneiss | | Granite | |
| | Keewatin | | Soudan Iron Formation; Ely Greenstone | | | Greenstone | Kitchi Schist; Mona Schist | Quinnesec Formation (Position Uncertain) | Quartzite & Schist; Quinnesec Formation and Hardwood Gneiss (Position Uncertain) | |

[1] After Goldich, et al. (19).

done on the middle and upper Precambrian rocks because locally they contain iron and coppers ores. Even with this important economic stimulus, much yet is to be learned concerning details of structure, stratigraphy, lithologic differences, facies changes, mineralogy, and geologic history. Detailed work is severely hampered by glacial deposits that blanket much of the region. Modern geological and geophysical studies are needed to extend the knowledge gained from limited outcrops into covered areas.

## Lower Precambrian Rocks

The lower Precambrian rocks of the Lake Superior Region occur in extensive areas north of the Mesabi Range in Minnesota, both north and south of the Marquette Range in Michigan, and in areas south of the Menominee and of the Gogebic Range in Wisconsin and Michigan. Detailed geological work in most of these areas has been limited. The lower Precambrian sequence includes rocks of Keewatin, Laurentian, Timiskamian, and Algoman ages.

KEEWATIN SYSTEM  The Keewatin characteristically is represented by a thick sequence of dominantly volcanic rocks, including common pillow lavas, with minor amounts of graywackes, slate, conglomerate, and iron formation. Paragneisses and schists, several thousand feet in thickness, occur in the Rainy Lake area in northern Minnesota and in Ontario and were termed Coutchiching by Lawson in 1888. These metasedimentary rocks are stratigraphically below greenstone flows. The age of the Coutchiching schists and paragneisses has been a subject of controversy since that time. There is a question as to age of the greenstones in the Rainy Lake area and their lateral relationship to the Keewatin greenstones in other areas. These greenstones may be in the lower part of the Keewatin system, which would require the paragneisses and schists to be of pre-Keewatin age, or the greenstones may be within the Keewatin sequence, so the metasedimentary rocks would be of Keewatin age.

Keewatin rocks in Minnesota are termed Ely greenstones and Soudan iron formation; on the Marquette Range, Michigan, the Mona and Kitchi schists; and on the Menominee Range in Michigan and Northern Wisconsin, the Quinnesec formation. Extensive areas of Keewatin rocks are chloritic, which resulted in the term greenstones being applied to these rocks. Where the Keewatin rocks are intruded

by Laurentian, Algoman, or Penokean granite, they may be metamorphosed into hornblende schists, hornblendites, biotite schists, gneisses, and hybrid rocks.

The Soudan formation, of the Vermilion iron range in Minnesota is in the upper part of the Ely greenstone, and consists of a number of lenses of thinly banded, cherty iron formation composed of interbedded fine-grained quartz-rich layers and hematite-rich layers. Much of the Soudan formation is a jaspilite composed of interbedded jasper and fine-grained gray hematite. The iron-formation lenses are from a few feet to a few hundred feet in stratigraphic thickness and are from a few hundred feet long to 3 to 4 miles in length. The bedding is straight to complexly folded. Locally, the iron formation is much thickened by folding. In the folded areas, the bedding is usually much crenulated.

LAURENTIAN INTRUSIVES  Laurentian rocks include granitic intrusives that cut Keewatin rocks and are pre-Timiskamian in age. The Saganaga granite, in northern Minnesota and adjacent Ontario, and the nearby Northern Light gneiss are of Laurentian age. The gneiss is composed of fine-grained, light-gray bands of feldspar, quartz, and biotite with coarser lenses of quartz and feldspar. The rock contains "eye-like" structures of quartz. The Saganaga granite is a pink to gray rock with large ovoids or "eyes" of quartz. These "quartz eye" granites and gneisses occur as pebbles and boulders in Knife Lake conglomerates (Timiskamian). In Wisconsin and Michigan, granites and gneisses that intrude Keewatin rocks and are pre-Huronian in age have been called Laurentian. Field relationships with Timiskamian rocks are unknown, so these granites and gneisses could be either Laurentian or Algoman in age.

TIMISKAMIAN SYSTEM  Rocks considered to be Timiskamian in age occur in Minnesota and may occur in Michigan. Rocks of the Knife Lake group in Minnesota are of post-Laurentian and pre-Huronian age and are correlated with the Timiskamian rocks of Canada. The Knife Lake group will be described briefly to illustrate the general characteristics of these rocks. It is possible that metamorphosed pre-Huronian sediments in Iron and Dickinson Counties in the Menominee Range area, Michigan, (the Dickinson Group of James, *et al.* (20)) and sedimentary rocks near Silver Lake, north of the Marquette Range, Michigan, may be of Timiskamian age. The pre-Huronian sedimentary rocks of Michigan, which are now often schists, gneisses, and amphibolites, will not be described further.

The Knife Lake group, in Northern Minnesota, is from 11,000 to 22,000 feet thick (Goldich, *et al.* (19)) and is found in a large area between the Vermilion Iron Range and the Canadian border and in a smaller area north of the Mesabi Range between Virginia and Gilbert, Minnesota. The Knife Lake rocks are largely slates, graywackes, and tuffs with conglomerates and agglomerates or volcanic breccias. The slates are gray to dark gray and green, laminated to well bedded, and massive. They are often interbedded with graywacke and locally have graded bedding. The graywackes are from light gray to greenish and dark gray, range from well bedded to massive, with grains ranging in size from silt to grit. In hand specimen, the graywackes often resemble medium, fine-grained, mafic igneous rocks. Locally extensive conglomerates occur. Clements (1) considered all conglomerates to be in the basal part of the sequence, but Gruner (6) showed that lenses and horizons of conglomerate occur throughout the Knife Lake group. Gruner reports the occurrence of a lens of conglomerate 4000 feet thick, north of Ogiskemunce Lake that has all but disappeared in 4 miles along the strike. At Cache Bay on Saganaga Lake, a thick conglomerate rests unconformably on the eroded Saganaga granite of Laurentian age. Tuffs are common but were mapped as slates by Clements. They usually are fine grained with subangular to angular grains commonly less than 4 mm in diameter. Tuffs may be associated with, or grade into, agglomerates. Some andesite porphyry occurs that may be a flow rock. No ellipsoidal greenstones, so characteristic of the Keewatin, were observed by Gruner. Volcanic materials in the Knife Lake group commonly are of andesite composition.

ALGOMAN INTRUSIVES    Pre-Huronian granitic rocks that intrude the Timiskamian and are pre-Huronian in age are correlated with the Algoman granites of Canada. These rocks include augite syenite, syenite, granite, gnessic granite, and gneiss. Radiometric dating of the Algoman granites indicates an average age of about 2.5 b.y. (Goldich, *et al.* (19)). In Minnesota, the granites in the Giants Range north of the Mesabi Range, near Lake Vermilion and at Snowbank Lake, are termed Algoman. Algoman granites may occur in Michigan and Wisconsin because pre-Huronian sedimentary rocks of possible Timiskamian age were intruded and metamorphosed in pre-Huronian time.

## Middle Precambrian Rocks

Huronian rocks of the middle Precambrian sequence are represented by thick metasediments and volcanics. The characteristic metasediments are quartzites, marbles, iron formation, and slates, indicating a mature weathering of the source rocks. The pre-Huronian metasedimentary rocks of the Lake Superior Region are characterized by materials derived from areas subjected to immature weathering.

There was a profound erosional period beteen the Algoman and Huronian, as is shown by a prominent unconformity at the base of the Huronian sequence throughout the region. The significance of this interval in time is not known. As pointed out by Goldich, *et al.* (19, P-74) the "radiometric dating method inherently makes periods of orogeny the logical basis for subdivision." Thus, between the Algoman and the Penokean orogenies, lies the Huronian. With a dating of about 2.5 b.y. for the Algoman and 1.7 b.y. for the Penokean, the Huronian has a maximum time interval of 800 million years.

Huronian rocks include the important iron formations of the Mesabi and Cuyuna Ranges in Minnesota and of the Michigan and Wisconsin iron ranges. The iron formations in the middle part of the Huronian appear to correlate between ranges and to tie the sedimentary sequences together. On the Mesabi, Cuyuna, and Gogebic Ranges, the Huronian (Animikie) sequence can be generalized as having (1) a pre-iron formation clastic unit, commonly from 50 to a few hundred feet thick, consisting of quartzite, quartz-argillite, and argillite and (2) a post-iron formation clastic unit several thousands of feet thick, consisting largely of argillite. This generalization cannot be made for the other Michigan iron ranges. The Michigan Huronian has the most complete and varied rock sequence in the Lake Superior region. It is described by Van Hise and Leith (2), James (17), Tyler and Twenhofel (9), James, *et al.* (20), and others. The accompanying reports in the Graton-Sales Volume, describing the Mesabi, Marquette and Menominee Ranges, summarize the geology of these areas and describe the Huronian (Animikie) rocks.

## Penokean Orogeny and Intrusives

The Penokean orogeny is a major mountain-building epoch in the Lake Superior Region. This orogenic belt extends through Central Minnesota, Northern Wisconsin, and the Upper Peninsula of Michigan. The axis of the

Penokean deformation appears generally to be south of the main Minnesota, Wisconsin, and Michigan iron ranges. Much of the metamorphism, and most of the structural features, found in the iron ranges may be related to the Penokean orogeny.

Granites, tonalites, granitic gneisses, and mafic intrusives (sills and dikes) are associated with the Penokean orogeny. The geographic distribution of these rocks is rather poorly known as their most extensive development is away from the important iron ranges, often in areas of heavy glacial cover.

Metamorphism, associated with the Penokean orogeny in Michigan, is described by James (14). Four nodes of high-grade metamorphism are recognized; (1) the northwestern area (Watersmeet Node), (2) south of the Menominee Range (Florence County Node), (3) Central Dickinson County (Peavey Node), and (4) South Marquette County (Republic Node). It seems probable that these nodes are satellitic to an extensive, high-grade, Penokean metamorphic belt that extends through northern Wisconsin and central Minnesota south of the Iron Ranges. Huronian rocks of a large part of the Minnesota, Michigan, and Wisconsin iron ranges are of low metamorphic rank.

The important "soft," natural iron ore deposits of the Lake Superior Region occur in the iron formations within low-rank metamorphic zones. Recrystallization of the iron formation with a change of cherty quartz to granular quartz and of stilpnomelane, minnesotaite, greenalite, and chlorite to pyroxenes and amphiboles, effectively stops the large-scale, secondary oxidation and leaching that is essential to the development of "soft" iron ores.

### Upper Precambrian Rocks

Upper Precambrian rocks in the Lake Superior Region record about 1 billion years of geologic time from about 1700 m.y. to the start of the Paleozoic era. This time span includes the erosion interval after the Penokean orogeny and the deposition of Keweenawan rocks both before and after the Grenville period of intrusion and deformation. The upper limit of the Precambrian is not clearly defined in this region.

KEWEENAWAN SYSTEM  Keweenawan rocks occur south and west of Lake Superior and appear to underline most of the Lake Superior basin. The Keweenawan rocks dip toward the basin with steeper dips around the outer margin and progressively flatter dips inward with essentially horizontal beds near the central part of the basin.

The Keweenawan sequence includes a basal sandstone and conglomerate, from a few feet to possibly 100 or more feet thick, a middle volcanic group with interbedded conglomerate and sandstone, and an upper group of conglomerate, shale, and sandstone. The volcanic group is up to about 20,000 feet thick, and the upper sedimentary group is from 17,000 to 18,000 feet thick. The Keweenawan rocks are described by Goldich, et al. (19), White (12), and Tyler, et al. (5).

Middle Keweenawan rocks are intruded by the large Duluth gabbro complex in Minnesota and smaller gabbroic bodies in Michigan and Wisconsin. A small granite body intrudes the Keweenawan flows near Mellen, Wisconsin. Goldich, et al. (19) dated these intrusive rocks by radiometric methods as 1000 to 1100 m.y. old. This age is comparable to that of the Grenville orogeny in the eastern United States and Canada. The middle Keweenawan flows were eroded and sediments deposited in the Lake Superior Basin during Upper Keweenawan time.

### STRUCTURE

The structural pattern of the Precambrian rocks of the Lake Superior Region is complex. It is the composite result of four widely separated periods of deformation, Laurentian, Algoman, Penokean, and Grenville.

The Laurentian orogenic period is imperfectly known as the effect of this early deformation is masked by the structural and metamorphic patterns of later orogenics.

The pre-Huronian rocks were strongly deformed by the Algoman orogeny. The Algoman period of deformation and intrusion established a general N60°E to N80°E strike for major folding of the Keewatin and Timiskamian rocks. This general regional strike also is the common direction for the Penokean folding of Huronian rocks and for the general axial direction of the Lake Superior Basin developed during Grenville time.

The Penokean period of deformation and intrusion appears to have developed most of the observed structures (faults and folds) of the Iron Ranges, with the exception of tilting and local faults on the Gogebic Range. The axis of the Penokean orogeny appears to extend across central Minnesota, south of the Cuyuna Range and across northern Wisconsin, south of the Gogebic and Menominee Ranges. The geology of the projected axis area is imperfectly known as much of the area is covered by thick glacial drift. The Mesabi Range, and its northeastward extension, the Gunflint Range, was less deformed and metamorphosed by the Penokean orogeny than are the other Iron Ranges. The Huronian rocks in this area may have been protected by the stable area of pre-Huronian rocks underlying and to the north of the Mesabi Range. Small faults and folds are common along the Mesabi Range. The Cuyuna Range, Minnesota, and the Michigan Iron Ranges were folded and faulted during Penokean time. These rocks are cut by post-folding dikes of Penokean or Keweenawan age.

The Lake Superior basin was formed in late Keweenawan time by a broad downwarping and faulting. Huronian rocks of the Gogebic Range are close to the edge of the basin and were locally faulted and tilted 55° to 65° northward during this deformation. Major faulting of Keweenawan rocks occurs along the south side of Lake Superior, i.e., the Keweenaw fault along the southeast side of the Keweenaw Peninsula and the Douglas thrust fault south of Superior and Ashland, Wisconsin (Figure 2). No important late Keweenawan faulting is recognized along the northwest side of the Lake Superior basin.

## LAKE SUPERIOR IRON ORES

Iron ores of the Lake Superior Region may be grouped into six types, based on mineral and chemical composition, ore structure, and geologic occurrence. These are: (1) soft ores, (2) hard ores, (3) conglomerate ore, (4) siliceous ore, (5) magnetite taconite ore, and (6) jasper ore. A seventh type of iron-bearing material, the hematitic and limonitic iron formations, also called oxidized iron formation, oxidized taconite, semi-taconite, or ferruginous chert, is a potential iron ore.

### Soft Iron Ore

Soft ores occur on all iron ranges and constitute the principal natural ore shipped from the Lake Superior Region. The soft ores are characteristically porous and are composed of hematite and limonite (goethite) with minor magnetite and manganese oxides. Quartz, clay minerals, and minor carbonate minerals are the common gangue materials. These ores range in texture and structure from compact to rubbly, friable, and sandy to fine grained

and semi-plastic. In most of the ore, the bedding, and some of the textural features of the iron formation are retained.

Soft ore bodies range in size from narrow ore stringers along a single fracture to deposits containing hundreds of millions of tons. The ores occur in trough, fissure, stratiform, and irregular orebodies. The most important ore bodies are in trough deposits that, on the Mesabi Range, are related to faults, folds, or zones of fracturing, on the Gogebic Range—to the intersection of dikes and impervious beds, on the Menominee Range—to fold troughs and on the Marquette Range—to a combination of troughs formed by folds, faults and dikes. Fissure ore bodies occur along faults and fractures cutting the iron formation. Stratiform ore bodies occur within certain favorable stratigraphic units of the iron formation. Ores in stratiform bodies may be, in part, related to impervious beds, fractured zones, or folds in which lithologically favorable iron formation was changed to ore. Irregularly shaped bodies, including the "blanket" or flat-lying ore bodies of the Mesabi Range, are related to fractured areas in the iron formation.

Soft ores occur from the rock surface to depths in excess of 4000 feet. On the Mesabi and Cuyuna Ranges in Minnesota, the bulk of the ore occurs within 400 to 500 feet of the present surface, with minor ore and oxidation and leaching to depths of about 1000 feet. Iron ores on the Gogebic and Marquette Ranges have been mined to a depth of over 3000 feet with oxidation and leaching known to over 4000 feet. Generally, soft ore bodies either extend to the present erosion surface, or are connected to the rock surface by oxidized and leached zones.

### Hard Iron Ore

Hard ores occur on the Marquette and Vermilion Ranges. The hard ores of the Marquette Range are composed of compact, fine grained, steel-gray hematite, specular hematite, magnetite, or martite. Ore bodies occur at the top of the Negaunee iron formation and are associated with jaspilite, unoxidized iron formation, and basic sills or dikes.

The hard ores of the Soudan Mine on the Vermilion Range are described by Klinger (15). They are composed of fine-grained, compact, steel-gray hematite associated with jaspilite and occur as narrow, lenticular, or somewhat sinuous bodies. A favored location for ore occurrence is at the ends of lenses of jaspilite, bounded by greenstone. The orebodies range from 50 to 1000 feet in strike length, up to 100 feet in thickness, and extend to depths in excess of 2500 feet. Ore contacts with greenstone usually are sharp but are often gradational with jaspilite. The hard ores of the Ely trough, Vermilion Range, are composed of a rubble of hard, angular fragments of steel-gray hematite in a fine-grained, soft-hematitic matrix, or of compact, steel-gray hematite composed of angular fragments of steely hematite in a gray hematite matrix, or of fragments of compact, steel-gray hematite in a red-brown, hematitic, calcite matrix. The contact between compact ore and calcitic ore is gradational. It seems likely that the ore with the soft hematitic matrix was derived from the calcitic ore by leaching of the calcite. The ores occur along the bottom of a faulted syncline of jaspilite. The jaspilite and ore occur above Ely greenstone. The ore contact with greenstone is sharp.

## Conglomerate Ore

Locally, conglomerates formed by weathering and erosion of iron formation and iron ores contain sufficient iron to be mined as ore. Conglomerates at the base of the upper Cretaceous sediments on the Mesabi Range, Minnesota, are mined locally. These ores consist of rounded to sub-angular fragments of iron ore and minor iron formation in a limonitic matrix. They usually are high in $Al_2O_3$, sulfur, and phosphorus. Some iron ore has been mined from conglomerates of the Goodrich formation at the base of the middle Huronian on the Marquette Range, Michigan. These conglomerates were derived from the Negaunee iron formation during a Precambrian erosion interval. The Goodrich conglomerate ore is a type of hard ore, and locally is associated with hard ore that occurs at the top of the Negaunee formation. It is possible that some of the iron in the Goodrich conglomerate was introduced at the time of the development of the hard ores.

## Siliceous Ore

A siliceous grade of iron ore containing over 18 per cent silica is shipped from the Lake Superior Region. The bulk of siliceous ore is obtained in Michigan from the Marquette Range; however, in past years, some siliceous ore also was mined from the Menominee Range. The common siliceous ores, mined in separate operations, contain from 35 to 40 per cent iron and 38 to 44 per cent silica.

The siliceous ores are obtained from iron-rich stratigraphic units or from areas where there has been some iron enrichment in the iron formation. The siliceous ores are an iron-rich ferruginous chert.

## Magnetite Taconite Ore

Magnetite taconite ore is the term applied to iron formations that contain sufficient magnetite to be processed commercially by fine grinding and magnetic separation to yield a high-grade iron-ore concentrate. Magnetite-bearing, cherty iron formations have an extensive occurrence on the Mesabi Range, the western end of the Gogebic Range, the Marquette Range, and a local occurrence on the Menominee Range. Only a part of the magnetite-bearing iron formation contains sufficient magnetite to be of ore grade. Taconite ores occur within the iron formation and are parts of magnetite-rich units that are of sufficient size to be mined by large-scale open-pit methods. Magnetite is the only ore mineral and commonly is associated with quartz or cherty quartz, silicates, carbonates, and some hematite and goethite. Magnetite taconites are mined extensively on the Mesabi Range and at the Empire Mine on the Marquette Range. The Mesabi and Marquette Range taconite ores are described in accompanying papers in the Graton-Sales Volume.

## Jasper Ore

Jasper ore is the term applied to iron formations composed of platy, crystalline, or specular hematite that is banded with fine-grained quartz and can be concentrated, after fine grinding, to yield a high-grade iron-ore product. Jasper ores commonly are red-brown to gray-white in color. They are a recrystallized, hematite-chert iron formation with little, if any, indication of enrichment in iron by natural processes. Hematite is the common ore mineral with some associated magnetite. The gangue is largely quartz with some silicates. Hematite is recovered commercially in spirals and by froth flotation. Jasper ores are being mined at the Humboldt and Republic mines on the Marquette Range and at the Groveland Mine on the Menominee Range, Michigan.

## ORIGIN OF THE LAKE SUPERIOR IRON ORES

Taconite ores and jasper ores occur as lithic units in cherty iron formations. The origin of these ores is the origin of particular facies of cherty iron formation, i.e., the deposition and the post-depositional history of the iron formations.

The geologic history of all types of ores starts with the deposition of cherty iron formation. The origin of the iron formations has been debated for many years without reaching agreement as to the source of the silica and iron or the environment and mode of deposition. There is general agreement that the cherty iron formations are chemical or biochemical sediments and that the iron and silica were derived either from the weathering of rocks or from volcanic sources. James (11), in 1954, discussed the origin of Precambrian iron formations and grouped them as oxide, silicate, carbonate, and sulfide facies. In 1966, James (26) summarized the extensive literature concerning the chemistry of iron-rich rocks, including Precambrian iron formations. A major part of the Lake Superior cherty iron formations appears to have been deposited as cherty carbonate-silicate iron formation. Iron formation believed to have been deposited as the oxide facies has a less widespread occurrence. There appears to have been a broad overlap of iron carbonate and iron silicate deposition to give a composite carbonate-silicate facies with a range from dominant carbonate to abundant silicate. There is doubt as to the abundance of primary iron-silicate deposition for the nature of the original iron-rich sediment is masked by post-depositional changes that are both diagenetic and metamorphic. The cherty iron formations characteristically are layered rocks with silica-rich layers from ⅛ inch to several inches thick alternating with thinner siliceous iron-rich layers. The alternating deposition of silica-rich and iron-rich layers is a feature to be explained in any theory of origin.

The two views commonly held for the source of iron and silica in Precambrian iron formation are: (1) from volcanic sources and (2) from the weathering processes. Van Hise and Leith (2) proposed a volcanic source for iron and silica and Lepp and Goldich (24) have brought up to date the concept of a source of iron and silica from weathering processes. Proponents of each view look to an immediate source for iron and silica in a contemporaneous sequence of supply and deposition.

Cherty iron formations are common in the upper part of the lower Precambrian and in the middle Precambrian throughout the world. Cherty iron formations are uncommon in the upper Precambrian and unknown after the Precambrian. The environmental conditions favorable to iron formation deposition were not unusual for an extensive period of geologic time; however, rock sequences in which evidence of abundant life can be found, do not contain cherty iron formations. This apparent relationship between cherty iron formation deposition and the development of life has led many workers to postulate that the earth's atmosphere had a different composition during the early and middle Precambrian time than it does now. Some, as Abelson (25), Holland (21), and Lepp and Goldich (24), postulate a reducing atmosphere in early Precambrian time with a gradual evolution to its present composition. A reducing Precambrian atmosphere would permit the transportation of silica to the seas as $H_4SiO_4$, and iron as bicarbonates. Also, under such atmospheric conditions, silica and iron may have been accumulated in sea water in larger amounts than it does in modern seas. If this situation existed, the Precambrian seas could have been the immediate source of silica and iron for the iron formations, much as the sea has been the immediate source of salts in later geological time. The iron and silica, derived over a long period of time from weathering and volcanic sources, would be deposited, as iron formation, in areas where the environment permitted deposition of silica and iron. In this concept, the silica and iron content of the seas would have decreased to the modern levels as the composition of the atmosphere changed. The important change in composition of silica and iron in the seas would be related to the period of change from a reducing to an oxygenated atmosphere, possibly in late middle Precambrian or in late Precambrian time. Red sandstones are common in the late Precambrian, indicating that oxygen was readily available, as was iron, at that time.

The alternation of silica-rich and iron-rich layers in iron formations is not easily explained. Some workers call upon annual climatic variations, cold and warm seasons; the literature concerning possible environments is extensive. Deposition has been suggested in barred basins, Woolnough (7), basins with restricted access to open oceans, James (11), in lakes, Hough (16) and Govett (27), and shallow epicontinental seas, White (13). The environment of iron formation deposition appears to require shallow water and, possibly, a restricted access to the open oceans. The granule-textured Biwabik iron formation of the Mesabi Range and the Ironwood iron formation of the Gogebic Range, have sedimentary characteristics of clastic sand-size sediments as noted by Mengel (22). This suggests environmental conditions in which iron minerals and chert formed as ovoid grains that may have been transported before final sedimentation.

The discovery of fossils in the Gunflint iron formation, Ontario, by Tyler and Barghoorn (12) in 1953, proved the existence of primitive life forms in the seas from which cherty iron formations were deposited. This is added evidence for the deposition of at least a part of the iron and silica as a biochemical sediment.

The post-depositional history of iron formations also is a matter of common debate as it is difficult to distinguish the diagenetic changes from those of metamorphic origin. All Lake Superior iron formations are more or less metamorphosed with recrystallization and the development of metamorphic minerals.

The magnetite-rich iron formations and the magnetite taconite ores contain magnetite as the dominant iron mineral with various amounts of siderite, and other carbonates, iron silicates, and cherty quartz or quartz. The magnetite-rich iron formation is believed to have been deposited as the carbonate-silicate facies with magnetite developed later by metamorphism. Evidence for the postdepositional origin of magnetite is presented by LaBerge (23). Gruner (8) and White ( 13) postulate a primary origin for the magnetite.

The jasper ores of Michigan may have developed from iron formations deposited in the oxide facies as primary hematite and chert. Under this concept, the hematite iron formation was later recrystallied, by metamorphism, to form a granular rock composed of hematite and quartz with minor iron silicates. An alternate origin is suggested by Van Hise and Leith (2), whereby the original carbonate-rich iron formation was oxidized during an erosion interval in middle Huronian time and then later recrystallized by metamorphism.

The soft iron ores of the Lake Superior Region were formed by the oxidation of the iron minerals in the iron formation to hematite and goethite and the leaching of the chert and the calcium, magnesium, and silica from the other gangue minerals. Soft ores of the Mesabi and Cuyuna Ranges, Minnesota, occur, within the iron formation, at shallow depths below the rock surface and are believed to have developed by the action of surface waters circulating in porous and permeable zones associated with faults, folds, and zones of fracturing. Gruner (8) postulated a modified system including hydrothermal activity. It seems, however, that the development of these ores can be explained adequately by the action of surface waters.

Some of the soft ores of the Gogebic and Marquette Ranges occur at depths in excess of 3000 feet in trough-shaped ore bodies. The configuration of these deposits is such that the "plumbing" needed to obtain a large volume of circulating surface waters is difficult to postulate. For these ores, some magmatic or hydrothermal assistance to induce circulation seems to be required. Even though some hydrothermal assistance seems necessary, the soft ore bodies are situated where their localization can be best explained by assuming downward moving waters, even to the maximum depths of known ore occurrence. This suggests a system of downward moving oxygenated surface waters that mingle with hydrothermal, magmatic waters at depth, with the warmed and hydrothermally influenced waters rising and escaping through as yet unrecognized channels. Making some basic assumptions, it is possible to postulate a system that would include the hard ores and soft ores, of the Marquette Range, in a single, unified system. This system would include downward moving surface waters that formed soft ore but at depth were heated and mixed with hydrothermal waters that ascended along available channelways to form hard ore bodies.

The hard ores of the Marquette and the Vermilion Ranges are believed to have formed by hydrothermal replacement of the silica, and possibly some silicate minerals, in the iron formation, by iron oxides. In both areas, the original iron formation replaced may have been a primary oxide facies of iron formation.

The hard ores occur only in the iron formation. This may be due to a relatively higher permeability of iron formation compared to slate or greenstone and to the more ready replacement of cherty quartz and certain silicate minerals than of the greenstone or slate. The contacts between iron ore and associated iron formation often are markedly gradational, whereas the contacts with greenstone or slate are sharp.

## SUMMARY

The Precambrian iron ores of the Lake Superior Region occur within cherty iron formations of Keewatin and Huronian age. These iron ores are classified by composition, ore structure, and geological occurrence into six ore types (1) soft ore, (2) hard ore, (3) conglomerate ore, (4) siliceous ore, (5) magnetite taconite ore, and (6) jasper ore. Oxidized iron formation, also called ferruginous chert, or oxidized taconite, or semi-taconite, is a potential ore that may become important through technological developments in ore dressing methods.

For almost 100 years, the natural iron ores have been mined on a large scale and are now approaching exhaustion. They are being replaced, and the lower grades of natural ores are being displaced, by high-grade iron ore, largely as pellets, produced from magnetite taconite ore and jasper ore.

The natural iron ores, including ores that are treated by washing, screening, and gravity concentration methods, were formed from the cherty iron formations where special conditions allowed the oxidation of the iron minerals and the leaching of silica. The soft ores are in-situ deposits. They represent a wide range of efficiency of leaching of the silica from slight leaching, sufficient partly to free the silica and iron oxide grains, to essentially complete leaching of the silica. Deformation of the iron formation, including both faulting and folding, developed porous and permeable zones within the iron formation that became channelways for solutions. Waters, either hot or cold, moved through the permeable zones, oxidizing the iron minerals to hematite and goethite and leaching the gangue minerals. The soft ores appear to have developed largely by the oxidizing and leaching action of downward moving surface waters, possibly aided by hydrothermal action to induce circulation, for the deep ores.

Hard ores are believed to have developed by hydrothermal replacement of the cherty iron formation. Iron was taken into solution, probably from another part of the iron formation, transported by hydrothermal waters to an area where silica was taken into solution and iron oxides deposited.

The siliceous ores, magnetic taconite ores, and jasper ores occur as lithologic units within the cherty iron formations. The history of ore development is the depositional and postdepositional history of the iron formation. Special local conditions may have had some effect on ore formation, but these special conditions also altered the adjacent, non-ore, iron formation.

## REFERENCES CITED

1. Clements, J. M., 1903, The Vermilion iron-bearing district of Minnesota: U.S. Geol. Surv. Mono. 45, 447 p.

2. Van Hise, C. R. and Leith, C. K., 1911, The Geology of the Lake Superior Region: U.S. Geol. Surv. Mono. 52, 641 p.

3. Wolff, J. F., 1915, Ore bodies of the Mesabi Range: Eng. and Min. Jour., v. 100, p. 89–94, 135–139, 178–185, 219–224.

4. Hotchkiss, W. O., 1919, Geology of the Gogebic Range and its relation to recent mining developments: Eng. and Min. Jour., v. 108, p. 443–501, 537–577.

5. Tyler, S. A., et al., 1940, Studies of the Lake Superior Pre-Cambrian by accessory mineral methods: Geol. Soc. Amer. Bull. v. 51, p. 1429–1538.

6. Gruner, J. W., 1941, Structural geology of the Knife Lake area of northeastern Minnesota: Geol. Soc. Amer. Bull. v. 52, p. 1577–1642.

7. Woolnough, W. G., 1941, Origin of banded iron deposits—a suggestion: Econ. Geol. v. 36, p. 465–489.

8. Gruner, J. W., 1946, The mineralogy and geology of the taconites and iron ores of the Mesabi Range, Minnesota: Iron Range Resources and Rehabilitation, St. Paul, 127 p.

9. Tyler, S. A. and Twenhofel, W. H., 1952, Sedimentation and stratigraphy of the Huronian of Upper Michigan: Amer Jour. Sci. v. 250, p. 1–27, 118–151.

10. Gruner, J. W., 1954 ,, A realistic look at taconite estimates: Min. Eng., v. 6, no. 3, p. 287–288.

11. James, H. L, 1954, Sedimentary facies of iron formation: Econ. Geol., v. 49, p. 235–291.

12. Tyler, S. A. and Barghoorn, E. S., 1954, Occurrence of structurally preserved plants in Precambrian rocks of the Canadian Shield: Science, v. 149, p. 606–608.

13. White, D. A., 1954, The stratigraphy and structure of the Mesabi Range, Minnesota: Minn. Geol. Surv. Bull. 28, 92 p.

14. James, H. L., 1955, Zones of regional metamorphism in the Precambrian of northern Michigan: Geol. Soc. Amer. Bull. v. 66, p. 1455–1488.

15. Klinger, F. L, 1956, Geology of the Soudan mine and vicinity: Geol. Soc. Amer. Guidebook for Field Trips, Field Trip No. 1, p. 120–134.

16. Hough, J. L., 1958, Fresh-water environment of deposition of Precambrian banded iron-formations: Jour. Sed. Petrol. v. 28, p. 414–430.

17. James, H. L., 1958, Stratigraphy of Pre-Keweenawan rocks in parts of northern Michigan: U.S. Geol. Surv. Prof. Paper 314-C, p. 27–44.

18. Carr, M. S. and Dutton, C. E., 1959, Iron-ore resources of the United States including Alaska and Puerto Rico, 1955: U.S. Geol. Surv. Bull. 1082-C, 109 p.

19. Goldich, S. S., et. al, 1961, The Precambrian

geology and geochronology of Minnesota: Minn. Geol. Surv. Bull. 41, 168 p.

20. James, H. L., *et al.,* 1961, Geology of central Dickinson County: U.S. Geol. Surv. Prof. Paper 310, 171 p.

21. Holland, H. D., 1962, Model for the evolution of the earth's atmosphere: *in* Engel, A. E. J., *et al., Editors, Petrologic studies: a volume in honor of A. F. Buddington,* Geol. Soc. Amer., p. 447–477.

22. Mengel, J. T., 1963, The cherts of the Lake Superior iron-bearing formations: Unpublished Ph.D. dissertation, Univ. Wisc.

23. La Berge, G. L., 1964, Development of magnetite in iron formations of the Lake Su-

perior Region: Econ. Geol., v. 59, p. 1313–1342.

24. Lepp, H. and Goldich, S. S., 1964, Origin of the Precambrian iron formations: Econ. Geol., v. 59, p. 1025–1060.

25. Abelson, P. H., 1966, Chemical events on the primitive earth: Nat. Acad. of Sciences Pr., v. 55, no. 6, p. 1365–1372.

26. James, H. L., 1966, Chemistry of the iron-rich sedimentary rocks: U.S. Geol. Surv. Prof. Paper 440-W, p. W1–W51.

27. Govett, G. J. S., 1966, Origin of banded iron formations: Geol. Soc. Amer. Bull. v. 77, p. 1191–1212.

# 24. The Marquette District, Michigan

GERALD J. ANDERSON*

## Contents

## Illustrations

## Tables

* The Cleveland-Cliffs Iron Company, Ishpeming, Michigan.

# Plates

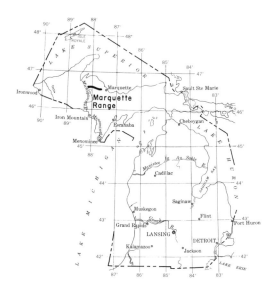

## ABSTRACT

The Marquette District of Central Northern Michigan is the oldest of the Lake Superior iron districts with a mining history dating from 1852 up to the present. The total production of all types of ore through 1965 was 338,120, 146 long tons. Emphasis up to the 1950's was on mining the direct shipping ores; however, demands for higher-grade and better-structured ores in the last decade has resulted in a complete shift in emphasis to beneficiating the low-grade Negaunee Iron-formation to produce high-grade iron ore pellets.

The Precambrian sediments of the Marquette Range have been correlated with the Animikie Series of the Thunder Bay District, Ontario, and are contained in a westward-plunging synclinorium that is approximately 33 miles in length and 3 to 6 miles in width.

The rock unit of major economic importance is the Negaunee Iron-formation that reaches a thickness of 2500 feet near the City of Negaunee. The primary iron content of the Negaunee Iron-formation averages 26 per cent, whereas the altered iron-formation approximates 31 per cent.

The Negaunee Iron-formation has been divided into three principal types: diagenetic, oxidized, and metamorphic. The ores produced to date fall into four categories: high-grade soft ores, high-grade hard ores, siliceous ores, and concentrates and agglomerates (pellets). The soft ores generally are found in synclines and in fault-trough structures, bounded by mafic dikes along at least one side, and are mainly found in the basal portion of the Negaunee. In contrast, the hard ores occur in the uppermost portion of the Negaunee and are controlled by structural features such as folds, faults, and dikes. Iron-formation that has undergone diagenetic and metamorphic alteration makes up the main beneficiating ore types.

Studies on the Negaunee Iron-formation have played an important role in the formulation of many of the theories on the origin of the Precambrian iron-formations.

## HISTORICAL

### Mining Operations in the Area

The Marquette District is situated in Marquette County and the east central part of Baraga County, Michigan. It was the first of the Lake Superior iron districts to be discovered (1845) and mined (1852). Iron mining continues to be the major mining activity. The early mining was in the hard ores found at or near surfaces; these ores originally were worked by open pit methods. As the mines went deeper, inclined skip roads were used to bring the ore to surface and, by 1880, most of the ore was produced from underground workings. Early mining methods were open stoping, room and pillar stoping, and square sets, followed by top slicing of the soft ores in 1890. Sub-level caving was used where conditions permitted, and large scale block caving was introduced in 1950. Production through the early 1950's was from the high-grade direct-shipping ores. However, since the mid-1950's the marketability of these ores has declined steadily, and the emphasis has shifted to beneficiating the low-grade Negaunee Iron-formation. The first commercial iron ore concentrating plant was opened at Humboldt in 1954 and, by the end of 1965, 73 per cent of the production from the district was in the form of beneficiated material and 27 per cent was direct-shipping ore. The direct-shipping ores continue to be mined underground, whereas the low-grade ores are being mined by open pit methods.

### Statistics of Mine Production

The total production of all types of ore from the Marquette Iron Range from 1852 through 1965 has been 338,120,146 long tons. A record was set in 1965 when 8,527,407 tons were produced. Table I shows average annual iron ore production for 5-year intervals since 1930

TABLE I.   *Iron Ore Production Marquette Iron Range*
*(Long Tons)*

| Production | Annual<br>Average | Cumulative<br>Total |
|---|---|---|
| 1852–1930 | | 170,560,033 |
| 1930–1934 | 2,211,588 | 181,617,973 |
| 1935–1939 | 3,855,689 | 200,896,418 |
| 1940–1944 | 5,464,542 | 228,219,128 |
| 1945–1949 | 4,468,985 | 250,564,053 |
| 1950–1954 | 5,151,652 | 276,322,313 |
| 1955–1959 | 5,079,787 | 301,721,248 |
| | Annual | |
| 1960 | 5,498,497 | 307,219,745 |
| 1961 | 4,240,268 | 311,460,013 |
| 1962 | 5,151,804 | 316,611,817 |
| 1963 | 5,820,771 | 322,432,588 |
| 1964 | 7,160,151 | 329,592,739 |
| 1965 | 8,527,407 | 338,120,146 |

and the total iron ore production on a yearly basis since 1960.

## PHYSIOGRAPHIC HISTORY AND PRESENT TOPOGRAPHY

Geographically, the Marquette District is on a topographic upland some 600 to 1200 feet above Lake Superior. The average ground elevation near the mines at Ishpeming and Negaunee is more than 1400 feet above sea level. The highest point—Fire Tower Hill in the Tilden Mine area—has an elevation of over 1875 feet. Locally, the topography is rugged with numerous lakes and swamps. The area drains both to Lake Superior and Lake Michigan.

## GEOLOGIC HISTORY

The principal rock units of the Marquette Range are Precambrian in age as shown on Figure 2. Structurally they are assigned to the Southern Province of the Canadian Shield (9). The thick sedimentary series in which the iron ore is found was called Huronian by Whitney in 1857, and the term was used for the next century. James (7) pointed out that the series differs considerably from the type section of the Huronian in Ontario and is more firmly correlated with the Animikie Series of the Thunder Bay District, Ontario; therefore, the United States Geological Survey uses the term Animikie.

The Animikie Series in the Marquette District is in a westward-plunging synclinorium that is approximately 33 miles in length and 3 to 6 miles in width (Figure 1). Near the Baraga County line, the south limb is then folded into a lesser downfold called the Republic Trough. The sedimentary series continues into Baraga, Dickinson, and Iron counties. Older (Archean) rocks are found to the north of the synclinorium, and a granite complex, thought to be related to the folding of the synclinorium, borders the southern edge.

The Cascade Range near Palmer, Michigan, on the southeast side of the Marquette Synclinorium is considered to be a faulted segment of the main structure. The Gwinn and Swanzy area, some 20 miles southeast of Negaunee, and the Dead River area, a few miles north of the Marquette Range, now are separate basins containing Animikie rocks but originally may have been within the main basin of Animikie deposition. The rock unit of major economic importance is the Negaunee Iron-

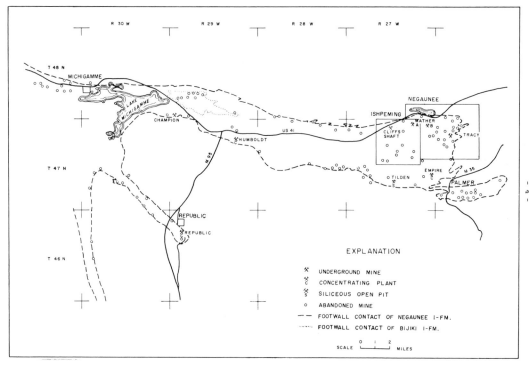

FIG. 1. *Generalized Map of the Marquette Range, showing locations of operating mines.*

formation. A rock unit of lesser economic importance is the Bijiki Iron-formation in the Upper Animikie. The Negaunee Iron-formation reaches a thickness of 2500 feet near the City of Negaunee, and recent work suggests an even greater thickness in the Palmer area. The entire district has been intruded by rocks of Post-Animikie and Pre-Keweenawan (Penokean) age. Cambrian and Ordovician sandstones and limestones are found to the south and southeast, and the entire district was glaciated extensively.

## ECONOMIC GEOLOGY

Iron ores of the Marquette Range are associated with the iron-formation. James (5) defined iron-formation as "a chemical sediment, typically thin-bedded or laminated, containing 15 per cent or more iron of sedimentary origin, commonly but not necessarily containing layers of chert." Such a definition is general but tends to encompass the different lithologies of the iron-formations. The iron-formations can be divided into sedimentary facies, zones of various degrees of oxidation, or zones of enrichment. The sedimentary divisions commonly are based on the dominant primary iron mineral

as proposed by James in 1954: sulfide, carbonate, oxide, and silicate; however, refinements of these divisions are often undertaken by mining companies when it becomes necessary to define the differences in iron content and metallurgical properties encountered during a mining operation.

The carbonate iron-formation is thought to have been the principal primary facies of the Negaunee Iron-formation. All of the known mineable ore bodies have been enriched through diagenetic replacement, more intensive metamorphism, and/or secondary oxidation accompanied by leaching. The primary iron content of Negaunee Iron-formation averages about 26 per cent, whereas the iron content of altered iron-formation approximates 31 per cent. Locally in the Negaunee area, the iron content is higher, averaging about 35 per cent.

Where enrichment has occurred, the chert commonly is replaced by iron-oxides, largely hematite and goethite but locally magnetite. Exploration in recent years indicates that secondary oxidation, leaching, and enrichment extend to far greater depths than thought earlier. Numerous drill holes have cut enriched iron-formation at depths of over 5000 feet from the present land surface. The deepest hole on

| | | | | | |
|---|---|---|---|---|---|
| | QUATERNARY | | | | GLACIAL DEPOSITS |
| PALEOZOIC | CAMBRIAN | | | JACOBSVILLE | SANDSTONE, CONGLOMERATE |
| PRE-CAMBRIAN | LATE | KEWEENAWAN | | | DIABASE INTRUSIVES, LAVA FLOWS, SANDSTONES |
| PRE-CAMBRIAN | MIDDLE | PENOKEAN | | | GRANITES<br>BASIC META-IGNEOUS MOSTLY INTRUSIVE |
| PRE-CAMBRIAN | MIDDLE | ANIMIKIE SERIES (FORMERLY CALLED HURONIAN) | PAINT RIVER GROUP | | FOUND IN THE SOUTHWEST OF THE MARQUETTE RANGE |
| PRE-CAMBRIAN | MIDDLE | ANIMIKIE SERIES (FORMERLY CALLED HURONIAN) | BARAGA GROUP | MICHIGAMME | UPPER ARGILLITE, GRAYWACKE<br>BIJIKI IRON FORMATION<br>MIDDLE ARGILLITE, GRAYWACKE<br>CLARKSBURG PYROCLASTICS<br>GREENWOOD MAGNETIC MEMBER<br>LOWER ARGILLITE, SLATE, GRAYWACKE |
| PRE-CAMBRIAN | MIDDLE | ANIMIKIE SERIES (FORMERLY CALLED HURONIAN) | BARAGA GROUP | GOODRICH | QUARTZITE, ARGILLITE, CONGLOMERATE |
| PRE-CAMBRIAN | MIDDLE | ANIMIKIE SERIES (FORMERLY CALLED HURONIAN) | MENOMINEE GROUP | NEGAUNEE | IRON FORMATION — UNOXIDIZED / OXIDIZED / ENRICHED / LOCALLY CLASTIC<br>(SIAMO-AJIBIK UNDIFFERENTIATED IN WESTERN PORTION) |
| PRE-CAMBRIAN | MIDDLE | ANIMIKIE SERIES (FORMERLY CALLED HURONIAN) | MENOMINEE GROUP | SIAMO | ARGILLITE, GRAYWACKE<br>SLATE — CONTAINS GOOSE LAKE IRON FORMATION |
| PRE-CAMBRIAN | MIDDLE | ANIMIKIE SERIES (FORMERLY CALLED HURONIAN) | MENOMINEE GROUP | AJIBIK | QUARTZITE — THIN TO THICK BEDDED |
| PRE-CAMBRIAN | MIDDLE | ANIMIKIE SERIES (FORMERLY CALLED HURONIAN) | CHOCOLAY GROUP | WEWE | GRAY SLATE — LOCALLY QUARTZITES, CONGLOMERATE |
| PRE-CAMBRIAN | MIDDLE | ANIMIKIE SERIES (FORMERLY CALLED HURONIAN) | CHOCOLAY GROUP | KONA | DOLOMITE — LOCALLY SILICIFIED, ALGAL STRUCTURES, MINOR QUARTZITE, SILTITE |
| PRE-CAMBRIAN | MIDDLE | ANIMIKIE SERIES (FORMERLY CALLED HURONIAN) | CHOCOLAY GROUP | MESNARD | QUARTZITE<br>CONGLOMERATE, GRAYWACKE, ARKOSE |
| PRE-CAMBRIAN | EARLY | | | | FELSITE PORPHYRY<br>TONALITE, GRANODIORITE<br>SCHISTS, METASEDIMENTS, GNEISSES<br>GREENSTONE, MASSIVE - ELLIPSOIDAL<br>FELSITE, METABASALT |

FIG. 2. *Geologic Column of the Principal Marquette District, modified from U.S. Geological Survey.*

the range, drilled to a depth of 6365 feet, encountered oxidation and enrichment to the bottom.

## Ore Occurrences

There are four general types of iron ore which have been produced and shipped from the Marquette Range. They are:

High-grade direct-shipping "soft" ores,
High-grade direct-shipping "hard" ores,
Siliceous ores,
Concentrates and agglomerates (pellets).

Traditionally in the Lake Superior Region, the direct-shipping ores were mined to provide an ore with a base natural (moisture included) iron content of 51.5 per cent. Recently, the direct-shipping ores have averaged 54 per cent Fe, natural, reflecting the competitive pressures of the high-grade foreign ores and iron ore pellets. A premium is paid for the hard lump Marquette Range ores that have a size range of more than 2 inches and less than 8 inches and average 61 per cent Fe natural. The siliceous type of ore, a specialty grade, constitutes small shipments of somewhat richer iron-formation that averages 38 per cent Fe natural and 39 per cent silica. Iron ore pellets produced on the Marquette Range contain from 61 to 65 per cent Fe natural.

In general, the soft ores are porous, friable, and earthy to semi-plastic and are made up chiefly of hematite and martite (locally with some magnetite), with minor amounts of goethite, chert, and occasionally silicates such as mica or chlorite. The hard ores are dense and compact, with low porosity, and are composed of magnetite, martite, dense compact hematite, and specularite. These ores form the lump product. A substantial amount of high-grade ore mined is made up of a mixture of hard and soft ores.

The amenability of the Negaunee Iron-formation to concentration results from its mineralogy and the distribution of the grains of iron minerals and has been classified by Anderson and Han (6) as is shown in Table II.

At the present time, the diagenetic and metamorphic types are being utilized and considerable research is being conducted on the oxidized type.

SOFT ORES Seventy per cent, or 236 million tons, of the total production from the Marquette District has been of the high-grade direct-shipping soft ore. The ores occurred, for the most part, in the basal portion of the

TABLE II.   *Types of Negaunee Iron-Formation*

**A. Diagenetic Iron-formation**
1. Magnetite-chert with some carbonate
2. Magnetite-silicates with carbonate-chert
3. Magnetite-silicates with clastics
4. Carbonate-chert with magnesium

**B. Oxidized Iron-formation**
1. Martite-chert
2. Martite-clastics
3. Goethite-hematite-chert
4. Goethite-chert

**C. Metamorphic Iron-formation**
1. Specular hematite-chert with or without sericite
2. Magnetite-chert with some chlorite and locally garnet
3. Grunerite-chert with magnetite or garnet

Negaunee Iron-formation in deposits up to 260 feet thick measured normal to the bedding and have been mined up to several thousand feet along structural troughs. The ore bodies are generally found in the synclines and in fault-trough structures bounded by mafic dikes along at least one side. Figure 3, a north-south section looking west passing through the eastern side of the City of Negaunee, shows the variety of ore occurrences present in this vicinity.

Another important type of soft ore occurrence is on large mafic sills in the upper part of the iron-formation. These were the first soft ores to be exploited in the Ishpeming area. The ore is irregular both as to its distribution and its thickness. These ore bodies had their principal development south of Ishpeming toward the axis of the Marquette synclinorium. At present, there are no mines operating in this type of ore although this type was an important ore source about 1900. The soft ores, lying on the metadiabase sheets, are limited by the same structural controls as the ores at the base of the Negaunee Formation. In general, these ores were high in iron and low in phosphorus and sulfur.

A sizable tonnage of soft ore in the Negaunee Formation is impregnated with gypsum, which results in sulfur analyses ranging from 0.2 to over 3.0 per cent in the ores. The gypsiferous ores occur at definite elevations, commonly bottoming near the present sea level and extending upward for several hundred feet. These ores appear to be structurally controlled.

At the present time, there are only two soft ore mines operating on the Marquette Range. The Mather Mine is located on the north limb of the syncline and produces ore from depths to 3500 feet; the Tracy Mine is located on the east-central end of the syncline and produces ore from depths to 1500 feet. The combined annual rated production of these two mines is around 3 million tons. (Plates I and

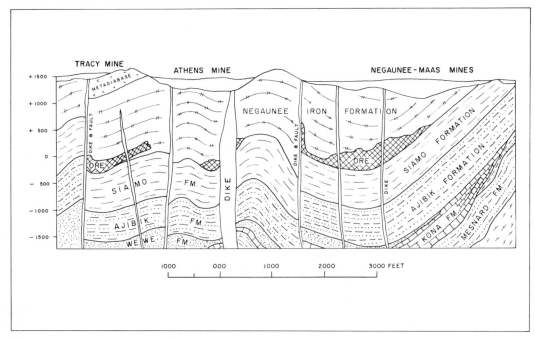

FIG. 3. *Generalized North-South Cross-Section 1 Mile East of West Line of R26W, looking west.*

PLATE I. *Photomicrograph of Mather Mine Soft Martite Ore. 63×. Martite, white; soft earthy hematite, gray.*

II show two types of soft ore from the Mather Mine.)

HARD ORES   Hard ores account for 20 per cent, or 67 million tons, of the total iron ore production (1852 to 1965) from the Marquette District. Most of the hard ores occur in the uppermost portion of the Negaunee Iron-formation immediately below the Goodrich Formation. The footwall of the ore may be unoxidized iron-formation, iron-formation of the "jaspilite" type, or a mafic sill. The outlines of hard ore bodies are frequently determined by structural features such as folds, faults, and dikes.

At present, there are two active hard ore mines: the Cliffs Shaft and the Champion. The production of hard ores is declining with a rapid depletion of available reserves.

CONCENTRATING ORES   Humboldt Mine—The first property on the Marquette Range and in Michigan to utilize fine grinding to obtain liberation and concentration of the iron minerals was the Humboldt Mine. This property has been producing high-grade iron ore concentrates since 1954 and pellets since 1960. The Humboldt deposit is located on the south limb of the Marquette Range synclinorium. Locally, the structural pattern consists of a north-south monocline dipping westward. Dips range from nearly vertical on the northern and southern portions to approximately 55 degrees in the central area.

The treatable iron-formation at Humboldt represents one of several known low-grade specularite iron-formation or jaspilite deposits found immediately below the Goodrich Formation in the upper part of the Negaunee Iron-formation. This portion of the iron-formation is a hard ore jasper (jaspilite) containing about 35 to 40 per cent iron. The hanging wall consists of a quartzite that is, in part, conglomeritic or argillaceous. An alternating series of mafic intrusives and unoxidized silicate iron-formation forms the mining footwall.

The ore section consists of a specular hematite-chert in the upper part and a specular hematite-magnetite-chert in the lower part. Visible subordinate minerals include sericite, muscovite, tourmaline, garnet, apatite, and chloritoid. Locally, sericite is a major mineral.

The mill flowsheet includes three stages of crushing and two stages of grinding in rod and ball mills. Liberation of the specular hematite is achieved at a minus 65 mesh grind and concentration is by froth flotation. The platy crystals of specular hematite are floated, and the chert is depressed. (Plate III illustrates a typical ore from the Humboldt Mine.)

PLATE III. *Photomicrograph of Humboldt Specular Hematite-Magnetite Ore; grid 115 microns. Specular hematite, white; magnetite, light gray; gangue (chert and sericite), gray and dark gray; pores, black.*

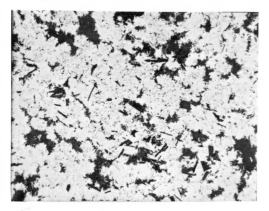

PLATE II. *Photomicrograph of Mather Mine Soft Hematite Ore. 310×. Hematite, white; pores and gangue, dark gray.*

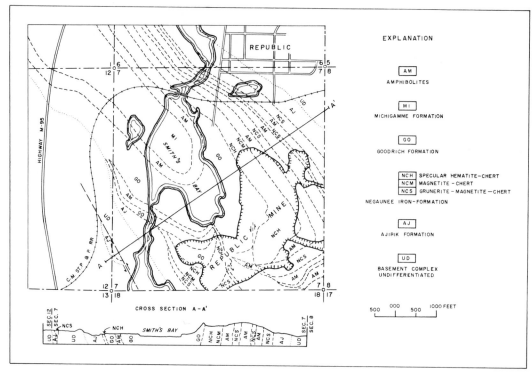

FIG. 4. *Geologic Map and Cross-Section of Republic Mine Area.*

Republic Mine—The largest concentrating ore property on the Marquette Range in 1965 was the Republic Mine with an annual output of over 3 million tons of iron ore pellets per year. Production started at this property in 1956 at 600,000 tons per year, and several expansions have brought it up to its present capacity.

The Republic deposit is located near the southern extremity of a tightly folded syncline known as the Republic Trough that extends southeasterly off the main Marquette Range synclinorium (see Figure 1). Like the Humboldt, the Republic ore body is located immediately below the Goodrich Formation (see Figure 4). The dip of the ore body ranges from 50 degrees near the plunging end of the trough to nearly vertical along the north limb of the trough. The ore body is a tabular mass 400 to 600 feet thick and is being mined along a strike length of 6000 feet. Information from underground workings of the "Old" Republic hard ore mine indicates that the specularite iron-formation extends down dip over 3000 feet. Economic conditions will dictate how deep the ore body may be mined.

The iron-formation at the Republic Mine is subdivided into four basic lithologic types according to major mineral constituents. These include a specular hematite-chert-muscovite conglomerate member at the base of the Goodrich Formation and, in descending order, a specular hematite-chert, a magnetite-chert and a magnetite-grunerite-chert. The upper three members constitute the ore feed for the Republic concentrator. The magnetite-grunerite-chert, although not being mined at present, is a potential ore reserve of fine-grained magnetite that may be utilized by a different flow scheme at some future date. The magnetite-grunerite iron-formation is cut by a series of sill-like amphibolite intrusives. At present, the mining footwall at Republic is either amphibolite intrusive or magnetite-grunerite iron-formation.

The flowsheet at the Republic Mine is essentially the same as at the Humboldt Mine except that, in addition to conventional flotation, the Republic mill employs a final hot-flotation process to upgrade the concentrates to produce a 65 per cent Fe pellet. (Plate IV illustrates a typical ore from the Republic Mine.)

Empire Mine—The Empire Mine and Plant is the first of its kind constructed to treat magnetite ores in Michigan. This property was placed in production in 1963 to concentrate a magnetite-rich portion of the Negaunee Iron-formation that occurs in the Palmer area (see Figures 1 and 5). The ore occurs in a monoclinal fold which dips about 35 degrees to the northwest. Dike-filled transverse faults cut the monocline into blocks. The apparent stratigraphic thickness of iron-formation in the Empire area is about 3000 to 3500 feet. Ore now is being mined from an area 4000 feet long and 1800 feet wide. Diamond drill hole information indicates good concentrating ores to a depth of 1100 feet.

The iron minerals in the Empire ore are magnetite, iron carbonates, iron silicates, martite, and earthy hematite with minor goethite and pyrite. The economic iron mineral is very fine-grained magnetite with a common size range of 400 to 800 mesh (0.037 mm to 0.019 mm). Geologic mapping and ore dressing tests indicate four types of ore from the hanging wall towards the footwall: 1) magnetite-chert-arenaceous iron-formation, 2) magnetite-chert-carbonate iron-formatin, 3) magnetite-chert-silicate iron-formation, 4) magnetite-chert-silicate-carbonate iron-formation (see Figure 5). The ore body is within the Negaunee Forma-

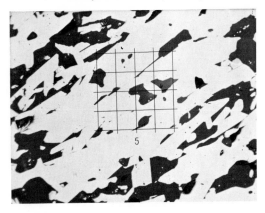

PLATE IV. *Photomicrograph of Republic Specular Hematite Ore; grid 40 microns. Specular hematite, white; gangue (chert), dark gray.*

tion. The hanging-wall rock stripping consists largely of lean sub-economic iron-formation that occasionally may contain ore material, and the footwall of the orebody is mainly a sillicate iron-formation. The actual mining limit differs according to the amenability of the iron-formation material to make a suitable

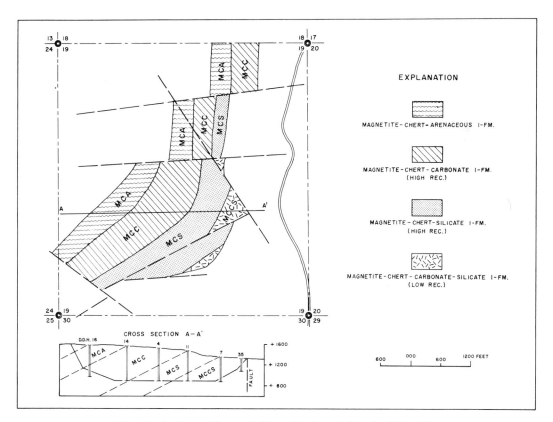

FIG. 5. *Geologic Map and Cross-Section of Empire Mine Pit.*

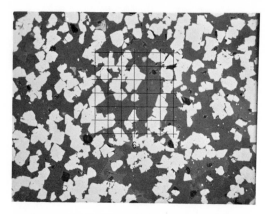

PLATE V. *Photomicrograph of Empire Mine Magnetite Ore; grid 29 microns. Magnetite, white; gangue (chert and carbonate), dark gray; pores, black.*

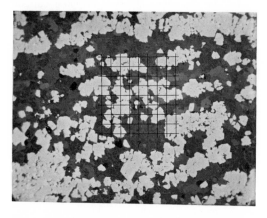

PLATE VI. *Photomicrograph of Empire Mine Magnetite Ore; grid 20 microns. Magnetite, white; gangue (carbonate, chert, and Silicate), gray to dark gray; pores, black.*

grade and give a satisfactory iron recovery. (Plates V and VI illustrate ores from the Empire Mine.) Liberation of the magnetite is achieved at a grind of 90 per cent minus 500 mesh, or about 0.032 mm. The ores are ground in an autogenous circuit and concentrated by magnetic separation and siphonsizers. Amine flotation is utilized for final upgrading to produce a plus 63 per cent Fe pellet.

## ORE GENESIS

The origin of Precambrian iron-formations has received much attention over the years. The work of Tyler and Twenhofel (4) and more recently the work of Lepp and Goldich

(10) and James (12) present discussions of the theories and mechanisms that have been proposed to explain the history of sedimentation, the structure and tectonic environment, and the geochemistry of iron-formations. It is significant to note that the Negaunee Iron-formation on the Marquette Range has played an important role in the formation of many of these theories.

There is no universal agreement on the origin of the high-grade ores of the Marquette Iron Range. There are also differences in opinion regarding both the method and the timing of the enrichment of the hard ores as opposed to the soft ores. The two hypotheses which have been applied to the soft ore deposits of the Lake Superior Region are the "cold water" hypothesis of Leith, *et al.* (2) and the "hydrothermal" hypothesis of J. W. Gruner (3). Both mechanisms can explain the observed secondary oxidation. There are proponents of both hypotheses for the origin of the soft ores found on the Marquette Range. The enrichment observed at great depths and some of the structural relationships of the soft ore bodies can be explained more easily by the hydrothermal hypothesis.

Monograph 52 (1, p. 278–279) outlines a possible origin and time sequence of hard ore formation. The theory described suggests that the upper portion of the Negaunee Iron-formation was exposed to weathering and concentration that was, in part at least, a mechanical classification to produce an iron-rich product. Burial by the Goodrich and later formations followed. After the deposition of the Michigamme Formation, the Animikie sediments were folded, faulted, and intruded. The metamorphism associated with the structural deformation formed the hard ores from the weathered ores. Later Post-Keweenawan erosion re-exposed the Negaunee and Bijiki Iron-formations. They were altered by ground-waters to form the soft ores in structurally favorable areas. Since the soft ores were formed after the dynamic metamorphism, they do not contain the specular hematite and display the other features of the hard ore.

The objection to this concept centers on the "weathered-surface" origin for all of the hard ores. It is necessary to postulate introduction of iron to explain both the geometry and detailed features of many hard ore bodies. This introduction of iron must have been accomplished before or during metamorphism. The hydrothermal school of thought would ascribe much of the hard ore to replacement of the iron-formation by high-temperature solution.

The iron-formation used for concentrating ores in modern processing plants can generally be assigned to 1) specular hematite-magnetite-chert, and 2) magnetite-carbonate-silicate-chert. Both are typically thin banded, reflecting original sedimentary stratification. However, the mineral assemblages appear to be related to the subsequent metamorphic changes. Working in the area of the Republic Mine, Han and Villar (11) found that regional metamorphism had erased most direct evidence relating to pre-metamorphic rock constituents. In other areas of the Marquette Range, where regional metamorphism has been much less intense, Han (8) has proposed a "diagenetic or low-grade metamorphic replacement of iron-rich carbonate by magnetite." Microscopic evidence suggests that "replacement" processes of silication, recrystallization, carbonatization, and silicification took place under static conditions, probably after the iron-formation was completely consolidated.

Many of the differences of opinion concerning the origin of the Negaunee Iron-formation rest in the nature of the environment of deposition and in the timing of diagenetic and metamorphic events. Greater knowledge of Precambrian geology will lead to a more complete understanding of ore genesis in the Marquette Range.

## REFERENCES CITED

1. Van Hise, C. R. and Leith, C. K., 1911, The geology of the Lake Superior Region: U.S. Geol. Surv. Mono. 52, 641 p.
2. Leith, C. K., *et al.,* 1935, Pre-Cambrian rocks of the Lake Superior Region: U.S. Geol. Surv. Prof. Paper 184, 34 p.
3. Gruner, J. W., 1937, Hydrothermal leaching of iron ores of the Lake Superior type—A modified theory: Econ. Geol., v. 32, p. 121–130.
4. Tyler, S. A. and Twenhofel, W. H., 1952, Sedimentation and stratigraphy of the Huronian of Upper Michigan: Amer. Jour. Sci., v. 250, p. 1–27, 118–151.
5. James, H. L., 1954, Sedimentary facies of iron-formation: Econ. Geol., v. 49, p. 235–293.
6. Anderson, G. J. and Han, T-M, 1956, The relationship of diagenesis, metamorphism, and secondary oxidation to the concentrating characteristics of the Negaunee Iron-formation of the Marquette Range: *in* Snelgrove, A. K., *Editor, Geological exploration,* Institute on Lake Superior Geology, Mich. Tech. Univ., p. 63–69.
7. James, H. L., 1958, Stratigraphy of Pre-Keweenawan rocks in parts of Northern Michigan: U.S. Geol. Surv. Prof. Paper 314-C, 44 p.
8. Han, T-M, 1962, Diagenetic replacement in ore of the Empire Mine of Northern Michigan, and its effect on metallurgical concentration (abs.): 8th Annual Institute on Lake Superior Geology, Mich. Tech. Univ., p. 7.
9. Leech, G. B., *et al.,* 1963, Age determinations and geologic studies: Geol. Surv. Canada, Paper 63–17, 140 p.
10. Lepp, H. and Goldich, S. S., 1964, Origin of Precambrian Iron-formations: Econ. Geol., v. 59, p. 1025–1060.
11. Han, T-M and Villar, J. W., 1965, Petrology of the silicate iron-formation in the Republic Mine area, Marquette County, Michigan (abs.): 11th Annual Institute on Lake Superior Geology, Univ. Minn., p. 17–18.
12. James, H. L., 1966, Chapter W: Chemistry of the iron-rich sedimentary rocks: *in Data of Geochemistry (6th ed.),* U.S. Geol. Surv. Prof. Paper 440-W, 61 p.

# 25. The Mesabi Iron Range, Minnesota

R. W. MARSDEN,* J. W. EMANUELSON,†
J. S. OWENS,‡ N. E. WALKER,§
R. F. WERNER‖

## Contents

## Illustrations

* United States Steel Corporation, Pittsburgh, Pennsylvania.
† Reserve Mining Company, Babbitt, Minnesota.
‡ The Hanna Mining Company, Nashwauk, Minnesota.
§ Pickands Mather and Company, Hibbing, Minnesota.
‖ Jones and Laughlin Steel Corporation, Virginia, Minnesota.

# Tables

## ABSTRACT

The iron ores of the Mesabi Range occur in a 340 to 750-foot thick, Precambrian cherty iron formation termed "taconite." For about 65 years, extensive natural iron ore bodies were mined, and the ores either shipped as a direct ore or processed by gravity methods to a shipping grade iron ore. Since 1956, magnetite taconite ores, occurring in magnetite-rich horizons in the Biwabik iron formation, have been commercially concentrated and agglomerated. The transition from natural ores to concentrating-grade ores is progressing rapidly so that, in the near future, a major percentage of the iron ore shipped from the Mesabi Range will be as iron ore pellets produced from taconite.

Natural iron ores of the Mesabi Range occur in the Biwabik formation as trough ore bodies in elongated channels and as irregular and tabular deposits in fractured areas associated with faults and folds. Ore bodies range in size from a few thousand tons to hundreds of millions of tons and may occur in any part of the iron formation. In some areas of strong ore development, ore was formed from the hanging-wall argillite to the footwall quartzite. The natural ores and beneficiating-grade natural crude ores were formed by the oxidation of the iron minerals, magnetite, greenalite, stilpnomelane, minnesotaite, and siderite, to hematite and goethite and by the removal of much of the silica in the cherty quartz and in the associated silicates by leaching. The ores are believed to have formed by the oxidizing and leaching action of surface waters.

Magnetite taconite ores occur as stratigraphic zones in the Biwabik formation where it contains sufficient magnetite to be concentrated profitably, by magnetic methods, to give a magnetite concentrate that is agglomerated into a high quality product. Unoxidized taconite has an extensive distribution away from and between channels of oxidation and leaching but only a part of the unoxidized taconite is of ore grade. Two general types of magnetite taconite ore occur: (1) magnetite taconite of the Main Mesabi Range, generally west of Mesaba, which is composed of magnetite associated with cherty quartz, greenalite, stilpnomelane, minnesotaite, and carbonates and (2) magnetite taconite of the Eastern Mesabi, generally east of Mesaba, composed of magnetite associated with quartz, cummingtonite, actinolite, ferrohypersthene, hedenbergite, fayalite, hornblende, pyroxene, garnet, and biotite. The Mesabi taconite is a metasediment derived from an iron- and silica-rich chemical sediment that has been regionally metamorphosed and, on the eastern end, additionally metamorphosed by the Duluth Gabbro Complex.

## INTRODUCTION

The Mesabi Iron Range of Northeastern Minnesota is a belt of iron formation extending from Birch Lake for about 110 miles southwesterly to a nebulous ending about 15 miles southwest of Grand Rapids. (Figure 1) The eastern end is at the intrusive contact with the Duluth Gabbro, whereas the western end disappears under a heavy cover of glacial and Cretaceous sediments. An extension of the Mesabi Range emerges from below the Duluth Gabbro about 40 miles northeast of Birch Lake as the Gunflint Iron Range of the Thunder Bay District of Ontario and adjacent Minnesota. The correlation seems positive since there are broad similarities in general stratigraphy and mineralogy. To the southwest of Pokegama Lake, the iron-formation is projected beneath glacial and Cretaceous material and is believed to follow a sinuous path to connect with the iron formation of the Cuyuna Range. The correlated segments of the Animikie iron formation, of which the Mesabi Range is a part, extend from near Schrieber, Ontario on the North Shore of Lake Superior to Brainerd, Minnesota, a distance of about 400 miles.

This paper has been edited by Ralph W. Marsden with contributions by John S. Owens, Ralph W. Marsden, Robert F. Werner, Neal E. Walker, and James W. Emanuelson. The identity of individual contributions is lost as information on specific areas has been used as required to maintain a logical presentation of information.

## HISTORY

The first recorded observation of iron-bearing rocks in the Mesabi Range was made in 1866 by Henry H. Eames, the first State Geologist of Minnesota. The early geological work of A. H. Chester and N. H. Winchell in the 1870's and 1880's and prospecting by Peter Mitchell in 1869–1872 revealed the occurrence of magnetic iron-bearing rocks that were of no economic interest at that time. In spite of the uneconomic nature of the outcropping iron formation, there was a continuing interest in the iron ore possibilities of the Mesabi Range after the late 1860's. The

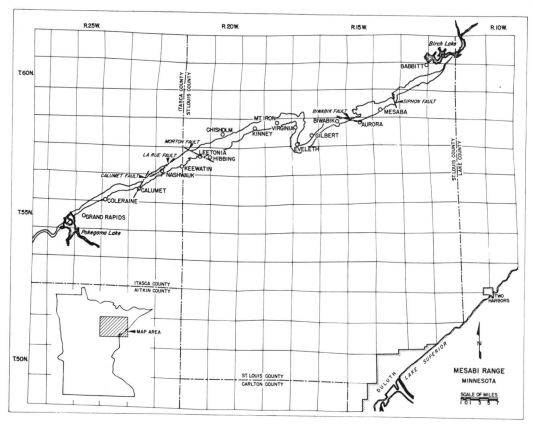

FIG. 1. *The Mesabi Range, Minnesota.*

persistence of prospectors was rewarded with the discovery of iron ore north of Mountain Iron on November 16, 1890 by J. A. Nichols, working for the Merritt Brothers. After the initial discovery, new iron ore deposits were found in rapid succession with the extent of the range known by 1900, at which time it had become a major source of iron ore.

Mining in the early years was conducted largely by underground methods. Open-pit mining, however, soon became the dominant manner of ore extraction; even by 1902, 47 per cent of the ore was mined from open pits. Underground mining continued on a small scale for about 60 years; the last underground mine, the Godfrey, closed in 1961.

Even though very large reserves of direct shipping ore were known, some iron ores were concentrated from lower-grade ore materials as early as 1907. The early beneficiation was done by simple washing of partly leached taconite; more advanced gravity concentration methods were introduced later to treat the more refractory ores. In 1915, a group headed by D. C. Jackling organized the Mesabi Syndicate to study Mesabi Range magnetite-bearing iron formation (commonly termed magnetite taconites). In 1919, the syndicate group formed the Mesabi Iron Company to concentrate magnetite taconite on the eastern Mesabi Range. Extensive research work was done from 1916 to 1924, and a processing plant was in operation from 1922 to 1924. This initial effort to produce iron ore concentrates from taconite, however, could not compete with the still abundant natural ores. In the late 1940's, there was a strong renewal of interest in the production of iron-ore concentrates from magnetite taconite. During the interval between 1924 and late 1940's, the Minnesota Mines Experiment Station continued research on the concentration of magnetite taconite and on the agglomeration of fine iron-ore concentrates. After World War II, the major mining companies established research facilities and soon developed full scale pilot operations. The Reserve Mining Company took over the lands of the Mesabi Iron Company and used the old Mesabi Iron Company plant at Babbitt, Minnesota for pilot operations. The Reserve Mining Company started large-scale commercial production of taconite pellets in 1956 and was followed by the Erie Mining Company in 1958 and the Eveleth Taconite Company in 1965. In 1965, pellets constituted about 36 per cent of the iron ore shipped from the Mesabi Range. By 1968, upon the completion of the taconite processing plants under construction or being expanded, pellets will constitute over 50 per cent of the iron ore shipped from the Mesabi Range. Extensive research work is being done on the concentration of the hematitic and limonitic iron formation, commonly called oxidized taconite or semi-taconite, and, in the future, this material may become a source of iron-ore concentrates.

## PRODUCTION

Iron ore produced and shipped from the Mesabi Range from 1892 to 1965 totaled 2,511,124,000 gross tons of which 687,129,000

TABLE I.   Iron Ore Production, Mesabi Range (gross tons)

|  | Direct Ore | Gravity Concentrates | Taconite[1] | Total |
|---|---|---|---|---|
| 1892–1900 | 31,390,000 | — | — | 31,390,000 |
| 1900–1910 | 192,828,000 | 668,000 | — | 193,496,000 |
| 1911–1920 | 297,777,000 | 35,151,000 | — | 332,928,000 |
| 1921–1930 | 278,712,000 | 53,241,000 | 219,000 | 331,953,000 |
| 1931–1940 | 184,293,000 | 46,589,000 | — | 230,882,000 |
| 1941–1950 | 452,386,000 | 145,589,000 | 78,000 | 598,117,000 |
| 1951–1960 | 334,715,000 | 190,686,000 | 42,151,000 | 567,552,000 |
| 1961 | 11,696,000 | 15,971,000 | 14,529,000 | 42,196,000 |
| 1962 | 10,724,000 | 17,181,000 | 14,257,000 | 42,162,000 |
| 1963 | 8,843,000 | 17,717,000 | 17,077,000 | 43,637,000 |
| 1964 | 9,933,000 | 18,357,000 | 19,372,000 | 47,662,000 |
| 1965 | 10,696,000 | 19,585,000 | 18,868,000 | 49,149,000 |
| Total | 1,823,995,000 | 560,578,000 | 126,551,000 | 2,511,124,000 |

Data: Univ. of Minnesota—Mining Directory (16).

[1] Includes magnetite taconite and oxidized taconite.

gross tons, or about 28 per cent, were concentrated from low-grade ore materials. Table I summarizes the production from the Mesabi Range.

## GRADES OF ORE

The average analyses of iron ores shipped from the Mesabi Range in 1965 are shown in Table II. For additional details concerning ore grades, the reader is referred to the American Iron Ore Association annual summaries. Shipments from the Mesabi Range are changing from natural ores containing about 51 to 57 per cent natural iron to pellets containing from 60 to 67 per cent iron. The high iron content and the uniform composition and size of Mesabi pellets make them very desirable at the blast furnace. There is a less favorable acceptance of the natural ores that differ markedly in quality and structure.

## PHYSIOGRAPHY

The iron-bearing rocks of the Mesabi Range follow the south side of the Giants Range. It is interesting that common practice has led to calling the topographic feature "The Giants Range" and the iron formation belt "The Mesabi Range." The word "Mesabi" in the Chippewa Indian language translates as "Giant." The Giants Range is a prominent topographic feature in the central and eastern part of the Mesabi area, but it disappears west of Hibbing where it merges into the low-lying glacial terrain. The general trend of the Giants Range and of the Mesabi Iron Range is east-north-east; however, in the vicinity of Virginia and Eveleth there is a pronounced bend in the Giants Range and in the iron range locally called "The Horn."

The headwaters of three major drainage basins converge in the Mesabi area with the Giants Range forming a part of the height of land that separates the Mississippi, Hudson Bay, and St. Lawrence watersheds. The elevation of the Mesabi area ranges from a ridge top elevation of about 1900 feet in the east to about 1300 feet in the west along the Mississippi River.

The general strike of the Mesabi Range is related to the east-north-east Precambrian structural trend, but the lesser features and the topography of the west end of the Mesabi Range are largely imposed by the Pleistocene glaciation and post-glacial erosion.

## GEOLOGIC HISTORY

The Precambrian sequence includes rocks classed by Goldich *et al.* (10), as Early, Middle and Late Precambrian. These are equivalent to Archean and Timiskamian, Huronian, and Keweenawan and have radiometric time ranges of plus 2500 m.y., 2500 to 1700 m.y. and 1700 to 600 m.y. respectively. These systems were interrupted by two major orogenies, the Algoman and Penokean (Table III). The Early Precambrian includes all rocks intruded, metamorphosed, and deformed during the Algoman and older orogenies. These form the "basement" rocks below the Animikie Group sediments of Huronian Age. Rocks of the Huronian system were intruded, metamorphosed and deformed by the Penokean orogeny and were eroded extensively before rocks of the Keweenawan system were deposited. Rocks of the Keweenawan system are considered to bridge the time between the Penokean Orogeny and the Paleozic.

There is a gap of almost a billion years in the geological record in the Mesabi area extending from Grenville time to the late Cretaceous. During this period, the natural iron ores formed and the Mesabi Range area was eroded to approximately its present bedrock topography. Marine Cretaceous sediments fill the low areas in the Precambrian surface and may occur in extensive areas south of the outcrop of the iron formation in the Western Mesabi area. In the Central and Eastern Mesabi, largely non-marine Cretaceous sediments are observed only as patches. Cretaceous rocks may have covered the Mesaba area; if

TABLE II.   *Average Grade and Tonnage for Iron Ores Shipped in 1965*

| Ore Class | Tons Shipped | Fe | Phos. | SiO$_2$ | Mn. | Al$_2$O$_3$ | Moisture |
|---|---|---|---|---|---|---|---|
| Bessemer | 1,372,479 | 57.44 | .036 | 8.55 | .58 | .43 | 5.85 |
| Non-Bessemer | 29,691,432 | 53.95 | .055 | 8.15 | .51 | 1.08 | 8.95 |
| Pellets | 18,868,041[1] | 61.17 | .021 | 8.53 | .29 | .37 | 2.82 |

Data: Iron Ore, 1965, American Iron Ore Association, (15,p. 95).

[1] Univ. of Minnesota—Mining Directory, 1965 (16).

TABLE III.　Stratigraphic Column and Chronology of the Mesabi Range and Adjacent Northern Minnesota[1]

| Era | ($10^9$ Years) | Period-System | Orogeny | Major Sequence | Formation | Thickness | Characteristic Rocks |
|---|---|---|---|---|---|---|---|
| Mesozoic | | Cretaceous | | | Coleraine | 0–200′+ | Conglomerate, sandstone and shales; locally the conglomerate constitutes an iron ore. |
| | (0.13 to 0.6 b.y.) | | — Unconformity — | | | | |
| Late Precambrian | (1.1 b.y.) | | | | Hinckley | 100′+ | Medium to coarse, yellow to salmon colored sandstone. |
| | | | | | Fond du Lac | 2000′+ | Sandstone with dark red to gray shale, commonly called "red clastics." |
| | | | Grenville | — Unconformity — | | | |
| | | Keweenawan | | North Shore Volcanic Group | Undivided | 25,000′+ | Duluth Gabbro complex, sills at Duluth, Beaver Bay complex, Logans intrusives and dikes, sills in the Eastern Mesabi Range—rocks are predominantly basic in composition. Thick lava flows with minor interbedded sediments. |
| | | | | | Puckwunge | 0–100′ | Sandstone and conglomerate. |
| | (1.7 b.y.) | | Penokean | — Unconformity — | | | Granites, granitic gneisses and tonalites. |
| Middle Precambrian | | Huronian | | Animikie Group | Virginia | 2000′+ | Argillite gray to black in color, fine grained, and thin bedded. |
| | | | | | Biwabik | 340–750′ | Iron formation or taconite which is a bedded ferruginous chert with a wide range of primary and metamorphic lithologies and in lithologies resulting from secondary oxidation and leaching, including extensive iron ore deposits. |
| | | | | | Pokegama | 0–200′ | Quartzite, which ranges from a fine grained quartz-mica argillite to a coarse grained micaceous quartzite. |
| | (2.5 b.y.) | | Algoman | — Unconformity — | | | Granites and gneisses. |
| | | Timiskamian | | Knife Lake Group | Undivided | ? | Slates, graywackes, arkoses, conglomerates, with flows, tuffs and pyroclastics. |
| | | | Laurentian | — Unconformity — | | | Granite. |
| Early Precambrian | (? b.y.) | Archean | | Keewatin | Soudan | 100′+ | Iron formation lenses of variable thickness and extent composed of chert or quartz, hematite and magnetite with some carbonates and silicates. It is commonly fine grained and distinctly banded. The bands may be straight or complexly folded. |
| | | | | | Ely | ? | Basaltic lavas or greenstones including pillow lavas, with tuffs, pyroclastics, clastic sediments and thin iron formations. |

[1] After Goldich, et al. with additions and modifications.

so, they were removed by post-Cretaceous erosion.

The Mesabi Range was glaciated during the Pleistocene and a mantle of glacial material deposited. The glacial drift is commonly from 20 to 100 feet thick but locally may exceed 300 feet.

### Stratigraphy

The Mesabi Range is made up of sedimentary rocks of the Animikie Group resting on a basement of older rocks and overlain by Cretaceous sediments and Pleistocene glacial deposits. The general stratigraphic column of the Mesabi Range and adjacent northeastern Minnesota together with a listing of the characteristic rocks in each formation and associated with each intrusive epoch is given in Table III.

The Animikie sediments include the Pokegama quartzite formation, the Biwabik iron-formation or taconite, and the Virginia argillite formation.

The Pokegama formation has a maximum thickness of 200 feet. It commonly consists of a basal conglomerate overlain by quartzite that ranges from fine quartzite or quartz-mica argillite to a coarse-grained micaceous quartzite. In the eastern Mesabi, the Pokegama is absent or represented only by a thin conglomerate.

The Biwabik formation is from 340 to 750 feet in thickness. It is divided into four members that can be traced across the entire Mesabi Range, Lower Cherty, Lower Slaty, Upper Cherty, and Upper Slaty. The members may be subdivided further into lesser lithologic units or zones of differing thicknesses and lateral extents. The iron-formation is termed "cherty" if it contains abundant cherty quartz, the bedding is wavy to irregular and massive, and the rock is generally lacking in fine clastics. Iron-formation is termed "slaty" if it contains some fine clastic material, the beds are thin and parallel, and there is a good bedding-plane parting. Lithologic and mineralogical differences within the iron-formation have an important bearing on the distribution and composition of natural iron ores and also on their amenability to concentration. In general, the cherty members are more favorable for concentration than the slaty members.

The cherty taconites of the Biwabik commonly consist of ovoid or ellipsoid grains or granules in fine to coarse sand size; these grains are comopsed of cherty quartz, iron silicates (greenalite, minnesotaite or stilpnome-

lane), iron carbonates, or iron oxides and are enclosed in a fine-grained matrix of cherty quartz, silicate, carbonate, or iron-oxide. The granules have no regular internal structure and may be composed of one mineral or a combination of minerals. Mengel (12) has studied the granules and finds that they appear to have been deposited as sand and have the associated sedimentary structures and textures of sandy sediments. The granules range in size from 0 03 mm to 4 mm but commonly are in the 0.25 mm to 1 mm size range. The granule cherts locally are cross bedded and may be associated with intraformational conglomerates. The cherty taconites may be wavy, straight, or irregularly bedded but tend to have thicker bedding layers than the slaty taconites. The cherty taconite horizons contain two algal chert zones, one at the base of the Lower Cherty member and another in the middle to upper part of the Upper Cherty member. The cherty taconites commonly contain about 28 to 33 per cent total iron.

Slaty taconite is a thin and even bedded, often laminated rock that commonly contains subordinate granule chert layers and beds and may contain a minor quantity of fine clastic material as shown by a small, yet significantly higher, $Al_2O_3$ content than the cherty taconites. The unoxidized slaty taconites generally are composed of different quantities of cherty quartz, minnesotaite, stilpnomelane, greenalite, iron carbonates, and magnetite. The granules and matrix of slaty taconites may be composed of any of the above listed minerals. The slaty taconites commonly contain from 20 to 30 per cent iron.

Gradational contacts occur between the adjacent members and the subunits of the Biwabik formation except for the contact of the Lower Cherty with the Lower Slaty member, which is rather sharp and can be defined within a few inches or a few feet. The gradational relationships between members and subunits result in a range of determinations of the member and unit boundaries and thicknesses. The four members, however, are recognized across the Mesabi as are some of the sub-units. The correlation of some sub-units between the Main Range area, west of Mesaba, and the Eastern Mesabi area is made difficult by the metamorphism and recrystallization of the Biwabik formation east of Mesaba. Table IV shows the general stratigraphic sequence and lithologies of the Biwabik formation. The lithologies described are representative of the Main Range area west of Mesaba.

The Virginia formation consists of fine-

*TABLE IV.*   *Generalized Stratigraphic Sequence and Lithology of the Biwabik Formation*

|  | Thickness in Feet |  | Thickness in Feet |
|---|---|---|---|

**Upper Slaty Member**—Laminated, silicate-carbonate, slaty taconite with interbedded lenses and layers of cherty, silicate-magnetite-carbonate taconite. A carbonate bed commonly occurs at the top of the member east of Nashwauk. West of Naswauk the upper contact is difficult to distinguish as the Upper Slaty has a ferruginous argillite zone containing 15 to 20 per cent iron at the top and a gradational contact with the Virginia formation.     *65–180*

**Upper Cherty Member**—Wavy to irregularly bedded, laminated to massive and mottled cherty granule taconite composed of cherty quartz, magnetite, stilpnomelane, minnesotaite, greenalite and carbonate. The chert is sometimes pinkish in color. The magnetite content ranges from minor to abundant. An algal chert zone from 5 to 10 feet thick occurs in the middle to upper part of the member east of Hibbing. East of Mesaba, the Upper Cherty member contains units rich in magnetite and is the important horizon being mined as taconite ore. West of Mesaba, the magnetite content differs widely and locally this member is of ore grade. With oxidation and leaching, the Upper Cherty is changed to iron ore, concentrating ore or oxidized taconite. Contacts with members above and below are gradational.     *80–300*

**Lower Slaty Member**—The Lower Slaty member is commonly 10 to 70 feet in thickness west, and 100 to 250 feet thick east of Hibbing. This member usually is divided into two units, an upper slaty taconite unit and a lower black silicate-carbonate unit termed the "Intermediate Slate." The upper unit is a laminated silicate-carbonate taconite interbedded with cherty granule taconite as nodules, layers, beds and zones. In the Mountain Iron to Eveleth area, a horizon up to 150 feet thick of cherty magnetite-rich taconite occurs between the upper unit and the Intermediate Slate. The Intermediate Slate unit is from 10 to 40 feet thick and occurs across the entire Mesabi Range. It is a black or dark green, slaty silicate-carbonate taconite with some associated carbonaceous and shaly material as shown by a common 2 to 6 per cent $Al_2O_3$ content. The magnetite content of the Lower Slaty commonly is low. By oxidation and leaching, the Lower Slaty is changed to iron ore, soft, fine grained, aluminous ore or "Painty Ore" and if siliceous, "Paint Rock," or oxidized taconite.     *10–250*

**Lower Cherty Member**—The Lower Cherty member is from 150 to 370 feet thick west of the Siphon Fault and from 0 to 50 feet thick to the east. This member characteristically is cherty taconite and is subdivided into three units that can be recognized across the Mesabi Range. The three sub-units may be further divided into several local units.     *0–370*

**Silicate-Taconite Unit**     *0–110*
This, the upper unit, is characterized by cherty silicate-carbonate taconite, often with coarse granules or pebbles. It is lower in iron than the average cherty taconite and is termed the "lean zone." It commonly contains little magnetite. The silicate taconite unit with oxidation and leaching forms yellow iron ore, sandy goethitic concentrating ore or yellow limonitic oxidized taconite.

**Magnetite-rich Cherty Taconite Unit**     *0–220*
The middle unit of the Lower Cherty member contains wavy to irregluar to massive bedded, sometimes mottled, cherty magnetite-carbonate-silicate taconite It characteristically has a granule texture with granules of cherty quartz, magnetite, stilpnomelane, minnesotaite, greenalite, and carbonate in a matrix of the same composition. This unit is an important taconite ore horizon where it contains sufficient magnetite and is of minable thickness. Upon oxidation and leaching, this unit may contain iron ore, concentrating ore, and oxidized taconite. The contacts with units above and below are gradational.

**Hematitic Taconite Unit**     *0–80*
This, the lower unit, has a varied lithology that is characterized by widely distributed but minor amount of primary hematite. It usually can be subdivided into three zones. The upper zone, often 5 to 10 feet thick, is even bedded, laminated slaty, silicate-carbonate taconite with differing amounts of magnetite. The middle zone is a cherty, magnetite-hematite carbonate taconite with some jaspery chert. It is wavy to even bedded and is sometimes mottled with a granule texture. The lower zone contains granular to jaspery chert with a chert pebble and quartz sand conglomerate at the base. Chert with algal structures occurs in the basal conglomerate zone across the Mesabi Range. Near Eveleth, a bed of chlorite cemented sandstone 8 to 10 feet thick occurs above the algal conglomerate.

                     Total     *340–750*

grained, thin-bedded, gray to black, locally carbonaceous argillite with layers of fine grained graywacke. It commonly is soft, but some phases are hard and siliceous. Some thin limestone beds and concretions occur in the lower part of the formation. The Virginia formation is conformable with the Biwabik formation.

The upper Cretaceous Coleraine formation rests on the eroded Precambrian surface. The lower portion consists of a conglomerate composed of angular to subangular pebbles and boulders of iron ore, taconite, porphyry, and argillite. Some of the iron ore pebbles have highly polished surfaces. The conglomerate grades upward into sands, silts, and clays, some of which are glauconitic. The Cretaceous sedi-

ments contain sharks' teeth, invertebrate shells, fish vertebrae, and scales, carbonized wood and plants, and a few ammonites. The conglomerate ranges in thickness from a few feet to about 100 feet. The total thickness of the Cretaceous is not known but probably exceeds 200 feet.

### Intrusives

Several diabase sills and dikes cut the Biwabik and Virginia formations. They are found from the eastern end of the range as far west as the center of Township 16 West and on the western Mesabi, near vertical, northwesterly trending, altered dikes have been

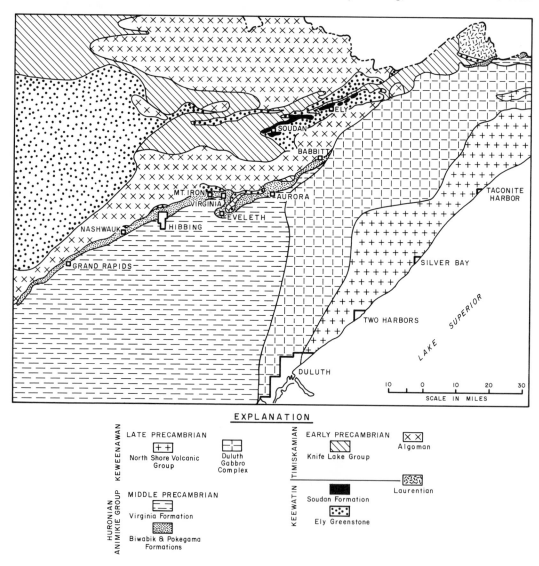

FIG. 2. *Geological Map of the Mesabi Range Area, Minnesota.*

observed in mines north and northeast of Keewatin. A medium-coarse grained, red syenite sill intrudes the Biwabik at Aurora. The Animikie formations are intruded by the Duluth Gabbro complex from south of Mesaba to Birch Lake where the Animikie rocks are truncated and the gabbro is in contact with the pre-Huronian rocks (Figure 2). In general, the dikes and sills have little, if any, metamorphic effect on the Biwabik. The large Duluth Gabbro mass, however, has had a profound effect on the Biwabik, resulting in strong metamorphic changes as far west as Mesaba.

## Structure

The broad structure of the Mesabi Range is a gently dipping monocline that strikes east-northeast and dips from 5° to 15°SE. The general strike is displaced by a large, gentle cross fold with a southwesterly plunging syncline at Virginia and a parallel anticline at Eveleth. Other prominent structural features (Figure 1) are the Siphon fault (or monocline) in the eastern part of T59N-R14W, the Aurora fold in T59N-R15W, the Biwabik fault

in T58N-R15 and 16W, a fault and fold system between Calumet and Leetonia, and the Sugar Lake anticline in T56N-R26W. The folds caused notable bends in the strike of the Animikie rocks and may have produced pronounced changes in the outcrop width of the iron formation. The Biwabik fault is a normal fault that dips steeply south with a vertical displacement of about 200 feet. It strikes about N80°W and can be traced for 4.5 miles. The offset along Biwabik fault causes a narrowing of the outcrop width of the iron formation. In the Calumet to Leetonia area, recent work shows a more extensive fault and fold pattern than was previously recognized (Figure 3). The cross faults at Calumet and Leetonia (the Calumet and Morton Faults, respectively) converge to the north and, in the intervening area, faults and monoclines fan out between these faults and dip toward the middle, forming broad steps into a graben in the central area. The northeast-trending La Rue reverse fault dips steeply northwesterly and extends through much of the area. It has a displacement of about 200 feet at the Galbraith Mine about 3 miles west of Keewatin.

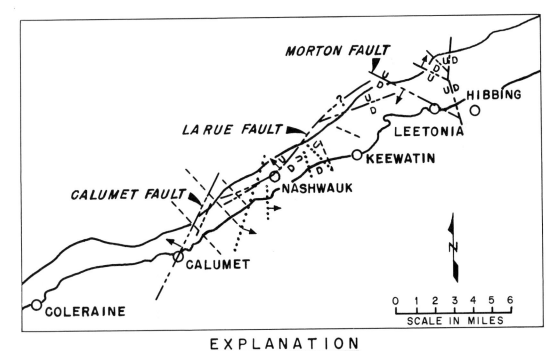

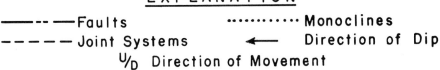

Fig. 3. *Structural Features in the Calumet-Leetonia Area.*

Vertical offsets of up to 100 feet are common in the Calumet to Leetonia area.

Minor local structural features are common along the entire Mesabi Range. These small folds and faults are accompanied by prominent jointing. Small folds and monoclines locally cause steep dips in the iron formation as those observed at the Gilbert Mine near Gilbert, Alpena Mine near Virginia, Grant Mine near Buhl, and Hull-Rust Mine at Hibbing. Faults along the Mesabi Range commonly strike N75°W or N20°W. Fault displacements commonly are less than 50 feet.

The iron formation and other Animikie rocks are prominently jointed with three main joint directions N10E, N45W, and E-W as noted by Gruner (4,9). The joint directions are well shown in some mined out natural ore pits where they appear as an etched pattern of slightly deeper oxidation and enrichment.

The minor structures, folds, monoclines, faults, and zones of strong jointing, are important as controls in the development of Mesabi natural iron ores. Associated with the natural ores are secondary structures related to the slumping. These structures include slump faults and folds that are the result of the compaction of the ore after extensive silica removal.

## MINERALOGY OF THE BIWABIK FORMATION AND IRON ORES

The mineralogy of the unoxidized taconite, which includes the magnetite-taconite ore, is described by Leith (1), Gruner (4), White (8), Gundersen and Schwartz (11), and others and will be reviewed but briefly in this report. The principal minerals in the unoxidized Biwabik formation in the Main Range, generally west of the middle of Range 14 West, are cherty quartz, magnetite, siderite, stilpnomelane, and minnesotaite with lesser amounts of hematite, calcite, dolomite, chamosite, greenalite, and chlorite. In the Aurora-Mesaba area, ferrodolomite or ankerite occurs in blotches and irregular areas. The principal minerals in the Biwabik formation, metamorphosed by the Duluth Gabbro east of Mesaba, are quartz, magnetite, amphiboles (commonly cummingtonite, actinolite, and hornblende), pyroxenes (commonly hedenbergite and ferrohypersthene with some diopside and clinohypersthene), andradite garnet, fayalite, and biotite.

The oxidized Biwabik taconite commonly is composed of cherty quartz, hematite, and goethite with minor amounts of magnetite, manganese oxides, and kaolinite. Martite is

common. East of Mesaba, oxidation and leaching of the Biwabik generally is lacking or is very minor and local.

The natural ores are composed of mixtures of hematite and goethite (limonite) with small amounts of cherty quartz, magnetite, manganese oxides, and kaolinite. The ores may be dominantly either hematite or goethite. Martitic ore or "blue ore" is developed from magnetite-rich horizons in the Upper and Lower Cherty members.

## MESABI RANGE IRON ORES

The iron ores of the Mesabi Range are of two types, natural ore and magnetite taconite ore. The magnetite taconite ore occurs in magnetite-rich horizons of the Biwabik formation that can be mined and concentrated by magnetic methods after fine grinding. The natural ores include the oxidized and leached iron-rich materials within the Biwabik formation that can be mined and shipped after crushing and sizing or can be beneficiated using screening and washing or gravity methods to yield a high-iron product. A small amount of natural ore occurs as conglomerate in the Cretaceous sediments derived from the Biwabik formation.

Oxidized cherty taconites are potential iron ores. They are composed of fine-grained cherty quartz, hematite, and goethite that cannot be concentrated commercially by gravity methods to yield an ore-grade product. The oxidized taconites are somewhat enriched in iron by slight to notable leaching of silica and oxidation of the iron carbonates and iron-silicate minerals. The oxidized taconites include materials ranging from friable cherty taconite termed "semi-taconite" to a hard massive rock. The textures and mineral grain sizes generally are similar to those found in the unoxidized taconites. The amenability of oxidized cherty taconites to concentration by various processes is being investigated by mining companies and government agencies. Because of their fine grain size, oxidized slaty taconites may never become significant sources of concentrates. Cherty oxidized materials, which not only are oxidized but also are partly leached and somewhat enriched in iron, show the most promise of making the transition from potential ore to ore in the near future.

### Magnetite Taconite Ore

Thick magnetite-rich horizons of the Biwabik formation, which can be concentrated by magnetic methods after fine grinding, are being mined in large tonnages as a concentrating-grade iron ore. The important factors in determining the commercial value of magnetite taconite deposits are size, magnetite content, grain size, and texture and stripping ratio of ore to waste. The deposits must be of a size and thickness that will permit mining by large-scale open-pit methods. The magnetite content must be sufficient to give a concentrating ratio of about 3 to 1 crude ore to concentrates. The grain size and texture should be so constituted that iron ore concentrates containing 60 to 67 per cent iron and less than 9 per cent silica can be obtained after grinding to 95 per cent minus 325 mesh (.043 mm). The stripping ratio of waste to ore must be low and commonly is less than one ton of waste per ton of iron ore concentrate. These factors are not fixed but show the general range of currently feasible mining conditions.

Mesabi taconite ores are considered to occur in two areas, the Main Range, west of Mesaba, and the Eastern Mesabi, east of Mesaba. This differentiation is both geographical and, more importantly, geological, the latter resulting from a superposition of a strong Grenville metamorphism, associated with the Duluth Gabbro, acting on an earlier, range-wide, regional, low-rank Penokean metamorphism.

The magnetite taconite ores in the Main Range area, generally west of Mesaba, commonly occur in the middle unit of the Lower Cherty and in the Upper Cherty members of the Biwabik formation. The cherty horizon of the Lower Slaty member in the Iron Mountain to Eveleth area also contains ore-grade material as do certain other units of the Lower Cherty. The taconite ore bodies of the Main Range are tabular stratigraphic units of the Biwabik. They may have assay boundaries caused by a decrease in magnetite content vertically and laterally or they may be terminated by oxidized zones. The only ore mineral is magnetite associated with cherty quartz, stilpnomelane, minnesotaite, greenalite, and carbonates. The ore bodies range widely in extent and thickness; some of the larger deposits are being mined in pits 2 to 4 miles long.

There are marked changes in the mineralogy and texture of the Biwabik formation in about the eastern 17 miles of the Eastern Mesabi area from near Mesaba to Birch Lake. The changes are usually gradational but range from gradational to abrupt. Among the variables that have influenced the mineralogic changes are: structure, permeability, original mineralogy, texture, and composition. These factors form a complexly interrelated system that lo-

cally influenced the encroaching metamorphic environment. Many of the details of the metamorphism are given by Gundersen and Schwartz (11), so only the broader aspects will be summarized here.

The metamorphic variables are observed most readily in mineralogic changes that caused the development of new mineral species or changes in the relative abundances of minerals. In general, there is a change from minerals stable at rather low temperatures in the Main Mesabi Range to those stable at higher temperature east of Mesaba. The highest-temperature mineral assemblage occurs where the iron-formation is in contact with the gabbro. The influence of the gabbro is observed westwardly from north of Aurora to Mesaba as blotchy areas and stringers of ferrodolomite or ankerite in the Lower Cherty member of the Biwabik. To the east and northeast of Mesaba, the taconite has been extensively silicated with the development of considerable minnesotaite and stilpnomelane and of some cummingtonite and actinolite. Farther to the east, cummingtonite becomes the most common silicate mineral in the taconite. Even farther eastward, cummingtonite and actinolite are dominant. Close to the gabbro, ferrohypersthene, actinolite, and hedenbergite are most important, but lesser amounts of fayalite garnet, diopside, biotite, and hornblende are present.

With the change in mineralogy, there is an accompanying change in the texture of the taconite from a common granule texture in a fine-grained matrix, to a rock with reconstituted granules in a felted, silicate-quartz matrix, to a rock with relic granules in a recrystallized rock, to a completely recrystallized, sandy textured, granular rock near the Duluth Gabbro. In the strong metamorphic zone of the eastern Mesabi, carbonate minerals and quartz were consumed in the formation of calcium-magnesium-iron silicates. It is possible that some magnetite may have formed from siderite, as the magnetite content of the Upper Cherty member is higher in the Eastern Mesabi than it is in the Aurora-Biwabik area a few miles to the west. Near the gabbro, the magnetite grain size increases as does the grain size of the associated quartz.

The determination that magnetite taconite is ore is made by laboratory testing of drill core and bulk samples. The magnetic-rich taconite horizons of the Biwabik are extensive and are locally of uniform thickness, yet they may change sufficiently in recoverable magnetite content laterally and vertically so that the material may be ore or waste. Some changes in magnetite content are facies changes and some are due to the development of martite by oxidation. In this report, the Biwabik formation is described by stratigraphic units and the term magnetite-rich taconite includes both ore and waste material. In generalizing, it also may include areas where the Biwabik formation is oxidized.

Commercial magnetite taconite processing plants are in production, under construction, or are being planned in most areas where substantial tonnages of ore-grade magnetite taconites are known. These areas of magnetite-taconite production or potential production are described in the following area reports.

CALUMET TO LEETONIA AREA   In the Calumet to Leetonia area, islands or tongues of unoxidized taconite, more or less unaffected by secondary oxidation and leaching, are located between channels of deeply oxidized rock. The continuity and the extent of the magnetite taconite bodies are limited by the oxidation patterns that are controlled largely by structure and to a lesser degree by stratigraphy and mineralogy. The principal oxidation zones are as follows:

(1) Along the northern edge of the Biwabik formation and extending down dip in the lower part of the iron formation.
(2) Along northwest-trending joint systems.
(3) Along cross-trending monoclines that may strike northwest, north, or northeast.
(4) Along faults, either strike faults or cross cutting faults that commonly strike N30°E to N80°E.

In general, the unoxidized taconite areas are more extensive in the southern part of the iron-formation belt where the lower part of the Biwabik formation is covered by the upper members and the Virginia formation. The contact between oxidized and unoxidized areas is fairly sharp. Iron silicates and carbonates oxidize more readily than magnetite, so some ore-grade material occurs in the partly oxidized halo around the unoxidized taconite blocks.

Magnetite occurs throughout the unoxidized Biwabik formation in amounts ranging from very minor to abundant. The Upper Slaty member contains some magnetite but cannot be concentrated satisfactorily. The Lower Slaty member and the upper and lower units of the Lower Cherty member contain little magnetite. Some magnetite taconite ore may occur in the

unoxidized Upper Cherty member; west of Nashwauk, it is largely oxidized. East of Keewatin, much of the Upper Cherty member is unoxidized but contains fine-grained, slaty taconite layers with generally marginal amounts of magnetite. Magnetite liberation is poor due to the intimate interlocking of fine silicates and carbonates. In the central part of the area, a part of the Upper Cherty member contains potentially economic magnetite taconite.

Magnetite-taconite processing plants, at Nashwauk and at Keewatin, will treat ore from the middle unit of the Lower Cherty member that ranges from 130 to 220 feet in thickness. The magnetically recoverable iron in the crude taconite ore ranges between 15 and 25 per cent, or from about 50 to 80 per cent of the total contained iron. The remaining 20 to 50 per cent of the iron in the unoxidized taconite occurs in primary hematite, siderite, minnesotaite, and stilpnomelane.

LEETONIA TO KINNEY AREA   The Leetonia to Kinney Area, about 11 miles long, has produced approximately one billion tons of natural ore or about 40 per cent of the total Mesabi Range iron ore production. Extensive areas of oxidation and leaching occur in structurally deformed areas. The largest natural ore deposit on the Mesabi Range, the Hull Rust-Mahoning ore body, occurred north of Hibbing in an area containing a number of small folds with associated minor faulting and extensive jointing. Another large ore body occurred south of Chisholm associated with an east-west trending structurally deformed zone on the Pillsbury-Monroe deposit. A third extensive area of iron-ore development occurred east of Chisholm along the south-dipping Buhl monocline that extends, with a general east-west trend, from Chisholm to Kinney. In general, the natural ore bodies are found in the middle part of the Biwabik formation and in the central part of the outcrop area, except near Kinney, where the Buhl monocline and associated ore bodies are in the upper part of the Biwabik formation near the south margin of the iron formation outcrop.

Unoxidized taconite occurs as areas along the north outcrop of the Biwabik, north of the natural ore bodies and associated oxidized zones, and as islands and tongues between channels of oxidation. The magnetite-rich stratigraphic unit of the Lower Cherty member extends throughout the Leetonia to Kinney area and represents a potential taconite ore horizon. The lower and upper units of the Lower Cherty member, and the Lower Slaty member, except for a thin cherty unit in the upper part of the Lower Slaty east of Chisholm, are low in magnetite content. The Upper Cherty member has a wide range in magnetite content and in some areas may be minable as taconite ore.

Gruner (4) presents information that shows the Lower Cherty member to be about 250 feet thick with a lower 55 to 80 foot unit low in magnetite, a central magnetite-rich unit 90 to 150 feet thick and an upper 40 to 100-foot unit with a low magnetite content. The Lower Slaty member is about 15 to 25 feet thick at Hibbing, 100 feet thick at Chisholm, and 155 feet thick at Kinney. Gruner (4) shows the Upper Cherty member to be from 180 to 200 feet thick with an algal horizon in the central part of the member east of Chisholm.

KINNEY TO GILBERT AREA  Important areas of unoxidized Biwabik formation, containing magnetite taconites are found in a 6-mile long area between Kinney and Mountain Iron and in approximately 3-mile long segments between Mountain Iron and Virginia, Virginia and Eveleth and Eveleth and Gilbert. Major tonnages of natural ores have been mined from ore bodies in this general area. These ores occurred along strongly oxidized and leached channels. The oxidized and leached zones often cut the entire iron formation and locally affected the lower part of the Virginia formation. The contact zone between unoxidized taconite and oxidized taconite (or natural ore) ranges from a few inches to many hundreds of feet in width.

The lower and upper units of the Lower Cherty the Lower Slaty, except for a magnetite-rich taconite horizon between Mountain Iron and Eveleth, and the Upper Slaty member, generally are low in magnetite. The Upper Cherty member has a wide range in magnetite content and locally contains commercial-quality taconite.

The middle unit of the Lower Cherty member, where unoxidized, contains important blocks of commercial-quality magnetite taconite in the Kinney-Gilbert area. These blocks are being mined near Mountain Iron and between Eveleth and Virginia. The magnetite taconite ore has a granule texture and is massive to irregularly and wavy bedded, and is mottled. It is composed of cherty quartz and magnetite with various amounts of minnesotaite, stilpnomelane, and carbonates. Magnetite generally is liberated from its gangue minerals if 85 to 95 per cent of the material is ground finer than 325 mesh (.043 mm). It

is practical to remove a magnetic-cobber tailing product after grinding to from 8 to 10 mesh (2.36 to 1.65 mm) so fine grinding is done on an enriched feed.

The Lower Slaty member in the Mountain Iron to Eveleth area contains a cherty taconite horizon ranging up to 150 feet in thickness that, where unoxidized, may be commercial quality magnetite taconite. The Lower Slaty taconite ore is similar to that in the Lower Cherty member in texture, magnetite content, and mineral composition but tends to be somewhat finer grained and to require a finer grind to yield a comparable-quality magnetite concentrate. The cherty-taconite horizon of the Lower Slaty member is usually overlain by a fine-grained, slaty taconite unit that has a lower magnetite content and is difficult to concentrate to a satisfactory product.

The Upper Cherty member ranges from 150 to 200 feet in thickness. Its magnetite content ranges from 10 to 35 per cent. The differences in magnetite are the result of the wide range in minnesotaite and stilpnomelane content and, to a lesser degree, of the variations in carbonate content. The Upper Cherty taconite is being processed by the Eveleth Taconite Company.

The magnetic iron content in the Biwabik formation in the Kinney to Gilbert area ranges from 5 per cent, or less, to about 25 per cent which is from 10 per cent, or less, to about 85 per cent of the total iron in the taconite.

GILBERT TO BIWABIK AREA  Magnetite taconite ore occurs in the unoxidized Biwabik formation between natural ore bodies and oxidized areas in the Gilbert to Biwabik area. The magnetic iron and soluble iron content of the Biwabik formation in a drill hole at Gilbert and in one at Biwabik are shown on Figure 4. The zones of high magnetic iron content are the potential ore zones. The low magnetic-iron zones, or non-ore horizons, are shown clearly. They correspond to the Upper Slaty and Lower Slaty members, the upper 40 to 50 feet of the Upper Cherty member, the upper 25 to 40 feet, and the lower 30 to 50 feet of the Lower Cherty member. The Upper Cherty member, below the algal zone, and the central unit of the Lower Cherty member may be magnetite taconite ore.

The Upper Cherty magnetite taconite is an irregular to wavy and straight bedded, laminated, layered or mottled, granule taconite with some slaty taconite. It is composed of cherty quartz and magnetite with various amounts of minnesotaite, stilpnomelane, and

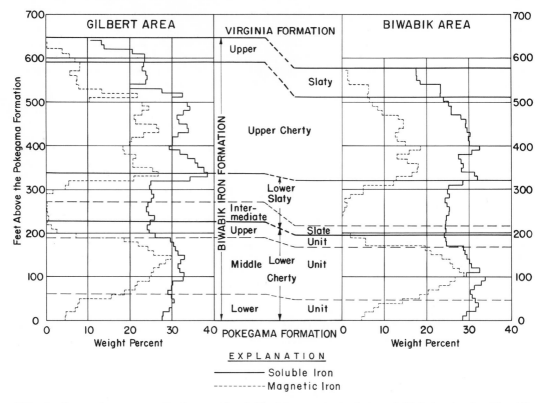

FIG. 4. *Comparison, magnetic iron and soluble iron content of two drill holes in the Biwabik iron formation in the Gilbert and Biwabik areas, Mesabi Range.*

carbonates. An algal zone occurs from 150 to 190 feet above the base of the unit. The algal zone is from 8 to 19 feet in thickness and contains chert with algal structures, conglomerate, and granule chert with magnetite in the matrix; in the Gilbert area, the zone includes a 6- to 8-foot slaty taconite zone that is absent in the Biwabik area. The magnetite-rich taconite shown in Figure 4 in the Upper Cherty member in the Gilbert area contains from 18 to 27 per cent magnetic iron which is from 55 to 85 per cent of the total iron; in the Biwabik area, on the other hand, the magnetic iron ranges from 13 to 18 per cent which is from 40 to 65 per cent of the total iron. This marked change in magnetic iron in the Upper Cherty member is found in two drill holes about 2 miles apart and represents a change from a good-quality magnetite taconite to a material of doubtful value.

The middle unit of the Lower Cherty member is 110 to 150 feet in thickness and is the main ore horizon. It is a massive, wavy to irregularly bedded, mottled to laminated and layered, granule to locally conglomeratic, cherty taconite with magnetite-rich layers and

layers with disseminated magnetite. The ore-grade material commonly contains from 18 to 25 per cent magnetic iron which is from 40 to 96 per cent of the total iron; the common range is between 70 and 80 per cent of the iron occurring in magnetite. The upper and lower contacts of the middle unit are gradational with the magnetite lean units.

AURORA TO MESABA AREA   Magnetite taconite of commercial value in the Aurora to Mesaba area is confined largely to the middle unit of the Lower Cherty member. The Upper Slaty, Upper Cherty and Lower Slaty members generally are low in magnetite content although, locally, zones within the Upper Cherty member may contain sufficient magnetite to be of ore grade. The unoxidized and unleached Biwabik taconite in this area is of low metamorphic rank and commonly is composed of various amounts of cherty quartz, magnetite, stilpnomelane, minnesotaite, greenalite, and carbonates with minor hematite. Magnetite taconite ore is mined by the Erie Mining Company where ore-grade material occurs in the middle unit of the Lower Cherty member.

The Lower Cherty member is about 140 feet thick near Aurora and thins eastward to about 115 feet north of Mesaba. It is subdivided into three units, as in the area to the west. The upper unit, from 20 to 40 feet thick, is a silicate taconite with granules and matrix composed of cherty quartz, stilpnomelane, minnesotaite, greenalite, and carbonates with minor magnetite. It contains more magnetite toward the base and has a gradational contact with the middle unit. The middle magnetite-rich unit is from 70 to 120 feet thick and contains an upper magnetite-silicate granule taconite zone with pink to brown ankerite mottling, a middle zone of irregularly to wavy bedded cherty-magnetite-silicate-carbonate taconite, and a lower zone that is commonly an even-bedded, magnetite-silicate taconite with some jaspery chert and with some thin argillaceous layers containing hematite. With an increase in argillaceous layers and hematite, the middle unit grades into the lower unit of the Lower Cherty member. This lower unit is from 15 to 30 feet thick and consists of hematitic-argillaceous layers in a magnetite-poor, granular cherty taconite with some jaspery chert. It grades downward into a greenish, silicate-rich argillaceous zone with some disseminated magnetite in a jaspery granule chert. The base of the Lower Cherty member commonly is marked by a thin conglomerate.

MESABA TO BABBITT AREA    From the vicinity of Mesaba eastward, the Biwabik formation is metamorphosed by the Duluth Gabbro that intrudes the Animikie rocks about 1.5 to 2 miles to the south of the Biwabik iron formation outcrop. The gabbro contact with the Virginia formation generally is parallel to the trend of the iron formation from Babbitt west to Mesaba where the contact turns sharply south away from the iron formation (Figure 2).

The Biwabik formation in the vicinity of Mesaba appears to be in a transition zone between a low-rank metamorphic area to the west and a medium-rank metamorphic area to the east. The Biwabik taconite at, and for about a mile east of, Mesaba contains stilpnomelane and minnesotaite as the common silicate minerals with which are associated some cummingtonite and actinolite. Where stilpnomelane and minnesotaite are present, some carbonate minerals commonly are found to occur associated with carbonate mottling and stringers. Farther to the east, where cummingtonite and actinolite are the common silicate minerals, carbonate minerals usually are absent except as late calcite. There also may be a marked decrease in quartz content of the taconite. A general increase in mineral grain size occurs with increase in metamorphic rank. Paradoxically, the magnetite in the lower-metamorphic-rank taconite west of Mesaba, liberates more easily, and at a somewhat coarser grind, than does the magnetite in the medium-rank metamorphic zone to the east of Mesaba. This difference may be due to the intimate magnetite-silicate fabric of the taconite in the Mesaba to Babbitt area.

In the Mesaba to Babbitt area, the Upper Cherty and the lower part of the Upper Slaty members are the most important sources of magnetite taconite ore. West of the Siphon fault, taconite ore occurs in the middle unit of the Lower Cherty member. This is an eastward extension of the magnetite-rich taconite zone west of Mesaba. The Lower Cherty is from 115 to 150 feet thick west of the Siphon fault, and about 30 to 50 feet in thickness east of the Siphon fault. Some commercial quality taconite occurs east of the fault in the Lower Cherty but it is only from 15 to 25 feet thick and is too thin to be of economic importance.

The Upper Slaty member is from 100 to 130 feet in thickness and is subdivided into seven submembers by Gundersen and Schwartz (11). In this report, the Lower Slaty is grouped into four units, termed Lower, Intermediate, Middle and Upper units. The upper unit of the Upper Slaty member (A and B of Gundersen and Schwartz) is from 18 to 28 feet thick. It is essentially a silicate rock with vaguely bedded quartzose taconite in the lower 10 to 20 feet, and a marble, 3 to 10 feet thick, at the top. The upper unit contains only minor magnetite. The silicate minerals are light-green diopside, hedenbergite, cummingtonite, hornblende and actinolite. The marble bed is a coarse-grained, white to gray calcite with some wollastonite, diopside, and andradite garnet. The contact with the overlying Virginia formation is distinct. The middle unit (C and D of Gundersen and Schwartz) is from 40 to 50 feet thick with laminated magnetite-quartz-cummingtonite-actinolite taconite with white quartz nodules and lenses in the lower 4 to 11 feet. This unit contains abundant fine-grained magnetite that requires very fine grinding to achieve liberation. It is not currently mined as magnetite taconite ore. The intermediate unit (E and F of Gundersen and Schwartz) is from 15 to 25 feet thick and includes rather massive, regularly bedded and interlayered fine-grained magnetite and cummingtonite taconite with thinner layers of fine-grained granular quartz. The bedding commonly ranges from 0.5 to 6 inches in thickness. Relic sedimentary structures are well preserved and include penecontemporaneous slumping, folding, faulting, and intraformational conglomerate. Magnetite is fined grained. It is poorly liberated after grinding to about 95 per cent minus 325 mesh (.043 mm). The top 2 to 5 feet of this unit contain "Septaria" structures with whitish quartz filling fractures in a granule quartz containing very fine-grained magnetite. This unit is not currently used as taconite ore. The lower unit (G submember of Gundersen and Schwartz) is from 20 to 35 feet thick. It is characteristically a massive, granule-textured magnetite-quartz-cummingtonite taconite with many thin magnetite-rich laminae. The magnetite is fine grained but can be processed to yield an acceptable grade of concentrate.

The Upper Cherty member is from 120 to 160 feet thick, averaging about 140 feet east of the Siphon fault. It is as much as 200 feet thick near Mesaba. The Upper Cherty is divided into 8 submembers by Gundersen and Schwartz; for this discussion, however, it is subdivided into 3 units. The upper unit (H and I of Gundersen and Schwartz) is from 10 to 20 feet thick. At the top is an 8- to 12-foot zone of irregularly bedded to laminated to massive silicate taconite containing fine-grained magnetite and quartz associated with yellowish to green amphiboles including cummingtonite, actinolite, and hornblende, and at the bottom, there is a 3- to 8-foot massive fine-grained quartz bed containing granule and algal structures. The middle unit (J, K, and L of Gundersen and Schwartz) is from 85 to 110 feet thick, averaging about 100 feet. It is thickest in the west and thins to the east. The taconite of this unit includes quartzose taconite, silicate taconite, and magnetite-rich taconite and, as a whole, contains moderate to abundant magnetite. This taconite is massive to laminated with relic granule and pebble structures. This unit has a wide range in mineralogy and texture and is described in detail by Gundersen and Schwartz. The common minerals are quartz, magnetite, cummingtonite, actinolite, hedenbergite, and, in the west near Mesaba, minnesotaite and stilpnomelane. This unit is the most important taconite ore horizon of the Eastern Mesabi Range.

The lower unit (M, N, and O of Gundersen and Schwartz) is from 35 to 40 feet thick east of the Siphon fault and up to about 90

feet thick to the west. It is composed largely of silicate taconite or quartz-silicate taconite and commonly contains cummingtonite with some actinolite and hornblende. Near Mesaba, minnesotaite and stilpnomelane are the most abundant silicates. Some higher temperature silicates occur south of Babbitt. This unit is magnetite poor.

The Lower Slaty member is 75 to 120 feet thick and includes an upper silicate taconite and quartz-silicate taconite, which is massive, mottled and silicate rich with only minor magnetite, and a lower zone, the intermediate slate, which is darker in color than the adjacent units. The dark color may be due to fine graphite associated with the abundant silicates and fine quartz.

The Lower Cherty member thins rapidly eastward, being about 115 feet thick at Mesaba but only about 30 to 50 feet in thickness in the area east of the Siphon fault. To the west of the Siphon fault, the magnetite-rich taconite in the middle unit of the Lower Cherty contains magnetite taconite ore, whereas to the east it contains only from 15 to 25 feet of the magnetite-rich taconite horizon and is too thin to be of economic importance.

BABBITT TO BIRCH LAKE AREA   The Biwabik formation is from 0 to 250 feet in thickness in the Babbitt to Birch Lake area. It was strongly metamorphosed by the Duluth Gabbro. The intrusive contact occurs from a few 100 to about 2000 feet south of the Biwabik formation outcrop in the area south of Babbitt to Section 10, T60N, R12W. In the area from northeast of Section 10 to Birch Lake, the gabbro intrudes the Biwabik formation and truncates it near Birch Lake.

The Biwabik formation in this area is recrystallized, and, in this process, coarse-grained quartz, magnetite, fayalite, ferrohypersthene, hedenbergite, hornblende, and cummingtonite were developed. The contact between the moderately metamorphosed zone to the west and strongly metamorphosed Biwabik to the east is gradational and occurs in the area generally south of Babbitt.

The stratigraphic units of the Biwabik, distinguished to the west of Babbitt, extend into this area, but the total thickness of the formation is much reduced. The Upper Slaty member is about 125 feet thick on the west edge of the area but is only about 55 feet thick about 3 miles to the east. The Upper Cherty member is from about 110 to 125 feet in thickness and may be subdivided into 3 units. The upper unit is from 10 to 20 feet in thickness and includes a 3- to 8-foot thick algal zone in which algal structures appear only as relic structures. This unit consists of discontinuous contorted layers or seams of magnetite in a dark granular quartz. The middle unit, from 60 to 100 feet thick, is magnetite-rich and corresponds to the magnetite-rich taconite zone to the west. It is composed of coarse-granular quartz, magnetite, ferrohypersthene, hedenbergite, and actinolite. The lower unit is silicate rich with granular quartz, ferrohypersthene, hedenbergite, fayalite, and magnetite.

The Lower Slaty is 65 to 75 feet in thickness and consists of a rather massive, dark-green to black taconite composed of quartz, ferrohypersthene, hedenbergite, and fayalite with small amounts of graphite and magnetite.

The Lower Cherty unit is about 30 feet thick in the western part of the area and is absent or present only as thin lenses in the area from about 2 miles east of Babbitt to Birch Lake. Where present, the Lower Cherty is a rather massive, coarse-grained, quartz-silicate and quartz-magnetite taconite. The silicates commonly are hedenbergite and fayalite with some garnet.

The structure of the taconite east of Babbitt superficially is simple, but in detail there are many faults, folds, and joints. The fault displacements commonly are only a few tens of feet. Small folds produce a local flattening of the dips and widening of the taconite outcrop. Much of the local structure may be related to the gabbro emplacement.

Commercial magnetite taconite occurs in the lower part of the Upper Slaty and in the Upper Cherty members.

## Natural Ores

The Mesabi Range natural ores have been well described by Leith (1), Wolff (3), Gruner (4), White (8) and others and will be described only briefly here. Commercial iron ore bodies occur as irregular deposits, within the Biwabik formation, where oxidation has changed the iron minerals of the primary unoxidized taconite to hematite and goethite and where leaching has removed sufficient silica so the material can be shipped to the steel furnaces after rather simple beneficiation processes. These processes include crushing, screening, washing, and gravity beneficiation methods. The natural ore deposits have a rough zoning in that essentially no deposits occurred in the eastern area where the taconite was metamorphosed by the Duluth Gabbro; whereas, the area between Mesaba and Hibbing contained a very high proportion of direct shipping ore with lesser amounts of partly leached, concentrating-grade material; and, west of Hibbing, the deposits had a preponderance of concentrating grade ore and rather small amounts of direct-shipping ore.

The natural ore deposits occur in a wide variety of shapes and sizes. In size, they contain from a few thousands of tons to hundreds of millions of tons. In shape, the deposits range from narrow fissures of ore following a single fracture, to channel-type deposits following a system of fractures, to blanket-type deposits where extensive areas in favorable stratigraphic horizons formed ore. In some deposits, ore was developed in essentially the entire Biwabik formation from the quartzite to the Virginia Slate.

All ore bodies are related to the Precambrian erosion surface, which generally, is the present bedrock surface. The ore bodies commonly lie directly below the glacial drift or, in some areas, below thin Cretaceous sediments. The ore extends to various depths, from a few feet to about 800 feet, but most ore is within 400 feet of the surface. Ore usually occurs in the outcrop area of the Biwabik formation although extensions of a few ore bodies reach a short distance under the Virginia slate, as at the Embarrass and St. James mines. In a few deposits, such as the Missabe Mountain and Auburn Mines, near Virginia, decomposed, kaolinized, and bleached Virginia argillite occurred in the central part of slump synclines over iron ore bodies. The ores may show a more widespread development in certain layers within the Biwabik formation than in others. One such horizon is at the top of the Lower Cherty member where a 15- to 30-foot portion of the upper silicate-taconite unit forms flat, strata-bound, yellow limonitic ore bodies. This type of ore body was mined extensively at the Godfrey underground mine east of Hibbing.

The mineralogy and chemistry of natural ore shows a direct relationship to the taconite from which it was derived. The slaty taconites commonly produce from slightly to highly aluminous ores containing from 1.0 to 6.0 per cent or more $Al_2O_3$. This ore is usually fine grained and red. It often is sticky, has a high moisture content and is termed "painty ore" or, if not of ore grade, "Paint Rock." The magnetite-rich middle unit of the Lower Cherty member commonly yields a blue, martitic, high-grade ore that is low in silica, alumina, and phosphorous. The silicate taconite at the top of the Lower Cherty member forms a yellow to brown ore, and the Upper Cherty member produces brown to blue ores that commonly are of non-Bessemer grade, ranging from 0.04 to 0.18 per cent phosphorus and from 0.5 to 1.5 per cent alumina. The silica content may be low, 2 to 8 per cent, in the cherty ores but tends to be higher, 5 to 10 per cent, in the painty ores. The ores shipped to the steel furnaces are blends of high- and low-silica ores, concentrated ores, and painty ores, designed to meet specific grade requirements.

The natural, direct-shipping ores occur in permeable areas where leaching action has been concentrated. There is a common correlation of ore bodies with faults, flexures, and folds that caused fracturing in the brittle taconite. White (8) concludes that about 75 per cent of the natural ore is related to various types of folds and flexures, 15 per cent to faults, and 10 per cent to stratigraphic and other factors.

It is believed that oxidation and leaching processes worked together, but that the oxidation of the ferrous iron minerals was easier to accomplish than the removal of the major quantities of silica required to produce high-grade ores. This is shown by the extensive areas of oxidation around some ore bodies.

The removal of 40 to 60 per cent of the taconite by the leaching of silica, with little, if any, introduction of material, created much pore space and the residual mass of iron oxide could not support the overlying ore and rock. The resulting slump often amounted to from 35 to 45 per cent of the original thickness of the taconite. In deep channel or trough ore bodies, prominent slump synclines developed and were accompanied by faulting and crumpling in the ore. In a general way, the depth of the ore body is indicated by the extent of slumping observed in its upper parts.

Quartz veins, some up to 15 to 20 feet wide and 0.5 miles long, are found in some ore bodies. The development of these veins may have had some influence on the ore genesis, but no direct relationship can be inferred.

## Cretaceous Ores

The Upper Cretaceous seas, acting on a shore line along the Mesabi Range, eroded and concentrated the iron-bearing materials of the Biwabik formation. Where the Biwabik being eroded, contained ore bodies, the conglomerates in the Cretaceous may be ore-grade deposits consisting of pebbles and grains of iron ore in a limonitic matrix. Some of these ore bodies resemble stream valley gravels, others

appear to have been formed as beach deposits, and some, as at the Enterprise mine near Virginia, appear to be accumulations of a crustal conglomerate, resembling "Canga," on the pre-Cretaceous surface. The Cretaceous ores are generally higher in sulphur, phosphorus, and alumina than ores formed in the Precambrian iron formation.

Cretaceous sediments are common in the Western Mesabi as far east as Hibbing and occur as isolated deposits at the Sherman mine near Chisholm, the Judson mine near Buhl, and the Enterprise mine at Virginia. In some areas, as the Judson mine, the Cretaceous ore conglomerate has been transported and the underlying Precambrian rock is the Virginia argillite.

## ORIGIN OF THE ORES

Much has been written on the origin of cherty iron formations stressing in particular the source of the iron and silica, the method of deposition, and the nature of the primary sedimentary materials. All agree as to the sedimentary origin of the Biwabik formation. There is not complete agreement on the nature of the primary sedimentary materials or the events that produced the variety of taconite lithologies observed. The reader is referred to Leith (1), Van Hise and Leith (2), Wolff (3), Gruner (4), White (8), Lepp and Goldich (14), Tyler (5,6), James (7,17), LaBerge (13) and others for a discussion of this very interesting subject.

The unoxidized taconites of the main Mesabi Range were developed from the primary sedimentary materials by diagenetic alterations and by low-rank metamorphism. In the Eastern Mesabi, the taconites were metamorphosed by the Duluth Gabbro Complex. This metamorphism produced various responses in the taconite that are related to the mineralogy, texture, and chemical composition of the taconite layers, their geological structures, and their distance from the gabbro contact. The magnetite taconite ores of the Mesabi Range are a part of the Biwabik iron formation that contain sufficient magnetite to allow profitable mining and processing.

The natural ores, and the concentrating-grade, natural ores, are the products of secondary oxidation and leaching of the slightly metamorphosed Biwabik taconite. The iron silicates, iron carbonates, and magnetite were oxidized to ferric oxides hematite and goethite and the calcium, magnesium, and much of the silica were removed by leaching.

These processes resulted in a residual product containing much of the iron and alumina and various amounts of the silica. There have been extensive discussions of the question as to source and nature of the oxidizing and leaching solutions that produced these secondary ores, but without agreement. Most, if not all, of the conditions required for development of Mesabi natural ore can be found in the oxidation and leaching action of surface waters, through usual weathering processes. Surface waters following permeable zones, such as faults, joints and fractures, acting over a long period of time, could have produced the observed configuration, distribution, textures, and composition of the Mesabi natural ores.

## SUMMARY

The natural iron ores and the natural iron-ore concentrates of the Mesabi Range are declining in importance and are being replaced by iron-ore pellets produced from magnetite taconite. Magnetite taconite ore, which can be mined commercially, occurs in unoxidized portions of certain stratigraphic horizons in the Biwabik iron-formation where it contains sufficient iron, as magnetite, with a grain size and texture which will produce a 60 to 67 per cent iron concentrate after fine grinding and magnetic separation.

The Mesabi Range taconites are of two general types, those of the Main Range area, from Mesaba westward, and those of the Eastern Mesabi from Mesaba east to Birch Lake. The magnetite taconites of the Main Range are composed of minerals of low metamorphic rank, with cherty quartz, magnetite, greenalite, minnesotaite, stilpnomelane, and carbonates as the principal minerals; these taconites have been affected only by the regional metamorphism associated with the Penokean orogeny. The Eastern Mesabi taconites have, in addition, been remetamorphosed and recrystallized by a Grenville metamorphism associated with the Keewanawan Duluth Gabbro. They are composed of minerals of medium to high metamorphic rank; the suite includes quartz, magnetite, cummingtonite, actinolite, hornblende, hedenbergite, ferrohypersthene, garnet, fayalite, and biotite. Commercial quality magnetite taconite occurs in the Upper Cherty, Lower Slaty (Mountain Iron to Eveleth area only), and Lower Cherty members of the Biwabik formation in the Main Mesabi area and in the Eastern Mesabi area, in the lower part of the Upper Slaty, and Upper Cherty

members, and west of the Siphon fault, in the Lower Cherty member. Production at an annual rate of about 30 million tons of taconite pellets is expected from the Mesabi Range in 1968.

## REFERENCES CITED

1. Leith, C. K., 1903, The Mesabi iron-bearing district of Minnesota: U.S. Geol. Surv. Mono. 43, 316 p.
2. Van Hise, C. R. and Leith, C. K., 1911, The geology of the Lake Superior region: U.S. Geol. Surv. Mono. 52, 641 p.
3. Wolff, J. F., 1915, The ore bodies of the Mesabi Range: Eng. and Min. Jour., v. 100, nos. 3, 4, 5, 6, p. 89–94, 135–139, 178–185, 219–224.
4. Gruner, J. W., 1946, Mineralogy and geology of the Mesabi Range: Iron Range Resources and Rehabilitation, St. Paul, 127 p.
5. Tyler, S. A., 1949, Development of Lake Superior soft iron ores from metamorphosed iron formation: Geol. Soc. Amer. Bull., v. 60, p. 1101–1124.
6. ———— 1950, Sedimentary iron deposits: p. 506–523 in Trask, P. D., *Editor, Applied sedimentation*, Wiley, N.Y., 707 p.
7. James, H. L., 1954, Sedimentary facies of iron-formation: Econ. Geol., v. 49, p. 235–291.
8. White, D. A., 1954, The stratigraphy and structure of the Mesabi Range, Minnesota: Minn. Geol. Surv. Bull. 38, 92 p.
9. Gruner, J. W., 1956, The Mesabi Range: Geol. Soc. Amer. Guidebook for Field Trips, Field Trip no. 1, p. 182–215.
10. Goldich, S. S., *et al.*, 1961, The Precambrian geology and geochronology of Minnesota: Minn. Geol. Surv. Bull. 41, 193 p.
11. Gundersen, J. N. and Schwartz, G. M., 1962, The geology of the metamorphosed Biwabik Iron-formation, Eastern Mesabi district, Minnesota: Minn. Geol. Survey Bull. 43, 139 p.
12. Mengel, J. T., 1963, The cherts of the Lake Superior iron-bearing formations: Univ. Wisc. Unpublished Ph.D. dissertation.
13. LaBerge, G. L., 1964, Development of magnetite in iron-formations of the Lake Superior region: Econ. Geol., v. 59, p. 1313–1342.
14. Lepp, H. and Goldich, S. S., 1964, Origin of the Precambrian iron-formations: Econ. Geol. v. 59, p. 1025–1060.
15. American Iron Ore Association, 1965, Iron ore: Cleveland, Ohio.
16. Alm, M. R., 1965, Mining Directory Issue, Minnesota, 1965: Univ. Minn. Bull. 68, 268 p.
17. James, H. L., 1966, Chemistry of the iron-rich sedimentary rocks: U.S. Geol. Surv., Prof. Paper 440-W, 60 p.

# 26. Iron Ore Deposits of the Menominee District, Michigan

CARL E. DUTTON,* PAUL W. ZIMMER†

## Contents

## Illustrations

## Tables

* U.S. Geological Survey, Madison, Wisconsin.
† The Hanna Mining Company, Iron River, Michigan.

morphism that concentration of magnetite and hematite is feasible, and pellets containing more than 60 per cent iron are being produced.

The older formation is also present in the southeastern part where it has been only slightly metamorphosed. Ore containing more than 50 per cent iron was formerly mined from local residual concentrations of iron oxides from which silica had been sufficiently leached, or replaced by iron oxide, or affected by combinations of these processes.

The younger iron-formation occurs in the western part of the district and, if unaltered, is interlayered siderite and chert. The ore in this area also contains more than 50 per cent iron and formed as local residual concentrations but in two stages. The siderite was converted to iron oxides (hematite and "limonite"), and silica was contemporaneously or subsequently removed by leaching, or replacement, or combinations of these processes.

## ABSTRACT

Iron ore in the Menominee district is mined from two iron-formations of middle Precambrian age. The older formation is present in the northeastern part; is composed mainly of hematite, magnetite, quartz, and minor silicates; and contains about 34 per cent iron. The rock has been so coarsened by meta-

## INTRODUCTION

The Menominee district is mainly in the southwestern part of the northern peninsula of Michigan but includes a small adjacent segment of northeastern Wisconsin (Figure 1). The part in Michigan consists of about two-thirds of Dickinson County and the southern third of Iron County, and the part in Wiscon-

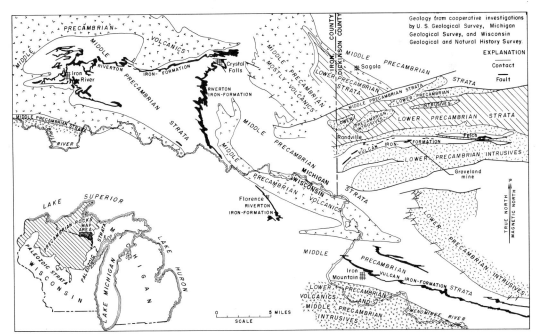

FIG. 1. *Map, showing generalized Precambrian geology of parts of northern Michigan and Wisconsin, after Dutton (8).*

sin is the northeastern quarter of Florence County. The district is approximately 50 miles from east to west; it is about 20 miles wide in Dickinson County, 5 miles wide in Florence County, and 10 miles wide in Iron County. Although the district is a geographic and geologic unit, its division into three areas—central Dickinson County, southern Dickinson County, and southern Iron County—arises from differences in lithology and structure of middle Precambrian formations and related iron deposits as will be described later.

Ore was discovered in southern Dickinson County in 1849, first shipments were made in 1877, and production ceased in 1959. Ore in southern Iron County was discovered in 1851, shipments began in 1882, and five mines are producing direct-shipping ore at present (1966). A small amount of direct-shipping ore was produced in central Dickinson County from 1882 to 1913, but mining and concentration of lean iron-ore formation into high grade pellets began in 1959 and have continued to the present. A few all-rail shipments are made from the district, but most ore is sent by train to Escanaba on Lake Michigan and transferred to boats for shipment to ports near Chicago, Illinois and Cleveland, Ohio.

There have been 96 producing mines in the Menominee district, 45 of which, or almost half, were in southern Iron County. There were 42 operating mines in 1918 but only six in 1964.

Total production of iron ore from the Menominee district through 1965 was 290,000,000 tons that had a value at the mines of about $1,200,000,000. Annual production reached a maximum of 6,569,413 tons in 1920, declined to 980,135 tons in 1938, but increased to 4,707,931 in 1951. High cost of operation caused many underground mines to close in the 1950's, and production dropped to 2,477,980 tons in 1959. As in all of the Lake Superior region, an increased demand for beneficiated iron ore has resulted in increased production from the Menominee district since 1959. This change is shown in the table of production of the Menominee district (Table I). Annual production of direct-shipping ore has been almost steady since 1959, and production of beneficiated ore will probably increase. The iron-formation in southern Dickinson County may also have possibilities for future beneficiation.

## GENERAL GEOLOGY

The Menominee district is part of the upland area between Lake Superior and Lake Michigan. The elevation ranges from 1100 to 1500 feet above sea level or approximately 500 to 900 feet above the level of Lake Superior.

Pleistocene glacial deposits are widespread; are mainly moraines, till plains, and outwash areas; and, except in the lowlands, are thin. Most of the upland areas are bedrock with a thin plaster of till and drift. The bedrock relief has thus been subdued by glacial deposits.

The abundance of outcrops ranges from

TABLE I.   Shipments of Iron Ore from Menominee District in Long Tons[1]

| Years | Southern Dickinson | | Central Dickinson | | Iron County[2] | | Total |
|---|---|---|---|---|---|---|---|
| | Direct Shipping | Bene-ficiated | Direct Shipping | Bene-ficiated | Direct Shipping | Bene-ficiated | |
| Through 1930 | 71,876,826 | — | 474,786 | — | 99,582,362 | — | 171,933,974 |
| 1931–1940 | 3,787,245 | — | — | — | 13,526,606 | — | 17,313,851 |
| 1941–1950 | 4,251,675 | — | — | — | 36,882,878 | — | 41,134,553 |
| 1951–1960 | 533,087 | — | — | 747,616 | 37,773,276 | 428,233[3] | 39,482,212 |
| 1961 | — | — | — | 775,938 | 3,109,964 | — | 3,885,902 |
| 1962 | — | — | — | 476,027 | 2,986,344 | — | 3,462,371 |
| 1963 | — | — | — | 1,100,864 | 3,203,330 | — | 4,304,194 |
| 1964 | — | — | — | 1,433,626 | 3,190,648 | — | 4,624,274 |
| TOTALS | 80,448,833 | — | 474,786 | 4,534,071 | 200,255,408 | 428,233 | 286,141,331 |

[1] Source: Lake Superior Iron Ores (2,4) and General Statistics covering Cost and Production of Michigan Iron Mines, Department of Conservation & Geological Survey, State of Michigan.

[2] Iron County totals include Iron River, Crystal Falls and a small production from Florence, Wisconsin.

[3] Book Mine Concentrator.

moderate in Dickinson County to generally sparse in southern Iron County and Florence County.

The sequence of main rock units in the Menominee district is shown in Table II.

The formations of middle Precambrian age are of special interest concerning iron deposits in the Menominee district. The strata are believed to be correlative with the Animikie Series (9) in other parts of the Lake Superior region but were included in the Huronian Series of former usage. Table II shows that two iron-formations are present. The Vulcan Iron-Formation is in the Dickinson County areas (Figure 1); this rock is composed mostly of iron oxide (hematite or magnetite) layers and interbedded quartz-rich (recrystallized chert) layers and is underlain in most places by thick dolomite and quartzite that accumulated during orogenically stable conditions. The Riverton Iron-Formation is in the Iron River-Crystal Falls area; this formation is characterized by alternate layers of siderite and chert* but is associated with predominantly fine- to medium-grained clastic materials. The two sequences that contain iron-formation are separated by a great thickness of metabasalt, some pyroclastic rock, graywacke, and slate that accumulated in an orogenically unstable environment. The iron formations are chemical deposits that formed in restricted basins where, according to James (6), normal free circulation was probably inhibited by offshore buckles that developed into volcanic island arcs during the evolution of a geosyncline.

The Precambrian stratified formations in the district are generally steeply dipping. Different structural patterns characterize each area—an elongate syncline in central Dickinson County, a faulted homocline in southern Dickinson County, and complex folds and faults in a synclinal basin in southern Iron County (Figure 1).

Intrusives into the iron-formation of the ore-bearing tracts are neither abundant nor large. Some granite is present in central Dickinson County, and some diabase and metadiorite in southern Dickinson and southern Iron Counties.

Rocks in the central Dickinson County area are at the staurolite or equivalent grade of metamorphism. Rocks are at the biotite or equivalent grade in the eastern part of the southern Dickinson County area, whereas those in the western part of that area and northwestward through Iron River-Crystal

Falls area are at chlorite or equivalent grade.

## CENTRAL DICKINSON COUNTY

Central Dickinson County is characterized by ridges and knobs that rise about 100 to 200 feet above intervening swampy lowlands. The Groveland mine (Figure 1) is on part of a ridge that is about 3500 feet long, 1000 feet wide, and rises 200 feet above the adjacent surface. The nearby areas of glacial drift are hummocky moraine in which relief is about 60 feet above the altitude of approximately 1200 feet.

The structure of principal interest in this part of the district is a long, narrow syncline containing four metamorphosed sedimentary formations. The Vulcan Iron-Formation is the youngest unit and occurs only in a series of small disconnected elongate doubly plunging east-trending synclines (Figure 1); the Groveland mine is in one of these small folds. The sedimentary sequence dips steeply and is in unconformable or fault contact with granite or with foliated granitic gneiss that contains layers of amphibolite.

The Vulcan Iron-Formation and all older rocks have been cut by late middle Precambrian granitic rocks and diabase. The iron-formation, as well as the other Precambrian rocks, are unconformably capped by remnants of relatively flat lying Cambrian sandstone. In the area of the iron-formation the base of this sandstone contains angular fragments of iron-formation.

The Vulcan Iron-Formation is a typical oxide facies of moderately high metamorphic rank as defined by James (7) and Cumberlidge and Stone (12). It is moderately coarse grained consisting mainly of hematite, magnetite, quartz, and minor amounts of tremolite-actinolite, cummingtonite, diopside, garnet, and carbonate. The quartz is gray to white metamorphosed chert or red jasper and minor granular quartz. Most of the carbonate is secondary and is either calcite or dolomite.

The metamorphosed Vulcan Iron-Formation is the iron ore of central Dickinson County. It is classed as a primary ore. At the Groveland mine of the Hanna Mining Company, the only operation in the area, almost the entire Vulcan Iron-Formation is being mined. It has not been significantly enriched or concentrated. Metamorphism has locally increased the grain size of the iron oxides and made some material more amenable to concentration

---

* Finely crystalline, non-clastic quartz.

*TABLE II.   Generalized Geologic Column for the Menominee District, Michigan (10, 11, and 14)*

| Geologic Age | | | | Name & Thickness of Formations | Lithology |
|---|---|---|---|---|---|
| Quaternary | | Pleistocene | | To 400 feet | Glacial deposits. Till, sand, gravel. |
| Lower Paleozoic | | Ordovician(?) and Upper Cambrian | | To 400 feet | Mainly sandstone, some dolomite. |
| Precambrian | Upper Precambrian | Keweenawan | | | Rare diabase dikes. |
| | | | | | Granite. |
| | | | | | Metagabbro and metadiabase. |
| | Middle Precambrian | Animikie Series | Paint River[1] Group | Fortune Lakes Slate Min. 4000 ft. | Dominantly slate and interbedded graywacke. |
| | | | | Stambaugh Formation about 50 ft. | Magnetic laminated flinty slate containing some graywacke and gray fissile slate. |
| | | | | Hiawatha Graywacke 50 ft. or less to 400 ft. | Mostly massive graywacke, arkosic near Iron River. Breccia of chert fragments in massive graywacke near the base or makes up entire formation near Crystal Falls |
| | | | | Riverton Iron-Formation 100–600 ft. | Mostly interbedded siderite and chert, where unoxidized. Interbedded dark siderite, nodular chert, and graphitic slate in upper third at Iron River. Hematite, "limonite," and chert where oxidized. |
| | | | | Dunn Creek Slate 400–800 ft. | Chiefly sericitic slate and siltstone but upper 20 to 50 feet are pyritic graphitic laminated slate and lower breccia of small pyritic graphitic slate fragments in matrix of same compostion. |
| | | | Baraga Group | Badwater Greenstone to 10,000 ft. or more | Massive to ellipsodal chloritized basaltic flows, in part agglomerates and tuffs. |
| | | | | Michigamme Slate about 5000 ft. | Mainly graywacke and slate or more metamorphosed equivalents. |
| | | | Menominee Group | Vulcan Iron-Formation about 300–800 ft. | Thin-bedded layers of magnetite, martite and specularite, alternate with quartz-rich layers; minor low-iron silicate minerals are present. |
| | | | | Felch Formation about 100–500 ft. | Mostly mica schist or phyllite containing thin quartzite layers. |
| | | | Chocolay Group | Randville Dolomite 500–2000 ft. | Mostly massive fine to coarsely crystalline dolomite |
| | | | | Sturgeon Quartzite 500–2000 ft. | Massive vitreous quartzite, locally schistose. |
| | Lower Precambrian | | | Pre-Sturgeon Formations Several thousand ft. | Metamorphic and intrusive rocks. |

[1] Descriptions of the Paint River and Baraga Groups refer to the Iron County area. Descriptions of the Menominee and Chocolay Groups refer to the Dickinson County areas.

than less metamorphosed iron-formation. The entire iron-formation averages 34 per cent iron; high grade pellets containing over 60 per cent iron are produced from this crude ore.

The structure of the ore body is a rather tight double plunging overturned syncline, with numerous smaller related folds within it (Figure 2). The axial plane of the syncline dips about 60°N. Mapping the structure is difficult because of several faults and the lack of distinctive stratigraphic units.

The main fault, which has strike-slip displacement, trends east roughly parallel to the axis of the syncline and dips steeply to the south. It is parallel to a number of other east-west faults in the region both to the north and south and may have a related origin. As important as this fault appears to be, the offset of several granitic dikes cut by the fault indicates a strike-slip (right-lateral) displacement of only about 150 feet. Other northwest-trending faults appear to show even less movement.

Several theories have dealt with the nature of the original iron-formation. The classic view developed by Van Hise and Leith (1) was that the primary iron-formation consisted of chemically precipitated iron carbonate and chert. James (6) reviewed these theories and suggested that primary iron-rich rocks can be divided into four major sedimentary facies: sulfide, carbonate, silicate, and oxide. From the following evidence, the iron-formation at Groveland appears to be an example of the oxide facies and as such is different from iron-formation in the southern Iron County area:

(1) Primary carbonates are rare and are not iron carbonates.

(2) Hematite and magnetite occur together. Boundaries are sharp between hematite-rich and magnetite-rich beds. This association is hard to explain if both were derived from an earlier carbonate.

(3) Some silicates are present, but high iron-bearing silicates, which might be expected from the metamorphism of iron carbonate, are not common.

Cumberlidge and Stone (12) divided the Vulcan Iron-Formation into four lithologic units called Basal, Lower, Middle, and Upper, but state "the subdivision is based more on lithology than on position in a stratigraphic sequence." The whole sequence may be equivalent to the Traders Member of the Vulcan Formation in southern Dickinson County.

The Basal unit is essentially a recrystallized chert or chemical precipitate that resembles a granular quartzite. It contains about 15 per cent iron, generally in the form of magnetite.

This unit has a maximum thickness of 110 feet and can be traced around the syncline even though the thickness on the north side is only 10 to 15 feet. Locally the unit is quite schistose. Because it has a low iron content, the unit is the only part of the iron-formation not considered as ore.

The Lower, Middle, and Upper units of the Vulcan Iron-Formation have been described by Cumberlidge and Stone (12) in terms of hematite and magnetite content and, to a lesser extent, of lithology. The Lower and Middle units are poorly bedded, hematite-rich, semi-clastic in appearance and, to some extent, contain less iron than the Upper unit. Interbedded magnetite-rich layers may be in the Lower or Middle unit.

The Upper unit of the Vulcan Formation is generally a magnetite-rich rock, distinctly bedded, generally a little higher in iron than the lower units and generally containing distinct silicate-rich layers. The upper part of the Upper unit contains hematite-rich layers.

Secondary changes have affected only two quite small parts of the Groveland ore body, but these constituted a significant part of the ore available during early mining. One of these zones, in which magnetite has been converted to hematite or martite, lies along the north side of the ore body beneath the Cambrian capping and can be traced down dip directly into magnetite-rich iron-formation. The other altered zone is adjacent to fractures in the northeast corner of the ore body where the normal iron-formation has been leached and the iron oxide converted to "limonite". The iron-formation has been shattered, stained with "limonite", and impregnated with calcite, pyrite, and very minor chalcopyrite. The Cambrian sandstone capping has likewise been stained and impregnated with "limonite", calcite and copper carbonate. Hydrothermal solutions may have caused this hydration and secondary deposition of carbonates and sulfides.

The ore body is within a metamorphic zone characterized by the presence of staurolite (7); the center of highest-temperature metamorphism is about 10 miles to the west. The principal effect of regional metamorphism on the iron-formation has been to increase the grain size by recrystallization, a change that has been of prime importance in making it possible to concentrate the ore at a reasonably coarse size. The quartz grain size has provided a fairly accurate index of the degree of metamorphism (7). The iron minerals have gone through a similar recrystallization. In the lower zone of metamorphism, the hematite, for example, measures about 0.01 mm whereas in the staurolite zone this increases to 0.07–0.3 mm; the average at Groveland is 0.1 mm. Metamorphism of the iron-formation is not related to the granitic dikes, and the dikes are not metamorphosed. Presumably, therefore, the metamorphism affecting the iron-formation preceded the granite intrusion in the area.

The sequence of geological events leading to the origin of the deposit at Groveland can be simply stated. It is generally believed that the iron-formation was deposited in shallow water and under oxidizing conditions in a restricted offshore basin. Numerous minor fluctuations in water depth probably gave rise to the variations in the ferrous-ferric iron ratio and hence in the hematite-magnetite ratio. Deposition may have been in a remnant basin of a much larger geosyncline, but at no time was this basin deep enough to be in the carbonate or sulfide zone (6).

Minor vulcanism to the north is evidenced by a thin persistent schist (metatuff) within the iron-formation on the north limb of the syncline. As described by James (6), strong regional structural disturbances, subsequent to burial, produced folds and faults. In late middle Precambrian time, intrusions of the basic dikes and sills were followed by granitic intrusions. Subsequent hydrothermal alteration and weathering only locally affected the iron-formation at Groveland.

Inasmuch as the iron-formation itself is the ore body at the Groveland mine, the age of "mineralization" in that part of the central Dickinson County area extends from the middle Precambrian, as based on the stratigraphic position of the Vulcan Iron-Formation, through the period of later metamorphism that increased the grain size. The iron-formation has not been significantly affected by any secondary enrichment process. It yields a salable product by concentration.

## SOUTHERN DICKINSON COUNTY

The prominent topographic features of southern Dickinson County are two discontinuous northwest trending bedrock ridges that are 200 to 300 feet above slightly irregular surfaces of morainic or outwash deposits.

The principal structures involving the iron

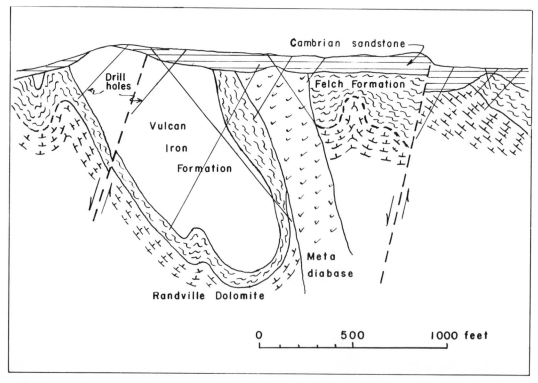

FIG. 2. *Generalized Geologic Section, Looking West, Groveland Mine Area. Line of section is about 3,000 feet east of the west margin of sec. 31, T.42N, R.29W.*

deposits in this area are two steeply southwestward dipping homoclines of the Vulcan Iron-Formation and associated strata from which ore was formerly produced at about 25 mines. This repetition is believed to be the result of vertical displacement of the more southwesterly of the homoclines along a major fault that parallels and lies between the segments. Each homocline is locally modified by northwest plunging right-lateral folds and by northeast-trending right-lateral faults.

Iron ore formerly mined from the Vulcan Iron-Formation in southern Dickinson County was of secondary origin. Iron oxides were concentrated through the action of subsurface water in leaching the chert or replacing it by the ore minerals. Ore bodies in this area are most common and best developed in iron-formation along the synclinal part of northwestward plunging right-lateral folds or at the southeast terminations against right-lateral faults. These structures probably provided very favorable conditions for iron enrichment by descending subsurface water. Significant enrichment probably occurred only after folding and subsequent erosion that exposed the iron-formations, so the age of "mineralization" has presumably continued since post-middle Precambrian time. Abundant ore-grade fragments in the basal strata of Upper Cambrian age in southern Dickinson County indicate that enrichment was well advanced at least locally, prior to that time.

## SOUTHERN IRON COUNTY

The topography in southern Iron County results largely from glacial deposits, as much as 400 feet thick, that are modified locally by post-glacial stream erosion. The most characteristic features are gentle slopes between broad hummocks that are about 150 feet above depressions and stream courses. Most of the depressions are occupied by lakes, ponds, or swamps. The few principal streams have irregular courses and shallow valleys. Exposures of bedrock are small—generally less than 500 square feet—and are moderately to widely scattered. Iron ore in southern Iron County has been mined principally around the cities of Iron River and Crystal Falls.

The Iron River-Crystal Falls area is part of a triangular-shaped basin with apices at each of the towns mentioned; the third apex is at Florence, Wisconsin. The most common structures within the basin are numerous steeply plunging folds, a few of which are locally overturned. The fold pattern is most complex near Iron River (Figure 1) where the general east trend is approximately parallel to the contact between the sedimentary sequence and the underlying volcanic rocks. Locally, fold axes are at right angles to the regional trend. Most of the folds at Crystal Falls plunge steeply eastward or westward. At least seven east trending faults have significantly displaced the Riverton Iron-Formation in the Iron River area. Similar faults are also present in the Crystal Falls area. Figures 3 and 4 show characteristic structural features near Iron River.

The complexities of stratigraphy and structure in the Iron River-Crystal Falls area have been deciphered primarily because two distinctive lithologic units, one a breccia and the other a magnetite-rich slate, are present in the rock sequence. Vertical sections through drag folds observed in the mine workings indicate that the Riverton Iron-Formation is stratigraphically underlain by a thin laminated graphitic slate, which, in turn, is underlain by a massive, finely granular graphitic rock that contains randomly oriented fragments of graphitic slate. This brecciated graphitic unit, which is easily recognized in mine exposures and drill cores, is widely distributed in the area. The other unit of equal or perhaps even greater general importance is the Stambaugh Slate, which is magnetic and can be traced under the extensive glacial drift by magnetometer surveys. This magnetic unit is stratigraphically above the Riverton but is separated from it by the Hiawatha Graywacke.

The Riverton Iron-Formation and associated strata in Florence County, Wisconsin, occur mainly in a complexly folded northwestward plunging syncline that is incomplete owing to removal of most of one limb by faulting and to absence of part of the other limb because of nondeposition or erosion.

Iron ore in the Iron River-Crystal Falls area is of secondary origin and resulted from at least three processes and related stages of enrichment. The first was the primary sedimentary process that produced layers of siderite, which with interbedded chert, comprise the Riverton Iron-Formation. The second process was alteration of siderite to one or more iron oxides (hematite or "limonite"). The final process was removal of sufficient chert to make the residual material a merchantable product.

The primary iron-formation is an alternation of siderite (48 per cent Fe) and chert in approximately equal amounts and in layers of approximately the same general range of thickness from 0.5 to 2 inches. The layers of siderite are gray and are commonly thinly bedded to laminated. The layers of chert are dark gray and are not internally layered. Thousands of partial analyses from drill records of the unaltered iron-formation average about 25 per cent metallic iron; all analyses were made on dried samples.

Analyses of samples, which seemed to be typical chert-siderite formation but proved to be slightly more ferruginous than the average, are shown in Table III. The estimated mineralogical composition based on analysis B is given in Table IV.

The iron-formation was enriched in oxide and residually concentrated by the percolation and chemical action of subsurface water that was predominantly or entirely of meteoric origin. Siderite has been converted to hematite (70 per cent Fe) and "limonite" (60 per cent Fe) in a wide range of ratios by removal of carbon dioxide, change to ferric iron, and some hydration. This oxidized iron-formation is an alternation of iron oxide layers and chert layers. It is common in the area and generally contains 35 to 45 per cent metallic iron. This iron content suggests that the alteration of siderite has not been accompanied by much decrease in chert due to leaching, replacement by iron oxides, or various combinations of these processes.

Locally, sufficient chert has been removed

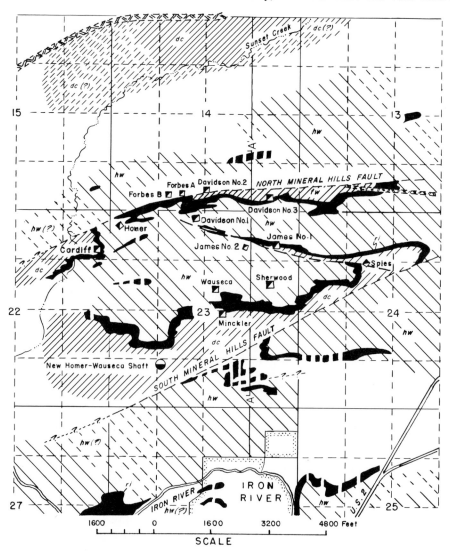

Fig. 3. *Generalized Geologic Map and Section of the Northern Part of the Iron River Area, after James and Dutton (8).*

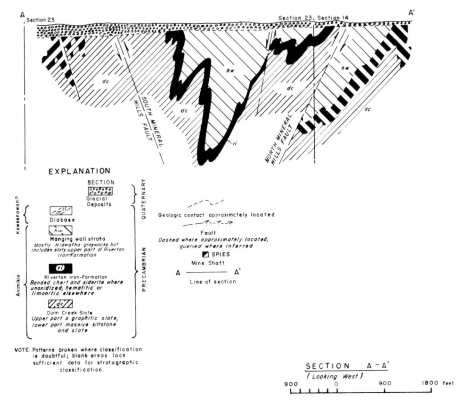

FIG. 3. (*Continued*)

so as to produce residual concentrations of iron oxides that are ore; that is, the iron content of large masses has been raised to an average of more than 50 per cent. These masses may retain part of the stratification of the original iron-formation, but collapse and slump due to leaching are commonly present on both large and small scales.

The paragenesis of the ore minerals is due to the conversion of siderite to iron oxides, as stated above. Locally the unoxidized iron-formation contained $MnCO_3$ that was similarly oxidized so as to yield special grades of ore in which manganese content averages about 3.36 per cent.[*]

The principal gangue material is finely crystalline nonclastic quartz derived from chert that has not been removed; the silica content averages about 4.96 per cent. Minor gangue constituents are primary and secondary calcite and pyrite; secondary gypsum is common. The ores contain an average of about 0.48 per cent $P_2O_5$ and about 2.82 per cent $Al_2O_3$. All the minerals containing phosphorus and aluminum

[*] This analysis and those in next paragraph are from reference 13, p. 94.

have not been determined fully, but some apatite has been recognized.

The physical controls affecting the formation of iron ore bodies in the Iron River-Crystal Falls area are stratigraphy and structure. The Riverton Iron-Formation is the only stratigraphic unit that contains a significant primary concentration of iron-bearing minerals, and all ore bodies are restricted to it. The formation ranges in thickness from 150 to 600 feet, but the upper one-third or less of the unit contains many interbedded layers and partings of "slate" (thinly bedded to laiminated argillite) that have inhibited oxidation, thus this upper part has no economic significance. The development of ore within the Riverton has been guided by the structure in directing the movement of subsurface water. Ore bodies are most common and generally best developed (1) in plunging synclinal structures, (2) along the axial part of these structures, and (3) in the upright rather than overturned limbs of these folds (Figure 3 and 4). The close relation of location and shape of ore bodies to structure suggests the probability that descending meteoric water has been the principal ore-forming agent. Evidences of possible hydrothermal ac-

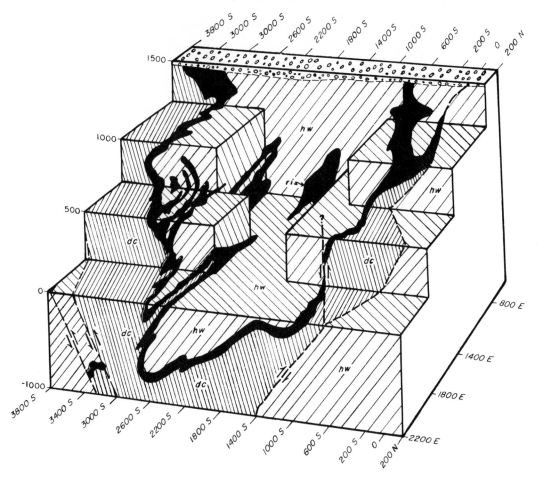

Fig. 4. *Diagram, showing structure of Riverton Iron-Formation in northern part of Iron River area. Origin of coordinates is northwest corner of sec. 23, T43N, R35W.*

tion are minor amounts of native copper and copper sulphides in some ores at Iron River.

A summary of the sequence of geologic events that produced iron ore in the Iron River-Crystal Falls area of southern Iron County begins with the deposition of carbonate facies of iron-formation because the ore has formed from, and is found only in, a stratigraphic unit characterized by interbedded siderite and chert. This type of iron-formation presumably accumulated in an environment that was below wave agitation and sufficiently reducing to yield ferrous compounds yet containing enough oxygen to destroy most organic material (6). The great predominance of siderite and chert and the virtual exclusion of clastic materials indicate deposition by chemical action. These prerequisites would most likely be present in a restricted basin formed by the rise of an offshore buckle during an orogeny.

Any associated pyrite may represent local or temporary existence of more thoroughly reducing conditions, as commonly would exist in deeper water. Associated iron oxides represent more thoroughly oxidizing conditions, as in shallow water. The iron-bearing silicates (stilpnomelane, minnesotaite, and chlorite) may have formed either as primary minerals or by alteration of earlier silicate and accumulated in a wide range of environments as indicated by associations ranging from sulphide to oxide. The chert interbedded with the siderite was also deposited by chemical action, but the chert-siderite association represents an alternation of critical chemical-physical relations in the transporting solutions, or within the basin.

The second stage in the formation of ore, as presented earlier, was conversion of siderite to hematite, or "limonite", or various proportions of these oxides.

TABLE III.   *Chemical Composition of Iron-Formation Rocks, Iron River Area (5)*

|  | A | B |
|---|---|---|
| $SiO_2$ | 24.25 | 32.2 |
| $Al_2O_3$ | 1.71 | 1.5 |
| $Fe_2O_3$ | 0.71 | 0.6 |
| FeO | 35.22 | 31.6 |
| MgO | 3.16 | 2.8 |
| CaO | 1.78 | 1.6 |
| $Na_2O$ | 0.04 | — |
| $K_2O$ | 0.20 | 0.2 |
| $H_2O$ | 0.21 | 0.2 |
| $H_2O+$ | 0.00 | 0.0 |
| $TiO_2$ | 0.00 | 0.0 |
| $P_2O_5$ | 0.91 | 0.8 |
| $CO_2$ | 27.60 | 24.8 |
| MnO | 2.11 | 1.9 |
| C (graphite) | 1.96 | 1.8 |
| S | N.D. | |
| Total | 99.86 | 100.0 |

A. Banded chert-carbonate iron formation. Beta drill hole 2, in SW¼ sec. 26, T. 43N, R. 35W. Analyst: Leonard Shapiro, U.S. Geol. Survey.

B. Analysis "A" recalculated to 25 per cent total iron. Each value, except that of silica, reduced proportionately. $SiO_2$ added to bring total to 100 per cent.

TABLE IV.   *Estimated Mineralogical Composition of Carbonate Iron-Formation (5)*

| Constituent | | Weight Per Cent |
|---|---|---|
| Carbonate | 81.0% $FeCO_3$<br>4.9% $MnCO_3$<br>9.5% $MgCO_3$<br>4.6% $CaCO_3$ | 62.4 |
| Chert | | 32.0 |
| Graphite | | 1.8 |
| Miscellaneous (phosphate, chlorite, iron oxide) | | 3.8 |
| | | 100.0 |

The final stage consisted of the removal of most of the chert by leaching, or by replacement with one or more iron oxides, or by combinations of these processes. This concen-tration probably has continued since post-middle Precambrian erosion exposed the folded iron-formation.

## REFERENCES CITED

1. Van Hise, C. R. and Leith, C. K., 1911, The geology of the Lake Superior region: U.S. Geol. Surv. Mono. 52, 641 p.
2. Lake Superior Iron Ore Association, 1938, Lake Superior iron ores: Cleveland, Ohio, 364 p.
3. James, H. L. and Dutton, C. E., 1951, Geology of the northern part of the Iron River district, Iron County, Michigan: U.S. Geol. Surv. Circ. 120, 12 p.
4. Lake Superior Iron Ore Association, 1952, Lake Superior iron ores: 2d ed., Cleveland, Ohio, 334 p.
5. James, H. L., 1951, Iron formation and associated rocks in the Iron River district, Michigan: Geol. Soc. Amer. Bull., v. 62, p. 251–266.
6. James, H. L., 1954, Sedimentary facies of iron formation: Econ. Geol., v. 49, p. 235–293.
7. James, H. L., 1955, Zones of regional metamorphism in the Precambrian of northern Michigan: Geol. Soc. Amer. Bull., v. 66, p. 1455–1487.
8. Dutton, C. E., 1958, Precambrian geology of parts of Dickinson and Iron Counties, Michigan: Mich. Basin Geol. Soc. Guidebook, 40 p.
9. James, H. L., 1958, Stratigraphy of pre-Keweenawan rocks in parts of northern Michigan: U.S. Geol. Surv. Prof. Paper 314-C, p. 27–44.
10. James, H. L., *et al.*, 1959, Bedrock geology, geologic map of the Iron River-Crystal Falls district, Iron County, Michigan: U.S. Geol. Survey Mineral Inv., Field Studies Map MF 225, sheet 3, 1"–2000'.
11. James, H. L., *et al.*, 1961, Geology of central Dickinson County, Michigan: U.S. Geol. Surv. Prof. Paper 310, 176 p.
12. Cumberlidge, J. T. and Stone, J. G., 1964, The Vulcan iron formation at the Groveland mine, Iron Mountain, Michigan: Econ. Geol., v. 59, p. 1094–1106.
13. American Iron Ore Association, 1964, Iron Ore 1964: Cleveland, Ohio, 133 p.
14. Bayley, R. W., *et al.*, 1966, Geology of the Menominee iron-bearing district Michigan and Wisconsin: U.S. Geol. Surv. Prof. Paper 513, 96 p.

# 27. Geologic Setting and Interrelationships of Mineral Deposits in the Mountain Province of Colorado and South-Central Wyoming[*]

### OGDEN TWETO[†]

## Contents

---

[*] Publication authorized by the Director, U.S. Geological Survey.
[†] U.S. Geological Survey, Washington, D.C.

## Illustrations

## Tables

## ABSTRACT

The classes of ore deposits in the mountain province of Colorado that have been the most productive in the past and that offer the greatest promise for the future are: (1) disseminated or stockwork molybdenum deposits associated with Tertiary stocks; (2) combination precious- and base-metal veins in Tertiary volcanic rocks; and (3) base-metal replacement deposits and veins in Paleozoic sedimentary rocks. Veins in Precambrian rocks, including precious metal-bearing sulfide veins, gold-silver telluride veins, and tungsten veins, have been mined extensively in the past, but their aggregate output has been far subordinate to that of the groups named above, and the outlook for major output from them in the future is poor. Principal ore deposits of south-central Wyoming are sedimentary iron ores of Precambrian age and uranium deposits in Tertiary sedimentary rocks. Precambrian deposits of several kinds are scattered through the mountains of both states. The most promising of these, based on present indications, are vanadiferous titaniferous magnetite in southern Wyoming and southwestern Colorado and thorium-niobium rare-earth deposits associated with alkalic intrusive complexes in Colorado.

Most of the mining districts of Colorado are in the Colorado mineral belt, a generally narrow but somewhat irregular belt that ex- tends northeastward across the north-trending mountain ranges and the north-northwest-trending geologic grain of the state. The belt is characterized by intrusive rocks of Laramide and younger age and by ore deposits, and it occupies environments ranging from Precambrian crystalline rocks through Paleozoic and Mesozoic sedimentary rocks to Tertiary volcanic rocks.

Two ages of intrusion and ore deposition recently have been recognized in the mineral belt. One is Laramide (late Cretaceous and early Tertiary), and the other is Oligocene. As interpreted herein, the major replacement deposits and most of the sulfide vein deposits in Precambrian and sedimentary rocks are Laramide in age, and the molybdenum deposits, the tungsten and the gold telluride veins in Precambrian rocks, and the precious- and base-metal veins in volcanic rocks are Oligocene in age. In general, the Laramide deposits are mesothermal in aspect, and the Oligocene deposits are principally epithermal.

The existence of two mineralization stages, in considerable part characterized by different suites of metals, has many applications in exploration.

## INTRODUCTION

Knowledge of the ore deposits of Colorado and southern Wyoming has advanced on many fronts since publication of the Lindgren Volume in 1933. Major advances in the knowledge of regional geology, of the chronology of geologic events, and of geologic processes have led to greatly improved understanding of the relationship of known deposits and districts to their geologic setting and provide many leads in the search for the unknown districts and deposits that are yet to be found.

The most significant departures from the state of knowledge in 1933—excellently summarized in the Lindgren Volume by Burbank and Lovering (13)—result from recognition that:

(1) Several of the post-Precambrian intrusive rocks with which ore deposits are associated are not Laramide (Late Cretaceous to early Eocene) in age but are of younger age, principally Oligocene.

(2) Consequently, that some ore deposits formerly classed as Laramide in fact are younger and hence that extensive parts of the region were mineralized in two major stages rather than in a single stage of Laramide age.

(3) That many faults formerly classed as Laramide in age in fact originated in Precambrian time and that many underwent major movement after Laramide time.

(4) That the San Juan volcanic pile of southwestern Colorado is a complex of caldera-related ash-flow tuff and lava piles, each with its own history, including, in some cases, an ore stage, rather than a single long-accumulating pile of lava flows that later was widely mineralized in a single ore-forming stage.

(5) That mid-Tertiary intrusive and extrusive igneous activity was not confined to the San Juan area but also was widespread elsewhere in the Colorado mountain region.

(6) That a north-trending basin-and-range fault system, younger than the Laramide fault systems and with its own distinctive mineralization features, extends deep into Colorado from the Rio Grande trough at the New Mexico line and, as a discontinuous feature, possibly extends all the way to Wyoming.

## History and Production

Mining began in Colorado in 1858 with the discovery of placer gold near the site of Denver. Many new gold and silver lode discoveries were made in following years in the mountains to the west, and, in the early 1870's, discoveries were made in the San Juan Mountains of southwestern Colorado. Significant production of base metals began with the opening

of the lead-carbonate deposits of Leadville in 1877, and, in the next few years, similar deposits were found in many other places, as at Aspen, Red Cliff (Gilman), and Kokomo. Last of the major precious- and base-metal districts to be discovered, in 1891, were the Cripple Creek gold district and the Creede silver district. Mining of tungsten, for which Boulder County was long famed, began in 1900. Mining of molybdenum at the great Climax deposit began in 1918 but has been continuous only since 1924. Production of fluorspar on a significant scale began in World War I, although a very minor production was made earlier. Production of uranium began as early as 1871 in the Central City district, but large-scale production of uranium did not start until the early 1950's.

Production of metals from the mountain province of Colorado is summarized in Table I. This tabulation does not include uranium and vanadium produced from the part of the State lying in the Colorado Plateau province, discussed separately in this volume by Fischer. The precious metals dominated Colorado metal output until 1940, when they were surpassed in annual value by molybdenum. Since 1945, base metals also have exceeded the precious metals in annual value (110, Figure 6).

In the interests of perspective, it should be noted that, of the total of $3.4 billion indicated in Table I, $2.6 billion, or 76 per cent, came from only five districts: Climax (28 per cent), Leadville (15 per cent), Ouray-Telluride-Silverton triangle (14 per cent), Cripple Creek (12 per cent), and Gilman (7 per cent).

TABLE I. *Value of Principal Metals Produced from the Mountain Province in Colorado, 1858 through 1964*

|  | Millions |
|---|---|
| Molybdenum | $ 952 |
| Gold | 920 |
| Silver | 610 |
| Zinc | 399 |
| Lead | 344 |
| Copper | 105 |
| Tungsten | 50 |
| Fluorspar | 39 |
| Uranium | 27 (est.) |
| Total | $3,446 |

Data from Koschmann (110) through 1962; updated through 1964 from Minerals Yearbook, U.S. Bureau of Mines.

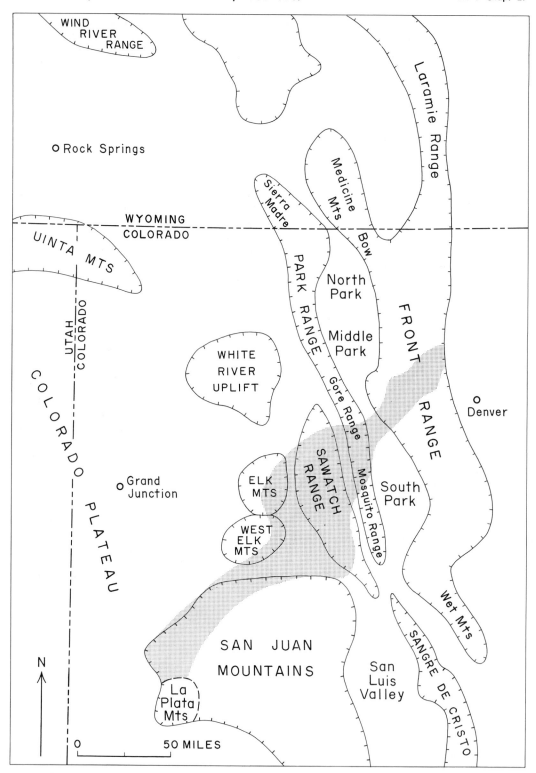

FIG. 1. *Principal Structural-Geomorphic Units in the Mountain Province of Colorado and South-Central Wyoming, and Outline of Colorado Mineral Belt.*

Metal production from south central Wyoming is mainly in uranium and iron, discussed in other reports in this volume by Harshman and by Bayley, respectively. Copper and gold produced from this area probably have not exceeded $5 million in total value.

## Physical Features and Geography

The mountainous part of Colorado and southern Wyoming has a sharp eastern border along which the mountains rise abruptly above the plains along a slightly sinuous line. In Colorado, the western border of the mountains is irregular and not sharply separable from the Colorado Plateau to the west, but, in Wyoming, the mountains are rather sharply bordered on the west by rolling and dissected plateau land. The mountainous region is made up of long subparallel ranges and subcircular mountain groups separated by large valleys or broad basins. Principal features of the region are shown in Figure 1.

Most of the metal mining districts of Colorado lie in the Colorado mineral belt, a generally narrow but somewhat irregular strip of ground that extends southwestward across the state from the mountain front near Boulder to the region of the San Juan Mountains. Location of the belt is shown in Figure 1. Location of mining districts in the belt and elsewhere in the mountain province of Colorado is shown in Figure 2. The mineral belt is in general an elevated zone. It contains many of the highest peaks of Colorado, and it contains or closely parallels the Continental Divide through much of its length.

## PRECAMBRIAN ROCKS AND EVENTS

Precambrian metamorphic and igneous rocks are widely exposed in the Front, Park, Sawatch, Sangre de Cristo, Uinta, Laramie, Medicine Bow, and Wind River ranges. In Colorado, the metamorphic rocks principally are strongly deformed high-grade gneisses and schists, but, in the west-central or Needle Mountains part of the San Juan Mountains and in the Uinta Mountains, they include younger units of less deformed and only moderately or slightly metamorphosed sedimentary rocks. Granites of three general age groups are intruded into the older metamorphic rocks, and at least one is intruded into the younger sequence. In south-central Wyoming, two major Precambrian sequences exist. One, consisting of very ancient gneisses and batholithic quartz diorite, apparently is much older than any of the rocks exposed in Colorado. An overlying sequence of metamorphosed sedimentary rocks and granites intrusive into them probably is approximately equivalent to the main or older sequence exposed in Colorado. Principal Precambrian events of the two areas are listed in Table II.

## Colorado

The older metamorphic rocks consist principally of biotitic quartzo-feldspathic gneisses, but they include many bodies of hornblendic or amphibolitic gneisses, as well as minor bodies of quartzite, marble, and calc-silicate gneiss. Most of these rocks seem to be metasedimentary, but some hornblende gneiss units might be metavolcanic, as are greenstones and presumed metarhyolites in southwestern Colorado. In the central Front Range, the older metamorphic rocks have been assigned in the past to the Idaho Springs Formation and in part to the Swandyke Gneiss, but these terms have been largely supplanted in recent years by a more explicit and stratigraphically meaningful lithologic terminology.

Structures in the old gneisses indicate three periods of deformation, here referred to as the early, the major, and the late deformations, in Precambrian time. Most of the visible structure in the rocks dates from the major deformation. In this deformation, the rocks were cast into folds ranging in size from regional down to microscopic and, in tightness, from open to isoclinal and overturned. Orientations of the fold axes differ from area to area but, over the Precambrian terrane as a whole, are predominantly northwest to west-northwest. The folding during the major deformation was accompanied by metamorphism of the gneisses to their present high grades and by intrusion of the oldest of the three main granitic groups.

The intense and widespread major deformation largely obliterated evidence of the early deformation, but structural and mineralogic vestiges of the early deformation have been widely recognized in recent years (124, p. 85; 153, p. 87; 168, 173, 177). These vestiges suggest open folding along northwest-trending axes, and metamorphism to at least the grade of mica schists.

The late deformation was characterized by cataclasis but included plastic folding in local areas as an accompaniment both of the cataclastic deformation and of intrusion of granites of the middle group (111,124). A belt of northeast-trending shear zones along the trend of what later came to be the Colorado mineral

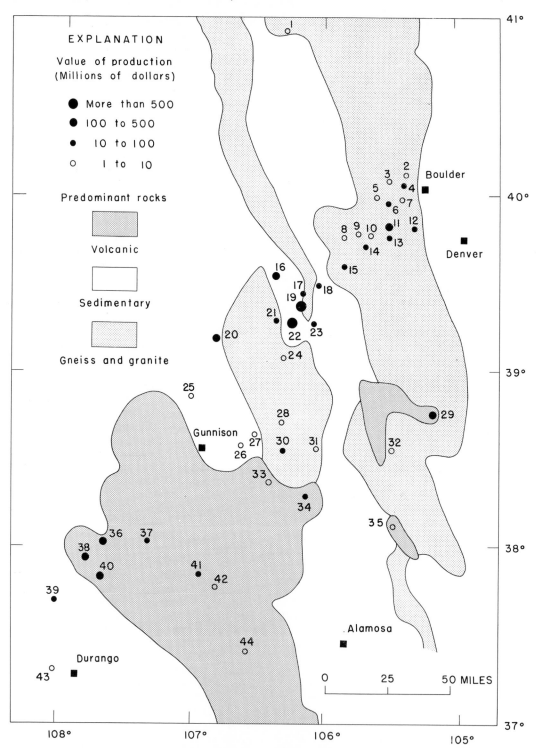

FIG. 2. *Principal Mining Districts in the Mountain Province of Colorado.*

EXPLANATION   *Figure 2*
*Mining Districts, Chief Products, and References*

| District or center | Chief products (In order of value) | References |
|---|---|---|
| 1. Northgate | $CaF_2$ | 90 |
| 2. Jamestown | Au, Ag, $CaF_2$ | 38; 52 |
| 3. Ward | Au, Ag | 19; 52 |
| 4. Gold Hill | Au, Ag | 26; 52 |
| 5. Caribou | Ag, Pb, U | 52; 61; 71 |
| 6. Nederland | W | 60 |
| 7. Magnolia | Au, Ag | 23; 52; 173 |
| 8. Urad | Mo | 84 |
| 9. Empire | Au, Ag | 52 |
| 10. Lawson-Dumont | Au, Ag | 52; 176 |
| 11. Central City-Blackhawk | Au, Ag, Cu, Pb, U | 52; 121; 122; 135 |
| 12. Ralston | U | 136; 144; 177 |
| 13. Idaho Springs | Au, Ag, Zn | 52; 64; 80; 161 |
| 14. Georgetown-Silver Plume | Ag, Pb | 52 |
| 15. Montezuma | Ag, Pb, Cu | 15 |
| 16. Gilman | Zn, Ag, Cu, Pb, Au | 33; 43; Radabaugh et al., this volume |
| 17. Kokomo | Zn, Pb, Ag, Au | 40 |
| 18. Breckenridge | Au, Ag, Pb, Zn | 14; 57 |
| 19. Climax | Mo, W | 41; 95; Wallace, this volume |
| 20. Aspen | Ag, Pb | 3; 16; 21 |
| 21. Sugar Loaf-St. Kevin | Ag, Au | 63; 78 |
| 22. Leadville | Ag, Zn, Pb, Au, Cu | 9; 59; Tweto, this volume |
| 23. Alma | Au, Ag, Pb | 25; 29; 53 |
| 24. Granite-Twin Lakes | Au, Ag | 7 |
| 25. Crested Butte | Ag, Pb, Zn | |
| 26. Gold Brick-Pitkin | Au, Ag, Pb | 6; 68 |
| 27. Tincup | Ag, Au, Pb | 17; 68 |
| 28. St. Elmo | Au, Ag | 68 |
| 29. Cripple Creek | Au | 18; 28; 49 |
| 30. Monarch-Tomichi | Zn, Ag, Pb, Cu | 68 |
| 31. Browns Canyon-Poncha | $CaF_2$ | 44 |
| 32. Tallahassee | U | 98 |
| 33. Cochetopa-Marshall Pass | U | 98 |
| 34. Bonanza | Ag, Pb, Cu | 11 |
| 35. Westcliff-Silver Cliff | Ag, Au, Pb | 2; 79 |
| 36. Ouray | Au, Ag, Pb, Cu | 24; 27; 39; 42 |
| 37. Lake City | Ag, Pb, Au | 42 |
| 38. Telluride | Au, Ag, Pb, Cu, Zn | 27; 70; 116; 174 |
| 39. Rico | Ag, Zn, Pb, Au, Cu | 34; 42 |
| 40. Silverton | Ag, Au, Pb, Zn, Cu | 12; 27; 55; 125; 175, Burbank and Luedke; this volume |
| 41. Creede | Ag, Pb, Zn | 92; 137; 151; Steven, this volume |
| 42. Wagonwheel Gap | $CaF_2$ | 44; Steven, this volume |
| 43. La Plata | Au, Ag | 48 |
| 44. Summitville | Au, Ag, Cu | 91 |

belt formed at this time (123), as did less persistent shear zones of this and other trends in many places in Colorado.

The granitic rocks in the Precambrian terrane have received many individual names, but they are classifiable in three general groups (152,169). Those of the oldest group are pre-dominantly granodioritic and gneissic, and they occur in generally concordant batholiths and phacoliths. The Boulder Creek Granite (or Granodiorite) of the Front and Park Ranges and the Denny Creek Granodiorite Gneiss of the Sawatch range are typical of this group. Rocks of the group have been classed as syn-

*TABLE II.   Summary of Precambrian Events*

| General age* | Central | Southwest | General age* | South-Central Wyoming |
|---|---|---|---|---|
| | COLORADO | | | SOUTH-CENTRAL WYOMING |
| | Erosion | | | |
| 1.0 | Intrusion of late granites, such as Pikes Peak | | | |
| | Erosion | Erosion | | Erosion |
| 1.35–1.45 | Late deformation; intrusion of granites of middle group, such as Silver Plume | Intrusion of granites of middle group<br>Deformation of Uncompahgre Fm.<br>Deposition of Uncompahgre Fm. | 1.35–1.45 | Intrusion of granites |
| | | | 1.6–1.9 | Deformation and metamorphism<br>Deposition of sediments |
| | Erosion | Erosion | | |
| 1.7–1.75 | Major deformation, metamorphism, and intrusion of early granites, such as Boulder Creek | do. | | |
| | | | | Erosion |
| | Early deformation | ? | | |
| >1.75 | Sedimentation and volcanism | do. | | |
| | | | 2.4–2.7 | Batholithic intrusion<br>Deformation and metamorphism |
| | | | >2.7 | Sedimentation, volcanism, and intrusion |

\* Age in billion years; references in text.

tectonic by many workers (80,135,169,173, 177), having been emplaced during the major period of folding and metamorphism of the old gneisses.

Granites of the middle group are only locally foliated and occur in both concordant and discordant batholiths and smaller plutons. The Silver Plume, Sherman, and related granites of the Front Range, the St. Kevin Granite of the Sawatch Range, and the Eolus Granite of the Needle Mountains are typical of the group. These granites generally are classed as late syntectonic to post-tectonic, and they are partly contemporaneous with the late, or cataclastic, period of deformation. Granites of the youngest group are represented principally by the Pikes Peak Granite, a coarse-grained massive granite in a large batholith that makes up a large part of the southern Front Range.

Dating of the various granites and other components of the Precambrian complex has been the subject of a large literature that records a gradual increase in some age assign-

ments as techniques and geologic understanding have improved. Recent reports (149, 154,162,169), which summarize much older work, place the age of the Boulder Creek and related granites at about 1.7 b.y., the Silver Plume and related granites at 1.35–1.45 b.y., and the Pikes Peak Granite at 1.0–1.05 b.y., based on Rb/Sr determinations. Metamorphism of the gneisses has been placed at about 1.75 m.y., based on Rb/Sr determinations (168,169), although one determination of 1.97 b.y. has been made (170). Age of the sediments parent to the gneisses is not established, but Hedge, *et al.* (168) have indicated an age only 100 m.y. greater than the major metamorphism would be consistent with the $Sr^{87}$ values. Such an age (approximately 1.9 b.y.) would be consistent with the oldest ages obtained from the younger sequence in southern Wyoming (Table II).

In central and northern Colorado, few if any rocks representing the time span between 1700 and 1450 m.y. are known, and differences in the depth-zone characteristics of the Boulder Creek and Silver Plume Granites suggest deep erosion as the principal event of this time. In southwestern Colorado, however, sediments of the Uncompahgre Formation were deposited on an eroded basement of old gneisses and syntectonic granites and then were folded and metamorphosed to slates, quartzites, and phyllites before intrusion of the Eolus Granite, a congener of the Silver Plume Granite.

Erosion must also have been the predominating event through most of the time from intrusion of the Silver Plume until the Cambrian, except during intrusion of the Pikes Peak Granite in the southern Front Range, as granites of all three age groups were widely exposed by the time of Cambrian sedimentation. Granitic dikes in the Gore Range have been dated as Pikes Peak in age by C. E. Hedge of the U.S. Geological Survey (oral communication, 1967), but other "young-looking" dikes, as of diabase and lamprophyre, yield ages closely similar to those of the middle group of granites (130,162). The essentially unmetamorphosed sandstone, quartzite, and shale of the Uinta Mountain Group in the Uinta Mountains, classed as younger Precambrian (146), possibly could be products of the time interval between the middle granites and Cambrian sedimentation, but alternatively they could be as old as the Uncomphagre Formation or even older, differing from other rocks of their age in being unintruded and unmetamorphosed.

## Wyoming

The older of the two metamorphic sequences of south-central Wyoming consists of schist, quartzite, metagraywacke, iron formation, greenstone, and meta-igneous intrusive rocks, all cut by batholithic quartz diorite and granite. The younger sequence is continuous with the sequence in Colorado but contains more hornblende gneiss and quartzite than in Colorado. Bayley's studies indicate two major deformations of the rocks of the older sequence in the Atlantic City area (this volume), and studies of Houston (148) in the Medicine Bow Mountains indicate two major deformations before deposition of the younger metasedimentary sequence, which is also deformed. A major northeast-trending shear zone that cuts both sequences in the Medicine Bow Mountains (148) has characteristics of the shear zones in the Colorado mineral belt and probably is a product of the third or late deformation recognized in Colorado (Table II).

The Precambrian igneous rocks of south-central Wyoming have not been so extensively studied as those of Colorado, but granitic bodies in the Sierra Madre, Medicine Bow, and Laramie ranges just north of the Colorado-Wyoming boundary in general are correlatable with the older and middle granites of Colorado. The Sherman Granite, which occupies a large area in the Laramie Range centered over the State boundary, long was correlated with the Pikes Peak Granite on lithologic grounds (52) but, with isotopic dating and more detailed geologic studies (149,130), is now known to be one of the middle granites. An anorthosite body that borders the Sherman Granite on the north is older than the Sherman (72,160, and Hagner, this volume) and may be a part of the older Precambrian sequence.

Ages of Wyoming Precambrian rocks shown in Table II are the products of many workers cited in a recent summary by Catanzaro (164); additional data on the Medicine Bow Mountains has been given by Hills, et al. (132) and Houston, et al. (148), and on the Sherman Granite by Peterman, et al. (149).

A Rb/Sr age of 2.32 b.y. for schist in the Red Creek Quartzite (147), the ancient basement rock beneath the Uinta Mountain Group in the Uinta Mountains, suggests that these mountains are in the province of the older metamorphic sequence of Wyoming.

## PALEOZOIC AND MESOZOIC ROCKS AND EVENTS

The Paleozoic-Mesozoic record in Colorado and south-central Wyoming is as much one of unconformity as of rocks. Sedimentation was repeatedly interrupted by uplift and erosion; consequently, many of the rock units exist only as remnants of the original bodies, and certain others were deposited only in restricted areas. In general, the region is characterized by a pre-Pennsylvanian Paleozoic sequence only 500 to 1500 feet thick and of rather restricted occurrence; a Pennsylvanian-Permian sequence as much as 10,000 feet thick in restricted basins; a Triassic-Jurassic sequence some hundreds of feet thick and of somewhat restricted occurrence; and a thick and widespread Cretaceous sequence. Thus the stratigraphic sequence differs markedly in various parts of the region. The sequences in representative areas in the mineral belt are shown in Table III.

From Cambrian through Mississippian time, the region was covered intermittently by shallow seas in which a sequence of carbonate and clastic rocks was layed down. This sequence is in general thinner in the Colorado-

TABLE III.   *Generalized Paleozoic and Mesozoic Stratigraphic Sequences*

| | Breckenridge | Leadville-Gilman Area | Aspen Area | Ouray-Telluride Area |
|---|---|---|---|---|
| Cretaceous | Pierre Shale 4000 ft; Niobrara Formation 300 ft; limestone and shale; Benton Shale 400 ft; Dakota Sandstone 200 ft | | Mesaverde Formation 1000+ ft; sandstone and shale; Mancos Shale 4000 ft; Dakota Sandstone 250 ft | Mesaverde Group 1000+ ft; Mancos Shale 3000+ ft; Dakota Sandstone 200 ft; Burro Canyon Formation 50 ft; sandstone |
| Jurassic | Morrison Formation 200 ft; shale and sandstone; Entrada Sandstone 50 ft | | Morrison Formation 400 ft; Entrada Sandstone 100 ft | Morrison Formation 700 ft; Wanakah Formation 100 ft; sandstone, siltstone, and limestone; Entrada Sandstone 50 ft |
| Triassic | Chinle Formation 200 ft; red shale | | Chinle Formation 300 ft | Dolores Formation 500 ft; red sandstone and siltstone |
| Permian | | | | Cutler Formation 1500 ft; red sandstone |
| Permian and Pennsylvanian | Maroon Formation 700 ft; red sandstone | Maroon Formation 2000 ft; red sandstone and conglomerate | Maroon Formation 8000 ft | Rico Formation 300 ft; red sandstone |
| Pennsylvanian | | Minturn Formation 6000 ft; red and gray sandstone and conglomerate; Belden Formation 200 ft; black shale and limestone | Minturn Formation 3000 ft; Belden Formation 700 ft | Hermosa Formation 1500–2000 ft; sandstone, shale, limestone, and gypsum; Molas Formation 50–100 ft; shale and regolith |
| Mississippian | | Leadville Dolomite 100–200 ft | Leadville Limestone 100–200 ft | Leadville Limestone 100–200 ft |
| Devonian | | Chaffee Formation 150 ft; dolomite and quartzite | Chaffee Formation 200 ft | Ouray Limestone 100 ft; Elbert Formation 100 ft; sandstone and shale |
| Ordovician | | Harding Sandstone, 75 ft; absent at Leadville; Manitou Dolomite 100 ft; absent at Gilman | Manitou Dolomite 300 ft | |
| Cambrian | | Peerless Formation 100 ft; dolomitic shale and sandstone; Sawatch Quartzite 100–200 ft | Peerless Formation 100 ft; Sawatch Quartzite 200 ft | Ignacio Quartzite 100–200 ft |
| | (Basement) | (Basement) | (Basement) | (Basement) |

Wyoming region than in bordering areas, indicating that the region had a tendency to stand slightly higher than adjoining parts of the stable continental platform.

Cambrian sediments were deposited on an essentially planar surface cut over Precambrian rocks, although relief of as much as 100 feet existed locally (123). As judged by absence of shore facies and the occurrence of outliers (including pebbles in younger conglomerates), these sediments may have been deposited over almost all of Colorado as well as adjoining Wyoming, but, in a period of erosion that preceded deposition of the Lower Ordovician Manitou Formation, they were stripped from large areas in the south-central and north-central parts of Colorado, and elsewhere they were beveled so as to thin eastward in the mountain province. Erosion after deposition of the Manitou removed this unit in the Gilman area—among others—and reduced it to only 20 feet in the Kokomo area (51). Similarly, the Middle Ordovician Harding Sandstone was extensively eroded before the Upper Ordovician Fremont Limestone was deposited, with the result that, except for an outlier in the Gilman area, the Harding is now largely restricted to the southeastern side of the mineral belt. The Fremont Limestone—an important host rock of ore deposits in the Garfield district—is even more restricted. However, limestone fragments containing fossils characteristic of the Fremont, and also fragments containing Silurian fossils, have been found in diatremes in the Precambrian terrane of the northern Front Range (100,142), suggesting, along with other indirect evidence, that both the Fremont and a Silurian limestone once may have been widespread in the region. These limestones, as well as older Paleozoic units in places, were stripped away in an extensive erosional episode that preceded deposition of the Upper Devonian Chaffee and Elbert Formations. This erosion produced the most widespread unconformity in the region, and, as a result of it, the Upper Devonian rocks lie at one place or another upon all of the units discussed above, as well as upon Precambrian rocks. The Devonian rocks in turn were irregularly eroded and thinned before deposition of the Middle Mississippian Leadville Limestone (or Dolomite). The Leadville, or its equivalents, evidently was deposited over the entire region, but in an erosional episode that preceded deposition of the Middle Pennsylvanian Belden-Minturn and Hermosa Formations, it was irregularly thinned by chemical weathering that produced a karst surface. The

Molas Formation, an irregular and pockety regolithic unit, consists largely of the insoluble residue from this weathering.

Pronounced uplift in Pennsylvanian time produced extensive highlands—the so-called Ancestral Rockies—separated by basins or narrow geosynclines in which many thousands of feet of Pennsylvanian and Permian clastic sediments and evaporites accumulated. Principal uplifts of this time were the Front Range highland, occupying most of the area of the present Front Range, Middle and North Parks, the northern Park Range, and adjoining parts of southern Wyoming; and the Uncompahgre-San Luis highland, occupying a broad northwest-trending belt extending from the present San Luis Valley through the northeastern half of the present San Juan Mountains and the Grand Junction area into Utah (Figure 2). All older sedimentary rocks were stripped from these highlands, exposing Precambrian rocks that became the source of most of the sediments in the adjoining basins. The Front Range and Uncompahgre-San Luis highlands persisted through the early Mesozoic, when alternating marine and continental clastic rocks were deposited in adjoining areas. Only with deposition of Late Jurassic units such as the widespread Morrison Formation, a fluviatile unit, were the highlands covered again, and, in succeeding Cretaceous time, they were deeply covered with marine sediments.

All these events have a bearing on exploration for ore deposits in the sedimentary rocks, for they produced many trends of thickening and thinning of units and many wedge edges in the units. Certain formations, particularly the thin pre-Pennsylvanian units, may prove to be unexpectedly absent in some areas, and, conversely, they may be present unexpectedly as outliers in others.

General compilations on the Paleozoic and Mesozoic stratigraphy of Colorado are contained in guidebooks of the Rocky Mountain Association of Geologists (74,81,96) and in reports by Johnson (35), Brill (32), and Tweto (51).

## LARAMIDE OROGENY

The Mesozoic Era closed and the Cenozoic Era opened with the Laramide orogeny, which produced extensive deposits of orogenic sediments, volcanic and intrusive rocks, ore deposits, and the principal structural features younger than Precambrian in the region. The orogeny began with uplift in the latter part of Cretaceous time and reached a structural climax in the late Paleocene and early Eocene. However, there is no well defined and generally recognized event to mark the end of the orogeny, and, as the pace of geologic events was rapid in the Cenozoic, the Laramide is visualized by some workers as extending into the Oligocene or even the Miocene. In this report, it is visualized as occupying the time span from the latter part of the Campanian stage of the Late Cretaceous to about the middle Eocene.

A first manifestation of the Laramide orogeny was uplift at the sites of most of the major ranges, forcing the Cretaceous seas to retreat eastward and allowing stripping of the sedimentary cover of the ranges to commence. Volcanism and intrusion followed closely, as early volcanic and intrusive rocks yield K/Ar ages of 65 to 70 m.y. (129,162,165), corresponding in time to the early Maestrichtian stage of the Upper Cretaceous, during which marine and overlying nonmarine sediments were being deposited on the plains (157). Concurrent with the uplift and volcanism, sediments consisting of materials derived from sedimentary, volcanic, and intrusive rocks of various ages accumulated in bordering basins. These sediments, which have been established paleontologically to range in age from late Cretaceous to early Eocene, are involved in the folds and faults that generally are identified as Laramide structural features, and hence the climax of Laramide deformation must have been at least as late as late Paleocene or early Eocene. Some of the deformation occurred earlier, however, as indicated by the angular unconformity between Paleocene and Cretaceous rocks in places (73), and between Paleocene and Eocene rocks (46, 82, 138, p. 96). Also, in the Leadville area, intrusive rocks younger than the major faults but older than many of the minor faults yield isotopic ages near the Cretaceous-Tertiary boundary (113).

The Wasatch Formation, of Paleocene and early Eocene age, is strongly deformed in many places, but the Lower and Middle Eocene Green River Formation is considerably less deformed (101) suggesting that tectonic movements were drawing to a close about middle Eocene time.

# CENOZOIC ROCKS AND EVENTS

## Sedimentary Rocks

Cenozoic sedimentary rocks are largely confined to the basins, plateaus, and plains. They are absent in many of the mining districts of Colorado, and they are mineralized only in a very few, but they are important to the geologic interpretation of many ore deposits because they constitute records of structural and igneous events to which the ores are related. In Wyoming, the Cenozoic sedimentary rocks contain major uranium deposits, as discussed by Harshman in this volume.

Because of their tectonic and volcanic origin, the sedimentary rock units are localized in occurrence, and they go by many different names. Principal units in Colorado are shown in Table IV. In general, the Paleocene units are predominantly fluviatile sandstones and conglomerates, many of which contain abundant volcanic debris. The Eocene units include lacustrine as well as fluviatile rocks and are predominantly siltstones, mudstones, and sandstones; the Green River Formation, widely known for its oil shale, is predominantly marlstone. The Oligocene units are typically tuffaceous and fine-grained. The Miocene units are widely varied and generally include much volcanic material, both as tuff and as erosional debris. The Pliocene units differ from those of the Miocene only in containing less tuff and, near the mineral belt, a preponderance of intrusive over volcanic rocks among its pebbles. Pleistocene deposits include tills of as many as nine glacial advances (107) as well as various gravels and alluviums.

Regional aspects of the Cenozoic stratigraphy, or parts of it, are discussed in reports by Brown (109), Siems (134), Scott (150), and Moore (88).

## Igneous Rocks

Cenozoic intrusive and volcanic rocks are widespread in the mountains of Colorado but rare in south-central Wyoming. The intrusive rocks, generally referred to as porphyries, are principally concentrated in the mineral belt, although they also occur in scattered areas throughout the mountain province in Colorado. Volcanic rocks are far more widespread, and, except in the San Juan Mountains, they lie principally outside of the mineral belt. Some of the volcanic rocks, such as basalts of northwestern Colorado, have little economic implication, but others of more felsic compositions are associated with ore deposits, as at Cripple Creek, Westcliff, and in the San Juan Mountains.

The porphyries of the mineral belt are predominantly quartz monzonites, although they range from diorite to granite in composition

TABLE IV.   *Principal Cenozoic Sedimentary Units*
*Queries indicate uncertainty as to age. Fm, Formation; Cgl, conglomerate; LB, lake beds; Vol, volcanics*

| | PLAINS AREA | HUERFANO PARK | NORTH AND MIDDLE PARKS | SOUTH PARK-ARKANSAS VALLEY |
|---|---|---|---|---|
| Pleistocene and Recent | Various glacial deposits, gravels, and alluviums | | | |
| Pliocene | Ogallala Fm | | | Dry Union Fm and Wagontongue Fm as used by DeVoto (128) |
| Miocene | Arikaree Fm | Devils Hole Fm (?) | North Park Fm and Troublesome Fm | |
| Oligocene | White River Fm; also Castle Rock Cgl | Farisita Cgl (?) | White River Fm and Rabbit Ears Vol | Antero Fm as used by DeVoto (128) and Florisant LB |
| Eocene | | Huerfano Fm Cuchara Fm | Coalmont Fm (upper part) | |
| Paleocene | Denver and Dawson Fms (upper part) | Poison Canyon Fm Raton Fm (upper part) | Coalmont Fm (lower part) and Middle Park Fm (upper part) | Denver Fm (upper part) |

TABLE IV.   (*Cont.*)

| | NORTHWEST COLORADO | SAN LUIS VALLEY-SAN JUAN MOUNTAINS | SOUTH SIDE OF SAN JUAN MOUNTAINS |
|---|---|---|---|
| Pleistocene and Recent | Various glacial deposits, gravels, and alluviums | | |
| Pliocene | | Santa Fe Fm (upper part) | |
| Miocene | Browns Park Fm | Santa Fe Fm (lower part) | |
| Oligocene | | Creede Fm | |
| Eocene | Green River Fm Wasatch Fm (upper part) | Telluride Fm (?) Blanco Basin Fm (?) | Wasatch Fm or San Jose Fm of Simpson (46) |
| Paleocene | Wasatch Fm (lower part) Ohio Creek Fm | | Nacimiento Fm Animas Fm (upper part) |

and also include alkalic and rare mafic facies. The porphyries have been described in detail in many of the district reports cited in Figure 2, and those of the Front Range have been described and correlated as a group by Lovering and Goddard (52) and by Wells (97). Aside from general studies made by Crawford (8) in the Sawatch and Mosquito Ranges more than 40 years ago, no regional studies of the porphyries southwest of the Front Range have been made, although excellent studies have been made locally. Studies of the mineral belt porphyries as a group are badly needed as a prerequisite to the complete understanding of this unique belt.

The intrusive rocks are typically hypabyssal in texture, structure, and mode of occurrence. To a considerable degree, the forms of the intrusive bodies differ with the environment. In Precambrian terranes, as in the Front and

Sawatch Ranges, the intrusive rocks occur in stocks and dikes. The Sawatch Range also includes a small batholith. In sedimentary terranes, as between Breckenridge and Leadville and west of the Sawatch Range, the intrusive rocks are principally in sills, but they occur also in scattered stocks, plugs, dikes, and laccoliths. The Elk Mountains are characterized by large irregular stocks, and the West Elk Mountains by large laccoliths.

With the exception of basalt fields, all of the larger piles of volcanic rocks in Colorado have been found in recent years to consist of ash flow tuffs and related lava flows and pyroclastic materials. By far the largest of the piles is that of the San Juan Mountains, which covers an area of 6000 square miles and was originally more than 5000 cubic miles in volume. The pile is a composite one, made up of several interfingering and overlapping individual piles, each characterized by a caldera center of eruption. Most such centers had geologically brief life histories, and a succession of them developed with passing time. Early ones were soon covered or partly destroyed, and they can now be identified only by painstaking geologic studies. Among the calderas thus far identified, the best known are the Silverton and Lake City calderas of the western San Juans, and the Creede caldera of the central San Juans, discussed by Burbank and Luedke and by Steven, respectively, in papers in this volume.

Stratigraphy of the many and complexly interrelated volcanic units is vastly complicated and will not be detailed here. Stratigraphic charts for the Silverton and Creede areas are included in the papers in this volume cited above, but a modern chart relating the units of these areas to each other and to all of the rest of the San Juan volcanic field is still to be realized. A general report on the San Juan field by Larsen and Cross (65), published in 1956 but based on work done much earlier, provides a wealth of data on the volcanic rocks, but the stratigraphic terminology and the concepts of volcanic mechanisms and history it presents have been greatly revised in recent years.

Principal felsic volcanic fields outside the San Juan Mountains are in the West Elk Mountains, southern South Park, and in the Wet Mountain Valley, between the Sangre de Cristo and Wet Mountains (Figure 1). Smaller ones exist on the crest of the Sawatch Range and in the Continental Divide area between North and Middle Parks. The West Elk volcanic field is largely unstudied but is known from the interfingering of volcanic units to be about the same age as the San Juan volcanic field. The volcanic field in South Park, known as the Thirtynine Mile field, is characterized by thin but widespread ash-flow tuff units and by caldera structure (127,166). Some ash flows from this center once extended across the site of the present canyon of the Arkansas southeast of Salida into Wet Mountain Valley, and ash flows in the Arkansas Valley north of Salida are related to them in time if not spatially. The small and isolated Cripple Creek volcanic center is evidently an outlier of the Thirtynine Mile field. The volcanic field in the Wet Mountain Valley was derived from eruptive centers in the Rosita Hills and Silver Cliff area a few miles east of Westcliff (171). Mineralized breccia pipes of the Westcliff district presumably are related to this volcanic activity.

## Age of Igneous Rocks

Until the advent of isotopic dating, the intrusive rocks of the mineral belt, as well as most similar intrusives outside the belt, were classed as Laramide in age (in the sense of Late Cretaceous-Early Tertiary). This classification, which included an assumption that all of the various intrusives were about the same age, was based originally on the general relationship of intrusive bodies to structural features that were classed as Laramide, with the knowledge that both intrusion and deformation post-dated the marine Cretaceous rocks. A later refinement was based on the occurrence of igneous pebbles in sedimentary units such as the Denver and Middle Park Formations, which were dated paleontologically as Late Cretaceous-Early Tertiary; then upon correlation of the pebbles with certain intrusive bodies; and then upon correlations among the intrusives based either on their interrelationships or on mutual relationships to structural features (22). Exceptions to the presumed early Laramide age were recognized in the West Elk Mountains and in the Spanish Peaks area of southern Colorado, where intrusive rocks cut the Eocene Wasatch and Huerfano Formations, respectively, and, of course, in the San Juan Mountains, where intrusive rocks cut the volcanics.

Most volcanics, except the basalt fields, were classed as Miocene, both because of the abundance of volcanic debris in Miocene sedimentary units and because the volcanics of the San Juan were tenuously dated paleobotanically as Miocene. The basalt fields were dated as late Tertiary or Quaternary, primarily on geomorphic grounds.

Isotopic dating has served to corroborate a Laramide age near the Cretaceous-Paleocene boundary for many intrusive rocks and for early volcanics. The Pando Porphyry, the earliest of many porphyries in the Leadville region, has a K/Ar age of 70 m.y., and the Lincoln Porphyry, younger in the porphyry sequence, has an age of 64 m.y. (113). The Eldora stock of the Front Range has yielded a Rb/Sr age of $60 \pm 9$ m.y. and a K/Ar age of about 54 m.y. (131). The nearby Audubon stock has been dated at about 67 m.y. and the Caribou stock at 68 m.y. (86,102). These stocks were regarded by Lovering and Goddard (22) as the earliest in the Front Range, and the Audubon stock was regarded by them as the source of igneous pebbles in the Upper Cretaceous and Paleocene Denver Formation. The stocks evidently were also the source of pebbles in the Middle Park Formation (73). A pumice stratum in the Denver Formation, 35 feet above the Cretaceous-Tertiary boundary as defined paleontologically, has a K/Ar age of 65 m.y. (129). In the San Juan Mountains, old volcanic rocks that may have been a source for igneous pebbles in the Upper Cretaceous and Paleocene Animas Formation have been dated by K/Ar at 66 m.y. (165).

These ages all cluster near the Cretaceous-Tertiary boundary fixed at 63 m.y. by Kulp (105) and at 70 m.y. by Holmes (87).

Existence of distinctly younger intrusive rocks in the mineral belt has been known or suspected on geologic grounds for some time. For example, the Montezuma stock of the Front Range cuts the Williams Range thrust fault, which is younger than the Middle Park Formation (73), and hence the stock must be younger than those that supplied pebbles to the Middle Park. Also, at Leadville (Tweto, this volume), as well as in many other places, some of the intrusive rocks have characteristics that suggest a shallower environment of intrusion and hence a younger age than for the typical mineral belt porphyries.

Although many igneous bodies are yet to be dated, accumulating data indicate a second, distinct, and widespread surge of intrusive and extrusive igneous activity 15 to 30 million years after the main period of Laramide igneous activity. This igneous stage is centered in the Oligocene (25 to 40 m.y. ages). Intrusive dates between 60 and 40 m.y., or roughly between Paleocene and Oligocene, are scarce. Hart (131) evidently favors a 54 m.y. K/Ar age for the Eldora stock, although he has made higher determinations by Rb/Sr; the Twin Lakes stock in the Sawatch Range has been dated as 51 and 56 m.y. by Moorbath (112,170), and a stock in another range has been dated as 46 m.y. (confidential information). Data on mid-Tertiary age determinations are summarized in Table V. Additional determinations in this age range are known but are confidential.

This widespread igneous activity in Oligocene time has important economic implications, as discussed in a following section.

TABLE V.    K/Ar Age Determinations on Post-Laramide Igneous Rocks

| Igneous body or locality | Age, m.y. | Reference |
|---|---|---|
| Basalt, Grand Mesa | 10 | U.S. Geol. Survey, 1966 (163) |
| Volcanic rocks, central San Juan Mountains | 26–28 | T. A. Steven, U.S. Geol. Survey, oral communication |
| Red Mountain plug, Urad | 27 | Taylor, et al. (172) and oral communication |
| Mt. Richthofen stock | 27–28 | Corbett (155) |
| Volcanics, Lulu Mountain | 27–28 | Corbett (155) |
| Volcanic-intrusive rocks, Nathrop | 28–29 | R. E. Van Alstine, U.S. Geol. Survey (oral communication) |
| Climax stock | 30–32 | Data from Amax Exploration, (with permission) |
| Rabbit Ears Volcanics, Middle Park | 33 | Izett (159) |
| Volcanics of Arkansas Valley near Salida | 34 | U.S. Geol. Survey, 1965 (153) |
| Volcanics of Thirty-nine Mile Field, South Park | 34–40 | Epis, et al. (166) |
| West Cliff-Rosita Hills | 38–40 | Siems (171) and oral communication |

## Structure

GENERAL FEATURES    The mountain ranges of Colorado and south-central Wyoming are, in the main, structurally simple upwarps, and so the structural geology of the region is far less complex than in areas farther west. The Front and Park Ranges and their northward extensions in Wyoming, and also the Sawatch Range, are broadly anticlinal, although modified by faulting along their flanks in many places. As anticlines, they have rather unconventional character, as they seem to have had broad flat tops rather than arching to a crest. In reality, they consist of paired, opposite-facing monoclines. This shape suggests that they are essentially uplifted blocks of basement rocks, and that the sedimentary rocks were folded principally by draping along their sides. Thrust faults and tight folds exist locally along the flanks of the mountains but probably mainly as a result of crimping in re-entrants, of expansion of the uplifted blocks, and of gravity sliding.

Although the mountain front and some of the main ranges trend generally north (Figure 1), the structural grain of the region is distinctly northwest, mainly at azimuths of N20°W to 35°W. Ranges such as the Sangre de Cristo, Wet, and Sawatch trend in this direction, and the Front and Park Ranges have many structural features that trend in this direction despite their northerly trend as ranges. The mineral belt, discussed separately below, trends northeastward across the trend of the ranges and the structural grain.

Most of the features that define the structural grain date from the Laramide, but some can be traced to the Precambrian, and some date at least from the late Paleozoic, if not earlier. In addition, post-Laramide faults and folds contribute to the northwest grain, but many of the faults may be reactivated Laramide faults.

SAN LUIS-UPPER ARKANSAS GRABEN  The largest post-Laramide structural feature of the mountain province is a great graben that extends generally northward through the San Luis and Upper Arkansas valleys. This is the northward continuation of the Rio Grande graben or trough of New Mexico and, thus, an element of basin-and-range structure. The graben comes to an end at about the Continental Divide near Leadville, but scattered en echelon north-trending faults possibly related to it continue northward, perhaps to the Wyoming line. The graben shows abundant evidence of recent fault movement, such as the great scarp of the Sangre de Cristo Range along the east side of the San Luis Valley, which is characterized by marked geomorphic unconformity, and Tertiary and Quaternary deposits faulted against Precambrian rocks at several places along the Upper Arkansas valley (Tweto, Leadville, this volume). However, the date of its origin is as yet unfixed. It is filled by sediments of the Miocene and Pliocene Santa Fe Formation in the San Luis Valley and of the Pliocene Dry Union Formation in the Arkansas Valley, but older deposits may lie beneath these units. Oligocene age of igneous rocks, apparently localized by valley border faults near Nathrop, and of volcanic rocks low on the valley side near Salida (Table V) suggests that the graben existed at least as early as Oligocene. On the other hand, it slices longitudinally through the east flank of the great Sawatch anticline and hence must be younger than Laramide.

FRONT RANGE  Geology and structure of the Front Range have been described in detail by Lovering and Goddard (52). The eastern flank of the range is essentially an abrupt monocline extensively modified by faulting and subsidiary folding. Flat-lying sediments of the Plains turn up steeply in the monocline and then flatten again, as indicated by scattered remnants of sedimentary rocks west of the mountain front. The monocline dips only moderately in the Laramie Range but steepens southward. From Fort Collins southward nearly to Boulder, in a segment of the mountain front that trends east of north, the monocline is rumpled by a series of en echelon northwest-trending anticlines and synclines, most of which are longitudinally faulted. Near Golden, the monocline is modified and partly cut out by the Golden fault, a west-dipping reverse fault that locally flattens to only a few degrees but that probably steepens with depth. Sedimentary rocks of the faulted monocline are vertical or overturned in the footwall of this fault. In a segment of the mountain front near Colorado Springs, faults similar to the Golden fault bring Precambrian rocks over the sediments, cutting out the monocline except for small remnants in the overriding block. The Wet Mountains, an offset southern extension of the Front Range, have a similar faulted front.

The Precambrian core of the Front Range, which makes up nearly the entire range, is broken by myriad fractures and faults. Largest and most persistent of these are NNE- to NNW-trending faults such as the Berthoud Pass fault of the Front Range and the Ilse fault of the Wet Mountains. These faults, as much as a half mile wide, are Precambrian shear zones that underwent much movement in later time. Far more abundant—and economically more significant—are northwest-trending faults generally known as breccia reefs. These faults also originated in Precambrian time and underwent extensive movement in later time (123). Many of them are hydrothermally altered and are mineralized with quartz and sparse hematite where they cross the mineral belt. Although barren in themselves in most of the mining districts, they were important in the localization of ore deposits in intersecting northeast-trending veins (52), and, in recent years, they have proved to be a principal habitat of uranium deposits, particularly on the fringes of, or outside, the mineral belt.

The western flank of the Front Range is faulted through much of its length, but remnants of a monoclinal structure—generally more gentle than on the east side of the range—remain in many places. Major faults along the flank include the Never Summer thrust in southeastern North Park; the Vasquez thrust near Granby and Fraser in Middle Park; the Williams Range thrust, which extends from the vicinity of Breckenridge north-northwest for 65 miles into west-central Middle Park; and the Elkhorn thrust, on the east side of South Park. None of these thrusts have very great horizontal displacements—generally only 1 or 2 miles—and they seem to be products both of lateral spreading of the vertically uplifted mountain block and of gravity sliding (73,104,117,133,155). Tight folding accompanies them only locally, as in the Vasquez Mountains re-entrant near Fraser (73). The thrusts all affect rocks as young as Paleocene, and two of them are cut or domed by stocks—the Never Summer by the Mt. Richthofen stock and the Williams Range by the Montezuma stock.

INTERMONTANE BASINS OR PARKS   The Front Range is separated from the Park Range to the west by a line of broad topographic and structural basins that includes North, Middle, and South Parks. These basins are of Laramide origin, as indicated by their content of sediments derived from the early Laramide uplifts, as in the Coalmont, Middle Park, and Denver Formations (Table IV). North and Middle Parks lie upon part of the old Front Range highland of late Paleozoic time, and so they have a rather thin sedimentary sequence that terminates downward with Triassic or Jurassic rocks lying upon the Precambrian. South Park, in contrast, was off the highland, and its sedimentary sequence extends downward to the Cambrian.

The floors of all three parks are disturbed by faults and folds, but Middle Park (73) considerably more than the other two. North Park is closed at its northern end by highlands consisting of Precambrian rocks that have been thrust southward over the sedimentary rocks along the west- to northwest-trending Independence Mountain thrust fault, and small east-dipping thrust faults border the northern part of its western side (145). In the center of the park, strata of the Miocene North Park Formation are sharply folded in a large westnorthwest trending syncline, indicating pronounced post-Miocene deformation. In Middle and South Parks, in contrast, the post-Eocene formations are only mildly warped, although faulted in places.

Middle and South Parks are separated by mountain spurs followed by the Continental Divide. The mineral belt occupies this same high ground and bordering parts of the two parks. Major faults of South Park, such as the Elkhorn thrust fault and, a few miles to the west of it, the South Park reverse fault, extend from the mineral belt southward to the Thirtynine Mile volcanic field, where they disappear beneath the volcanics and associated Oligocene sedimentary rocks (50).

PARK RANGE   As shown in Figure 1, the Park Range as an overall unit contains segments known by other names, although the segment from the Colorado River to the Wyoming boundary also is known specifically as the Park Range. Each of the three major segments, i.e., Park, Gore, and Mosquito, has characteristic geology and structure. The northern, or Park, segment closely resembles the Front Range, being essentially monoclinal on its eastern flank and extensively faulted (but also monoclinal) along its western flank. This segment of the range lies entirely upon a part of the late Paleozoic Front Range highland. The Gore Range, in contrast, is at the site of the border of the old highland. The Gore fault, which borders the high or Precambrian-rock part of the range on its west side, was a bordering scarp of the highland in Pennsylvanian time, as indicated by the abrupt thinning of strata of the Minturn Formation along the line of the fault. Various evidence indicates that the Gore fault was a Precambrian fault that was reactivated in Pennsylvanian time (123), and it was reactivated again in Laramide and perhaps later time. The eastern flank of the Gore Range is monoclinal and locally faulted, just as the Park Range segment to the north.

The Mosquito Range, including a short northern segment known as the Tenmile Range, is a fault-block range carved from the eastern flank of the huge Sawatch anticline, the core of which constitutes the Sawatch Range, the next range west of the Mosquito Range (Figure 1). These ranges are in the basin of late Paleozoic sedimentation south and west of the old Front Range highland.

In general, the west side of the Mosquito range is a modified fault scarp, and the east side dips monoclinally into the South Park basin. The faults that border the range on the west are in part Laramide faults, in part younger basin-and-range faults related to the Arkansas valley graben, and in part Laramide faults that also served as graben faults. The principal fault in the northern part of the range is the Mosquito fault, a west-dipping normal fault with a displacement (including that on subsidiary faults) of about 10,000 feet. The projected intersection of the north-trending Mosquito fault and the S30°E-trending Gore fault, a few miles northeast of Kokomo, is occupied by the early Tertiary Humbug stock. North of the stock, the Mosquito fault shows evidence of Precambrian origin (123). South of the stock such evidence has not been noted, but the fault, or a zone along it, was the locus of intrusion of Laramide porphyries, indicating that the fault existed at least by early Laramide. From the latitude of Kokomo or Climax southward beyond Leadville, the Mosquito and related faults show evidence of post-Laramide movement, such as displacement of the Pliocene Dry Union Formation and even of glacial units, indicating that the Mosquito fault zone served as part of the younger Tertiary graben fault system. Midway between Leadville and Climax, the Mosquito fault is intersected on its footwall side by the north-northwest trending London reverse fault, a Laramide fault that is the major structural feature of the Alma mining district. This fault extends far to the southeast in South Park, where it is marked by scattered small intrusive bodies.

At the southern edge of the Leadville district, the main line of Laramide displacement turns off from the Mosquito fault and onto the north-northwest-trending and nearly vertical Weston fault (Tweto, Leadville, this volume). The Weston fault has small stocks scattered along it, and, at Buffalo Peaks, 20 miles southeast of Leadville, it is covered for a short distance by volcanic rocks that may correlate with those of the Thirtynine Mile field. About 12 miles north of Salida, the fault is cut out by the large Whitehorn stock. At about this latitude, an en echelon fault of similar trend, the Badger Creek fault, takes up a few miles to the east and extends southward across the Arkansas River at Coaldale.

SANGRE DE CRISTO RANGE   This range, only about half of which lies in Colorado, is structurally continuous with the Mosquito Range. It is characterized by greater amounts of sedimentary rocks than the ranges discussed above and also by somewhat greater deformation of

these rocks, but it does not differ greatly from the other ranges in its fundamental features. The east side of the range is essentially a monocline modified by longitudinal folds and many small thrust faults. The west side is a fault scarp bordering the graben of the San Luis Valley. This scarp is mainly in Precambrian rocks, although sedimentary rocks cap the crest of the range in many places. The northern end of the range is, like the Mosquito Range, a part of the faulted flank of the great Sawatch anticline. Most of the folding and thrusting in the range presumably date from the Laramide, but the graben faults along the west side of the range, and possibly some part of the fold structure, are younger. Structural geology of parts of the range has been described by Burbank and Goddard (20), Litsey 76), and Gableman (58).

SAWATCH RANGE   The Sawatch Range consists principally of Precambrian rocks in the core of a great anticline 90 miles long and 40 miles wide. Upturned sedimentary rocks wrap around the northern end of the range and are more or less continuous along its western flank. The eastern flank of the range is bounded by the San Luis-Arkansas fault graben through most of its length and hence is primarily in Precambrian rocks. The sedimentary rocks of the east flank of the anticline—as contrasted to the flank of the range—are mainly in the ranges east of the graben, as the Mosquito, northern Sangre de Cristo, and southern Gore ranges.

On the northeastern flank of the range, from Tennessee Pass northwestward through Gilman to Minturn, the sedimentary rocks dip monoclinally off the range and anticline at angles of 12° to 15° (33). Along the west side of the range, dips locally are much steeper, and the monocline is rumpled by folds. Small thrust faults occur along parts of the western flank. The largest of these, the Tincup-Morning Glim fault, is cut out for several miles by the Tertiary Mt. Princeton batholith (68). Near Aspen, the sedimentary rocks of the flank of the range are broken by many steep faults, largest of which is the Castle Creek fault, a north-northwest trending steep reverse fault that turns at Aspen and extends northwest down the Roaring Fork valley for several miles and then passes into a monocline (16).

The southern part of the Sawatch Range is characterized by scattered infolds of sedimentary rocks in a matrix of Precambrian rocks. The Monarch and Tomichi mining districts are located in such infolded Paleozoic rocks. West of these districts, the range is joined from the west by the Gunnison uplift, a structural high occupying part of the site of the old Uncompahgre-San Luis highland, where Paleozoic rocks are absent, and the Jurassic Morrison Formation lies upon the Precambrian. South of Monarch-Tomichi, near Bonanza (Figure 2), rocks and structures of the Sawatch Range disappear beneath volcanics of the San Juan field.

Precambrian rocks of the interior of the range—like those of the Front Range—are broken by many faults, some of which have existed since the Precambrian. Many faults along the east side of the range, however, seem to be part of en echelon series of graben faults that project from the Arkansas valley border into the flank of the range. A major feature of the northern part of the range is the Homestake shear zone, a Precambrian shear zone several miles wide that extends southwestward across the range from near Gilman to near Aspen (123). Other and smaller shear zones of similar trend are spaced through much of the length of the range.

ELK MOUNTAINS AND WHITE RIVER PLATEAU
The Elk Mountains can be viewed as a broad structural terrace in the west flank of the Sawatch anticline or, more precisely, as a very asymmetric northwestward-plunging anticline branching from the west-central part of the Sawatch anticline. No major structural or topographic feature separates the Elk Mountains from the Sawatch Range. The principal structural feature of the Elk Mountains, on their west side, is a faulted monocline, which has been described in some detail by Vanderwilt (21). In the southern part of the mountains, the monocline is overturned and broken by thrust faults. Farther north, it is vertical and broken by steep normal faults, and still farther north it becomes a simple, if steep, monocline. The monocline extends sinuously northwestward far beyond the limits of the Elk Mountains. As the Grand Hogback, it forms the southwestern and western boundary of the White River uplift (Figure 1) and continues northwestward to merge with the Uinta Mountains fold (21). This monocline is generally taken as the boundary between the mountain province on the east and the Colorado Plateau province on the west. Strata of the Paleocene and early Eocene Wasatch Formation are involved in it, as, to a lesser degree, are strata of the overlying Green River Formation (101); hence it must be of post-early Eocene age.

Large stocks in the Elk Mountains seem to have been controlled by the monocline and related faults (21), and one stock cuts out the main fault for several miles. The Treasury Mountain dome, a prominent feature of the Elk Mountains, is an intrusive dome west of the monocline (21). Precambrian and Paleozoic rocks are exposed in the dome, which is surrounded by Cretaceous rocks. The small mining districts of Marble and Crystal are located in the Paleozoic sedimentary rocks on the flanks.

The White River Plateau is a large and structurally simple flat dome that was extensively eroded before a cover of late Tertiary basalt flows forming a higher plateau called the Flat Tops was formed on it. The White River Plateau is broken by numerous faults in its southern part (118), and it is connected with the Sawatch Range to the southeast by a sagging anticlinal axis.

WEST ELK MOUNTAINS     The West Elk Mountains differ from the Elk Mountains, into which they grade, chiefly in being characterized by large laccoliths and sills in gently dipping Cretaceous and Wasatch rocks and in containing extensive volcanic rocks of the San Juan Mountains type in their southern part. Aside from local bending near intrusive bodies, the sedimentary rocks are only very gently folded, but, in parts of the mountains, they are shattered by many faults and dikes. A concentration of dikes and small plutons holds up a northeastern ridge of the mountains known as the Ruby Range. Large parts of the mountains, including almost the entire volcanic part, never have been studied in detail, but maps of the northeastern part have been published recently (156,167).

SAN JUAN MOUNTAINS     The San Juan Mountains are largely on the site of the Uncompahgre-San Luis highland of late Paleozoic and early Mesozoic time, but the southwestern part of the mountains lies in the border zone of the sedimentary basin that lay to the southwest of the highland. Consequently, no sedimentary rocks older than Jurassic are likely to exist beneath the volcanics that make up most of the mountains except in a fringe area along the west and southwest side of the volcanic field, and in the extreme northeast, near Bonanza.

In Laramide time, much of the area occupied by the present mountains was deformed in a great dome, as has been described by Burbank (42). Intrusion and early volcanism accompanied this deformation. The dome later was beveled by erosion, and then the Eocene Telluride Formation was deposited over the eroded surface, at least in the western part of the mountains. The Needle Mountains, an area of Precambrian rocks some 25 miles in diameter in the southwestern part of the present San Juan Mountains, represent a part of the apical portion of the early dome. The Precambrian rocks of these mountains are bordered on the south and west by upturned Paleozoic sedimentary rocks, and, on the north and east, they are covered by volcanic rocks.

Burbank (24,42) has identified a northeast-trending monoclinal hinge line, resulting from the Laramide deformation, extending through Ouray and Rico (Figure 2). Laramide intrusive rocks and associated ore deposits at Ouray and the mineralized structural dome that characterizes the Rico district are features of this hinge line, which is regarded as the southwest-extension of the main Colorado mineral belt (123). The La Plata Mountains intrusive center, south of Rico, also may be a part of the line of Laramide deformation and intrusion, as it seems to have been a source of igneous materials in the Late Cretaceous and Paleocene Animas Formation (67).

Northeast-trending Precambrian shear zones exist in the Precambrian rocks at the northern edge of the volcanic field (123). They project toward the Ouray-Telluride area but have not been identified there. The Cimarron fault, a major west-northwest-trending fault that cuts the northern flank of the mountains also may be as old as Precambrian (146). If so, it was reactivated in Laramide time, as it displaces the Cretaceous rocks as much as 2500 feet. It is overlain in places by unbroken volcanic rocks.

Structure of the great volcanic field of the San Juans is only in the process of being worked out. In the western mountains, Burbank and Luedke (this volume) have recognized a major early volcano-tectonic depression some 15 by 30 miles in size. Nested within this depression are the Silverton and Lake City calderas (or cauldrons), each about 10 miles in diameter. The Silverton caldera, the older, more dissected, and better studied, is characterized by complex radial and concentric fracture and dike systems, many of which are mineralized, and by intrusive stocks and pipes along its margin (Burbank and Luedke, Figure 3, this volume; also 42). In the central San Juan Mountains, the Creede caldera, the youngest in a group that only now is coming to light,

is associated with a north-northwest trending graben fault system that contains the productive veins of the Creede district (Steven, this volume; also 92). Other calderas, such as a probable one in the Bonanza area, exist elsewhere in the San Juan Mountains but have not yet been studied.

## Mineral Belt

Most of the mining districts of Colorado lie in the Colorado mineral belt, a generally narrow but somewhat irregular strip of ground that extends southwestward across the mountain province from the mountain front near Boulder to the region of the San Juan Mountains (Figure 1). The belt is characterized geologically by intrusive igneous rocks and associated ore deposits of Laramide and younger age and, in some places, by fissures and veins of northeasterly trend. The intrusive rocks typically are hypabyssal porphyries in stocks, laccoliths, sills, and dikes. The associated ore deposits are mesothermal and epithermal veins, replacement bodies, and stockworks. The fissures and veins are generally short and occur principally in clotlike swarms in the Front Range part of the belt. These features are scattered and discontinuous, and no Laramide structural feature runs through the belt to connect them. Parts of the belt, however, are interconnected by northeast-trending Precambrian shear zones, and the belt as whole coincides approximately with a swarm of such shear zones (123).

Aside from the shear zones and the Tertiary igneous and mineralization features, the geology within the mineral belt is the same as in the bordering areas, and, in most places, this background geology dominates over the features that define the belt. Thus, the mineral belt seems to be a weakly defined geologic unit characterized principally by magmatic rather than tectonic features, but, despite this apparent weakness, it cuts indiscriminately across the geologic grain of the region. It extends diagonally across the major mountain ranges, crosses major faults and regional folds, lies athwart the ancient highlands or positive areas, and occupies geologic environments ranging from Precambrian crystalline rocks through Paleozoic and Mesozoic sedimentary rocks to Tertiary volcanic rocks.

Tweto and Sims (123) concluded, from this behavior and the family resemblances of the igneous rocks throughout the belt, that the mineral belt must be an expression of an underlying batholithic body, as suggested previously by others (8,22). Pronounced negative gravity anomalies along the belt (114,140) support this conclusion. Presumably, the zone of crustal weakness defined by the ancient Precambrian shear zones of the mineral belt allowed deep-lying magma to rise high into the crust, forming a batholith, or series of them, along the belt. The hypabyssal intrusives that define the belt are interpreted as cupolas and apophyses in the roof of the batholith. In rising toward the surface, the hypabyssal intrusives were guided by whatever local structures were present and suitably oriented. Consequently, although the intrusives are a feature of the mineral belt, many of them are closely related to structural features that are not confined to the belt or related to it in origin.

As it has no sharp boundaries, the mineral belt can be defined in many ways. Defined so as to include the principal mining districts, most of the intrusive bodies, and, in the San Juan Mountains, the area containing Laramide intrusive centers, it has the shape shown in Figure 1. So defined, it is 10 to 15 miles wide in the Front Range and widens abruptly along the Mosquito fault to about 30 miles in the Park Range. In the Sawatch Range, it widens even more, as intrusive rocks and ore deposits are scattered through the 90-mile length of the range, but, as these are separated in places by large tracts that contain few features typical of the belt, the mineral belt also is more diffuse here than in ranges to the northeast. In the Elk and West Elk Mountains, intrusive rocks are scattered over a wide area, but ore deposits are confined to the eastern part of the mountains. From Gunnison southwestward across the old Uncompahgre-San Luis highland to Ouray, a distance of 50 miles, the mineral belt is only faintly expressed. From Ouray southwestward to the La Plata Mountains, it is only a narrow belt as defined by exposed Laramide intrusive centers, but parts of it might have been destroyed by the mid-Tertiary volcanism.

## MINERAL DEPOSITS

The mineral deposits of Colorado are great in number and varied both in composition and in geologic character. The 44 main mining districts shown in Figure 2 constitute only a fraction of the mineralized localites in the mountain province. A map by Fischer, *et al.* (37) shows some hundreds of metallic deposits distinguishable at a scale 1:1,000,000, and individual commodity maps of the state (*see* 139) show many additional localities. Wyoming is far less endowed in numbers and variety of metallic deposits, but deposits of its two chief metallic products, uranium and iron, are outstanding in size.

Individual discussion of so many districts and localities is beyond the scope of this report and volume. Almost all have been described somewhere, and references in this report will lead to most of them. References listed in Figure 2 and throughout this report emphasize recent works, especially since publication of the Lindgren Volume, and most of them will in turn lead to an older literature not cited in this report.

In the text that follows, the ore deposits are discussed as genetic, geologic, and compositional groups rather than by rank or geography. They are also discussed within a general age framework, not for the sake of systematism but because age carries with it a connotation of a set of conditions, and this set of conditions serves both to put limits on the characteristics and distribution of groups of deposits and to suggest new localities for consideration.

As indicated in Table I, the principal metals produced from the mountain province of Colorado are molybdenum, gold, silver, lead, and zinc. Other important commodities are copper, tungsten, uranium, and fluorspar. Lesser amounts of several others have been produced, and a potential exists for production of thorium, rare earths, niobium, titanium, vanadium, and perhaps nickel and cobalt. Compared with many other western states, Colorado is a copper-poor province, although byproduct copper is obtained from many of its districts. The state also is poor in mafic and ultramafic rocks and hence in metals characteristically associated with such rocks. Of the leading deposits, the great majority are of Laramide or Tertiary age, but Precambrian deposits of many kinds are scattered throughout the area, including adjoining Wyoming. The principal iron, thorium-rare earth, and vanadiferous titanium deposits in the two states are in the Precambrian, as are the few examples of copper-nickel ores of mafic association. In addition, beryllium and several nonmetallic minerals are obtained from the Precambrian.

General summaries of the mineral deposits of Colorado have been made under the guidance of Vanderwilt (45), Del Rio (85), and the U.S. Geological Survey (139). Nonmetallic minerals have been summarized by Argall

(47). Summaries for Wyoming have been made by Birch (62), Osterwald, *et al.* (83), and the U.S. Geological Survey (94). These reports contain extensive bibliographies on individual mineral commodities.

## PRECAMBRIAN ORE DEPOSITS

The Precambrian ore deposits include sedimentary, magmatic, and regional metamorphic deposits as well as epigenetic contact metamorphic, replacement, and vein deposits. The epigenetic deposits, valuable chiefly for copper and gold, seem to represent local mineralization at different times in different places rather than distinct epochs of widespread metallization in the Precambrian. Most of the Precambrian deposits, except of iron, are small, but they are of interest because they may eventually point the way to other and larger deposits in the extensive Precambrian terranes of the region.

### Sedimentary Iron

Important deposits of iron in Precambrian sedimentary rocks are exposed in three main localities in southern Wyoming. These localities are in a slightly bowed east-west line extending from the southern end of the Wind River Range (Figure 1) to the Hartville uplift east of the Laramie Range. The three deposits are in the older Precambrian sequence of Wyoming (more than 2 b.y., Table II), but their stratigraphic relation to each other is not established. Whether or not they are a part of a single stratigraphic zone, their wide spacing leads to a speculation that other deposits might be concealed in the large areas covered by younger sedimentary rocks. Iron formation is not a component of the younger Precambrian sequence and hence is absent in the Precambrian of southernmost Wyoming and Colorado.

The most productive of the Wyoming iron deposits is the Sunrise deposit in the Hartville uplift east of the Laramie Range. This deposit has been a principal source of iron ore for the furnaces at Pueblo, Colorado, since 1901, and it has thus far yielded about 30 million tons of direct-shipping hematite ore from surface and underground workings (94). The hematite occurs as beds or lenses in tightly folded schist. The deposit has been described briefly by Carter (119) and Harrer (158).

Taconite deposits at Atlantic City, at the southeastern end of the Wind River Range, are described by Bayley in this volume. Large-scale mining of these deposits began in 1962, and, considering the size of the deposits, production presumably will eventually outstrip that of Sunrise.

Iron deposits of the Seminoe Mountains, about 80 miles east-southeast of the Atlantic City area, have been described by Lovering (10) and Harrer (158) and mapped by Bayley (141). The deposits are tightly folded taconite and accompanying lenses of replacement hematite associated with greenstone schists in the hanging wall of an overthrust fault. Although one body of taconite contains about 100 million tons (158), the deposits on the whole consist of many small bodies scattered through an area of about 10 square miles (141), and they have not been worked.

### Magmatic Titaniferous Iron

Titaniferous and vanadiferous magnetite occurs in several bodies in the anorthosite complex of the Laramie Range, Wyoming, and in a stock in the Powderhorn area, Colorado. The largest body in the Laramie Range is that of Iron Mountain, described by Hagner in this Volume. Of the several grades of ore recognized, the best contains 40 to 50 per cent Fe, 16 to 23 per cent $TiO_2$, and 0.5 to 0.7 per cent $V_2O_5$ (94). The Union Pacific Railroad estimated 178 million tons of all classes in Iron Mountain and vicinity (158). The deposits have not yet been worked for their contained metals, but they have been mined extensively for use as heavy aggregate (158). The anorthosite complex that contains the titaniferous magnetite lenses and dikes has itself been investigated as a source of alumina (56). The pure anorthosite contains as much as 29 per cent $Al_2O_3$; it occurs in thick layers interstratified with noritic anorthosite of somewhat lower alumina content (56,72,160).

The deposit in the Powderhorn area has been described by Rose and Shannon (89). The alkalic stock of which it is a part is discussed below, under Thorium.

### Thorium and Rare Earths

Thorium and related rare-earth deposits are associated with alkalic intrusive rocks that are thought to be of late Precambrian age in two localities in Colorado, the Powderhorn area in Gunnison County, 20 miles southwest of Gunnison (Figure 2), and in the McKinley Mountain area in the Wet Mountains, 10 miles northeast of Westcliff (no. 35, Figure 2).

The complex Iron Hill stock of the Powderhorn area consists of carbonatite, alkalic pyroxenite, melilite-rich rock (uncompahgrite), ijolite, and nepheline and soda syenites, described by Larsen (31). Thorium, accompanied by rare earths and niobium, occurs in carbonatite dikes, in veins, and in dikes and segregations of magnetite-ilmenite-perofskite rock (66,103). The thorium-bearing dikes are mainly within the stock, and the veins are mainly in the bordering Precambrian granites and gneisses. The veins have a complex mineralogy highlighted by orthoclase, hematite, barite, fluorite, carbonates, biotite, and sulfides as well as by thorite and thorogummite, and they contain as much as 4 per cent $ThO_2$. Somewhat vanadiferous titaniferous magnetite also occurs in this alkalic complex.

Thorium veins of the Wet Mountains are described by Christman, et al. (79) who considered them to be related genetically to an albite syenite stock of late Precambrian age. However, two Precambrian alkalic complexes containing carbonatite dikes, characterized by thorium, rare earths, and niobium, later were discovered a few miles north and northwest of the McKinley Mountain area by Parker and Hildebrand (120), who relate the thorite veins to these complexes. Phair (106), however, has suggested a Tertiary origin by a process of lateral secretion. More than 400 thorium-bearing veins have been mapped in the area, most of them less than 5 feet wide and between 100 and 1000 feet long. They are in Precambrian gneisses and are characterized by iron oxides, carbonate minerals, barite, quartz, and minor sulfide minerals as well as by thorium minerals. Thorium oxide content of channel samples ranges from 0.02 to nearly 5 per cent (79).

The rare-earth minerals, xenotime and monazite, containing some thorium, occur in exceptional concentrations in places in the gneisses of the Central City area. Gneiss lenses as much as 5 feet thick and several hundred feet long contain 1 to 5 per cent xenotime-monazite by volume. Young and Sims (108) ascribe the concentrations to a process of metamorphic segregation.

## Copper-Nickel in Mafic Associations

The most productive of the Precambrian base- and precious-metal deposits are copper deposits associated with mafic rocks. The two largest deposits of this kind are in the Encampment district on the east flank of the Sierra Madre Mountains in Wyoming, nearly on the Colorado boundary, and the Sedalia deposit near the south end of the Mosquito Range in Colorado (Figure 1). Both deposits were worked near the turn of the century and were described, respectively, by Spencer (4) and by Lindgren (5). The Encampment deposits, which are credited with most of Wyoming's total copper output of about 17,000 tons, are in hornblende schists that have been intruded by noritic diorite-gabbro and granite. Pyrrhotite-chalcopyrite ore containing some gold and, in places, nickel and cobalt and traces of platinum, occurs: (1) disseminated in hornblende schist, (2) in tactitelike garnet-epidote streaks in the schist, (3) as bedding replacement bodies in quartzitic layers in the schist, (4) in hornblendic contact metamorphic zones bordering the diorite-gabbro, and (5) in quartz veins and pegmatites. Minor galena-sphalerite-pyrite ore, containing some silver, is of similar types of occurrence, and some quartz veins contain gold. Spencer found the diorite-gabbro to be cupiferous and thought it to be the source of most of the copper, although he classed that in the tactitelike zones as metamorphic and hence older.

In the Sedalia deposit, chalcopyrite, locally accompanied by sphalerite, galena, and silver, is disseminated in hornblende gneiss or schist just as in some of the Encampment ore. The deposit yielded 60,000 to 75,000 tons of ore containing 5 per cent copper and $1.00 to $2.50 in gold and silver per ton. The Sedalia deposit is the largest of several in the Cleora district, at the southern tip of the Mosquito Range. This district shows a remarkable parallelism with the Encampment district, as it contains all of the Encampment ore types in close association with dark diorite or gabbro, but except for the Sedalia mine, it has not produced much. Its copper ores contain some tungsten (93) but apparently have not been tested for nickel.

A third deposit, having these same characteristics but containing considerable nickel and some cobalt, is in the Gold Hill district in the Front range (no. 4, Figure 2). In this deposit, described by Goddard and Lovering (30), chalcopyrite-pyrrhotite-pyrite ore, containing polydymite and other nickel minerals, occurs as a replacement in amphibolite adjoining the upper terminus of a dikelike body of noritic gabbro. The ore body contains as much as 4 per cent nickel, as well as copper and cobalt, but it has not been extensively worked due to lack of treatment facilities.

A deposit at the Malachite mine, in the Front Range just west of Denver, may also belong to this group. Some of the ore, consisting of chalcopyrite, sphalerite, and pyrrhotite, was in an amphibolite interpreted by Lindgren (5) to be a metagabbro, but most of the ore evidently occurred as a replacement of hornblendic gneisses along a fracture zone.

### Skarn and Calc-Silicate Replacement Deposits

Base metals and tungsten occur as a metamorphic component or an early postmetamorphic replacement of skarns, calc-silicate gneisses, and amphibolites in many localities. Few of the deposits have produced much, but the abundance of rocks of these kinds encourages the idea that bigger and better deposits will be found.

A leading example is the Cotopaxi deposit, on the Arkansas River 20 miles southeast of the Cleora district discussed above. In this deposit (5,52), amphibolite in lenses enclosed in gneissic granite contans chalcopyrite, sphalerite, and a little galena. A modest production of copper ore has been made from the deposit. At the Copper King mine, in the Front Range only a few miles south of the Wyoming boundary, anthophyllite-cummingtonite skarn has been partly replaced by magnetite, pyrrhotite, pyrite, chalcopyrite, and sphalerite (77). The skarn, along with gneisses, occurs as inclusions in granite. A pitchblende vein occupys a fault that cuts the skarn, granite, and other rocks. Sims, et al. (77) related the sulfide minerals in the skarn to the enclosing granite, which is of Precambrian age, but they concluded that the pitchblende vein was Laramide in age.

Other base-metal replacement deposits include the St. Louis or Milliken deposit in the Front Range a few miles northwest of Urad (no. 8, Figure 2), where sphalerite and galena are disseminated in diopsidic calc-silicate gneiss (52); the Slavonia prospects on the western side of the northern Park Range, where ore similar to that of the St. Louis deposit occurs; and the Pearl district on the eastern side of the northern Park Range, where chalcopyrite and sphalerite occur in small bodies in both biotitic and hornblendic gneisses (145).

Tungsten, in the form of scheelite or scheelite-powellite, is of rather widespread occurrence in the calcic gneisses. A report by Tweto (93) describes 21 occurrences of Precambrian scheelite in Colorado and Wyoming. Some of the scheelite seems to date from the time of regional metamorphism, and some seems to be related in origin to Precambrian granitic and pegmatitic rocks that invaded the

gneisses. None of the deposits is large, although a little tungsten ore has been produced from several of them.

### Gold and Copper Veins

The leading Precambrian vein deposits are those of Atlantic City, Wyoming, described by Bayley in this Volume and credited with an output of about $2,000,000. Other vein deposits are widespread but have made only a minor output. Examples are copper veins in the Laramie Range (83), copper-tungsten veins in the Cleora district and southern South Park (5,93), and scattered veins in the northern Front Range (52). Most of the veins are characterized by glassy quartz and contain silicate gangue minerals that give them a high-temperature aspect.

## AGE AND CHARACTER OF LARAMIDE AND TERTIARY DEPOSITS

The ore deposits of the mineral belt generally have been regarded in the past as being Laramide in age and as having had an ultimate source in a widespread igneous body that was parent to the hypabyssal porphyries as well as to the ore deposits. Thus viewed, the belt presents some perplexing genetic problems because of marked differences in character among closely spaced districts. Thus, for example, the Gold Hill and Central City-Blackhawk precious-metal districts (nos. 4 and 11, Figure 2) are separated by the Boulder (or Nederland) tungsten district characterized by quartz-ferberite veins almost to the exclusion of any other. The Urad deposit (no. 8, Figure 2) contains only molybdenum whereas its close neighbors are pyritic gold and lead-silver districts. The great Climax deposit (no. 19, Figure 2), containing only molybdenum and accessory tungsten, is closely surrounded by the base- and precious-metal districts of Gilman, Kokomo, Alma, and Leadville. Similarly, low-grade molybdenum deposits in the central Sawatch Range, thus far unworked, are surrounded by base- and precious-metal districts. Among the precious-metal districts, some, such as Central City-Blackhawk, are characterized by sulfide ores; some, such as Magnolia, by telluride ores; and some, such as Gold Hill and Jamestown, by intermingled sulfide deposits and telluride deposits.

No pattern of regional zoning emerges from this distribution, although zoning is well defined in some of the local areas, such as the Central City district and environs (115).

Lovering and Goddard (22,52) attributed major ore types, such as lead-silver, pyritic-gold, molybdenum, tungsten, and telluride, to separate groups of igneous rocks. These groups, according to Lovering and Goddard, reflected both: (1) the timing of local intrusion with respect to the differentiational history of a widespread parent magma body, and (2) differing evolutional histories of local bodies of magma once separated from the source body. Widespread early lead-silver and pyritic-gold ore types were regarded as products of a widespread source body and more localized and younger tungsten and telluride ores as products of more restricted magma bodies with independent differentiation histories. The Climax molybdenum deposit was classed as a product of an early but local magma. Although intrusion and mineralization were visualized as coming in surges, the overall process was visualized as having been progressive and essentially continuous and having been accomplished entirely during the Laramide.

Certain features of the mineral belt have long seemed inconsistent with a single major stage of mineralization of Laramide age. Deposits in some districts, such as Central City, Kokomo, and Leadville, are distinctly mesothermal in their mineralogic and textural characteristics, whereas in others such as Nederland, Magnolia, and Gold Hill and Jamestown in part, they are distinctly epithermal. Some, such as several in the Front Range, are closely associated with deeply eroded stocks that are thought to have had volcanic edifices that contributed sediments to the Laramide Denver and related formations, but others, such as Urad, the Sawatch Range molybdenum deposits, and some deposits of the Mt. Princeton batholith complex in the Sawatch Range, are associated with volcanic or very shallow intrusive rocks. Many districts contain "sport" deposits atypical of the district. For example, the Ward district (no. 3, Figure 2), is characterized by pyritic copper-gold ore in coarse-grained quartz veins containing some wolframite, but a deposit of vuggy galena-barite-calcite-silver ore in the White Raven mine has quite a different genetic connotation. Wahlstrom (19) has long argued a distinctly younger age for the White Raven deposit than for the other ores, as he also has for telluride deposits in Boulder County (54).

The establishment of a post-Laramide or Oligocene age of certain intrusive rocks by isotopic dating, and the fact that some of these, notably at Climax and Urad, are mineralized, makes clear that there were two major mineralization events in the mineral belt, not one. Although much more dating is yet to be done, the thesis is advanced here that the principal base- and precious-metal sulfide deposits in nonvolcanic areas, generally of mesothermal aspect, are Laramide; and that the major molybdenum, tungsten, and telluride deposits as well as precious-metal veins in volcanic rocks, all of epithermal aspect, are products of the later or Oligocene epoch of igneous activity. Such a two-fold division of intrusion and mineralization is compatible with the age relations between various igneous rocks and between various ores, and it would account satisfactorily for the marked differences between the ores of districts that not only are associated spatially but also are structurally related.

For many years, the Laramide age of the base- and precious-metal sulfide ores, as deduced from geologic evidence, was supported by lead-uranium age determinations on pitchblende in the veins of Central City. Although the values obtained were disparate, many fell in the 55 to 70 m.y. range, and an average of $59 \pm 5$ m.y. proposed by Eckelman and Kulp (69) as a best value was widely adopted and was used as a basis for fixing the age of the Laramide (87,105). However, Banks and Silver (126) later demonstrated inhomogeneity and a complex history of isotopic disturbance in the pitchblende ores, and they thus questioned the validity of the age assignments. The pitchblende of Central City is paragenetically older than the early pyrite of the sulfide ores (122,136), but it occurs in close association with sulfides in the same veins. The ores in these veins, which apparently formed at temperatures as high as 600°C (115), are a part of the older, mesothermal suite of the Front Range mineral belt, which on geologic grounds must be considerably older than beginning Oligocene, or 40 m.y. Thus the pitchblende ages, although not exact, may be approximate minimum ages.

## LARAMIDE DEPOSITS

### Veins in Precambrian Rocks

The ore deposits of many districts in the Front Range are sulfide veins valuable principally for precious metals. Other types of occurrence, such as stockworks or pipes are rare. The veins are in Precambrian rocks or locally in or bordering Laramide intrusive rocks. The veins occupy fissures or small faults of Laramide and Precambrian ages that trend between north-northeast and east. Most of the faults are only a few hundred feet long, and very few exceed a mile. Most of them show displacements of only a few or several feet. The mineralized faults, or veins, are clustered in swarms, each of which constitutes a mining district or center. In Boulder County, the vein swarms are almost continuous from district to district, but elsewhere they are scattered, poorly aligned, and apparently isolated from each other.

The ore deposits in such veins are in ore shoots that have thicknesses of several inches to a few feet, stope lengths of a few tens to a few hundreds of feet, and pitch lengths from a few tens to 1000 feet. The ore bodies thus are small in tonnage, and many of them could be worked only because of their rather high grade, which in many cases was 1 to 10 ounces of gold or 50 to 500 ounces of silver per ton. Since the veins are discontinuous and the ore bodies small, the outlook for future production from deposits of this kind is poor, primarily because the cost of discovery is high in proportion to the value of an ore body.

Districts with veins of this character include Jamestown in part, Gold Hill in part, Ward, Caribou, Empire, Lawson-Dumont, Central City-Blackhawk, Idaho Springs, Georgetown-Silver Plume, Montezuma, and Breckenridge in part (Figure 2). These districts are all described by Lovering and Goddard (52).

Veins in Precambrian rocks in other ranges have the same general characteristics as those of the Front Range, except that a few have somewhat greater size or persistence. Such veins are scattered through the Precambrian terrane of the Park Range (within the mineral belt) and are minor features of the Alma, Leadville, Gilman, and Monarch-Tomichi districts. They also characterize the Sugar Loaf-St. Kevin district, the Granite and Twin Lakes districts, and, in part, the Gold Brick and Pitkin districts (Figure 2).

## Replacement and Vein Deposits in Sedimentary Rocks

Replacement and vein deposits in sedimentary rocks are one of the most productive classes of ore deposits in Colorado, having accounted for 25 to 30 per cent of the value of the State output. Included in this group are the principal deposits at Leadville, Gilman, Aspen, Tincup, Monarch-Tomichi, Alma, Kokomo, and Rico; and some ores of this type also are obtained at Breckenridge, Crested Butte, Pitkin, and Ouray, as well as in several minor districts. The deposits in all these districts are characterized principally by silver-lead-zinc or simple lead-zinc ores, but some of the ores also contain copper and gold. The deposits of the various districts have many features in common, and, although each district or deposit has its distinguishing characteristics, the deposits described in the reports on Leadville and Gilman in this volume serve to exemplify the group as a whole.

Of all host rocks of the replacement deposits, the most productive by far is the Leadville Dolomite, and second most productive is the underlying Dyer Dolomite member of the Chaffee Formation or, in the San Juan Mountains, the Ouray Limestone. The principal deposits of the Leadville, Gilman, Aspen, Alma, and Tincup districts are in these units, as are some of the deposits in the Monarch-Tomichi, Pitkin, Ouray, and Rico districts. The Lower Ordovician Manitou Dolomite is an important host rock in the Leadville and Monarch-Tomichi districts, and the Upper Ordovician Fremont Limestone also is a host rock in districts in the southern Sawatch Range, the only area within the mineral belt in which it is preserved. Sawatch Quartzite and the Peerless Formation contain replacement deposits at Leadville, Gilman, and Alma. Such deposits generally are richer in gold and copper than the deposits in the overlying dolomites.

In many districts in which the Leadville is mineralized, the overlying Pennsylvanian rocks are only weakly mineralized or are barren, but, in a few districts, the Pennsylvanian rocks are the principal host rocks of the ore deposits. The main replacement deposits of Kokomo (40) are in limestone or dolomite units within the Minturn Formation, and the principal deposits at Rico (42) are in gypsum and limestone of the Hermosa Formation. At Ouray, some ore is in the Hermosa, but greater amounts are in the limy Wanakah Formation and in the Dakota Sandstone (42). In the La Plata district (no. 43, Figure 2), where sedimentary rocks have been reduced to isolated plates and blocks by porphyry intrusion, ore occurs in one place or another in all units of the sedimentary sequence from the Hermosa Formation to the Mancos Shale (48). At Breckenridge, where pre-Pennsylvanian Paleozoic rocks are absent, replacement deposits occur in redbeds of the Maroon(?) Formation and in the Dakota Sandstone (14).

In every district characterized by replacement deposits, some part of the ore occurs in veins in the sedimentary rocks. The relative importance of the veins as sources of ore ranges widely from district to district. In the Gilman and Aspen districts, the production from veins has been small in proportion to that from the bedding replacement deposits. In many other districts, including Leadville, Alma, Monarch-Tomichi, and Tincup, veins have been a larger source of ore, although subordinate to the replacement deposits. In still others, as Ouray and Breckenridge, veins dominate over the replacement deposits. In most districts, the veins and replacement deposits are closely associated, and, in many places, replacement deposits simply branch from veins at sites where the veins intersect "favorable" host rocks. Many large replacement bodies, however, have no recognized connection with veins, as indicated in the reports on Leadville and Gilman in this volume.

Although the primary ores of the deposits in sedimentary rocks consist principally of pyrite, sphalerite, and galena, they differ in various ways from district to district. Some, as at Aspen and Ouray, are rich in silver whereas others, such as the large manto bodies at Gilman, have a very low silver content. The primary ores at Leadville, generally containing a few ounces of silver per ton, are perhaps the most typical. The sphalerite in most of the major zinc deposits is marmatitic, but the sphalerite of lead-silver ores, as of Aspen, is much purer. The sphalerite of Kokomo, which occurs principally as shells around bodies of massive pyrrhotite, has a zinc content of only 45 to 50 per cent. Replacement deposits at Leadville and Gilman are characterized by a rather high content of manganese, which occurs principally as manganosiderite, but the deposits in most other districts do not have this characteristic.

Many of the replacement deposits contain some chalcopyrite and traces of gold, but the chief sources of copper and gold, and also of high-grade primary silver ore, are either: (1) veins, (2) replacement deposits in siliceous rocks, or (3) bodies of paragenetically late copper-silver-gold minerals impregnating or replacing earlier pyrite-sphalerite replacement ore. The veins commonly are richest in copper, silver, and gold where they traverse quartzites, sandstones, or porphyries. In such settings, they are quartz veins that commonly contain free gold, argentiferous tetrahedrite or tennantite, and, in places, ruby silver minerals, as well as the more ubiquitous chalcopyrite, pyrite, sphalerite, and galena. The London and other veins in the Alma district, described by Singewald (25,29), are typical examples; other

examples are in the Breckenridge district (14), Monarch-Tomichi district (68), Leadville district (this volume), and Ouray district (42).

Since veins of this type commonly change to lead-zinc veins upon intersecting carbonate rocks and then spread into replacement bodies, the veins almost certainly are of the same age as the blanket replacement ores. Whether ore bodies characterized by paragenetically young copper, silver, and gold minerals *within* larger pyrite or pyrite-sphalerite replacement bodies are of the same age as the veins and blanket replacement bodies is another matter. As indicated in the reports on Gilman and Leadville in this volume, the richest ores of Gilman and some of the richest primary ores of Leadville are of this type. These ores contain gold-silver telluride minerals, free gold in association with the telluride minerals, and a suite of bismuth-bearing lead, silver, and copper sulfosalt and sulfide minerals. At Gilman, where ore of this type is best seen and most studied, these minerals coat vugs and fractures in, and partly replace, early massive pyrite that occurs principally in chimneys at the lower ends of manto ore bodies. The young ore minerals are eratically distributed, and they enriched only parts of the early massive pyrite bodies (*see* 43, Figure 7). Radabaugh, *et al.* (this Volume) evidently regard the copper-silver-gold minerals as late products of a one-stage mineralization process, and as such, to be localized in the chimneys as the "source" of the manto ores.

An alternative view is that the copper-silver-gold minerals are a product of a separate and younger mineralization, presumably the Oligocene mineralization stage. Paragenetic relations indicate only that these minerals are younger than all of the other ore and gangue minerals, but these relations do not establish how much younger. Other features, however, suggest a separate mineralization event: (1) the lead-copper-silver-gold-bismuth-antimony minerals in the younger ore contrast markedly in composition with the pyrite-marmatite-manganosiderite assemblages in the early ore; (2) much pyrite in the so-called chimney ore is a granular aggregate of crystals, in contrast to massive fine-grained pyrite in the early ore, suggesting extensive recrystallization and reorganization of the early pyrite; (3) temperature connotation of the young minerals is distinctly lower than for the pyrite-marmatite assemblage, which also includes a little pyrrhotite and magnetite; and (4) texturally, the drusy and encrusting young minerals are epithermal in habit. None of these arguments is conclusive, but, from the point of view of understanding the mineralization process, and of exploration, much more is to be gained by considering the possibility of a second mineralization than by ignoring the idea.

In a similar way, the long-debated issue of the origin of the dolomite in the Leadville Dolomite/Limestone also is pertinent to an understanding of the ore-forming processes and ore occurrence, and hence to exploration. The Leadville is, by and large, a dolomite within the mineral belt and a limestone outside the belt, and this suggests that the dolomite facies is somehow related to the belt. Emmons (1), the earliest worker on replacement deposits in the region, found the Leadville to be entirely dolomite in the Leadville district, and he considered the dolomite to be sedimentary in origin. At Aspen, however, Spurr (3) found both dolomite and limestone in the Leadville, and he ascribed the dolomite to hydrothermal replacement along presumed bedding plane faults. Vanderwilt (16) dismissed Spurr's replacement theory and classed the alternating limestone and dolomite as a "normal stratigraphic sequence," implying that the dolomite was sedimentary or diagenetic. In the Gilman-Leadville region, Lovering and Tweto (33) and Tweto (51), impressed by an abrupt change in the Leadville from 100 per cent limestone to 100 per cent dolomite along a line 3 miles north of Gilman, classed the early dark fine-grained facies of the dolomite as hydrothermal and regarded it as the product of the earliest of several alteration stages that preceded ore deposition. Such an origin was supported by the finding of Engel, et al. (75) that, based on oxygen isotope ratios, the dark dolomite at Gilman seems to be of distinctly higher temperature origin than the limestone facies of the Leadville. In the report on Gilman in the present volume, Radabaugh, et al. dispute some of the conclusions of Engel, et al. and class the dark fine-grained dolomite as sedimentary or diagenetic.

Dolomite is interbedded with limestone in the Leadville in many places besides Aspen, from the White River Plateau (118,143) to the southern Sawatch Range (68) and the Sangre de Cristo Range (76). There can be little doubt that primary dolomite exists in the Leadville, even though much of the dolomite in localities as distant from the mineral belt as the White River Plateau is secondary and classed as hydrothermal (143). The question is: Is the dark fine-grained dolomite of the mineral belt, as in the Gilman-Leadville-Alma region, to be assigned to the primary or to

the secondary dolomites? As indicated in the report on Leadville in this volume, the dolomite predates the Pando Porphyry, the earliest of the porphyries, which in turn considerably predates the ores. Thus, if the dolomite is hydrothermal, it is very early and cannot be so much related to individual areas and deposits as to some premonitory regional change along the mineral belt as a whole. The vast quantities of magnesium required to dolomitize so much limestone might have been liberated from basement rocks as they were assimilated or metamorphosed by the mineral belt magma as it first started to rise into the crust. If, on the other hand, the dolomite is primary, then its great concentration within the mineral belt as compared with areas outside the belt would indicate that the mineral belt was somehow a geologic entity as early as Mississippian time, long before it acquired its identifying characteristics of intrusive rocks and ore deposits.

Most of the ore bodies in the Leadville are in dolomite as contrasted with limestone, and they generally are in areas where the Leadville is thin, although not all areas of thin Leadville are mineralized. As a result of pre-Pennsylvanian erosion, the Leadville ranges in thickness from only a few feet to more than 500 feet. In almost all mineralized areas, it is less than 200 feet thick, and, within these areas, ore bodies seem to be concentrated where it is thinnest (Leadville, this Volume). The pre-Pennsylvanian erosion was largely chemical, producing a karst surface. Presumably, the Leadville was most thinned in areas where karst channel and cave development was greatest. As has been widely recognized (Leadville and Gilman, this volume), the karst channel system played an important role in the mineralization process, particularly in guiding the pre-ore dolomite-altering and leaching solutions. Thus a rough correlation between ore occurrence and a thin Leadville might be expectable.

Considering that every district with ores in the pre-Pennsylvanian Paleozoic rocks is in an area where these rocks are exposed at the surface and that these rocks are covered by Pennsylvanian and younger rocks over wide areas in the mineral belt, the outlook for the occurrence of additional major deposits or districts is good. The discovery of such deposits may be difficult, however, because the Pennsylvanian rocks are thick, and, to judge by the Gilman district, they may show little sign of the ore bodies beneath them. In other areas, however, as at Kokomo, they are themselves mineralized.

## POST-LARAMIDE DEPOSITS

### Tungsten and Telluride Deposits

The ferberite tungsten veins and gold-silver telluride veins of the northeastern end of the mineral belt in the Front Range have many features in common. The ferberite veins occur in the Nederland (or Boulder) tungsten district and the telluride veins in the Magnolia, Gold Hill, and Jamestown districts, but the two types occur together in the eastern end of the tungsten district and adjoining parts of the Gold Hill and Magnolia districts. In some veins, the ferberite seems to be paragenetically younger than the telluride minerals, but in others the opposite is true, and the two ores thus are judged to be of about the same age. The tungsten and telluride veins clearly are younger than the precious- and base-metal sulfide veins, since they cut sulfide veins, occur within reopened sulfide veins, occupy faults that displace sulfide veins, and cut porphyry dikes that cut sulfide veins. Both the tungsten and the telluride veins are characterized by a dense chalcedonic quartz known as horn quartz and by open spaces and a drusy, encrusting structure that indicates epithermal character and is in contrast with the much tighter structure of the mesothermal sulfide veins. Many of the tungsten and telluride veins are small—only inches wide—but they have been workable because of their high grade. They have been described in detail by Lovering and Tweto (60), Lovering and Goddard (52), and Goddard (26).

Lovering and Goddard (22,52) recognized the tungsten and telluride veins to be the youngest in the Front Range, but they assigned them to the Laramide on the assumption that all ore formation and porphyry intrusion were Laramide. With the recognition of post-Laramide intrusion and mineralization elsewhere in the mineral belt, the tungsten and telluride ores more logically are assigned to the post-Laramide mineralization stage than to the Laramide, although precise dating is not yet available. The marked compositional contrast between the tungsten-telluride veins and the earlier sulfide veins suggests a lack of any direct relationship. More importantly, the difference in character of the veins, expressed as the difference between epithermal and mesothermal, suggests that much of the cover that existed when the early sulfide veins formed was removed before the later tungsten and telluride veins were formed. The geologic record indi-cates very extensive erosion in the Front Range between Laramide and Oligocene time; the erosion that reduced the cover between the two stages of ore deposition is better assigned to this long period of erosion than to a brief interval within the Laramide.

The tungsten and telluride veins have together produced no more than $60 million during nearly 100 years of mining. As the veins are small and scattered and the ore shoots are generally under 1000 tons and have poor persistence in depth, the outlook for significant production from deposits of this class in the future is poor.

### Molybdenum Deposits

Molybdenum is Colorado's greatest metallic mineral resource. The Climax molybdenum deposit alone has accounted for 28 per cent of the mineral output recorded in Table I, and figures published in annual reports of the Climax Company indicate 35 to 40 years reserves even at the present high rate of production. In addition, the Urad deposit (no. 8, Figure 2), now being developed for production by the Climax Division of American-Metals Climax, will greatly increase the molybdenum output in the future; a reserve of 236 million tons of 0.45 per cent $MoS_2$ has been announced, and exploration is continuing. Still other molybdenum potential exists, as in the Sawatch Range, where low-grade molybdenum-bearing rock evidently occurs in large volume, and in smaller deposits, as at Apex in the Central City district and at Cumberland Pass in the Tincup district. The outlook for further discoveries is good, since the molybdenum mineralization is associated with post-Laramide intrusive bodies, and such bodies not only are widespread but in general are only slightly eroded or, possibly, not yet unroofed.

The Climax deposit is described by Wallace in this Volume. This great deposit overlies a stock that has been reached only in deep exploration. The stock is in a zone of intense intrusion a few miles wide that extends along the Mosquito fault from the southern part of the Leadville district northward through Climax and the Kokomo district to about the intersection of the Mosquito and Gore faults. Thus, the Climax deposit, along with those of the Leadville and Kokomo districts, evidently is genetically related to the Mosquito fault zone (Leadville, this volume), but the deposit also is cut and extensively displaced by the Mosquito fault, as indicated in the report by Wallace in this volume. The large post-ore displacement raises intriguing questions on the possibilities of ore deposits beneath the Pennsylvanian rocks on the west side of the fault throughout the area from the Kokomo district to the Leadville district.

The Urad deposit has not yet been described in light of intensive recent exploration, but a brief description based on earlier information has been given by Carpenter (84). In general, the ore body is in a small porphyry stock of Oligocene age and in bordering Precambrian rocks. As at Climax, molybdenite occurs in myriad veinlets in shattered silicified and sercitized rock.

At Apex, in the northwestern part of the Central City district, molybdenite occurs in brecciated zones or pipes in a quartz monzonite stock. Although molybdenum production has been small, the deposit is of special interest because copper occurs in dikes that are presumed to be offshoots of the stock. These dikes have been mined in the nearby Evergreen mine, described by Lovering and Goddard (52).

## Precious- and Base-Metal Deposits in Volcanic Rocks

Precious- and base-metal veins and pipes in volcanic rocks have thus far been the most productive class of ore deposits in Colorado, having accounted for about one third of the value of the total output, but they doubtless will lose their position to molybdenum in just a few years. The high value of the output from deposits of this group reflects the dominance of precious metals in the ores, for these deposits are the principal source of gold in Colorado and a major source of silver. Base-metal output has been appreciable from some of these deposits, as in the San Juan Mountains, but subordinate to the precious metals.

Mining districts in volcanic rocks include Cripple Creek, Westcliff-Silver Cliff, Bonanza, Ouray (in part), Lake City, Telluride, Silverton, Creede, Wagonwheel Gap, and Summitville, shown in Figure 2, as well as many minor districts. Cripple Creek, in the southern Front Range far outside the mineral belt, is the most productive single district of the group, and it has outproduced all other districts in the Front Range combined.

The Cripple Creek district coincides with a small area of phonolitic volcanic rocks occupying a steep-walled basin in the Precambrian Pikes Peak Granite. The volcanic rocks and basin were long interpreted as a volcanic

crater, but Loughlin and Koschmann (18) and Koschmann (28,49) established that the basin is a fault-bounded subsidence feature and that it is filled largely with stratified waterlaid sediments. These sediments, which have a thickness of more than 3000 feet, consist of a lower unit of arkose, conglomerate, and siltstones made of nonvolcanic materials and an upper unit of crudely sorted and bedded phonolitic volcanic debris generally referred to as breccia. These rocks are cut by dikes and irregular bodies of various alkalic igneous rocks but not by any igneous body identified as the source of the volcanic debris that fills the upper part of the basin. The floor of the basin is made uneven by fault blocks or so-called "islands" of Precambrian rocks. Vein systems in the rocks of the basin are aligned along the trend of the fault blocks in the basin floor, and along the lines of projection of basin-border faults. The veins are characterized by gold-telluride ores, and gold has been essentially the only product of the district. Mining has been conducted to depths of more than 3000 feet, but tenor of the veins decreases at such depths. Since the entire basin, about 2 by 4 miles, has been extensively explored, the future of the district would seem to rest largely on the possibility that large-volume operations, as contrasted to vein mining, might prove feasible in places. As indicated above, the Cripple Creek area seems to be an outlier of the extensive Thirtynine Mile volcanic field to the west. This field and also many deeply down-faulted blocks in Precambrian rocks between it and Cripple Creek are worthy of investigation for other deposits of the Cripple Creek type.

In the San Juan volcanic field, the most heavily mineralized area by far is the vicinity of Ouray, Telluride, and Silverton, in the western San Juan Mountains. In this area, veins and subordinate mineralized pipes occur throughout the Silverton caldera, 8 to 10 miles in diameter, and through a border zone 3 to 4 miles wide, making an essentially continuous mineralized district 18 miles in diameter. Many individual mining districts or areas are recognized within this large area, and, for convenience, these often are grouped together into three main units, the Ouray, Telluride, and Silverton districts, but even this division is artificial, reflecting mainly the apportioning of credit for production to three different counties.

Geology of the area is summarized in the report by Burbank and Luedke in this Volume, and additional information on the ore deposits is included in an earlier summary by Burbank, Eckel, and Varnes (42) as well as in many reports on parts of the area (12,27,39,55,-116,125,175).

The Silverton caldera is a slightly elliptical area of subsidence bounded by intra- and post-volcanic ring faults with aggregate displacements in the thousands of feet. The interior of the caldera contains many short veins of various orientations and, in places near its margins, pipes of richly mineralized breccia. Most of the producton from the area has come from systems of long veins that radiate outward from the caldera, particularly in its northern half (Figure 3 *in* Burbank and Luedke (this volume, and 42, plate 28). Many of the veins follow dikes, which follow early faults and are followed by late faults. Some of the vein structures are 5 miles long and have a vertical extent of as much as 5000 feet; typical widths are 3 to 5 feet. Among hundreds of productive veins of this kind are expecially famed ones such as the Smuggler-Union east of Telluride, the Camp Bird southwest of Ouray, the Sunnyside northeast of Silverton, and the Shenandoah-Dives southeast of Silverton. Since World War II, the largest mining operation in the San Juan region has been at the Idarado mine, on the Black Bear, Argentine, Montana, and other veins southeast of Telluride. Workings extend 6 miles through the mountain from the Telluride to the Ouray side. The mine and particularly its veins and ores were nicely described as of 1957 by Hillebrand (70), but replacement deposits found since then in sedimentary rocks beneath the volcanics have not been described. The Idarado operations are mainly at considerable depth beneath old surface mines, and, as the gold and silver content of the veins in general decreases downward, the output now is primarily base metals but with an important contribution of gold and silver.

Burbank (42) has noted that the Smuggler vein was stoped through a length of more than 9000 feet and to a height of 2500 feet, and the Idarado operation has shown that mineralization extended through a minimum vertical range of almost 4000 feet. Burbank (42) has attributed the continuity of the veins to uniformity of the massive tuff-breccia of the San Juan Formation, the lowest unit in the volcanic sequence, and has noted that both the vein structures and the intensity of mineralization weaken upward above this unit. The San Juan is as much as 3000 feet thick, and it underlies large areas on the caldera borders. Considering the great numbers and persistence of veins

in the Silverton caldera area and the continuity of mineralization in them, this area would seem to offer a good chance of continued major production, although costs of finding and developing major new ore bodies likely would be high.

In general, the Silverton caldera area is characterized by older volcanic suites and deeper erosion than in most of the rest of the San Juan Mountains. The Creede caldera, about 30 miles to the east, contains younger volcanic rocks and is far less dissected. Indeed, the principal topographic features are volvanic and tectonic in origin, rather than erosional. The Creede caldera and its ore deposits are described by Steven in this volume, and ore deposits of the Creede and the Summitville districts have been described in greater detail by Steven and Ratté (91,151).

The ore deposits at Creede, valuable principally for silver and lead, are in veins that occupy faults of a graben system that began to form during volcanism and that were recurrently active until some time after volcanism had ceased. The mineralization of the faults is postvolcanic, but, as recent exploration of the Bulldog vein has shown, veins may be stronger in older volcanic units than the young ones due to the prolonged history of fault movement. At Summitville (91) mineralization occurred during the time span of eruption of the Fisher Quartz Latite, the youngest unit of the volcanic sequence. Other flows and related shallow intrusive rocks of the Fisher were altered and mineralized before younger flows of the Fisher were extruded.

Thus, it is evident that there may be more deposits covered by younger and unaltered volcanic flows in this region. More importantly, as Steven (137, this Volume) has emphasized in writings and speech, the ore deposits of the Creede area are so young and, in many places, erosion since their formation has been so slight, that only the original surface manifestations of deposits are exposed. In the Silverton caldera area as in most mining districts, erosion has gone far enough to bring the present surface to the level of ore deposition, but, in the Creede and some other areas in the San Juans, the present surface is essentially the one that existed at the time of ore deposition and thus may be above the level of ore deposition.

Considering these circumstances, the opportunity for discovery of additional major deposits or districts in this region seems to be excellent, but it will require painstaking unraveling of a complex volcanic-tectonic history as well as development of criteria for recognizing ore deposits from a level above their level of formation. Mineralogic studies of the Creede veins being made by P. M. Bethke, P. B. Barton, Jr., and E. M. Roedder of the U.S. Geological Survey will supply some of the criteria, but alteration and geochemical studies also are needed.

Possibilities for the occurrence of ore deposits beneath the volcanics of the San Juan exist in some places but are poor in many others. In calderas, prevolcanic basement rocks probably do not exist, the volcanic rocks being most likely underlain by intrusive rocks. In a large part of the San Juan field, the volcanic rocks probably lie on Precambrian rocks of the old Uncompahgre-San Luis highland, and so the possibilities for replacement bodies in sedimentary rocks are nil. Sedimentary rocks do exist along the western and southwestern sides of the volcanic field and at the northeast corner, near Bonanza. Replacement bodies could exist in these areas, as in the Idarado on the west side of the Silverton caldera, but, on the east side of the caldera, the sedimentary rocks are gone. Near Ouray, it is perhaps possible that ore deposits of the Laramide stage of mineralization might be covered by volcanic rocks, but a few miles to the east and southeast, such deposits, if they existed, probably would have been destroyed in the San Juan super-caldera. Uranium deposits of the Cochetopa district (no. 33, Figure 2), in Precambrian, Jurassic, and Cretaceous rocks at the northern edge of the volcanic field (98), probably formed beneath the volcanics. A few miles farther south on Cochetopa Creek, the only mercury prospect in Colorado is in the Dakota Sandstone immediately beneath the volcanics (36).

## Mineral Deposits of San Luis-Arkansas Graben

Hot springs, some with a high metal content and some with extensive rock alteration, are scattered the length of the San Luis-Arkansas graben. The fluorspar deposits of Browns Canyon and Poncha Springs (no. 31, Figure 2) seem to be related to this fault system. Perhaps it is only coincidence, but, much farther north, the fluorspar deposits of the Northgate district (no. 1, Figure 2) are in north-trending veins cutting Precambrian and Tertiary sedimentary rocks (90), on the line of projection of the graben fault system. North-trending faults on the west side of North Park contain fluorspar and uranium (145), and they also might be

part of the system. Most faults of the graben system are poorly exposed or are concealed beneath surficial deposits in the upper Arkansas and San Luis valleys, but they seem promising targets for investigation for deposits of the hot spring or shallow epithermal type.

The graben fault system intersects the mineral belt in the Leadville-Climax area, the most productive part of the entire belt. In this area, north-trending structures predominate, suggesting influence of the graben system. The Mosquito fault, on which both the Leadville and Climax districts are located, trends this direction, as do many subsidiary faults at Leadville and trunk veins and vitrophyric young dikes in the Sugar Loaf-St. Kevin district (78). However, these faults are all proved to have existed in Laramide time, although reactivated later, and the mineralization at Leadville probably is principally Laramide. To what extent, if any, the Oligocene mineralization of Climax and possibly of the Sugar Loaf-St. Kevin district represents a magma of the graben system as contrasted to one of the mineral belt is not yet known. This question could have an important bearing on exploration, particularly for molybdenum.

## REFERENCES CITED

1. Emmons, S. F., 1886, Geology and mining industry of Leadville, Colorado: U.S. Geol. Surv. Mon. 12, 770 p.
2. Emmons, S. F., 1896, The mines of Custer County, Colorado: U.S. Geol. Surv. 17th Ann. Rept., pt. 2, p. 405–472.
3. Spurr, J. E., 1898, Geology of the Aspen mining district, Colorado: U.S. Geol. Surv. Mon. 31, 260 p.
4. Spencer, A. C., 1904, The copper deposits of the Encampment district, Wyoming: U.S. Geol. Surv. Prof. Paper 25, 107 p.
5. Lindgren, W., 1908, Notes on copper deposits in Chaffee, Fremont, and Jefferson Counties, Colorado: U.S. Geol. Surv. Bull. 340, p. 157–174.
6. Crawford, R. D. and Worcester, P. G., 1916, Geology and ore deposits of the Gold Brick district, Colorado: Colo. Geol. Surv. Bull. 10, 116 p.
7. Howell, J. V., 1919, Twin Lakes district of Colorado: Colo. Geol. Surv. Bull. 17, 74 p.
8. Crawford, R. D., 1924, A contribution to the igneous geology of central Colorado: Amer. Jour. Sci., 5th ser., v. 7, p. 365–388.
9. Emmons, S. F., et al., 1927, Geology and ore deposits of the Leadville mining district, Colorado: U.S. Geol. Surv. Prof. Paper 148, 368 p.
10. Lovering, T. S., 1929, The Rawlins, Shirley,

and Seminoe iron-ore deposits, Carbon County, Wyoming: U.S. Geol. Surv. Bull. 811-D, p. 203–235.

11. Burbank, W. S., 1932, Geology and ore deposits of the Bonanza mining district, Colorado: U.S. Geol. Surv. Prof. Paper 169, 166 p .

12. Burbank, W. S., 1933, Vein systems of the Arrastre Basin and regional geologic structure in the Silverton and Telluride quadrangles, Colorado: Colo. Sci. Sco. Pr., v. 13, p. 136–214.

13. Burbank, W. S. and Lovering, T. S., 1933, Relation of stratigraphy, structure, and igneous activity to ore deposition of Colorado and southern Wyoming: *in Ore deposits of the Western States* (Lindgren Volume), A.I.M.E., N.Y., p. 272–316.

14. Lovering, T. S., 1934, Geology and ore deposits of the Breckenridge mining district, Colorado: U.S. Geol. Surv. Prof. Paper 176, 64 p.

15. Lovering, T. S., 1935, Geology and ore deposits of the Montezuma quadrangle, Colorado: U.S. Geol. Surv. Prof. Paper 178, 119 p.

16. Vanderwilt, J. W., 1935, Revision of structure and stratigraphy of the Aspen district, Colorado, and its bearing on the ore deposits: Econ. Geol., v. 30, p. 223–241.

17. Goddard, E. N., 1936, Geology and ore deposits of the Tincup mining district, Gunnison County, Colorado: Colo. Sci. Soc. Pr., v. 13, p. 551–595.

18. Loughlin, G. F.. and Koschmann, A. H., 1936, Geology and ore deposits of the Cripple Creek district, Colorado: Colo. Sci. Soc. Pr., v. 13, p. 217–435.

19. Wahlstrom, E. E., 1936, The age relations of the Ward ores, Boulder County, Colorado: Econ. Geol., v. 31, p. 104–114.

20. Burbank, W. S. and Goddard, E. N., 1937, Thrusting in Huerfano Park, Colorado, and related problems of orogeny in the Sangre de Cristo Mountains: Geol. Soc. Amer. Bull., v. 48, p. 931–976.

21. Vanderwilt, J. W., 1937, Geology and mineral deposits of the Snowmass Mountain area, Gunnison County, Colorado: U.S. Geol. Surv. Bull. 884, 184 p.

22. Lovering, T. S. and Goddard, E. N., 1938, Laramide igneous sequence and differentiation in the Front Range, Colorado: Geol. Soc. Amer. Bull., v. 49, p. 35–68.

23. Wilkerson, A. S., 1939, Geology and ore deposits of the Magnolia mining district, Boulder County, Colorado: Colo. Sci. Soc. Pr., v. 14, p. 81–101.

24. Burbank, W. S., 1940, Structural control of ore deposition in the Uncompahgre district, Ouray County, Colorado: U.S. Geol. Surv. Bull. 906-E., p. 189–265.

25. Butler, R. D. and Singewald, Q. D., 1940, Zonal mineralization and silicification in the Horseshoe and Sacramento districts, Colorado: Econ. Geol., v. 35, p. 793–838.

26. Goddard, E. N., 1940, Preliminary report on the Gold Hill mining district, Boulder County, Colorado: Colo. Sci. Soc. Pr., v. 14, p. 103–139.

27. Burbank, W. S., 1941, Structural control of ore deposition in the Red Mountain, Sneffels, and Telluride districts of the San Juan Mountains, Colorado: Colo. Sci. Soc. Pr., v. 14, p. 141–261.

28. Koschmann, A. H., 1941, New light on the geology of the Cripple Creek district, Colorado: Colo. Mining Assn., Denver, 28 p.

29. Singewald, Q. D. and Butler, B. S., 1941, Ore deposits in the vicinity of the London fault, Colorado: U.S. Geol. Surv. Bull. 911, 74 p.

30. Goddard, E. N. and Lovering, T. S., 1942, Nickel deposit near Gold Hill, Boulder County, Colorado: U.S. Geol. Surv. Bull. 931-0, p. 349–362.

31. Larsen, E. S., 1942, Alkalic rocks of Iron Hill, Gunnison County, Colorado: U.S. Geol. Surv. Prof. Paper 197-A, 64 p.

32. Brill, K. G., Jr., 1944, Late Paleozoic stratigraphy, west-central and northwestern Colorado: Geol. Soc. Amer. Bull., v. 55, p. 621–656.

33. Lovering, T. S. and Tweto, O. L., 1944, Preliminary report on geology and ore deposits of the Minturn quadrangle, Colorado: U.S. Geol. Surv. Open File Rept., 115 p.

34. Varnes, D. J., 1944, Preliminary report on the geology of a part of the Rico dome, Dolores County, Colorado: U.S. Geol. Surv. Strategic Mineral Inv. Rept.

35. Johnson, J. H., 1945, A resume of the Paleozoic stratigraphy of Colorado: Colo. Sch. Mines Quart., v. 40, no. 3, 109 p.

36. Tweto, O. and Yates, R. G., 1945, Memorandum report on the Cochetopa Creek quicksilver deposit, Saguache County, Colorado: U.S. Geol. Surv. Strategic Mineral Inv. Rept. 3-189, 4 p.

37. Fischer, R. P., *et al.* 1946, Metallic mineral deposits of Colorado: U.S. Geol. Surv. Missouri Basin Studies No. 8 with text, tables, and maps, 1:1,000,000.

38. Goddard, E. N., 1946, Fluorspar deposits of the Jamestown district, Boulder County, Colorado: Colo. Sci. Soc. Pr., v. 15, p. 1–47.

39. Kelley, V. C., 1946, Geology, ore deposits, and mines of the Mineral Point, Poughkeepsie, and Upper Uncompahgre districts, Ouray, San Juan, and Hinsdale Counties, Colorado: Colo. Sci. Soc. Pr., v. 14, p. 287–466.

40. Koschmann, A. H. and Wells, F. G., 1946, Preliminary report on the Kokomo mining district, Colorado: Colo. Sci. Soc. Pr., v. 15, p. 49–112.

41. Vanderwilt, J. W. and King, R. U., 1946,

Geology of the Climax ore body: Min. and Met., v. 27, p. 299–302.

42. Burbank, W. S. *et al.,* 1947, the San Juan region: *in* Vanderwilt, J. W., *Editor, Mineral Resources of Colorado,* State of Colo., Mineral Res. Bd., p. 396–446.

43. Tweto, O. and Lovering, T. S., 1947, The Giiman district, Eagle County (Colo.), *in* Vanderwilt, J. W., *Editor, Mineral Resources of Colorado,* State of Colo., Mineral Res. Bd., p. 378–387.

44. Van Alstine, R. E., 1947, Fluorspar investigations, *in* Vanderwilt, J. W., *Editor, Mineral Resources of Colorado,* State of Colo., Mineral Res. Bd., p. 457–465.

45. Vanderwilt, J. W., 1947, *Editor, Mineral Resources of Colorado:* State of Colo., Mineral Res. Bd., 547 p.

46. Simpson, G. G., 1948, The Eocene of the San Juan basin, New Mexico: Amer. Jour. Sci., v. 246, p. 257–282, 363–385.

47. Argall, G. O., 1949, Industrial minerals of Colorado: Colo. Sch. Mines Quart., v. 44, no. 2, 477 p.

48. Eckel, E. B., 1949, Geology and ore deposits of the La Plata district, Colorado: U.S. Geol. Surv. Prof. Paper 219, 179 p.

49. Koschmann, A. H., 1949, Structural control of the gold deposits in the Cripple Creek district, Teller County, Colorado: U.S. Geol. Surv. Bull. 955-B, p. 19–60.

50. Stark, J. T., *et al.,* 1949, Geology and origin of South Park, Colorado: Geol. Soc. Amer. Mem. 33, 188 p.

51. Tweto, O., 1949, Stratigraphy of the Pando area, Eagle County, Colorado: Colo. Sci. Soc. Pr., v. 15, p. 149–235.

52. Lovering, T. S. and Goddard, E. N., 1950, Geology and ore deposits of the Front Range, Colorado: U.S. Geol. Surv. Prof. Paper 223, 319 p.

53. Singewald, Q. D., 1950, Gold placers and their geologic environment in northwestern Park County, Colorado: U.S. Geol. Surv. Bull. 955-D, p. 103–172.

54. Wahlstrom, E. E., 1950, Melonite in Boulder County, Colorado: Amer. Mineral., v. 35, p. 948–953.

55. Burbank, W. S., 1951, The Sunnyside, Ross Basin, and Bonita fault systems and their associated ore deposits, San Juan County, Colorado: Colo. Sci. Soc. Pr., v. 15, p. 285–304.

56. Hagner, A. F., 1951, Anorthosite of the Laramie Range, Albany County, Wyoming, as a possible source of alumina: Wyoming Geol. Surv. Bull. 43, 15 p.

57. Singewald, Q. D., 1951, Geology and ore deposits of the upper Blue River area, Summit County, Colorado: U.S. Geol. Surv. Bull. 970, 72 p.

58. Gabelman, J. W., 1952, Structure and origin of northern Sangre de Cristo Range, Colorado: Amer. Assoc. Petrol. Geols. Bull., v. 36, p. 1547–1612.

59. Behre, C. H., Jr., 1953, Geology and ore deposits of the west slope of the Mosquito Range: U.S. Geol. Surv. Prof. Paper 235, 176 p.

60. Lovering, T. S. and Tweto, O., 1953, Geology and ore deposits of the Boulder County tungsten district, Colorado: U.S. Geol. Surv. Prof. Paper 245, 199 p.

61. Wright, H. D., 1954, Mineralogy of a uranium deposit at Caribou, Colorado: Econ. Geol., v. 49, p. 129–174.

62. Birch, R. W., 1955, Wyoming's mineral resources: Wyoming Nat. Res. Bd., 166 p.

63. Singewald, Q. D., 1955, Sugar Loaf and St. Kevin mining districts, Lake County, Colorado: U.S. Geol. Surv. Bull. 1027-E, p. 251–299.

64. Harrison, J. E. and Wells, J. D., 1956, Geology and ore deposits of the Freeland-Lamartine district, Clear Creek County, Colorado: U.S. Geol. Surv. Bull. 1037-B, p. 33–127.

65. Larsen, E. S., Jr. and Cross, W., 1956, Geology and petrology of the San Juan region, southwestern Colorado: U.S. Geol. Surv. Prof. Paper 258, 303 p.

66. Olson, J. C. and Wallace, S. R., 1956, Thorium and rare-earth minerals in Powderhorn district, Gunnison County, Colorado: U.S. Geol. Surv. Bull. 1027-O, p .693–721.

67. Shoemaker, E. M., 1956, Structural features of the central Colorado Plateau and their relation to uranium deposits: U.S. Geol. Surv. Prof. Paper 300, p. 155–170.

68. Dings, M. G. and Robinson, C. S., 1957, Geology and ore deposits of the Garfield quadrangle, Colorado: U.S. Geol. Surv. Prof. Paper 289, 110 p.

69. Eckelman, W. R. and Kulp. J. L., 1957, Uranium-lead method of age determination. Part II: North American localities: Geol. Soc. Amer. Bull., v. 68, p. 1117–1140.

70. Hillebrand, J. R., 1957, The Idarado mine: New Mex. Geol. Soc. 8th Ann. Field Conf. Guidebook, p. 176–188.

71. Moore, F. B., *et al.,* 1957, Geology and uranium deposits of the Caribou area, Boulder County, Colorado: U.S. Geol. Surv. Bull. 1030-N, p. 517–552.

72. Newhouse, W. H. and Hagner, A. F., 1957, Geologic map of anorthosite areas, southern part of Laramie Range, Wyoming: U.S. Geol. Surv. Mineral Investigations, Field Studies Map MF 119, 1:63,360.

73. Tweto, O., 1957, Geologic sketch of southern Middle Park, Colorado: *in* Finch, W. C., *Editor, Guidebook to the geology of North and Middle Park basins, Colorado,* Rocky Mtn. Assoc. Geols., p. 18–31.

74. Curtis, B. F., *Editor,* 1958, Symposium on Pennsylvanian rocks of Colorado and adjacent areas: Rocky Mtn. Assoc. Geols., Denver, 184 p.

75. Engel, A. E. J., *et al.,* 1958, Variations in

isotopic composition of oxygen and carbon in Leadville Limestone (Mississippian, Colorado) and in its hydrothermal and metamorphic phases: Jour. Geol., v. 66, p. 374–393.

76. Litsey, L. R., 1958, Stratigraphy and structure of the northern Sangre de Cristo Mountains, Colorado: Geol. Soc. Amer. Bull., v. 69, p. 1143–1178.

77. Sims, P. K., *et al.*, 1958, Geology of the Copper King uranium mine, Larimer County, Colorado: U.S. Geol. Surv. Bull. 1032-D, p. 171–221.

78. Tweto, O. and Pearson, R. C., 1958, Geologic map of the southwestern part of the Holy Cross quadrangle, Colorado: U.S. Geol. Surv. Open File Map.

79. Christman, R. A., *et al.*, 1959, Geology and thorium deposits of the Wet Mountains, Colorado: U.S. Geol. Surv. Bull. 1072-H, p. 491–535.

80. Harrison, J. E. and Wells, J. D., 1959, Geology and ore deposits of the Chicago Creek area, Clear Creek County, Colorado: U.S. Geol. Surv. Prof. Paper 319, 92 p.

81. Haun, J. D. and Weimer, R. J., *Editors,* 1959, Symposium on Cretaceous rocks of Colorado and adjacent areas: Rocky Mtn. Assoc. Geols. Guidebook 11th Field Conf., Denver, 210 p.

82. Johnson, R. B., 1959, Geology of the Huerfano Park area, Huerfano and Custer Counties, Colorado: U.S. Geol. Surv. Bull. 1071, p. 87–119.

83. Osterwald, F. W. *et al.*, 1959, Wyoming mineral resources: Wyoming Geol. Surv. Bull. 50, 259 p.

84. Carpenter, R. H., 1960, Molybdenum: *in* Del Rio, S. M., *Editor, Mineral Resources of Colorado,* First sequel: Colo. Mineral Res. Bd., p. 317–325.

85. Del Rio, S. M., *Editor,* 1960, Mineral Resources of Colorado, First sequel: Colo. Mineral Res. Bd., Denver, 764 p.

86. Hart, S. R., 1960, A study of mineral ages in a contact metamorphic zone: *in Variation in isotopic abundances of strontium, calcium, and argon and related topics,* Mass. Inst. Tech., 8th Ann. Prog. Rept. for 1960, U.S. Atomic Energy Comm., NYO-3941, p. 131–154.

87. Holmes, A., 1960, A revised geologic time scale: Edinburgh Geol. Soc. Trans., v. 17, pt. 3, p. 204.

88. Moore, F. E., 1960, Summary of Cenozoic history, southern Laramie Range, Wyoming and Colorado: *in* R. J. Weimer and J. D. Haun, *Editors,* Geol. Soc. Amer. Guidebook for Field Trips (Guide to the geology of Colorado), p. 217–222.

89. Rose, C. K. and Shannon, S. S., Jr., 1960, Cebolla Creek titaniferous iron deposits, Gunnison County, Colorado: U.S. Bur. Mines R.I. 5679, 30 p.

90. Steven, T. A., 1960, Geology and fluorspar deposits of the Northgate district, Colorado: U.S. Geol. Surv. Bull. 1082-F, p. 323–422.

91. Steven, T. A. and Ratté, J. C., 1960, Geology and ore deposits of the Summitville district, San Juan Mountains, Colorado: U.S. Geol. Surv. Prof. Paper 343, 70 p.

92. Steven, T. A., and Ratté, J. C., 1960, Relation of mineralization to caldera subsidence in the Creede district, San Juan Mountains, Colorado: U.S. Geol. Surv. Prof. Paper 400-B, p. 14–17.

93. Tweto, O., 1960, Scheelite in the Precambrian gneisses of Colorado: Econ. Geol., v. 55, p. 1406–1428.

94. U.S. Geological Survey, 1960, Mineral and water resources of Wyoming: U.S. Senate Doc. 76, 86th Cong., 2d ses., 40 p.

95. Wallace, S. R., *et al.*, 1960, Geology of the Climax molybdenite deposit: A progress report: *in* Weimer, R. J., and Haun, J. D., *Editors,* Geol. Soc. Amer. Guidebook for Field Trips, (*Guide to the geology of Colorado*) Field Trip B-3, p. 238–252.

96. Weimer, R. J. and Haun, J. D., *Editors,* 1960, Guide to the geology of Colorado: Geol. Soc., Amer. Guidebook for Field Trips, 310 p.

97. Wells, J. D., 1960, Petrography of radioactive Tertiary igneous rocks, Front Range mineral belt, Colorado: U.S. Geol. Surv. Bull. 1032, p. 223–272.

98. Wright, R. J. and Everhart, D. L., 1960, Uranium: *in* Del Rio, S. M., *Editor, Mineral resources of Colorado,* first sequel. Colo. Mineral Res. Bd., Denver, p. 329–365.

99. Berg, R. R. and Rold, J. W., *Editors,* 1961, Lower and Middle Paleozoic rocks of Colorado: Rocky Mtn. Assoc. Geols., Guidebook 12th Field Conf., Denver, 236 p.

100. Chronic, J. and Ferris, C. S., 1961, Paleozoic outliers in southeastern Wyoming: *in* R. R. Berg and J. W. Rold, *Editors, Lower and Middle Paleozoic rocks of Colorado:* Rocky Mtn. Assoc. Geols., Guidebook 12th Field Conf., p. 143–146.

101. Donnell, J. R., 1961, Tertiary geology and oil-shale resources of the Piceance Creek basin between the Colorado and White Rivers, northwestern Colorado: U.S. Geol. Surv. Bull. 1082-L, p. 835–891.

102. Hart, S. R., 1961, Mineral ages and metamorphism, *in* J. L. Kulp, *Editor, Geochronology of rock systems:* N.Y. Acad. Sci. Ann., v. 91, art. 2, p. 192–197.

103. Hedlund, D. C. and Olson, J. C., 1961, Four environments of thorium-, niobium-, and rare-earth-bearing minerals in the Powderhorn district of southwestern Colorado: U.S. Geol. Surv. Prof. Paper 424-B, p. 283–286.

104. Holt, H. E., 1961, Geology of the lower Blue River area, Summit and Grand Coun-

ties, Colorado: Unpublished Ph.D. thesis, Univ. Colo.

105. Kulp, J. L., 1961, Geologic time scale: Science, v. 133, no. 3459, p. 1111.

106. Phair, G. and Fisher, F. G., 1961, Potassic feldspathization and thorium deposition in the Wet Mountains, Colorado: U.S. Geol. Surv. Prof. Paper 424-D, p. 1–2.

107. Tweto, O., 1961, Late Cenozoic events of the Leadville district and Upper Arkansas Valley, Colorado: U.S. Geol. Surv. Prof. Paper 424-B, p. 133–135

108. Young, E. J. and Sims, P. K., 1961, Petrography and origin of xenotime and monazite concentrations, Central City district, Colorado: U.S. Geol. Surv. Bull. 1032-F, p. 273–297.

109. Brown, R. W., 1962, Paleocene flora of the Rocky Mountains and Great Plains: U.S. Geol. Surv. Prof. Paper 375, 119 p.

110. Koschmann, A. H., 1962, The historical pattern of mineral exploitation in Colorado: Colo. Sch. Mines Quart., v. 57, no. 4, p. 7–25.

111. Moench, R. H., et al., 1962, Precambrian folding in the Idaho Springs-Central City area, Front Range, Colorado: Geol. Soc. Amer. Bull., v. 73, p. 35–58.

112. Moorbath, S., 1962, Evidence for the origin and absolute age of mineralized Tertiary intrusives in the southwestern U.S. from strontium isotope ratios: Mass. Inst. Tech., 10th Ann. Progress Rept. NYO 3943, p. 81–83.

113. Pearson, R. C., et al., 1962, Age of Laramide porphyries near Leadville, Colorado: U.S. Geol. Surv. Prof. Paper 450-C, p. 78–80.

114. Qureshy, M. N., 1962, Gravimetric-isostatic studies in Colorado: Jour. Geophys. Res., v. 67, no. 6, p. 2459–2467.

115. Sims, P. K. and Barton, P. B., Jr., 1962, Hypogene zoning and ore genesis, Central City district, Colorado: Geol. Soc. Amer., Buddington Volume, p. 373–396.

116. Vhay, J. S., 1962, Geology and mineral deposits of the area south of Telluride, Colorado: U.S. Geol. Surv. Bull. 1112-G, p. 209–310.

117. Wahlstrom, E. E. and Hornback, V. Q, 1962, Geology of the Harold D. Roberts Tunnel, Colorado: West portal to Station 468+49: Geol. Soc. Amer. Bull., v. 73, p. 1477–1498.

118. Bass, N. W. and Northrop, S. A., 1963, Geology of the Glenwood Springs quadrangle and vicinity, northwestern Colorado: U.S. Geol. Surv. Bull. 1142-J, 74 p.

119. Carter, D. A., 1963, Sunrise iron mine: in Guidebook to the geology of the northern Denver basin and adjacent uplifts, Rocky Mtn. Assoc. Geols., p. 264–266.

120. Parker, R. L., and Hildebrand, F. A., 1963, Preliminary report on alkalic intrusive rocks in the northern Wet Mountains, Colorado: U.S. Geol. Surv. Prof. Paper 450-E, p. 8–10.

121. Sims, P. K., et al., 1963, Economic geology of the Central City district, Gilpin County, Colorado: U.S. Geol. Surv. Prof. Paper 359, 231 p.

122. Sims, P. K., et al., 1963, Geology of uranium and associated ore deposits, central part of the Front Range mineral belt, Colorado: U.S. Geol. Surv. Prof. Paper 371, 119 p.

123. Tweto, O. and Sims, P. K., 1963, Precambrian ancestry of the Colorado mineral belt: Geol. Soc. Amer. Bull., v. 74, p. 991–1014.

124. U.S. Geological Survey, 1963, Geological Survey Research, 1963: U.S. Geol. Survey Prof. Paper 475-A, 299 p.

125. Varnes, D. J., 1963, Geology and ore deposits of the South Silverton mining area, San Juan County, Colorado: U.S. Geol. Surv. Prof. Paper 378-A, 56 p.

126. Banks, P. O. and Silver, L. T., 1964, Reexamination of isotopic relationships in Colorado Front Range uranium ores: Geol. Soc. Amer. Bull., v. 75, p. 469–476.

127. Chapin, C. E. and Epis, R. C., 1964, Some stratigraphic and structural features of the Thirtynine Mile volcanic field, central Colorado: Mtn. Geologist, v. 1, p. 145–160.

128. DeVoto, R. H., 1964, Stratigraphy and structure of Tertiary rocks in southwestern South Park: Mtn. Geologist, v. 1, p. 117–126.

129. Everenden, J. F., et al., 1964, Potassium-argon dates and the Cenozoic mammalian chronology of North America: Amer. Jour. Sci., v. 262, p. 145–198.

130. Ferris, C. S. and Krueger, H. W., 1964, New radiogenic dates on gneiss rocks from the southern Laramie Range, Wyoming: Geol. Soc. Amer. Bull., v. 75, p. 1051–1054.

131. Hart, S. R., 1964, The petrology and isotopic-mineral age relations of a contact zone in the Front Range, Colorado: Jour. Geol., v. 72, p. 493–525.

132. Hills, A., et al., 1964, Chronology of some Precambrian igneous and metamorphic events of the Medicine Bow Mountains, Wyoming (abs.): Geol. Soc. Amer. Spec. Paper 82, p. 92.

133. Sawatsky, D. L., 1964, Structural geology of southeastern South Park, Park County, Colorado: Mtn. Geologist, v. 1, p. 133–139.

134. Siems, P. L., 1964, Correlation of Tertiary strata in mountain basins, southern Colorado and northern New Mexico: Mtn. Geologist, v. 1, p. 161–180.

135. Sims, P. K. and Gable, D. J., 1964, Geology of Precambrian rocks, Central City district, Colorado: U.S. Geol. Surv. Prof. Paper 474-C, 52 p.

136. Sims, P. K. and Sheridan, D. M., 1964, Geology of uranium deposits in the Front

Range, Colorado: U.S. Geol. Surv. Bull. 1159, 116 p.

137. Steven, T. A., 1964, Geologic setting of the Spar City district, San Juan Mountains, Colorado: U.S. Geol. Surv. Prof. Paper 475-D, p. 123–125.

138. U.S. Geological Survey, 1964, Geological Survey Research, 1964: U.S. Geol. Surv Prof. Paper 501-A, 367 p.

139. U.S. Geological Survey, 1964, Mineral and water resources of Colorado: U.S. Senate Interior and Insular Affairs Comm., Committee Print, 88th Cong., 2d ses., 302 p.

140. Woolard, G. P. and Joesting, H. R., *Chmn.,* 1964, Bouger gravity anomaly map of the U.S.: U.S. Geol. Surv. Spec. Series Map.

141. Bayley, R. W., 1965, Preliminary geologic map of the Bradley Peak quadrangle, Wyoming: U.S. Geol. Surv. Open File Map.

142. Chronic, J., *et al.,* 1965, Lower Paleozoic rocks in diatremes in southern Wyoming and northern Colorado (abs.): Geol. Soc. Amer. Spec. Paper 87, p. 280–281.

143. Conley, C. D., 1965, Petrology of the Leadville Limestone (Mississippian), White River Plateau, Colorado: Mtn. Geologist, v. 2, p. 181–182.

144. Downs, G. R. and Bird, A. G., 1965, The Schwartzwalder uranium mine, Jefferson County, Colorado: Mtn. Geologist, v. 2, p. 183–192.

145. Hail, W. J., Jr., 1965, Geology of northwestern North Park, Colorado: U.S. Geol. Surv. Bull. 1188, 133 p.

146. Hansen, W. R., 1965, The Black Canyon of the Gunnison, today and yesterday: U.S. Geol. Surv. Bull. 1191, 76 p.

147. ——, 1965, Geology of the Flaming Gorge area, Utah-Colorado-Wyoming: U.S. Geol. Surv. Prof. Paper 490, 196 p.

148. Houston, R. S., *et al.,* 1965, Regional aspects of structure and age of rocks of the Medicine Bow Mountains, Wyoming (abs.): Geol. Soc. Amer. Spec. Paper 87, p. 287–288.

149. Peterman, Z. E., *et al.,* 1965, Precambrian geochronology of the northeastern Front Range, Colorado (abs.): Geol. Soc. Amer. Spec. Paper 87, p. 127.

150. Scott, G. R., 1965, Nonglacial Quaternary geology of the southern and middle Rocky Mountains, *in The Quaternary of the United States:* Princeton Univ. Press, Princeton, p. 243–254.

151. Steven, T. A. and Ratté, J. C., 1965, Geology and structural control of ore deposition in the Creede district, San Juan Mountains, Colorado: U.S. Geol. Surv. Prof. Paper 487, 90 p.

152. Tweto, O., 1965, Regional features of Precambrian rocks in north-central Colorado (abs.): Geol. Soc. Amer. Spec. Paper 87, p. 304.

153. U.S. Geological Survey, 1965, Geological Survey Research, 1965: U.S. Geol. Survey Prof. Paper 525-A, 376 p.

154. Wetherill, G. W. and Bickford, M. E., 1965, Primary and metamorphic Rb-Sr chronology in central Colorado: Jour. Geophys. Res., v. 70, no. 18, p. 4669–4686.

155. Corbett, M. K., 1966, The geology and structure of the Mt. Richthofen-Iron Mountain region, north-central Colorado: Mtn. Geologist, v. 3, p. 3–21.

156. Gaskill, D. L. and Godwin, L. H., 1966, Geologic map of the Marcellina Mountain quadrangle, Gunnison County, Colorado: U.S. Geol. Surv. Map GQ-511.

157. Gill, J. R. and Cobban, W. A., 1966, The Red Bird section of the Upper Cretaceous Pierre Shale in Wyoming: U.S. Geol. Surv. Prof. Paper 373-A, 73 p.

158. Harrer, C. M., 1966, Wyoming iron ore deposits: U.S. Bur. Mines I.C. 8315, 114 p.

159. Izett, G. A., 1966, Tertiary extrusive volcanic rocks in Middle Park, Grand County, Colorado: U.S. Geol. Surv. Prof. Paper 550-B, p. 42–46.

160. Klugman, M. A., 1966, Resumé of the geology of the Laramie anorthosite mass: Mtn. Geologist, v. 3, p. 75–84.

161. Moench, R. H. and Drake, A. A., 1966, Economic geology of the Idaho Springs district, Clear Creek and Gilpin Counties, Colorado: U.S. Geol. Surv. Bull. 1208, 91 p.

162. Pearson, R. C. *et al.,* 1966, Geochronology of the St. Kevin Granite and neighboring Precambrian rocks, northern Sawatch Range, Colorado: Geol. Soc. Amer. Bull., v. 77, p. 1109–1120.

163. U.S. Geological Survey, 1966, Geological Survey research, 1966: U.S. Geol. Surv. Prof. Paper 550-A, 385 p.

164. Catanzaro, E. J., 1967, Correlation of some Precambrian rocks and metamorphic events in parts of Wyoming and Montana: Mtn. Geologist, v. 4, p. 9–21.

165. Dickinson, R. G., *et al.,* 1967, Late Cretaceous uplift and volcanism on the north flank of the San Juan Mountains, Colorado (abs.): Geol. Soc. Amer. Rocky Mtn. Sect., Prog. 1967 Ann. Mtng., p. 31.

166. Epis, R. C., *et al.,* 1967, Geologic history of the Thirtynine Mile volcanic field, central Colorado (abs.): Geol. Soc. Amer. Rocky Mtn. Sect., Prog. 1967 Ann. Mtng., p. 34–35.

167. Gaskill, D. L., *et al.,* 1967, Geologic map of the Oh-Be-Joyful quadrangle, Gunnison County, Colorado: U.S. Geol. Survey Map GQ-578.

168. Hedge, C. E., *et al.,* 1967, Age of the major Precambrian regional metamorphism in the northern Front Range, Colorado: Geol. Soc. Amer. Bull., v. 78, p. 551–558.

169. Hutchinson, R. M. and Hedge, C. E., 1967, Precambrian basement rocks of the central

Colorado Front Range and its 700 million year history: Geol. Soc. Amer. Rocky Mtn. Sect., 20th Ann. Mtng. Guidebook, 51 p.

170. Moorbath, S., *et al.,* 1967, Evidence for the origin and age of some mineralized Laramide intrusives in the southwestern United States from strontium isotope and rubidium-strontium measurements: Econ. Geol., v. 62, p. 228–236.

171. Siems, P. L., 1967, Volcanic geology of the Rosita Hills-Silver Cliff district, Custer County, Colorado (abs.): Geol. Soc. Amer. Rocky Mtn. Sect.. Prog. 1967 Ann. Mtng., p. 62.

172. Taylor, R. B., *et al.,* 1967, Mid-Tertiary volcanism in the central Front Range, Colorado (abs.): Geol. Soc. Amer. Rocky Mtn. Sect., Prog. 1967 Ann. Mtng., p. 68.

173. Wells, J. D., 1967, Geology of the Eldorado

Springs quadrangle, Boulder and Jefferson Counties, Colorado: U.S. Geol. Surv. Bull. 1221-D, 85 p.

174. Bromfield, C. S. (*In press*), Geology of the Mount Wilson quadrangle, western San Juan Mountains, Colorado: U.S. Geol. Surv. Bull. 1227.

175. Burbank, W. S. and Luedke, R. G., (*In press*), Geology and ore deposits in the Eureka mining district of the Silverton caldera, southwestern Colorado: U.S. Geol. Surv. Prof. Paper.

176. Hawley, C. C., *et al.,* (*In press*), Geology and ore deposits of the Lawson-Dumont-Fall River district, Clear Creek County, Colorado: U.S. Geol. Surv. Bull. 1231.

177. Sheridan, D. M., *et al.,* (*In press*), Geology and uranium deposits of the Ralston Buttes district, Jefferson County, Colorado: U.S. Geol. Surv. Prof. Paper 520.

# 28. Ore Deposits of the Atlantic City District, Fremont County, Wyoming[*]

RICHARD W. BAYLEY[†]

## Contents

* Publication authorized by the Director, U.S. Geological Survey.
† U.S. Geological Survey, Menlo Park, California.

## Illustrations

## Tables

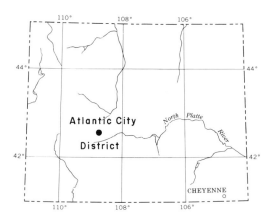

on the same lodes have not been explored. They appear to hold the only hope for future gold mining.

Sedimentary iron-formation, similar to the Lake Superior taconite, occurs in the northern part of the district. Locally, where the iron-formation is thickened by folds and faults, sufficient tonnages are present to be minable, as at Iron Mountain, where the U.S. Steel Corporation produces about 1.3 million tons of iron-ore pellets annually. Exploration of the iron deposits began in 1954, and the first iron-ore pellets were shipped to Provo, Utah in August 1962. Indicated reserves of iron-formation (about 30 per cent Fe) are reported to be about 300 million tons.

## ABSTRACT

The Atlantic City district encompasses several districts and has been previously called by different names, e.g., Atlantic gold district, Atlantic City-South Pass mining district, and Sweetwater mining district. The district is located in a terrane of folded and faulted early Precambrian metamorphic rocks which lies near the south end of the Wind River Range in Fremont County, Wyoming, Historically it is a gold district of little production, but today is an iron district of considerable importance. Gold was discovered in placer deposits about 1842 and in quartz-arsenopyrite veins about 1867, after which there occurred a mining boom lasting a few years. Minor, intermittent gold production from placers and lodes has continued to date, but the total production has probably not exceeded a few millions of dollars. Past lode mining was generally shallow (in the oxidized zone), the deepest about 360 feet. The deeper, less oxidized and sulfide ores

## INTRODUCTION

The Atlantic City district straddles the south part of the Wind River Range in Fremont County, Wyoming about 28 miles south of Lander (Figure 1). It is a gold district historically but an iron district of considerable importance today.

The earliest geologic accounts of the district, dating from about 1870, are fragmentary and concern mainly the gold deposits; the first geologic mapping of note was done in 1914 by A. C. Spencer (4), who recognized the major rock units and established the basic geologic framework. Unpublished reports and maps by Armstrong (8), De Laguna (5), and King (9) contributed additional geologic detail for various parts of the district. Armstrong's maps of the underground workings of the Carissa mine represent the only available subsurface data.

The writer mapped the four 7½-minute quadrangles that cover the district in five field seasons, 1958 to 1962. The iron deposits were mapped in detail in the first season (17), and the remainder of the district and environs in the field seasons following (19,20,21,22). By exceptional good fortune, the geologic mapping of the iron deposits preceded mining and coincided with a very thorough diamond-drilling exploration of them by the U.S. Steel Corporation and affiliates. Maps and drill-hole data were given freely, permitting an accuracy in geologic mapping seldom attainable. For data and personal courtesies I am especially grateful to R. H. B. Jones, Columbia-Geneva Steel Division; F. M. Galbraith and V. Thompkins, Columbia Iron Mining Company; and the U.S. Steel Corporation.

## HISTORICAL BACKGROUND

### Gold Mining

The history of gold mining in the district has been related by Knight (2) and Nickerson (1). The two accounts differ in many particulars, but agree in a general way. It seems that placer gold was found in the streams of this area about 1842 by emigrants to California, who worked the placers in groups—to guard against hostile Indians—at intervals until about 1867, when the Carissa Lode was discovered, and the district was stampeded. In 1869, there was reported to be 2000 people in the district. By 1871, all of the major lode deposits had been found, and many of the placers had been worked in detail, but the boom was short lived, and 1875 found most of the mines idle and the district nearly deserted.

Occasional placer and lode mining continued after 1886 to the present. Numerous unsuccessful attempts to rehabilitate and operate existing mines sums up the lode-mining activity in general. A highlight in the period was the dredging of Rock Creek below Atlantic City, which produced $400,000 in gold. As yet no attempt has been made to explore the deep ores of the district, even though the value of the ore has proved continuous to the greatest depth yet mined, and on geologic grounds a great vertical extent seems certain for the major veins. The deepest mine, the Carissa, is bottomed in ore at only 360 feet.

### Gold Production

No reliable records exist for the early, and probably the most productive, period of gold mining in the district. On the basis of pub-

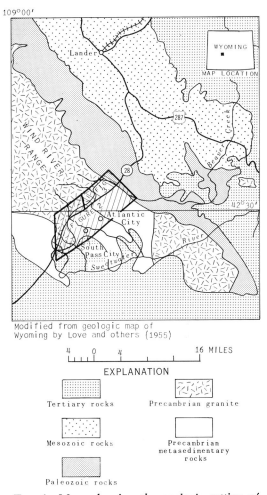

Modified from geologic map of Wyoming by Love and others (1955)

4    0    4    16 MILES

#### EXPLANATION

Tertiary rocks

Precambrian granite

Mesozoic rocks

Precambrian metasedimentary rocks

Paleozoic rocks

FIG. 1. *Map, showing the geologic setting of the Atlantic City District, Fremont County, Wyoming, and the locations of Figures 2 and 3.*

lished total gold production figures for the State of Wyoming, Spencer (4, p. 27–28) concluded that the value of gold produced in the district to 1913 was less than $1,000,000. However, estimates of individual mine outputs made in 1911 by Jamison (3) greatly exceed Spencer's estimated total. Jamison's estimates are shown in the Table on page 592.

Spencer regarded the Jamison figures as high. He conceded, however, that the total value probably exceeded that indicated by the State's statistics, but not by millions. Reasoning that production before and after 1875 was probably about equal, he arrived at the figure $1,500,000, which he considered a fair estimate of total gold production to 1916. De

| Lode Mines | | Placers | |
|---|---|---|---|
| Miners Delight | $1,200,000 | Meadow Gulch | $1,000,000 |
| Carissa | 1,000,000 | Yankee Gulch | 500,000 |
| Carabou | 500,000 | Spring Gulch | 30,000 |
| Garfield | 400,000 | Promise Gulch | 30,000 |
| Victoria Regina | 350,000 | Smith Gulch | 20,000 |
| Franklin | 300,000 | Red Canyon | 20,000 |
| Mary Ellen | 125,000 | Atlantic Gulch | 15,000 |
| Lone Star | 40,000 | Beaver Creek | 10,000 |
| Carrie Shields | 35,000 | Others | 140,000 |
| Other mines | 187,000 | | |
| | | | $1,725,000 |
| | $4,137,000 | | |

Laguna (5) accepted Spencer's 1916 figure as a starting point and estimated the gold production to 1938 at about $2,000,000. The total had not changed significantly by 1945, according to Armstrong (8, p. 37), nor has it changed significantly to the present day.

## Iron Mining

It is not known when or by whom the iron deposits were first discovered, but their location and extent has been a matter of public record since 1916 (4). Because they are composed of low grade magnetic iron-formation (taconite), their exploitation necessarily had to await the development of modern taconite technology and a favorable economy. In August 1962, the first iron-ore pellets were shipped to Provo, Utah from the U.S. Steel Corporation's Atlantic City Project (Figure 2), thus successfully concluding nearly 8 years of exploration and development begun in 1954.

## Iron Production

The amounts of iron-formation (taconite) mined from the U.S. Steel Corporation's open-pit mine at Iron Mountain are as follows (18):

1962—1,043,111 gross tons
1963—3,782,488 gross tons
1964—3,656,062 gross tons

These figures indicate that the mine operated at about its reported intended capacity in 1963 and 1964 (10, p. 84); they indicate also that the mill operated at close to intended capacity for the same period and manufactured about 1.3 million gross tons per year of iron ore pellets [iron-formation to pellet ratio about 2.64 (24, p. 84)].

## PRESENT TOPOGRAPHY AND PHYSIOGRAPHIC HISTORY

The Atlantic City District is high, semi-desert country above 7500 feet in elevation. The southwest part, to about the latitude of South Pass City, is a sage- and grass-covered plain of low relief—a Tertiary pediment—cut by steep-sided gullies. Beyond the plain, to the northwest limit of the district, are mature hills, with approximately accordant tops, as much as 700 feet high. The hills and plains are composed of deeply dissected Precambrian rocks of various kinds. High cliffs of easterly dipping Paleozoic rocks mark the northeast limit of the district.

The development of the present topography began with the uplift of the Wind River Range during the Laramide Orogeny. The uplifted range may be viewed simply as a northeast-tilted block more than 100 miles long, uplifted along a major fault that bounds the range on the southwest. Erosion during Tertiary time removed a full complement of Mesozoic and Paleozoic layered sedimentary rocks from the highest portions of the block and laid the Precambrian surface bare. The debris from the range was deposited in the adjacent Green River and Wind River Basins.

By about Miocene time, a broad pediment surface had advanced into the range to about the present 8200-foot elevation, and the range crest was skirted in its own debris to slightly above the same level. In reaction to a subsequent regional uplift, the streams removed vast quantities of the original Tertiary deposits until only remnants remained. The present topography on the Precambrian rocks then, represents a partly denuded Tertiary erosion surface, and an exhumed and dissected Precambrian surface. Outliers of Cambrian rocks indicate that relief on the Precambrian surface was about 400 feet; it has not been greatly changed since except in the glaciated high parts of the range to the north of the district.

## GEOLOGIC HISTORY

The readable geologic record in the Atlantic City district is but a fraction of the total history. Figuratively, we have been permitted to examine volume 2 and the last half of volume 5 of a 5-volume geologic history, each volume representing about 1 billion years, the earliest being number 1. The more recent events leading to the present topography have been briefly touched upon; our chief concern here will be with the ancient Precambrian rocks.

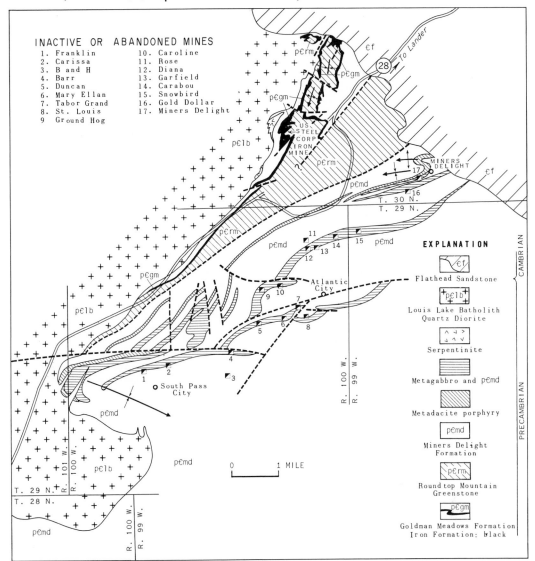

FIG. 2. *Generalized Geologic Map of the Atlantic City District, Fremont County, Wyoming. Tertiary and Quaternary deposits omitted. Heavy dashed lines indicate major faults.*

## Stratigraphy

Layered metasedimentary and metavolcanic rocks make up most of the Precambrian terrane. These were deposited on a gneiss and mafic volcanic basement in a geosyncline that probably extended northeastward to the limits of the State. Sedimentation began with a thin, basal, transgressive sand layer, followed by muds and chemically precipitated iron-formation. Thus, extraordinary tectonic stability and remoteness from any source of coarse clastic materials characterized the early phase of sedi-

mentation. In contrast, the later and final phase of sedimentation was characterized by tectonic instability, by submarine volcanism, and by great abundance of clastic materials, deposited mainly from turbidity currents. The total sequence of layered rock units deposited is as follows:

Miners Delight Formation 5000 to 10,000 feet — Graywacke (turbidites), schist, conglomerate, ellipsoidal greenstone (andesite)

Roundtop Moun-    Ellipsoidal greenstone
   tain Greenstone    (basalt) and green
   5000 feet    schist

Goldman Meadows    Schist
   Formation 700 to    Iron-formation
   800 feet    Schist
   Iron-formation
   Schist
   Quartzite

## Metamorphosed Intrusive Igneous Rocks

Intrusive igneous rocks of several kinds cut the above formations; the approximate order of their emplacement is as follows:

(1) Serpentinite (metaperidotite)

(2) Metadacite porphyry sills or flows (?)

(3) Metagabbro (diabase) dikes and sills

(4) Metaleucodacite and metatonalite dikes and stocks

(5) Metagabbro (diabase) dikes

(6) Louis Lake batholith, quartz diorite mainly, with pegmatitic and porphyritic granite phases

(7) Diabase dikes

## Early Deformation and Metamorphism

After the emplacement of the igneous rocks (1 through 5), but partly contemporaneous with the last intrusions of diabase, all the rocks of the district were tightly folded on north-easterly axes, regionally metamorphosed to a moderate degree, and displaced by major northeast-trending faults.

During this early deformation, all known gold-bearing quartz veins were emplaced in minor fault zones, in major shear zones between structurally competent and structurally incompetent rocks, and in fractures in intrusive igneous rocks, mainly metagabbro.

## Later Deformation and Batholith Emplacement

The early folding and metamorphism of the rocks was followed after some interval by the upwelling of vast quantities of quartz dioritic magma that regionally engulfed and destroyed all but a small fraction of the older rocks. The time of this event has been approximately determined by mineral dating as 2.5 to 3.3 billion years ago (23). Severe deformation attended the intrusion of the batholithic rocks; the marginal parts of the igneous bodies were strongly deformed, and the already-folded

country rocks, particularly west of South Pass, were refolded. Deformation had ended by the time the granitic rocks were hard enough to fracture and permit the venting of gases and fluids to form a suite of post-deformation pegmatites.

## Fresh Diabasic Gabbro

Regional uplift and a protracted period of erosion probably followed the intrusion of the granitic rocks. During this time a swarm of diabase dikes was emplaced in northeast-trending joints of regional extent.

## Late Faulting

Subsequent to the emplacement of diabase dikes, the rocks of the district were cut by major east- and northeast-trending faults that displaced some rock units as much as 2 miles. This faulting is believed to be of Precambrian age, but Laramide movement is known to have taken place on the eastern extension of at least one of the faults. Several of the late faults are marked by zones of crushed and bleached rock in which are minor quartz veins bearing chalcopyrite or its oxidation minerals. These minor veins are copper-rich in places and carry a trace of gold but are generally thin, discontinuous, and few in number.

## Lost Interval

Except for uplift and erosion, there is no record of geologic events in the district between the time of the late faulting and the time of the first transgression of the Paleozoic sea. The hiatus at the Middle Cambrian overlap represents at least 1 billion and probably as much as 2.5 billion years of Middle and Late Precambrian time and part of Early Cambrian time.

From Middle Cambrian to Early Paleocene time, nearly 3 miles of sedimentary rocks were deposited on the site of the Atlantic City District; thereafter, the Wind River Range was on the rise, and, by Eocene time, the Precambrian rocks were already exposed much as they are today.

## STRUCTURE OF PRECAMBRIAN ROCKS

Although the district covers more than 100 square miles, only secondary and lesser order structures are in evidence. The major part of the controlling first-order structure lies buried beneath Paleozoic and younger rocks east of the district. This master structure is inferred, from the exposed deformed rocks, to be a southwest-pitching synclinorium; the adjacent anticlinorium axis lies northwest of Iron Mountain in the area now underlain by the Louis Lake batholith.

The generalized structure of the district is shown on Figure 2. Some features important to the understanding of a structure but not explicit in that figure are as follows:

Most all rock units dip steeply, commonly within 20 degrees of the vertical.

The long northeast-trending fault on Figure 2 is a major fault and an early one. It formed after the first folding but before the intrusion of the batholith and the second folding. The Roundtop Mountain Greenstone and the Miners Delight Formation face each other across the fault, that is, up section in each is toward the fault. The northeast-trending fault between the Goldman Meadows Formation and the Roundtop Mountain Greenstone is overlain by Tertiary deposits. Its location, trend, and apparent movement have been inferred on the basis of a magnetic survey that appears to show an abrupt truncation of the iron-formation of the Goldman Meadows Formation where it bends around the nose of an anticline crossfold. Faults, which were contemporaneous with the early folding and are older than those discussed above, cut the several formations but are generally of small displacement. The metadiabase-filled faults shown on Figure 3 are of this older set, as are the quartz vein-bearing shear zones on either side of the metagabbro dike that defined the gold belt on Figure 2. The faults that displace the Miners Delight Formation and the included gold belt metagabbro and other dikes are all quite young, much younger than the major northeast faults. A few faults of this younger set cut the iron-bearing rocks at Iron Mountain and vicinity (Figure 3), discussed below.

Northwest of the gold belt or southernmost belt of metagabbro (Figure 2), the rock section seems fairly simple. Although beds are extremely plicated locally, there seems to be no notable repetition of beds by folding—except for folds in the northeast that have disrupted the iron-bearing rocks (Figures 2 and 3). But south of the metagabbro belt referred to, tight, closely-spaced, isoclinal folds, which trend generally east and northeast, (21, 22), are characteristic. These isoclinal folds are recumbent in the vicinity of South Pass City (Figure 2) and are refolded just west of there into a broad east-pitching synform and an adjacent domical antiform with a syntectonic granite core (Figure 2).

The structure at Iron Mountain and vicinity is rather complicated and deserves special consideration. Despite the large size, the basic structure is steeply pitching drag fold, the anticlinal part of which was halved by faulting, as noted above (Figure 2). The limbs of the fold, as defined on the basis of the key iron-formation, are nearly vertical and are even slightly overturned at a few places. The iron-formation is intensely plicated internally. The plications and larger fold elements plunge 70° to 90°SW, and a considerable extension of the limbs in the plunge direction is indicated by mineral lineations, by elongated basalt pillows, and by boudinage structures. The syncline was compressed and shortened, in the longitudinal sense, probably after it had attained its form and probably after the limbs were nearly vertical. There are several indicators of the shortening: (1) East to northeast cross folds have affected opposite limbs in parallel and thus have caused parallel sinuosities in the axial plane of the syncline. The structure at Iron Mountain is dominated by folds of the above class that have parallel counterparts in the limb opposite (Figure 2). All such folds on the west limb of the main syncline are turned, in a sense, opposite to that expected if genetically related to the formation of the original syncline; consequently, the plunging crests and troughs are upside down. (2) Imbrication and telescoping of limbs along what appear to be lateral and reverse faults have caused shortening of the limbs of the syncline at a number of places. Nearly all of these fault zones were invaded by basaltic magma, probably while the movement was taking place. (3) Shortening and thickening of the iron-formation as a result of internal folding and plication has taken place at many places but most particularly at Iron Mountain. The plications are systematic, generally southwest plunging, and indicate a drag-fold-type coupling of the iron-formation that at most places resulted in stress release by faulting.

The enormous concentration of iron-formation at Iron Mountain, which makes up the U.S. Steel Corporation's iron ore body, is structural in origin, as indicated above, and not due to any stratigraphic thickening of the iron-formation. Note that, where the iron-formation enters and emerges from the Iron Mountain structure, it is of average thickness, about 150 feet (Figure 3). Between the points of entry and egress, an anticline-syncline cross

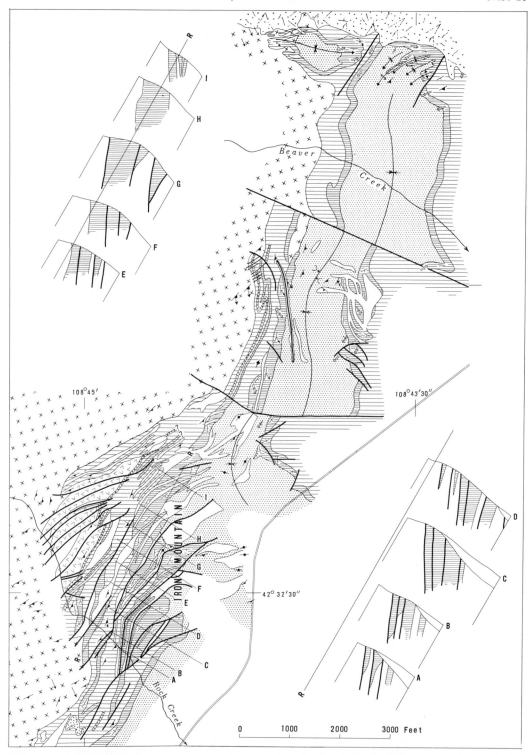

FIG. 3. *Geologic Map and Cross Sections of Precambrian Iron Deposits near Atlantic City. Tertiary and Quaternary deposits omitted. Cross sections spaced arbitrarily along reference line R. Note, in the explanation, that the metasedimentary rocks are not divided on the basis of formations. The upper schist includes some rocks of the Goldman Meadows and Roundtop Mountain Formations. The granite includes mainly quartz diorite, as well as some granite, of the Louis Lake batholith.*

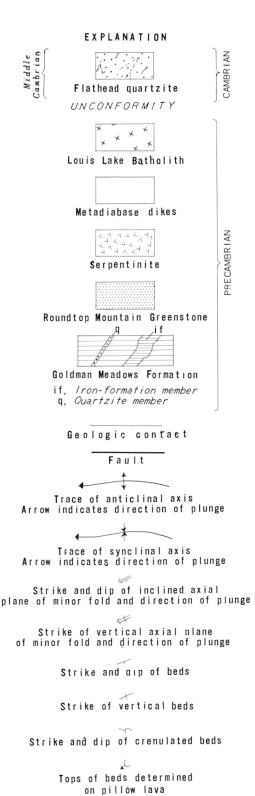

EXPLANATION

Middle Cambrian

Flathead quartzite

*UNCONFORMITY*

CAMBRIAN

Louis Lake Batholith

Metadiabase dikes

Serpentinite

PRECAMBRIAN

Roundtop Mountain Greenstone

q        if

Goldman Meadows Formation

if, *Iron-formation member*
q, *Quartzite member*

Geologic contact

Fault

Trace of anticlinal axis
Arrow indicates direction of plunge

Trace of synclinal axis
Arrow indicates direction of plunge

Strike and dip of inclined axial
plane of minor fold and direction of plunge

Strike of vertical axial plane
of minor fold and direction of plunge

Strike and dip of beds

Strike of vertical beds

Strike and dip of crenulated beds

Tops of beds determined
on pillow lava

Strike and dip of beds and
direction of plunging lineation

Strike and dip of foliation

Strike of vertical foliation

Plunge direction of lineation

Strike and dip of foliation
and direction of plunging lineation

fold is superposed on the original syncline. Compaction and thickening of the iron-formation in this fold zone took place in several ways: First, and probably most important, was by movement, probably under high compression, of iron-formation into the crest area of the anticlinal part of the fold couplet. Inasmuch as the crest of the anticline was already occupied by iron-formation, the movement took place by internal accordion-type folding that, in effect, shortened the strike length of the iron-formation and thickened it about fourfold. The second way in which the iron-formation was compacted and thickened was second in sequence as well. Axial-plane cleavage developed across the already-folded iron-formation. The iron-formation appears to have been repeated by upward and southwestward movement along some cleavage planes, particularly in the north and south parts of the Iron Mountain structure. Cross sections A, B, and I (Figure 3) indicate how effectively the reverse faults on cleavage planes have amassed iron-formation at the present surface. Nearly equal-spaced cleavage planes were invaded by basaltic magma, now metagabbro, probably at the time of movement. The magma issued up through the breached trough of the original syncline, where it formed massive dikes in the upper schist member of the Goldman Meadows Formation and the greenstones of the Roundtop Mountain formation. The dikes are 200 to 400 feet thick in the schist and greenstone; they neck down to only 10 to 25 feet thick where they cross the somewhat more competent iron-formation; a few flare out and thicken again in the schist on the footwall side of the iron-formation.

Reverse faults, of much younger vintage than the cleavage faults just described but of similar strike and dip, provide the third way in which the bulk of the iron-formation at Iron Mountain may have been increased. Particularly in the south part of the ore body, large blocks of iron-formation appear to have moved up on these faults; however, it has not been possible to determine whether or not this faulting actually resulted in a net gain of iron-formation near the surface. The vastly different ages of the early and late faults at Iron Mountain are indicated by the fact that the younger faults cut granitic rocks west of Iron Mountain that were not intruded until after the basic Iron Mountain structure had formed. Further, it is known from age-dating of biotites from the granitic rocks deformed by the younger faults and from granitic rocks away from the faults and not so affected, that nearly 1.6 billion years elapsed between the time the granitic batholith cooled and the time faulting occurred.

## ECONOMIC GEOLOGY—IRON

### Primary Iron Ore

The iron-formation members of the Goldman Meadows Formation are the only source of iron ore in the district. These iron-formations, so-called "taconites," are metasedimentary rocks, are rather precisely fixed stratigraphically, and react in a characteristic way to deformation and metamorphism.

The distribution of iron-formation is shown on Figure 2 and more exactly on Figure 3. For more detail, there is a published map at 1 inch = 500 feet (17).

The iron-formation is a hard, dense, laminated rock of dark color, commonly gray and black. Alternate layers are iron-rich (dark) and quartz-rich (light) and generally less than 1 cm thick. Quartz and magnetite make up 90 per cent of the rock, and amphibole (grunerite or actinolite), chlorite, and garnet are sparingly present. The iron-formation is very fine grained; the microtexture is granular or hornfelsic. Individual quartz crystals, averaging about 0.2 mm across, are about twice the size of the magnetite crystals, which average about 0.12 mm. The average iron content of the iron-formation as indicated by 150 analyses is 33.5 per cent; the silica averages about 50 per cent. Minor elements, such as Ti, P, and S, which determine the quality of the ore, average 0.025, 0.046, and 0.011 per cent respectively. The average specific gravity of 10 representative samples is 3.4. A chemical analysis of a sample chipped from resistant layers across 200 feet of outcrop on Iron Mountain (Figure 3) is presented below. An analysis representing the weighted average of

TABLE I.  *Chemical Analyses of Iron-Formation*

|                  | A      | B            |
|------------------|--------|--------------|
| $SiO_2$          | 56.23  | 47.32        |
| $Al_2O_3$        | 0.45   | .24          |
| $Fe_2O_3$        | 34.96  | 26.72        |
| FeO              | 5.67   | 14.87        |
| MgO              | 1.13   | 2.16         |
| CaO              | 0.81   | 1.26         |
| $Na_2O$          | 0.15   | n.d.         |
| $K_2O$           | 0.12   | n.d.         |
| $+H_2O$          | 0.52   | 1.39         |
| $TiO_2$          | 0.02   | n.d.         |
| $P_2O_5$         | 0.05   | .067         |
| MnO              | 0.07   | .4 estimated |
| $CO_2$           | 0.06   | 4.18         |
| Cl               | —      | n.d.         |
| F                | —      | n.d.         |
| S                | —      | n.d.         |
| C                | n.d.   | 0.05         |
|                  | 100.24 | 98.66        |

A. Analysis of iron-formation from sec. 23, T. 30 N., R. 100 W., Fremont County, Wyoming, by Dorothy F. Powers, U.S. Geological Survey.

B. Weighted average of 12 analyses of banded magnetite iron-formation, from Biwabik Formation, Mesabi District, Minnesota, after Gruner (7, Tables 5, 7, and 8).

12 analyses of banded magnetite iron-formation from the Precambrian Biwabik Formation, Mesabi district, Minnesota, given by Gruner (7, Tables 5, 7, and 8), is shown for comparison.

The rocks represented by the above analyses are approximately equivalent iron-formation facies (magnetite-banded iron-formation of James, 11, p. 261), but the Mesabi analyses indicate that the magnetite-banded rocks of the Biwabik probably contain more iron silicates than the Wyoming example, and, probably primary carbonate, whereas the Wyoming example does not.

### Origin of Iron Ore

Granted that the iron-formation is a product of chemical sedimentation, the primary or diagenetic mineralogy, which, at sedimentation, is closely controlled by the sea-bottom chemical environment (14), will very precisely control what minerals form during thermal metamorphism (13). Actually, an infinite variety of starting mixtures of chert, magnetite, hematite, iron silicate, and iron-carbonate could, upon metamorphism, form banded magnetic iron-formation; but the great preponderance of quartz and magnetite in the metamorphosed Wyoming iron-formation and the lack of any granular or oolitic structures suggest strongly that the original starting material was nearly all chert and diagenetic magnetite. The metamorphism has merely coarsened these chief constituents and formed new silicates (amphibole, chlorite, garnet) from an original silicate (possibly chamosite) and probably some iron carbonate.

### Iron Ore Bodies

The main ore body of the district lies along the north trend of Iron Mountain, which crests about 500 feet above the valley of Rock Creek and is eminently well situated topographically for open-pit mining. The structural origin of the body has been described (p. 16), and the body is depicted in three dimensions on Figure 3. There are three separate and structurally different parts to the ore body: (1) an arcuate north ore body (north of cross section E, Figure 3), which, though faulted, seems to have maintained its structural continuity, and therefore may be expected to continue downward to great depth; (2) a small faulted-in ore body of shallow depth (east half of cross section G, Figure 3); and (3) a south ore body which is separated from the others by faults and a belt of schist which is faulted internally, and is of great but limited depth (cross section A to D, Figure 3). The several ore bodies are ribbed by metagabbro dikes, as indicated in the figure. Also, there are many minor schist bodies sliced in by faults and infolded with the iron-formation close to its upper and lower contacts. These waste rock inclusions in the ore bodies necessitate close, daily supervision at the mining face to maintain a uniform mill feed.

Published tonnage and grade figures for this mine area are as follows:

Mining World (16, p. 27):

121 million tons (25 to 32 percent Fe) proved reserves
300 million tons (22 to 35 percent Fe) indicated ore

Engineering and Mining Journal (24, p. 84):

111 million tons (30 percent Fe) proved reserves

Several areas of iron-formation, which lie northeast of Iron Mountain (Figure 3) and which, because of their small bulk, would not have been considered ore before the development at Iron Mountain, may now be classed

as ore because of that development. These potential mine areas are believed to be owned by the U.S. Steel Corporation. The longer haul from these deposits to the present mill south of Rock Creek would be the only added cost. The larger of these areas of iron-formation outcrop are on hills and are very suitable for open-pit mining. They will produce only millions of tons of iron-formation in contrast to the tens of millions of tons expected to be mined at Iron Mountain, but properly utilized they will lengthen the life of open-pit mining in this district by 3 to 5 years at least, assuming 3 million tons per year volume, and longer at a reduced rate.

### Summary of Mining and Milling at Iron Mountain

The iron-formation is mined from an open pit benched each 75 feet to maintain an overall 1:1 slope. Electrically powered drills are used for blast holes, and electrically powered 6-yard shovels load 45-ton, Diesel-powered trucks which haul about 4000 feet to the primary crusher plant. Cone crushers, rod mills, and ball mills reduce the ore to about 90 per cent −270 mesh. A straight magnetic separation is made in two stages over barrel-type magnetic cobbers that utilize permanent magnets. The magnetic concentrate, then about 65 per cent iron and 9 to 12 per cent silica, is pelletized and roasted. See Engineering and Mining Journal (24) for mill details.

### ECONOMIC GEOLOGY—GOLD

The gold mined in this district has come from quartz veins and from placer deposits derived from them.

### Quartz-Gold Veins

The quartz veins generally follow the grain of the wall rocks and occupy zones of shearing and very commonly are extremely sheared themselves. The major shear zones dip steeply at most places, as do the enclosing strata. With very few exceptions, the gold-bearing veins are restricted to a narrow, interrupted belt of sill-like metagabbro bodies that trends northeast across the district (see mine locations 1 to 15, Figure 2). This close spatial relationship of veins to the metagabbro belt is not genetic but structural. Shearing during folding, between the metagabbro and adjacent graphitic and micaceous schists and along schist inclusions in the metagabbro, provided the accessways for the quartz veins. The preference of the metagabbro for this particular belt of rocks seems to lie in the fact that the rocks of the belt, namely graphite and other schist, basaltic flows and agglomerate, and graywacke conglomerate, provided easier access to gabbro magma than the surrounding ever-bedded graywacke. At other places, however, particularly in the vicinity of the Gold Dollar and Miners Delight mines (locations 16 and 17, Figure 2), the gabbro intruded the graywacke. The point to be stressed is that most veins occupy shear zones formed within the structurally very competent metagabbro belt or between the metagabbro and the less competent schist and graywacke on its flanks. These shear zones are, at most places, demonstrably older than the major faulting, which caused extensive shearing and brecciation but contain only very local and minor copper-gold mineralization. Locally, as at locations 1, 3, and 11 (Figure 2) gold-bearing quartz veins occur in the graywacke away from any metagabbro.

The quartz veins have been classified as to shape and attitude by several people in various ways; I will not add to the already-complicated terminology. Geologic mapping has shown that almost all veins of any economic interest are strike veins; that is, they conform in strike to the trend of the country rocks. Moreover, they occur in mappable zones of shearing, faults, and joints, on the surface and underground.

The veins generally do not represent fracture fillings; they are not concretionary or zoned. Although shearing subsequent to emplacement has almost everywhere obscured their original relations to the country rocks, replacement of country rocks by quartz is indicated in a few places, particularly where the veins cut metagabbro. The main production of gold has been from veins or leads composed of numerous veins 2 to 7 feet thick. Large veins, up to 60 feet wide occur, particularly along the north margin of the general gold belt, but these large veins have proved gold poor and, although extensively prospected, have not been mined.

## Mineralogy of Veins

Although hundreds of veins were examined during geologic mapping, nothing significant can be added to previous descriptions of their mineralogy. Surface exposures show mainly quartz, some of it massive and clean but more often interlayered with splits of weathered country rock of various kinds (mainly ferruginous or graphitic dirt) and the whole

very much iron stained. Some quartz is cellular and vuggy; some shows limonite-filled pits, the sites of oxidized sulphide crystals. During sampling the amount of iron oxide present was thought indicative of the amount of sulphides once present, and hence the amount of gold present, but the results of 100 assays indicate no close relationship between the amount of iron oxide in surface exposures and the amount of gold present. Quite unoxidized specimens of vein material from stock piles and mine dumps show white or gray or black quartz, pyrite, arsenopyrite, calcite in some areas, and uncommonly chalcopyrite or green copper oxides. Most of the ore specimens examined have been exposed to the weather for many years. They smell of arsenic and show a dusty yellow efflorescence, that is presumed to be an arsenic compound. A few samples showing this efflorescence were assayed and found to contain gold; the gold in one was valued at $50 per ton. According to Spencer (1) pyrrhotite and galena are subordiate minerals in the ore; however, he observed galena only at the Mary Ellen mine. Cannon (23) found a little galena at the Snowbird mine. Neither pyrrhotite nor galena was found during this investigation, and spectrographic analyses of 100 vein samples indicate no extraordinary concentrations of lead. Though generally microscopic in size, tourmaline is a common mineral of the gold veins and wall rocks close to veins and is abundant enough locally be blacken the vein quartz.

## Chemical Composition of Veins

Amounts of major elements in the quartz leads sampled differ extremely because of contamination of wall rocks; that the ores have, however, a rather distinctive signature with respect to the concentrations of certain minor elements is indicated by combined spectrographic analyses, colorimetric analyses (As), and fire assays (Au, Ag) for 17 samples found to contain gold. The signature is given below and is expressed in terms of observed ranges of enrichment of certain elements with resepct to the amounts of those elements normally found in silicic igneous rocks (12).

## Nature of Ore-Bringing Solutions

All of the elements listed above, except Au, As, Ag, B, Bi, Cu, and Mo, could have been present in the maximum concentrations shown in the metagabbro and other mafic country

TABLE II.    *Minor Element Signature of Atlantic City Gold Ores*

| Element | Acid Igneous Rocks (average per cent) | 18 Gold-Bearing Veins (minimum and maximum times average per cent acid igneous rocks) |
|---|---|---|
| Au | .000001 | 60x—4600x |
| As | .00015 | <2x—10,000x |
| Ag | .000015 | <2x—>70x |
| B | .0015 | 2x—100x |
| Bi[1] | .0002 | ? —10x |
| Co | .00051 | >1x—5x |
| Cr | .0025 | >1x—5x |
| Cu | .0031 | 5x—10x |
| Mo | .0002 | >1x—10x |
| Ni | .0008 | 10x—20x |
| Sc | .0007 | >1x—10x |
| V | .004 | >1x—5x |

[1] Four samples from area of Carissa mine.

Other minor elements such as Pb, Sr, Y, Yb, and Zr occur consistently in the ores in amounts not exceeding those normally found in acid igneous rocks.

rocks; it is not necessary to assume, therefore, that they were imported. Thus, it is possible to arrive at a simplified hypothetical ore-bringing solution consisting mainly of water vapor plus other gases, a few major elements, namely Si, Fe, and K, and the minor elements Au, As, Ag, Bi, B, Cu, and Mo. The major components of this medium suggest that the whole was derived from a differentiating primary igneous source. Because all of the igneous rocks of the district except for Louis Lake batholith and some late diabase dikes are pre-ore and pre-metamorphism, it is assumed that the batholith, at some stage of development prior to intrusion, caused the metamorphism and supplied the mineralizing solutions; but either or both could have been caused by some subterranean body not exposed anywhere.

It should be noted here, for the purpose of comparing this district with others, that the leucodacite dikes and stocks, mentioned earlier, are spatially closely related to a few of the gold deposits and are locally mineralized themselves. Some of the leucodacite is quartz-oligoclase porphyry, and the soda content is unusually high (see analysis below).

Probably these porphyries correspond to the quartz-albite porphyries commonly found in other similar gold districts (6,15). They are undoubtedly pre-ore in this district, though probably the last material intruded before the gold mineralization.

## Wall-Rock Alteration

Wall-rock alteration is slight and not megascopically apparent. Thin sections of wallrocks from several veins show mainly either sericitization or replacement of all preexisting feldspar by untwinned microcline, and some rock replacement by quartz, calcite, and tourmaline. At the B & H mine, where the wall rocks were sampled in detail, there is total replacement of oligoclase by microcline in the metagraywacke close to the quartz lead. The alteration decreases outward and disappears about 6 feet from the quartz lead, which is itself 1 to 6 feet wide.

Thin sections from many leads and their wallrocks show that generally the wall-rock septa within the leads are mineralogically similar to the wallrock adjacent to the leads except that the in-vein metamorphic minerals, mainly biotite, feldspar, and hornblende, are much larger and show better cystal forms.

TABLE III.    *Chemical Analyses of Leucodacite Porphyry and Leucotonalite*

| | A | B |
|---|---|---|
| $SiO_2$ | 69.64 | 72.9 |
| $Al_2O_3$ | 15.43 | 14.7 |
| $Fe_2O_3$ | .92 | .8 |
| FeO | .89 | 1.3 |
| MgO | .63 | 1.0 |
| CaO | 1.89 | 1.1 |
| $Na_2O$ | 5.12 | 5.7 |
| $K_2O$ | 2.21 | 1.0 |
| $H_2O$ | 1.11 | .88 |
| $TiO_2$ | .35 | .31 |
| $P_2O_5$ | .10 | .11 |
| MnO | .03 | .04 |
| $CO_2$ | 1.18 | <.05 |
| Cl | — | n.d. |
| F | — | n.d. |
| S | .02 | n.d. |
| | 99.52 | 100. |

A. Leucodacite porphyry, veined by quartz and containing arsenopyrite crystals. Assays .01 oz. per ton Au and .01 oz. per ton Ag. Sec. 36, T. 30 N., R. 100 W., Fremont County, Wyoming. Dorothy F. Powers, analyst, U.S. Geological Survey.

B. Leucotonalite. Sec. 7, T. 29 N., R. 99 W., Fremont County, Wyoming. Rapid analysis by P. L. D. Elmore, I. H. Barlow, S. D. Botts, and Gillison Chloe, U.S. Geological Survey.

## Temperature of Ore Formation

The wall-rock alteration suggests that the temperature of the mineralizing solutions was quite high, but not significantly higher than the temperature of the contemporaneous middle-grade regional metamorphism that affected all the rocks. The mineralogy of the metagray-wackes suggests that they recrystallized near the low-temperature limit of the amphibolite facies, about 400°C (10); the in-vein minerals suggest no higher temperature, but the environment was doubtless more fluid, hence the larger metamorphic minerals.

## Age of Gold Deposits

Geologically the gold veins were emplaced early in the deformational history of the district and very probably pre-Louis Lake batholith, even though the batholith may have been the source of the gold. A galena sample from the Snowbird lead (loc. 15, Figure 2), analyzed by Cannon (23) gave the oldest model lead age yet found in the United States, 2.75 billion years. The Louis Lake batholith, which is postkinematic with respect to the early folding and synkinematic with respect to the second folding and definitely younger than the gold mineralization has been dated by various methods at 2.2 to 3.3 billion years old. The gold deposits therefore cannot be younger than early Precambrian.

## Tenor of Gold Veins

As in most lode gold districts of this type, the Atlantic City lodes are commonly lean but locally contained rich ore shoots. It is generally accepted that the oxidized near-surface ores were somewhat enriched. The early production records would seem to indicate extreme enrichment of the oxidized shallow ores, but it is suspected that much of the early mining was done by very selective gophering on rich leads—thus, the apparent value of the surface ores is too high. Spencer's (4) comments of the gold content of the ores and the persistence of veins seem as valid today as they were in 1916 and are repeated as follows.

"Gold content of the ores. Very little definite information can be given concerning the amount of gold carried by the ores that have been mined from the veins of the district. Notes in Raymond's reports indicate that the ores mined prior to 1872 returned as a rule from $20 to $40 a ton under treatment by stamp milling and simple amalgamation. Some ores were worked that yielded only $15 a ton, and occasional lots yielded as high as $200 a ton. 'The average yield of several thousand tons of ore from different lodes has been from $30 to $40 per ton.'[1] Among the mines mentioned by Raymond in 1870 are the Carrie Shields, Young America, Carissa, Golden Gate, Wild Irishman, Gold Hunter, Calhoun, Duncan, Mary Ellen, Barnaba, Buckeye State, Soules & Perkins, Caribou, Miners Delight, and Bennet Line.

"Between July 20 and November 1, 1868, the original Hermit mill, near South Pass City, treated 1,040 tons of ore yielding an average of $36 a ton; and between April 20 and July 1, 1869, it crushed 480 tons of ore, averaging $47 a ton.[1] [Raymond, R. W., Statistics of mines and mining in the States and Territories west of the Rocky Mountains for 1869, p. 330, 1870].

"Ore from the Miners Delight vein is said to have yielded from $16 to $200 a ton, the average having been about $40.

"Perusal of old reports relating to the district leaves an impression that many of the lodes occurring in the district might be expected to yield ores carrying (sic) on the average as much as $20 a ton. It is believed, however, that as a general rule this expectation will not be realized, though with little doubt shoots of rich ore will be found. It is more likely that assays will show averages of $6 to, say, $15 a ton.

"The records of production in possession of the Geological Survey, which are held as confidential with respect to individual mines, indicate that 11,105 tons of ore produced by eight mines since 1902 yielded an average of $8.15 a ton. (See list of assays, p. 40.) On the face of the returns, if a few tons of exceptionally high-grade ore are left out of consideration, the yield appears to have ranged from $5 to $9.18 a ton. The figures given represent the metal recovered, and there is no way of ascertaining the actual gold content of the ores as mined. In so far as gold and silver yields are both given the ratio of gold to silver is found to vary from 5.03 to 10.02, the weighted mean ratio for 8,958 tons of ore being 6.79, which corresponds to a fineness in gold of 0.871."

"Probable persistence of the veins. From a practical man's point of view, the first importance naturally attaches to the question whether or not the veins and other gold-bearing deposits of the district will be found to persist to great depth, and if so whether they will continue to carry about the same amounts of gold as near the surface.

"In regard to physical persistence, the conclusions may be drawn that these deposits are of deep-seated origin, that the present topographic surface is a chance surface due to erosion, without significant relation to the ore deposits, and that on the whole the deposits must be abundant at any depth that might be chosen for consideration as they are at the existing surface. Although these general conclusions are fully warranted by the geologic features of the district, they should not be taken as a guaranty that all the veins of the district will be found to be continuous to indefinite depths. It is probable that the lodes showing long outcrops, like the Carissa and Miners Delight, persist to great depths, whereas it would not be surprising if lodes that can be traced at the surface for very short distances are found to pinch out at correspondingly moderate depths. On the whole the writer is inclined to believe that in this district strike veins, if well defined, are likely to prove more persistent than cross veins.

"As to the downward continuance of gold content, though it is likely that there has been some enrichment through solution and redeposition in the oxidized portions of the lodes, it is not believed that any really large proportion of the gold in the upper parts of the veins has been secondarily precipitated from surface solutions; and no hesitation is felt in stating the conclusion that the occurrence of valuable ores is not limited to a shallow zone. It may be expected that here, as is the rule in other districts, different parts of the same lode will be found to carry varying amounts of gold, or, in other words, that in any vein the best ore will be found in the form of shoots.

"The foregoing conclusions and suggestions indicate the writer's belief that the district is worthy of further development."

## Copper-Gold Veins

Thin quartz-calcite veins, bearing chalcopyrite occur in shear zones related to the major faults. The faults displace the gold-quartz veins described above, and therefore the copper-gold veins in the fault zones represent a separate and later epoch of mineralization. The veins are commonly only a few inches thick. They have been extensively prospected by test pitting, but nothing has been produced from them. Assayed samples from three test pits show traces of gold, and spectrographic analyses show copper up to 5 per cent. The copper is in chalcopyrite and malachite. Limonite stains most exposures and most chalcopyrite seen is partly altered to limonite. The wallrocks adjacent to this set of veins are mostly sheared and brecciated sericite-altered graywacke. These rather soft and bleached wallrocks are found in zones up to 100 feet wide along certain of the major east-trending faults.

## Future of Gold Mining

It is difficult to find any sound basis for evaluating the future of gold mining in this district. The record is incomplete and unimpressive, in comparison with important gold districts. What is known about the cost of mining and, in exceptional instances, the cost of acquisition of claims and the cost of construction of mills and other paraphernalia (and there were a great many mills) leads one to suspect that the total expense may have exceeded the value of the total gold production.

The recent geological mapping indicates that most of the exposed lodes have been tested and that the possibility of finding new ones is remote. Also, on a district-wide basis, most exposed veins are too lean to be of economic interest. However, the available production records suggest that the major vein systems, i.e., the Carissa, Carabou-Diana, and Miners Delight, were quite rich in gold. Each of these systems is 0.25 to 0.5 miles long, as indicated by the geologic mapping; each may be expected to extend to great depth; but none has been mined below 360 feet. Necessarily then, any future gold mining in the district must be based on recovery of the partly oxidized and sulphide ores from the deeper parts of these vein systems, and possibly some others that seem less promising. Below about 200 feet, the district is still in its virginal state. Whether the deep ores will warrant mining remains to be learned. Exploration by drilling seems the most practical and direct way to the answer, but as yet (1966) it has not been tried.

## REFERENCES CITED

1. Nickerson, H. G., 1886, Early history of Fremont County: State of Wyoming, Historical Dept. Quart. Bull., v. 2, no. 1, (1924).
2. Knight, W. C., 1901. The Sweetwater mining district, Fremont County, Wyoming: Wyo. Univ., School Mines, Univ. Geol. Surv. Bull. 5, 35 p.

3. Jamison, C. E., 1911, Geology and mineral resources of a portion of Fremont County, Wyoming: Wyo. Geol. Surv. Bull. 2, 90 p.

4. Spencer, A. C., 1916, The Atlantic gold district and the North Laramie Mountains, Fremont, Converse, and Albany counties, Wyoming: U.S. Geol. Surv. Bull. 626, 85 p.

5. De Laguna, W., 1938, Geology of the Atlantic City district, Wyoming: unpublished Ph.D. thesis, Harvard University, n.p.

6. Gallagher, D., 1940, Albite and gold: Econ. Geol., v. 35, p. 698–736.

7. Gruner, J. W., 1946, The mineralogy and geology of the taconites and iron ores of the Mesabi Range Minnesota: St. Paul, Minn., Office of Commissioner of the Iron Resources and Rehabilitation and Minn. Geol. Surv., 127 p.

8. Armstrong, F. C., 1948, Preliminary report on the geology of the Atlantic City-South Pass mining district, Wyoming: unpublished Ph.D. dissertation, Univ. Wash.; U.S. Geol. Surv. open-file report, 65 no 1. [Washington, D.C.].

9. King, R. H., 1949, Iron deposits near Atlantic City, Wyoming: unpublished report, Natural Resources Research Institute, Univ. Wyo., Wyo. Geol. Surv., 10 p.

10. Rosenquist, I. Th., 1952. The metamorphic facies and the feldspar minerals: Bergen, Museum, Årbok 1952, Naturvidenskapelig Rekke, no. 4, 116 p.

11. James, H. L., 1954, Sedimentary facies of iron-formation: Econ. Geol., v. 49, p. 235–293.

12. Vinogradov, A. P., 1956, Regularity of distribution of chemical elements in the earth's crust: Geokhimiya, 1956, v. 1 p. 6–52.

13. Yoder, H. S., Jr., 1956–1957. Isograd problems in the metamorphosed iron-rich sediments: in Ann. Rept. Dir. Geophys. Lab., Year Book 56, Carnegie Inst. Washington, p. 232–237.

14. Huber, N. K., 1958, The environmental control of sedimentary iron minerals: Econ. Geol., v. 53, p. 123–140.

15. Ward, H. J., 1958, Albite porphyries as a guide to gold ore: Econ. Geol., v. 53, p. 754–756.

16. Anon, 1960, Wyoming taconite project to get underway: Min. World, v. 22, no. 8, p. 27.

17. Bayley, R. W., 1963, A preliminary report on the Precambrian iron deposits near Atlantic City, Wyoming: U.S. Geol. Surv. Bull. 1142-C, 23 p.

18. State Board of Equalization of Wyoming, Ad Valorem Tax Department, 1963–64, Twenty-third Biennial Report of the State Board of Equalization.

19. Bayley, R. W., 1965, Geologic map of the Miners Delight quadrangle, Fremont County, Wyoming: U.S. Geol. Surv. GQ-460, 1:24,000.

20. ———— 1965, Geologic map of the Louis Lake quadrangle, Fremont County, Wyoming: U.S. Geol. Surv., GQ-461, 1:24,000.

21. ———— 1965, Geologic map of the Atlantic City quadrangle, Fremont County, Wyoming: U.S. Geol. Surv., GQ-459, 1:24,000.

22. ———— 1965, Geologic map of the South Pass City quadrangle, Fremont County, Wyoming: U.S. Geol. Surv., GQ-458, 1:24,000.

23. Cannon, R. S., Jr., et al., 1965, Ancient rocks and ores in south-central Wyoming (Abs.): Rocky Mountain Section. Geol. Soc. Amer., 8th Ann. Meeting, Fort Collins, Colorado, May 1965, p. 27.

24. Anon., 1965, U.S. Steel's Atlantic City Ore Mine, first taconite producer in the west: Eng. and Min. Jour., v. 166, no. 3 p. 73–92.

# 29. Multiple Intrusion and Mineralization at Climax, Colorado

STEWART R. WALLACE,* NEIL K. MUNCASTER,†
DAVID C. JONSON,* W. BRUCE MACKENZIE,‡
ARTHUR A. BOOKSTROM,† VAUGHN E. SURFACE†

## Contents

\* Climax Molybdenum Company, Golden, Colorado.
† Climax Molybdenum Company, Climax, Colorado.
‡ Climax Molybdenum Company, Urad, Colorado.

# Illustrations

# Tables

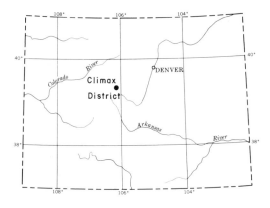

## ABSTRACT

In mid-Tertiary time a wet silici-alkalic magma penetrated the Precambrian rocks of what is now the Tenmile Range of Central Colorado and formed the Climax Stock. The stock is a composite one and was intruded in four separate irruptions, each giving rise to a major hydrothermal event. The first three of these produced large stockwork ore bodies that cap and flank the tops of their respective source intrusives; the fourth, and last, yielded almost no commercial mineralization.

The ore bodies and their related zones of alteration are circular to ring-shaped in plan and arcuate in section. The different phases of the stock are so positioned with respect to each other that many of the circular and arcuate zones of mineralization and alteration overlap or are juxtaposed. Evidence of a time separation between the different ore bodies is found in certain dikes that were intruded between periods of mineralization and are well mineralized in one ore body but are barren in another.

If it is assumed that the parent magma at Climax contained 10 per cent "water," that the "water" contained 0.1 grams per liter of molybdenum, and that the processes of extraction were 100 per cent efficient, then approximately 10 cubic miles of magma would be required for the formation of each ore body. The volume requirements, taken in conjunction with the cross-sectional areas of the different phases of the stock to explored depths, suggest to the authors that each magma chamber was columnar, extended to a depth of thousands of feet, and was connected with a much larger master reservoir. The repetition of similar intrusive, structural, and hydrothermal events, indicates periodic tapping of this reservoir. A regular chemical evolution of the parent magma is suggested by the systematic change in metal ratios and differences in the sites of deposition, relative to the source intrusive, in each of the four separate igneous-hydrothermal stages.

The preparation of ground was due to changes in magmatic and/or hydrothermal pressures within the igneous-hydrothermal columns; small, repeated, up and down movements of the rock enclosing the tops of the columns formed the stockwork fractures into which the mineralizers penetrated.

The forceful emplacement of each successive phase of the Climax Stock resulted in a slight doming and rotation to the westward of the earlier-formed geologic features. Post-ore normal displacement along the Mosquito fault is estimated at about 9000 feet. Part of one ore body was cut by the fault and lies at depth below the Paleozoic sedimentary strata on the hanging wall. On the footwall side of the fault, subaerial erosion and glaciation removed a large part of one ore body and uncovered the top of a second.

## INTRODUCTION

The Climax molybdenum deposit is unique not because of its mineralogy, its general type, the tenor of its ore, or the size of its reserves; it is the combination of these features that places it among the truly unusual ore deposits of the world, and, as such, it has attracted the attention and aroused the interest of economic geologists and mining men for nearly half a century.

The deposit is a large mineralized stockwork, generally similar in overall aspect to many of the porphyry-copper deposits. In spite of this similarity, the geologist at Climax has two distinct advantages over most of his porphyry-copper counterparts in attempting detailed geologic analysis:

(1) Because of its physical setting, Climax is an underground mine rather than an open pit, and underground openings lend themselves more readily to detailed geologic mapping than do pit benches. Also, they last longer and can be re-examined from time to time as geologic concepts change and develop. For example, within the last eight years a number of workings dating back to the early 1920's have been re-mapped. These drifts are on the upper levels of the mine and can be restudied in light of geologic relations now exposed at depth.

(2) By all odds the greater part of the molybdenum content at Climax is present in the primary sulfide form. The mineralization thus exhibits its original shape, and the patterns of grade distribution can be studied in

detail and related to primary geologic causes or events.

During the last 10 years the mine has undergone a rapid and continuing expansion and development program, and a wealth of new geologic data has become available. Study of this information, coupled with a re-examination of earlier records, has led to the development of the concepts presented here. Ideas have changed repeatedly and no doubt will continue to do so as knowledge expands with continued study; hopefully, these changes will not require major revision of the present conclusions.

Since the first comprehensive work of B. S. Butler and J. W. Vanderwilt in 1934, (4) more than 25 geologists have engaged for various periods of time in the collection and interpretation of geologic data at Climax. These workers have contributed much, and though they are not mentioned individually, their efforts are gratefully acknowledged. The authors also wish to express their appreciation to the company management for permission to publish the information contained herein.

### Location and Physical Features

Climax is located midway along the Colorado mineral belt on the west slope of the Tenmile Range* in central Colorado (Figure 1). The camp is at an altitude of about 11,500 feet and lies astride the Continental Divide above Fremont Pass. Tenmile Creek, which heads in a cirque just above the mine, flows northward from the pass to the Colorado River drainage; the East Fork of the Arkansas River drains the south side of the Divide.

Development of present physiographic features probably started in the Middle Tertiary, following major uplift and faulting, and, by the close of the Pliocene epoch, the major drainages and mountain ranges had been established. During Pleistocene time, existing valleys and ridges were further sculptured and steepened by glacial ice. Present local relief is about 3500 feet.

### History of Production

A gray metallic mineral, variously identified as galena, graphite, and silver, was discovered in 1879 by Charles J. Senter amid the prominent iron-stained bluffs on the slopes of Bart-

---

* The Tenmile Range, not specifically shown on Figure 1, is a subsidiary range of the Park Range.

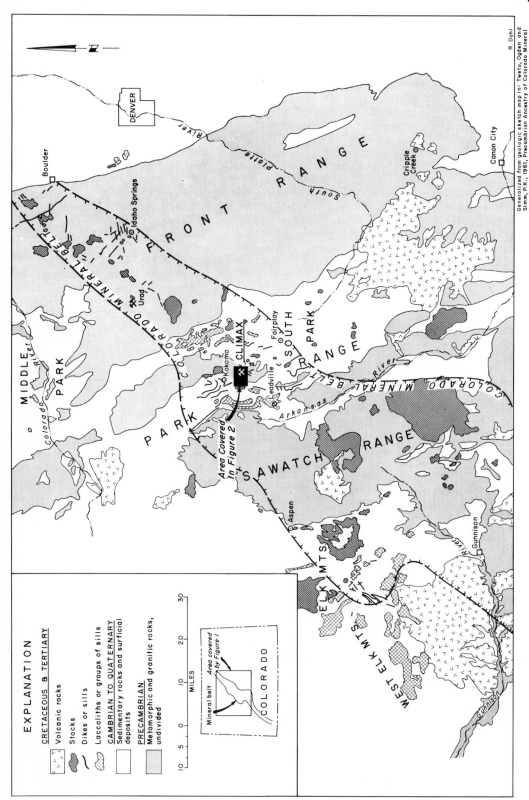

Generalized from geologic sketch map in: Tweto, Ogden and Simm, P.K., 1961, Precambrian Ancestry of Colorado Mineral Belt: Geol. Soc. of America Bull., vol 74, pp. 991–1014.

Fig. 1. *Relationship of Climax to the Colorado Mineral Belt.*

lett Mountain between the then active mining camps of Leadville and Kokomo. Sixteen years later, in 1895, the mineral was finally identified correctly as molybdenite by Professor George of the Colorado School of Mines. This knowledge did not, however, spur great activity. Although Bartlett Mountain undoubtedly contained large amounts of molybdenite, an effective method of separating the mineral from the rock had not been developed. Besides, who wanted molybdenum?

By 1912, the superior qualities that molybdenum imparted to special steels were well known to the tool-steel industry. This market was a small one, however, and, in 1913, it was estimated that the annual world market could be supplied by the molybdenum contained in a single ton of concentrates; the price was negotiable, in the neighborhood of $1.75 per pound. World War I greatly stimulated interest in molybdenum as a toughener for steel, and, from 1915 through 1917, several shipments were made from the showings on Bartlett Mountain to a custom mill in Leadville where trial runs were made using the new flo-

million tons per year until 1933. The last 25 years have seen a steady improvement in mining efficiency, and this, coupled with a strong demand for molybdenum, has led to an increase in production and a lowering of the economic cut-off. Today, the mill handles a daily average of more than 43,000 tons, grading something less than 0.4 per cent $MoS_2$. In 1957, approximately 40 years after initial production, the first hundred millionth ton of ore was processed. In January 1966, nine years later, the second hundred millionth ton was treated, and, during the year, the plant shipped the billionth pound of contained molybdenum. Present reserves are in excess of 400 million tons, and the ore bodies have by no means been "drilled out."

In 1948, a by-product plant was built to treat the tailings from the rougher molybdenite circuit. Since that time, about 18,000,000 pounds of $WO_3$, 800,000 pounds of tin and 400,000 tons of pyrite have been produced; monazite has been recovered in small amounts as required by demand. Production in 1966 is tabulated below.

TABLE I.    Production of Molybdenum and By-products, Climax, 1966

| Pounds Mo Contained | | Pounds $WO_3$ in Huebnerite Concentrate | Pounds Tin in Cassiterite Concentrate | Tons of Pyrite | Pounds of Monazite |
|---|---|---|---|---|---|
| In $MoS_2$ concentrate | 55,779,063 | 1,623,085 | 98,804 | 35,601 | 58,905 |
| In $MoO_3$ | 486,622* | | | | |
| | 56,265,685 | | | | |

* In August 1966, Climax started operation of a plant designed to recover molybdenum in the form of $MoO_3$ from oxidized ores. "Mixed" ore, containing both sulfide and oxide molybdenum as mined, is being produced from the near surface parts of two ore bodies; about 5700 tons per day are processed in the new oxide-treatment plant.

tation process as a method of recovery. The first real production muck was taken from the ground in February 1918, and the mine operated until March 1919 when it shut down for lack of a market. For the next five years, company research metallurgists worked steadily to find peacetime uses of molybdenum, and, by 1924, a small market had been developed in the steel industry. The mine re-opened in August of that year, and, except for two work stoppages, it has been in continuous operation ever since.

In the early years of operation, the mill feed averaged nearly 0.9 per cent $MoS_2$, and the tonnage treated was small—well under 0.5

## GEOLOGIC SETTING

### General Features

The Mosquito fault, a major break in the earth's crust, lies along the west side of the Tenmile Range and separates Precambrian crystalline rocks on the east from Paleozoic sedimentary rocks on the west (Figure 2). In the vicinity of Climax, the fault strikes northerly and dips steeply to the west; normal displacement, largely post-ore, is thought to be about 9000 feet. Early Tertiary porphyries of the type common to the Colorado Mineral Belt and generally ranging in composition from

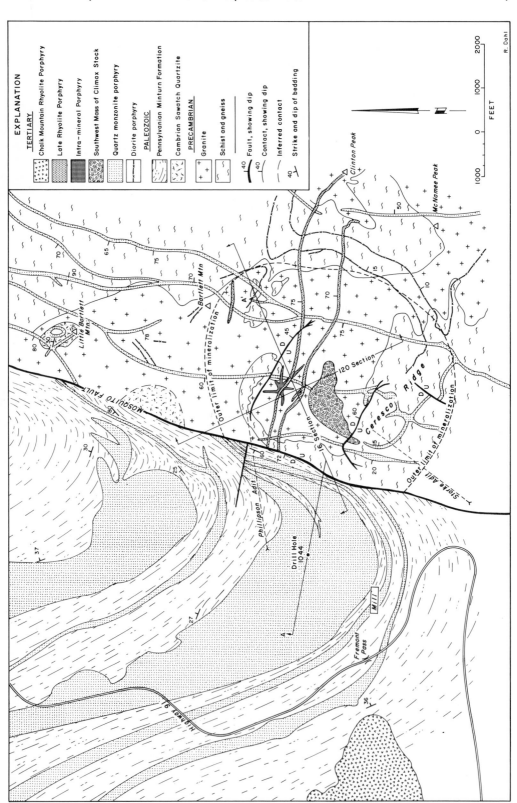

Fig. 2. *Generalized Bedrock Geology of the Climax Area.*

diorite to quartz monzonite have intruded rocks on both sides of the fault. The molybdenite mineralization at Climax is related to a composite stock of rhyolitic porphyry. This stock and its related dikes are mid-Tertiary in age and intrude the Precambrian crystalline rocks and the earlier Tertiary porphyries on the footwall side of the Mosquito fault. Molybdenite is present as three separate stockwork ore bodies, each genetically related to a different phase of the Climax Stock. Part of one body has been displaced by post-ore movement along the Mosquito fault and lies at depth beneath the Paleozoic cover on the hanging wall.

## Pre-Ore Rocks and Events

The oldest rocks of the area are Precambrian biotite-quartz-feldspar schists and gneisses of the Idaho Springs Formation. These rocks, in places, have been partly to completely transformed into masses of granitic rock by metasomatism, then injected with scattered mafic dikes (now metalamprophyre), and finally intruded by three separate, but closely related, pulses of Silver Plume Granite, all in Precambrian time. Prior to the development of the metasomatic granite, the schist and gneiss were thrown into a series of north-northeast-trending folds. Granitization of the metasedimentary rocks is thought to be late syntectonic with respect to this period of folding. Both the metasomatic granite, and the metalamprophyre dikes that cut it, were involved in a younger, and much less intense, period of deformation; minor fold axes and crinkles in the mine area trend northeasterly at angles of from 20° to 50° to trends of the older folds.* This deformation was followed by intrusion of the three phases of Silver Plume Granite. South of Climax, the intrusive granites form a stock; in, and immediately adjacent to, the Tenmile Cirque, the three successive phases are represented by small irregular masses and by dikes and flat- to gently dipping sheets.

A. H. Koschman and M. H. Bergendahl (11) mapped a series of major east-west trending folds in the Precambrian rocks of the northern Tenmile Range. Available evidence now suggests that this fold system continues

* Precise analysis of linear elements in the Climax area is impossible because of later disruption by granite intrusion, re-orientations of unknown extent and magnitude by intrusive doming in mid-Tertiary time, and still later flexing along the Mosquito fault.

southward to the vicinity of Climax and beyond; Climax appears to lie on a major synclinal axis of this fold system. The age of this period of folding is unknown.

The sedimentary record of central Colorado displays ample evidence of rapid lateral changes in thickness of formations and of erosion or nondeposition of strata during Paleozoic and Mesozoic time. The Paleozoic section at Climax is reasonably complete (Table II) though many of the formations appear to be thinner than they are to the south in the Leadville district or to the west in the Sawatch Range. Northward toward Kokomo, the Lower Paleozoic rocks thin rapidly, and the lower part of the Pennsylvanian-Permian Minturn Formation is missing. These features reflect the presence of a generally positive area during much of Early Paleozoic time and a pronounced highland during the Early Pennsylvanian. (Ogden Tweto, 5)

Mesozoic sedimentary rocks are exposed to the north in the Tenmile, Gore, and Williams River ranges and are present in South Park east of Climax. Presumably, they blanketed the area, but none is preserved at Climax, and their thickness at the time of ore formation is conjectural.

Intrusive centers of the Colorado mineral belt surround Climax on all sides. Dikes, sills, and sheets of porphyry within the Climax area are similar in appearance and probably related to intrusives of the Leadville district to the south, the Buckskin Gulch and Mount-Lincoln areas to the southeast, possibly the Breckenridge district to the northeast, the Kokomo district to the north, and the Pando and Chicago-Ridge areas to the west and southwest. The abundance in the Climax area of rocks from the various locales noted above is dependent upon the proximity of the source, the strength of the igneous event, and the availability of suitable structures at the time of intrusion.

Gently dipping sheets of diorite porphyry crop out around the headwall of the Tenmile Cirque and mark the earliest Tertiary igneous event in the immediate mine area. The source of these is unknown, but they extend several miles south of the mine area to the head of the Arkansas Valley and bear no direct relationship to the igneous complex centered at Climax. The diorite is cut by Elk Mountain porphyry that, in turn, is cut by Lincoln porphyry. Both the Elk Mountain and Lincoln porphyries are quartz monzonites and were named years ago by S. F. Emmons (2) and Whitman Cross (1) for major exposures on

TABLE II.  Tabular Summary of General Geologic History of the Climax Area

| ERA | Period | Rock Name | Thickness | Rock Description | Event |
|---|---|---|---|---|---|
| Cenozoic | Recent | | | Alluvium, talus, rock glaciers | Minor movement on Mosquito fault; continued oxidation of ore. |
| | Quaternary | | | Moraine and glaciofluvial deposits | Pleistocene glaciation; sculpturing of the landscape; oxidation of ore; minor movement on Mosquito fault. Uplift of Tenmile Range by major normal movement along Mosquito fault; major subaerial erosion and development of physical features. |
| | Tertiary | Climax group of porphyries | | Rhyolitic rocks | Intrusion of composite stock and dike swarms with doming and fracturing of country rocks accompanied by three productive stages of mineralization. |
| | | Lincoln Porphyry, Elk Mountain Porphyry, Pando Porphyry | | Quartz monzonitic rocks | Intrusion of sheets, sills, and north-northeast-trending dikes; dikes intruded along through-going fractures that strike parallel with the Mosquito fault but show no displacement. |
| | | Diorite Porphyry | | Diorite porphyry | Intrusion of narrow dikes and gently-dipping sheets. |
| Mesozoic | Cretaceous | Laramie Formation (10,18) | 300–500' | Sandstone, shale and coal | |
| | | Fox Hills Formation (10,18) | 100–300' | Sandstone | |
| | | | Removed by late Cretaceous erosion | | |
| | | | Removed by Cenozoic erosion (Section shown is a composite derived from typical South Park, Blue River, and Eagle Valley sections.) | | Intermittent deposition and erosion throughout Paleozoic and Mesozoic time. |
| | | Pierre Formation (10,18) | 2500' | Shale with minor sandstone | |
| | | Niobrara Formation (18) | 500' | Shale and limestone | |
| | | Benton Formation (18) | 400' | Black shale | |
| | | Dakota Formation (18) | 250' | Sandstone and shale | |
| | Jurassic | Morrison Formation (18) | 250' | Marlstone with lenses of sandstone, mudstone, and limestone. | |
| | | Entrada (?) Formation (6) | 50' | Sandstone | |
| | Triassic | Lykins (?) Formation (6,12) | 200' | Red, nonmicaceous shale | |
| | Permian | Maroon Formation (5) | 2000' | Arkosic sandstone and shale | |

| Era | System | Formation | Thickness | Lithology | Remarks |
|---|---|---|---|---|---|
| Paleozoic | Pennsylvanian | Minturn Formation | 3200' | Arkosic grits and conglomerates with several thin carbonate marker beds | Intermittent deposition and erosion throughout Paleozoic and Mesozoic time. |
| | | Belden Formation | ? | Black shale, sandstone, and limestone | |
| | Mississippian | Leadville Formation | 65' | Leadville Dolomite member | |
| | | | 8' | Gilman Sandstone member | |
| | Devonian | Chaffee Formation | 77' | Dyer Dolomite member | |
| | | | 45' | Parting Quartzite member | |
| | Silurian | | | | |
| | Ordovician | Manitou Formation | 8' | Dolomite | |
| | | Peerless Formation | ? | Shale and dolomite | |
| | Cambrian | Sawatch Formation | 55' | Quartzite | |
| Precambrian | | Silver Plume Granite (three separate but closely-related phases) | | Medium-grained granite and pegmatite | Intrusion of stocks, dikes and sheets at Climax and to south of Climax; development of large, east-west-trending folds in Tenmile Range (age uncertain). |
| | | | | Fine-grained to medium-grained equigranular granite | |
| | | | | Coarse-grained trachytoid granite | |
| | | Metalamprophyre | | Metalamprophyre | Intrusion of dikes and small irregular bodies minor folding along northeasterly-trending axes. |
| | | Granite gneiss and pegmatite | | Granite gneiss and pegmatite | Development of granitic rocks in situ. Most contacts with metasediments are concordant and gradational. |
| | | | | | Major folding along north-northeast trending axes. |
| | | | | | Folds have steep west-facing limbs and gentle east-facing limbs. |
| | | Idaho Springs Formation (6) | | Biotite schist and gneiss | Development of foliated metasedimentary rocks by deep burial. |
| | | | | Banded gneiss, white granulite (15) | |

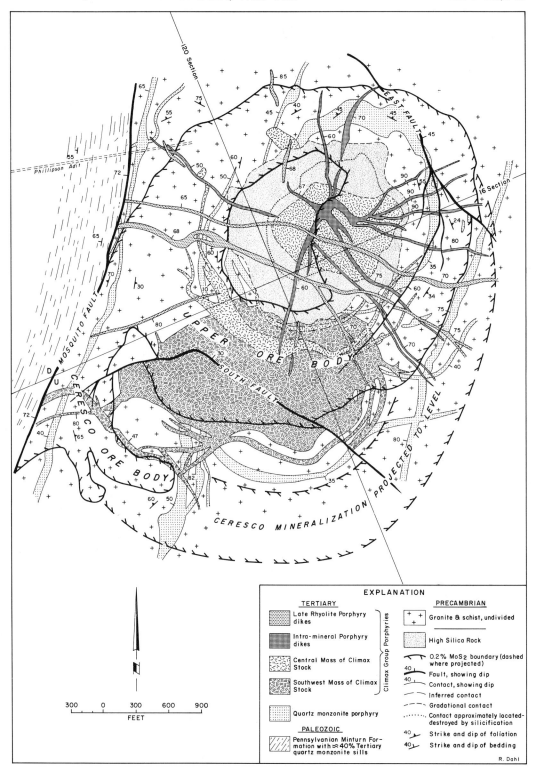

FIG. 3. *Phillipson Level, showing generalized geology and ore zones.*

Elk Mountain in the northwestern part of the Kokomo district and Mount Lincoln in the Tenmile Range east-southeast of Climax, respectively. Rocks from both centers intermingle in the Climax area where they form a reticulated set of gently dipping sheets and steep, northeasterly trending feeder dikes. In places, the two types of quartz monzonite porphyry occupy the same structures, forming composite dikes. Where these have been sheared by differential movement of the walls or affected by hydrothermal alteration associated with the Climax ore bodies, it is difficult to distinguish one from the other. Both rocks are shown by the same pattern in the illustrations that accompany this report (Figures 2, 3, 4). Sills of Pando porphyry (identified by Ogden Tweto of the U.S. Geological Survey) and of a rock resembling Chalk Mountain Rhyolite Porphyry have been intersected by deep drill holes on the hanging wall side of the Mosquito fault. Neither of these rocks has been recognized in the mine workings or in core from drill holes in the mine area. Dikes of rhyolite porphyry, similar to the later white rhyolite of the Leadville district, occur in Platte Gulch just east of the Tenmile Cirque and south of the mine at the head of the Arkansas Valley.

## ORE RELATED INTRUSIVE AND STRUCTURAL EVENTS—THE CLIMAX STOCK

### General Statement

The pre-ore geologic events, briefly reviewed above, served primarily to establish the geologic environment into which the Climax Stock was intruded. Undoubtedly the pre-ore geologic framework that evolved during Precambrian, Paleozoic, Mesozoic, and Laramide time influenced the localization of the Climax Stock. The exact nature, however, of the various possible controls, their inter-relationships and their importance relative to each other is unknown; certainly, their relationship to mineralization is indirect at best. The Climax Stock and the mechanics of its emplacement, on the other hand, have had a direct bearing on the development and distribution of the ore and, as such, deserve special attention. Paramount to this thesis is the geologic evidence of a direct and immediate relationship between hydrothermal and igneous events.

In Oligocene time, a wet silici-alkalic magma penetrated the crust near the axis of a northeasterly trending Precambrian anticline and formed an intrusive complex referred to as the Climax Stock. The exact age of the initial intrusion is unknown; K/Ar determinations place the last major thermal event, presumably hydrothermal, at approximately 30 m.y.

The stock is a composite one and was intruded as four main bodies or phases, each with its own suite of hydrothermal products—related to it genetically and in time and space (Figure 3, 4). The emplacement of each successive phase was thus accompanied and followed by a hydrothermal event. The four principal igneous-hydrothermal stages, in order of their occurrence, are tabulated below.

*TABLE III.   Stages of Intrusion and Mineralization*

| Stage | Igneous Event (phase of Climax Stock) | Hydrothermal Event (ore body or alteration zone) |
|---|---|---|
| I | Southwest Mass of the Climax Stock and an associated irregular and anastomosing dike swarm. | Ceresco Ore Body |
| II | Central Mass of the Climax Stock and related arcuate dikes and sheets. | Upper Ore Body |
| III | Aplitic Porphyry Phase of the Climax Stock and Intramineral Porphyry dikes (radial dike swarm). | Lower Ore Body |
| IV | Porphyritic Granite Phase of the Climax Stock and Late Rhyolite Porphyry dikes. | Late "Barren" Stage of mineralization—predominantly pyritic alteration—essentially noc ommercial mineralization. |

The Climax Stock penetrated the crust both by stoping and by shouldering aside and doming the country rock. The mechanics of emplacement were somewhat different for the different phases of the stock, but the end product in each case was an intrusive body, the top and flanks of which were surrounded by zones of fractured rock. Fluids introduced into these roughly hemispherical zones of broken ground altered the rocks, deposited the minerals, and formed the ore bodies.

The ore bodies are remarkably similar, but the intensity and distribution of mineralization and alteration do differ somewhat with each.

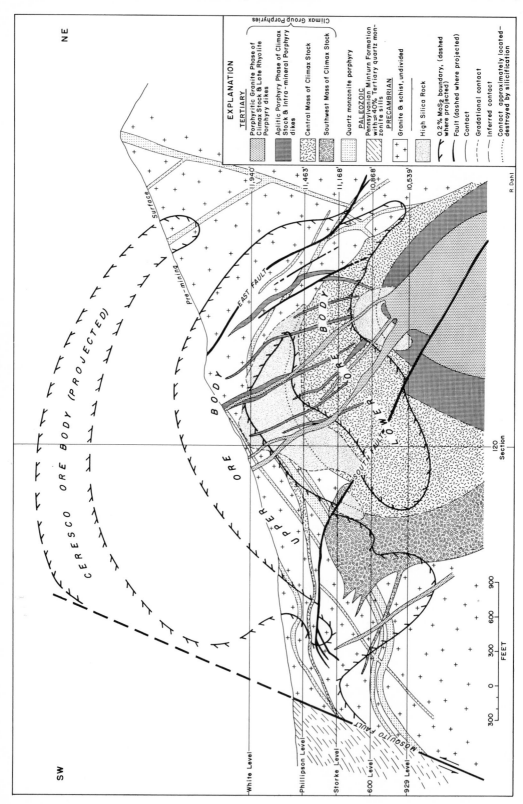

FIG. 4. *16 Section, showing generalized geology and ore zones.*

In general, a zone of introduced silica, referred to as High-Silica Rock, lies on and across the upper contact of each productive phase of the Climax Stock. The molybdenite ore bodies cap and flank these zones of silicification; zones of pyrite-tungsten mineralization lie along the hanging walls* of the molybdenite ore bodies.

Each of the principal igneous-hydrothermal stages, though closely related in space, indeed overlapping, was distinct and separate in time. One major sub-stage has been provisionally identified (see section on "Dual Nature of the Upper Ore Body"), and minor intrusive pulses were common during at least two of the principal ore-forming stages. The minor igneous pulses produced small pegmatite and aplite dikes containing disseminated molybdenite and huebnerite and a few porphyry dikes and dikelets that both cut, and are cut by, quartz-molybdenite veinlets. These rocks were contemporaneous with, or marked only a slight interruption in, the hydrothermal event and were expelled from essentially the same source as that which generated the hydrothermal solutions.

As nearly as can be determined, the original chemical and mineralogical composition of the various phases of the Climax Stock was essentially identical: quartz, orthoclase, albite, and biotite. Because of this and because in many places these rocks are strongly altered, the valid and consistent distinction between various types of Climax porphyry has been an obdurate problem. Indeed, the identification of porphyry types and the division of these into a logical and internally consistent classification has been the single most difficult geologic problem at Climax. Differences in type and intensity of alteration and mineralization, in mode of occurrence, and in texture have been used as a basis for subdivision into the major units; complete and detailed classification of all rock types and subtypes, however, has not been possible. The solution presented in this paper has gradually evolved over the past twelve years and represents the consensus of current thought. This in no way guarantees its correctness.

The interpretation of geologic features within the lower and deeper portions of the

Climax Stock is based in large measure on information obtained from diamond drill holes and must be considered tentative. The final solution to many of the remaining geologic puzzles must await the development of the deeper levels of the mine.

## The Southwest Mass

DISTRIBUTION AND HABIT   On the Phillipson level* the Southwest Mass of the Climax Stock forms an elliptical body which measures about 1800 feet by 1100 feet (Figure 3). It is elongated in an east-west direction and is some 350 feet distant from the Central Mass of the stock at its nearest point. Below the level, the Southwest Mass gradually assumes a more equidimensional outline, decreases in size and plunges northeastward toward the Central Mass. On the Storke level, the two masses come together. Here they barely touch, but on the 600 level they are well joined and on the 929 level the two parts of the stock have merged and appear as a single elliptical body, the long axis of which trends northeastward. The maximum known development of the Southwest Mass is between 100 and 200 feet above the Phillipson level. Above this elevation it necks down and leans northeastward over the top of the Central Mass. Incomplete evidence suggests that the Southwest Mass may have flared out again at its distal end. Subaerial erosion and mining subsidence have removed much of the direct evidence however, and the true configuration of the top of the Southwest Mass is based largely on projection and the distribution of related mineralization.

From the level at which the Southwest Mass diverges from the Central Mass, it rises in discordant fashion across the southwest limb of the Precambrian anticline that generally flanks and encloses the Climax Stock. Precambrian structural trends are sharply truncated and inclusions of country rock are not uncommon. Above the Phillipson level, the Southwest Mass, though partly discordant in plan, generally follows in section the schistosity of the metamorphic rocks. A highly irregular swarm of dikes and dikelets surrounds the parent intrusive.

Although the magma exerted enough pres-

---

* The nature of the mineralization and structure at Climax is such that the terms hanging wall and footwall cannot be applied according to strict definition when referring to ore. As used herein (and at Climax), the term hanging wall refers to the upper and outer surface of an ore body or zone of mineralization or alteration; footwall refers to the lower and inner surface.

* The Phillipson level is established as the zero or reference level. The Grizzly level, White level, and Leal level are respectively 60, 500, and 700 feet above the Phillipson level. The Storke level is 300 feet below the Phillipson; the 500, 600 and 929 levels are the corresponding number of feet below the Phillipson.

sure on its walls to fracture them, the habit of the intrusive and its related dikes points surely to stoping, rather than forceful intrusion, as the dominant method of emplacement.

PETROGRAPHY AND MINERALIZATION   In hand specimen, much of the porphyry of the Southwest Mass is reasonably fresh-looking, even where mineralized. Subhedral phenocrysts of quartz and orthoclase averaging 2 mm to 3 mm in size are set in a fine-grained matrix of the same minerals. The quartz phenocrysts commonly exhibit coarse aggregate extinction, but numerous crystal outlines are unmistakably those of free-formed crystals. Albite (An₆) is common at depth but is sparingly present on the 500 level, rare on the Storke level, and absent on the Phillipson level and above. Along the southwestern contact of the porphyry on the Storke level, alteration is rather weak, and in places islands of nearly fresh rock remain between veinlets. Here the porphyry is flecked with numerous small ragged crystals of primary biotite, and it seems probable that biotite was present throughout much of the Southwest Mass prior to hydrothermal alteration and the development of sericite and/or orthoclase. Biotite and/or sericite define a primary flow structure that can be observed in many exposures of the Southwest Mass. It generally is weak and is best developed near the contacts.

A petrographic oddity of the Southwest Mass is the presence of scattered grains of potash feldspar that exhibit extinction patterns reminiscent of microcline. The patterns are very fine and generally quite faint; crystals showing this type of extinction have been labeled pseudomicrocline (possibly anorthoclase). The significance of this feature is unknown. Pseudomicrocline is not present in all specimens from the Southwest Mass but is completely lacking in other parts of the Climax Stock.

The Ceresco Ore Body lies well outside the boundary of the Southwest Mass and the porphyry is therefore not affected by this mineralization (Figure 3). Dikes from the Southwest Mass do extend into the Ceresco Ore Body workings and in most exposures are well mineralized. At a few places, however, dikes both cut off and are cut by quartz-molybdenite veinlets. These dikes were derived from the same general, though perhaps not specific, source as the hydrothermal fluids and were essentially contemporaneous with the mineralization.

One porphyry dike cuts off nearly all molybdenite-bearing veinlets. It includes numerous fragments of mineralized wall rock and is clearly younger than most, if not all, of the Ceresco Ore Body mineralization. The few stray veinlets that have been observed to cut it are in an area adjacent to the Upper Ore Body, and the sparse mineralization is attributed to that source. Molybdenite mineralization of both the Upper and Lower ore bodies is imposed upon the Southwest Mass of the Climax Stock.

## The Central Mass

DISTRIBUTION AND HABIT   The Central Mass of the Climax Stock is so designated because of its position in relation to the patterns of alteration and mineralization on the Phillipson level (Figure 3, 5). It is a vertical plug with a domal upper surface that tops out some 200 feet above the Phillipson level. The outline of the plug is roughly circular in plan and enlarges downward to observable depths; its diameter on the Phillipson level is 1200 feet, on the Storke level 1800 feet, and on the 500 level 2100 feet. Below the 500 level, it merges rapidly with the Southwest Mass and loses its identity. The symmetry and habit of the Central Mass thus are distinctly different than those of the Southwest Mass. This contrast reflects a difference in the method of emplacement.

The Central Mass was forcefully intruded along the axial plane of a Precambrian anticline. This fold, not shown in the illustrations that accompany this report, was modified to a doubly plunging anticline and finally developed into a dome draped over the top of the intrusive. On 16 Section (Figure 4) and on the Phillipson level (Figure 3), the pattern of the dome is shown by the trace of the quartz-monzonite dikes, sheets and sills that lie close to the Central Mass and were deflected by its emplacement. Note that, in general, they wrap around the Central Mass but are truncated by the Southwest Mass. Major contacts of the Central Mass are, as a rule, concordant and inclusions are almost unknown, further attesting to the forceful mode of emplacement. Additional evidence of doming is found in the texture of the quartz-monzonite porphyries. Secondary foliation is almost universally present in those bodies that lie directly above and on the flanks of the Central Mass. The foliation is ascribed to shearing of the dikes and sills by differential movement of their walls in response to doming. The foliation is particularly well developed just to the east of the eastern contact of the

Central Mass, and in places the quartz-monzonite porphyries have been transformed into a rock resembling porphyroblastic gneiss.* Dikes and sills are attentuated and in places pinch out entirely.

Dikes of Climax porphyry related to the Central Mass are peripheral to and above the Central Mass and are arcuate in plan and section (Figures 3, 4). They may have been emplaced as flat sheets early in the intrusive cycle and then deformed by doming as a result of continued intrusion of the Central Mass, or the dikes may have followed arcuate tension fractures related to doming as postulated by Anderson (3). Certainly they are less abundant and far more regular than those associated with the Southwest Mass and give much less evidence of stoping action of the magma.

The habit of the Central Mass and its related dikes thus indicates that stoping was considerably less important as a mechanism of emplacement than it was for the Southwest Mass and that forceful intrusion was dominant, at least above the 500 level. The density of fractures associated with the Central Mass is greater than that related to the Southwest Mass, and the pattern is more uniformly developed.

PETROGRAPHY AND MINERALIZATION Petrographically, the Central Mass of the Climax Stock is similar to the Southwest Mass. It is somewhat coarser grained and is considerably more altered. In thin section, orthoclase crystals are notably ragged, and the quartz phenocrysts are circular-to-oval-shaped aggregates of irregular interlocking grains. Both plagioclase and biotite are completely lacking above the Storke level. Minor amounts of secondary (?) biotote are found on the 500 level,† and plagioclase is observed in specimens from the 600 level and below; pseudo-microcline is unknown.

Primary flow structures, similar to those found in the Southwest Mass, are common throughout the Central Mass. They generally

* Not all of this foliation resulted from the forceful intrusion of the Central Mass. Later phases of the Climax Stock were similarly emplaced and accentuated the foliation. Orientation of forces was similar in all cases, but movement was concentrated in the eastern part of the mine during the intrusion of the Aplitic Porphyry and Porphyritic Granite.

† Here it occurs as small clots suggestive of segregations formed by alteration. Thin discontinuous stringers of biotite cut silicified porphyry and are secondary.

are quite weak and are best developed along and near the contacts. In the eastern part of the Central Mass, the porphyry exhibits a very strong foliation with steep dip, generally parallel to subparallel with the primary flow structure. This foliation is expressed by the elongation of quartz phenocrysts, and, in places, the long dimension of the phenocrysts (now quartz aggregates) is five to six times their thickness. This feature of the porphyry is clearly not the result of orientation by liquid flow. It is interpreted as a secondary foliation produced by shearing of the porphyry mass in response to near vertical forces from below.

The margins of the porphyry mass are, in places, appreciably finer grained than the interior; the fine-grained rocks are thought to be a chilled margin. In some places, however, no such border zone exists and coarse-grained porphyry is in sharp contact with the wall rock. Vague, ill-defined inclusions of fine-grained porphyry are noted occasionally within the coarse-grained contact zone. These are interpreted as fragments of a chilled margin that were stripped from the walls of the magma chamber by renewed intrusive activity and incorporated in the coarser-grained rock.

The Upper Ore Body generally lies above and outside the Central Mass of the Climax Stock, but, in places, footwall mineralization does extend into the porphyry. The Central Mass has been strongly affected by hydrothermal events related to the Lower Ore Body.

## The Aplitic Porphyry Phase and Intra-mineral Porphyry Dikes

DISTRIBUTION AND HABIT The Aplitic Porphyry Phase is the third major unit of the Climax Stock. It was intruded as a plug into the Central Mass and is similar to it in size and shape. Its apex lies just above the 600 level and slightly to the east of the axis of the Central Mass.

The Aplitic Porphyry was forcefully emplaced and accentuated the dome developed during the intrusion of the Central Mass. Concurrently, the upper part of the Central Mass was itself deformed and broken by a system of radial tension fractures. These fractures were filled by a series of discontinuous dikes that, for reasons given elsewhere in this paper, are referred to as Intra-mineral Porphyry dikes. These dikes form a crude, star-shaped mass on the White level and a radial pattern on the Phillipson, Storke, and 500 levels. The "center" of the swarm on the White level is not, so far as is known, the top of a pipe-like mass or chimney that acted as a source. Rather, it represents merely the coalescence of dikes in an area of maximum distension near the top of the dome. The dikes display a center of sorts on the Phillipson level (Figure 3), but centers are completely lacking on the Storke and 500 levels. The Intra-mineral Porphyry dikes have been traced as far down as the 600 level, but their exact source has not been determined. They appear to be related to the Aplitic Porphyry Phase, and the interior of this mass is the presumed source. The axis of the dike swarm plunges steeply to the east and individual dikes, except those that strike east-west, have steep easterly dips; either the force causing the radial fractures was not oriented vertically, or the dike swarm (at least the fracture system) has been slightly tilted since its formation.

PETROGRAPHY AND MINERALIZATION The Aplitic Porphyry Phase is mineralogically similar to the other units of the Climax Stock. Texturally, as its name implies, it is a fine-grained rock containing small sparse phenocrysts of quartz set in a fine granular matrix of quartz and alkali feldspar; some specimens contain tiny flakes of biotite, and more than one rock type may be present. Near the contact, coarse-grained to pegmatitic dikes, pods, and segregations contain large irregular clots of coarse-grained biotite, fluorite, and potash feldspar.

Three textural varieties of rocks are found in the Intra-mineral Porphyry dike swarm. Two of these, a fine-grained pyritic rhyolite and a biotitic porphyry with sparse phenocrysts of quartz and orthoclase, are of minor importance and are found in only a few dikes. The preponderance of Intra-mineral Porphyry dikes have a somewhat coarser-grained aspect than the Aplitic Porphyry and contain more abundant and larger phenocrysts of quartz and orthoclase. Crystals of albite, though uncommon, have been observed in specimens from many of these dikes, particularly above the Phillipson level. Many of the quartz phenocrysts (xenocrysts ?) are broken, and angular fragments, and shards are common. The groundmass is very fine grained, "dirty," and packed with finely-divided fragmental material. Breccia zones of broken wall rock commonly are associated with these dikes.

The Lower Ore Body lies just above the upper contact of the Aplitic Porphyry Phase, and footwall molybdenite mineralization commonly is present in the upper and outer parts of this unit of the Climax Stock. Lower Ore

Body mineralization also affects the Intra-mineral Porphyry dikes. These dikes cut cleanly across Upper Ore Body mineralization, however, and in many places include mineralized fragments. Systematic sampling of Intra-mineral Porphyry dikes on the White, Grizzly, Phillipson, and Storke levels shows conclusively that, as the Lower Ore Body is approached both vertically and horizontally, the molybdenite content of the dikes increases; within the Lower Ore Body proper, the dikes are well mineralized and yield assays just as high as those of wall rock samples. Intra-mineral Porphyry dikes cut the High-Silica Rock on the footwall of the Upper Ore Body but are affected by silicification related to tungsten mineralization on the hanging wall of the Lower Ore Body; in many areas the two zones of silicification are coincident.

## The Porphyritic Granite Phase and Late Rhyolite Porphyry Dikes

DISTRIBUTION AND HABIT The Porphyritic Granite Phase of the Climax Stock lies at depth within and below the Aplitic Porphyry Phase, and very little is yet known about it. Its size and shape appear generally to be similar to that displayed by the two preceding phases of the stock, but this is based almost entirely on scattered drill hole information. Accurate delineation of this unit must await development of the 929 level. On the basis of present information, the high point of the Porphyritic Granite Phase appears to lie slightly to the northeast of, and somewhat below, the apex of the Aplitic Porphyry Phase (Figure 4).

Late Rhyolite Porphyry is prominently expressed by two persistent dikes that strike east-southeast across the mine area from the Mosquito fault on the west to the limit of mineralization on the east (Figure 3). On the surface (Figure 2), they have been followed still farther eastward, across the floor of the Tenmile Cirque and up the headwall; one has been traced over the Divide and beyond, a distance of well over a mile from the Mosquito fault. Both dikes dip steeply northward.

In the mine area, other Late Rhyolite Porphyry dikes, together with the two just described, form an imperfect radial pattern generally similar to that displayed by the dikes of Intra-mineral Porphyry. Indeed, the Late Rhyolite Porphyry and the Intra-mineral Porphyry occupy many of the same fractures, and, in places, the dikes are composite, both along strike or down dip and from wall to wall.

It seems likely, however, that at least some of the Late Rhyolite Porphyry dikes fill fractures formed by forces related to intrusive events that post-date the Aplitic Porphyry Phase. The exact source of the dikes is unknown, but present evidence indicates that they are related to the Porphyritic Granite Phase.

PETROGRAPHY AND MINERALIZATION The Porphyritic Granite is a medium- to coarse-grained xenomorphic-granular aggregate of orthoclase, albite, and quartz in which are set clusters of quartz grains that stand out as "eyes." These "eyes" have aggregate extinction and commonly exhibit somewhat ragged circular to oval outlines. A small number, however, have subhedral to euhedral form and their origin as phenocrysts seems clear. Biotite is present as scattered flakes and small clots. Replacement textures are common and microscopic examination suggests that albite ($An_2$) both replaces and is replaced by orthoclase. The Late Rhyolite Porphyry dikes display considerable variety in texture and aspect, but all bear the unmistakable stamp of Climax rocks.

One heading on the 929 level recently intersected the Porphyritic Granite Phase. Here, pegmatitic pods and irregular dikes of coarse quartz, potash feldspar, fluorite, and rhodochrosite lie along and above the contact; disseminated pyrite, chalcopyrite, sphalerite, and molybdenite are minor constituents of the pegmatites. Various minerals of the Late "Barren" Stage of mineralization are present as veinlets, disseminations, and segregations in the upper part of the Porphyritic Granite Phase.

The Late Rhyolite Porphyry dikes are well mineralized with pyrite and, in places, contain appreciably more than the wall rocks; possible explanations are:
(1) late solutions rich in iron and sulfur may have found preferred channelways along dike-related fractures and (2) biotite in the dikes may have acted as nuclei of preferential deposition.

Single veinlets of quartz and molybdenite cutting Late Rhyolite Porphyry have been found at perhaps two or three places on the 500 level and above. On the 600, 629, and 929 levels such occurrences, though rare, are far more common. Although tungsten minerals have not been identified in the Late Rhyolite Porphyry dikes, numerous special samples show beyond doubt that tungsten is present in abnormal concentrations; tungsten assays of dike samples are universally less than those of wall rock but are far too high to be accounted for by mineralized inclusions. In

places, the dikes are strongly affected by argillic alteration or by quartz-topaz-sericite replacement.

## MAJOR FAULTS

### East Fault

The East fault, so named because of its exposure in workings in the eastern part of the mine, is a normal fault that dips about 45° to 50°E. Where it crosses the northernmost drift on the east side of the Phillipson level, it is a single fault. Southward, it divides into a stranded fault with several branching and interconnecting breaks of different strengths and widths spread across a zone several hundred feet thick (Figure 3, 4). The fault system is arcuate in strike and roughly follows the eastern limb of both the Upper Ore Body and the Lower Ore Body. It displaces both ore bodies, but, because its movement is distributed along several strands that are subparallel with the ore bodies, both in strike and in dip, the exact amount of displacement is not known. Total movement across the zone is probably less than 350 feet. To the north the fault extends beyond the limit of information; southward, it dies out in several strands and has not been traced as far as the Ceresco Ore Body.

Crushed rock, gouge, and fine-grained chalcedonic white and gray quartz are the most abundant constituents within the East fault zone. The fault also contains abnormal concentrations of coarse fluorite, rhodochrosite, and pyrite, and fine-grained sphalerite, chalcopyrite, and galena. These minerals belong to the Late "Barren" Stage of mineralization, and the formation of the fault therefore pre-dates that stage. The fault is thought to have formed in response to intrusive forces developed during the emplacement of the Porphyritic Granite Phase of the Climax Stock.

### South Fault

The South fault is a normal fault with about 200 to 300 feet of post-ore movement. In the mine the fault averages between 10 and 20 feet in thickness and dips about 30°NE. It steepens up-dip and on Ceresco Ridge has dips of between 50° and 60°. It has a somewhat arcuate strike and in plan is convex to the north. The fault is commonly filled with light- to dark-gray chalcedonic quartz, in many places brecciated and cemented by fine crystalline quartz; vugs coated with small quartz crystals are common. Coarse fluorite is present in places.

The South fault dies out to the southeast, and, where its projection crosses the crest of Ceresco Ridge, the typical mineralization is not observed; if the fault is present, the mineral content is not. To the west, as it approaches the Mosquito fault, it breaks up into a number of minor structures. The origin of the South fault is not yet satisfactorily explained. Its displacement of the ore and its mineral content indicate that it formed late in the sequence of events at Climax.

### Mosquito Fault

The Mosquito fault is a major break in the earth's crust. It has been traced for many miles both north and south of Climax and is not local in origin. In the vicinity of Climax, the average strike is about N10°E, but, where it passes along the western edge of the Upper Ore Body, it swings eastward to about N20°E. The significance, if any, of this change in strike is unknown.

The fault dips 70° to the west at Climax and has normal displacement of about 9000 feet; the hanging wall appears to have moved southward, perhaps as much as 1500 feet. Where cut by mine workings, the Mosquito fault is a zone several hundred feet wide of fractured and broken rock, and, in places, crushed rock and gouge are 25 to 50 feet thick. White barren quartz, slightly vuggy and with minor fluorite, fills much of the fault where intersected by deep drill holes.

Deep drill holes have also intersected anomalous concentrations of molybdenite and tungsten on the hanging wall of the fault. This is presumed to be mineralization belonging to the Ceresco Ore Body (Figure 8) and most of the observed movement along the fault is therefore thought to be post-ore.

The pre-ore history of the fault is not fully known. Large through-going dikes of quartz-monzonite porphyry fill fissures that parallel the Mosquito fault (Figure 2). These dikes are Laramide in age and place the development of the present Mosquito fault fracture system at least as far back as early Tertiary. At various times during the Paleozoic era, the fault well may have been active in response to epeirogenic movements of the Ancestral Rockies. Ogden Tweto and P. K. Sims (17) have noted its Precambrian ancestry, and, in many places, the present fault trace follows or parallels a major Precambrian shear zone;

strike slip may have been a prominent direction of movement during the Precambrian.

### THE ORE BODIES

#### General Statement

Molybdenite at Climax occurs in three distinct but overlapping ore bodies, each related spatially, temporally, and genetically to one of the three productive phases of the Climax composite stock. The three ore bodies in order of decreasing age are: the Ceresco Ore Body, the Upper Ore Body, and the Lower Ore Body. The Ceresco Ore Body is actually the uppermost of the three but has been largely stripped away by post-mineral erosion and only the roots remain; if reconstructed as a complete ore body it would lie above the Upper Ore Body (Figure 4). It was not discovered,* however, until after the terms Upper and Lower Ore Body (8) had been established, and their names have, therefore, been retained.

The three ore-related phases of the Climax Stock are closely grouped, and the resulting overlap of zones of alteration and mineralization has led to a confusing and paradoxical array of geologic evidence. It seems best, therefore, theoretically to isolate one ore body and to discuss it as though the other ore bodies did not exist, before introducing the complexities arising from the superposition of mineral products of one hydrothermal event upon those of another. Of the three ore bodies, the Upper Ore Body is by far the best known. It has been much more thoroughly drilled and developed and has yielded nearly all the ore thus far produced. It is the largest of the three ore bodies,† contains the highest grade ore, and exhibits the most intense zones of hydrothermal alteration.

#### The Upper Ore Body

SIZE, SHAPE AND ATTITUDE   The Upper Ore Body is circular to ring shaped in plan and arcuate in section (Figure 3, 4). On the Leal level, 700 feet above the Phillipson level, the area of mineralization prior to mining formed a crude half circle, the other half of which

---

* Anomalous mineralization was first recognized on Ceresco Ridge by prospectors in 1916; the ore body was not "discovered" until 1956, and evidence of its full pre-erosion extent and configuration was not fully developed until 1964.

† The Ceresco Ore Body may have been larger prior to erosion.

had been scraped away by glacial abrasion. The preglacial diameter of the ore zone at this elevation was approximately 2000 feet; projection of the ore zones on section shows the apex of the ore body about 500 feet above the Leal level. On the White level 200 feet below, the assay plan shows a ring-shaped ore zone with inner and outer diameters of 700 and 2200 feet respectively. The Upper Ore Body reaches its maximum development 500 feet below the White level. Here, on the Phillipson level, the ore body is slightly elliptical and has minor and major axes from hanging wall to hanging wall of 2700 and 3600 feet; the width across the ore averages about 800 feet.

The general attitude of the Upper Ore Body is not horizontal, i.e., the base of the hemispherical zone described above does not lie on a horizontal plane. The roots of the ore body extend to much greater depths along the western and northwestern sectors than they do elsewhere. The westerly component of this tilt is shown on 16 Section (Figure 4), the northerly on 120 Section (Figure 5). The net effect of this attitude is that, on the Storke level and below, the Upper Ore Body describes an arc of about 180 degrees convex to the northwest, rather than a complete ring. Ore-grade mineralization of the northwest limb of the Upper Ore Body extends downward to the elevation of what would be the 1200 level, but very little is known of its distribution at this depth. The lack of "horizontality" of the Upper Ore Body is attributed in large part to differential uplift and tilting resulting from the forceful intrusion of the Aplitic Porphyry Phase and the Porphyritic Granite Phase into the eastern part of the Central Mass of the Climax Stock (Figure 4).

Hydrothermal events that preceded, accompanied, and followed the deposition of molybdenite, produced zones of feldspathic alteration, pyrite-tungsten mineralization, and intense quartz replacement that repeat the shape of the ore zone and lie, respectively, within, above, and below the ore body.

DISTRIBUTION AND ORIGIN OF FRACTURES
The mineralized fractures that constitute the stockwork ore body described above are not completely random. Random fractures are everywhere present, to be sure, and, in some few places, no preferred orientation of fractures can be observed. In most underground workings, however, one or more of four fracture sets are well developed, and these contain essentially all of the non-random veinlets in the Upper Ore Body. The four fracture sets can be classified into three types according to origin:

(1) Vertical to steeply dipping* fractures arranged in a radial pattern about the Central Mass of the Climax Stock; these are thought to be radial tension fractures; (2) radially arranged fractures with moderate dips; two sets are (at 60° to 70°) present and equally developed in some places; in most places one set is weakly developed or absent; these are classed as shear fractures; and (3) Flat or gently dipping fractures; these are the least common of the three types; interpreted as tension fractures.

Many specimens show several generations of veinlets, one offsetting another, and apparently fracturing continued throughout the mineralizing epoch. Most of the mineralized fractures that show preferred orientation do not fit any regional stress pattern and appear to be entirely local in origin. Small repeated vertical movements, both up and down, of the Central Mass are thought to have generated the stresses that produced the fractures that preceded and accompanied mineralization. The origin of the forces may have been magmatic or hydrothermal.

POTASSIC ALTERATION    Potash metasomatism, expressed primarily by the development of feldspar, was a prominent feature of the alteration process at Climax. Paragenetically, orthoclase is a persistent mineral and was introduced prior to, throughout, and following the ore-forming period; the most intense feldspathic alteration preceded ore deposition.

Feldspathization is especially strong on and immediately above the contact of the Central Mass, and, in some areas, the surrounding biotite schist has been transformed for hundreds of feet from the contact into a fine granular aggregate of tan to faintly pinkish orthoclase containing minor quartz; sparse remnants of altered and bleached schist, greatly elongate and in subparallel alignment, serve to identify the original rock. In other places, the aspect of the feldspathized schist is somewhat different; layered rocks of fine alternating laminae of clear quartz and white porcellaneous potash feldspar grade imperceptibly through hundreds of feet into strongly altered but recognizable schist.

In all but the most extreme cases of feld-

---

* Where non-vertical the dips commonly have an easterly component in keeping with the westerly tilt of the ore body.

spathization, the textures of the Precambrian granites are preserved, but the mineralogy is not. Microscopic examination shows that, within the ore zone, oligoclase crystals have been replaced by potash feldspar, and microcline, so common in the fresh granites, has been transformed to orthoclase. Within the Climax porphyry, orthoclase forms ragged pseudomorphs after albite and replaces the groundmass as irregular interconnecting clots and patches.

Although the areas of strongest feldspathic alteration correspond generally with those of best mineralization, the two are not directly related in time. Quartz-molybdenite veinlets commonly cut cleanly through the feldspathized rocks of the ore zone, and this potassic alteration in large part pre-dates the mineralization.[*] The initial invasion by potassium-rich fluids of the fractured rocks overlying the Central Mass is thought to have accompanied and followed very closely the final upward movement of the magma. Presumably, fluids boiled off by an adiabatic decrease in pressure spread outward along fractures and soaked through the intervening rocks, resulting in the pervasive alteration and mass replacement. The whereabouts of the sodium and calcium that were displaced are unknown.

As noted previously, orthoclase formed throughout the entire hydrothermal sequence. Potash feldspar was deposited with the ore, and thin stringers and anhedral masses are common in quartz-molybdenite veinlets. Orthoclase and adularia also are present in some of the quartz-pyrite-tungsten veinlets, although sericite is the more common potassic mineral here.

THE ORE    All molybdenite is finely crystalline and generally is so intimately intergrown with, embedded in, and enclosed by quartz and other minerals that it will not smear or soil the hands when rubbed. It is present in the following types of occurrence:

(1) In quartz-filled fractures generally less than 0.25 inches in thickness;

(2) In quartz-filled tabular structures large enough to be classed as veins;

(3) Disseminated in pegmatitic pods and aplite dikes;

(4) As inclusions or fragments in younger intrusive rocks and breccia;

(5) As irregular clots of sparsely scattered crystals in High-Silica Rock; and

(6) As "paint" on tight "dry" fractures and joint surfaces.

The quartz-molybdenite veinlets (category 1) are by far the most important and contain perhaps 95 to 97 per cent of all the molybdenum in the Upper Ore Body. The fractures along which these veinlets formed already have been described.

Quartz is by far the most abundant mineral in the veinlets and in many thin sections is the only gangue mineral observed. It is fine grained and forms a tight interlocking mosaic with sutured grain boundaries. Orthoclase and fluorite are common and generally, though not consistently, occur as discontinuous septa or scattered grains along the medial parts of veinlets. Sericite is a minor constituent and, where present, generally forms a discontinuous selvage. The molybdenite is present as tiny hexagonal plates embedded in the quartz. These may be scattered uniformly throughout the gangue or concentrated in clots and streaks. By far the most common distribution is as concentrations along one or both walls.

Pyrite and topaz are not uncommon within quartz-molybdenite veinlets but are generally restricted to, or concentrated near, intersections with quartz-sericite-topaz-pyrite veinlets. Composite veinlets are common, and most, if not essentially all, of the pyrite and topaz found in the quartz-molybdenite veinlets was introduced later.[*] Genetically, the pyrite and topaz belong to a different generation of veinlets.

Molybdenite and its associated gangue minerals were deposited both by open-space filling and by replacement. Evidence of deposition in open spaces is found in banded veins and "tight," smooth-walled fractures coated with films of fine molybdenite. These features are minor, however, and account for only a very small portion of the material deposited. Although some open-space filling undoubtedly took place in many of the quartz-molybdenite veinlets, the authors believe that replacement was the dominant process in veinlet formation. Fluid inclusions are rare (E. Roedder, personal

---

[*] In some specimens, potash feldspar clearly displaces molybdenite. In places, this feldspar belongs to the Lower-Ore-Body period of alteration; in others it results from alteration related to the late phase of the Central Mass (see section on "Dual Nature of the Upper Ore Body"). Where feldspathization was weak, orthoclase developed mainly in the inter-veinlet areas, and the quartz-molybdenite veinlets themselves were little affected.

[*] See description of pyrite and tungsten in a following section of this paper.

communication, 1966) and comb structure, crustification, and drusy vugs such as might be expected in open-space filling are unknown. On the other hand, contacts between vein material and wall rock are irregular and gradational, and veinlets that intersect at acute angles commonly show no offset from dilation. These features clearly indicate the importance of replacement in veinlet formation. Finally the expansion required to accommodate the quartz-molybdenite veinlets alone presents a serious obstacle to large scale open-space filling.

The quartz-molybdenite veinlets exhibit considerable variation in width. If they were formed primarily by replacement rather than by open-space filling, the difference in separation of veinlet walls cannot be appealed to, and some other explanation is required. The explanation preferred, with some reservations, by the authors is that the thicker veinlets were formed by more reactive solutions.

The ore veinlets show a wide range in the ratio of molybdenite to quartz. Those that are richer in molybdenite are thinner, on the average, than are the highly siliceous veinlets and generally are straighter and sharper walled; many of the quartz-rich veinlets have irregular swellings and, in places, spread out into diffuse patches of replacement silica. Two other relationships between veinlet types should be pointed out here:

(1) Veinlets rich in molybdenite are more common, relative to quartz-rich veinlets, in the hanging-wall portion of the ore body; highly siliceous veinlets increase in a general way toward the footwall and

(2) Veinlets rich in quartz, as a rule, cut those containing a higher percentage of molybdenite.

The conclusions to be drawn from the observations noted above are these:

(1) The hanging wall area of the ore body was the site of initial precipitation of molybdenite.[*]

(2) Early formed veinlets were rich in molybdenite relative to later-formed veinlets. From this it is inferred that, with the passage of time, solutions became progressively richer in silica and poorer in molybdenum, and vein material was deposited closer to the source. As the "front" of deposition gradually retreated, ore solutions reacted more vigorously with the wall rocks. On the footwall of the

---

[*] This would not necessarily follow if the silica introduction was entirely separate and later; the authors do not believe that it was.

---

Upper Ore Body, silica, virtually lacking in molybdenum, diffused outward from the fractures, completely replaced the intervening rock and formed the zone of High-Silica Rock that underlies the ore.

HIGH-SILICA ROCK  The High-Silica Rock is a finely crystalline, white-to-light-gray hydrothermal quartz. It underlies the Upper Ore Body and forms a zone about 1500 feet in diameter and from 300 to 600 feet thick that mimics the shape of the ore body. Within this zone, replacement by silica has been so complete that only a few, small scattered remnants of the original rocks have escaped destruction. Most of these, although they can be recognized as relics of earlier rocks, cannot be specifically identified, and, in some places, altered "inclusions" of any kind are completely lacking for scores of feet.

The rocks thus replaced consisted essentially of quartz and orthoclase prior to silicification and rather small amounts of silica were required for conversion. The displaced potassium and aluminum are thought to have migrated into the ore zone and beyond where they formed sericite, hydromica, and allophane.

The intense silicification in the hanging wall of the High-Silica Rock has encroached on the footwall of the Upper Ore Body in many places. Here the original rock has been essentially destroyed, and the contained molybdenite dispersed. The former presence of quartz-molybdenite veinlets is evidenced by vague, elongate, discontinuous concentrations of molybdenite within the quartz and by weak "ghost" veinlets. Much of the displaced molybdenite apparently did not travel far but remained as diffuse concentrations of finely disseminated molybdenite in the High-Silica Rock.

Beyond the border of the High-Silica Rock the intensity of alteration grades through strongly silicified to moderately and weakly silicified rocks.[*] The footwall of the High-Silica Rock is quite sharp, but strong silicification extends for several hundred feet beyond the hanging wall. Above and peripheral to the High-Silica Rock, the ore thus is cut by numerous veinlets of quartz, many containing little or no molybdenite. In places, the quartz apparently followed the earlier molybdenite veinlets, and many veinlets thus appear to be composite.

The preceding section of this paper described the general increase in the silica con-

---

[*] Almost all rocks at Climax show some silicification.

tent of the quartz-molybdenite veinlets as mineralization receded from the hanging wall to the footwall and thence into the High-Silica Rock; the process was viewed as a gradual, continuous one. It is still held to be so, but there can be little doubt that a younger quartz-rich phase of alteration advanced far beyond the hanging wall of the High-Silica Rock and that many of the quartz-rich veinlets in the ore belong to this phase.

Pyrite and Tungsten Pyrite averages between 2 and 3 per cent in the mill feed and is thus by far the most abundant metallic mineral at Climax. It is of little economic importance, however, and, as a consequence, its distribution is poorly known; this is particularly true of the pyrite that is genetically related to the Upper Ore Body. Available information, nevertheless, shows that the Upper Ore Body pyrite is concentrated in a broad zone generally above and peripheral to the molybdenite ore; samples from this zone generally contain from 3 to 6 per cent pyrite.

Pyrite commonly is found in a family of veinlets referred to as quartz-pyrite-sericite veinlets. In most specimens, quartz is predominant but mineral content is not constant, and the relative amounts of quartz, pyrite and sericite differ widely. Topaz and fluorite are common and, in places, abundant in these veinlets.

The hanging wall of the Upper Ore Body is mantled by a zone of tungsten mineralization (Figure 5) that is roughly coincident with the pyritic zone just described. The tungsten mineralization, though very low grade, is distinctly anomalous and forms a well-defined zone about 4000 feet in diameter and 300 to 500 feet thick. Tungsten occurs as the black oxide and has been identified as both huebnerite and wolframite. Quite possibly both are present, but, for the purpose of discussion, the tungsten mineral will hereafter be referred to as huebnerite. Huebnerite is commonly found as tiny grains scattered irregularly along the pyrite-bearing quartz-sericite veinlets described above. Its distribution is very erratic and even within the more consistent and better-grade parts of the tungsten zone, assays from successive ten-foot intervals of core may show as much as a ten-fold difference. The higher grade parts of the ore zone average about 0.06 per cent $WO_3$.

Although the tungsten zone generally is peripheral to the molybdenite, there are many places along the hanging wall of the Upper Ore Body where the two types of mineralization overlap. Here, huebnerite-bearing veinlets invariably cut the quartz-molybdenite veinlets,

and, at any one place, the tungsten mineralization is consistently younger. The observed relations suggest two possibilities:

(1) The solutions changed at the source and traveled, with little or no evidence of their passage, through the molybdenite ore along new fractures and deposited their mineral on the hanging wall of the ore body or

(2) Molybdenite and huebnerite were carried by the same solutions and deposition was essentially contemporaneous but took place at different sites. Huebnerite (and pyrite) traveled farther and dropped out of solution well beyond the point at which molybdenite was forming. Both minerals (molybdenite and huebnerite) were precipitating on the hanging wall of their respective zones. With the passage of time, the loci of deposition shrank inward, and eventually both minerals were forming in the footwall parts of their own zones.

The authors prefer the second alternative.

Tin and Other Minor Constituents Tin is present in trace amounts in the Climax ore. The concentration is so small that, in most samples, it cannot be detected by standard chemical analysis and consequently very little is known about its distribution. Small crystals of cassiterite set in vugs or embedded in pockets of sericite have been found in a few places in the quartz-pyrite-sericite family of veinlets, and specimens of vein material from these localities contain tin in assay amounts. The common habit is as sparse irregular patches or loose clusters of tiny dark-brown grains with oval outlines; in most thin sections, cassiterite is not observed. What little spectrographic work that has been done suggests that tin occurs in anomalously high concentrations in the tungsten zone. This is in accord with the presence of tin in assay amounts in several of the quartz-pyrite-sericite (topaz-fluorite) veinlets, and the general affinity of tin for tungsten and topaz. Tin, therefore, has been provisionally grouped with the tungsten "stage" and zone of mineralization.

Brannerite and ilmenorutile (7) have been identified in the tungsten and tin concentrates produced in the by-product plant. The appearance of brannerite and ilmenorutile in these particular concentrates has no genetic significance, but they are geochemically compatible with the huebnerite and cassiterite and are tentatively grouped with them. The by-product monazite probably is present as an accessory mineral in the Precambrian igneous and metamorphic rocks.

Chlorite and epidote are almost conspicuous by their paucity. Fine-grained flakes and aggregates of these minerals replace some of the biotite, hornblende, and plagioclase on the hanging wall of the ore body and beyond.

Significant quantitative data relating to the chemistry of alteration and mineralization are all but lacking; geologic studies only recently have progressed to the point at which meaningful samples can be taken.

DUAL NATURE OF THE UPPER ORE BODY    The description of the Upper Ore Body given above is accurate, to the best of our knowledge, so far as it goes. It is not, however, complete. In the preceding explanation, it seemed best to sacrifice completeness in the interest of brevity and clarity and certain aspects of the Upper Ore Body that now warrant consideration were omitted.

The distribution of ore in Figure 3 shows the Upper Ore Body as a reasonably uniform ring. The Ceresco Ore Body merges with it on the west, to be sure, and the top of the Lower Ore Body appears along the footwall of the east limb and modifies the pattern of mineralization in that quadrant. The Upper Ore Body itself, however, exhibits no significant irregularities.

In Figure 5, the distribution of molybdenite again is shown on the Phillipson level plan, but the grade lines are drawn at a higher cut-off. Viewed in this manner the Upper Ore Body displays a striking peculiarity—an elongate, slightly arcuate zone of lower-grade mineralization lies within the south limb of the ore body. If an appropriate section is selected and the ore zones again drawn using the higher cut-off, the south limb of the Upper Ore Body shows a distinct bi-lobate root (Figure 5). The two root zones lead upward into higher-grade strands that merge near the apex of the ore body into a zone of very high-grade ore. From the top of the ore body, this zone spirals counterclockwise down the north flank of the ore body to the west limb. Southward along the west limb the high-grade zone bifurcates into separate strands that extend eastward and grade into the two root zones of the south limb.

This pattern of grade distribution is peculiar, yet well developed, and leads immediately to speculation regarding its origin and significance. Study of the pattern suggests that it may be the result of a dual mineralization. The bi-lobate root zone of the south limb seems far too well developed to be explained simply as a random irregularity, and each lobe is thought to be the root of a separate ore body. These merge up-dip into a single ore body containing a high-grade zone with a peculiar spiral configuration. This geometry is thought to be the result of the partial coincidence of two hemispherical "surfaces" of similar diameter, i.e., two Climax ore bodies.

The Upper Ore Body tungsten zone, also on Figure 5, displays a pattern similar to, and consistent with, that traced by the molybdenite. The similarity of two different zones of mineralization ($MoS_2$ and $WO_3$), each exhibiting the same peculiarities, is too systematic to be ascribed to chance alone, and the authors accept the geometric features noted above as evidence of the dual nature of the Upper Ore Body. The two phases are so nearly one that physically they may be so considered; genetically they seem to be two.

The general theory of Climax geology developed thus far is that each major hydrothermal event is related to a different intrusive parent. If this is valid, and if the Upper Ore Body consists of two juxtaposed but genetically distinct ore bodies, then there should have been two separate pulses or phases of the Central Mass of the Climax Stock. Theory also predicts that these two phases should be very closely related in space and similar in habit. The description of the Central Mass of the stock given in an earlier section of this paper applies to the first or early phase; a rock that may qualify as the second or late phase is found in the eastern part of the Central Mass on the Phillipson level (Figure 5). It is similar to the porphyry of the first phase but has a finer-grained groundmass and is somewhat less altered. When first recognized in drill holes, it was thought that it might possibly be the source of the Intra-mineral Porphyry dikes. It lacks the fragmental texture of the Intra-mineral Porphyry however, and exposures in recent drifts show it to be cut by these dikes; it is also somewhat more altered and mineralized than the dike rocks. Its extent is imperfectly known, particularly at depth. It is thought to be the fine-grained margin of a second phase of the Central Mass that grades downward into coarser-grained rock. Both phases of the Central Mass are affected by Lower Ore Body alteration and mineralization, and, at depth, the two phases have not been distinguished one from the other. The second phase is provisionally designated as that part of the Central Mass that gave rise to the inner part of the Upper Ore Body.

The argument for dual mineralization and for overlap is based largely on the pattern of ore distribution and only incidentally on rock type and degree of alteration. No paradoxical relationships have been identified, and no really striking geologic anomalies, other than

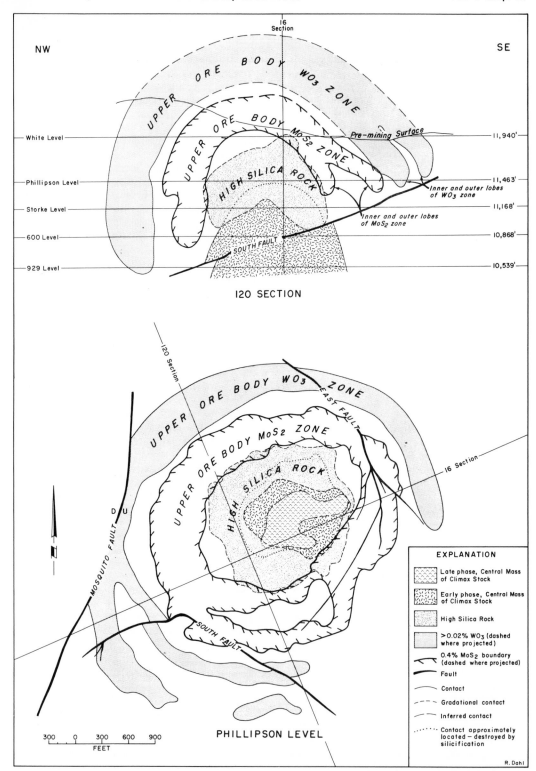

FIG. 5. *Generalized Geology and Ore Zones, showing dual nature of the Upper Ore Body.*

the pattern of mineralization itself, have been recognized. Certainly, the Upper Ore Body could not have been presented, as it has been in this paper, as the product of two closely related but separate igneous-hydrothermal stages without firm evidence in other parts of the mine of overlapping zones of alterations and mineralization. Evidence of some of these is presented in a later section of this paper.

## The Ceresco Ore Body

SIZE, SHAPE AND ATTITUDE    Knowledge of the Ceresco Ore Body that has been accumulated to date suggests strongly that it was very similar to the Upper Ore Body. On the upthrown side of the Mosquito fault, the top of the ore body has been completely stripped away by erosion and only its roots remain as evidence of its general size, shape, and attitude (Figure 3, 4). The roots form a zone of low-grade ore and anomalous mineralization about 4500 feet in diameter that is generally peripheral to the Upper Ore Body. It is not, however, perfectly centered with respect to the Upper Ore Body but is somewhat eccentric to it to the southeast (Figure 3). On the Phillipson level, the Ceresco Ore Body merges with the western limb of the Upper Ore Body just north of 16 Section. Northward from this point two things happen:

(1) In plan, the degree of overlap increases and,

(2) The bottom of the Ceresco Ore Body rises.

Where the resultant combined mineralization lies above the Phillipson adit (Figure 3), the presence of the Ceresco Ore Body is shown only by a distortion in the shape of the hanging wall, and, eastward from this point, the Ceresco mineralization gradually loses its identity. Its projection carries it across the glory hole to an area on the slopes of Bartlett Mountain where soil sampling has shown the presence of a very strong geochemical anomaly. From this area, anomalous mineralization has been traced by diamond drilling southward across the floor of the Tenmile Cirque into commercial ore underlying Ceresco Ridge. All in all, the roots of the Ceresco Ore Body have been traced for about 300° of arc,* and only that portion across the northern part of the glory hole is missing. Its pre-mining presence here is suspected on the basis of production

---

* Enough ore is preserved beneath the present erosion surface to permit mining for about one-half of this 300° of arc; the strike length of minable ore is about 4900 feet.

records* although it may have been quite weak.

The distribution of known mineralization shows that the ore body has a marked tilt to the southwest. It is thus quite possible that weakly mineralized ground rather than ore may have existed at the pre-mining erosion surface along the missing segment. Tilting is attributed to doming and deformation caused initially by the forceful intrusion of the Central Mass of the Climax Stock to the northeast of the center of the Ceresco Ore Body. Additional tilt, this to the northwest, resulted from the intrusion of the Aplitic Porphyry and the Porphyritic Granite phases of the stock (Figure 4, 7).

In part, the distribution of molybdenite may reflect the irregular development of the Southwest Mass. As noted earlier, the apex of the Southwest Mass has been removed by erosion. Where the Southwest Mass intersects the surface, it is rising to the northeast and is thought to have extended above what is now the floor of the cirque. Presuming that molybdenite deposition was roughly equidistant from the porphyry contact, this would explain the Ceresco mineralization on Bartlett Mountain. The shape and position of the Southwest Mass and the relationship of mineralization to the intrusive contact may also have contributed to the asymmetry and tilted attitude of the ore body (Figure 7).

Near the Mosquito fault, the ore zone boundaries have been deformed by drag (Figure 3, 4). The Ceresco Ore Body has been cut by the fault, and what is presumed to be the displaced segment of the ore has been intersected in deep drill holes (Figure 8).

SIGNIFICANT   DIFFERENCES—UPPER   AND CERESCO ORE BODIES   The similarities between the Ceresco Ore Body and the Upper Ore Body are far more striking than are their differences. In a way, the fact that the two ore bodies are so alike is not surprising; they were formed by similar processes in similar environments. There are some notable differences however, and the significance of these is enhanced because they stand out from the general background of sameness.

---

* Drill hole information is almost completely lacking in the area above the north limb of the Upper Ore Body, and the former presence here of the Ceresco Ore Body can only be inferred. The ore was mined by caving, and some of the old draw points in this area produced more ore than was anticipated on the basis of tonnage assigned to the Upper Ore Body.

Ceresco Ore Body mineralization is farther from the contact of the Southwest Mass than is the Upper Ore Body from the contact of the Central Mass (about 400 to 500 feet compared with 200 to 300 feet). The fracture pattern is less well developed than in the Upper Ore Body; fracture density is less and the distribution is irregular. Mineralization is, therefore, generally lower grade and more erratic.

Very little data have been collected on the distribution of pyrite and tungsten related to the Ceresco Ore Body. Pyrite is certainly much more abundant on the hanging wall of the molybdenite zone, and anomalous, though weak and erratic, tungsten mineralization appears to be concentrated there as well. The strength and magnitude of the tungsten zone related to the Ceresco Ore Body is much less, as measured by existing sample data, than that belonging to the Upper Ore Body; it also appears to be farther beyond the molybdenite hanging wall.

Many of the differences noted above appear to be related to differences in the parent intrusive; when considered together, they seem to fit into a grouping that may have some genetic significance.

Fracture patterns at Climax primarily are local in origin and related to the mechanism of stock emplacement. The somewhat sinuous shape of the Southwest Mass, its discordant contacts, and the greater abundance of inclusions indicate that stoping, as opposed to forceful intrusion, was a dominant process in its emplacement. The paucity and irregularity of fractures (and veinlets) in the Ceresco Ore Body relative to those in the Upper Ore Body are consistent with this mode of intrusion and are so interpreted. The fracture pattern accounts, in large part, for the erratic distribution of $MoS_2$ and the rather low grade of the ore body.

The importance of stoping in its emplacement suggests that the magma of the Southwest Mass was more fluid than that of the Central Mass. Such a fluid magma is envisioned as a copious source of somewhat dilute hydrothermal solutions. In contrast with those of the Central Mass, the ore solutions from which the Ceresco Ore Body was derived probably contained a lower concentration of metal ions or complexes, and this might account for Ceresco ore deposition having taken place so far from the intrusive contact.

## The Lower Ore Body

SIZE, SHAPE AND ATTITUDE   The Lower Ore Body lies 100 to 200 feet above the upper contact of the Aplitic Porphyry Phase and is smaller and slightly lower grade than the Upper Ore Body (Figure 4). It lies below the Upper Ore Body and is slightly eccentric to it toward the east. The two ore bodies join on the eastern part of the Phillipson level. The ore configuration at depth is incompletely known but the roots rest more nearly on a horizontal plane than do those of either the Upper Ore Body or the Ceresco Ore Body. The Lower Ore Body does, however, appear to be canted slightly to the west. Deformation is attributed to the intrusion of the Porphyritic Granite Phase below, and slightly to the east of, the center of the Lower Ore Body.

DISTRIBUTION AND ORIGIN OF FRACTURES
The general pattern of mineralized fractures is very similar to that of the Upper Ore Body. The origin of the fractures is attributed to small but repeated vertical movements of the Aplitic Porphyry Phase of the stock.

In the eastern part of the mine near the top of the Lower Ore Body, several quartz-molybdenite veins as much as one foot thick have been traced for many tens of feet. These veins dip gently and are filled with irregularly alternating bands of quartz and molybdenite. These veins cut almost all of the veinlets and are believed to have formed by open space filling of subsidence fractures opened late in the mineralizing epoch. Similar veins well may have been present in the other ore bodies.

POTASSIC ALTERATION   Potassic alteration related to Lower Ore Body events has been imposed on various types of Upper Ore Body alteration, and its strength and extent are, therefore, difficult to assess. In places, it overlaps the highly silicified rock beneath the Upper Ore Body. This is an inhospitable host rock, and the effects of feldspathization are quite weak; in other areas, it reinforces the Upper Ore Body feldspathic alteration. All in all, Lower Ore Body feldspathization is thought to be less widespread and somewhat weaker than that related to the Upper Ore Body.

Biotite, mixed with potash feldspar, quartz, and fluorite is found in pegmatitic segregations near the upper contact of the Aplitic Porphyry Phase. Biotite also is present well above the contact. On the 500 level, some of the strongly silicified rocks marginal to the High-Silica Rock contain small clots and stringers of fine biotite. The clots appear to have formed by segregation of very finely-disseminated biotite originally present as an accessory in the porphyry. Biotite in the stringers is thought to have a similar origin, but the mineral constituents migrated into tabular zones.

THE ORE    Specimens of ore from the Lower Ore Body generally are indistinguishable from those of the Upper Ore Body, but there are exceptions. On the eastern part of the Phillipson level, the hanging wall mineralization of the Lower Ore Body overlaps the footwall mineralization of the Upper Ore Body in the High Silica Rock (Figure 4). Here, Upper Ore Body molybdenite is present as ghost veinlets and diffuse clots, whereas molybdenite of the Lower Ore Body is in sharp-walled veinlets that cut cleanly through the High-Silica Rock; nowhere do Upper Ore Body quartz-molybdenite veinlets cut High-Silica Rock.*

HIGH-SILICA ROCK    The body of High-Silica Rock underlying the Lower Ore Body is much smaller than that on the footwall of the Upper Ore Body. This is not to say, however, that the Lower Ore Body silicification is weaker. Within the zone beneath the Lower Ore Body (Figure 4), the alteration is every bit as strong as it is on the footwall of the Upper Ore Body. Replacement is essentially complete and what few fragments of rock remain generally can not be identified.

TUNGSTEN    A major difference between the Upper Ore Body and the Lower Ore Body is the development of their respective tungsten zones. The Upper Ore Body tungsten zone is 300 to 500 feet thick and, as will be remembered, lies generally outside and peripheral to the Upper Ore Body molybdenite. The tungsten zone belonging to the Lower Ore Body, on the other hand, is not only well developed above and outward from the molybdenite, it is present well down into the Lower Ore Body and, at several points, extends to the footwall (Figure 8). The average thickness of the zone is about 700 feet, and, in places, it is more than 1000 feet. Almost all the by-product tungsten thus far produced has come from this zone, and it is by far the most well-studied and understood.

Huebnerite is present in the same type of quartz-pyrite-sericite veinlets previously described. In many places, quartz replacement

has spread out from the veinlets, and, where these are closely spaced, pervasive silicification results. Such silicified areas commonly contain sparse disseminated pyrite and minor amounts of sericite and fluorite; they are somewhat different in aspect than the High-Silica Rock that underlies the ore bodies.*

## The Late "Barren" Stage

Mineralization belonging to the Late "Barren" Stage is present partly as stockwork veinlets, partly as stringers and disseminations in pegmatitic pods, partly as segregations in silicified zones, and partly as irregular veins filling the East and South faults. Quartz, pyrite, sericite, topaze, fluorite, montmorillonite, kaolinite, rhodochrosite, chalcopyrite, sphalerite, galena, huebnerite, and molybdenite are all represented in various habits and forms. The distribution of these several constituents is widespread but spotty, and the pattern is sketchily known.

On the 929 level, recent drifting has intersected the Porphyritic Granite. Diamond drill hole information, combined with direct observation of the new exposures, suggests that a weakly developed zone of stockwork molybdenite lies astride the upper contact of the Porphyritic Granite Phase. It appears to repeat the shape of the contact and extends from about 50 feet within the intrusive to perhaps 100 to 150 feet beyond the contact. Below and inside this stockwork is a zone 20 to 30 feet thick in which the Porphyritic Granite contains disseminated molybdenite. Central to this zone, the intrusive is barren.

The Late Rhyolite Porphyry dikes, believed to have their source within the Porphyritic Granite Phase of the stock, are cut and veined by nearly all of the quartz-molybdenite veinlets within the Late "Barren" Stage molybdenite stockwork zone noted above. The few veinlets that stop against dike contacts are thought to be footwall mineralization of the Lower Ore Body.

Within the High-Silica Rock that underlies the Lower Ore Body, much of the molybdenite is present as ghost veinlets similar to those that are found in the silicified zone below the

---

* Upper Ore Body veinlets may have cut a zone of silicified rock underlying the Ceresco Ore Body. Fragmentary evidence indicates that such a zone did exist but it has been almost completely removed by erosion. Where it remains, it is beyond the influence of Upper Ore Body mineralization.

* The quartz-pyrite-sericite type of silicification related to, and on the hanging wall, of the Lower Ore Body has been imposed upon the zone of silicification related to, and on the footwall, of the Upper Ore Body. The overlap or reinforcement of one zone of silicification upon another may partly explain the unusual size of the High-Silica Rock beneath the Upper Ore Body.

Upper Ore Body. In contrast, quartz-molybdenite veinlets of the Late "Barren" Stage cut sharply through the Lower Ore Body High-Silica Rock.

Both the Late Rhyolite Porphyry dikes and the Porphyritic Granite Phase are younger than all of the Lower-Ore-Body-footwall silicification, but, in places, they are almost completely replaced by pyritic silica. The silicified zones, in addition to pyrite, topaz, sericite, and fluorite, contain irregular patchy concentrations, as much as several feet across, of fine, disseminated black sphalerite.

Pegmatite dikes on, and just above, the contact consist of feldspar and quartz as crystals and growths several inches in size mixed with blotches and seams of coarse-grained purple and green fluorite and rhodochrosite. Pyrite and chalcopyrite are the common sulfides in the dikes, but sphalerite, galena, and sparse disseminated molybdenite have also been observed.

The concentration in the East fault of minerals characteristic of the Late "Barren" Stage of mineralization has been noted in an earlier section of this report; similar mineralization is found in a number of small discontinuous breaks throughout the ore body, and some Late "Barren" Stage mineralization apparently spread along preferred channelways to the far reaches of the mineralized area.

Whether or not the Late Rhyolite Porphyry dikes acted as favored channels for mineralizing solutions is uncertain. It is known, however, that these dikes on the 500, Storke, Phillipson, and White levels contain abnormal concentrations of tungsten. Assays of special samples show the dikes to contain much less than the wall rocks (generally 65 to 85 per cent less) but far more than can be explained on the basis of inclusions. The localities at which the special dike samples were taken are all within the Lower-Ore-Body tungsten zone, and the wall rocks are mineralized by quartz-pyrite-sericite-huebnerite veinlets of that event; the Late Rhyolite Porphyry dikes are not. The dikes generally are well pyritized and at least some of the pyritic alteration of the Late "Barren" Stage spread as far as 1800 feet from the source. The tungsten mineral has not been recognized in the Late Rhyolite Porphyry dikes but is assumed to be present in the pyrite-bearing veinlets. Tungsten mineralization of the Late "Barren" Stage appears to have been even more extensive (though lower grade) than that related to the Lower Ore Body. On the basis of sample data, it overlaps the Lower Ore Body tungsten zone and probably reinforces it throughout.

## Recurrent Mineralization

A view frequently expressed by geologists who visit Climax, as well as by some of those who have worked here, is that the various types of mineralization and alteration are the result of a single mineralizing epoch and that the observed temporal and spatial relationships can best be explained on the basis of continuous hydrothermal activity along retreating fronts. Thus, different types of mineralization and alteration take place simultaneously at different places, the various loci gradually shifting inward in response to declining thermal environment and changes in chemical activity. Contradictory relations and local anomalies are easily explained by temporary reversals of the processes at work and a consequent overlap of the resultant products. The present authors generally accept this thesis as it applies to a single ore body but believe that the three ore bodies are indeed separate, and that each is the result of a distinct series of events, similar in nature but separated from the others by significant intervals of time. Complications arise, however, because the ore bodies, their related zones of alteration, and their parent intrusives, though separated in time, are superimposed in space.

Although the overlap in space of one feature upon another introduces geologic complexities, it also provides the proof of separation in time. The geologic feature that best illustrates the division in time is the Intra-mineral Porphyry dike swarm (Figure 3, 4).

The habit of the Intra-mineral Porphyry dikes is such that they are present in considerable abundance in both the Upper and Lower ore bodies, and many have been traced "continuously" by exposure in drift and by drill hole from one ore body to the other. These dikes, therefore, provide an opportunity, unequaled at Climax, to examine in a systematic way, differences in the intensity of mineralization as a function of position and time.

Examination and sampling of the Intra-mineral Porphyry dikes and their walls show that within the Lower Ore Body they are every bit as well mineralized as the rocks that enclose them; in the Upper Ore Body they are essentially barren of molybdenite mineralization.* All Intra-mineral Porphyry dikes exhibit this behavior toward mineralization.

The single dike that perhaps best illustrates

---

* The dikes perhaps more properly should be labeled Inter-mineral rather than Intra-mineral. The term Intra-mineral was used in the 1960 report by Wallace, *et al.* (14) and is here retained to avoid confusion.

this is shown on Figure 6. This particular dike is quite persistent within the area sampled and has been exposed in mine workings at several points on a number of different levels. It is in the northeast quadrant of the mine area and, on the Phillipson plan, has an outline reminiscent of a scimitar (Figure 3).

Study of the assay* values of dike and wall rock samples collected at various points (Figure 6) reveals the following:

### Molybdenite Mineralization

(1) Where the Intra-mineral Porphyry dike is within the Lower Ore Body, the molybdenite content of the dike is high, as high as that of the wall rock (sample locality #1). The dike is well mineralized and cuts off no quartz-molybdenite veinlets.

(2) On the hanging wall of the Lower Ore Body, the molybdenite content of the dike, although considerably less, is still appreciable (sample localities #2, #3, and #4). At locality #2, assays of dike and wall rock are about equal; all the molybdenite present is Lower Ore Body-related. At localities #3 and #4, the dike contains roughly one-half as much molybdenite as the wall rock. The molybdenite is in part Lower Ore Body-hanging-wall mineralization and in part Upper Ore Body-footwall mineralization; dike samples within the zone of influence of both ore bodies, e.g., sample locality #3, both cut off and are cut by quartz-molybdenite veinlets.

(3) Samples of dike rock from localities still farther beyond the hanging wall of the Lower Ore Body contain even less molybdenite. At sample locality #5, only a very few quartz-molybdenite veinlets pass through the dike; at locality #6, the dike cuts off all molybdenite-bearing veinlets. The molybdenite in the wall rock samples is due almost entirely to Upper Ore Body mineralization. Sample locality #5 is on the footwall of the Upper Ore Body, and locality #6 is well within it; assay values of wall rock samples reflect their positions relative to each other. At many places within the Upper Ore Body, e.g., sample locality #6, the dike contains inclusions of mineralized wall rock.

(4) On the hanging wall of the Upper Ore Body, the concentration of molybdenite in both dike and wall rock is very low (sample locality #7).

### Huebnerite Mineralization

(1) As noted on previous pages of this paper, the distribution of huebnerite is erratic

---

* The assays shown on Figure 6 usually represent an average of the values derived from three or four samples.

in detail. This habit is illustrated by the assay results shown on Figure 6. So far as is known, the inconsistent relationships of tungsten concentration in dike and wall rock at sample localities #1 through #6 have no particular significance and are thought to be random. The tungsten mineralization at these localities is that belonging to the Lower Ore Body, reinforced by low-grade mineralization of the Late "Barren" Stage.

(2) Sample locality #7 is well out on the hanging wall of the Lower Ore Body tungsten zone, and the assay of the Intra-mineral Porphyry dike is correspondingly low. The concentration of $WO_3$ in the wall rock, however, is three times that of the dike and is thought to reflect the presence of the Upper Ore Body tungsten zone.

To the best of our knowledge, the Intra-mineral Porphyry dikes exhibit no exception to the pattern of mineralization summarized above and illustrated in Figure 6.* They were intruded prior to the formation of the Lower Ore Body but followed the development of the Upper Ore Body and mark a definite time break in the sequence of mineralization at Climax. Similar, though less well documented, structural and intrusive interruptions have been noted between the Ceresco and Upper Ore Body periods of mineralization and between the formation of the Lower Ore Body and the Late "Barren" Stage of mineralization. The Ceresco Ore Body is the oldest, followed in succession by the Upper Ore Body, the Lower Ore Body, and the Late "Barren" stage of mineralization.

### Summary Statement—The Ore Bodies

A graphic summary of the main igneous, hydrothermal and structural events at Climax is shown on Figure 7. The diagram is reasonably self-explanatory and requires little comment in the text. Each of the first three major intrusive phases was accompanied and followed by the development of a molybdenite ore body (the dual nature of the Upper Ore

---

* Composite dikes, though not abundant at Climax, have been recognized. The dikes may be composite wall to wall, down dip, or along strike. Two discontinuous, north-south striking dikes of Intra-mineral Porphyry contain segments of Late Rhyolite Porphyry on the Phillipson level (Figure 3). In evaluating evidence displayed by dikes, either for or against any particular concept, identity of the dike is critical. Fortunately, most of the Intra-mineral Porphyry dikes have a diagnostic fragmental texture not shared by other dikes of the Climax suite of porphyries.

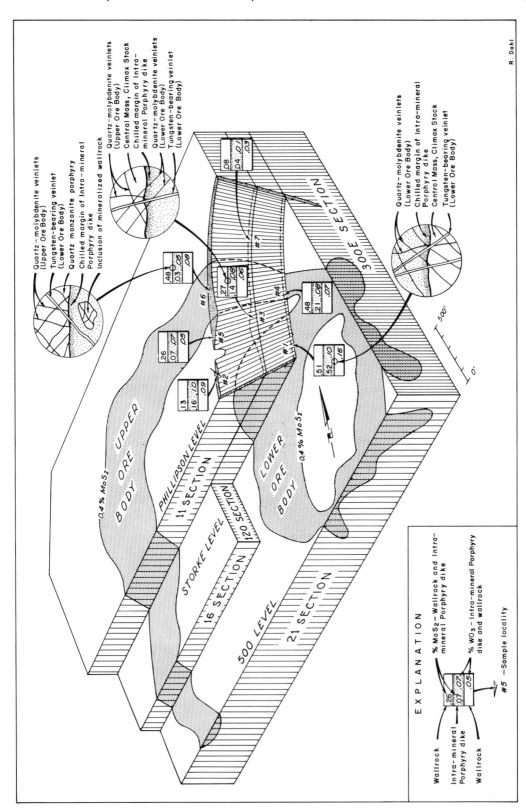

Fig. 6. *Block Diagram, looking N68°W and showing an Intra-mineral Porphyry dike cutting Upper Ore Body mineralization and cut by Lower Ore Body mineralization.*

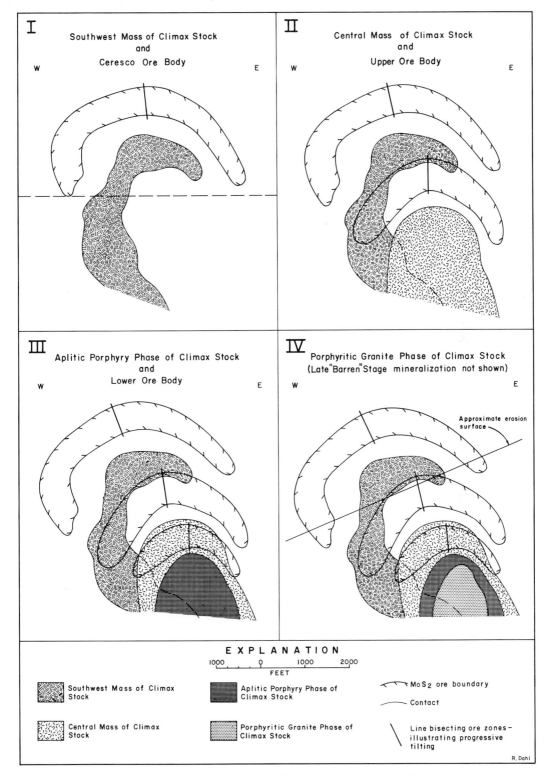

Fɪɢ. 7. *Diagrammatic Sections, showing multiple intrusion and mineralization and progressive tilting.*

Body is not illustrated in this figure); the last intrusive phase yielded almost no commercial mineralization.

Each phase of the Climax Stock was emplaced somewhat to the eastward of its predecessor. Uplift and doming that accompanied the forceful intrusion of each of the three youngest phases imparted a slight (but cumulative) westward rotation to the pre-existing geologic features. During the intrusion of the Porphyritic Granite Phase of the Climax Stock much of the movement took place along the East fault system. At the time the Central Mass and the Aplitic Porphyry Phase were emplaced, the East fault did not exist and rotational movement was accomplished by a cumulative slippage across wide zones adjacent to the eastern contacts of the intrusives. Evidence of this movement is shown by the intense secondary foliation in the eastern part of the Central Mass and in the nearby quartz monzonite dikes and sills.

Each successive ore body is smaller and formed closer to the upper contact of its source intrusive. The weak molybdenite mineralization of the Late "Barren" Stage, though not shown, is the youngest, the smallest, and the closest to its contact.

With each major hydrothermal stage in sequence, the footwall of the tungsten zone formed closer to the upper contact of its source intrusive. The respective hanging walls did not, however, recede by corresponding amounts, and the thickness of the zones of mineralization thus increased with each stage; the distance between related zones of molybdenite and huebnerite deposition decreased (Figure 8). Another, and perhaps related, feature of the huebnerite mineralization is that, relative to molybdenum, tungsten seems to be quantitatively* more important with each succeeding period of mineralization, i.e., the ratio of tungsten to molybdenum increased with each hydrothermal event including the Late "Barren" Stage of mineralization.

The orderly change with time and hydrothermal stage in:

(1) Distance between site of molybdenite deposition and related contact;

(2) Distance between zones of related molybdenite and huebnerite mineralization; and

(3) Ratio of molybdenum to tungsten (and Zn, Mn, Cu, and Pb)

suggests the periodic tapping of a master igneous-hydrothermal system undergoing a gradual and regular chemical evolution.

* Exact quantitative data to support this are not available.

## THE CYCLIC MECHANISM

The pattern at Climax of repeated igneous-hydrothermal events is far too well developed to dismiss as due simply to a normal circumstance in the process of ore formation; it requires a causative cyclic mechanism. One such mechanism is discussed below.

If the assumption is made that the parent magma of the Climax ore bodies contained 10 per cent water and that this water was all given off as a hydrothermal fluid containing 0.1 grams per liter* of molybdenum, an estimated 25 to 30 cubic miles of magma would be needed to yield the amount of molybdenum present at Climax. If the three ore bodies are arbitrarily considered to be equal, about 10 cubic miles of magma would thus be required for each. This volume is far greater than that of any of the units of the Climax Stock to observable depth, and much of the source magma, therefore, must have existed at greater depths and supplied molybdenum during the course of mineralization to those parts of the igneous-hydrothermal systems where the ore solutions were being concentrated.

The horizontal dimensions of the various productive units of the Climax Stock give areas of between $\frac{1}{8}$ and $\frac{1}{15}$ square miles. These well may enlarge somewhat downward, but the authors envision them as cylindrical bodies, the vertical dimensions of which are much greater than their horizontal ones; the magma chamber during each igneous-hydrothermal stage is considered to have been a column thousands of feet in length, connecting at depth with a much larger reservoir.

Small, but repeated, vertical movements of magma in the column fractured the rocks at the top of the chamber and allowed the more easily driven fluids to escape. The concomitant drop in confining pressure reduced the solubility of the hyperfusibles, and the dense

* This concentration is the maximum suggested by Roedder (13) based on his 1960 study of fluid inclusions. More recent analyses (9,16) by neutron activation have shown concentrations of metal (zinc) in certain fluid inclusions to be several orders of magnitude greater. Whether or not this would apply to molybdenum at Climax is unknown. In any case, in estimating the requisite volume of solution the absolute concentration of metal is not critical; what is important is the difference in metal concentration between solutions entering the zone of precipitation and those leaving it. In the present example at Climax, the authors have assumed a 100 per cent efficiency in the extraction of "water" from the magma and of molybdenum from the "water."

water-rich gas, streaming up within the column, dissolved metallic constituents from the magma and carried them to the top of the chamber and into the fractured country rock where they formed the ore.

The rate of extraction of volatile constituents from the magma by this process exceeded the rate of their generation by crystallization. Thus, there was a net loss of water in the residual magmatic fluid and a consequent increase in the freezing point. As the process of mineralization continued, the system lost energy, the temperature dropped, and finally, the magma congealed, presumably at some constriction well down in the column. The magmatic source was thus sealed off from the hydrothermal part of the system, and the process of ore formation gradually ceased.

At depth, magma in the master reservoir continued to crystallize, and the concentration of mineralizers and dissolved metal ions or complexes increased. The recharged magma rose in the crust, following the existing zones of weakness, until the weight of the column plus the resistance of the country rock to further intrusion equaled the magmatic driving force. The chamber roof was fractured, fluids escaped, pressure dropped and the ore forming process re-initiated.

At Climax the cycle was repeated four times. The authors know of no reason, save one, why the cycle may not have been repeated again. Possibly a change in the chemistry of the system marked by the appearance of zinc, manganese, copper, and lead as the major metallic elements, other than iron, presaged the final exhaustion of the master reservoir.

## POST-ORE GEOLOGIC HISTORY

At some time following the formation of the ore bodies, the Mosquito fault was reactivated, and major post-ore displacement initiated. As the footwall block rose, erosion stripped away the sedimentary cover and, eventually, exposed the Precambrian rocks and the top of the Ceresco Ore Body contained within them. By Pleistocene time, the Ceresco Ore Body had been largely removed by the ancestral Tenmile Creek, but parts of the north and south limbs were preserved on the slopes of Bartlett Mountain and beneath Ceresco Ridge, respectively. It seems probable that the valley floor had been deepened sufficiently to expose the apex of the Upper Ore Body, although this is not definitely known. Certainly, the water table was below the top of the ore, and the upper part of the ore body was oxidized. In the inter-stream areas, oxidation stayed well ahead of erosion and on Ceresco Ridge locally extended to depths of more than 400 feet below the surface.

Glacial abrasion and transport at the headwaters of the Tenmile Creek sculptured the amphitheater and removed all but the bare roots of the north limb of the Ceresco Ore Body on Bartlett Mountain. In the floor of the Tenmile valley, the ice uncovered the top of the Upper Ore Body and removed much of the oxide capping that had developed prior to glaciation. The Tenmile glacier also scoured the north side of Ceresco Ridge and ice flowing in the Arkansas Valley attacked the base of the ridge on the south side. The crest of Ceresco Ridge, however, was unaffected by glacial action, and the pre-glacial oxidized material was undisturbed. Downstream from the nose of Ceresco Ridge in the lower part of camp, moraine consists of at least two layers. An upper layer, about 10 to 15 feet thick, contains abundant fragments of mineralized rock from the Tenmile Cirque. This material is underlain by moraine characterized by unaltered rocks typical of those found in the Arkansas Valley. A thin, discontinuous layer of material quite rich in organic debris separates the two moraines.

Glaciation was followed by renewed oxidation but probably at a much reduced rate. The average annual temperature is below freezing and presumably has been throughout the Recent epoch. Fossil ice is buried beneath ablation moraine and talus in the Tenmile Cirque, permafrost is common in the area, and most precipitation comes as snow. Certainly, the rate of oxidation since Pleistocene time has been very slow and most of the oxidation should be classified as pre-glacial.

The exact form of the oxide species of molybdenum minerals is not well known. The yellow molybdic ochre, ferrimolybdite, is not uncommon, but the great preponderance of molybdenum in oxidized minerals is found in sulfates and in what appear to be amorphous iron oxides. Thin films with a faint bluish cast that coat a few of the fracture surfaces are presumed to be ilsemannite.

Collecting truly representative samples of oxidized "ore"* is difficult. At Climax, trench-

---

* The material is ore only if it contains enough sulfide molybdenum to warrant extraction and handling in the customary fashion. If this material also contains sufficient molybdenum in the oxidized form, the tailings from the rougher sulfide circuit are sent to the oxide recovery plant for further treatment.

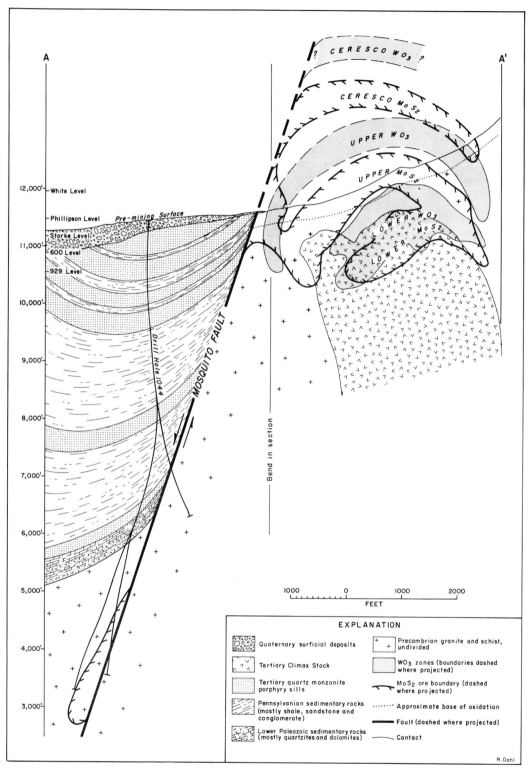

FIG. 8. *Generalized Section A–A′ (Figure 2) through the Climax Area.*

## SUMMARY OF GEOLOGIC EVENTS

(Refer to Figure 7, 8)

| Event | Remarks |
|---|---|
| (1) Precambrian, Paleozoic, Mesozoic and pre-Oligocene Tertiary events. | Influence on generation and localization of ore unknown—indirect at best. |
| (2) Intrusion of Southwest Mass of the Climax Stock. | Magma very wet, and stoping, rather than forceful intrusion, was dominant process of emplacement; wall rocks peripheral to contact fractured but density of fractures is rather low and pattern somewhat irregular. |
| (3) Introduction of hydrothermal solutions into fractured rock surrounding porphyry and deposition of minerals to form Ceresco Ore Body; quartz-molybdenite stockwork ore body with zone of quartz-pyrite-huebnerite veinlets peripheral to it. | Solutions probably abundant but quite dilute; molybdenite precipitated 500 feet or more from contact; fracture controlled mineralization generally weak and erratic; tungsten zone very weak; feldspathic and siliceous zones possibly once present, now largely removed by erosion. |
| (4) Intrusion of Central Mass of Climax Stock (early phase). | Magma wet, but forceful intrusion and doming, rather than stoping, were dominant processes of emplacement at levels of observation; wall-rock fracture pattern dense and well developed; Ceresco Ore Body somewhat deformed and tilted to the southwest. |
| (5) Introduction of potash to form feldspathized zone; deposition of outer lobes of Upper Ore Body molybdenite and huebnerite; formation of some of the High-Silica Rock below ore zone. | Feldspathic alteration especially strong on and above upper contact of Central Mass; hydrothermal solutions more concentrated than during Ceresco ore formation and molybdenite deposited closer to contact (about 300 feet); dense stockwork with high-grade ore; tungsten zone of mineralization on hanging wall of molybdenite ore body well developed; northwest quadrant of Upper Ore Body overlaps on the northwest quadrant of the Ceresco Ore Body. |
| (6) Intrusion of Central Mass of Climax Stock (late phase). (Figure 5.) | Little structural disruption of the early phase of the Central Mass. |
| (7) Introduction of hydrothermal products as in (5); formation of inner lobes of mineral zones. | Juxtaposition and reinforcement of zones of mineralization and alteration noted in (5) above. |
| (8) Intrusion of Aplitic Porphyry Phase of Climax Stock and radial dike swarm of Intra-mineral Porphyry. | Magma wet but forceful intrusion dominant; shear foliation and radial tension fractures developed in Central Mass of Climax Stock; stockwork fractures well developed; Upper Ore Body and Ceresco Ore Body tilted slightly to the northwest. |
| (9) Formation of Lower Ore Body. | Solutions more concentrated than earlier ones and molybdenite deposition close to source intrusive (100 to 200 feet above upper contact of Aplitic Porphyry Phase); zone of tungsten mineralization is particularly large and strong; east limb of Lower Ore Body superimposed on east limb of Upper Ore Body. |
| (10) Intrusion of Porphyritic Granite Phase of the Climax Stock and Late Rhyolite Porphyry dikes; formation of East fault zone. | East fault opened; normal movement of 200 to 300 feet offsets Upper and Lower ore bodies; Lower Ore Body tilted slightly to west. |
| (11) Late "Barren" Stage mineralization. | Introduction of quartz, pyrite, sericite, fluorite, topaz, rhodochrosite, chalcopyrite, sphalerite, galena, huebnerite, and molybdenite; plagioclase partly altered to orthoclase, kaolinite and montmorillonite; quartz and pyrite locally abundant, other mineralization weak or erratic—distribution generally unknown; deposition of late-stage minerals in East fault. |
| (12) South fault displaces all ore bodies. | Ore bodies offset 200 to 400 feet by normal movement; brecciated chalcedonic quartz and finely crystalline vuggy quartz fill fault; fault contains some coarse fluorite and rhodochrosite and probably originated before close of Late "Barren" Stage mineralization. |
| (13) Major normal displacement along Mosquito fault. | Post-ore displacement estimated at 8500 to 9500 feet; part of Ceresco Ore Body cut off by fault and preserved on the hanging wall. |
| (14) Subaerial erosion and oxidation. | Rapid erosion of upthrown block; deep oxidation of ore in inter-stream area of Ceresco Ridge. |
| (15) Glaciation. | Minor adjustments along Mosquito fault. |
| (16) Renewed subaerial erosion and oxidation. | Minor adjustments along Mosquito fault (personal communication, Ogden Tweto). |

ing, followed by both bulk and channel sampling and three different types of drilling, have been tried in an effort to obtain a fair sample. As nearly as can be determined on the basis of present information, oxidation of the ores resulted in neither significant enrichment in, nor loss of, molybdenum.

## REFERENCES CITED

1. Cross, W., 1886, *in* Emmons, S. F., *Geology and mining industry of Leadville, Colorado:* U.S. Geol. Surv. Mon. 12, 770 p.
2. Emmons, S. F., 1898, Tenmile district special, Colorado: U.S. Geol. Surv. Geol. Atlas, Folio 48, 6 p.
3. Anderson, E. M., 1924, *in* Bailey, E. B., *et. al., The Tertiary and post-Tertiary geology of Mull, Loch Aline, and Oban:* Geol. Surv. Scotland Mem.
4. Butler, B. S. and Vanderwilt, J. W., 1931, The Climax molybdenum deposits of Colorado: Colo. Sci. Soc. Pr., v. 12, no. 10, p. 309–353.
5. Tweto, O., 1949, Stratigraphy of the Pando area, Eagle County, Colorado: Colo. Sci. Soc. Pr., v. 15, no. 4, p. 149–235.
6. Singewald, Q. D., 1951, Geology and ore deposits of the Upper Blue River area, Summit County, Colorado: U.S. Geol. Surv. Bull. 970, 74 p.
7. Vanderwilt, J. W. and King, R. U., 1955, Hydrothermal alteration at the Climax molybdenite deposit: A.I.M.E. Tr., v. 202, p. 41–53.
8. Wallace, S. R., *et al.,* 1957, Ring-fracture intrusion and mineralization at Climax, Colorado: a preliminary report (abs.): Geol. Soc. Amer. Bull., v. 68, no. 12, p. 1809–1810.
9. Czamanske, G. K., 1959, Sulfide solubility in aqueous solutions: Econ. Geol., v. 54, p. 57–63.
10. Haun, J. D. and Weimer, R. J., 1960, Cretaceous stratigraphy of Colorado: Geol. Soc. Amer. Guidebook for Field Trips, (*Guide to the geology of* Colorado), p. 59–65.
11. Koschmann, A. H. and Bergendahl, M. H., 1960, Stratigraphy and structure of the Precambrian metamorphic rocks in the Tenmile Range: U.S. Geol. Surv. Prof. Paper 400B, p. B249–B252.
12. Oriel, S. S. and Craig, L. C., 1960, Lower Mesozoic rocks in Colorado: Geol. Soc. Amer. Guidebook to Field Trips, (*Guide to the geology of* Colorado), p. 43–56.
13. Roedder, E., 1960, Fluid inclusions as samples of the ore-forming fluids: 21st Int. Geol. Cong. Rept., pt. 16, p. 218–229.
14. Wallace, S. R., *et al.,* 1960, Geology of the Climax molybdenum deposit: a progress report: Geol. Soc. Amer. Guidebook for Field Trips, (*Guide to the geology of* Colorado), Field Trip B3, p. 238–252.
15. Bergendahl, M. H., 1963, Geology of the northern part of the Tenmile Range, Summit County, Colorado: U.S. Geol. Surv. Bull. 1162-D, p. D1–D19.
16. Czamanske, G. K., *et al.,* 1963, Neutron activation analysis of inclusions for copper, manganese, and zinc: Science, v. 140, no. 3565, p. 401–403.
17. Tweto, O. and Sims, P. K., 1963, Precambrian ancestry of the Colorado mineral belt: Geol. Soc. Amer. Bull., v. 74, p. 991–1014.
18. Leroy, L. W., 1964, Generalized composite stratigraphic section, South Park, Colorado: The Mountain Geologist, v. 1, no. 3, p. 115.

# 30. Geology and Ore Deposits of the Gilman (Red Cliff, Battle Mountain) District, Eagle County, Colorado

R. E. RADABAUGH,* J. S. MERCHANT,† J. M. BROWN,†

## Contents

* The New Jersey Zinc Co., Tucson, Arizona.
† The New Jersey Zinc Co., Gilman, Colorado.

# Illustrations

# Tables

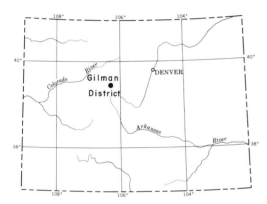

## ABSTRACT

The Gilman district is on the northeast flank of the Sawatch Range in central Colorado. It has yielded a total of 10,000,000 tons of ore having a value of over $250,000,000. Paleozoic sediments intruded by a Tertiary quartz latite sill and unconformably overlying Precambrian intrusives and metasediments comprise the country rock of the area.

The sediments strike northwesterly and dip 8° to 12° northeasterly. Structures in the Precambrian are related to the Homestake shear zone of the Colorado mineral belt. Only minor folding and faulting occur in the sediments.

The principal ore bodies are massive sulfide replacement deposits in carbonate rocks. They consist of long, pipe-like mantos in the completely dolomitized Mississippian Leadville Limestone and funnel-shaped chimneys of copper-silver ore cutting across the Mississippian and Devonian strata. Mineralization is continuous from the chimneys into the mantos. The principal ore bodies are roughly in the shape of a four-tined fork which points up dip. Smaller manto deposits occur in the Cambrian Sawatch Quartzite, and fissure veins are present in the Precambrian rocks.

The principal minerals in the mantos are marmatite and galena in a gangue of pyrite and manganosiderite. The chimneys consist of a central core of pyrite with erratically distributed chalcopyrite, tetrahedrite, freibergite, and occasional galena with a shell of zinc ore on the outer margin. Both the chimneys and the mantos are surrounded by imperfect shells of manganosiderite and sanded dolomite. The mineralization is of Laramide age.

Hydrothermal alteration affects all rock units to varying degrees. Clay mineral alteration halos occur around the ore bodies. Hydrothermal solutions have developed sanded rock, rubble filled channels, and banded zebra textures, principally in the Leadville Limestone.

Structural control of the ore bodies is not well defined. Both the location of the district and the chimneys are probably related to basement structures. The mantos are largely controlled by stratigraphic factors and a pre-Pennsylvanian karst topography in the Leadville Limestone, which were modified by hydrothermal activity, and by weak zones of northeasterly trending faults.

## INTRODUCTION

The Gilman district, also known as the Red Cliff or Battle Mountain district, is in central

Colorado approximately 75 miles west-southwest of Denver and about 20 miles north-northwest of Leadville. The town of Red Cliff is near the southern end of the district, and Gilman, the present-day center of mining activity, is near the middle at an elevation of 9,000 feet.

L. C. Graton (4, 6, 7) made one of the early geologic studies of the Eagle mine. He recognized the potential of the area and worked out many of the basic geologic features which have withstood the test of time with remarkable accuracy.

A. W. Pinger, one of Graton's assistants, was employed by The New Jersey Zinc Company and made many important contributions to the knowledge and understanding of the ore deposits. Over the years, a number of other geologists working for The New Jersey Zinc Company have also made significant contributions, and unpublished reports by Adams (10), W. H. Brown (14, 15), Callahan (17), Jerome (18), O'Neill (30), and Snively (33) have been particularly helpful. It is not feasible to acknowledge each individual contribution, but data compiled by many workers have been freely drawn upon in the preparation of this paper. Published reports on the geology of the Gilman district include those by Means (2), Crawford and Gibson (9), Lovering and Behre (13), Lovering and Tweto (20, 23), and Tweto (24, 26).

In addition, other investigators have made studies of particular phases of the geology in the Gilman district, and The New Jersey Zinc Company has made detailed investigations of certain features to gain a better understanding of the deposits and to assist with exploration. T. G. Lovering (38) investigated the temperatures and depth of formation of the deposits, and Silverman (45) has recently published a study of base metal diffusion at Gilman. Davidson (28) studied hydrothermal dolomitization and other alteration effects and their relation to metallization.

Roach (40) made a study of thermoluminescence and porosity of the host rocks in the Eagle mine. The results of the preliminary investigation suggested that thermoluminescence might be a useful exploration tool. Additional work was done by The New Jersey Zinc Company, and glow curves were made of a large number of samples from a variety of geologic environments. It was determined that thermoluminescence is highly erratic and is strongly influenced by many factors not related to mineralization.

Several studies have been made of the trace metal distribution in the district and in the surrounding area. The New Jersey Zinc Company investigated the distribution of copper, lead, zinc, and mercury. The trace metal halos are erratic and narrowly restricted.

A wide variety of geophysical methods have been tested. These include magnetics, applied potential, electromagnetic surveys, and gravity. Underground torsion balance surveys, because of the geometry of mine workings with respect to the favorable horizons, have some promise of being a useful exploration tool, although the range is very limited. None of the other geophysical techniques are applicable under the existing conditions.

The writers are grateful to The New Jersey Zinc Company for permission to prepare and publish this paper and in particular to S. S. Goodwin, Vice President, W. H. Callahan, Manager of Exploration, and W. L. Jude, Superintendent of the Eagle Mine. J. W. Johnson prepared many of the maps and illustrations used in the paper.

Much of the following discussion pertains principally to the Eagle mine of The New Jersey Zinc Company at Gilman. This mine has the most extensive workings and has been studied in the most detail. Many of the smaller mines are inaccessible in whole or in part, and the available information on them is sparse.

## HISTORICAL INFORMATION

### History

Ore was discovered in the Gilman-Red Cliff area in late 1878 or early 1879 by prospectors from Leadville who recognized the significance of the iron stained cliffs along the Eagle River and certain other similarities in the geology of the two districts. A rush ensued, and a number of discoveries were soon made resulting in the organization of the Battle Mountain district in June 1879. Construction of a smelter was started in the fall of 1879 and was completed in 1880. However, it was abandoned about a year later when the Denver and Rio Grande Railway reached Red Cliff permitting the ore to be shipped to the smelters at Leadville.

The first production consisted of silver-bearing lead-carbonate ores from the Leadville Limestone. The fissure vein deposits in the Precambrian rocks were discovered shortly thereafter, but the high grade gold ores in the quartzite were not recognized until several years later. As the mines reached greater

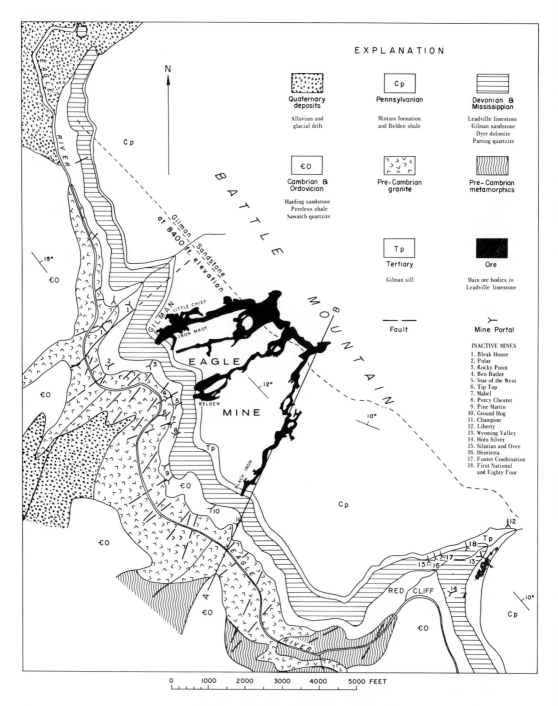

Fig. 1. *Geologic Map of the Gilman District, Eagle County, Colorado, showing location of principal ore bodies. Compiled from maps by The New Jersey Zinc Company and Lovering and Tweto (20).*

depths sulfide ores were encountered. These were of lower grade and contained much zinc which was not acceptable to the smelters. A roaster and magnetic separator were constructed in 1905 to treat the complex sulfide ores, and the production of zinc gradually assumed major proportions.

The New Jersey Zinc Company entered the district in 1912 and over a period of years acquired and consolidated the principal mines into one operating unit, the Eagle mine. Construction of an underground, selective flotation mill to treat the zinc ore was started in 1928 and completed in 1929. The mill was operated until late 1931, when because of the low metal prices then prevailing, zinc production was discontinued. During the period from 1932 through 1940, production was entirely from the copper-silver bearing chimney ore bodies. The copper-silver ore was shipped directly to the smelters. Zinc ore production was resumed in 1941 and has constituted the principal tonnage since that time. There has been continuous production from the district from 1879 to the present, a period of 88 years, although during several periods of low metal prices only relatively small tonnages of the higher grade ores were mined.

## Mining Operations

The Eagle mine of The New Jersey Zinc Company is the principal mine in the district (Figure 1) and is the only operating property at the present time. It includes what were originally the Little Chief, Iron Mask, Belden, and Black Iron mines in the limestone and the Bleak House, Polar-Rocky Point, and other mines in the quartzites. All these mines have been interconnected by common workings. The mine is developed by a series of inclines more or less on the dip of the beds and by strike drifts with interconnecting crosscuts. Development and exploration headings extend for about 9000 feet along the strike of the beds and 8500 feet down dip from the outcrops of the ore bodies. The deepest workings are approximately 2800 feet vertically below the upper slopes of Battle Mountain. Much of the ground is heavy and requires timber support, especially in the ore. The mine and mill have a capacity of 1200 tons of ore per day. The Eagle mine has produced about 90 per cent of the total production of the district.

The Polar-Rocky Point, Ground Hog, Percy Chester, Champion, Pine Martin and other mines (Figure 1) in the quartzite have in the past produced substantial tonnages of ore.

Many of these ore bodies were only a few feet thick, which resulted in the development of gently inclined, sheet-like stopes on the dip of the bedding, some barely high enough for a man to crawl through.

The Mabel, Ben Butler, and Tip Top mines (Figure 1) were the principal mines developed on fissure veins in the Precambrian rocks. The veins are narrow and discontinuous and yielded only a small aggregate tonnage. The Horn Silver, Wyoming Valley, and a number of other small mines are in the limestone near Red Cliff (Figure 1).

## Production

The Gilman district has produced a total of approximately 10,000,000 tons of ore. Of this total, the Eagle mine has produced 90 per cent, consisting of 6,000,000 tons of zinc ore and 3,000,000 tons of copper-silver ore. For many years the Eagle mine has been one of the leading zinc producers in Colorado and during the nineteen-thirties produced about 85 per cent of the copper and 65 per cent of the silver output of the state.

Statistics in terms of recovered metals and values compiled by the United States Bureau of Mines on the production from Eagle County for the period from 1880 through 1964 are summarized below. These statistics include a small amount of production from other districts, but approximately 99 per cent has come from the Gilman district.

|  |  | Value |
|---|---|---|
| Gold (lode) | 348,253 ounces | $ 11,045,758 |
| Silver | 64,108,793 ounces | 47,416,901 |
| Copper | 206,190,473 pounds | 27,336,738 |
| Lead | 255,658,692 pounds | 22,354,143 |
| Zinc | 1,293,639,629 pounds | 146,538,953 |
| Total |  | $254,692,493 |

In addition to the other ores, approximately 210,000 tons of manganese ore has been mined from the oxidized zone. According to Umpleby (3), the ore averaged 15 per cent manganese and 38 per cent iron.

## TOPOGRAPHY AND PHYSIOGRAPHIC HISTORY

The Gilman district is in an area of steep mountainous topography between the Sawatch

Range on the west and the Gore Range on the east. Both ranges have a number of peaks with elevations ranging from 13,000 to 14,000 feet.

The Eagle River has cut a northerly trending, narrow, V-shaped canyon through the district. The area has a relief of some 2600 feet. The bottom of the Eagle River Canyon has an average elevation of 8400 feet, and the top of Battle Mountain is about 11,000 feet. The town of Gilman, where the principal mines are located, is perched on a shoulder of Battle Mountain at the top of a precipitous cliff, some 600 feet almost vertically above the river. Many of the older mine workings were driven in the steep walls of the canyon.

According to Behre (12), the region was uplifted during the latter part of the Laramide revolution. This was followed by a long period of erosion, probably during the Oligocene, that reduced the area to a peneplain as evidenced by the concordant summits of the Sawatch Range. A later uplift with subsequent degradation to a mature surface ensued. This stage is represented by hills with elevations from 11,000 to 12,000 feet. This was followed by an uplift which probably corresponds to the late Pliocene deformation of the Front Range. The present topography is largely the result of dissection by both glacial and stream erosion resulting from this uplift, although a slight later uplift caused trenching by the Eagle river.

## STRATIGRAPHY

Rocks ranging from Precambrian to Quaternary in age are present in the Gilman district. The Precambrian is represented by metamorphics and intrusives, mainly granite. Cambrian and Ordovician formations are principally clastics. There are no Silurian rocks in the district. Carbonate rocks with minor interbedded clastics comprise the Devonian and Mississippian strata. All the formations from the Cambrian through the Mississippian are relatively thin having a total thickness of only 560 feet. The Pennsylvanian strata consists of a great thickness (about 6000 feet) of shales, sandstones, and arkosic conglomerates with a few thin carbonate beds in the upper third of the formation. There are no Mesozoic rocks, although strata of this age are present a few miles distant. A quartz latite sill of Laramide age was intruded just above the base of the Pennsylvanian strata. The Quaternary deposits consist of glacial drift, landslides, and local alluvium. Most of the rock units are well exposed in

the steep northeast wall of the Eagle River Canyon (Figure 1).

The stratigraphic section is summarized in Table I. The nomenclature used by the U.S. Geological Survey and others has not been strictly followed because the subdivisions of the Cambrian, Devonian, and Mississippian strata are here referred to as formations rather than members. Because of space limitations, only certain of the stratigraphic units will be discussed.

### Sawatch Quartzite—Rocky Point Horizon

A zone from a few inches to 10 feet in thickness, named the Rocky Point horizon, occurs about 10 to 15 feet below the top of the Cambrian Sawatch Quartzite. This zone is widely mineralized and locally ore bearing and has been an important source of ore. It varies greatly in thickness, composition, and structure. At places it is represented by a few thin shale partings and slight breaking of the quartzite. At other places, the zone consists of strongly broken quartzite, referred to as a crackle breccia, or it may consist of lenticular beds of reddish-brown sandy dolomite up to 6 or 7 feet in thickness. It is more porous and permeable than the typical Sawatch quartzite.

### Gilman Sandstone

The Gilman Sandstone of Middle Mississippian age is a thin but distinctive formation, averaging 15 feet in thickness. However, because of unconformities at both the top and the bottom, the thickness is highly variable, ranging from a minimum of 10 feet to a maximum of 50 feet. The greater thicknesses occur where channels cut into the underlying Devonian Dyer Dolomite.

The Gilman Sandstone is composed of quartzite and impure sandstone, with interbedded dolomite and minor shale. A coarse breccia of subangular sandstone blocks in a matrix of sandy dolomite or greasy, black clay is a conspicuous feature of the formation. The dimensions of the quartzite blocks range from a few inches to several feet, and the breccia bed may comprise more than half the thickness of the formation.

The origin of the breccia facies has been a subject of speculation. It has been interpreted as the result of supergene solution, cave formation, and collapse (20, p. 27–36; 27, p. 181–184), probably prior to the deposition of the Leadville Limestone. It is more likely,

TABLE I.　*Stratigraphic Section of the Gilman District*

| Age | Formation | Thickness Feet [1] | Character |
|---|---|---|---|
| Middle Pennsylvanian | Minturn Formation | 6,000 | Gray and maroon shale, sandstone, and coarse arkosic grits with thin beds of limestone in upper half. |
| | Belden Shale | 125 | Thin bedded, carbonaceous shale with thin beds of limestone and a few beds of sandstone. Grades upward into Minturn formation. |
| | *Intrusive Contact* | | |
| Laramide | Gilman Sill | 53 - 150 (80) | Fine grained, quartz latite porphyry with quartz and plagioclase phenocrysts. |
| | *Intrusive Contact* | | |
| Middle Pennsylvanian | Belden Shale | 5 - 50 (12) | Thin bedded, carbonaceous shale with thin beds of limestone and a few beds of sandstone. Regolithic shaly material at base. |
| | *Unconformity* | | |
| Middle Mississippian | Leadville Limestone | 90 - 170 (140) | Completely dolomitized, interbedded, locally cherty, dark gray, fine grained and light gray, saccharoidal beds in lower 95 feet. Light gray, coarsely crystalline, vuggy, massive beds with zebra structure in upper 45 feet. Several sedimentary breccias. Karst erosion surface at top. Main ore bearing horizon with limestone mantos and chimneys. |
| | *Unconformity* | | |
| | Gilman Sandstone | 10 - 50 (15) | Variable assemblage of arkosic and dolomitic sandstone, quartzite, dark gray dolomite, and sedimentary breccia consisting of blocks of sandstone in dark gray, clay shale. Tan, dense, waxy bed at top. Chimney ore. Bottom of smaller chimneys. |
| | *Unconformity* | | |
| Upper Devonian | Dyer Dolomite | 40 - 90 (70) | Medium and dark gray, medium bedded dolomite in lower 31 feet. Dark gray, thin bedded dolomite in upper 39 feet. 1 to 6 inch sand grain marker bed 40 feet above base. Chimney ore. |
| | Parting Quartzite | 35 - 40 (37) | Light gray, coarse, uneven grained quartzite. Local lenticular conglomerate. Commonly cross bedded. Several chimneys bottom in this formation. |
| | *Unconformity* | | |
| Middle Ordovician | Harding Sandstone | 25 - 60 (45) | Variable assemblage of quartzite, argillaceous sandstone, and clay shale. Varicolored gray, brown, and apple green. Two chimneys bottom in this formation. |
| | *Unconformity* | | |
| Upper Cambrian | Peerless Shale | 65 - 70 (68) | Thin bedded, sandy dolomites and shales with a few thin quartzite beds. Reddish, purplish, and greenish gray. |
| | Sawatch Quartzite | 130 - 215 (190) | Light gray, uniform, medium grained quartzite with thin beds of sandy dolomite in upper half. Rocky Point horizon consisting of brecciated quartzite and lenticular beds of pink, sandy dolomite 10 to 15 feet below top is host for quartzite manto ore bodies. |
| | *Unconformity* | | |
| Precambrian | Granite Diorite Schist | | Strongly foliated, metasedimentary, quartz-biotite schist intruded by minor diorite and by massive, locally porphyritic, gray to pink granite with abundant pegmatite. Local injection gneiss. Fissure veins. |

[1] Figures in parentheses are average thicknesses in the Eagle mine.

however, that this facies is an intraformational breccia resulting from wave action and tidal currents on a shallow sea bottom. A breccia facies, the lithology of which varies from place to place, occurs in the basal Mississippian strata (Gilman Sandstone) at numerous places in the Sawatch Range (19, p. 331–332, 348–376), and the lithologic variations suggest penecontemporaneous erosion and deposition. The lack of collapse structures in the Gilman sandstone strata overlying the breccia facies also argues against collapse.

The greasy, black clay-like material in the breccia zone is known only in the vicinity of the Eagle Mine. It is most abundant near the ore bodies, and, although its lateral extent is greater than that of the mineralized area, it gradually disappears away from the mine. This clay is a product of hydrothermal alteration associated with the formation of the ore bodies.

## Leadville Limestone

The Middle Mississippian Leadville Limestone is completely dolomitized throughout both the Gilman district and the area to the south to, and beyond, Leadville, a distance of more than 30 miles. The dolomitization extends about 3 miles north of Gilman. At other places in central and western Colorado, the Leadville consists of interbedded limestones and dolomites with limestone predominating.

The dolomitized Leadville Limestone varies from finely to coarsely crystalline and from light to dark gray. The darker beds are generally fine-grained, and the lighter beds are usually moderately to coarsely crystalline. Chert occurs in some of the strata both as thin lenticular beds and as angular to subangular fragments. The lower two-thirds of the formation consists of alternating dark gray fine-grained and light gray coarsely crystalline strata. The upper one third, the Discontinuous Banded zone, is a massive, coarsely crystalline, medium to light, pearly gray rock in which bedding is only rarely preserved. These two units are separated by the Pink Breccia horizon.

Patches of vuggy, white, recrystalline dolomite are common in the Leadville Limestone. Some patches consist entirely of white, coarsely crystalline dolomite, but most are made up of alternating layers of white, coarsely crystalline dolomite and dark gray, fine-grained rock, the zebra rock of the miners. The individual bands range from $\frac{1}{16}$ to $\frac{1}{2}$ inch in thickness and from 1 to 6 inches long and are usually roughly parallel to the bedding.

The zebra banding occurs throughout the formation but is particularly abundant in the Discontinuous Banded zone.

The Pink Breccia is from 1 inch to 10 feet thick and is characterized by angular to ellipsoidal fragments of fine-grained, medium gray to pink dolomite in a shaly matrix. Locally in the Eagle mine, the Pink Breccia consists of interstratified sandstone and shale lenses with irregularly distributed fragments of fine-grained dolomite. At places the contacts are very irregular, particularly at the base, with channels cutting several feet into the underlying strata. The pink color of many of the dolomite fragments is due to microscopic inclusions of ferric oxide resulting from hydrothermal alteration. A greasy clay, locally present in the Pink Breccia particularly near ore, also results from hydrothermal alteration.

Several hypotheses have been advanced for the origin of the Pink Breccia. Lovering and Tweto (20, p. 38, 82–85) attribute it to bedding plane faulting. However, the lithologic variations and the unconformable contacts indicate it is an intraformational breccia of sedimentary origin.

The thickness of the Leadville Limestone in the Gilman district averages approximately 140 feet but ranges from 90 to 170 feet. Much of the variation is due to a karst topography with sink holes and cavernous ground developed during a pre-Pennsylvanian period of erosion. The local thinning of the formation has been accentuated by extensive solution by hydrothermal activity with the resultant sagging and collapse of the higher beds.

The sink holes in the karst surface are filled with a greasy medium gray to gray-green, very fine-grained clay or silt-like material composed largely of quartz and illite with scattered nodules of chert. This is known as "shaly lime," although it contains very little calcareous material. Most of the "shaly lime" is strongly contorted with many slickenside surfaces, probably caused by slumping rather than tectonic movement. Many depressions filled with "shaly lime" penetrate the Leadville limestone as much as 50 feet and in a few places extend to the Gilman sandstone, a distance of approximately 140 feet.

The "shaly lime" represents an accumulation of the residual products of the chemical weathering of the limestone during the development of the karst topography. Residual soils are present between the Mississippian and Pennsylvanian strata at many places in western Colorado and, where well developed, are known as the Molas formation.

## Gilman Sill

The Gilman Sill is intruded near the base of the Belden Shale, usually 10 to 15 feet above the top of the Leadville limestone, but locally as much as 50 feet. It has been widely, and locally strongly, altered by hydrothermal solutions, but the sill itself caused only very minor metamorphism of the enclosing Belden Shale. The extensive hydrothermal alteration indicates that the sill is older than the ore. It has been correlated with the Pando Porphyry of the Pando area (32, p. 510), which is also present in the Leadville district (41, p. B11). Potassium-argon age determinations on Pando Porphyry from Leadville indicate an age of 70 million years (42), thus placing it in the currently accepted age range of the Laramide.

## STRUCTURE

### Regional Structure

The Gilman district is on the northeastern flank of the Sawatch Range, a broad elliptical-shaped domal uplift, the long axis of which trends north-northwest. Precambrian basement and Tertiary intrusive rocks are exposed in the core of the range with quaquaversally dipping sediments on the flanks. Between the Sawatch Range and the Gore Range to the east, Paleozoic sediments occupy a broad, northwesterly trending, asymmetrical syncline formed by the uplift of the two flanking mountain ranges. The district is in the wide, gently dipping, western limb of this syncline. The narrow, steeply dipping, eastern limb is truncated by the Gore fault which forms the western boundary of that range. This fault brings Precambrian crystalline rocks into contact with Pennsylvanian sediments and is a branch of the Mosquito fault of the Leadville district.

The district is in the Homestake shear zone (43, p. 1001–1005), one of the elements of the Colorado mineral belt. The mineral belt, 250 miles long and 10 to 35 miles wide, cuts diagonally across the northerly trending mountain ranges and is characterized by intrusive porphyries and associated mineral deposits of Laramide age (43). In the northern part of the Sawatch Range, the northwestern edge of the mineral belt is marked by the Homestake shear zone. This zone, 7 or 8 miles wide and approximately 20 miles long, is made up of a large number of individual shears and faults. It is a zone of weakness developed early in Precambrian time along which there has been minor recurring movement during many subsequent periods resulting in thickness variations and facies changes in the Paleozoic sedimentary rocks.

### District Structure

The district is in an area of simple structure characterized by minor flexures, discontinuous, high-angle faults of small displacement, and minor bedding plane movement. This is in marked contrast to the complex structures of the nearby Leadville district.

The beds dip from 8° to 12°NE and have an average strike of N43°W. The district is near the center of a broad shallow syncline approximately 200 feet in depth trending down the dip of the beds. Two broad, northerly trending flexures are superimposed on the southeast limb of the district syncline, and a gentle, northwesterly trending terrace formed by a local flattening of the dip occurs at the down-dip edge of the main ore bodies (Figure 2). Numerous minor, northeasterly trending crenulations are superimposed on the larger structures and are most pronounced in the vicinity of the manto ore bodies.

There are two sets of steeply dipping faults and fractures, a northeasterly trending set and a northwesterly trending one. All the faults are discontinuous and of small displacement. Faults occur in both the Precambrian rocks and in the Paleozoic sediments, but only rarely do individual structures cut both types of rock.

The faults of the northeast set are the strongest and most abundant. In the Precambrian rocks, they are better developed in the granite than in the metasediments and are more numerous and closely spaced in the area of the main ore bodies. The apparent displacements are small, and the movements are largely strike slip. The faults in the Precambrian rocks are elements of the Homestake shear zone.

Many of the faults in the Precambrian rocks stop at the contact with the overlying sediments and others flatten and turn into bedding at the unconformity. A few of the structures cross the contact, and these steepen where they pass from the crystalline rocks into the sediments. Most of these die out in the lower part of the quartzite, but one, the Bleak House Fault, has been traced from the granite through the Sawatch Quartzite into the base of the Peerless Shale. No fault is known to extend from the Precambrian to the Devonian and Mississippian formations.

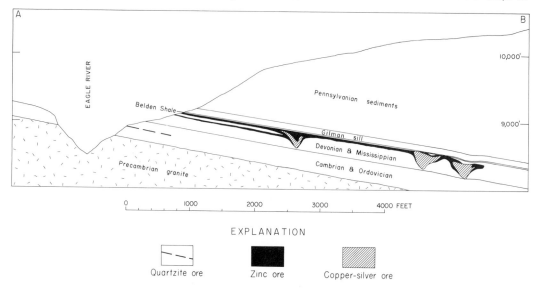

FIG. 2. *Geologic Section, showing chimney, limestone manto, and quartzite manto ore bodies and looking northwest along line A–B of Figure 1.*

A number of high-angle, northeasterly trending faults of small displacement are present in the carbonate rocks in the Eagle mine. Most have a displacement of only a few feet with a maximum of approximately 40 feet. They are discontinuous and can be traced only short distances both vertically and horizontally. Many turn abruptly into bedding movement, especially at the contacts of beds of somewhat different lithology. Others feather-out into a number of small breaks and disappear completely in a short distance. Viewed broadly, they form poorly defined zones of small, discontinuous breaks which are somewhat better developed in the vicinity of ore bodies. Some of these structures are the result of minor tectonic movement, but many are the result of differential solution, probably due to hydrothermal activity, and the consequent slumping and adjustment of the beds.

The northwesterly structures consist largely of fractures and joints and are very discontinuous. Locally in the mine workings, they are relatively abundant, but they are irregularly distributed and cannot be traced for more than short distances.

Some bedding-plane and low-angle faulting is present in the district. The most definite bedding-plane structure is in the Harding Sandstone and is indicated by a gouge and crushed zone and by a local thickening of the formation. Other low-angle faults are present in the Sawatch Quartzite, and small adjustments are

common along bedding in the carbonate rocks. Certain of the features previously attributed to bedding-plane movement probably are disconformities and diastems of sedimentary origin.

The folding and faulting in the district represent minor adjustments of the type that occurs in any series of rocks of varying lithology and competency when gently folded as by the Sawatch uplift. None of the known structures has sufficient continuity to have served as throughgoing channelways for mineralizing solutions. In fact, many of the structures in the carbonate rocks appear to be effects rather than causes.

## AGE OF MINERALIZATION

The age of mineralization in the Gilman district can only be approached through indirect lines of evidence and by comparison with the nearby Leadville district with which it shares enough points in common to suggest a similar age. The basal Pennsylvanian rocks are slightly mineralized, and the ore deposits are clearly younger than these strata. The Gilman Sill of Laramide age is the only post-Pennsylvanian rock in the district. Hydrothermal alteration associated with the ore bodies affects the Gilman Sill, and, locally, minor mineralization is present in it indicating the sill is older than the ore.

The timing of the porphyry intrusion and ore deposition with respect to the regional folding cannot be conclusively demonstrated. However, the form of the ore bodies—long, thin mantos extending up the dip from massive, crosscutting chimneys, the axes of which are inclined slightly up dip—strongly suggests that the strata were tilted to their present attitude prior to the deposition of the ore. The tilting of the beds in the Gilman district and the faulting and tilting at Leadville are related to the Sawatch uplift and are of approximately the same age. Tweto (41) has shown that the faulting at Leadville is pre-ore, and this suggests the beds were tilted before the ore was emplaced.

Isotopic analyses have been made of lead from the Battle Mountain district. Data from two sources (Table III) are in close agreement, but all the lead is anomalous, and widely differing ages may be derived from them.

The age of the mineralization in the Battle Mountain district is clearly post-porphyry and somewhat less definitely post-tilting. Exactly how much younger it may be is not known. Presumably, the mineralization is related to the late stages of Laramide orogeny and intrusion.

## HYDROTHERMAL ALTERATION

Hydrothermal alteration affects to varying degrees all rock formations in the Battle Mountain district. O'Neill (30) made an exhaustive study of the alteration, and the results of this investigation are briefly summarized below. The sequence is essentially the same in all rock types, and the stages are listed in their paragenetic order.

(1) An early weak stage characterized by the chloritization of biotite and hornblende where present, and partial replacement of feldspar by dolomite.

(2) Three stages of argillic alteration consisting of the development of kaolinite and dickite followed successively by two previously undescribed hydromicas classified according to their birefringence as hydromica (0.015) and hydromica (0.025). Silicification persisted throughout these three stages but was strongest in the hydromica (0.015) stage during which jasperoid was developed.

(3) Two stages both contemporaneous with ore deposition, the first characterized by sericite and the second by a lath-like, low birefringent hydromica.

(4) Two stages which may be partly retrogressive consisting of a siderite-kaolinite stage followed by gypsum and hydromica (0.010). These two stages partly overlap each other

and are either very late ore or post-ore in age.

The alteration through the sericite stage forms roughly concentric shells, modified by the type of wall rock, around the major ore bodies. The later stages are restricted in their distribution. O'Neill interprets the hydrothermal alteration as a continuous process with the various zones moving outward from, and up dip along, the channels carrying the hydrothermal solutions which gradually increased in temperature and acidity through the sericite stage.

Clay mineral alteration is most pronounced in the Gilman Sill and is also strongly developed in the Precambrian granite and the siliceous sediments, particularly where they are feldspathic. Clay minerals, although present, are only sparsely developed in the carbonate rocks, where other types of alteration are more pronounced.

Hydrothermal alteration of the carbonate rocks is most strongly developed in the Leadville Limestone, but it also affects the other carbonate strata to varying degrees. The Leadville formation consists entirely of dolomite in the Gilman district and to the south beyond Leadville, a distance of 30 miles. The dolomitization is generally interpreted as being the result of hydrothermal activity, but this is open to question as there is evidence to suggest it may be of sedimentary or diagenetic origin.

Several lines of evidence may be used to argue for a hydrothermal origin of the dolomite in the Gilman and Leadville districts. In some of the smaller mining districts of central Colorado such as Aspen, Tin Cup, and Alma, dolomitization transects bedding along ore contacts and faults and is closely related to sulfide mineralization. Both the ore deposits and the halo of dolomite are of small to moderate size. In the Gilman and Leadville districts, the extensive dolomitization may be interpreted as reflecting the large size of the ore deposits.

The dolomitization is roughly coextensive with a zone of porphyry intrusives, and O'Neill (30, p. 78–79) relates the dolomitization of the limestone to the early chlorite stage of hydrothermal alteration. Engel, Clayton, and Epstein (37) studied the isotopic composition of the oxygen in the Leadville Limestone and concluded that the dolomite is hydrothermal. However, an analysis of their data indicates that only certain phases of the dolomite are related to hydrothermal activity, but that other phases, even near ore, are not.

The factors favoring a sedimentary or diagenetic origin are discussed briefly below. Inter-

TABLE II.  *Composition of Dolomite from Gilman*
*and Leadville Districts*

| District | Weight Per Cent | | | | |
|---|---|---|---|---|---|
| | CaCO₃ | MgCO₃ | FeCO₃ | MnCO₃ | Insol |
| Gilman[1] | 53.9 | 44.1 | 0.7 | 0.2 | 1.1 |
| Leadville[2] | 52.9 | 44.3 | 0.5 | 0.1 | 2.3 |

[1] Average of 106 analyses.
[2] Average of 4 analyses (11, p. 35 and 34, p. 36–37).

bedded dolomite is common in the Leadville Limestone outside of mineralized areas, and Johnson (19, p. 338) believes these dolomites "are original rather than secondary deposits". The size of the dolomitized area and the uniform chemical composition of the rock argue in favor of a sedimentary or diagenetic origin. A comparison of the chemical composition of the dolomite from the Gilman and the Leadville districts is given in Table II. The individual samples vary only slightly from the averages, and the composition shows no evidence of influence by ore deposits or structures. The dolomite, both near and distant from ore, and the unaltered limestone have a remarkably uniform trace element content of 3 ppm. copper, 3 ppm. lead, and 10 ppm. zinc (35, p. 1692; 37, p. 393), and this is more indicative of a sedimentary than a hydrothermal origin. The isotopic data presented by Engel, Clayton, and Epstein (37) show that the $O^{18}/O^{16}$ composition of the dark gray dense dolomite does not change significantly with distance from ore and is similar to that of unaltered rocks outside the mining district.

The writers believe there is, therefore, strong, although not conclusive, evidence indicating that the great mass of dolomite in the Gilman-Leadville area is of sedimentary or diagenetic origin. It probably is a dolomite facies upon which hydrothermal alteration was later superimposed in restricted areas near the main conduits.

Recrystallization of the dolomite produced the zebra structure and related, irregular patches of coarsely crystalline white dolomite that are a common feature of the Leadville formation and characterize the upper third of that formation in the Gilman district. Recrystallization in part began along minute joints and fractures, but sedimentary features were the primary control of the zebra banding. The banding is more or less parallel to bedding, and finely laminated dolomite, grading into zebra structure in which the original bedding has been completely destroyed, has been observed in the Eagle mine. Zebra rock is widespread in dolomitized Leadville Limestone but, in general, is more abundant near ore. It was only sparingly developed in the Dyer Dolomite and is always close to ore.

Jasperoid occurs in several large masses in the Leadville Limestone and the Dyer Dolomite. None of the jasperoid occurs with ore, and the nearest occurrence is 3000 feet from the Eagle mine. However, the up-dip ends of the mantos have been removed by erosion, and it is not known whether jasperoid was at one time present in the system. O'Neill (30) related the jasperoid to hydrothermal alteration, and believes most of it was formed before the deposition of the sulfides.

Sanding is one of the more prominent types of hydrothermal alteration in the carbonate rocks. It results when the bond between the individual dolomite grains has been destroyed by hydrothermal solutions. All degrees of sanding occur. Weak sanding results in a slight increase in porosity, and the rock is slightly friable. Stronger sanding produces different degrees of friability with the most intense resulting in a loose, freely running, dolomite sand.

Shells of sanded dolomite a few inches to several tens of feet thick surround the manto and chimney ore bodies. Sanding is more abundant close to ore but is not confined to the immediate vicinity. Large pockets of intensely sanded dolomite occur as much as 1000 feet from the nearest known ore.

Sanding is strongest and most widespread in the Leadville Limestone, but also occurs in the Dyer Dolomite and in the dolomitic portions of the Gilman Sandstone and the Sawatch Quartzite. Strong sanding in the Dyer, Gilman, and Sawatch formations occurs only in close proximity to ore.

Sanded dolomite in the Leadville district has been ascribed to the oxidation of the sulfide ore (11, p. 36; 34, p. 37). However, in the Gilman district, it is attributed to hydrothermal activity as it is present below the zone of oxidation and contains widely distributed sericite and other hydrothermal alteration products. The sanding appears to be contemporaneous with the deposition of sulfides. However, in any one chimney-manto system, sanding developed up dip and on the margins while sulfides were still being deposited in other parts of the system, thus locally giving it a pre-ore aspect where replaced by ore minerals.

Extensive solution also resulted from the hydrothermal activity. Channels and caves were widely developed in the Leadville limestone. Some are filled with stratified dolomite sand, and others contain a rubble resulting from the collapse of the overlying rocks into the open space. The rubble consists of a heterogeneous mixture of fragments of Gilman Sill, Belden Shale, Leadville Limestone, and locally ore. Where unmineralized, the rubble is referred to as cap collapse breccia, and, where mineralized, as rubble ore.

Several types of clay seams and selvages from one inch to about 2 feet in thickness occur in, near, and around ore. They are most abundant near the ore bodies. Most probably resulted from hydrothermal alteration, although some of those parallel to bedding may have been primarily of sedimentary origin with superimposed hydrothermal alteration.

## ECONOMIC GEOLOGY

Large deposits of zinc and copper-silver ores, containing many millions of tons in all, occur in the district. Smaller deposits, mainly valuable for their precious metal content, are also present. Two principal types of deposits are represented, and these may be divided into four sub-types. All have certain features in common but vary in form, stratigraphic relations, and proportions of the various minerals. They may be classified as follows:

Fissure vein deposits—

(1) Veins cutting the Precambrian crystalline rocks and Cambrian sediments and principally valuable for their gold content.

Replacement deposits—

(1) Quartzite mantos—thin bedding-replacement deposits in the Sawatch Quartzite, locally with high precious metal content.

(2) Limestone mantos—long, pencil-like replacement bodies with roughly elliptical cross section and more or less oriented on the bedding in the Leadville limestone. Most of the zinc ore occurs in these mantos.

(3) Chimneys—roughly funnel-shaped, replacement deposits cutting across the bedding, frequently through two or more formations. Most of the copper-silver ore occurs in the chimneys.

Ore minerals have been found in the rocks of all the stratigraphic units from the Precambrian through the Pennsylvanian. The largest and most economically important ore bodies are in the Mississippian Leadville Limestone, and some of these extend downward through the Devonian strata. The ore bodies in the

Cambrian Sawatch Quartzite are appreciably smaller but, none the less, have contributed significantly to the value of district production because of their precious metal content. The occurrences in the Precambrian rocks are of minor economic importance. Scattered mineralization of the same general type as that in the major ore bodies occurs in the other formations, but only rarely is it of sufficient size and grade to be mined. The limestone mantos and the chimneys have together contributed about 95 per cent of the total tonnage, two-thirds of which came from the mantos and one-third from the chimneys.

The larger ore bodies are in a small area of slightly less than 1 square mile. A composite plan of the ore bodies (Figure 3) resembles an irregular four-tined fork with the open end pointing up the dip of the beds to the southwest. The tines are formed by the long, narrow, manto ore bodies, and the base consists of three major chimneys connected by strike mantos. Because the extent and continuity of the two interior mantos had not been fully determined during the early days of development, the pattern of the then-known ore bodies resembled and was called the "horseshoe." It has long been referred to as such.

### Mineralogy

The major primary ore bodies are essentially massive sulfide deposits. Pyrite is the most abundant mineral, marmatitic sphalerite is the most abundant ore mineral, and siderite is the principal nonmetallic constituent. These three minerals comprise about 95 per cent of the hypogene mineralization. Chalcopyrite and galena make up the bulk of the remainder, with all the other minerals together comprising no more than 1 or 2 per cent of the ore.

On a mineralogical basis, the ores may be classified as (1) pyrite-zinc ore, (2) siderite-zinc ore, and (3) copper-silver ore. In addition, there are large masses of essentially barren pyrite and a few small bodies of barren siderite.

The silver minerals, although present only in small amounts, account for much of the value of certain of the ores. The most abundant is freibergite. Others that have been identified are hessite, petzite, dyscrasite, pearceite, polybasite, pyrargyrite, proustite (?), schapbachite (matildite), stromeyerite, and argentite.

Other primary metallic minerals include arsenopyrite, pyrrhotite, tetrahedrite-tennantite, and native gold. Bournonite, beegerite,

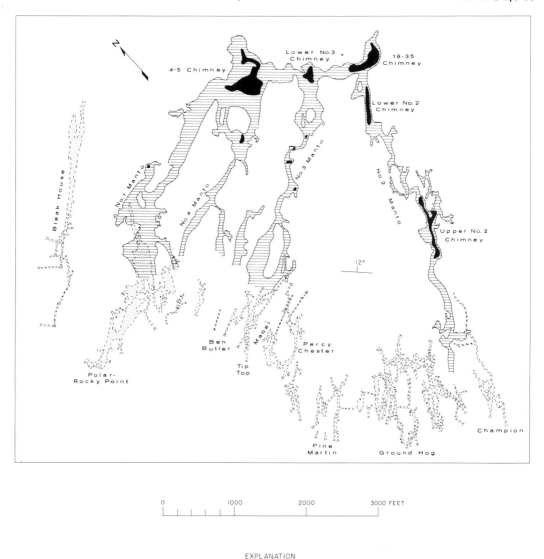

Fig. 3. *Composite Plan Map of the Principal Ore Bodies in the Eagle Mine and Adjacent Area.*

and cubanite (?) have also been reported. Magnetite and hematite are found locally in the Rocky Point horizon of the Sawatch Quartzite. A botryoidal form of sphalerite, schalenblende, occurs at one locality in the Rocky Point horizon.

In addition to siderite, the common but less abundant non-metallic minerals, are barite, dolomite, quartz, rhodochrosite, and apatite. A calcium-magnesium carbonate mineral containing 3 to 4 per cent each of iron and man-

agnese belonging to the dolomite-ankerite-kutnahorite group occurs in large rhombic crystals up to two inches long in certain parts of the Eagle mine.

Cerussite, anglesite, smithsonite, cerargyrite, and gold are the principal minerals in the oxidized zone. Native sulfur, native copper, and the secondary copper sulfides, chalcocite, covellite, and bornite are sparsely developed. Hair-like crystals of goslarite as much as 10 inches long coat the walls of many of the old mine workings, and stalactites of chalcanthite occur locally. Epsomite is also common as a post-mining mineral.

The sphalerite is the black high-iron variety, marmatite. Minor amounts of lighter-colored, brown to reddish-brown sphalerite occur locally. Typical Gilman marmatite averages about 54 per cent zinc, 11 per cent iron, 0.3 per cent copper, 0.2 per cent manganese, 0.2 per cent cadmium, and 0.2 per cent lead. The iron content of individual samples ranges from a low of 6.5 per cent to a high of 11.8 per cent. Some of the copper is in solid solution, but much of it occurs as discrete, microscopic inclusions of chalopyrite.

In general, marmatite from the up-dip portions of the mantos contains less iron than that from the down-dip ends and from the chimneys, suggesting a lower temperature of formation with distance from the source. However, the iron content may vary betwen individual samples from the same area, and some of the crystals have a zonal growth. Different zones from the same crystal may vary by as much as 6 per cent iron (39, p. 4).

All the siderite is the manganiferous variety, manganosiderite. It averages 28 per cent iron, 10 per cent manganese, and 8 per cent magnesium oxide. Individual samples range from 18.9 to 35.7 per cent iron, 6.5 to 16.5 per cent manganese, and 5.3 to 12.1 per cent magnesia. The composition of the manganosiderite is apparently influenced by the amount of sulfide minerals present in the ore (31, p. 18). Where the content of sulfide minerals is high, the manganese content of the manganosiderite is low.

## Paragenesis

The paragenesis of the ore in the Gilman district has been studied by Graton and Short (5), Pinger (8), Callahan (17), Lovering and Tweto (20), and Dykstra (21). Most of the investigations have been confined to ore in the carbonate rocks, although Lovering and Tweto (20) worked out the paragenetic sequence in several of the quartzite manto and Precambrian fissure ore bodies. The interpretations based on these studies vary somewhat in detail, but there is general agreement on most of the major points. The paragenesis of the principal minerals is given in Figure 4.

An early stage of iron deposition it characteristic of all types of ore in the district. Most of the siderite was deposited first, replacing the carbonate rocks prior to the deposition of the sulfides. This was probably accompanied

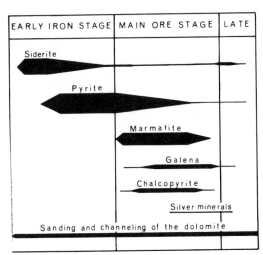

Fig. 4. *Paragenesis of the Primary Ore Minerals.*

by magnetite and hematite in the Rocky Point horizon. The solutions producing siderite, magnetite, and hematite were almost completely lacking in sulfur. The first sulfide mineral to form was pyrite, which was developed on an extensive scale. It replaced the siderite to different degrees, in some places completely and in other places only slightly, and it also replaced dolomite. Pyrite is always the first sulfide mineral to form at any particular place. Minor arsenopyrite and pyrrhotite were deposited with the early pyrite. Although the bulk of the pyrite was deposited early, it continued to form throughout the whole of the mineralizing sequence. Pyrite is the most abundant mineral and comprises about 75 per cent of the sulfide minerals.

Marmatite, galena, and chalcopyrite were the next sulfide minerals to be deposited. The sequence of deposition of these minerals cannot be clearly separated, and they overlap and replace one another. All three minerals replace siderite and pyrite. Some chalcopyrite is always present in the marmatite of both the mantos and the chimneys as microscopic inclusions and in solid solution, and the two were probably deposited simultaneously. The sulfosalts of copper and silver are largely confined to the chimneys. They started to form during the period of chalcopyrite deposition but are in large part slightly younger. A late stage of galena deposition also occurs in the chimneys and locally fills open spaces and slightly replaces the earlier metallic minerals. This galena is always rich in silver as it contains inclusions of hessite and petzite and locally a little native gold.

The last minerals to form occur as open space fillings, commonly as well developed crystals lining vugs in both the ore and the country rock. Included are all the sulfide minerals together with dolomite, quartz, siderite, apatite, barite, and rhodochrosite.

The mineralization was a dynamic and continuing process. The early minerals were replaced to different degrees by the later minerals, and the mineral or minerals being replaced were taken into solution and carried up dip or toward the margins where they were redeposited. At any one place, there is a more or less definite sequence of deposition, but when the whole mineralizing system is considered, many of the minerals were being deposited simultaneously.

A sequence similar to that described for the ore in the carbonate rocks in the Eagle mine is reported by Lovering and Tweto (20, p. 91) in the quartzite ore. This sequence consists of an early barren pyrite stage which enlarged openings in the quartzite and a younger stage that deposited the ore minerals and the gangue minerals along with more pyrite.

## Ore Structures

Three distinct types of structure are recognized in the ore: (1) bedrock ore, (2) sand ore, and (3) rubble ore. There is some intergradation of the types, but commonly each is distinctive and readily recognized.

Bedrock ore results from the volume-for-volume replacement of the host rock with the preservation of pre-ore bedrock features such as bedding, jointing, and faulting. Textures and structures are exceptionally well preserved in some of the bedrock ore.

Sand ore consists of loosely bonded sulfide and more rarely siderite grains. Sand ore is more or less friable and where most strongly developed is a loose, freely running sulfide sand. It is fine-grained, but the individual grains are larger than those of the dolomite sand. Mineral banding is common in the sand ore and may in part represent relict stratification of hydrothermally developed dolomite sand. It is more commonly liesegang-type banding resulting from a diffusion phenomenon. Close to waste, the banding conforms to the ore-waste contact, but in the interior of the ore body the banding is usually intricately contorted. Some of the sand ore may have been a replacement of previously sanded dolomite. Part of it, however, appears to us to have resulted from incomplete replacement

of dolomite host rock by sulfides from which the unreplaced dolomite was later leached.

Rubble ore consists of a heterogeneous mixture of sub-angular blocks of varying size and varying composition. The blocks range from a fraction of an inch to several feet in dimensions. Locally the blocks of rubble all have a similar composition, but much of the rubble ore consists of a heterogeneous mixture of various types of ore and may include Gilman Porphyry Sill and Belden Shale cap rock. The spaces between the blocks are open and sometimes of considerable size. There is no filling or development of drusy crystals in the voids between the blocks. The blocks are, however, slightly cemented by sulfides, usually colloform pyrite, where they are in contact with each other. Lovering and Tweto (20, p. 98) interpret the rubble as having formed prior to the deposition of the sulfides with subsequent replacement by sulfides of the individual blocks in the rubble. However, it appears more likely that the collapse took place after the deposition of the sulfides. The heterogeneous composition of the blocks, the lack of drusy coatings and filling of the voids, and the angularity of the blocks suggest a post-ore origin.

## Fissure Veins

The fissure veins are mostly in the Precambrian rocks, where they occur in high-angle northeasterly trending faults. They are most numerous and strongest in the massive competent granite in the area of the "horseshoe" (Figures 1 and 3). The dips of the veins range from 55° to vertical.

The fissure veins range in width from a few inches to a maximum of about 10 feet. Much of the vein material consists of bleached and altered rock, all more or less sheared and brecciated. The sulfide mineralization in the veins is thin and discontinuous, but the controlling structures and alteration are stronger and more persistent. Most of the sulfide minerals occur in thin stringers ranging from less than one inch to a maximum of about 6 inches in width, but these may thicken at intersections of fractures of slightly divergent strike. Disseminated sulfides occur only locally. Some of the veins spread out and are more strongly mineralized immediately below the contact with the Sawatch Quartzite, and in a few the mineralization follows the contact for a short distance. Some of the veins extend a short distance into the Sawatch Quartzite, and one, the Bleak House, extends from the granite through the quartzite into the base of the Peerless Shale.

The ore from the fissure veins is principally

of value for its gold content, although silver is locally important. It is siliceous and pyritic, but commonly some chalcopyrite, marmatite, and galena are present. Gold tellurides have been reported from several of the mines (1, p. 451). The mineralogy of the ore in the veins is very similar to that in the other types of ore deposits. There are a number of prospects on the fissure veins but only a few small mines; the Mabel, Ben Butler, Tip Top, Alpine, and Bleak House being the principal ones. The production from the fissure veins has been small, although of high dollar value because of the precious metal content of the ore.

## Quartzite Mantos

The quartzite mantos are principally in the Rocky Point horizon of the Sawatch Quartzite. Sulfide mineralization also occurs in a few thin beds below the Rocky Point horizon but is rarely extensive enough to be mined. The ore in the Rocky Point horizon ranges from a few inches to a maximum of about 8 feet thick. At some places, the mineralization is confined to one bed, and at other places the sulfides occur in two, or rarely three, beds separated by barren quartzite. Most of the ore shoots are from 10 to 50 feet wide, but one, the Bleak House, has a maximum width of 160 feet. In some areas, such as the Ground Hog and Rocky Point mines (Figure 3), a number of narrow, anastomosing ore shoots occur in a zone 1500 feet wide along the strike of the strata. Several ore bodies have been mined for 1500 to 2000 feet down the dip of the beds, and uneconomic sulfide mineralization continues beyond the limits of mining.

The mineralization in the Rocky Point horizon differs with the type of rock in which it occurs. In the crackle breccia, most of the ore minerals are confined to the fractures, although locally the quartzite fragments are more or less completely replaced. Replacement of the host rock is usually more complete in the sandy dolomite facies, and much of this ore consists of massive sulfides. At some places, solution channels and caves occur in the quartzite. Some of these are partially filled with sulfide minerals, but others are barren.

Pyrite is the most abundant primary sulfide mineral in the Rocky Point ore bodies. Marmatite, chalcopyrite, galena, and sparse argentite and gold and silver tellurides occur with the pyrite. Manganosiderite, rhodochrosite, barite, and quartz are the most common gangue minerals. Magnetite, hematite, and apatite are present in one area and may have been deposited during an early low sulfur stage

of the mineralization during which siderite was deposited in the carbonate rocks.

The ore in the near-surface portions of the quartzite mantos has been completely oxidized. The depth of oxidation varies considerably from place to place and is reported to have extended 700 feet down the dip of the beds in the Ground Hog mine (9, p. 74). The oxidized ore is largely exhausted, and none has been mined for many years. It consisted of a spongy, porous mass of limonite and some manganese oxides carrying native gold. The richest values were usually found in a "joint clay" near the top of the ore body, although local depressions in the quartzite floor also contained rich pockets of gold. The silver occurred as cerargyrite with possibly some native silver. Small quantities of secondarily enriched copper ore containing chalcocite, covellite, and bornite have been produced.

The principal quartzite manto mines are the Bleak House, Polar-Rocky Point group, Percy Chester, Pine Martin, Ground Hog, Champion, and the quartzite workings of the Ben Butler (Figures 1 and 3). The ore bodies are much smaller than those in the overlying carbonate rocks, but because of the high precious metal content of much of the ore, the quartzite mantos have contributed approximately 10 per cent of the total value of the production of the district. Most of the production came from the oxidized portions as the primary sulfide mineralization in many of the ore bodies is too low grade to be economic. However, the Bleak House mine has yielded a substantial tonnage of sulfide zinc ore, and some primary copper ore has been produced from the Rocky Point mine.

## Limestone Mantos

The limestone mantos are long, irregular, pipe-like ore bodies more or less in the plane of the bedding. They are related to, and a continuation of, the chimney ore bodies. The mantos consist mainly of zinc ores, whereas the chimneys contain most of the copper-silver ore. The limestone manto ore bodies together with the chimneys comprise the major part of the "horseshoe" (Figure 3).

The four longer manto orebodies follow a somewhat sinuous course obliquely up the dip of the beds. Two shorter mantos are on the strike of the beds connecting three major chimneys at the toe of the "horseshoe." The No. 1 manto (Figure 3) on the northwest side is approximately 2500 feet in length, and the No. 2 manto on the southeast has continuous mineralization for almost 4000 feet, although it is a composite of several manto-chimney systems. The up dip ends of the No. 1 and No. 2 mantos are 4500 feet apart where they crop out at the surface. The width of the mantos varies from approximately 50 feet to over 400 feet and the thickness from 2 to 145 feet. They are roughly elliptical in cross section, except at the up dip ends where several of them flare out into thin, wide, blanket-like bodies.

With but two possible exceptions, the limestone mantos are confined to the Leadville Limestone. Graton (7, p. 31–32) described two small manto-like bodies in the Dyer Dolomite near the up-dip end of the No. 1 manto. One is 15 feet above the Parting Quartzite and the other just below the Gilman Sandstone. Both are thin, 2 to 8 feet, and of small size. They are highly pyritic and more like the copper-silver ore than the zinc ore. The area is no longer accessible, and the writers have no first hand knowledge of these ore bodies.

The zinc mantos vary in stratigraphic position within the Leadville limestone (Figure 2). Some are in the uppermost beds immediately below the Belden shale, but others are lower in the section and are capped by dolomite. The lower part of the No. 1 manto replaces the full thickness of the Leadville Limestone, but further up dip it is confined to that portion below the Pink Breccia. Locally, some of the thinner mantos cut across bedding at a low angle from one horizon to another. At the lower end of the No. 3 manto there are two distinct layers of ore, but these coalesce up dip into one layer.

Pyrite, marmatite, and siderite are the most abundant minerals in the manto ore bodies. The relative proportions of these minerals vary widely from place to place. Some small areas consist largely of marmatite with small amounts of pyrite or siderite. Other parts of the ore bodies may be mainly pyrite or siderite with a small amount of marmatite. Galena is always present in small amounts. W. H. Brown (14, p. 16) investigated mineral zoning and determined that the proportion of lead with respect to zinc gradually increases up the dip of the mantos. Chalcopyrite is also always present in small amounts, but silver minerals and gold are sparse. At the down-dip ends of several of the mantos adjacent to the chimneys, the cores consist almost entirely of pyrite which is referred to as "non-commercial iron."

Bedrock, sand, and rubble ore all occur in the zinc manto ore bodies. Sand and rubble ore are more common in the down-dip portions, but are present locally throughout the

ore bodies. The upper parts of the mantos are largely bedrock ore, sometimes with a channel of sand and rubble ore in, along the side, or at the top.

The mantos are usually surrounded by an irregular shell of siderite, and this in turn is surrounded by a shell of sanded dolomite. These shells are somewhat discontinuous and vary from a few inches to a number of feet in thickness. The contacts between the zones are sharp, with little or no intermixing of the various constituents. Where the mantos lie immediately below the capping shale, the upper surface is commonly very irregular with many pockets and channels filled with "shaly lime" or cap collapse extending down into the ore, some for many feet.

The up-dip ends of those mantos that reach the surface are oxidized. The depth of oxidation varies from place to place, but locally it extends 1200 feet down dip. The mines encountered water and sulfide ores at approximately the same elevation, indicating the water table before mining was near the base of the oxidized zone. The oxidized ores are largely exhausted, and none has been mined for many years.

Silver and lead constituted the main metals of economic value in the oxidized zone. The principal ore minerals were cerargarite, cerussite, and anglesite. The principal gangue minerals were limonite, jarosite, melanterite, and a mixture of manganese oxides. Only small amounts of zinc carbonate were found along the sides and bottoms of the ore bodies, and, based on the composition of the unoxidized portions of the mantos, appreciable zinc must have been carried away in solution. The oxidation process, however, effected a concentration of silver and gold as the content of these two metals is five to ten times higher in the oxidized ore than in the primary manto ore.

Where the primary mineralization is predominantly manganosiderite, the oxidized ore is rich in manganese but low in lead and precious metals. In the No. 2 manto ore body (Figure 3) the manganese ore occurs in a separate bed approximately in the middle of the Leadville Limestone and about 40 feet below the oxidized sulfide ore, which is just under the Belden Shale.

The principal limestone mantos are in the Eagle mine (Figure 3). A small limestone manto is present in the Wyoming Valley and Horn Silver mines (Figure 1) near the southern edge of the district. This ore body occurs in the upper part of the Leadville Limestone just below the Belden Shale. The ore ranges from 18 inches to 4 feet thick, and the width varies greatly, with a maximum of 300 feet. The ore extended from surface approximately 1000 feet down dip. Replacement was not complete and the ore body contains a number of horses of poorly mineralized rock.

Several other small ore bodies in the Leadville Limestone occurred in the Red Cliff area. These include the Silurian, Ovee, Henrietta, Foster Combination, First National, and Eighty Four mines (Figure 1). The ore bodies were small and irregular, but the limestone host rock is strongly sanded The ore was completely oxidized and was chiefly valuable for its silver content, although some lead and zinc have been reported.

## Chimney Ore Bodies

The chimneys are irregular, funnel shaped masses that cut across the bedding (Figures 2 and 5). There are 11 chimneys in all. The three larger ones are at the down-dip ends of the three larger limestone mantos at the toe of the "horseshoe". The other chimneys are irregularly distributed along the trends of the mantos (Figure 3).

A typical chimney (Figure 5) consists of a pyritic core surrounded by irregular and incomplete shells of zinc ore, siderite, and sanded dolomite. The core is 90 to 95 per cent pyrite with varying amounts of chalcopyrite, tetrahedrite, freibergite and other silver minerals, galena, and gold. The bulk of the core comprises copper-silver ore, but the distribution of valuable metals is highly erratic. Small masses of very high grade ore containing 5 or more per cent copper, several hundred to several thousand ounces per ton of silver, and several ounces per ton of gold can change in a few feet to essentially barren pyrite called non-commercial iron.

The zinc and siderite shells are best developed in the Leadville Limestone, where they may be several tens of feet thick. These shells are thinner or may be entirely missing in the lower formations. The zinc shell of the chimneys is continuous with and mineralogically similar to the zinc ore in the mantos.

The individual chimneys differ widely in size and shape. Some, especially the smaller ones, are almost circular in plan, others are roughly elliptical, and a few are elongated along the trend of the mantos. All mushroom out at the top into irregular shapes, spreading out more on the up-dip than on the down-dip side. The largest chimney is 400 feet in diameter at the top, and the smallest is about 40 feet. All chimneys taper downward, and the axes

pitch about 70° northeast. Two chimneys extend from the top of the Leadville limestone down into the top of the Harding Sandstone, a vertical distance of about 275 feet. Some bottom in the Parting Quartzite and others in the Gilman Sandstone. The carbonate rocks are completely replaced by sulfides, but the Gilman Sandstone and Parting quartzite in places are only partly replaced. Remnants of the Gilman Sandstone, where traceable through a chimney, sag slightly toward the center. Mineralization in the base of a chimney dies out abruptly and only extends a short distance along joints and fractures below the massive sulfides.

Rubble ore is common in the chimneys, especially in the interior and the upper part of the down-dip side. Commonly a core of rubble ore will extend through much of that portion of the chimney in the Leadville Limestone. The lower parts of the chimneys below Gilman Sandstone contain more bedrock ore. Sand ore is abundant, especially near the tops of the chimneys. Much of the chimney ore is very vuggy and porous.

The tops of the chimney are very irregular due to "shaly lime" pendants and channels of cap collapse (Figure 5), which have been found as deep as the Gilman Sandstone. All the chimneys are below the zone of oxidation.

## Grade of Ore

The average grade of the primary zinc ore is approximately 12.0 per cent zinc, 2.0 per cent lead, 1.0 ounces of silver, and 0.02 ounces of gold per ton. The average grade mined for yearly periods has varied from 11.7 to 17.0 per cent zinc and from 1.2 to 2.4 per cent lead. The copper-silver ore from the chimneys averages 18.7 ounces of silver and 0.08 ounces of gold per ton and 3 5 per cent copper. Small tonnages of copper-silver ore have been mined that have yielded several thousand ounces of silver and several tens of ounces of gold per ton.

Accurate data are not available for the grade of the oxidized ores. There are numerous legends of small, fabulously rich pockets, but, according to the more sober and reliable reports, the oxidized ore from the quartzite mantos ranged from 10 to 80 ounces of silver and 0.1 to 4.8 ounces of gold per ton. The oxidized ore from the limestone mantos ranged from 8 to 21 ounces per ton of silver, 0.2 to 0 8 ounces per ton of gold, and 8 to 15 per cent lead. The oxidized manganese ore averaged 15 per cent manganese, 38 per cent iron, and 1 to 2 per cent silica (3).

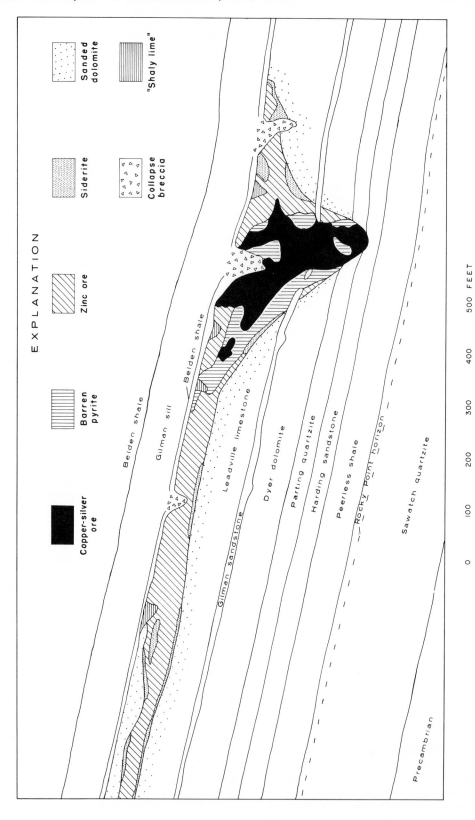

Fig. 5. *Diagramatic Section of Typical Chimney and Limestone Manto Ore Bodies, looking northwest.*

## Factors Controlling Form and Location of Ore Bodies

The controls for the ore bodies in the district are very subtle and are not completely known. The location of the district as a whole is probably related to the basement structures of the Homestake shear zone of the Colorado mineral belt. The district is underlain by Precambrian granite, and this rock, because of its greater competency in contrast to that of the inclosing metasediments, has had an important influence on the type and continuity of the structures developed in this part of the shear zone.

The granite is moderately faulted and fractured over a wide area and some of these structures are mineralized. Both the breaks and fissure veins are more numerous and more closely spaced in the area of the "horseshoe," and trend northeasterly under the ore bodies. Only a few of the fissure veins persist from the granite into the quartzite, and most of these die out upward within a short distance. The Bleak House vein, the only known exception, cuts the granite, the full thickness of the Sawatch Quartzite, and the lower part of the Peerless Shale. It is mineralized for this entire interval. Another fissure vein, the Doddridge winze ore body (9, p. 73) in the Ground Hog mine, may have a continuity similar to that of the Bleak House but has long been inaccessible for study. However, none of the structures in the granite and quartzite continues into the Devonian and Mississippian strata where the major ore deposits occur.

Although no throughgoing conduits have been found, the vertical stacking of the deposits in the various rock units suggests a high-angle control for the ore bodies. The mantos in the limestone lie more or less vertically above the larger mantos in the quartzite, and these in turn occur above the areas of greatest fracturing in the Precambrian granite (Figure 3).

The ore shoots in the quartzite are localized by both stratigraphic factors and by fracturing. The Bleak House manto has been formed in a dolomite lens in the Rocky Point horizon on both sides of the Bleak House vein, and several smaller mantos occur along the same vein in lower dolomitic beds in the Sawatch Quartzite. In the Ground Hog mine, the ore occurs in the quartzite facies of the Rocky Point horizon in a zone of abundant northeasterly faults and fractures.

In the Leadville Limestone, the distribution of ore depends largely on the location of the chimneys that served as the major inlets for the hydrothermal solutions. However, it is not possible to relate the chimneys to any specific structural control. There are mine workings directly under the bottoms of three of the larger chimneys, and no feeding structures have been found. The bottoms of several of the chimneys have been completely mined out, and only thin veinlets of sulfide minerals were found in minor fractures immediately below the massive sulfide ore, and these die out in a few feet. The only direct evidence that the mineralizing solutions moved upward through the rock below the chimneys consists of bleaching and slight argillic alteration of some of the lower strata and scattered small patches and stringers of pyritic mineralization in the Sawatch, Peerless, and Harding formations. However, upward movement of the solutions through the chimneys into the mantos is indicated by both the mineral zoning in the chimney manto systems and by solution movement studies based on Newhouse's criteria (16).

Three of the major chimneys lie along a northwesterly trending line at the toe of the "horseshoe" (Figure 3) and are connected by strike mantos. This linearity has prompted speculation about possible basement structures such as a northwest-trending fault zone or a contact between contrasting rock types, but not enough is known about the Precambrian below the chimneys to arrive at definite conclusions. In the sediments there is a small northwesterly trending terrace parallel to the line of the chimneys. Various theories have been proposed relating the chimneys to the intersection of this terrace with northeasterly fault zones (15, p. 15), small northeasterly synclinal flexures (25, p. 14), and the thrust fault in the Harding Sandstone (20, p. 114). None of the theories, however, adequately explain the structural control of the chimneys. The writers believe the primary control is in the Precambrian basement rocks but have few facts with which to support this hypothesis.

Stratigraphic and small structural features exercised important secondary controls on the form and distribution of ore in the carbonate rocks. Almost all the manto ore is in the Leadville Limestone, and only a few small bodies of sulfides occur in the Dyer Dolomite outside of the chimneys. The concentration of manto ore in the Leadville Limestone is in part due to the impermeable Belden Shale cap rock and in part due to stratigraphic features in the Leadville. The impervious shale deflected the upward migration of the solutions rising through the chimneys, resulting in the formation of the manto ore bodies in the underlying Leadville Limestone.

The manto ore bodies vary in stratigraphic

position in the Leadville Limestone. Much of the ore is in the upper part of the formation immediately below the capping Belden Shale. However, several large mantos occur in lower beds and are capped by carbonate rocks, and other mantos change from one bed to another. Variations in porosity and permeability related to sedimentation and diagenesis, no doubt, influenced subsequent events and the localization of the manto ore bodies. Such variations may account for the greater favorability of the Leadville Limestone compared to that of the Dyer Dolomite as well as for the stratigraphic position of the ore in the Leadville Limestone itself. Rove (22, p. 72–73), Ohle (29, p. 893), and Roach (40) have investigated porosity and permeability of various stratigraphic units in the Eagle mine, but because of the limited scope of the investigations, it is not feasible to draw firm conclusions from the results.

A karst topography was developed in the Leadville Limestone prior to the deposition of the Pennsylvanian strata, and its development was, no doubt, influenced by sedimentary and diagenetic features of the rock. This created zones of porosity and permeability which were followed by the ore depositing solutions as indicated by the greater abundance of sink holes filled with "shaly lime" in the vicinity of the ore bodies than away from them. Hydrothermal solutions associated with the ore-forming stage enlarged and accentuated the zones of porosity and permeability producing sanding, channeling, and collapse breccias. The most pronounced thinning in the Leadville Limestone is associated with the ore bodies.

The zones of weak, discontinuous, northeasterly trending faults and fractures also influenced the localization of the limestone mantos. These structures are more numerous in the vicinity of the ore bodies, and at several places in the Eagle mine, ore follows along small faults for short distances. Some of these structures may have developed prior to Pennsylvanian time as the result of intermittent movement on the Homestake shear zone (43, p. 1008) and, thus, influenced the development of the karst zones. Although some of the northeasterly trending faults and fractures were factors in the localization of the mantos, many of them are the result of adjustments in the strata due to reduction in volume by both karst and hydrothermal solution activity.

The limestone mantos are associated with shallow synclinal structures upon which minor second-order folds or crenulations have been superimposed. These structures have been interpreted (15, p. 15; 25, p. 14) as being responsible for ground preparation and localization of the manto ore bodies. Although these folds are spatially related to ore, they, like many of the faults, are largely the result of reduction in volume.

The structural features responsible for the chimneys, although imperfectly understood, are the most important primary controls for the localization of the ore. The solutions that formed the limestone manto ore bodies utilized the best of the available pathways, whatever their nature. Callahan (44, p. 239–241) interprets the karst topography as being the fundamental control for the limestone manto ore bodies, and the writers are in accord with this interpretation. The complete replacement of the rock by massive sulfides and the hydrothermal alteration destroyed many of the original features responsible for the localization of the ore. The formation of the ore bodies was a dynamic and continuing process which strongly modified the initial controls and produced additional features, thus making it difficult to distinguish between cause and effect.

### Source of the Ore

The source of the mineralizing solutions is not known. Other than for the Gilman sill, the nearest known intrusive is a small pluton of granodiorite of Laramide age (42, p. C78–C79) approximately 11 miles southwest of the district. Neither the sill nor this stock are the source of the ore. Presumably the solu-

TABLE III. Isotopic Composition of Lead from the Gilman District

| Sample | Lead Isotopes, Atomic Per Cent | | | |
|--------|------|-------|-------|-------|
|        | 204  | 206   | 207   | 208   |
| 1      | 1.36 | 24.35 | 21.37 | 52.92 |
| 2      | 1.35 | 24.48 | 21.35 | 52.81 |
| 3      | 1.36 | 24.52 | 21.22 | 52.90 |
| 4      | 1.37 | 24.37 | 21.33 | 52.93 |
| 5      | 1.38 | 24.78 | 21.31 | 52.50 |
| 6      | 1.26 | 28.58 | 20.08 | 50.06 |
| 7      | 1.29 | 27.43 | 20.45 | 50.81 |

Samples 1 to 3—Analyses by Isotopes, Inc.: 1, manto and chimney ore composite, Eagle mine; 2, Bleak House vein at Rocky Point horizon, Sawatch Quartzite; 3, veins in Precambrian granite, Eagle mill.

Samples 4 to 7—Atomic per cent calculated from ratios from Engel and Patterson (36): 4, ore lead; 5, dolomite at ore contact; 6, dolomite near edge of dolomitized area approximately 2 miles from Eagle mine; 7, unaltered Leadville Limestone.

tions moved upward, perhaps by a circuitous route, from a deeper source that could be the parent magma for both the ore and the known intrusives.

The isotopic composition of lead from ore occurrences in host rocks of widely differing ages is essentially uniform (Table III, Samples 1–4), indicating that all the ores in the district were derived from a common source. The different isotopic composition of indigenous trace lead in the Leadville Limestone (Sample 6 and 7) implies that the ore lead was not derived from that formation. No lead isotope data are available with regard to the other sedimentary formations of the district.

## SUMMARY OF GEOLOGIC EVENTS

The first recorded event was the deposition of a thick series of sediments that were subsequently strongly folded, faulted, and metamorphosed to form the Idaho Springs Schist. The basic structures of the Colorado mineral belt were developed during this diastrophism. Diorite and granite were later intruded, the area was uplifted, and a long period of erosion and peneplanation followed.

The Paleozoic was characterized by repeated periods of submergence and deposition interspersed with periods of uplift and erosion resulting in the deposition of a series of clastic and carbonate rocks, many of the individual units of which are separated by erosional unconformities. Minor movements probably occurred at several times along the elements of the Homestake shear zone. The most important Paleozoic events from the standpoint of the formation of the ore deposits were the deposition of the Leadville Limestone and the subsequent period of uplift and erosion during which a karst topography was developed.

There are no Mesozoic rocks in the district. However, the entire area was probably covered by a thick series of Mesozoic sediments prior to the Sawatch uplift and the subsequent erosion.

Starting in late Cretaceous and continuing into early Tertiary time, there was a long period of uplift, faulting, and intrusion. The Sawatch uplift formed at this time giving the strata their northeasterly dip. The Gilman sill was intruded during the Laramide orogeny, but it is not known whether this occurred before or after the uplift. Hydrothermal activity from an unknown source followed the intrusion resulting in the formation of the primary ore deposits and the alteration of the wall rocks.

A long period of stream and glacial erosion followed the Laramide orogeny and the formation of the ore deposits resulting in the development of the present mountainous topography. The upper ends of the manto ore bodies were eroded away and the near surface portions of the ore bodies were oxidized.

## REFERENCES CITED

1. Pearce, R., 1890, The association of gold with other metals in the West: A.I.M.E. Tr., v. 18, p. 447–457.
2. Means, A. H., 1915, Geology and ore deposits of Red Cliff, Colorado: Econ. Geol., v. 10, p. 1–27.
3. Umpleby, J. B., 1917, Manganiferous iron ore occurrences at Red Cliff, Colorado: Eng. and Min. Jour., v. 104, p. 1, 140–141.
4. Graton, L. C., 1922, Geological report No. 1 on Eagle mines, Empire Zinc Company, Gilman, Colorado: The New Jersey Zinc Company, Private Rept., 33 p.
5. Graton, L. C. and Short, M. N., 1922, Description of specimens from the Eagle mine, Colorado: The New Jersey Zinc Company, Private Rept., 9 p.
6. Graton, L. C. and Ettlinger, I. A., 1922, Geological report No. 2 on Eagle mines, The Empire Zinc Company, Gilman, Colorado: The New Jersey Zinc Company, Private Rept., 30 p.
7. Graton, L. C., 1923, Geological report No. 3 on Eagle mines. Empire Zinc Company, Gilman, Colorado: The New Jersey Zinc Company, Private Rept., 71 p.
8. Pinger, A. W., 1923, Mineralogy of the ores: in Graton, L. C., *Geological report No. 3 on Eagle mines, Empire Zinc Company, Gilman, Colorado:* The New Jersey Zinc Company, Private Rept., p. 72–85.
9. Crawford, R. D. and Gibson, R., 1925, Geology and ore deposits of the Red Cliff district, Colorado: Colo. Geol. Surv. Bull. 30, 89 p.
10. Adams, S. F., 1927, Geological information at Gilman to August 24, 1927: The New Jersey Zinc Company, Private Rept., 18 p.
11. Emmons, S. F., *et al.*, 1927, Geology and ore deposits of the Leadville mining district, Colorado: U.S. Geol. Surv. Prof. Paper 148, 368 p.
12. Behre, C. H. Jr., 1933, Physiographic history of the upper Arkansas and Eagle Rivers, Colorado: Jour. Geol., v. 41, p. 785–814.
13. Lovering, T. S. and Behre, C. H. Jr., 1933, Battle Mountain (Red Cliff, Gilman) mining district (Colorado): 16th Int. Geol. Cong., Guidebook 19, p. 69–76.
14. Brown, W. H., 1937, Geologic report on the Eagle mine of the Empire Zinc Company: The New Jersey Zinc Company, Private Rept., 24 p.

15. Brown, W. H., 1938, Geologic report on the Eagle mine of the Empire Zinc Company: The New Jersey Zinc Company, Private Rept., 22 p.

16. Newhouse, W. H., 1941, The direction of flow of mineralizing solutions: Econ. Geol., v. 36, p. 612–629.

17. Callahan, W. H., 1942, Geology, geophysics, and exploration at Gilman, Colorado: The New Jersey Zinc Company, Private Rept., 199 p.

18. Jerome, S. E., 1943, Results of geological conference at Gilman, Colorado: The New Jersey Zinc Company, Private Rept., 79 p.

19. Johnson, J. H., 1944, Paleozoic stratigraphy of the Sawatch Range, Colorado: Geol. Soc. Amer. Bull., v. 55, p. 303–378.

20. Lovering, T. S. and Tweto, O. L., 1944, Preliminary report on geology and ore deposits of the Minturn Quadrangle, Colorado: U.S. Geol. Surv. Open File Rept., 115 p.

21. Dykstra, F. R., 1947, Paragenesis of the ore mineralization at the Eagle mine, Gilman, Colorado: Columbia Univ., Unpublished Masters Thesis, 20 p.

22. Rove, O. N., 1947, Some physical characteristics of certain favorable and unfavorable ore horizons: Econ. Geol., v. 42, p. 57–77, 161–193.

23. Tweto, O. L. and Lovering, T. S., 1947, The Gilman district, Eagle County; *in* Vanderwilt, J. W., *Editor, Mineral Resources of Colorado:* State of Colo., Mineral Res. Bd., p. 378–387.

24. Tweto, O. L., 1947, Gilman mining district: Rocky Mtn. Assoc. Geols. Guidebook, Field Conf. June 1947, p. 55–59.

25. Fowler, G. M. and Hernon, R. M., 1948, The Eagle mine, Gilman, Colorado: The New Jersey Zinc Company, Private Rept., 26 p.

26. Tweto, O. L., 1948, Gilman mining district, Eagle County, Colorado; *in* Vanderwilt, J. W., *Editor, Guide to the geology of central Colorado:* Colo. Sch. Mines Quart., v. 43, p. 122–127.

27. ———— 1949, Stratigraphy of the Pando area, Eagle County, Colorado: Colo. Sci. Soc. Proc., v. 15, p. 149–235.

28. Davidson, R. N., 1950, Hydrothermal alteration effects in the Leadville (Colorado) limestone and their relation to metallization (abs.): Geol. Soc. Amer. Bull., v. 61, p. 1551.

29. Ohle, E. L., Jr., 1951, The influence of permeability on ore distribution in limestone and dolomite: Econ. Geol. v. 46, p. 667–706, 871–908.

30. O'Neill, T. F., 1951, Hydrothermal alteration in the Battle Mountain mining district, Eagle County, Colorado: The New Jersey Zinc Company, Private Rept., 294 p.

31. Rodda, J. L., 1951, Composition studies of Gilman sphalerite and manganosiderite: The New Jersey Zinc Company, Private Rept., 19 p.

32. Tweto, O. L., 1951, Form and structure of sills near Pando, Colorado: Geol. Soc. Amer. Bull., v. 62, p. 507–531.

33. Snively, N., 1952, The controls of Gilman ore deposition: The New Jersey Zinc Company, Private Rept., 46 p.

34. Behre, C. H., 1953, Geology and ore deposits of the west slope of the Mosquito Range (Colorado): U.S. Geol. Surv. Prof. Paper 235, 176 p.

35. Engel, A. E. J. and Engel, C. G., 1956, Distribution of copper, lead, and zinc in hydrothermal dolomites associated with sulphide ore in the Leadville limestone (Mississippian, Colorado) (abs.): Geol. Soc. Amer. Bull., v. 67, p. 1692.

36. Engel, A. E. J. and Patterson, C. C., 1957, Isotopic composition of lead in Leadville limestone, hydrothermal dolomite, and associated ore (Colorado) (abs.): Geol. Soc. Amer. Bull., v. 68, p. 1723.

37. Engel, A. E. J., *et al.,* 1958, Variations in the isotopic composition of oxygen and carbon in Leadville limestone (Mississippian, Colorado) and its hydrothermal and metamorphic phases: Jour. Geol., v. 66, p. 374–393.

38. Lovering, T. G., 1958, Temperatures and depth of formation of sulfide ore deposits at Gilman, Colorado: Econ. Geol., v. 53, p. 689–707.

39. Rodda, J. L., 1958, Zinc and iron contents of Gilman sphalerite: The New Jersey Zinc Company, Private Rept., 9 p.

40. Roach, C. H., 1960, Thermoluminescence and porosity of host rocks at the Eagle Mine, Gilman, Colorado: U.S. Geol. Surv. Prof. Paper 400-B, p. B107–B111.

41. Tweto, O. L., 1960, Pre-ore age of faults at Leadville, Colorado: U.S. Geol. Surv. Prof. Paper 400-B, p. B10–B11.

42. Pearson, R. C., *et al.,* 1962, Age of Laramide porphyries near Leadville, Colorado: U.S. Geol. Surv. Prof. Paper 450-C, p. C78–C80.

43. Tweto, O. L., and Sims, P. K., 1963, Precambrian ancestry of the Colorado mineral belt: Geol. Soc. Amer. Bull., v. 74, p. 991–1014.

44. Callahan, W. H., 1964, Paleophysiographic premises for prospecting for strata bound base metal mineral deposits in carbonate rocks: *Cento Symposium on Mining Geology and Base Metals,* Istanbul, p. 191–248.

45. Silverman, A., 1965, Studies of base metal diffusion in experimental and natural systems; Part II, Diffusion systems at Gilman, Colorado: Econ. Geol., v. 60, p. 325–349.

# 31. The Titaniferous Magnetite Deposit at Iron Mountain, Wyoming[*]

ARTHUR F. HAGNER[†]

## Contents

## Illustrations

[*] Publication authorized by the Director, U.S. Geological Survey and the State Geologist, Geological Survey of Wyoming.
[†] University of Illinois, Urbana, Illinois.

# Tables

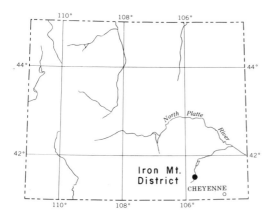

## ABSTRACT

The titaniferous magnetite deposit at Iron Mountain, Wyoming, is in Precambrian anorthosite. Individual ore bodies are lenses, commonly arranged en echelon, conformable to the platy crystal structure and compositional layering of the anorthosite; a few transect these features at various angles. The ore bodies are on the east flank of the principal anticline in the anorthosite mass.

Magnetite and ilmenite, accompanied by spinel, are the ore minerals. The principal gangue minerals are olivine, which was introduced with the ore, and plagioclase, the chief constituent of the anorthosite.

The anticlinal region appears to have been a structurally weak zone, and westward-bowing curves of the axis were especially favorable sites for localization of the two largest massive ore bodies. These bodies are synclinal and plunge southeast.

The proportion of magnetite-ilmenite to olivine and to anorthosite appears to be related to structural features. Where magnetite-ilmenite is present with olivine, the magnetite-ilmenite is encountered along the most pronounced parts of a fold where anorthite content is highest, whereas olivine is associated principally with more gently folded portions.

During granulation and recrystallization of the anorthosite mass, temperature and pressure changes furnished the energy required to release and mobilize iron and other elements that migrated to areas of low pressure, the ore zone. Some iron and magnesium may have combined with silica from partly dissociated plagioclase to form olivine.

The ore formed by replacement of anorthosite as indicated by: (1) the marked changes in mineralogy along strike, dip, and plunge of an ore body; (2) the concentration of massive ore in folds and olivine on limbs; (3) the halo of low-grade mineralized rock that transects the structure of the anorthosite in the hanging wall; and (4) the layering in the large southern ore body, that strikes more easterly than the trend of the body.

## HISTORICAL INFORMATION

The titaniferous magnetite deposit at Iron Mountain, Wyoming, has been known for more than a hundred years, and reconnaissance work dates back to the surveys of Stansbury (1), Hayden (2), and King (3). Brief studies of the deposit were made by Knight (4), Kemp (5, 7), Lindgren (6), and Ball (8) and this work was summarized in 1913 by Singewald (10) who also added new information. In 1941, Diemer (12) discussed the titaniferous magnetites of the Laramie Range, Wyoming, and included the first generalized map of the Iron Mountain deposit, other than a sketch map published by Kemp in 1905 and included in the bulletin by Singewald.

In 1944, a detailed geologic map of Iron Mountain was made by Newhouse and Hagner (19), assisted by M. L. Troyer, for the United States Geological Survey and the Geological Survey of Wyoming. In their report on the deposit, Newhouse and Hagner discuss the geology, structure, grades and types of ore, and include a geologic map and sections. Much of the substance of the present paper has been taken from that open file report. During their mapping, Newhouse and Hagner spotted drilling sites for the United States Bureau of Mines, and 17 holes were drilled (14). The results of this work prompted the Union Pacific Railroad Company, owner of a substantial part of the property, to begin an exploratory program in 1951. The program consisted of geologic mapping, geophysical surveying, diamond drilling, and surface trenching; field studies have been continued during subsequent seasons. Metallurgical tests have been made by the Union Pacific Railroad Company, the United States Bureau of Mines, and other organizations to determine the most efficient and economical process for concentrating low-grade ores and for further reduction of higher-grade ores and concentrates into marketable products. In recent years, the Pacific States Cast Iron Pipe Company of Provo, Utah, the Anaconda Copper Company, and the Magnetites Products Company of Florence, Colorado, have been engaged in exploration work in the area.

Since 1960, Magnetites Products Company has mined 400,000 tons of high-grade ore at Iron Mountain that has been shipped to New Orleans and used as an aggregate for heavy concrete to coat submerged pipelines to off-shore wells in the Gulf Coast region. Before shipping, the ore is crushed, sized, and passed over a magnetic pulley to produce a uniform, high-gravity (4.5) product of minus quarter-inch size.

To date, over 21,000 tons of ore have been mined and shipped for metallurgical pilot plant tests, but no iron, titanium, or vanadium has been produced commercially.

In 1944, Newhouse and Hagner (19) estimated that the Iron Mountain deposit contained 9,150,000 tons of indicated and inferred ore. On the basis of additional drilling, trenching, and magnetometer surveying, the Union Pacific Railroad Company estimated the ore reserves to be 23,056,800 tons (29). This figure includes the low-grade "mineralized anorthosite" of Newhouse and Hagner, most of which was not included in their ore estimate.

## REGIONAL SETTING

### General Geology

Iron Mountain is 47 miles northeast of Laramie by road in sections 22, 23, 26, and 27 of T19N, R71W, Albany County, Wyoming. The nearest railroad station is at Farthing on the Colorado and Southern Railroad 9 miles to the southeast. Bosler, on the Union Pacific Railroad, is about 31 miles west. Two improved county roads and several ranch roads provide access to the area.

Iron Mountain is a rugged northward-trending ridge in the Laramie Range, breached near the southern end by North Chugwater Creek. The ridge, which is near the eastern margin of an anorthosite mass, rises about 660 feet above the creek level to an elevation of approximately 7450 feet. Titaniferous magnetite ore crops out principally along the ridge crest. Where the ore consisted of massive magnetite-ilmenite with little olivine and anorthosite, it stood above the surrounding anorthosite in prominent outcrops. Mining operations during the past few years have reduced or covered many of these outcrops, but parts of some are now visible in pits. The major ore zone consists of a series of ore bodies that extend about 5000 feet along and near the crest of the ridge.

The Laramie Range is part of the northern portion of the Southern Rockies and constitutes the eastern limb of the Rocky Mountain Front Range in the area; the western limb is the Medicine Bow Mountains, and the two ranges are separated by the Laramie Basin. The Laramie Range is an asymmetric anticline that trends north-south. Paleozoic rocks along the west flank dip gently west whereas on the east flank these rocks dip steeply east or are overturned and thrust faulted (Figure 1*). The core of the Range has been truncated by a gently rolling surface that slopes to the east and is cut by numerous streams. The physiographic history of the Laramie Range has been discussed by Blackwelder (9) and Moore (27).

The oldest rocks in the Laramie Range are Precambrian quartzite, marble, hornblende schist, biotite schist, and gneisses. Next in age are the anorthositic rocks, the host rocks of the titaniferous magnetite deposits. Anorthosite has been intruded by masses and dikes of norite. After the norite crystallized, quartz syenite, gneisses, granites, pegmatite, and "lamprophyre" dikes were formed. The anorthosite series and its relationship to other rocks of the Laramie Range is shown on the *Geologic Map of Anorthosite Areas, Southern Part of Laramie Range, Wyoming,* by Newhouse and Hagner (24). A simplified and slightly modified version of this map is presented by Klugman (26).

* Figure 1 is a highly simplified and reduced copy of U.S. Geol. Surv. Min. Invest. Field Studies Map 119.

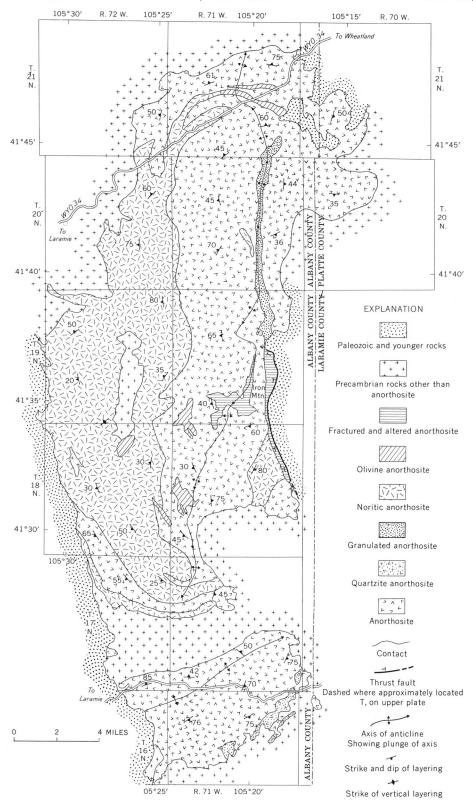

EXPLANATION

Paleozoic and younger rocks

Precambrian rocks other than anorthosite

Fractured and altered anorthosite

Olivine anorthosite

Noritic anorthosite

Granulated anorthosite

Quartzite anorthosite

Anorthosite

Contact

Thrust fault
Dashed where approximately located
T, on upper plate

Axis of anticline
Showing plunge of axis

Strike and dip of layering

Strike of vertical layering

FIG. 1. *Generalized Geologic Map of the Laramie Range Anorthosite Area.*

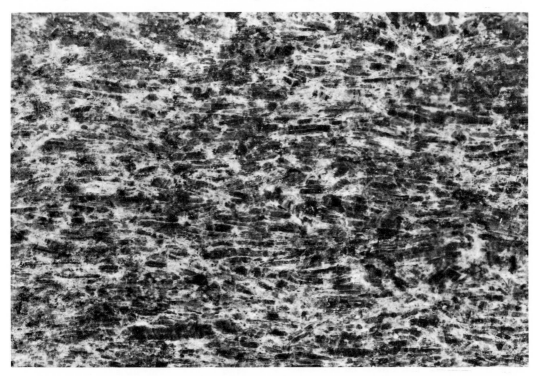

FIG. 2. *Platy Crystal Structure in Anorthosite.*

## Anorthosite of the Laramie Range

The anorthosite series ranges in composition from rocks that are almost entirely plagioclase to those containing 50 per cent or more dark minerals. Anorthosite contains 90 per cent or more plagioclase with the remainder ferromagnesian silicates and minor magnetite-ilmenite. Noritic anorthosite contains, for the most part, 10 to 20 per cent orthopyroxene, with magnetite-ilmenite as an accessory mineral. Locally, orthopyroxene constitutes up to 50 per cent of the noritic anorthosite. In places, olivine forms 5 or more per cent of the anorthosite and may reach over 50 per cent; orthopyroxene is commonly, but not everywhere, present. Thus, the rock is an olivine anorthosite, troctolite, or olivine norite depending upon the percentages of olivine and pyroxene it contains.

Anorthosite generally forms angular outcrops and exhibits a foliation brought out by parallelism of plagioclase crystals, called platy crystal structure (Figure 2). Pronounced layering, due to different proportions of minerals or to different grain size, is often displayed; this layering is parallel to the platy crystal structure. Noritic and olivine anorthosite are more massive, platy structure is poorly developed or absent, and outcrops are rounded. The

anorthosite series has been described by Fowler (11), Hagner (17), and Klugman (26).

A complete section of the anorthosite series is not exposed. The part that is exposed has the form of a folded lens or tabular-shaped mass.

The northern mass consists of 4 major layers of anorthosite and noritic anorthosite, each several thousand feet thick (Figure 1). The lowermost of these layers is anorthosite exposed near the domal part of a major anticline in the eastern part of the mass and extending southwest along its crest. Above this unit is a noritic anorthosite layer with a great thickness on the western flank of the anticline. A third layer upward is anorthosite that is exposed on the extreme south and north ends of the anticline. The fourth and uppermost layer is noritic anorthosite that is exposed only in the extreme southern part of the northern mass (13).

The major structure of the anorthosite is the sharply defined anticline, or antiform,* in the eastern part which trends north-south for 25 miles (Figure 1). An offset of this structure

* This term probably is preferable to anticline, but anticline has been used in several publications on the area and will be used in this paper.

is present as a southern anorthosite mass. In this mass, the anticline trends N60°E and is about 7 miles long. In the northern area, the anorthositic rocks dip, in general, to the west except where modified by the major anticline. Dips of platy crystal structure and layering on the northern anticline are from 20 to 60 degrees, the majority being from 40 to 50 degrees. This structure is modified by minor folds.

A zone of granulated anorthosite, 9 miles long and from 1000 to 3000 feet wide, is present north of Iron Mountain along or near the anticlinal axis (Figure 1). Here the anorthosite is light gray, granulated, and in general finer grained than elsewhere. This zone continues southward as a fractured and altered zone about 8 miles long. Another, more irregular, fractured and altered zone extends for about 4 miles west and south of Iron Mountain along the anticlinal crest where it changes direction.

With the exception of a few small outcrops, the titaniferous magnetite ore bodies in anorthosite are within a mile of the major anticlinal axis in the northern mass. The major deposits and accompanying mineralization occur for the most part where the anticline changes direction, in particular where the axis is convex to the east. Nearly all these deposits are in the lowermost major layer of anorthosite.

### Place of Mineralization in the Geologic History

The titaniferous magnetite ore bodies of the Laramie Range formed after the anorthosite rocks and before the intrusion of granite dikes. Age relationships of the ore bodies to syenite, quartz syenite, norite, and post-anorthosite gneisses were not found during the mapping. The ore bodies are, however, a part of the Precambrian activity of the region and are believed to have formed during the period of granulation and recrystallization of the anorthosite. The sequence of events is summarized below:

(1) Early Precambrian sedimentation, possibly with associated mafic flows that are now represented by hornblende schist.

(2) Deformation and metamorphism of the rocks in (1) to quartzite, marble, schists, and gneisses.

(3) Introduction of the anorthositic rocks followed by granulation and recrystallization along the axis of the major anticline. The granulated, sheared, and altered zones on and near the anticline in the northern anorthosite mass

may have developed at this time, along with widespread shearing and fracturing.

(4) Displacement of the anorthosite body, to form the northern and southern masses, accompanied by introduction of norite as large intrusive masses and dikes; in places these masses and dikes transect the structure of the anorthosite.

(5) Introduction of quartz syenite as replacement bodies followed by the development of gneisses between the northern and southern anorthosite masses as replacements of norite and quartz syenite. Locally cordierite bodies formed in the norite and gneiss as part of this replacement process. (16)

(6) Development of the Sherman Granite pluton, in large part by replacement of pre-existing rocks. (18) Locally granite and lamprophyric, mafic, and ultramafic dikes cut the granite pluton.

(7) Peneplanation.

## ECONOMIC GEOLOGY

### Form of the Ore Bodies

Titaniferous magnetite ore bodies large enough to be of commercial interest occur at Iron Mountain (Figure 3). The largest of these is about 1200 feet long and up to 80 feet thick; locally faulting has increased the thickness to over 100 feet. This ore body occurs near the southern part of the deposit where North Chugwater Creek crosses the ore zone. In the northern part of the deposit where Iron Mountain reaches its maximum elevation, another rather large ore body reaches a length of about 375 feet and a thickness of up to 70 feet; faulting has increased this thickness to over 100 feet near the north end. Both of these bodies are arcuate in form and plunge to the southeast. The remainder of the ore bodies at Iron Mountain are considerably smaller. Massive ore is known to be present at over 30 places in the Laramie Range but most of these other exposures are small; only a few lenses are as much as several hundred feet long and up to 20 feet thick.

Individual ore bodies at Iron Mountain are lenses or tabular masses commonly arranged in discontinuous and overlapping, or en echelon, bodies. These bodies are, with rare exceptions, conformable to the platy crystal structure and compositional layering of the anorthosite host rock.

The Iron Mountain deposit is on the east flank of the major anticline near the eastern border of the anorthosite mass where it is over-lapped by Paleozoic and Mesozoic sedimentary rocks. Two minor layers of noritic anorthosite, each several hundred feet wide, are present within the major anorthosite layer in the Iron Mountain map area (Figure 3). The massive magnetite-ilmenite ore, however, is restricted to the anorthosite.

Many outcrops of anorthositic rocks are sheared, and, where large pits or trenches provide excellent exposures, the widespread occurrence of closely-spaced shears, fractures, and minor faults is evident. Shear zones may be parallel to the platy structure and compositional layering or transect them. In places, minerals are aligned along shear planes where the anorthosite has recrystallized during shearing.

At several places in the Iron Mountain area, titaniferous magnetite ore bodies are offset by faults that appear to have only a few tens of feet of displacement. Numerous small slick-ensided surfaces are present along and near the footwall of the massive ore. Some ore bodies have been cut by granite dikes. Fractures in the anorthosite and ore may be filled with carbonate.

## Mineralogy of the Deposit

The mineralogy of the ore is simple. Magnetite and ilmenite, accompanied by spinel, are the ore minerals; rarely, grains of pyrrhotite may be seen in specimens of drill core. The principal gangue minerals are olivine, which is widespread at Iron Mountain, and plagioclase, the chief constituent of the host rock anorthosite; orthopyroxene is present where ore occurs in noritic anorthosite. A small amount of hematite is found as an alteration product of magnetite.

Magnetite and ilmenite commonly occur as anhedral, roughly equidimensional grains, but they may also be elongated parallel to the platy crystal structure of the host rock. The ore ranges from fine- to coarse-grained, individual crystals ranging in diameter from 0.2 cm to 1 cm. In massive ore, magnetite is more abundant than ilmenite, whereas, in disseminated ore and in mineralized anorthosite, ilmenite is more abundant than magnetite (15).

In thin section, some of the magnetite-ilmenite is rimmed by biotite and green, fibrous hornblende. Commonly, these coronas consist of a narrow, incomplete inner rim of biotite and an outer rim of hornblende. Generally the biotite rim is only a small fraction of a millimeter thick; it may contain granular mag-netite-ilmenite. The hornblende rim is more continuous than the biotite rim and is usually 0.1 to 0.2 mm wide. Where biotite is absent, hornblende may be in direct contact with magnetite-ilmenite. Locally, the outer rim contains as much biotite as hornblende; both rims may contain a very small amount of green spinel and may have a symplectic texture. The rims appear to have formed at the expense of plagioclase, not at olivine-ore contacts.

Spinel is ubiquitous in the ore and occurs as exsolved blebs and lenses in magnetite and in ilmenite exsolved from magnetite and as small grains scattered through these minerals. The grains are somewhat elongated masses along the boundaries of magnetite and ilmenite that range in size from 0.2 mm to 0.3 mm. Spinel is more commonly associated with ilmenite than with magnetite.

Olivine ranges in diameter from a fraction of a millimeter to several centimeters. Grains are commonly rounded and show alteration, principally along fractures, to granular magnetite and serpentine. The fayalite content ranges from 27 to 58 per cent, the great majority of olivines containing between 40 and 56 per cent fayalite. Small amounts of talc, chlorite, biotite, and an unidentified brown mineral may occur as alteration products. Olivine appears to have crystallized, in part, earlier than magnetite-ilmenite because, in places, the latter cuts across, surrounds, and replaces olivine grains. Olivine has replaced minerals of the anorthosite.

The plagioclase of the host rock anorthosite ranges in size from a fraction of a centimeter to occasional crystals that are many centimeters long and 5 cm across; most of the grains fall between the limits of 3 mm to 13 mm across and 5 mm to 25 mm long. The ratio of length to width ranges from about 2:1 in some places to as much as 15:1 elsewhere outside the Iron Mountain area. This ratio also changes with composition, being less in the calcic varieties. The composition of plagioclase in anorthosite ranges from 38 to 56 per cent anorthite; in noritic anorthosite this range is from 46 to 58 per cent. Plagioclase contains opaque, crystallographically oriented rods, needles, and minute particles that probably are exsolved material. DeVore (23) has tentatively identified the needles as being titaniferous hematite that has exsolved some titanium as ilmenite or rutile. He points out that "Along borders of the plagioclase crystals, along zones of strain, microfaulting, and bending, and in the recrystallized plagioclases, the exsolved bodies either are greatly reduced in

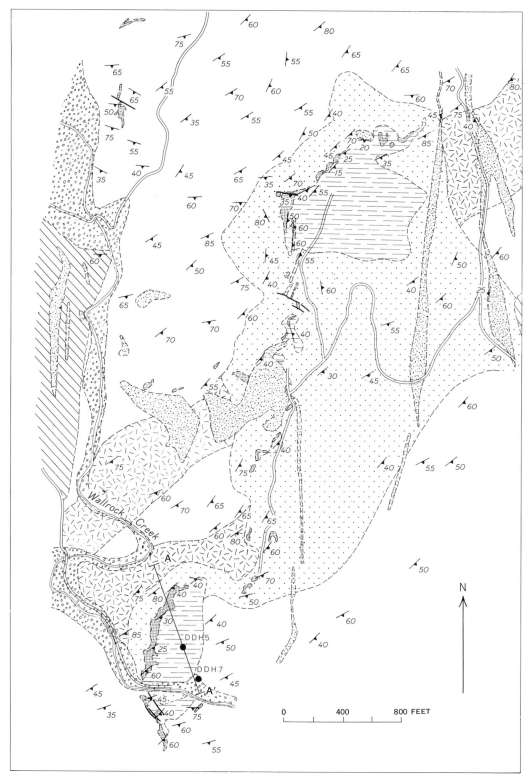

FIG. 3. *Generalized Geologic Map of the Iron Mountain Titaniferous Magnetite Deposit. Section A—A' is shown in Figure 5.*

EXPLANATION

Tertiary and younger rocks, undivided

Sherman "Granite"
Pink or gray, fine- to medium-grained, massive dikes

Magnetite-ilmenite, Grade 1
Magnetite-ilmenite containing from 0 to 35 percent silicates by volume, mostly olivine; massive to layered; lenticular.   TiO$_2$ 16 to 23 percent

Mineralized anorthosite, Type 1
Contains grades 2 and 3 "ore" and olivine in high concentrations but the bulk of the rock is 85 to 90 percent plagioclase with olivine and magnetite-ilmenite

Mineralized anorthosite, Type 2
About 75 percent of the rock contains 5 to 15 percent olivine and 2 to 7 percent magnetite-ilmenite; the remainder is anorthosite; foliated and layered

Fractured and altered anorthosite
Gray, medium-grained plagioclase with disseminated olivine and local magnetite-ilmenite; massive to foliated and layered; fractured and altered

Noritic anorthosite
Gray, coarse-grained plagioclase with 10 to 20 percent orthopyroxene and magnetite-ilmenite; massive to foliated and layered

Anorthosite
Gray- to blue-gray, medium- to coarse-grained plagioclase with 0 to 10 percent orthopyroxene and magnetite-ilmenite; massive to foliated and layered

– – – – – – –

Contact approximately located

⤢ 30

Strike and dip of layering

―――

Fault

number or are absent." Most plagioclase is fresh, but some shows incipient alteration, particularly along fractures and twin lamellae, to sericite, carbonate, and clay. The mineral differs in composition with the kind of anorthosite, with structural features, and with proximity to ore bodies.

Drill cores through noritic anorthosite at Iron Mountain contain orthopyroxene as a common mineral, but these cores generally contain little or no high-grade ore. If ore is present, it generally is of the low-grade disseminated type. Orthopyroxene, averaging 68 per cent ferrosilite, is commonly altered in part to one or more of the following minerals: hornblende, actinolite, biotite, epidote, or chlorite.

The Iron Mountain titaniferous magnetite deposit consists of two chief mineralogic types. One contains massive magnetite-ilmenite with minor spinel. A second type contains olivine and in places plagioclase in addition to the minerals of the first type. A third type, not found at Iron Mountain, consists of magnetite-ilmenite, spinel, and apatite. Locally these minerals are arranged in bands producing a marked compositional layering of the ore.

The first two types of ore commonly are associated in the same ore body, and one type may grade into the other along strike and down dip. Within an ore body, olivine frequently is more abundant in some layers than in others; thin layers may consist almost entirely of olivine and these adjoin other layers containing almost none. The layering at Iron Mountain commonly is parallel to the plane of platy crystal structure and layering in the adjoining anorthosite.

Massive magnetite-ilmenite containing only a small percentage of silicate is resistant to weathering and forms prominent outcrops. Olivine, the most common silicate in the ore, is largely removed from outcrops by weathering to give pitted surfaces (Figure 4). Where olivine exceeds 20 per cent, the ore disintegrates and generally does not crop out. Of the minerals in ore taken from outcrops, magnetite may show alteration effects whereas ilmenite and spinel do not. In these specimens, hematite is present in magnetite, principally along cracks and grain boundaries and around the margins of exsolved ilmenite and spinel. In places, magnetite is completely altered to hematite.

Both anorthosite and noritic anorthosite have been profoundly modified in the central part of the map area by the introduction of olivine and magnetite-ilmenite. The hanging

Fɪɢ. 4. *Layering of Magnetite-Ilmenite and Olivine, as brought out by olivine pits.*

wall and footwall of the major ore bodies are mineralized anorthosite that differs considerably in width. Width is related to the size of ore bodies, to strike and dip of layering, and to topography.

TABLE I.  *Anorthite Content of Plagioclase in Different Environments, Iron Mountain, Wyoming*

| Environment of Specimens | Average Anorthite % | Number of Specimens | Range |
|---|---|---|---|
| Anorthosite not adjoining ore | 48.9 | 20 | 40–56 |
| Mineralized anorthosite, types 1 and 2 | 50.0 | 26 | 45–55 |
| Anorthosite in and adjoining (within 2 ft.) grades 2 and 3 ore | 49.1 | 23 | 38–56 |
| Anorthosite adjoining (within 15.5 ft. or less) grade 1 ore | 54.1 | 6 | 51–56 |

The mineralized rock has been divided into two types. Type I contains grades 2 and 3 "ore" and olivine, but the bulk of the rock consists of 85 to 95 per cent plagioclase with olivine and magnetite-ilmenite distributed throughout. This type could not be subdivided because of insufficient exposures. Type II has less abundant mineralization; about 75 per cent of the rock contains 5 to 15 per cent olivine and 2 to 7 per cent magnetite-ilmenite, the remainder is anorthosite. One type may pass into the other or into ore along strike or down dip. The extent of these types is shown in Figure 3. This halo of low-grade mineralization developed principally along the hanging wall of the ore zone; it transects and replaces the host anorthosite.

Studies by W. H. Newhouse and the writer established that the anorthite content of plagioclase adjacent to massive ore was several per cent higher than that some distance away. H. S. Brown (21) assembled the data on which this generalization was based and added many anorthite determinations of his own. Table I

contains data on anorthite content of plagioclase in different rock environments at Iron
Mountain. It will be noted that the anorthite
content of plagioclase within 1 inch to 16.5
feet of grade I ore is about 5 per cent higher
than that that some distance away, and is also
higher than that adjoining "ore" grades 2 and
3 and the mineralized anorthosite.

Turner (15) has discussed the complex series of exsolution intergrowths and overgrowths in the magnetite-ilmenite ores of the
Laramie Range. He describes the following
forms in some detail:

(1) Ilmenite intergrowths in magnetite;
(2) Ilmenite-spinel intergrowths within magnetite grains;
    Major lamellae;
    Minor lamellae;
(3) Spinel intergrowths in magnetite;
    Disseminated blebs;
    Strings of blebs;
(4) Ilmenite-spinel frames about magnetite
grains;
(5) Magnetite-spinel intergrowths in ilmenite;
(6) Ilmenite-spinel overgrowths on ilmenite;
(7) Ilmenite overgrowths on spinel.

The following description of mineral textures has been taken, in large part, from
Turner. The ore minerals occur as separate
anhedral grains of ilmenite and as intimate
intergrowths of ilmenite and spinel in magnetite. Ilmenite grains of the massive ore are
generally free from inclusions but occasional
irregular masses of spinel and exsolution lenses
of magnetite and spinal may be present. In
disseminated ore, the ilmenite grains "are
usually bordered by an overgrowth of ilmenite
containing abundant blebs of spinel which has
been exsolved from the adjacent magnetite."
(15, p. 8) The intergrowths show ilmenite
forming a network of fine lamellae along
octahedral partings of magnetite. In places,
ilmenite is present as incomplete rims and
patches around magnetite and as veinlets and
fracture fillings in it.

Magnetite contains several types of ilmenite,
ilmenite-spinel, and spinel exsolution forms.
In ore with "less than 5 to 10 per cent silicates
the magnetite grains contain only exsolved
blebs and lenses of spinel and an almost ul-
tra microscopic network of exsolved ilmenite. . . . The ilmenite network may be
found all through the magnetite, but the exsolved spinel is found only in the central portions of the grains." In lean ores and along
borders of thick ore lenses, where silicates
form over 5 to 10 per cent of the ore, "ilmen-

ite-spinel exsolution features appear within and
about the borders of the magnetite grains in
addition to the forms mentioned above. These
magnetite grains also have margins free from
intergrowths except the fine network," however, "certain large, well-developed exsolved
ilmenite-spinel lamellae may extend to the
intergrain boundaries." (15, p. 8–9)

Where the ore contains less than 5 per cent
silicates, exsolved ilmenite laths are abundant.
Ilmenite laths and grains "always contain exsolved spinel blebs. Where exsolved ilmenite
laths are missing the ilmenite grains in the
ore never contain exsolved spinel blebs. . . .
Spherical spinel blebs occur in magnetite in
all ore samples, whether exsolved ilmenite laths
are present or not." However, where the
laths occur, the number of blebs seems "to
be less than where the laths are missing." (15,
p. 33)

## Grade of Ore

The Iron Mountain ore bodies have been
divided into 3 grades on the basis of volume
percentage of silicates present. The $TiO_2$ percentage of these grades was obtained from assays made by the United States Bureau of
Mines.

Grade 1 consists of massive magnetite-ilmenite with from 0 to 35 per cent silicates,
mostly olivine and minor spinel. The $TiO_2$ percentage ranges from 16 to 23.

Grade 2 consists of magnetite-ilmenite with
35 to 65 per cent silicates, chiefly olivine and
minor spinel, plagioclase, and orthopyroxene.
The $TiO_2$ percentage ranges from 10 to 16.
"Ore" of this grade rarely crops out.

Grade 3 consists of magnetite-ilmenite with
65 to 85 per cent silicates, largely plagioclase
with minor olivine and orthopyroxene. The
$TiO_2$ percentage ranges from 5 to 10. Grade
3 "ore" does not form outcrops.

A chemical analysis of grade 1 ore is given
in Table II and for comparison analyses of
massive ore from two other deposits in the
Laramie Range are included. Partial chemical
analyses of silicate "ore" are shown in Table
III. Many partial analyses of the ore at Iron
Mountain were made by the United States Bureau of Mines (14).

The percentage of ore minerals and gangue
may change laterally and in depth so that individual ore bodies consisting of massive magnetite-ilmenite grade into disseminated ore and
vice versa. This is best seen in geologic sections
depicting an ore body that has been penetrated
by one or more drill cores thus illustrating

TABLE II.   *Chemical Analyses of Massive Ore*

| | (1) Iron Mountain | (2) Shanton | (3) Taylor |
|---|---|---|---|
| $SiO_2$ | none | .26 | .36 |
| $Al_2O_3$ | 8.12 | 5.13 | 3.01 |
| $Fe_2O_3$ | 26.26 | 38.34 | 26.15 |
| FeO | 41.80 | 32.66 | 36.31 |
| MgO | 3.62 | 2.93 | 1.85 |
| CaO | none | .10 | .21 |
| $Na_2O$ | nd | .54 | .20 |
| $K_2O$ | nd | .40 | .22 |
| $H_2O-$ | nd | .02 | .03 |
| $H_2O+$ | .58 | .34 | .43 |
| $TiO_2$ | 19.30 | 19.56 | 30.84 |
| MnO | .09 | .14 | .08 |
| S | nd | .03 | .01 |
| $P_2O_5$ | nd | <.01 | <.01 |
| $V_2O_5$ | nd | .40 | .27 |
| | 99.77 | 100.86 | 99.98 |

(1) Fresh grade 1 ore from drill core at Iron Mountain Analyst, F. A. Gonyer.

(2), (3) Slightly weathered magnetite-ilmenite. Magnetite partly replaced by hematite. Analysts, Norman Davidson (2) and I. Warshaw (3).

TABLE III.   *Partial Chemical Analyses of Silicate "Ore" and Mineralized Anorthosite, Iron Mountain, Wyoming (after drying)*
Analyst: Ledoux & Co., Inc.

| Sample | Total Iron | $(TiO_2)$ | $(V_2O_5)$ | $(SiO_2)$ | $(P_2O_5)$ |
|---|---|---|---|---|---|
| Wy-40 | 37.80 | 10.75 | 0.07 | 17.64 | none |
| Wy-19 | 56.70 | 9.26 | 0.29 | 3.24 | none |
| Wy-20 | 49.46 | 17.26 | 0.23 | 3.60 | 0.02 |

Wy-40.   D.D. Hole 9 at 75 feet. Magnetite-ilmenite 40%, olivine 40%, and plagioclase 20%. Grade 2 "ore."

Wy-19.   D.D. Hole 17 at 202–205 feet. Magnetite-ilmenite concentrated from core containing 10% magnetite-ilmenite, 60% olivine, and 30% plagioclase to a concentrate with an estimated 70–90% magnetite-ilmenite. Mineralized anorthosite, Type 1.

Wy-20.   D.D. Hole 14 at 150 feet. Magnetite-ilmenite concentrated from core containing 20% magnetite-ilmenite, 30% olivine, and 50% plagioclase to a concentrate having 80–90% magnetite-ilmenite. Mineralized anorthosite, Type 1.

the changes that take place down dip (Figure 5). Since grades 2 and 3 do not drop out, lateral changes can be observed at the surface only where trenches or pits have exposed the lower grades of "ore."

## ORE GENESIS

The ores are considered to have formed by replacement of anorthosite. This hypothesis is supported by the marked change in mineralogy along strike, dip, and plunge of an ore body and its individual layers, whether a fraction of an inch or tens of feet thick. These variations in mineralization, which range from almost pure olivine to massive magnetite-ilmenite, are difficult to explain in any manner than by replacement. In addition, the concentration of massive ore in the folds and of olivine on the limbs, and the halo of low-grade mineralized rock that locally transects the structure of the anorthostie in the hanging wall, are suggestive of introduction and replacement. Hand specimens, microscopic sections, and occasional outcrops show features that are indicative of replacement of anorthosite by magnetite-ilmenite and olivine. Perhaps the most convincing criterion of replacement in this kind of deposit is the layering, in the large southern ore body, that strikes more easterly than the trend of the body (Figure 3). It is believed that magmatic segregation in place, or injection of a magmatic body, could not produce such a feature.

The anticlinal region appears to have been a structurally weak zone, and westward-bowing curves of the axis where it is convex to the east were especially favorable sites for the localization of the two largest massive ore bodies at Iron Mountain.

The en echelon arrangement of ore bodies is believed to be due to the introduction of ore along zones of shear fracturing that deviate from the direction of compositional layering and platy crystal structure. The layering in the large southern ore body that strikes more easterly than the trend of the ore lens thus bears the same relation to the lens as a whole as do the en echelon lenses. The ore is considered to have formed by complete replacement along a zone of en echelon fractures. In contrast with this, the layering in most ore bodies is essentially parallel to the major plane of the lens.

Most of the massive ore bodies are in anorthosite; massive ore bodies are small and uncommon in noritic anorthosite. This is believed to be due to the more brittle nature of anortho-

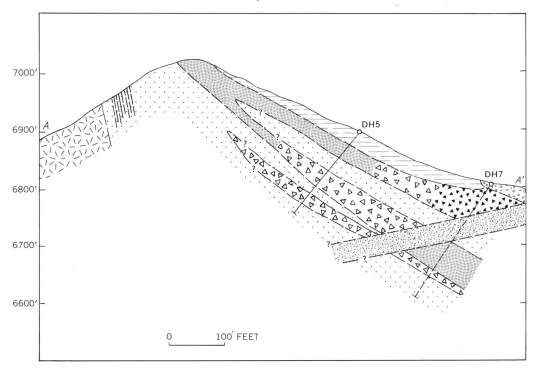

F IG . 5. *Geologic Section, illustrating change in grade of ore with depth. Section along D. D. Hole 5. D. D. Hole 7 is projected to the plane of the section. The patterns are the same as those given in the explanation for Figure 3 with addition of large open triangles for magnetite-ilmentite, grade 2, and small solid triangles for magnetite-ilmenite, grade 3; parallel dashes in mineralized anorthosit type 2, indicate platy crystal structure.*

site, to the massive character of noritic anorthosite, and to the influence of these physical factors on the fractures that admitted the ore mineralization.

The proportion of magnetite-ilmentite to olivine, as well as the proportion of these to the anorthosite, appears to be related to structural features. Where magnetite-ilmenite is present with olivine, the former two minerals are along the most pronounced parts of a fold, whereas olivine is concentrated in the limbs and the more gently folded portions. This, coupled with anorthite data (Table I), suggests that the anorthite content is highest in the most pronounced parts of folds.

As the principal ore minerals are magnetite and ilmenite, accompanied by spinel, much of the material introduced during the formation of ore bodies was iron, titanium, magnesium, and minor vanadium. The ore minerals, spinel, and olivine are considered to be indicative of high-temperature conditions. Brown (21) has suggested that if the temperature of the host rock had been raised sufficiently during deformation, plagioclase of lower anor-

thite content might become unstable in relation to plagioclase with higher anorthite. Partial dissociation of plagioclase would release silica that could combine with iron and magnesium to form olivine. If this happened, magnetite-ilmenite that formed in plagioclase of successively lower anorthite content should be accompanied by more and more olivine. This seems to be the case at Iron Mountain, but the suggestion does not contain any clue as to the possible source of the iron and magnesium with which the silica combined.

DeVore believes that granulation and recrystallization released titanium from the exsolved rods, needles, and fine particles of titaniferous hematite in the plagioclase of the anorthosite, and that this titanium "joined Fe and Mg in the dispersed state and migrated to the sites of deposition." (23, p. 187) He estimates that the volume of these needles would account for the titanium, iron, and magnesium in the plagioclase. DeVore's spectrographic analyses "show a decrease of Ti and, in some specimens, of Fe and Mg from the core of the crystals to the inclusion-free border or granulated matrix." The average $TiO_2$ content of 23 analyzed plagioclase cores from the Laramie anorthosite is 0.13 per cent, whereas the borders of these specimens average 0.08 per cent $TiO_2$. Nineteen nongranulated plagioclases from the Laramie anorthosite contain an average of 0.11 per cent $TiO_2$ compared with a 0.06 per cent average for 9 granulated plagioclases. DeVore calculated that from 2 cubic kilometers of granulated anorthosite "enough $TiO_2$ would be lost from the rock to form 30 million tons of titaniferous magnetite containing 15 per cent $TiO_2$." He concludes that granulation and recrystallization occurred in the presence of iron and magnesium, which increase in the newly formed plagioclase, but that these elements were not derived from the recrystallization of plagioclase. The titanium of the ore was derived from plagioclase "which has already joined Fe and Mg in the dispersed state and migrated to the sites of deposition." This hypothesis would account for the source of the titanium in the ore but not for the iron and magnesium.

At the present stage of investigation, it seems logical to believe that during granulation and recrystallization of anorthosite in the anticlinal region, plagioclase high in the anorthite molecule relative to that in the generality of Iron Mountain plagioclase formed in the most pronounced parts of folds, and olivine developed in the limbs. Temperature and pressure changes furnished the energy required to move iron and other elements from an unknown source to areas of low pressure, i.e., to the ore zone. The high concentration of oxide minerals within the ore zone is evidence that oxidizing conditions existed there or, at least, that oxide minerals were more stable there than elsewhere.

Geochemical work by Reinking (30) indicates that granulation and recrystallization has resulted in a loss of an average of 0.15 weight per cent Fe from the titanohematite needles in the plagioclase. His calculations show that if only 0.10 per cent Fe was released, the granulation of 3.4 cubic kilometers of anorthosite would have released enough iron to form 23 million tons of high grade ore (44% Fe). The analyses by Reinking indicate that the plagioclase in the anorthosite immediately adjacent to the ore is Fe deficient, thereby providing a ready source for the ore-forming material. Titanium analyses by Reinking show a loss with recrystallization and thus support DeVore's findings.

Trace element analyses of mineral fractions by Reinking support the suggested metamorphic origin. The relative abundances of Ni, Cr, and V in the Iron Mountain ores, show by analogy with studies at the Skaergaard (20,22), that the ores fluids could not be residual from a normal basic melt such as may have formed the norite of the anorthosite complex. The abundance of chromium in the anorthosite strongly argues against the ore-forming fluids being residual to the crystallization of the anorthosite or being early liquid segregations from an anorthositic melt that later might have replaced the anorthosite. Had an iron-rich segregation formed, either early or late, it would undoubtedly have contained most of the chromium, leaving little to be dispersed in the feldspar. Reinking also suggests that geochemical evidence does not support the possibility that the ores fluids represent mobilized preexisting ore or that they were derived from the mafic silicates in the anorthosite.

If iron and other metals moved as ions, they combined with oxygen, perhaps from replaced plagioclase in the ore zone, to form the ore minerals and spinel. If these metals moved as oxide molecules or complexes, there is no need for a source of oxygen to form magnetite-ilmenite in the ore zone. Some iron and magnesium may have combined with silica from partly dissociated plagioclase to form olivine. A somewhat similar process has been proposed for the source and development of magnetite ore at the Scott mine in the Sterling Lake district of New York (28). Some accessory magnetite-ilmenite may also have migrated as complexes in a manner akin to that which occurred in the Ausable district of New York (25).

The ore bodies contain only elements and minerals found in the anorthositic country rocks, and these rocks may have been the source of the ore-forming material. Another possible source of the ore constituents, suggested by several previous investigators, is a magma at depth. A geochemical study underway, of host-rock minerals and ore, will yield more detailed and precise information on which to base a hypothesis of ore formation and source of ore material.

## CONCLUSIONS

During Precambrian time the following events took place prior to, and contemporaneous with, the formation of the Iron Mountain ore deposit:

(1) Granulation, recrystallization, and fracturing of the anorthosite mass along the axis of the major anticline in the northern body.

(2) Movement of iron, titanium, and other elements to low pressure areas (the ore zone) on the east limb of the anticline in the lowermost major layer of anorthosite.

(3) Replacement of plagioclase by ore and gangue minerals with movement of material along platy crystal structure and along en echelon fractures that transect this structure. The largest massive ore lenses formed in westward-bowing curves of the anticlinal axis in the most pronounced parts of folds where anorthite content was highest. Lower grade, disseminated ore containing appreciable olivine moved to the limbs of folds where anorthite content was lower.

(4) The presence of corona structures around magnetite-ilmenite grains within the ore, and of fractures and faults that transect the ore bodies, suggest that the ore formed before the close of deformation.

## ACKNOWLEDGMENTS

The writer wishes to thank P. W. Guild, T. P. Thayer, and Norman Herz of the U.S. Geological Survey for carefully reading the manuscript and for suggestions that considerably improved its presentation. Thanks are also due James A. Marsh, R. H. Godbe, and Richard Chojnacki of the Union Pacific Railroad Company for information pertaining to the exploratory program of the Company at Iron Mountain. Paul Brenton furnished information on the production, processing, and use of magnetite for aggregate by the Magnetites Products Company.

## REFERENCES CITED

1. Stansbury, H., 1853, Exploration and survey of the Valley of the Great Salt Lake of Utah: U.S. 32nd Cong. Spec. Sess., Senate Executive Doc. 3, 487 p.

2. Hayden, F. V., 1871, Preliminary report (fourth annual) of the United States Geological Survey of Wyoming and portions of contiguous Territories: 2nd Ann. Rept. Progress, Washington, p. 14.

3. King, C., 1876–78, Geological exploration of the 40th parallel, etc.: U.S. (War Dept.) Chief Eng. Ann. Repts., v. 1, p. 27; v. 2, p. 13–16; v. 3, p. 107–109.

4. Knight, W. C., 1893, Geology of the Wyoming experiment farms, and notes on the mineral resources of the state: Univ. Wyo. Exp. Sta. Bull. 14, p. 177.

5. Kemp, J. F., 1899, A brief review of the titaniferous magnetites: (Columbia) Sch. of Mines Quart., v. 20, p. 352–355.

6. Lindgren, W., 1902, A deposit of titanic iron ore from Wyoming: Science, v. 16, p. 984–985.

7. Kemp, J. F., 1905, Die Lagerstätten titanhaltigen Eisenerzes im Laramie Range, Wyoming, Vereinegten Staaten: Zeit. f. prakt. Geol., Jg. 13, H. 21, S. 71–80.

8. Ball, S. H., 1907, Titaniferous iron ore of Iron Mountain, Wyo.: U.S. Geol. Surv. Bull. 315, p. 206–212.

9. Blackwelder, E., 1909, A Cenozoic history of the Laramie region: Jour. Geol. v. 17, p. 429–444.

10. Singewald, J. T. Jr., 1913, The titaniferous iron ores in the United States: U.S. Bur. Mines Bull. 64, p. 111–125.

11. Fowler, K. S., 1930, The anorthosite area of the Laramie Mountains, Wyoming: Amer. Jour. Sci., 5th ser., v. 19, p. 305–315, 373–403.

12. Diemer, R. A., 1941, Titaniferous magnetite deposits of the Laramie Range, Wyoming: Geol. Surv. Wyo. Bull. 31, 23 p.

13. Newhouse, W. H. and Hagner, A. F., 1945, Structure of the Laramie Range anorthosite, Wyoming (abs.): Geol. Soc. Amer. Bull., v. 56, p. 1184–1185.

14. Frey, E. 1946, Exploration of Iron Mountain titaniferous magnetite deposits, Albany County, Wyo.: U.S. Bur. Mines, R.I. 3968, 37 p.

15. Turner, G. L., 1947, Textures of the iron-titanium minerals of the Laramie Range, Wyoming: Unpublished M.S. thesis, Univ. Chicago, 31 p.

16. Newhouse, W. H. and Hagner, A. F., 1949, Cordierite deposits of the Laramie Range, Albany County, Wyoming: Geol. Surv. Wyo., Bull. 41, 18 p.

17. Hagner, A. F., 1951, Anorthosite of the Laramie Range, Albany County, Wyoming, as a possible source of alumina: Geol. Surv. Wyo., Bull. 43, 15 p.

18. Harrison, J. E., 1951, Relationship between structure and mineralogy of the Sherman granite, southern part of the Laramie Range, Wyoming-Colorado: Unpublished Ph.D. thesis, Univ. Ill., 79 p.

19. Newhouse, W. H. and Hagner, A. F., 1951, Preliminary report on the titaniferous iron deposits of the Laramie Range, Wyoming: U.S. Geol. Surv. Open File Rept., 17 p.

20. Wager, L. R. and Mitchell, R. L., 1951, The distribution of trace elements during strong fractionation of basic magma—a further study of the Skaergaard intrusion, East Greenland: Geochim. et Cosmochim. Acta, v. 1, p. 129.

21. Brown, H. S., 1954, Anorthosite variations around titaniferous iron deposits, Laramie Range, Wyoming: Unpublished M. S. thesis, Univ. Ill., 30 p.

22. Vincent, E. A. and Phillips, R., 1954, Iron titanium oxide minerals in layered gabbros

of the Skaergaard intrusion, East Greenland: Geochim. et Cosmochim. Acta, v.
6, p. 1–26.

23. DeVore, G. W., 1955, The role of adsorption
in the fractionation and distribution of elements: Jour. Geol. v. 63, p. 159–190.

24. Newhouse, W. H. and Hagner, A. F., 1957,
Geologic map of anorthosite areas, southern
part of Laramie Range, Wyoming: U.S.
Geol. Surv. Min. Invest. Field Studies Map
MF 119, 1 : 63,360.

25. Hagner, A. F. and Collins, L. G., 1959, Host
rock as a source of iron, Ausable Forks
Magnetite district, New York: Geol. Soc.
Amer. Bull., v. 70, p. 1613–1614.

26. Klugman, M. A., 1960, Laramie anorthosite:
Geol. Soc. Amer. Guidebook for Field
Trips (Guide to the geology of Colorado),
Field Trip B-2, p. 223–227.

27. Moore, F. E., 1960, Summary of Cenozoic
history, Southern Laramie Range, Wyoming
and Colorado: Geol. Soc. Amer. Guidebook for Field Trips (Guide to the geology
of Colorado), Field Trip B-2, p. 217–222.

28. Hagner, A. F., *et al.,* 1963, Host rock as
a source of magnetite ore, Scott Mine, Sterling Lake, N.Y.: Econ. Geol. v. 58, p.
730–768.

29. Chojnacki, R., 1965, Personal Communication.

30. Reinking, R. L., 1967, Geochemistry of the
Iron Mountain magnetite deposit, Albany
County, Wyoming: Unpublished Ph.D. the-
sis, Univ. Ill.

# 32. Leadville District, Colorado[*]

OGDEN TWETO[†]

## Contents

[*] Publication authorized by the Director, U.S. Geological Survey.
[†] U.S. Geological Survey, Washington, D.C.

# Illustrations

# Tables

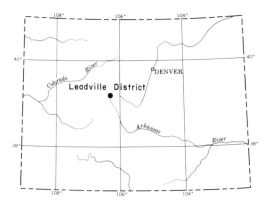

## ABSTRACT

The Leadville district, on the west flank of the Mosquito Range in central Colorado, has produced silver, zinc, lead, gold, and minor metals valued at $512,000,000. The ore deposits are in a sequence of dolomites and quartzites, Cambrian through Mississippian in age and about 500 feet thick, which is extensively intruded by porphyry sills, dikes, and plugs of Tertiary age. The sedimentary rocks and sills dip about 15°E and are broken by many faults, which are predominantly of near-north trend and downthrown to the west.

The ore deposits are principally blanket or manto replacement deposits, but in the eastern part of the district many veins occur also. The replacement deposits are largely confined to three dolomite units in the stratigraphic sequence. Of these, the uppermost or Leadville Dolomite is the most productive. Most ore bodies are on the underside of porphyry sills and, particularly, beneath sills that occupy unconformities. Many replacement bodies have vein roots or branch from veins. The veins occupy faults and are productive mainly in the section of quartzites and dolomites and included sills.

The primary ores consist principally of pyrite, marmatitic sphalerite, and galena but locally contain chalcopyrite, silver minerals, bismuth minerals, and gold. Principal gangue minerals are manganosiderite and jasperoid. The ore deposits have a crude zonal pattern centered around an intrusive center at Breece Hill, which evidently influenced ore deposition thermally.

The ore deposits were oxidized to depths of several hundred feet in Miocene time. Bonanza ores mined in the early days were characterized by cerussite, cerargyrite, and embolite. Later, zinc carbonate ores were identified and mined on a large scale.

The Leadville district is a complexly faulted and intruded area at the intersection of major fault systems trending N20°W and N15°E in the Mosquito Range. The intensely fractured area evidently localized igneous intrusion and subsequent mineralization. Ore solutions rose on many of the faults in the area, presumably from the same source as the porphyry magmas. Movement on these faults continued into the Pleistocene, displacing ore bodies, oxidation zones, Pliocene sediments, and glacial deposits.

## INTRODUCTION

The Leadville district, in Lake County, Colorado, was one of the first of the limestone-replacement and "carbonate ore" mining districts discovered in the Western United States, and even yet remains among the largest of these districts in total output. It owes its leading position among districts of its kind to the high grade of some of its ores, to the great number of ore bodies in an intensively mineralized area of about 8 square miles, and to the variety of its mineral products, which include silver, zinc, lead, gold, copper, manganese, iron, bismuth, and pyrite. The wide variety in metal content of the ores, the division of all ores into oxidized and unoxidized classes, and a wide range in the forms and geologic occurrence of the ore bodies combine to make the ore deposits highly varied geologically, mineralogically, and metallurgically.

### Historical Summary

Mining in the Leadville district began with the discovery of placer gold in California Gulch in 1860. A town, Oro City, was quickly established, and mining activity boomed for a year but then declined gradually through more than a decade. Discovery of lode gold at the Printer Boy mine in 1868 helped extend the life of the gold camp but did not reverse the trend of declining production. Gold output in the period 1860–1874 is estimated at about $6 million by Henderson (7).

In 1874, a heavy brown material that had hampered the gold placer operations was identified as argentiferous cerussite, and by 1876 several lode discoveries had been made. In 1877, the town of Leadville was founded and quickly grew to a population of more than 15,000. By 1880, twelve smelters were in operation, and annual production had risen to 10,000,000 ounces of silver and 66,000,000 pounds of lead—in 140,000 tons of ore.

Ore of the bonanza period was oxidized or so-called carbonate ore, mostly from depths of less than 500 feet. By the late 1880's, many mines had reached sulfide ores. Although many sulfide ore bodies were worked successfully by selective mining and sorting to eliminate the undesirable zinc, production declined slowly until about 1893, when gold-rich ores were discovered in the Breece Hill area, in the eastern part of the district. These ores, together with growing recovery of zinc sulfide from 1899 onward, sustained the district until the depression of 1907, when output dropped

drastically. In 1909, however, the wall rocks in many of the old workings were discovered to be zinc carbonate, and, from 1910 to 1925, when mining of zinc carbonate ended, zinc was the principal product of the district.

Following 10 years of declining productivity, the district was rejuvenated in 1935 with organization of the Resurrection Mining Company. This company found vein ores beneath the old replacement-body stopes in the eastern part of the district and for 20 years made a substantial output of complex sulfide ores from these and other sources. Exploration after World War II led to discovery of ore in a deeply downfaulted block more than a square mile in area in the eastern part of the district, and this area was brought into production in the early 1950's. By 1957 costs were equaling returns, and operations ended, placing the Leadville district in dormancy for the first time in 97 years. Late in 1965, however, reopening of the Irene mine in the downfaulted block was announced by the American Smelting and Refining Company.

### Production

Total production of gold, silver, copper, lead, and zinc from Lake County is summarized in Table I. Although the Lake County figures include output from outlying districts, at least 97 per cent of the total is attributable to the Leadville district proper. In addition to the output indicated in Table I, mines of the Leadville district produced 936,000 long tons of metallurgical manganese-iron ore containing 14 to 45 per cent Mn, and about 2,500,000 tons of manganiferous fluxing ore, much of it of metallurgical grade, through 1939 (13). An unknown but at least equal

TABLE I.  Production of Gold, Silver, Copper, Lead, and Zinc in Lake County, Colorado, 1859–1963, in terms of recoverable metals
Based on figures reported by Henderson (7) and in Minerals Yearbook and other Publications of the U.S. Bureau of Mines.

|        | Unit        | Quantity    | Value        |
|--------|-------------|-------------|--------------|
| Gold   | Troy ounces | 2,985,776   | $ 66,942,109 |
| Silver | Troy ounces | 240,055,514 | 196,259,750  |
| Copper | Short tons  | 53,109      | 15,521,319   |
| Lead   | Short tons  | 1,088,204   | 110,542,249  |
| Zinc   | Short tons  | 785,380     | 118,222,469  |
|        |             |             | $507,487,896 |

amount of low-manganese iron fluxing ore was also produced, as well as an unknown amount of metallurgical magnetite-hematite ore. The fluxing ores, used in the smelting of ores from many other districts as well as those of Leadville, generally contained a few ounces of silver per ton and/or a few per cent of lead. In effect, they were submarginal silver-lead ores made workable by their high content of iron and manganese. Large amounts of pyrite for manufacture of sulfuric acid were shipped from the district from 1906 through 1921, but records are available only for the period 1918–1921, when a total of about 37,000 tons was shipped (7). Bismuth ores, generally containing 5 to 16 per cent Bi, were produced at times from a small group of mines on the north slope of Breece Hill and from other sources, but production records are fragmentary. Smelter recoveries of byproduct bismuth, though unknown, may have been substantial, as the sulfide ores in many parts of the Leadville district are bismuthiferous.

Assuming a value of $20,000,000 for the manganese and other minor products, and assigning 97 per cent of Lake County base- and precious-metal production to Leadville, total value of production from the Leadville district through 1963 is $512,000,000. Of this value, about 37 per cent was in silver, 22 per cent in zinc, 21 per cent in lead, 13 per cent in gold, 4 per cent in minor products, and 3 per cent in copper. A graph by Henderson (7) shows that gold dominated the output from 1860 to 1877, silver dominated from 1877 to 1904, and zinc dominated from 1904 to 1920. Since 1920, lead and zinc have shared the leading position.

## Investigations and Status of Knowledge

Study of the geology and ore deposits of Leadville has been underway off and on since 1879, when S. F. Emmons (1) began the work leading to U.S. Geological Survey Monograph 12, a major foundationstone of the science of economic geology. Much has been published on Leadville since then, including a brief but important supplement to the monograph in 1907 by Emmons and Irving (2) and a general revision in 1927 by Emmons, Irving, and Loughlin (6), but even so, many aspects of the geology and ore deposits are poorly known or even have not been studied. The basic bedrock geology is unsatisfactorily known in many parts of the district, in part because surficial deposits and vast dumps prevent surface observations, and in part because many mine

workings, and especially, thousands of prospect shafts, did not get examined during the frenzy of digging. Similarly, although Emmons and successors made invaluable observations of the ores and ore bodies in many places, they studied only a part of the total. No record remains even of the locations of many ore bodies, much less of their geologic occurrence and character. Short of driving new workings, no hope can be entertained of ever recovering these data, for the ground at Leadville is notably weak, and caving has obliterated a large proportion of the former workings, including an estimated 75 per cent of all shafts.

With the exception of a few specimens studied by Chapman (14) and Loughlin (6) and some unpublished work by G. M. Schwartz, the ores of Leadville have never been studied microscopically, let alone by more advanced techniques. Similarly, although most of the rocks in the district are altered, the subject of rock alteration remains at the empirical stage, partly because adequate underground access has not existed since this field of economic geology has flowered and, partly, because of the complexity and the enormity of the job.

Despite the limitations, much has been learned in the last two or three decades of the geology, the setting, and the geologic development of Leadville. These subjects and their economic implications constitute what is new in the text that follows; the rest leans heavily on earlier reports, notably that of Emmons, Irving, and Loughlin (6), which remains the most current general source on the ores.

## PHYSIOGRAPHIC AND STRUCTURAL SETTING

### Physical Features

The town of Leadville, at an altitude of 10,200 feet, is on the western slope of the Mosquito Range, at the head of pedimented surfaces that slope westward about 3 miles to the axis of the upper Arkansas Valley. The mining district extends eastward from the town, across hilly terrain that rises and steepens eastward toward the crest of the range, which is at altitudes of 13,000 to 14,000 feet. From Leadville and most points in the district, one looks westward across the broad valley of the upper Arkansas to the peaks of the Sawatch Range, which exceed 14,000 feet in altitude. Both the Mosquito and the Sawatch

Ranges were heavily glaciated; they are scarred by cirques and deep canyons in their upper parts and strewn with moraines in the lower parts.

Most of the Leadville mining district lies between two of the glaciated canyons on the side of the Mosquito Range, Evans Gulch on the north and Iowa Gulch on the south, although some mining was done southward from Iowa Gulch to the next canyon, Empire Gulch (Figure 1). Between Iowa and Evans Gulches is California Gulch, a short and unglaciated drainage, and to the north of it, a still smaller drainage, Stray Horse Gulch. Between these

drainages are named hills of various degrees of eminence that long have been used as a basis for geographic classification, and in a general way, even for classification of geology, mines, and ore deposits, as each has its own set of geologic characteristcs. These landmark hills are shown in Figure 1.

Upper Tertiary alluvial and Pleistocene glacial deposits as much as 600 feet thick cover the bedrock through most of the district. Although these deposits contain no ore, except locally some placer gold, they are economic factors in the district because they mask the geology, are difficult and costly to work or

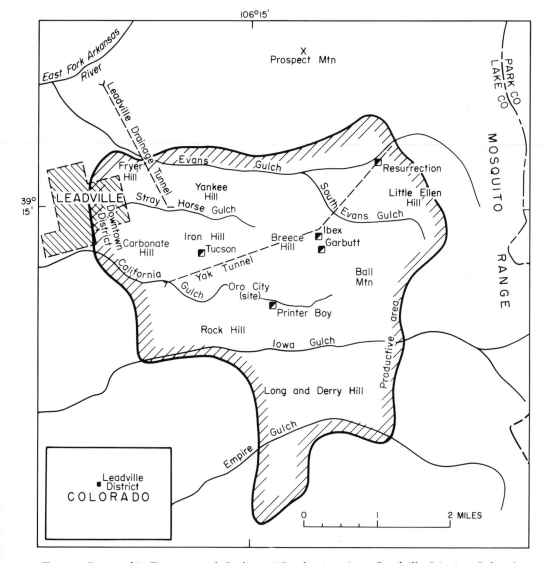

FIG. 1. *Geographic Features and Outline of Productive Area, Leadville District, Colorado.*

explore through, and are reservoirs of troublesome water. Detailed study of these deposits in recent years has led to recognition of a fault history extending into the Pleisocene.

## Historical Development

Physiographic development of the Leadville area is inseparably tied to a history of repetitive fault movements through most of Cenozoic time. As discussed in another report in this volume, the Sawatch Range is the core of a huge anticline broken longitudinally on its eastern flank by graben faults of the upper Arkansas valley. The Mosquito Range, also a part of the anticline, is essentially a fault-block range east of the graben. The graben was first outlined by faults that formed in early Laramide (latest Cretaceous or Paleocene) time, but it was markedly accentuated later, from Eocene or Oligocene time to the Pleistocene.

Possibly by early Miocene a mature topography had developed on the slopes of the mountains bordering the upper Arkansas valley, and widespread pediments had been cut over the lower slopes. By the close of the Miocene, these old surfaces, particularly the pediments, had been deeply dissected, presumably as a result of renewed large-scale fault movements. Although this old topography is now largely buried beneath younger deposits, exploration data indicate a southwest-trending canyon beneath the town of Leadville that is cut 500 to 600 feet below the old pediment surface, and seismic data (15) indicate that an ancestral Iowa Gulch was cut 800 feet below the general bedrock surface. In such canyons, great volumes of sedimentary rock generally favorable to the occurrence of ore deposits have been removed, leaving only the generally unfavorable Precambrian rocks.

In the early Pliocene, thick clastic sediments were deposited in the valley and high onto its sides, covering the old topography. Subsequently, as the mountains rose or the valley sank farther, these sediments were faulted against bedrock in many places. The already irregular bedrock surface beneath the sediments thereby became more irregular. The Pendery fault on the eastern outskirts of Leadville, for example, is marked in the subsurface by a post-sediment bedrock scarp 200 feet high, considerably affecting exploration and mining, and also the water problems in mining. Similarly, in the largely unexplored pediment area west of Leadville, depths to bedrock evidently range through hundreds of feet as a result of the combination of rough pre-Pliocene topography and post-Pliocene faulting.

By the beginning of the Pleistocene, the Pliocene sediments had been extensively dissected and stripped from the higher slopes as well as faulted. Glaciers of the first of several glacial episodes were widespread, taking the form of discontinuous icecaps in the mountains and broad, far-going ice streams in the valleys. During a long interval that preceded the next glacial episode, the early drift was deeply weathered and dissected. The weathering produced a tough red-brown gumbo that is a familiar sight in dumps in the western part of the district, but it probably did little to alter the ores, as these had already been oxidized to greath depth, as discussed below. Following this period of weathering, and probably in response to renewed fault movement, a flood of gravel derived principally from the Mosquito Range was deposited over the weathered and dissected surface. The deposition ended with the formation of a second set of pediments, which now are conspicuous features of the upper Arkansas valley. After this, glaciers of six successive glacial advances cut the cirques and canyons so evident in the Mosquito and Sawatch ranges and formed the moraines and outwash deposits that give Leadville its characteristic gravelly apperance. Effects of the successive glaciations were to expose some ore deposits and to cover others, but mainly to remove and destroy great quantities of ore, for many an ore body is truncated at the glacial canyon walls.

Fault movement continued through most if not all of the Pleistocene. Drift of the oldest glacial episode is displaced tens of feet in many places, and successively younger deposits are displaced correspondingly smaller amounts.

## Oxidation

Although present water levels in the Leadville district are everywhere at shallow depth, the ore deposits in the unglaciated parts of the district are oxidized to depths of several hundred feet. This oxidation, which so greatly enhanced the value of the early-day ore, evidently was a product of a dry climate and a ground-water base-level greatly lowered by graben-faulting in Miocene time. Deep canyon cutting in the valley sides presumably furthered the lowering of ground-water levels, allowing oxidation to considerable depths. The oxidation now recorded in the district must have been almost solely a product of this time, for oxidation levels appear completely independent of the presence or absence of Pliocene deposits or other surficial materials. Ground-water levels probably began rising as the Arkansas Valley filled with Pliocene sediments, and continued to rise as the valley was further filled with glacial deposits in the Pleistocene.

## GEOLOGY

### Stratigraphy

Stratigraphic interest at Leadville centers on pre-Pennsylvanian Paleozoic rocks, as these are the main host rocks of the ore deposits. These rocks form a sequence only about 500 feet thick that lies on Precambrian granitic rocks and is overlain by a thick sequence of Pennsylvanian rocks. All the sedimentary rocks are cut to ribbons by Tertiary and Quaternary(?) intrusive prophyries, of which there are many varieties. The porphyries are largely in sills or slightly discordant sheets, but they occur also in dikes, pluglike bodies, and irregular compound bodies not readily classifiable as to form. They expand the Paleozoic sedimentary sequence in places to two or three times its normal thickness and segment it into islands surrounded by porphyry. The Precambrian rocks are cut by dikes and pluglike bodies but are far less extensively intruded than the sedimentary rocks.

Essential features of the sedimentary and igneous rocks are summarized in Tables II and III. Terminology of these rocks has evolved through several decades from a miner's to a standard stratigraphic nomenclature. Since a large literature and a great volume of mine records are built around the older nomenclatures, these also are indicated in Tables II and III.

Sedimentary and igneous rocks of the district have been described by Emmons (1), Emmons, Irving, and Loughlin (6), and Behre (19). Paleozoic sedimentary rocks have been described in terms of current nomenclature by Tweto (18). Some of the lower Tertiary igneous rocks have been dated radiometrically by Pearson and others (26). Pliocene and Pleistocene deposits and history have been described briefly by Tweto (25).

### Structure

The structural geology of Leadville is simple in its broad features but complex in detail. The sedimentary rocks are on the flank of the great Sawatch anticline, and they dip homoclinally eastward at about 15°; the only departures from this simple arrangement are local, as near some faults and in blocks rotated by igneous intrusion. The tilted slab of sedimentary rocks and its included, generally concordant, porphyry bodies is broken by a series of generally northtrending faults, most of which are upthrown to the east (Figure 2). Thus the east-dipping rocks are repeatedly stepped up to the east, with the result that the productive zone rises with the topography and remains more or less continuously within working distance of the surface for 4 miles in the downdip direction. To a considerable extent, Leadville owes its productivity to this rather remarkable structural coincidence, for without it the ore zones would be hopelessly deep going eastward, and with too much fault movement, they would be eroded, or "in the air."

Simple though the broad structural arrangement may be, the details are moderately complicated due to complex relations between intrusions and faults and to complex histories of faulting. With the exception of a few early faults, most of the faults were created during the stage of Laramide porphyry intrusion (Figure 3). Many underwent repeated movements as successive porphyry magmas were emplaced (24), and many were reactivated later in conjunction with the Arkansas Valley graben faulting, which continued at least into the Pleistocene. Faults displace different porphyries in different degrees. Some faults contain dikes or serve as boundaries of porphyry bodies; some cut certain varieties of porphyry and are cut by other varieties. Many faults mark abrupt changes in thickness and/or stratigraphic position of sills. In addition, some contain breccias whose compositions are inconsistent with the present displacement, as for example, basement-rock fragments between walls of Paleozoic rocks or porphyry, and some are associated with tight drag features that are geometrically inconsistent with the movements implied by the present displacements. Further, some faults alternate between normal and reverse character along strike, owing to changes either in direction of displacement or in direction of dip. Faults of constant dip may alternate between being apparently normal and apparently reverse in the downdip direction, owing to differences in the thicknesses and stratigraphic positions of sills on each side.

These characteristics, together with map relations of the rocks, indicate that many of the major faults had complex histories of movement. The faults probably formed origi-

Table II
Stratigraphic Units in Leadville District, Exclusive of Cenozoic Igneous Rocks

| Age | Unit | Thickness (feet) | Former nomenclature | Character |
|---|---|---|---|---|
| Pleistocene | Drifts of six glacial episodes | 0–300 | Younger glacial drift | Till and outwash gravel |
|  | — Unconformity — | | | |
|  | Malta Gravel | 0–300 | High terrace gravel | Crudely bedded cobble gravel |
|  | — Unconformity — | | | |
|  | Ancient glacial drift | 0–150 | Earlier glacial drift | Till; in many places, 5–40 ft of tough clayey red-brown gumbo at top |
|  | — Unconformity — | | | |
| Pliocene | Dry Union Formation | 0–600 | Lake beds | Brown pebbly sandy silt and interbedded sand and gravel |
|  | — Unconformity — | | | |
| Pennsylvanian | Minturn Formation | 1000 ft preserved 6000 ft in region | Weber grits (Weber(?) Formation) | Upper part, lenticular feldspathic sandstones and conglomerates; lower 500 ft, even-bedded black to white coarse-grained micaceous quartzite and gray to black shale |
|  | Belden Formation | 150–400 (Thickens southeastward across district) | Weber shales (Weber(?) Formation) | Black carbonaceous shale and interbedded thin-bedded dark-gray limestone and sandstone; local beds of bony coal |
|  | — Unconformity — | | | |
|  | Molas Formation | 0–40 (pockety) | Generally misidentified as oxidized mineralized matter | Structureless red and yellow siltstone and mudstone containing abundant chert fragments |
|  | — Unconformity — | | | |
| Mississippian | Dolomite member *(Leadville Dolomite — Chaffee Formation)* | 0–190 Generally 60–100 | Blue Limestone also Leadville Limestone | Massive dark-gray to blue-black dolomite; irregularly cherty; normally fine-grained but widely altered to coarsely crystalline facies |
|  | Gilman Sandstone Member | 5–25 Generally 10–15 | | Sandy dolomite with chert fragments grading downward into coarse-grained yellow-gray sandstone |
|  | — Unconformity — | | | |
| Devonian | Dyer Dolomite Member | 70–105 Generally 80 | | Thin-bedded gray to black fine-grained dolomite with characteristic hackly fracture |
|  | Parting Quartzite Member | 25–45 Generally 30 | Parting Quartzite (Yule Limestone — Parting Quartzite Member) | White to tan crossbedded coarse-grained quartzite and quartz-pebble conglomerate; in places, 5–15 ft of buff, green, and maroon shale at base |
|  | — Unconformity — | | | |
| Ordovician | Manitou Dolomite | 90–120 Generally 115 | White Limestone (Yule Limestone — White Limestone Member) | Medium-bedded white to gray crystalline dolomite; characterized by white chert |
|  | — Unconformity — | | | |
| Cambrian | Peerless Formation | 60–80 Generally 65–75 | Transition shale; also in part White Limestone and in part Sawatch Quartzite and Peerless Shale Member of Sawatch | Thin-bedded maroon, buff, and green dolomite and dolomitic shale, grading downward into brown glauconitic sandstone; generally gray to black in mine workings |
|  | Sawatch Quartzite | 95–115 Generally 100 | Lower white quartzite or Cambrian quartzite | Medium-bedded fine-grained white quartzite; pinkish near top and gray to tan at base |
|  | — Unconformity — | | | |
| Precambrian | St. Kevin Granite, unnamed older granite and local biotite gneiss | | "Archean"; Silver Plume(?) and Pikes Peak Granites as applied by Behre (1953) | St. Kevin Granite: equigranular to porphyritic biotite-muscovite granite. Older granite: coarse-grained, gneissic, locally porphyritic |

Table III
Cenozoic Igneous Rocks, Leadville District

| Age | Unit | Former Nomenclature | Character |
|---|---|---|---|
| Late Tertiary and Quaternary(?) | Rhyolite and rhyolitic explosion breccia | Rhyolitic agglomerate | Flow- and chill-banded dense intrusive rhyolite and vitrophyre, grading into explosion breccias consisting largely of foreign rock fragments |
| | Rhyolite porphyry | Late White Porphyry | Slightly porphyritic white intrusive rhyolite |
| | Little Union Quartz Latite | Trachyte | Intrusive brownish-gray quartz latite with conspicuous biotite phenocrysts |
| Early Tertiary | Younger porphyries, Breece Hill-Ball Mountain area; of at least 6 varieties, including Iowa Gulch Porphyry | Gray Porphyry or Gray Porphyry Group | Quartz monzonite and quartz latite porphyries; some with fluidal banding and abundant inclusions |
| | Lincoln Porphyry | | Bluish- to greenish-gray quartz monzonite porphyry with abundant 1/2-1 in K-feldspar and 1/4-1/2 in quartz phenocrysts |
| | Johnson Gulch Porphyry | | Gray quartz monzonite porphyry with scattered 1-in pink K-feldspar phenocrysts |
| | Evans Gulch Porphyry | | Gray-green, very finely porphyritic, almost equigranular quartz monzonite porphyry |
| | Sacramento Porphyry | | Gray to greenish-gray quartz monzonite porphyry; plagioclase the most prominent phenocryst |
| | Pando Porphyry | White Porphyry, Mount Zion Porphyry Early White Porphyry | Fine-grained light-gray to white quartz latite porphyry, generally sericitized |

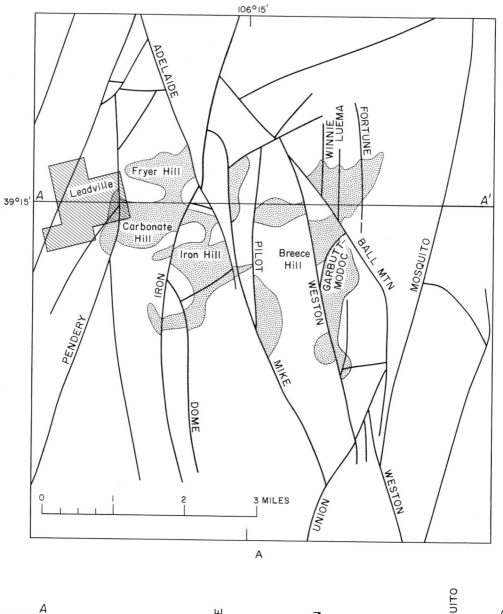

Fig. 2. *A. Principal Faults and Areas of Principal Ore Bodies, Leadville District, Colorado. B. Generalized Cross-section A-A'. Dot Pattern, Pliocene and Pleistocene Deposits; Hachured Line, Top of Basement.*

nally as parts of orderly regional tectonic fracture systems, but they evidently were successively modified on a local basis by magma movements and the room-making activities of intrusive bodies. The comparatively early (Laramide) origins of the faults, the long histories of movement, and the differing histories of different faults or even segments of the same fault invalidate or vastly complicate the simple classifications based on pre-ore or post-ore age and normal or reverse displacement long used at Leadville.

Viewed with respect to the regional fault pattern, the Leadville district is an intensely faulted and intruded area at the intersection of two major fault systems that meet at an acute angle. South of the district for many miles, the principal longitudinal faults of the Mosquito Range trend about N20°W. The Weston fault, the largest, is a near-vertical fault, locally normal and locally reverse, along which the west side is displaced downward 3,000–5,000 feet (Figure 4). North of the district, the principal longitudinal faults of the range trend about N15°E. The Mosquito fault, the largest, is a west-dipping normal fault with more than 5,000 feet of displacement. Faults of both systems date from early Laramide, but those of N15°E trend, much more than those of N20°W trend, served also as graben faults later in Cenozoic time.

The major faults within the district are principally of these two trends. The north-northwest-trending faults die out near the northern edge of the district, but the north-northeast-trending faults continue southward beyond the district (Figure 4). Along faults of both systems, the downthrown sides are predominantly to the west, giving a total stratigraphic throw of about 10,000 feet across the entire broad Mosquito-Weston fault zone. Among lesser faults in the district are many of northeast trend, a few of east-west trend, and some randomly oriented faults attributed to intrusion.

In general, the faults become simpler and more sharply defined from east to west across the district, and they also show increasing post-Laramide movement. In the eastern part of the district, the faults were complexly intruded after undergoing large displacement, and they also underwent movements during and following intrusion. As a result, large faults were almost obliterated, leaving as their only visible remnants the post-porphyry fractures, displacements on which are tiny in proportion to those on the original faults. At the Garbutt mine, for example, an early Garbutt-Modoc fault displaced the sedimentary rocks

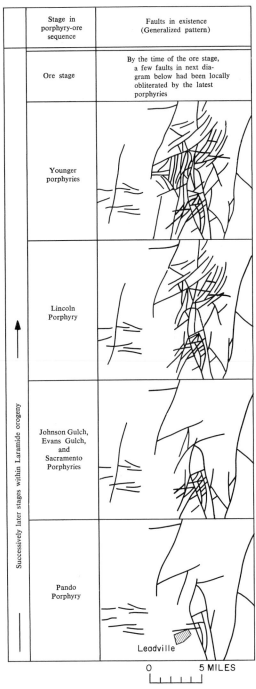

FIG. 3. *Fault Patterns in the Leadville Area, Colorado, at Successive Stages Defined by Sequence of Porphyries.*

at least 1000 feet. The fault was then invaded by successive porphyries in irregular ramifying bodies, which in places destroyed the fault as

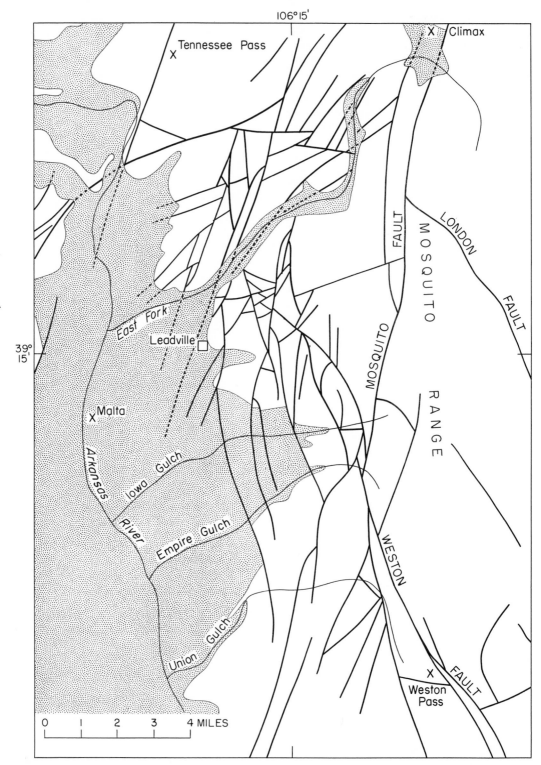

FIG. 4. *Fault Structure of the Mosquito Range in Leadville Region, Colorado. Shaded Area, Pliocene and Pleistocene Deposits.*

a physical feature and in other places pushed segments of the physical fault hundreds of feet out of position. During post-porphyry movement the Garbutt fissure formed, in places following remnants of the old fault surface, in places cutting into one wall or the other, and in places cutting only porphyry. Consequently, along some segments of the Garbutt vein, the walls are displaced only a few feet or even a few inches, whereas along others they are displaced hundreds of feet.

Similarly, a segment of the Weston fault on Breece Hill was obliterated by intrusion of a small pluton of porphyry along it, and another lengthy segment served as the sidewall of a sill of porphyry 400 to 600 feet thick. As a result, a major strand of the Weston fault, which here is dying out northward, is a reverse fault upthrown to the west as measured in sedimentary rocks above the sill, but a normal fault downthrown to the west as measured in strata beneath the sill. Two miles farther south, in a complexly intruded and faulted area described by Linn (28), an old strand of the Weston fault dips steeply west and is downthrown to the west in part of its course but in another part dips steeply east and is downthrown to the east; a new strand dips east and is downthrown to the east but has drag features and breccia that suggest that downthrow was originally to the west.

Other faults, such as the Pilot and Mike, not only had complications caused by intrusion of the porphyries but were later intruded by pipes, dikes, and irregular bodies of rhyolite and rhyolitic explosion breccia. These materials invaded veins and replacement ore bodies also, destroying much ore. The Pilot and Mike faults had post-Laramide movement, as shown by displacement of the Dry Union Formation south of the Leadville district. Other faults in the eastern part of the district probably had such movement also, but the Dry Union Formation is not present to prove it.

The faults of the western part of the district, such as the Adelaide, Iron, and Pendery, were not extensively intruded by porphyries, but differences in the thickness, sequence, and stratigraphic positions of sills on their two sides clearly indicate that the faults existed at the time of porphyry intrusion, and, moreover, that movements occurred on them between successive intrusions. Since they are earlier than the porphyries, which in turn are earlier than the ore, these faults are pre-ore, although they have been long classed as post-ore. They had extensive post-ore movement, however,

and they displace ore bodies, Dry Union Formation, and the earlier glacial deposits.

## Age of Intrusion and Mineralization

The ore deposits of Leadville, along with those of the Colorado mineral belt generally, have long been regarded (a) as Laramide (Late Cretaceous-early Tertiary) in age; (b) as being somehow related to the abundant intrusive igneous rocks, or porphyries, that accompany them; and (c) as representing the last major event in a sequence of events that, in simplified form, included (1) folding and faulting, (2) igneous intrusion, and (3) mineralization (4,6,8,10,12,29). However, age of mineralization in this generalized picture is only a working hypothesis. As discussed in the article on Colorado in this volume, much depends on the amount of time encompassed by "Laramide," on whether the intrusive rocks are all of the same age or include two or more distinct age groups, on the time span between intrusion and mineralization, and on whether mineralization occurred in just one stage, or in more than one.

Regional structural and stratigraphic relations strongly suggest that deformation in the Leadville district, along with the rest of the southern Rocky Mountain province, began in Late Cretaceous time. Until the advent of isotopic dating, these regional relationships were the only basis for assigning a Laramide age to the older structural features, the porphyries, and the ore deposits of Leadville. Isotopic ages of 70 and 64 million years for the Pando and Lincoln Porphyries, respectively, (26) corroborate a Laramide age very close to the Cretaceous-Paleocene boundary for at least some of the porphyries. However, there are several porphyries younger than the Lincoln, and most of these have intrusive habits that suggest a shallower environment of emplacement, which in turn suggests a distinctly younger age. Because of their pervasively altered character, these porphyries have not been dated. If there are two ages of porphyry, at least some ore is later than the younger porphyries, but whether all ore is of the same age is not certain. Marked differences in the character of the ores in various parts of the district, indications of multiple source channelways, and the paragenetic relations in many of the ores, all are as compatible with two stages of mineralization as with one.

Whether in one or more than one stage, mineralization was separated from porphyry intrusion by a time gap that is of unknown extent but that is generally assumed to have been small. Direct dating of the ore has not yet been accomplished. Linn (28) reported a K/Ar age of 60 m.y. for sericite in the Pando ("White") Porphyry. He regarded the sericite as probably deuteric, but the low age as compared with the 70 m.y. figure for fresh Pando (26) suggests either that the sericite is hydrothermal and thus possibly close in age to the ore, or that it is a mixture of deuteric and hydrothermal sericites giving an intermediate age value. Work on ore leads from the district has shown that the leads differ widely in isotopic composition (31,34). The leads evidently had a complicated geologic history and their isotopic composition thus does not serve to fix the age of mineralization.

In summary, mineralization at Leadville probably is in some part correctly referred to the "Laramide," or more specifically to the Paleocene or early Tertiary. It is not proved to be of this age, however, and wholly or in part, it could be appreciably younger. Available geologic evidence requires only that it be of pre-Miocene age.

## ORE DEPOSITS

### Ore Bodies

The ore bodies at Leadville can be broadly classed as blanket, manto, or replacement bodies and as veins, but many intermediate types exist, as well as pipes, stockworks, and irregular bodies. In the western part of the district, blanket bodies or variants of them are by far the predominant type. In the eastern part of the district, veins and blanket bodies are both common, and in many places these two kinds are linked. Throughout the district, ore bodies differ in various ways from area to area, and in a given area, they differ also with stratigraphic level.

The blanket bodies were known in the heyday of Leadville mining as "contacts" because they are located in dolomite on the underside of porphyry sills or quartzite beds and hence were encountered with a change in rock as shafts were sunk. The contacts were numbered in the order encountered in the shafts, and they thus differed in terminology from area to area. Some mines, in highly mineralized and extensively intruded areas, had as many as eleven "contacts." Where mineralization was intense, all carbonate rock layers between the

porphyry sills were completely replaced by ore and gangue minerals, so that superposed "contacts" or blanket bodies in reality became one large ore body containing sheets (sills) of porphyry waste. In most places, however, the ore bodies formed distinct, if lumpy, layers separated from one another by unmineralized dolomite as well as by porphyry. Among the ore bodies, the uppermost or "first contact," generally located in the upper part of the Leadville Dolomite and under a sill of Pando ("White") Porphyry, was most widespread, and successively lower "contacts" were less extensive.

These features are well illustrated at Carbonate Hill and in the so-called Downtown district, where sedimentary rocks and porphyry sills are in a generally ordered arrangement. Through most of this area there were five "contacts," each 5 to 50 feet thick, spaced through a vertical distance of about 450 feet, and of these the uppermost was by far the most extensive. In places on the north end of Carbonate Hill, however, some of the "contacts" fused, making the ore as much as 200 feet thick, and, along a vein, 400 feet thick. Ore bodies throughout the Carbonate Hill-Downtown area ranged from a few feet in width and height and a few tens of feet in length to tens of feet in cross-section dimensions and hundreds of feet in length. They also varied widely in form, depending in part on stratigraphic position. The stratigraphically higher ore bodies, generally of lead carbonate-silver ore, were chiefly channel-shaped and moderately sharply defined; in many places they were set in a more or less continuous blanket of manganese or iron oxide fluxing ore, and locally they were bordered by bodies of zinc carbonate ore. Sulfide ore bodies, at lower stratigraphic levels, were shaped more like irregular blankets, and many had one or more keels along veins. Zinc carbonate ore bodies were very irregular in form and occurrence and in their relations to lead carbonate and sulfide ore bodies, but they were principally concentrated at the third and fourth "contacts," near the stratigraphic level of the Parting Quartzite Member.

On Fryer Hill, in contrast, porphyry intrusion reduced the dolomite sequence above the Parting to isolated plates up to several acres in size. In the heart of the hill, all such plates were completely mineralized, although not all material in them was high-grade ore after oxidation. Consequently, there was no orderly sequence of "contacts," and the sizes, shapes, and locations of ore bodies depended principally upon how the porphyry had sundered the dolomite.

The largest and best defined ore bodies in Leadville were those of Iron Hill and Rock Hill. These bodies were long, somewhat irregular, branching and anastomosing shoots trending northeast down the dip and were spaced 200–500 feet apart in the strike direction. They lay at the "first contact" (under Pando Porphyry) in the northern part of Iron Hill, and at the "second contact" (under Johnson Gulch(?) Porphyry) in the southern part of the area. According to a tabulation by Emmons, *et al.* (6, p. 191), the larger ore shoots were 1100 to 3100 feet long, 200 to 520 feet wide, and 50 to 120 feet thick. In places they were underlain by other ore bodies at stratigraphically lower levels and also by fissure-vein and stockwork deposits of the Cord, Whitecap, and Tucson mines, which have been described and illustrated in detail by Loughlin (5, 6).

Blanket deposits of the eastern part of the district, although numerous, are generally smaller and more irregular than those of the western part, due principally to the greater abundance and discordance of porphyry bodies and to greater density of faults. Many of these replacement bodies have vein roots, or branch out along the bedding at successive favorable horizons from a central vein.

As shown by diagrams of Emmons, *et al.,* (6, fig. 51, 52; pl. 56, 57), the veins of the district have a predominant near-north trend and individually are rather short. The Winnie-Luema and Fortune veins, the longest and most productive in the district, have been traced 4000 and 5000 feet, respectively. The Winnie-Luema was stoped more or less continuously through a length of about 3000 feet and a depth of about 500 feet, and was as much as 40 feet wide. Branching veins and nearby replacement bodies greatly increased the tonnage of its gold-silver-copper-lead ores. The Fortune was originally worked principally for its bordering replacement bodies or mantos in the dolomite units (Dyer and Leadville) above the Parting Quartzite. Fissure and associated replacement deposits on the Fortune and intersecting northeast-trending veins in the sedimentary section beneath the Parting (Figure 5) were the basis for the Resurrection operation from the mid-'30's to the mid-'50's. These deposits yielded 2 million tons of complex sulfide milling ore having a gross value of $40 million (32).

Although smaller, many other veins in the eastern part of the district were highly productive. In some places on Breece Hill, swarms

FIG. 5. *Distribution of Ore (black) below Parting Quartzite Member on Fortune and Satellite Veins, Resurrection Mine, Leadville District. Manto (shaded) is above Parting Quartzite. After Walker (32).*

of intersecting small veins made stockworks of pyritic gold ore.

## Stratigraphic Relations

As Leadville is primarily a replacement-ore district, "favorable beds" and their stratigraphic relations are of paramount interest. Although precise data are unavailable, probably at least 90 per cent of the output from the district has come from the stratigraphic zone between the top of the Sawatch Quartzite and the base of the Molas or Belden, a zone only 350 to 400 feet thick (Table II). Within this zone, the dolomite units have been by far the most productive host rocks. Although modern stratigraphic usage recognizes three such dolomite units—the Leadville Dolomite, the Dyer Dolomite Member of the Chaffee Formation, and the Manitou Dolomite—old usage recognized only a "Leadville" or "Blue" limestone that included the present Dyer, and a "White" limestone. In this usage, which applied through Leadville's most productive era, the "Leadville" far exceeded the "White" or Manitou in productivity. The Leadville of present usage probably exceeds the Dyer and Manitou in productivity, but by no means to the degree that is commonly credited to it. Recent work has shown that, in some parts of the district, the Leadville is severely reduced in thickness beneath an unconformity, and that much ore previously assumed to have been in the Leadville was actually in the Dyer.

The unconformity at the top of the Leadville is a product of erosion under karst conditions prior to deposition of the Pennsylvanian Belden Formation. The Molas Formation, an irregular and pockety unit, is a product of the erosion and is largely the slightly reworked insoluble residue from chemical weathering of the Leadville in pre-Belden time. Recognition of the Molas at the top of the Leadville, and of the Gilman Sandstone Member as a persistent unit at the base, establishes that the dolomite member of the Leadville varies widely in thickness and, indeed, is absent in some parts of the district. Unfortunately, this was not known in the days of active prospecting and mining, and as a result money and effort were spent fruitlessly in searching for missing thicknesses of carbonate rock. Lamentably, the Molas Formation too was a source of wasted effort. Being an iron-stained clayey unit somewhat resembling oxidized ore, it was generally identified as "contact matter" and then, after assaying, as "barren contact." Many a prospect shaft was sunk through glacial drift and por-phyry to a bottom in the Molas, leaving the Leadville untested, and in places extensive workings were driven in the Molas in search of the nonexistent place where the "barren contact" would become productive. Lack of knowledge of the Gilman also was costly, because in the frenzy of the bonanza days it was often misidentified as Parting, and shaft sinking was stopped in it, leaving the underlying Dyer unexplored.

In several heavily mineralized areas, replacement ore was obtained in appreciable quantities from the thin-bedded and slightly shaly dolomite and sandstone of the Peerless Formation. Quartzites in the Sawatch and Parting are generally poor in replacement ore except in intensely mineralized areas, where they are mineralized in a blotchy pattern in leached sandy areas in the otherwise glassy quartzite. Many veins, however, such as those of the Winnie-Luema, Resurrection, Cord, and Tucson mines, had good ore in the quartzites.

Precambrian rocks have not been productive except along a few veins, notably the Winnie-Luema. Similarly, the Pennsylvanian rocks have been productive principally along veins, although they contain small replacement bodies in limy beds intersected by veins in the Ibex area on Breece Hill. Porphyry of all kinds is barren of ore through most of the district, but veins in the eastern part contained ore in porphyry, and small veins in porphyry in the Ibex mines were the source of rich gold ores. The Antioch pyritic gold stockwork on Breece Hill is in porphyry, and the nearby South Ibex stockwork is in mixed porphyry and metamorphosed grits of the Minturn Formation.

## Wall-Rock Alteration

With the exception of some quartzite, all the rocks of the Leadville district are altered in some degree, and many are altered intensely. Alteration of the sedimentary rocks began with dolomitization, which transformed the original limestone of the Leadville to a dark finely crystalline dolomite, not only in the district but throughout this part of the Colorado mineral belt.* Dolomitization of the Manitou, which outside the mineral belt is a dolomitic limestone, presumably was completed at this time also. The Dyer and Peerless were already dolo-

---

* See article on Colorado in this volume for discussion of hydrothermal versus syngenetic origin of dolomite in the Leadville Limestone/Dolomite in the Colorado mineral belt.

mites and evidently were not affected. Following the initial pervasive dolomitization, the Leadville of many areas underwent successive recrystallizations, first to a banded dark-gray and white rock known as "zebra," and then to a coarser, gray-tan phase, known as the "coarse pearly" or "discontinuous banded," characterized by destruction of the zebra structure. Engel, et al. (22) concluded from a study of oxygen isotopes that these various dolomite facies were formed at temperatures above those of limestone in the Leadville outside the Colorado mineral belt and at progressively higher temperatures as ore is approached. Their method was later questioned (27) but more recently (33) was found to have yielded generally valid results. From a study of fluid inclusions, T. G. Lovering (23) concluded that the dolomite of the "zebra" and younger facies formed at temperatures of 200° to 300°C.

Time relations between these changes involving dolomite and the emplacement of the porphyries are not fully established. The earliest, or "dark dense," variety of dolomite existed before the porphyries were intruded, since this dolomite occurs as inclusions in the Pando Porphyry, the earliest of the porphyries and, in places, was metamorphosed by the porphyry. The younger varieties of dolomite have not been observed with certainty as inclusions and although in places they show a slight color metamorphism at igneous contacts, this could have been inherited. Probably, the younger varieties of dolomite are of post-porphyry age. On Breece Hill, which is extensively intruded by porphyry, dolomites of the Leadville and Dyer were metamorphosed to magnetite-diopside-forsterite(?) rock, which subsequently altered to magnetite-serpentine. Whether any of the metamorphosed Leadville had zebra structure is unknown.

All the porphyries were deuterically altered after they were emplaced, and then they were widely altered hydrothermally. Reactions at the hydrothermal stage led predominantly to sericite, resulting in intensely sericitized rocks.

Presumably at about the time of porphyry alteration, dolomites were locally leached and disaggregated, or "sanded", and in places open channels or "water courses" developed in them. Similarly, quartzite was locally leached to a porous, cellular texture. Mineralization followed and produced further wall-rock changes as well as widespread replacement of dolomitic rocks by sulfides. In the western half of the district, dolomite was widely replaced by manganosiderite and, in many places, by gray jasperoid. Throughout the district but particularly in the eastern part, porphyries were weakly to moderately pyritized, as were the Pennsylvanian rocks in the Breece Hill-Ball Mountain area. At this same time, normally reddish or yellowish rocks such as those of the Peerless and Molas Formations were reduced and finely pyritized, so that underground they are now generally gray to black, except where later re-oxidized. Finally, oxidation produced widespread alteration of the already altered rocks, as well as of the ores. Pyritic rocks, including many porphyries, were leached and iron-stained and probably underwent some replacement by clays. Iron and manganese oxides and ferruginous and manganiferous brown and black jasperoid replaced dolomite as well as manganosiderite and ore. Dolomites underwent post-ore "sanding" and channeling, as did some quartzite. In many places, rocks of all kinds were reduced to a ferruginous clayey(?) sulfate-rich muck in which the parent materials are barely or not at all recognizable.

As indicated previously, no comprehensive study has been made of any of these kinds of alteration, and particularly of alteration in the porphyries. Emmons, et al. (6) briefly considered alteration in dolomite, the origin of jasperoid, and alteration related to oxidation. Contrary to conclusions stated here, they classed zebra rock as a product of spent ore solutions, "sanding" as an oxidation feature, and jasperoid as mainly hypogene. However, at Leadville as at nearby Gilman (16,17), zebra rock and its recrystallization products clearly were replaced by ore, and a dolomite sanding stage preceded deposition of the ores, as shown by replacement of sanded rock and of resedimented dolomite sand. At Leadville, minor sanding may have occurred again at the oxidation stage, as suggested by the occurrence of sanded zones descending into dolomite from the bottoms of oxidized ore bodies. That at least some of the jasperoid is supergene is shown by the close association of yellow-brown to brown-black iron- and manganese-rich jasperoids with oxidized ore bodies. Indeed, there is a complete gradation from oxidized fluxing ores containing $Fe + Mn > Si$ to jasperoids containing only a few per cent Fe plus Mn.

## Primary Ores

The primary ores of the district are mixed sulfide ores typically containing more than one metal of value and differing in composition both with location in the district and with kind of occurrence. They have been described and classified in considerable detail by Emmons, *et al.* (6). The commonest variety, characteristic of the blanket deposits, is a mixture of pyrite, sphalerite, and galena, with a subordinate gangue of quartz (mostly as jasperoid) and manganosiderite or ankerite. Most of this ore is almost free of copper minerals, but locally it contains chalcopyrite. The three main sulfide minerals occur in all proportions from roughly equal mixtures to almost monomineralic concentrations of one or the other. Regardless of these proportions, the ores generally contain a little silver, principally as argentite. Correlation between silver content and galena content of the ores in poor (6, Fig. 58). Except for late vug crystals, the sphalerite in such ores is a dark-brown to black marmatitic variety; three analyses show 12.1 to 17.8 per cent Fe + Mn, with the Mn ranging from 1.3 to 3.7 per cent (6, p. 157).

Some of the mixed sulfide ore is very fine grained and nonporous, and some is granular and finely vuggy. The ore generally shows bedding and other structures of the original dolomite, and much of it also exhibits planar, curved, and concentric rhythmic banding, as well as a clotty structure reflecting differences both in mineralogy and grain size. Paragenetically, pyrite is invariably the earliest of the sulfide minerals and was succeeded by sphalerite and galena in overlapping relations, although galena deposition persisted after that of sphalerite. Minor ore minerals, as chalcopyrite and argentite, followed the galena. Widespread replacement of dolomite by manganosiderite preceded the sulfide stage, and quartz was deposited principally in a stage preceding and accompanying pyrite. Minor gangue minerals such as barite, rhodochrosite, and dolomite are paragenetically late. All the ore and gangue minerals occur in minor quantity as drusy crystals in vugs, presumably as a result of minor solution and reprecipitation late in the mineralization process.

Grade of sulfide ores such as these varies widely, depending principally on the proportion of galena-marmatite to pyrite, and on the silver content. In recent decades, 12 to 15 per cent combined lead and zinc and 2 to 4 ounces silver per ton has been a common mining grade. Due to the weak ground and many other factors, high mining costs have always characterized the Leadville district, and though of an attractive grade by the standards of many districts, ore of the stated composition has been near the economic limit of mining existing at the time. Ore of this general value has been the mainstay of the district through much of its history, and although Leadville has the reputation of being a bonanza camp, it owes a considerable part of its output to the prosaic sulfide milling ores.

Although the vein ores are varied in composition, the bulk of them are mixed sulfide ores differing from the blanket ores principally in containing appreciable copper, gold, and silver. Most of them are highly pyritic and have a quartz or jasperoid gangue. Some veins are almost pure pyrite, which in places contains notable amounts of gold and silver. In other veins, chalcopyrite accompanies the pyrite, occurring either as veinlets cutting the pyrite or as an inconspicuous interstitial filling between pyrite grains. Sphalerite generally is a conspicuous component of the veins, occurring either in intergrowths with pyrite and quartz or in distinct seams, but only in recent decades has zinc been recovered from vein ores, notably from the Resurrection mine. Galena occurs principally in intergrowths with quartz and the other sulfides, and in some veins, such as the Winnie-Luema, in rich pods or seams. Silver occurs principally as argentite but also as pyrargyrite and as argentiferous tetrahedrite ("freibergite"). In some rich silver ores, argentite is in fine intergrowths with bismuthinite and galena; such mixtures have long been called "lillianite," "kobellite," and "schapbachite" in the district. Insofar as known, the gold is all in native form, although the precise nature and occurrence of some of it are unknown. Some, however, is readily visible, and considerable amounts of sacked "high grade" containing several per cent gold were obtained from small vens in the Ibex mines. Some of this gold was closely associated with sphalerite, occurring as intergrowths, inclusions, veinlets, and coatings in or on the sphalerite.

The vein ores are in part fissure fillings but in greater part seem to have replaced the fractured rocks in and along faults and fissures. They characteristically widen where they intersect dolomite and certain quartzite and grit beds, and they generally narrow in shale and porphyry. They also change in character with change in wallrocks, changing from siliceous to pyritic in passing from grit or quartzite to porphyry, and generally containing a higher proportion of sphalerite and galena where the walls are dolomite.

Thanks to their silver, gold, and copper content, the vein ores are generally higher in grade than the blanket sulfide ores, but they range widely in tenor. As with the blanket ores, much of the output has been at grade levels near the economic limit of mining existing at the time. Most vein ore mined in recent decades has had a sale value of approximately $20 per ton.

Variants of the common mixed-sulfide blanket and vein ores include (1) contact metamorphic magnetite-specularite-siderite ores, which in general have been productive only where they have been enriched by addition of paragenetically younger gold, pyrite, sphalerite, and chalcopyrite; (2) high-bismuth ores; (3) tungsten-bearing siliceous pyritic gold ores, characterized by paragenetically old wolframite and young scheelite; and (4) manganosiderite, protore of the oxidized manganese ores. These ores have been described by Emmons, *et al.* (6). High-bismuth silver-gold ore (but not bismuth ore *per se*) has been described by Chapman (14), who identified in it galenobismutite, aikinite, and alaskaite, as well as bismuth- and silver-bearing galena (9). These bismuth minerals are accompanied by argentite, chalcopyrite, tennantite, altaite, hessite, and native gold. The gold is principally associated with the hessite, as is true also in the rich silver ores of the nearby Gilman district (17). Chapman (14) assigned all these minerals to the paragenetically late chalcopyrite stage, which followed that of galena.

## Secondary Ores

The secondary ores include both the oxidized or "carbonate" ores and enriched sulfide ores. These two classes of ores have complex overlapping relationships to each other and to the primary ores, and they occur through extensive vertical ranges, reflecting both the effects of fluctuating and gradually rising water levels since mid-Tertiary time and the effect of post-mineral fault movements on the circulation system. In glaciated canyons, the oxidized zone extends only to shallow depth or is even absent, but in most other parts of the district it extends to depths of 400 to 600 feet and, locally, to more than 900 feet (6). Although the oxidized and sulfide zones are sharply separated in places, more typically they overlap. In places, partly oxidized ore extends through a vertical range of as much as 600 feet, a situation that has considerably complicated ore treatment or beneficiation at Leadville. The sulfide ore in the zone of overlap and immediately beneath it is in some places secondarily enriched but is only primary ore in others.

The enriched sulfide zone is far better developed and extends to much greater depths in the veins than in the blanket deposits. It is characterized particularly by chalcocite and minor accompanying bornite and covellite, which have made copper-rich ore bodies in many veins that otherwise are lean in copper. Increased silver content also characterizes the enriched zone. Much of the silver is in an unseen and unknown form, but some is native and some is in the sooty coatings known to the miners as "sulphurettes"; Chapman (14) concluded that most argentite is secondary and that the chief primary silver mineral is hessite. Above-average gold values also are common in the enriched zone; Emmons, *et al.* (6) considered that there had been much solution and redeposition of gold, although recognizing that many gold ores had been created by residual concentration.

Among the oxidized ores, the most productive were the argentiferous lead carbonate ores mined during the first few decades of Leadville's history. The lead in these ores was principally in the form of the carbonate, cerussite, but some was in other forms, such as residual galena and anglesite, minimum, massicot, wulfenite, vanadinite, pyromorphite, and plumbojarosite. Silver accompanied the lead in widely ranging concentration and occurred also in distinct silver ore bodies, poor in lead. The silver was principally in the form of halides, notably as embolite and cerargyrite, but occurred also in native form and as secondary argentite, part of which was in inclusions that colored some cerussite gray to black. Residual galena was characteristically greatly enriched in silver, in many places containing hundreds, and even thousands, of ounces of silver per ton.

The oxidized lead-silver ores were of four main kinds but included many intermediate types. So-called "sand carbonate" ore consisted principally of granular, crumbly cerussite, either white or stained brown to black by iron and manganese oxides; it formed extensive, sharply defined ore bodies in the western part of the district. So-called "hard carbonate," which was far more widespread and abundant, was essentially a ferruginous jasperoid containing small to microscopic crystals of cerussite and silver minerals. A third variety, containing either lead-silver or silver alone, was a pasty to soupy mixture of iron sulfates and hydroxides. Its value could be determined only by assay, since much material of similar appearance was—and is—barren. The fourth kind was siliceous ore that generally contained either silver or lead but not both. Some very rich silver ores were of this type. They consisted essentially of jasperoid that was coated and seamed by silver halides. Some such ore was gray jasperoid coated by translucent green-gray embolite, but much of it was yellow-brown or brown-black due to admixtures of iron and manganese oxides in both the jasperoids and the coatings. Lead-silver ore bodies of all these kinds generally were bordered by, and graded into, iron-manganese fluxing ores or iron-manganese jasperoids. The jasperoids, plus porphyry waste, constitute the bulk of the dumps which now remain as almost the only visible evidence of the old carbonate workings.

The oxidized zinc ores in part accompanied and bordered the oxidized lead-silver bodies, but in general they were at lower depths, and some were isolated in dolomite considerably below the oxide-sulfide boundary as expressed in ores of other kinds. The oxidized zinc ores have been described by Loughlin (3,6), and a recent summary of them has been made by Heyl (30). The chief mineral of the oxidized zinc deposits was the carbonate smithsonite, but calamine, chalcophanite, hetaerolite, and zinciferous clays, as well as more rare species, also were present. The zinc minerals replaced dolomite or manganosiderite, in places almost duplicating the original rock in appearance. More commonly, however, the zinc minerals were accompanied by iron and manganese oxides, or by dark jasperoids, which colored the ore brown to black. Many large bodies of zinc carbonate contained 30 to 40 per cent Zn and, for many years, 18 to 20 per cent Zn was the cutoff grade both for mining and for metallurgical treatment, although bodies existed at various grades down to a few per cent. Unlike the low-grade oxidized lead-silver ores, however, the low-grade oxidized zinc ores had no value either as flux or as manganese ore.

Secondary ores in the veins were in general characterized by cerussite, native gold, and in places by native copper, chalcanthite, and copper carbonates, as well as by iron oxides Bismuth ore bodies, all of which were oxidized, were characterized by the carbonate bismutite, which presumably was derived from primary bismuthinite and other bismuth-bearing minerals.

## Zonal Relations

A zonal pattern of ore deposits centered around a so-called stock at Breece Hill was recognized by Emmons, *et al.* (6) and was further described by Loughlin and Behre (11).

This pattern, though somewhat irregular in detail, is characterized by a change from contact-metamorphic or hypothermal deposits at the center through typical mesothermal deposits in an intermediate zone to local epithermal deposits on the borders of the district. Loughlin and Behre regarded the zonal pattern as an expression of combined space-time relations in the mineralization process and also emphasized the importance of geologic structure in distorting the pattern. In terms of space, they considered the Breece Hill "stock" as the main center of mineralization, although recognizing another, smaller, one to the south, on Printer Boy Hill. They classed the area of magnetite-serpentine deposits with associated sulfides on Breece Hill as pyrometasomatic and hypothermal; the manganosiderite-pyrite-marmatite assemblages typical of the central part of the district as intermediate mesothermal; the jasperoid-pyrite-sphalerite-galena deposits farther west as intermediate mesothermal; the barite-galena-silver-sphalerite deposits of western, southern, and northern localities in the district as cooler mesothermal; and outlying dolomite-resinous sphalerite-galena deposits to the south of the district as epithermal. In terms of time, pyrometasomatic pyroxene and olivine (parent minerals of the serpentine) were followed successively at Breece Hill by hypothermal magnetite, hotter mesothermal siderite and manganosiderite, and intermediate and cooler mesothermal sulfide minerals and gold. Similarly, in the intermediate zone, hotter mesothermal manganosiderite-marmatite was followed in time by cooler mesothermal galena, silver and bismuth minerals, chalcopyrite, and gold.

The so-called stock of Breece Hill is not a stock in the conventional sense but is a composite of many different intrusions in a tenuous matrix of sedimentary rocks. Early large intrusions were basically concordant, with the result that sedimentary strata, although greatly distended and in places enveloped by porphyry, maintain stratigraphic relations and rough continuity through and under the "stock." In reality, the "stock" is a pile of thick sills, which locally are fused to make a Christmas-tree laccolith. The pile has as a root a thick dike (or elliptical pipe?) along the Weston fault and its branches, as well as smaller dikes or irregular intrusive bodies along the Pilot and other faults. The pile is pierced by a pipelike body of younger porphyry or microintrusion breccia and also is cut by dikes and pipes of other porphyries, including several of the youngest known at Leadville. All these rocks, and ore deposits as well, are locally cut by rhyolitic dikes and intrusion breccias. The entire hill is strongly altered and widely mineralized; unquestionably, it was a center of intrusion and hydrothermal activity. The crude zonal pattern strongly suggests that this intrusive center thermally influenced the mineralization process through much of the district, but as is discussed in following sections, the mineralizing solutions themselves probably originated in a deeper and larger source than this single localized center, and they rose along many deepgoing faults, including some that localized the Breece Hill intrusive center.

## LOCALIZATION OF ORE BODIES

Ore bodies of the Leadville district are localized by combined structural, stratigraphic, and intrusive features. In general, except for the veins, structural features were of lesser importance than stratigraphic and intrusive features in the localization of individual ore bodies, but they had important influence on the distribution of ore bodies. The ore bodies most characteristic of the district are blanket bodies of large area but generally moderate thickness lying in one of the dolomite units beneath a porphyry sill or sheet or beneath quartzite or shale. Some of these ore bodies have small veins as roots, and a few branch from through-going veins or fissures, but most have no evident direct fracture control. Of all such ore bodies, the largest and most extensive are just beneath the unconformity marked by the Molas Formation and just beneath a thick porphyry sill also at this horizon.

### Role of Stratigraphic Features

In relation to the underlying rocks, the unconformity beneath the Molas has a wide range of stratigraphic levels. It is generally at some horizon between 50 and 200 feet above the base of the dolomite member of the Leadville but, in places on Fryer and Yankee Hills and the north end of Breece Hill, it is at the level of the Gilman Sandstone Member; and locally on Fryer Hill, it is in the Dyer Member of the Chaffee Formation as low as the top of the Parting Quartzite Member. This old karst or weathered zone had porosity that clearly fostered mineralization. The porosity results: (1) from channels and small caves within the dolomite; (2) locally,

from an extremely uneven top surface of the dolomite, with steep pinnacles separated by pits and upward-opening channelways; and (3) from fine-scale or intergranular openings that give some of the rocks a punky character. Although most of the channels and depressions in the dolomite are filled with silt and chert fragments of the Molas Formation, they obviously remained more permeable than the dolomite or the overlying porphyry or the black shale of the Belden Formation, as indicated by the concentration of alteration features along them. Many of the channels and depressions are bordered by sanded dolomite or even by younger open channelways produced by pre-ore dolomite-leaching solutions. Along many, the dolomite is farther advanced in the evolutionary sequence—(1) dense-dark dolomite, (2) zebra dolomite, (3) coarse pearly dolomite—than is that nearby.

Thus a porosity or plumbing system inherited from the period of karst weathering in pre-Belden time evidently was accentuated by the activities of pre-ore altering and leaching solutions and then served to guide ore solutions and localize some ore deposits. Some heavily mineralized areas, such as Fryer Hill, seem to correlate with an exceptionally porous and thin Leadville. In this area, the Leadville was mineralized almost continuously, so that there were no ore shoots except as defined by porphyry cutting the Leadville. The well-defined ore shoots in dolomite of areas such as Iron Hill may have had a similar control, as suggested by the erratic thinning of the Leadville in many places on Iron Hill (6, plates 22–26). The widespread occurrence of material from the Molas on dumps of the old workings on Iron Hill indicates that the thinning must be due to unconformity and not to transfer of some of the Leadville to the upper side of the capping sill. However, a few of the ore shoots in Iron Hill are aligned along small faults, so ancient channeling may not have been the only control of the ore shoots.

A second favored horizon of mineralization was at the level of the Gilman Sandstone Member which also was a porous zone, apparently as a result of ground-water leaching at the time the karst surface was developing (18). A third favored horizon was in the upper part of the Manitou Dolomite, in thin-bedded and slightly shaly strata just beneath either the tight shale or glassy quartzite of the Parting Member. Inasmuch as the top of the Manitou marks an unconformity of considerable magnitude, this zone may also be somewhat porous, although it does not have the conspicuous porosity of the top of the Leadville.

## Role of Intrusive Features

Aside from the three stratigraphic zones just named, the occurrence of ore in dolomite is principally determined by porphyry contacts, and the three zones themselves contain such contacts in many places. A chart showing the positions of blanket ore bodies in various parts of the Leadville district (6, plate 59) clearly shows the close correlation between porphyry sills and ore bodies, the ore occurring on the under side of the sill regardless of the stratigraphic position of the sill in a dolomite unit.

The sills evidently served to deflect the ore solutions so that they spread widely from whatever channelways or fractures provided their access. Permeable zones that allowed solutions to cross the sills probably existed only locally, because most fractures in the sills are tight and gougey as a consequence of the hydrothermally altered character of the porphyries. Of all the sills, the most persistent and thickest is at or near the base of the Belden. This widespread soft shaly zone, marking the greatest discontinuity in physical properies in the stratigraphic column, evidently was the most attractive of all sites for sill intrusion, and as a result it is occupied in most places by the earliest of the porphyries, the Pando ("White") Porphyry, in a sill as much as 1000 feet thick. Where the Pando is absent, Johnson Gulch Porphyry generally occupies this horizon.

Many ore bodies are localized in dolomite at the intersections of dikes with the undersides of sills, generally, but not exclusively, on the downdip side of the dike as measured in the dolomite. Some such dikes are feeders of the sills, and others are of younger porphyries that cut the sills. Most of the dikes in the district being Johnson Gulch Porphyry and other varieties that have been loosely grouped with Johnson Gulch as "Gray porphyry," this type of ore occurrence led in the early days to the widely held notion that "Gray porphyry" was the source of the ores. The relationship is only geometric, however, as indicated by the fact that keels or sharp rolls in sill contacts produce the same effect and that ore may be localized at the intersection regardless of the identity of either the dike or the sill. Probably, in some cases the dike impeded solutions moving laterally under influence of the sill, and, in others, the sill impeded solutions rising along the dike.

## Role of Faults

Faults had a major effect on distribution of ore bodies within the district, and they also localized many individual ore bodies, as in veins and many of the smaller replacement bodies. Distribution of areas of principal ore bodies with respect to major faults (Figure 2) strongly suggests that faults such as the Ball Mountain, Weston, Pilot, Mike, Adelaide, and Iron were channelways for mineralizing solutions. Ore deposits along these faults and many lesser faults connected with them are characterized by mineral assemblages with higher-temperature connotation than in intervening areas, and they thus constitute anomalies in the broad zonal pattern. In general, the ores are most marmatitic, pyritic, and sideritic along such faults, and they are also richest in copper, gold, and bismuth. Most major faults are devoid of recognizable veins, but, in one place or another, almost all contain fragments of ore. In some places such fragments represent ore derived from blanket deposits dragged along the faults during post-ore movement. In many others, however, the fragments are not correlatable with blanket deposits and probably represent former veins.

From the presumed principal source channelway faults, the mineralizing solutions evidently spread widely in the network of lesser faults and were deflected long distances beneath sills. Abundant evidence exists of the role of minor faults in the mineralizing process, as for example, in the Resurrection mine in the eastern part of the district (Figure 5). An outstanding example in the western part of the district is the Tucson-Maid reverse fault beneath Iron Hill and the north part of Carbonate Hill. In its upper part, this northwest-trending fault dips gently northeastward almost parallel to the strata, but it steepens downward to a dip of 45° to 60°NE. Where steep, it has a maximum stratigraphic throw of about 150 feet, but where gentle it has a throw of only a few feet. Although the fault itself is only slightly mineralized, it is bordered along its course through Iron and Carbonate hills by many blanket ore bodies and branching mineralized fissures, as shown in detail by Loughlin (5, Plates 6, 7; 6, Figures 18–20). The fault clearly was a major ore control at all stratigraphic levels except the "first contact" in these highly productive areas. The Tucson-Maid reaches the surface only west of the crest of Carbonate Hill. From the northwest slope of this hill, it has been traced northwestward through a maze of fault blocks to the Pendery fault at Stray Horse Gulch. Thus, although it was not recognized during the early-day mining on Fryer Hill, it probably underlies this hill and may have been a factor in the extensive mineralization there.

Another kind of fault control of ore deposits in the western part of the district is exemplified by the Cord vein, off the Yak Tunnel in southern Iron Hill, which has been described in detail by Loughlin (6, p. 69, 287). The vein occupies a short steep northeast-trending fault with a displacement of only a few feet, and replacement ore bodies make out from it where it crosses dolomite layers sandwiched between porphyry sills or quartzite. Northeast-trending faults coinciding with the long axes of two of the long manto ore bodies on Iron Hill have been classed as post-ore (6, p. 94), but their coincidence with the manto bodies and indications of mineralization in underlying workings on at least one of them suggest a genetic relation between the manto bodies and the faults. In the Leadville Drainage Tunnel beneath Fryer Hill, a swarm of small northeast-trending veins in a broad sheeted zone cuts quartzite beneath the level of old stopes in the dolomite, suggesting fracture feeding from below of at least some of the replacement deposits.

## LOCALIZATION OF DISTRICT AND FORMATION OF ORE DEPOSITS

Regionally, the Leadville district is located at the intersection of the northeast-trending Colorado mineral belt (29) with the near-north-trending Mosquito-Weston fault system, which in turn coincides in part with the basin-and-range Arkansas Valley graben fault system. These relations are discussed in the report on Colorado in this volume.

On a more local basis, the district is a complexly faulted and intruded area at the intersection of the N20°W-trending Weston fault system with the N15°E-trending Mosquito fault system. The district, on the west side of the intersection and in the hanging wall of the Mosquito fault, contains many faults of both systems, as well as cross-fault systems of two or three orientations. The overall pattern suggests two fault systems fighting their way through each other, with the Weston system dying out just north of the district.

Although many of the faults other than the main Mosquito and Weston faults formed concurrently with intrusion and had complicated histories of independent movement caused by intrusion, their conformance with regional fault trends indicates that they are not solely the products of intrusion. Rather, they localized and fostered intrusion that occurred at the same time as the faulting. The Mosquito and Weston faults are characterized by stocks and other intrusive centers at several places along their combined courses of nearly 100 miles, and the Mosquito fault especially is characterized by great variety and abundance of porphyries in a hanging-wall belt a few miles wide extending from the Leadville area northward for several miles. Evidently, the fault served as a major conduit for porphyry magmas that rose from the batholith that is generally visualized as underlying the Colorado mineral belt (29). Because of its structural location the Leadville area was more susceptible to fracturing than neighboring areas to the north and south, and it therefore became one of the principal sites of intrusion along the Mosquito-Weston fault system.

That the mineralized area at Leadville closely coincides with a complexly faulted and intruded area can scarcely be an accident. A "hot" and localized source of ore-depositing solutions and deposition under influence of a marked temperature gradient are indicated by abundant evidence, such as the range from pyrometasomatic to epithermal in the characteristics of the deposits, the rough zonal pattern, the widespread occurrence of highly marmatitic sphalerite, the intense sericitic alteration of all rocks except carbonates and quartzites, and isotopic evidence (22) that the dolomite alteration facies in the Leadville-Gilman area formed at temperatures distinctly higher than did the unaltered carbonate rocks of the surrounding region. These features, together with the close association of ore deposits with intrusive rocks, make the Leadville district an ideal example of an igneous-related "hydrothermal" district. Obviously, however, the source of its ore was not in any of the exposed rocks or intrusive bodies, for with the exception of the late rhyolitic intrusion breccia, these rocks all predate the ore. Presumably, the ore solutions either had their source in the underlying batholith or were in part mobilized out of other rocks under influence of the batholith, and just as the porphyry bodies before them, they rose in this area because structural conditions here were most favorable for their doing so. The question of the ultimate source of the metals will not be debated here. Isotopic data suggest a crustal source and complex history for the lead in the Leadville deposits (34). Sulfur isotope determinations on four samples showed a $S^{30}/S^{34}$ ratio of 22.10

to 22.35, very near meteoritic sulfur composition (21, p. 146). A uniform Cu, Pb, and Zn content of 3, 3, and 10 ppm, respectively, in carbonate rocks of the Leadville both in and outside the mineral belt (20) indicates that there was no mobilization of trace metals from these rocks.

The hydrothermal ore solutions evidently reached the sites of deposition by many devious routes, but in general they rose through the basement along many of the faults in the fault complex and then were in part deflected into subhorizontal paths in the stratified section of sedimentary rocks and sills. In the stratified section, they migrated widely beneath the thick sills and, especially, beneath sills at the horizons of unconformities. They replaced the dolomite beneath the sills to form the widely extensive blanket deposits and, throughout the eastern part of the district, they also formed veins.

## POST-DEPOSITIONAL HISTORY OF ORE DEPOSITS

Once formed, the ore deposits were modified by faulting, oxidation, intrusion of rhyolite or rhyolitic breccia, and erosion. Post-ore fault movement—almost entirely by reactivation of pre-ore faults—probably occurred at many different times, but most of it followed oxidation, as shown by displacement of oxidation levels. On Iron Hill, for example, on the east side of the Iron and related faults, the top of the sulfide zone is in Manitou Dolomite and at an elevation of about 10,300 feet, whereas on the west side it is 600 feet lower, in porphyry intruded into the upper part of the Leadville Dolomite. The top of the basement, in contrast, is displaced about 1500 feet by the same faults. Oxidation evidently occurred principally in Tertiary time and largely before deposition of the Lower Pliocene Dry Union Formation, for oxidation levels are independent of the presence or absence of Dry Union Formation or any other surficial deposits. Presumably, the oxidation occurred in the period of pedimentation, graben faulting, and later canyon cutting in Miocene time, before the groundwater level of the Arkansas Valley was raised by Dry Union sedimentation.

Emplacement of rhyolitic intrusion or explosion breccia in pipes and dikes followed oxidation and preceded glaciation. Most of the breccia or rhyolite bodies are in Evans and South Evans Gulches, where the oxidized zone has been destroyed by glacial erosion and hence the relation of rhyolitic breccia to oxidized ore cannot be determined. An elliptical pipe on the northwest flank of Breece Hill, however, is in part in an unglaciated area, and its fresh, unweathered character in contrast to adjoining strongly oxidized ore deposits indicates a post-oxidation age. The rhyolitic pipes and dikes reamed through many an ore body, and they probably destroyed much ore. At least a minor mineralization (or a remobilization of earlier ore?) accompanied their emplacement, for small sulfide veinlets cut the breccia in places, and some pebbles of basement rocks abraded to roundness in the explosive intrusive process are coated by drusy sulfides.

Since some ore bodies subcrop beneath the Dry Union Formation, erosion had reached the level of the Leadville ore bodies at least by the beginning of the Pliocene, and probably had reached them by the Miocene. As the Dry Union Formation was gradually stripped, the deposits were again exposed to erosion, and then, in the Pleistocene, they were deeply eroded along the courses of the glacial canyons. How much ore may have been removed from the district by erosion is an academic question, but whether any of it was redeposited in any concentrated form in the sediment-laden Arkansas Valley is a matter for investigation.

## REFERENCES CITED

1. Emmons, S. F., 1886, Geology and mining industry of Leadville, Colorado: U.S. Geol. Survey Mon. 12, 770 p. and atlas of 35 sheets.
2. Emmons, S. F. and Irving, J. D., 1907, The Downtown district of Leadville, Colorado: U.S. Geol. Survey Bull. 320, 75 p.
3. Loughin, G. F., 1918, The oxidized zinc ores of Leadville, Colorado: U.S. Geol. Survey Bull. 681, 91 p.
4 Crawford, R. D., 1924, A contribution to the igneous geology of central Colorado: Am. Jour. Sci., 5th ser., v. 7, p. 365–388.
5. Loughlin, G. F., 1926, Guides to ore in the Leadville district, Colorado: U.S. Geol. Survey Bull. 779, 37 p.
6. Emmons, S. F., et al., 1927, Geology and ore deposits of the Leadville mining district, Colorado: U.S. Geol. Survey Prof. Paper 148, 368 p.
7. Henderson, C. W., 1927, Production, history, and mine development (of Leadville district), p. 109–144, in Emmons, S. F., et al., Geology and ore deposits of the Leadville mining district, Colorado: U.S. Geol. Survey Prof. Paper 148, 378 p.
8. Burbank, W. S., 1933, Relation of Paleozoic and Mesozoic sedimentation to Cretaceous-

Tertiary igneous activity and the development of tectonic features in Colorado, *in Ore deposits of the Western States* (Lindgren Volume): New York, A.I.M.E., p. 277–301.

9. Chapman, E. P. and Stevens, R. E., 1933, Silver and bismuth-bearing galena, Leadville. Econ. Geol., v. 28, p. 678–685.

10. Lovering, T. S., 1933, The structural relations of the porphyries and metalliferous deposits of the northeastern part of the Colorado Mineral Belt, *in Ore deposits of the Western States* (Lindgren Volume): New York, A.I.M.E., p. 301–307.

11. Loughlin, G. F., and Behre, C. H., Jr., 1934, Zoning of ore deposits in and adjoining the Leadville district, Colorado: Econ. Geol., v. 29, p. 215–254.

12. Lovering, T. S. and Goddard, E. N., 1938, Laramide igneous sequence and differentiation in the Front Range, Colorado: Geol. Soc. America Bull., v. 49, p. 35–68.

13. Hedges, J. H., 1940, Possibilities of manganese production at Leadville, Colorado: U.S. Bur. Mines I.C. 7125, 23 p.

14. Chapman, E. P., 1941, Newly recognized features of mineral paragenesis at Leadville, Colorado: A.I.M.E. Tr., v. 144, p. 264–275.

15. Swartz, J. H., *et al.*, 1943, Report of a geophysical survey of proposed drainage tunnel routes in the Leadville mining district, Leadville, Colorado: U.S. Bur. Mines, mimeo. report, 18 p.

16. Lovering, T. S., and Tweto, O. L., 1944, Preliminary report on geology and ore deposits of the Minturn quadrangle, Colorado: U.S. Geol. Survey open-file rept., 115 p.

17. Tweto, O. and Lovering, T. S., 1947, The Gilman district, Eagle County (Colo.): p. 378–387 in Vanderwilt, J. W., Editor, *Mineral resources of Colorado,* State of Colo., Mineral Res. Bd., 547 p.

18. Tweto, O., 1949, Stratigraphy of the Pando area, Eagle County, Colorado: Colorado Sci. Soc. Proc., v. 15, p. 149–235.

19. Behre, C. H., Jr., 1953, Geology and ore deposits of the west slope of the Mosquito Range: U.S. Geol. Survey Prof. Paper 235, 176 p.

20. Engel, A. E. J. and Engel, C. G., 1956, Distribution of copper, lead, and zinc in hydrothermal dolomites associated with sulfide ore in the Leadville Limestone (Mississippian, Colorado): (Abs.) Geol. Soc. Amer. Bull., v. 67, p. 1692.

21. Kulp, J. L., *et al.*, 1956, Sulfur isotope abundances in sulfide minerals: Econ. Geol., v. 51, p. 139–149.

22. Engel, A. E. J., *et al.*, 1958, Variations in isotopic composition of oxygen and carbon in Leadville Limestone (Mississippian, Colorado) and in its hydrothermal and metamorphic phases: Jour. Geol., v. 66, p. 374–393.

23. Lovering, T. G., 1958, Temperatures and depth of formation of sulfide ore deposits at Gilman, Colorado: Econ. Geol., v. 53, p. 689–707.

24. Tweto, O., 1960, Pre-ore age of faults at Leadville, Colorado; *in* Short papers in the geological sciences: U.S. Geol. Survey Prof. Paper 400-B, p. 10–11.

25. Tweto, O., 1961, Late Cenozoic events of the Leadville district and upper Arkansas Valley, Colorado; *in* Short papers in the geologic and hydrologic sciences: U.S. Geol. Survey Prof. Paper 424-B, p. 133–135.

26. Pearson, R. C., *et al.*, 1962, Age of Laramide porphyries near Leadville, Colorado; *in* Short papers in geology and hydrology: U.S. Geol. Survey Prof. Paper 450-C, p. 78–80.

27. Friedman, I. and Hall, W. E., Fractionation of $O^{18}/O^{16}$ between coexisting calcite and dolomite: Jour. Geol., v. 71, p. 238–243.

28. Linn, K. O., 1963, Geology of the Hellena mine area, Leadville, Colorado: Unpublished Ph.D. thesis, Harvard Univ., 157 p.

29. Tweto, O. and Sims, P. K., 1963, Precambrian ancestry of the Colorado Mineral Belt: Geol. Soc. America Bull., v. 74, p. 991–1014.

30. Heyl, A. V., 1964, Oxidized zinc deposits of the United States, Part 3. Colorado: U.S. Geol. Survey Bull. 1135-C, 98 p.

31. Pierce, A. P., 1965, Isotopic nature of ore-leads in the Colorado mineral belt, (abs.) *in* Program of Rocky Mtn. Sect., Geol. Soc. America.

32. Walker, W. J., 1965, Structural control of the ore deposits of the Resurrection mine, Leadville mining district, Colorado. Unpublished report, 22 p.

33. O'Neil, J. R. and Epstein, S., 1966, Oxygen isotope fractionation in the system dolomite-calcite-carbon dioxide: Science, v. 152, p. 198–200.

34. Pierce, A. P., *et al.*, (*in press*) Isotopic nature of ore-leads in the Colorado mineral belt: Manuscript, 52 p.

# 33. Ore Deposits in the Central San Juan Mountains, Colorado[*]

THOMAS A. STEVEN[†]

## Contents

## Illustration

## Table

[*] Publication authorized by the Director, U.S. Geological Survey.
[†] U.S. Geological Survey, Denver, Colo.

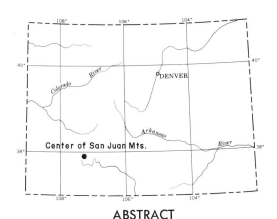

## ABSTRACT

Most mineralized areas in the central San Juan Mountains, Colorado, are associated with the youngest subsidence structures in a large volcanic cauldron complex that formed concurrently with eruption of the surrounding ash-flow field. Two local types have been recognized: (1) fissure-vein deposits along faults characterized by relatively late movement, and (2) hydrothermally altered and mineralized areas associated with postsubsidence Fisher Quartz Latite centers. The first type has provided most production in the past; the potential of the second type has yet to be adequately tested.

## INTRODUCTION

Ore deposits in the central San Juan Mountains, Colorado, are mostly within a large area of complexly overlapping volcanic subsidence structures that subsided repeatedly in response to voluminous ash-flow eruptions in middle Tertiary time (Figure 1). Except for the nearly circular calderas, these structures will be referred to by the general term "cauldrons". Detailed and reconnaissance studies (3,4,5,6,8) have shown that mineralization was generally late in the local sequence of volcanic and related geologic events and that most known ore deposits are localized around the youngest subsidence structures in the complex. These conclusions have already successfully guided exploration for concealed ore deposits in the Creede mining district (4,7) and should be helpful in judging the potential of other areas nearby.

Comprehensive discussions of the geology and ore deposits in the Creede and Summitville districts, the main producing areas in the cen-

tral San Juans, have been published recently (3,8), as has a brief account of the geologic setting of the Spar City district (6). Many of the details in these reports need not be repeated here, where emphasis will be given instead to regional considerations that bear on the localization of mineralization.

## GEOLOGIC SETTING

The central San Juan cauldron complex is surrounded by a sequence of related volcanic rocks, largely of ash-flow deposits with associated lava and breccia units, that accumulated concurrently with subsidence. This sequence is one of the youngest in the great volcanic pile that constitutes most of the San Juan Mountains in southwestern Colorado; only the widespread late basaltic lavas of the Hinsdale Formation are younger. The source area for the central San Juan sequence is marked by a complex of overlapping volcanic subsidence structures that formed by collapse of the roofs of local magma chambers after they had been partly evacuated by recurrent eruption of great quantities of ash. Four individual collapse structures have been reported (8), but recent work has established the existence of several more, and undoubtedly others remain undetected. In all, a north-trending area at least 15 to 20 miles wide and 35 miles long, centered a few miles south of the town of Creede, shows evidence of episodic collapse during the period of volcanic activity (Figure 1).

The youngest of these collapse structures are: (1) the Creede caldera near the center of the cauldron complex, (2) a poorly defined area of subsidence immediately south of the Creede caldera, and (3) the San Luis Peak caldera near the north end of the complex; the first two appear to have formed about contemporaneously near the end of the related period of volcanism and the last a little earlier. Most of the known ore deposits and larger hydrothermally altered and mineralized areas in the central San Juan Mountains are associated in space and time of origin with one or another of these three local subsidence structures. The most significant exception is the Summitville district (3) near the southeast margin of the cauldron complex.

The general stratigraphic relations of the rock units associated with the Creede and San Luis Peak calderas within the central San Juan cauldron complex are summarized in Table 1. Major units from other centers intertongue marginally with these units, but a general summary of the complex volcanic stratigraphy of

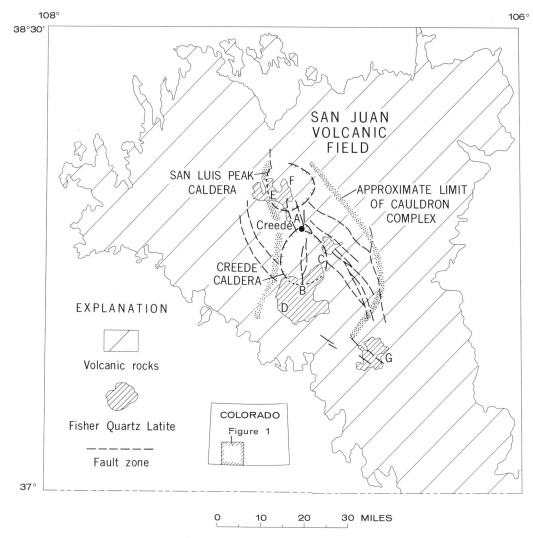

FIG. 1. *San Juan Volcanic Fields.*

the central San Juans is beyond the scope of this article.

These rock units can be divided with respect to recurrent subsidence. Ash-flow eruptions responsible for the Outlet Tunnel Member of the La Garita Quartz Latite culminated in collapse of the La Garita cauldron along a series of north- to northwest-trending normal faults in the area north of the Creede caldera and east of the San Luis Peak caldera. Most of the collapsed area is covered by younger rocks and the original form or extent of the subsided area is poorly known.

The rhyolite of Miners Creek, Bachelor Mountain Rhyolite, Shallow Creek Quartz Latite, and Phoenix Park Member of the La Garita Quartz Latite were erupted, at least

in part, onto the irregular surface formed by the earlier collapse. The rhyolite of Miners Creek is a local steep-sided accumulation of rhyolite flow and pyroclastic rocks. This is overlapped by a thick mass of pumiceous ash-flow material that constitutes the Bachelor Mountain Rhyolite. To the west, the upper part of the Bachelor Mountain Rhyolite intertongues with quartz latitic lava flows and breccias of Shallow Creek Quartz Latite. To the east, the Bachelor Mountain rocks intertongue with and are overlain by crystal-rich welded tuffs of the Phoenix Park Member of the La Garita Quartz Latite.

Eruptions of the Bachelor Mountain Rhyolite culminated in cauldron collapse of the probable vent area near the present town of

TABLE 1.   *Volcanic Stratigraphy near the Creede and San Luis Peak Calderas*

| Fisher Quartz Latite | Creede Formation |
|---|---|

↑ RECURRENT SUBSIDENCE ↓  
OF  
CREEDE CALDERA

RECURRENT SUBSIDENCE OF CREEDE CALDERA

DOMING OF CREEDE CALDERA

Snowshoe Mountain Quartz Latite

Nelson Mountain Quartz Latite  ↑ SUBSIDENCE OF SAN LUIS  
Rat Creek Quartz Latite  ↓    PEAK CALDERA

Huerto Formation { Wason Park Rhyolite / Mammoth Mountain Rhyolite / Farmers Creek Rhyolite } Quartz latite and rhyolite lava flows

SUBSIDENCE OF BACHELOR MOUNTAIN CAULDRON

Shallow Creek Quartz Latite { Bachelor Mountain Rhyolite / Windy Gulch Member / Campbell Mountain Member / Willow Creek Member } La Garita Quartz Latite / Phoenix Park Member

Rhyolite of Miners Creek

SUBSIDENCE OF LA GARITA CAULDRON  
Outlet Tunnel Member  
Modified from Steven and Ratté (1964)

Creede and in the development of a series of northwestward-trending step faults that initiated faulting in the area later to become the Creede graben.

Succeeding eruptions deposited first a local accumulation of pumiceous tuffs and welded tuffs east of Creede that constitutes the Farmers Creek Rhyolite. This was followed in turn by widespread thick welded tuffs comprising the Mammoth Mountain and Wason Park Rhyolites, as well as other major units in adjacent areas. Concurrent eruptions of dacitic lava and pyroclastic rocks southwest of Creede deposited the Huerto Formation; these dacitic rocks intertongue with the widespread welded tuffs. Indirect evidence suggests that the area now occupied by the Creede caldera may have subsided repeatedly concurrently with the ash-flow eruptions.

Crystal-rich quartz latitic ash was erupted later in intracaldera time to form the tuffs and welded tuffs of the Rat Creek and Nelson Mountain Quartz Latites, and the San Luis Peak caldera subsided in several stages in response to these eruptions. The last major eruptions in intracaldera time resulted in the accumulation of more than 6000 feet of crystal-rich welded tuff of the Snowshoe Mountain Quartz Latite in the core of the Creede caldera; the caldera sank concurrently with and following accumulation of the Snowshoe Mountain Quartz Latite.

Minor eruptions during intracaldera time re-sulted in local volcanic necks and flows that cut and intertongue with the more widespread ash-flow deposits.

Final subsidence of both the San Luis Peak and Creede calderas was followed by strong doming of the caldera cores. Local flows, domes, and pyroclastic breccias of Fisher Quartz Latite were then erupted along broken zones associated with both calderas. Concurrently, lake and stream deposits, volcanic ash, and travertine were deposited in the structural moat around the margin of the Creede caldera to form the Creede Formation.

The last main period of faulting followed deposition of the Creede Formation. Several older faults were reactivated at this time.

## CREEDE CALDERA AND ASSOCIATED MINERAL DEPOSITS

The Creede caldera is a nearly circular subsided block 10 to 12 miles in diameter. Caldera subsidence apparently continued over an extended period, because evidence for faulting concurrent with eruption of the Farmers Creek Rhyolite and succeeding formations can be seen locally around the caldera margin. The extent of this early subsidence or the form of the subsided block cannot be told from present exposures. The last subsidence accompanied and followed eruption of the Snowshoe Mountain Quartz Latite and involved all of the caldera core as now known.

Several faults in the Creede graben extending north to northwest from the caldera are known to have been active during subsidence, and tangential grabens intersecting the northeastern and southwestern margins of the caldera clearly formed during or shortly following final collapse of the caldera. Following final subsidence, the core of the caldera was domed so that the highest parts near the center stood about 4000 feet above the moatlike margin.

Post-caldera faulting, indicated by offset Creede Formation beds and Fisher Quartz Latite lavas, is evident at places around the caldera margin and in grabens extending from the caldera.

Three areas around the Creede caldera have had sufficient production or mining activity to qualify as mining districts (Figure 1). The Creede district along the northern margin of the caldera is by far the largest and most productive, and from the beginning of mining in 1889 until 1964 it had produced about $66,400,000 worth of metals, largely silver ($39,500,000), lead ($15,600,00), and zinc ($7,000,000), with lesser amounts from copper ($980,000) and gold ($3,350,000). The Spar City district along the southern margin of the caldera was prospected in the early part of the 1900's and several small mines were developed, but actual production was small. The Wagon Wheel Gap fluorspar district along the eastern margin of the caldera was active from about 1911 to 1950, and an estimated 190,000 tons of fluorspar was produced.

## Creede District

The ore deposits in the Creede district are associated principally with faults in a complex graben that extends radially north-northwest from the Creede caldera (8). Displacement in the graben was concentrated largely along four main fault zones, the Alpha-Corsair and Bulldog Mountain fault zones on the west side, and the Amethyst and Solomon-Holy Moses fault zones on the east side of the graben. Although best displayed in the extensive mine workings along the Solomon-Holy Moses and Amethyst fault zones, ore deposits are known in all four zones, and many minor faults are altered and mineralized as well. Some ore has been mined from the basal part of the Creede Formation in the Monon Hill area, where sedimentary beds overlying the broken caldera margin are impregnated with sulfide minerals.

Mineralization appears to have taken place toward the end of the late faulting, as indicated by veins that cut the Creede Formation and

Fisher Quartz Latite and by ore minerals that impregnate parts of the Creede Formation. Sufficient displacement had taken place along some faults to form extensive systems of hanging-wall fractures before mineralization, but the main veins commonly show repeated brecciation and rehealing during vein deposition, and in many places they are cut by post-ore fractures. The late movement was not sufficient to affect the hanging-wall fractures nearly as thoroughly as it affected the main veins, and most smaller veins show no post-ore movement.

Although mineralization followed or was in part concurrent with the youngest faulting, many of the faults formed earlier and were reactivated just prior to mineralization. The clearest evidence is along the Amethyst fault zone, where the late mineralized fault locally deviates as much as several hundred feet from the trend of an ancestral fault formed during subsidence of the Bachelor Mountain cauldron (8). The Bulldog Mountain fault zone also appears to be near the trend of an older fault formed during subsidence of the Bachelor Mountain cauldron, but the earlier displacement was opposite that on the younger fault. The Solomon-Holy Moses fault zone shows some brecciation and silicification earlier than the late faulting, but the evidence is obscure. The Alpha-Corsair fault zone apparently was active only during the last main period of movement.

Veins along the Amethyst fault zone and related hanging-wall fissures have supplied more than 95 per cent of production from the Creede district. Most production prior to 1920 came from the Amethyst vein proper, but since then most production has come from hanging-wall veins. Since 1940 the great preponderance of ore mined in the Creede district has come from the O H vein, the largest and most persistent of the many hanging-wall veins that have been developed. A detailed account of the structures along the Amethyst fault zone, and their control on ore deposition, is given by Steven and Ratté (8).

The Solomon-Holy Moses vein zone has been the second most productive vein zone in the Creede district, but in all it has supplied only a few per cent of the metals produced. These have come largely from the southern part of the vein zone where the Solomon, Ridge, and nearby properties were active during the 1890's and early 1900's, and from the northern part where the Phoenix mine was a major producer during the 1950's.

The Alpha-Corsair fault zone has not been

explored extensively, although it has the largest displacement of any fault zone in the district, hydrothermally altered and mineralized rocks are apparent at many places along its trend, and ore valued at about $600,000 has been produced from the Corsair mine near its southern end.

The Bulldog Mountain fault zone was virtually unexplored until recently; the geologic environment seemed favorable for the occurrence of ore deposits, however, and an exploration program based on recommendations by the Geological Survey (4) was underway when this report was prepared. Preliminary results reported by Cox (7) are encouraging.

Primary ore in the Creede district, as exemplified by that along the Amethyst and related hanging-wall veins, is largely mineralized fault breccia consisting of fragments of country rock in a variable mixture of clay, chlorite, amethystine to white quartz, and sulfide minerals. Open-space filling appears to have been the dominant process in vein formation. The mineral assemblage along the O H vein, as reported by Bethke *et al.* (2, p. 1826), consists of galena, sphalerite, chalcopyrite, pyrite, quartz, hematite, and chlorite, and minor amounts of fluorite, barite, and ankerite.

### Spar City District

The Spar City district (6) is along the south flank of the Creede caldera near the intersection of the southward-extending core graben and the broken zone along the caldera margin (Figure 1). The faults related to caldera subsidence and resurgence were buried by Fisher Quartz Latite and Creede Formation, and the whole assemblage rebroken before mineralization. Veins containing galena, sphalerite, barite, manganese oxides, and quartz cut both the Creede Formation and Fisher Quartz Latite and fill fissures that trend both easterly and northerly, approximately parallel to older trends. Detailed relations are obscured by surficial deposits and by dense forest, but the deposits are similar in mineralogy, appearance, and age to those in the Creede district.

### Wagon Wheel Gap Fluorspar District

The Wagon Wheel Gap fluorspar district occurs along the east margin of the Creede caldera at the intersection of the ring-fracture zone and the southernmost fault in a broad, generally tangential graben that extends at least 15 miles southeast. The ring-fracture zone shows evidence of repeated faulting, beginning

concurrently with eruptions of Farmers Creek Rhyolite, and ending after the Creede formation and Fisher Quartz Latite accumulated. Evidence for faulting at any given time is generally local, and in most places the different increments of displacement are difficult to separate. Most subsidence appears to have taken place during Farmers Creek time or during final subsidence of the Creede caldera; displacement during the last movement, following the Fisher eruptions and Creede alluviation, appears to have been minor and perhaps local.

The east-southeast-trending graben fault that intersects the ring-fracture zone formed after the Wason Park Rhyolite was erupted, but more specific dating depends largely on analogy. Other faults in the graben are younger than the Nelson Mountain Quartz Latite and older than Fisher Quartz Latite, and thus were active during the same general time span as the final subsidence of the Creede caldera. Most displacement on the graben fault in the Wagon Wheel Gap district probably took place during this same period, but later movements may have occurred in post-Fisher time when other nearby faults are known to have been active.

The fluorspar deposits and two of the three associated hot springs in the Wagon Wheel Gap district are along the east-northeast-trending graben fault (1, p. 238). The close association of the fluorspar vein and the hot springs suggests that the thermal episode related to mineralization may still be active; actual mineralization, however, predated the end of Pleistocene as fragments of the vein have been found in high gravels that are older than the youngest glacial deposits.

## DEPRESSED AREA SOUTH OF THE CREEDE CALDERA

Sparse reconnaissance south of the Creede caldera suggests that, in an irregular area 8 to 10 miles across and largely covered by Fisher Quartz Latite, pre-Fisher units are structurally depressed relative to the adjacent areas. Limited data suggest that this depression was relatively late, probably about concurrent with final subsidence of the Creede caldera. The succeeding Fisher eruptions filled the depressed area with a great pile of viscous quartz latitic to rhyolitic flows, domes, and associated breccias.

Brief reconnaissance has indicated local hydrothermally altered and mineralized rocks within the Fisher lavas, particularly near local eruptive centers. At the best known of these, near the southwestern margin of the Fisher field at the head of Red Mountain Creek, rocks in and around a volcanic neck are highly altered, impregnated with pyrite, and veined by quartz. Gold in amounts of interest are reported from old diggings in this area.

Two older periods of alteration and mineralization are apparent in the area just west of upper Red Mountain Creek. Native sulfur was deposited in highly altered rocks near a local Huerto center along Trout Creek, and a later period of hydrothermal activity altered other rocks nearby. The economic potential of these areas is unknown.

## SAN LUIS PEAK CALDERA

The San Luis Peak caldera near the northern end of the central San Juan cauldron complex is marked by a nearly circular subsided block about 10 miles across that is filled by an abnormally thick accumulation of Nelson Mountain quartz latite. Only the southern and southwestern parts of the caldera have been mapped in detail; the other parts are still little known.

Initial collapse probably resulted from eruption of Rat Creek tuffs, for the older tuffs in the succeeding Nelson Mountain Quartz Latite were deposited in a preexisting depression. Where seen along the southern margin of the caldera, the edge of this depression was a steep north-dipping slope against which the younger tuffs abut. In places the adjacent Rat Creek tuffs were deeply channelled by north flowing streams prior to the Nelson Mountain eruptions. Accumulation of the Nelson Mountain Quartz Latite was accompanied by local steep faulting of the caldera margin. The youngest Nelson Mountain ash flows were deposited across the western margin of the caldera without significant change in thickness or attitude, indicating no late subsidence in this area.

The core of the San Luis Peak caldera shows some evidence of repeated uplift or resurgence during Nelson Mountain eruptions, but most doming seems to postdate these eruptions. Late resurgence is most apparent in the eastern part of the caldera, and the highest part of the dome is near San Luis Peak.

The western part of the San Luis Peak caldera is covered by quartz latitic to rhyolitic flows and breccias of Fisher quartz latite that were erupted from local vents after subsidence and resurgent doming of the caldera. Preliminary reconnaissance has indicated two general areas of hydrothermal alteration and mineralization associated with the San Luis Peak caldera; these investigations are incomplete, however, and other similar areas may exist. Many rocks in and near the local Fisher vents in the western part of the San Luis Peak caldera are highly altered, and lead and silver assays of ore grade or near ore grade have been reported. Little systematic prospecting has been done in the past, and present knowledge is too incomplete to judge the mineral potential.

The other area of known mineralization is the Bondholder district near the confluence of Cascade and Spring creeks, northwest of San Luis Peak. Personal observations have been limited to a single vein near the mouth of Cascade Creek, where galena, sphalerite, pyrite, chlorite, and amethystine quartz fill an east-trending fissure. The general appearance of the ore is identical with much of that in the Creede district farther south. Ore-grade assays have been reported reliably from several other veins in the same area. The host rock in lower Cascade Creek in Nelson Mountain Quartz Latite, so mineralization postdates filling of the San Luis Peak caldera.

## SUMMITVILLE DISTRICT

The Summitville district (3) shows many features in common with other central San Juan mineralized areas, although it is not closely associated with any of the known late subsidence structures in the central San Juan cauldron complex. The district is within a local pile of Fisher Quartz Latite adjacent to the southeast margin of the cauldron complex (Figure 1). Although the regional setting is not well understood, the district seems localized along a projection of the tangential graben extending southeast from the Creede caldera in the Wagon Wheel Gap area, and is at the intersection of this trend with a discontinuous set of more westerly trending faults that have been mapped near the southern end of the cauldron complex. The age of these Fisher Quartz Latites relative to other accumulations of Fisher rocks has not been established, but it appears to be about the same. The viscous flows and domes at Summitville represent local summit volcanoes that are younger than any other units nearby except the Hinsdale Formation, and they are localized along the projected trend of a major fault zone that was active at the time of final subsidence of the Creede caldera.

The hydrothermal alteration and mineralization at Summitville clearly took place during Fisher Quartz Latite time (3, p. 38–39), as highly altered and mineralized lower lavas of the Fisher are overlain unconformably by un-altered upper lavas of the Fisher. The ore de-posits are within a quartz latitic volcanic dome and are believed to be localized above one margin of the underlying neck (3, p. 40). Solfataric activity altered large volumes of rock near the postulated vent and locally devel-oped quartz-alunite replacement veins contain-ing pyrite, enargite, and gold. From 1873 to 1950, nearly $7.5 million worth of metals was produced. Of this, about $7 million worth was from gold, $350,000 from silver, $75,000 from copper, and $6,000 from lead (3, p. 6).

An older age of hydrothermal alteration and mineralization is apparent along the margin of an older quartz monzonitic intrusive south of the Summitville district (3, p. 38). Some small mines and prospects were developed in this area in the early days, but production has been negligible.

## GEOLOGIC CONTROLS

Most known mineralized areas in the central San Juan Mountains, ranging from the highly productive Creede district to obscure areas of hydrothermally altered rock, formed late in the local sequence of volcanic and related geo-logic events. Two general types of structural setting are recognized:

(1) Fissure veins along, or disseminated de-posits near, faults the last movement on which just preceded or in part coincided with min-eralization. (Creede and Spar City districts; possibly Wagon Wheel Gap and Bondholder districts).

(2) Hydrothermally altered areas in and near Fisher Quartz Latite volcanic centers (western part of San Luis Peak caldera; de-pressed area south of the Creede caldera; Sum-mitville district).

In all fissure-vein districts except the Bond-holder, faulting was recurrent throughout vol-canism, but mineralization generally was con-centrated along the latest faults active. In the Creede and Spar City districts, this faulting followed accumulation of the Creede Forma-tion and Fisher Quartz Latite. Some faulting took place concurrently in the Wagon Wheel Gap district, but is not known to have affected the mineralized fault. The fissure vein districts are not closely associated with centers of Fisher volcanism; mineralized faults near Fisher lavas appear younger than, and inde-pendent in origin from, the flows, and the pro-ductive Creede district is not near any major Fisher centers. The hydrothermal solutions ap-parently rose along the then recently active permeable fault zones, probably from cooling magma not necessarily closely associated with the then recent volcanic accumulations.

The hydrothermally altered and mineralized areas closely associated with accumulations of Fisher Quartz Latite, on the other hand, are grouped in and around volcanic centers, and hydrothermal activity appears closely tied in time and space to the volcanic activity. In many places, altered and mineralized lower lavas of the Fisher are covered by younger, unaltered Fisher lavas. Faulting appears minor around many Fisher centers, and cursory ex-aminations have disclosed little fracture control of alteration and mineralization. Inasmuch as this hydrothermal activity appears roughly co-incident with Fisher volcanism, it appears somewhat older than the fissure-vein mineral-ization elsewhere in the area. The only produc-tive mines developed so far in this environment are in the Summitville district, but several other similar areas appear worthy of additional prospecting.

## REFERENCES CITED

1. Emmons, W. H. and Larsen, E. S., 1913, The hot springs and mineral deposits of Wagon Wheel Gap, Colorado: Econ. Geol., v. 8, p. 235–246.
2. Bethke, P. M., *et. al.*, 1960, Time space rela-tions of the ores at Creede, Colorado (abs): Geol. Soc. Amer. Bull., v. 71, p. 1825–1826.
3. Steven, T. A. and Ratté, J. C., 1960, Geology and ore deposits of the Summitville district, San Juan Mountains, Colorado. U.S. Geol. Surv. Prof. Paper 343, 70 p.
4. ———— 1960, Relation of mineralization to cal-dera subsidence in the Creede district, San Juan Mountains, Colorado: *in Short papers in the geological sciences:* U.S. Geol. Surv. Prof. Paper 400-B, p. B14–B17.
5. Steven, T. A., 1964, Geologic setting in the Spar City district, San Juan Mountains, Colorado: *in Short papers in geology and hydrology:* U.S. Geol. Surv. Prof. Paper 475-D, p. D123–D127.
6. Steven, T. A., and Ratté, J. C., 1964, Revised Tertiary volcanic sequence in the central San Juan Mountains, Colorado: *in Short papers in geology and hydrology:* U.S. Geol. Surv. Prof. Paper 475-D, p. D54–D63.
7. Cox, M. W., 1965, Geologic theory leads to ore at Creede (letter to the editor): Eng. and Min. Jour., v. 166, no. 9, p. 6.
8. Steven, T. A., and Ratté, J. C., 1965, Geology and structural control of ore deposition in the Creede district, San Juan Mountains, Colorado: U.S. Geol. Surv. Prof. Paper 487, 90 p.

# 34. Geology and Ore Deposits of the Western San Juan Mountains, Colorado[*]

WILBUR S. BURBANK,[†] ROBERT G. LUEDKE[‡]

## Contents

## Illustrations

## Tables

[*] Publication authorized by the Director, U.S. Geological Survey.
[†] U.S. Geological Survey, Exeter, New Hampshire.
[‡] U.S. Geological Survey, Washington, D.C.

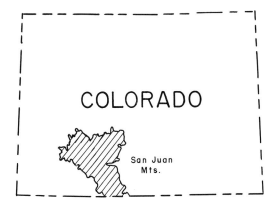

COLORADO

San Juan
Mts.

## ABSTRACT

The impressive western San Juan Mountains of Colorado were carved by Pleistocene and Recent erosion from a thick blanket of Tertiary volcanic rocks that rests upon a basement of metamorphic, sedimentary, and igneous rocks ranging in age from Precambrian to Tertiary. These rocks record a long, complex, and fairly complete sequence of geologic events that have affected the region through much of geologic time. The geologic history of the region includes several periods of deformation, erosion, and igneous activity; closely associated with two of these periods were two periods of mineralization. The older of these was related to Late Cretaceous-early Tertiary (Laramide) intrusive activity, and its deposits occur in the older sedimentary units. The younger, of later Tertiary age and related to the widespread middle Tertiary extrusive and intrusive activity, is the chief concern of this paper.

Tertiary volcanism built a great plateau surrounding the San Juan volcanic depression, a complex of subsided crustal blocks in the vent areas. The Silverton and Lake City cauldrons, nested within the larger depression, were domed by resurgent magma that created keystone grabens along the distended crests of the domed floors and many postcauldron radial and concentric fractures, dikes, and intrusive plutons within and marginal to the cauldron sites.

The postvolcanic evolution of large quantities of gases, chiefly carbon dioxide and water, propylitically altered the rocks in and around the volcanic depression prior to the younger period of ore deposition. The ore deposits, localized mainly in the radial and concentric fractures about the cauldrons and in the associated graben faults, include fissure veins, chimneys, replacements, and disseminations in volcanic and underlying basement rocks. Fissure veins, constituting the major economic source, yielded precious- and base-metal ores. Most veins are compound with sphalerite, galena, and chalcopyrite; some veins contain silver-bearing arsenical sulfosalts and free gold. These minerals occur in predominantly quartz gangue. However, gangues composed of several manganese silicates and carbonates and relatively barren of metals are abundant locally and probably originated by alteration of rocks at roots of faults and fissures during evolution of carbonatic solutions. The bulk of the ores are hypogene and probably had their source in deep parent magmas of the shallow eruptive rocks.

Total production since the early 1870's has exceeded half a billion dollars in recovered metals, with a ratio of precious- to base-metal values of about 3 to 2. Since 1930, improvements in metallurgical treatment and mining techniques have increased the base-metal yield.

## INTRODUCTON

The San Juan region, as roughly outlined by the Tertiary volcanic field occupies about 6000 square miles in southwestern Colorado, but covers 8000 to 10,000 square miles when defined more broadly to include marginal areas of other geologic character. It is a mountainous region with a total relief of about 8000 feet; altitudes range from about 6000 to over 14,000 feet. The Continental Divide, swinging in a westerly looping fashion, separates the region's drainage into the so-called western (Pacific) slope, drained by the Colorado River system, and the eastern (Atlantic) slope, drained via the Rio Grande. The region receives moderate precipitation and is noted for its long and often severe winters.

The few small settlements within the San Juan region, founded principally as mining-towns, also serve as centers to the stock raising, lumbering, and tourist industries. Many of these centers formerly were connected by narrow-gauge railroad lines, now removed, but are readily accessible by all-weather roads. Several larger towns at the edges of the region do have both rail and air facilities.

The approximate western one-third of the San Juan region, the area of concern of this paper (Figure 1), is mostly rugged alpine terrain containing 12 of the region's 14 peaks that reach 14,000 feet. Consequently, accessibility and climate have been important to the development of the area. The moderate to heavy snowfall has temporarily curtailed mining activities at different properties in the past.

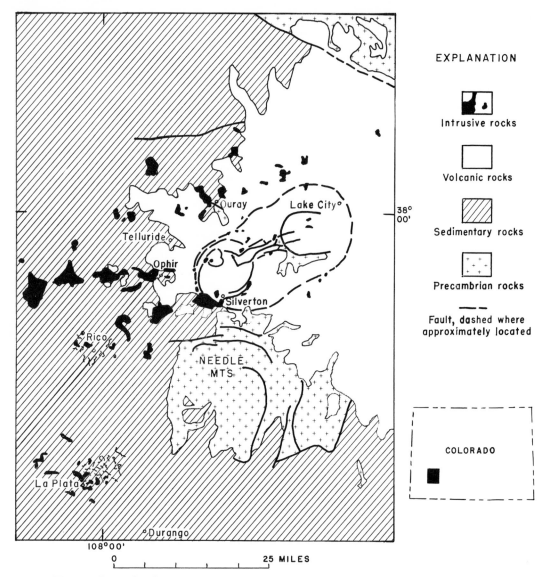

**EXPLANATION**

Intrusive rocks

Volcanic rocks

Sedimentary rocks

Precambrian rocks

Fault, dashed where
approximately located

COLORADO

FIG. 1. *Generalized Geologic Map of the Western San Juan Mountains, Colorado.*

The mining industry in the western San Juan Mountains, although very much dependent upon the prevailing methods of mining, milling, and marketing, has had neither spectacular boom nor prolonged slump periods to date and has been a major contributor to the economy of the State.

This paper is concerned principally with the geologic setting and ore deposits, particularly of later Tertiary age, of the western San Juan Mountains. The following discussion is summarized from the many papers concerning the geology and ore deposits of the western San Juans, and from our present investigations in cooperation with the Colorado State Mining Industrial Development Board and a predecessor agency, the Colorado State Metal Mining Fund Board.

## HISTORY OF MINING

Prospecting for gold and silver in the western San Juan Mountains began more than a century ago. Gold reportedly was discovered in 1848 near the future site of Lake City by a member of an exploring party under the command of John C. Fremont, but this discovery went unheralded. The first real prospecting in the area was in 1860, when a party under the leadership of John Baker* penetrated the mountains to where Silverton now stands, but heavy winter snows and Indian harassment discouraged this and further attempts in this part of the mountains for at least 10 years. Historical accounts, however, do mention limited prospecting along the Dolores River near the present Rico townsite (Figure 1) in 1861, 1866, and 1869. The first gold ore produced in the western San Juan Mountains was in 1870 near the Silverton townsite. Prospecting and claim location of some soon-to-be prominent mines in the area occurred during the early 70's; the majority of these locations were silver lodes. However, not until the spring of 1874, upon completion of negotiations of a new treaty between the Federal Government and the Ute Indians, was the region officially thrown open to settlement. Exploration in earnest, rapid growth, and real mining then began. During the 30 years, 1870–1900, the richest deposits were discovered and mined.

Early mineral production in the several different mining districts composing the western San Juans was principally of silver and some gold, copper, and lead. High costs of equipment and the expenses and difficulties of treating and transporting ores resulted in slow development of the new discoveries. The building of smelters and the construction of roads and narrow-gauge railroad lines in the 80's and 90's all tended to reduce the costs of ore production. About 1890, the silver-bearing low-grade base-metal ores of the area were first treated successfully by concentration and amalgamation at both the Sunnyside and Silver Lake mills near Silverton. Then, in 1917, the first commerical lead-zinc selective flotation plant in North America was built at the Sunnyside mine to improve further the treating of the low-grade complex ores. However, not until 1945 was the three-way separation of copper, lead, and zinc concentrates of the ores accomplished, beginning with those of the Black Bear mine near Telluride. The mining industry of the area generally has kept pace with economic trends, and now, through consolidation of properties in parts of some districts, is doing deeper exploration and mining.

## GEOLOGIC HISTORY

The complex geologic history of the San Juan Mountains region as a whole is as yet

* The correct Christian name, John, Jim, or Charles, is in doubt.

too imperfectly known to make a thorough discussion possible. However, a general review of the geologic development of the western San Juans reflects many aspects, particularly pre-Tertiary, of the geologic history of the entire region (21) and thus establishes a regional geologic setting for this and companion papers.

Many geologic events have affected the western San Juan Mountains throughout geologic time and are recorded by the great variety of metamorphic, sedimentary, and igneous rocks ranging in age from Precambrian to Recent. Exposures of Precambrian rocks are somewhat limited, except in the Needle Mountains; hence relations between different rock types in the different locales as yet are not fully understood. The Paleozoic and Mesozoic rocks of the western San Juans are in almost all respects representative of those of the San Juan region in general. The Tertiary volcanic history likewise is similar throughout the San Juan region but differs in detail within each of the several principal eruptive centers.

Outcrops of Precambrian rocks in the western San Juans (Figure 1) are confined to a few small isolated masses of a few square miles each, except for the much larger mass in the Gunnison River area at the northeast corner of Figure 1 and the mass about 600 square miles in extent that makes up the Needle Mountains. Metasedimentary, metavolcanic, and igneous rocks in these exposures represent a great variety of environmental conditions and are roughly divisible into two age groups (Fred Barker, 1966, written communication). The presumed oldest and most widespread group consists of intensely metamorphosed and deformed mica schist, hornblende schist, amphibolite, gneiss, and some quartzite; it is exposed in both the Gunnison River area and the Needle Mountains. In both areas, but particularly the Needle Mountains, these high-rank metamorphosed rocks were intruded, in part, synkinematically by mostly discordant igneous plutons of intermediate to silicic composition and of various sizes. The younger group consists of less metamorphosed and deformed conglomerate, quartzite, argillite, and slate; it is exposed in the Needle Mountains and the canyon south of Ouray. Near Ouray, alternating thick layers of quartzite and slate total over 8000 feet in thickness (26). This younger group also was intruded by small to large igneous plutons having a variety of compositions (mostly silicic). The Precambrian rocks southwest of the Lake City subsided block (Figure 1) are granites intruded by diabase dikes. The Precambrian terrane was

TABLE I.　*Geologic Units of the Western San Juan Mountains*

| Era | System | Unit | Rock or Character |
|---|---|---|---|
| Cenozoic | Quaternary | | Alluvium, talus, rock glaciers, landslide deposits, glacial drift, etc. |
| | Tertiary | Hinsdale Formation | Basalt flows and some Rhyolite tuffs. |
| | | Intrusive rocks | Gabbro to rhyolite stocks, dikes, and plutons. |
| | | Potosi Volcanic Group | Quartz latitic to rhyolitic welded ash-flow tuffs; divisible into Gilpin Peak Tuff, Sunshine Peak Rhyolite, and several unnamed units. |
| | | Silverton Volcanic Group | Andesitic to rhyodacitic flows, breccias, and tuffs, and quartz latitic welded ash-flow tuffs; divided into four units: Picayune Formation at base, Eureka Tuff, Burns Formation, and Henson Formation at top. |
| | | San Juan Formation | Predominantly rhyodacitic tuff breccia with some flows and tuffs. |
| | | Lake Fork Formation | Rhyodacite flows, breccias, and tuffs. |
| | | Telluride Conglomerate | Gray to red-brown conglomerate with some sandstone, siltstone, and shale. |
| | | —Unconformity— | |
| Mesozoic | Cretaceous | Intrusive rocks | Granodioritic stocks, laccoliths, dikes, and sills; some clastic dikes. |
| | | Mesaverde Group | Yellow sandstone and gray shale; subdivided into the Cliff House Sandstone, Menefee Formation, and Point Lookout Sandstone. |
| | | Mancos Shale | Gray fissile shale with few thin beds of sandstone and limestone. |
| | | Dakota Sandstone | Gray to yellow quartzitic sandstone with some conglomerate and shale. |
| | | —Unconformity— | |
| | Jurassic | Morrison Formation | Upper part, Brushy Basin Shale Member, variegated mudstone with some sandstone. Lower part, Salt Wash Sandstone Member, sandstone with some mudstone and limestone. |
| | | Junction Creek Sandstone | Yellow mostly crossbedded sandstone, locally quartzitic. |
| | | Wanakah Formation | Upper part, mudstone, shale, and siltstone with some limestone; in center, Bilk Creek Sandstone Member, sandstone; at base, Pony Express Limestone Member, gray bituminous limestone with some shale and gypsum. |
| | | Entrada Sandstone | Light-colored crossbedded sandstone. |
| | | —Unconformity— | |
| | Triassic | Dolores Formation | Red shale, siltstone and sandstone with some limestone and conglomerate near base. |

*TABLE I.   Geologic Units of the Western San Juan Mountains (Continued)*

| Era | System | | Unit | Rock or Character |
|---|---|---|---|---|
| | | | —Unconformity— | |
| | Permian | | Cutler Formation | Red locally arkosic shale, siltstone, sandstone, and conglomerate. |
| | ? | | | |
| | | Pennsylvanian and Permian(?) | Rico Formation | Red to brown sandstone and conglomerate with some shale and limestone. |
| Paleozoic | Carboniferous | Pennsylvanian | Hermosa Formation | Red arkosic sandstone and fossiliferous green shale, sandstone, and limestone. |
| | | | Molas Formation | Red limy shale, sandstone, and conglomerate. |
| | | | —Unconformity— | |
| | | Mississippian | Leadville Limestone | Massive gray dense limestone with few sandy and chert layers. |
| | Devonian | | Ouray Limstone | Gray, buff, and white dense dolomite and limestone. |
| | | | Elbert Formation | Buff dolomitic limestone with interbedded shale and sandstone. |
| | | | —Unconformity— | |
| | Cambrian | | Ignacio Quartzite | White to pink quartzite and quartzitic conglomerate. |
| | | | —Unconformity— | |
| Precambrian | | | | Gneiss, schist, amphibolite, quartzite, and slate assigned to Irving Greenstone, Vallecito Conglomerate, Uncompahgre Formation, and unnamed units; locally intruded by mafic to silicic dikes and sills, and by mostly granite stocks and batholiths. |

extensively eroded to a gently rolling surface and was deeply weathered locally before submergence by the sea in Late Cambrian time.

The first of a succession of Paleozoic and Mesozoic sedimentary strata (Table I) was deposited during Late Cambrian time with a pronounced angular unconformity to the beveled Precambrian rocks. The Paleozoic rocks are divisible into two general sequences. The lower or older sequence comprises thin widespread quartzites, dolomites, and limestones of Cambrian through Mississippian ages that were deposited under shallow marine conditions. Several disconformities exist in the stratigraphic sequence. After deposition of a few tens of feet to about 200 feet of thin-bedded quartzites and conglomerates of the Ignacio Quartzite of Late Cambrian age, epeirogenic uplift of the region resulted in withdrawal of the sea, subsequent erosion, and removal locally of the Ignacio. In the northeastern part of the San Juan region, Cambrian rocks are missing, probably owing to erosion rather than nondeposition, and Ordovician rocks directly overlie Precambrian rocks. No Ordovician, Silurian, or Lower and Middle Devonian beds are recognized in the western San Juans, so that the thin-bedded quartzites, sandstones, and shales of the Elbert Formation of Late Devonian age disconformably overlie the Ignacio. However, at Ouray, the lowermost Paleozoic unit is lacking, and the Elbert lies with angular unconformity upon the Precambrian rocks. The Elbert Formation, 100 feet or less thick, grades upward into several tens of feet of limestone beds with some shaly partings of the Ouray Limestone of Late Devonian age. Disconformably overlying the Ouray are 250 feet or less of massive to thick-bedded marine dolomites and limestones of the Leadville Limestone of Mississippian age. Again there was epeirogenic uplift and a withdrawal of the sea; subsequent deep weathering developed a karst topography on the surface of the Mississippian limestones.

In contrast to the older sequence, the upper or younger sequence comprises several thousand feet of conglomerates, sandstones, siltstones, limestones, salt, and gypsum of Pennsylvanian and Permian ages that were deposited marginally to an ancient major landmass in the San Juan region. This "Ancestral

Rockies" area, the Uncompahgre-San Luis Highland, extended from west-central Colorado southeasterly through the San Juan region into north-central New Mexico and served as a source of sediments during late Paleozoic and part of Mesozoic time. The Molas Formation of Pennsylvanian age, consisting in part of residual materials and in part of shales and sandstones, accumulated on the weathered Mississippian surface mostly before advancement of the sea, is the basal unit of a thick redbed sequence. With the gradual and continuous uplifting of the highland, there was a transition from marine deposition to continental deposition as indicated successively by the sedimentary rocks of the Hermosa Formation of Pennsylvanian age, Rico Formation of Pennsylvanian and Permian(?) age, and Cutler Formation of Permian age.

Continental deposition continued through much of Triassic and Jurassic time with apparent conformity everywhere except in the vicinity of Ouray. Here, a pronounced angular unconformity separates the Cutler and older units from the Dolores Formation of Late Triassic age. The extensive erosion, strong monoclinal folding with subsidiary folds axial to the mountain core, and faulting indicate local deformation at the end of the Paleozoic. Rocks of Triassic and Jurassic age generally are about 1500 feet thick and include beds of sandstone, siltstone, shale, mudstone, and some limestone and gypsum. These rocks are widespread and represent deposition in both terrestrial and lagoonal or near-shore environments. The rocks of Cretaceous age, in contrast, are characteristic of mostly marine, near-shore, and coastal swamp deposition, and consist of a sequence nearly 9000 feet thick of sandstone, locally coal bearing, and shale with minor shaly limestone. Most of the Cretaceous section is missing north and west of the mountains but is complete near Durango on the south (Figure 1), where it grades upwards into the early Tertiary rocks (21).

In Late Cretaceous and early Tertiary time, renewed domal uplift (Laramide orogeny) of the western San Juan region again was accompanied by erosion, monoclinal folding, and faulting, but was accentuated by considerable igneous activity. This extrusive and intrusive igneous activity, together with the genetically related ore deposition, occurred at Ouray and probably at Rico and La Plata. Erosion by streams heading generally in the Needle Mountains vicinity gradually reduced the western San Juan region to one of fairly low relief dotted with occasional residual hills formed by the more resistant rocks such as the laccoliths in the vicinity of Ouray.

A variety of materials derived from early Tertiary erosion accumulated locally on this surface to form the Telluride Conglomerate of Oligocene(?) age. A wide volcanic plateau of eruptive materials with related volcano-tectonic subsidence structures was built conformably on the Telluride, where present, or on the surface of the pre-Tertiary rocks, which are mainly Precambrian. This layered volcanic rock succession, aggregating nearly 1.5 miles in thickness and having a volume greater than 1000 cubic miles, consists of many rock types and depositional types and may be divided into three principal stratigraphic sequences (28).

The oldest sequence comprises the Lake Fork Formation and the San Juan Formation; these were erupted from a cluster of central-vent volcanoes extending from near Lake City southwestward to near Silverton (Figure 1). The fragmental debris from these explosive eruptions was deposited mainly as mud flows and lava flows over hundreds of square miles and to a thickness of over 3000 feet near Ouray. Sometime late during this eruptive period, without an appreciable volcanic quiescence and erosion, eruptive and structural activities changed, and the large oval-shaped San Juan volcanic depression began to form, perhaps by engulfment of the vents. This subsidence structure is about 15 miles wide, 30 miles long, and covers more than 400 square miles.

Within and mostly confined to the San Juan depression were the eruptions of the second or middle stratigraphic sequence, the Silverton Volcanic Group, which is divisible into four principal units. The oldest of the four, the Picayune Formation, consists mostly of mafic lava flows, tuffs, and breccias that probably were erupted from central-vent volcanoes; some of the lavas overflowed and spread several miles beyond the northwest rim of the volcanic depression. The next unit, the silicic Eureka Tuff, consists of thick welded ash-flow tuffs. Gradual deepening of the basin accompanied these lava and ash eruptions. Following this early subsidence, there was resurgence, arching, and faulting of the depression floor. Eruptions of the compositionally intermediate Burns Formation next occurred as domes and volcanic piles along the rims of the depression and as domes and thick flows on the basin floor. At this time, ashy and pumiceous freshwater limestones and shales were deposited locally on the basin floor. Continuing eruptions of lava and ash, but of more mafic composition, resulted in the Henson Formation, the youngest of the middle sequence.

The youngest sequence, the Potosi Volcanic Group, consists almost entirely of silicic welded ash-flow tuffs. Eruptions of these tuffs resulted in the formation and subsidence of the younger and nearly circular comagmatic Silverton and Lake City cauldrons. These cauldrons are each about 10 miles in diameter and are within the San Juan volcanic depression (Figure 1).

Postvolcanic faulting, fissuring, and intrusive igneous activity in part was accompanied and in part was followed by pervasive rock alteration and the formation of the major late Tertiary ore deposits.

Late Tertiary erosion locally was accompanied by basaltic and rhyolitic eruptions of the Hinsdale Formation northeast, east, and southeast of Lake City. The present rugged mountainous topography was developed by several stages of Pleistocene glaciation and Recent stream and mass-wasting activities.

## STRUCTURE

The geologic history has outlined briefly the long and complex structural record of the western San Juan region. Four, and probably more, major periods of deformation are recorded in the rocks; the structural pattern of each period controlled to some degree, and was obscured partially by, the pattern of each succeeding period. Nevertheless, each major period differs in certain respects from the others, and each coincides generally with the closing phase of a major division of geologic time. Except for Precambrian time, the intensity of deformation generally decreased as igneous activity increased through the ages. The deformation near the end of the Paleozoic and Mesozoic was characterized by structures associated with mountain building. Also, each of the major periods of deformation was followed by extensive erosion prior to deposition of the overlying strata. Crustal disturbances between the major deformations are indicated by erosional unconformities and by gaps in the stratigraphic sequence.

Only the Precambrian rocks are intensely metamorphosed, folded, and faulted. Folds range from broad and open in the Needle Mountains just south of Silverton to close and locally overturned in the Uncompahgre River canyon south of Ouray. The folds, faults, and minor structures generally strike through an arc of a few degrees north of west in the

western part of the area to a few degrees north of east in the eastern part. Near Ouray this deformation was accompanied by minor intrusive activity (26,31), but in the Needle Mountains and perhaps between Silverton and Lake City, granitic masses of batholitic proportions were emplaced.

The absence of strata of different ages in the western San Juans indicates periods of no deposition or of uplift and erosion during both the Paleozoic and Mesozoic eras. Apparently, this epeirogenic activity was not accompanied by faulting or folding. However, the late Paleozoic domal uplift of the region produced monoclinal folds and faults with little or no igneous activity. Related to the monoclinal folds are outward-plunging anticlinal folds along axes of uplift radial to the mountain core. The stratigraphic section was tilted gently to steeply westward and northward near the Precambrian core of the mountains and was locally eroded, as indicated by the angular unconformity between Paleozoic and Mesozoic strata at Ouray (26).

Deformation in late Mesozoic and early Cenozoic time (Laramide orogeny) is indicated by a renewal of domal uplift with further monoclinal and radial folding and faulting that in part followed previously established trends. The monocline near Ouray in part coincides with that formed earlier and in part is displaced farther from the mountain core. This Laramide deformation was accompanied by igneous intrusion (concordant and discordant plutons at Ouray, and perhaps at Rico and in the La Plata Mountains) and by an early period of mineralization (11).

Tertiary time was distinguished by widespread volcanic activity, intrusion, faulting, and formation of the cauldron complex. These structural features largely controlled the distribution and localization of later ore deposition. Most important are the Silverton and Lake City cauldrons, each about 10 miles in diameter, that lie within the northeast-trending older and larger San Juan volcanic depression, which is about 15 by 30 miles in size (Figure 1). These major structures are the result of successive and repetitive stages of eruption, intermittent subsidence of some blocks, partial engulfment of volcanic piles, and resurgent magma that caused local uplift and radial and concentric fracturing. The northeasterly alignment of some of these major Tertiary volcanic structures reflect tectonic trends as old as Precambrian, as suggested by Tweto and Sims (29, p. 1006). The Silverton cauldron site, particularly, is at the intersection of this north-

FIG. 2. *Structural Map of Silverton Cauldron and Vicinity. Larger intrusive bodies, V-pattern; volcanic pipes, stippled pattern; faults, heavy lines; dikes and veins, light lines; locations of mines, x; mines: 1. Idarado, 2. Camp Bird, 3. Sunnyside, 4. Gold King, 5. American Tunnel, 6. Terry Tunnel, 7. Shenandoah, 8. Silver Lake.*

easterly trend and the late Paleozoic Uncompahgre-San Luis Highland northwesterly trend.

In response to repeated eruptions of ash flows of the Potosi Volcanic Group, the nearly circular crustal block of the Silverton cauldron subsided 1000 to 2500 feet along closely spaced vertical to steeply inward-dipping normal faults composing the ring-fault zone. Late magmatic resurgences formed the principal systems of radial and concentric fissures, faults, and dikes marginal to the cauldron (Figure 2). Doming and distention of the subsided block resulted in renewal of faulting along some earlier established faults and in the principal formation of the northeast-trending Eureka graben (19) that connects the Silverton and Lake City cauldrons.

## ECONOMIC GEOLOGY

### Introduction

The principal ore deposits of the western San Juan Mountains formed during two major metallogenetic epochs; the older is associated with Late Cretaceous-early Tertiary laccolithic centers chiefly in Paleozoic and Mesozoic sedimentary rocks, and the younger is associated with later Tertiary volcanic centers. A few scattered deposits of possible Precambrian age occur on both the northern and southern flanks of the mountains, but little is known about these.

Ore deposits of the earlier major epoch of mineralization are related to the intrusive activity at the laccolithic centers at Ouray (7), Rico, and La Plata (14). These three centers (Figure 1) are believed to be along an ancient, broad northeast-trending zone of weakness, and are located where the zone is intersected by northwest- and/or west-trending structural lines. The base- and precious-metal ore deposits occur in both fissures and associated blanket replacements. At Ouray, ores and intrusives in the sedimentary rocks definitely are older than, and were partly eroded before, the middle Tertiary eruptive activity. The Rico and La Plata centers are related to the Ouray center by analogy of ore types and structural features (11, p. 399). The deposits of this early period of mineralization have been described adequately several times in the literature (11) and will not be discussed further in this paper.

The later epoch of mineralization, which yielded the large mineral output from ore deposits in the Tertiary volcanic rocks, is associated with subsidence and related structures and outlying intrusives at the Silverton and Lake City eruptive centers (Figures 1 and 2). The following discussion is concerned with this younger period.

### Mining Operations

The major mining operations of the western San Juans have been primarily in fissure veins in the Tertiary volcanic rocks, which include some of the longest and deepest veins of the State. The veins extend downward into the pre-Tertiary rocks where small replacement deposits in beds of limestone, shale, and conglomerate have been mined locally, but most of these deposits have been of minor economic value. However, recent discoveries of substantial replacement deposits lateral to productive veins cutting sedimentary beds immediately underlying the volcanic cover may qualify earlier conclusions (11, p. 401–402) regarding the future potential value of such deposits.

The bulk of the ores mined in the western region have been those of gold, silver, copper, lead, and zinc. A few deposits of tungsten (20), manganese, and fluorspar have been exploited. There also has been interest in the exploitation of rhodonite, a prominent gangue of some veins. However, the remaining potential ores are primarily of the complex base-metal (zinc-lead-copper) type with local substantial silver and gold.

Most mining operations within the districts and/or camps (5,11) during the last 30 to 40 years in Hinsdale, Ouray, San Juan, and San Miguel counties (Figure 3) have been small and sporadic. During World War II, Government metal premiums and exploration grants led to many small operations few of which survived postwar economic conditions and the difficulties of operation in this rugged mountain area. However, since the early 1930's, mining development has tended toward property consolidation, several of the larger operations being periodically successful. From 1927 to 1945, the operation of the Shenandoah-Dives Mining Co. in the Animas district (J, Figure 3) south of Silverton, which was the mainstay of the Silverton area for much of this period, was discontinued for lack of reserves above operating levels. In 1930–31, the operations of the Sunnyside Mining and Milling Co. at the Sunnyside property (Figure 2) were discontinued after mining the principal development ore bodies above the Terry Tunnel level (18); operations were resumed in 1937 but again discontinued within a year. The Camp Bird mine (Figure 2) near Ouray was successfully operated by lessees from 1925 to the mid-1950's. In 1959, the American Tunnel at Gladstone (Figures 2 and 3), originally driven many years ago to develop the Gold King vein system, was extended by the Standard Metals Corporation to a position about 300 feet beneath the Sunnyside mine workings and connected to the shaft from the Terry Tunnel level on the Washington vein. The most significant mining development in the western San Juan's mineralized area since the mid-1940's has been the consolidation and operation of several of the older producers near

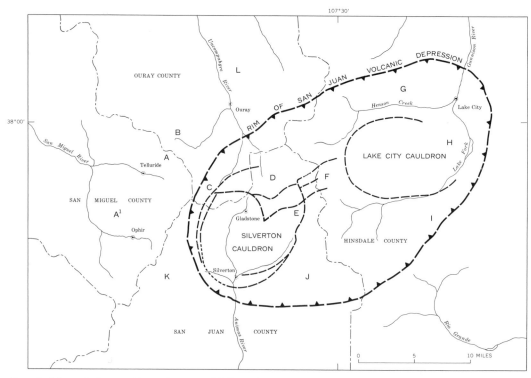

Fig. 3. *Map of County Boundaries and Mineral Districts and Camps, showing relations to the San Juan Volcanic Depression and the Silverton and Lake City Cauldrons. A, Upper San Miguel (Telluride) district; A', Ophir and South Telluride districts; B, Sneffels (Imogene Basin) district; C, Red Mountain district; D, Upper Uncompahgre (Poughkeepsie Gulch and Mineral Point) district; E, Eureka district; F, White Cross (Burrows Park) district; G, Galena (Henson Creek) district; H, Lake (Lake San Cristobal) district; I, Carson Camp; J, Animas district; K, Ice Lake Basin district; L, Ouray (Uncompahgre) district—chiefly early Tertiary mineral deposits.*

Telluride by the Idarado Mining Co. (22), through a deep tunnel from the mill site near Telluride in the Upper San Miguel district (A, Figure 3) and through the Treasury Tunnel in the Red Mountain district (C, Figure 4). Details of some of this operation and their results are given by Hillebrand (22) and Varnes (13).

### Statistics of Mine Production

To date, the output from lode mining in the western San Juan Mountains has yielded several hundred million dollars from gold, silver, copper, lead, and zinc (Table II); the total output from placer mining probably is insignificant. The bulk of the production has come from ores of later Tertiary ore deposits in the four listed counties; the total value for Ouray County, however, includes an inseparable 12 to 14 per cent from early Tertiary o e. Since the beginning of mining the production from ore deposits of later Tertiary age has

yielded about 22 per cent of the all-time State production recorded since the 1850's (Table II); production only since 1930 has yielded about 41 per cent of the all-time four-county total recorded since the 1870's (Table III). However, in San Miguel County, production amounting to about 135 million dollars since 1930 exceeded the recorded production attained from 1875 to 1930. Improvements in zinc recovery from the complex base-metal ores and increases in copper production since 1930 are reflected especially in both Ouray and San Miguel Counties (Table III).

As shown in Figure 4, gold and silver accounted for much of the total value of the western San Juan's combined totals from the beginning of mining to the minor post-World War I depression in the early 1920's; from then until the present, except for the major depression period in the 1930's, the combined base metals (copper-lead-zinc) account for most of the total. For the period 1875 to the silver panic of the mid-1890's, silver accounted

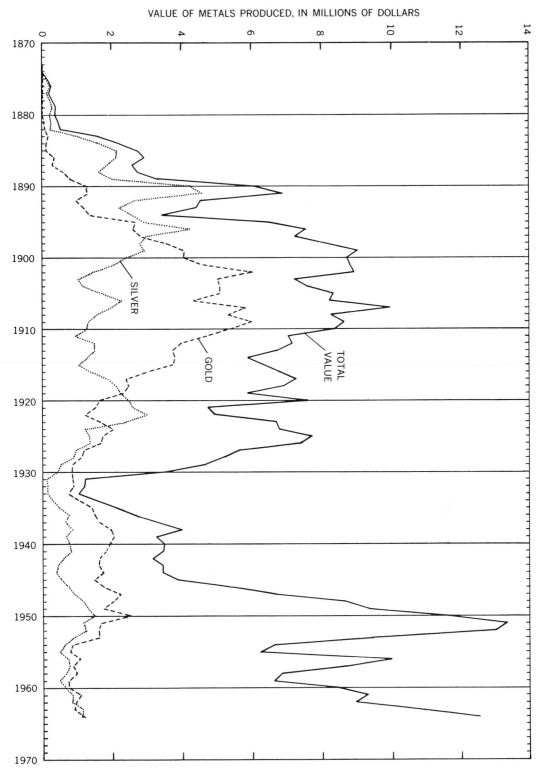

VALUE OF METALS PRODUCED, IN MILLIONS OF DOLLARS

FIG. 4. *Graph of Values of Metal Production in the Western San Juan Mountains Region (Hinsdale, Ouray, San Juan, and San Miguel Counties), Colorado.*

TABLE II. Production of Gold, Silver, Copper, Lead, and Zinc in the Western San Juan Region to 1964, inclusive, in terms of recovered metals

| County (and reporting period) | Ore (Short tons) | Gold (Troy ounces) | Gold (Value) | Silver (Troy ounces) | Silver (Value) | Copper (Short tons) | Copper (Value) | Lead (Short tons) | Lead (Value) | Zinc (Short tons) | Zinc (Value) | Total Value |
|---|---|---|---|---|---|---|---|---|---|---|---|---|
| Hinsdale (1875–1964) | 114,667 | 71,558 | $ 1,491,411 | 5,831,910 | $ 4,723,594 | 1,506 | $ 427,383 | 50,594 | $ 4,424,084 | 867 | 138,033 | $ 11,204,505 |
| Ouray (1878–1964) | 3,720,175 | 1,966,937 | 43,397,979 | 47,359,441 | 36,910,745 | 20,274 | 7,643,486 | 116,692 | 15,048,022 | 33,483 | 8,128,864 | 111,129,096[1] |
| San Juan (1873–1964) | 11,009,486 | 1,668,796 | 39,609,967 | 44,455,794 | 31,173,918 | 42,286 | 13,039,245 | 289,058 | 38,380,396 | 139,966 | 22,516,491 | 144,720,017 |
| San Miguel (1875–1964) | 19,023,046 | 3,926,484 | 91,934,321 | 62,742,885 | 48,008,537 | 45,214 | 23,503,815 | 244,345 | 46,248,840 | 166,916 | 41,535,902 | 251,231,264 |
| Total | | 7,633,775 | 176,433,678 | 160,390,030 | 120,816,794 | 109,280 | 44,613,929 | 700,689 | 104,101,342 | 341,232 | 72,319,290 | 518,284,882[2] |

Data: Compiled from U.S. Geological Survey Mineral Resources of the United States and the U.S. Bureau of Mines Minerals Yearbook(s).

[1] About 12–14 per cent of this is from production of early Tertiary ore deposits.

[2] This 4-county total value represents about 22 per cent of the all-time Colorado total which for the reporting period 1858–1964, inclusive, is $2,382,759,011.

for 40 to 80 per cent of the dollar value of the ore produced. Similarly, from mid-1890's to World War I, gold with 40 to 70 per cent, accounted for the principal dollar value of the ore produced.

The veins of the northwest sector in San Miguel and Ouray Counties (A and B, Figure 3) have yielded a major part of the production value, totaling nearly 350 million dollars (Table II). A small part of this value, probably about 4 million dollars, came from veins of the western sector around Ophir and south of Telluride (A', Figure 3). Also, a little over 10 million dollars is allotted to chimney deposits of the Red Mountain district (C, Figure 3) in Ouray County.

The metal production of San Juan County, with a total value of almost 145 million dollars, came principally from the Eureka (E, Figure 3) and Animas (J, Figure 3) districts; a few older camps of the Upper Uncompahgre district (D, Figure 3) contributed a small amount. The largest producers in this county were the Sunnyside, Shenandoah-Dives, and Silver Lake mines (Figure 2). Smaller production from numerous veins came from the Animas district south and east of the cauldron rim and from a few veins in the Eureka district near Gladstone in the central part of the cauldron. Of these, the Gold King was a principal gold producer.

The mining areas of Hinsdale County generally are divided into the Galena district (G, Figure 3) along the north rim of the Lake City cauldron and the Lake district (H, Figure 3) along the east rim. Ore valued at a few hundred thousand dollars was produced from outlying districts such as Carson Camp (I, Figure 3) on the south rim of the San Juan volcanic depression and from the White Cross district (F, Figure 3) in the graben area between the two cauldrons. Since 1930, the mines of Hinsdale County have yielded about $550,000 in metals, a minor fraction of the county's total value recorded since 1875.

The detailed distribution of mines within the western region as a whole is shown in the Mineral Resources of Colorado (11, plates 4, 28, and 31, respectively). However, a few of the principal mines and vein systems, particularly of the Silverton cauldron, are given in Figure 2 of this report.

For more comprehensive accounts of the production in the western San Juan Mountains the reader is referred to Burbank (11), Henderson (4), King and Allsman (17), and Vanderwilt (12). Somewhat more detailed county and district records for the period 1932 to 1957 also are given in the annual volumes (Minerals Yearbook) of the U.S. Bureau of Mines.

## Forms and Regional Localization of the Ore Bodies

The major ore bodies of the area are fissure veins, but a few occur as chimneys and tabular replacements. Some but not all veins follow the walls of dikes, including both igneous and clastic types. Most of the longer and more productive veins, such as those northwest of the Silverton cauldron, occupy tensional fractures more or less radial to the cauldron rim; others have diagonal and concentric trends possibly related to cone fracturing or to shearing stresses during subsidence. Larger faults and fractures associated with the Eureka graben also were sites of some of the widest veins.

Some of the longest veins have been stoped nearly continuously for 9000 feet, and some of the greatest vertical developments on productive veins exceed 2500 feet. Veins are as wide as 50 feet but probably average 3 to 6 feet. Most veins dip steeply, generally more than 70°, but some so-called "flat veins" have dips as low as 45° to 50°.

Most "chimneys," of the miner's terminology are essentially vertical pipes, but some consist of ramifying channels and associated replacement deposits of little regularity in form (8). They are confined mainly to the ring-fault zone of the Silverton cauldron and are localized commonly by so-called "ore breaks" (mineralized fissures) related to the ring faults or to intersecting fractures of various trends. A few chimney deposits are located on the rims of, or within, volcanic pipes.

Most replacement deposits are irregularly shaped disseminations of ore minerals in fractured volcanic rocks, or are approximately tabular bodies in bedded shales, limestones, sandstones, conglomerates, and tuffs; the deposits occur both within the volcanic rocks as well as in underlying sedimentary rocks. Few replacement deposits have proved economically important, although recently discovered tabular bodies in the underlying sedimentary rocks in the Telluride district (A, Figure 3) may be of substantial value. A few replacement deposits in Paleozoic limestones at igneous contacts or where penetrated by pipelike mineralized bodies have been prospected in the Red Mountain district (C, Figure 3), but none has yielded appreciable ore.

The more productive districts of the north-

TABLE III. Production of Gold, Silver, Copper, Lead, and Zinc in the Western San Juan Region, 1932 to 1964, inclusive, in terms of recovered metals

| County | Gold (Troy ounces) | Gold (Per cent of all-time production) | Silver (Troy ounces) | Silver (Per cent of all-time production) | Copper (Short tons) | Copper (Per cent of all-time production) | Lead (Short tons) | Lead (Per cent of all-time production) | Zinc (Short tons) | Zinc (Per cent of all-time production) | Total Value | Per cent of all-time county value |
|---|---|---|---|---|---|---|---|---|---|---|---|---|
| Hinsdale | 929 | 1.3 | 95,353 | 1.6 | 53 | 3.5 | 1,445 | 2.9 | 226 | 26.1 | $ 546,212 | 4.9 |
| Ouray | 213,054 | 10.8 | 5,103,863 | 10.8 | 8,459 | 41.7 | 34,426 | 29.5 | 32,733 | 97.8 | 31,483,858 | 27.8 |
| San Juan | 408,876 | 24.5 | 9,512,903 | 21.4 | 10,683 | 25.2 | 66,467 | 22.9 | 37,272 | 26.6 | 46,746,259 | 32.3 |
| San Miguel | 758,827 | 19.3 | 14,988,417 | 23.9 | 36,185 | 80.0 | 132,590 | 54.2 | 157,143 | 94.1 | 134,200,210 | 53.5 |
| Total | 1,381,686 | 18.1 | 29,700,536 | 18.5 | 55,380 | 50.6 | 234,928 | 33.6 | 227,374 | 66.6 | 212,976,539 | 41.1 |

Data: compiled from Minerals Yearbooks, U.S. Bureau of Mines, and Mineral Resources of Colorado (11).

western (8,31,32) and western (27) sectors to the Silverton cauldron area (Figure 2), where radial veins and dikes have their greatest continuity, demonstrate that doming and tensional fissuring, influenced locally by outlying intrusive bodies, were the prime factors in localizing the deposits. The radial veins and dikes converge toward the Mt. Sneffels intrusive body lying about 6 miles northwest of the cauldron rim (Figure 2). As light subsidence or sagging of the central part of this northwesterly swarm of fissures also may have contributed to repeated openings of the fissures. A few fissures have diagonal trends and curve in spiral-like or arcuate form to the cauldron rim; an example is the Camp Bird vein (Figure 2, no. 2) and its western continuation. Some of these fissures may be cone fractures formed as the result of interaction of shearing and concentric forces generated by subsidence of the cauldron block.

South of Silverton, in the Animas district, Varnes (30, p. A23-A33) concluded that there were three systems of fractures within a triangular area of 3 or 4 miles on a side: (1) a system approximately concentric, extending to 2.5 miles south of the cauldron rim; (2) a system of shear fractures and related tension fractures in the western part of the area; and (3) a system of shear fractures in the eastern part of the area. Thus, Varnes places much more emphasis upon shearing action generated by the subsiding cauldron block than upon stresses of uplift and radial tension. Actually, most fissures of this area are somewhat diagonal to the cauldron rim as compared with those of the northwest sector.

The fractures of the northern and northeastern (10,15) sectors of the Silverton cauldron are related to a north-trending horst and graben system and to the large northeast-trending Eureka graben that connects the Silverton and Lake City cauldrons (Figures 1 and 2). Fissures more or less concentric to the cauldron ring-fault zone cross these trends. The veins tend to switch from one set to the other, and form complex local patterns. The open fissures of the concentric set probably developed as a result of subsidence of the central cauldron block (Figure 3).

The Eureka graben and paralleled northeast-trending veins of the adjacent Mineral Point area (Figure 2) probably represent an initial distention of the crust, followed by intermittent subsidence of the main graben block; as a result, beds in the volcanic units on either side of the main graben tilt away from the graben axis. Intermittent reopening of these fissures

during mineralization resulted in the formation of compound vein structures. The highly productive veins of the Sunnyside group lie on the north side of the Eureka graben near the intersection of the main northeast fault zone and a northwest-trending fault that bounds a downfaulted block of the central cauldron (Figure 2). Other veins within the central part of the cauldron are of diverse trend and for the most part are not readily interpreted in terms of various stresses that might have been generated during uplift or subsidence.

A strong north-trending vein set along the west side of the cauldron evidently is related to the western rim faults. This pronounced trend, which extends throughout the western part of the cauldron and beyond it to the north, also may have been influenced by a buried structural line. This structural line in general represents an erosional boundary between the Precambrian rocks and the westerly dipping Paleozoic and Mesozoic sedimentary rocks in the uplifted basement underlying the Tertiary volcanic rocks. Here the entire width of the cauldron ring zone, ranging from 2 to 3 miles across, was subjected to intense solfataric alteration. Many of the "chimney" ore bodies lie within this zone.

In the Lake City cauldron area, more study will be required to understand vein localization. Reconnaissance observations indicate that, unlike the vein system of the Silverton cauldron, the vein system here is more restricted to the borders of the cauldron. This restricted vein distribution possibly may be related to the paucity of both peripheral and outlying intrusive bodies around this cauldron.

## Pre-Ore Propylitization

Many cubic miles of the volcanic rocks within the San Juan volcanic depression and for some miles beyond its rim have been propylitically altered to varying degrees (23). This alteration intensifies with depth. The altered rocks range from low-grade quartz-chlorite-calcite types to albite-epidote-chlorite types analogous to hornfels.

The principal chemical changes during propylitization are addition of carbon dioxide, water, and locally sulfur; sulfur, however, does not appear to have been as essential a component as in the later more localized solfataric alteration. Chemical data are inadequate to assess fully the effects of propylitic alteration at depth and laterally, in part because of differences in initial compositions of the rocks and in part because of the magnitude of the sampling problem. In general, a few tenths to 5 per cent $CO_2$ has been added and about half this quantity of water. Where local changes can be compared between rocks of closely similar composition, some iron and magnesia appear to have been added in proportion to added $CO_2$. Veining of the rocks and microscopic cracks show that the products of rock decompositions have migrated locally, but evidence is lacking for any regional or zonal migration.

Propylitic alteration preceded vein formation and the general introduction of sulfur and metals. The widespread distribution of the effects of this alteration in obvious relation to the major volcanic structural elements and the lack of direct relation to fissures and exposed intrusive bodies indicates that the causative agents were of deep-seated origin, related to postvolcanic evolution of the parent magmas.

## Altered Rocks of the Ore Deposits

Altered rocks genetically related to mineralizing processes commonly are restricted to within a few feet, rarely a few tens of feet from the margins of deposits. Pyrite, quartz, sericite, clay minerals, calcite, and chlorite are typical alteration products more or less directly associated with the introduction of certain ores and gangues in the vein deposits. Pyrite, chlorite, and clay minerals are also intergrown at places with vein quartz.

In the walls of manganiferous vein gangues, rhodonite and rhodochrosite have replaced wallrock minerals and can be correlated with this particular stage of vein formation. But, in many instances, the superposition of alteration products in the walls of veins that repeatedly have been opened and mineralized results in numerous mineral changes, some of which are difficult to correlate with particular stages of vein formation.

In areas of solfataric alteration, the mineral suites generally are those in which large proportions of the rock bases have been removed and silica and sulfur added in various proportions. Silicification is particularly characteristic of the walls of some chimney ore deposits. These features are discussed further in the description and genesis of chimney deposits.

## Structure and Mineralogy of the Deposits

Structure and mineralogy of the deposits are best considered together, as the paragenesis of the minerals cannot be considered inde-

pendently of vein structure as a whole. The vein deposits nearly everywhere have a compound structure, consisting of repetitions of similar associations of ore and gangue as well as of strikingly dissimilar vein matter (Figure 5). Textures of the mineral filling also differ widely from one stage to another, ranging from uniform intergrowths of ore and gangue to banded and vuggy textures. Fine- to medium-textured intergrowths of ore and gangue are on the whole more characteristic of the

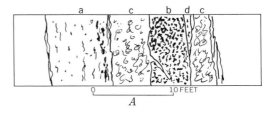

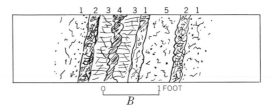

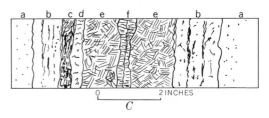

Fig. 5. *Sketches of Structures of Compound Veins, Silverton Cauldron. A. No-Name vein, Eureka district, northeast graben area: a, coarse quartz containing pyrite and lesser sphalerite; b, mainly sphalerite, chalcopyrite, and galena intergrown with quartz (50 to 70 per cent sulfides by weight); c, rhodonite ribs and patches of tephroite, friedelite, and carbonate; d, quartz. B. Mendota vein, Telluride district, outer northwest radial vein system (after Purington, 1): 1, country rock, pyritized and sericitized near walls; 2 (left), sphalerite with calcite; 2 (right), spalerite and galena; 3, white quartz; 4, rhodochrosite; 5, bluish quartz with finely disseminated sulfides and sulfosalts. C. London vein, Upper Uncompahgre district, northeast radial vein system (after Ransome, 2): a, country rock; b, quartz and chalcopyrite; c, tetrahedrite; d, quartz; e, galena; f, quartz.*

earlier stages of base-metal ore, but locally the sulfides may be segregated in lenses or layers more or less devoid of gangue. Finely banded textures generally are more characteristic of later stages of vein filling, especially in some gold-quartz veins.

The common sulfides of the base-metal ores are pyrite, sphalerite, galena, chalcopyrite, and tennantite. In the higher grade silver-bearing ores, argentiferous tennantite, tetrahedrite, and proustite constitute larger proportions of the sulfides, which usually include the common sulfides as well. The principal source of silver in the ore shoots was argentiferous tennantite, and the appearance of much proustite was considered by the miners as an indication of approach to the ore shoot limits. Arsenopyrite is a constituent of some gold-silver veins. Tellurides of gold and silver are very sparse, or wanting altogether, but have been identified in microscopic quantities. Some veins, notably those of Poughkeepsie Gulch in the northern sector of the Silverton cauldron (D, Figure 3), were noted for pockets of rich silver ore containing sulfobismuthites of doubtful identity. Those more positively identified include matildite, aikinite, and argentiferous cosalite (25, p. 34–35).

Free gold in association with quartz and locally adularia generally was introduced late, but microscopic blebs of gold have been identified in the sulfides of early base-metal ores that generally contain small quantities of gold as well as silver.

The common gangues of the veins include quartz, barite, calcite or locally ankerite, and fluorite, all directly associated with sulfide deposition. The complex manganiferous gangues of some areas (24), which include rhodonite, rhodochrosite, tephroite, alleghanyite, and friedelite, are not associated as directly with major sulfide deposition, and occur mostly in relatively barren bodies. The reconstitution of these gangues to simpler aggregates of rhodonite and quartz or rhodochrosite resulted in intermixing with later silver- or gold-bearing solutions, so that not all bodies of these minerals are devoid of appreciable amounts of sulfides.

Sequences of deposition in compound veins include some or all of the following gangue-sulfide associations: (1) sparse early and somewhat ribbony quartz with partings of chloritic substances and a little pyrite; (2) common base-metal sulfides in a dominant quartz gangue; (3) rhodonite and other manganese silicates of lower silica content, with very sparse sulfides and including traces of alabandite; (4) quartz or barite, or both, with silver ores and additional quantities of common sulfides and locally rhodochrosite, or ankeritic carbonate; and (5) quartz with free gold and locally associated sulfides, adularia, calcite, and fluorite. Barren vuggy quartz and locally barren carbonate are found in some veins. The common base-metal sulfides are found in all stages of mineralization but in lesser quantities in later vein stages and, with the exception noted, in certain bodies of mixed manganese silicates.

Paragenetic relations in the manganese silicate state are particularly interesting with regard to the obscure origin of these bodies, some of which are very large in the Sunnyside fault and associated vein system. The earliest minerals of the suite were fine microscopic intergrowths of friedelite and subordinate tephroite, suggestive of a coagulated precipitate from solutions that had been abruptly supersaturated. Most of these earlier aggregates, except remnant patches, were converted to coarser crystallized alleghanyite and finally to rhodonite along cracks and around vuggy patches, indicating the addition of silica from a continued flow of solutions. Rhodochrosite also formed by local addition of $CO_2$. The silicates low in silica, tephroite, alleghanyite, and friedelite appear to have been deposited only in open spaces and never are observed as replacements of quartz or wallrocks. Rhondonite, however, is commonly seen as a replacement of wallrock silicates or of quartz. These relations indicate that the mixed manganese silicate gangues of early crystallization were deposited from solutions relatively undersaturated with silica as compared with manganese. This condition represents a very marked change from the preceding quartz-rich sulfide deposition and also from following vein stages of quartz and sulfide. In many veins, especially those of average widths, only rhondonite or rhodochrosite are associated commonly with siliceous gangues; if these occurrences had a similar evolution, it must have taken place below levels now exposed, or earlier stages were completely obliterated. It is also possible, if not likely, that these earlier stages were confined to certain areas.

The ores of the chimney deposits have entirely different internal relations than those of the typical fissure veins. Sulfides in the chimney deposits tend to be segregated in bodies devoid of intergrown gangues other than a little barite and fluorite or products of rock decomposition such as sericite and clay minerals. Quartz is rarely intergrown with the sulfide bodies. Enargite is a common copper ore as compared with its rarity in vein deposits. Sulfosalts of copper and silver are common minerals, and stromeyerite is of local occurrence.

The environment of the chimney ores also contrasts with that of the veins. Chimneys commonly lie in or near strongly leached and solfatarically altered volcanic rocks of the cauldron rim zones. Over large areas, the rocks are converted to aggregates of quartz, dickite, and other clay minerals, pyrophyllite, zunyite, diaspore, and alunite. Essentially all the original iron has been converted to pyrite. The cauldron rim zones also are perforated in some places with large numbers of siliceous volcanic intrusive and breccia pipes, indicating that ring intrusions probably underlie the cauldron rims at shallow depths. The environmental conditions are suggestive of those that might underlie modern fumarolic and hot-spring areas.

The stronger alteration associated with the chimneys has resulted in leaching of ramifying channels and chimneylike openings in the shallower rocks. Ore bodies in these structures formed both by filling of the previously leached spaces and by replacement of the altered walls of channels. The silica leached from the rocks locally caps the sites of chimneys. The redeposited quartz and other materials of rock decomposition tended to seal the vents of the altering solutions; the building up of pressures in underlying channels eventually allowed ore-bearing solutions to bleed into the leached spaces and deposit sulfides by more gradual loss of pressure, temperature, and gaseous constituents.

## Genesis of the Mineral Deposits

Some features of the environment, mineral products, and paragenetic sequences of the mineral deposits are of interest in relation to later stages of volcanic evolution, and these relationships may point to possible sources of mineralizing and altering agents (16). The end stages of volcanism may have had a bearing on the earliest hydrothermal activity that preceded ore-forming activity. (1) The later volcanic eruptions culminated in violent ash-flow eruptions, which could have depleted the shallower magmatic reservoirs of the more siliceous differentiates charged with their energizing volatiles. (2) Any residual parent magmas of more mafic composition, such as those that formed the exposed dioritic and gabbroic stocks, (24,32), would tend to hold carbon dioxide in solution in greater proportion than

normally present in the more siliceous differentiates that had been erupted (23). (3) Cooling and gradual crystallization of these parent magmas under the fractured cover of volcanic rocks may well have involved osmotic conditions under which their volatiles high in $CO_2$ were squeezed out of the reservoirs. Pressures of dissolved gases under such conditions are known to be very high, and this could account for the earlier clastic injections of basement rock fragments that invaded fissures following igneous dikes and preceding vein formation. (4) The sealing and choking of early fissures by such injections would aid in maintaining pressures of the gases at depth and cause their widespread penetration of the volcanic cover. (5) Possible carbonatization of basement country rocks might have desilicated them locally and released sufficient silica to account for the early barren quartz of the veins and perhaps for much of the quartz gangue.

Although reopening of fissures and repeated veining took place even during base-metal deposition, the strongest reopening of fissures commonly followed the quartz-base metal stage. This action was particularly pronounced in parts of the Eureka district adjacent to the major graben faults. In these localities, the manganese silicates also attained their greatest development. The influx of manganiferous solutions low in silica at this stage conceivably may be attributed to fractionation of the residual solutions of the early propylitic and carbonatic stages of rock alteration. Abrupt release of pressure at the roots of faults and fissures may have resulted in an influx of solutions from wallrocks and upper parts of consolidated intrusive bodies, analogous to, but much less violent than, the earlier influx of clastic debris. Concentration of manganese in such laterally secreted solutions could result from several physical and chemical factors. (1) Release of gases would tend to fractionate any residual solutions as they moved toward fissured and broken ground. In carbonatic solutions manganese could become enriched in fluids from mixed carbonates by loss of $CO_2$ pressure. (2) The formation of epidote and chlorites, as well as iron oxides, would have tended to fix iron and magnesium. (3) Prior desilication of some rock bodies by carbonatization at depth might also account in part for the low-silica content of the manganiferous solutions. Such an origin of these solutions would explain in large part the extremely low sulfide content of material deposited at shallower depths. The main sources of sulfur and metals may have been in the mafic parent magma at depth. The ensuring influx of these metal-bearing solutions followed with further and minor refissuring of the manganiferous gangues.

## Hypogene and Supergene Enrichment

Within observable depths, some silver-copper ore shoots appear to have been concentrated from underlying low-grade base-metal veins. The question arises whether these are examples of simple zonal deposition, supergene enrichment, or hypogene enrichment by fractionation and recrystallization of underlying bodies of ore. Microscopic, textural, and field studies of the ores by Bastin (3), Kelley (10), Moehlman (6), and Burbank (8) indicate that the ores have been extensively recrystallized and replaced, and are primarily of hypogene origin. There is general agreement that most of the larger ore bodies enriched in silver sulfosalts and in silver-bearing copper minerals are of hypogene rather than supergene origin. The positions of some of these enriched shoots above zinc-lead ore bodies of very low silver and copper content tend to favor a hypothesis of hypogene recrystallization of primary ore and selective leaching and fractional resolution of vein minerals. Reopening of the veins and consequent passage of mineralizing solutions over previously deposited vein matter is conductive to recrystallization along the vein. Recrystallization at lowering temperatures and pressures also would be conducive to fractionation (and ultimate redeposition) of the copper and silver in solution complexes. Some evidence that sphalerites of the veins have been recrystallized is afforded by analyses of their iron and minor metal contents. Extreme differences in iron content, seemingly without regard to association and environment, were noted in a number of randomly collected sphalerite samples. Hillebrand (22, p. 185–186) reports similar results based upon much more systematic sampling of the Black Bear vein in the Idarado mine. He suggests possible recrystallization of the sphalerites during declining energy conditions as one possible cause of the unsystematic variation of iron content.

As much of the quartz and other gangue minerals, such as the manganese silicates, illustrate repeated veining and recrystallization as well as successive stages of replacement, it would appear unlikely that common sulfides, supposedly more readily recrystallized than many gangues, would have escaped effects of the repeated passage of mineralizing solutions.

Supergene enrichment of veins in some areas has been established, but this appears to be confined to very shallow depths. Erosion generally has kept pace with oxidation and leaching processes. Small quantities of native silver, argentite, sooty chalcocite, and locally carbonates and sulfates of the lead-zinc-copper ores are minerals of this origin.

## Geochemical Prospecting

Preliminary tests have indicated that probable usefulness of arsenic as an indicator in geochemical prospecting in areas where direct surface observations are somewhat obscure. In areas of strong solfataric alteration, the more soluble base metals may be leached and carried away in acid-sulfate waters derived from oxidation of pyrite, but the arsenic, lead, and, locally, silver tend to remain in some oxidized residues (9). Most of the sulfosalts of the solfatarically altered area are arsenical, so that in general arsenic was a common constituent of the ore-forming solutions. Arsenic tests also might prove useful in prospecting some gold-silver veins containing arsenopyrite.

## REFERENCES CITED

1. Purington, C. W., 1898, Preliminary report on the mining industries of the Telluride quadrangle, Colorado: U.S. Geol. Surv. 18th Ann. Rept., pt. 3, p. 745–850.
2. Ransome, F. L., 1901, A report on the economic geology of the Silverton quadrangle: U.S. Geol. Surv. Bull. 182, 265 p.
3. Bastin, E. S., 1923, Silver enrichment in the San Juan Mountains, Colorado: U.S. Geol. Surv. Bull. 735, p. 65–129.
4. Henderson, C. W., 1926, Mining in Colorado: U.S. Geol. Surv. Prof. Paper 138, 263 p.
5. Burbank, W. S., 1935, Camps of the San Juans: Eng. and Min. Jour., v. 136, p. 386–398.
6. Moehlman, R. S., 1936, Ore deposition south of Ouray, Colorado: Econ. Geol., v. 31, p. 377–397, 488–504.
7. Burbank, W. S., 1940, Structural control of ore deposition in the Uncompahgre district, Ouray County, Colorado: U.S. Geol. Surv. Bull. 906-E, p. 189–265.
8. ———— 1941, Structural control of ore deposition in the Red Mountain, Sneffels, and Telluride districts of the San Juan Mountains, Colorado: Colo. Sci. Soc. Pr., v. 14, no. 5, p. 141–261.
9. Varnes, D. J. and Burbank, W. S., 1945, Lark mine, Cement Creek area, San Juan County, Colorado: Colo. Mining Assoc. Yearbook, p. 36–37.
10. Kelley, V. C., 1946, Ore deposits and mines of the Mineral Point, Poughkeepsie, and

Upper Uncompahgre districts, Ouray, San Juan, Hinsdale Counties, Colorado: Colo. Sci. Soc. Pr., v. 14, no. 7, p. 287–466.

11. Burbank, W. S., *et al.,* 1947, The San Juan region, *in* Vanderwilt, J. W., *Editor, Mineral resources of Colorado:* State of Colo. Mineral Res. Bd., p. 396–446.

12. Vanderwilt, J. W., 1947, Metals: gold, silver, copper, lead, and zinc districts by counties, *in* Vanderwilt, J. W., *Editor,* Mineral resources of Colorado: State of Colo. Mineral Res. Bd., p. 23–228.

13. Varnes, D. J., 1947, Recent developments in the Black Bear vein, San Miguel County, Colorado: Colo. Sci. Soc. Pr., v. 15, p. 135–146.

14. Eckel, E. B., 1949, Geology and ore deposits of the La Plata district, Colorado: U.S. Geol. Surv. Prof. Paper 219, 179 p.

15. Hazen, S. W., Jr., 1949, Lead-zinc-silver in the Poughkeepsie district and part of the Uncompahgre and Mineral Point districts, Ouray and San Juan Counties, Colorado: U.S. Bur. Mines R.I. 4508, 110 p.

16. Burbank, W. S., 1950, Problems of wall-rock alteration in shallow volcanic environments, *in Applied geology, a symposium:* Colo. School Mines Quart., v. 45, no. 1B, p. 287–319.

17. King, W. H. and Allsman, P. T., 1950, Reconnaissance of metal mining in the San Juan region, Ouray, San Juan, and San Miguel Counties, Colorado: U.S. Bur. Mines I.C. 7554, 109 p.

18. Kuryla, M. A., 1950, Sunnyside mine: U.S. Bur. Mines I.C. 7554, p. 70–71.

19. Burbank, W. S., 1951, The Sunnyside, Ross Basin, and Bonita fault systems and their associated ore deposits: Colo. Sci. Soc. Pr., v. 15, no. 7, 285–304.

20. Belser, C., 1956, Tungsten potential in the San Juan area, Ouray, San Juan, and San Miguel Counties, Colorado: U.S. Bur. Mines I.C. 7731, 18 p.

21. Larsen, E. S., Jr. and Cross, W., 1956, Geology and petrology of the San Juan region, southwestern Colorado: U.S. Geol. Surv. Prof. Paper 258, 303 p.

22. Hillebrand, J. R., 1957, The Idarado mine: New Mexico Geol. Soc., 8th Field Conf., Guidebook, p. 176–188.

23. Burbank, W. S., 1960, Pre-ore propylitization, Silverton caldera, Colorado: U.S. Geol. Surv. Prof. Paper 400-B, p. B12-B13.

24. Burbank, W. S. and Luedke, R. G., 1961, Origin and evolution of ore and gangue-forming solutions, Silverton caldera, San Juan Mountains, Colorado: U.S. Geol. Surv. Prof. Paper 424-C, p. C7-C11.

25. Eckel, E. B., 1961, Minerals of Colorado—A 100-year record: U.S. Geol. Surv. Bull. 1114, 399 p.

26. Luedke, R. G. and Burbank, W. S., 1962, Geology of the Ouray quadrangle, Colorado: U.S. Geol. Surv. Geol. Quad. Map GQ-152, 1:24,000.

27. Vhay, J. S., 1962, Geology and mineral deposits of the area south of Telluride, Colorado: U.S. Geol. Surv. Bull. 1112-G, p. 209–310.

28. Luedke, R. G. and Burbank, W. S., 1963, Tertiary volcanic stratigraphy in the western San Juan Mountains, Colorado: U.S. Geol. Surv. Prof. Paper 475-C, p. C39–C44.

29. Tweto, O. and Sims, P. K., 1963, Precambrian ancestry of the Colorado mineral belt: Geol. Soc. Amer. Bull., v. 74, p. 991–1014.

30. Varnes, D. J., 1963, Geology and ore deposits of the South Silverton mining area, San Juan County, Colorado: U.S. Geol. Surv. Prof. Paper 378-A, 56 p.

31. Burbank, W. S. and Luedke, R. G., 1964, Geology of the Ironton quadrangle, Colorado: U.S. Geol. Surv. Geol. Quad. Map GQ-291, 1:24,000.

32. ———— 1966, Geology of the Telluride quadrangle, Colorado: U.S. Geol. Surv. Geol. Quad. Map GQ-504, 1:24,000.

# 35. The Uranium and Vanadium Deposits of the Colorado Plateau Region[*]

R. P. FISCHER[†]

## Contents

## Illustration

## Tables

* Publication approved by the Director, U.S. Geological Survey.
† U.S. Geological Survey, Denver, Colo.

## ABSTRACT

The Colorado Plateau region has been the principal domestic source of uranium, vanadium, and radium. The value of these commodities produced from the region through 1964 slightly exceeds $2 billion.

Most of the deposits occur in streamlaid lenses of sandstone in the Chinle and Morrison Formations. The ore minerals below the zone of oxidation are low-valent oxides and silicates of uranium and vanadium; all of these, except the quite stable vanadium silicates, oxidize readily to a variety of high-valent uranium and vanadium minerals. The ore minerals partly replace fossil wood but mainly impregnate sandstone, typically forming tabular ore bodies that are nearly concordant to the bedding of the host rock. The deposits range in size from those containing only a few tons of ore to those containing more than 1 million tons. Stratigraphic, lithologic, and sedimentary structural factors obviously influenced the localization of these deposits, but igneous activity, hydrothermal mineralization, and tectonic structures show no similarly consistent influence on localization. The deposits are epigenetic, but the source of the ore metals, the nature of the ore-bearing solutions, and the time of ore emplacement are uncertain.

A few uranium deposits in the region are dominantly discordant to bedding and occur in collapsed breccia pipes or along linear fractures. Their origins are also uncertain.

## INTRODUCTION

The general distribution and types of uranium deposits in the Colorado Plateau region are indicated in Figure 1. Typically the deposits are in sandstone and form tabular ore bodies that are nearly concordant (peneconcordant) with bedding. Most of these are in beds of the Chinle Formation of Triassic age or in the Morrison Formation of Jurassic age. Some deposits occur in other formations, however, and many of these are also in sandstone, although a group of deposits in New Mexico occurs in limestone. A few deposits are in collapsed breccia pipes, and a few form veins along fractures. Accumulations of uranium have also been found in beds of shale and coal and in igneous rocks in the Colorado Plateau region, but deposits of these types are not shown on Figure 1 as they have been virtually unproductive.

The geology of the Colorado Plateau region and the habits of the principal types of uranium deposits are briefly described below. This general section is followed by four separate sections that describe in greater detail the ore deposits and the application of geology to orefinding and development in selected mining districts. These districts are outlined on Figure 1.

## Production Data and History of Mining

Though 1964, the uranium-bearing deposits of the Colorado Plateau region have yielded about 70 per cent of the uranium recovered from domestic ores, about 98 per cent of the recovered vanadium, and nearly all of the recovered radium. Production data by significant periods of operation are shown in Table I. In addition, about $8,000,000 worth of silver was obtained from ore bodies in the Silver Reef area, Utah (5), before the associated uranium was economically recoverable, and some copper has been recovered from uranium-bearing deposits at other places. Consideration has been given to attempting to recover molybdenum and selenium from some of the uranium ores.

Vanadium is more abundant than uranium in most of the deposits in the Morrison Formation in Colorado, Utah, and Arizona and also in some of the deposits in the Chinle Formation in Utah and Arizona (1). These deposits have yielded about 80 per cent of the total vanadium production shown in Table 1. Nearly all of the rest has come from deposits in the Entrada Sandstone along the west side of the San Juan Mountains, Colorado (10,2), and from the Navajo(?) and Entrada Sandstones near Rifle, Colorado (21). These deposits are similar geologically to the uranium and uranium-vanadium deposits in the Morrison and Chinle Formations, but their uranium content is generally so low that it can only be recovered as a byproduct. These deposits were worked for vanadium alone from 1910 to 1920, 1925 to 1932, and during World War II; ore that has been mined since 1948 has been blended with uranium-rich ores to yield both vanadium and uranium.

The first uranium ore in sandstone was found in the Morrison Formation in Montrose County, Colorado, in 1898. Similar deposits were soon found in the adjoining parts of Colorado and Utah, including some deposits in Utah in beds now classed as Chinle, but only a small amount of high-grade ore was mined, mainly for export, through 1909. During the 1910–23 period, the deposits in the Morrison and Chinle formations yielded about 67,000 tons of high-grade ore, mainly for its radium

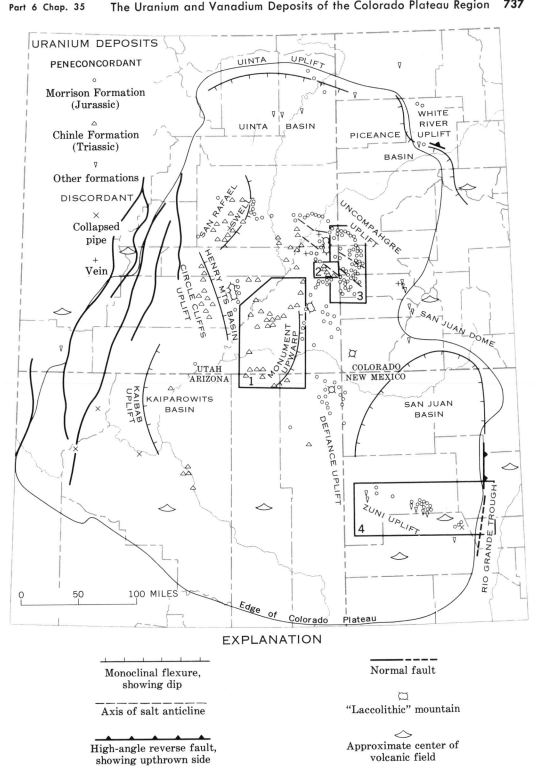

FIG. 1. *Principal Uranium Deposits and Major Structural and Igneous Features in the Colorado Plateau Region. Blocks show approximate areas covered by the following district reports: 1, Monument Valley-White Canyon; 2, Lisbon Valley; 3, Uravan Mineral Belt; 4, Zuni Uplift (Grants and Laguna.)*

TABLE I.   *Production and Approximate Value of Uranium, Radium, and Vanadium from the Colorado Plateau Region*

| Period of operation | Ore mined and shipped (short tons) | Mill yield, expressed as metal content and estimated value of product obtained | | | | | |
| --- | --- | --- | --- | --- | --- | --- | --- |
| | | Uranium | | Radium | | Vanadium | |
| | | Short tons | Millions of $ | Grams | Millions of $ | Short tons | Millions of $ |
| 1898–1909 | 9,000[1] | | | | | few(?) | (?) |
| 1910–1923 | 267,000[2] | some | (?) | 202 | 20 | 2,084 | 15 |
| 1924–1935 | 312,000[3] | some(?) | (?) | few(?) | (?) | 2,420 | 14 |
| 1936–1947 | 1,318,000 | 1,260[4] | 24 | | | 9,126 | 32 |
| 1948–1964 | 44,000,000 | 88,000 | 1,900 | | | 54,200 | 190 |
| Totals (rounded) | 46,000,000 | 89,000 | 1,900 | 202 | 20 | 68,000 | 250 |

[1] Most of this ore apparently was shipped to Europe.

[2] Includes about 200,000 tons of ore from Placerville, Colorado, yielding about 1550 tons of vanadium but no uranium or radium.

[3] Includes about 304,000 tons of ore from Rifle, Colorado, yielding about 2350 tons of vanadium but no uranium or radium.

[4] Most of this uranium was recovered from tailings from vanadium mills.

Source: Published and unpublished data of the U.S. Bureau of Mines and the U.S. Atomic Energy Commission.

content, but some vanadium and a little uranium were recovered as byproducts. Only about 8000 tons of uranium-vanadium ore was mined from these deposits from 1924 to 1935. In 1936, aggressive mining was resumed in Colorado and then spread into Utah and northern Arizona during the war years. Vanadium was the desired metal, but some byproduct uranium was recovered from the vanadium-mill tailings. Mining declined to a negligible rate in the mid-1940's.

Following the 1948 announcement of the U.S. Atomic Energy Commission's uranium ore-buying program, ore production increased rapidly from reactivated mines. During the next seven years, intensive exploration, aided by modern techniques and geologic guidance, found many new ore bodies in the established mining districts as well as in areas remote from known deposits. Initial discoveries in the following major mining districts were made during this period: White Canyon, Utah, 1948; Grants, New Mexico, deposits in the Todilto Limestone, 1950, deposits in the Morrison Formation, 1951; Laguna, New Mexico, 1951; Lisbon Valley, Utah, deposits in the Chinle Formation, 1952; Cameron, Arizona, 1952; and Ambrosia Lake, New Mexico, 1955. Uranium ore production in the Colorado Plateau region increased from 73,000 tons in 1948 to about 1 million tons in 1954 and to nearly 6 million tons in 1960; it declined to about 3.5 million tons in 1964, due to adjustment of uranium purchase contracts.

## GEOLOGY

### Geologic History

Precambrian igneous, metamorphic, and sedimentary rocks crop out in only a few places in the Colorado Plateau region. By the beginning of Paleozoic time, erosion had apparently reduced this region to a moderately flat surface, and since then the Plateau has been quite stable structurally, at least in comparison with adjoining regions.

Known Paleozoic beds of pre-Pennsylvanian age consist of marine carbonates and clastics and total a few hundred feet thick. Although these rocks are exposed only in a few places, they are assumed to be fairly extensive and uniform in thickness in the Plateau region and to have been deposited in shallow seas on a stable platform. In late Paleozoic time, crustal stability in the eastern part of the Plateau was interrupted by the development of the Paradox Basin in southwestern Colorado and southeastern Utah and uplift in the adjoining Uncompahgre area and in the Zuni and Defiance areas in New Mexico and Arizona (Figure 1). A thick sequence of marine shale and limestone, with abundant salt and sulfate evaporites, accumulated in the Paradox Basin in Pennsylvanian time. These beds grade upward into beds of Permian age; the Permian beds consist dominantly of continental clastics derived from the Uncompahgre Highland.

The upward movement of the evaporites in Pennsylvanian time along fractures that probably developed with recurrent uplift of the Uncompahgre initiated the so-called "salt anticlines" of the Paradox Basin; the flow of the evaporites continued into the Mesozoic.

During the Mesozoic Era, the Colorado Plateau region remained quite stable except for a gradual lowering that kept it close to sea level. A few thousand feet of beds, mainly nonmarine sandstone and shale, were deposited during the Triassic, Jurassic, and Early Cretaceous. The principal source of these sediments shifted during Triassic time from the east to the south and west, where highlands developed that yielded mainly sand, mud, and volcanic debris. During the early part of the Late Cretaceous, the Plateau and large areas to the east and north sank below sea level, and the thick and uniform Mancos Shale accumulated. Mancos deposition ended, and the Mancos sea was displaced, as sands and muds that formed the beds of the Mesaverde Group encroached eastward from rapidly rising mountains west of the Plateau.

During the Laramide orogeny in latest Cretaceous and early Tertiary time, the Plateau province was outlined and some of the major structural components in it were born, but the Plateau still acted as a moderately stable unit relative to the more intensively deformed regions surrounding it. The sedimentary units that formed in the Plateau region during and after the Laramide orogeny are mostly of early Tertiary age and consist mainly of sands and muds deposited by streams and lakes in basins adjoining highlands (12).

## Structure and Igneous Activity

In most of the Colorado Plateau region the Precambrian basement rocks are covered by a veneer of sedimentary beds, 1 to 3 miles thick. Typically, these beds lie nearly flat, but in places they have been disturbed by folds and faults, some of which displace the basement, and in places they have been intruded or covered by magma. Figure 1 shows the principal structural and igneous features (8).

The most conspicuous structural features are uplifts. The largest of these are crustal blocks, as much as 100 miles long, that are tilted like partly raised trap doors, with thousands of feet of structural displacement. These blocks are probably bounded by faults in the basement rocks, but in general the sedimentary cover drapes over the block edges without faulting, forming the prominent monoclinal flexures of the region. Some uplifts are coupled with basins formed by blocks tilted downward; the three largest basins—the San Juan, Piceance, and Uinta—are coupled with uplifts outside the Plateau proper. The "laccolithic" mountains of the central part of the Plateau are domal uplifts of several thousand feet relief and the order of 10 miles across, raised by the invasion of magma into the sedimentary cover. The so-called "salt anticlines," which resulted from arching of beds due to the upward flowage of evaporites along linear structures, have been breached by erosion and now form spectacular valleys a few miles across and 10 to 30 miles long.

Faults are prominent structural features in parts of the Plateau region. High-angle reverse faults occur in places along the eastern margin of the Plateau, but all other exposed faults are of the normal type. Normal faults of large displacement bound large elongate blocks along the west side of the Plateau, and a complex series of normal faults bounds the Rio Grande trough along the southeast edge of the Plateau. Faults of small to moderate displacement occur in places in the surface rocks along the monoclines and the salt anticlines; it is probable that these flexures are associated with faults of larger displacement in the basement rocks. Minor faults are common in many parts of the Plateau region, as are joint systems in the more competent beds that are mainly sandstone.

Collapsed pipes are common in the Laguna area, New Mexico, and near Moab, Utah, and a few occur in the Grand Canyon area, Arizona, and in the San Rafael Swell, Utah. The larger ones are a few hundred feet in diameter, and their cores are displaced downward a hundred feet or more and are rather intensively brecciated. A few of these pipes are uranium-bearing and form deposits having unique characteristics.

Igneous activity in the Plateau is minor relative to that in adjoining areas to the east, south, and west. In the central part of the Plateau, intrusive bodies of small to moderate size form the several "laccolithic" mountains. Their age has not been definitely established though probably they formed during early or mid-Tertiary time. A little mineralization with typically hydrothermal characteristics is associated with these intrusives. Some volcanic activity occurred in places in the southern part of the Plateau; it ranges in age from Middle Tertiary to Quaternary.

## Stratigraphy of Uranium-Bearing Rocks

Table II is a generalized stratigraphic section of the principal Permian and Mesozoic rocks in much of the Colorado Plateau region; these are the beds that crop out in most of the uranium-mining districts, and they contain nearly all of the peneconcordant uranium deposits in the region. Brief descriptions of the lithology of the Chinle and Morrison Formations are given below, as these two units have yielded most of the ore. In many respects these two units are lithologically similar to other units that contain peneconcordant uranium deposits in the Plateau region and also to other formations that contain like deposits in Wyoming, South Dakota, and elsewhere.

The Chinle Formation (19) is widespread in the Colorado Plateau region. It is about 2000 feet thick in southwestern Utah, and from there it thins rather gradually to several hundred feet in the eastern and northern parts of the Plateau. Beds of similar lithologic character extend to the west, north, and southeast

of the Plateau. The Chinle is composed mainly of brightly colored mudstone and siltstone, dominantly red but in places varicolored. Some sandstone, generally light-colored, is present nearly everywhere, and in places it predominates; conglomeratic sandstone is common, particularly at the base of the formation. Clay minerals derived by alteration of volcanic debris are abundant, especially in some of the finer-grained beds. Although the Chinle Formation is divided into several members in the southern and western parts of the Plateau, conditions of deposition of the Chinle were virtually the same throughout the region—the sediments probably accumulated in an environment of an alluvial plain with meandering streams, short-lived lakes, and extensive mud flats. Highlands south of the Plateau region probably were the chief sources of these sediments, though parts of the ancestral Uncompahgre may have contributed some material to the Chinle in western Colorado and southeastern Utah and to equivalent strata in north-central New Mexico. Known uranium

TABLE II.   *Generalized Stratigraphic Column of Upper Paleozoic and Mesozoic Formations in the Colorado Plateau Region*

| System | Major units and equivalents | General distribution and approximate average thickness or range (feet) | | | | Principal characteristics and origin |
|---|---|---|---|---|---|---|
| | | Utah | Ariz. | Colo. | N. Mex. | |
| Cretaceous | Mesaverde Group | 3000 | 1000 | 2000 | 1000 | Sandstone and shale; continental and marine |
| | Mancos Shale | 3000 | 1000 | 3000 | 1000 | Shale, dark-colored; marine |
| | Dakota Sandstone | 100 | 100 | 100 | 100 | Sandstone and shale; continental |
| | Burro Canyon Formation | 200 | 0 | 0–200 | 100 | Sandstone and mudstone; continental |
| Jurassic | Morrison Formation | 800 | 800 | 600 | 500 | Sandstone and mudstone; continental |
| | Bluff and Junction Creek Sandstones | 0–300 | 0–60 | 0–200 | 100 | Sandstone; continental (eolian) |
| | Summerville Formation | 300 | 100 | 100 | 100 | Shale; marginal marine |
| | Curtis Formation | 200 | 0 | 0 | 0 | Sandstone and shale; marine |
| | Todilto Limestone | 0 | 0 | 0 | 20–100 | Limestone and gypsum; lagoonal(?) |
| | Entrada Sandstone | 500 | 200 | 200 | 150 | Sandstone; continental (eolian) |
| | Carmel Formation | 400 | 0–100 | 0–100 | 0 | Sandstone, shale, limestone, and gypsum; marginal marine |
| TR ? | Navajo Sandstone | 1000 | 500 | 100 | 0 | Sandstone; continental (eolian) –?– |
| | Kayenta Formation | 200 | 200 | 200 | 0 | Sandstone; continental |
| Triassic | Wingate Sandstone | 500 | 500 | 0–300 | 100 | Sandstone; continental |
| | Chinle Formation | 700 | 1000 | 500 | 1000 | Shale, sandstone, and conglomerate; continental |
| | Moenkopi Formation | 600 | 400 | 400 | ? | Shale and sandstone; marginal marine |
| Permian | Cutler Formation | 1000 | 1000 | 1000 | 1000 | Shale and arkosic sandstone; continental |

deposits in the Chinle are virtually restricted to Arizona and Utah. In northeastern Arizona and southeastern Utah, most of the deposits occur in whichever member locally forms the base of the formation. Typically these deposits occur in the thicker parts of conglomeratic sandstone lenses, many of which occupy channels or scours cut into the underlying formation. Carbonized fossil wood is common in these lenses, especially in the ore-bearing parts.

The Morrison Formation (7) was deposited over the entire Plateau region except the southern and southwestern edges, and it extends far to the east and north. In the Plateau region it comprises four members, each of somewhat different lithologic character and distribution; locally they intertongue. These members all accumulated in stream and flood-plain environments. The Morrison has yielded most of the uranium ore from the Plateau region, and it contains the bulk of the known reserves in the United States.

The Salt Wash Member is composed of material derived from a source area southwest of the Plateau. Its maximum thickness is about 600 feet where the Colorado River crosses the Utah-Arizona boundary, and there it is composed almost wholly of sandstone, in part conglomeratic. It spread as a fan eastward and northeastward, thinning gradually and decreasing in grain size. In eastern Utah, western Colorado, and the adjoining parts of Arizona and New Mexico, it forms a broad facies zone that is favorable for uranium-vanadium deposits. In this zone, layers of sandstone and mudstone are interbedded and occur in about equal proportions. The sandstone is fine- to medium-grained and light-colored, either pale red or pale gray; it occurs in lenses up to 100 feet thick and commonly 1 to 5 miles wide. The mudstone is poorly bedded, somewhat silty, and generally red. East and north of this ore-bearing facies the Salt Wash loses its identity as the sandstone layers become more evenly bedded and decrease in amount.

The Recapture Member and the overlying Westwater Canyon Member cover northwestern New Mexico, but each barely extends into the adjoining states. Both were derived from highlands south of the Plateau, and each has a maximum thickness of a few hundred feet in the vicinity of Gallup, New Mexico, where they are composed mainly of conglomeratic sandstone. As is true of the Salt Wash, the Recapture and Westwater Canyon spread as fans eastward and northward from their points of maximum thickness, decreasing in grain size and consisting of interbedded sandstone and

claystone. The claystone in the Recapture is reddish, that in the Westwater Canyon is gray. The sandstone beds in both members are dominantly gray, medium-grained, but poorly sorted; sandstone in the Recapture is thin-bedded, argillaceous, and soft, whereas sandstone in the Westwater Canyon occurs in thick lenses and is arkosic and moderately hard. Only a few small uranium deposits have been found in the Recapture, but the Westwater Canyon contains many large deposits in the Grants area, New Mexico.

The Brushy Basin Member is widespread; originally it may have covered all of the Plateau, but it was removed by pre-Dakota erosion from the southern and southwestern parts. Where present it commonly ranges from 200 to 400 feet in thickness. It is composed predominantly of poorly bedded mudstone in which clay minerals derived from the alteration of volcanic ash are abundant. The mudstone beds are variegated in shades of red, purple, green, and gray. Thin lenses of dense gray limestone are common and thin to thick lenses of light-colored, coarse-grained to conglomeratic sandstone occur in places. In the Laguna area, New Mexico, a large sandstone lens, locally called the Jackpile sandstone, contains a few large deposits. A few small deposits occur in sandstone lenses in the Brushy Basin in Colorado and Utah.

## ORE DEPOSITS

### Mineralogy and Elemental Composition of the Ore

The ore minerals (20) below the zone of oxidation are low-valent oxides and silicates of uranium and vanadium: uraninite [$UO_{2+}$], coffinite [$U(SiO_4)_{1-x}(OH)_{4x}$], montroseite [$VO(OH)$], and vanadium-bearing mica, chlorite, and clay. The common copper sulfides are widespread in sparse amounts and abundant enough in a few places to be ore minerals. Accessory minerals are mainly sulfides; pyrite and marcasite are common but generally not abundant, and trace amounts of galena and sphalerite are widespread. Minerals containing molybdenum, selenium, chromium, nickel, cobalt, and silver are also present, but only in a few places are some of these minerals abundant enough to be recognized. Introduced gangue minerals other than those that commonly cement sandstone are inconspicuous or absent.

In peneconcordant deposits, some of the ore

minerals are disseminated in quite clean sandstone, but generally they are most concentrated by and near carbonaceous material. This material occurs in two general forms. One is the so-called asphaltic material that impregnates masses of sandstone, coating sand grains, in the same manner as does "dead oil," but the exact nature and origin of this material is not known. Asphaltic material is reorganized in only a few places, but in these places it is abundant; in the Grants and Laguna areas, New Mexico, it carries disseminated coffinite, whereas at Temple Mountain in the San Rafael area, Utah, it carries mainly uraninite. The second form of carbonaceous material is present in almost all deposits. It consists of recognizable remains of plants, in part macerated leafy material but mostly uncrushed woody material occuring as fragments, branches, and even whole tree trunks. The plant fossils with which the ore minerals are associated are carbonized; plant fossils that were replaced by silica or carbonate minerals generally do not host ore minerals. Although some uraninite and coffinite are disseminated in sandstone without recognizable carbonaceous matter, principally they replace carbonized fossil wood, commonly preserving cell structure, and mineralize the adjacent sandstone, filling pore spaces and partly replacing sand grains. Montroseite and the copper sulfides likewise mainly replace fossil wood and the sandstone adjacent to it, but they also are disseminated in sandstone without fossil wood. The micaceous vanadium minerals mainly occupy the pore spaces in sandstone, coating the sand grains, whether or not the host sandstone contains fossil wood.

In the uranium-bearing collapsed pipes and veins, sulfides are more abundantly associated with uranium minerals. These minerals coat fracture and breccia surfaces and impregnate the porous rock adjacent to fractures. Vanadium minerals are sparse or absent in these deposits.

The vanadium silicates are virtually stable in the zone of oxidation, but montroseite and the two primary uranium minerals oxidize readily, and the contained metals are converted to a soluble higher valent state. Where sufficient vanadium is available, however, uranium is fixed almost in place as carnotite $[K_2(UO_2)_2V_2O_7\cdot3H_2O]$ or tyuyamunite $[Ca-(UO_2)_2V_2O_7\cdot5-8\frac{1}{2}H_2O]$, both of which are quite stable in the zone of oxidation. Hence, in the vanadium-rich uranium deposits, the mineralogy in the zone of oxidation is simple,

and there is little or no migration of either uranium or vanadium to cause enrichment or impoverishment of the deposit at and near the surface. On the other hand, in deposits that are low in vanadium, the uranium is mobilized, or it forms a variety of secondary minerals, some of which are not very stable themselves, so there is apt to be migration of uranium and impoverishment at the outcrop, perhaps with some secondary enrichment at some place behind the outcrop.

## Ore Bodies and Their Localization

The peneconcordant uranium deposits of the Colorado Plateau region are rather uniform in grade and consistent in habits, though they vary greatly in size. The ore produced has averaged about 0.25 per cent $U_3O_8$, and most mines yield ore of about this grade. The edges of most ore bodies are defined by a rather abrupt drop in grade; layers of ore-grade material too thin to mine commonly extend beyond the stope walls, but large halos of low-grade material are not common. Typically the deposits are tabular or lenticular layers that are nearly concordant to bedding of the host rock but that do not follow the bedding in detail. In places, two or three individual ore layers, separated by several feet of barren sandstone, partly overlap. Some ore layers split into two or more partly overlapping tongues. The ore layers average only a few feet thick, but they range in thickness from a few inches to a few tens of feet. The ore bodies range in size from small masses several feet across, containing only a few tons of ore-bearing rock, to those hundreds of feet across, containing a million or more tons of ore. Some ore bodies are roundish in plan, but most are elongate, many conspicuously so; nearby bodies are generally elongate in the same direction. Elongate pods of ore thicker and richer than average occur in many ore bodies, especially in vanadium-rich ores. The term "rolls" is often applied to these pods because of their well-defined and smoothly rounded surfaces. Generally the rolls in a single ore body trend parallel to the long axis of that body.

Stratigraphic, lithologic, and sedimentary structural factors are significant in the localization of these deposits—most of the deposits occur in a few stratigraphic zones, most are in rather porous sandstone that contains carbonaceous material, and most are in the thicker central parts of sandstone lenses and are generally elongate parallel to the courses of the streams that deposited these lenses. In contrast, igneous activity, hydrothermal mineralization, and tectonic structure show no similarly consistent influence on localization of deposits. Some peneconcordant deposits occur near intrusive or extrusive igneous rocks, but they do not differ in character or size from those many miles remote from known igneous activity. Deposits of typically hydrothermal character are sparse in the Plateau region, but there are some veins that carry base and precious metals, mostly copper and silver. Where these veins are close to peneconcordant uranium deposits, some geologists have suggested a genetic association whereas others assume that two types of mineralization happen to be superposed. Some of the deposits shown on Figure 1 certainly are along the flanks of some of the major flexures in the region, and many other deposits are similarly distributed along structures too small to be shown on this map, but this spatial association may be due to a happenstance of erosion, exposure, and discovery of deposits at and near outcrops rather than to a genetically controlling relationship. Certainly, in recent years, subsurface exploration has found many deposits remote from outcrops and in all types of structural environments. Not many through-going fractures in the region are mineralized, and most faults that cut peneconcordant ore bodies merely displace them. In some places near typical ore bodies, however, secondary ore minerals accumulate along fault planes or in the sandstone adjacent to them. Highly productive ore bodies of this type have pronounced vertical and horizontal extensions along some fracture zones in the Grants area, New Mexico. These bodies are called "stacked" or "postfault" bodies in contrast to the term "prefault" ore applied to the adjacent typical peneconcordant ore bodies (23) that have been even more productive.

Several collapsed pipes in the Colorado Plateau region are uranium-bearing (Figure 1), but only the Woodrow pipe (29) in the Laguna area, New Mexico, and the Orphan Lode (24, p. A7-A12) in the Grand Canyon area, Arizona, have been productive. The best ore is along the outer fracture, or ring fault, of each structure, but ore also occurs in fractured rock within the pipes, especially where the fractured rock is porous enough to be impregnated by uranium minerals.

The few uranium-bearing vein deposits in the Plateau have yielded insignificant amounts of ore. These deposits occur along faults of small displacement, and the uranium minerals coat and impregnate the fractured rock.

## Rock Alteration

Most of the peneconcordant uranium deposits in the Plateau region are surrounded by masses of slightly altered sandstone and mudstone that extend a few hundred to a few thousand feet beyond ore. Unaltered sandstone is pale red due to hematite paint, whereas the altered sandstone is pale gray and contains finely disseminated pyrite below the zone of oxidation but is speckled pale yellow-brown at and near the surface due to the weathering of the pyrite. Mudstone beds associated with unaltered sandstone are dominantly red, but those in contact with altered sandstone are greenish-gray and commonly contain finely disseminated pyrite. The alteration probably reduced ferric iron to ferrous. Keller (18) found no differences in clay mineralogy between unaltered and altered rock near many deposits, but Austin (27) reports widespread development of kaolinite in the Grants area, New Mexico. The nature of the altering solutions and the paragenetic relations between alteration and ore deposition have not been clearly established. Nevertheless, because the altered rock surrounds many deposits and is easily recognized, it has proved to be a useful tool in exploration (3, 4).

More intense alteration, mainly argillic, has been described at several places in the region by Professor Kerr and his students (13, 26, 30), who relate this type of alteration to the effects of hydrothermal solutions.

## Age of Mineralization

On the basis of geologic evidence alone, the peneconcordant uranium deposits can be dated only vaguely. Because the ore-bearing layers do not follow the bedding in detail, the ore minerals obviously were introduced at some time after the host sandstone beds accumulated. Because most or all of the faults that cut peneconcordant ore bodies seemingly displace them, the ore probably was emplaced before faulting. Although not many faults in the Plateau region can be dated accurately, on the basis of the geologic history of the region it is reasonable to assume that not much faulting occurred before Laramide time; many faults, however, may be younger. If the major flexures that formed during Laramide time did not influence the localization of deposits—as suggested above, but for which there is no positive evidence—then the deposits probably were emplaced before Laramide deformation. The replacement by ore minerals of uncrushed cell structures in plant fossils suggests mineralization before deep burial, either by the ore minerals themselves or by some mineral that was later replaced by the ore minerals.

Isotopic dating of the ore has not yielded results that appear to be more definitive than the geologic evidence alone; in fact, the possible age range can be even greater, depending upon what interpretations are made. Most isotopic age determinations are strikingly discordant. $Pb^{207}:Pb^{206}$ ratios give impossible ages—almost all ages are older than the ore-bearing rocks, most are Precambrian. Calculated ages from $Pb^{206}:U^{238}$ and $Pb^{207}:U^{235}$ ratios are commonly fairly consistent for individual samples, but they differ so much among samples from a single district, or even between samples from a single mine, that the selection of an age depends upon the interpretations applied. Stieff *et al.* (6) interpret $Pb^{206}:U$ ratios to give a mean age of 73 m.y. for 4 samples from the Chinle Formation (range, 65–80 m.y.) and a mean of 72 m.y. for 34 samples from the Morrison (range 15–140 m.y.); they prefer an age of mineralization about 10 m.y. younger than the calculated mean. These ages would coincide with Laramide time. Miller and Kulp, on the other hand, have given two different interpretations. In 1958, they gave a $Pb^{206}:U^{238}$ "apparent age" range of 84 to 294 m.y. for 10 samples from the Chinle, saying (17, p. 937–938): "It is possible from the isotopic data to have all deposition occurring within the last five million years but it (sic) does not preclude other periods of deposition such as in Laramide time. The isotopic ages are apparent ages only and bear no direct relation to the time of deposition." In 1963, they interpreted the $Pb^{207}:U^{235}$ ratios from these same samples and others to mean (28, p. 627): "The time of primary uranium mineralization, as indicated by the $U^{235}-Pb^{207}$ isotopic ages, is during or soon after the formation of the enclosing rocks in Lisbon Valley, Utah, and Cameron, Arizona—i.e., about 210 m.y. ago. In the Uravan, Temple Mountain, and Monument Valley districts the deposition possibly occurred in mid-Cretaceous time represented by the highest $U^{235}-Pb^{207}$ age for each area—i.e., about 110 m.y. ago." For whatever significance the reader may apply, he should have in mind that most of the deposits in the Lisbon Valley and Cameron areas

are in the Chinle Formation, those in the Uravan area are in the Morrison, and those in the Temple Mountain and Monument Valley areas are in the Chinle.

## Origin

The peneconcordant uranium deposits have several characteristics that are genetically significant but puzzling to interpret. The deposits are widespread geographically, but most are restricted to a few stratigraphic zones containing beds of similar lithology. The ore minerals mainly impregnate sandstone and replace plant fossils; mineralization was of low intensity, was associated with mild alteration, and was virtually without gangue minerals. The ore bodies are tabular masses that are nearly parallel to bedding and that correlate in shape and location with sedimentary structures; tectonic structures and igneous and hydrothermal activity have not conspicuously influenced the localization of these bodies.

Various theories have been proposed that attempt to incorporate most or all of these factors into a likely explanation of origin and source of ore metals. Syngenesis has been proposed but never strongly supported; because the ore bodies are discordant to bedding, at least in detail, this process obviously is not applicable in its strict sense. Epigenesis is requisite but is divided into two major concepts—hypogene and supergene—relating to the nature of the ore-bearing solutions and source of the ore metals. The hypogene concept supposes that hydrothermal solutions carrying ore metals were derived by magmatic activity, probably deep-seated. These solutions would have had to enter the ore-bearing beds at various places and travel along them to sites where conditions caused precipitation of the ore minerals. In the process these solutions probably mixed with ground water and lost heat and pressure, thereby yielding low-temperature deposits that lack many features common among other types of hydrothermal deposits (9,16). Deposition during or after Laramide time is generally suggested. The supergene concept assumes the solutions were moving ground waters that leached trace amounts of metals from the ore-bearing beds and associated strata. Solutions that are weakly alkaline and moderately reducing, with a high carbonate content, are thought to be geochemically capable of carrying the ore metals in their high-valent states; precipitation of the low-valent ore minerals could be induced by the reducing action of carbonaceous material or hydrogen sulfide (25,11). Isotopic ratios of sulfur from minerals associated with the deposits range widely, suggesting a source of sulfur from hydrogen sulfide formed by anerobic bacteria rather than a magmatic source (15). Ore formation could have occurred whenever and wherever these geochemical processes were operative; the movement of solutions and bacterial activity may have been greater before deep burial of the ore-bearing beds in Cretaceous time.

Data regarding the uranium-bearing collapsed pipes in the Plateau region are insufficient to justify generalizations. Features characteristic of hydrothermal activity that are associated with the pipes in the Temple Mountain area, Utah, have been described by Kerr *et al.* (14). A possible igneous affiliation for the Woodrow pipe, Laguna area, New Mexico, is suggested by Wylie (29), whereas Hilpert and Moench (22) suggest that this pipe is an intraformational slump structure in which the ore accumulated from ground waters.

## Resources and Outlook

Reserves of indicated and inferred ore in the Colorado Plateau region totaled about 36,000,000 tons, averaging about 0.24 per cent $U_3O_8$, at the end of 1964.* This represents about 55 per cent of the U.S. total reserves of ore of this general grade. Of the Plateau total reserves, about 95 per cent are in the Morrison Formation and 4 per cent in the Chinle. The Grants and Laguna areas contain about 85 per cent of the Plateau total, the Uravan Mineral belt about 7 per cent, the Lisbon Valley area about 3½ per cent, and the Monument Valley—White Canyon areas nearly 1 per cent.

Resources in undiscovered deposits of similar character probably exceed the known reserves. Likely much of this material will be found in the vicinity of known deposits.

## REFERENCES CITED

1. Fischer, R. P., 1942, Vanadium deposits of Colorado and Utah: U.S. Geol. Surv. Bull. 936-P, p. 363–394.
2. Fischer, R. P., *et al.*, 1947, Vanadium deposits near Placerville, San Miguel County, Colo-

rado: Colo. Sci. Soc. Proc. v. 15, p. 117–134.
3. Fischer, R. P., 1949, Federal exploration for carnotite ore: Spec. Pub., Colorado Mining Association, Denver, Colo., 14 p.
4. Weir, D. B., 1952, Geologic guides to prospecting for carnotite deposits on the Colorado Plateau: U.S. Geol. Surv. Bull. 988-B, p. 15–27.
5. Proctor, P. D., 1953, Geology of the Silver Reef (Harrisburg) mining district, Washington County, Utah: Utah Geol. and Mineral. Surv. Bull. 44, 169 p.
6. Stieff, L. R., *et al.*, 1953, A preliminary determination of the age of some uranium ores of the Colorado Plateaus by the lead-uranium method: U.S. Geol. Surv. Circ. 271, 19 p.
7. Craig, L. C., *et al.*, 1955, Stratigraphy of the Morrison and related formations, Colorado Plateau region—a preliminary report: U.S. Geol. Surv. Bull. 1009-E, p. 125–168.
8. Kelley, V. C., 1955, Regional tectonics of the Colorado Plateau and relationship to origin and distribution of uranium: Univ. N. Mex. Pubs. in Geol., no. 5, 120 p.
9. McKelvey, V. E., *et al.*, 1955, Origin of uranium deposits: *in Fiftieth Anniversary Volume*, Part I, Econ. Geol., p. 464–533.
10. Bush, A. L., 1956, Vanadium-uranium deposits in the Entrada Sandstone, western San Juan Mountains, Colorado (abs.): Geol. Soc. Amer. Bull., v. 67, p. 1678.
11. Gruner, J. W., 1956, Concentration of uranium in sediments by multiple migration-accretion: Econ. Geol., v. 51, p. 495–520.
12. Hunt, C. B., 1956, Cenozoic geology of the Colorado Plateau: U.S. Geol. Surv. Prof. Paper 279, 99 p.
13. Kelley, D. R. and Kerr, P. F., 1957, Clay alteration and ore, Temple Mountain, Utah: Geol. Soc. Amer. Bull., v. 68, p. 1101–1116.
14. Kerr, P. F., *et al.*, 1957, Collapse features, Temple Mountain uranium area, Utah: Geol. Soc. Amer. Bull. v. 68, p. 933–982.
15. Jensen, M. L., 1958, Sulfur isotopes and the origin of sandstone-type uranium deposits: Econ. Geol., v. 53, p. 598–616.
16. Kerr, P. F., 1958, Uranium emplacement in the Colorado Plateau: Geol. Soc. Amer. Bull., v. 69, p. 1075–1112.
17. Miller, D. S. and Kulp, J. L., 1958, Isotopic study of some Colorado Plateau ores: Econ. Geol., v. 53, p. 937–948.
18. Keller, W. D., 1959, Clay minerals in the mudstones of the ore-bearing formations: *in Geochemistry and mineralogy of the Colorado Plateau uranium ores*. U.S. Geol. Surv. Prof. Paper 320, p. 113–119.
19. Stewart, J. H., *et al.*, 1959, Stratigraphy of Triassic and associated formations in parts of the Colorado Plateau region: U.S. Geol. Surv. Bull. 1046-Q, p. 487–576.

* Based on ore reserve estimates provided by the U.S. Atomic Energy Commission, Grand Junction Office, 1965.

20. Weeks, A. D., *et al.*, 1959, Summary of the ore mineralogy: *in Geochemistry and mineralogy of the Colorado Plateau uranium ores,* U.S. Geol. Surv. Prof. Paper 320, p. 65–79.

21. Fischer, R. P., 1960, Vanadium-uranium deposits of the Rifle Creek area, Garfield County, Colorado: U.S. Geol. Surv. Bull. 1101, 52 p.

22. Hilpert, L. S. and Moench, R. H., 1960, Uranium deposits of the southern part of the San Juan Basin, New Mexico: Econ. Geol., v. 55, p. 429–464.

23. Granger, H. C., *et al.*, 1961, Sandstone-type uranium deposits at Ambrosia Lake, New Mexico—an interim report: Econ. Geol. v. 56, p. 1179–1210.

24. Granger, H. C. and Raup, R. B., 1962, Reconnaissance study of uranium deposits in Arizona: U.S. Geol. Surv. Bull. 1147-A, p. A1-A54.

25. Hostetler, P. B. and Garrels, R. M., 1962, Transportation and precipitation of uranium at low temperatures with special reference to sandstone-type uranium deposits: Econ. Geol., v. 57, p. 137–167.

26. Abdel-Gawad, A. M. and Kerr, P. F., 1963, Alteration of Chinle siltstone and uranium emplacement, Arizona and Utah: Geol. Soc. Amer. Bull., v. 74, p. 23–46.

27. Austin, S. R., 1963, Alteration of Morrison sandstone: *in Geology and technology of the Grants uranium region,* N. Mex. Bur. Mines and Mineral Resources, Mem. 15, p. 38–44.

28. Miller, D. S. and Kulp. J. L., 1963, Isotopic evidence on the origin of the Colorado Plateau uranium ores: Geol. Soc. Amer. Bull., v. 74, p. 609–630.

29. Wylie, E. T., 1963, Geology of the Woodrow breccia pipe: *in Geology and technology of the Grants uranium region,* N. Mex. Bur. Mines and Mineral Resources, Mem. 15, p. 177–181.

30. Jacobs, M. B. and Kerr, P. F., 1965, Hydrothermal alteration along the Lisbon Valley fault zone, San Juan County, Utah: Geol. Soc. Amer. Bull., v. 76, p. 423–440.

# 36. Uranium Deposits of the Grants Region

VINCENT C. KELLEY,* DALE F. KITTEL,† PAUL E. MELANCON‡

## Contents

## Illustrations

## Table

* University of New Mexico, Albuquerque, New Mexico
† The Anaconda Company, Grants, New Mexico
‡ Homestake-Sapin Partners, Grants, New Mexico

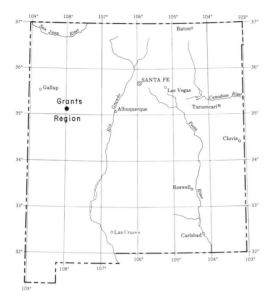

## ABSTRACT

Uranium of the Grants region occurs predominantly in continental sandstones of the upper part of the Jurassic Morrison Formation, but significant lesser deposits are found in limestone of the Jurassic Todilto Formation and in black shale of the Cretaceous Dakota Formation. The deposits are disseminations that form runs ranging from a few hundred tons to several million tons. The ore consists mainly of uraninite, uraniferous carbonaceous material, coffinite, and such secondary oxidized minerals at tyuyamunite, carnotite, and uranophane. In sandstone, the ore runs were localized by mudstone, interstitial carbonaceous material, and primary sand channel trends. In limestone, ore deposition was related to folds and fracture zones.

The great bulk of production has come from the Ambrosia Lake mines in the Grants district and the Jackpile and Paguate mines in the Laguna district. The ores, which generally have been 0.20 to 0.30 per cent $U_3O_8$, are treated at mills operated by Kerr-McGee Corporation, Homestake-Sapin Partners, and The Anaconda Company, all in the Grants district. In 14 years after discovery in 1950, the deposits had yielded concentrates valued in excess of $800,000,000.

Since deposition of the Jurassic host rocks, a geologic history involving tilting, faulting, erosion, changing ground-water environments, and oxidation has considerably influenced the existence and character of the deposits. It is evident that their origin is not simple. However, a degree of agreement appears probable for a pre-Dakota age for the original ore runs, and, at Ambrosia Lake, two principal stages of ore formation, separated by Laramide faulting, are recognized. All earlier ores have been modified by Quaternary oxidation, solution, and enrichment during which time much secondary tyuyamunite, metatyuyamunite, carnotite were formed.

## INTRODUCTION

The Grants uranium region includes a belt of deposits in Jurassic and Cretaceous sediments stretching some 85 miles along the south side of the San Juan Basin of northwestern New Mexico. The principal districts have been referred to as Gallup, Grants, and Laguna (16, p. 1). Within these are the Gallup, Churchrock, Smith Lake, Ambrosia Lake, Grants, North Laguna, and South Laguna areas. Because the Ambrosia Lake mines and the Jackpile and Paguate mines are so dominant, these names are used by some in a district sense.

The belt of districts is separated physiographically into two parts by Mount Taylor, a late Tertiary volcano rising to 11,389 feet atop a basalt-capped plateau (Figure 1). The Laguna districts, which include Anaconda's great open-pit Jackpile and Paguate mines, lie to the east of the plateau in a valley at about 6000 feet elevation.

Ambrosia Lake and other areas of mining in the Grants and Gallup districts lie to the west of Mount Taylor along a series of southward-facing cliffs, cuestas, mesas, and intervening soft-rock valleys. The Ambrosia Lake area is located in a valley at about 7000 feet, some 3 miles wide and 12 miles long, eroded in Cretaceous Mancos Shale. Production has come from some 100 mines in the Ambrosia Lake and adjoining areas. The ores are treated at mills operated by Kerr-McGee Corporations, Homestake-Sapin Partners, and The Anaconda Company.

## DISCOVERY, EXPLORATION, AND MINING

Uranium minerals have been known to occur in sedimentary outcrops of the Grants region since 1920 (16, p. 1), and were mapped in 1948 (2, p. 23), but it was not until late in 1950 after a discovery by Paddy Martinez, a Navajo Indian prospector and sheepherder, was publicized that small tonnages of uranium-bearing sandstone and limestone were

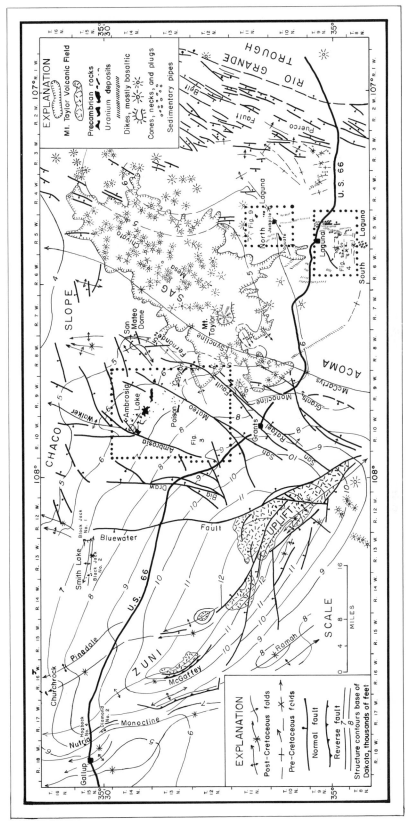

Fig. 1. *Geologic features of the Grants Region.*

mined directly from the outcrops and sold to the United States Atomic Energy Commission buying station at Monticello, Utah (Ingles M. Gay, personal communication).

This initial mining was followed by a period that was chiefly devoted to intense exploration and development. Ore bodies were developed in the Todilto Limestone and "Poison Canyon" sandstone tongue* north and west of Grants and in the Todilto Limestone and Jackpile sandstone on Laguna Pueblo lands. Methods used to explore and define the deposits of the districts included prospecting with and without radiation detectors, radiometric traverses, geobotanical sampling, airborne radiometric reconnaissance using fixed-wing aircraft and helicopters, test-pitting, trenching, rim-stripping, wagon drilling with percussion drills, diamond-core drilling, rotary noncore drilling used in conjunction with natural radioactivity and electric logging, and extensive long-hole drilling from the underground workings (18, p. 3).

Mining activities resumed in 1952 and increased sharply in 1953, reflecting the opening of the United States Atomic Energy Commission uranium ore-buying station at Bluewater, New Mexico, and the completion by The Anaconda Company of a mill designed to treat the limestone ores. Most of the ore produced during 1953 and 1954 was from open-pit operations in the Todilto Limestone and "Poison Canyon" sandstone. Daily production from these open-pit mines ranged from 300 tons per day at the Haystack mine, where crawler loading equipment was used, to 5 tons per day in small pits where the broken rock was sorted and mucked by hand.

The first underground mines in the Grants region were small operations, 20 to 200 tons per day, that were developed from adits driven into ore-bearing outcrops. Inclines and small two-compartment rectangular vertical shafts up to 300 feet in depth were used to reach ore that had been discovered behind the rims with percussion wagon drills and noncore rotary drills.

Room and pillar mining methods were used to develop and extract the ore. In most cases the ore was hand trammed from the working face to the portal or shaft. Air-powered slushers soon replaced hand mucking, and jacklegs completely replaced hand-held rock drills. A few pieces of diesel-powered crawler-type mucking, loading, and tramming equip-

* Not to be confused with the Poison Canyon Formation of southeastern Colorado, named by R. C. Hill in 1888.

ment were also used in some of these early underground operations. Stulls and open-timber cribbing provided most of the back support. A few roof bolts were installed, but it was not until the Ambrosia Lake mines opened that their use became widespread.

One of the first large-scale underground mining operations utilizing diesel equipment was conducted by The Anaconda Company to check surface drilling information and to obtain large samples from the Jackpile ore body for amenability tests. The successful completion of this program opened the way for stripping operations and mining of what was to become the world's largest open-pit uranium mine.

Prior to the opening of the Ambrosia Lake mines in 1955, ore had been mined and sold from some 60 properties, with total production from any one individual mine generally not exceeding 25,000 tons.

A wildcat drilling operation conducted by Louis Lothman was responsible for the discovery of uranium in the Ambrosia Lake area in the spring of 1955 (4, p. 26). This discovery became the Dysart No. 1 mine, the first and for many years the largest in the Ambrosia Lake area. Lothman's discovery resulted in an exploration boom that reached its peak in 1956 when some 70 drilling rigs were in operation in what is now the Ambrosia Lake area. Major ore bodies were also discovered on the Quinta properties in the Churchrock area, just north and east of Gallup, and on the Black Jack properties in the Smith Lake area. The last substantial new ore body was delineated by drilling in the northeast part of the Churchrock area in 1960.

The uranium ore discoveries from the combined exploration activities in the Grants region were a major factor in changing the position of the United States from a nation dependent upon foreign suppliers to a nation with a current surplus of known uranium reserves. These exploration activities were almost all financed with private capital provided by several entities including individuals, partnerships, hastily formed corporations, and long-established mining and oil corporations. The reserves were eventually consolidated by one means or another into groups that justified the construction of four concentrators for Ambrosia Lake ores and the building of a new section on the Anaconda mill to treat the sandstone ores developed on the Laguna Pueblo lands.

Phillips Petroleum Company and Kermac Nuclear Fuels Corporation built their mills in the heart of the Ambrosia Lake mining area in secs. 28 and 31, respectively, of T14N, R9W. Homestake-New Mexico Partners and Homestake-Sapin Partners built mills side by side on New Mexico Highway 53 some 6 miles north of U.S. Highway 66 and 18 miles south of the mining area in sec. 26, T12N, R10W.

The importance of uranium production from the Grants region is best demonstrated by the fact that through December 31, 1964, a period of less than 14 years, the gross value of concentrates produced from mills in the districts has exceeded $800,000,000. These ores have been mined from the following host rocks:

Rock bolts and heavy woven wire fencing are used widely for roof and rib support in main haulageways and development drifts and crosscuts.

The room and pillar mining method generally is used to extract ores that are less than 20 feet thick. During development, it is common practice to drill vertical holes with a jack-leg drill up and down at regular intervals along the drifts and crosscuts as well as fans of long-holes from crosscut faces that have been stopped in protore or barren rock. This supplementary drilling developed substantial tonnages of previously unknown ore above and below,

| | | |
|---|---|---|
| Morrison Sandstone (Jackpile, "Poison Canyon", and Westwater) | 22,130,000 tons @ 0.22% | $U_3O_8$ |
| Other sandstone | 150,000 tons @ 0.20% | $U_3O_8$ |
| Todilto Limestone | 974,000 tons @ 0.22% | $U_3O_8$ |
| Total: Grants districts | 23,250,000 tons @ 0.22% | $U_3O_8$ |

Of this total production 11,920,000 tons @ 0.22 per cent $U_3O_8$ has been mined from T14N, R9 and 10W (the Ambrosia Lake area).*

With the exception of the Dysart No. 1 mine, initial mining operations in the Ambrosia Lake area were plagued with problems and disappointments. The problems resulted from inexperience in mining uranium ore below the water table and were magnified by trying to accelerate operations to meet deadlines imposed by short-term concentrate selling contracts. Early disappointments were caused by improper evaluation of the drilling results and mining capabilities, the change in underground position and configuration of the ore bodies from their projected locations and shapes based on surface drill holes, the sharp decrease in mined grade from the sampled grade because of wet mining conditions, and the inability to maintain passable haulageways for track-less equipment in the wet ground. Most of the problems were overcome with experience and ingenuity.

Sublevel track haulage systems that were installed in the wet mines proved to be effective for ore transportation and predraining overlying ore bodies. The use of diesel-powered mucking and hauling equipment is now generally limited to dry mines and more or less dry areas of wet mines. In these cases, diesel equipment has proved to be most efficient.

* This information has been provided by the Grand Junction Office of the United States Atomic Energy Commission.

as well as on, the mining level. During retreat stoping, the pillars are commonly halved or quartered before being taken by slabbing. Caving is controlled by using stulls, open timber cribbing, hydraulic stulls, or concrete posts that are poured from the surface through drill holes.

Several methods have been used to recover thick ore. The area to be stoped generally is prepared by cutting the ore into pillars on one or more levels. A slot is then established, and the ore is prepared for blasting by ring drilling or longhole drilling parallel to the slot. The ore is blasted with nitrocarbonitrate and removed from the slot by slushers. Cut and fill, scrams and ring drilling, top slicing, and square-set timbering are other methods that are, or have been, used to recover the thick ores.

Back filling of stoped areas with wet or dry sandfill is common practice in Ambrosia Lake mines. Alluvial blow sand is used for the dry fill, and the wet fill is prepared from mill tailings or by classifying alluvial material. Filling stabilizes the adjoining ground for future mining, reduces the area of the mine that must be ventilated or in which ventilation must be controlled, and, in the deeper mines, prevents excessive caving that might rupture overlying aquifers and cause a sudden inflow of large quantities of water.

## STRATIGRAPHY

The sedimentary rocks of the Grants region range from Pennsylvanian to Cretaceous. Pre-

TABLE I.   *Stratigraphic Nomenclature in the Grants District*

| System | Units | Thickness (Feet) |
|---|---|---|
| | Eroded | |
| Cretaceous: | Point Lookout Formation (Hosta Sandstone) | 150+ |
| | Crevasse Canyon Formation | 800–1000 |
| | Mancos Shale | 800–1000 |
| | Dakota Sandstone (sandstone, shale, coal) | 10–150 |
| Jurassic: | Morrison Formation | 100–600 |
| | Brushy Basin Shale Member including Jackpile sandstone | 20–350 |
| | Westwater Canyon Sandstone Member including "Poison Canyon" tongue | 0–300 |
| | Recapture Shale Member | 50–175 |
| | Zuni Sandstone (Bluff tongues) | 130–400 |
| | Summerville Formation (mudstone, sandstone) | 50–220 |
| | Todilto Formation (gypsum, limestone) | 0–125 |
| | Entrada Sandstone | 150–250 |

cambrian gneiss, schist, and granite are exposed in the core of the Zuni Mountains, and overlying this in the flanks are the Pennsylvanian Magdalena Formation (125 feet), Permian Abo, Yeso, and San Andres formations (659, 900, 300 feet), and Triassic Docum Group (1600 feet) (5). Uranium deposits of consequence are almost entirely in the Jurassic, but a few small deposits have been mined from the lowermost beds of the Cretaceous. The currently prevailing stratigraphic nomenclature of the significant units in the Grants districts is shown in Table I.

Small amounts of ore have been mined from the Dakota, Recapture, Summerville, and Entrada; considerable ore has been produced from limestone of the Todilto; the great bulk, however, has been and is being produced from Westwater Canyon and Brushy Basin sandstones.

In the Laguna district, the Todilto is up to 125 feet thick and consists of 5 to 35 feet of fetid gray laminated limestone overlain by as much as 90 feet of gypsum and anhydrite. The basal limestone consists of laminated to massive beds, and the upper part is massive. The gypsum zone pinches out southward and is only locally present south of U.S. Highway 66. In the Grants and Ambrosia Lake areas, only the limestone is present, and this commonly is a tripartite unit 15 to 30 feet thick consisting of parallel laminated limestone and some siltstone at the base, a middle unit of crinkled thin limestone beds, and an upper massive, often coarsely crystalline, limestone. The upper unit is locally absent.

The basal platy zone is up to 15 feet thick, thin-bedded, and light gray to grayish-brown. Thin carbonaceous partings commonly occur along bedding planes, and recrystallization has occurred along joint planes and in folds. The medial limestone zone is up to 6 feet thick, extremely crenulated, usually recrystallized, and of the same colors as the platy part. Because of these crenulated bedding planes, this middle unit is locally referred to as the "crinkly" zone. The upper massive limestone is usually assigned to the Todilto although some assign it to a transition zone between the Todilto and Summerville formations.

The Westwater Canyon Sandstone Member is quite irregular in thickness owing to lensing and lateral gradations into Brushy Basin- or Recapture-type mudstone and claystone. The "Poison Canyon" sandstone, which is up to 85 feet thick, is a tongue of Westwater Canyon sandstone in the Brushy Basin Member. The Westwater averages about 150 feet in thickness in the Ambrosia Lake area, but locally it may reach 300 feet or thin and disappear into zones of arenaceous mudstone. Its sandstone beds vary from parallel-bedded to irregularly cross-bedded. The color may be light-gray, yellow-brown, red-brown, or grayish-black, depending on proximity to the surface, mineralization, or content of organic material. The sand is fine to coarse, poorly sorted, and locally conglomeratic. The composition is feldspathic with considerable sanidine (9, p. 41). Locally, the quartz fraction contains slightly corroded, high-temperature, bipyramidal phenocrysts. In general, the composition indicates an acidic tuffaceous derivation for many of the beds.

The Brushy Basin resembles the Recapture and consists of greenish-gray and reddish-brown mudstones with numerous Westwater-type sandstone lenses, channel fills, or arenaceous zones. It is as much as 500 feet thick, but may be much less, owing partly to intertonguing with the underlying Westwater Canyon and Zuni sandstone tongues but principally to plication at the extensive pre-Dakota erosion surface.

In the Laguna district, the Westwater and Recapture units are thought to have thinned markedly and are both relegated by Hilpert (14, p. 14–15) to about the lower 100 feet of the Morrison (Figure 2). On the other hand, the overlying Brushy Basin is thought to have thickened to about 500 feet (6, p. 435). In the vicinity of the Jackpile mine, as much as 220 feet of the uppermost part of this thickness consists of a large sandstone channel deposit (Jackpile sandstone) preserved in a broad pre-Dakota structural downwarp that was more or less parallel to the northeasterly depositional trend.

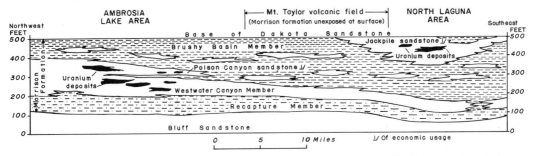

FIG. 2. *Stratigraphy of the Grants Region, showing Morrison stratigraphy between the Ambrosia Lake and North Laguna areas. (Figure 8 shows details of ore zones in the Ambrosia Lake area.)*

## IGNEOUS ROCKS

A variety of shallow intrusive rocks and rather extensive basaltic flows constitute the principal igneous rocks. Diabase dikes and sills are common in the Laguna district; most of these are less than 10 feet wide, but a few that branch from plugs are several tens of feet wide. Dike trends are principally northerly and northwesterly; the maximum known length is about 10 miles. Sills are found in many stratigraphic horizons in the Laguna district. The most notable of them is a sheet as much as 75 feet thick in the Entrada-Summerville interval east and southeast of Laguna. This sill metamorphosed the uranium deposits and locally produced a vanadium-rich garnet (19, p. 161).

Numerous basaltic necks occur in or adjacent to the Mount Taylor volcanic field. Most of these fed volcanos, and several sections of partly dissected cones and feeder necks are exposed along the edge of Mesa Chivato and lesser volcanic-capped mesas. These necks may consist of solid lava, lava breccia, or lava-sedimentary mixed breccias.

About 7 miles northeast of Grants, in what is referred to as East Grants Ridge, there is an elliptically shaped dome 1.0 to 1.5 miles in diameter. The central part is aphanitic, lithophysaical, flow-banded rhyolite surrounded by a peripheral-chilled sheath of obsidian and perlite.

## STRUCTURE

The principal regional structures of the area are the Zuni uplift and the Acoma sag. The Zuni uplift is a broad northwest-trending upwarp asymmetrical to the southwest, and the Acoma sag is a broad, flat, little deformed downwarp that slopes very gently northward between the Zuni uplift on the west and the Lucero uplift and Puerco fault belt margins to the Rio Grande trough (Figure 1). McCartys syncline near the margin of the Zuni uplift forms the axis of the sag. Deposits at Ambrosia Lake lie along the northeastern corner of the Zuni uplift in the Chaco slope into the San Juan Basin. The Laguna deposits lie along the eastern side of the Acoma sag. The eastern and northeastern parts of the Zuni uplift are broken by numerous faults. One, a westerly to northwesterly trending system, considerably modifies the crest and southwesterly limb of the uplift. Another, forming a rough fan between north and northeast, modifies the northeastern sector of the uplift and affects to some extent the distribution, if not the localization, of the Ambrosia Lake and Grants deposits. Mapping by Thaden and Santos (23, p. 20) has more clearly delineated the pattern of these faults and in particular the San Rafael fault zone and the Grants monocline. The irregular structural declivity formed by the Grants monocline, San Rafael fault zone, and the Fernandez monocline lets the Acoma sag abruptly down 1000 to 1500 feet against the Zuni block.

The Zuni uplift and McCartys syncline, together with the associated major faults and folds such as the Ambrosia anticline, are considered to be Laramide. The Puerco fault belt is probably largely Pliocene, and its development considerably modified the eastern side of the Acoma sag during subsidence of the Rio Grande trough. Some faults that dislocate the volcanics capping Mesa Chivato around Mount Taylor may be late Pliocene or early Pleistocene. There is little evidence that any of the Laramide or younger structures served as localizers of the principal ore bodies.

Structures of most importance to ore localization consist of (1) fractures, (2) folds of the Todilto Limestone, (3) broad gentle anticlines, synclines, and bends of post-Morrison-pre-Dakota age, and (4) small subsidence pipes in the Jurassic beds.

Both tectonic and gravity flow folds are present in the Todilto Limestone. Together with fractures, they seem to have localized ore deposits in numerous places. The tectonic folds affect the adjacent formations and are in general larger than the minor disharmonic flow folds. As a consequence, somewhat larger deposits are found associated with the tectonic folds. The flow folds are intraformational and, in many places, range down to minute crenulations. In the Ambrosia Lake and South Laguna areas, most of the folds trend either northerly or westerly, and commonly follow the flanks and troughs of the pre-Dakota folds. A second group of folds is late Jurassic or early Cretaceous and probably is related to northward regional tilt of a broad east-west upwarp which extended across central New Mexico and Arizona during and following Morrison time. The folds of this disturbance are best known and displayed in the Acoma sag, although their presence is suggested in the Ambrosia Lake area.

In the Laguna district, there are two principal groups; (1) westerly trending curved warps as much as 5 to 10 miles long, 1 to 2 miles wide, and a few hundred feet in amplitude and (2) a north-northwest trending lesser group 2 to 3 miles long, 0 5 miles wide, and up to 100 feet in relief. The former set is most common south and southwest of Laguna and just beyond the southern margin of Figure 1. The lesser group lies around the Jackpile mine and some 10 miles to the southeast (6, p. 439). The Laramide Ambrosia anticline is thought also by some to show subsurface evidence of slight pre-Dakota warping.

About 300 collapse structures in the Bluff and Summerville formations have been mapped in the South Laguna area. They range from a few inches to 200 feet in diameter and from a few feet to possibly 300 feet in height. None of these South Laguna structures is known to penetrate into either the Morrison or the Todilto.

## OCCURRENCES

Host rocks for the ore bodies in the Grants districts are limestone, sandstone, and carbonaceous shale, all of continental origin. The deposits in the limestone are replacement, disseminations, and fracture coatings that form tabular to elongate bodies ranging from a few hundred to more than 200,000 tons. The deposits in the sandstone are grain coatings, in-terstitial fillings, replacements, and fracture coatings that form runs (1, p. 153) ranging from a few hundred to several million tons. The deposits in the Dakota carbonaceous shale are submicroscopic disseminations that form thin runs of only a few thousand tons. Although about 20 uranium minerals have been identified, coffinite, $U(SiO_4)_{1-x}$ $(OH)_x$; uraninite, $UO_2$; tyuyamunite, $Ca(UO_2)_2(VO_4)_2 \cdot$ $5\text{-}8H_2O$; and uraniferous carbonaceous material are predominant.

## Limestone Deposits

Discoveries in the Todilto Limestone resulted in well over 100 workable uranium deposits of different sizes, all occurring within the major belt of uranium mineralization that borders the southern edge of the San Juan Basin (Figure 1). Although these occurrences have provided the only appreciable production from a limestone host in the United States, it is insignificant when compared to that from the Morrison sandstones. By the end of 1962, the limestone production amounted to less than one per cent of the total from the Grants region.

The Todilto uranium deposits lie principally in two areas; those along the Todilto bench from the Haystack mine area to the F-33 mine (Figure 3), and those in the South Laguna area. A few small deposits have been also mined in an area a short distance east of Grants and 3 or 4 miles south of the F-33 mine.

The Todilto bench is transected by three fault sets; north-south, east-west, and N20°E. Displacements of the F-33 deposit indicate that the N20°E set is postore.

Folds in the Todilto are both harmonic and disharmonic with respect to the adjacent beds. The harmonic folds are part of the regional structure, and their distribution is incompletely known. The disharmonic intraformational folds are small, irregular, and may be confined to any unit of the Todilto or lower Summerville. They range from gentle arches to tight and recumbent forms. Considerable fracturing and occasional brecciation are associated with these intraformational folds. Both types of folds have consistently localized uranium deposits.

Very few of the Todilto uranium deposits have yielded more than 5000 tons of ore. An exception, which produced more than 100,000 tons of ore before depletion in 1961, was the Haystack Mountain Development Company's open-pit mine in sec. 19, T13N, R10

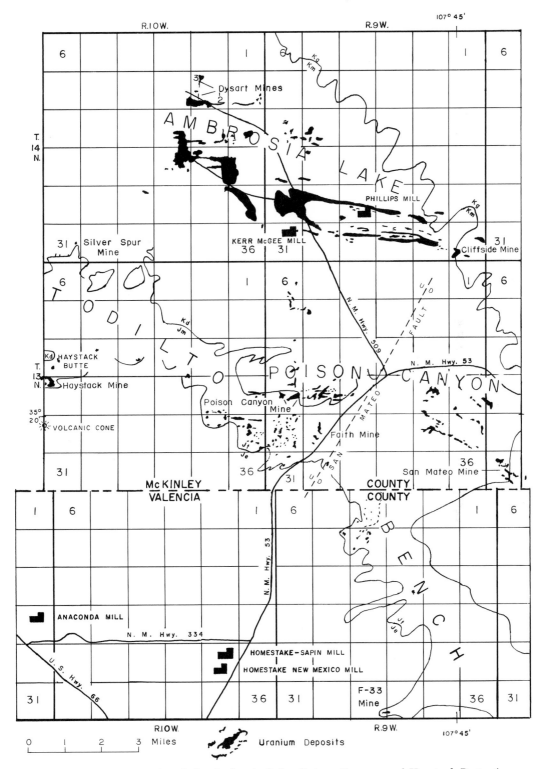

FIG. 3. *Uranium Deposits of the Ambrosia Lake, Poison Canyon, and Haystack Butte Areas.*

W. It was in a northwest-trending ore body about 1150 feet long and 100 to 500 feet wide. The deposit was in a harmonic northwesterly plunging syncline, and the ore occurred chiefly in the platy member of the Todilto as disseminated yellow secondary uranium vanadates.

Another large deposit in secs. 33 and 34, T12N, R9W has been worked from the underground F-33 mine. The ore body trends N70°E for a developed distance of about 1500 feet with an average width of about 70 feet. The massive unit forms a bulbous lens immediately above the crinkly zone and is up to 20 feet thick (Perry West, personal communication). Most of the uranium is concentrated in intraformational folds in this thickened lens of the massive unit. Primary uraninite in blebs and disseminations constitutes nearly all the commercial ore in the F-33 mine. With it are very minor amounts of pyrite, fluorite, and barite. Some areas in the mine contain numerous solution channels and cavities in which the uranium vanadates, tyuyamunite, $Ca(UO_2)_2(VO_4)_2 \cdot 5\text{-}8H_2O$ and carnotite, $K_2(UO_2)_2(VO_4)_2 \cdot 3\text{-}5H_2O$; the vanadium oxides, haggite, $V_2O_3 \cdot V_2O_4 \cdot 3H_2O$ and paramontroseite, $V_2O_4$; and the rare calcium vanadate, metahewettite have been deposited. Substantial reserves of uranium ore remain at the F-33 mine, where underground operations were suspended in 1959 because of the curtailment of the uranium market.

The Faith mine in sec. 29, T13N, R9W has also produced more than 50,000 tons of uranium ore. Its shaft was sunk for Food Machinery Corporation to below the 450-foot level, where the ore was mined from a series of about 30 disconnected ore shoots that ranged up to 250 feet long, 60 feet wide, and 15 feet thick. Mining was carried out until 1963, when the ore was depleted. Most of the ore shoots were oriented north-south along intraformational folds, but an abrupt change to an east-west orientation occurs in the extreme southern part of the trend. The largest ore shoot occurred on a broad intraformational fold upon which were several superimposed closed folds. The major uranium mineral was uraninite, and with it were gummite, fluorite, barite, pyrite, calcite, iron oxides, and manganese oxides.

Section 25, T13N, R10W, and section 30, T13N, R9W, together contain approximately one hundred uranium deposits that have produced more than two million pounds of $U_3O_8$ from both open-pit and underground operations. Although the majority of the deposits are depleted, some of them were still being mined at the end of 1965. The indi-vidual deposits differed considerably in size and shape. For the most part, they comprised a few hundred to a few thousand tons of randomly oriented localizations of disseminated uraninite distributed in any or all of the limestone units. In the northern part of section 25, where some quite large ore bodies occurred, an easterly trend is evident. In section 30, many of the large ore bodies were oriented north-south. Apparent structural ore controls were chiefly intraformational folds and related joints. The supergene ore minerals included tyuyamunite, carnotite, gummite and uranophane, and calcite, pyrite, hematite, fluorite, and barite as gangue.

Another area from which more than 50,000 tons of Todilto ore were produced is in secs. 4 and 9, T12N, R9W, where 50 small deposits, ranging in size up to 2000 tons, were mined by open-pit methods and from adits driven into the walls of the pits. Control of the deposits by folds was often evident during mining. The ore occurred in all Todilto units and consisted mostly of uraninite in association with minor amounts of yellow uranium vanadates and uranophane, and with barite, calcite, fluorite, and iron oxides. Locally fluorite massively replaced the limestone.

The South Laguna area, about 30 miles east-southeast of Grants, comprises T8N R4W; T8 and 9N, R5 and 6W; and the eastern part of T8 and 9, R7W, (Figure 4), all of which is Laguna Pueblo land.

The dominant structural features in the normally flat-lying beds of the South Laguna area are a series of broad and gentle pre-Dakota folds which trend both westerly and northerly across the general area (25). Innumerable small harmonic and intraformational Todilto folds are inconsistently oriented. In conjunction with joints, they often exerted significant controls on the emplacement of the uranium mineralization. However, numerous intraformational folds are barren of uranium mineralization. Only a few faults of much displacement are known to occur in the South Laguna area. Most of them are in the eastern half of T9N, R5W, and are oriented generally north-south. No relationship has been noted between faults and uranium mineralization.

Extending for about 15 miles from the southwest corner of the area in a northeasterly direction is an outcrop zone of Todilto Limestone in which numerous radiometric anomalies were discovered in the mid-1950's by scintillation-equipment surveys and conventional foot prospecting. Although much close-spaced shallow drilling was done in the anomalous areas by The Anaconda Company, only

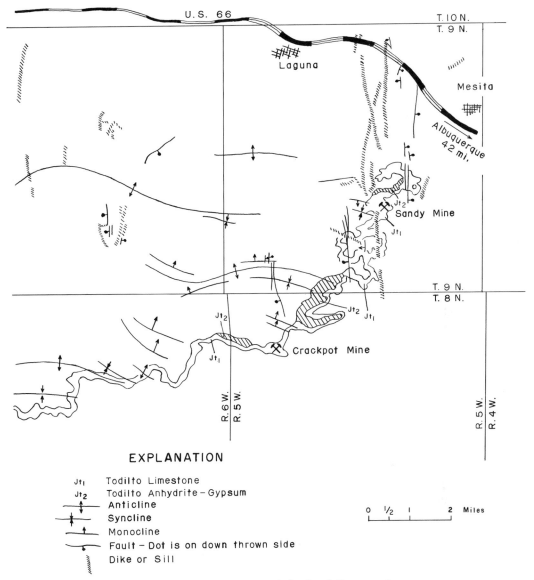

FIG. 4. *Uranium Deposits of the South Laguna Area.*

EXPLANATION

| | |
|---|---|
| $Jt_1$ | Todilto Limestone |
| $Jt_2$ | Todilto Anhydrite–Gypsum |
| | Anticline |
| | Syncline |
| | Monocline |
| | Fault – Dot is on down thrown side |
| | Dike or Sill |

0    1/2    1    2  Miles

two anomalous areas were sufficiently mineral-ized to warrant mining operations. One was the Crackpot mine near the center of the northwest quarter of sec. 8, T8N, R5W; the other was the Sandy mine in the north-central part of sec. 27, T9N, R5W.

Nearly all Crackpot ore occurred in the crinkly and platy zones of the Todilto; the upper massive unit was virtually barren of ura-nium. Intraformational folds were the chief ore control. The minerals consisted chiefly of uraninite and yellow uranium vanadates that were irregularly disseminated in a small fold.

Accessory minerals other than calcite and limonite were not in evidence.

The Sandy deposit consisted of several small pods in Todilto Limestone and upper bleached Entrada Sandstone on a gentle south-facing monocline. Several small harmonic and intra-formational folds provided some ore control, but the ore cutting across the tilted Todilto-Entrada contact indicated that folds were not the only control. The ore was similar mineral-ogically to that in the Crackpot deposit. Less than 5000 tons of ore were produced from the Sandy and Crackpot open-pit mines, and no ore has been produced from either mine since 1955.

## Sandstone Deposits

Uranium deposits have been found in the Entrada Sandstone, Morrison Formation, and the Dakota Sandstone. The deposits in the Morrison are among the largest and most continuous known. Those in the Entrada and Dakota are small, isolated, and widely scattered.

ENTRADA SANDSTONE  Deposits in the Entrada occur just below the Todilto Limestone contact in the bleached part of the upper sandy member. This member is 80 to 250 feet thick, reddish orange to light gray, moderately well-cemented and well-sorted fine- to medium-grained quartz sandstone. The upper 5 to 30 feet is generally bleached to light gray; this is thought to be the result of weathering of the pre-Todilto surface.

The principal Entrada uranium deposits were at the Haystack mine, sec. 19 T13N, R10W; the Zia mine sec. 15, T12N, R9W; and the Sandy mine in sec 27, T9N, R5W, in the the South Laguna area. These and all other known deposits in the districts appear to be supergene occurrences derived from overlying Todilto ore bodies. Field observations of the present terrain indicate that the overlying impermeable formations were eroded away making it possible for meteoric waters to reach the limestone. Tension fractures associated with ore-localizing intraformational folds provided the permeability necessary to allow the waters to leach the uranium from the limestone. The uranium was reprecipitated in the Entrada in the form of uraninite, coffinite, and tyuyamunite. The resulting small ore bodies were slightly elliptical in shape.

MORRISON FORMATION  More than 95 per cent of the uranium ore produced in the Grants region has been mined from the Westwater Canyon and Brushy Basin members. The Westwater Canyon is the principal host rock west of Mount Taylor. This part of the district includes the Ambrosia Lake, Smith Lake, Churchrock, and Poison Canyon areas (Figure 5). In the Laguna district, the Brushy Basin member including the Jackpile sandstone is the principal host.

Special terms are in use locally to describe recognizable variations in the occurrences of black Westwater deposits west of Mount Taylor.

| Older Ores | Younger Ores |
|---|---|
| prefault | postfault |
| trend | redistributed |
| primary | secondary |
| roll | stack |

Some terms apply to both younger and older ores. The term stack is intended to refer to a thick postfault, redistributed occurrence, but the superposition of two or more runs has the appearance of, and is sometimes erroneously termed, a stack. Redistribution may be only lateral and not form a stack.

Prefault or trend ore bodies are properly termed runs. They usually are elongate in a west-northwesterly direction. In the lower Westwater, they are generally a part of a larger, elongate, thin layer of mineralization in which the grade, thickness, and width of mineralization have been observed to increase at predictable intervals into runs.

In places, the ore changes elevation abruptly, crosses the bedding, and thickens to form a roll (11, p. 67). It has been observed that when this change in elevation occurs across the west-northwesterly trend direction, the higher elevation is consistently to the north and up-section (Figure 6).

In the upper Westwater, the mineralization is not as widespread, and the deposit generally forms a single wide run that is lenticular in cross section and may have small satellitic prefault ore bodies along its north and south edges. Here again the northern ore bodies have a higher stratigraphic elevation than those to the south.

A direct relationship has not been established between prefault ores and specific structural features. Faults observed in the older ore bodies throughout the district indicate postore movement inasmuch as the ore and bedding planes are displaced the same amount (8, p. 1191). An indirect relationship exists between the uranium deposits and regional deformation that controlled the original linear features in the Morrison sandstone host beds.

Redistributed or postfault uranium deposits generally are associated with, and quite often engulf, an older prefault deposit. These deposits tend to be more equidimensional horizontally and considerably thicker (up to 100 feet) than the prefault occurrences. Their direction of elongation is commonly related to an increase in fracture density along which the redistribution has taken place. Changes in grain size and sorting, the presence or absence

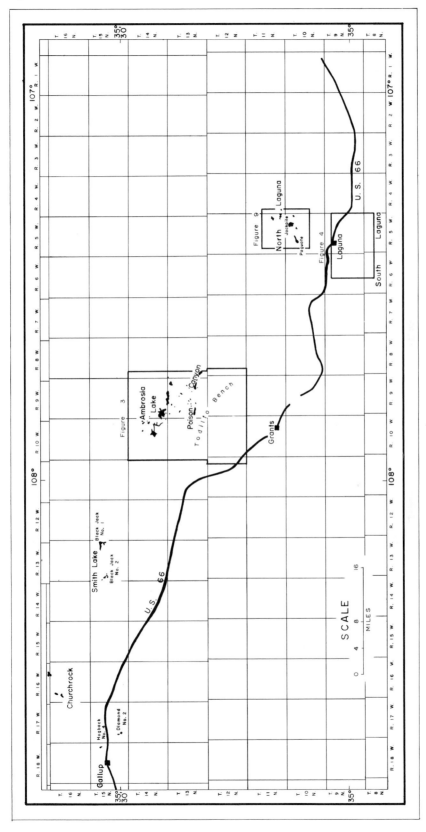

FIG. 5. *Spatial Relationships of the Areas and Deposits in the Grants Region.*

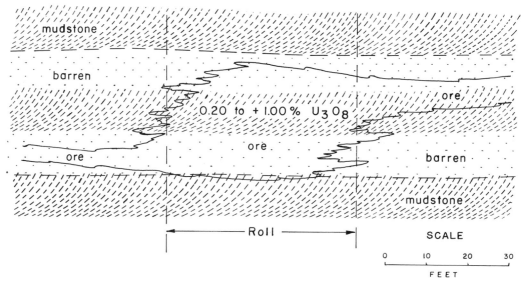

FIG. 6. *A Typical Ore Roll in the Ambrosia Lake Area.*

of intraformational mudstone layers, the presence or absence of gouge in fault zones, and the number of generations of fracturing are all features that have influenced the shape, position, and grade of the redistributed deposits (Metzger, oral communication).

The redistribution process, which has apparently been in progress since the earliest Laramide tectonic activity, changed the hydrodynamics and set formation waters in motion. Redistribution is accomplished by circulation of aerated ground water through an area of older mineralization. These solutions change the uranium from a plus four to a plus six valance state that can form complexes with bicarbonate and sulfate ions present in the waters. These uranium-rich solutions proceed down dip until their oxidizing capability lessens to the point where precipitation takes place in a favorable environment. Oxidation, solution, and reprecipitation are well illustrated by the presence of high-grade redistributed ore in a zone along the contact between oxidized and unoxidized sandstone. From this contact zone the grade decreases and finally feathers out into the unoxidized sandstone. Remnants of prefault ore that may have provided the uranium for redistribution have been observed in oxidized sandstone through which redistributing solutions have passed. Figure 7 shows a stack that occurs in Homestake-Sapin Partners' Section 23 mine. This is a striking example of redistribution along the receding interface between the oxidized and unoxidized environments. Here, the bleached remnants of prefault ore are shown down dip because they are on the south limb of the Ambrosia dome. However, they are toward the regionally updip source of the moving ground waters. Elsewhere in the Ambrosia Lake area, these remnants are normally found updip from the stacks. While the interface zone may or may not be enriched, the grade gradually decreases to the north, and the stack splits into tabular deposits of low-grade ore.

The uranium minerals that make up the deposits have been chemically precipitated as sand grain coatings, interstitial cement, fracture fillings, and replacements of carbonaceous material. The minute grain size makes megascopic mineral identification nearly impossible and microscopic study difficult. The most reliable identifications have been made using X-ray equipment.

Coffinite, uraninite, and uraniferous carbonaceous material are the principal uranium-bearing materials found throughout the districts in the unoxidized deposits. Granger (12, p. 22) describes the carbonaceous material as . . . "an authigenic organic matter that seems to have been introduced in a fluid state and has remained as a precipitate or residue to form grain coatings, interstitial cement, and fracture fillings." It is coextensive with prefault ore and seems to be the matrix or gangue in which many of the prefault ore metals occur. The carbonaceous material is generally uraniferous and contains coffinite, but where oxidized, as in shallow deposits exposed at the surface, the coffinite may be destroyed, and the uranium leached. Coalified fossil trunks, limbs, and fragments of wood and grasses may also be ore-bearing, but they are scarce relative to the vast quantities of authigenic carbonaceous material.

The most common secondary minerals are tyuyamunite, metatuyamunite, $Ca(UO_2)_2$-$(VO_4)_2 \cdot 3\text{-}5H_2O$, and carnotite, Zippeite, $2UO_3 \cdot SO_3 \cdot 5H_2O$, andersonite, $Na_2Ca(UO_2)(CO_3)_3 \cdot 6H_2O$, and bayleyite, $Mg_2(UO_2)(CO_2)_2 \cdot 18H_2O$, are commonly found as postmine efflorescent deposits on the mine walls.

Commonly occurring interstitial gangue minerals are pyrite, marcasite, calcite, jordisite, ilsemannite, and ferroselite ($FeSe_2$). Native selenium, barite, calcite, and pyrite occur as fracture fillings, and pascoite is a common postmine efflorescent occurrence.

The most significant deposits in the Westwater Canyon Member are found in T14N, R9 and 10W, the Ambrosia Lake area. The minable uranium ore bodies, those containing more than 2 pounds of $U_3O_8$ per ton, form a nearly continuous west-northwesterly striking deposit that extends some 8 miles from the southeast corner of T14N, R9W to the northeast quarter of R10W. Along the southern edge of the Ambrosia Lake trend these

more or less stratified deposits occur throughout the entire thickness of the member, including the "Poison Canyon" tongue. Santos (21, p. 55) noted that this stratigraphic range gradually lessens to the north as follows: "Through the center of the belt, ore deposits occur from near the middle to the top of the Westwater Canyon, and along the northern margin of the belt, they occur at the top only."

The subdivision of the 250-foot thick Westwater into distinguishable sandstone units that are separated by discontinuous mudstone layers and beds has been a practical necessity for correlating the ore bodies (Figure 8). These ore-bearing units range in thickness from 12 to 60 feet (13, p. 102). The ore ranges in thickness from less than 2 feet in runs to more than 100 feet in redistributed stacks.

The localization of the uranium deposits within the various sandstone units occurred in at least two phases. The first deposition occurred in reducing environments created when areas of dense vegetation were rapidly buried and later converted into carbonaceous material. During this time, sedimentary features controlling porosity and permeability had their most important effect. Interstitial and interbedded mudstones controlled the flow of and trapped the carbonaceous fluids and uranium-bearing solutions. Large areas of unoxidized clean sandstone that contain little

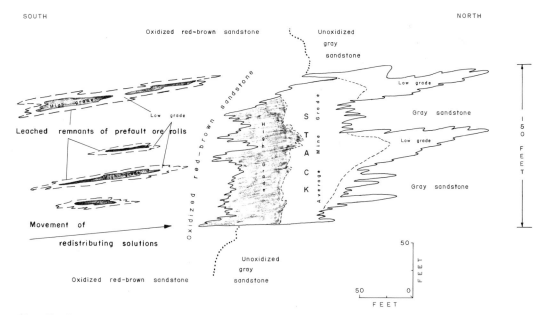

FIG. 7. *Cross section, showing the relationship between leached remnants of prefault runs and redistributed stack on the south flank of the Ambrosia dome.*

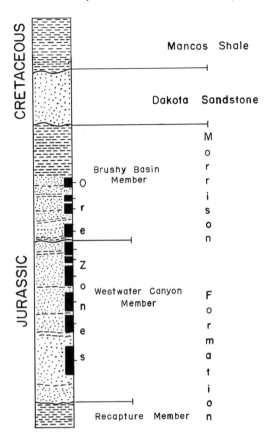

FIG. 8. *Ore Zones of the Westwater Canyon and Brushy Basin Members in the Ambrosia Lake Area.*

interstitial mudstone are barren even though natural radioactivity logs suggest that uranium-bearing solutions probably passed through the sandstones.

The importance of carbonaceous material as an ore localizer and the sedimentary features that controlled ore emplacement are well illustrated by the "A" and "B" ore bodies in the Section 30 mine. The "A" ore body and sandstone are at the top of the Westwater; they overlie and are separated from the "B" ore body and sandstone by the "K" shale.

"Deposition of the "A" sand was interrupted by intermittent folding and faulting which disturbed the surface enough to cause erosion through the "K" shale. . . ."

\*\*\*

"Fossil vegetal matter is abundant in the "A" sandstone and is sparse in the "B" sandstone. Some areas containing vegetal matter are mineralized whereas other areas are barren. However, the ore is always associated with carbonaceous

material. The "B" sandstone did not contain enough vegetal matter to produce the amount of carbonaceous material present." (10, p. 78).

The additional carbonaceous material moved from the "A" sandstone into the "B" sandstone through the hole eroded in the "K" shale. This abundant carbonaceous material and its controlled distribution were responsible for the localization and superimposition of two of the largest prefault deposits in the Ambrosia Lake area.

More typical trend or prefault ore bodies occur in the lower Westwater sandstones in secs. 10, 11, 14, 15, 22, 23, and 25, T14N, R10W, and in secs. 29, 30, 32, 33, 34, 35, and 36, R9W. The elongation of these deposits is more or less parallel to the direction of sedimentation and roughly coincident with thickening of the lower sandstone units. However, evidence of their control by specific sandstone lenses or channels is obscure and inconclusive. Recognizable carbonaceous trash (stems, twigs, leaves, etc.) is rare in these deposits. In contrast the major prefault ore bodies found in the upper Westwater in secs. 17, 18, 20, 26, 27, 28, 30, and 34, T14N, R9W, appear to be localized within specific sandstone lenses or channels that occur in the thicker parts of the upper Westwater units.. Carbonaceous trash is common in these deposits.

Molybdenum, in the mineral jordisite $(MoS_2)$, is a common accessory in both the upper and lower deposits where it is found as a halo or fringe on the edges as well as within the ore (22, p. 100). The mineral montroseite is also an included accessory mineral.

The second and possible succeeding phases of localization occurred when the uranium in older prefault runs was redistributed by moving ground waters. The effectiveness of the redistribution process along zones of increased permeability is well demonstrated in the Poison Canyon trend where prefault uranium deposits in the San Mateo fault zone and associated syncline have been completely removed. This uranium may or may not have been reprecipitated. However, it may be significant that two of the highest-grade deposits in the Grants district, the Hogan and the Cliffside, are down dip from the area of removal (20, p. 131).

The permeability that allowed deposits to be destroyed was also necessary for the formation of the characteristic Ambrosia Lake stack ore bodies. Examples of stack ore bodies that have been localized in fault zones and areas of increased fracture density are found in secs. 10, 11, 22, 23, 24, and 25, T14N, R10

W, and secs. 29, 30, 35, and 36, R9W. Carbonaceous material is generally present but less abundant in redistributed deposits.

Production from individual Ambrosia Lake mines has ranged from less than 100 tons to more than 1000 tons per day. At the end of 1965, production from the larger mines averaged about 600 tons per day. This restriction in production is the result of curtailed purchasing by the AEC.

The first discoveries in the Westwater were made in runs that form the "Poison Canyon Ore Trend" (20, p. 122). The occurrences in this trend also span some 8 miles from the Blue Peak mine, sec. 24, T13N, R10W, where uranium ore occurs at the surface, southeasterly to the San Mateo mine, where ore lies at a depth of about 1400 feet (Figure 3).

Most of the ore occurs in typical upper Westwater runs that are found where "Poison Canyon" sandstone exceeds 40 feet in thickness. Although most of the ore is localized near the base of the unit, the changes in thickness do not reflect channeling or thickening at the base but are the result of the accumulation of stray sandstone lenses at the top of the unit (20, p. 126).

The southern edge of the "Poison Canyon" trend is marked by a straight sharp boundary between ore and barren rock that more or less parallels the long axis of the sandstone lenses. In comparison, the northern edge is irregular, and the gradation from ore to waste is gradual; this irregularity is the result of redistribution down dip along northeasterly striking fracture systems. Uranium in the ore bodies in sec. 7 and 8, T13N, R9W, was almost certainly dissolved from runs in the trend and carried more than 2 miles down dip along a fault system before being redeposited. The San Mateo fault zone is an area where removal of pre-existing ore bodies was complete.

All production from the "Poison Canyon" trend has been from underground mines except for that from the rim at the Blue Peak mine, and from the Poison Canyon open pit. Daily production from individual mines along the trend varied from a few tons to several hundred tons.

The Smith Lake area is about 20 miles north-northwest of Ambrosia Lake in T15N, R13W (Figure 5). The Homestake-Sapin Partners' Black Jack No. 1 and No. 2 mines are the only producers and deposits in this area.

The Black Jack No. 1 deposit is a J-shaped easterly trending ore body that hooks south and west, and lies on the east end of the Mariano Lake anticline. This deposit of more than one million tons occurs in the middle and upper Westwater. There are seven ore zones (7, p. 29) in the Black Jack deposit. The lower three intertongue and form an east-west prefault run some 3600 feet long. Shorter northeasterly prefault runs that have been partly removed are found in the southern part of the deposit. In this ore body, the sharp ore-waste cutoff occurs along the north edge of the run. The upper four zones appear to be redistributed deposits that have formed in areas of increased fracture density that occur along both normal dip-slip and strike-slip faults (17, p. 47). The apparent redistribution from north to south is puzzling. The mine has produced more than 1000 tons per day and averages about 750 tons per day.

The Black Jack No. 2 deposit is in the north-central part of sec. 18, T14N, R13W (Figure 5). It lies at the southwest end of the Mariano Lake anticline and occurs in an upper Westwater sandstone that is locally called the "Poison Canyon tongue" (15, p. 49). This sandstone lens is a very local occurrence and ranges in thickness from 18 to 60 feet in the mine area. More than 200,000 tons of ore were produced from the deposit before it was mined out in 1964. Daily production ranged from 100 to 200 tons. There were three zones of ore; the lowermost zone near the base of the sandstone was the most prevalent. Observations in the mine indicate that the ore was prefault and was deposited in a sand-filled channel where shapes of the ore pods were controlled by changes in permeability related to interbedded mudstone.

The westernmost Westwater occurrences have been found in T15, 16, and 17N, R16W, in the Churchrock area (Figure 5). Ore production from the area has been limited. A few small mines along the rim produced a few thousand tons each, and the Churchrock mine operated by United Nuclear Corporation produced less than 50,000 tons from the Westwater before it was closed. A substantial tonnage of low-grade reserves remains.

In the Churchrock area, uranium mineralization has been found throughout the Westwater, with the bulk occurring in the middle of the member. Deposition and apparently latter leaching of the deposit occurred along a northeasterly striking fracture zone. Natural radioactivity logs of widely spaced drill holes in the above three townships indicate that uranium-carrying solutions ranged widely in the Churchrock area.

A new deposit has been discovered north-northeast of the Churchrock mine. Drill holes in this Northeast Churchrock deposit have cut minable thicknesses of ore grade mineralization throughout the member. The best concentrations are near the base and in the upper part of the member. The configuration of the deposits suggests that the lower ores are pre-fault and the upper ores redistributed.

The North Laguna area is about 30 miles east of Grants and includes the Jackpile, Paguate (pronounced Pah-wah'-tee), St. Anthony, and Woodrow ore bodies (Figure 9). The Jackpile and Woodrow deposits were discovered in November, 1951, with airborne scintillation equipment; an outcrop of the St. Anthony deposit was discovered in early 1954; and the Paguate and L-Bar deposits, neither of which is exposed at the surface, were discovered in 1956 as a result of systematic exploration drilling.

The Jackpile sandstone is the host rock for all economically important uranium deposits in the North Laguna area. This unit was derived from the southwest and deposited in a northeasterly-trending downwarp known to be at least 35 miles long, 15 miles wide, and 220 feet deep. The Jackpile and Paguate ore bodies are in the thicker part of the Jackpile sandstone.

There are practically no faults in the area, and those that are present are small and have no apparent control over primary ore localization. Mineralization was influenced in different degrees by such controlling factors as cross-bedding, carbonaceous material, mudstone layers and lenses, bedding planes, lithologic changes, weak intraformational faults, and thickness of host rock.

The Jackpile deposit has a known length of about 1.5 miles, and an average width of about 0.5 miles. The Paguate deposit has a known length of over 2 miles and an average width of a few hundred feet. In the Jackpile mine, nearly all the uranium mineralization occurs in the lower half of the host sandstone. The ore ranges in thickness from 20 to 50 feet and is locally separated by various thicknesses of barren sandstone of similar lithology. Uranium mineralization in the eastern part of the Paguate deposit normally occurs in the upper one-third of the Jackpile sandstone; locally, mineralized in the Jackpile lenses have been truncated by the overlying Dakota Sandstone. In the western part of the deposit, ore normally occurs in the lower two-thirds of the

host sandstone. To date the following minerals from the Jackpile and Paguate mines have been identified:

Autunite, $Ca(UO_2)_2(PO_4)_2 \cdot 10-12H_2O$
Becquerelite, $7UO_3 \cdot 11H_2O$
Carnotite, $K_2(UO_2)_2(VO_4)_2 \cdot H_2O$
Coffinite, $U(SiO_4)_{1-x}(OH)_{4x}$
Hydrogen-autunite, $HUO_2PO_4 \cdot 4H_2O$
Metatorbernite, $Cu(UO_2)_2(PO_4)_2 \cdot nH_2O_n = 4$ to 8
Metatyuyamunite, $Ca(UO_2)_2(VO_4)_2 \cdot 5-7H_2O$
Phosphuranylite,    $Ca(UO_2)_4(PO_4)_2(OH_4 \cdot 7H_2O$
Schoepite, $4UO_3 \cdot 9H_2O$
Sklodowskite, $Mg(UO_2)_2(SiO_3)_2(OH)_2 \cdot 6H_2O$
Soddyite, $(UO_2)_5(SiO_4)_2(OH)_2 \cdot 5H_2O$
Tyuyamunite, $Ca(UO_2)_2(VO_4)_2 \cdot 7-10.5H_2O$
Uraninite, $UO_2$
Uranophane, $Ca(UO_2)_2(SiO_3)_2(OH_2) \cdot 5H_2O)$

Noncommercial quantities of selenium, molybdenum, and vanadium also occur in association with the uranium minerals. Metallurgical studies show the normal mill feed from the Jackpile and Paguate deposits to be about as follows:

| | |
|---|---|
| Black oxidized uranium complexes | 80% |
| Uraninite | 15% |
| Black organo-uranium complexes | 3% |
| Coffinite | 2% |

Coffinite and uraninite are somewhat more abundant in the high-grade ores. The uranium minerals are intimately mixed with and replace carbonaceous material or occur as the cement in the sandstone.

Production from the Jackpile and Paguate mines averaged more than 4000 tons per day prior to May, 1959, when the AEC curtailed its purchases of uranium concentrate. Since then production has been reduced to about 1400 tons per day.

The St. Anthony deposit is about 2 miles northeast of the Jackpile mine. It occurred in the Jackpile sandstone unit and was mined from the 250-level of a vertical shaft by Climax Uranium (a subsidiary of Climax Molybdenum) during the period 1957–60. The ore body was about 1000 feet long, 50 to 300 feet wide, and up to 30 feet thick. The mineralogy of the deposit was essentially the same as that of the Jackpile and Paguate ore bodies, being for the most part a complex of black uranium oxides in which the higher-grade portions comprised coffinite to a large extent, with some associated uraninite. Normal production amounted to somewhat less than 200 tons per

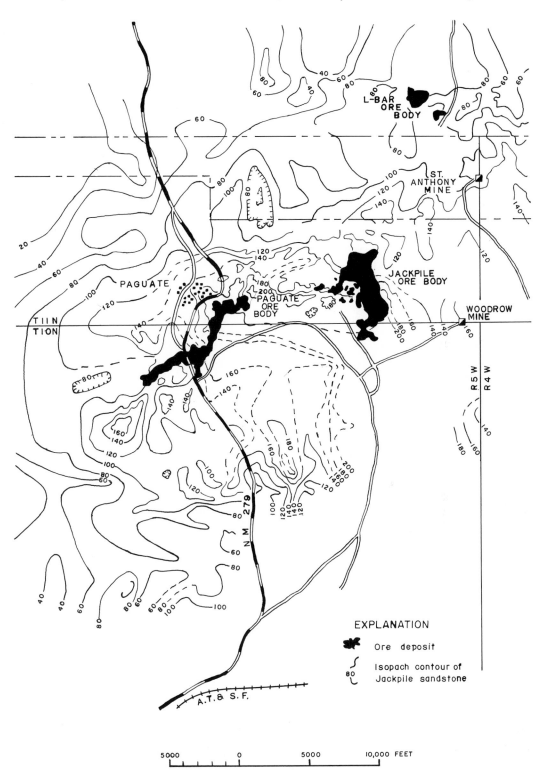

FIG. 9. *Map, showing the relationship of the North Laguna ore bodies to the thickness of the Jackpile sandstone.*

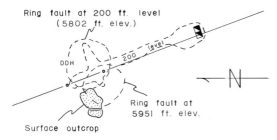

Surface outcrop

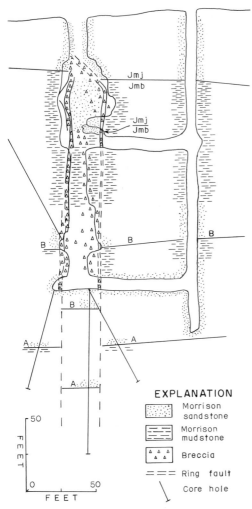

EXPLANATION

Morrison sandstone

Morrison mudstone

Breccia

=== Ring fault

Core hole

FIG. 10. *Plan and Section of the Woodrow Mine, North Laguna Area.*

day of ore with a grade of about 0.20 per cent $U_3O_8$, most of it mined by a pillar-retreat stoping.

The L-Bar deposit is a partially developed ore body located about a mile northwest of the St. Anthony deposit on land belonging to the L-Bar Cattle Company. Its shape as developed to date is roughly a square about 2000 feet on a side, but drill-hole data indicate that the mineralization may extend for another 2000 feet southeasterly. The deposit is made up of numerous, fairly discontinuous thin lenses of uranium mineralization that occur throughout the entire section of Jackpile sandstone. The ore composition is the same as that found in the Jackpile and Paguate mines.

The Woodrow mine is about a mile east of the Jackpile mine. The deposit occurred in a nearly vertical breccia pipe of Jackpile sandstone. The pipe is, for the most part, bounded by strong ring faults that penetrate into Brushy Basin mudstone near the surface and continues downward into the Morrison Formation beyond a known depth of 272 feet. The core of the pipe has been downthrown approximately 40 feet, as shown on Figure 10, and quite thick, black clay-gouge was formed along the fault. Most of the mineralization occurred in the core, but, in the interval from 31 to 51 feet below the surface, small quantities of ore assaying as high as 20 per cent $U_3O_8$ were found within the ring fault. In this same interval the mineralization extended up to 10 feet outside the ring fault along minor fractures (24, p. 177). The following minerals have been identified from the Woodrow pipe:

*Autunite, $Ca(UO_2)_2(PO_4)_2 \cdot 10\text{-}12H_2O$
Barite, $BaSO_4$
* Becquerelite, $7UO_3 \cdot 11H_2O$
Chalcopyrite, $CuFeS_2$
*Coffinite, $U(SiO_4)_{1-x}(OH)_{4x}$
Galena, $PbS$
Jarosite, $KFe_3(OH)_6(SO_4)_2$
Johannite, $Cu(UO_2)_2(SO_4)_2(OH)_2 \cdot 6H_2O$
Marcasite, $FeS_2$
* Meta-autunite, $Ca(UO_2)_2(PO_4)_2 \cdot 8H_2O$
* Metatorbernite, $Cu(UO_2)_2(PO_4)_2 \cdot nH_2O_n =$ 4 to 8
* Pyrite, $FeS_2$
Sabugalite, $Ha1 (UO_2)_4(PO_4)_4 \cdot 16H_2O$
Torbernite, $Cu(UO_2)_2(PO_4)_2 \cdot 12H_2O$
* Uraninite, $UO_2$
Uranopilite, $(UO_2)_6(SO_4)(OH)_{10} \cdot 12H_2O$
* known to occur also in the Jackpile deposit

The Woodrow pipe is without question a unique uranium host that has certain geologic characteristics otherwise unknown except at the Orphan pipe on the south rim of Grand

Canyon. The strong ring faults, the breccia, and sulfides suggest that the mineralization in the pipe may be hydrothermal.

The Woodrow deposit was mined from the 100-foot level to the surface in 1954, and from the 200-foot to the 100-foot level in 1956. The first mining phase produced ore with an average grade of 1.53 per cent $U_3O_8$ and 0.05 per cent $V_2O_5$, whereas the second phase produced ore with an average grade of 0.32 per cent $U_3O_8$ and 0.03 per cent $V_2O_5$.

DAKOTA SANDSTONE   Uranium occurs in basal interbedded sandstones and shales of the Dakota. The deposits have been found in the Gallup area along the hogback formed by the Nutria monocline, in the Churchrock area, and in the southwestern corner of the Ambrosia Lake area. The first mine in the Dakota sandstone, and one of the first to be brought into production, was the small open-pit Silver Spur mine in sec. 31, T14N, R10W (Figure 3); it was also one of the first producing mines in the Grants area. The largest producers, about 40,000 tons each, have been the Diamond No. 2 mine south of U.S. Highway 66 on the Nutria monocline (Figure 1) and the Dakota level of the Churchrock mine. At the end of 1965, the Diamond No. 2 was still producing a few hundred tons per month and was the only Dakota deposit being worked. The total production from the Dakota deposits has been about 100,000 tons containing four pounds of $U_3O_8$ per ton.

The uranium has been concentrated in fine- to coarse-grained sandstone and interbedded carbonaceous shale that in places contain enough carbonized vegetal remains to form thick peat beds. The Hogback No. 4 deposits occurred entirely within a shaly carbonaceous bed.

Primary uranium minerals, uraninite and coffinite, are found in deposits that lie below the water table. On the outcrop and in near surface deposits, these have been oxidized to form carnotite and other secondary minerals. These primary and secondary minerals impregnate the host rock as sand grain coatings and interstitial fillings, and the secondary minerals commonly fill fractures.

The deposits are nearly equidimensional in the horizontal plane, and their thickness has been limited to less than 20 feet by the thickness of the sandstone units in which they occur. The elongation of the deposits is generally parallel to northerly trending fractures associated with the ore bodies.

The occurrence of several of the ore bodies in folded, faulted, and fractured areas suggests that the uranium in the Dakota deposits may be post-Laramide and may have been derived from eroded updip or underlying Morrison deposits and redistributed along the fractures where the favorable carbonaceous environment was encountered.

## ORIGIN

Theories of the origin of the Grants deposits, as with most others of the Colorado Plateau, have fallen into the following genetic types:

(1) Syngenetic-sandstone, mudstone, tuff, carbonaceous shale
(2) Ground water
   (a) Lateral secretions
   (b) Supergene
(3) Hydrothermal

Some theories are combinations of the above or are variations of one of them, such as a penesyngenetic category. To evaluate the genesis of the deposits properly, it is absolutely essential to know as nearly as possible the geologic history beginning with the source and deposition of the sedimentary materials. It is therefore fitting to begin with the nature and source of the Morrison beds.

The Morrison sediments are markedly tuffaceous and clearly indicate a volcanic provenance in which there may have been flow, pyroclastic, geyser, and hot spring eruptions. Significant quantities of uranium may have been associated with these eruptions. If this were true, the uranium could have been brought to the sites of Morrison deposition by surface and subsurface waters, fluvial debris, and fallout. The surface and subsurface waters moving toward the Morrison depositional sites could have continued to extract uranium from unstable rock debris during transportation. Associated with the rocky sediment was also considerable organic debris or trash which accumulated in irregular deposits that eventually became instrumental in localizing many of the ore deposits. Diagenetic redistribution of uranium may have continued during burial, compaction, and cementation. These were the conditions and environments during the earliest stage of possible uranium accumulation by sedimentary syngenetic or penesyngenetic processes.

The sedimentary stage was brought to a halt in most of the area in late Jurassic time by the rise of the Mogollon highland with its broad gentle tilt to the north (3, p. 83). Some gentle open folding accompanied the tilting while widespread stripping reworked the upper sediments to the north. Channel sands such as the Jackpile and other post-Brushy Basin deposits farther to the north may have formed during this episode. Water tables, ground-water dynamics, and water chemistry were undoubtedly modified considerably during this rejuvenation. Previous uranium deposits could have been weathered, leached, eroded mechanically, and redeposited several times. To many geologists, this has been a favored time for formation of the older runs west of Mount Taylor and for the Jackpile and Paguate deposits.

This second stage was terminated with marine encroachments which accompanied regional subsidence of subcontinental dimensions and eventually buried the Grants region nearly to 5000 feet in late Cretaceous and possibly to 7000 by Paleocene time. Although temperatures and pressures were increased during this stage, ground water circulation essentially ceased and modifications of deposits would have been local and slight.

The next stage of potential ore formation began with Laramide deformation and continued through the Tertiary. For the Grants region, the rise of the Zuni uplift and the subsidence of the San Juan Basin and its subsidiary Gallup and Acoma embayments were the predominating tectonic influences. In addition to the Zuni dome, all the folds and most of the faults that are so prevalent in the vicinity of Ambrosia Lake were formed during this time. Metallizing porphyry intrusions may have occurred at this time in the more than 300 square mile area between Ambrosia Lake and Laguna that is covered by the Mount Taylor volcanic field. The obvious effects upon the deposits beyond deformation included general decline of pressure and temperature and increased mobility of the ground water. Additionally, hydrothermal alteration and introduction of uranium is a possibility at this stage.

Continued regional uplift probably resulted in first re-exposure of the Morrison along the Zuni uplift by middle Tertiary. By the end of the period the uranium-bearing horizons had been stripped and truncated well off the crest of the uplift where the Precambrian was exposed in a wide erosion surface. During all this time and into the present, oxidation and solution by meteoric water progressively modified and "worked" the deposits down dip to their present positions, especially to the north and northeast of the uplift. Some of this downward migration of the uranium-bearing solu-

tions is thought also to have crossed the formations downward to be precipitated in the Todilto Limestone.

The history of exposure and modification at Laguna was quite different, as these great deposits lay in the Acoma embayment well removed from the energizing effects of the Zuni uplift. Exposure came later, slower, and along a much broader area owing to the low dips. As a result, supergene buildup was not shifted so much down the dip as in the redistributed stacks which characterize the Ambrosia Lake and Gallup areas.

The foregoing outline of the geologic history since late Jurassic furnishes a framework of conditions that must have controlled the formation of the deposits regardless of the specifics of their origin. The source of the metal, whether from sedimentary debris, groundwater introductions, or hydrothermal additions is the principal genetic problem along with timing of deposition. To decide upon these problems, petrographic, spectrographic, paragenetic, geochemical, isotopic, and radiometric studies have been pursued. These appear to raise additional subsidiary problems such as the chemistry of the transporting fluids, of precipitation, or of concentration, and the significance of the widespread alteration in the Morrison sands. Nevertheless, some conclusions appear to have reached near unanimity. One of these is recognition at Ambrosia Lake of two stages (prefault and postfault) of orebody development associated with down-dip oxidation, solution, and enrichment. Secondly, as the result of the work of Moench and his associates, there appears to be near unanimity on a pre-Dakota age for the primary ore runs, even though as outlined above, some minor differences of age and derivation may be involved. If an early derivation is agreed upon, then later modification and redistribution, energized by the Laramide disturbances and erosion, become more or less logical, corollary, sequential steps that would fit into the known geologic history.

## REFERENCES CITED

1. Emmons, W. H., 1940, The principles of economic geology: McGraw-Hill, New York, 529 p.
2. Smith, C. T., 1954, Geology of the Thoreau quadrangle, McKinley and Valencia counties, New Mexico: N. Mex. Bur. Mines and Min. Res. Bull. 31, 36 p.
3. Kelley, V. C., 1955, Regional tectonics of the Colorado Plateau and relationship to the origin and distribution of uranium:

Univ. New Mexico Pub. in Geol., No. 5, 120 p.

4. Birdseye, H. S., 1957, The relation of the Ambrosia Lake uranium deposits to a pre-existing oil pool: p. 26–29, *in Geology of the southwestern San Juan Basin,* Four Corners Geol. Soc. Second Field Conf., 198 p.

5. Smith, C. T., 1957, Geology of the Zuni Mountains, Valencia and McKinley counties, New Mexico: p. 53–61, *in Geology of the southwestern San Juan Basin,* Four Corners Geol. Soc., Second Field Conf., 198 p.

6. Hilpert, L. S. and Moench, R. H., 1960, Uranium deposits of the southern part of the San Juan Basin, New Mexico: Econ. Geol., v. 55, p. 429–464.

7. Fitch, R., 1961, Let's look at Lance's uranium mines: Mining World, v. 23, no. 11, p. 23–27.

8. Granger, H. C., *et al.,* 1961, Sandstone-type uranium deposits at Ambrosia Lake, New Mexico—an interim report: Econ. Geol., v. 56, p. 1179–1210.

9. Austin, S. R., 1963, Alteration of Morrison sandstone: p. 38–44, *in* Kelley, V. C., *Editor, Geology and technology of the Grants uranium region,* N. Mex. Bur. Mines and Min. Res. Mem. 15, 277 p.

10. Clary, T. A., *et al.,* 1963, Geologic setting of an anomalous ore deposit in the Section 30 mine, Ambrosia Lake area: p. 72–79, *in* Kelley, V. C., *Editor, Geology and technology of the Grants uranium region,* N. Mex. Bur. Mines and Min. Res. Mem. 15, 277 p.

11. Gould, W., *et al.,* 1963, Geology of the Homestake-Sapin uranium deposits, Ambrosia Lake area: p. 66–71, *in* Kelley, V. C., *Editor, Geology and technology of the Grants uranium region,* N. Mex. Bur. Mines and Min. Res. Mem. 15, 277 p.

12. Granger, H. C., 1963, Mineralogy: p. 21–37, *in* Kelley, V. C., *Editor, Geology and technology of the Grants uranium region,* N. Mex. Bur. Mines and Min. Res. Mem. 15, 277 p.

13. Harmon, G. F. and Taylor, P. S., 1963, Geology and ore deposits of the Sandstone mine, southeastern Ambrosia Lake area: p. 102–107, *in* Kelley, V. C., *Editor, Geology and technology of the Grants uranium region,* N. Mex. Bur. Mines and Min. Res. Mem. 15, 277 p.

14. Hilpert, L. S., 1963, Regional and local stratigraphy of uranium-bearing rocks: p. 6–18, *in* Kelley, V. C., *Editor, Geology and technology of the Grants uranium region,* N. Mex. Bur. Mines and Min. Res. Mem. 15, 277 p.

15. Hoskins, W. G., 1963, Geology of the Black Jack No. 2 mine, Smith Lake area: p. 49–52, *in* Kelley, V. C., *Editor, Geology and technology of the Grants uranium region,* N. Mex. Bur. Mines and Min. Res. Mem. 15, 277 p.

16. Kelley, V. C. (*Editor*), 1963, Geology and technology of the Grants uranium region: N. Mex. Bur. Mines and Min. Res. Mem. 15, 277 p. Contains papers by numerous authors which are cited individually.

17. McRae, M. E., 1963, Geology of the Black Jack No. 1 mine, Smith Lake area: p. 45–48, *in* Kelley, V. C., *Editor, Geology and technology of the Grants uranium region,* N. Mex. Bur. Mines and Min. Res. Mem. 15, 277 p.

18. Melancon, P. E., 1963, History of exploration: p. 3–5, *in* Kelley, V. C., *Editor, Geology and technology of the Grants uranium region,* N. Mex. Bur. Mines and Min. Res. Mem. 15, 277 p.

19. Moench, R. H., 1963, Geologic limitations on the age of uranium deposits in the Laguna district: p. 157–166, *in* Kelley, V. C., *Editor, Geology and technology of the Grants uranium region,* N. Mex. Bur. Mines and Min. Res. Mem. 15, 277 p.

20. Rapaport, I., 1963, Uranium deposits of the Poison Canyon ore trend, Grants district: p. 122–135, *in* Kelley, V. C., *Editor, Geology and technology of the Grants uranium region,* N. Mex. Bur. Mines and Min. Res. Mem. 15, 277 p.

21. Santos, E. S., 1963, Relation of ore deposits to the stratigraphy of the Ambrosia Lake area: p. 53–59, *in* Kelley, V. C., *Editor, Geology and technology of the Grants uranium region,* N. Mex. Bur. Mines and Min. Res. Mem. 15, 277 p.

22. Squyres, J. B., 1963, Geology and ore deposits of the Ann Lee mine, Ambrosia Lake Area: p. 90–101, *in* Kelley, V. C., *Editor, Geology and technology of the Grants uranium region,* N. Mex. Bur. Mines and Min. Res. Mem. 15, 277 p.

23. Thaden, R. E. and Santos, E. S., 1963, Map showing the general structural features of the Grants district and the areal distribution of the known uranium ore bodies in the Morrison: p. 20, *in* Kelley, V. C., *Editor, Geology and technology of the Grants uranium region,* N. Mex. Bur. Mines and Min. Res. Mem. 15, 277 p.

24. Wylie, E. T., 1963, Geology of the Woodrow breccia pipe: p. 177–181, *in* Kelley, V. C., *Editor, Geology and technology of the Grants uranium region,* N. Mex. Bur. Mines and Min. Res. Mem. 15, 277 p.

25. Moench, R. H., 1964, Geology of the Dough Mountain quadrangle, New Mexico: U.S. Geol. Survey Geol. Quad. Map, GQ-354, 1:24,000.

# 37. Geology and Exploitation of Uranium Deposits in the Lisbon Valley Area, Utah

HIRAM B. WOOD*

## Contents

## Illustrations

## Table

* U.S. Atomic Energy Commission, Grand Junction, Colo.

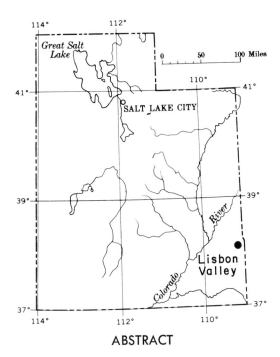

## ABSTRACT

Uranium ore deposits in the Lisbon Valley area are in an arcuate belt, 15 miles long by one-half-mile wide, on the southwest flank of the Lisbon Valley anticline. They range in size from 500 to 1,500,000 tons and by mid-1965 had yielded 6,147,000 tons containing 48,530,000 pounds of $U_3O_8$. Ore bodies average 6 feet thick, are tabular, amoeba-shaped masses, concordant to the bedding, and are in the thickest, lowest sandstone unit of the Moss Back Member of the Chinle Formation of Triassic age. The host rock is predominantly a fluviatile, calcareous, fine-grained to conglomeratic sandstone. Uraninite, the principal ore mineral, fills the pores in the sandstone and partly replaces sand grains and fossil wood fragments.

During Late Permian and Early Triassic time pre-Chinle Formations were arched to form the ancestral Permian anticline. Moss Back streams truncated Moenkopi and Cutler beds which capped the ancestral anticline and then covered it with fluviatile clastics, which were subsequently covered by thousands of feet of Triassic, Jurassic, and Cretaceous sediments. Uranium, probably derived from the Chinle Formation by diagenetic processes, was transported in connate ground waters, was moved by compaction or hydrostatic forces, and was deposited under reducing conditions. Uranium was emplaced around the crest of the ancestral anticline, prior to the Laramide orogeny. During the orogeny, the Tertiary-Lisbon valley anticline was superinduced on the Permian anticline and, penecontemporaneous with uplift, was faulted parallel to its longitudinal axis. The Big Indian ore belt on the southwest flank footwall block was elevated to approximately its present position, and the northeast flank hanging wall block was elevated only slightly.

An extension of the Big Indian ore belt around the northeast flank of the ancestral Permian anticline may exist in the Moss Back sandstone in the downthrown block northeast of the Lisbon Valley fault.

## INTRODUCTION

This paper reviews the exploration and mining of the uranium deposits in the Big Indian ore belt and discusses the geology important to the deposits. The ore belt is on the southwest flank of the Lisbon Valley anticline, near the center of the Colorado Plateau (Figure 1 and see Fischer Figure 1, this volume), and is about 40 miles southeast of Moab, Utah.

The ore deposits described herein are in the Moss Back Member of the Chinle Formation of Triassic age. The uranium-vanadium deposits in the Salt Wash Member of the Morrison Formation of Jurassic age at the northwest end of the Lisbon Valley anticline and also in Dry Valley south of the anticline and the copper deposits in the Dakota Group sandstone beds of Cretaceous age along the Lisbon Valley fault are not described in this paper (Figure 1).

Geologists and mine operators of Atlas Corporation, Homestake Mining Company, Standard Metals Corporation, Utex Exploration Company, Continental Materials Corporation, and Humeca Exploration Company have supplied mine maps, drill hole logs, and some geologic maps, without which this study would not have been possible. Information on mining methods and cost was supplied by Lochinvar B. Birch of the U.S. Atomic Energy Commission. Critical reviews and constructive suggestions by R. P. Fischer of the United States Geological Survey and by R. A. Laverty of the United States Atomic Energy Commission were very helpful. Unpublished geologic reports on the Big Indian mining district, prepared prior to 1958, by M. A. Lekas, N. E. Salo, John Volgamore, P. L. Grubaugh, and Robert Schoen, all former U.S. Atomic Energy Commission geologists, were used extensively for back-up information.

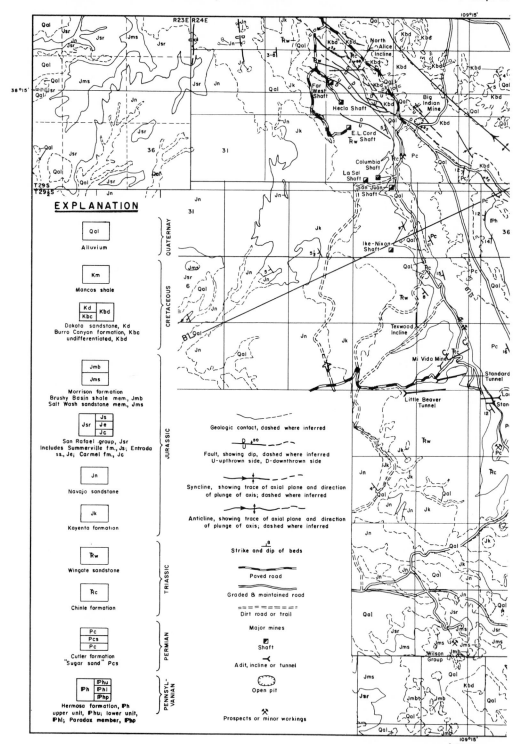

**EXPLANATION**

QUATERNARY

| Qal |
Alluvium

CRETACEOUS

| Km |
Mancos shale

| Kd |
| Kbc | Kbd |
Dakota sandstone, Kd
Burro Canyon formation, Kbc
undifferentiated, Kbd

JURASSIC

| Jmb |
| Jms |
Morrison formation
Brushy Basin shale mem., Jmb
Salt Wash sandstone mem., Jms

| Js |
| Jsr | Je |
| Jc |
San Rafael group, Jsr
Includes Summerville fm., Js; Entrada
ss., Je; Carmel fm., Jc

| Jn |
Navajo sandstone

| Jk |
Kayenta formation

TRIASSIC

| Ɍw |
Wingate sandstone

| Ɍc |
Chinle formation

PERMIAN

| Pc |
| Pcs |
| Pc |
Cutler formation
"Sugar sand" Pcs

PENNSYL-VANIAN

| Ɋhu |
| Ɋh | Ɋhl |
| Ɋhp |
Hermosa formation, Ɋh
upper unit, Ɋhu; lower unit,
Ɋhl; Paradox member, Ɋhp

Geologic contact, dashed where inferred

Fault, showing dip, dashed where inferred
U-upthrown side, D-downthrown side

Syncline, showing trace of axial plane and direction
of plunge of axis; dashed where inferred

Anticline, showing trace of axial plane and direction
of plunge of axis; dashed where inferred

Strike and dip of beds

Paved road

Graded & maintained road

Dirt road or trail

Major mines

Shaft

Adit, incline or tunnel

Open pit

Prospects or minor workings

FIG. 1. *Geologic Map of Lisbon*

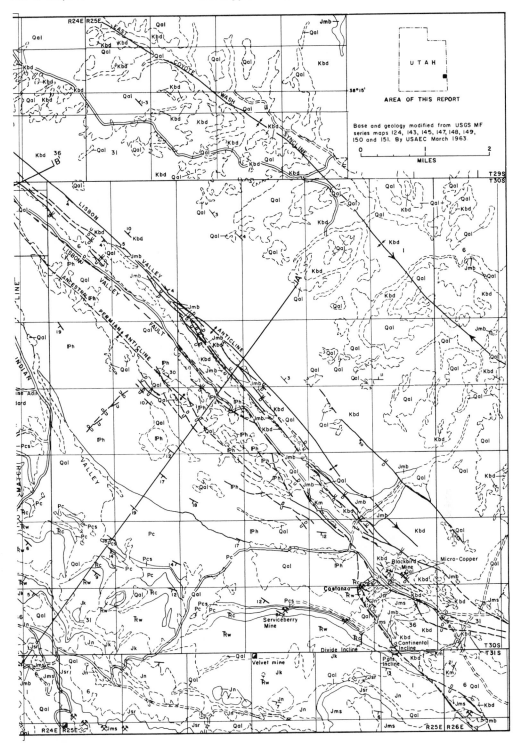

*Valley Area, San Juan County, Utah.*

## EXPLORATION AND MINING HISTORY

### Exploration History and Costs

The first discovery of uranium-vanadium ore on the Lisbon Valley anticline was made in 1913, at the south end of the anticline, on outcrops of basal Chinle sandstone. In 1948, low-grade uranium deposits were discovered and developed in upper Cutler sandstone outcrops along the center of the southwest flank of the anticline. These were the deposits that attracted Charles Steen to the area. In July 1952, Steen drilled the famous 70-foot deep discovery hole on the Mi Vida claim just off the Big Buck claims and down dip from the mines in the Cutler Formation. He encountered about 13 feet of uraninite ore in the Moss Back sandstones, about 100 feet higher stratigraphically than anticipated (1, p. 5) (2, p. 1). Following this discovery, claim staking and exploration drilling progressed rapidly to the northwest and southeast and continued with intensity through 1956. During this ensuing period, the following deposits in the Chinle were discovered by drilling in the north half of the Big Indian ore belt: Standard (Big Buck), Little Beaver, Louise, Texwood-Stinko, Ike-Nixon, La Sal, Columbia, San Juan, Cord (Jen), Radon (Hecla), Far West, and North Alice (Figure 2). A peak in exploration was reached in 1955, when 647,000 feet of drilling was reported. By mid-1956, exploration drilling began to taper off. By the end of 1964, over 4500 holes totalling about 2,200,000 feet had been drilled in the search for uranium on the anticline, and over 3000 holes, spaced 200 to 500 feet apart, had been drilled within the delineated ore belt (Figure 2). This intensity of drilling argues against the existence of any undiscovered large ore deposits in the drilled areas, although a number of small ore deposits may remain undiscovered.

The discovery in 1962 and the mining of uranium ore in 1964 at the Costanza mine (Figure 1) in sections 26 and 35, T30S, R25E established the existence of uranium ore on the northeast side of a hinge fault that has more than 2000 feet of displacement. This high angle fault is one of the main bifurcating faults at the south end of the Lisbon Valley fault. Between May 1964 and July 1965, Humeca Exploration Company drilled five deep holes (2500 feet ±), in the center of section 21, and in the southwest corner of section 22, T29S, R24E. Interpretation of Century Geophysical Company gamma ray logs of these holes indicated that two holes penetrated

up to 33 feet of Moss Back sandstone and 2 to 8 feet of uranium ore. This discovery established the occurrence of uranium ore in the downthrown block northeast of the main Lisbon Valley fault at the north end of the anticline.

Present direct drilling costs in the district are about $1.00 to $1.25 per foot for non-core rotary drilling and about $5.00 per foot for core drilling to depths of about 500 feet. Usually coring is limited to an interval of about 40 feet, which includes the gray zone of the basal Moss Back sandstone and the upper few feet of the red Cutler sandstone. Radiometric logging costs 12 to 15 cents per foot. Costs were considerably higher during the period of major exploration than now. Assuming an average cost of $3.50 per foot for drilling, including direct and indirect drilling costs, about $7,700,000 has been spent for drilling in the area. The average discovery and development rate, based on production plus ore reserves, has been about 25 pounds $U_3O_8$ per foot of drilling, at a cost of about 14 cents per pound $U_3O_8$ developed.

### Mining History and Costs

Vanadium ore production from the Chinle Formation at the Divide and Serviceberry mines in south Lisbon Valley was reported in 1917, 1940, and 1941. These same mines were reopened in 1948 (1) for their uranium content. Also in 1948, the Big Buck mines in the Cutler Formation in Big Indian Valley were mined for uranium. Intermittent production from these small deposits continued until 1952. In December 1952, Steen shipped the first ore from the Mi Vida mine (2, p. 3).

Production from the Moss Back sandstone beds has ranged from two to six million pounds of $U_3O_8$ per year and reached a peak production of over 6,377,000 pounds in Fiscal Year 1958, Table I. Due to the exhaustion of a few of the major ore deposits, the ore production rate started dropping in 1960 and has leveled off to about 4,000,000 pounds per year.

In the central and southern deposits, the vanadium content is high enough for that metal to be extracted economically. During the 1948–56 period, vanadium was extracted from ore that was shipped from these mines to some of the processing mills on the Colorado Plateau. However, most of the ore has been processed at the Atlas Corporation alkaline leach mill at Moab, Utah, and to date this mill has recovered only a small amount

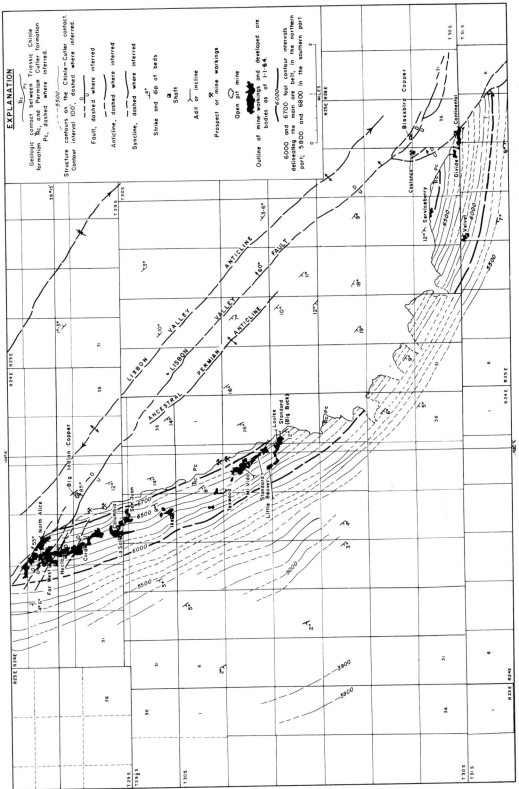

FIG. 2. *Structure Map of the Lisbon Valley Anticline, showing delineated Big Indian Ore Belt with ore bodies outlined.*

TABLE I. *Annual Chinle—Cutler Ore Production 1948 through F. Y. 1965*

|  | Tons | Pounds U$_3$O$_8$ | Pounds V$_2$O$_5$* |
|---|---|---|---|
| 1948 to July 1, 1953 | 15,288 | 143,093 | 361,862 |
| F. Y. Ending |  |  |  |
| July 1, 1954 | 71,391 | 742,452 | 1,749,724 |
| July 1, 1955 | 208,781 | 1,998,764 | 3,271,362 |
| July 1, 1956 | 394,713 | 2,795,701 | 2,643,571 |
| July 1, 1957 | 592,304 | 4,707,193 | 3,409,168 |
| July 1, 1958 | 773,042 | 6,377,746 | 2,401,720 |
| July 1, 1959 | 771,229 | 6,349,628 | 3,821,721 |
| Subtotal | (2,898,748) | (23,114,582) | (17,659,128) |
| July 1, 1960 | 760,585 | 5,649,143 | Incomplete |
| July 1, 1961 | 640,536 | 4,179,223 | vanadium |
| July 1, 1962 | 590,319 | 3,819,600 | assay |
| July 1, 1963 | 418,477 | 4,309,452 | records. |
| July 1, 1964 | 507,398 | 4,650,565 |  |
| July 1, 1965 | 330,748 | 2,806,691 |  |
| TOTAL | 6,146,811 | 48,529,256 |  |

*Note: The above production was compiled from U.S. Atomic Energy Commission ore receipts and includes ore shipments to numerous uranium processing mills on the Colorado Plateau. From November 1956, when the Moab mill went on stream, to 1960, the mill assayed the ore for vanadium, but after 1960 they discontinued the practice. Post-1959 vanadium assays are available on ore shipments to other plateau mills, but these shipments were sporadic

of vanadium on an experimental basis. Molybdenum assays ranging up to 0.25 per cent are common but very spotty in the Big Indian ores. Those deposits with anomalous molybdenum, such as the North Alice, South Almar, Mi Vida, Standard, and Velvet, normally assay only 0.03 to 0.07 per cent molybdenum.

Up to July 1, 1965, about 6,147,000 tons of ore at a grade of 0.39 per cent U$_3$O$_8$ containing about 48,530,000 pounds of U$_3$O$_8$, which is over 8 per cent of the U.S. total, had been mined from deposits in the Big Indian ore belt, including production from the Cutler Formation, Table I. Continuation of operations to 1971 should result in a total production of about 57 million pounds of uranium oxide worth about $450,000,000.

In the Big Indian belt, all underground mining methods allow full caving, and rarely is a stoped area accessible after the ore body has been depleted. The mining method selected was often determined by the adequacy of development drilling, by the knowledge of sub-

surface ground conditions, and by the kind of mining equipment the company had available. The mining equipment in use depends on the thickness of the ore and varies from jacklegs to jumbos for drilling; slushers and front-end-diesel-motor and air-motor loaders for mucking; battery or trolley motors with cars on track, Young-Shuttle Buggies, or Koehring's Dumpters for haulage.

Many of the major mines are accessible by shafts 400 to 800 feet deep, but the Mi Vida, Big Buck (Standard), Louise, and North Alice are entered by inclines or adits. Direct mining costs, plus haulage to mill of about 4.5 cents per ton mile, ranged from about $6.40 per ton for ore 15 to 45 feet thick, using the room and pillar method, to about $13.10 per ton for ore 3 to 8 feet thick, using the long-wall retreat method. The mines were dry, the ore bodies were fairly uniform in thickness, the dip of the bedded host rock was usually less than 10°, fracturing was not excessive, and the average grade was high for this type of bedded sandstone ore.

## PHYSIOGRAPHY

Lisbon Valley is one of the many northwest-trending, subsequent stream valleys formed along breached salt anticlines in the Paradox Basin of the Colorado Plateau (20) (23) (28). The Cane Creek, Moab, Lisbon Valley, and Dolores (Slick Rock) anticlines are on the west edge of the deeper part of the basin; they all have Triassic and younger formations exposed; they all have uranium mineralization in basal Chinle sandstone beds.

The Lisbon Valley anticline is a compound structure formed by folding during both the Permian and Tertiary periods. The two anticlines have separate but nearly parallel axes; however, the younger Tertiary anticline forms the present physiographic structure. Weir, *et al.* (31) named the smaller, ancestral Permian anticline, the axis of which is west of the fault, the Lisbon Valley anticline, and the larger Tertiary anticline, the axis of which is east of the fault, the Lisbon Canyon anticline. The writer prefers to consider the smaller Permian anticline the ancestral Permian anticline and the larger Tertiary anticline the Lisbon Valley anticline (Figures 1 and 2).

The Lisbon Valley anticline covers an area about 21 miles long and 9 miles wide. Altitudes range from about 6000 to 7200 feet. Many high sandstone-capped cuestas, cut by canyons or gorges 200 to 500 feet deep, characterize this area, which is typical of the Colorado Plateau. Good access roads follow the valley floors. Except for a few drill roads, most of the hog-back ridges are inaccessible by motor vehicle.

## GENERAL GEOLOGY

### Igneous Rocks

No igneous rocks are exposed on the anticline, and none has been encountered in the numerous oil and gas test wells that have penetrated over 11,000 feet of sediments on the anticline. The nearest igneous intrusives are in the La Sal Mountains (South Mountain), 7 air miles north of the North Alice mine (Figure 1). During the Tertiary period, the diorite, monzonite, and syenite porphyrys of the La Sal laccoliths, dikes, sills, and plugs were intruded into and through at least 9000 feet of sediments (Pennsylvanian through Cretaceous in age) (15). No physiographic, structural, or mineralogic evidence appears to relate the La Sal igneous intrusives to the Lisbon Valley uranium deposits; however, some geologists postulate a relationship based on geographic considerations, on the Tertiary age of the major faulting, and on a few $U^{238}/Pb^{206}$ isotope age determinations that indicate early Tertiary ages.

### Stratigraphy

A generalized stratigraphic section at the anticline is shown in Figure 3, which symbolizes the lithology and gives approximate thicknesses of formations. Only the ore-bearing formations and those important in explaining the characteristics and the genesis of the Big Indian ore belt are described in the text.

CUTLER FORMATION  The Cutler Formation at the Lisbon Valley anticline exemplifies the red bed facies (10, 11, 26) and ranges from 900 to 1800 feet in thickness. It thins gradually to the north and to the west and thickens to the east and northeast toward the Uncompahgre uplift. Throughout the salt anticline area, at the top of the Cutler, there is a notable erosional unconformity. Erosional thinning of the Cutler over the anticlines indicates the Permian ancestry of most of the anticlines (24).

Where exposed in Big Indian Valley (Figure 1), the fluviatile upper part of the Cutler consists of alternating beds or lenses of light pink, orange, and buff mudstone, calcareous siltstone, and arkosic sandstone. The sandstone beds are well-sorted, are fine- to medium-grained, have a saccharoidal texture, and are as much as 50 feet thick. The sandstone is composed of quartz, feldspar, and biotite, with clay as the predominant binder, but locally calcite may be the main cement.

A few uranium ore pods in Cutler sandstone crop out along the west escarpment of Big Indian Valley about 100 feet stratigraphically below the Permian-Triassic nonconformity and 1000 to 1500 feet up dip and northeast of the eastern limit of the ore belt. Other ore pods are found in these massive sandstone units where they subcrop under the Moss Back ore deposits. Some ore pods are as much as 40 feet below the nonconformity, but most are within 6 feet of it. The Cutler sandstone beds, where exposed in the mine workings, appear to be more extensively bleached than they are on the rim outcrops. The thickness of this bleached zone below the unconformity does not appear to be directly related to the size or position of the overlying ore bodies but does appear to be related to the thicker and more porous of the Cutler sandstone beds.

MOENKOPI FORMATION  The Moenkopi Formation, widespread throughout the Paradox Basin, does not crop out on the anticline (12) (27), but it is penetrated in oil and gas test wells drilled low on the flanks of the anticline (22). At the nearest outcrops on the Colorado River, the Moenkopi consists of interbedded dark-red to chocolate-brown, laminated, micaceous, ripple-marked shales and siltstones, containing a few thin, well-sorted, fine-grained sandstone beds. It is unconformably overlain by the Chinle Formation.

CHINLE FORMATION  The Chinle Formation consists of fluviatile and lacustrine sediments and averages about 400 feet thick along the ore belt. The lower part is the Moss Back Member (type section in White Canyon), which ranges from 10 to 80 feet in thickness and was deposited on the Cutler erosion surface by streams flowing westerly and northwesterly. The upper part of the Chinle, which ranges from about 275 to 400 feet in thickness in this area, has not been precisely correlated, but it is probably equivalent to the Church Rock Member in White Canyon (11) (32).

The Moss Back Member, the uranium host rock in the Big Indian belt, is predominantly a fluviatile, cross-bedded, calcareous, fine- to coarse-grained, arkosic, poorly-sorted sandstone with interbedded lenses of mudstone and calcarenite conglomerates. Sparse to abundant coalified plant material, mostly as woody trash,

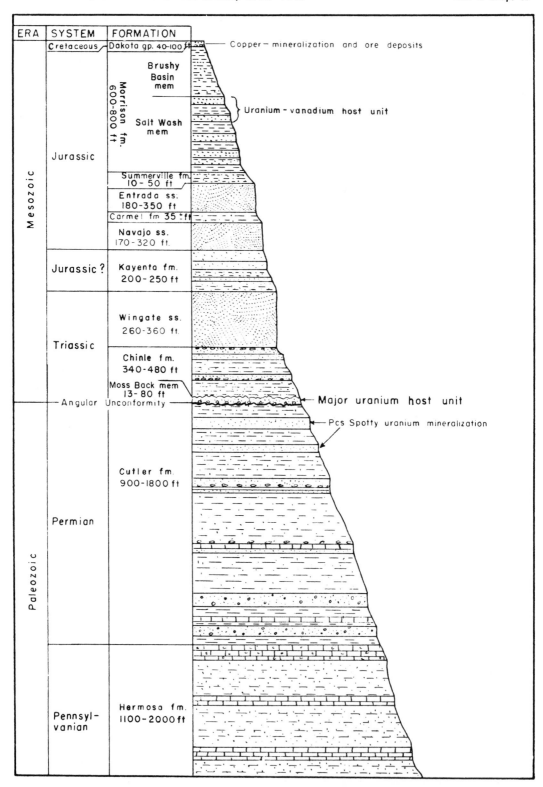

Fig. 3. *Generalized Stratigraphic Section in the Lisbon Valley Area.*

occurs in sandy lenses and pockets in and above the basal part of the ore sandstone or in highly coalified mudstone beds above the ore sandstone. The upper facies of the Moss Back Member contains proportionably more calcarenite conglomerate and micaceous mudstone beds. The Moss Back is gray-green to dark brown to greenish-gray and light gray in contrast to the variegated red color of the overlying upper Chinle beds. The Moss Back, in the northern and central part of the anticline, averages 45 feet thick, and in the southern part only 13 feet thick. The Moss Back grades lithologically upward into the variegated reddish-brown, purple, and lavender thin-bedded siltstones and claystones of the upper Chinle. Many of the claystones are lacustrine, and some of the clay beds are bentonitic, having been derived from volcanic ash (14) (38).

Isopach maps of the Chinle Formation indicate that this formation is over 100 feet thicker to the southwest and down dip from the central group of ore bodies than it is over these ore bodies, and, in contrast, it is about 100 feet thinner to the west or down dip from the northern group of ore bodies than it is over them. This notable thickening and thinning, 480 to 340 feet, is characteristic of fluviatile sediments in this region.

CUTLER-CHINLE NONCONFORMITY The angle of discordance between the beds of the Cutler and Chinle Formations at the northwest end of the ore belt is less than 2°; at the Standard-Big Buck mine near the central part of the ore belt, it is about 6°; and at the southeast end of the belt it is 3° to 4° (Figure 4). The paleotopography of the nonconformity varies from a fairly smooth, undulating surface with less than 5 feet of relief to a deeply scoured surface with over 30 feet of relief. The paleotopographic relief is greater in the area underlying the ore bodies in the central part of the ore trend where the greatest Permo-Triassic uplift and erosion occurred. Elongated scours or troughs are common, but no pattern or common orientation is evident. Distribution of the ore bodies is erratic with respect to the paleo lows and highs (Figure 5). Some large scours parallel the strike of the Cutler subcrops, and other scours trend down the dip of the Cutler beds, but there are no persistent or well-defined channels, such as the Shinarump (basal Chinle sandstone) channels of the White Canyon district (R. C. Malan, this volume). From general observation in the mines, it appears that the variable topography of the surface of nonconformity produced turbulent Moss Back

streams and caused shifting of stream courses. This resulted in many facies changes and the deposition of intraformational conglomerates and thick sandstone lenses favorable for uranium emplacement.

Conspicuous bleaching, manifested by color changes of the rock from darker to lighter colors, has taken place in the upper Cutler and lower Chinle beds, predominantly along the nonconformity. The bleaching is noticeable throughout the Big Indian ore belt, occurring in, above, and below ore and extending laterally away from the ore deposits, although it is less conspicuous in areas remote from known ore. The greater intensity of bleaching indicates that strong reducing conditions, caused either by migrating humic acids derived from diagenetic processes, by sulfur-bearing waters rising from salt anticlines or oil fields, or by uranium mineralizing solutions, existed in and near the ore.

In the vicinity of ore deposits, the contact between the Cutler and Chinle beds is often difficult to identify, particularly where the basal Chinle sandstone contains an abundance of reworked Cutler arkosic sands or where the beds on both sides of the contact are bleached to light gray, are calcareous, are uraniferous, and are otherwise lithologically similar. The following mineralogic, lithologic, and color variances may be used to locate the contact. The Moss Back has an abundance of carbon specks or woody fragments, an abundance of red or gray subangular chert pebbles and concretions, and some muscovite and chlorite. It is poorly sorted and is more cross-bedded and more coarse-grained than the Cutler. The Moss Back in the vicinity of ore is light to dark-gray, greenish-gray, or buff with some limonite specks. The Cutler sandstone, on the other hand, has an abundance of fresh and altered feldspar and of biotite altered to chlorite and has a few thin gray and red chert beds or lenses. It is normally friable and fairly well sorted. Where overlain by ore, it is mottled-gray, tan, brown, and pink. The mottled coloration grades downward into a normal rusty-red or brown (12) (18).

## Structure of the Anticline and Fault Zone

The Lisbon Valley anticline is a faulted asymmetric anticline, formed by flowage of salt and gypsum in the Paradox Member of the Hermosa Formation (Figures 1, 2, 3, 4). Over 6000 feet of salt and anhydrite beds have been penetrated in a few of the oil and gas test wells. According to Budd (22, p. 121),

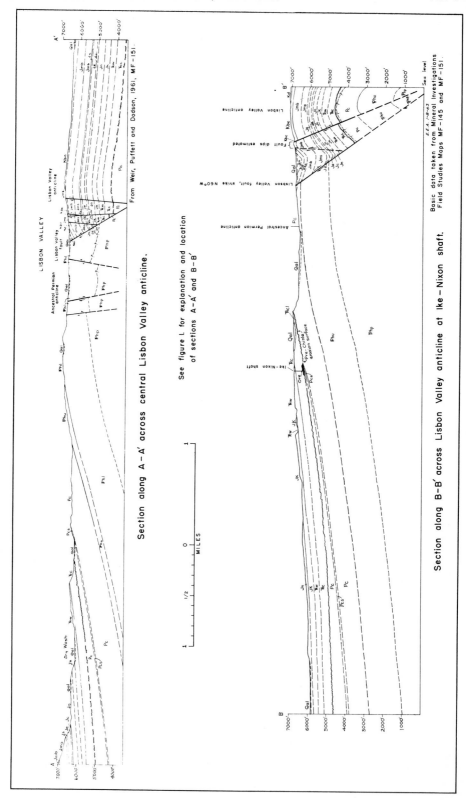

FIG. 4. *Geologic Sections across the Lisbon Valley Anticline.*

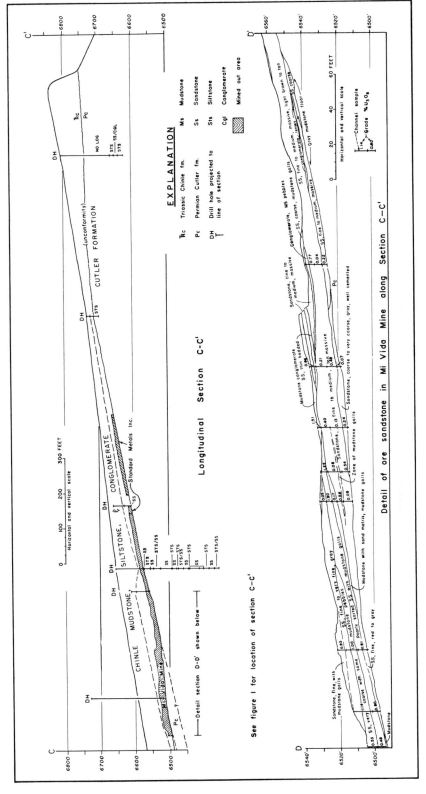

FIG. 5. *Geologic Section across the Ore Body at the Utex (Mi Vida) and Standard (Big Buck) Mines.*

the anticline has 3000 feet of vertical closure at the surface.

The anticline is faulted along its longitudinal axis by the Lisbon Valley normal fault. The fault plane has dips ranging from 50° to 85°NE. The surface trace of the fault is about 21 miles long but probably has a greater length at depth. The fault zone contains gouge, shear bands, and closely spaced step faults with a variety of dips, some reversed. Adjacent beds are often broken by drag and gash fracturing. At the crest of the anticline, the displacement is about 4000 feet, and Hermosa beds butt against Dakota beds; 8 miles northwest at the North Alice mine, the displacement is about 2000 feet, and Chinle beds butt against Morrison beds. Displacement decreases toward the northwestern and southeastern ends of the anticline, where the fault splits into a number of smaller hinge-type branches. These branching faults have different dips and strikes, which complicate interpretation of the structural geology. Horst and graben blocks are common along the fault but are most prevalent at each end of the anticline.

Almost all branch faults and major fractures are parallel or subparallel to the Lisbon Valley fault and the long axis of the anticline. At the Cord, Far West, and North Alice mines, subsidiary normal faults dip to the northeast and southwest and displace the ore bodies (18). Small strike slip displacements are measurable where the faults are exposed underground (Figure 2). The intensity of fracturing or the strike of the fractures does not appear to control the location or shape of any observed ore body. However, the longitudinal fractures, which are usually open, have caused mining problems and unexpected caving.

On the southwest flank, the Chinle beds dip 10° to 15°SW, and the Cutler beds dip as much as 20°. On the northeast flank, the Jurassic and Cretaceous beds dip 5° or less to the northeast. All bedding dips decrease toward the ends of the anticline (Figures 1, 4).

The difference in the shape of the Permian anticline from the shape of the Tertiary anticline is indicated by: (1) the approximate 15° angle between the strike of the Cutler beds and the strike of the Chinle beds in the northern part of the ore belt; (2) by the flattening of the dip in the Cutler beds on the crest of the ancestral Permian anticline; and (3) by the difference in the discordant angle of dip between the Chinle and Cutler beds throughout the anticline. The present location of the ancestral Permian anticline, now a minor anticline or flexure on the southwest

flank of the Tertiary anticline, is shown on Figures 1, 2, 4.

## Geologic History of the Anticline

In Pennsylvanian time, a deep basin developed southwest of the northwest-trending Uncompahgre Highland. A thick sequence of marine shale, limestone, and evaporites accumulated in this basin; the Lisbon Valley area is on the southwestern edge of the deeper part of this basin. Plastic flow of the evaporites into northwest-trending zones of weakness, folds or grabens, caused by fracturing and faulting, initiated the formation of the salt anticlines of the region (24,35).

Uplift of the ancestral anticline started in Late Permian time and probably continued into Late Triassic time. During this period, a local topographically high area developed that was exposed to extensive erosion. Hundreds of feet of Cutler and Moenkopi sediments were denuded from the crest of the ancestral arch, prior to deposition of fluviatile sediments over the structure by westward and northwestward flowing Moss Back streams. Mild uplift probably continued during deposition of the Moss Back sediments, as is evidenced by the occurrence of intraformational conglomerates that contain mudstone, calcarenite, and chert pebbles or fragments derived from erosion of Permian and Triassic sediments.

Emplacement of uranium minerals to form the Big Indian ore belt started soon after deposition of Chinle sediments and probably continued until movement of connate waters, caused by compaction of sediments or by regional folding, became negligible. During the ensuing Triassic, Jurassic, and Cretaceous periods, the Paradox Basin continued to subside and thousands of feet of eolian sandstone, fluviatile sandstone and mudstone, marginal marine, marine, lake sediments, and volcanic ash beds were deposited over the region. This depositional sequence was interrupted by brief hiatuses, which probably coincided with some regional uplifting. Proof of continuous or intermittent salt flowage during this period is not available, but it is inferred, from the variations in formation thicknesses and from the hiatuses in the overlying sediments, that mild salt flowage could have occurred intermittently (19,24,27).

Major uplift of the present Lisbon Valley anticline, which was superinduced on the ancestral Permian anticline, and the major offset along the Lisbon Valley fault occurred during the Laramide orogeny. Salt and gypsum flowed into the footwall of the fault, thrusting the footwall block upward and southwestward. Tilting of the southwest flank accentuated the arcuate bend of the ore belt. In the hanging wall, away from the fault zone, the beds were elevated only slightly. Rejuvenated flowage of evaporites and accompanying movement along the fault probably continued into Oligocene or Miocene time and still may be active. The unoxidized and unaltered condition of the brecciated fault zones along some of the small subsidiary faults, which displace uranium ore bodies at the north end of the belt, indicates that some of this faulting may be fairly recent.

Extensive erosion of the Lisbon Valley anticline started in Late Eocene time (35), removed approximately 5000 feet of sediments, and exposed Pennsylvanian sediments on the crest of the anticline.

## ECONOMIC GEOLOGY

### Size, Shape, Grade, and Distribution of the Uranium Deposits

The 15-mile-long by half-mile wide Big Indian ore belt consists of numerous intermittent bodies ranging in size from 500 to 1,500,000 tons of ore. A five-mile stretch of the south-central part of the ore belt has been removed by erosion, leaving about 6 miles of coalesced or separate ore deposits in the northern part of the belt and about 4 miles of scattered smaller deposits in the southern quarter. Possible evidence that ore bodies have been eroded, between the Standard and Serviceberry mines, is indicated by the anomalous radioactivity in the alluvium of Big Indian Valley in the vicinity of Big Indian Rock (7), and by the existence of uranium mineralization on a point of Moss Back cropping out in section 30, T30S, R25E.

The ore belt contains two large groups of nearly coalesced deposits from which 74 per cent of the districts ore has come as of July 1, 1965. Approximately 20,000,000 pounds of $U_3O_8$ has been mined from the north group of deposits and 16,000,000 pounds from the central group. Three intermediate-size deposits lie between the two larger groups and production from them ranges from 2,800,000 to 3,500,000 pounds of $U_3O_8$. Eight smaller ore bodies, ranging in size from 500 to 90,000 tons or ore, are scattered throughout the ore belt (Figure 2). In general, the ore deposits range from a few inches to 45 feet in thickness, with an average thickness of 6 feet. The ore has an average grade of 0.39 per cent $U_3O_8$, contains about 15 per cent $CaCO_3$, and has a dry bulk density factor of about 13 cubic feet per ton. The northern 6 miles of the big Indian ore belt contains about 2600 acres, of which about 300 acres is underlain by ore. All the deposits are irregular, amoeba-shaped masses that lie concordantly to the bedding. Within each deposit, there is a notable variability in thickness and grade, but in general, the grade of uranium ore drops sharply at the edges of the ore bodies. This sharp cutoff precludes the probable existence of any large reserves of low-grade uranium ore.

### Mineralogy and Geochemistry of the Big Indian Ore

Uraninite, $UO_2 \cdot UO_3$, is the principal uranium ore mineral. Intimately associated are small amounts of coffinite, $U(SiO_4)_{1-x}(OH)_{4-x}$, and the vanadium minerals montroseite, $VO(OH)$, doloresite, $V_2O_4 \cdot nH_2O$, and vanadium clay or vanadium hydromica. Secondary uranium and vanadium minerals, such as metatyuyamunite, $Ca(UO_2)_2(VO_4)_2 \cdot 5\text{-}7H_2O$, pascoite, $Ca_3V_{10}O_{28} \cdot 16H_2O$, and corvusite, $V_2O_4 \cdot 6V_2O_5 \cdot nH_2O$ are found in areas of oxidation, but they are of no quantitative importance (8). Beautiful high-grade uraninite ore specimens, usually replacements of coalified wood, have been collected from many of the mines; some samples assay about 80 per cent $U_3O_8$. Ore minerals dominantly fill the interstices between grains of sand, but they also replace calcite, carbonaceous plant remains, and, to a lesser extent, detrital quartz and feldspar grains and other accessory minerals.

The principal detrital constituents of the quartzose sandstone host rock are quartz, limy rock fragments (calcarenites), feldspar, mica, chert, clay, and coalified plant remains. Minerals commonly associated in anomalous quantities with the ore deposits are barite, celestite, galena, pyrite, chalcopyrite, chalcocite, molybdenite, chalcedony, greenockite, and sericite (8) (31).

Paragenetic studies (8) indicate that detrital minerals were deposited and cemented by calcite; silica and barium sulfate solutions precipitated authigenic silica and barite; local deformation of the mudstones and some of the less competent sandstone beds resulted in fracturing of quartz, feldspar, and calcite; and ore solutions penetrated the sandstone, replaced some minerals and plant remains, and precipitated in the voids. In places, there was also fracturing of the detrital grains after uraninite was formed, but before introduction of some of the galena and greenockite. Calcite was probably deposited, dissolved, and redeposited many times; consequently, calcite is replaced by uraninite and other minerals and in turn replaces them (13, p. 548).

Geochemical and mineralogic studies (17, 31) show that: (1) lead correlates with uranium, (2) molybdenum occurs as a halo above some of the central and southern ore bodies, (3) vanadium ranges from one-third to three times the per cent of uranium in the Mi Vida mine and in other deposits to the south, and (4) is negligible in the Ike-Nixon mine and in other deposits to the north. Most of the ore from the Mi Vida mine on the south may be classified as uranium-vanadium or vanadiferous uranium ore because the ore will usually assay over 0.40 per cent $V_2O_5$. Other extrinsic elements, anomalous in or around the ore bodies, are arsenic, barium, cadmium, copper, iron, strontium, and zirconium and sometimes selenium, cobalt, zinc, yttrium, and columbium.

Mill heads average about 15 per cent $CaCO_3$, but the $CaCO_3$ content differs considerably from deposit to deposit within the ore belt and also from place to place in the individual ore bodies. The calcium carbonate concentration is highest where it occurs as a cement and as detrital grains. Rarely does high-grade ore assay over 20 per cent $CaCO_3$. Commonly the $CaCO_3$ has been recrystallized to form visible sand crystals showing a poikilitic texture (8). Although these deposits are hundreds of feet above the present water table, the low permeability, due to cementation by calcite, has protected them from oxidation and leaching by vadose waters.

Coalified wood fragments are common, but show little spatial relation or quantitative correlation to the ore deposits. Coaly material is usually in the clayey beds immediately above the ore-bearing sandstone, where it is concentrated along bedding planes and is disseminated in mudstone conglomerates. Because the coaly layer is usually barren and is embedded in mudstone, it is not broken into during mining, which may account for the low carbon content of the mill heads. Within the ore bed, some coalified logs are partially replaced by uraninite, but others are not.

## Isotope Analyses and Geochronology of the Big Indian Ore

Uraninite ore samples collected from the North Alice and South Almar (Cord) mines (Figure 1) by the author were submitted to John Rosholt (U.S.G.S.) for isotope analysis. The results of isotope fractionation studies on $U^{238}$, $U^{235}$, and $U^{234}$, as given in J. N. Rosholt's letter on May 13, 1964, for these samples, are summarized in this quotation: "The Lisbon Valley samples studied show extremely good examples of radioactive equilibrium between $U^{238}$ and $U^{234}$. No other area in sandstone-type ore deposits, yet analyzed, has indicated such an environment of consistently stable uranium mineralization. The deviation from the equilibrium $U^{234}/U^{235}$ ratio is within $\pm 1$ per cent for all six samples, thus within our experimental error. There is no evidence to indicate that the uranium mineralization in the fault zone (at the mines mentioned above) occurred at a more recent time than that some distance from the fault zone."

Uranium-lead isotope age determinations range from 85 to 295 m.y. and disagreements on interpretation and value of the age determinations are acknowledged by many (16, 34,36). Most of the $U^{235}/Pb^{207}$ apparent age determinations range between 150 and 210 m.y. and imply that the ore deposits are Jurassic to Triassic in age.

## Favorable Lithology of the Host Rock

Lithologically the most favorable host rock is a gray, poorly-sorted, fine- to coarse-grained, calcareous, arkosic quartzose sandstone containing some interbedded mudstone and limestone pebble conglomerates and some mudstone and siltstone lenses, all poorly sorted and of variable permeability. The highest-grade ore is in semi-permeable, fine-grained, sandy lenses that contain less than 30 per cent calcium carbonate as cement or as clastic limestone grains. Jasper, smoky quartz, pyrite, and a reddish-brown or pink calcite, colored by hematitic inclusions, are common in the higher-grade deposits. There is an abundance of mudstone pebbles and coalified wood-trash either in or directly overlying the host rock.

## Stratigraphic Position of the Ore Bodies

The Chinle Formation thickens and thins notably, 480 to 340 feet in short distances. There is no apparent relation of the thickness of the Chinle Formation or of the Moss Back Member to the position of the ore belt.

Usually, within the ore belt, the larger ore deposits are where the Moss Back Member is thicker, such as across the north-central part of the anticline, but the higher-grade deposits in the northern ore bodies are in thin basal sandstone units. Although thinner and lithologically less favorable Moss Back sandstone is spread over the southern part of the anticline, even this less favorable host rock, within the ore belt, contains numerous small but minable ore deposits. Persistent paleostream channels, which were scoured into the Cutler paleotopography and later filled with thick lenses of favorable Moss Back sandstone, do not occur under the Big Indian ore belt. Elongated sandstone-filled scours, containing uranium ore, do occur at some of the larger ore deposits, but their control of any ore distribution is local.

Prevalently the lowest sandstone or conglomerate unit of the Moss Back Member is the uranium host rock, and it rests on or within a few feet of the Cutler-Chinle nonconformity. Within a host unit, the lithology and the sedimentary structural features, which directed the course of the ore fluids, control the localization of the ore. The ore bodies may cut across bedding planes within the host and may occur above and below a mudstone seam, but they do not cross through intercalated mudstone beds.

Lekas and Dahl (12) in 1956 and Puffett and Weir (21) in 1959 recognized the spatial relation of the Moss Back ore deposits to the subcrop, at the nonconformity, of a Cutler sandstone unit, called locally the "sugar sand." Where this unit crops out at the Standard mine portal, it is a 30-foot-thick, friable, porous, saccharoidal, fine-grained, light-gray sandstone. Underground it can be traced from the portal, in a northwesterly direction, to the north end of the Mi Vida mine. This "sugar sand" has not been definitely traced north of the Mi Vida mine. It is believed that the northern ore bodies are underlain by different, stratigraphically lower, but physically similar Cutler sandstone units. The common occurrence of this sandstone unit or its physical equivalent under the ore belt has been a useful guide in exploration. Lekas and Dahl suggest that the subcrop of this porous Cutler sandstone unit acted as a conduit for ore solutions, thus localizing the ore deposits within the delineated belt, and Noble (37) suggests that a pressure change occurred at the intersection of the two (Cutler and Chinle) acquifers. It is suggested by the writer that the Cutler porous sandstone units may have acted as a conduit for rising gases or solutions that precipitated the uranium minerals or changed the pH or Eh sufficiently to permit precipitation.

It has been suggested by some geologists (6,12) that the position of the erosional pinch-out of the Moenkopi Formation, around the southwest flank of Lisbon Valley anticline, was influential in positioning the ore belt. They theorized that it acted as an impervious cap to rising uraniferous solutions, until the pinch-out was reached or that the uraniferous solutions, migrating laterally in the fairly permeable Moss Back beds, were confined between the quite impermeable Moenkopi and upper Chinle beds. The validity of this hypothesis depends on the location of the Moenkopi pinch-out, which has not been determined.

## Structural Position of the Ore Bodies

A structure map, (Figure 2), showing contours drawn on the nonconformity at the base of the Chinle Formation, shows that the 6-mile-long northern half of the ore belt lies between the 6000- and 6700-foot contours and the southern one-quarter of the ore belt is between the 5800- and 6800-foot contours. Furthermore, over 90 per cent of the ore occurs between the 6200- and 6700-foot contours in the northern half of the ore belt. Exploration drilling to date has failed to discover large ore bodies on the southwest flank of the anticline outside of this delineated belt. This vertical limitation was first referred to in 1955 as the "magic contour interval"(6).

The horizontal width of the ore bodies or ore body aggregates along the trend of the ore belt ranges from 800 to 3000 feet in the northern section. At the northwest and southeast ends, where the ore belt is intersected by the Lisbon Valley fault, the ore bodies are spread over a wider area, 3000 feet to over 3600 feet across the trend, giving the appearance that the ore is spread out along the Lisbon Valley fault. This spread of the ore belt may be explained by a flattening of the angle of nonconformity between the Cutler and Chinle beds from 6° at the crest to 3° at the ends, and by a gentle flattening of the dip of these beds at each end of the anticline.

The limitation of the ore belt within the "magic contour interval" may imply a water table or a water-oil-gas interface as a control for positioning the Big Indian uranium deposits (30). Another interpretation is that the structural position of the ore belt may represent an ancestral water table where the descending vadose water encountered connate water.

## Summary of Geologic Characteristics of the Big Indian Ore Belt

(1) Most of the ore in the 15-mile-long by half-mile wide Big Indian ore belt is within a 500-foot elevation interval.

(2) The large ore bodies are hundreds of feet above the present water table, but more than 95 per cent of the ore is unoxidized uraninite.

(3) The spread of the ore bodies across the belt is appreciably wider at the northwest and southeast ends of the anticline, where the beds have a more gentle dip and the Cutler-Chinle angle of nonconformity is less than it is along the central part of the southwest flank.

(4) The bedded ore deposits are displaced, by Tertiary normal faulting, at the North Alice, Far West, Cord, and Continental (Section 36) mines.

(5) The main ore deposits are usually confined to the lowest sandstone bed in the Moss Back Member of the Chinle Formation.

(6) Uranium deposits appear to be more persistent or to occur more often where porous, friable, fine-grained Cutler sandstone units, bleached to light gray, subcrop at the nonconformity under lithologically favorable Moss Back sandstone.

(7) Small deposits of uraninite ore are in the Cutler sandstone adjacent to the Moss Back ore deposits. Carnotite deposits also occur about 500 feet laterally up-dip and at approximately the same elevation, but lower stratigraphically, in the Cutler section than the ore occurrence subjacent to the Moss Back ore.

(8) Persistent bleaching, resulting from the leaching of ferric oxide, is unmistakable in the Chinle and Cutler Formations in and around uranium ore and particularly along the nonconformity.

(9) Abundant coalified plant material occurs as lenses or as aggregations in and above the basal ore sandstone or as carbon trash disseminated in carbonaceous mudstone beds above the ore sandstone.

(10) Zoning of vanadium within the ore belt is manifested by the notable increase of the V:U ratio in the Mi Vida and other ore bodies to the south and by some differences, although erratic, in the up dip and down dip vanadium to uranium ratio within the ore bodies.

(11) Anomalous copper is spotty in Moss Back uranium deposits and in nearby barren Moss Back sandstone. In contrast, copper ore containing only traces of uranium, is common in the Dakota Group sandstones at many places along the Lisbon Valley fault.

(12) Molybdenum occurs as an anomalous halo around some of the southern ore bodies and is anomalous in samples collected from ore deposits adjacent to the subsidiary faults in the northern part of the ore belt.

(13) No anomalous radioactivity has been found along the trace of the Lisbon Valley fault or along the fault zone at depth, except where the Moss Back, containing uranium mineralization, butts against the fault, and ore bodies do not appear to have been localized or shaped by fractures.

## Ore Genesis and Mode of Deposition

Except for slight variations in their theory, Steen (2), Isachsen (6), Loring (18), Weir and Puffett (29), and Jacobs and Kerr (39), postulate a hydrothermal origin, advocating access of rising uranium, vanadium, and copper solutions by way of the Lisbon Valley fault or by way of some unnamed conduit. Lekas and Dahl (12) discuss a variety of hypotheses of origin and localizing features. Kennedy (25,31), and Noble (37), each postulate, but in a slightly different manner, that connate waters, which were expelled from fluviatile sediments during compaction, may have been the ore-forming and ore-transporting fluid. Waters and Granger (3) and Schultz (14, 38), discuss how the extrinsic elements may have been derived from Chinle volcanic ash or tuff beds.

The possibility that uranium and other extrinsic elements in the deposits could have been precipitated by reduction from telethermal, connate, or meteoric waters leaves the genesis open to many hypotheses. The hypothesis that the ore solutions rose through the thick series of Pennsylvanian gypsum, halite, shale and limestone beds, and then used the Cutler porous sandstone beds as conduits to known points of deposition, seems improbable. Also, the absence of supporting evidence of hydrothermal activity and the absence of any ore deposits localized along fractures or faults that could have acted as conduits for ascending thermal solutions makes it improbable that these deposits are of hydrothermal-hypogene origin.

In general, the geologic features of these deposits appear to be congruous with the theories postulated by Kennedy, Noble, and the writer, and with some of the theories suggested by Lekas and Dahl.

The writer believes that the uranium was indigenous to the Chinle Formation; was mobilized or released from the siliceous glass by diagenetic processes, which started soon after deposition of sediments; was moved laterally or outward from the least permeable sediments into more permeable sandstones, as consolidation and compaction continued; and was probably comingled with vadose ground waters. Under hydrostatic forces or water of compaction forces (30,37), the solutions moved laterally through the permeable Moss Back sandstone toward points of expulsion or precipitation. At times, movement may have been rejuvenated by ground water flushing the aquifer.

It is reasonable to assume that the uraniferous solutions contained some natural gases, such as methane and dense carbon dioxide; that they were mixed with ground waters; that the uranium was transported as a nearly neutral, highly stable, uranyl dicarbonate or tricarbonate complex; and that precipitation resulted from reduction of hexavalent uranium to form uraninite (33). As explained by Garrels (5), the part played by liquid or gaseous $CO_2$ in the dissolution and transportation of uranium is not fully understood, but, because large quantities of $CO_2$ are associated with the Mississippian oil reservoir under the northwest flank of the Lisbon Valley anticline, this probable relationship should be considered and studied. Upon reaching a favorable depositional site, precipitation may have been caused by change in the pH and Eh, or by reduction in pressure, or by encountering hydrogen sulfide produced by bacterial action on carbonaceous material (34). The presence of an abundance of coalified plant material within the host rock and the probable leakage into the Cutler-Chinle nonconformity of natural gas or sulfur-solutions from the Mississippian oil and gas reservoirs beneath the ore belt, assures the existence of sufficient reducing agents (16).

The mechanism suggested, if effective, implies that these deposits were emplaced during the fairly static period following deposition of the fluviatile Chinle Formation; that emplacement continued during burial under thousands of feet of Triassic and Jurassic sediments; but that emplacement ended prior to the start of the Laramide orogeny. The ore deposits were positioned within the ore belt by either a water-gas interface (30) or by a connate-vadose water interface near the crest of the ancestral Permian anticline, or by subcrops of either the porous upper Cutler sandstones or the Moenkopi Formation at the nonconformity around the flanks of the ancestral anticline. The multiple oxidation-solution-migration-accretion theory of Gruner (9), does not explain some of the geologic conditions described above, but it explains enough of these conditions to make this theory worthy of consideration.

## Ore Guides and Potential of the Area

Future exploration along the Big Indian ore belt may be guided by the stratigraphic and structural ideas and favorable host rock characteristics described herein and by giving due consideration to the probable shape and size of the target. Favorable Moss Back beds are spread over the northern part of the Permian anticline, with the thickest sandstone beds on the crest of the anticline. Less favorable sandstone beds cover the southern part of the anticline.

The more irregular the relief of the Cutler paleotopography, the better are the chances of finding lithologically favorable Moss Back sandstones. The scours trending parallel to the strike of Cutler beds are deeper and commonly contain uranium deposits on the up dip side of the scour.

The favorable characteristics of the Moss Back on the northeast side of the Lisbon Valley fault have not been thoroughly evaluated. A Moss Back thickness of about 33 feet has been interpreted from a gamma log of a deep hole drilled northeast of the fault. Geologic logs of other deep test wells drilled northeast of the fault, and still on the structure, indicate a variable thickness of Moss Back sandstone and probably the occurrence of Moenkopi sandstone under the Moss Back. Unless more rapid changes exist in the character of the Chinle streams than has been observed, it appears logical to assume that favorable Moss Back sandstone beds do exist across the Lisbon Valley fault.

In summary, it is believed that the original ore belt encircled the crest of the Permian anticline as a band. The southwest flank of the anticline has been thoroughly explored for large deposits. An extension of the Big Indian ore belt, similar in size and grade to the known ore belt, probably occurs in the downthrown block northeast of the Lisbon Valley fault at depths of 2400 to 2700 feet beneath the Dakota-capped surface. As of the fall of 1965, only nine holes had been drilled into the Moss Back sandstone northeast of the fault, which leaves this area as the most favorable unexplored area remaining on the Lisbon Valley anticline.

## REFERENCES CITED

1. Dix, G. P., Jr., 1953, The uranium deposits of Big Indian Wash, San Juan County, Utah: U.S. Atomic Energy Comm. RME-4022, 15 p.

2. Steen, C. A., *et al.*, 1953, Uranium mining operations of the Utex Exploration Co. in the Big Indian district, San Juan County, Utah: U.S. Bur. Mines I. C. 7669, 13 p.

3. Waters, A. C. and Granger, H. C., 1953, Volcanic debris in uraniferous sandstones and its possible bearing on the origin and precipitation of uranium: U.S. Geol. Surv. Circ. 224, 26 p.

4. Gruner, J. W., *et al.*, 1954, The mineralogy of the Mi Vida uranium ore deposit of the Utex Exploration Co. in the Indian Wash. area, Utah: U.S. Atomic Energy Comm. RME-3094, p. 15–27.

5. Garrels, R. M. and Richter, D. H., 1955, Is carbon dioxide an ore forming fluid under shallow earth conditions?: Econ. Geol., v. 50, p. 447–458.

6. Isachsen, Y. W., 1955, Uranium deposits of the Big Indian Wash-Lisbon Valley mining district, San Juan County, Utah: Nuclear Eng. and Science Cong. Cleveland, Ohio, Dec. 12–16, Preprint 281, Am. Inst. Chem. Engrs., 13 p.

7. Research, Inc.—Dallas, Texas (Tripp, R. M.), May 1955, Research and development of geophysical and geochemical techniques for uranium exploration on the Colorado Plateau: U.S. Atomic Energy Comm., Final Contract Rept. RME-3111, 94 p.

8. Gross, E. B., 1956, Mineralogy and paragenesis of the uranium ore, Mi Vida mine, San Juan County, Utah: Econ. Geol., v. 51, p. 632–648.

9. Gruner, J. W., 1956, Concentration of uranium in sediments by multiple migration-accretion: Econ. Geol., v. 51, p. 495–520.

10. Herman, G. and Sharps, S. L. 1956, Pennsylvanian and Permian stratigraphy of the Paradox salt embayment: *in Geology and economic deposits of east central Utah,* Intermountain Assoc. Petrol. Geols. 7th Ann. Field Conf. Guidebook, p. 77–84.

11. Johnson, H. R., Jr. and Thordarson, W., 1956, Regional synthesis studies, Utah and Arizona: *in Geologic investigations of radioactive deposits:* U.S. Geol. Surv. TEI-640, p. 188–191.

12. Lekas, M. A. and Dahl, H. M., 1956, The geology and uranium deposits of the Lisbon Valley anticline, San Juan County, Utah: *in Geology and economic deposits of east central Utah,* Intermountain Assoc. Petrol. Geols. 7th Ann. Field Conf. Guidebook, p. 161–168.

13. Holland, H. D., *et al.*, 1957, The use of leachable uranium in geochemical prospecting on the Colorado Plateau. I. The distribution of leachable uranium in core samples adjacent to the Homestake ore body, Big Indian Wash, San Juan County, Utah: Econ. Geol., v. 53, p. 546–569.

14. Schultz, L. G., 1957, Studies of clay in Triassic rocks: *in Book 2, Geologic investiga-*tions of radioactive deposits: U.S. Geol. Surv. TEI-690, p. 497–504.

15. Hunt, C. B. and Waters, A. C., 1958, Structural and igneous geology of the La Sal Mountains, Utah: U.S. Geol. Surv. Prof. Paper 294-I p. 305–364.

16. Jensen, M. L., 1958, Sulfur isotopes and the origin of sandstone-type uranium deposits: Econ. Geol., v. 53, p. 598–616.

17. Kennedy, V. C., 1958, Geochemical studies in the Lisbon Valley area: *in Geologic investigations of radioactive deposits:* U.S. Geol. Surv. TEI-740, p. 29–36.

18. Loring, W. B., 1958, Geology and ore deposits of the northern part of the Big Indian district, San Juan County, Utah: Unpublished Ph.D. dissertation, Univ. Ariz., 75 p.

19. Shoemaker, E. M., *et al.*, 1958, Salt anticlines of the Paradox Basin: *in Guidebook to the geology of the Paradox Basin,* Intermountain Assoc. Petrol. Geols. 9th Ann. Field Conf. p. 39–59.

20. Byerly, P. E. and Joesting, H. R., 1959, Regional geophysical investigations of the Lisbon Valley area, Utah and Colorado: U.S. Geol. Surv. Prof. Paper 316-C, p. 39–50.

21. Puffett, W. P. and Weir, G. W., 1959, Geologic mapping, Lisbon Valley, Utah-Colorado: U.S. Geol. Surv. TEI-752, p. 13–18.

22. Budd, H., 1960, Notes on the Pure Oil Company discovery at northwest Lisbon: *in Guidebook—Geology of the Paradox Basin fold and fault belt,* Four Corners Geol. Soc. 3rd Field Conf., p. 121–124.

23. Buss, W. R., 1960, Physiography and geomorphic history of part of southeastern Utah and adjacent Colorado: *in Guidebook—Geology of the Paradox Basin fold and fault belt,* Four Corners Geol. Soc. 3rd Field Conf., p. 31–32.

24. Elston, D. P. and Landis, E. R., 1960, Cutler unconformities and the early growth of the Paradox Valley and Gypsum Valley salt anticlines: U.S. Geo. Surv. Prof. Paper 400, p. B261.

25. Kennedy, V. C., 1960, Origin of uranium-vanadium deposits in the Lisbon Valley area, San Juan County, Utah (abs): Geol. Soc. Amer. Bull., v. 71, p. 1904.

26. Kunkel, R. P., 1960, Permian stratigraphy in the salt anticline region of western Colorado and eastern Utah: *in Guidebook—Geology of the Paradox Basin fold and fault belt,* Four Corners Geol. Soc., 3rd Field Conf. p. 91–97.

27. Stewart, J. H. and Wilson, R. F., 1960, Triassic strata of the salt anticline region, Utah and Colorado: *in Guidebook—Geology of the Paradox Basin fold and fault belt,* Four Corners Geol. Soc., 3rd Field Conf., p. 98–106.

28. Weir, G. W., *et al.*, 1960, Preliminary geologic map and section of the Mount Peale 2 SE quadrangle, San Juan County, Utah:

U.S. Geol. Surv., Mineral Investigations, Field Studies Map MF-143, 1:

29. Weir, G. W. and Puffett, W. P., 1960, Similarities of uranium-vanadium and copper deposits in the Lisbon Valley area, Utah and Colorado: U.S.A.: 21st Int. Geol. Cong. Rept., pt. 15, p. 133–148.

30. Germanov, A. I., 1961, Geochemical and hydrodynamic conditions of epigenetic uranium mineralization in petroleum-water zones: Geochemistry, no. 2, p. 107–120 (translated from Geokhimiya).

31. Kennedy, V. C., 1961, Geochemical studies of mineral deposits in the Lisbon Valley area, San Juan County, Utah: U.S. Geol. Surv. Open File Rept., p. 156.

32. Weir, G. W., *et al.*, 1961, Preliminary geologic map and section of the Mount Peale 4 NW quadrangle, San Juan County, Utah: U.S. Geol. Surv. Mineral Investigations, Field Studies Map MF-151, 1:

33. Hostetler, P. B. and Garrels, R. M., 1962, The transportation and precipitation of uranium and vanadium at low temperatures, with special reference to sandstone-type uranium deposits: Econ. Geol., v. 57, p. 137–167.

34. Adler, H. H., 1963, Concepts of genesis of sandstone-type uranium ore deposits: Econ. Geol., v. 58, p. 839–852.

35. Case, J. E., *et al.*, 1963, Regional geophysical investigations in the La Sal Mountains area, Utah and Colorado: U.S. Geol. Surv. Prof. Paper 316-F, p. 106.

36. Miller, D. S. and Kulp, J. L., 1963, Isotopic evidence on the origin of the Colorado Plateau uranium ores: Geol. Soc. Amer. Bull., v. 74, p. 609–630.

37. Noble, E. A., 1963, Formation of ore deposits by water of compaction: Econ. Geol., v. 58, p. 1145–1156.

38. Schultz, L. G., 1963, Clay minerals in Triassic rocks of the Colorado Plateau: U.S. Geol. Surv. Bull. 1147-C, p. C35-C65.

39. Jacobs, M. B. and Kerr, P. F., 1965, Hydrothermal alteration along the Lisbon Valley fault zone, San Juan County, Utah: Geol. Soc. Amer. Bull., v. 76, p. 423–440.

# 38. The Uranium Mining Industry and Geology of the Monument Valley and White Canyon Districts, Arizona and Utah

ROGER C. MALAN*

## Contents

## Illustrations

* U.S. Atomic Energy Commission, Grand Junction, Colo.

# Tables

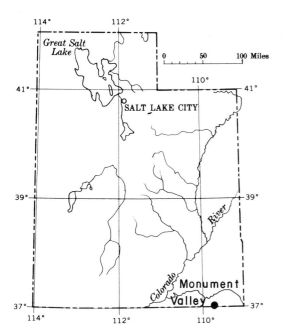

Most of the deposits are in an arcuate belt, convex to the west. This belt, 3 to 12 miles wide, extends from Monument Valley northward nearly 130 miles. It is along the western flank of the ancient Monument Valley-Monticello upland, an area that was slightly uplifted at the beginning and again at the end of deposition of the Shinarump. Erosion and subsequent reworking of the Shinarump sediments in the vicinity of this upland are postulated to have made possible the transportation, in solution, of the uranium contained in these sediments. The soluble uranium probably was carried by migrating ground water into sites favorable for precipitation in Shinarump beds bordering the upland. Thus, the original very small amounts of dispersed uranium in the early Shinarump sediments were accreted in the favorable belt where the accumulations of carbonaceous plant remains and their decay products provided a persistent reducing environment.

## INTRODUCTION

The Monument Valley and White Canyon districts cover about 3000 square miles in the central portion of the Colorado Plateau in northeastern Arizona and southeastern Utah (Figure 1). In these districts, deep canyons dissect a high tableland. Altitudes range from 4000 to 9000 feet. Most of the uranium deposits in these districts are in an arcuate belt which is 130 miles long and between 3 and 12 miles wide.

A study by the U.S. Atomic Energy Commission of the uranium resources in the Monument Valley and White Canyon districts was started in 1962 and completed in 1964; it is the basis of this paper. This project, originally assigned to R. G. Young and E. A. Noble, was completed by the present author. Many of the original stratigraphic interpretations, ideas on the origin of the uranium, and determination of environmental favorability, developed during the time when R. G. Young was in charge of the project, were later reported by Young (23). As a result of subsequent investigations, some of the earlier ideas have been modified by the writer in this report.

## ABSTRACT

The Monument Valley and White Canyon districts are in northeastern Arizona and southeastern Utah. Exploration and mining for uranium has been conducted in these districts since the late 1940's. In July 1965, ore reserves plus ore production at 174 properties were approximately 3.3 million tons of ore containing about 19 million pounds $U_3O_8$, of which about 10 per cent remained in reserves. The two largest mines, Monument No. 2 and Happy Jack, together account for about 45 per cent of the sum of production and reserves. Approximately half the deposits in these districts contain less than 1000 tons of ore.

Nearly 5000 feet of Permian, Triassic, and Jurassic sedimentary rocks, mainly continental in origin, are exposed in these districts. All the important uranium deposits are in the Shinarump, the basal member of the Chinle Formation of Triassic age. The Shinarump lies on a widespread unconformity. It is composed of fluvial sediments, generally less than 100 feet thick, that were deposited by streams flowing from a source to the south.

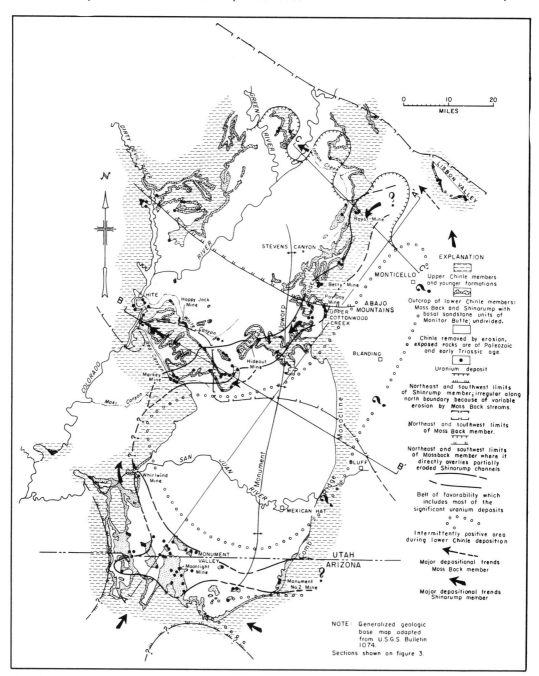

FIG. 1. *Generalized Geologic Map of the Monument Valley and White Canyon Districts, Arizona and Utah.*

About 174 properties with recorded production and/or reserves and about 100 additional occurrences without production or reserves were investigated. Stratigraphic sections of the lower Chinle Formation were measured at many localities and samples were collected for petrographic study. Fluvial channels at the base of the Chinle Formation of Triassic age, which are the sites for nearly all the uranium deposits, were mapped on aerial photographs.

The channel segments were correlated insofar as was possible, and, by projections, a complex paleodrainage system was reconstructed.

The helpful suggestions and the critical review of this paper by E. W. Grutt, Jr., U.S. Atomic Energy Commission, and R. P. Fischer, U.S. Geological Survey, are gratefully acknowledged. The author also wishes to express his appreciation for the helpful cooperation given him by mine operators and owners on numerous occasions.

## URANIUM INDUSTRY

### History

WHITE CANYON   The copper, which is associated with many of the uranium deposits in the White Canyon district, was first discovered in the 1880's. B. S. Butler (1) of the U.S. Geological Survey identified uranium minerals in the White Canyon area in 1920 at what is now the Happy Jack mine. The first recorded uranium production in the area was from the Fry 4 claim in 1946. Uranium mining at the Happy Jack mine, which subsequently was developed into the largest mine in the district, began in 1949. During that year, Vanadium Corporation of America constructed a small uranium mill on the Colorado River near Hite and operated it until 1953. Prospecting was intense from 1948 to 1951 in the White Canyon, Red Canyon, and Deer Flat portions of the district, and as a result many claims were staked. Drilling programs by the U.S. Atomic Energy Commission and U.S. Geological Survey in the early 1950's stimulated a new wave of prospecting and discovery in 1953 and 1954, extending the area of known deposits northeastward into the Elk Ridge locality.

Although the number of producing mines increased from 6 in 1952 to 47 in 1956, only 13 mines were active in July 1965. A total of 113 mines have contributed to the total production. Production for a single year was greatest in 1958.

MONUMENT VALLEY   The Monument No. 2 mine, discovered by a Navajo in 1942, was developed into the first producing uranium mine in Monument Valley in 1948.

In the late 1940's and early 1950's, many deposits, small to medium in size, were discovered in paleochannel exposures at rim outcrops. In 1955 and 1956, a cluster of important deposits including the Moonlight mine was discovered by Industrial Uranium Corporation

in buried channels at moderate depths in the central portion of the Monument Valley district. Production in Monument Valley reached a peak in 1955, when 14 mines were operating.

Most of the ore that has been produced from the Monument No. 2 mine has been beneficiated in an upgrader located at the mine site. From 1948 to 1957, ore from other mines in Monument Valley was shipped to mills on the Colorado Plateau outside Monument Valley. A mill at Mexican Hat, Utah, constructed by Texas Zinc Minerals Corporation in 1957, was operated until March 1965. While it was operating, this mill processed most of the ore produced in Monument Valley except that from the Monument No. 2 mine. It also processed most of the ore produced in the White Canyon district. Since March 1965, most of the ores formerly treated by the Texas Zinc Minerals mill have been sent to the Atlas Minerals Corporation mill at Moab, Utah, for processing.

The Monument No. 2 mine has produced about 60 per cent of the ore mined in the district. Fifty-three properties in all have produced uranium, but only 4 mines were operating in July 1965.

### Production and Reserves

There are 54 properties in Monument Valley and 120 in White Canyon with either available reserves or reserves and recorded production. As of July 1, 1965, the sum of the production and the remaining available reserves for the two districts combined was about 3,300,000 tons averaging 0.29 per cent $U_3O_8$ or about 19,000,000 pounds $U_3O_8$; 49 per cent of this total was credited to the Monument Valley district and 51 per cent to the White Canyon district. In July 1965, ore reserves in Monument Valley and in White Canyon were about 96,000 and 217,000 tons respectively.

### Exploration

In Monument Valley, about 1.1 million feet of drilling, resulting in the discovery of about 1.448 million tons of ore, has been done for an estimated total cost of $1.5 million. The estimated average drilling costs per ton of ore and per pound $U_3O_8$ have been $1.04 and $0.16 respectively. The tons of ore developed per foot of drilling has been about 1.3.

In the White Canyon district, about 1.2 million feet of drilling, resulting in the discovery of about 1.864 million tons of ore, has been done for an estimated total cost of about $2.76 million. The estimated average costs per ton of ore and per pound U₃O₈ have been $1.48 and $0.28 respectively. The tons of ore developed per foot of drilling has been about 1.6. The drilling cost per ton of ore developed ranges from $0.25 to over $5.00.

In places where the depths to the base of the Shinarump do not exceed 300 feet, holes are drilled in an irregular pattern of fences, and holes are usually spaced 25 to 100 feet apart. Where hole depths are between 300 and 500 feet, the spacing of drill holes is usually between 100 and 500 feet. Nearly all drill holes are less than 500 feet in depth; however, 11 holes in the vicinity of the Happy Jack mine in White Canyon were drilled to an average depth of 1100 feet.

An estimated 95 per cent of all the drilling in the Monument Valley and White Canyon districts has been by rotary non-core methods. Current drilling cost is about $1.50 per foot of hole.

## Mining

Nearly all mining in the two districts is underground. The only significant exception is at the Monument No. 2 mine in Monument Valley where open-pit mining is used. Underground mining methods vary from regularly spaced pillars at the Happy Jack mine in White Canyon to random rooms and pillars in medium size mines, to "drift mining" in the narrow, small deposits.

In about 90 per cent of the underground mines, adit or shallow incline access is possible from canyon rims. Shafts or steep inclines are used for access to deposits that are more remote from such rims.

## GEOLOGY

### Geologic History

About 5000 feet of Permian and Triassic sediments, mainly continental in origin with minor marine interruptions, were deposited in the area of study (Figure 2). The following simplified stratigraphic history of the Cutler and Chinle Formations is based on AEC observations and on the studies of McKee (16) and Stewart (17).

The Cutler Formation of Permian age was deposited by westward-flowing streams heading in the ancient Uncompahgre highland in southwest Colorado and emptying into a slowly subsiding basin. The facies grade from a conglomeratic arkose near the ancient highland to mainly red siltstone in the basin. During

periods of isostatic adjustment in the basin, the eolian Cedar Mesa and De Chelly Sandstone Members of the Cutler Formation were deposited.

Following Cutler deposition, a shallow sea advanced from the northwest into the subsiding basin. During this time, red beds of the Moenkopi Formation of Triassic age were deposited in tidal flats in advance of the shallow sea. Streams continued to flow from the Uncompahgre highland westward into the basin.

Deposition of the Moenkopi stopped with regional uplift of the basin. This uplift started a period of erosion during which channels were cut into the Moenkopi sediments by streams flowing generally northward from a highland area in southern or central Arizona and southern New Mexico. This highland, where granitic and volcanic rocks were exposed, was the main source of the sediments that formed the Chinle Formation of Triassic age. The earliest Chinle sediments, those that formed the Shinarump Member, were carried by the northward-flowing streams and were deposited mainly in stream channels over a wide area in northern Arizona and southern Utah. In the area northeast of Monument Valley and southeast of White Canyon, however, these sediments were eroded shortly after deposition as a result of minor uplift in that area. Volcanic activity increased in the Arizona and New Mexico highlands during the time of deposition of the upper members of the Chinle Formation. These upper members are composed of as much as 900 feet of tuffaceous sediments.

### Stratigraphy

All the important uranium deposits in the Monument Valley and White Canyon districts are in the rather thin Shinarump Member of the Chinle Formation. In a few cases, ore extends downward a few feet into the underlying Moenkopi Formation. The following stratigraphic descriptions are restricted to the lower members of the Chinle Formation.

The Chinle Formation unconformably overlies the Moenkopi and crops out throughout much of the area (Figure 1). The Chinle, which ranges in thickness from 500 to 1200 feet, has been subdivided by Stewart (17, p. 500) into seven members including, in ascending order, the Temple Mountain, Shinarump, Monitor Butte, Moss Back, Petrified Forest, Owl Rock, and Church Rock Members. The Temple Mountain Member, which may be a facies of the Shinarump, is a thin unit re-

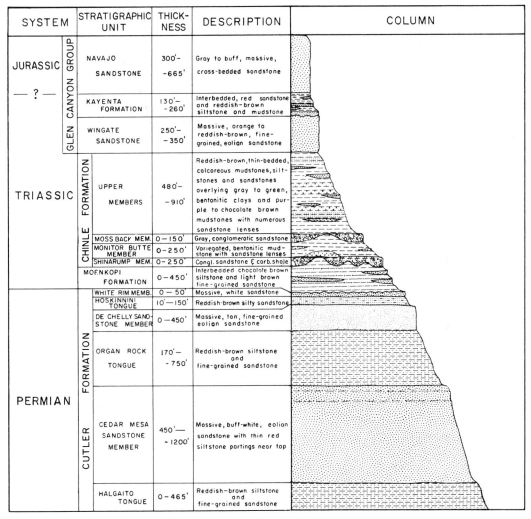

| SYSTEM | STRATIGRAPHIC UNIT | THICK-NESS | DESCRIPTION | COLUMN |
|---|---|---|---|---|
| JURASSIC | GLEN CANYON GROUP — NAVAJO SANDSTONE | 300'- -665' | Gray to buff, massive, cross-bedded sandstone | |
| — ? — | KAYENTA FORMATION | 130'- -260' | Interbedded, red sandstone and reddish-brown siltstone and mudstone | |
| | WINGATE SANDSTONE | 250'- -350' | Massive, orange to reddish-brown, fine-grained, eolian sandstone | |
| TRIASSIC | CHINLE FORMATION — UPPER MEMBERS | 480'- -910' | Reddish-brown, thin-bedded, calcareous mudstones, silt-stones and sandstones overlying gray to green, bentonitic clays and pur-ple to chocolate brown mudstones with numerous sandstone lenses | |
| | MOSS BACK MEM. | 0-150' | Gray, conglomeratic sandstone | |
| | MONITOR BUTTE MEMBER | 0-250' | Variegated, bentonitic mud-stone with sandstone lenses | |
| | SHINARUMP MEM. | 0-250' | Congl. sandstone & carb. shale | |
| | MOENKOPI FORMATION | 0-450' | Interbedded chocolate brown siltstone and light brown fine-grained sandstone | |
| PERMIAN | CUTLER FORMATION — WHITE RIM MEMB. | 0-50' | Massive, white sandstone | |
| | HOSKINNINI TONGUE | 10'-150' | Reddish-brown silty sandstone | |
| | DE CHELLY SAND-STONE MEMBER | 0-450' | Massive, tan, fine-grained eolian sandstone | |
| | ORGAN ROCK TONGUE | 170'- -750' | Reddish-brown siltstone and fine-grained sandstone | |
| | CEDAR MESA SANDSTONE MEMBER | 450'- -1200' | Massive, buff-white, eolian sandstone with thin red siltstone partings near top | |
| | HALGAITO TONGUE | 0-465' | Reddish-brown siltstone and fine-grained sandstone | |

FIG. 2. *Generalized Stratigraphic Column of the Monument Valley and White Canyon Districts, Arizona and Utah.*

stricted to the San Rafael Swell northwest of the White Canyon district.

The Shinarump, the lowermost member of the Chinle, consists of fluvial sediments which were deposited in stream channels and flood plains. These fluvial sediments are composed of lenticular beds of sandstone, conglomerate, siltstone, and mudstone; they contain abundant fragments of carbonized wood and minor amounts of silicified wood. The carbonaceous debris is partially replaced by ore minerals in the uranium deposits. Individual beds range from a few inches to 40 feet in thickness. The sandstone is commonly light buff and me-dium- to coarse-grained and is usually con-glomeratic at the base. The pebbles are pre-dominately quartzite, quartz, and chert with

some limestone, sandstone, siltstone, and mud-stone. Calcite is the most common cementing material in the sandstone and conglomerate. Depositional features include longitudinal bars, which fill scours, and torrential cross-bedding. The thickness of the Shinarump ranges from about 10 feet to nearly 250 feet. Preceeding the deposition of the overlying Monitor Butte Member, Shinarump flood plain and channel sediments, in areas peripheral to uplands, were thinned by erosion; channel sediments in the uplands were either truncated or completely removed.

The Monitor Butte Member of the Chinle Formation, which unconformably overlies the Shinarump Member (23), is composed of a thin, discontinuous basal sandstone unit overlain by a thick mudstone unit. This sandstone unit has been included in the Shinarump by other investigators (20). The areal distribution of the basal sandstone unit of the Monitor Butte and the distribution of the Shinarump are similar. This basal sandstone unit is a buff to cream, conglomeratic, variably carbonaceous, quartzose sandstone cemented with calcite or clay. Pebbles in the conglomerate are gray and pink quartzite and quartz and are as large as 4 inches in diameter. The basal sandstone unit changes from a nearly continuous blanket of sandstone in Monument Valley to less continuous lenticular beds of sandstone which interfinger with beds of mudstone in White Canyon. In the White Canyon area, the zone of transition from the basal sandstone unit to the overlying mudstone unit of the Monitor Butte is characterized by a gradual decrease in the number of sandstone lenses. The mudstone unit is composed of grayish-green, micaceous mudstone and claystone beds and a few thin conglomeratic sandstone beds.

The Monitor Butte thins from about 250 feet in Monument Valley to a featheredge northward along the north part of Elk Ridge (Figures 1, 3). North of Elk Ridge, the Moss Back Member of the Chinle rests unconformably either on erosional remnants of the Shinarump or on the Moenkopi. The northward thinning of the Monitor Butte is believed to be a result of truncation during the period of erosion which preceded deposition of the Moss Back Member.

The Moss Back Member of the Chinle Formation, which is the host rock for the large uranium deposits in the Lisbon Valley area, is comprised of buff to gray sandstone and conglomeratic sandstone with some conglomerate, siltstone and mudstone. Quartzite and chert pebbles are abundant in the conglomerates. In places, limestone pebbles predominate. The Moss Back is present only in the northern portion of the area (Figure 1). It extends from near White Canyon on the southwest to somewhat beyond the Lisbon Valley area on the northeast. From the northeast to the southwest, the Moss Back overlaps a succession of truncated sediments including the Cutler Formation, Moenkopi Formation, and the Shinarump and Monitor Butte Members of the Chinle Formation (Figure 3). The predominant regional trend of streams that deposited the Moss Back was northwesterly; however, local southwesterly trends are present in the upper Indian Creek area and along Elk Ridge (Figure 1). The Moss Back ranges in thickness from a featheredge to 150 feet. It interfingers with mudstone of the overlying Petrified Forest Member of the Chinle.

The upper members of the Chinle including the Petrified Forest, Owl Rock, and Church Rock are comprised of 500 to 1000 feet of varicolored mudstone and siltstone with minor sandstone, conglomerate, and limestone. They contain no known uranium deposits in the Monument Valley and White Canyon districts.

## Shinarump Channel Systems

The recognition of Shinarump channels and channel patterns (Figure 4) is important, because all of the significant uranium deposits in the Monument Valley and White Canyon districts are in these channels. In the 1950's, several investigators including Reinhardt (3), Grundy and Oertell (9), Larsen and Schoen (11), Evensen and Gray (7), Johnson and Thordarson (14), Lewis and Trimble (15), Thaden, *et al.* (21), Lewis and Campbell (24), and Witkind and Thaden (20), studied channels and mapped channel segments on the basis of outcrops and available data from drill holes. Their work provided a background of information useful in this study.

As used in this report, Shinarump channels are the courses of paleostreams which were incised into the Moenkopi and which were filled with fluvial sediments. Scours are the discontinuous, stream-incised, cut-and-fill components within the channels. These scours developed at stages during the lateral shifting of the main stream channel. Sediments in scours in the lower portions of channels are the hosts for the uranium deposits. Channels in Monument Valley are U-shaped in cross section, contain mainly sandstone and conglomerate, are quite narrow, and commonly contain only one ore-bearing scour. Channels in White Canyon are broader; carbonaceous mudstone and siltstone are more abundant, and some channels contain as many as three separate subparallel ore-bearing scours.

In unexplored ground, channels can be projected with reasonable accuracy up to distances of about one mile between exposures on opposite sides of a mesa. Lithologic similarities, the common trends of longitudinal bars, and the common attitude of cross-laminations aid in correlating between two exposures. Over distances greater than a mile, projections for some channels can be made on the basis of geologic features in the Monitor Butte and Moss Back Members. Slumping in the Monitor Butte often occurred above the courses of

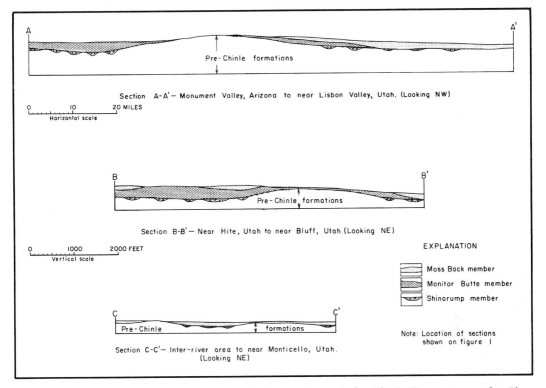

A — A'

Pre-Chinle formations

Section A-A'— Monument Valley, Arizona to near Lisbon Valley, Utah. (Looking NW)

0     10     20 MILES
Horizontal scale

B — B'

Pre-Chinle formations

Section B-B'— Near Hite, Utah to near Bluff, Utah.(Looking NE)

0     1000     2000 FEET
Vertical scale

EXPLANATION

Moss Back member

Monitor Butte member

Shinarump member

C — C'

Pre-Chinle formations

Section C-C'— Inter-river area to near Monticello, Utah.
(Looking NE)

Note: Location of sections
shown on figure I

FIG. 3. *Generalized Geologic Sections of the Lower Part of the Chinle Formation at the Close of Deposition of the Moss Back Member.*

those Shinarump channels that are filled with mudstone, and, at these places, the Moss Back is also locally thicker.

Where the Moss Back and Monitor Butte are removed by erosion, Shinarump channels commonly yield a characteristic topographic expression. Channels filled with carbonaceous mudstone commonly form low ridges, for the Shinarump mudstone is generally more resistant to erosion than is the Moenkopi Formation. Shinarump sandstone in sandstone-filled channels, on the other hand, is commonly highly jointed, and hence these channels are more easily eroded than the Moenkopi; many present day stream courses and valleys follow these sandstone-filled channels.

Sandstone and conglomerate were deposited in places where the channels were narrow and had quite high gradients, whereas, carbonaceous mudstone was deposited in places where the channels were broad and meandering and had low gradients. The type of sediments deposited in the channels was determined by the position of the channels relative to an upland that was elevated at the beginning of deposition of the Shinarump. This upland, named the Monument Valley-Monticello up-

land in this report, most probably extended from Monument Valley to the vicinity of Monticello, Utah (Figure 1). In the White Canyon district, the facies of the Shinarump changes from dominant sandstone along the western flanks of the Monument Valley-Monticello upland to dominant mudstone in an adjacent lowland to the west. The favorable belt in which nearly all of the important uranium deposits in White Canyon are located is this zone of transition between sandstone on the flanks of the upland and mudstone in the adjacent lowland.

Variations in gravel size, orientation of dips in cross-strata, and diagnostic fossils in gravel indicate that the Shinarump was deposited by streams flowing northward from a source area in southern New Mexico and southern or central Arizona (16, p. 24; 17, p. 506, 523). The positions of the eastern and northern margins of the ancient upland (Figure 1), which extended from Monument Valley to Monticello, are inferred from subsurface data from a few widely-spaced oil tests. Fluvial sandstones of the basal Chinle, which were penetrated by these holes drilled east of the ancient upland, may be either the Shinarump or the

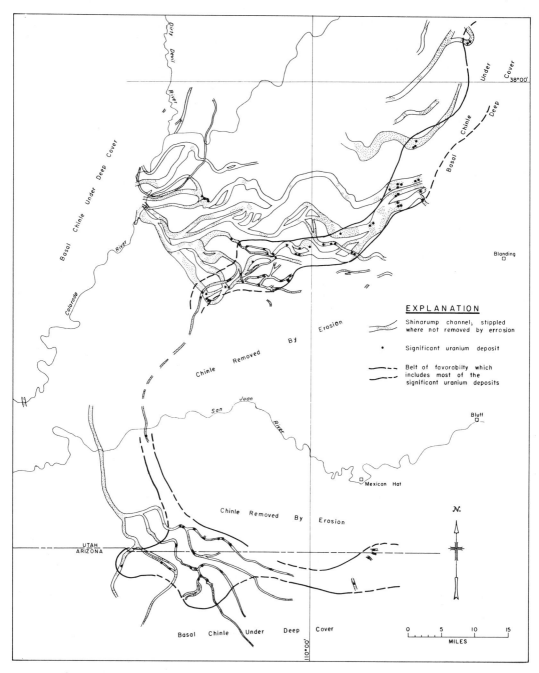

FIG. 4. *Shinarump Channel Systems in the Monument Valley and White Canyon Districts, Arizona and Utah.*

Moss Back, or perhaps both of these members. These fluvial sandstones east of the ancient upland may have been deposited by streams flowing northerly or northwesterly from the highland in southern or central Arizona and southern New Mexico. Streams flowing to the northwest may have been diverted to the northeast along the east margin of the barrier formed by the Monument Valley-Monticello upland. Johnson and Thordarson (14) suggest that streams depositing the Shinarump sediments in the White Canyon and Elk Ridge localities may have originated in the ancestral Uncompahgre highland to the northeast. However, the development of such drainage patterns between the ancient Uncompahgre highland and the White Canyon district might have been poor because of the barriers formed by the northwest-trending salt anticlines in the Paradox Basin. The Chinle internally thins and nearly pinches out over the core of the Paradox Valley anticline (12, p. 50). The author favors a source to the south or southeast for the Shinarump in the Monument Valley and White Canyon districts and for the Shinarump (?) east of the ancient upland extending from Monument Valley to Monticello.

The Monument Valley-Monticello upland was elevated in the early stages of development of the Shinarump drainage system. Evidence for the existence of the upland during the development of the Shinarump drainage system is the parallelism of the courses of many larger channels to the western margin of the upland in the northern portion of the White Canyon district (Figures 1 and 4). Farther south, the streams flowed westerly away from the upland in response to the regional gradient. The sediments deposited by smaller tributary streams draining this earlier upland and part of the sediments deposited in the large channels in the adjacent lowland were later eroded during a period of renewed uplift of the Monument Valley-Monticello upland. This uplift probably did not exceed 500 feet (Figure 3). The basal sandstone unit of the Monitor Butte Member consists partly of Shinarump sediments that were reworked during this period.

Evidence of meander cutoffs, shifting of channels, and anastomosing streams are present in the Shinarump drainage system (Figure 4). In parts of the White Canyon area, there are two rather distinct generations of channels in the Shinarump. The younger channels commonly cut through the older channels and locally form quite different directional patterns. Where these two systems can be distinguished, the more important uranium deposits are always in the younger system.

## Structural Geology

The Monument Valley and White Canyon districts are located in the southern portion of the Monument Upwarp, a large asymmetric, north-trending anticline of probable Laramide age (Figure 1). The western flank of the upwarp dips 2° to 4° west into the Kaiparowits and Henry Mountains Basins in central Utah; the upwarp is bounded on the east and south-

east by the Comb Monocline which dips steeply to the east. Several north- and northeast-trending asymmetric anticlines and associated synclines are superimposed on the upwarp. Small, high-angle, normal faults are present throughout the region, but they are most common in the northern portion. West-to southwest-trending faults bound well-defined grabens in the area west of the Abajo Mountains.

## URANIUM DEPOSITS

### Distribution and Habit

With the exception of the Happy Jack mine and its satellite deposits, all important deposits in the Monument Valley and White Canyon districts are restricted to an arcuate belt of favorable sandstone 3 to 12 miles wide and about 130 miles long. This favorable belt extends from Monument Valley on the south, through White Canyon and Elk Ridge, to Indian Creek on the north (Figure 1). The only place where its continuity has not, as yet, been established is the unexplored, deeply buried segment of Chinle sedimentary rocks 15 miles long between the San Juan River and Red Canyon. Finch (13) named the area southeast of this buried segment the Monument Valley Belt and the area to the northeast the East White Canyon Belt. Young (23) has proposed the name Monument Mineral Belt to include the entire belt. The belt is postulated to coincide with the distribution of favorable sandstones of the Shinarump along the west margin of the Monument Valley-Monticello upland.

Uranium deposits are primarily restricted to favorable carbonaceous sandstone and conglomerate beds in the lower part of the Shinarump Member of the Chinle Formation; however, in a few mines such as the Moonlight and Happy Jack, ore extends downward as much as 15 feet into the siltstone of the underlying Moenkopi. The channel at the Monument No. 2 mine locally scoured through the underlying Moenkopi Formation and the Hoskinnini Tongue of the Cutler Formation into the De Chelly Sandstone Member of the Cutler Formation. At that mine, vanadium ore extends downward into the De Chelly for 10 to 20 feet.

As viewed in plan, ore deposits are linear, non-linear, and curvilinear in outline. In the linear deposits, the ratio of length to width is commonly at least 5 to 1 and may reach 50 to 1. Most of the deposits in the Monument Valley and White Canyon districts are linear.

Typical examples are the Monument No. 2 and Markey mines. Non-linear deposits are irregular amoeba-shaped or somewhat eliptical in outline. Length to width ratios range from 1 to 1 to 5 to 1. There are only a few non-linear deposits; the Happy Jack and Betty mines are of this type. Curvilinear deposits are strongly elongated and broadly curving with parallel to subparallel scours controlling the local ore trends. The ratio of length to width is about the same as in linear deposits. Meander loops in Shinarump channels were favorable sites for ore deposition. An example is the Hideout mine where several separate, subparallel, curving ore trends occupy scours formed at different stages in the development of a major meander.

Ore bodies consist of closely-spaced, lenticular ore pods which are generally concordant with bedding. Single ore pods range from a few feet to a few hundred feet in length and from less than one foot to 12 feet in thickness. The average length ranges from about five to ten times the average width. Mineable ore is continuous in one mine for 7000 feet. Deposits range in size from a few tons to more than 800,000 tons. About half of the 174 deposits are smaller than 1000 tons in size and 96 per cent are smaller than 50,000 tons (Table I); only six deposits contain more than 50,000 tons each. The largest deposit contains nearly one-third of the sum of total production and remaining available reserves in the Monument Valley and White Canyon districts. The ten largest deposits contain about 70 per cent of the combined total production and reserves.

Most of the ore now being mined is unoxidized. Except for the large Monument No. 2 deposit, nearly all the surficial oxidized deposits have been mined.

## Copper and Vanadium

The uranium deposits contain variable amounts of copper and vanadium. Ores from the Monument No. 2 mine in the eastern part of the Monument Valley district contain average amounts of 1.40 per cent $V_2O_5$ and nil copper. In the other Monument Valley deposits for which some data are available, vanadium ranges from 0.22 per cent to 0.81 per cent and copper ranges from 0.29 per cent to 2.50 per cent; weighted averages are 0.60 per cent $V_2O_5$ and 0.71 per cent copper. In White Conyon, the vanadium content of those deposits for which some data are available ranges from 0.02 per cent to 1.20 per cent $V_2O_5$; the weighted average is 0.23 per cent $V_2O_5$. In the White Canyon deposits, copper ranges from 0.12 per cent to 1.30 per cent and averages 0.69 per cent. These averages for each district are not representative, because they are based solely on production from mines for which the vanadium and copper content was recorded. Although incomplete, the above data are indicative of variations and relative orders of magnitude.

In general, the vanadium content of ores in Monument Valley decreases from east to west, but copper increases from east to west. The copper content in the White Canyon dis-

TABLE I.   *Distribution of Deposits by Locality and Size, Monument Valley and White Canyon Districts*

| Locality | Number of Deposits [1] | Tons [2] | Distribution of Deposits by Size | | | | | |
|---|---|---|---|---|---|---|---|---|
| | | | Less than 1,000 T. | 1,000 to 10,000 T. | 10,000 to 50,000 T. | 50,000 to 100,000 T. | 100,000 to 500,000 T. | Greater than 500,000 T. |
| Monument Valley | 54 | 1,448,000 | 29 | 11 | 12 | 0 | 1 | 1 |
| White Canyon | 46 | 713,000 | 28 | 13 | 4 | 0 | 0 | 1 |
| Red Canyon | 29 | 492,000 | 13 | 8 | 6 | 1 | 1 | 0 |
| Deer Flat | 11 | 279,000 | 2 | 6 | 2 | 0 | 1 | 0 |
| Elk Ridge—upper Cottonwood | 25 | 233,000 | 8 | 8 | 9 | 0 | 0 | 0 |
| Stevens Canyon—upper Indian Creek | 9 | 146,000 | 4 | 0 | 5 | 0 | 0 | 0 |
| Total | 174 | 3,311,000 | 84 | 46 | 38 | 1 | 3 | 2 |
| Per cent of total | | | 48% | 26% | 22% | .6% | 2% | 1% |

[1] Includes deposits with production and/or available reserves as of 7/1/65.

[2] 7/1/65 available reserves plus production.

trict increases from east to west, but no pattern of distribution is evident for vanadium.

## Mineralogy

In the unoxidized parts of the Monument No. 2 mine, uraninite and coffinite are associated with vanadium minerals such as montroseite, corvusite, doloresite, and vanadium hydromica. Sulfides of iron, copper, and lead are also present. Oxidized ore minerals from this mine are tyuyamunite, carnotite, hewettite, and navajoite. All these minerals are associated with oxides of iron.

In other mines in Monument Valley and White Canyon, the suite of unoxidized minerals is the same as that at the Monument No. 2 mine, but copper sulfide minerals are more abundant, and montroseite is less abundant.

The uranium minerals, torbernite, uranophane, uranopilite, betazippeite, and johannite have been identified in samples from oxidized deposits. Malachite, azurite, and hydrous copper and iron sulfates are common accessory minerals.

Calcium carbonate is present in ore mostly as cementing material in the sandstone host rock. In Monument Valley mines, calcium carbonate ranges from 1.4 per cent to 10.3 per cent and averages 4.6 per cent. Calcium carbonate content generally seems to be inversely proportional to vanadium content in Monument Valley deposits. Analyses for calcium carbonate in White Canyon mines range from 1.3 per cent to 8.0 per cent and average 2.4 per cent; this is about half the average in Monument Valley mines. No relationship between calcium carbonate and vanadium content is evident in White Canyon mines. Calcium carbonate cannot be correlated with copper in either district. There is a quite high calcium carbonate content in the Royal mine near Indian Creek at the extreme northeastern end of the White Canyon district. Abundant calcium carbonate is present in lime-pebble conglomerate and calcareous sandstone in the Moss Back Member which overlies the Shinarump host rocks in this locality.

## Ore Controls

The uranium deposits are most commonly localized in the more deeply scoured portions of Shinarump channels. These deeper scours occur on the outside of bends and in relatively straight portions of channels in those places where the less resistant beds of the Moenkopi

were present. The deep scours were subsequently filled with longitudinal sand and gravel bars and carbonaceous debris. When the scours were filled and lower stream gradients were attained, the sandstone and conglomerate beds were covered by layers of silt and carbonaceous mudstones. In all probability, the transmissive coarse sediments provided the main pathways for uranium-bearing ground waters, and the carbonaceous debris in the sandstone, conglomerate, and overlying mudstone created the reducing environment necessary for precipitation of the uranium.

The Happy Jack deposit in White Canyon is outside the favorable belt within which nearly all the other significant deposits occur (Figure 1). The main Happy Jack deposit and its associated cluster of smaller deposits are nonlinear types. These deposits are confined to an area in which a sharp meander of a younger Shinarump channel crosses and scours into a broad bend in an underlying Shinarump channel which is filled with carbonaceous mudstone. These geologic conditions permitted large quantities of uraniferous ground water from the two channels to enter a single favorable environment. As evidenced by the large size of the Happy Jack deposit, the ore forming process must have been very efficient.

The Monument No. 2 mine, in eastern Monument Valley is a strongly elongated, northwest-trending, linear deposit in a channel situated along the northern flank of an upland which was uplifted at the close of deposition of the Shinarump (Figure 1). The southeastward segment of the channel, where it crosses the upland, was removed by erosion following the uplift. The original northwest inclination of the northwest segment of the channel was reversed to the southeast by uplift during Laramide time of the Gypsum Creek Dome which is an element of the larger Monument Upwarp. Most of this northwest segment was subsequently removed by erosion.

## Genesis

The uranium in the deposits may have been derived from: (1) erosion and leaching of large masses of granitic rocks, arkose, and tuffaceous sediments or (2) from hypogene fluids generated by magmatic activity. The author favors the first theory.

LEACHING OF GRANITIC, ARKOSIC, AND TUFFACEOUS ROCKS  The uranium deposits of the Monument Valley and White Canyon districts may have been formed through multiple migration-accretion (5). Small amounts of uranium, vanadium, and copper which were leached from rocks actively undergoing erosion were introduced into the Shinarump channel systems by surface and ground waters during or soon after deposition of the channel sands. Although no definitive evidence is on hand, the uranium may have been obtained from the weathering and leaching of granitic igneous rocks, arkosic sandstones, and tuff beds containing trace amounts of uranium in the source areas of the sediments. It is postulated that uranium, as the uranyl ion, moved northward in ground water through the permeable channel sediments until it was fixed by reduction in the vicinity of accumulations of organic debris. This organic material could have been the reducing agent, or it could have supplied the energy source for anaerobic bacteria which generate hydrogen sulfide, a powerful reducing agent capable of reducing the water-soluble uranyl ion to the insoluble uranous state (10). Minor and dispersed concentrations of uranium, vanadium, and copper, few of which may have been large or rich enough to mine, thus were formed soon after sediments accumulated in the Shinarump channel systems.

Following deposition of the Shinarump, renewed upwarp of large areas that coincided in part with the already existing uplands took place north and southeast of Monument Valley and east of the White Canyon district (Figure 1). The minor and dispersed occurrences of uranium deposited initially in Shinarump sediments were solubilized by oxidation during the erosion of channels from higher parts of the newly uplifted areas and the partial erosion of numerous large channels in the adjacent prevailing lowland. This remobilized uranium was transported by ground waters into reducing environments in the channels that remained along the lower flanks of the rejuvenated uplifts. In the White Canyon district, the transition zone between the upland and lowland is characterized by a gradual change from permeable sands in channels on the flanks of the uplifts to impervious carbonaceous mudstone in these same channels in the lowlands. In Monument Valley, the most favorable places for uranium deposition were the heterogeneous channel sandstones that were deposited where anastomosing streams converged (Figure 4).

HYPOGENE FLUIDS GENERATED BY MAGMATIC PROCESSES   The presence of a few conspicuous Tertiary plugs and many small dikes of lamprophyric rock (20, p. 51) in Monument Valley and laccolithic intrusives of horn-

blende andesite porphyry (8, p. 144) in the Abajo Mountains, a few miles east of the northern portion of the favorable belt has suggested to some workers a possible magmatic origin during Tertiary time for uranium in the Monument Valley and White Canyon districts. Williams (2, p. 148) believes the Monument Valley intrusives to be of middle to late Pliocene age. Hunt (6, p. 82) suggests that the Abajo intrusives are of middle Miocene age, and Witkind (22, p. 104) provisionally proposes a Miocene or Pliocene age. Finnel (18, p. 52) suggests that the ore solutions moved upward along buried faults from a deep source during the Laramide orogeny. All these ages are much younger than the age of 180 m y. for the uranium as proposed by Young (23) and favored by the author.

A number of isotopic age determinations on samples from Monument Valley and White Canyon have been made. $Pb^{206}/U^{238}$ age determinations of only slightly altered uraninite samples from the Monument No. 2 mine range from 60 to 100 m.y. and average 78 m.y. (20, p. 96). The ages of samples from the Happy Jack mine are within this same range (4, p. 15). Ages of samples from basal Chinle deposits in other areas are much older. Age determinations on samples from Lisbon Valley, Utah, and Cameron, Arizona, average 150 m.y. and 175 m.y. respectively (19). As commented upon by Young (23, p. 872), the isotopic ratios that result in wide differences in age may reflect the complicated redistribution of uranium during the deposition of the lower Chinle and perhaps during the Laramide orogeny.

To the author, the most compelling evidence against a magmatic hydrothermal origin of the uranium deposits is that the favorable belt is transverse to the Monument Upwarp of Laramide age and the distribution of deposits is not spatially related to intrusives or to patterns of faults, fractures, and folds within the upwarp. If the deposits were Tertiary in age, some better correlation with these structures would be expected.

## GUIDES TO PROSPECTING

Certain geologic factors seem to have influenced the localization of uranium deposits in the Monument Valley and White Canyon districts. An understanding of the geologic events is helpful in prospecting for new deposits in these districts. Some important considerations are:

1. The belt of important deposits in the

White Canyon district is marginal to the ancient Monument Valley-Monticello upland. The possible genetic relationship between deposits in the White Canyon district and the large deposits in the Lisbon Valley area, 20 miles northeast of the Royal mine near Indian Creek, warrants consideration. The Lisbon Valley area is also situated on the flank of an ancient upland, the ancestral Lisbon Valley anticline. The host rock is the Moss Back Member of the Chinle which overlies the widespread unconformity at the base of the Chinle in that area.

2. In the White Canyon district, nearly all the important deposits are situated within a belt that coincides with the facies change in the Shinarump from predominately sandstone to predominately carbonaceous mudstone.

3. Uranium deposits occur only in the lower portions of Shinarump channels which are trough-shaped depressions filled with sandstone, conglomerate and variable amounts of carbonaceous mudstone.

4. Uranium deposits may exist in channels which are obscured by overburden at the rims of mesas within the belt of favorability. In localities such as Elk Ridge, where overburden is extensive, more exploratory rim stripping may be warranted.

5. Undiscovered uranium deposits may exist in Shinarump channels in portions of the belt in which the Shinarump is deeply buried by younger members of the Chinle.

6. The probable courses of buried Shinarump channels can often be inferred by correlating the channel segments which have been established through exploration and reconstructing the channel patterns.

## REFERENCES CITED

1. Butler, B. S., 1920, White Canyon region: p. 619–622 in The ore deposits of Utah: U.S. Geol. Surv. Prof. Paper 111, 672 p.
2. Williams, H., 1936, Pliocene volcanics of the Navajo-Hopi Country: Geol. Soc. Amer. Bull., v. 47, p. 111–171.
3. Reinhardt, E. V., 1952, Uranium-copper deposits near Copper Canyon, Navajo Indian Reservation, Arizona: U.S. Atomic Energy Comm. RMO-902, 11 p.
4. Stieff, L. R., et al., 1953, A preliminary determination of the age of some uranium ores of the Colorado Plateau by the uranium-lead method: U.S. Geol. Surv. Circ. 271, 19 p.
5. Gruner, J. W., 1956, Concentration of uranium in sediments by multiple migration-accretion: Econ. Geol., v. 51, p. 495–520.
6. Hunt, C. B., 1956, Cenozoic geology of the Colorado Plateau: U.S. Geol. Surv. Prof. Paper 279, 99 p.
7. Evensen, C. G. and Gray, I. B., 1957, Geology of Monument Valley uranium deposits, Arizona and Utah: U.S. Atomic Energy Comm. RME-95.
8. Witkind, I. J., 1957, Abajo Mountains, Utah: p. 143–147 in Geologic investigations of radioactive deposits, Semiannual Progress Rept. December 1, 1956, to May 31, 1957, U.S. Geol. Surv. TEI-690, issued by U.S. Atomic Energy Comm., Oak Ridge.
9. Grundy, W. D. and Oertell, E. W., 1958, Uranium deposits in the White Canyon and Monument Valley mining districts, San Juan County, Utah and Navajo and Apache Counties, Arizona: in Guidebook to the geology of the Paradox Basin, Intermountain Assoc. Petrol. Geols., 9th Ann. Field Conf., p. 197–207.
10. Jensen, M. L., 1958, Sulfur isotopes and the origin of sandstone-type uranium deposits: Econ. Geol., v. 53, p. 598–616.
11. Larsen, R. N. and Schoen, Robert, 1958, Shinarump-filled channels and their relationship to uranium in the White Canyon mining district, San Juan County, Utah: U.S. Atomic Energy Comm., unpub. RME.
12. Shoemaker, E. M., et al., 1958, Salt anticlines of the Paradox Basin: in Guidebook to the geology of the Paradox Basin, Intermountain Assoc. Petrol. Geols., 9th Ann. Field Conf., p. 39–59.
13. Finch, W. I., 1959, Geology of uranium deposits in Triassic rocks of the Colorado Plateau: U.S. Geol. Surv. Bull. 1074-D, p. 125–164.
14. Johnson, H. S., Jr. and Thordarson, William, 1959, The Elk Ridge-White Canyon channel system, San Juan County, Utah: Its effect on uranium distribution: Econ. Geol., v. 54, p. 119–129.
15. Lewis, R. Q., Sr. and Trimble, D. E., 1958, Geology and uranium deposits of Monument Valley, San Juan County, Utah: U.S. Geol. Surv. Bull. 1087-D, p. 105–131.
16. McKee, E. D., et al., 1959, Paleotectonic maps, Triassic system: U.S. Geol. Surv. Misc. Geol. Invest. Map 1–300.
17. Stewart, J. H., et al., 1959, Stratigraphy of Triassic and associated formations in part of the Colorado Plateau region: U.S. Geol. Surv. Bull. 1046-Q, p. 487–576.
18. Finnell, T. L., et al., 1963, Geology, ore deposits, and exploratory drilling in the Deer Flat area, White Canyon district, San Juan County, Utah: U.S. Geol. Surv. Bull. 1132, 114 p.
19. Miller, D. S. and Kulp, J. L., 1963, Isotopic evidence on the origin of the Colorado Plateau uranium ores: Geol. Soc. Amer. Bull., v. 74, p. 609–630.
20. Witkind, I. J. and Thaden, R. E., 1963, Geology and uranium-vanadium deposits of the

Monument Valley area, Apache and Navajo Counties, Arizona, with a section on serpentine at Garnet Ridge by E. E. Malde and R. E. Thaden: U.S. Geol. Surv. Bull. 1103, 171 p.

21. Thaden, R. E., *et al.*, 1964, Geology and ore deposits of the White Canyon area, San Juan and Garfield Counties, Utah: U.S. Geol. Surv. Bull. 1125, 166 p.

22. Witkind, I. J., 1964, Age of the grabens in southeastern Utah: Geol. Soc. Amer. Bull., v. 75, p. 99–106.

23. Young, R. G., 1964, Distribution of uranium deposits in the White Canyon-Monument Valley districts, Utah-Arizona: Econ. Geol., v. 59, p. 850–873.

24. Lewis, R. Q., Sr. and Campbell, R. H., 1965, Geology and uranium deposits of Elk Ridge and vicinity, San Juan County, Utah: U.S. Geol. Surv. Prof. Paper 474-B, p. B1–B69.

# 39. Geology and Uranium-Vanadium Deposits in the Uravan Mineral Belt, Southwestern Colorado

J. E. MOTICA*

## Contents

## Illustrations

* Union Carbide Corporation, Grand Junction, Colo.

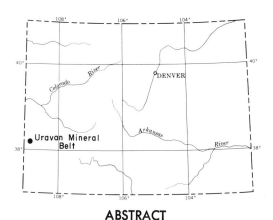

Exploratory and development drilling from the surface is conducted in three phases: first, wide-spaced drilling to find and outline favorable channels; second, moderate-spaced drilling in favorable ground to locate ore deposits; and third, close-spaced drilling to block-out ore. Long-hole drilling from underground stations is used to find and delineate extensions of ore bodies being mined.

Due to intense production since 1948 from shallow deposits, best potential for new deposits exists in areas where the ore-bearing unit is 600 to 800 feet deep.

## ABSTRACT

Ores containing uranium and vanadium minerals have been mined from the Salt Wash Member of the Morrison Formation from many localities in the Colorado Plateau region since about 1900. The most productive deposits are in a relatively small area in Southwestern Colorado referred to as the Uravan mineral belt.

The mineral belt is a narrow elongate area in which uranium-vanadium deposits in the Salt Wash have a closer spacing, larger size, and higher grade than those in adjoining areas. Pre-1948 ore production from the mineral belt was 655,000 tons averaging 1.91 per cent $V_2O_5$ and 0.28 per cent $U_3O_8$. Production during the 1948 to 1964 period was 7,900,000 tons of ore averaging 1.46 per cent $V_2O_5$ and 0.27 per cent $U_3O_8$.

The ore deposits occur principally in the uppermost sandstone unit of the Salt Wash. This unit consists of sandstone lenses formed by a system of aggrading braided streams flowing in an easterly direction generally normal to the mineral belt. These lenses, referred to as channels, are as much as one mile in width and average about 50 feet in thickness and can be traced for several miles along their courses. Areas favorable for one deposits are recognized principally by the following criteria: (1) a host sandstone thickness of over 30 feet, (2) the presence of carbonaceous material in the host sandstone, and (3) gray mudstones and clays associated with the ore-bearing sandstone. The ore minerals are believed to have been precipitated from laterally migrating solutions; there is no apparent genetic relationship between ore deposits and tectonic structural features.

## INTRODUCTION

Uranium-vanadium deposits in the Salt Wash Member of the Morrison Formation occur at many locations in the Colorado Plateau region. In general, the more productive occurrences are in southwestern Colorado and adjacent portions of southeastern Utah in an area referred to as the Uravan mineral belt.

The mineral belt extends in an arcuate pattern from Polar Mesa in southeastern Utah to the Slick Rock district in southwestern Colorado as indicated by Figure 1. As it is known at present, the Uravan mineral belt is approximately 70 miles long and 2 to 8 miles wide. The boundaries of the belt are indistinct and are subject to personal interpretation.

The idea of a mineral belt was first conceived in 1943 by R. P. Fischer (1) of the U.S. Geological Survey, who defined it as a narrow elongate area in which the uranium-vanadium deposits in the Salt Wash generally have closer spacing, larger size, and higher grade than those in adjoining areas and the region as a whole. Exploration work within and adjacent to the area postulated by Fischer to be a mineral belt has verified Fischer's original concepts of it.

Even though total uranium ore production from other uranium districts may exceed that from the Uravan mineral belt, the area of this belt has special significance in the uranium industry due to its long and colorful history and to the nature of the hundreds of ore occurrences in it. Some of the world's earliest recorded production of uranium minerals came from outcrops in the Uravan mineral belt. The belt also has the distinction of being the world's leading producer of vanadium from 1948 to the present, as a result of coproduction with uranium.

Due to the strategic importance of the ore deposits within the mineral belt, the geology

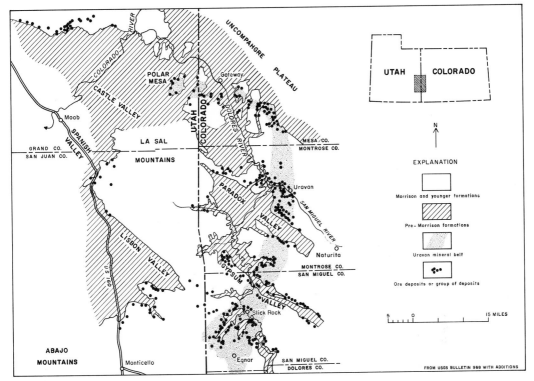

FIG. 1. *Map, showing the location of the Uravan Mineral Belt.*

has been intensely studied and results reported in publications issued by the U.S. Geological Survey, the U.S. Atomic Energy Commission, and geologists working with private concerns. Much has been learned about the geology of the ore deposits in the Salt Wash; however, important questions remain unanswered. The purpose of this paper is to summarize known pertinent geologic features relative to the uranium-vanadium ore deposits in the Uravan mineral belt as well as to outline exploration techniques in common use throughout the area.

## History and Production

Interest in the deposits in the Uravan mineral belt varied from the time of discovery in 1898 to the present, depending on the demand for a particular metal. Initial interest in the deposits was by the French who recovered radium from the ores in the early 1900's. Domestic recovery of radium from these ores prevailed from 1910 to 1923, when the uranium deposits in the Belgian Congo were brought into production and this source sup-

plied the demand for radium. From 1915 to 1923, however, some by-product vanadium and a little by-product uranium were recovered from the radium ore of the mineral belt ores. There was virtually no production from the mineral belt from 1923 until 1937, when the mines were reactivated in response to increased demands for vanadium. Mining for vanadium continued to 1944, but some uranium was recovered as a by-product under the Manhattan Project during World War II years. Most mines were inactive from 1945 to 1948, when the Atomic Energy Commission initiated the domestic uranium procurement program. Production was greatly expanded under this program and has been intensive from 1948 to the present, vanadium being recovered from the ores as well as uranium. Some of the operators in the mineral belt are participating in the AEC's uranium stretch-out program and production is assured through 1970.

Most of the ore mined in the Uravan mineral belt has been from underground mines. Production records furnished by the U.S. Atomic Energy Commission show the following for the mineral belt:

| Period | Tons of Ore | Grade | |
|---|---|---|---|
| | | $V_2O_5$ | $U_3O_8$ |
| Pre-1948 | 655,000 | 1.91% | 0.28% |
| 1948–thru 1964 | 7,900,000 | 1.46% | 0.27% |

Principal operators with milling facilities under the AEC program include Union Carbide Corporation, Vanadium Corporation of America, and Climax Uranium Company. In addition to ores processed from properties controlled by these milling companies, significant quantities of ore are treated from properties operated by independents.

## GEOLOGY

### Stratigraphy of the Salt Wash

The Salt Wash Member of the Morrison Formation has been studied extensively by Craig (3), who concluded that the member was formed as a large alluvial fan by an aggrading system of braided streams diverging to the north and east from an apex in south-central Utah. The entire Salt Wash in the Uravan mineral belt area averages about 250 feet thick and is composed of interbedded units of sandstone and mudstone.

The sandstone units within the Salt Wash range in color from red to brown to light gray. The mudstone units vary from red-brown to gray-green. The colors of the units within the Salt Wash will be further discussed under the section of this paper which describes guides to ore deposits.

The Salt Wash conformably overlies the Summerville Formation, which consists of approximately 100 feet of marine shale, mudstone, and thin-bedded sandstone. Overlying the Salt Wash is the Brushy Basin Member of the Morrison Formation; it is composed predominately of variegated mudstone, with lenses of siltstone, sandstone, and conglomerate, principally in the lower portion of the member. The Brushy Basin varies in thickness from 400 to 500 feet throughout the mineral belt. The contact between the Salt Wash and the Brushy Basin is not a distinct boundary but rather one that is picked somewhat arbitrarily. Some geologists place the contact at the top of the uppermost prominent sandstone in the Salt Wash. Craig (3) describes the contact as at the base of a widespread chert pebble conglomerate in the lower portion of the Brushy Basin.

Commonly the Salt Wash consists of three prominent stratigraphic units of sandstone separated by layers of mudstone and thin sandstone lenses. The three sandstone units are similar in lithology and thickness. Occasionally a unit will split to give more than three sandstone layers or a unit may be entirely missing and only two prominent sandstone layers will be present. The uppermost unit is by far the most important from the standpoint of ore production and reserves, but some ore has been mined from deposits in the other two Salt Wash units as well as from sandstone beds in the lower part of the Brushy Basin.

The uppermost sandstone unit in the Salt Wash represents several separate sandstone deposits formed by several separate aggrading braided stream systems at the same stratigraphic level. Many geologists working with the Salt Wash refer to these fluvial sandstone deposits as channels.

The character of the individual ore-bearing sandstone channels is remarkably similar throughout the Uravan mineral belt. The channels are as much as one mile or more in width and can be traced for several miles along their courses. The depositional direction is easterly, generally normal to the Uravan mineral belt. The edges of the sandstone channels are indistinct and are usually represented by a gradual thinning of sandstone and lateral gradation to mudstone. The sandstone thickness within a channel ranges from 25 feet to 100 feet with an average of about 50 feet. The abnormally thick portions of a channel occur randomly and generally represent local conditions which are not continuous over large areas.

### Lithology of the Salt Wash

The ore-bearing sandstone is fine to medium grained and is composed predominately of quartz with minor amounts of accessory minerals consisting of feldspar, chert, zircon, tourmaline, garnet, staurolite, rutile, apatite, and magnetite-ilmenite. In light brown and light gray portions of the host rock, iron occurs as limonite in the oxidized zones and as pyrite and marcasite in the unoxidized zone. Nondetrital components are mainly silica, calcite, dolomite, and gypsum.

Clay minerals are important constituents of the Salt Wash ore-bearing sandstones, occurring as films coating the sand grains, as galls and pebbles, and as lenses of claystone and mudstone. The dominant clay minerals as reported by Keller (7) are illite and chlorite with subordinate amounts of kaolin.

Carbonaceous materials occur in restricted

portions of the ore-bearing sandstone. The carbonaceous occurrences represent the coalified remains of vegetal matter enclosed within the sandstone at the time of formation. The carbonized plant remains vary from minute flakes to entire trees and occur in amounts ranging from sparse to abundant, generally in portions of sandstone containing abundant clay. The carbon-bearing section of the sandstone averages only a few feet in thickness.

Sedimentary structures in the ore-bearing sandstone include crossbedding, current lineation, festoons, ripple marks, and scour and fill features. Typically, these structures are present principally in the middle or lower portion of the ore sandstone and the upper portion is generally horizontally bedded or structureless. The sandstone is typically a hard competent rock that stands well during mining.

## Structural Features

The most conspicuous tectonic structural features in the Uravan mineral belt area are two northwest-trending salt anticlines which are traversed by the belt. The sedimetary beds along the flanks of these salt anticlines dip at angles ranging from about 5° to 15°. The sediments throughout the balance of the mineral belt are essentially flat-lying or dip at low angles.

The only significant faulting in the mineral belt area is that associated with collapse of portions of the salt anticlines. Faults parallel to the anticlinal axes were developed along the margins of the collapsed areas. Ore deposits in the host sandstone have been displaced by some of these faults, indicating that the faults are post-mineral.

## ORE DEPOSITS

Principal metals recovered from the Uravan mineral belt ores are uranium and vanadium. Radium was recovered early in the 1900's but no substantial amounts have been recovered since about 1923. Trace amounts of other metals have been detected in the ores, but none of these occurs in quantities sufficient to warrant recovery. Metals detected other than those found in common rock forming minerals include molybdenum, copper, silver, selenium, chromium, nickel, cobalt, rare earths, and manganese.

Principal uranium minerals in the unoxidized ores are uraninite and coffinite while the main uranium mineral in the oxidized ores is carnotite with minor amount of tyuyamunite. Vanadium-bearing clays, consisting of chlorite and hydromica, are the main vanadium minerals in both the oxidized and unoxidized ores. Montroseite, a low-valent vanadium oxide, occurs in significant amounts in the unoxidized ores. The vanadium in the vanadiferous clays is firmly fixed and is relatively unaffected during oxidation; however, montroseite oxidizes readily first to an intermediate mineral, corvusite, which imparts a blue-black color to the ore. Upon further oxidation the vanadium forms a series of vanadates consisting of carnotite, hewettite, metahewettite, pascoite, rauvite, fervanite, and hummerite. These secondary vanadates do not account for a large proportion of total vanadium in the ores. The mineralogy of the Salt Wash ores is well summarized in a paper by Weeks *et al.* (6).

The uranium-vanadium mineralization imparts a distinct coloration to the host rock, and grades can often be estimated relatively closely on the basis of color. The vanadiferous clays vary in color from a light gray to almost black, depending on the vanadium content. The reduced uranium and vanadium oxide minerals are black, and carnotite is a bright yellow.

Studies show that there has been very little, if any, migration of uranium and vanadium during oxidation. A thin film of carnotite will occasionally be found along fractures, particularly in the fault zone bordering the salt anticlines, but the amount of mineralization found under these conditions rarely is sufficient to make ore.

The uranium and vanadium minerals almost always occur intimately mixed together regardless of the environment containing the mineralization or type of mineral association. Occurrences of one metal without the other in large amounts are virtually unknown in the Uravan mineral belt area. Generally the amount of vanadium exceeds the uranium in ratios ranging from 3:1 to 10:1 for large quantities. This ratio will tend to vary throughout individual deposits as well as between channels. Controls affecting the vanadium-uranium ratio are not known at this time.

The nature of the uranium-vanadium mineralization suggests that the ore minerals were precipitated from laterally migrating solutions inasmuch as the minerals are not related to fractures or through-going feeders. The ore minerals impregnate the host sandstone and occur principally as interstitial cement with lesser amounts occurring as replacement of vegetal material and with clay galls and thin clay seams.

The Salt Wash ore deposits in the Uravan

mineral belt area are relatively small, tabular or podlike bodies, the long dimensions of which are parallel to the bedding. The deposits are irregularly shaped, both in section and in plan, and generally consist of a single ore layer which may split into two or more tongues. In places, more than one ore layer may be present in the principal ore-bearing sandstone, and the ore layers are separated by unmineralized rock. Commonly the ore deposits are elongate in a direction parallel to the long axis of the channel. The deposits can be almost anywhere within the boundaries of the sandstone channel but they are generally in areas with average or greater sandstone thickness. The ore deposits favor the portion of the sandstone containing carbonaceous material although not all carbon-bearing portions are mineralized.

Typically the ore deposits are nearly concordant to the bedding of the host rock but frequently the ore layers cut sharply across the bedding for no obvious reason, forming deposits known as rolls. Shawe (4) describes roll ore bodies as layered deposits that cut across sandstone bedding in sharply curving

forms. The roll ore bodies are typically C or S shaped in cross-section (Figure 2) and may be elongated for several hundred feet generally parallel to the long axis of the sandstone channel. The rolls are a few feet wide at their widest portion and up to 10 feet in height with an average of about 5 feet. A significant feature of the roll ore bodies is the sharp boundary along the edges between mineralized and barren rock. Roll ore deposits are found in both oxidized and unoxidized portions of of the host sandstone. Shawe (4) postulates that the roll ore bodies were formed by precipitation of minerals at the interface between solutions of different composition and density. Ore production from roll deposits is relatively small compared to the total production from other Salt Wash deposits.

The ore deposits vary considerably in size, ranging from deposits consisting of a single mineralized fossil log containing a few tons of ore to those that may be continuous for several hundred feet containing many thousand tons of ore. Wood and Lekas (5) studied 666 typical Salt Wash ore deposits that ranged in size from a few hundred to 150,000 tons. Results of the study showed that 70 per cent of the deposits contained less than 3000 tons each. The ore thickness ranges from a few inches to about 25 feet, but the average is only 3 to 4 feet. Sub-ore grade mineralization is associated with the ore deposits, but large halos of low grade mineralization are not common. The edges of the ore deposits are very irregular and are defined by either abrupt or gradual changes in grade or thickness.

The intensity of uranium-vanadium mineralization varies considerably throughout a typical deposit ranging from weakly mineralized rock to ore containing several per cent $U_3O_8$ and $V_2O_5$. Average shipping grades of typical mines are 0.2 to 0.3 per cent $U_3O_8$ and 1 to 2 per cent $V_2O_5$.

A typical mine area (Figure 3) generally consists of one or more clusters of individual ore bodies which may or may not be connected by low grade mineralization. Economics of mining generally determine the lateral extent of mine workings from a given mine opening. Production from individual mines ranges from a few hundred to a few hundred thousand tons.

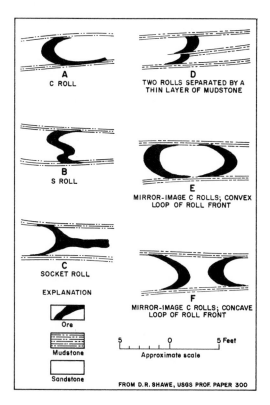

Fig. 2. *Cross-sectional Forms of Roll Ore Bodies.*

## GUIDES TO ORE

Principal criteria used in exploring for uranium-vanadium deposits in the Salt Wash throughout the Uravan mineral belt are: (1)

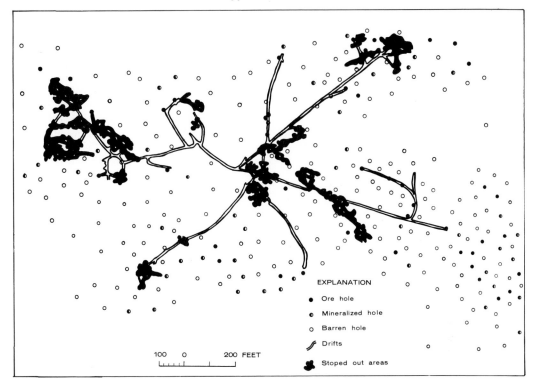

FIG 3. *Plan Map, showing a Typical Salt Wash mine.*

thickness of the host sandstone, (2) quantity of carbonaceous matter within the ore-bearing sandstone, (3) color of the host sandstone and associated clays and mudstones, (4) lithology and sedimentary structure of the host sandstone, and (5) radioactivity. Weir (2) describes a statistical study of some of these features as seen in drill core and suggests a scheme of numerical rating to appraise the favorableness of the rock penetrated by a drill hole.

Although ore deposits are known to occur almost anywhere within the channels, regardless of sandstone thickness, the deposits favor the thicker portions of the sandstone unit. Generally the minimum favorable sandstone thickness is about 30 to 35 feet and the most favorable thickness is from 50 to 90 feet. In areas where the sandstone unit splits into more than one layer, ore deposits frequently occur in each layer.

Organic matter appears to have greatly influenced the emplacement of the ore minerals since, although not all carbon-bearing zones are mineralized, almost all ore deposits are associated either directly with or are adjacent to carbonaceous materials. Rich concentrations of ore minerals frequently occur with local

accumulations of coalified plant remains, sometimes referred to as "trash piles," and fossil logs almost completely replaced by ore minerals are common throughout the mineral belt. However, the size and grade of an ore deposit is not always governed by the amount of carbonaceous material present. Large ore deposits are known to occur with only moderate amounts of carbon and unmineralized carbon-bearing zones commonly occur adjacent to ore deposits. The permeability of the host sandstone does not seem to be a key factor in determining whether or not a carbonaceous zone will be mineralized. The association of ore deposits with coalified plant remains is not fully understood.

The color of the ore-bearing sandstone and associated clays and mudstones differs between areas favorable and unfavorable for ore deposits in the Salt Wash. In the favorable areas, the host sandstone is light brown in the oxidized zone and a light gray where unoxidized, and the clays and mudstones are some shade of gray in both the oxidized and unoxidized zones. In unfavorable areas, the sandstone and associated clays and mudstones are principally red colored in both the oxidized and unoxidized zones. There appears to be a close spatial relationship between the distribution of gray clays and mudstones with carbonaceous matter, indicating that the red ferric oxide has either been reduced to the ferrous state or has been removed by solvents originating from the decay of vegetal matter.

Lithology and sedimentary structures within the host sandstone serve to indicate potential carbon-bearing zones. The carbonaceous material generally is located in lithologic zones containing abundant clay galls or seams and in thin bedded sandstone.

Radioactivity is helpful in defining areas favorable for ore deposits. Only a small proportion of the carbon-bearing areas contain ore deposits, and a significant amount of the carbonaceous material is not even weakly mineralized. All drill holes are generally probed for radioactivity, and carbonaceous zones showing weak radioactivity are considered to be more favorable than those that are completely barren.

There is no apparent genetic relationship between regional structures and ore deposits in the Salt Wash in the Uravan mineral belt, although the salt anticlines may have influenced the Salt Wash drainage pattern. Ore deposits are found in sandstone channels located in synclines as well as those associated with anticlines. The major portion of the mineral belt is free of faults, and the only significant faulting is that associated with collapse features in the salt anticlines. Where faults intersect ore deposits, the deposits are displaced, indicating that the particular faults are post-mineral.

Typically an upper Salt Wash sandstone channel within the Uravan mineral belt exhibits favorable characteristics in varying degrees over a large area; however, estimates are that less than 10 per cent of the favorable area actually contains ore deposits. Studies to date have failed to reveal the specific features that localized the ore deposits.

## EXPLORATION METHODS

Exploration procedures generally in use are based on the favorable criteria outlined above. Initially, efforts are made to outline the boundaries of the sandstone channels containing the ore deposits. This information is compiled from outcrop studies and from drilling data. In unexplored areas, wide-spaced core drilling is used for the first drilling phase. Holes are placed on about 1000 foot centers in fences approximately normal to the projected direction of stream deposition. Information from this phase of exploration not only delineates

the channels but serves to outline broad areas of favorability.

The second drilling stage is conducted within the favorable areas indicated by phase I and the purpose is to attempt to obtain ore intersections. Core drills are generally used during this phase and holes are placed at random on 200 to 600 foot centers or as offsets to any mineralized holes drilled during phase I.

The third drilling phase is the ore-blocking-out phase which is designed to outline the ore deposits. Normally, ore or mineralized holes from stage II are offset on a square grid. Due to the erratic nature of the Salt Wash ore bodies, holes should be drilled on about 25 foot centers if the deposits are to be outlined with a high degree of accuracy. Since this is impractical because of the high cost of sufficient drilling, drill hole centers are expanded to different distances depending on the depth to the ore horizon. At shallow depths of 150 feet or less, drill holes are on about 50 foot centers. Block-out centers are expanded as depths increase to where holes are drilled on about 200 foot centers where the ore-bearing sandstone is 600 to 800 feet below the surface. Only a small proportion of the total ore reserves is intersected by holes on these centers, but sufficient data are collected to compute reserves over relatively large areas. The ratio of ore holes to total holes drilled serves as a basis for computing ore reserves, and the computations are greatly influenced by the favorability of the ore-bearing sandstone.

Core drills are used generally for phases I and II, and for block-out drilling in deep areas, but in all of these only the ore-bearing sandstone is cored. Wagon drills and rotary plug drills are used for blocking out ore in relatively shallow areas. Generally all holes are probed for radioactivity which gives a reliable estimate of uranium grade since the uranium and decay products are in good equilibrium in both the oxidized and unoxidized zones.

Underground long-hole drilling, employing diamond drills or percussion drills, is used extensively to supplement surface drilling, principally in deep mines where final block-out surface drill holes are on centers of 100 feet or greater. Stations for drill sites to be located in haulage drifts are generally predetermined and are included in the primary mine development plans. In planning the location of the underground drill sites, efforts are made to position the stations so as to obtain a large angle of intersection with the target zone; however, where this is difficult, raises are driven for the drill position, thus increasing the angle of intersection.

Several holes, in a radiating pattern, are drilled from each drill station to intersect favorable target zones within the ore-bearing sandstone. Maximum effective length of diamond drill holes is about 200 feet and of holes drilled with percussion machines about 75 feet. Holes are drilled with water as a circulating medium, and cores or drill cuttings generally are not saved, but the holes are probed for radioactivity.

Geologic maps showing ore trends and favorability of the host sandstone are constructed for those areas of interest as geologic information becomes available. These maps are used to plan successive drilling programs and also provide information which is useful for planning primary mine development. A common method used in constructing these maps is to assign numerical ratings to each drill hole and then to contour the values. The numerical ratings represent a summation of values assigned individual geologic guides used to evaluate favorability. Data acquired from underground geologic mapping are added to the maps to provide additional information that may indicate favorable trends.

Drilling depths to the ore horizon throughout the Uravan mineral belt vary considerably. The Salt Wash ore sandstone is exposed along numerous drainages within the belt, and where benches have developed on top of the Salt Wash, drilling depths are less than 250 feet. Most of the ore produced to date in the Uravan mineral belt has come from these areas with relatively shallow depths to ore. There are large areas in the mineral belt, however, where the surface is formed by the Dakota Formation and depths to the Salt Wash ore horizon are approximately 700 feet. A considerable amount of exploration for Salt Wash ore deposits has been done in these deeper areas, and a significant amount of production has resulted from successful exploration programs.

## OUTLOOK FOR FUTURE DISCOVERY

Most of the lands throughout the Uravan mineral belt area consist of public domain, and mineral rights are held under mining claims. Large areas within the mineral belt were withdrawn from entry during the early years of the AEC uranium procurement program for exclusive exploration by government agencies. Much of the land has been restored to the public domain as unfavorable for ore deposits; however, significant favorable areas are still withdrawn.

The relatively shallow areas in the mineral belt have been intensively explored and resulting ore deposits heavily produced since 1948 under the AEC uranium program. The reserves will continue to be depleted since many of the properties will be produced through 1970 under the AEC stretch-out program.

Best estimates are that most of the ore reserves in the shallow areas in the mineral belt will have been mined by 1971 except those reserves on withdrawn ground, and the remaining significant reserves will be in areas where the ore horizon is 600 to 800 feet deep. There are large areas in the Uravan mineral belt with these depths that remain to be explored.

## REFERENCES CITED

1. Fischer, R. P. and Hilpert, L. S., 1952, Geology of the Uravan mineral belt: U.S. Geol. Surv. Bull. 988-A, 13 p.
2. Weir, D. B., 1952. Geologic guides to prospecting for carnotite deposits on Colorado Plateau: U.S. Geol. Surv. Bull. 988-B, p. 15–27.
3. Craig, L. C., *et al.,* 1955, Stratigraphy of the Morrison and related formations, Colorado Plateau region—a preliminary report: U.S. Geol. Surv. Bull. 1009-E, p. 125–168.
4. Shawe, D. R., 1956, Significance of roll ore bodies in genesis of uranium-vanadium deposits on the Colorado Plateau: *in Contributions to the geology of uranium and thorium,* U.S. Geol. Surv. Prof. Paper 300, p. 239–241.
5. Wood, H. B. and Lekas, M. A., 1958, Uranium deposits of the Uravan mineral belt: *in Guidebook to the geology of the Paradox Basin,* Intermountain Assoc. Petrol. Geols. 9th Am. Annual Field Conf., p. 208–215.
6. Weeks, A. D., *et al.,* 1959, Geochemistry and mineralogy of the Colorado Plateau uranium ores; summary of the ore mineralogy: *in Geochemistry and mineralogy of the Colorado Plateau uranium ores,* U.S. Geol. Surv. Prof. Paper 320, p. 65–79.
7. Keller, W. D., 1962, Clay minerals in the Morrison formation of the Colorado Plateau: U.S. Geol. Surv. Bull. 1150, 90 p.

# 40. Uranium Deposits of Wyoming and South Dakota*

E. N. HARSHMAN†

## Contents

## Illustrations

## Tables

* Publication authorized by the Director, U.S. Geological Survey.
† U.S. Geological Survey, Denver, Colo.

## ABSTRACT

Uranium mining and milling are rather new industries in Wyoming and South Dakota, but, in the past 15 years, ore valued at more than $400,000,000 has been mined and milled in the region. The major uranium deposits are in structural and/or erosional basins or on the flanks of uplifts. The principal host rocks are medium- to coarse-grained sandstones that were deposited in continental or brackish-water environments and were derived in large part from the granitic cores of the uplifts and the sedimentary rocks that flank the cores. The ages of the host rocks range from Cretaceous to Miocene, but more than 95 per cent of the known reserves in the region are in rocks of early Eocene age. The bulk of the ore in the region is unoxidized and lies below the water table; some oxidized ore has been mined in the Powder River Basin and Black Hills districts. The unoxidized ore comprises pyrite, uraninite and/or coffinite, marcasite, hematite, jordisite(?), and selenium. Minerals in the oxidized ore include hydrous uranium carbonates, phosphates, and sulfates and hydrous uranium and/or vanadium silicates.

The deposits are in the form of ore rolls that contain from a few hundred to a few hundred thousand tons of ore, ranging in grade from 0.1 per cent to 1.0 per cent $U_3O_8$. The rolls are at the margins of large tabular bodies of sandstone that has been altered by the ore-bearing solutions. The deposits are believed to have resulted from the interaction of an alkaline, oxidizing, ore-bearing ground water and a reducing environment.

Exploration guides include the spatial relations of ore to altered sandstone, favorable lithology in the host rock, the content of certain elements and/or compounds in ground water, and the relation of deposits to major paleodrainages.

## INTRODUCTION

Uranium mining and milling are rather new industries in Wyoming and South Dakota, but they are making significant contributions to the economy of the region. Ore valued at about $400,000,000 (at $8 00 per pound of contained $U_3O_8$) was mined and, for the most part, milled in the area during 1951 through 1965.

During its growth, the industry passed through periods of extensive prospecting, intensive exploration, and rapid mine development—a remarkably successful "crash" program in which geology played an important role. This paper discusses briefly the broad aspects of the geology and ore deposits of the Wyoming-South Dakota uranium-bearing region and of several of the smaller mining areas, genetic concepts, and exploration guides. It is followed by three short papers, the authors of which are geologists well qualified to discuss the geology and ore deposits of the major uranium-mining districts in the region.

### History of Discovery and Development

Uranium mineral occurrences were known in the Wyoming-South Dakota region as early as 1935, but the first discovery of uranium in economically significant amounts was made in June 1951 along the east side of Craven Canyon about 10 miles north of Edgemont, South Dakota (18). Subsequently, important discoveries in Wyoming were made in the Powder River Basin in 1951, in the Gas Hills, Crooks Gap, and Baggs areas in 1953, at Copper Mountain in 1954, and in the Pryor Mountains and Shirley Basin in 1955. In most areas, the discoveries were followed almost immediately by exploration, development, and mining activities.

Ore-buying stations were established by the United States Atomic Energy Commission at Edgemont, South Dakota, in late 1952, and at Riverton, Wyoming, in early 1955. When it was apparent that the Wyoming-South Dakota region contained large reserves of uranium ore, plans were made for construction of several mills. The first mill was completed in July 1956 by Mines Development, Inc., at Edgemont, South Dakota. Its rated capacity was 300 tons per day. The Western Nuclear mill, with a capacity of 400 tons per day, was completed at Jeffrey City, Wyoming, in July 1957, and the Lucky Mc mill, rated at 750 tons per day, was completed in the Gas Hills area in February 1958. Mills with a capacity of about 500 tons per day were completed by Fremont Minerals at Riverton, Wyoming, in November 1958; by Globe Mining Co. and Federal-Gas Hills in the Gas Hills, in late 1959; and by Petrotomics Co., in the Shirley Basin, in October 1962. The capacities of most of these processing plants have been increased above their original tonnages.

From a very modest start of about 1600 pounds of $U_3O_8$ in 1951, annual production in Wyoming rose to slightly more than 7,000,000 pounds of $U_3O_8$ in 1961, and, among the States, Wyoming became the second largest producer of uranium. South Dakota, production reached a maximum of slightly more than 460,000 pounds of $U_3O_8$ in 1963. About 21 per cent of the 1965 domestic production of uranium oxide came from the Wyoming-South Dakota region. A sharp decline in production of $U_3O_8$ in 1965 resulted from agreements between most of the uranium mining companies and the Atomic Energy Commission whereby production was curtailed—in the so-called "stretchout program."

Exploration activities did not parallel the rapid upswing in production of uranium. By 1958, the combined efforts of industry and government proved so successful in the discovery of uranium ore that production threatened to exceed domestic requirements, and it was necessary for the government to limit purchases. This limitation had an adverse effect on exploration and development activities, and only a few of the more aggressive mining companies continued to explore for new deposits. The Shirley Basin, discovered in 1955 and developed in 1957–60, was the last major new district brought into production in this country.

In 1965, there was renewed interest in exploration for uranium, due principally to the anticipated rapid increase in world requirements for nuclear-fueled power plants. It is forecast that uranium for power to be generated in the 1980's will have to come from deposits and/or mining districts yet to be discovered. By the spring of 1966, established uranium mining companies, as well as companies new in the industry, were actively exploring for uranium deposits in the Wyoming-South Dakota region.

### Production and Reserves

Uranium production for the States of Wyoming and South Dakota, compiled from figures furnished by the United States Atomic Energy Commission and the United States Bureau of Mines, was as shown in Table I.

As of October 1, 1965, the United States Atomic Energy Commission estimated the ore reserves in the Wyoming-South Dakota region to be:

| State | Ore (tons) | $U_3O_8$ (lbs) |
|---|---|---|
| Wyoming | 22,000,000 | 110,000,000 |
| South Dakota | 200,000 | 1,300,000 |

Most of the reserves in Wyoming are in continental clastic rocks of early Tertiary age;

**TABLE I.   Uranium Production and Reserves— Wyoming and South Dakota**

| Year | Wyoming | | South Dakota | |
|---|---|---|---|---|
| | Ore (tons) | $U_3O_8$ (lbs) | Ore (tons) | $U_3O_8$ (lbs) |
| 1951 | 268 | 1,608 | | |
| 1952 | 40 | 360 | 4,228 | 22,831 |
| 1953 | 4,629 | 29,672 | 11,491 | 49,892 |
| 1954 | 13,538 | 88,826 | 18,493 | 65,273 |
| 1955 | 51,663 | 230,281 | 24,088 | 98,189 |
| 1956 | 156,509 | 702,882 | 35,302 | 129,178 |
| 1957 | 247,699 | 1,189,947 | 69,800 | 231,215 |
| 1958 | 651,790 | 3,282,698 | 35,489 | 142,375 |
| 1959 | 864,582 | 4,337,433 | 45,734 | 171,449 |
| 1960 | 1,357,225 | 6,740,180 | 41,104 | 156,180 |
| 1961 | 1,521,064 | 7,091,427 | 43,588 | 148,688 |
| 1962 | 1,301,784 | 6,386,210 | 29,452 | 104,390 |
| 1963 | 1,173,420 | 5,892,238 | 72,088 | 462,015 |
| 1964 | 1,490,353 | 6,950,847 | 110,147 | 423,195 |
| 1965 | 1,048,176 | 4,560,625 | 44,738 | 104,500 |
| TOTAL | 9,882,740 | 47,485,234 | 585,742 | 2,309,370 |

those in South Dakota are in continental sandstones of Early Cretaceous age and in lignites of Paleocene age. These reserves are about 37 per cent of the total United States reserves.

## GEOLOGY

### Geologic History

The Precambrian rocks of central Wyoming and western South Dakota crop out only in the cores of the mountain ranges. They are part of an ancient terrain that was folded, metamorphosed, and subsequently invaded by batholiths, generally granitic in composition, by pegmatites, and by mafic dikes. After this complex history of deposition, metamorphism, and intrusion, the region was eroded and, by the end of Proterozoic time, was reduced to a quite flat surface.

Throughout the Paleozoic Era and much of the Mesozoic, the central Wyoming-western South Dakota region was a stable foreland that lay along the east side of the Cordilleran geosyncline. Epicontinental seas repeatedly transgressed the area from the west and northwest, and in them were deposited sediments representing most geologic systems. The formations generally are thicker in the western part of the region than in the eastern part, and some formations disappear eastward because of erosion or nondeposition.

Near the end of the Jurassic Period, highlands began to develop in the geosynclinal area to the west of Wyoming, and the western shores of the Cretaceous seaways migrated progressively eastward. Great amounts of clastic debris were shed into the Cretaceous seas from highlands to the west and to the east.

Near the end of the Cretaceous Period, the seas withdrew from the region, and the first movements of the Laramide orogeny began to deform the featureless landscape. Uplift started in what is now western South Dakota, as well as along the west side of the old Cretaceous seaway in central and western Wyoming. Folding continued through the Paleocene, and, by the end of that time, the major mountain uplifts and the basins were well outlined. Paleocene sediments were deposited at the foot of steep mountain fronts, across broad fluvial plains, and in lakes that formed in many of the basins.

Folding of the mountain masses, accompanied by reverse faulting, continued through early Eocene time, and, in central Wyoming, thick continental sediments accumulated in the subsiding intermontane basins. In eastern Wyoming and western South Dakota, the debris eroded from the Black Hills uplift apparently was transported from the area by streams. Basin subsidence and mountain uplift virtually ceased by middle Eocene time, but debris from the mountains, augmented by material of volcanic origin, continued to fill the basins throughout Oligocene and Miocene time. Uplift in late Miocene time resulted in renewed erosion and, near the mountain fronts, deposition of coarse-grained debris on the truncated edges of the earlier Tertiary rocks.

Regional uplift toward the end of Pliocene time initiated the cycle of erosion that continues today, a cycle that has exhumed land forms that were carved initially in Paleocene and/or Eocene time.

The geologic history of the area is more fully discussed by Keefer (26), Thomas (8), and Gries (20).

## Physical Setting

Most of the Wyoming-South Dakota uranium-mining region is in the Wyoming Basin and the Great Plains physiographic provinces of Fenneman (1), although the northern part of it lies in the Middle Rocky Mountain province. The Wyoming Basin province is continuous with the Great Plains province to the east through a gap between the Laramie and Bighorn Mountains, and it connects with the Colorado Plateau province to the south through a gap between the Uinta Mountains and the Park Range.

The region is dominated by broad basins, and it forms a topographic discontinuity between the southern and the central Rocky Mountains. Structural continuity, however, is shown by the low mountains now buried or partly buried by the Tertiary basin fill.

In the western part of the region, the basin surfaces are 6500 to 7500 feet above sea level; in the eastern part altitudes approximate 3000 to 3500 feet. The highest peaks in the rugged mountains that surround the region range from about 13,000 to 14,000 feet above sea level.

All but the southwest corner of the region is in the Missouri River drainage basin; the southwest corner is drained by the Green River and its tributaries. It is noteworthy that most of the major streams draining the region pass through, not around, the bounding mountain ranges. Superimposed drainage is clearly evident.

## Structure

The principal structural elements in the Wyoming-South Dakota uranium-bearing region are shown on Figure 1. Most of the northwesterly trending mountain ranges are asymmetric anticlinal folds with one gently dipping limb and one steeply dipping, overturned, and/or thrust-faulted limb. Thrusting is most common in the western part of the region and is less common, or absent, in the eastern part.

The intermontane basins are structural features formed by simple downwarps or downwarps modified by folds and/or faults. Many of the basins are asymmetrical with their deepest parts near an adjacent uplift. The basins are filled with a considerable thickness of clastic continental rocks of Tertiary age, derived in large part from the granitic cores of the uplifts and from the sedimentary rocks that flanked the cores.

## Igneous Activity

Cenozoic igenous activity, shown on Figure 1, is minimal in the Wyoming-South Dakota uranium-bearing region, although directly to the northwest is the major volcanic field of the Absaroka Mountains. Intrusive rocks of Laramide age crop out as sills, dikes, and laccoliths in the northern part of the Black Hills area. In the Rattlesnake Hills of central Wyo-

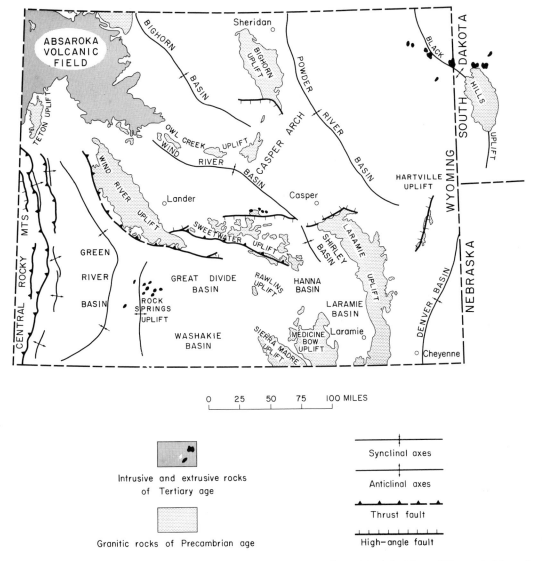

FIG. 1. *Tectonic Map of Wyoming and Western South Dakota. Modified from "Tectonic Map of the United States," U.S. Geol. Surv. and Amer. Assoc. of Petro. Geols., 1961.*

ming, several volcanic necks of middle Eocene age bear testimony to volcanic activity in that area. There was Tertiary extrusive activity in the Leucite Hills area, north of the Rock Springs uplift, where erosional remnants of cinder cones and lava flows are preserved in small buttes and mesas.

## Stratigraphy

The principal units present in the uranium mining districts of Wyoming and western South Dakota are shown in the generalized stratigraphic column of Table II. Rocks of Paleozoic age crop out on the flanks of many of the mountain ranges and underlie the Wyoming basins, generally at considerable depth. They were deposited in epicontinental seas and comprise limestones, dolomites, sandstones, and shales; in some areas they are phosphatic and in others they contain anhydrite. Abrupt facies changes are common, and the Paleozoic stratigraphic section thins from about 3000 feet in the western part of the area to about 2000 feet in the Black Hills (8) (20) (26).

Mesozoic rocks crop out low on the flanks

TABLE II.   *Principal Stratigraphic Units in Uranium-Mining Areas*

| Geologic time units | | | Central and southeastern Wyoming area | | Black Hills area | |
|---|---|---|---|---|---|---|
| Cenozoic | Quaternary | Recent and Pleistocene | Stream alluvium and terrace gravels | | | |
| | Tertiary | Pliocene | Ogallala Fm | | Ogallala Fm | |
| | | Miocene(?) | Browns Park Fm | | (Absent) | |
| | | Miocene | Arikaree Fm | | Arikaree Fm | |
| | | Oligocene | White River Fm | | White River Fm | |
| | | Eocene | Wagon Bed Fm | | (Absent) | |
| | | | Wind River* and Wasatch* Fms | | | |
| | | Paleocene | Fort Union Fm | | Fort Union* Fm | |
| Mesozoic | Cretaceous | Upper | Lance Fm | | Hell Creek Fm | |
| | | | | | Fox Hills Ss | |
| | | | Lewis Sh | | Pierre Sh | |
| | | | Mesaverde Fm | | | |
| | | | Steele and Cody Shs | | | |
| | | | | | Niobrara Fm | |
| | | | Frontier Fm | | Carlile Sh, Greenhorn Fm, and Belle Fourche Sh | |
| | | Lower | Mowry Sh | | Mowry Sh | |
| | | | Thermopolis Sh | | Newcastle Ss and Skull Creek Sh | |
| | | | Cloverly Fm | | Fall River and Lakota Fms | Inyan Kara* Gr. |
| | Jurassic | | Morrison Fm | | Morrison Fm | |
| | | | Sundance Fm | | Sundance Fm | |
| | | | Gypsum Spring Fm | | Gypsum Spring Fm | |
| | | | Nugget Ss | | (Absent) | |
| | Triassic | | Chugwater Fm | Jelm and Chugwater Fms | Spearfish Fm | |
| | | | Dinwoody Fm | Goose Egg Fm | | |
| Paleozoic | Permian | | Phosphoria and Park City Fms | | Minnekahta Ls | |
| | | | | | Opeche Fm | |
| | Pennsylvanian | | Tensleep Ss | Casper Fm | Minnelusa Fm | |
| | | | Amsden Fm | | | |
| | Mississippian | | Madison Ls | | Pahasapa Ls | |
| | Devonian | | (Absent) | | Englewood Fm | |
| | Ordovician | | Bighorn Dol | | Whitewood Ls | |
| | | | | | Winnipeg Fm | |
| | Cambrian | | Gallatin Ls | | Deadwood Fm | |
| | | | Gros Ventre Fm | | | |
| | | | Flathead Ss | | (Absent) | |
| Precambrian | | | Granitic and metamorphic rocks | | | |

[Shows general ages and succession of units; correlation between columns is only approximate.  Fm, formation; Ss, sandstone; Sh, shale; Ls, limestone; Gr, group; Dol, dolomite.  Asterisk (*) indicates principal uranium-bearing unit] of the mountain ranges, and, in most basins, they form the eroded surface upon which the post-Cretaceous rocks were deposited. They comprise a thick sequence of sandstones, mudstones, and shales and include red beds in the lower part of the sequence. Rocks of Mesozoic age in western Wyoming aggregate as much as 30,000 feet in thickness, but, in the Black Hills area, they thin to about 6000 feet (8).

Sedimentary rocks of Tertiary age are probably as much as 20,000 feet thick in the deepest parts of some of the Wyoming basins, but they are virtually absent from the Black Hills. However, remnants of rocks of Miocene and/or Pliocene age on the flanks of the Black Hills indicate that at one time rocks of middle and late Tertiary age may have extended across the area and at least partially buried the Black Hills uplift. The Tertiary rocks are tuffaceous and clastic and are of fluviatile, lacustrine, and paludal origin.

Paleocene rocks are principally fine grained and, in large part, were derived from the sedimentary rocks on the flanks of newly formed uplifts. However, by early Eocene time, the granitic cores of these uplifts were exposed by erosion; the rocks of Eocene age are composed principally of arkosic material carried into the basins by torrential streams. Low-rank coal and lignite occur in rocks of Paleocene and early Eocene age; materials of volcanic origin are sparse.

Mid-Tertiary volcanic activity in the Absaroka Mountains-Yellowstone Park area produced large amounts of airborne volcanic ash, and the Oligocene, Miocene, and Pliocene rocks in Wyoming and South Dakota contain this ash as a major constituent.

The principal host rocks for uranium deposits in the Wyoming-South Dakota region are the Inyan Kara Group of Early Cretaceous age, the Wind River and Wasatch Formations of early Eocene age, and the Fort Union Formation of Paleocene age. In the uranium-bearing areas of western South Dakota and northeastern Wyoming, the Inyan Kara Group ranges in thickness from about 325 to 650 feet and includes the Lakota Formation and the overlying Fall River Formation. The Lakota Formation is nonmarine, ranges in thickness from 200 to 500 feet, has a poorly defined basal contact, and differs greatly in lithology from place to place. The formation thickens southward and eastward from the northwestern part of the Black Hills. The rocks comprise a sequence of fine- to medium-grained sandstone, siltstone, and clay, deposited under fluvial, lacustrine, and paludal conditions. Most of the sandstone occurs as poorly bedded lenses, some of which are of considerable lateral extent. Carbonaceous shale is present in the lower part of the formation in some areas.

The Fall River Formation overlies the Lakota with a marked disconformable contact. It is 125 to 150 feet thick and comprises mostly fine-grained sandstone with some interbedded siltstone. The Fall River was deposited in a marginal marine environment, and its thin-bedded character contrasts with the more massive character of the nonmarine Lakota.

The Wind River Formation and its partial equivalent, the Wasatch Formation, are the principal host rocks for uranium deposits in Wyoming. In the uranium-mining areas, they are early Eocene in age and comprise a sequence of interbedded arkosic sandstones, conglomerates, siltstones, and shales that in some areas contain coal and/or lignite. The sediments were derived from the mountain ranges surrounding the basins and were deposited in alluvial fans, stream channels, swamps, and in low flat areas between streams. Correlation of gross lithologic units can be made over distances of several miles, but there is little continuity of individual coarse-grained beds. Continuity of individual fine-grained and carbonaceous beds is much better.

Either the Wind River or the Wasatch is present in all the major Wyoming basins, but neither is present in western South Dakota. The formations are thin at the margins of the basins but reach thicknesses of several thousand feet in the centers of some of the basins (27).

On a regional basis, the Wind River (or Wasatch) is divisible into two distinctive lithologic units—a mountainward facies, comprising coarse clastic rocks with boulders as much as 25 feet in diameter, and a basinward facies, comprising fine-grained clastic sediments with some interbedded pebble, cobble, and boulder beds. The two facies interfinger, the extent of either having been determined by the vigor of the streams transporting the eroded debris toward the centers of the basins. The coarse-grained facies may extend basinward from the mountain flank as much as 20 miles.

The Wind River Formation in the Wind River Basin has been divided according to lithologic and faunal differences into an upper unit—the Lost Cabin Member, and a lower unit—the Lysite Member. The Lost Cabin tends to contain more coarse-grained conglomeratic material and less fine-grained silty material than does the Lysite. This twofold division cannot be made in many areas of Wyoming.

The Fort Union Formation in Wyoming comprises a nonmarine sequence of inter-bedded conglomerates, sandstones, mudstones, and shales, generally carbonaceous and drab. On a regional basis, the formation may be divided into two distinctive units: a lower one, composed of interbedded carbonaceous shale, shale, sandstone, and minor amounts of con-glomerate and an upper one, generally more fine grained and composed of clayey siltstone, shale, ironstone, and coal.

The thickness of the Fort Union Formation ranges from 0 to as much as 8000 feet in this region. In the Great Divide Basin, the entire thickness is not exposed, but it is esti-mated to be about 1100 feet; in the Powder River Basin, the thickness ranges from about 2000 to 3000 feet. The formation may be as much as 8000 feet thick in the deepest part of the Wind River Basin (27). Here the upper unit is divisible into two members—the Shot-gun Member, comprising fine- to medium-grained clastic rocks deposited in areas margi-nal to a large body of water known as Walt-man Lake, and the Waltman Shale Member, deposited contemporaneously in the lake.

The Fort Union Formation is not present in the Black Hills areas, but, in northwestern South Dakota, it is exposed in several flat-topped buttes and ridges that rise as much as 500 feet above the surrounding country (28). It consists of interbedded sandstones, shales, claystones, and coal. The sediments were deposited in flood plains, in swamps, and, to a lesser extent, in brackish water or in shallow seas. The formation ranges in thick-ness from about 700 feet (28) to about 1200 feet (12).

The Fort Union Formation in Wyoming is not known to contain significant deposits of uranium ore, but, in South Dakota, it is ore bearing.

## ORE DEPOSITS

The locations of the principal mining dis-tricts in the Wyoming-South Dakota region are shown on Figure 2. The districts, for the most part, are in axial portions or on the flanks of basins filled with Tertiary rocks. Most of the uranium deposits are in clastic rocks of continental origin and of Cretaceous or early Tertiary age. The bulk of the ore produced in the two-state area has been composed of low-valent uranium oxides and silicates, with

a vanadium-to-uranium ratio of less than 1 to 1; however, some ore has been mined from oxidized deposits in the Powder River Basin and the Black Hills districts.

The deposits generally are concordant with the bedding of the host rocks, and many de-posits are in the form of rolls (29) (14) (32). They contain from a few hundred to a few hundred thousand tons of ore, and the grade, as mined, has averaged about 0.24 per cent $U_3O_8$ in Wyoming and about 0.20 per cent $U_3O_8$ in South Dakota.

Olin M. Hart, in a companion chapter in this section of the Volume, discusses the ura-nium deposits of the Black Hills area; a general description of the Wyoming deposits follows.

## Mineralogy of the Wyoming Deposits

Uraninite, $(U^4_{1-x}, U^6_x)O_{2+x}$, and coffinite, $U(SiO_4)_{1-x}(OH)_{4x}$, are the principal low-valent uranium minerals in the unoxidized sandstone-type deposits of Wyoming. Pyrite, marcasite, hematite, molybdenum (probably as jordisite), and selenium (in part as ferro-selite) are associated with the uranium min-erals; all are considered to be epigenetic and to have been deposited during the ore-forming processes. Calcite cement is present in and near the ore bodies; it is genetically related to ore.

Minerals of high-valent (oxidized) uranium and/or uranium-vanadium are present above the water table in some of the Wyoming ura-nium-mining districts. The most common min-erals are hydrous uranium carbonates, phos-phates, and sulfates and hydrous uranium and/or vanadium silicates. Other common oxi-dization products are ilsemanite and "selenium bloom." In many areas, minerals of oxidized uranium are sparse, for there is insufficient vanadium and/or phosphorus to form con-spicuous amounts of the insoluble uranium vanadate or phosphate minerals, and, once in solution, uranium remains mobile and is leached from the zone of oxidation.

In unoxidized ore, uranium and sulfide min-erals coat sand grains, fill interstices between grains, and, to a minor extent, replace the grains. Carbonized plant debris in some de-posits contains considerable uranium and/or pyrite, but, in many deposits, there is no direct relation between the organic carbon content of the ore and its uranium content. Minerals in oxidized ore fill interstices in sandstone and form efflorescences on natural outcrops, pit walls, and mine openings.

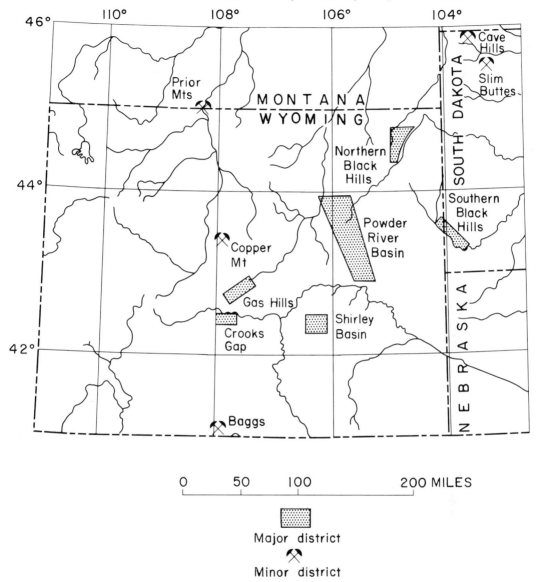

FIG. 2. *Uranium-Mining Districts of Wyoming and South Dakota.*

## Character and Size of the Wyoming Deposits

The principal uranium deposits in Wyoming are in arkosic sandstones and conglomerates of continental origin and of early Eocene age. The ore-bearing sandstone units range in thickness from a few tens of feet to several hundred feet. Generally these units contain interbedded siltstones or shales, and in many districts they are underlain and overlain by quite thick sequences of fine-grained rocks. Most of the deposits are found in the more permeable portions of the sandstone beds, but some have been emplaced at or near the contact between sandstone and an underlying siltstone or shale. Some deposits occupy the entire sandstone interval between two essentially impermeable beds; others occupy only a portion of that interval and are suspended within it. Carbonaceous plant remains are present in most of the ore-bearing sandstones or in the shales interbedded with them.

Most low-valent uranium ore occurs below

the ground-water table in roll-type or blanket-type deposits. The terms "roll" and "blanket" describe the forms of the ore bodies and have no genetic significance. Roll-type ore bodies, which extend laterally for as much as several thousand feet, are commonly crescent- or C-shaped in cross section, although other curved forms have been described (6). Blanket deposits, as the name implies, are thin tabular ore bodies, generally conformable with the flat bedding of the sedimentary rocks. The roll-type ore deposits are economically far more important than are the blanket type; it is probable that many of the deposits previously described as blanket type were actually the upper or lower limbs of roll-type deposits.

Typically, ore in a roll-type deposit is in sharp contact with the host sandstone on the concave side of the ore body, but on the convex size the uranium content of the ore decreases gradually to trace amounts. The highest uranium content may be along the inner, or concave, margin of the ore body, or it may be at some distance from it. The highest selenium content generally is found in a narrow zone astride the inner or concave contact of ore and altered host rock. The highest molybdenum content is on the convex side of ore in the transition zone between ore and mineralized sandstone. The larger roll-type deposits contain several hundred thousand tons of ore ranging in grade from a few hundredths to more than 1 per cent $U_3O_8$.

In some parts of roll-type ore bodies, sedimentary structures appear to have exercised considerable control on ore deposition, and the location of the ore was influenced by such features as crossbedding; in other parts, little or no control is apparent, and the ore cuts indiscriminantly across sedimentary structures. A sound generalization is that major sedimentary features have controlled the position of ore bodies, while minor sedimentary features have, to a limited extent, controlled the intricacies of their form.

Some low-valent ore has formed by the redeposition of uranium leached from older deposits and carried by percolating ground water into a reducing environment. This ore tends to be of the blanket type and to form in sandstone near shale contacts or in the shale itself. Economically, it is unimportant.

Deposits of high-valent, or oxidized, uranium ore are found in areas of Wyoming where vanadium is present in the unoxidized ore in quantities sufficient to form insoluble uranium vanadate minerals. Oxidation destroys

deposits of low-vanadium content. Oxidized ores generally tend to be pseudomorphous with the unoxidized deposits from which they formed, and, depending on the extent of oxidation, they consist of oxidized minerals surrounding an unoxidized core or they consist entirely of the oxidized progeny of unoxidized minerals. Deposits of this type have been mined in the Powder River Basin (22) and in shallow surface workings in other districts. The larger oxidized deposits contain a few tens of thousands of tons of ore ranging in grade from a few hundredths to a few tenths of a per cent $U_3O_8$.

## Relation to Alteration

The roll-type ore deposits in Wyoming are related to large tabular bodies or tongues of altered sandstone. In this discussion, the term "altered" is used in a restricted sense to denote certain physical and chemical changes, excluding ore deposition, effected by the ore-bearing solution on the medium through which it passed. The largest ore bodies lie along the margins of the altered sandstone tongues and extend outward for distances of a few to as much as a few hundred feet. Smaller ore bodies are emplaced along the top and bottom surfaces of the tongues. The tongues in different districts are similar in their size and shape, their position in permeable sandstone beds, and their spatial relation to ore. They are not necessarily similar in the elemental and mineralogical composition that distinguishes them from unaltered rock, although similarities do exist.

Altered tongues are as much as 5 to 6 miles long, 1 to 3 miles wide, and a few tens of feet to as much as 200 feet thick. In some areas, they consist entirely of medium- to coarse-grained sandstone; in other areas they contain siltstone on claystone beds or lenses. Similarities between altered sandstone tongues in the major uranium mining districts include: (1) a distinctive color, although not the same in all districts; (2) a selenium content of about 10 ppm in contrast to about 1 ppm in unaltered sand; (3) an eU/U ratio generally higher than in ore or unaltered sand; (4) calcium carbonate, organic carbon, and sulfate contents generally lower than in ore or unaltered sand; (5) a partial or complete destruction of some of the heavy minerals; and (6) similar form of some heavy minerals, particularly pyrite. Altered sand tongues and the character of the alteration in various uranium mining districts are discussed in companion papers in this volume and by Bailey (23), Harshman (14,31), King and Austin (32), and Melin (21).

## Origin of the Deposits

The sandstone-type uranium deposits in Wyoming have many broad similarities, and it seems probable that the genetic processes responsible for their origin were similar but not necessarily identical.

Any acceptable concept of origin must explain the wide geographic and restricted stratigraphic distribution of the deposits, the spatial relation of ore to large tongues of altered sandstone, the elemental and mineralogic composition of ore and altered sandstone, and the physical parameters of the ore bodies. Completely satisfactory concepts have not been advanced, but significant progress in that direction has been made in the last decade.

A syngenetic theory of origin for the major Wyoming deposits is untenable, for although most ore bodies are peneconcordant in a broad sense, more often than not they transect cross-bedding and other sedimentary structures; for this reason there have been few advocates of syngenesis.

Epigenetic concepts of origin of sandstone-type deposits can be divided into two broad categories based on the source of the uranium and the character of the solutions in which it was transported to the site of original deposition.

Advocates of the magmatic hydrothermal or hypogene concept (13) (19) (11) propose that uranium and the associated elements were expelled from a magma in fluids that passed upward into zones of progressively lower temperatures and pressures. These fluids entered gently dipping permeable sedimentary rocks, comingled with ground water to a greater or lesser extent, and eventually reached a place where temperature and pressure conditions caused deposition. Some geologists propose that a redistribution of these hypogene ore deposits (or concentrations) by ground water formed the deposits now being exploited. The major distinctions between the hypogene concept and others disappear with the hypothesis of excessive redistribution of uranium by ground water.

The hypogene concept of origin has been rejected by most geologists working on the Wyoming deposits. There is no evidence of Phanerozoic igneous activity in several of the uranium mining districts nor evidence that would relate any of the deposits to unknown igneous sources, but there is a considerable amount of evidence to support the supergene concept of origin.

Proponents of the supergene concept advocate that: (1) percolating vadose water leached uranium and associated elements from various source rocks and transported them basinward through the transmissive ore-bearing formations and (2) deposition occurred when these solutions entered an environment conducive to precipitation of the transported elements. These events probably took place below the water table in most areas but may have done so at the water table in a few areas. There is no unanimity of opinion as to the specific source of the uranium. Some investigators believe that it came from the arkosic sediments that are the principal host rocks for the deposits (21,32); others believe that uranium was released to ground water by devitrification of glass in the tuffaceous rocks which overlie, or once overlay, the uranium mining areas (2,12). It is entirely possible that the source was multiple, for uranium is a mobile element, present in small amounts in rocks of many types.

The chemical character of the ore-bearing ground water is speculative, although, if the present is a key to the past, it was slightly alkaline and oxidizing. Hostetler and Garrels (15) have shown that moderately alkaline slightly reducing ground water containing sulfate and carbonate ions is capable of dissolving uranium and transporting it, as a carbonate complex, to the site of deposition. Harshman (31) has shown that slightly alkaline moderately oxidizing ground water may have transported uranium and the elements associated with it to form the Shirley Basin deposits, while Melin (21), who studied the same area, believes that the transporting ground water was acid (pH $4\pm$) and reducing so far as uranium was concerned.

Proposed mechanisms for precipitation of the ore minerals are similarly diverse. Sulfur isotope studies on pyrite, associated with uranium in some of the major Wyoming deposits, suggest that uranium was precipitated by the reducing action of hydrogen sulfide of biogenic origin (10,30). Other investigators propose that precipitation was caused by the reducing action of natural gas (7) or of carbonaceous material associated with most deposits (16). Still others believe that pH changes (21) or that Eh and pH changes (31,23) were of paramount importance is causing deposition.

Additional field and laboratory research is needed to determine the principal factor or combination of factors that caused precipitation.

Most geologists believe that some mechanism of concentration, perhaps similar to that postulated in the multiple migration-accretion hypothesis of Gruner (5), is necessary to form high-grade uranium ore bodies from ore-bearing solutions thought to contain, at the most, only a few parts per million of uranium. The so-called "primary black ores" may contain uranium that has passed through many cycles of oxidation, migration, and reduction before reaching its present resting place.

### Age of the Deposits

The Wyoming deposits, occurring as they do in rocks of early Eocene age, are less than 55 million years old; how much less is less certainly known. The Powder River Basin deposits, as interpreted from lead-lead and lead-uranium ratios, range from 7 to 13 million years in age (22). Studies on the distribution of $Th^{230}$ and $Pa^{231}$ and on the fractionation of $U^{234}$ (25) suggest a minimum age of 250,000 years for deposits in the Shirley Basin. It is possible that certain daughter products of uranium were introduced at the time the deposits were formed or that certain daughters have been leached differently since that time; in either event, the age determinations would be considerably in error.

## EXPLORATION GUIDES

There are no infallible exploration guides to buried sandstone-type uranium deposits in Wyoming and South Daokta. However, there are a number of general guides that, when combined with broad geologic and hydrologic studies, should lead to more successful and less costly exploration programs.

First consideration must be given to the regional guides, for they help define potential uranium-bearing areas. Data gathered over the last decade have shown that the major uranium deposits are in or on the flanks of structural and/or erosional basins that are filled with clastic sedimentary rocks of continental origin. These basins are adjacent to the granitic cores of major uplifts, and most of them are, or once were, filled with Tertiary rocks containing considerable volcanic ash.

Within favorable areas, certain lithologic, stratigraphic, and paleodrainage factors appear to have favored the transportation and deposition of uranium. Medium- to very coarse-grained, arkosic, crossbedded sandstones of continental fluvial and lacustrine origin and of high transmissivity are the favored host rocks in Wyoming, and physically similar rocks, that contain little feldspar, are the hosts for the South Dakota deposits. Other favorable characteristics of a potential host rock include the presence of coalified woody material or other forms of organic carbon and the presence of natural gas or any other reductant capable of precipitating uranium from solution, either directly or indirectly.

A favorable host sandstone generally is associated with fine-grained rocks. If it is a few tens of feet thick, it may be overlain and/or underlain by shales or siltstones, but, if it is several hundred feet thick, shales and siltstones generally are interbedded with it. Restricted transmissivity, due to the presence of impervious beds, is often cited as the favorable factor, but it is probable that an even more important factor is the carbonaceous trash that is almost always associated with the fine-grained rocks.

There is some evidence that ore deposits, within favorable areas, will be found in or near paleostream channels cut by the major streams that transported and deposited the host rock sediments. The favorability factor in this instance would be the character of the sediments deposited by these streams and the subsequent funneling of uranium-bearing solutions through these sediments.

Available data indicate that the uranium content of ground water may be used as a guide to uranium-bearing areas (4), but that it may fail as a guide to the deposits themselves. A knowledge of the rate and depth of oxidation is a necessary adjunct to the use of this guide, for they govern, to a large extent, the amount of uranium in solution. As a general guide, 5 ppb or less uranium in ground water suggests that an area is quite barren, while 10 ppb or more suggests that it may contain uranium deposits. The use and limitations of this guide are illustrated by ground-water samples from the Shirley Basin, Wyoming. All samples contain from 10 to 25 ppb uranium, indicating a potential uranium-bearing area, but there was no significant difference in the uranium content of ground water from the ore-bearing sandstone and that of water from springs in the generally barren White River Formation.

A sulfate content of 450 ppm or more in ground water has been suggested as indicative of uranium-bearing areas (30). Here again, oxidation is an important factor, for it controls the rate at which the sulfur in pyrite accompanying uranium is converted to sulfate and enters the ground water. The sulfate content of ground water from uranium-bearing areas in the Shirley Basin that are not undergoing oxidation is within the 80–150 ppm range, but, in areas undergoing oxidation, it is as much as 500 ppm.

Altered sandstone has proved to be an excellent guide to ore in the major uranium mining districts of Wyoming, and it is probable that the genetic and spatial relations of ore to alteration are similar for deposits as yet undiscovered. Altered sandstone tongues are large targets easily delineated by widely spaced drilling. Once a tongue is delineated, maximum exploration efforts can be concentrated along its margins, where the largest ore bodies commonly are found. Criteria for recognizing altered sand tongues were discussed earlier in this paper.

The exploration guides discussed above must be used with caution, for uranium deposits are the result of a fortuitous combination of many chemical and physical factors, and, while the broad genetic controls for ore deposition may have been similar for most deposits, local controls must have differed from area to area. For this reason, guides applicable to one area may fail in another, and guides not yet recognized may be of major importance in exploring new areas.

## DESCRIPTION OF MISCELLANEOUS DISTRICTS

Three major uranium mining districts in Wyoming and South Dakota are described in papers that follow this regional discussion. The following districts, however, not all of lesser importance,* deserve mention:

### Copper Mountain District

This small uranium mining district is on the north flank of the Wind River Basin in the central part of Wyoming (Figure 2). The host rock for the deposits is a very coarse-grained, arkosic sandstone and boulder conglomerate that was derived from the Precambrian granite upon which it was deposited. The

* *Editor's Note:* Papers on the Gas Hills and Crooks Gap uranium mining districts of Wyoming, scheduled for inclusion in this Volume, were not submitted. Brief descriptions of these two districts appear in this chapter.

host rock is probably Eocene in age. Near the steep slope of Copper Mountain, it laps onto the eroded surface of the granite or fills old stream channels in the granite. Uranium, in the form of coffinite and various yellow oxides, coats sand grains, fills interstices between grains, and, in many places, forms rinds on boulders. Minor amounts of uranium have been found in tuffaceous rocks interbedded with the arkosic material.

Radioactive anomalies and a few low-grade concentrations of uranium exist in the granite, and it is possible that the Copper Mountain deposits represent uranium weathered from the granite, transported in ground water, and deposited in the arkosic sedimentary rocks.

### Poison Basin District

This district is on the southeast flank of the Washakie Basin about 6 miles west of Baggs, Wyoming (Figure 1). The Browns Park Formation, of Miocene(?) age, is the host rock for the uranium deposits. The formation, which generally is gray to buff, is of continental origin and is at least partly aeolian; it' comprises a thick series of fine- to medium-grained, crossbedded, tuffaceous sandstone, tuff, and thin quartzite. Vine and Prichard (3) reported the formation to be about 300 feet thick in the Poison Basin area.

The low-valent or reduced gray ore consists of uraninite and coffinite that coat sand grains and fill interstices between grains. Considerable pyrite is associated with the ore minerals. Selenium and molybdenum, elements characteristic of many of the Wyoming deposits, are present in appreciable amounts. The upper parts of the deposits have been oxidized, and the brown sandstone contains uranophane, meta-autunite, schroeckingerite, and other high-valent uranium minerals. The contact between oxidized and unoxidized sandstone is generally very sharp.

The ore bodies occur in the more permeable parts of the sandstone, and, to some extent, their location appears to have been influenced by faults that are common in the area. Some of the deposits are gently dipping tabular bodies that follow bedding, and, according to R. Rackley (oral communication, 1966), some are roll-type deposits in which oxidation has destroyed the upper limb.

The origin of the deposits has been attributed to deposition from circulating ground water, but the source of the uranium is not definitely known. The uranium may have originated in the volcanic ash common in the rocks present in the Poison Basin area. The lack of carbonaceous material in the host sandstone and the presence of natural gas in the Browns Park Formation led Grutt (7) to propose that the deposits were formed by the precipitation of iron and uranium from circulating ground water by reaction with $H_2S$ in the natural gas.

## Lignite of South Dakota

Although uranium associated with lignite occurs over a wide area along the southwestern part of the Williston Basin, ore-grade material is confined mainly to the northwest corner of South Dakota and the southwest corner of North Dakota. According to Denson and Gill (24), deposits of minable grade occur in the continental Fort Union Formation, of Paleocene age. It comprises a thick sequence of flat-lying lignite-bearing sandstone, shale, and claystone. The carbonaceous host rocks range in thickness from 6 inches to more than 2 feet and are characterized by high ash contents and quite high permeabilities. Denson and Gill believe that volcanic ash in the White River and Arikaree Formations, which once overlay the area, was the source of the uranium, that ground water was the transporting medium, and that the uranium was fixed by the reducing action of the lignite. Uranium deposition was controlled by the proximity of the lignite to the unconformity marking the base of the overlying rocks, the permeability of the beds directly overlying the host lignite, the presence of shallow local troughlike folds, the absence of impervious rocks above the host lignite, and by the permeability of the host material.

Most of the uranium occurs as a disseminated amorphous organo-uranium compound, but uraninite and a number of yellow secondary uranium minerals have been identified in the deposits. The grade of the lignite ore is about 0.33 per cent $U_3O_8$, somewhat higher than that of the sandstone-type ore in South Dakota.

## Pryor Mountain-Little Mountain District

This district lies on the east flank of the Bighorn Basin along the border between north-central Wyoming and south-central Montana (Figure 2). The host rock for the deposits is the upper part of the Madison Limestone, the Mississippian age. More specifically, the deposits are in partly filled caverns and solution channels formed in the limestone during Late Mississippian or Early Pennsyl-

vanian time and perhaps modified in Tertiary time.

The fill material consists of limestone blocks, fallen from the walls and backs of the caverns, chert nodules and other insoluble constituents of the limestone, and sand, silt, and clay washed in from the surface or from strata that overlay the Madison. In some of the caves there is considerable siliceous cement. The fine-grained debris may be well bedded and cemented, or it may be poorly bedded and unconsolidated.

According to Hart (9), the principal ore minerals are tyuyamunite and metatyuyamunite that are associated with calcite, hematite, gypsum, barite, opal, fluorite, celestite, and pyrite. The ore minerals are disseminated in the fine-grained cavern fill, coat fractures in the limestone, and form crusts on cavern walls, limestone blocks, calcite crystals, and chert nodules. Some caverns contain no uranium minerals.

The individual deposits range in size from sparse coatings on the walls of a single fracture or cavern to deposits in large caverns or connected caverns that may extend several hundred feet horizontally and contain as much as 10,000 tons of ore.

The source of the uranium in the deposits is not known, although it does not appear to have been the Madison Limestone. Bell (17) stated: "The two most likely methods of emplacement seem to be: (1) the deposits are supergene enrichments of uranium leached from formations that are younger than the Madison Limestone and which formerly covered the area, and (2) the tyuyamunite is the oxidized residue of hydrothermal deposits." Sufficient data are not at hand to prove or disprove either postulate.

## Gas Hills District

The Gas Hills, one of the major uranium mining districts in the United States, is in Fremont and Natrona counties, Wyoming, about 60 miles east of Lander (Figure 2). The total production from the district, through the year 1966, is about 34,674,800 pounds of $U_3O_8$, and ore reserves on January 1, 1967, were estimated to be about 9,782,000 tons with a grade of 0.20 percent $U_3O_8$ (data from U.S. Atomic Energy Commission, written communication, 1967).

The district is on the south flank of the Wind River Basin, a major structural and topographic depression, bounded by the Owl Creek, uplift on the north and the Sweetwater uplift on the south (Figures 1, 2). The Beaver Divide, a northward facing escarpment, is near the southern edge of the district and is the most prominent topographic feature in the area. Surface drainage is northward from the divide.

Soister (33) shows the Gas Hills district to be in an area where a trough-like depression, underlain by rocks of Paleozoic and Mesozoic age, is filled with a sequence of continental rocks of fluvial origin and of Tertiary age. The older rocks dip 10° to 15°N; the Tertiary rocks dip a few degrees to the south.

The uranium deposits are in the upper part of the Wind River Formation of early Eocene age. The ore-bearing beds in the Gas Hills, as in the Shirley and Powder River Basins, are coarse-grained, conglomeratic arkoses, a basin-border facies of the Wind River Formation that contrasts markedly with a fine-grained facies nearer the center of the basin.

Roll-type ore bodies are by far the most important source of uranium in the Gas Hills, although blanket-type bodies of reduced black ore, as well as near-surface oxidized ore bodies, have been mined. The deposits are in three belts, trending northerly and about 3 miles apart. Two of the belts are aligned with channels, exposed in the Beaver Divide escarpment, that were eroded through the Wagon Bed Formation and subsequently filled with rocks of the White River Formation (Table II). Genetic implications have been based on this alignment.

In the Gas Hills, as in many other Wyoming districts, the ore deposits are related to tongues of altered conglomeratic sand produced by the ore-bearing soutions. Most of the ore occurs along the margins of the altered tongues, in elongate bodies that are C shaped in cross section. Some ore is emplaced on the top and bottom surfaces of the tongues. Some of the rolls are suspended entirely within a sandstone interval; however, many rolls terminate, top and bottom, at impermeable siltstone beds.

Epigenetic minerals in the ore bodies include uraninite, coffinite, pyrite, marcasite, calcite, jordisite, and one or more selenium minerals. Selenium is most abundant near the contact between ore and altered sand, and molybdenum generally is found on the convex side of the roll in a zone between ore and mineralized ground. The ore minerals coat sand grains and fill the interstices between the grains.

Most geologists working in the area conclude that the deposits are the result of deposition from ground water moving through the most transmissive parts of the Wind River Formation. Opinion is diverse regarding the source of the uranium, the hydrodynamics of the ore-bearing solutions, and the factors causing deposition. Recent investigations by Armstrong (written communication, 1965) and Cheney and Jensen (30) indicate that the three belts in the Gas Hills district may all be related to a single large tongue of altered sand, irregular in plan and extending across the district. The deposits have been estimated to be more than 500,000 years old (30).

Exploration guides for the Gas Hills area have been described by King and Austin (32).

## Crooks Gap District

The Crooks Gap uranium mining district is in the southeastern part of Fremont County, Wyoming, about 55 miles southeast of Lander and 25 miles south of the western part of the Gas Hills district (Figure 2). From the time of its discovery, in 1953, through 1966, the district produced about 4,866,000 pounds of $U_3O_8$ (data from U.S. Atomic Energy Commission, written communication, 1967).

The known deposits are on the north and west flanks of Sheep Mountain, a few miles north of the drainage divide between the Great Divide Basin and the Sweetwater uplift (Figure 1). The area is structurally complex. Rocks of pre-Tertiary age are intensely deformed by folding, normal faulting, and overthrusting. Rocks of Tertiary age are only moderately deformed.

The host rock for the uranium deposits is the Battle Spring Formation, a sequence of coarse-grained, friable, arkosic sandstones, conglomerates, conglomeratic sandstones, and thin interbeds of carbonaceous siltstone. The formation is 2000 to 3000 feet thick in the mining area. These rocks are of early Eocene age and are partly equivalent to the coarse-grained facies of the Wasatch and Wind River Formations in other Wyoming basins (Table II). The Battle Spring Formation rests either on the eroded surface of the Fort Union Formation of Paleocene age or on the eroded surface of the older rocks.

According to Eric Newman of Western Nuclear, Inc. (oral communication, 1967), the uranium deposits in the Crooks Gap area are in the lower 800 feet of the Battle Spring Formation. On a regional basis, the ore-bearing zones are concordant with the beds, which dip 20°SE; within the zones, ore may crosscut sedimentary features. Roll-type uranium deposits, associated with slightly altered sand, are thought to be present in the Crooks Gap area, but they are poorly defined and difficult to recognize. Much of the ore appears to be in irregular and/or blanket-type deposits. The complex structural and stratigraphic setting for the Crooks Gap deposits contrasts sharply with the more simple settings in the Gas Hills and Shirley Basin. A "plumbing" system, complicated by faulting and impermeable siltstone beds, is thought to account for the irregularity of the deposits. The ore-bearing solution is thought to have been ground water; the shape and position of some of the ore bodies suggest that it moved from south to north. Unoxidized ore is black and contains uraninite, coffinite, and pyrite; calcite, selenium, and molybdenum (jordisite) are associated with uranium in the deposits. Near surface ore bodies, of minor economic importance, contained uranium phosphates, silicates, sulfates, and vanadates.

## REFERENCES CITED

1. Fenneman, N. M., 1931, Physiography of western United States: McGraw-Hill, New York, 534 p.
2. Love, J. D., 1952, Preliminary report on uranium deposits in the Pumpkin Buttes area, Powder River Basin, Wyoming: U.S. Geol. Surv. Circ. 176, 37 p.
3. Vine, J. D. and Prichard, G. E., 1954, Uranium in the Poison Basin area, Carbon County, Wyoming: U.S. Geol. Surv. Circ. 344, 8 p.
4. Denson, N. M., *et al.,* 1956, Water sampling as a guide in the search for uranium deposits and its use in evaluating widespread volcanic units as potential source beds for uranium [S. Dak., Wyo.] (revised): p. 673–680, *in Contributions to the geology of uranium and thorium,* U.S. Geol. Surv. Prof. Paper 300, 739 p.
5. Gruner, J. W., 1956, Concentration of uranium in sediments by multiple-migration-accretion: Econ. Geol., v. 51, p. 495–520.
6. Shawe, D. R., 1956, Significance of roll ore bodies in genesis of uranium-vanadium deposits on the Colorado Plateau (slightly revised) p. 239–241, *in Contributions to the geology of uranium and thorium,* U.S. Geol. Surv. Prof. Paper 300, 739 p.
7. Grutt, E. W., Jr., 1957, Environment of some Wyoming uranium deposits: Advances in nuclear engineering: 2d Nuclear Eng. and Sci. Conf., Philadelphia, Pr., v. 2, p. 313–323.
8. Thomas, H. D., 1957, Geologic history and structure of Wyoming: Wyo. Geol. Surv., Contr., Reprint Ser., no. 18, 11 p.
9. Hart, O. M., 1958, Uranium deposits in the Pryor-Big Horn Mountains, Carbon County, Montana, and Big Horn County, Wyoming: 2d UN Int. Conf. on the Peaceful Uses of Atomic Energy (Geneva) Pr., v. 2, p. 523–526.
10. Jensen, M. L., 1958, Sulfur isotopes and the origin of sandstone-type uranium deposits [Colorado Plateau and Wyoming]: Econ. Geol. v. 53, p. 598–616.
11. Kerr, P. F., 1958, Criteria of hydrothermal emplacement in [Colorado] Plateau uranium strata, 2d UN Int. Conf. on the Peaceful Uses of Atomic Energy (Geneva), Pr. v. 2, p. 330–334.
12. Denson, N. M., *et al.,* 1959, Uranium-bearing lignite in northwestern South Dakota and adjacent states [Montana-North Dakota]: U.S. Geol. Surv. Bull. 1055-B, p. 11–57.
13. Page, L. R., 1960, The source of uranium in ore deposits: 21st Int. Geol. Cong. Rept., pt. 15, p. 149–164.
14. Harshman, E. N., 1962, Alteration as a guide to uranium ore, Shirley Basin, Wyoming: U.S. Geol. Surv. Prof. Paper 450-D, p. D8–D10.
15. Hostetler, P. B. and Garrels, R. M, 1962, Transportation and precipitation of uranium and vanadium at low temperatures, with special reference to standstone-type uranium deposits: Econ. Geol., v. 57, p. 137–167.
16. Adler, H. H., 1963, Concepts of genesis of sandstone-type uranium ore deposits: Econ. Geol., v. 58, p. 839–852.
17. Bell, K. G., 1963, Uranium in carbonate rocks: U.S. Geol. Surv. Prof. Paper 474-A, p. A1–A29.
18. Gott, G. B. and Schnabel, R. W, 1963, Geology of the Edgement NE quadrangle, Fall River and Custer Counties, South Dakota: U.S. Geol. Surv. Bull. 1063-E, p. 127–190.
19. Gabelman, J. W. and Krusiewski, S. V, 1964, Zonal distribution of uranium deposits in Wyoming, U.S.A.: 22d Int. Geol. Cong., Repts. (in press).
20. Gries, J. P., 1964 Geology—Geologic history of the Black Hills: *in Mineral and water resources of South Dakota:* U.S. 88th Cong., 2d sess., Comm. Print, p. 23–28.
21. Melin, R. E., 1964, Description and origin of uranium deposits in Shirley Basin, Wyoming: Econ. Geol., v. 59, p. 835–849.
22. Sharp, W. N. and Gibbons, A. B., 1964, Geology and uranium deposits of the southern part of the Powder River Basin, Wyoming: U.S. Geol. Surv. Bull. 1147-D, p. D1–D60.
23. Bailey, R. V., 1965, Applied geology in the Shirley Basin uranium district, Wyoming: Univ. Wyo. Contr. Geol., v. 4, no. 1, p. 27–35.
24. Denson, N. M. and Gill, J. R, 1965, Uranium-bearing lignite and carbonaceous shale in the southwestern part of the Williston Basin—A regional study, *with a section on* Heavy minerals in Cretaceous and Tertiary rocks associated with uranium occurrences, by W. A. Chisholm: U.S. Geol. Surv. Prof. Paper 463, 75 p.
25. Dooley, J. R., *et al.,* 1965, Radioactive dis-

equilibrium studies of roll features, Shirley Basin, Wyoming: Econ. Geol. v. 59, p. 586–595.

26. Keefer, W. R., 1965, Geologic history of Wind River Basin, central Wyoming: Amer. Assoc. Petrol. Geols. Bull., v. 49, p. 1878–1892.

27. Keefer, W. R., 1965, Stratigraphy and geologic history of the uppermost Cretaceous, Paleocene, and lower Eocene rocks in the Wind River Basin, Wyoming: U.S. Geol. Surv. Prof. Paper 495-A, p. A1–A77.

28. Pipiringos, G. N., *et al.*, 1965, Geology and uranium deposits in the Cave Hills area, Harding County, South Dakota: U.S. Geol. Surv. Prof. Paper 476-A, p. A1-A64.

29. Rosholt, J. N., *et al.*, 1965, Isotopic geology of the Gas Hills, Wyoming, uranium in sandstone, Powder River Basin, Wyoming, and Slick Rock district, Colorado: Econ. Geol., v. 60, p. 199–213.

30. Cheney, E. S. and Jensen, M. L., 1966, Stable isotopic geology of the Gas Hills, Wyoming, uranium district: Econ. Geol., v. 61, p. 44–71.

31. Harshman, E. N., 1966, Genetic implications of some elements associated with uranium deposits, Shirley Basin, Wyoming: U.S. Geol. Surv. Prof. Paper 550-C, p. C-167–C173.

32. King, J. W. and Austin, S. R., 1966, Some characteristics of roll-type uranium deposits at Gas Hills, Wyoming: Min. Eng., v. 18, no. 5, p. 73–80.

33. Soister, P. E., 1967, Geologic map of the Coyote Springs quadrangle, Fremont County, Wyoming: U.S. Geol. Surv. Misc. Geol. Investigations Map I-481, 1:24,000.

# 41. Uranium in the Black Hills

OLIN M. HART*

## Contents

## Illustrations

## Table

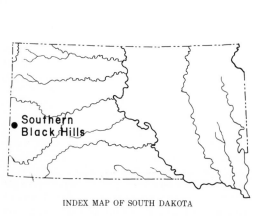

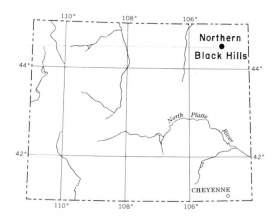

INDEX MAP OF SOUTH DAKOTA

* Homestake Mining Company, Lead, South Dakota.

## ABSTRACT

Uranium ores occur in the Lower Cretaceous Inyan Kara group of heterogeneously stratified fluvial and fluvial-marine sandstones in the Black Hills of western South Dakota and northeastern Wyoming. There are three principal mining districts, each in the ancestral and present drainage area of a major river partially responsible for stripping the region to its present topographic form.

Primary minerals in the deposits are coffinite and uraninite with minor amounts of paramontroseite and haggite. The ore minerals coat sand grains and fill interstices of complexly cross-stratified sandstone along solution fronts similar to "roll" type deposits of the other districts. Minerals of oxidized deposits are typically carnotite and tyuyamunite with different proportions of secondary vanadium accessory minerals. Ground water was the transporting medium and deposition of primary uranium and vanadium minerals occurred in reducing environments produced and controlled by physio-chemical characteristics of the sedimentary rocks.

## INTRODUCTION

In June 1951, uranium ores were discovered in the southern Black Hills near the town of Edgemont, in Fall River County, South Dakota. Numerous deposits were quickly developed in this area, and, by 1952, additional deposits had been discovered in the northern Black Hills, in Crook County, Wyoming. These deposits were the first economically significant discoveries of uranium ores in sandstone to be found beyond the boundaries of the Colorado Plateau region in the United States. In December 1952, the United States Atomic Energy Commission established an ore-buying station at Edgemont, and in July 1956, the mill of Mines Development, Incorporated was put in operation to process ores from the Black Hills and adjacent areas.

Uranium mining centered in three principal areas; namely, the Edgemont district along the southwestern flank of the Black Hills in Fall River County, South Dakota, and in the Carlile and Hulett Creek districts along the northwestern flank of the Black Hills in Crook County, Wyoming. The Edgemont district was the major producing area through 1958, largely from oxidized or partially oxidized ores developed from outcrops and by shallow surface drilling. By 1959, unoxidized deposits became the major source of ore and they account for

about 75 per cent of all production and reserves.

All production has been from Lower Cretaceous fluvial or fluvial-marine sandstones mined with conventional trackless equipment, either by rim-stripping and open pits or by underground room and pillar methods from adits, inclines or vertical shafts. Anomalous radioactivity and/or uranium minerals have been found in numerous formations in the Black Hills area, but occurrences in other than the Lower Cretaceous or Inyan Kara group sediments have proved to be of no commercial interest to date.

## PRODUCTION

*TABLE I.   Black Hills Uranium Production*

| Year | Southern Hills | | Northern Hills | |
|------|------|------|------|------|
| | Tons | Pounds $U_3O_8$ | Tons | Pounds $U_3O_8$ |
| 1951–52 | 4,007 | 19,186 | — | — |
| 1953 | 11,331 | 50,011 | 4,885 | 34,048 |
| 1954 | 17,160 | 60,256 | 7,777 | 42,369 |
| 1955 | 22,635 | 90,160 | 7,301 | 33,480 |
| 1956 | 34,355 | 126,176 | 20,951 | 79,502 |
| 1957 | 69,151 | 229,910 | 11,222 | 42,643 |
| 1958 | 35,332 | 141,691 | 10,141 | 43,303 |
| 1959 | 45,734 | 171,449 | 45,607 | 200,876 |
| 1960 | 40,833 | 157,252 | 84,019 | 377,231 |
| 1961 | 43,591 | 148,691 | 75,865 | 352,201 |
| 1962 | 29,365 | 103,980 | 93,343 | 415,669 |
| 1963 | 29,963 | 88,531 | 73,453 | 322,631 |
| 1964 | 70,595 | 131,512 | 93,188 | 379,927 |
| Total | 454,052 | 1,518,805 | 527,752 | 2,323,880 |

Source: Unpublished data, United States Atomic Energy Commission, written communication, E. A. Youngberg, 1965.

## PHYSIOGRAPHY

The Black Hills region is a broad domal shaped uplift of Laramide age composing the easternmost outer range of the Rocky Mountains in western South Dakota and northeastern Wyoming (1). The domal area trends northwesterly and is approximately 160 miles long and 50 miles wide. Erosion has exposed a Precambrian core of crystalline and metamorphic rocks that trends north-south, flanked by upturned truncated Paleozoic and Mesozoic strata. The outermost geomorphic expression of the Black Hills structure is a bold inward-facing hogback of sandstones in the Inyan Kara group. The northwestern and southern ends of the uplift respectively have plunges of a few degrees to the northwest and south-east, while the east and west flanks have steep dips.

## STRATIGRAPHY

The Inyan Kara group has been subdivided by Waagé (3) into two formations. The lower, or Lakota formation, is dominantly sandy sediments of non-uniform continental facies lithogenetically related to the underlying Morrison formation, in part Upper Jurassic in age. The upper part of the group, the Fall River formation, is dominantly sandy sediments of marginal marine facies related lithogenetically to, and gradational with, the overlying marine Skull Creek shales. (Figure 1)

The Lakota formation ranges in thickness from 250 to 500 feet and is composed of fine- to coarse-grained lenses of buff to white sandstone and conglomeratic sandstone, irregularly interbedded with varicolored claystone. The coarse fractions are chert, quartzite, and quartz. Lithology and succession differ so appreciably from one place to another that unqualified statements as to the nature of the lithic sequence, applicable to the whole outcrop area, cannot be made. There is no obvious

stratigraphic break that, regionally, marks the contact of the Lakota with the underlying Morrison formation, and, at many places, arbitrary local features must be used to separate the two units. Most of the Lakota sandstones occur as massive lenses. Some lenses are broad and sheetlike, others are elongated and of limited lateral extent. Massive sandstones are largely concentrated in the southern Black Hills outcrop area where the succession differs from that of the northwestern area of the Black Hills through addition of beds emplaced progressively at the base of the formation as it thickens eastward and southeastward towards the source area. In the northwestern part of the Black Hills, the sandstone lenses thin appreciably, resulting in a lesser total thickness of the formation compared with the southern Black Hills succession. The sandstones range from fine-grained and silty to coarse-grained and conglomeratic. They are generally poorly bedded and massive. Some are compact and well cemented, some are sugary and friable, others may have much interstitial clay. Claystones range from semiplastic clay to hard silty claystone with blocky fractures, and many are bentonitic. (Figure 2)

The Fall River formation is about 150 feet thick. It consists chiefly of fine-grained, thin-bedded, laminated to tabular cross-laminated, buff to brown sandstone, interbedded with gray to black shale and siltstone. Sandstones and thinly interbedded siltstones and shales commonly contain ripple marks, and trails, casts, and borings of soft-bodied wormlike animals. Thin bedding and lamination characterize the Fall River sequence and contrast strongly with the generally lenticular, massive, cross-laminated beds of the underlying Lakota formation.

## ORE DEPOSITS

The black unoxidized ore minerals of the Black Hills uranium deposits are uraninite and coffinite with associated vanadium minerals, paramontroseite and haggite ($V_2O_3 \cdot V_2O_4 \cdot 3H_2O$). These minerals impregnate the interstices of sandstones and are intimately associated with pyrite, marcasite, calcite, and rarely jordisite and ilsemannite. Corrosion and embayment of detrital quartz grains by ore minerals are common. Vanadium to uranium ratios of unoxidized or black ores average about 1.5:1.0.

Oxidized ores are of two general types: (1) those that are predominantly yellow and (2) those that are mainly purplish black to reddish brown. Yellow ores have carnotite and

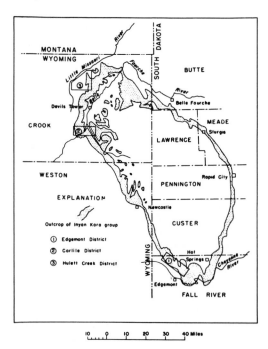

<FIG. 1. *Outcrop map of the Inyan Kara group in the Black Hills, showing location of the principal uranium districts and major streams.*>

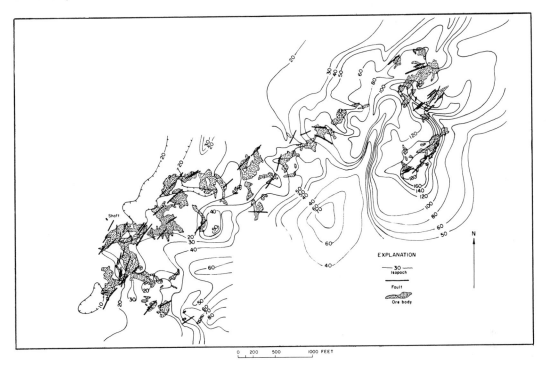

FIG. 2. *Isopach Map of the Basal Lakota Sandstone and Distribution of Ore Bodies at the Hauber Mine.*

tyuyamunite as the principal minerals; autunite and uranophane have only minor and local significance. The ore minerals are commonly associated with differing amounts of calcite, iron oxide, carbonaceous material, and clay minerals, interstitital to detrital quartz grains that are commonly embayed by the ore minerals. Vanadium to uranium ratios of the yellow ores average about 0.4:1.0. Yellow ores are the prevalent oxidized deposits in the Lakota sandstones.

The purplish black to reddish brown oxidized ores contain carnotite and tyuyamunite with the vanadium minerals corvusite and rauvite in sufficient quantity to impart the darker coloration. These minerals are commonly associated with small amounts of calcite, iron oxide, and occasional remnants of pyrite. The uranium and vanadium minerals occur interstitially to the detrital quartz grains of the sandstone. Secondary rims of quartz overgrowth on quartz grains are common, and the secondary rims and the quartz grains are normally corroded and embayed by the ore minerals. Vanadium to uranium ratios average about 2.0:1.0. The purplish black to reddish brown ores are the common oxidized deposits in the Fall River formation.

A pink hematite staining accompanies most of the uranium deposits in the Black Hills area. It borders or partly encloses the deposits and terminates abruptly against the zone containing visible uranium and/or vanadium minerals. The hematite impregnates interstitial clay and coats sand grains. The persistent association of hematite staining and ore is an excellent guide in surface and subsurface exploration.

The deposits are elongated tabular lenses and pods in sandstones ranging from coarsely conglomeratic cross-bedded sandstone with varying amounts of carbonaceous content to fine-grained intricately laminated and cross-laminated, highly carbonaceous sandstone and claystone. The host sandstones are the aquifers of the local lithic succession, or they are spatially related to sandstones that serve as channelways permitting ground water entry and movement through the host horizons. Carbonized plant fragments in the host sandstone are of prime importance in ore deposition. The carbonized plant fragments are believed responsible for the reducing environment necessary to precipitate uranium and other metals carried in ground water.

Sedimentation structures such as channels, lenticular beds, and cross-bedding had the effect of channeling circulating waters through the host formations. Compositional variation of the host rocks, sedimentary structures, lithic variation, and chemical favorability are major controlling features localizing ore deposits. Minor tectonic features do not appear to have contributed to localization of ore deposits.

## Hauber Ore Deposit

The Hauber ore body is the largest in the Black Hills and contains about 600,000 tons as a total reserve averaging 0.23 per cent $U_3O_8$. The deposit is in the Hulett Creek district in northeastern Crook County, Wyoming, on the gently dipping terrace of the Black Hills monocline which forms the western edge of the uplift. Ores occur in the basal sandstone unit of the Lakota formation which is complexly cross-bedded, interstratified with clays and mudstones, and contains sparse to abundant macerated carbonaceous material with a few scattered small carbonized logs and limb fragments. The sandstone is poorly sorted, ranging from fine well-rounded to sharp angular quartz grains to conglomeratic sandstones with chert and quartz pebbles up to ½ inch in diameter. The Lakota formation near the Hauber mine is about 250 feet thick. (Figure 2)

Dips on the terrace of the monocline are 2° to 5°N to NW with local reversals of 1° to 2° caused by minor domal-form flexures common in the region. A series of pre-ore normal faults transect the deposit ranging from 50 to 500 feet in length within the mine area. Faults adjacent to the deposit are as much as 5000 feet long. Nearly all faults strike N45°E to N60°E, with steep dips mostly northwest, a few to the southeast; the northwest sides of the faults are commonly downthrown from inches to a maximum of 50 feet. Spatial relations of the faults to a structural change of trend and split of the Black Hills monocline, suggest the faulting is the result of tensional stress induced in the local sedimentary succession by folding of the monocline during uplift (4).

Detailed studies of mineralogy on the deposit have not been made, but Gruner (personal communication) identified coffinite as the principal ore mineral. Uraninite, as well as the black vanadium minerals paramontroseite and haggite are presumed to be present in the ores based on similarities of field appearance to other unoxidized deposits where mineralogic relations have been determined. The uranium and vanadium minerals impregnate the interstices of the sandstone and are intimately as-

sociated with pyrite, marcasite, and different amounts of calcite.

The deposit is a complex of rudely tabular lenses and pods in heterogeneously stratified, cross-bedded, conglomeratic sandstones with interstratified clay and mudstones. The ore minerals impregnate sandstones along "chemical fronts" typical in all respects to the common "roll" deposits in other districts. Because of wide variations in permeability of the host rocks, mineralizing solutions could not follow a simple path, but were required to flow through zones of widely differing permeability and composition where mineral deposition took place at multiple interconnecting horizons.

The many irregular elongated ore pods are rudely aligned in a N60°E trend, approximately parallel to a facies boundary where the basal Lakota sandstone succession thins rapidly and pinches out through lateral grading into a thick claystone-mudstone succession lying to the immediate north and west of the mine area (Figure 2). Mineralization is confined essentially to one horizon in sandstones less than 30 feet thick; it occurs in multiple horizons in sandstones 30 to 180 feet thick. In the thicker sandstones, up to three mineable horizons are distributed through a vertical range of 50 to 75 feet above the Morrison-Lakota contact.

Average thickness of a deposit is 4 to 6 feet, though thicknesses up to 12 feet are not uncommon. Occasionally thicknesses of mineable ore up to 30 feet have occured where multiple horizons have assumed a stacked position of one ore bed over another.

Hematite staining of sand grains and interstitial clays is a persistent feature on the general up-dip or southeast side of the ore trend. The staining may completely surround ore pods, particularly those on the southeast side of the main ore trend. Staining stops abruptly at zones containing visible uranium and/or vanadium minerals.

Tenor of mineralization differs considerably and low grade or barren material may be found within mineable ore bodies. Higher grade ores of 0.50 per cent to more than 1.00 per cent $U_3O_8$ are usually more prevalent on the southeastern side facing the hematite staining, while the northwestern side grades to an assay cutoff. However, because of lithic heterogeneity, an irregular distribution is most common.

Faulting has had little or no control on ore deposition other than causing local offsetting of favorable host horizons and producing local permeability barriers. Many ore pods pass directly through faults and mine observation indicates the faults to be pre-mineral.

## ORIGIN

There is scant evidence bearing upon the source of uranium in the Black Hills, but much evidence indicating ground water as the principal distributing agent. The most likely source was Oligocene and younger sediments that were present over much of Wyoming, eastern Montana, and western North and South Dakota and overlay much of the northern and southern parts of the present Black Hills. These sediments contain much devitrified volcanic ash and tuffaceous sandstones and were reported by Gill and Moore (2) to be more uraniferous (0.0015 per cent U) than average sedimentary rocks. The tuffaceous ash had its source in explosive volcanism centered in the Absoraka volcanic field in northwestern Wyoming.

It is postulated that, during the erosion of Oligocene and younger sediments from the western flanks of the Black Hills and adjacent regions, ground water leached the available mineral from the tuffaceous source beds and contributed uranium to Inyan Kara sediment through recharge and migration of uranium-bearing solutions toward the three major river channels. It is readily apparent that each of the three major uranium districts (Figure 1) is directly related to one of the three major rivers responsible for denuding the sedimentary succession and producing present topography.

The uranium and vanadium minerals were deposited from ground water as primary oxides in a reducing environment produced and controlled by physiochemical characteristics of the Inyan Kara host formations but only at spatially favorable locales. There were undoubtedly numerous phases of solution, migra-tion, and redeposition of uranium and vanadium minerals from source to present sites of deposition, and the reader is referred to the abundance of literature treating the subject of uranium transport and deposition in aqueous solutions for review of this process.

## USE OF GEOLOGY IN EXPLORATION, DEVELOPMENT, AND MINING

Adequate geology is most important in successful discovery and development of any new and significant uranium deposits in the Black Hills. Exploration will be far removed from outcrops and will reach depths of 200 to 600 feet. Exploratory drilling will require interpretation and evaluation of the lithic succession as a host environment, the presence or absence of mineralization effects and indicators, local and regional stratigraphic studies, and the evaluation and projection of sedimentary structures. Continued geologic surveillance is particularly required in production from multiple horizon deposits for maximum ore recovery and planning mining programs.

## REFERENCES CITED

1. Darton, N. H. and Paige, S., 1925, Description of the central Black Hills quadrangles: U.S. Geol. Surv. Geol. Atlas, Folio 219, 34 p.
2. Gill, J. R. and Moore, G. W., 1956, Carnotite-bearing sandstone in Cedar Canyon, Slim Buttes, Harding County, South Dakota: U.S. Geol. Surv. Bull. 1009-I, p. 249–264.
3. Waagé, K. M., 1959, Stratigraphy of the Inyan Kara group in the Black Hills: U.S. Geol. Surv. Bull. 1081-B, 90 p.
4. Robinson, C. S., et al., 1964, Stratigraphy and structure of the northern and western flanks of the Black Hills uplift, Wyoming, Montana, and South Dakota: U.S. Geol. Surv. Prof. Paper 404, 134 p.

# 42. Uranium Deposits in the Eocene Sandstones of the Powder River Basin, Wyoming

VERNON A. MRAK*

## Contents

## Illustrations

## Table

* Consultant, Boulder, Wyoming.

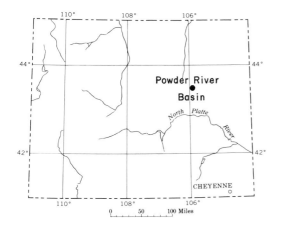

## ABSTRACT

The Powder River Basin of northeast Wyoming was the first area in the state to receive attention during the early days of uranium exploration. Although the uranium occurrences are many and widespread, most of the known ore deposits are quite small and difficult to mine. As a result of this, detailed regional exploration in the Powder River Basin has been neglected.

The uranium ore deposits occur in that part of the Wasatch Formation that is of earliest Eocene age, and they are scattered over an area of about 400 square miles. All uranium ore deposits are associated with red beds or altered sandstones within the Wasatch Formation.

Uranium ore was first shipped from the region in 1953. Total production through June of 1965 is 506,257 tons of uranium ore containing 1,922,527 pounds of $U_3O_8$. The average grade of ore is about 0.19 per cent $U_3O_8$. The major producing area of the Powder River Basin is the Monument Hill district where about 91 per cent of the basin's total uranium ore has been mined.

The source of uranium is presumed to have been the Oligocene, Miocene, and perhaps Pliocene tuffaceous rocks that once overlay the area. Joints, fractures, and permeable beds or zones were the passageways for the migration of uraniferous waters from the overlying beds to the Eocene host rocks.

## INTRODUCTION

This paper describes the distribution and general geologic features of the Tertiary sandstone-type uranium deposits found in the Powder River Basin, Wyoming (Figure 1), in an area extending from the Pumpkin Buttes south to the vicinity of Douglas.

The discovery of commercial uranium deposits in the Pumpkin Buttes area in October, 1951, was made by J. D. Love of the United States Geological Survey during a ground check of radioactivity anomalies located by an airborne radiometric reconnaissance flown by the United States Geological Survey upon the recommendation of Love. Love's find foreshadowed a series of similar statewide discoveries that initiated a uranium exploration and development program that now has placed Wyoming second to New Mexico in total known uranium reserves.

The search for radioactive minerals in the Powder River Basin, which followed the announcement of discovery of uranium (2), has uncovered hundreds of radiometric anomalies and uranium occurrences throughout the region. The principal areas of mineralization are located near the Pumpkin Buttes in Campbell and Johnson Counties, at Turnercrest, south of the Pumpkin Buttes, and at Monument Hill and Box Creek in Converse County (Figure 1).

Uranium ore was first shipped from the Powder River Basin in July, 1953, and, as of June, 1965, 506,257 tons of uranium ore containing 1,922,527 pounds of $U_3O_8$ had been mined from 80 individual deposits. Mining and development have progressed slowly but continuously and this trend probably will persist because of the rather small size and scattered nature of the ore deposits known at present. However, the objective of most exploration has been the near-surface ore deposits. Considering the great areal extent and favorable thicknesses of host rock along with the widespread nature of uranium occurrences, new discoveries probably will be made when intensive exploration is renewed. By applying the pattern of "ore front" exploration, new deposits of significant size may be found at greater depths without excessive random drilling.

## DISTRICT GEOLOGY

### Physiography

The Powder River Basin encompasses an area about 12,000 square miles in northeastern Wyoming (Figure 1). It is a broad structural and topographic basin, open to the north and terminated on the west by the Bighorn Mountains and the Great Pine Ridge Escarpment, on the south by the Laramie Mountains and

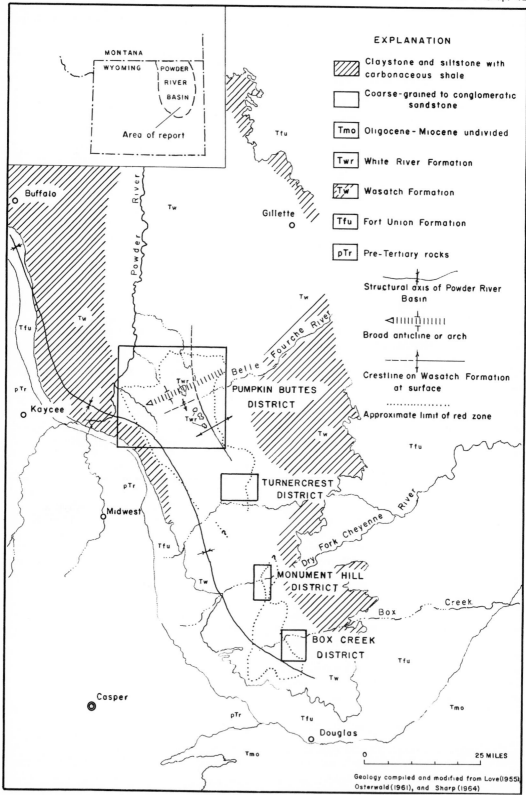

FIG. 1. *Geologic Map of the Powder River Basin, Wyoming, showing distribution of uranium mining districts.*

the Hartville Uplift, and on the east by the Black Hills and related structures. The region is one of moderate relief consisting of flat-top highlands with small playa lakes separated by sagebrush-covered rolling hills and wide gentle drainages. Elevations range from 4500 feet to about 5600 feet, but the Pumpkin Buttes, which are a prominent landmark in the central part of the basin, rise to an elevation of 6000 feet, or about 1000 feet above the surrounding countryside.

The northern part of the basin is drained to the north and northeast of the Powder River and Belle Fourche River. Throughout most of the mineralized area south of the Pumpkin Buttes, the drainage by the Cheyenne River and its tributaries is eastward, transverse to the regional structural trend.

## Stratigraphy

All of the known uranium deposits in the Powder River Basin are in the Wasatch Formation of earliest Eocene age. The Wasatch Formation is underlain by the Fort Union Formation of Paleocene age and overlain by the White River Formation of Oligocene age. The White River is present as an erosional remnant capping the Pumpkin Buttes, where it is in unconformable contact with the underlying Wasatch, and overlaps the older rocks to the west in the Bighorn Mountains, to the east in the Black Hills, and to the south on the margin of the Powder River Basin.

FORT UNION FORMATION  The Fort Union Formation is composed of 1900 to 3200 feet of buff to gray fine-grained sandstone, dark-gray siltstone, carbonaceous shale, and coal beds. The Fort Union is divided into three units: the Tullock, the Lebo shale, and the Tongue River members. The Roland coal bed, which is the thickest and most extensive coal bed in Wyoming (1), occurs near the top of the Tongue River member and usually is considered the stratigraphic unit marking the top of the Fort Union Formation. This coal bed has been recognized in the Great Pine Ridge southwest of the Pumpkin Buttes area, and it is exposed on the banks of the Dry Fork of the Cheyenne River immediately east of the Monument Hill uranium district. Underlying the Monument Hill district the coal bed is not present and appears to have been removed by pre-Eocene erosion.

WASATCH FORMATION  The Wasatch Formation, all but the upper 200 feet of which is

of earliest Eocene age (9), has a maximum thickness of 3500 feet (1) in the northern part of the Powder River Basin. The formation is of fluvial origin deposited in rapidly aggrading northward-flowing streams. It wedges out laterally and to the south, and it is exposed over about 7000 square miles of the Powder River Basin.

The Wasatch Formation is composed of drab, tan to brown claystone and siltstone with some coal seams, interbedded with mostly tan to buff, cross-bedded, arkosic, fine- to coarse-grained sandstone lenses. In some areas, the sandstone lenses, which normally are drab, are either red or, to a lesser extent, light gray to white. The sandstone of the Wasatch Formation is coarser grained near the southern and central part of the basin.

The red facies of the Wasatch Formation are important because all of the known uranium ore deposits are at or near the contact of red and drab sandstone, in a zone in which some degree of red coloration is present. The colors range from red-purple to red to pink and orange-red. A sandstone lens may be entirely red or it may contain thin red stringers and nodules of red interspersed throughout its entire thickness. The color change is a cross-cutting feature having no apparent lithologic control. The red color usually terminates as lobes or stringers projecting into the non-red beds. The color contact is extremely irregular in both cross-section and plan. Deep drilling has shown that no red color is present below the permanent water table.

The light-gray to white sandstones of the Wasatch Formation may also have some relationship to uranium mineralization. In the Monument Hill district, uranium ore bodies occur several hundreds of feet down dip from the predominant red coloration, and the area between the red sandstone and the ore bodies is generally light gray to white. These light-colored intervals appear to be highly oxidized altered tongues. Immediately adjacent to some ore bodies, the white coloration is attributed to an abundance of montmorillonite (7) that occurs as minute pellets interspersed with the sand grains.

Calcite-cemented sandstone concretions are common and usually take the form of elongate oval masses having average cross-sectional areas of about 15 square feet and maximum lengths of about 70 feet. Concretions are generally elongated to the north and many times are found in groups lying roughly en echelon. Less frequently the concretionary masses are spherical.

WHITE RIVER FORMATION  The White River Formation of Oligocene age, which originally may have blanketed the Powder River Basin, was recognized by Love (2) as the protective cap preserving the Pumpkin Buttes. It is comprised of three facies: a soft basal coarse-grained sandstone unit, from 100 to 200 feet thick, overlain by a caprock composed of silica-cemented conglomeratic sandstone about 50 feet thick. This silica is fluorescent and slightly radioactive. The caprock series is overlain by small remnants of white blocky claystone typical of the lower part of the White River Formation.

## Structure

The Powder River Basin is structurally a broad, gentle syncline overlying the rather stable foreland region of the Cordilleran System. The axis of the syncline lies west of the center of the basin and bears about N30°W, approximately parallel to the Bighorn Mountain-Great Pine Ridge alignment (Figure 1). Dips are usually less than 3° on the east side of the basin, but are very steep on the southwest and west sides near the basin margin.

Evidence of structure rarely is discernible in the Tertiary rocks within the central part of the Powder River Basin. Regional mapping in the Pumpkin Buttes area (7) indicates the presence of a broad north-trending low-amplitude anticline lying just east of that area that is transected by a broad west-plunging arch (8) (Figure 1).

A trace of the basin trough is shown to exist (8) at the surface in the Wasatch Formation immediately west of the Monument Hill mineralized area. Small-scale faulting is present in the Monument Hill and Box Creek areas. A few small faults are also reported (7) in the vicinity of the Pumpkin Buttes. Only one small uranium deposit observed by the writer appeared to have definite structural control.

Although there is little surface expression of structure, Osterwald (6) believes that gradual, long-continued growth along the structural trends of Precambrian age underlying the Powder River Basin has caused the Wasatch to become slightly warped and jointed and that these later structures are the fundamental controls for the localization of uranium.

## ORE DEPOSITS

More than 300 uranium occurrences are known in the Powder River Basin in an area extending from north of the Pumpkin Buttes

southeasterly to a point about 17 miles north of Douglas, Wyoming.

The region is divided into four mining districts (Figure 1) separated by areas devoid of uranium deposits. In the Pumpkin Buttes district, located on the north end of the uranium mineralization, the deposits are distributed irregularly over an area of approximately 350 square miles. Uranium ore has been produced from 55 individual ore deposits; the largest located northeast of the Pumpkin Buttes has produced in excess of 5000 tons of ore. Three mines have produced over 1000 tons and the remaining 51 mines have each produced 500 tons or less with the majority producing less than 100 tons. Most of the mines of the district are located west of the Pumpkin Buttes in a region somewhat more severely dissected by erosion.

The Turnercrest district is located south of the Pumpkin Buttes. Uranium has been produced from seven deposits. Two mines have produced over 1000 tons, and the remaining five mines have produced less than 100 tons.

The Monument Hill district lies south of, and is separated from, the Turnercrest area by a rather large barren area drained by eastward-flowing streams that may have stripped away most of the favorable host rocks. In this barren region, the red-colored sandstones are essentially absent. Although only a few deposits have been mined in the Monument Hill district, most of those found are among the largest in the Powder River Basin. This district has produced about 91 per cent of the uranium ore shipped from the basin. The major production (Figure 2) has come from properties owned or leased by Western Nuclear, Incorporated, which has a total production of 212,100 tons containing 495,700 pounds of $U_3O_8$; B and H Mines, Incorporated, which has a total production of 140,900 tons containing 593,400 pounds of $U_3O_8$; and Mrak Mining Company whose properties have produced 109,500 tons containing 538,800 pounds of $U_3O_8$.

The southernmost area of uranium deposits is the Box Creek district from which 3500 tons of ore have been produced from two mines.

The total uranium production through June, 1965, from the Powder River Basin is shown in Table I.

The exploitable uranium deposits in the Wasatch Formation have several characteristics in common. Uranium and vanadium are the dominant elements of the oxidized ore. The deposits are, almost without exception, associated with areas of red coloration. They are

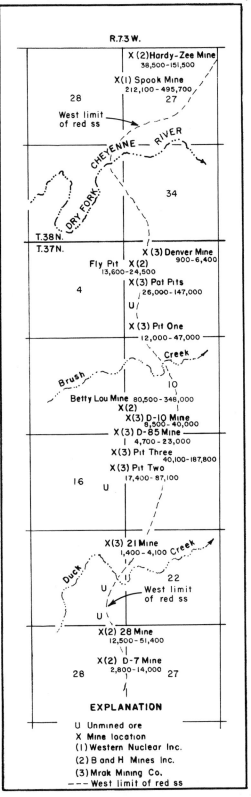

TABLE I.    *Uranium Production in the Powder River Basin, Wyoming, through June 30, 1965*

| Year | Tons | Pounds of $U_3O_8$ |
|---|---|---|
| 1953 | 602 | 3,195 |
| 1954 | 2,436 | 20,365 |
| 1955 | 6,970 | 42,915 |
| 1956 | 24,011 | 111,369 |
| 1957 | 48,953 | 222,491 |
| 1958 | 39,336 | 188,841 |
| 1959 | 31,077 | 124,596 |
| 1960 | 55,016 | 235,810 |
| 1961 | 30,370 | 137,833 |
| 1962 | 74,340 | 206,327 |
| 1963 | 110,459 | 363,107 |
| 1964 | 64,164 | 194,426 |
| 1965 (First $\frac{1}{2}$) | 18,523 | 71,252 |
| Totals | 506,257 | 1,922,527 |

Data: U.S. Atomic Energy Commission.

generally located east of the synclinal axis of the Powder River Basin and are found in the middle or predominantly lenticular sandstone sequence of the Wasatch Formation. They all occur within 165 feet or less below the surface of the ground and have not been found more than a few feet below the water table. With one exception, all deposits have been leached and modified by surface oxidation. The deposits lie at elevations ranging from about 4500 feet to more than 5400 feet.

The uranium deposits may be broadly classified into three general types on the basis of size and grade, mineralogy, and their association with red sandstone (Figure 3).

The first type (Figure 3-A) comprises small high-grade concretionary or roughly tabular deposits usually containing less than 35 tons of uranium ore with an average grade of 1 per cent or more. The minerals are uraninite associated with manganese or iron oxides and surrounded by uranophane and/or secondary uranium vanadates. These isolated high-grade pods are found within the red beds and generally are considered of little economic importance. The writer believes that the majority of the mines in the Pumpkin Buttes area were developed from this type of deposit.

The second type (Figure 3-B) is composed

FIG. 2. *Map of the Distribution and Align-* *ment of Uranium Ore Bodies in the Monument Hill Mining District in the Powder River Basin, Converse County, Wyoming. Figures give total production in tons of ore and pounds of $U_3O_8$ for individual mines through June 30, 1965.*

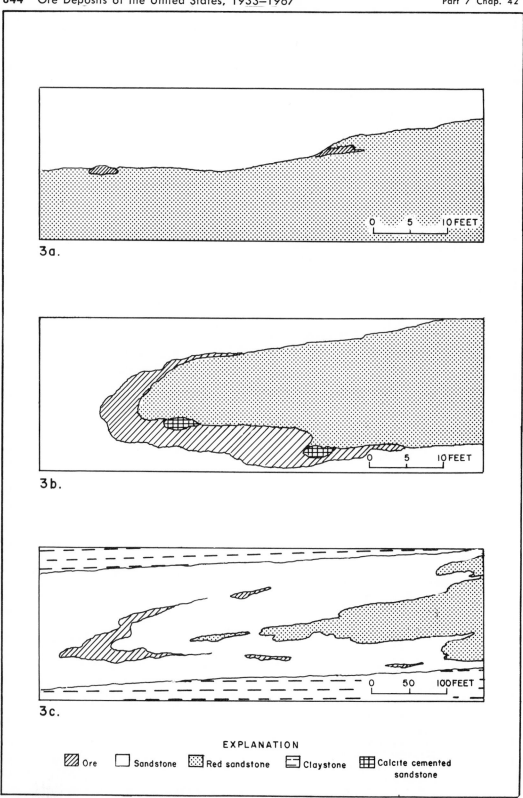

FIG. 3. *Diagrammatic Cross-Sections, showing the three types of uranium deposits found in the Powder River Basin, Wyoming.*

of intermediate size deposits containing as much as 5000 tons of ore averaging about 0.40 per cent $U_3O_8$. This type of deposit lies adjacent to the buff-colored sandstone at the contact with the pink to red sandstone. Uranium minerals occur as a shell, several inches to several feet thick, coinciding with the maximum penetration of red coloration and transecting all sedimentary features. Where the contact is most irregular, the uranium minerals may be disseminated throughout the areas between the red-colored stringers and may attain a thickness of 8 to 12 feet. Uranium vanadates are the dominant ore minerals; however, small high-grade uraninite pods associated with pyrite are found in the buff sands at or near the contact, and, to a lesser extent, the uraninite-manganese-iron oxide pods are found in the red sands at the contact. Calcite is the prevalent gangue mineral.

The third type of deposit (Figure 3-C) is characteristic of the Monument Hill district. More than 150,000 tons of ore with grades averaging between 0.20 and 0.25 per cent $U_3O_8$ have been mined from single oxidized deposits that appear to be skeletal remnants of formerly much larger unoxidized ore bodies. Uranium deposits of this type are similar to the second or intermediate type deposit with the notable exception that the ore bodies are located several hundreds of feet down-dip from the predominantly red-colored zone. The interval between the red beds and ore bodies is occupied by leached or altered sandstones having a light-gray to white coloration. Vestiges of red are present in the ore bodies and in the altered interval in the form of red clay galls, thin pale-pink sandstone stringers, and rarely thin concentric banding surrounding high-grade uranium concentrations or calcite-cemented concretions.

The ore bodies are generally irregularly "C" shaped in cross-section, convex down-dip away from the red beds, and sinuous in plan with their overall trend roughly parallel to the margins of the red sandstones. Ore deposition appears to be localized at or near the terminus of a mineralized alteration front with the altered or leached sandstone forming the interior of the altered tongue. These Powder River Basin deposits appear to be similar to those in the Gas Hills and Shirley Basin described in companion papers for other districts in the Wyoming-South Dakota region.

The ore bodies occur intermittently for a distance of almost 7 miles along the mineralized trend in the Monument Hill district. The areas devoid of uranium ore deposits along the mineralized trend may be the result of non-deposition but more often appear to be caused by present or Quaternary stream erosion. Quaternary erosion has cut channels as much as 70 feet deep into the Wasatch Formation and may be observed as brown sandy-silt channel fills on the open-pit mine walls. Often, where a Quaternary channel crosses an ore body, the ore will either have been removed by erosion or it will have been leached by the large amounts of fresh water that percolated through the host rock adjacent to the channels during the period of down cutting. Generally this is shown by the extreme radiometric deficiency of uranium nearest the channels.

The mineralogy of the uranium ore is based almost entirely on the oxidized deposits. Gruner (3,5) and Sharp (10,11) indicate that the uranium vanadates, metatyuyamunite, and carnotite are the most prevalent minerals. The list of minerals identified include uranophane, uraninite, coffinite, and the uranium carbonates; liebigite, bayleyite, and zellerite. Several vanadium minerals and manganese minerals, in addition to pyrite and native selenium, have also been identified. The common gangue minerals are calcite, gypsum, pyrite, hydrated iron oxides, and barite.

Recently uraninite and coffinite were identified by Sharp (Butler, written communication, 1965) as the ore minerals of the unoxidized deposit in Pit 2 presently being mined in the Monument Hill district (Figures 2 and 4). These primary uranium minerals probably represent the mineralogy of most deposits prior to oxidation.

## ORIGIN

W. N. Sharp (10,11) concludes that the uranium deposits are the result of a redistribution of elements within the Wasatch Formation during a period of excessive heat and pressure developed by sedimentary loading and anticlinal folding. In this environment, a dense $CO_2$ phase developed, "which provided a means for the migration and accumulation of metals and the reddening of the sandstone."

The writer, however, believes, as do Love (2) and Denson (4), that the deposits are epigenetic and that the source of uranium was the tuffaceous Oligocene and Miocene rocks that overlie or once overlay the area. Circulating ground water was the ore-bearing medium. Uranium was taken into solution by oxidizing surface waters and perculated downward through joints and fractures in the rocks over-

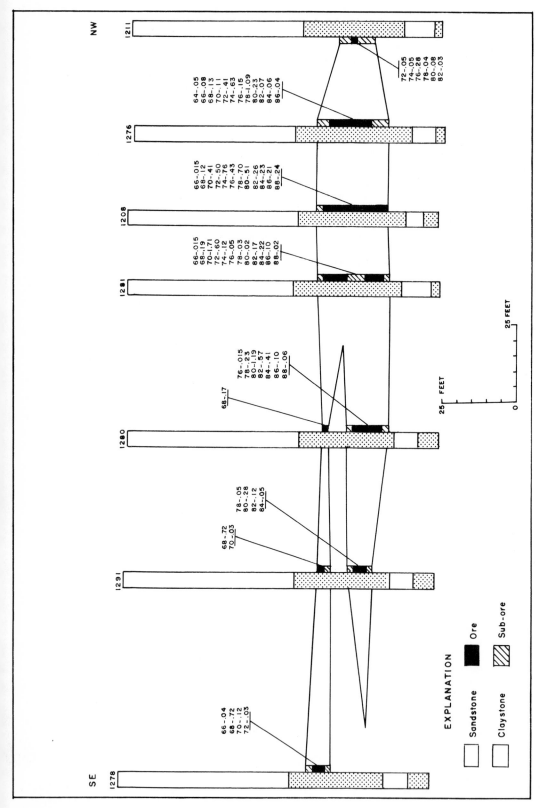

FIG. 4. *Cross-Section of the Mrak No. 2 Pit Located in the Monument Hill District, showing depth of ore and chemical assay value (in per cent U₃O₈) of ore as sampled on 2-foot intervals.*

lying the ore-bearing Wasatch Formation. These fractures were probably more numerous in areas of structural deformation, such as those described by Osterwald (8) and Sharp (10,11). Some of the downward percolating water may have entered and moved laterally through quite flat-lying permeable zones such as the sandstone beds in the base of the White River Formation or the unconformity at the top of the Wasatch Formation.

The mechanism for deposition of the first type of uranium deposit, that is, the small uranium-manganese pods within the red beds, is insufficiently understood to be described by the writer; however, the deposition of the large ore bodies adjacent to or down-dip from the red beds appear to be the result of chemical change at an interface between uranium-bearing oxidizing waters and the almost stagnant waters contained in the sediments below the basin spill point. This is similar to the theory advanced by Harshman (12) for the accumulation of uranium in the Shirley Basin of Wyoming.

## USE OF GEOLOGY IN EXPLORATION, DEVELOPMENT, AND MINING

### Exploration Method

The first phase of uranium exploration in the Powder River Basin was accomplished by airborne radiometric surveys and was followed by ground location and surface examination of points of anomalous radioactivity. Many small ore bodies found at the surface were delimited and mined by mechanized front-end loaders. The more sophisticated operators used rotary drills to explore the anomalous areas with holes spaced 25 feet apart. Depth of drilling was usually controlled by the depth of the first claystone encountered, whether it was at a depth of 10 feet or 40 feet. With few exceptions this was the extent and method of exploration and development of almost all of the radioactive anomalies found within the Powder River Basin. The Atomic Energy Commission did, however, drill a number of wide-spaced holes to depths exceeding 400 feet.

When it was found that ore was present at considerable depth in the Monument Hill district, drilling depths were increased and most holes were bottomed at the top of the permanent water table. Only a few holes drilled in the Monument Hill district or elsewhere in the Powder River Basin exceed 200 feet in depth. It is estimated that about

1,200,000 feet of exploratory and mine-control drilling has been done in the Powder River Basin.

As a result of drilling most of the Monument Hill district on a 300-foot by 600-foot diamond grid, the writer was able to detect the sandstone beds that served as conduits for mineralizing solutions. This was determined by radiometric probing of the drill holes. Slight radioactive anomalies were observed at the upper and lower sandstone-claystone contacts for distances in excess of 1000 feet updip from the ore bodies or the terminus of the alteration front. The anomalous radioactivity may suggest that some uranium or radium from the migrating solutions was adsorbed by the enclosing claystone.

Many of the Powder River Basin uranium deposits are at the contact of red and drab sandstones, and this spacial relation of ore to a color boundary is an excellent exploration guide. However, in the Monument Hill area, and possibly to the south, the deposits are related to altered sand tongues that extend several hundred feet down dip from the red-drab sandstone contact. In this area, exploration should be continued down dip from the red sandstone as long as the altered sandstone has even a slight amount of anomalous radioactivity.

## REFERENCES CITED

1. Berryhill, H. L., Jr., *et al.*, 1950, Coal resources of Wyoming: U.S. Geol. Surv. Cir. 81, 78 p.
2. Love, J. D., 1952, Preliminary report on uranium deposits in the Pumpkin Buttes area, Powder River Basin, Wyoming: U.S. Geol. Surv Circ. 176, 37 p.
3. Gruner, J. W. and Smith, D. K., Jr., 1955, Tenth progress report for period April 1, to October 1, 1955: U.S. Atomic Energy Comm. RME-3125.
4. Denson, N. M., *et al.*, 1956, Water sampling as a guide in the research for uranium deposits and its use in evaluating widespread volcanic units as potential source beds for uranium [S. Dak., Wyo.] (revised): U.S. Geol. Surv. Prof. Paper 300, p. 673–680.
5. Gruner, J. W., *et al.*, 1956, Annual report for April 1, 1955, to March 31, 1956, part 2: U.S. Atomic Energy Comm. RME-3137.
6. Osterwald, F. W., 1956, Relation of tectonic elements in the Precambrian rocks to uranium deposits in the Cordilleran foreland of the western United States (slightly revised): p. 329–335, *in Contributions to the geology of uranium and thorium:* U.S. Geol. Surv. Prof. Paper 300, 739 p.

7. Sharp, W. N., 1956, Geologic investigations of radioactive deposits: U.S. Geol. Surv. TEI 620, Semiann. Prog. Rept. Dec. 1, 1955, to May 31, 1956, p. 181–183.

8. Osterwald, F. W. and Dean, B. G., 1961, Relation of uranium deposits to tectonic pattern of the central Cordilleran foreland: U.S. Geol. Surv. Bull. 1087-I, p. 337–390.

9. Love, J. D., et al., 1963, Relationship of latest Cretaceous and Tertiary deposition and deformation to oil and gas in Wyoming: in Backbone of the Americas, Amer. Assoc. Petrol. Geols., Mem. 2, p. 196–208.

10. Sharp, W. N., et al., 1964, Geology and uranium deposits of the Pumpkin Buttes area, of the Powder River Basin, Wyoming: U.S. Geol. Surv. Bull. 1107-H, p. 541–638.

11. Sharp, W. N. and Gibbons, H. B., 1964, Geology and uranium deposits of the southern part of the Powder River Basin, Wyoming: U.S. Geol. Surv. Bull. 1147-D, 60 p.

12. Harshman, E. N., 1966, Genetic implications of some elements associated with uranium deposits, Shirley Basin, Wyoming: U.S. Geol. Surv. Prof. Paper 550-C, p. C167–C173.

# 43. Uranium Deposits of the Shirley Basin, Wyoming*

E. N. HARSHMAN†

## Contents

## Illustrations

* Publication authorized by the Director, U.S. Geological Survey.
† U.S. Geological Survey, Denver Colorado.

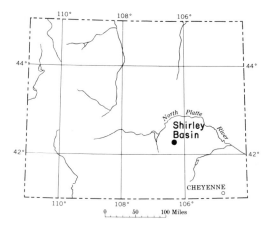

## ABSTRACT

The Wind River Formation of Eocene age is the host rock for large high-grade uranium deposits in the Shirley Basin. The major deposits are in a northwest-trending belt of sandstones that were deposited in stream channels and that were lithologically favorable for uranium accumulation. Paleotopography influenced the character and position of the belt.

The major deposits within the belt are at, or near, the margins of large tongues of altered sandstone formed in the most transmissive parts of thick sandstone beds. At least two tongues of altered sandstone are present in the belt, one of which is 5 miles long, 3 miles wide, and 70 feet thick; the other tongue is somewhat smaller.

Ore bodies consist of a few hundred tons to several hundred thousand tons of material containing from 0.10 to about 2.00 per cent $U_3O_8$. The ore mineral is uraninite associated with pyrite, marcasite, hematite, and calcite. Deposition is believed to have taken place when neutral to slightly alkaline, oxidizing, uranium-bearing ground water, on entering the basin, encountered a reducing environment. The reductant may have been $H_2S$ of biogenic origin.

Geology played an important role in discovery, exploration, and development of ore in the Shirley Basin.

## INTRODUCTION

### Location and History

The Shirley Basin is in southeastern Wyoming about 35 miles airline south of Casper (Figure 1 in Harshman's regional paper on uranium in Wyoming and South Dakota). Uranium was discovered in the area late in 1955 when Teton Exploration and Drilling Company located and explored claims in the northeast part of the basin. The discovery went almost unnoticed until July 1957 when several thousand claims were located in an area of about 150 square miles. A period of intensive exploration followed, and, by the spring of 1959, it was apparent that the Shirley Basin contained one of the nation's major reserves of uranium ore. The first Shirley Basin ore was mined early in 1960 by the Utah Construction and Mining Company from the 330-foot level of their underground mine. In July 1960, the Petrotomics Company began stripping 3,000,000 yards of overburden in preparation for open-pit mining. The first ore was mined on this property in December 1960.

Although ore was found by drilling on other Shirley Basin properties, no new mines were activated until August 1964 when the Centennial Development Company completed a shaft and prepared to mine an ore body on the Nall lease of the Homestake Mining Company.

As of December 1965, all three mines were operating, although Utah Construction and Mining Company has abandoned underground operations in favor of solution mining from drilled wells.

### Production and Reserves

The following production data from the Shirley Basin were compiled from unpublished figures of the U.S. Atomic Energy Commission and reflect production allocations established by that Agency:

| Year | Tons of Ore | Lbs of $U_3O_8$ | Value at $8 per lb |
|---|---|---|---|
| 1960 | 22,286 | 248,956 | 1,991,648 |
| 1961 | 235,273 | 1,715,009 | 13,720,072 |
| 1962 | 138,660 | 982,074 | 7,856,592 |
| 1963 | 181,827 | 1,266,101 | 10,128,808 |
| 1964 | 143,406 | 799,555 | 6,396,440 |
| Total | 721,452 | 5,011,695 | 40,093,560 |

The annual production figure for pounds of $U_3O_8$ includes uranium produced by solution mining (in situ leaching) as well as uranium contained in mined crude ore; the average grade of the Shirley Basin ore, therefore, cannot be calculated from tons of ore and pounds of U₃O₈ produced.

As of October 1, 1965, the U.S. Atomic Energy Commission estimated the ore reserves in the Shirley Basin to be about 50,000,000 pounds of U₃O₈.

## Acknowledgments

The cooperation of the mining companies and claim owners in the Shirley Basin is gratefully acknowledged. Many of the basic data in this paper were obtained with their permission or directly from them.

## GEOLOGY

The Shirley Basin is a structural and topographic feature of low to moderate relief, drained by the southward-flowing Little Medicine Bow River and bounded by the Laramie Mountains on the northeast and the Shirley Mountains on the southwest. Altitudes range from about 6800 feet along the river in the southeastern part of the basin to about 7600 feet along its northern rim.

### Stratigraphy

The oldest rocks in the area are gneiss, schist, and granite that are cut by diabase dikes; all are of Precambrian age. The Precambrian rocks underlie the basin at depths of as much as 5000 feet and crop out in the cores of the Laramie and Shirley mountains. They are overlain by a series of continental and marine sedimentary rocks ranging in age from Mississippian to Cretaceous, and a series of continental clastic and pyroclastic rocks ranging from Eocene to Miocene in age (see Table II in Harshman's paper in this Volume on uranium in Wyoming and South Dakota). The Tertiary formations, because of their relation to the ore deposits, will be described in the following paragraphs.

The oldest rocks of Tertiary age in the Shirley Basin are the fluviatile deposits of the Wind River Formation of early Eocene age. They consist of interbedded and intertonguing beds of sandstone, conglomerate, siltstone, and claystone, all compacted but poorly cemented. The sediments, predominantly arkosic, were derived principally from granitic outcrops west and/or southwest of the basin. They were deposited in a northwesterly trending depression eroded in rocks of Cretaceous and older age.

The Wind River Formation has a maximum known thickness of about 500 feet, and it wedges out against the older rocks on the flanks of the basin.

Overlying the Wind River Formation conformably is a 150-foot thick series of interbedded arkosic sandstones and silicified claystones, generally ledge-forming and of a pale greenish hue. On the basis of stratigraphic position, lithology, and the presence of middle Eocene vertebrate fossils in the basal beds of the series, these rocks are correlated with the Wagon Bed Formation of middle and late Eocene age described by Van Houten (4) in the Beaver Rim area of Wyoming 80 miles to the northwest. In the central part of the Shirley Basin, the Wagon Bed Formation is not present, apparently having been removed by erosion in late Eocene time.

The White River Formation of Oligocene age unconformably overlies the Wagon Bed Formation or the Wind River Formation where the Wagon Bed has been removed by erosion. It is divided into two distinctive mappable units. The lower unit consists of claystone and tuffaceous siltstone with subordinate amounts of tuff; the upper unit is boulder conglomerate interbedded with tuffaceous siltstone. The maximum known thickness of the White River in the Shirley Basin area is about 700 feet. The White River overlaps the older Tertiary formations and wedges out against the pre-Tertiary rocks on the flanks of the Laramie Mountains. At one time, it may have buried all but the higher peaks in the Laramie Range.

Capping several prominent ridges north of the Shirley Basin is a series of tuffaceous siltstone and sandstone beds as much as 120 feet thick. These beds are correlated with the Arikaree Formation of Miocene age, and they rest on the White River Formation with apparent conformity.

No rocks of Tertiary age younger than the lower part of the Arikaree Formation are present in the area, but rocks of Miocene and Pliocene age as much as 500 feet in thickness may have been removed by erosion.

### Structure

Structural features in the Shirley Basin are simple and reflect at least two periods of deformation. In Laramide time, the pre-Tertiary rocks were folded into a broad syncline, the axis of which strikes northwesterly and lies about 8 miles southwest of the ore deposits. In the Laramie Mountains, rocks in the east limb of the syncline dip from 4° to 8°SW;

in the west limb near the Shirley Mountains, there are comparable northeasterly dips. Minor faulting accompanied folding.

The Tertiary rocks generally have a northerly dip of 1° to 2°, due largely to a northward tilting of the area. Local variations in dip are in part due to differential compaction and in part due to original depositional inclination. Faulting accompanying this period of folding was unimportant. These structures affect rocks as young as Miocene in age; they probably originated in early or middle Pliocene time.

## Ground Water

The permanent ground-water table is at depths of a few feet to 300 feet below the ground surface. The ground-water gradient is southward at 10 to 30 feet per mile.

The principal anions in the ground water are bicarbonate (200 ± ppm), and sulfate (100 ± ppm) as shown by analyses of 25 water samples from the Wind River Formation. Radioelements include uranium (20 ± ppb), radium (10 pc/1), and radon (as much as 200,000 pc/1). Field and laboratory measurements of pH range from 6.6 to 8.3; most fall in the range 7.5 to 8.0.

## ORE DEPOSITS

### Paleotopographic Control

The major ore deposits are in a well-defined belt that, for the most part, lies west of a prominent ridge on the pre-Wind River erosion surface. This ridge is formed by the Wall Creek Sandstone Member of the Frontier Formation of late Cretaceous age and is due to the erosional resistance of these sandstone beds. Near the ore deposits, it is buried by 125 to 350 feet of Tertiary rocks, but a few miles to the southeast, where the Tertiary rocks have been stripped away, it crops out and forms a conspicuous topographic feature. Relief on the buried ridge is about 400 feet in the northern part of the area and about 100 feet in the southern part. The ridge is breached in several places by ancient stream channels.

The buried ridge is believed to have controlled the position of a belt of Wind River sediments that had characteristics favorable for ore deposition at a later time. Streams originating west of the area and flowing easterly were diverted to the northwest by the ridge, and, because of a sudden decrease in stream gradi-

ent, sediment was deposited along the west toe of the ridge. Throughout the period of Wind River deposition, swampy flood-plain conditions alternated with periods of stream cutting and deposition. As a result, Wind River strata just west of the ridge consisted of permeable channel sandstones interbedded with silty claystones and organic trash, an ideal environment for the deposition of uranium.

Sediments favorable for subsequent metallization were deposited locally in some of the tributary stream channels 3 to 10 miles west of the buried ridge. Uranium deposits have been found in some of these rocks, but, compared with the deposits in the mineral belt, they are of minor significance. They will not be discussed in this report.

## Size and Character

Individual uranium deposits in the mineralized belt contain from a few hundred tons to several hundred thousand tons of ore. Grades generally range from several hundredths to about 2 per cent $U_3O_8$, although ore containing as much as 20 per cent $U_3O_8$ has been mined. Ore as mined contains from about 0.20 to 0.60 per cent $U_3O_8$; the grade is dependent on the type and cost of mining. The ore is unoxidized and lies just below the water table to as much as 400 feet below it.

The ore mineral is uraninite. It is fine-grained or sooty and associated with pyrite, marcasite, calcite, hematite, and organic material. Uraninite fills interstices between sand grains, coats grains, and, locally, fills fractures in grains. In some places, it replaces, or is disseminated in, carbonized fossil plant remains. Pyrite was deposited before, during, and after the deposition of uraninite. Marcasite is intergrown with pyrite, and it is probably younger than uraninite. Calcite is both younger and older than uraninite, but it probably was precipitated from the uranium-bearing solutions. Hematite generally is associated with calcite; its age relations with uraninite are not known.

Metallic elements, other than uranium, are scant in the ore. High-grade ore generally contains less than 0.1 per cent V; less than 0.01 per cent Pb, As, Mn, Cr; less than 0.001 per cent Cu, Ni, Be; and less than 0.0005 per cent Mo.

## Relation to Alteration

The ore deposits in the mineral belt are genetically and spatially related to large tabular bodies or tongues of greenish-yellow altered* sandstone. Figure 1 shows the general outlines and relative positions, in plan view, of the two principal tongues. Each is in an extensive sandy interval separated stratigraphically by 50 to 75 feet of siltstone and silty claystone. A third tongue probably is present, but data concerning it are meager, and it will not be considered in this discussion.

Detailed study and sampling have been possible only along the western edge of the upper tongue where mining operations have exposed large high-grade ore bodies. This discussion is based on data obtained from exposures in the upper tongue. Drilling data show that the two tongues, and the ore deposits related to them, are similar.

The upper tongue is 5 miles long in a northwesterly direction and about 3 miles wide. It is at depths of 100 to 450 feet beneath the surface. The thickness of altered sandstone ranges from a few feet near the edge of the tongue to about 70 feet at the center. The upper and lower surfaces of the altered sandstone tongue generally are comformable with the regional dip of the Wind River Formation. The position and configuration of the altered-unaltered sandstone contact in some places is controlled by crossbedding and other sedimentary structures, but, in other places, the structures appear to exercise no control on the character of the contact. In the Utah mine area, a 15-foot thick silty claystone bed splits the upper altered sandstone tongue horizontally into two lesser tongues; both are ore-bearing at their margins.

Figure 2, a vertical section across the west edge of the upper altered sandstone tongue, illustrates the spatial relations of altered sandstone, ore, unaltered sandstone, and calcite cement as well as the configuration of the edge of the tongue. Large high-grade ore bodies are found at the edge of the tongue, and smaller ore bodies are found along the top and bottom surfaces. Ore bodies near the edge are commonly C-shaped in cross section, a shape to which the term "roll" has been applied, and they are elongate parallel to the margins of the altered sandstone tongue. Ore bodies are as much as 50 feet wide and 2500 feet long. Irregularities in the margins of the

---

* In this report the term "altered" is used in a restricted sense to refer to those physical and chemical changes, excluding mineralization, effected by an ore-bearing solution on the medium through which it passed. "Unaltered" denotes the lack of such changes.

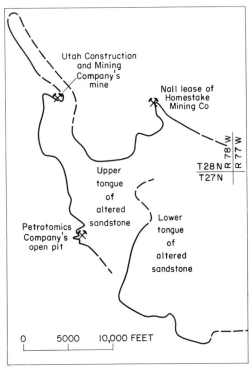

Fig. 1. *Plan Map, showing outline of upper and lower altered sandstone tongues, Shirley Basin, Wyoming. Lines indicate edges of altered sandstone tongues; dashed where approximately located.*

altered sand tongues, both in plan and in section, generally are loci of the major ore bodies (Figures 1, 2).

The curved inner boundary between altered sandstone and ore is generally very sharp, whereas the outer boundary between ore and unaltered unmineralized sandstone is gradational. Calcite cement, containing streaks and irregular masses of hematite, is found along the contact between altered sandstone and ore, in ore, and in weakly mineralized ground near ore; it is rarely found in altered sandstone.

An altered sandstone tongue may terminate laterally in a compound roll, as illustrated in Figure 2, consisting of a main roll with minor rolls on the top or bottom surfaces, or it may terminate in a simple or single roll. Each part of a compound roll has the features of a single roll.

The Wind River Formation in the lower part of the basin typically contains considerable pyrite, abundant carbonized plant remains, and calcite in concretionary masses. In the altered tongues, alteration has destroyed most of the pyrite, carbonaceous material, and

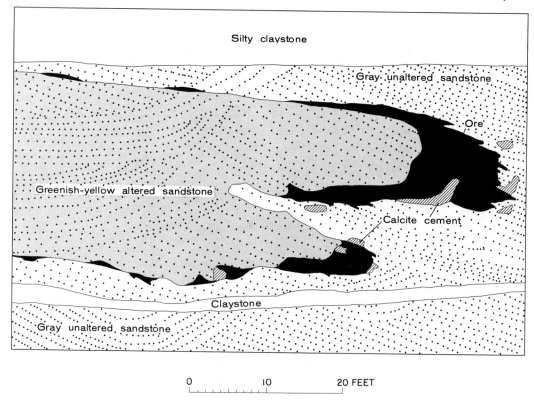

FIG. 2. *Section Normal to West Edge of Altered Sandstone Tongue, showing relation of ore and calcite cement to altered sand (1, Fig. 122.1).*

calcite, and has increased the iron content of the montmorillonite clay. The yellowish green of the altered sandstone as contrasted with the gray of the unaltered sandstone, the most obvious difference between the two, is probably due to a redistribution of iron from pyrite to the clay.

## Origin

There has been considerable speculation on the geochemistry of the ore-bearing solutions in the Shirley Basin. Harshman(1) proposed that uranium was transported in weakly acid, moderately oxidizing ground water and deposited under reducing conditions in the lower part of the basin. Melin (3) proposed that uranium, and some of the elements associated with it in the ore were transported in acid (pH 4–4.5), slightly reducing ground water and that precipitation was due entirely to neutralization of the acid by the alkaline ground through which the water moved. Bailey (5) suggested that the ore-bearing ground water was weakly alkaline and slightly oxidizing and

that precipitation was a result of reduction, possibly by $H_2S$ of biogenic origin.

Recent studies by Harshman (6) on the distribution and geochemistry of elements associated with uranium in the Shirley Basin ore bodies suggest that the ore-bearing medium was ground water, neutral to weakly alkaline and oxidizing, and that uranium was deposited as a result of Eh and pH changes in the ore-bearing solution as it moved down the flank of the basin and into the reducing environment in the lower part of the basin. The zone of deposition appears to have been a dynamic feature, migrating basinward (northwesterly) by oxidation and solution on the concave (up-dip) side and reduction and deposition on the convex (downdip) side. A sharp but temporary drop in pH, due to oxidation of pyrite, probably occurred near the updip side of the zone. Transportation of uranium as a carbonate complex and reduction by $H_2S$ of biogenic origin are suggested. This genetic concept is compatible with the alkaline environment of the general area, the known geochemistry of the ore elements, and the alteration associated with the deposits. The degree to which it is valid will be determined only by additional field and laboratory research in this complex phase of economic geology.

Available data suggest a multiple source for the uranium now concentrated in the ore deposits. Analyses show that ground water issuing from the tuffaceous beds of the White River Formation contains 10 to 20 ppb uranium and establish this material as a possible source of uranium. Tests made on the granite that served as a source for much of the sandstone and conglomerate in the Wind River Formation demonstrate that uranium can be leached from this material by a solution approximating the composition and pH of present-day Shirley Basin ground water. There is no evidence of igneous or hydrothermal activity of Tertiary age in or near the Shirley Basin, and it seems improbable that magmatic hydrothermal solutions contributed uranium directly to the deposits or to the ground water from which they are believed to have formed.

## Age

The Shirley Basin uranium depostis are epigenetic and thus they are younger than the 55,000,000-year-old Wind River Formation in which they occur. How much younger is not known. Total lead-uranium ratios, corrected for "background" lead in the Wind River rocks prior to ore deposition, suggest ages of 10,000,000 to 15,000,000 years. However, considerable error in these ages may have been introduced by assuming that all radiogenic lead in the deposits resulted from the decay of uranium in the deposits, and the assumption that no common lead was introduced with the uranium.

Studies on the distribution of $Th^{230}$ and $Pa^{231}$ and on the fractionation of $U^{234}$ by Dooley et al. (2) indicate a minimum age of more than 250,000 years.

## GEOLOGY IN EXPLORATION AND DEVELOPMENT

Geology played an important role in the discovery of uranium in the Shirley Basin and in exploration, development, and mining activities that followed. The original drilling by Teton Exploration and Drilling Company was undertaken as a result of geologic studies that showed similarities between the Shirley Basin and other basins in Wyoming known to contain uranium

The general concept that ore bodies would be found in old stream channels, particularly associated with large meanders, guided exploration for several years after discovery of ore in the Shirley Basin. Drill holes, generally drilled by non-coring methods, were located on a grid pattern, and drill-hole data, including gamma-ray, resistivity, and lithologic logs were used to prepare isorad, isopach, and paleotopographic maps. These general ore-guides proved successful, and the uranium mineral belt was discovered and roughly delineated before the relation of ore to altered rock, the more direct guide to ore, was recognized.

Altered sand associated with ore was first noted by geologists of the Utah Construction and Mining Company in drill cuttings and mine development headings. Shortly thereafter, the alteration-ore association was established in the Petrotomics pit, and it became apparent that the spatial relations of ore to altered sandstone were characteristic of the entire mineral belt. Here, then, was a direct guide to the large high-grade ore bodies along the margins of the altered sandstone tongues and a less direct guide to the somewhat smaller ore bodies on their top and bottom surfaces.

The exploration geologists became adept at recognizing altered sandstone even where color distinctions were made difficult by dilution and mixing of drill cuttings. In addition to color differences, altered sandstone was found to be characterized by a lack of carbonaceous trash, calcium carbonate cement, and pyrite, by a gross radioactivity slightly greater than that of unaltered sandstone, and by a radiometric peak at its contact with unaltered sandstone. On the basis of these criteria, grid drilling on wide-spaced centers was employed to bracket the margins of altered sandtsone tongues, and fence drilling, on centers as close as 10 feet in the fences, was used to locate ore bodies along the margins. Ore bodies on the top and bottom surfaces were located by either fence or grid drilling on centers ranging from 10 to 50 feet.

Successful mining operations, and a better understanding of the origin of uranium deposits in sandstone attest to the efforts and abilities of those geologists responsible for the discovery, exploration, and development of the Shirley Basin uranium deposits.

## REFERENCES CITED

1. Harshman, E. N., 1962, Alteration as a guide to uranium ore, Shirley Basin, Wyoming:

Art. 122 in U.S. Geol. Surv. Prof. Paper 450-D, p. D8-D10.

2. Dooley, J. R., Jr., *et al.*, 1964, Radioactive disequilibrium studies of roll features, Shirley Basin, Wyoming: Econ. Geol., v. 59, p. 586–595.

3. Melin, R. E., 1964, Description and origin of uranium deposits in Shirley Basin, Wyoming: Econ. Geol., v. 59, p. 835–849.

4. Van Houten, F. B., 1964, Tertiary geology of the Beaver Rim area, Fremont and Natrona Counties, Wyoming: U.S. Geol. Surv. Bull. 1164, 99 p.

5. Bailey, R. V., 1965, Applied geology in the Shirley Basin uranium district, Wyoming: Univ. Wyo., Dept. Geol., Contribs. to Geol., v. 4, no. 1, p. 27–35.

6. Harshman, E. N., 1966, Genetic implications of some elements associated with uranium deposits, Shirley Basin, Wyoming: U.S. Geol. Surv. Prof. Paper 550-C, p. C167–C173.

# 44. Western Utah, Eastern and Central Nevada

## WILLIAM PAXTON HEWITT*

## Contents

* Utah Geological Survey and University of Utah, Salt Lake City, Utah.

# Illustrations

# Tables

## ABSTRACT

Mineral deposits of western Utah and eastern and central Nevada have produced in excess of $8,500,000,000 since 1871. Through 1965, Bingham Canyon had produced over $4,600,000,000 and seven other camps over $100,000,000 each, principally from base metals. In addition, they yield iron and uranium ores; contain the world's largest beryllium reserves; and have produced the United States' most important gold discovery in 25 years.

The area is famous for its silver production and its bonanza-type precious metal deposits, but such production has become unimportant. The area's greatest producers of metallic wealth are base-metal deposits in limestones and porphyry-type copper ores.

In contrast to silver's decline, gold production has been rising—the result of by-product operations at Bingham Canyon. The area also contains gold deposits in which the gold is so finely disseminated it is not recoverable by panning. The area's gold ores frequently are associated with realgar, cinnabar, and carbon.

The ore deposits generally are attributed to Tertiary intrusives,* but, at Mountain City, they may be related to Paleozoic volcanism. Mountain City has yielded rich chalcocite from quartz-pyrite-chalcopyrite bodies in siliceous sediments.

Regional stratigraphy and structure are difficult to interpret, 69 per cent of the area being obscured by alluvium or volcanics. In general, central Nevada consists of chert, shale, silty sandstones, lavas, and pyroclastics; eastern Nevada and western Utah of many alternations of limestone, shale, and sandstone. Central Nevada has been structurally active since late Devonian. Although granitic stocks may have intruded eastern Nevada and western Utah prior to the Cretaceous, these areas remained quite stable until affected by the Laramide Revolution.

Precambrian exposures, occupying 3 per cent of the area, have not been important hosts. However, the question is raised as to whether Precambrian carbonates, properly situated with respect to invading mineralizers, might yet prove to be important.

This paper describes 16 precious metal camps, 6 limestone replacement districts, 2 silver-base metal vein deposits, and 9 miscellaneous types and discusses briefly the 11

major camps that are described in detail in this Volume.

## INTRODUCTION

This paper, discussing western Utah and eastern and central Nevada, synthesizes the mining geology of most of the Great Basin, specifically that portion from 117°30'W, west of Tonopah and Battle Mountain, Nevada, to the Wasatch Mountains east of Salt Lake City, Utah (Figure 1). Included are eight districts with production in excess of $100,000,000: Bingham Canyon, Ely, Park City, Tintic, Iron Springs, Tonopah, Pioche, and Goldfield. Bingham Canyon, with a production of over $4,600,000,000, ranks among the great camps of the world. A ninth camp, Eureka, Nevada, may have produced in excess of $120,000,000 (Nolan and Hunt, this volume) although its recorded production is only $58,000,000. Outside the area of this report are the Virginia City and Yerington districts of Nevada; Nevada's iron, diatomite, and magnesite production; and Utah's uranium deposits on the Colorado Plateau.

Cumulative production figures for the whole area of this report are difficult to obtain, particularly for camps that operated shortly after the Civil War. However, total recorded production of gold, silver, copper, lead, and zinc in Utah and Nevada, from 1871 through 1965, exceeds $8,495,000,000. Since Virginia City's heyday was prior to this period, and most of the metallic and non-metallic yield of Utah and Nevada is from the area of this report, production can be referred with little modification to the total two-stage figure.

First described by Fremont (1) in 1845, the Great Basin, a desert with interior drainage, is bounded on the east by the Wasatch Range, on the west by the Sierra Nevada. Its topography, aptly described as an "army of caterpillars crawling toward Mexico,"† is composed of playas and isolated ranges rising starkly to forested summits.

American prospectors, scouring the Basin for gold during the decade that preceded the Civil War, deaf to Mexican miners who recognized "Mucha plata! Mucha buena plata!" in the annoying blue mud of the Comstock (10), failed to discover its silver ores until 1859 when Virginia City became one of the giants of the mining world. Following Virginia City, the Basin's bonanza camps were rapidly

---

* Some ores, as at Ely, are associated with intrusives of Cretaceous age.

† Ascribed to Dutton by Thomas B. Nolan (19).

discovered, but "The cream was soon skimmed and the catastrophic fall in production which coincided with the panic of 1879 was merely the accentuation of an inevitable decline" that reached its ebb in the decade following 1890 (20).

The discoveries of Tonopah in 1900 and of Goldfield in 1903 initiated a second surge in precious-metal production that lasted for a quarter of a century, with peak production occurring in 1913 (Figure 2). Silver reached its peak in 1922, gold in 1910. Silver production has declined steadily ever since; but even though precious metal bonanza camps no longer operate, gold production has been rising since 1930. This upward trend is partially explained by the increased price in 1934, and, since 1945, by the tremendous daily tonnages from low-grade porphyry copper mines wherein gold is an important by-product (Figures 2, 4).

With the early discovery of free-milling precious metal vein deposits came the discov-

ery of silver ores associated with lead, which required the construction of smelters. Among them, Eureka, Nevada, has contributed significantly to the art of lead smelting.* Lead production reached its peak in 1930 (Figure 3). Another contribution of the silver-lead camps has been the production of zinc, mainly from lead-zinc sulfide ores. Peak zinc production occurred in 1951.

Exploration within two of the silver-lead camps has led to two of the great porphyry-copper deposits of North America: Bingham Canyon, Utah, and Ely, Nevada. Moreover, the open-pit operation at Bingham Canyon was the first to demonstrate that profits can be won from large tonnage operations on low-grade disseminated ores. Today the average observer, overawed by this outpouring of

* Eureka also is famous for its early mining litigations that resulted in Supreme Court decisions that substantially modified the American mining code (Nolan and Hunt, this Volume).

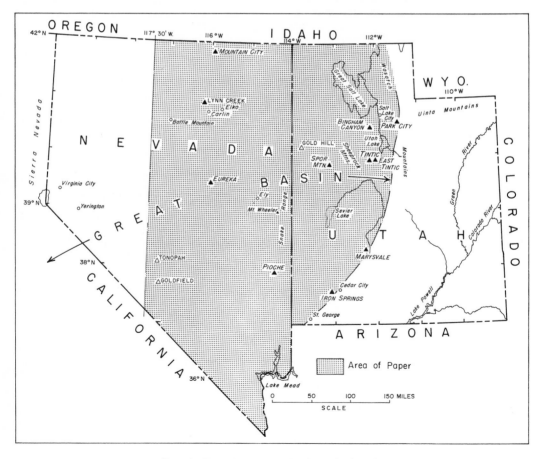

FIG. 1. *Location Map: Utah and Nevada.*

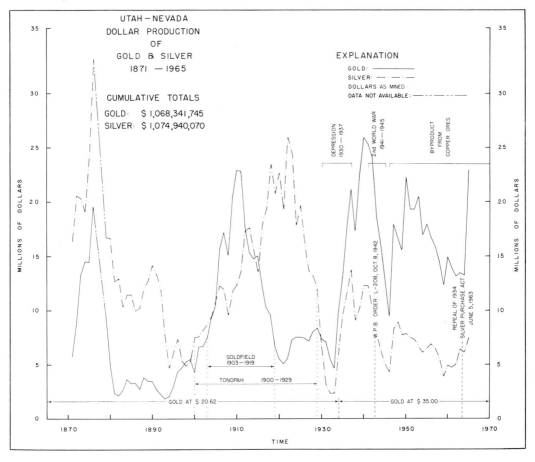

FIG. 2. *Gold and Silver Production: Utah–Nevada, 1871–1965.*

open-pit material, forgets that the open-pit operation is but one phase of Bingham Canyon where underground mines have yielded over $777,000,000; 2,377,000 ounces of gold; 136,125,000 ounces of silver; 816,490,000 pounds of copper; 4,180,279,000 pounds of lead; and nearly 2,000,000,000 pounds of Zinc from nearly 44,000,000 short tons of crude ore (Rubright and Hart, this volume) (Figure 4).

The Great Basin continues its outpouring of mineral wealth as old camps continue to operate and new discoveries are added. In 1949, uranium discoveries at Marysvale, Utah led to the development of one of North America's important uranium-bearing vein deposits (Kerr, this volume). The 1960's witnessed the discovery of the World's largest known beryllium deposits at Spor Mountain and at Gold Hill. And, in 1964, the first important gold discovery in the United States in over a quarter of a century was brought into

production at Lynn Creek near Carlin, Nevada.

## STRATIGRAPHIC AND STRUCTURAL HISTORY

Within the area of this paper 69 per cent of the bed rock is obscured beneath great depths of alluvium or volcanics, and the complex history of the Great Basin is understood only in broad outline. The attached paleostratigraphic cross section by R. L. Armstrong (77), selected from a paper being published by the Utah Geological Survey, illustrates the stratigraphic complexities in the northeastern portion of the Great Basin (Figure 5).

### Precambrian Outcrops

Precambrian outcrops account for less than 3 per cent of the area's surface, or possibly

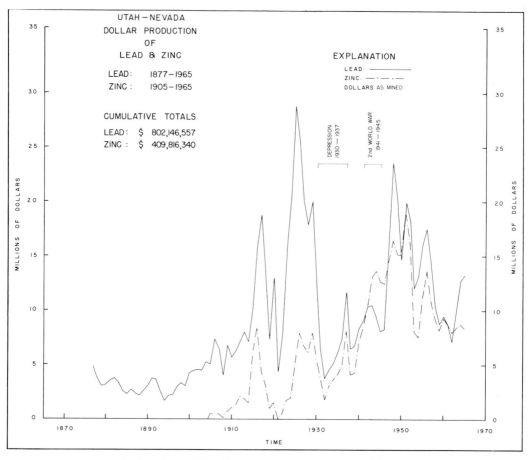

FIG. 3. *Lead and Zinc Production: Utah–Nevada, 1877–1965.*

9 per cent of the non-volcanic, non-alluvium exposures. They occur, widely scattered, in two general areas: a central group, generally of older rocks, extending from the vicinity of Ely, Nevada, easterly to the Deep Creek Mountains in western Utah; and, absent only to the west, a peripheral group at a radial distance of 100 to 200 miles (Figure 6). The older members, folded and recrystallized, contain schists, marbles, and granite gneiss (62). The younger members, particularly to the east, only slightly metamorphosed, contain indurated shales and sandstones that become slates and quartzites locally, and infrequent exposures of dolomite that are weakly recrystallized.

There has been little regional range-to-range correlation. It is presumed that Precambrian sediments were deposited in a northeasterly trending seaway that occupied western Utah and much of Nevada, with continental facies to the east. In general, Precambrian beds are separated from the Cambrian by an angular

unconformity, but some exposures conformably underlie Cambrian sediments (40,38).

Precambrian rocks contain glacial deposits (38,48,55), and preserve the history of Precambrian orogenies (54). Exposures in the Grouse Creek Mountains in northwestern Utah are among the oldest dated rocks in North America (69). Detailed structural studies may determine whether they contain North America's oldest carbonate sediments (78).

## Mineral Deposits in Precambrian Rocks

Precambrian rocks of the Great Basin, essentially non-productive and largely of scientific interest, may contain deposits that are convertible into producing mines. In extreme southern Nevada, Precambrian schists and gneisses contain three mineralized areas. In the Bunkerville district subcommercial deposits of copper, nickel, cobalt, and platinum occur within small ultrabasic dikes as pods and

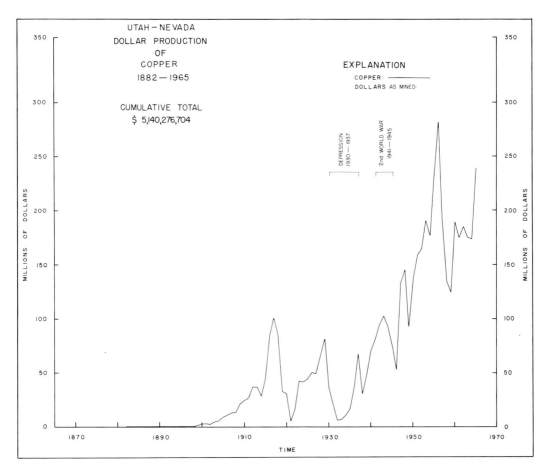

FIG. 4. *Copper Production: Utah–Nevada, 1882–1965.*

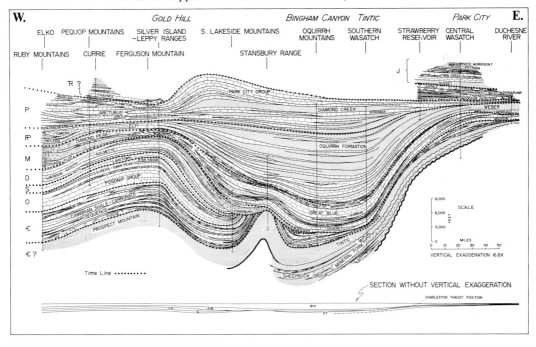

FIG. 5. *W-E Paleostratigraphic Cross-Section Facing Northerly; Northeast Portion of the Great Basin: Elko to Park City.*

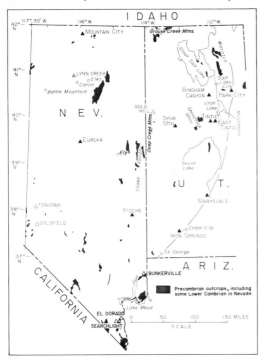

FIG. 6. *Precambrian Outcrops: Western Utah and Eastern Nevada.*

disseminated knots and grains of chalcopyrite, pyrite, pyrrhotite, and magnetite (45,64). At El Dorado and Searchlight, each commercially productive, veins, probably of Tertiary age but within presumed Precambrian host rocks, have yielded gold and silver.

Elsewhere, building stones have been produced from these ancient rocks. These rocks also contain high-content alumina minerals, but only in central Utah in the vicinity of Salt Lake City have these minerals been mined for their ceramic properties (47).

## Paleozoic Events

In the early Cambrian, a narrow seaway, advancing on Precambrian rocks, occupied a narrow trough in western Utah and eastern Nevada. By late Cambrian, the sea had spread westward to California and eastward into central Utah. Great thicknesses of sands, now quartzites, were deposited along the shoreline. Throughout the Paleozoic, western Utah and eastern Nevada were covered by shallow seas, the eastern shorelines of which fluctuated across central Utah.

These seas overlay a rather quiet area on which localized but pronounced subsidence occurred. The Oquirrh Basin, one such area of local subsidence, extended across northwestern Utah into eastern Nevada (67). Possibly forming in Mississippian time, it accumulated over 25,000 feet of Pennsylvanian and Permian calcareous and arenaceous sediments (67,34). There were areas of uplift, notably in Ordovician and Devonian times (67,43,28,32) but there seems to have been no orogenic disturbance. Limestone is the most common sediment, minor shale and sandstone beds occur; graywackes and pyroclastics are absent.

In western Nevada, in the western portion of the area encompassed by this paper, early Paleozoic eastern Nevada assemblages of limestone, shale, and sandstone grade into chert, shale, silty sandstone, lavas, and pyroclastics. Moreover, western Nevada underwent a succession of orogenies that raised geanticlines, split the seaways, and folded the sediments. The first of these, the Antler, occurred from late Devonian into early Pennsylvanian when most of western Nevada stood above the geosyncline (32). It shed great thicknesses of clastic sediments. During the late Mississippian, over 7000 feet of conglomerate, thinning eastward into sandstone and shale, covered east central Nevada, and over 10,000 feet of dirty sandstone accumulated to the west. Associated with this orogeny is the Roberts Mountains thrust zone of indefinite age. Evidence for it is preserved in isolated ranges from the vicinity of Mountain City in north central Nevada, southwesterly toward Eureka. In the opinion of R. J. Roberts (56), it carried cherty western rocks eastward, as much as 90 miles, and over-rode the carbonate rocks of central Nevada.

Following the Antler orogeny, Pennsylvanian conglomerates were laid down on the bevelled edges of distorted strata, and shallow seas drowned western and central Nevada. Then, in late Permian, an imperfectly documented orogeny, the Sonoma, occurred in north central Nevada. (63.)

## Mesozoic and Cenozoic Events

Siltstones, calcareous shales, limestones, and gypsum were deposited by a Triassic sea, the last sea to cover the entire area (61). These sediments intertongue with, and are overlain by, continental deposits that were laid down during a period of widespread eolian deposition that was followed during the middle Jurassic by seas that appeared in central Utah. These deposited 3000 feet of muddy limestone and gypsum, which are the last marine strata deposited in the area of this report.*

The late Jurassic marked the beginning of a third mountain-building period, the Nevadan orogeny, which continued through early Cretaceous. Folding, thrusting, and intrusion of batholiths were most intense in western Nevada, but isolated stocks of granitic material were intruded throughout the region.

During late Jurassic and early Cretaceous, most of Nevada and western Utah stood as highlands, coarse clastics were shed into central Utah and into interior basins in Nevada, and there was an outpouring of volcanics. Cretaceous rocks consist of clastics, isolated patches of volcanics, and lacustrine sandstones (61).

The Laramide Revolution, commencing in the late Cretaceous and continuing into the Tertiary, seized eastern Nevada and western and central Utah. It was a period of folding and thrusting; of the widespread ejection of great thicknesses of pyroclastics; of the accumulation of shales and marlstones in large lacustrine basins; and, like the Nevadan Orogeny, of the intrustion of numerous stocks. This revolution continued into late Miocene or early Pliocene, a conclusion substantiated by absolute ages of 13 m.y. from a granitic intrusive in western Utah's Sheeprock Range (37), 12 and 16 m.y. from quartz monzonite and granodiorite intrusions in southern Nevada (70), and 10 to 13 m.y. dates from pitchblende at Marysvale (Kerr, this volume).

In a classic sense, the Laramide Revolution represents the fourth great orogenic period since Precambrian time, but, as quoted by Bick (71), Nolan (19) pointed out that there seems to have been "a long epoch of crustal activity that spasmodically affected the Great Basin throughout later Mesozoic and early Tertiary time," and Bick (71) refers to the Nevadan-Laramide events as occurring within a single, long-continuing orogeny.

## Emplacement of Granitic Rocks

In western Nevada, strongly folded Mesozoic sediments cut by granitic rocks suggest a connection between these intrusives and the Sierra Nevada batholith of presumed Jurassic age. In contrast, in western Utah and eastern Nevada granitic intrusives are largely Tertiary. Ferguson (20) pondered whether "the locus

---

* Seas, of course, reoccupied the geosyncline, but only to the east of the area; there great thicknesses of Cretaceous sediments, largely shales, were deposited.

of intrusion moved gradually eastward from the Sierra Nevada to the Rocky Mountain region"; whether "areas of granitic rocks, intermediate in position between the Sierra Nevada batholyths and the Tertiary batholyths to the east, are also intermediate in time"; or "whether there may have been two distinct and sharply separated episodes of granitic intrusions." He preferred the latter, a preference partially substantiated by Schilling (70), who, on the basis of a compilation of isotopic age determinations, has concluded that "intrusive activity in the Great Basin is not distributed randomly in time"; that "more intrusives were emplaced at certain times than at others"; and that "these maximas (sic) of igneous activity appear to have moved . . . in pulses or waves."

## Basin and Range Orogeny

In the late Tertiary a fifth mountain-building period affected the entire area. Still in progress, it has split the surface without regard to pre-existing structures into northerly trending fault-blocks. Raising some, often twisting or tilting them, it has formed isolated ranges that rise above down-dropped blocks now buried beneath thousands of feet of alluvium. Vertical displacements in excess of two miles are not uncommon. The bed-rock floor of the intermontane valleys, in some cases, has been depressed beneath sea level; valleys are choked with as much as 7000, possibly even 10,000 feet, of sediment (65), and individual ranges rise from a few thousand feet to a mile or more above surrounding plains.

## GENETIC RELATIONS OF GREAT BASIN ORE DEPOSITS

### Associated Intrusives

Most Great Basin ore deposits appear to be genetically related to intrusive porphyritic igneous rocks, especially to aphanitic porphyry. Stringham (29) observed that 86 per cent of 107 mineral districts within the Basin and Range Province are spatially associated with porphyritic intrusive rocks; 79 per cent with aphanitic porphyries. He also observed that base-metal and precious-metal deposits showed a preference relationship with porphyritic types; tungsten, with granitoid types. Unfortunately, not all intrusive porphyritic rocks contain ore deposits. Within the Basin and Range are many barren intrusives, and

Stringham (33) observed from an analysis of 207 mining districts in the western United States, 39 of them being from the area of the present study, that "Productive acid porphyry bodies are generally of fair size," as contrasted to small dikes, "and seem to cross-cut enclosing rock structures with very little or no disturbance due to intrusive action," that "barren porphyries may be of various sizes but on the whole have a tendency either to be concordant with wall rock structures, or, where cross-cutting relations exist, show obvious intrusive structural effects such as brecciation, folding, or up arching." Of the 39 mining districts within the area of the present report, 90 per cent are associated with large masses of intrusive porphyry that show little or no disturbance associated with intrusive action.

### Age of Intrusions

J. A. Whelan has made an extensive study of age determinations on rocks associated with Utah ore deposits. He has reported (78) a possible connection between copper mineralization and Miocene or older intrusives; and between beryllium mineralization and Pliocene or younger intrusives. Schilling's (70) study of Nevada intrusives indicated no relationship between age of intrusive and the following types of deposits: tungsten contact-metasomatic deposits, lead-zinc-silver limestone replacement bodies, copper contact deposits, lead-silver veins, or porphyry copper deposits. In Nevada (70) "no more ore can be expected around an intrusive body of one particular age than around those of any other age" and those "periods of maximum igneous activity, those having the largest number of intrusive bodies, are the periods of maximum ore genesis."

### Host Rocks

Exposed sedimentary rocks in western Utah and eastern Nevada are predominantly limestones and shales, in western Nevada largely clastics. These differences are reflected in types of ore deposits. Thus, limestone replacements, absent in western Nevada, predominate in eastern Nevada and western Utah, whereas vein types, common to both areas, are the prevalent type in western and west-central Nevada.

Paleozoic carbonates are noted for the replacement deposits of Bingham Canyon, Park City, Tintic, Pioche, and Eureka. Paleozoic clastic members contain the pyritic-chalcopyrite deposits of Mountain City, the disseminated gold deposits of Lynn Creek (Carlin)

and Gold Acres, and numerous important veins, among them Delamar.

Mesozoic carbonates are the principal host at Iron Springs in southern Utah and contain replacement bodies at Park City. In contrast, Mesozoic clastic members contain only one important deposit, Silver Reef, and no major deposits.

Cenozoic rocks, largely volcanics, contain the bonanza Tertiary vein deposits of west central Nevada, the uranium deposits of Marysvale, and high-grade disseminated beryllium deposits and the Yellow Chief uranium occurrence—both in the vicinity of Spor Mountain in western Utah.

### Structure

There seems to be no relation between ore deposits and any particular type of structure. Although windows within postulated major thrusts have been the object of recent intensive exploration activity, more than an overthrust is involved in the production of ore bodies. The area of this report contains no recognized overthrust that post dates an intrusive stock, and the role of overthrusts in ore deposition seems to be more indirect than direct. In places, they have produced a breccia porosity that localized invading solutions, as at Goodsprings, Nevada, where ore bodies are in bedded breccias related to flat thrust faults (7), and at Eureka, Nevada (Nolan and Hunt, this volume), where brecciated zones hundreds of feet above a major thrust are an ore host. Thrusts have disrupted potential ore hosts, as at Eureka, Nevada, where folded pre-Laramide thrusts have determined the distribution of favorable rock formations (36), and apparently have prepared them for replacement; and thrusts may conceal ore zones by over-riding favorable horizons with unfavorable rocks. There also are important examples of local direct control, as at Bingham Canyon where segments of the North Fault have carried ore bodies (Rubright and Hart, this volume).

Whereas, there is general consensus as to the importance of thrust faults in the role of ore deposition, there is no consensus as to the magnitude of the thrust faults within the Basin. There is no question as to the recognition of major thrusts. At Bingham Canyon, movements of over 5000 feet are thoroughly documented (Rubright and Hart, this volume). But suggested movements of 90 miles (63), a figure commonly invoked in the Roberts Mountains thrust zone, bring a counter claim that the same evidence might be explained by more modest movements associated with ir-regular paleogeographic outlines and rapidly changing sedimentary facies. As emphasized, evidence consists of a few exposures in a vast informational vacuum.

## ORE DEPOSITS

### Distribution of Production

The Great Basin's boundaries are indefinite, and within the area of this discussion southern Nevada and southwestern Utah lie within the drainage of the Colorado River. Therefore, the following figures dealing with the per cent of surface area occupied by various rock types relate not to the Basin proper but to all of Nevada east of the 117°30′ meridian, and to all of Utah west of a line drawn arbitrarily through Park City, Marysvale, Cedar City, and St. George, Utah (Figure 1). Within this area, the surface is occupied by the following rock types in the following proportions:

| | |
|---|---|
| Alluvium | 47% |
| Tertiary volcanics | 22% |
| Intrusives | 1% |
| Post Precambrian sedimentary rocks | 27% |
| Precambrian exposures | 3% |
| | 100% |

The subject of this paper is confined largely to the 50 per cent occupied by intrusives, post-Precambrian sedimentary rocks, and Tertiary volcanic cover (Figure 7). Within these two areas are some 244 mining districts, of which 76 per cent lie in post-Precambrian intrusive or sedimentary rocks and 20 per cent in Tertiary volcanics. Their estimated production surpasses 6600 million dollars, of which 69 per cent is from open pit and underground mines at Bingham Canyon, 29 per cent from the combined production of Ely, Park City, Pioche, Tintic, Tonopah, Ophir-Rush Valley, and Goldfield, and 2 per cent from Eureka, Cottonwoods, Mercur, Round Mountain, Manhattan, Cortez, and Mountain City. In the remaining 1 per cent are some 41 districts with a production of less than 10 million dollars each.

### Gold Statistics

The Great Basin has yielded over 33,400,000 ounces of gold, 68 per cent as a by-product from base-metal operations, 25 per cent from strictly gold producers, and 7 per cent from operations that produced gold and silver.

Seventy per cent of the gold from precious metal mines has come from Tertiary volcanics. In contrast, there is not a single producer from Tertiary volcanics among the by-product operations. Fifty-eight per cent of all by-product gold has come from the two large "porphyry copper" camps: Bingham Canyon, Utah and Ely, Nevada, where production is derived from ores disseminated in porphyry intrusives and associated sediments and from replacement deposits in adjacent limestone. Another 40 per cent of the by-product gold has come from other limestone replacement camps.

It is interesting that the 62 per cent of the gold-producing mines yielded only 32 per cent of the metal. In contrast, the 38 per cent, which are by-product operations, yielded 68 per cent of the total gold production. These figures are summarized in Table I.

The important gold producers, except for Goldfield, lie in the eastern Great Basin within carbonate rocks. This same geographic-geologic relationship holds true when districts are compared from a standpoint of total monetary yield. Excluding Virginia City, outside the area of this discussion, only two camps producing in excess of $100,000,000 (dollars received at time of ore production) are in Tertiary volcanic rocks in the western Great Basin. These are Tonopah, a silver-gold producer, and Goldfield, a gold producer. In contrast, six: Bingham Canyon, Ely, Park City, Tintic, Pioche and Iron Springs—and probably a seventh, Eureka, Nevada, lie in carbonate rocks in the eastern part of the Basin. Of these seven, six are base metal producers; one is a producer of iron ores.

## GREAT BASIN DISTRICTS DESCRIBED IN THIS VOLUME

Of the 11 districts described in detail, seven—Bingham Canyon, Park City, Tintic, East Tintic, Iron Springs, Pioche, and Eureka, Nevada—lie in miogeosynclinal host rocks. Ignoring the porphyry copper ores at Bingham, major production in all seven has been mined from limestone-replacement ore bodies of many types: among them mantos, chimneys, and bedding replacement along fissures. In some camps, replacement has occurred throughout great lengths of the geologic column, but, except at Park City where depth conditions are insufficiently known, the first important limestone horizon encountered by

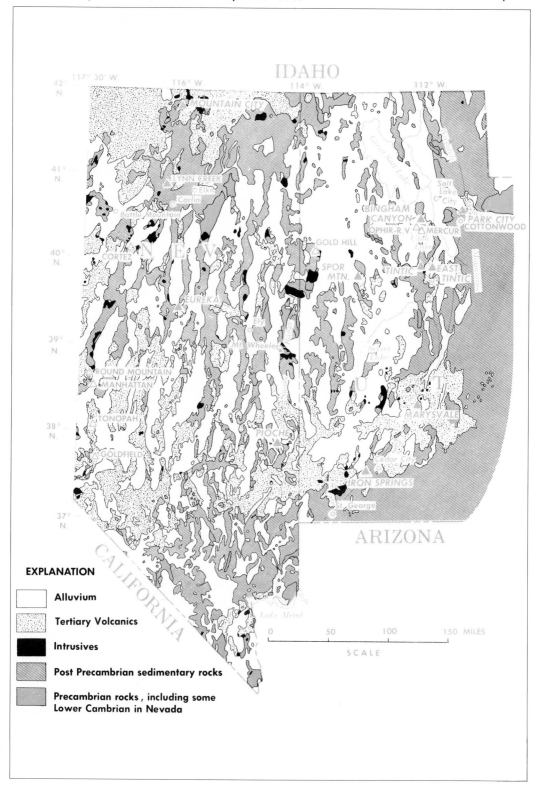

Fɪɢ. 7. *Generalized Geology: western Utah and eastern Nevada.*

TABLE I.   *Precious Metal Production from the Great Basin Segregated According to Type of Production and Type of Host Rock*

| Precious Metal Producers | Mines | | Gold | |
|---|---|---|---|---|
| | No. | % | Oz. | % |
| *Gold Producers* | | | | |
| Tertiary Volcanics | 10 | 20% | 5,186,165 | 15.5% |
| Intrusives, Schist, Gneiss | 3 | 6% | 407,541 | 1.2% |
| Sedimentary Rocks: Replacements | 2 | 4% | 1,601,764 | 4.8% |
| Sedimentary Rocks: Veins | 5 | 10% | 816,700 | 2.5% |
| Sedimentary Rocks: Disseminations & Breccias | 4 | 8% | 332,700 | 1.0% |
| | 24 | 48% | 8,344,870 | 25.0% |
| *Gold and Silver Producers* | | | | |
| Tertiary Volcanics | 4 | 8% | 2,269,560 | 6.8% |
| Intrusives | 1 | 2% | 10,000 | |
| Sedimentary Rocks: Replacements | 2 | 4% | 52,000 | .2% |
| | 7 | 14% | 2,331,560 | 7.0% |
| *By-Product* | | | | |
| Tertiary Volcanics | (none) | | | |
| Intrusives | (none) | | | |
| Intrusives & Sediments | 2 | 4% | 13,264,442 | 39.7% |
| Sedimentary Rocks: Replacements | 12 | 24% | 9,227,335 | 27.6% |
| Sedimentary Rocks: Veins | 5 | 10% | 237,400 | .7% |
| | 19 | 38% | 22,729,177 | 68.0% |
| Total | 50 | 100% | 33,405,607 | 100.0% |

invading mineralizers has been abundantly replaced. At Pioche and East Tintic, these horizons were the oldest known Paleozoic limestones—interbeds in the lower Middle Cambrian Pioche and Ophir Shales; at Eureka, Nevada, the Middle Cambrian El Dorado Dolomite; at Bingham Canyon, the calcareous interbeds in Pennsylvanian quartzites; and, at Iron Springs, the Homestake limestone member of the Jurassic Carmel Formation. At Park City, nearly 50 per cent of the camp's production has come from the Jenney horizon of the Pennsylvanian-Permian Park City Formation, but great tonnages have been produced from ore shoots along various fissures, notably the Crescent, usually where the fissures cut carbonate rocks, but also where they have intersected the Weber Sandstone in some areas and Tertiary volcanics in others.

The intermixture of favorable limestones with quartzites at Bingham Canyon, their contact with a calcareous quartzite at Park City, and their interbedding with shales at Pioche and East Tintic suggest ore-control importance for limestone-clastic interfaces. It may be merely interesting that at Pioche, East Tintic, Eureka, and Park City the great ore horizons overlie thousands of feet of quartzites, whereas at Bingham Canyon they are interbedded with them. But it may be significant that important production at Pioche and East Tintic has been mined from basal Cambrian quartzites. If these basal Cambrian horizons have been mineralized, what then is the possibility that somewhere carbonate horizons of the underlying Precambrian rocks might yet prove to be important hosts? Seemingly, these Precambrian carbonates should be tested where they might overlie an area into which mineralizers had been introduced.

Eureka, Nevada, and East Tintic are each involved with the complexities of thrust faulting. At East Tintic favorable horizons have been overridden and concealed. At both sites, stresses attendant to these thrusts have been major factors in the preparation of ground that localized important ore deposits. At Tintic, the opposite extreme, structural controls for many of the replacement ore bodies either are locally obscure or are not apparent.

In all seven districts, mineralization is attributed to Tertiary intrusions. Yet in few of the districts has there been sufficient erosion or underground probing to expose the complete pattern of intrusives or to trace the paths of the mineralizers that are judged to have been dependent upon them.

Of these seven districts, Bingham Canyon is unique for its almost classic example of metallic zoning. Bingham Canyon, richly mineralized throughout an area 3.5 miles in diameter, has been mined down dip in excess of one mile. In a central core, occupying an area of about 0.4 square mile, mineralization— dominately pyrite with chalcopyrite, bornite, and molybdenite—occurs along fractures and as disseminated grains within a silicified, sericitized, biotitized, and kaolinitized granite porphyry. Outward from this disseminated mineralization and adjacent to the porphyry is a zone of substantial width containing massive chalcopyrite-bearing pyrite bodies that have replaced silicated limestone horizons within steeply dipping calcareous quartzites. In the vicinity of these bodies, the quartzites and silicated limestones contain disseminated pyrite and chalcopyrite. As this zone rises along the porphyry contact, the massive sulfides change from pyrite-chalcopyrite bodies to silver-bearing galena-sphalerite mixtures with minor copper, and on the periphery, lead-zinc ores with a higher silver content. Locally the zone widens and the silicated limestones give way to banded silicates which, in turn, give way to marble grading into limestone. Primary sulfides within this surrounding envelope, irregularly concentrated in veins and replacement deposits, occur on three sides of the porphyry intrusive but are absent to the north. However, base metals are not homogeneously distributed, ores on the east side of the intrusive being predominantly lead-zinc types and on the west, rich in copper.

At Tintic also, there is a zoning pattern. For a mile north of the Silver City stock, copper and gold ores predominate. One to two miles further on they are succeeded by lead-silver ores. Within these zones are alternations to the pattern. Vertically, pattern is nonexistent. At Eureka, Nevada, and Park City patterns are either absent or, at best, anomalous when related to exposed stocks.

Iron Springs has clarified the importance of laccoliths, and of all the districts has furnished the most penetrating insight to a probable process of ore genesis. Here there is strong evidence that these replacement deposits are consequent upon metals "sweated out" of the crystal lattice of iron-rich silicate minerals, under deuteric attack, after emplacement of the laccoliths.

As is true of many limestone camps, deep developments at Tintic, East Tintic, Park City, and Eureka, Nevada are plagued with water problems. The problem at East Tintic is compounded by corrosive thermal waters similar in composition to hot springs in western Utah. Temperatures of 147°F have been measured.

The discovery of gold at Lynn Creek in lower Paleozoic siltstones and calcareous shales, exposed in a window surrounded by overthrust sediments, has drawn major attention to this type of occurrence. The gold is widely distributed in rocks that are not obviously altered and is too fine to be recovered by panning. Here Newmont's open-pit Carlin gold project, based on an ore body with an estimated reserve of 11 million tons of ore that average 0.32 ounces of gold per ton (66) dramatizes the United States' first significant gold discovery in 25 years. Fault zones attributed to the Roberts Mountains thrust (57, p. 7) are exposed in the walls of the open pit, but neither their genetic nor geometric relation to the ore body has been detailed. As described by McQuiston (66), the gold occurs in submicroscopic discrete grains (the largest only 7 by 2 microns); has an average fineness in excess of 900 parts per thousand; and is disseminated irregularly throughout host rocks that are minutely fractured and in places silicified. Sulfides are seldom seen, but pyrite, sphalerite, galena, chalcopyrite, cinnabar, native arsenic, and barite have been identified In its lack of visible gold it resembles the abandoned camps of Mercur and Gold Acres.

Of the 11 districts, only Mountain City lies in eugeosynclinal host rocks. Its history, intimately involved with the structural and orogenic complexities of the Great Basin, is the least understood. Its replacement ore bodies, intimate simple mixtures of quartz, pyrite, and chalcopyrite—unique in western Utah and eastern and central Nevada—are restricted to an intensely deformed stratigraphic horizon of black to grey Ordovician phyllites and minor quartzites in the upper plate of the Roberts Mountains thrust. At Mountain City, there are no recognizable paths along which the mineralizers might have entered. Within the general region a Cretaceous quartz monzonite intrusion occurs; a few narrow altered porphyry dikes of unknown age are in the mine area; and in the associated sediments are several thick series of andesitic to basaltic sills, flows, and pillow lavas—converted to greenstones—

of Ordovician and post-Mississippian Paleozoic ages. Speculation centers around the possibility that the ores are related to Pennsylvanian volcanism. If they are, areas of Paleozoic-early Mesozoic volcanism within Nevada's eugeosynclinal sequences are speculative prospecting targets. Although Mountain City lies in the upper plate of the Roberts Mountains thrust, there are no data as to the age or magnitude of thrusting. The district is best known for its exceptionally rich chalcocite ores.

Pioche has produced over $100,000,000; 102,707 ounces of gold, 17,116,451 ounces of silver; 2,930 short tons of copper; 142,085 short tons of lead; and 296,108 short tons of zinc.* Discovered in 1868 or 1869, its initial production yielded a bonanza of silver ores from breccia veins within basal Cambrian Prospect Mountain Quartzite. Healed with silver minerals, largely chlorides with a little lead, these veins by 1872 were second only to the Comstock in their yield of silver (8).

In addition to veins in quartzite, the district has yielded silver-lead-zinc ores from a shattered and mineralized, highly sericitized rhyolite porphyry dike, and silver-lead-zinc-copper ores from replacement bodies in Middle and Upper Cambrian limestones. However, it was not until 1932 that the large tonnage silver-lead-zinc replacement bodies, deposited in limestone horizons interbedded within the Lower Cambrian Pioche Shale, were discovered and brought into production. Pioche's Cambrian carbonates are productive through a vertical range of 5500 feet that extends from the Lower through the Upper Cambrian, but, as pointed out by Westgate and Knopf (8), there is no one mine in which replacement ores occur throughout this range. The maximum vertical extent of mineralization at any single property is 1500 feet.

In addition to rich silver ores, the district produces iron-manganese ores with a low content of silver, lead, and zinc from the same limestone horizons that yield the silver-lead ores; and, as a mineralogic curiosity, tabular crystals of wolframite in a quartz gangue associated, in some occurrences, with galena, sphalerite, and pyrite.

The district's highly productive interbedded limestones within the Lower Cambrian Pioche Shale are similar in setting to those at Ophir. The productive veins in basal Cambrian quartzite and its productive mineralized dike are similar to the gold occurrences at Delamar. The district differs from Delamar in that the

* J. H. Schilling, personal communication.

latter has lacked production from interbedded limestones in the overlying Pioche Shale.

From the neighboring Jack Rabbit district, largely from the Bristol Silver mine, there has been a production of 5628 ounces of gold; 5,291,623 ounces of silver; 11,126 short tons of copper; 18,346 short tons lead; and 1958 short tons of zinc* from replacement bodies, largely chimneys, in Cambrian limestone.

Marysvale is the only district of the 11 described that lies in Tertiary volcanic terrain. It has yielded minor quantities of precious and base metal ores, but significant amounts of bonanza-grade uranium have been mined from veins and, to a lesser extent, from disseminations.

Spor Mountain, the last of the group, has yet to produce its first ton of ore. It contains the world's largest recognized reserve of what is hoped to be a space-age cinderella, beryllium. Confined to replacements of Tertiary carbonate gravels within a specific tuff horizon, Spor Mountain deposits indicate the importance of paleotopography as an ore control and point to a specific rhyolite as the ore source.

Physiographically, Park City and Marysvale are the only districts among the 11 that do not lie within the Basin proper but are within its drainage area immediately to the east. Park City is in the folded structures of the Rocky Mountains. Marysvale is in a transition zone between the High Plateaus to the east and the Great Basin to the west. Furthermore, both districts are associated with Tertiary volcanics, but whether the volcanics are of the same age is undetermined.

## OTHER GREAT BASIN DISTRICTS

### Gold Deposits

The Great Basin, renowned for bonanza-grade precious metal veins, has had virtually no production from this source for 30 years. Only the Mayflower area of the Park City district now produces from such deposits (Figure 8). Current production predominantly is a by-product of lead-zinc-copper producers, principally from the large tonnage low-grade open-pit porphyry copper ores of Bingham Canyon, which now ranks as the second most important gold producer in the United States.

An ore that is mined, but not mined in the first 60 years of mineral production, is an extremely fine-grained low-grade dissemination

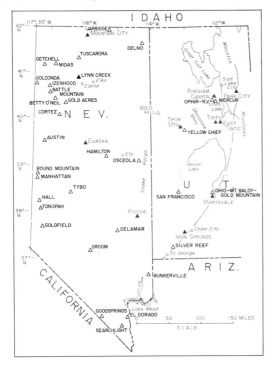

FIG. 8. *Location map: Great Basin Mining Districts in western Utah and eastern Nevada.*

of gold within a sedimentary host in close regional association with an overthrust. The Gold Acres deposit at Tenabo, operating from 1936–1961 and producing 120,000 ounces of gold and 6500 ounces of silver,* was an example. The newly discovered Lynn Creek deposit north of Carlin, which went into operation in 1965, is another. Both are unusual in that gold, not recoverable by panning, is observed under extreme magnification. Finely dispersed gold likewise is typical of the Getchell ores, although visible gold also occurs.

GETCHELL   The Getchell vein, 7000 feet long, averaging 40 feet in width but ranging from a few feet to more than 200 feet, has been mined to depths of 800 feet. It cuts granodiorite, dacite and andesite porphyry dikes, Paleozoic sediments, and Tertiary basalts and rhyolite tuffs. According to Joralemon (23), its ore shoots represent a wholesale non-selective replacement of all sheared rock types by sulfides and gangue minerals. Between ore shoots, replacement has been limited and selective. The most important gold-bearer is a hydrothermal replacement of the wall rock by a porous gumbo of minute quartz crystals in

* J. H. Schilling, personal communication.

a submicroscopic intergrowth of quartz and amorphous carbon that concentrated gold from later gold-bearing solutions. Associated sulfide minerals—pyrite, pyrrhotite, arsenopyrite, and marcasite—occur throughout the vein. The most abundant sulfides, orpiment and realgar, are restricted to areas with substantial gold content. The presence of realgar is a guide to good gold ore. Despite this relationship between arsenic sulfides and high-grade gold, Joralemon reports the only minerals strikingly associated with gold are carbon and magnetite. The vein contains restricted pockets of ilsemannite and stibnite, and Joralemon remarks on the similarity of mineralogy at Getchell with that of the mud deposited at Steamboat Springs. Observing that the ores at Getchell and the mud at Steamboat Springs are post-andesite, he compares them with the gold-realgar ores at Manhattan and Round Mountain—which are post-rhyolite. Joralemon, noting mineralogic and textural features at Getchell suggesting a genetic relation to cinnabar deposits, theorizes that ores at both Getchell and Manhattan "represent a gradation between the earlier gold ores, such as Goldfield, and more recent cinnabar deposits."

The Getchell mine began operation in 1938, became idle in 1945 with the exhaustion of the high-grade shoots, resumed operation on low-grade open-pit material in 1962, and again ceased operation in 1967. Through 1965, it had been credited by Bergendahl (46) with a production in excess of 436,000 ounces of gold. Getchell ore, although not unique, is not typical of Great Basin gold deposits in that it contains virtually no silver.

MANHATTAN   Manhattan, mined for gold both in placer and lode deposits, produced $10,360,000 through 1949 (24, p. 115). Placer mining, begun in 1905, was "beheaded" a year later by the San Francisco fire but was resumed from 1938–1946. Total placer yield has approached $4,600,000. Lode mining, begun in 1918, had produced $5,760,000 through 1946.

Manhattan veins occur in Paleozoic rocks, intrusives into them, and in Tertiary lavas. The most important lodes have been mined from Cambrian limestones, slates, and schists. Shallow veins, reportedly, were the most productive but veins in the White Caps mine, stoped to widths of 5 to 7 feet, have persisted in the Cambrian White Caps Limestone to depths of 1300 feet where heavy ground forced the abandonment of deep mining. As at the Getchell, Manhattan mineralogy is complex and Kral (24, p. 118) records realgar, orpiment, stibnite, pyrite, arsenopyrite, and free gold in a gangue of fine-grained quartz, coarsely crystalline calcite, fluorite, sericite, leverrierite ($Al_2O_3 \cdot 2SiO_2 \cdot 2\frac{1}{2}H_2O$), and adularia. The district is noted for its realgar and stibnite and Kral reports that lenses of both realgar and of pure stibnite, up to 20 feet long by 4 feet wide, apparently were bypassed in operations seeking milling ore (24, p. 124). Ferguson (6, p. 84–86) states that in the White Caps mine cinnabar, reported from above the 200-foot level, in 1923 was found in appreciable quantity on the 980-foot level; that dark jasper-like quartz that forms the best ore contains dustlike specs of carbonaceous matter; and that, although realgar and orpiment are auriferous, visible gold is lacking, even in oxidized ore (6, p. 84). In the Manhattan district, only the White Caps mine has failed to yield free gold. The White Caps, like the Getchell, has a low silver content, its bullion yielding 17 parts of gold to 1 part of silver in contrast to a more usual Manhattan ratio of 2 or 2.5 to 1.

ROUND MOUNTAIN This district, discovered in 1906, is credited with a production of some 540,000 fine ounces of gold, one-third of which may have been from placers (46). Underground mining ceased in 1935, placer mining in 1959. As described by Ferguson (4) free gold, minor pyrite, and rarely realgar occur in a gangue of quartz with accessory adularia and alunite and rarely fluorite, in flatly dipping veins. These, seldom more than a few inches wide but strongly persistent, have been mined to a maximum vertical depth of 350 feet. The fineness of the bullion varied from 574 to 696 parts per thousand.

The presence of realgar suggests a genetic similarity with Getchell and Manhattan; that of alunite, a similarity with Goldfield.

Ferguson believes there have been two periods of vein formation. In the first, the veins received primary gold and auriferous pyrite and are associated with silicification of wall rocks and with sericitization. In the second, the veins received supergene gold and manganese that were derived from oxidation of auriferous pyrite, lack quartz, and are associated with kaolinitic alteration.

According to Kral (24, p. 151), the Nevada Porphyry Gold mines, attempting to develop a large low-grade mine based on pyritized rhyolite, blocked out 13 million tons with a reasonably assured grade of $1.50 in gold; and, after the increase in the price of gold in 1934, milled 6,000 tons of better-grade material and recovered $2.53 a ton. He further reports the

A. O. Smith Company explored and sampled the Round Mountain-Manhattan area from 1935–1936, but no operation ensued.

MERCUR Another important deposit, the mineralogy of which strongly resembles the Getchell vein, is Mercur, Utah. Operated from 1870 to 1950, it yielded approximately $26,000,000 from gold ore with a bullion fineness of 833 parts per thousand.* According to Butler and Heikes (3, p. 382–395), from 1870 to 1913, it operated on $3.50 to $5.72 ore yielding over $19,000,000 from a series of silicified horizons throughout a 300-foot thickness of Mississippian limestone. Stopes ranged from 5 to 27 feet in height, with maximum down-dip development of 1500 feet beneath the outcrop. Although the bulk of the ore came from a few hundred feet beneath the surface, at all horizons there was a rapid change from ore grade to non-ore material. Pyrite, realgar with some orpiment, and cinnabar occurred in a gangue of cherty quartz, calcite, barite, and sericite. Gold was never visible, and, whereas realgar and cinnabar indicated good ore, there was much good ore that resembled waste, and the gold appeared to be associated with carbon rather than with sulfides. Like the Getchell ore, silver was sparsely associated with the gold.

Mercur ores contain thallium. According to unpublished records of the U.S. Bureau of Mines, during 1945 to 1946, sampling of dumps and of sand and slime tailings from the Manning and Mercur mines showed a thallium content that varied from .023 to .059 per cent. The thallium carrier was not identified as the tailings had been roasted to remove the carbon, then finely ground and cyanided.†

GOLDFIELD Goldfield is another unique precious-metal district. Here the gold is associated with alunite and kaolin, as well as with flinty quartz. Discovered in 1903 and exhausted by 1938, it produced over $100,000,000 in gold, old price (22), with minor silver, from flat, easterly dipping veins. These, mined to a vertical depth of 1000 feet, were enclosed in a series of Tertiary andesites and associated volcanics, the most important of which, a prominant dacite, has been interpreted as an intrusive by Ransome but as a flow by Searls. Searls (22) reports that over 70 per cent of the production came from this dacite; that no vein persisted into the pre-Tertiary complex of

---

* U.S. Bureau of Mines, Minerals Yearbook and related publications.

† Carl Schack, personal communication.

Cambrian shales and alaskite; and that the lower levels "were essentially contours of the Tertiary surface."

Discovery was delayed because free gold occurred not in jaspery outcrops but in adjacent earthy mixtures of kaolin, alunite, gypsum, iron oxides, and fragments of porous quartz. Only a small portion of the ore filled fractures, although the deposits are intimately related to fissuring. Instead, the deposits are "irregular masses of altered and mineralized rock traversed by multitudes of small irregularly intersecting fractures, such fracturing passing in many places into thorough brecciation" (2). The brecciated rock may either be loose and associated with interstitial limonite and kaolin or cemented by quartz, limonite, or hematite. These brecciated or altered deposits contain the ore shoots. Although the two are equally irregular and are structurally indistinguishable, they are not coextensive. Moreover, the ore shoots have assay walls, although the change from high grade to waste may occur over short distances.

Unoxidized ore consists of native gold, goldfieldite, pyrite, bismuthinite, arsenical famatinite ($3Cu_2S \cdot Sb_2S_5$), and minor sphalerite within a flinty quartz gangue associated with mixtures of alunite and kaolinite. In depth, the ores contained abundant famatinite, and Searls (22) reports ore-grade figures from the Tertiary, pre-Tertiary contact horizon of

| Oz. | | % |
|-----|-----|-----|
| Au | Ag | Cu |
| 1.07 | 5.58 | 4.21 |
| .31 | | 1.13 |
| 1.35 | 4.41 | 2.79 |

Tolman and Ambrose (11) recognized marcasite, wurtzite, tennantite, sylvanite, hessite, petzite, yellow gold carrying silver, and red gold low in silver. Their studies indicate that alunitization and probably kaolinitization were completed prior to the introduction of the ore minerals.

A minor tin content in the copper-rich ores,[*] although it was not recognized during mining, was reported by the U.S. Geological Survey in February, 1941.

The ore-bearing andesite series is partially overlain by lake beds, and these, in turn, by

[*] Searls gives as a reference: Department of the Interior, Information Service, Feb. 20, 1941.

basalts and rhyolites. Searls (22) states in regard to age of mineralization there can "be no question about the fact that the Goldfield veins were formed, partially oxidized, and attacked by erosion before there was substantial accumulation of sediments or tuffs in the upper Miocene lake."

The source of the ore is not known, but Searls considers "that whatever the original source of the gold in this bonanza camp—perhaps a rhyolite vent or plug or breccia pipe—its distribution was accomplished by the Columbia Mountain fault," the most prominent mineralized feature in the district.

Gold Acres   The Gold Acres open pit gold property somewhat resembles the Lynn Creek deposit. In 1939, Vanderburg (16, p. 39–46) described the principal ore body as brecciated iron-stained, silicified limestone, dipping flatly southwest between a porphyry hanging wall and a blue limestone foot wall. Brecciation is attributed to the Roberts Mountains thrust (46, p. 92). As at Lynn Creek, thrust planes are exposed in the walls of the pit, and the gold, irregularly distributed, is so fine-grained it cannot be recovered by panning.

Gold Acres, operating from 1936 to 1961, produced some 120,000 ounces of gold and 6500 ounces of silver[†] from ore containing as little as $4.00 per ton in gold (16, p. 39).

Placer Gold   At the present time, there is no formal placer mining within the Great Basin, but formerly it was important. Bingham Canyon, prior to the development of its lode mines, may have yielded $1,000,000 in placer gold from 1868 to 1871 (Rubright and Hart, this volume). Osceola is said to have produced $2,000,000 to $5,000,000 in gold prior to 1907, 90 per cent from placers. The source of the Osceola placers appears to have been shattered masses of Cambrian quartzite (5, p. 253).

More recently, Manhattan, Round Mountain, and Battle Mountain have witnessed dredging operations based on bajada-type placers typical of the arid West. Only the dredging operation at Manhattan has proved successful. Between 1938 and 1946 the Manhattan Gold Dredging Company produced $4,596,427 from 21-cent gravel (24, p. 113–131), and Bergendahl (46) credits the district with a production of 206,000 ounces of placer gold.

In 1946, the Manhattan dredge was moved

[†] J. H. Schilling, personal communication.

to Copper Canyon at Battle Mountain, where the source appears to have been mineralized conglomerates (Pennsylvanian Battle Formation) and graywackes (Cambrian Harmony Formation). At Round Mountain, the Round Mountain Gold Dredging Corporation carried out large-scale operations from 1950 to 1959 with a production of 150,000 ounces of placer gold (46). Both operations are considered to have been unprofitable or marginal at best. The source area at Copper Canyon is being developed into an open-pit porphyry copper mine by the Duval Corporation.

GOLD-CARBON-CINNABAR-REALGAR RELATION The association of realgar, cinnabar, carbon, and gold appears to be common in this Utah-Nevada area. Even within the massive sulfides at Bingham Canyon, small showings of stibnite, realgar, orpiment, cinnabar, and pyrite have been discovered from a later sequence of mineralization that cut across the first stage of lead-zinc deposition (Rubright and Hart, this volume). Associated with this later sequence is an erratic higher-grade gold content.*

### Silver-Gold Deposits

In precious metal operations throughout the Great Basin, silver, by weight, usually exceeds gold. These, the "typical" Tertiary veins in volcanic host rocks, rarely produce $10,000,000, but among the group is Tonopah, the larger of two precious metal camps within the area of this paper to have produced over $100,000,000. Lesser districts are Jarbidge, Delamar, Midas, and Tuscarora.

TONOPAH Tonopah, discovered in 1900 and essentially exhausted by 1948, still produces a small yearly tonnage. It has yielded over $150,000,000 from quartz veins in Tertiary tuffs, breccias, trachytes, and rhyolites. Thus (13, p. 70–71):

|  | Ounces | | |
|---|---|---|---|
| Total Tons | Gold | Silver | Total $ |
| (Thru 1932) | | | |
| 8,262,944 | 1,790,961* | 168,525,219* | 148,206,139$ |

\* According to personal communication from John H. Schilling, Nevada Bureau of Mines, production through 1965 had reached 1,861,200 ounces of gold and 174,152,628 ounces of silver.

\* R. D. Rubright, personal communication.

According to Nolan (12) as quoted by Kral (24, p. 169–174), ore structures, replacement veins following faults and minor fractures, frequently are ill-defined and generally have assay walls. On the Mizpah Vein they have been stoped for lengths of 1500 feet. Electrum, argentite, polybasite, and arsenical pyrargyrite, as well as other silver and base metal sulfides, occur in a gangue of quartz, fine-grained pinkish carbonate, barite, and altered wall rock.

JARBIDGE Jarbidge operated from 1910–1949. As described by Granger (26, p. 83–101), it produced over $10,100,000 in gold and silver from some 40 northerly trending quartz veins. These veins, maintaining sharp contacts with country rock, occupy a belt 10 miles long by 4.5 miles wide within rhyolite extrusives, range individually from 1000 feet to 6 to 7 miles long and from 1 to 30 feet wide, and contain inclusions of rhyolite, sedimentary rocks, and seams of gouge or clay. Ore shoots have been mined through lengths of up to 700 feet and to depths of possibly 600 feet. Ore minerals consist of gold, silver, electrum, argentite, cerargyrite, the silver-lead selenide—naumannite $(Ag,Pb)Se$, chalcopyrite, pyrargyrite, and pyrite within a gangue of quartz, with associated adularia and calcite. The gold, fine-grained and rarely seen, averaged about 0.5 ounces per ton. Alteration features, extending back from the veins for as much as 100 feet, consist of an envelope of devitrified and silicified rhyolite within which the feldspars are sericitized. The rhyolite, pyritized, is cut by veinlets of quartz and adularia.

DELAMAR Delamar, unique among precious metal deposits because production has been from veins within basal Cambrian Prospect Mountain Quartzite, operated from 1902 to 1934. Seventy-five per cent was obtained from 1903 to 1909. As described by Callaghan (14), its production through 1934 of

|  | Oz. | | |
|---|---|---|---|
| Total Tons | Gold | Silver | Total $ |
| 806,287 | 190,987 | 333,466 | $4,166,244 |

came from three types of ore shoots:
(1) Bodies of quartzite breccia cemented and partly replaced by cherty quartz and cut

by vugs and veinlets of comb quartz, with sparsely distributed fine-grained sulfide;

(2) Narrow veins of cherty, fine-grained quartz with local concentrations of free gold and sulfides;

(3) Free gold along bedding planes within bedded quartzite.

Most of the tonnage has come from brecciated quartzite healed with cherty quartz. Sulfides, restricted to cherty quartz or to vugs lined with quartz crystals, occur as scattered grains of pyrite, arsenical tetrahedrite, and minor chalcopyrite. Bornite, sphalerite, and galena have been reported. Tellurides occur, but the mineralogy has not been reported (14; 52, p. 32). Bismuth has been recorded from assays.

Sedimentary wall rocks have not been significantly altered, although some calcareous horizons have been silicified. Rhyolite dikes in contrast have been severely altered to sericite and clays.

MIDAS  Midas operated from 1908 to 1942. Country rock is a series of rhyolite and andesite flows cut by rhyolite and andesite dikes. As described by Granger (26, p. 64–72) its production of

|              | Oz.     |           |            |
|--------------|---------|-----------|------------|
| Total Tons   | Gold    | Silver    | Total $    |
| 401,752      | 126,726 | 1,630,268 | $4,137,417 |

came from narrow veins, traceable 1500 to 3500 feet, within rhyolite and along shattered rhyolite-andesite contacts. At the Elko Prince mine, the ore shoot, 15 inches wide by 600 to 800 feet long, topped 270 feet beneath the surface. Elsewhere ore shoots have been less than 200 feet long, and, in the Rex and Gold Crown veins, there were stope widths of 5 to 25 feet. Primary ore minerals are pyrite, gold, and argentite associated with quartz. Alteration, intense in the vicinity of ore bodies, has introduced pyrite, quartz, chlorite, and sericite in the rhyolite; has sericitized feldspars; and has carbonated and chloritized augites in the andesites.

TUSCARORA  Tuscarora operated from 1877 to 1930 and from 1937 to 1939. As described by Granger (26, p. 150–166), its recorded production of 424,617 tons of gold and silver ore worth $10,711,872 came from a series of altered pyroclastics and volcanics. Ore deposits were of two types: gold-bearing stockworks with a gold to silver ratio of 1 to 4 or 1

to 5 and narrow low-tonnage bonanza silver lodes with a gold to silver ratio of 1 to 150. There may have been some 18 silver veins. Ore minerals were silver, horn silver, argentite, stephanite, proustite, and pyrargyrite associated with pyrite, enargite, arsenopyrite, bornite, chalcopyrite, sphalerite, and galena in a gangue of quartz, calcite, and altered wall rock.

Most of the mines that were essentially silver producers were exhausted prior to 1900, and none is operating. Examples include Austin, Hamilton, Cortez, and Betty O'Neil. Austin, Hamilton, and Cortez have yielded over $10,000,000 each.

## Silver Deposits

AUSTIN  Discovered in 1862 and largely exhausted by 1903, Austin produced high-grade silver ores from narrow quartz fissures within a quartz monzonite. The veins varied from a few inches wide to as much as three feet, averaging about 15 inches. Vanderburg (16, p. 68–78) estimates they have yielded over $26,000,000 with 75 per cent of the production prior to 1887. He states that the surface ores, down to the water table at a depth of 75 feet, apparently assayed $100 to $400 a ton with silver chloride the principal ore mineral; that below water table the ore shoots carried ruby silver (both dark and light), polybasite, enargite, stephanite, silver glance, galena, sphalerite, copper glance, marcasite, arsenopyrite, tetrahedrite, and chalcocite.

HAMILTON  The near-surface bonanza silver ores of Mount Hamilton, discovered in 1867, were confined to Treasure Hill. They were soon exhausted, yet Lincoln (5) credits them with a production of possibly $22,000,000 between 1868 to 1887. According to Humphrey (30), Treasure Hill ores, almost entirely cerargyrite, were mined from a quartz-and-calcite-healed breccia zone in the Devonian Nevada Limestone. This zone, less than one mile long by half a mile wide, extends from the surface to a depth of less than 50 feet. Replacement bodies developed where brecciation was less intense, and manto-like stopes, 100 to 150 feet wide, have been mined for lengths of 1700 feet. The ore formed directly beneath a shale capping and does not persist in depth. Ore has not been found in the Mississippian Joana Limestone, although the latter is replaced by jasperoid.

In addition to the exhausted bonanza ores, the Hamilton district has yielded rich lead-silver ores from small mines on veins and re-placement bodies located in the lead-silver belt to the west of Treasure Hill. Lincoln (5) records a lead-silver belt production of 145,000 tons worth $6,000,000 up to 1909, most of which must have been produced prior to 1902 because he quotes U.S. Geological Survey figures of 17,638 tons worth $1,003,947 for the period from 1902 to 1921. Since then there has been a desultory produc-tion. Schilling (personal communication) cred-its the Hamilton district with 4718 ounces of gold; 5,932,495 ounces of silver; 196.46 short tons of copper; 103,165.2 short tons of lead; and 165.45 short tons of zinc.

CORTEZ   Cortez is a limestone camp in which the replacement ores occur along bedding planes, in limestone and quartzite in and near a dike fissure, and in fissures parallel to dikes. The last have been the most productive. Ore bodies have been mined through a vertical range of 1600 feet beneath the surface. The largest, 630 feet long, 5 to 15 feet wide, had a vertical extent of 200 feet. According to Roberts (57, p. 5) most of the ore has oc-curred in Cambrian Hamburg Dolomite just below the Eureka Quartzite.

Much of the production was prior to 1900, but the district was a steady producer until 1930 and an intermittent producer until 1952.* Vanderburg (15, p. 22–28) credits Cortez with production of $13,000,000 through 1936. The new U.S.G.S. gold discovery is in the Cortez area.

BETTY O'NEIL   The Betty O'Neil, a vein-like replacement deposit in limestone and slate, was discovered in 1880, flooded in 1882, reopened in 1920, and ceased formal operations in 1941. As described by Vanderburg (16, p. 59–63), the mine, by 1936, had produced over $2,900,000 from some 4,200,000 ounces of silver. Developed to a depth of 1600 feet along dip, its ore, chiefly silver with minor copper and lead, contained tetrahedrite, stephanite, argentite, polybasite, galena, pyrite, and sphal-erite in a mixture of silicified limestone and slate that has been fractured and recemented by calcite and quartz.

## OTHER MINOR DISTRICTS

### Limestone Replacement Camps

In addition to limestone camps described individually in this volume, others are worthy

\* J. H. Schilling, personal communication.

of comment. Among them, in the eastern part of the Great Basin, are Ophir and Rush Valley; San Francisco and its satellite districts near Milford; Groom; and Gold Hill. In the west is Goodsprings. These, and other minor camps, are plotted in Figure 7.

OPHIR AND RUSH VALLEY   Ophir and Rush Valley, herein considered as a single district, was discovered in 1864 or 1865 (3, p. 362–382). Like Bingham Canyon and Mercur, it lies in the Oquirrh Mountains and yields rich silver-lead ores, with zinc and copper, from replacement deposits in Paleozoic lime-stones. Unlike Bingham Canyon, the produc-tion of which comes from Pennsylvanian lime-stones and, unlike Mercur, where the ore comes from Mississippian limestones, much of Ophir production comes from limestones interbedded with the Cambrian Ophir Shale. In this respect, it resembles Pioche and East Tintic. The district, through 1964, has pro-duced over $95,000,000; 100,000 ounces of gold; 50,000,000 ounces of silver; 23,000 short tons of copper; 360,000 short tons of lead; and 18,000 short tons of zinc.* As a mineral-ogical curiosity, it has produced rich concen-trations of scheelite associated with silver-bear-ing galena ores and wolframite as a late replacement of the scheelite (27).

SAN FRANCISCO AREA   Including the nearby Preuse, Star, North Star, Beaver Lake and Rocky districts, San Francisco has produced over $40,000,000; 38,000 ounces of gold; 19,000,000 ounces of silver; 26,000 short tons of copper; 200,000 short tons of lead; and 21,000 short tons of zinc.* These were mined from replacement bodies in Permian, Car-boniferous, and Cambrian limestones; from contact metamorphic deposits; and from brec-cia pipes that occur both in quartz monzonite intrusives and Tertiary ignimbrites. Ore bodies, except in the Cactus and the Horn Silver mines, have been small to medium in size. In the Cactus, ore occurs in a copper-bearing breccia pipe. In the Horn Silver, the lead-sil-ver-copper chimney, which may be a breccia pipe, lies between a Cambrian limestone foot-wall and a hanging wall of Tertiary volcanics. Ore, first discovered in the 1860's, has been produced intermittently to the present time; the period of greatest production extended from 1905 into the 1920's.

\* U.S. Bureau of Mines, Minerals Yearbook and related publications.

The San Francisco-Milford area is typical of those districts within the Great Basin that are intriguing for their speculative prospecting possibilites. With its numerous ore shoots, its variety of mineral deposits and their host rocks, and the large percentage of its surface that consists of alluvium or volcanics that conceal pediments, it resembles the great Pima-Mission-Esperanza area of Arizona.

GROOM   The Groom district, a small producer of argentiferous galena ores, is plagued with fault problems. As at Pioche, production has come from limestones interbedded with the Lower Cambrian Pioche Shale (21).

GOLD HILL   Gold Hill, reportedly discovered in 1858 (3, p. 475–484), has operated intermittently. Its production record of $2,800,000; 25,000 ounces of gold; 832,000 ounces of silver; 1,700 short tons of copper* fails to convey the uniqueness of the district. Whereas most of the production has come from limestone replacement bodies mined for the above metals, these are rich in arsenical minerals, and the district has produced 36,227 short tons of arsenic (73). It is the principal potential source for additional arsenic supplies within the Great Basin.

The district also produces from contact metamorphic deposits, from pegmatites, and from quartz veins within quartz monzonite. It has produced scheelite from these deposits, and, in 1964, Griffitts (53) believed it contained the second largest known deposit of beryllium. The beryllium occurs as fine-grained bertrandite within quartz-calcite veins that cut an epidotized quartz monzonite.

GOODSPRINGS   Goodsprings is the only district in the Great Basin within which ores have been mainly those of zinc.

Discovered in 1856, it has produced $31,660,221; 90,508 ounces of gold; 2,085,051 ounces of silver; 2463 short tons of copper; 46,349 short tons of lead; 107,082 short tons of zinc; 20 short tons of molybdenum; 5.5 short tons of cobalt; 4.1 short tons of vanadium; and 506 pounds of platinum.† According to Hewett (7) the most productive bodies have been replacements in dolomitized Lower Mississippian limestone. Although 95 per cent of the lead-zinc ores have been mined from a zone 600 feet thick, the ores occur throughout 4000 feet of Devonian, Mississippian, and Lower Pennsylvanian limestone.

Bedded premineral breccias contain the largest ore bodies, and the district is noted for its thrust faults. Hewett believes the major thrusts have determined the local distribution of intrusive rocks; that centers of distribution of these intrusives have controlled the general location of the ore bodies. He also notes most of the ore deposits occur in breccias along thrust faults, but the major thrust faults seldom contain ore deposits. Callahan (72) speculates these premineral breccias might be collapse features related to karst development in Mississippian time, and Mills (74) cites the presence of unconformities within the Paleozoic sequence and postulates these may have guided invading mineralizers.

Hewett observed a close coincidence between dolomitized limestone and ore deposits, as well as large areas of widespread dolomitization within which no ore bodies are known. He concluded the coincidence of ore deposits and dolomitization is of a general nature rather than a precise relationship.

In addition to lead-zinc-copper ores from replacement bodies, the district has produced gold and silver from highly altered intrusive porphyries. Cobalt has been produced. In four localities Hewett reports that wherever copper is present cobalt likewise occurs.

### Silver-Base-Metal Vein Deposits

The Great Basin, noted for its precious metal Tertiary vein deposits and its yield of base metals from limestone replacement deposits, has produced few important veins that yielded either base metals or silver and base metals combined. Important exceptions occur at Bingham Canyon and at Park City. Other exceptions are Tybo, and the Ohio-Mount Baldy-Gold Mountain District.

TYBO   Tybo, an important producer from 1865–1879 and from 1929–1937, has yielded $9,789,281 from 596,040 tons of silver-lead ore (24, p. 189–195). Country rock, a series of Cambro-Ordovician-Silurian limestones and shales with minor quartzite and chert, is cut by porphyritic quartz-latite dikes and is overlain by Tertiary lavas. The important ore shoots carry pyrite, about equal amounts of sphalerite and argentiferous galena, minor chalcopyrite, pyrrhotite, and arsenopyrite, all within a gangue of fine-grained quartz and coarse calcite. According to Ferguson (9 p. 8), these ore shoots have replaced quartz latite

---

* U.S. Bureau of Mines, Minerals Yearbook and related publications.

† J. H. Schilling, personal communication.

porphyry dikes within a prominent fault. Kral states the fault-vein has been explored 2500 feet horizontally and 1310 feet vertically (24).

OHIO, MOUNT BALDY, AND GOLD MOUNTAIN
The discovery of small gold-bearing ore shoots within strong persistent quartz-carbonate veins in Bullion Canyon in the late 1860's led to the establishment of three gold producing districts: Ohio, Mount Baldy, and Gold Mountain. Production has been derived from two types of deposits: veins in Tertiary Bullion Canyon volcanics, which extend downward into underlying pre-Tertiary sediments, and from manto replacements of Permian Kaibab Limestone. The Ohio and Mount Baldy production was largely from gold-silver-lead replacements in the Kaibab Limestone; Gold Mountain production (the ores were virtually exhausted by 1917) was from veins in the volcanics. Vein ores were of two types (3, p. 536–558): gold-silver ores, frequently with a little pyrite and tetrahedrite, and lead-copper-zinc ores with gold and silver in which the ore minerals were pyrite, chalcopyrite, tetrahedrite, galena, and sphalerite in a gangue of quartz and iron-manganese calcite. The lead-copper-zinc vein deposits were closely associated with sedimentary rocks, however, Butler and Heikes (3) were undecided whether this represented a mineralogic change with depth or a wall rock control. Estimated production through 1923, according to Callaghan and Parker (35), amounted to

| District | $ | Ounces | |
|---|---|---|---|
| | | Gold | Silver |
| Gold Mountain | $3,241,583 | 143,561 | 460,228 |
| Ohio and Mount Baldy | $2,000,000 | | |

Since 1923, production from the precious metal veins has been insignificant. However, from 1960 to 1964* replacement deposits in the Mount Baldy district have produced 1086 ounces of gold; 63,022 ounces of silver; 18.4 short tons of copper; 396.5 short tons of lead; and 833.8 short tons of zinc.

In addition to base metal ores from both replacement and vein deposits, the Ohio-Mount Baldy district has yielded other minerals: mercury selenide from small replacements in lime-

stone (3, p. 551); large deposits of alunite (3, p. 546); and, since 1950, it has been one of the nation's important sources of uranium (Kerr, P. F., this Volume). These uranium ores occur as veins in granitic intrusives and as fracture fillings and disseminations in volcanic rocks; whereas, in the western United States, most uranium ores occur as disseminations in sedimentary host rocks. Scientifically, the district is of interest because of the K/Ar dates of 10 to 13 m.y. obtained from Marysvale uraninite and 12 m.y. from sericite in the Deer Trail mine tunnel (Kerr, P. F., this volume).

## "Sandstone" Deposits

Sandstone-type deposits, although well-known for copper, vanadium and uranium content, seldom have proved economically profitable when exploited for copper. In contrast, they have become renowned for their uranium production. Less well-known are "sandstone" deposits that yield silver.

Within the Great Basin, sandstone-type vanadium or vanadium-copper deposits of commercial importance are lacking; sandstone-type uranium deposits are of secondary importance. However, two "sandstone" deposits are of interest: Silver Reef, for its past silver production, and the Yellow Chief, an uranium producer, for its location in Miocene-Pliocene tuffaceous sandstones and its areal proximity to the fluorspar and the beryllium deposits of Spor Mountain.

SILVER REEF  Silver Reef, with a recorded production of over $9,700,000,* is the only Great Basin area yielding a significant return from continental rocks. Broadly similar to the "sandstone"-type deposits in the Mesozoic sandstones of the Colorado Plateau, its important commercial mineralization has been confined to silver minerals, largely silver chloride, rather than to uranium, vanadium, and copper salts. Moreover mineralization at Silver Reef occurs in the middle of the Upper Triassic Chinle Formation rather than in basal sandstones of the Chinle or the closely associated Shinarump Conglomerate, the host rocks of Colorado Plateau ores. Proctor (25) reports minor associated amounts of copper, selenium, vanadium, and uranium. The district's heyday was from discovery in 1875 to 1888. Silver production prior to 1910 was valued at

---

* U.S. Bureau of Mines, Minerals Yearbook and related publications.

$7,822,900. It has yielded a few minor shipments of uranium.

YELLOW CHIEF   The Yellow Chief is one of the few significant uranium producers of the Great Basin. According to Bowyer (39) it ranks with mines containing over 100,000 tons of ore with a grade in excess of 0.2 per cent $U_3O_8$. The host, a tuffaceous sandstone derived from Tertiary volcanics, is presumed to be late Miocene or early Pliocene. Ore occurs in lenses of various sizes; is beta-uranophane, a secondary uranyl silicate, to the virtual exclusion of other uranium species; lacks carbonaceous material or iron sulfides; but contains small amounts of manganese oxide, beryllium, and nodules of chert-like material.

It has been suggested the ore is the result of oxidation of primary tetravalent uranium minerals. Another theory is that the uranium-bearing fluorite pipes at Spor Mountain, immediately west of the Yellow Chief, were eroded and the uranium derived from the pipes went into solution and subsequently was precipitated in the Yellow Chief sandstones (39). A genetic relationship between these spatially related deposits of uranium, fluorite, and beryllium at Spor Mountain and at Yellow Chief

is suggested by the presence of beryllium in the Yellow Chief ores; the presence of both beryllium and uranium in the fluorspar pipes at Spor Mountain; and the presence of fluorite in deposits of bertrandite ($4BeO \cdot 2SiO_2 \cdot H_2O$) in the Tertiary volcanics that form pediments to Spor Mountain. Without implying genetic relationship, there is a direct age association between the postulated late Miocene-early Pliocene age of the Yellow Chief deposits and the 10 to 13 m.y. K/Ar dates from Marysvale pitchblende.

## Miscellaneous Deposits

FLUORSPAR   The Great Basin contains numerous, though usually small, fluorspar deposits in various host rocks. In Spor Mountain, the one district within the area of this discussion where a sizeable tonnage has been produced,* the ore occurs in high-grade pulverulent masses that have defied all attempts to make either acid or ceramic grade concentrates.

BERYLLIUM   Beryllium's widespread occurrence and abundance in the eastern part of the Great Basin was not recognized until the Spor Mountain and Gold Hill ores were discovered in 1960 and 1962 (53), and it was not until 1963 that Williams (44) and Cohenour (41,42) focused attention on the deposits. From the Sheeprock Mountains westward to Eureka, Nevada, northward to Gold Hill, and southward to the Mineral Range are many different types of beryllium deposits, not all of them of ore grade. There is disseminated beryl in granite in the Sheeprock Mountains; beryl within quartz-muscovite veins in the Deep Creek Mountains; beryl within pegmatites at Granite Peak; beryl with phenacite, bertrandite, fluorite, scheelite assemblages in lime-silicate gangue minerals at Mount Wheeler; bertrandite within quartz-carbonate-adularia veins at Gold Hill; disseminated bertrandite in Tertiary tuffs at Spor Mountain and Honeycomb Hills; and helvite in a contact metamorphic aureole in the Mineral Range (Figure 9).

At Spor Mountain, the ore is associated with fluorite and occurs as a replacement of limestone pebbles within a Tertiary tuff. At Mount Wheeler, it occurs in one of the limestone interbeds of the lower Cambrian Pioche Shale, a stratigraphic setting similar to that of lead-

* 144,000 tons of metallurgical grade fluorspar were produced from 1943 to 1962 from a series of replacement chimneys within Ordovician and Silurian dolomites (51).

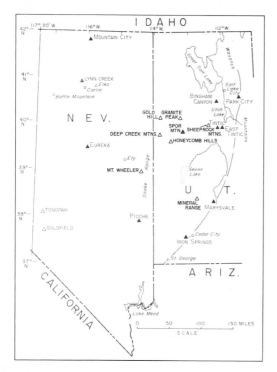

FIG. 9. *Location Map: Beryllium deposits in western Utah and eastern Nevada.*

silver ores at Pioche, Ophir, and East Tintic. In the Deep Creek Mountains and at Gold Hill, the veins lie in granitic rocks, as do the pegmatites at Granite Peak. Veins in the Deep Creek Mountains also occur in dolomite and in schist.

K/Ar age determinations associated with beryllium mineralization are recorded by Whelan (78) and by Park (76) in Table II.

Some of the ores carry significant percentages of other elements. At Mount Wheeler, the deposits were prospected for scheelite. At Spor Mountain, Shawe (60) reports a 0.22 per cent $LiO_2$ content from 18 samples and at Honeycomb Hills a 0.09 per cent $LiO_2$ content from 14 samples.

At Gold Hill the BeO content of the quartz-carbonate-adularia veins is both erratically distributed and directly related to the presence of adularia (79). Beryllium minerals have not been identified, and El Shatoury (79) believes the BeO content is isomorphously contained within late stage feldspar. Compared with the bertrandite deposits at Spor Mountain, the Gold Hill ores are low grade.

Bertrandite, $4BeO \cdot 2SiO_2 \cdot H_2O$, contains roughly 40 per cent BeO. At Spor Mountain and at Honeycomb Hills, it is finely disseminated. Griffitts (53), who at one time attributed the Gold Hill BeO content to bertrandite, regards the Spor Mountain and Gold Hill districts as capable of yielding 15 million tons with a BeO grade of 0.5 per cent or better. The large proportion of the reserves are at Spor Mountain. The ore grade of 0.5 per cent BeO appears low compared with the 15 to 12 per cent BeO obtained from beryl concentrates, but it is high grade compared with the crude-rock grade of a beryl deposit. Bertrandite requires an extraction process different from that for beryl, and the development of these Great Basin ores awaits the expansion of beryllium product markets (44,53).

Spherobertrandite, $Be_5Si_2O_7(OH)_4$, based on Stanley's postulated structure, has been formed in the laboratory at temperatures below 225°C (75). Since spherobertrandite has not been reported from Spor Mountain, it is inferred by Stanley* that the Spor Mountain ores may have formed above this temperature, although possibly at temperatures not much higher.

**Tungsten** Tungsten occurs widely throughout the Basin, most often as scheelite disseminated throughout contact-type deposits of

* D. A. Stanley, personal communication.

TABLE II. *K/Ar Age Determinations Associated with Beryllium Mineralization in Utah*

| Locality | Million Years | Remarks |
|---|---|---|
| Sheeprock Mountains | 13 to 22 | Granite |
| Spor Mountain | 16.2 ($\pm$1.5) | Upper topaz mountain rhyolite |
| Deep Creek Mountains | 17.7 ($\pm$0.9) | Quartz-muscovite-beryl veins in Granite Canyon |
| Mineral Range | 15.5 ($\pm$1.5) | Granite dike in Miller mine |
| Gold Hill | 8.0 ($\pm$0.8) | Quartz-carbonate-adularia veins in Rodenhouse Wash |
| Granite Peak | 30.0 ($\pm$2.0) | Pegmatite veins |

lime-silicate minerals. Two other types of deposits are the scheelite-galena ores at Ophir (27) and a complex of tungsten, manganese oxide, and iron oxide at Golconda. At Golconda, the tungsten deposit occurs as cementing material and as veins within gravel (17). There is no readily available record of tungsten production from the Great Basin. Schilling (59) states that United States production amounts to 13 per cent of the World's production and that Nevada has produced roughly 30 per cent of that total. Although high prices bring out many producers, there were no tungsten operations as of 1966.

**Tin** Tin occurs as cassiterite at three places—Goldfield, Delno, and Izenhood. None is commercially significant. These deposits are of interest because of their diverse geologic settings, none of which resembles the "classic" Cornwall picture. At Goldfield, copper-rich concentrates have carried cassiterite. At Delno, a manto system in Paleozoic limestone has been mined down dip for possibly 3000 feet over widths of 80 to 125 feet. Primary mineralization is essentially argentiferous galena within a siliceous gangue, but adjacent to the stopes are silicified unmined areas low in lead in which an unidentified tin mineral occurs as thin, yellow coatings on fractures. Olsen (31,68) considers it an amorphous mixture of cryptocrystalline tin oxide and silica. There may be several minerals because specimens were at one time identified by U.S. Bureau of Mines personnel† as arandisite, $5SnO_2 \cdot 3SiO_2 \cdot 4H_2O$. At Izenhood, 22 miles north of Battle Mountain, cassiterite-bearing veinlets occur in rhyolite flows. Fries (18)

† S. R. Wilson, personal communication.

reports occasional ore shoots 15 to 20 feet long by 4 to 6 feet wide but no concentrations of commercial significance. Along the surface at lower elevations are minor rivulets with concentrations of wood tin. These feed into a gravel-choked valley in which the upper 40 feet have been tested by shafts. The deeper gravels apparently have not been tested, but Vanderburg (16, p. 56–57) observed that in Nevada the occurrence of placer tin probably resembles that of placer gold. He further noted that a uniform concentration from surface to bedrock cannot be expected because of cloud-burst action, and the richer pockets probably will occur near the bedrock surface.

IRON    The Great Basin contains numerous magnetite and hematite deposits, both as veins and as replacements in limestones or volcanic rocks. Most are small, but important deposits occur in two areas: one, west of the area of this paper, where a series of deposits has been exploited intensely since the Korean War with much of the production shipped to Japan; the other, the Iron Springs district (Mackin, Bourret, and Jones, this volume), near the Basin's southeastern edge, where a series of replacement deposits in the Homestake limestone member of the Jurassic Carmel Formation have sustained Utah blast furnaces for some 40 years. These Iron Springs deposits have yielded over 69,000,000 tons of ore with a gross value exceeding $400,000,000.

MANGANESE    Manganese production, quite unimportant, has resulted from the war-time stimulus of small mines. The deposits are of several types: oxidized enrichments of manganese carbonate replacements in limestones; veins in igneous rocks; deposits from recent hot-spring activity, as at Golconda that was mined for tungsten; and limestone-replacement halos extending outward from base-metal ore bodies, as at Pioche (49,50). Although reportedly amenable to concentration, ore grades have been in the 20 to 30 per cent range.

MOLYBDENUM    Molybdenum deposits in the Great Basin, extremely important, are of various types. Large quantities of molybdenum are derived as a by-product from the great open-pit porphyry-copper deposits at Bingham Canyon, Utah, and Ruth, Nevada. Over 500,000,000 pounds of $MoS_2$ have been produced at Bingham Canyon (7,3); and some $4,000,000 in molybdenite concentrates at Ruth (58). Molybdenum occurs as ilsemannite

at Marysvale, Utah, where Kerr (this volume) reports that "pound for pound, almost as much molybdenum has been mined as uranium;" but none has been recovered. At the Hall property, a porphyry-type molybdenum deposit has been under exploration since 1957 (58). Elsewhere quartz veins and contact metamorphic deposits contain numerous occurrences, usually associated with scheelite, but none has proved significant.

## ACKNOWLEDGMENTS

The Nevada Bureau of Mines, the United States Bureau of Mines, and the United States Geological Survey have cooperated wholeheartedly in supplying production data for the preparation of this regional summary. The author is deeply indebted to their personnel; to all contributors of regional papers, to Richard Lee Armstrong of Yale for the courtesy of releasing a restored stratigraphic section; to John H. Schilling of the Nevada Bureau of Mines for his numerous responses for assistance; and to all members of the staff of the Utah Geological Survey. If errors of fact or judgment have occurred, the author accepts them on his shoulders.

## REFERENCES CITED

1. Fremont, J. C., 1845, Report of the exploring expedition to the Rocky Mountains in the year 1842, and to Oregon and north California in the years 1843–1844; 28th Congress, 2d Session, Senate Doc. 174–177, Washington, 174 p.
2. Ransome, F. L., et al., 1909, The geology and ore deposits of Goldfield, Nevada: U.S. Geol. Surv. Prof. Paper 66, 258 p.
3. Butler, B. S., et al., 1920, The ore deposits of Utah: U.S. Geol. Surv. Prof. Paper 111, 671 p.
4. Ferguson, H. G., 1921, The Round Mountain district, Nevada: U.S. Geol. Surv. Bull. 725, p. 383–406.
5. Lincoln, F. C., 1923, Mining districts and mineral resources of Nevada: Nevada Newsletter Pub. Co., Reno, 295 p.
6. Ferguson, H. G., 1924, Geology and ore deposits of the Manhattan district, Nevada: U.S. Geol. Surv. Bull. 723, 163 p.
7. Hewett, D. F., 1931, Geology and ore deposits of the Goodsprings quadrangle, Nevada: U.S. Geol. Surv. Prof. Paper 162, 172 p.
8. Westgate, L. G. and Knopf, A., 1932, Geology and ore deposits of the Pioche district, Nevada: U.S. Geol. Surv. Prof. Paper 171, 77 p.

9. Ferguson, H. G., 1933, Geology of the Tybo district, Nevada: Univ. Nev. Bull., v. 27, no. 3, 61 p.

10. Lyman, G. D., 1934, The saga of the Comstock Lode: Charles Scribner's Sons, N.Y., 399 p.

11. Tolman, C. F. and Ambrose, J. W., 1934, The rich ores of Goldfield, Nevada: Econ. Geol., v. 29, p. 255–279.

12. Nolan, T. B., 1935, The underground geology of the Tonopah mining district: Univ. Nev. Bull., v. 29, no. 5, 49 p.

13. Hewett, D. F., *et al.,* 1936, Mineral resources of the region around Boulder Dam: U.S. Geol. Surv. Bull. 871, 197 p.

14. Callaghan, E., 1937, Geology of the Delamar district, Lincoln County, Nevada: Univ. Nev. Bull., v. 31, no. 5, 69 p.

15. Vanderburg, W. O., 1938, Reconnaissance of mining districts in Eureka County, Nevada: U.S. Bur. Mines I. C. 7022, 66 p.

16. ———— 1939, Reconnaissance of mining districts in Lander County, Nevada: U.S. Bur. Mines I. C. 7043, 83 p.

17. Kerr, P. F., 1940, Tungsten-bearing manganese deposit at Golconda, Nevada: Geol. Soc. Amer. Bull., v. 51, p. 1359–1389.

18. Fries, C., 1941, Tin deposits of northern Lander County, Nevada: U.S. Geol. Surv. Bull. 931, p. 279–294.

19. Nolan, T. B., 1943, The basin and range province in Utah, Nevada, and California: U.S. Geol. Surv. Prof. Paper 197-D, p. 141–196.

20. Ferguson, H. G., 1944, The mining districts of Nevada: Univ. Nev. Bull., v. 38, no. 4, (Geol. and Min. ser. no. 40), 108 p.

21. Humphrey, F. L., 1945, Geology of the Groom district, Lincoln County, Nevada: Univ. Nev. Bull., v. 39, no. 5, (Geol. and Min. ser. no. 42), 53 p.

22. Searls, F. A., Jr., 1948, A contribution to the published information on the geology and ore deposits of Goldfield, Nevada: Univ. Nev. Bull., v. 42, no. 5, (Geol. and Min. ser. no. 48), 24 p.

23. Joralemon, P., 1951, The occurrence of gold at the Getchell mine, Nevada: Econ. Geol., v. 46, p. 267–310.

24. Kral, V. E., 1951, Mineral resources of Nye County, Nevada: Univ. Nev. Bull., v. 45, no. 3, (Geol. and Min. ser. no. 50), 223 p.

25. Proctor, P. D., 1953, Geology of the Silver Reef (Harrisburg) mining district, Washington County, Utah Geol. and Mineral. Surv. Bull. 44, 164 p.

26. Granger, A. E., *et al.,* 1957, Geology and mineral resources of Elko County, Nevada: Nev. Bur. Mines Bull. 54, 190 p.

27. Weintraub, J., 1957, Mineral paragenesis and wall rock alteration at the Ophir mine, Tooele County, Utah: Unpublished M.S. thesis, Univ. Utah, 44 p.

28. Osmond, J. C., Jr., 1958, Tectonic history of the basin and range province in Utah and Nevada (abs.): Min. Eng., v. 10, no. 11, p. 1132, 1134.

29. Stringham, B., 1958, Relationship of ore to porphyry in the basin and range province, U.S.A.: Econ. Geol., v. 53, p. 806–822.

30. Humphrey, F. L., 1960, Geology of White Pine mining district, White Pine County, Nevada: Nev. Bur. Mines Bull. 57, 119 p.

31. Olsen, D. R., 1960, Geology and mineralogy of the Delno mining district and vicinity, Elko County, Nevada: Unpublished Ph.D. thesis, Univ. Utah, 96 p.

32. Osmond, J. C., Jr., 1960, Tectonic history of the basin and range province in Utah and Nevada: Min. Eng., v. 12, no. 3, p. 251–265.

33. Stringham, B., 1960, Differences between barren and productive intrusive porphyry: Econ. Geol., v. 55, p. 1622–1630.

34. Welsh, J. E. and James, A. H., 1961, Pennsylvanian and Permian stratigraphy of Central Oquirrh Mountains, Utah: p. 1–16, *in* Cook, D. R., *Editor, Geology of the Bingham mining district and northern Oquirrh Mountains,* Utah Geol. Soc., Guidebook to the geology of Utah, no. 16, 145 p.

35. Callaghan, E. and Parker, R. L., 1962, Geology of the Delano Peak quadrangle, Utah: U.S. Geol. Surv. Geol. Quad. Map GQ-153, 1:5208.

36. Nolan, T. B., 1962, The Eureka mining district, Nevada: U.S. Geol. Surv. Prof. Paper 406, 78 p.

37. Odekirk, J. R., 1962, Lead-alpha age determinations of five Utah rocks: Unpublished M.S. thesis, Univ. Utah, 31 p.

38. Blackwelder, E., *et al.,* 1963, Precambrian rocks of Utah: p. 39–44, pt. 2, paper 5, *in* Crawford, A. L., *Editor, Surface, structure and stratigraphy of Utah,* Utah Geol. and Mineral Surv. Bull. 54a, 175 p.

39. Bowyer, B., 1963, Yellow Chief uranium mine, Juab County, Utah: p. 15–22, *in* Sharp, B. J., and Williams, N. C., *Editors, Beryllium and uranium mineralization in western Juab County, Utah:* Utah Geol. Soc. Guidebook to the Geology of Utah, no. 17, 59 p.

40. Christiansen, F. W., 1963, Cambrian rocks of Utah: p. 45–50, pt. 2, paper 6, *in* Crawford, A. L., *Editor, Surface, structure and stratigraphy of Utah,* Utah Geol. and Mineral. Surv. Bull. 54a, 175 p.

41. Cohenour, R. E., 1963, The beryllium belt of western Utah: p. 4–7, *in* Sharp, B. J., and Williams, N. C., *Editors, Beryllium and uranium mineralization in western Juab County, Utah,* Utah Geol. Soc., Guidebook to the geology of Utah, no. 17, 59 p.

42. ———— 1963, Beryllium and associated min-

eralization in Sheeprock Mountains, p. 8–13, *in* Sharp, B. J., and Williams, N. C., *Editors, Beryllium, and uranium mineralization in western Juab County, Utah:* Utah Geol. Soc., Guidebook to the geology of Utah, no. 17, 59 p.

43. Hintze, L. F., 1963, Summary of Ordovician stratigraphy of Utah: p. 51–62, pt. 2, paper 7, *in* Crawford, A. L., *Editor, Surface, structure and stratigraphy of Utah,* Utah Geol. and Mineral. Surv. Bull. 54a, 175 p.

44. Williams, N. C., 1963, Beryllium deposits, Spor Mountain, Utah: p. 36–59, *in* Sharp, B. J., and Williams, N. C., *Editors, Beryllium and uranium mineralization in western Juab County,* Utah, Utah Geol. Soc., Guidebook to the geology of Utah, no. 17, 59 p.

45. Beal, L. H., Cobalt and nickel: p. 78–81, *in Mineral and water resources of Nevada,* Nev. Bur. Mines Bull. 65, 314 p.

46. Bergendahl, M. H., 1964, Gold, p. 87–100, *in Mineral and water resources of Nevada,* Nev. Bur. Mines Bull. 65, 314 p.

47. Crawford, A. L. and Tuttle, C. E., 1964, Non-metallic mining and processing: p. 125–138, *in* Crawford, A. L., *Editor, Geology of Salt Lake County,* Utah Geol. and Mineral. Surv. Bull. 69, 192 p.

48. Crittenden, M. D., Jr., 1964, General geology of Salt Lake County: p. 11–45, *in* Crawford, A. L., *Editor, Geology of Salt Lake County,* Utah Geol. and Mineral. Surv. Bull. 69, 192 p.

49. ——— 1964, Manganese: p. 113–119, *in Mineral and water resources of Nevada,* Nev. Bur. Mines Bull. 65, 314 p.

50. ——— 1964, Manganese: p. 103–108, *in Mineral and water resources of Utah,* Utah Geol. and Mineral. Surv. Bull. 73, 275 p.

51. Dasch, M. D., 1964, Fluorine: p. 162–168, *in Mineral and water resources of Utah,* Utah Geol. and Mineral. Surv. Bull. 73, 275 p.

52. Everett, F. D., 1964, Reconnaissance of tellurium resources in Arizona, Colorado, New Mexico, and Utah; including selected data from other western states and Mexico: U.S. Bur. Mines R. I. 6350, 38 p.

53. Griffitts, W. R., 1964, Beryllium: p. 71–75, *in Mineral and water resources of Utah,* Utah Geol. and Mineral. Surv. Bull. 73, 275 p.

54. Hashed, A. H., 1964, Geochronological studies in the central Wasatch Mountains, Utah: Unpublished Ph.D. thesis, Univ. Utah, 98 p.

55. Marsell, R. E., 1964, Glaciation: p. 55–68, sec. 3, *in* Crawford, A. L., *Editor, Geology of Salt Lake County,* Utah Geol. and Mineral. Surv. Bull. 69, 192 p.

56. Roberts, R. J., 1964, Paleozoic rocks: p. 22–25, *in Mineral and water resources of Nevada,* Nev. Bur. Mines Bull. 65, 314 p.

57. ——— 1964, Exploration targets in north-central Nevada: U.S. Geol. Surv. Open-file report, 10 p.

58. Schilling, J. H., 1964, Molybdenum: p. 124–132, *in Mineral and water resources of Nevada,* Nev. Bur. Mines, Bull. 65, 314 p.

59. ——— 1964, Tungsten: p. 155–161, *in Mineral and water resources of Nevada,* Nev. Bur. Mines, Bull. 65, 314 p.

60. Shawe, D. R. and Duke, W., 1964, Lithium associated with beryllium in rhyolitic tuff at Spor Mountain, western Juab County, Utah: U.S. Geol. Surv. Prof. Paper 501-C, p. C86–C87.

61. Silberling, N. J., 1964, Mesozoic rocks: p. 27–30, *in Mineral and water resources of Nevada,* Nev. Bur. Mines Bull. 65, 314 p.

62. Stewart, J. H., 1964, Precambrian and lower Cambrian rocks: p. 21, *in Mineral and water resources of Nevada,* Nev. Bur. Mines Bull. 65, 314 p.

63. Wallace, R. E., 1964, Structural Evolution: p. 32–39, *in Mineral and water resources of Nevada,* Nev. Bur. Mines Bull. 65, 314 p.

64. Longwell, C. R., *et al.,* 1965, Geology and mineral deposits of Clark County, Nevada: Univ. Nev. Bull. 62, 218 p.

65. Maybey, D. R., 1965, Gravity and aeromagnetic surveys: p. 105–111, *in Geology of the Cortez quadrangle, Nevada,* U.S. Geol. Surv. Bull. 1175, 117 p.

66. McQuiston, F. W., Jr., and Hernlund, R. W., 1965, Newmont's Carlin gold project: Min. Cong. Jour., v. 51, no. 11, p. 26–30, 32, 38–39.

67. Roberts, R. J., *et al.,* 1965, Pennsylvanian and Permian basins in Northwestern Utah, northeastern Nevada and south-central Idaho: Amer. Assoc. Petrol. Geols. Bull., v. 49, p. 1926–1956.

68. Olsen, D. R., 1965, New variety of cassiterite from Nevada (abs.): Geol. Soc. Amer. Special Paper 82, p. 340.

69. Sayyah, T. A., 1965, Geochronological studies of the Kinsley stock, Nevada, and the Raft River Range, Utah: Unpublished Ph.D. thesis, Univ. Utah, 99 p.

70. Schilling, J. H., 1965, Isotopic age determinations of Nevada rocks: Nev. Bur. Mines Rept. 10, 79 p.

71. Bick, K. F., 1966, Geology of the Deep Creek Mountains: Utah Geol. and Mineral. Surv. Bull. 77, 120 p.

72. Callahan, W. C., 1966, Paleophysiographic premises for prospecting for strata-bound base metal mineral deposits in carbonate rocks: Nev. Bur. Mines Rept. 13, p. 5–50.

73. Mardirosian, C. A., 1966, Mining districts and mineral deposits of Utah: map, 1 sheet, privately printed (521—5th Ave., Salt Lake City, Utah), 1:750,000.

74. Mills, J. W., 1966, The role of unconformities in the localization of epigenetic mineral

deposits in the United States and Canada: A.I.M.E., New York Meeting, March 3, 1966.

75. Stanley, D. A., 1966, A preliminary investigation of the low-sodium portion of the system $BeO-Na_2O-SiO_2-H_2O$: Unpublished Ph.D. thesis, Univ. Utah, 65 p.

76. Park, G. M., 1967, Some geochemical and geochronological studies of the beryllium deposits in western Utah: Unpublished M.S. thesis, Univ. Utah.

77. Armstrong, R. L., in press, Cordilleran miogeosyncline: Utah Geol. and Mineral. Surv.

78. Whelan, J. A., in press, Radioactive and isotopic age determinations of Utah rocks: Utah Geol. and Mineral. Surv.

79. El Shatoury, H., 1967, Mineralization and alteration studies in the Gold Hill mining district, Tooele County, Utah: Unpublished Ph.D. thesis, Univ. Utah.

# 45. Non-Porphyry Ores of the Bingham District, Utah

R D. RUBRIGHT,* OWEN J. HART*

## Contents

* United States Smelting and Mining Company, Lark, Utah.

## Illustrations

## Tables

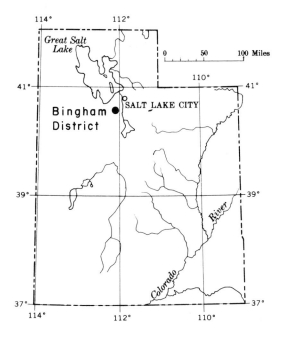

## ABSTRACT

In the Bingham district over a span of more than 90 years, 43,947,104 tons of "non-porphyry" copper, lead, zinc, gold, and silver ore have been mined from a folded and faulted alternating series of Pennsylvanian and Permian limestones and sandstones and from some Tertiary intrusives. This ore came from six major mine areas located east, south, and west of the Bingham stock, site of Kennecott's porphyry copper deposit. The sediments were affected by varying degrees and types of metamorphism within 2000 feet or more of the Bingham stock. The intrusives consisting of the mineralized Bingham and the unmineralized Last Chance stocks and numerous dikes and sills of granite and monzonite were intruded into the asymmetrical northwest-trending and plunging Bingham syncline. Extrusives, although present, contain no ore. The bulk of the non-porphyry ore in the district was found above the Midas thrust.

The primary ores are found in three major types of deposits: concordant bedding fissures, crosscutting fissures, and replacement deposits. These deposits consist of both copper and lead-zinc ore bodies that contain various amounts of gold and silver. These are in places roughly zoned both laterally and vertically and have been mined for up to 10,000 feet outward from the center of the Bingham stock and up to 5000 feet vertically below the surface.

Secondary ore was of minor importance and was found in and under the oxidized zone at shallow depths.

Genetically the Bingham stock, the ores, and related phenomena have had a common source.

## INTRODUCTION

The Bingham mining district located in the Oquirrh Mountains about 20 miles southwest of Salt Lake City, Utah, is well known as the site of the Kennecott Copper Corporation's porphyry copper deposit situated in and close to the Utah Copper or Bingham stock. Less well known are the ore deposits of gold, silver, copper, lead, and zinc which are adjacent to the stock and extend to a distance of 10,000 feet from its center. These deposits have been developed by hundreds of miles of mine workings to a vertical depth of 5000 feet below the surface. This article deals with these "non-porphyry" ore deposits.

Beginning at a point northeast of the stock and continuing around clockwise to a position west of it are the mines of the major non-porphyry deposits. These are in order: the Lark section of the United States and Lark Mine, the Combined Metals Mine, the U.S. section of the United States and Lark Mine, the Armstrong or old Boston Consolidated Mine, the Utah Metals and Tunnel Company Mine, the Highland Boy Mine, and the Utah Apex

Mine. The remainder of the mines, part of which are north of the stock, are of lesser importance (Figure 1).

A comprehensive review of the entire district is somewhat restricted by the absence of information on, or the inaccessibility of, some of the older workings.

## HISTORICAL INFORMATION

### Mining Operations

Mining in the Bingham district began in 1863, following discovery of galena ore in

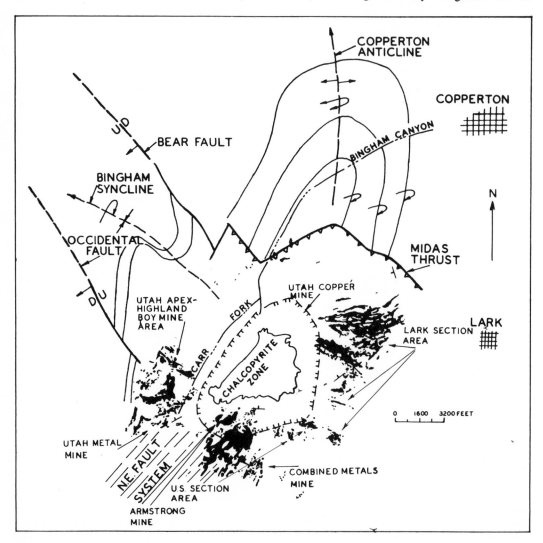

FIG. 1. *Principal Structures, Metal Zones, and Composite Stoping. Modified from 1961 Kennecott map.*

Galena Gulch, a tributary to Bingham Canyon. The first shipment of ore is reported to have been made by the Walker brothers in 1868 to Baltimore (1). Development was slight until the summer of 1870 except for placer gold, $1,000,000 worth of which was produced up to 1871. In 1873, a railroad was built to Bingham. During 1891 to 1892, there were 21 producing mines in the district. The first shipment of high grade copper ore was made from the Highland Boy Mine in 1896 (2). Until 1897, Bingham production consisted largely of silver, lead, and gold in that order of importance, reaching a total value of $39,000,000 and only $1,075,000 in copper (3).

At the turn of the century, several major companies were formed by consolidation of smaller mines. By 1924, the major mines were those of the Utah Consolidated Mining Company, the Utah Apex Company, the Utah Metal and Tunnel Company, the United States Smelting Refining and Mining Company, the Montana-Bingham Consolidated Mining Company, and the Bingham Mines Company (3).

Utah Copper Company began production in 1904 by underground methods that continued until 1906. At this time, surface waste stripping was begun, in addition to the underground work, preparatory to open pit mining that started in March, 1907. In March, 1914, underground mining was suspended. Since that time all the porphyry copper ore production has been from open pit operations (12). Utah Copper Company became a division of Kennecott in 1936.

Of the non-porphyry mines, the United States Smelting Refining and Mining Company, having acquired the Montana-Bingham and Bingham Mines properties, is the only company producing ore at this time. This property consists of two units, the U.S. and Lark sections.

## PHYSIOGRAPHIC HISTORY AND PRESENT TOPOGRAPHY

The physiographic history of the Bingham district logically begins with the deposition of 25,000 feet of Pennsylvanian and Permian sediments which were predominantly fine-grained sandstones and interbedded limestones. The upper 10,000 feet or more is exposed in the Bingham district. At the end of Cretaceous or in early Tertiary time, a series of large north-northwest-trending asymmetrical folds, which plunge toward the northwest, were formed. Subsequent erosion first carved the folded rocks into a mountainous topography,

TABLE I.    *Statistics of Mine Production*

Total Production of Non-Porphry Ores of the Bingham District Through 1964

| | |
|---|---|
| Crude Ore | 43,947,104 short tons |
| Metals Recovered at Smelters: | |
| Gold | 2,377,418 ounces |
| Silver | 136,125,378 ounces |
| Copper | 816,498,957 pounds |
| Lead | 4,180,279,900 pounds |
| Zinc | 1,709,091,700 pounds |
| Total Value | $777,138,213 |

The value is in terms of current dollars in the year of production. The total amount of zinc produced is greater than that shown as recovered since considerable zinc was lost in the past in smelting.

Source: U.S. Bureau of Mines.

then to a mature surface. Renewed orogeny as a result of Basin and Range faulting in late Miocene or Pliocene time, coupled with erosion, has produced the present terrain. Gilluly (5, p. 91) states, "Some of the faults which are the cause of the topographic prominence of the present range were probably rejuvenations of earlier breaks dating from the early Tertiary but others were probably newly formed." This faulting has continued to the present day.

The Bingham district lies on the east slope of the Oquirrh Range and is one of high relief, ranging from an elevation of 5500 feet at the edge of the Great Salt Lake valley to over 9000 feet on the crest of the ridge which limits the district on the west. It covers parts of three drainage system: Bingham Canyon with its upper tributaries Carr Fork, Bear Gulch, and Galena Gulch, a small portion of the Butterfield Canyon drainage that lies to the south, and numerous small gulches draining into the Great Salt Lake valley east of the ridge that bounds one portion of the Bear Gulch drainage.

Although the relief is great, the outline of the mountains is smooth and rounded. Few ledges or cliffs are formed because the preponderant quartzite and sandstone break into small fragments upon weathering.

A striking feature of the present topography and one seen by thousands of tourists yearly is the stupendous open pit of Kennecott Copper Corporation's porphyry-copper mine which is over 2000 feet deep and 10,000 feet across at its widest point.

TABLE II.   *Stratigraphic Column of the Bingham District*

| Age and Formation | Member | Thickness in Feet |
|---|---|---|
| **Tertiary** | | |
| Extrusives | | |
| Intrusives | | |
| **Permian** | | |
| Curry | Midas limestone | 290–560 |
| | Quartzite | 200–300 |
| | Bemis limestone | 100–360 |
| | Sandstones and siltstones | 1580 |
| | Unconformity ——————— | |
| **Pennsylvanian** | | |
| Oquirrh | Dixon shales and limestones | 5–20 |
| | Winamuck shales and limestones | 40 |
| | Quartzite (predominantly) | 3000+ |
| | Congor limestone | 10–25 |
| | Quartzite | 480 |
| **Tertiary** | | |
| Intrusive | Fortuna sill, porphyritic monzonite | 70 |
| **Pennsylvanian** | | |
| Oquirrh | Quartzite | 60 |
| | Mayflower limestones | 110 |
| | (Unknown gap between Petro and Mayflower limestones—may be same members) ? | |
| | Petro limestone | 6–10 |
| | Quartzite | 310 |
| | Parnell limestone | 30 |
| | Quartzite | 800 |
| | Commercial (Yampa) limestone | 120–200+ |
| | Quartzite | 100 |
| | Lark limestone | 10 |
| | Quartzite | 255–320 |
| | Jordan (Highland Boy) limestone | 140–230 |
| | Unconformity ——————— | |
| | Quartzite | 400–1630 |
| | Lower Intermediate (Utah Metal) limestone | 45–70 |
| | Quartzite | 260–350 |
| | Highland ("14" and "850") limestones | 85–130 |
| | Quartzite | 210–350 |
| | "A" (Washington) limestone | 50–225 |
| | Quartzite | 110–380 |
| | "B" (Levant) limestone | 45–90 |
| | Quartzite | 200 |
| | "C" (Fern) limestone | 30 |
| | Quartzite | 125 |
| | "D" (Queen) limestone | 200–250 |
| | Quartzite | 280 |
| | "E" limestone | 320 |

# GEOLOGIC HISTORY

## Stratigraphic Column

Tertiary intrusives and extrusives cover a big portion of the Bingham district. The intrusives, consisting of stocks, dikes, and sills, cut the entire sedimentary series (Figure 2). The intrusives consist of granite, quartz monzonite, monzonite, granodiorite, quartz diorite, granite porphyry, quartz monzonite porphyry, and biotite quartz latite porphyry (10). The extru-

sives, according to Smith (12), consist of andesite, andesite porphyry, andesite breccia, agglomerates, ash deposits, and rarely basalt.

All the sediments within a distance of roughly 2000 feet from the Bingham stock show different degrees of metamorphism as described in more detail later.

The Curry formation, Permian in age, lies at the northern part of the district. The larger part of it is under the Midas thrust. It is 2800 feet thick and is composed of two rock types, dark gray silty, platy limestones and fine-grained calcareous, thin-bedded sandstones and siltstones. Small tonnages of ore have been produced from fissures in the Bemis and Midas limestone members, which occur close to the top of the Curry formation, under the Midas thrust as shown in Figure 3. The Midas is a dark gray to black limestone and is more calcareous than the Bemis that consists of alternating layers of sandstone and siltstone with limestone. There is an unconformity at the base of the Curry.

The Pennsylvanian Oquirrh formation contains the rest of the section to be described. Over 30 limestones and interbedded sandstones make up the upper 6500 feet of the Oquirrh formation down to the base of the Jordan limestone. The interbedded sandstones, which are quartzitic in the productive area, are light colored, cross bedded, coarse- to fine-grained, and feldspathic. These will not be described individually.

The Dixon and Winamuck limestones, which are close to the top of the Oquirrh formation, are found under the Midas thrust in the Lower Bingham area. They are composed of black shales and some light to dark colored limestones.

The Congor limestone, which was ore bearing, ranges from a black limestone to a calcareous quartzite and at times lies above a porphyry sill 12 to 15 feet thick.

The Fortuna sill, which is porphyritic monzonite, has a 3- to 5-foot calcareous shale bed under it in places. Ore has been produced from veins on both the hanging wall and footwall of the sill.

The Mayflower limestones consist of two members separated by 50 feet of quartzite. The unweathered underground exposures show a peculiar greenish to brownish mottling of impure sandy limestone. The upper member was ore bearing in places.

On the east side of the Bingham stock, from the footwall of the Mayflower limestones to the hanging wall of the Commercial limestone, there is 1185 feet of quartzite which contains

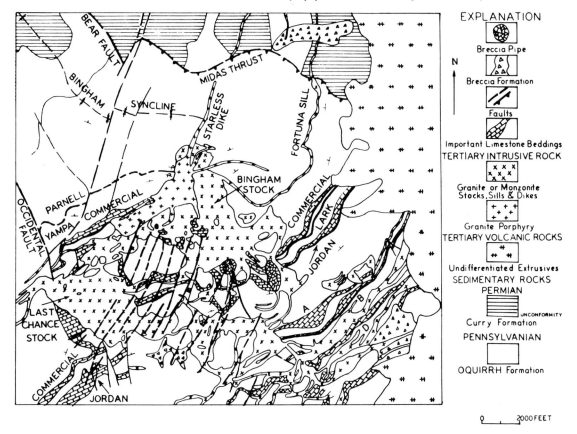

FIG. 2. *Surface Geology Map of the Bingham district. Modified from 1961 Kennecott map.*

several thin, unnamed, non-productive in the large part, sandy limestones. On the west side of the Bingham stock in the Utah Apex-Highland Boy Mine area, in the same interval above the Commercial limestone, are found two productive limestone members, the Petro and the Parnell (Figure 4). The Petro is a very dark gray to black limestone. The Parnell is also dark gray to black but is more sandy.

The 800-foot interval between the Parnell and the top of the Commercial limestone shows calcareous quartzite with interbedded thin, sandy to silty limestones, from two of which a little ore has been produced. Correlation of any of these thin limestones from one side of the Bingham stock to the other is difficult.

The Commercial (Yampa) limestone, except where it is cut out by intrusives, is found throughout the district to lie above the Midas thrust. Where little altered, it is dark gray to black and changes over short distances from cherty and silty to quite pure. It is said to contain two quartzite lenses 20 and 90 feet above the base of the Utah Apex-Highland

Boy mine area. It is one of the major ore-producing limestones.

The Lark limestone is a cherty, sandy limestone and, where only slightly altered, is black. It is a major ore-bearing limestone in the Lark section only.

The Jordan (Highland Boy) limestone is another major ore-bearing limestone found throughout the district in the hanging wall of the Midas thrust. In the western part of the district, it appears as a gray to dark gray limestone which is silty, cherty, and sandy (4). In the eastern part of the district, beyond the metamorphic zone, it is black and locally contains a quartzite member near the hanging wall. According to Welsh (12, p. 8), the base of the limestone is marked by an angular unconformity dividing the upper Pennsylvanian from the lower Pennsylvanian.

The next unit is predominantly quartzite and ranges from 400 feet in thickness on the west side of the district to over 1600 feet on the east side and contains numerous thin, sandy, in part cherty, lenticular limestones known as

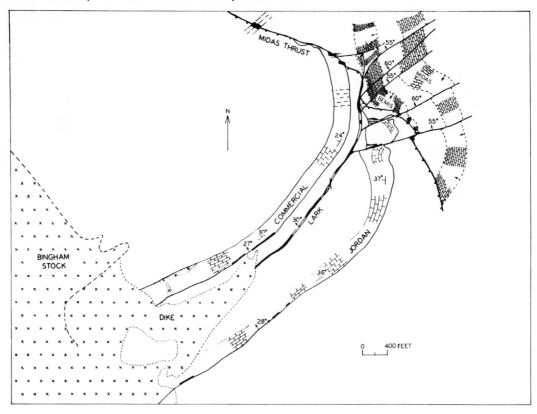

FIG. 3. *Geology of a Lark Section Level. Bemis and Midas limestones shown under the Midas thrust are overturned and Permian in age.*

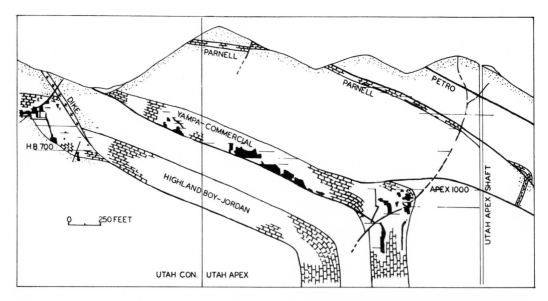

FIG. 4. *Vertical North-South Section through the Highland Bay, Yampa, and Utah Apex Mines, showing the four productive limestones. Modified from 1924 map by R. N. Hunt. Black areas are stopes.*

the Sub-Jordan series which with a few exceptions were not ore bearing.

The Lower Intermediate or Utah Metal limestone can be correlated across the district and consists of sandy limestone with some interbedded quartzite. It has been productive in the Lark section where it has also been called the St. Joe limestone.

The Highland or "14" and "850" limestones consist of three beds in the U.S and Lark mine where two of them have been ore bearing. These beds when unmetamorphosed are dark gray and sandy. In the western part of the district they are called the "14" and "850" limestones and were also ore bearing.

The "A" (Washington) limestone shows a thickness of from 50 to 225 feet, the differences being due either to lensing or to conflicting interpretations. Although it has been described as a pure dark gray limestone in the western part of the district, where seen in the U.S. and Lark Mine, it is much more quartzitic. Little ore production has come from this member.

The "B" (Levant) limestone is one of the major ore sources in the U.S. section where the productive portion is in the metamorphosed zone. The limestone is almost completely silicated here.

The "C" (Fern) limestone is a highly silicated bed in the U.S. section and is only sparingly mineralized.

The "D" limestone is another major ore-bearing limestone in the U.S. section. It also is silicated in the productive zone. On the surface away from the metamorphosed zone, this limestone is grayish-black, thin bedded, and fine grained with abundant chert nodules.

The "E" limestone, where seen underground in the U.S. section, is intensely silicated and is found in contact with the Last Chance stock. It is similar lithologically to the "D", but has less chert.

## Structure

EARLY STEEP NORMAL FAULTS  The structural geologic history of the Bingham district began with a series of steep, normal faults to which Rollin Farmin (7) refers in the Utah Delaware and Utah Metals Mines as the South and Top faults. This is probably the same system called the Roll fault in the U.S. section of the U.S. and Lark Mine. These structures, which have from 800 to 1000 feet of offset, were pre-folding and pre-igneous intrusion. They are offset by all bedding faulting and also by the steep northeast-striking fissure system that is so well

developed in the western part of the district (Figure 1). There is an igneous intrusive along one of the major breaks and some mineralization. There are similar faults with smaller offsets buried well down on the west limb of the asymmetrical Bingham syncline in the U.S. section. In the U.S. section, all of these structures strike northwestward and dip on an average 50°SW; all other faults that cut this early series offset it.

FOLDING  The late Cretaceous or early·Tertiary folding that produced the Bingham syncline and the Copperton anticline created the very important bedding-plane fault pattern that has such a significant bearing on the location of the concordant ore bodies. This folding in the Bingham district is part of the series of northwest-southeast trending Tertiary folds that affected all the southern Oquirrh Mountains. These plunge to the northwest and are in part asymmetrical, with steep to overturned northeast flanks.

BEDDING-PLANE FAULTING  This faulting has measured offsets as great as 1500 feet; although it was initiated by folding, it is not thought that all of the movement occurred at the time of folding. The footwall contact of many of the major limestones with quartzite is the location of the greater porportion of the movement along bedding-planes, but some of the bedding movement occurred within the limestones. This intrabedding faulting usually took place along a lithologic break such as at the contact of a sandy portion with a more limey bed. As noted by Winchell (5) in 1924, "The Highland Boy limestone as a formation locally consists of as much as 80 per cent quartz; petrographically such parts must be considered to be sandstone."

Some bedding-plane faulting is, in places, very pronounced and contains wide gouge zones; in others, it is so completely healed as to be difficult to detect. In the Lark section of the U.S. and Lark Mine, there is considerable brecciation of the quartzite under some bedding faulting that forms a favorable zone for mineral deposition, and large tonnages of ore have been mined from the quartzites so broken. Late post-mineral bedding faulting has affected the ore and wall rock.

Normal fault slump toward the Bingham stock went unnoticed for a long time because of its association with bedding movement. The early pre-folding steep normal faults such as the Roll, Top, and South faults are dropped normally along the bedding structures as much

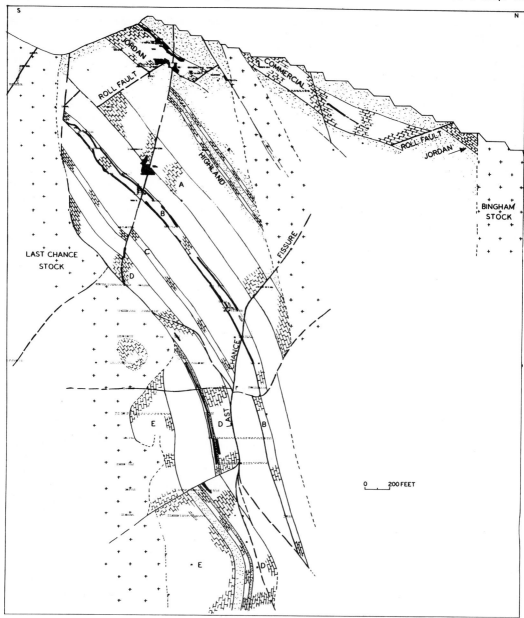

FIG. 5. *Verical North-South Section of the U.S. Section. Black areas as stopes.*

as 1500 feet (See Figure 5). That all of these offsets were not made during folding is suggested by bedding-fault offsets of porphyry masses, breccia pipes, and post-mineral faulting in the ore and faulting of the North fault or Midas thrust by bedding faults. There have been several stages of movement along these bedding faults, both pre- and post-ore deposition.

THRUSTING The Midas thrust, which is exposed on the surface in the northern part of the district and underground in the Lark section of the U.S. and Lark Mine, where it is known as the North fault, was in part postfolding since it cuts through the overturned Copperton anticline and places Pennsylvanian Oquirrh formation over Permian rocks. This thrust strikes northwestward and dips 20° to

40°SW. C. L. Thornburg of the United States Smelting Refining and Mining Company staff, who was the first to work on the thrusting aspect of this fault, says that the stratigraphic separation is in excess of 5000 feet. This large structure helped fold and offset the Commercial, Lark, and Jordan limestones of the Lark section (Figure 3). The visible fault plane may be only inches wide, but the quartzites and sandy limestones found in its footwall side are at times brecciated for dozens of feet and partly recemented with silica. Although it has been said that the Midas thrust (or North) fault limits the known significant lead-zinc ore in the eastern portion of the Bingham district, there are both ore-bearing fissures and concordant bedding ore bodies found under it in the Copperton anticline, some of which are producing ore at the present time. It is thought that the high-angle reverse fault, the Bear fault, which is so closely associated with the Midas thrust, may also have been formed at this time (Figure 1).

NORTHEAST-STRIKING FAULTS  This fault system is found in the western part of the district. Billingsley and Locke (9) considered part of this system to be tear faults. While some of this faulting may have been initiated by the Midas thrust, part of it must also have been associated with folding, and the close association along the strike of the Bingham stock and the Last Chance stock with this fault pattern suggests that common tectonic forces may have been responsible for the location of both of these features.

The major northeast-striking structures are well mineralized, and large tonnages of ore have been produced from them. Post-mineral faulting has produced mullions with dips of 40° to 50° downward toward the Bingham stock. Some of this movement probably reflects the slump adjustment of bedding faults, which the northeast ones cut at almost right angles in the U.S. section. Offsetting of the Last Chance porphyry stock attests to late movement along this system.

NORTHWEST-STRIKING FAULTS  The Occidental fault is a large, normal Basin and Range fault with a 1500-foot vertical displacement. It roughly bounds the Bingham district on the west and was formed prior to igneous intrusion since, according to Rollin Farmin (7), it was stoped out during the emplacement of the Last Chance stock and contains dikes intruded along it. The Occidental fault itself is sparingly

mineralized and strikes on an average N25°W and dips 62°SW.

Although it has been said that the Occidental fault limits lead-zinc mineralization to the west, there has been lead-zinc ore mined both from concordant bedding fissures and northeast-striking fissures west of this large structure. More recent movement is shown by some offsetting features.

BASIN AND RANGE FAULTING  In late Miocene or Pliocene time, renewed uplift along Basin and Range faults resulted in the present topographic prominence of the Oquirrh mountains. This faulting, which is probably confined to the west side of the Oquirrhs, resulted in tilting of the range. This, of course, renewed movement on the Occidental fault which is a Basin and Range structure and likewise renewed movement on most of the smaller faults in the district.

## Igneous Intrusion

The emplacement of the Bingham and Last Chance stocks in a northeastward direction, roughly coinciding with the fault system of the same strike (as was mentioned above), suggests the same tectonic force might have governed the development of both of these features. These stocks are composed principally of granite, granite porphyry, and monzonite and are each more than a square mile in area. The Bingham stock is mineralized, but the Last Chance stock, as such, is barren.

The area to the east and south of the Bingham stock contains large irregular dikes and sill-like masses of monzonite (Figure 2). The area northeast of the Bingham stock is cut by long tabular dikes and sills.

Although there is evidence of some dilation accompanying many of the small dikes and sill-like bodies, the larger intrusives furnish good evidence of having stoped their way into place.

Stringham (10) mentioned smaller bodies of biotite-quartz latite porphyry, actinolite syenite, granodiorite, and quartz diorite as being found in the Bingham stock area. Several stages of igneous activity are indicated by both surface and underground observation. Current work on age determination of the intrusive igneous rocks is said to show only small variations in age from one rock type to another.

## Breccia Pipes

Breccia pipes are found both to the south and east of the Bingham stock. These differ widely in shape as well as size—the larger ones being up to 600 feet in diameter and over 3000 feet in vertical extent. They are composed of rounded to angular fragments of most rocks found in the district including silicated and partially altered limestones depending on the location of the pipe. Some of the pipes provide evidence of more than one stage of igneous activity since they include fragmental igneous rock and are cut by small dikes. Although they were formed before the ores were deposited, they are generally barren. In one rare case, the fringe remnant of one pipe was the location of a lead-zinc replacement ore body. The matrix of some is silica—hydrothermally derived; others, however, have a porphyry matrix. A penetration over 4000 feet vertically below the surface in the U S. section under an irregular breccia pipe shows solid porphyry. This same pipe, from all available evidence, appears to cone out at less than 1000 feet below the surface. This particular pipe would tend to demonstrate the collapse theory of formation. That these pipes have an intimate relationship to igneous activity is clearly evident. The fine matrix and the pronounced rounding of some of the fragments within certain local areas of some pipes suggests attrition caused by gaseous movement.

## Igneous Extrusives

The extrusive rocks are located on the extreme eastern flank of the Oquirrh mountains in the Bingham district and extend out into the Great Salt Lake valley (Figure 2). In the Lark area, the flow material has been dropped normally by a major mineralized bedding fault. Smith (12, p. 105) says, "The rock of the volcanic area consists of flows, volcanic breccia, agglomerates, and ash deposits. The flows are andesites and latites which rest on a surface of moderate relief. The surface resulting from the deposition of the volcanic flows was eroded to moderate relief, and then later deeply covered by volcanic agglomerates. After this the dark porphyry dikes and perhaps some dikes originating in the light porphyry intrusives were intruded into the volcanic rocks. Ash deposits near Bingham Canyon interbedded with later gravels indicate some very late volcanic activity."

## Age of Mineralization

The mineralization and metallization followed the igneous intrusion closely and as such are probably early Tertiary in age.

## ECONOMIC GEOLOGY—PRIMARY ORE

### Forms of the Ore Bodies

Three specific forms of non-porphyry ore deposits are found in the Bingham district. These are concordant bedding-fissure deposits, crosscutting fissure deposits, and replacement deposits.

### Concordant Deposits

STRATIGRAPHIC RELATIONS Concordant ore deposits were those formed in and along bedding or near-bedding faults which originated during folding of the Bingham syncline. These bedding faults, with their subsequent major ore deposits, are found in the Petro, Parnell, Commercial (Yampa), Lark, Jordan (Highland Boy), Lower Intermediate (St. Joe and Utah Metal), Highland ("14" and "850"), "B" (Levant), and "D" limestones (Figure 5). Smaller production has come from other bedding structures. Most of the concordant ore is associated with strong footwall bedding faults; however, the Commercial, Jordan, "B", and "D" limestones had two productive bedding-fault ore zones—one on or close to the footwall and the other up within the limestone. These ore deposits have been called "mantos," and, if the term is used for tabular, bedded deposits, then most of them fall into this category. The dips may be from 10° to vertical or even overturned. The ore deposits range in size from small tabular shoots to that of the Lark vein, which has been mined for up to one mile on strike and more than one mile down-dip from the surface. They grade into replacement deposits where the host rock for any of a number of reasons became more susceptible to ore deposition. Brecciation caused by crossfaulting or folding or simple fissure intersections prepared areas for replacement that break the continuity of the concordant deposits. The controls that change concordant ore to a replacement type of deposit are not always apparent.

In the Utah Apex-Highland Boy mine area, concordant ore was mined from the Petro, Parnell, Commercial (Yampa), Jordan (Highland Boy), and the "14" and "850" limestones (Figure 4). In the Utah Metals, Armstrong, and U.S. section Mines, the Commercial, Jordan, Highland ("14" and "850"), "B", "C", and "D" limestones were productive. In the Lark section, the Commercial, Lark, Jordan, and Lower Intermediate limestones have furnished the bulk of the ore. Lesser production has come from the Congor limestone, from veins on both the hanging wall and footwall of the Fortuna sill (the Sierra Grande and Fortuna veins, respectively), and the May-flower vein—all of which are stratigraphically higher than the Commercial limestone in the Lark section. The Dixon, Winamuck, and Tie-waukee veins in Lower Bingham in the foot-wall of the Midas thrust have been mined to a limited extent, mainly for lead-zinc silver ores. Some production has come from bedding faults enclosed in quartzite.

Mineralogy   Like other forms of ore de-posits in the Bingham district, the concordant bodies comprised both copper and lead-zinc ore. Hunt (3) describes the copper deposits in the western part of the district as massive coarse-grained sulfide bodies, vein-like, located along fissures, bedding planes, and intrusive contacts and consisting largely of pyrite and chalcopyrite with very little quartz or other gangue minerals and a little gold and silver. Although this type of deposit is normally not large, those along the footwall of the Commer-cial (Yampa) and Parnell limestones contained extensive shoots of both copper and lead ore. In places, lead-zinc and copper ore are in-timately intermixed in the same ore shoot. The lead-zinc ore bodies are likewise massive coarse-grained sulfides of galena, sphalerite, much of which is marmatitic with varying amounts of pyrite and with some copper, gold, and silver. While the galena itself contains silver, much of the silver is found as argentif-erous tetrahedrite-tennantite.

Gangue minerals may be: (1) silica as jas-peroid, quartz, opal, or chalcedony; (2) cal-cite; (3) saponite; (4) montmorillonoid clay; (5) talc; (6) chlorite; (7) wollastonite, diop-side; and (8) mica. Calcite and quartz are by far the most common.

The near-surface concordant copper deposits in the Jordan, Commercial, and Mayflower limestones as well as the Fortuna sill had small mineable bodies of carbonates and oxides, but in places only 50 feet below the surface, these turned into only slightly oxidized pyritic cop-per ore. Chalcocite was found below many of the oxidized zones. Secondary bornite, chal-cocite, covellite, cuprite, and native copper are found as small showings to a depth of 3600 feet down-dip in the Lark section. In the same area, according to Boutwell (2), the oxidized zone in the old Brooklyn Mine in the Jordan limestone extended to a depth of 1450 feet. The mineral changes in depth in the con-cordant ore bodies are about the same as in

some of the fissure ore bodies; sphalerite be-comes marmatitic (Rose, 13), pyrite becomes more dominant, and silica increases. Total amount of mineralization decreases and tends to funnel downward. This tendency was noted by Billingsley and Locke before 1937 (9). Er-ratic, high-grade gold and silver has been found at considerable depths in some of the mines.

Wall-rock alterations, associated with con-cordant bedding ore deposits in metamor-phosed limestones in the U.S. section, Arm-strong Mine, and part of the Lark section, differ a great deal but locally may contain some or all of the following: talc, saponite, green trioctahedral montmorillonoid, chlorite, fine-grained trioctahedral and dioctahedral mica, vermiculite minerals, and mixtures of cal-cite and dolomite. In one instance in the Lark section, a lead-zinc ore body contained abundant streaks of kaolinite-dickite. Halloysite is present near opalized limestone in several places at the surface. Deep in the U.S. section, the limestone above one bedding fissure was silicified (opal) and argillized (dioctahedral montmorillonoid) (Hunt, 11). In the rest of the district, outside of the zone of intense meta-morphism, millions of tons of lead-zinc ore have been mined from concordant deposits in little altered black limestone. Small amounts of quartz and calcite were common in the wall-rocks of these ore bodies.

Vertical differences in grade can be shown for the U.S. and Lark sections. These are the only properties from which statistics can be readily compiled.

The following table shows average grades

*TABLE III.    Average Grade by Level for a Major Concordant Ore Body*

| Levels | Au | Ag | Cu | Pb | Zn |
|---|---|---|---|---|---|
| +200 to surface | .057 | 9.33 | | 39.23 | |
| 0 + 200 | .012 | 2.74 | .22 | 9.43 | 6.14 |
| 0 | .017 | 2.26 | .37 | 9.47 | 8.50 |
| 200 | .014 | 3.30 | .46 | 10.82 | 8.31 |
| 500 | .015 | 3.54 | .59 | 11.88 | 9.10 |
| 800 | .016 | 3.92 | 1.03 | 12.79 | 6.78 |
| 1200 | .026 | 4.34 | 1.42 | 10.73 | 3.43 |
| 1600 | .019 | 3.41 | .77 | 17.53 | 11.21 |
| 2050 | .034 | 2.84 | .86 | 10.40 | 7.42 |
| 2500 | .030 | 5.12 | 1.27 | 17.75 | 5.15 |
| 2750 | .031 | 3.83 | .88 | 10.91 | 8.27 |
| 3000 | .033 | 4.08 | .82 | 9.11 | 7.72 |
| 3250 | .024 | 4.66 | .77 | 12.46 | 8.43 |

by level for one of the major concordant ore bodies in the Lark section which extends through a vertical range of 2700 feet or 4500 feet down-dip. Boutwell (2) in describing the old upper part of this ore shoot gave the average assays shown on the first line of the table. These represent over 50,000 tons of ore mined for a dip distance of 1600 feet from the surface. The grade of this ore indicates selective, high grade mining, as was subsequently proved by assays of ore left in stopes and later mined. Other major concordant deposits over a comparable dip distance show only minor changes in grade with depth.

FACTORS CONTROLLING FORM AND LOCATION OF ORE BODIES   The shape of the asymmetrical Bingham syncline determines the attitude of the concordant ore deposits in the district. Most of the major concordant ore bodies plunge toward the Bingham stock, which was intruded into the axis of the syncline.

Beginning in the Lark section, where the limestone beds are cut off by the Midas thrust, they strike roughly north-south with an average dip of 22°W. There is a progressive change in strike and dip as the beds are followed from this point to the U.S. section where the strike is northwestward and the dip steepens to vertical and even overturns in the bottom of the mine (Figure 5). Continuing farther along strike, past an apophysis of the Bingham stock, into the Utah Apex-Highland Boy mine area, the beds assume an east-west strike and finally strike slightly southwestward as they are cut by the Occidental fault. In this area, the beds change from a low dip high in the two mines to vertical and even overturned at depth.

In the Lark section, there are six major ore-bearing concordant bedding structures on or near the footwall contacts of the Fortuna sill, Mayflower limestone, Commercial limestone, Lark limestone, Jordan limestone, and the Lower Intermediate limestone. Much less production has come from nine other bedding deposits in quartzite and thin sandy limestones. In the Lark section, the quantity of ore produced from the concordant deposits has greatly exceeded that produced from crosscutting fissures and replacement bodies, whereas, in the rest of the district the opposite is true.

The control for the emplacement of the concordant lead-zinc ore is both structural and chemical. Structurally controlled ore is found in and along the bedding fault zones. There is no doubt that the footwall bedding fissure of the Lark limestone was the conduit for the ore solutions, with ore having been deposited

locally in broken quartzite below the limestone, yet in places the entire width of the limestone was replaced. In the latter case, the ore may be from 8 to 15 feet thick. The Lark limestone is, for the most part, black and only little altered. This is in contrast to the major concordant lead-zinc ore body in the Commercial limestone in the Lark section that was emplaced along a weak bedding structure in bleached, altered limestone. This concordant deposit grades upward into replacement bodies which may be as much as 30 feet thick and in one case are separated into three distinct layers.

The Commercial, Lark, and Jordan concordant ore bodies at times show a similar sequence of deposition as shown by layering of mineral assemblages, controlled possibly in part by renewed faulting at the time of ore deposition. The following description of a stope in the Jordan limestone illustrates these facts. Pyrite and quartz mineralized a zone six feet thick that included the bedding fissue and part of the limestone above it; this zone remains essentially unbroken. However, a 6- to 8-foot layer directly above this zone contains banded pyrite, galena, and sphalerite and in places grades upward into a brecciated collapse zone that was replaced in part by galena, sphalerite, and pyrite. Large amounts of ore have commonly been found in the broken quartzite under the Lark and Jordan limestones usually as galena and pyrite.

In the upper part of the present workings of the Lark section, some lead-zinc ore was mined from a thin, sandy limestone layer 10 to 15 feet below the footwall of the Jordan limestone. This structure did not persist laterally or in depth.

A few restricted areas in the Lark section have had concordant copper deposits intimately associated with pyrite and a few that have graded laterally into lead-zinc ores. These, for the most part, have contained less than 5 per cent copper.

An example of the localization of concordant copper deposits within a favorable bed is given by Boutwell (2) in which he described thin lenses of calcareous carbonaceous shale between the quartzite and the footwall of the Fortuna sill with ore having formed along this contact in lenticular pencils or pod-shaped shoots.

In the Lark section, the Commercial, Lark, and Jordan limestones are very favorable host rocks for ore deposition. When replaced by sulfides, these beds may contain nodules and layers of unreplaced syngenetic chert; such chert is absent in massive high-grade ore.

Sills and dike-like masses of monzonite are associated with some of the concordant bedding deposits; the main effect of these intrusions was detrimental to ore deposition in that they cut out parts of the productive zone and today create difficulties due to their expansion when exposed during mining operations. In specific cases, however, they have acted as damming structures that localized small replacement ore bodies below them.

In the U.S. section and the Armstrong Mine, the concordant bedding structures and the northeast-striking fissures form a near perfect intersecting pattern. Although there are ten major bedding fissures in six separate limestones, only six of these fissures contain definite concordant ore bodies. Because of the intersecting pattern formed by the northeast-striking fissure system with the bedding fissures, replacement ore bodies in the Jordan and Commercial limestones were formed that extended not only along the bedding fissure zone, but also along the crosscutting fissures. Although these are in part concordant bodies, they probably should be classed as replacements and will be described under that heading.

The Highland ("14" and "850") limestones, consisting of two members, have had concordant deposits along their footwalls. Another concordant ore-bearing limestone known as the Lower Intermediate (St. Joe) in the Lark section is not ore bearing in the U.S. section or in the Utah Metals Mine where it is known as the Utah Metal limestone. The "B" limestone, also known as the Levant in the Utah Metals Mine, has supplied over 1.25 million tons of lead-zinc ore in the U.S. section from two concordant ore bodies, one on the footwall and one near the hanging wall. This limestone has been productive from where it is cut off by the Roll fault down-dip for over 4000 feet. The "D" limestone also has two concordant ore zones which are within the limestone and which together have produced over 1.5 million tons of lead-zinc ore. The "D" limestone has been productive for over a dip length of 3400 feet. The "B" and "D" limestones and their contained ore deposits did not outcrop in the U.S. section (Figure 5).

All the limestones in the U.S. section, lying as they do between the Bingham stock and the Last Chance stock, are metamorphosed, and those of the alphabetical series in the heart of the productive area are silicated. An important feature influencing the deposition of the concordant ore deposits in the U.S. section, and not seen elsewhere in the district, has been

the introduction of large tabular bodies of coarsely crystalline white to blueish white calcite along the bedding faults. This carbonatization followed silicatization and as such is found enclosed by walls of silicated limestone or with one wall of quartzite and the other of silicated limestone. The first stage of igneous intrusion silicated all the alphabetical limestones. A second stage of igneous or higher-intensity hydrothermal activity is indicated as following the introduction of the calcite on bedding faults since locally a portion of such bedding close to igneous rock is now composed of very coarse-grained wollastonite, diopside, and garnet. This was followed by deposition of lead-zinc ore which replaced much of the calcite. The unreplaced calcite in places forms almost a halo around the ore bodies. In other places, it was only partially replaced and remains as gangue. Seven different concordant bedding zones in five separate limestones have this pre-ore-deposited calcite along strong bedding faults in almost completely silicated limestones. Six out of the seven zones contain major ore bodies. The location of these concordant ore bodies was dependent not only on the permeability of the coarse-grained calcite, but also on the repeated movement along bedding-plane faults. This same process of carbonatization is thought to have introduced calcite for hundreds of feet out into the crosscutting fissures that intersect the bedding structures at almost right angles.

Post-Ore Metamorphism  In one instance, where portions of a concordant ore body were once believed to have been brecciated by bedding faulting, it is now thought to have undergone post-ore leaching that caused collapse of the ore and some of the wall-rock; the breccia was subsequently cemented by silica and calcite. The ore fragments, which at times were large, were also angular. Later deposition of realgar, orpiment, stibnite, and pyrite cuts across this breccia. Unbroken ore is found at both ends of this collapse zone.

Post-mineral faulting along concordant bedding structures has resulted in deformation by pressure, causing a pseudoflow structure in galena that is seen frequently throughout the district.

## Crosscutting Fissure Deposits

Stratigraphic Relations  The crosscutting fissures usually cut the bedding fissures or veins of the limestone units and the interbedded quartzites at high angles and in some cases have cut part of both the Last Chance and Bingham stocks. The major crosscutting fissures of the northeast-striking fissure system are located on the south and southwest side of the Bingham stock and are found in the Utah Apex, Highland Boy, Utah Metals, Armstrong, U.S. section, and Combined Metals Mines (Figure 1). Across 10,000 feet of this northeast-striking fissure zone, there are thirty named fissures and numerous unnamed ones that have produced ore. Some have been explored over a vertical range of 5000 feet. Contrary to an earlier and frequently quoted opinion that fissures do not contain ore except in the vicinity of limestone beds, the major part of the tonnage of the northeast system has been produced from within porphyry and quartzite. In many cases, structure has proven to be more important than host rock in effecting ore deposition. The fissures are usually steep and most dip eastward; some have reverse dips, and many flatten with depth. Most of them show normal faulting offsets; however, farther to the southwest, part of this northeast-striking system shows high-angle reverse faulting (4).

In the Lark section, more than a million tons of lead-zinc ore have been mined from eight named crosscutting fissures. Smaller production has come from a few unnamed ones. Of the named fissures, five join with the Lark vein on its footwall side and strike northeastward through quartzite and limestone beds and dip toward the northwest (Figure 3). These have been mined up to 900 feet laterally under the Lark vein and over a vertical range of 1700 feet. Fissure ore in the Lark section has come from both limestone and quartzite and varies in width from 1 to 10 feet.

In the Utah Apex and Highland Boy Mine area, crosscutting fissures have also produced lead-zinc ore in both limestones and quartzites. The Leadville fissure lead-zinc ore bodies were 2500 feet long and were mined to a depth of 2000 feet (12).

MINERALOGY Three main types of ore have been mined from these fissures. In order of importance there are lead or lead-zinc, pyritic-copper, and silicious gold-silver. In the lead or lead-zinc fissures, the sulfides consist of pyrite, sphalerite, and galena with minor amounts of chalcopyrite, silver-bearing tetrahedrite, and tennantite; rarely, native gold has occurred. The sulfides are usually coarse grained and massive. Bands of alternating sulfides and gangue are common in these crosscutting fissures, which range from inches to over 20

feet in width. Quartz and calcite are the common gangue minerals. Much calcite was introduced into some of the wider parts of the major fissures in the Last Chance stock prior to ore deposition and at times remains as unreplaced masses with widths as great as 12 feet. Rhodonite and rhodochrosite are frequently found as gangue near the outer limit of mineral deposition. The gold and silver content of the ore usually increased toward this outer limit along the northeast-striking fissure system. As the Bingham stock is approached, pyrite becomes the predominant mineral. James, Smith, and Bray (12, p. 93) in describing the mineralogy of the Bingham stock area say: "during mineralization some iron appears to have been removed from the central zone and deposited peripherally." This iron, as pyrite, forms a zone that is neither uniform nor continuous, such pyrite being found in many of the fissures as they are followed toward the Bingham stock. In the Combined Metals, U.S., Armstrong, and part of the Utah Metals Mines, the following zonal pattern has been observed (but only at comparatively shallow depths) outward from the center of the Bingham stock: from 4000 to 5000 feet from the stock, a pyrite and cupriferous pyrite zone; from 5000 to 7000 feet, a lead and lead-zinc zone; and from 7000 to 9000 feet plus, a zone of less lead-zinc but more gold and silver. On the Last Chance fissure, which is one of the major northeast-striking fissures, the grade diminished at between 4000 and 5000 feet below the surface due to the decrease in the size of the fissure, and pyrite became predominant over lead-zinc. However, the total amount of mineralization also decreased. A small amount of pyrrhotite has been found on the lowest penetration. Large distorted subhedral pyrite crystals, up to six inches across, are characteristic of this area at depth, and the fissure gangue becomes dominantly quartz. Erratic high-grade gold and silver were mined from some of the lead-zinc ore shoots on the lower levels. Considerable hydrothermal leaching occurred near the fissures in silicated limestone, and some areas are honeycombed with openings. Fissures, where tighter, exhibit highly polished slickensides.

In the Lark section, since the crosscutting fissures are located thousands of feet from the Bingham stock, no zonal pattern was developed. The Midas thrust or North fault in the Lark section is considered as one of the crosscutting fissures and contains several ore shoots. For the most part, these ore shoots were deposited at the intersection of the Midas thrust and weaker crosscutting faults. Because of this restricted nature of the ore shoots, no zonal pattern is seen either laterally or in depth.

Although no zonal arrangement is described laterally for the Leadville vein in the northwest part of the district in the Utah Apex and Highland Boy Mines, a mineralogical change in depth was described by Lindgren (8, p. 121) "In the Leadville vein at Bingham a low-temperature lead-zinc deposit changes at a depth of about 2000 feet into tennantite and chalcopyrite ore."

Siliceous gold-silver fissures have been found throughout the district, usually high in the mines. Although referred to as gold-silver-bearing fissures, they contained appreciable copper, lead, and zinc, and in places graded into lead-zinc deposits. According to Boutwell (2), the common gangue was calcite, quartz, barite, and rhodochrosite; these minerals were associated with pyrite, galena, sphalerite, and occasionally arsenopyrite. These same fissures in the oxide zone had the oxides and carbonates of some of the above elements. The silver content ranged from a few ounces to over 100 ounces per ton. Some of the major lead-zinc fissures of the northeast-striking system were mined for their gold and silver content close to the outer and upper limit of ore deposition.

Pyritic copper has been mined from fissures in the western part of the district as described by Hunt (3). These are included in the description of concordant ore bodies in this paper. Old records show that high-grade copper ore was frequently mined in association with lead-zinc, gold, and silver. None of the fissures in the district has been mined specifically for its copper content in recent years.

Wall-rock alteration depends not only on host rock but in part on distance from the igneous intrusive.

The sedimentary series of the Armstrong and U.S. section, lying as it does between the Bingham stock and the Last Chance stock, was intensely metamorphosed by silicatization, silicification, and marmorization. In the Lark section, the effects of metamorphism are confined to the western part of the mine, reaching a point 3800 feet from the center of the Bingham stock. The processes involved here were silicatization, silicification, and bleaching. Metallization occurred for an additional 5500 feet beyond the limit of metamorphism into black, slightly altered limestone for a total mineralized length of 9300 feet from the center of the Bingham stock. In the Apex and Highland Boy Mine area, metamorphism of

the sediments extended from 1500 to 2000 feet from the west edge of the Bingham stock.

Hunt (11, p. 110–112) describes a typical wall-rock alteration along one of the principal fissures in the Last Chance stock in the U.S. section, "The sulfides in the vein consist of coarse galena, brown sphalerite, and pyrite with a few small grains of chalcopyrite and numerous chalcopyrite blebs in sphalerite. The gangue is massive quartz and carbonate with a few tufts of interstitial white mica. . . . The features described from this well-zoned alteration envelope bordering the fissures can be summarized as a series of mineralogical changes progressing toward the vein: (1) plagioclase is partially broken down to montmorillonoid clay and brown biotite develops local blotches of green color; (2) sericite develops at the expense of montmorillonoid clay, plagioclase and alkali feldspar; (3) chlorite appears in place of biotite; (4) chlorite disappears and there is a marked increase in the abundance of sericite. . . . Bordering the vein, the porphyry has been strongly altered in a zone over two inches in width."

Wall-rock alterations along fissures in quartzite and silicated limestones have also been well described by Hunt (11). The following sulfides were found in the fissure zone: pyrite, sphalerite, and galena with minor amounts of chalcopyrite and tetrahedrite-tennantite and some stibnite. The sulfides were coarse-grained and massive in texture. Where the fissure cuts through quartzite, the only alteration observed in the quartzite at the contact with the lead-zinc ore was a general increase in the amount of interstitial calcite. According to Hunt (11, p. 123), "At the contact between lead-zinc ore and silicated limestone, a soft zone extended several inches away from the ore. This altered limestone consisted mainly of a talc-saponite mixture with quartz and calcite, and trace amounts of magnesium bearing kaolin-like material. Optically, minor quantities of epidote, diopsidic pyroxene and irregular grains of either idocrase or zoisite were observed within the talc-saponite matrix." At other places within the silicated limestone, he identified light colored clays as being dominantly montmorillonoid. He also determined some green and gray clays as being composed of the following: chlorite, trioctahedral mica, green talc, and a green montmorillonoid clay. At times gypsum and some of the vermiculite minerals are present in small amounts.

In the Lark section, the fissures are outside the zone of metamorphism in only slightly altered limestones and quartzites. As such, the only obvious accompanying alteration products are quartz and calcite.

The Utah Apex and Highland Boy mine area probably has the same type of wall-rock alteration along its crosscutting lead-zinc fissures as that seen in similar host rock in the U.S. section, Armstrong, and Lark Mines. This assumption has been made since little has been published on this subject for the Utah Apex and Highland Boy Mines, and they are no longer accessible.

FACTORS CONTROLLING FORM AND LOCATION OF ORE BODIES  The northeast-striking fissure system in the southern and western part of the district forms for the most part a simple parallel fault pattern in the heart of its productive zone. Some of the structures, as they were explored southwestward toward the outer limit of ore deposition, changed from simple tension fractures to an en-echelon pattern and finally horsetailed. A near perfect intersecting system is developed where the northeast-striking system cuts the bedding structures at almost right angles. Slumping along bedding-plane faults toward the Bingham stock caused associated movement along the crosscutting northeast-striking fissures as shows by mullions with dips of 40° to 50° downward toward the Bingham stock. Unlike fissure intersections that have formed large replacement pipe-like ore bodies in other parts of the district, the intersections of the northeast-striking fissures and bedding structures deep in the U.S. section had no such influence on ore deposition. The low porosity of the silicated host rock may have been responsible for this. Winchell (5) shows that completely silicated limestones of Bingham have a porosity of only 3.3 per cent whereas partially silicated rock averages 5.9 per cent. Another controlling factor of ore deposition, in addition to openings produced by renewed faulting, has been the introduction of coarse-grained calcite in both the crosscutting fissures and along bedding structures that, with their greater porosity, captured the bulk of the ore solutions. Proof of this is shown by the quantity of ore found in crosscutting fissures irrespective of host rock. The lateral extent of ore deposition on the northeast-striking fissures may be as much as 3000 feet greater near the surface than at depth. One of the obvious reasons for this is the more open nature of the fissures near the surface regardless of host rock. On the lowest mine level in the district, one of the major northeast-striking fissures cut silicated limestone that was honeycombed by hydrothermal leaching. These near-barren

openings appear to have been channels for mineral solutions that deposited ore in the fissures above.

In the Lark section, five of the crosscutting fissures form a roughly parallel pattern, striking northeastward from their junctions with the Lark vein (Figure 3). The Lark vein was probably the main conduit for the ore deposited in this fissure system. The bulk of this ore was produced from fissure intersections, with pipe-like replacement ore shoots being formed in black, slightly altered Jordan limestone. One such ore shoot has been mined for 3480 feet down its plunge.

Two fissure zones in the Lark section, which have made replacement ore deposits at their junctions with the Midas thrust, have had only limited ore away from these junctions. Both the strength of the fissure zones and the accompanying ore diminished up and down dip. One of these fissure-controlled ore bodies in quartzite terminated up dip mineralogically in that the lead-zinc ore changed to pyrite. The other ore body, formed in limestone, terminated up dip as the fissure zone died out. Both end down dip where the fissure zones weaken. The dip distance mined on these two ore bodies was from 1400 to 2000 feet.

POST ORE METAMORPHISM  As seen at times in lead-zinc replacement ore deposits, there has been strong corrosion and pitting of some of the fissure minerals and the addition of post-ore silica and calcite. Pseudoflow structure in galena has developed along many of the fissures by post-mineral faulting.

## Replacement Ore Bodies

STRATIGRAPHIC RELATIONS  The major non-porphyry mines of the Bingham district have had large replacement ore bodies of both copper and lead-zinc. In the Utah Apex-Highland Boy mine area, the large copper ore bodies were found in the Commercial and Jordan limestones within 1500 to 2000 feet of the Bingham stock. The lead-zinc deposits were likewise found in these two limestones beyond the copper deposits and up to 4000 feet from the stock. These limestones are 200 feet or more in thickness and locally were replaced almost from footwall to hanging wall by ore. The host rocks for the copper ores were light colored, highly metamorphosed, silicated limestones consisting of garnet, wollastonite, diopside, tremolite, asbestos, and specularite (3). The lead-zinc ores were found in marmorized to only slightly altered dark-colored limestones.

Hunt (3) says that ores were found down to 2500 feet vertically or 5000 feet on the dip of the limestones from the surface.

In the old Boston Consolidated or Armstrong Mine area, as well as in the U.S. section of the U.S. and Lark Mine, the Jordan and Commercial limestone ore bodies were oxidized near the surface; these ore bodies extended down dip for 2000 feet. The limestones are likewise in the zone of intense metamorphism and are silicated and marmorized in the western part, becoming both marmorized and silicified to the east. Deep in the U.S. section is a lead-zinc ore body that replaces part of the marginal remnant of a breccia pipe. This pipe is composed of angular to rounded fragments of silicated limestone, quartzite, and porphyry in a porphyry matrix. Farther to the east in the U.S. section is a large area of the Jordan limestone that is almost completely silicified with opaline silica and contains irregular replacements of lead-zinc, pyrite, and chalcopyrite.

In the upper part of the Lark section in the Jordan and Commercial limestones, smaller replacement ore bodies were found that were similar to those found in those same limestones high in the U.S. section in that they contained both copper and lead-zinc in both sulfide and oxide minerals. These ore bodies also extended roughly 2000 feet down dip from the surface. At lower elevations in the Lark section, irregular replacement lead-zinc deposits extend upward from the concordant footwall bedding-fissure ore for dozens of feet into the Jordan and Commercial limestones. Other replacement ore bodies in the Lark section are found in these two limestones where they are cut by strong crosscutting fissures; these have been discussed under crosscutting fissures. Other occurrences of ore found in different host rocks are the replacement deposits found in highly fractured quartzite above and below the Lark limestone and below the Jordan limestone. The intersection of a fissure zone with the Midas thrust has formed a large pipe-like replacement ore body in quartzite. This was also discussed under crosscutting fissures.

In the Lower Bingham area, Boutwell (2) describes irregular sinuous pipes of ore in the hanging wall as replacements from fissures. Although these were high-grade in gold, silver, and lead, they were of minor economic importance because of their limited size.

MINERALOGY  The mineralogy of the near surface oxidized and enriched zone throughout the district is similar enough for all the occur-rences to be described together. This zone, however, differs greatly in depth over the district. In this zone, the copper ores in limestone consisted of azurite, malachite with cuprite, and rarely native copper and were usually underlain by supergene chalcocite. At lower depths, the chalcocite appeared only as a coating on pyrite and chalcopyrite. Undoubtedly, many other copper minerals occurred that are not described here, but since these areas have been mined out, covered by dumps, or for other reasons are inaccessible, nothing more is known about them.

Cores of galena enclosed by anglesite and cerussite are typical occurrences in the oxidized zone. Minor amounts of the lead oxide, massicot, have been found in the same area. Gold and silver, although very erratic in occurrence, locally provided high-grade siliceous ore in association with both lead-zinc and copper in the oxidized zone. Boutwell (2), in describing a zone of sulfide enrichment, mentions the occurrence of gold and silver tellurides.

With the exception of the enriched portions of the copper-replacement ore bodies, the primary ore is essentially chalcopyrite and pyrite with small amounts of gold and silver. The Highland Boy ore is reported to have contained over 75 per cent pyrite. The sulphides of this mine were coarse-grained, although frequently every structural and textural feature of the limestone was preserved. Orrin Peterson (4, p. 912), in describing the Highland Boy copper ore shoot, says, "This zone probably contained more than 50 million tons of heavily mineralized ground, a considerable portion of which was too lean to mine." Several million tons were mined from this Highland Boy copper ore shoot that, for a long time, averaged 2.5 to 3.0 per cent copper. Lower-grade pyritic copper ores were found at depth.

According to Boutwell (2), in the rest of the major mines in the district, the primary copper replacement ore was essentially chalcopyrite and pyrite with minor gold and silver in a siliceous gangue. The copper content had an approximate mean of between 3 and 4 per cent, the accessory gold averaging from ten cents to $1 and silver averaging from 2 to 5 ounces per ton. No replacement-copper ore bodies have been mined in recent years.

Primary lead-zinc-replacement ore bodies differ mineralogically depending on host rock, depth, and distance from the Bingham stock. A description of the alteration and associated minerals found in the Jordan and Commercial limestones in the Utah Apex and Highland Boy mine area outward along strike from the

Bingham stock tells of the following zones being crossed. First, the intensely metamorphosed white to pale green lime-silicate zone with abundant garnet is crossed, then a crystalline carbonate zone with all gradations from silicated rock to pure marble, then still farther out the silicate rocks and marble become streaked with gray to black bands of apparently unaltered limestone, and finally grade into black limestone. Most of the replacement copper ore came from the altered white limestone. The lead-zinc-replacement ore bodies were found in the outer margin of the altered white limestone or in the large area of black or apparently unaltered limestone. In this same fringe area, lead ore was intermingled with copper ore, but inward from this zone would be found copper with only minor lead or outward lead with mere traces of copper (3).

The lead ores consisted of galena, pyrite, sphalerite, chalcopyrite, gold, and silver. Locally within the district, some later age stibnite is found with the lead ore. The zinc, copper, and the silver content differs considerably from one area to another. Quartz and calcite are the common gangue minerals. In some of the large replacements in the metamorphosed limestones and in one breccia remnant, calcite, chlorite, and specularite were abundant gangue minerals. The texture of the lead-zinc-replacement ore is for the most part coarse-grained.

In the Lark section, the lead-zinc replacements in little altered black limestones and in quartzites have little associated wall-rock alteration and not much gangue. Unreplaced limestone and quartzite are the only obvious diluents. In the limestone replacement bodies, the ore boundaries are usually sharp as compared to those in quartzite, which are frequently gradational and must be determined by assaying. In the Lark section, one occurrence of a replacement ore body in quartzite was mined on dip for a distance of 1400 feet where the lead-zinc ore passed into pyrite that extended for several hundred feet farther upward.

No generalization can be made concerning grade of the lead-zinc replacement ore bodies in any host rock, as is shown by the following table of the four major replacement ore shoots in the Lark section.

Number 1 is only 600 feet away from number 2, and both make junctions with the Midas thrust. However, as shown above, number 1 is a replacement in quartzite while number 2 is in limestone. A comparison of numbers 2 and 3, both of which were formed in limestone, shows sizeable differences in grade also.

TABLE IV.   *Comparison of Grade of Replacement Ore Shoots in the Lark Section*

| Host Rock | Tonnage Mined | Average Grade | | | | |
|---|---|---|---|---|---|---|
| | | Au | Ag | Cu | Pb | Zn |
| #1 Quartzite | 134,000 | .038 | 5.50 | .98 | 15.30 | 1.95 |
| #2 Limestone | 263,000 | .019 | 2.90 | .31 | 10.36 | 14.18 |
| #3 Limestone | 76,000 | .090 | 5.19 | .67 | 14.13 | 5.62 |
| #4 Limestone | 40,000 | .022 | 8.89 | .42 | 9.70 | 8.18 |

Number 4, likewise formed in limestone, shows still a different grade. No reason, as yet, is apparent for these differences.

Replacement lead-zinc deposits found in the opaline silicified Jordan limestone in the eastern part of the U.S. section formed pipe-like bodies at the junctions of bedding fissures and cross fissures. At times these were in the form of bulges extending upward from concordant bedding-fissure ore bodies. Copper and lead-zinc deposits were intermingled in the stopes. One old famous stope is reported to have carried 45 per cent lead, 18 ounces of silver, and 1.5 per cent copper. Most of these replacements became pyritic downdip as they approached porphyry. Pyrite was scattered through large areas of the footwall quartzite. A sample of the barren, altered material within the ore bodies assayed 88.25 per cent $SiO_2$ (2).

FACTORS CONTROLLING FORM AND LOCATION OF ORE BODIES   The contact metamorphic copper-replacement ore deposits of the Utah Apex-Highland Boy mines were formed adjacent to the Bingham stock in the Jordan and Commercial limestones, probably just after silicatization. These were evidently the first ores formed in the district. Fissures, fractures, and bedding planes seem to have facilitated alteration, and these features, together with dikes, have, in some cases, determined the local distribution of ores (3). Part of the closely spaced parallel pattern of the northeast-striking fissure system high in the U.S. and Armstrong mines was responsible for the large replacement lead-zinc and copper ore bodies in the Jordan and Commercial limestones of these two mines. These crosscutting fissures contained good ore bodies for thousands of feet below the replacements. The junction of several major fissures at or near the breccia-pipe remnant deep in the U.S. section is probably responsible for this replacement ore body. The intersection of the Midas thrust with cross fissures as well as cross fissure intersections with favorable limestone beds have controlled the localization of some of the major lead-zinc replacement deposits in the Lark section.

POST-ORE METAMORPHISM In the western part of the district, the near-surface lead-zinc- and copper-replacement ore bodies were broken by post-mineral cross faulting which facilitated oxidation of these deposits. In the deeper part of the Lark section, the replacement ore bodies show little effect of post-mineral faulting.

Higher in the Lark section, a small late-stage deposit of cinnabar was found associated with a lead-zinc-replacement ore body in fractured quartzite above the Lark limestone. There was no apparent effect on the earlier lead-zinc sulfides. The close association of the cinnabar with just the lead and zinc sulfides suggests that these may have acted as precipitating agents for the cinnabar.

At certain localities within the district, there has been strong corrosion and pitting of galena, which apparently was the last mineral deposited in the first stage of metallization. No correlation can be made between this observation and renewed igneous activity, although there were several stages of the latter. Some of the galena removed during the process of corrosion may be what was deposited on other sulphides to give the appearance of overlap.

## Mineral Paragenesis of Primary Ores

The paragenetic sequence of the major minerals is: (1) quartz, (2) calcite, (3) pyrite, (4) sphalerite, (5) chalcopyrite, (6) tetrahedrite, (7) galena. There is ample evidence of a weaker repeated deposition of the sequence or part of it as is often seen on corroded and pitted galena and sphalerite. Some observations indicate overlap. Peterson (14) suggested there may be even a third set of sulfides on the other two in the Highland Boy and Apex Mines. Enargite is rare, and its position in the sequence is known only from a few crystal specimens coating pyrite. Silver is found in with both galena and tetrahedrite.

Small showings of stibnite, realgar, orpiment, cinnabar, and pyrite form a later sequence that cut across the first stage of sulfides locally in the U.S. section. Locally in the Lark section, cinnabar is a later mineral, apparently precipitated by the earlier galena and sphalerite that formed a small replacement in quartzite. The cinnabar of the U.S. section and the deposit in the Lark section are 11,000 feet apart, but both were formed at almost the same elevation. The U.S. section showing was almost 3000 feet below the surface. Neither occurrence was commercial.

Quartz, calcite, and some pyrite continued through the sequence. Native gold which is very rare was probably deposited late.

## Geologic Sequence for the Formation of Primary Ores

(1) Deposition of an alternating series of sandstone and limestone.

(2) Folding of the Bingham syncline and the contemporaneous formation of bedding or near-bedding faults, thrust faults, and some of the cross-cutting fissures.

(3) Igneous intrusion and the subsequent silication, silicification, and marmorization of the limestones in a large part of the district.

(4) The contact-metamorphic copper ore of the Utah Apex-Highland Boy Mine area, according to Lindgren (8) was formed following silication.

(5) Introduction of the pre-ore silica and later calcite along the bedding faults and cross-cutting fissures in much of the district.

(6) Leaching of portions of the limestones in the Lark section and formation of some collapse breccias.

(7) Renewed faulting before and during ore deposition.

(8) Primary ore deposition with some repeated deposition and overlap in certan places, followed by late-stage deposition of stibnite, pyrite, realgar, orpiment, and cinnabar locally.

Note: Stage 4 may actually have been the first portion of stage 8.

# ECONOMIC GEOLOGY—SECONDARY ORE

## Supergene Sulfide Enrichment

Supergene sulfide enrichment has not been very important in the nonporphyry ore deposits of the Bingham district. Oxidation in the district was dependent not only on the location of the primary ore deposit but also on numerous factors such as type of host rock, faulting in the host rock, amount and kinds of minerals, rate of erosion, attitude of the host rock, and depth of the water table.

The zone of oxidation is quite shallow, due in part to rapid erosion. Sulfides were found outcropping. In certain areas of the Commercial and Jordan limestones, the zone of oxidation is underlain by large masses of pyrite, with some chalcopyrite, that locally were coated with chalcocite. Hunt (3) states that supergene chalcocite was an important ore mineral in the upper 300 feet of many of the Highland Boy Mine stopes. As mentioned earlier under concordant ore bodies, the Brooklyn area of the Jordan limestone shows the greatest depth of oxidation, 1450 feet down the dip. This area overlies the deepest occurrence of secondary copper and silver which reached a maximum distance of 3600 feet below the surface on the dip of the vein. Boutwell (2) in describing the silver found in the carbonate lead-zinc ores of the Telegraph Mine area says that, in over 10,000 tons of ore, the silver content averaged 57 ounces per ton. In numerous other mines, the high silver content was closely associated with carbonate ore. In a few instances silver, probably as argentite in supergene chalcocite, has been found at the lowest level reached by secondary sulfides. Assays of over 80 ounces of silver have been recently obtained locally. This is over ten times the norm for silver in the enclosing sulfides.

## Oxide Enrichment or Residual Concentration

In the oxidized zone, in addition to the usual occurrence of azurite and malachite, cuprite, some native copper, and chrysocolla, galena was found enclosed by layers of anglesite and cerussite. Although rare, massicot has been found with the carbonates. Gold ore in a honeycombed siliceous gangue with some pyrite was found in part of the near-surface oxidized zone. Rapid erosion of such zones would account for the nearby placer gold.

## Controls of Supergene Ores

The deepest occurrence of secondary copper and silver, found in a restricted zone, shows both physical and chemical controls and is explained by the footwall bedding-fault zone During the deposition of primary ore, this fault zone and several feet of the limestone were replaced by quartz and pyrite. Post-mineral faulting fractured part of the silicified zone and the quartzite beneath it. Copper and silver-rich solutions migrating downward through this unreactive siliceous environment replaced primary chalcopyrite, pyrite, and galena which lay adjacent to the siliceous zone. This process reached a maximum depth of 3600 feet below the surface. At times, the secondary silver-bearing chalcocite, with covellite, digenite,

cuprite, and native copper, formed thin-bedded deposits within the primary sulfides. These secondary deposits, because of their limited size, did not make ore by themselves. Similar occurrences have been associated with the Lark vein, but these were not as large.

The bulk of the secondary ore mined in the district came from partly within and below oxidized zones of the large replacement deposits in the thicker limestones. These were confined to comparatively shallow depths because of the precipitating action of the massive sulfides.

## Paragenesis of Secondary Ore

Boutwell (2), in describing the secondary enrichment in the higher portions of the old mines, noted that the paragenetic sequence varied throughout the district. In the Highland Boy mine, it was characterized by chalcopyrite being coated with bornite, whereas in another property covellite coated chalcopyrite. Other occurrences showed chalcopyrite coating pyrite, which in turn was coated with black copper sulfides, which were later determined to be chalcocite, tenorite, melaconite, and some tetrahedrite. An analysis of some selected samples of the black sulfides yielded 3.8 ounces of gold, 58.6 ounces of silver, and 42.3 per cent copper, plus some tellurium. The position of silver in the paragenetic sequence is still not known, other than it is intimately associated with the chalcocite and may replace it. Recent observation of secondary sulfides taken from a narrow fissure zone suggests the following sequence: bornite, covellite, digenite, then chalcocite.

## Geologic Sequence for the Formation of Secondary Ores

(1) Deposition of primary ore.

(2) Erosion to place the proper sulfide close to the surface.

(3) Rocks made permeable, in part by fracturing due to faulting.

(4) Oxidation of iron sulfides to yield metals in solution.

(5) Non-reactive host rock to allow solutions to migrate downward to the water table.

(6) Proper sulfides below the water table in an oxygen-free environment for replacement by metals from the solutions.

## ORE GENESIS

The intimate relationship between the Bingham stock and the ore, together with certain features of the ore, indicate a common source. As shown in Figure 1, the chalcopyrite zone, located for the most part in the Bingham stock, is the locus of Kennecott's disseminated porphyry copper deposit and is also the center of the "non-porphyry" ore deposits of the district. Another geographical association is the 2000 foot wide zone of metamorphism that surrounds most of this intrusive. Although the sediments show the effects of hydrothermal alteration in the vicinity of most of the intrusives in the district, those adjacent to the Bingham stock show the greatest degree of alteration, which is roughly zoned outward from the intrusive. The above geographical association and the zoning of the metamorphics, together with an irregular zoning and rake of the non-porphyry ores, substantiate the concept of a common source for the Bingham stock, the hydrothermal solutions, and the ores.

## ACKNOWLEDGMENTS

The writers are indebted to the officials of the United States Smelting Refining and Mining Company for permission to publish this paper and for allowing time spent in its preparation. We would also like to thank Mr. C. L. Thornburg for comments and suggestions. Kennecott geologists and R. N. Hunt furnished, from previous papers, portions of the illustrations used, and to them we offer our sincere appreciation.

## REFERENCES CITED

1. Bancroft, H. H., 1890, History of Utah, p. 741.

2. Boutwell, J. M., 1905, Economic geology of the Bingham district, Utah: U.S. Geol. Surv. Prof. Paper 38, 413 p.

3. Hunt, R. N., 1924, The ore in the limestones at Bingham, Utah: A.I.M.E. Tr., v. 70, p. 857–883.

4. Peterson, O., 1924, Some geological features and court decisions of the Utah Apex—Utah Consolidated Controversy, Bingham district: A.I.M.E. Tr., v. 70, p. 904–932.

5. Winchell, A. N., 1924, Petrographic studies of limestone alterations at Bingham: A.I.M.E. Tr., v. 70, p. 884–899.

6. Gilluly, J., 1932, Geology and ore deposits of the Stockton and Fairfield quadrangles, Utah: U.S. Geol. Surv. Prof. Paper 173, 171 p.

7. Farmin, R., 1933, Influence of basin range faulting in mines at Bingham, Utah: Econ. Geol., v. 28, p. 601–606.

8. Lindgren, W., 1933, Mineral deposits, 3d ed., McGraw-Hill, N.Y., 930 p.

9. Billingsley, P. and Locke, A., 1937, Structure of ore districts in the continental framework: A.I.M.E. Tr., v. 144, p. 9–64.

10. Stringham, B., 1953, Granitization and hydrothermal alteration at Bingham, Utah: Geol. Soc. Amer. Bull., v. 64, p. 945–991.

11. Hunt, J. P., 1957, Rock alteration, mica, and clay minerals in certain areas in the United States and Lark Mines, Bingham, Utah: Ph.D. dissertation, Univ. Calif., 321 p.

12. Cook, D. R., Editor, 1961, Geology of the Bingham mining district and northern Oquirrh Mountains: Utah Geol. Soc. Guidebook to the geology of Utah, no. 16, 145 p.

13. Rose, A. W., 1966, The iron content of sphalerite from the Central district, New Mexico and the Bingham district, Utah: Econ. Geol., v. 56, p. 1363–1384.

# 46. Fine Gold Occurrence at Carlin, Nevada

DONALD M. HAUSEN,* PAUL F. KERR†

## Contents

* Newmont Exploration Ltd., Danbury, Connecticut.
† Columbia University, New York, New York.

## Illustrations

# Tables

fault. The ore body is generally stratiform and is more or less conformable to altered beds near the top of the formation, underlying Devonian limestones.

Two sequences of mineralization are recognized: (1) an earlier base metal-barite assemblage related to early Cretaceous intrusives ($121 \pm 5$ m.y.), and (2) a later low temperature Au-As-Hg-Sb assemblage of near surface emplacement. The earlier sequence consisting of sparse galena and sphalerite in barite with anomalous amounts of zinc, lead, nickel and copper, associated with dikes of dacitic composition, is of little economic importance. The later sequence of gold, realgar, cinnabar, and stibnite associated with extensive silicification and argillic alteration of limestone beds, has resulted in important deposits of gold.

Argillic alteration of hydrothermally leached carbonate strata has provided the environment in which the most prominent gold deposition took place. Carbonate minerals in the limestone host rock have been replaced by microcrystalline quartz and chalcedony to form stratiform silica masses and recrystallized lenses of euhedral quartz. Zones of porous silicification are light gray to white, elliptical in shape, and more or less follow bedding. Silicification is bordered by argillic alteration and pyritization. Deposition of gold usually lies in a zonal pattern that encircles chimney areas of silicification.

## ABSTRACT

Fine colloidal gold near Carlin, Nevada is disseminated in leached carbonate strata of the Roberts Mountains Formation in the Lynn "window" of the Roberts Mountains thrust

Ore textures, mineral assemblages, and alteration criteria in the Carlin ore body all favor late stage epithermal mineralization. Gold introduction is attributed to late hydrothermal solutions rising along elliptical conduits controlled by permeability of select horizons in the Roberts Mountains Formation. Precipitation of gold has occurred mostly in illitic clays, organic matter, pyrite, and microcrystalline quartz.

## INTRODUCTION

### Location

The Carlin mine is located in north central Nevada, in the Lynn Mining District, in northeastern Eureka County, 40 miles northwest of Elko and 20 miles north of Carlin, (Figure 1). Near-surface ore is being mined from an open pit at an elevation of about 6300 feet above sea level in the Tuscarora Mountains. Gold occurs in Silurian rocks along the eastern edge of the Lynn Window of the Roberts Mountain thrust fault. The host rock belongs to the eastern assemblage in the lower plate group of the Silurian Roberts Mountains Formation (Roberts, *et al.* 1958).

### Acknowledgments

This paper includes much hitherto unpublished information from the files of Newmont Exploration, Ltd., and has benefitted from correspondence and conversations with many at the Carlin mine, the Danbury laboratory, and the New York offices of Newmont Mining Corporation. In particular we thank R. B. Fulton, Vice President and Manager of Exploration; F. W. McQuiston, Jr., Vice President of Metallurgy; D. J. Christie, Metallurgist; R. D. MacDonald, Director of the Danbury Laboratory; and R. W. Hernlund and W. C. Hellyer.

We express our appreciation to J. B. McBeth, P. Loncar, R. L. Akright, and Byron Hardie of the Carlin Gold Mining Company who, provided assistance in making observations at the deposit.

Fire assays for gold and silver were provided by H. Treweek of the Carlin mill and G. V. Larsen of the Danbury Laboratory. Chemical analyses were provided by W. Hart and G. V. Larsen under the supervision of J. D. Crozier, Chief Chemist at the Danbury Laboratory. Sample preparation was conducted by L. DeVries and A. Bochnia. Thin and polished sections were prepared by L. DeVries. We thank all for their generous assistance.

We are grateful to T. I. Taylor of the Chemistry Department and to E. Hamburg of the Physical Metallurgy Department at Columbia University for guidance in the use of the electron microscope. We appreciate the cooperation of P. J. M. Ypma and R. J. Holmes in making suggestions and for their critical reading of the manuscript. We thank Mary Hausen for aid in manuscript assembly.

### Techniques

X-ray diffraction analyses were performed on a General Electric Diffraction Unit, Model XRD-5, using a wide range goniometer and Ni filtered copper $K\alpha$ radiation ($\lambda = 1.5418$). Identification of clay minerals was facilitated by the use of sedimented slides. X-ray diffraction patterns were compared from: (1) untreated samples; (2) glycolated samples; and (3) heated samples taken to 550°C for one hour, in accordance with procedures outlined by Molloy and Kerr (25).

Emission spectographic analyses were obtained from the American Spectrographic laboratories in San Francisco, and from direct use of the Baird spectrograph at Columbia University. X-ray fluorescence analyses were performed semiquantitatively on finely ground Carlin samples by scanning from 10° to 90°, $2\theta$, on a wide range goniometer, utilizing a General Electric Spectrometer, Model XRD-5.

Differential thermal analyses of Carlin ore samples were run on a Brinkman Differential Thermo-Analyzer, Model DDTA IV. This instrument utilizes a thermocouple-amplifier-recorder arrangement to produce a low noise, high gain, differential thermal curve and temperature tracing on a standard Honeywell chart recorder. Gases given off by differential thermal analyses of carbonaceous and pyritic samples were analyzed by their infrared spectra using a Perkin Elmer Model 21 Infrared Spectrophotometer.

Selected hand specimens were inspected visually under the binocular microscope. Thin and polished sections were examined at higher magnifications by means of a Zeiss Universal Microscope with a split beam light source. Gold in the Carlin ore is finely disseminated through poorly consolidated argillaceous silty rocks that require careful impregnation with epoxy plastic prior to cutting, grinding, and polishing.

Electron micrographs were taken of submicroscopic gold from the main ore body. Microscopic equipment included an RCA electron microscope, model EMU-2B and a Siemens electron microscope.

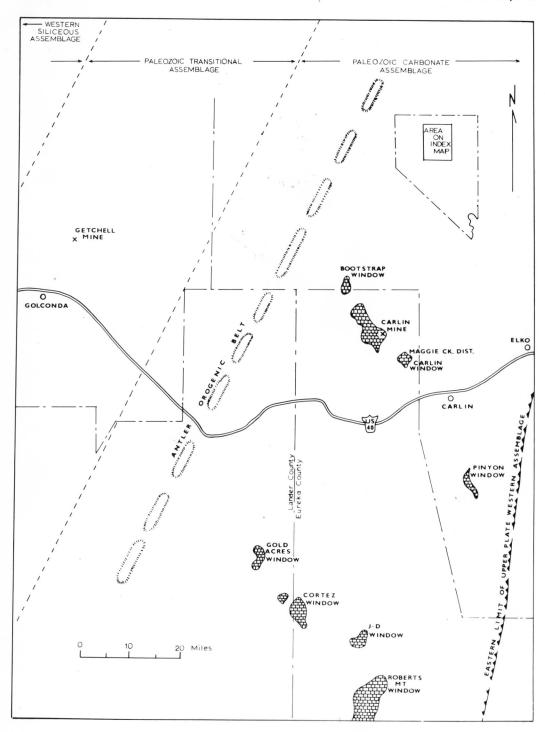

FIG. 1. *Index Map of the Carlin Mine.*

## HISTORICAL INFORMATION

Visible gold was discovered in placer deposits in the Lynn district in April, 1907. This was followed by sporadic production from shallow surface and underground workings, mostly from eroded upper plate rocks of the Ordovician Vinini Formation. In lower plate rocks in recent pit operations, visible gold has not been observed.

The Carlin property is operated by the Carlin Gold Mining Company owned by the Newmont Mining Corporation. Exploration was first prompted by a U.S. Geological Survey report by R. J. Roberts (23), describing the alignment of mining districts in north central Nevada. Mapping and claim location were followed by drilling in 1962, with significant gold being found in the third drill hole. Estimated gold reserves in early 1965 were approximately 11 million tons, averaging 0.32 ounces of gold per ton (32).

Construction of a cyanidation mill was started in June, 1964, and gold production commenced in April, 1965.

Mineralogic studies were initiated during the summer of 1964 at the Newmont Research and Development Laboratory in Danbury, Connecticut, in support of metallurgical and geological investigations both at Carlin and Danbury. Several trips were made to the Carlin property in 1965 and 1966.

The Carlin pit was large enough in the late summer of 1966 to expose the altered lower-Paleozoic rocks that serve as a host for mineralization. Although megascopic gold has not been observed in the deposit, alteration features associated with gold content have been inspected in the field and samples collected for detailed laboratory study.

White to light gray, porous, silicified circular to elliptical chimneys, with brown pyritic halos, have been observed near high angle pre-ore faults and in lenses along sedimentary beds. In general, these bleached zones are more or less elongate parallel to bedding, representing zones of high permeability through which solutions apparently migrated upward.

## GENERAL GEOLOGY

### Regional Features

The Carlin mine is situated near the eastern edge of the Lynn window of the Roberts Mountains thrust fault. According to Roberts *et al.* (21), thrusting is related to the Pre-Carboniferous Antler orogeny and occurred along a regional thrust plane that brought western eugeosynclinal clastic rocks into contact with eastern miogeosynclinal carbonate rocks of correlative age. Roberts presumes a single broad continuous sheet extended over a wide region in North Central Nevada; he has designated this as the Roberts Mountains thrust fault.

Other investigators, however, believe that the fault is not a single thrust sheet but rather that multiple planes of thrusting can be distinguished in intricate thrust slices both in the Seetoya Mountains north of Elko (26) and in the Torquina Range south of Cortez (30).

Nevertheless, the term, Roberts Mountains thrust has been assigned by Roberts *et al.* (21) to faults bordering the Lynn and Carlin windows north of Carlin, Nevada, and, to avoid extended structural analysis beyond the scope of this paper, the name will be retained in this discussion.

The effect of thrust faulting as an ore control feature at the Carlin deposit is limited, but using the distribution of windows in the over-thrust sheet as a guide to loci of mineralization has merit. The Lynn window is typical of other exposures of lower-plate carbonate rocks including the Bootstrap window to the north as well as the Carlin, Pinyon, Gold Acres, Cortez, J-D, and Roberts Mountains windows to the south (Figure 1).

According to Roberts (23), the windows have been formed as the result of erosion of upper-plate rocks near local areas of post-thrust uplift and doming. Roberts further points out that the principal mining districts are located in and around eroded windows of lower-plate carbonate rocks and that the alignment of windows indicates zones of structural weakness along which igneous rocks and related ore-bearing fluids have penetrated. Windows are aligned along zones or belts trending to the northwest. The Carlin gold deposit is situated near the north end of the Lynn-Pinyon belt, which extends from the Pinyon window northward through the Carlin and Lynn windows to the Bootstrap window. The Shoshone-Eureka belt southwest of the Lynn-Pinyon belt trends northwestward from the Eureka window through the J-D, Cortez, Gold Acres, and Goat Ridge windows toward the Golconda and Getchell mines. Lower-plate rocks at the north end of this belt consist of the transitional assemblage that may be difficult to distinguish locally from upper-plate rocks of the western assemblages.

A geochemical association of arsenic, mercury, and antimony with gold has been recognized at several localities including Getchell, Bootstrap, Carlin, and Gold Acres in north-central Nevada (Erickson *et al.* 24,29,37). Anomalous association of arsenic, antimony, tungsten, and mercury with gold in the Cortez window also is reported by Erickson *et al.* (37) south of a major northeast fault in silicified Devonian limestones in the lower plate of the Roberts Mountains fault. Similar metallic assemblages have been reported in hot spring deposits, notably from Steamboat Springs, Washoe County, Nevada.

### Structure of the Deposit

The Carlin ore body consists of irregularly stratiform masses of mineralized Roberts Mountains Formation that strike northeast and dip northwest conformably beneath Devonian limestones. The stratigraphy of the district is discussed in the following section, headed "Sedimentary Rocks," and is shown in tabular form in Figure 2.) In plan, the ore body is elongate to the northeast along the strike of the beds, and forms three recognizable mine units (Figures 3, 4): (1) the main ore body, (2) the easterly extension, and (3) the west ore body.

In general, faults are of two types, (1) the older low-angle Roberts Mountains thrust and (2) more or less vertical faults which may intersect the earlier thrust fault. Between are tilted segments of strata. Although the general aspects of the structural pattern are beginning to emerge, the fault structures in the pits are

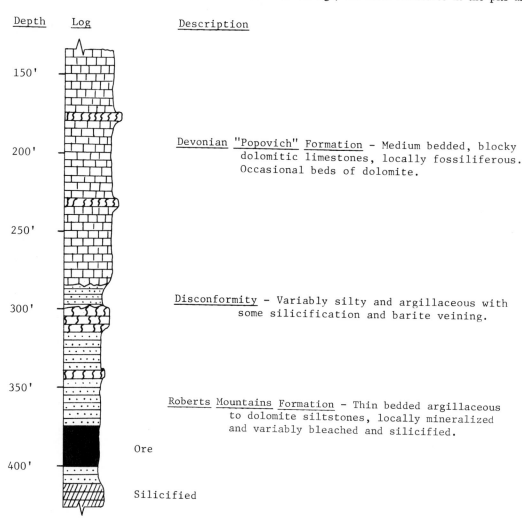

FIG. 2. *Lithologic Log of Drill Hole Based upon X-ray Diffraction.*

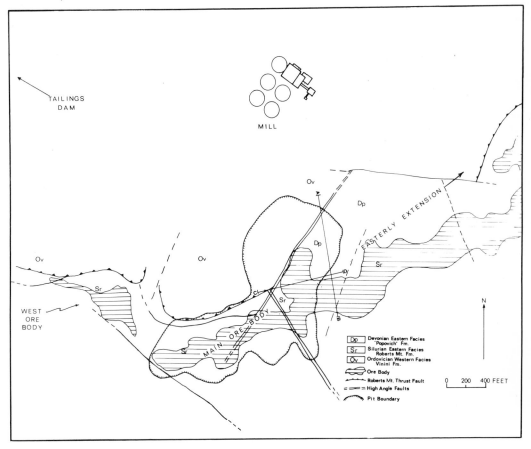

FIG. 3. *Schematic Map of Carlin Mine.*

not yet adequately exposed to complete the structural interpretation.

The west ore body is separated from the main ore body with an intervening northeast-trending vertical fault that marks in places the western boundary of the main ore body (Figures 3, 4). This fault can be followed intermittently along the northwestern edge of the main ore body and to the northeast where it continues to the north of the easterly extension. The Roberts Mountains thrust fault is intersected by the northeast fault along the northwestern edge of the main pit. A considerable vertical displacement of upper plate rocks of the Vinini Formation apparently has occurred along the northeast fault near the point of intersection with the thrust.

A vertical northwest fault is exposed in the east face of the main pit and strikes about N35°W (Figure 3). This fault appears to be pre-mineral and has been intruded locally by a quartz-porphyry dike and barite veins. The dike and adjacent wall rock are highly altered

and numerous silicified "chimneys" occur nearby.

A schematic diagram of the east face of the main pit is shown in Figure 5. The Roberts Mountains Formation is overlain by Devonian limestones with the local name "Popovich" Formation (31). The underlying limestones show argillic alteration, silicifications, and mineralization . These beds are slightly off-set by a steep northwest dual fault pattern that has produced slight differences in strike by tilting. The fault pattern is intruded by a quartz-porphyry dike ranging from about 2 to 6 feet wide. Steep dikes of similar composition in the Roberts Mountains Formation extend to discontinuous outcrop northwesterly to the major northeast fault. Upper-plate rocks of the Vinini Formation are exposed on the northwest side of the major northeast fault at the north end of the pit (Figures 3, 4, 5).

The Roberts Mountains thrust fault is projected in the foreground of the sketch, Figure 6. The intersection of the thrust sheet with

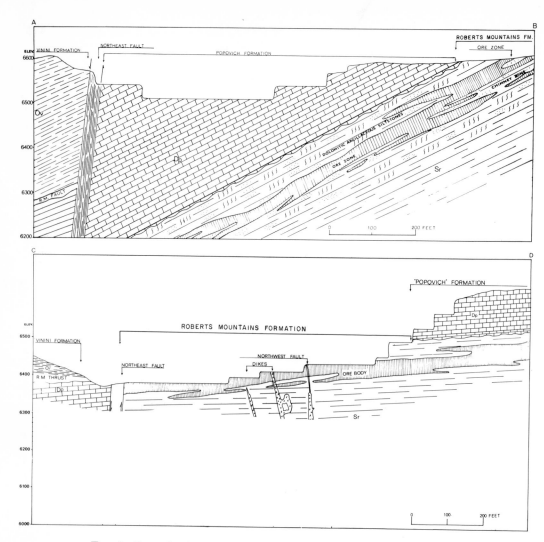

FIG. 4. *Cross Sections of Carlin Ore Body (A-B above, C-D below).*

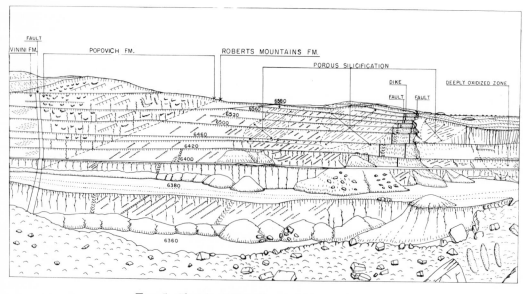

FIG. 5. *Sketch of East Face of Main Ore Body.*

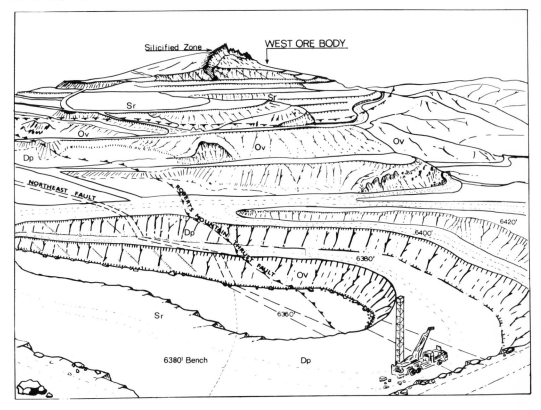

FIG. 6. *Sketch of West Ore Body.*

the vertical northeast fault occurs near this location. In cross-section, the thrust zone consists of about 10 to 20 feet of alternating layers of bleached, brecciated, and iron-stained rock rubble, above which lie nearly horizontal beds of cherts and carbonaceous shales of the Vinini Formation. Sediments immediately below the fault are highly altered beds of the "Popovich" limestone. Argillic alteration is prevalent along the thrust zone.

The thrust zone between the Vinini and Roberts Mountains Formations contains major montmorillonite with lesser illitic clays and quartz. Carbonate minerals are absent from the fault zone and from adjacent units of Vinini and Roberts Mountains beds. Small amounts of alunite were identified locally in the Vinini Formation near the fault.

## SEDIMENTARY ROCKS

### Roberts Mountain Formation

Platy silty limestones of the lower Silurian Roberts Mountains Formation have been de-

scribed by Roberts *et al.* (21) from the Lynn and Carlin windows. In addition, unaltered samples of Roberts Mountains Formation were obtained for comparative study from road cuts along Maggie Creek Canyon within the Carlin window about 6 miles southeast of the mine. Two vertical sections were sampled, ranging from 10 to 20 feet thick. Samples consist of medium to hard, dense, dark-gray dolomitic limestone, silty to sandy in composition and thin-bedded. The rock is platy to blocky and closely jointed. Joints are filled locally with calcite. Chert is thin bedded and sparse to absent.

In the pit area, the Roberts Mountains Formation has been altered extensively with local bleaching and iron staining, but bedding features are often preserved, displaying a platy, thin-bedded character. However, carbonate minerals have been largely removed, leaving porous, low-density rocks that range in composition from argillaceous to dolomitic siltstones. Late introduction of chalcedonic silica has resulted locally in silicified seams or lenses that frequently parallel bedding.

On close study, the mineral composition and

*TABLE I. Mineralogic Composition of Unaltered and Altered Roberts Mountains Carbonate Rocks*

| Mineral Constituents | Unaltered Roberts Mountain at Maggie Creek | | Altered Roberts Mountain (Main Ore Body Composite) Per Cent |
|---|---|---|---|
| | 20' Section Per Cent | 10' Section Per Cent | |
| Calcite | 50 | 40 | 4 |
| Dolomite | 15 | 10 | 10 |
| Quartz | 20 | 20 | 40 |
| Illite | 5–10 | 5–10 | 10–20 |
| Montmorillonite | 5–10 | 10–20 | 10–20 |
| Kaolinite | 2 | 2 | 2–5 |
| Pyrite | tr | tr | tr |
| Carbonaceous Material | tr | tr | tr |

textures of Carlin ore appear to be more closely related to alteration effects than to primary sedimentary composition. However, primary bedding outlines are retained and reflect the general pattern of the original unaltered Roberts Mountains limestone.

The Roberts Mountains Formation as exposed below the thrust fault consists mostly of argillaceous siltstones, with quartz and small to moderate amounts of montmorillonite, illite, and traces of kaolinite. Dolomite and barite increase locally to major proportions. The for-

mation was much higher in carbonate content and lower in silica prior to mineralization (Tables I, II). Calcite is the major component of the unaltered rocks with small to moderate amounts of dolomite. In leaching, most of the calcite was removed, but the dolomite largely remained. Illite, montmorillonite, kaolinite, and detrital quartz were moderately abundant in unaltered rocks. All but montmorillonite were increased by major amounts after calcite was leached by hydrothermal solutions. Original angular grains of quartz are poorly sorted

*TABLE II. Chemical Composition of Unaltered and Altered Roberts Mountains Carbonate Rocks*

| Chemical Composition | Unaltered 20' Section of Dol. Ls. Maggie Creek Per Cent | Altered Roberts Mountain Main Ore Body Composite Per Cent | Method of Analysis |
|---|---|---|---|
| $SiO_2$ | 26.46 | 69.14 | Chemical Analysis |
| $Al_2O_3$ | 9.06 | 11.18 | " " |
| $Fe_2O_3$ | 1.47 | 3.23 | " " |
| CaO | 30.96 | 5.12 | " " |
| MgO | 3.98 | 2.45 | " " |
| $K_2O$ | 1.25 | 1.75 | Spectrographic |
| $Na_2O$ | 0.05 | <0.01 | " |
| $TiO_2$ | 0.2 | 0.6 | " |
| $MnO_2$ | 0.01 | 0.02 | " |
| BaO | 0.015 | 0.5 | " |
| SrO | 0.015 | 0.015 | " |
| ZnO | <0.005 | 0.01 | " |
| CuO | 0.004 | 0.008 | " |
| PbO | — | 0.003 | " |
| $As_2O_5$ | — | 0.25 | " |
| HgO | — | 0.01 | " |
| $Sb_2O_5$ | — | 0.02 | " |
| NiO | 0.003 | 0.01 | " |
| $ZrO_2$ | 0.03 | 0.04 | " |
| $CO_2$ | Most of remainder | | |
| Au | <0.005 oz/ton | 0.49 oz/ton | Fire Assay |

and randomly scattered through a dolomitic matrix containing disseminated clays. Individual grains range from 15 to 70 μ in size, and show relatively few inclusions, cavities or strain shadows.

Dolomite rhombs up to 60 μ are randomly distributed through a fine grained carbonate matrix. Small amounts of pyrite and carbonaceous matter in the fresh rock were probably syngenetic but, during alteration, were remobilized locally and redistributed. The dark gray of the unaltered limestone is attributed to carbonaceous matter.

Unaltered Roberts Mountains limestones from Maggie Creek have bulk densities near 2.63, compared with values ranging from 1.95 to 2.40 for ore-bearing replaced samples from the Carlin pit.

## "Popovich" Formation

The "Popovich" formation (Devonian) overlies the Roberts Mountains Formation and consists of gray, fossiliferous medium to thin-bedded limestone, locally dolomitic and silty near the base. This, the youngest formation in the area, was named "Popovich," by Hardie (31), after an early prospector in the district. Fossil evidence has dated the "Popovich" formation at the base of the upper Devonian, stratigraphically equivalent to the Devils Gate Limestone. The "Popovich" unconformably overlies the Roberts Mountains Formation, but some movement appears to have occurred along the contact.

The "Popovich" has been altered hydrothermally along its base and along nearby joints and fractures. Altered rock has been leached of its carbonate content and replaced by clay and iron oxide. Altered samples usually are reddish brown, soft, porous, and low in density. Select zones in the "Popovich" have been silicified similarly to the Roberts Mountains, notably near the northeast fault at the north end of the ore body.

Unaltered "Popovich" usually is fine to medium-grained dolomitic limestone similar in composition to the Roberts Mountains but with less silt and clay. The mineralogic and chemical compositions of altered and unaltered "Popovich" are compared in Tables III and IV. Small amounts of carbonaceous matter and pyrite are distributed along bedding planes, resulting in a dark-gray rock. Most samples are also fossiliferous, containing a variety of microfossils that resemble spicules, crinoid stems, and types of foraminifera. Microfossils have been recrystallized and replaced by microcrystalline quartz.

Dolomitic limestones grade into massive dolomite near the base. Samples of altered and unaltered dolomite were analyzed from near the base of the "Popovich" along a road cut north of the east pit (Table IV). A comparison of analyses for the two samples indicates major removal of both dolomite and calcite during alteration. Micropore space in the rock is indicated by low-bulk densities (near 1.60) as compared with 2.75 for nearby unaltered dolomite.

## Vinini Formation

The Vinini Formation (middle Ordovician) occurs in upper-plate rocks along the Roberts Mountains thrust fault. This formation consists mostly of thin-bedded siliceous cherts, organic shales, minor quartzites, and lenses of limestone. The bedded cherts and black organic

TABLE III. *Mineralogic Composition of Unaltered and Altered Samples from the Popovich Formation*

| Mineral Constituents | Unaltered Popovich Sample No. 1 | Unaltered Popovich Sample No. 2 | Altered Popovich Sample No. 3 |
|---|---|---|---|
| Calcite | 45 Per Cent | 5 Per Cent | <2 Per Cent |
| Dolomite | 15 | 60 | <2 |
| Quartz | 15 | 20 | 60 |
| Illite | 5–10 | 2–5 | 10–20 |
| Montmorillonite | 2–5 | 2–5 | 5–10 |
| Kaolinite | 2–5 | <2 | 2–5 |
| Pyrite | Tr | Tr | — |
| Carbonaceous Material | Tr | Tr | — |

Sample 1 represents an average of analyses for 125' of drill cuttings from Popovich.
Sample 2 is from a dolomite bed near base of Popovich.
Sample 3 is altered material adjacent to Sample 2 of fresh dolomite.

*TABLE IV.  Chemical Composition of Unaltered and Altered Samples from the Popovich Formation*

| Chemical Constituents | Unaltered Popovich Sample No. 2 | Altered Popovich Sample No. 3 | Method of Analysis |
|---|---|---|---|
| $SiO_2$ | 32.04 | 81.00 | Chemical Analysis |
| $Al_2O_3$ | 10.82 | 12.34 | "        " |
| $Fe_2O_3$ | 1.90 | 1.61 | "        " |
| $CaO$ | 18.66 | 1.40 | "        " |
| $MgO$ | 9.10 | 0.58 | "        " |
| $K_2O$ | 1.0 | <1.0 | Spectrographic |
| $Na_2O$ | 0.07 | <0.01 | " |
| $TiO_2$ | 0.30 | 0.35 | " |
| $MnO_2$ | 0.01 | 0.03 | " |
| $BaO$ | 0.03 | 0.08 | " |
| $SrO_2$ | 0.008 | 0.008 | " |
| $ZnO$ | 0.002 | 0.003 | " |
| $CuO$ | 0.010 | 0.015 | " |
| $PbO$ | <0.001 | 0.002 | " |
| $As_2O_5$ | 0.01 | 0.02 | " |
| $HgO$ | <0.01 | <0.01 | " |
| $Sb_2O_5$ | <0.01 | <0.01 | " |
| $NiO$ | 0.003 | 0.003 | " |
| $ZrO_2$ | 0.02 | 0.02 | " |
| $CO_2$ | Most of remainder | | |
| $Au$ | <0.005 oz/ton | 0.04 oz/ton | Fire Assay |

shales are clearly of the western lithologic assemblage. According to Roberts *et al.* (21), the most complete stratigraphic section of the Vinini Formation can be seen in the Tuscarora Mountains in northern Eureka County. Here the Vinini Formation is locally mineralized, and gold has been mined from small workings, but no large ore bodies have been found. However, near the Carlin pit, traces of gold have been detected in Vinini outcrops.

At Carlin, the Vinini Formation consists of siliceous shales in two groups. An upper unit of undetermined thickness contains moderate to major amounts of well-crystallized illite and small to trace amounts of K-feldspar in addition to major quartz. A lower unit contains major amounts of mixed layered illite-montmorillonite in addition to major quartz. Both Vinini units contain little to no kaolinite and montmorillonite and are separated by a manganese zone of wad containing anomalously high zinc, arsenic, barium, and nickel.

## Lithologic Interpretation by X-ray Diffraction Analysis

Numerous samples from across the Carlin ore body have been examined by x-ray diffraction. Analyses are semiquantitative and based on relative intensities of characteristic reflections for major minerals, including quartz, calcite, dolomite, kaolinite, illite, and barite. Measured intensities were compared with intensities calibrated from standard samples. An example is provided by the lithologic section in Figure 2 that was interpreted from x-ray diffraction analyses of cuttings from a drill hole located east of the main ore body.

Estimated percentages have provided a rapid semi-quantitative method for evaluating lithologic trends that otherwise might escape detection by visual methods of logging. Microscopic examination of samples in grain mounts also has been used to supplement x-ray diffraction data in distinguishing fine massive quartz from granular detrital quartz.

Observations drawn from x-ray and microscopic examination of drill cuttings and outcrop samples are summarized below:

(1) Major lithologic types of gold-bearing samples collected across the ore body are: (a) limestones, (b) dolomitic limestones, (c) argillaceous siltstones, and (d) silicified rocks. Samples may be silty, argillaceous, dolomitic, calcareous, siliceous, carbonaceous, pyritic, and baritic in composition. Rock units are gradational with few sharp differences in mineralogic composition.

(2) Calcite has been removed selectively from mineralized zones leaving an altered rock matrix consisting mostly of detrital quartz, clays, and relic dolomite. Illite is the dominant clay mineral associated with mineralization, comprising as much as 50 per cent of some ore samples. Quartz may range from 25 per cent in highly argillized samples to 75 per cent in silicified ore zones. Relic dolomite may range from nil to as much as 40 per cent in a few samples. Calcite is normally sparse but has been enriched locally in some portions of the ore, apparently as a late epigenetic mineral.

(3) Unaltered "Popovich" limestones are highly calcareous (>70 per cent calcite) and low in detrital quartz, except near the base, where dolomite and silty quartz increase significantly. Here, alteration is accompanied by removal of carbonates, mostly calcite, accompanied by argillization and local silicification.

(4) Several different rock units appear in drill cuttings. An upper calcareous unit detected in cuttings from the northeast portion of the ore body apparently is "Popovich" limestone. The underlying Roberts Mountains Formation consists of alternate zones of argillaceous siltstones, dolomites, and chalcedonic strata of highly varied composition. Massive silicified zones may occur locally in altered portions of either formation, accompanied by an increase in quartz. Other differences may be detected by x-ray diffraction, including increases of: (a) kaolinite along altered dikes of porphyry, (b) montmorillonite along fault zones, and (c) barite along veins and replacements.

(5) Gold largely is confined to altered rock units of the Roberts Mountains Formation. Larger amounts occur locally in argillaceous siltstones. In porous light gray chimney-like forms with fine crystalline quartz, gold usually is low. In general, small amounts of gold are disseminated through silicified rocks including chalcedonic lenses, altered dikes, and even occasionally in zones of barite replacement.

(6) In general, argillaceous and dolomitic siltstones are prevalent in ore samples from the main ore body; the rock becomes more dolomitic, chalcedonic, and carbonaceous across the east ore body.

## IGNEOUS ROCKS

Igneous rocks examined from the area include quartz porphyry dikes in the pit and a large quartz diorite intrusive exposed about 3 miles to the north.

### Quartz Porphyry Dikes

Quartz porphyry dikes that cut the "Popovich" limestone and the Roberts Mountains Formation in the Carlin pit dip steeply northeast and strike northwest, following a system of northwest faults and associated fractures. The dikes range from a few inches to 10 feet in thickness and may differ widely in thickness over short distances. Barite veins and sparse base-metal mineralization appear to be related closely to the dike intrusions, as shown by mutual association along the northwest fault on the east face of the main ore body. Subsequent low-temperature alteration of dikes and wall rock apparently occurred at a much later date associated with gold mineralization.

Borders of most dikes for a distance of about one-half inch on both sides are extremely fine grained and are zones of chilled contact against host rocks. Central zones are more crystalline in thin section but have been replaced completely by secondary clay minerals and quartz. Most samples are porphyritic, containing phenocrysts of embayed quartz and relics of amphiboles and feldspars. Apparent relics of hornblende are euhedral and typically pseudo-hexagonal to rhombic in cross-section (Figure 7). Subhedral phenocrysts of plagioclase have been replaced completely by microcrystalline quartz of chalcedonic texture. Quartz phenocrysts are only slightly altered with the exception of reaction rims along external borders. Reaction rims are attributed to primary processes of interaction between phenocrysts and the cooling magma. Relic groundmass textures range from pilitic along dike margins to intersertal in interior portions. Felty needles of feldspar in the groundmass have undergone pseudomorphic replacement

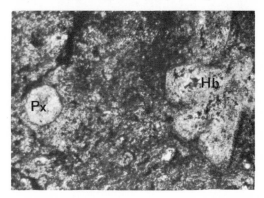

Fig. 7. *Euhedral Outlines of Hornblende (Hb) and Pyroxene (Px) in Altered Dike from Main Pit. ×18, Transmitted light.*

by microcrystalline quartz and kaolinite. Microscopic textures indicate that the dikes were either dacite or quartz latite in composition, but the general field term "quartz porphyry" is used in this paper.

X-ray diffraction analyses of dike samples show major quartz and kaolinite with minor dickite, illite, montmorillonite, and hydrous iron oxides. Feldspars and mafic silicates were not detected. Argillization of dikes and surrounding sediments has resulted from the same processes of alteration since the same assemblages of clay minerals are widely distributed through both.

Gold ore continues into the dikes in the vicinity of ore bodies, but, in places, gold may be essentially absent.

## Quartz Diorite

A small plutonic intrusion of quartz diorite is exposed in the Gold Strike claims in section 30 about 3 miles north of the Carlin mine. At the surface, the intrusive is moderately to highly altered and weathered. Nearly fresh cuttings from a drill hole at about 195 to 200 feet depth are gray to dark gray and medium to fine grained in texture. X-ray diffraction scans show major constituents to be plagioclase, mica, hornblende, quartz, and chlorite. Thin sections reveal a hypidiomorphic granular texture consisting mostly of anhedral plagioclase and quartz with moderate amounts of biotite and hornblende. Accessory amounts of sphene were also observed (Figure 8). Potash feldspars were not detected.

Plagioclase, near sodic andesine in composi-

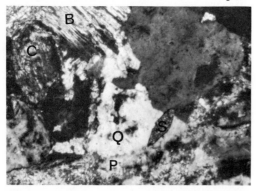

Fig. 8. *Photomicrograph of Quartz Diorite 4 Miles North of Carlin Mine. Alteration of plagioclase (P) to sericite and clays. Biotite (B) is partially altered to chlorite (C), and late quartz (Q) contains sphene. ×18. Crossed nicols.*

**TABLE V.**   *Analytical Data on K/Ar Age Determination of Biotite from Quartz Diorite*

| Argon Analyses | | Potassium Analyses | |
|---|---|---|---|
| $Ar^{40*}$ ppm | 0.0333 | $K_{(per\ cent)}$ | 3.53 |
| | 0.0315 | | 3.74 |
| $Ar^{40*}/Total\ Ar^{40}$ | 0.775 | $K_{Average\ (per\ cent)}$ | 3.65 |
| | 0.860 | | |
| Average $Ar^{40*}$ | 0.0324 | $K^{40}$ ppm | 4.44 |
| | Ratio | | |
| $Ar^{40*}/K^{40*} = 0.0073$ | | | |

Age $= 121(\pm 5) \times 10^6$ years (middle to early Cretaceous)

tion, is locally sericitized and argillized. Hornblende and biotite are altered partially to chlorite. Pyrite locally is abundant in altered samples and is associated with alteration of feldspars and mafic silicates. In weathered samples, pyrite is oxidized to hydrous iron oxides. Small amounts of gold (0.02 oz/ton) were detected in altered samples.

An age of 121 ± 5 m.y.) (early Cretaceous) was determined for biotite separated from the intrusive using the K/Ar method. The sample contained less than 10 per cent impurities most of which were hornblende that did not affect the accuracy of the determination. Data were reported by Harold W. Krueger, technical director of Geochron Laboratories, Inc. (Table V). Because of petrographic similarity, the dikes in the Carlin pit may be of the same age. The close proximity and similarity in composition of the quartz diorite and the quartz porphyry dikes suggest a common magmatic source.

The potassium-argon dates of Nevada and Utah intrusives apparently provide reliable ages for the crystallization of these rocks (27). The ages of these intrusives fall broadly into two groups (12–64 m.y. and 93–182 m.y.) with periods of maximum ore genesis believed to coincide in general with periods of igneous activity (33). An age of 121 m.y. for intrusives near the Carlin mine is compatible with determinations from the Cortez mine in northern Eureka County that range from 124 to 150 m.y. A number of intrusives of apparent Cretaceous age are distributed along a northwesterly trend through the Lynn and Bootstrap windows.

## BARITE MINERALIZATION

Barite deposits are of widespread occurrence in Nevada (Gianella, 1941) and often are associated with precious and base-metal mineralization. Barite occurs in several places in the Carlin ore body and in the overlying "Popovich" formation. Deposits occur as: (1) veins and irregular replacement bodies in limestones; (2) injections along high-angle faults and dike contacts; and (3) irregular replacements that follow bedding, notably along the disconformity between the "Popovich" formation and the Roberts Mountains. Most of the veins dip steeply and strike northwesterly.

Barite and dike material are associated mutually along a northwest fault exposed in the east face of the main ore body. Most dikes also contain anomalously high barium on chemical analysis. Anomalously high lead, zinc, nickel, and copper are also found both in dikes and barite veins.

Galena and sphalerite have been identified microscopically in barite samples from the main pit. Barite is replaced locally by microcrystalline quartz veinlets that contain finely crystalline pyrite. Realgar, stibnite, and carbonaceous matter have been identified also as late fracture fillings in barite. Galena is altered locally to jordanite, while sphalerite is partially replaced in places, apparently by tennantite. The latter replacements probably occurred at a later age of mineralization, possibly associated with late gold-arsenic solutions.

## GOLD DEPOSITS

### General Features of the Ore Body

Drilling to date indicates that the Carlin main ore body is an irregular inclined deposit (Figure 4), located near the top of the Roberts Mountains Formation and underlying the "Popovich" formation. Thickness may range from a few feet to nearly 100 feet with gradational assay boundaries. The upper ore boundary occurs just below the "Popovich," ranging stratigraphically from a few feet to a few tens of feet below the contact. The lower contact grades into light gray zones of porous silicification, locally penetrated by dark-gray stratiform, but tapering, chalcedonic masses.

The ore mass follows the inclination of strata and continues downward from the surface until intersected by the northeast fault. In horizontal section, the ore body is elongate, extending to the northeast along strike.

### Lithologic Influence on Ore Deposition

Permeable horizons of silty dolomitic limestones in the Roberts Mountains Formation

have provided a favorable host environment for mineralizing solutions.

As exposed, nearly 200 feet of uppermost Roberts Mountains strata have been leached and locally silicified, presumably by ascending solutions associated with mineralization. Gold in this interval may range from a few ounces per ton to virtually nil (<0.002 oz/ton). In outcrops, low-grade to barren materials are not readily distinguished from those that are highly mineralized, hence closely spaced samples and numerous assays are required to delineate the ore body.

Mineralized portions of the ore body usually range from replaced argillaceous to dolomitic siltstones that in places may be carbonaceous, calcareous, or silicified. Only a few consistently recognizable rock types may be associated visually with barren rock or ore. Highly bleached, rounded, porous, silicified zones in the Roberts Mountains usually contain negligible amounts of gold. On the other hand, a peculiar spotted rock (Figure 9), that may overlie silicified zones in the deposit often contains significant amounts of gold. "Spots" appear to be initial patches of alteration that form small patches of argillitization parallel to bedding. Seams of argillic alteration in "spotted" rock commonly range from a few tenths of a millimeter to several centimeters thick. Smaller seamlets are usually lenticular in cross section and follow continuous paths along bedding planes or into fractures, resembling "worm borings" (Figure 10). Larger seams of alteration are more continuous in cross section, usually follow bedding planes, and locally are irregular and sinuous in distribution. Spotted features are a light-grayish

FIG. 9. *"Spotted" Ore Concentrations; in argillic alteration in Roberts Mountains Formation these yield a "spotted rock" often high in gold. From 6400 foot bench, main ore body.*

FIG. 10. *Argillic Alteration Seamlets along Bedding Planes. Bleached seamlets resembled "worm borings" in mineralized Roberts Mountains Formation. From 6400' bench, main ore body.*

alteration in contrast to a residual dark-gray rock matrix in unweathered areas. On weathering "spots" are stained with ferruginous oxide and appear as dark-brown areas in a light-reddish-buff matrix. This type has been observed and sampled from several locations; each location was highly mineralized but was bordered by a major silicified zone. Under the microscope, "spots" reveal clay-mineral orientation. Seamlets of alteration display a fabric of epigenetic clays, mostly illite flakes, that are aligned more or less parallel to bedding. Under crossed nicols, the extinction angles lie on the plane of lineation. Visible gold is detected in the matrix of oriented clay seamlets, whereas no gold has been detected in unoriented clay seamlets.

In a sequence of highly mineralized samples from the 6400-foot bench (Figure 11), a correlation may be established between gold content and the abundance of oriented clay seamlets in the ore. In lower grade samples, e.g., sample No. 9 containing 0.19 oz/ton Au, oriented seamlets are less abundant but occur along multiple bedding planes (Figure 12). In high grade samples, above 2 oz/ton, oriented seamlets interfinger and coalesce into a continuous fabric of oriented clays (Figure 13). Dark reddish brown to opaque matter associated with oriented clay seamlets is amor-

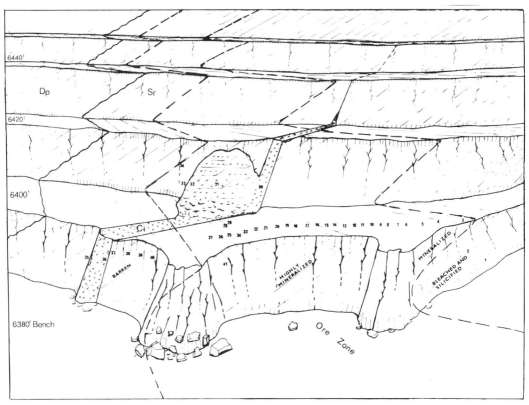

FIG. 11. *Sketch of Ore Zone, showing location of samples taken along 6400-foot bench. The ore zone lies near the top of the Silurian Roberts Mountains (Sr) underlying Devonian "Popovich" (Dp), and is intruded by Cretaceous dike (Ci) along NW fault.*

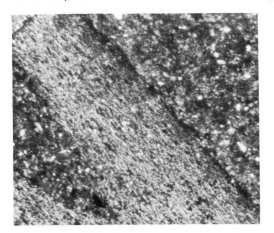

FIG. 12. *Photomicrograph of Argillization Seamlets in Dolomitic Siltstone (Sample No. 9. Figure 12 and Table V), showing oriented clay fabric in contrast to random fabric of surrounding area.* ×22, *partially crossed nicols.*

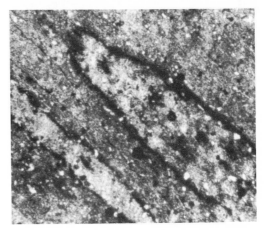

FIG. 13. *"Island Relics" of Unaltered Dolomitic Sandstone in Matrix of Coalescing Alteration Seamlets (Sample No. 5, Figure 12 and Table V).*×21. *partially crossed nicols.*

phous and appears to be a mixture of ferruginous oxides.

Oriented clay seamlets are believed to be epigenetic alteration channels along which gold was introduced as the carbonate was leached from the rock. Carbonate minerals are largely absent in oriented clay seamlets. The lamellar texture of the clay fabric may be explained either by flow orientation from solution movement or by the collapse of local rock structure

along zones of carbonate removal. The latter explanation seems more plausible from the view point of rock deformation.

Deposition of gold appears to have occurred in the capillaries of argillaceous siltstones, associated closely with the removal of carbonate along bedding planes. Certain minor differences in lithology have noticeable effects on ore grade. (Table VI). Small siltstone-sandstone seams, up to several inches thick, usually are less mineralized than adjacent finer-grained

*TABLE VI.   Metal Content Across Ore Zone on 6400-foot Bench of the Main Pit, Carlin Ore Body*[1]

| Rock Type and Sample Numbers (Fig. 21) | Average Fire Assay Au/Oz. Ton | Average Semiquantitative Spectrographic Analyses Per Cent | | | | | | | | | | |
|---|---|---|---|---|---|---|---|---|---|---|---|---|
| | | As | Sb | Hg | Zn | Cu | Fe | Ba | Ni | Zr | Sr | Rb |
| Dolomitic argillaceous siltstone (8, 9, 10, 11, 12, 13, 15, 17, 19, 20, 21, 22, 25, 34, 38, 39, 40, 41) | 2.66 | 0.22 | .01 | .01 | .020 | .005 | 3.0 | 0.1 | .004 | .030 | .008 | .007 |
| Argillaceous siltstone (4, 23, 29) | 2.19 | 0.22 | .01 | .01 | .044 | .008 | 4.7 | 0.3 | .003 | .020 | .013 | .010 |
| Siltstone-sandstone (14, 16, 24, 26) | 1.26 | 0.28 | <.01 | <.01 | .015 | .004 | 4.3 | 0.4 | .002 | .026 | .005 | .005 |
| Calcified argillaceous siltstone | 0.84 | 0.17 | <.01 | <.01 | .01 | .003 | 4.0 | 0.3 | .002 | .045 | .006 | <.005 |
| Silty argillaceous dolomite (1, 5, 7, 27) | 0.805 | 0.12 | <.01 | <.01 | .020 | .006 | 3.0 | 0.2 | .003 | .046 | .009 | .009 |
| Silty calcareous dolomite (18) | 0.19 | 0.12 | <.01 | .01 | .020 | .005 | 3.0 | 0.1 | .004 | .030 | .008 | .007 |
| Altered porphyry dike (30, 31, 32, 36) | 0.153 | 0.59 | <.01 | <.01 | .15 | .016 | 12.5 | 0.8 | .050 | .050 | .021 | <.005 |
| Calcareous siltstone near dike (33 above, 35, 37 below) | 0.052 | 0.09 | <.01 | <.01 | .12 | .006 | 2.8 | 0.2 | .024 | .025 | .008 | .005 |
| Bleached argillaceous siltstone (2, 3) | 0.023 | 0.025 | <.01 | <.01 | <.002 | .003 | 1.25 | <.1 | .002 | .042 | .013 | <.005 |

[1] Sample numbers are located on sketch map shown in Figure 12.

TABLE VII. *Dilution of Gold Content by Impregnations of Silica and Calcite*

| | Silica Impregnation | Calcite Impregnation |
|---|---|---|
| Sample Location | 6440-foot Bench (Fig. 22) | 6550-foot Bench (Fig. 23) |
| Au Oz/Ton | | |
| Porous | 0.06 oz/ton | 0.49 oz/ton |
| Impregnated | 0.03 " | 0.20 " |
| Bulk Density | | |
| Porous | 1.91 Sp. G. | 2.21 Sp. G. |
| Impregnated | 2.58 " | 2.53 " |

argillaceous siltstones. Sandy and fossiliferous beds are more permeable, are commonly silicified, and carry lower amounts of gold.

Where quartz porphyry dikes carry gold, they usually are less mineralized than adjacent sediments. An example is provided by a quartz porphyry dike that transects the ore zone between the 6380-foot and 6420-foot benches (Figs. 5, and 11). Samples of the dike from the ore zone between the 6400-foot and 6420-foot benches range from 0.003 to 0.48 oz/ton of gold, compared with 0.19 to 5.67 oz/ton in adjacent altered sediments. Samples of the dike, where it intrudes sparsely mineralized sediments between the 6380-foot and the 6400-foot benches, range from nil to 0.002 oz/ton Au, compared with 0.003 to 0.045 oz/ton in adjacent sediments. Gold contents and descriptions of rock units shown in Figure 11 are compared in Table VI.

FIG. 14. *Silicified Rock from 6440-foot Bench along East Face of Main Ore Body. Silica has penetrated white permeable beds, resulting in dark gray chalcedonic rock of higher bulk density.*

FIG. 15. *Calcite Vein (white) Penetrating Light-Gray Leached Rock. The vein has been accompanied by calcitic permeation of the border rock with a change to dark gray. From 6400' bench along east face of main ore body.*

Ores usually are of lower grade in rocks of chalcedonic or calcareous composition. Apparently microcrystalline quartz and calcite have been added to local rock units after the gold was deposited with dilution of gold values (Table VII). Pore spaces in highly porous siltstones have been filled locally in places with silica (Figure 14) and also with calcite (Figure 15), resulting in slightly lower gold content and higher bulk density. The gold also is more difficult to extract from these ore types.

## Relationship of Structure to Gold Deposition

The Roberts Mountains overthrust per se has little apparent structural control over the distribution of ore. The thrust plane undoubtedly has served as a conduit for migrating solutions over a long period but appears of little importance as a site for ore deposition. However, ore concentrations of importance occur adjacent to portions of the northwest faults.

Porous "white" silicification is most prevalent along northwest faults that cut the main and the west ore bodies. Fault zones, however, are sparsely mineralized, carrying quite low amounts of gold. A number of samples of gouge and bleached rock adjacent to northwest and northeast faults have been sampled and analyzed (Table VIII), without finding evidence of significant gold mineralization. Samples of gouge usually contain illite and montmorillonite. Kaolinite and dickite become locally abundant as alteration products of quartz porphyry dikes that intrude high-grade faults in the area.

*TABLE VIII.   Gold Content in Fault Gouge and Silicified Rocks*

Northeast Fault, 6360-foot and 6380-foot Benches, Main Ore Body
|  |  |
|---|---|
| Gouge along fault | <0.005 oz/ton Au |
| Vinini near fault | 0.005 " |

Northwest Fault, Top Bench, West Ore Body
|  |  |
|---|---|
| Bleached shear zone in fault | <0.005 oz/ton Au |
| Argillaceous stiltstone, foot wall of fault | 0.02 " |
| "          "       hanging wall of fault | 0.32 " |
| Barite in fault | 0.05 " |

Northwest Fault, 6240-foot Bench, East Face of Main Ore Body
|  |  |
|---|---|
| Barite in fault | 0.07 oz/ton Au |
| Bleached dike in fault | 0.19 " |

Highly altered fault zones may have served as conduits for mineralizing solutions but have not become significantly mineralized. Gouge from northwest faults usually contains from a trace to 0.05 oz/ton Au.

Original minor fracturing and jointing have been somewhat obscured in the Roberts Mountains Formation by intense argillizing alteration and local silicification. The alteration, as inferred from silicification, is a general permeation and cuts across jointing and fracturing.

The unconformity between the "Popovich" and the Roberts Mountains has been highly silicified and is sparsely mineralized in places. Alteration of the "Popovich" appears to be localized largely along vertical joints. Altered "Popovich" samples usually contain only traces of gold (0.005 to 0.03 oz/ton).

## ROLE OF COLLOIDAL GOLD IN ORE FORMATION

### Microscopic Gold

Not even specks of gold are visible in Carlin ore to the unaided eye, but with the microscope at 1000x or more, gold from the main ore body may be detected in a variety of argillaceous rocks with a range in carbonate content. Textural associations of gold in the 5 to 0.5 $\mu$ range are frequently visible under the microscope, but submicron sizes (0.2–0.5 $\mu$) may be barely resolved at high magnifications, while particles below the limits of resolution are barely visible but can not be resolved. Particles much less than 0.5 $\mu$ in diameter must be observed with the electron microscope. Gold often borders detrital quartz grains (Figure 16), and may be included within fractures in detrital quartz or disseminated in matrix clays. Gold particles do not appear to have been transported mechanically.

Larger gold particles (5 to 1 $\mu$) are irregular but generally subrounded to oval in outline. Reflected colors range from pale to deep yellow in polished sections. The low hardness and slight negative relief aid in distinguishing between gold and pyrite. It is estimated that possibly 90 per cent or more of the gold may be submicroscopic (<0.2 $\mu$).

Samples containing gold microscopically visible at high magnifications often show pore spaces where calcite has been selectively removed. Gold is common along friable zones, but impregnation with epoxy plastic is required to retain it in situ during polishing.

Microscopic gold has been observed locally in association with late quartz veinlets in silicified siltstones (Figure 17), and with barite and along external boundaries of pyrite. Oxidation of the pyrite to hematite has liberated smaller inclusions of gold.

FIG. 16. *Particle of Gold (Au) Attached to Detrital Quartz (Q) Grain. Sample from main ore body.* ×900, *Incident light.*

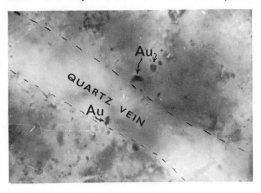

FIG. 17. *Mineralized Quartz Veinlet in Silicified Siltstone. Gold (opaque) is distributed within the veinlet and through the adjacent silicified host rock. ×720, Transmitted light.*

*TABLE IX.   Gold Content in Screen Fractions (6400' Bench)*

| Screen Fraction | Per Cent of Sample by Wt. | Au Oz/Ton | Per Cent of Total Au |
|---|---|---|---|
| +200 mesh (74 μ) | 5.0 | 2.29 | 5.9 |
| +400 mesh (28 μ) | 14.2 | 1.13 | 8.3 |
| −400 slimes (5–28 μ) | 43.3 | 1.57 | 35.3 |
| Decanted slimes (−5 μ) | 37.5 | 2.58 | 50.5 |

## Submicroscopic Gold

The oval outline of the gold particles is preserved in electron micrographs at magnifications up to 120,000× (Figure 18). Individual particles may range down to less than 0.005 μ. Their identification has been confirmed by limited area electron diffraction on clusters of rather high gold concentration. Earlier identifications were made in late 1965 with the assistance of Emil Hamburg, graduate student in Physical Metallurgy, Columbia School of Mines, utilizing a Siemans Electron Microscope.

The fine dimensions of Carlin gold are confirmed by analyses of screen fractions from a high-grade composite sample from the 6400-foot bench (Table IX).

In the analysis, 85.8 per cent of the gold is concentrated in the −400 slimes and decanted slimes where half of the gold is associated with illitic clays.

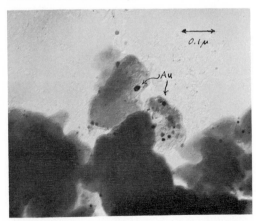

FIG. 18. *Electron Micrograph of Native Gold Clusters in Clay Matrix.*

Attempts were made to dissolve silicate minerals selectively and leave the gold, using various solvents and fluxes including hydrofluoric acid, caustic soda, borax, sodium carbonate, potassium carbonate, and sodium acid phosphate. Solubility tests were conducted at temperatures ranging from room temperature to about 100°C; fluxing was conducted from 750° to 900°C. In each instance, significant amounts of gold were dissolved along with the silicates. A method was not found to separate colloidal-sized gold from the silicates by physical treatment.

## Precipitation of Colloidal-Sized Gold

The mechanics of colloidal gold precipitation have been investigated by many workers, Butler (4); Haycock (7); Smith (13); Stillwell and Edwards (17); W. H. White (14); Van Aubel (9); and Joralemon (19). Most investigators agree that hydrothermal gold is carried in true solution and not as colloidal particles. This conclusion is supported by textural studies of vein gold (14) and solubility studies of metallic gold in alkali sulfide solutions (13).

Precipitation of gold from alkali sulfide solutions may depend on several mechanisms including: (1) sulfide ion concentration; (2) pH of the solution; (3) the temperature and pressure system; and (4) the nature of mineral surfaces for adsorption or nucleation of gold. Metallic gold crystals may be precipitated from alkali sulfide solutions by lowering the sulfide ion concentration (13). The sulfide ion concentration may be lowered by the escape of $H_2S$ under low pressure or by oxidizing the sulfur to sulfates. The latter lowers the pH and enhances precipitation.

Haycock (7) states that submicroscopic gold tends to occur when deposition is contemporaneous with that of the host minerals, whereas, coarser gold tends to form when deposition is later than the host. Essentially contemporaneous deposition would appear to prevail at Carlin. Here various clay minerals, carbonaceous bitumen, pyrite, arsenic sulfides, and silica have accompanied gold deposition. Gold has been adsorbed selectively by, or nucleated onto, surfaces of illitic clays, carbonaceous matter, iron sulfides, and quartz. Submicroscopic colloidal-sized gold may be distributed over a range in particle size from about 0.2 $\mu$ down to the size of a gold atom (1.45 Å or 0.000145 $\mu$). Haycock (7) restricts colloidal-sized gold to a limited range between about 6 Å (0.0006 $\mu$), and 75 Å (0.0075 $\mu$). Colloidal-sized, gold, as defined by Haycock, may account for as much as half of the gold in the deposit, as indicated by the optical and electron microscopes and reference to Haycock's grain size curves (7, p. 409). The rapid solubility of the Carlin gold in cyanide solution also indicates the distribution of fine gold through a permeable rock matrix.

### Influence of Host-Rock Permeability

The deposition of gold may be regarded as a subsidiary phase in the sequence of leaching and silicification of sedimentary limestones influenced greatly by permeability. More permeable beds served as conduits for solution movement, from which auriferous fluids diffused into minute interstices and tiny fractures of less permeable beds. Gold deposition occurred mostly along the finer capillaries and not along the main avenues of fluid transport. These observations agree with those of a number of investigators including Mawdsley (8). Johnston (11), Pardee and Park (16), W. H. White (14), and Chase (18), who have noted the importance of fine fractures and microcavities in minerals, notably quartz, in admitting gold to a site of emplacement.

The Roberts Mountains limestone with accompanying thin clay streaks becomes readily permeable to ore solutions after the calcium carbonate has been removed. Distribution of gold appears to depend more on micro-permeability than macro-permeable characteristics, Highly permeable silty strata, porous silicified zones, and drusy chalcedonic seams are inherently low in gold compared with nearby argillaceous beds with pore dimensions that are submicroscopic but provide large surface areas.

### Gold Deposition in Clay

The association of gold with kaolinite and fine micas has long been noted (3). The possibility exists that, following the precipitation of fine gold from true solution, positive charges on the edges of colloidal particles in the Carlin matrix clay may have attracted the negatively charged colloidal-sized particles of gold. H. van Olphen (28) has reproduced an electron micrograph showing crystals of kaolinite more or less translucent in the electron beam, with small opaque spheres of gold attached to the edges of almost hexagonal flakes of kaolinite. The kaolinite crystals measure about .5 to 1.0 $\mu$ while the gold particles measure about .005 to .01 $\mu$.

In discussion, H. van Olphen points out that, in contrast to the large flat surfaces of clay mineral crystals where negative charges prevail, the edges of the crystals expose edges of the layer lattice structure where the charge is positive. Thus, the positive double layer on the edge of the clay crystal flake would attract gold while the flat negative double layers would repel gold. This may account for the absence of gold particles on the flat surfaces of the kaolinite flakes in contrast to the concentration that takes place around the rims of the crystals.

This colloidal phenomenon is of particular interest in view of the natural concentration of gold which appears to take place in the small streamers of clay at Carlin. It would appear that colloidal gold, once precipitated from solution, would be apt to become entrapped by such a mechanism. Thus the positive charges of the mono-layers on the edges of clay crystals may have provided a source of attraction for the gold particles that, in turn, resulted in the concentration of more than normal amounts of gold in the clay.

### Organic Material and Fine Gold

Gold and organic material, at times resembling asphaltite, locally are associated in ores from the easterly extension of the main ore body. Also concentrates rich in organic matter show an increase in gold content, and the mutual association of some gold and organic matter is indicated by metallurgical test work. Gold insoluble in cyanide solution has been noted to increase with carbon content, (Table X), although most of the gold readily becomes soluble after the hydrocarbon is removed in roasting.

Organic matter is a known precipitant for gold, and precipitation of gold from cyanide solutions apparently is accomplished by physical adsorption of the gold onto surfaces of some forms of carbon. Such a phenomenon also might be suggested to have occurred between the organic matter in the Roberts Mountains limestone and the ore solutions, except that large sections of organic limestones in

*TABLE X.    Insoluble Gold Content in Carbonaceous and Pyritic Ores*

| Locality | Carbon Per Cent | Pyrite Per Cent | Insoluble Gold | |
|---|---|---|---|---|
| Mercur District, Utah[1] | | | | |
|    Oxidized ore | 0.105 | N.R. | $0.40 | $/Ton |
|    Raw Base (Arsenic) | 0.358 | N.R. | 0.90 | " |
|    Pyritic base | 0.450 | N.R. | All but trace | " |
| Carlin Ore Body, Nevada[2] | | | | |
|    Oxidized ore, Main Ore Body | N.D. | 0.2 | 8 | per cent of total Au |
|    Pyritic ore, Northeasterly extension | N.D. | 20.0 | 31.6 | " |
|    Carbonaceous ore, Northeasterly extension | 0.18 | 1.0 | 79.7 | " |
|    Carbonaceous-pyritic ore, Northeasterly extension | 0.45 | 3.0 | 87.7 | " |
|    Carbonaceous-pyritic ore, Northeasterly extension | 0.53 | 3.0 | 87.9 | " |

[1] Association of carbon and insoluble gold in Mercur ores described by B. S. Butler, et al (1920), p. 394.

[2] Investigations into treatments of Carlin refractory gold from metallurgical files of Newmont Explortion Limited, Danbury, Connecticut. It should be noted that refractory carbonaceous and pyritic ores are relatively minor in occurrence in the Carlin ore body, but are worthy of note for comparison with other carbonaceous gold deposits in Nevada and Utah.

the vicinity of Carlin are virtually unmineralized.

The main ore body at the Carlin mine has been partly oxidized by recent weathering, resulting in removal of much of the pyrite and carbon. Some unoxidized ore has been encountered in the main pit that is dark gray to black and deleterious to gold recovery in the mill.

Unweathered limestones of the "Popovich" and Roberts Mountains formations are slightly organic and pyritic through a stratigraphic interval of hundreds of feet. The amount of organic carbon by assay mostly ranges from a few tenths to several per cent. Some redis-

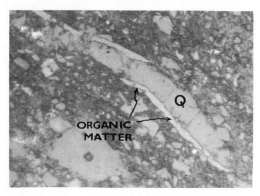

Fig. 19. *Organic Residues, Resembling Asphaltic Bitumen, along Borders of Quartz Veinlet (Q) in Dolomitic Siltstone. Organic matter appears to have been mobilized and emplaced epigenetically after the quartz veinlet. ×110, Incident light.*

tribution has occurred during periods of mineralization. The migratory nature of organic matter appears upon thin section study. It is such as might result from petroliferous origin. The latter, however, has not been verified by other studies.

The organic matter is medium to dark gray, slightly anistropic in incident light, opaque except on thinnest edges, and takes a poor polish. Bituminous residues have accumulated within a variety of accessible spaces including: (1) passageways along silica veinlets (Figure 19); (2) fractures in sedimentary rocks and in barite veins; (3) bedding planes; and (4) interstitial pores in argillaceous siltstones. Most of the bituminous matter occurs as micron-size disseminations in interstitial clays.

Paragenetic features indicate that much of the bitumen in carbonaceous ore was remobilized and redistributed during and after the introduction of gold and hydrothermal silica. Highly carbonaceous ores appear to have been literally soaked in "petroleum" that subsequently was polymerized in situ by thermal solutions.

Organic matter in the ore is amorphous as indicated by x-ray diffraction patterns, but qualitative estimates of the relative amounts and types of the material can be made by differential thermal analysis. D.T.A. curves were run in air to temperatures as high as 1000°C. Oxidation temperatures were determined and reaction gases examined both for organic matter and pyrite from a number of samples. Pyrite oxidizes exothermally in air between 440° and 575°C, accompanied by the evolution of SO₂. Organic matter begins to oxidize slowly at about 350°C and reaches exothermic maxima at two temperature intervals, near 490° and 570°C, respectively. Strong endothermic reactions near 760° and 850° are attributed to the decomposition of carbonate minerals. The relative amounts of organic matter were roughly estimated in a number of samples from the intensity of oxidation exotherms.

Infrared spectra were found useful in the analysis of gases from differential thermal analyses. Surges of oxidation products such as sulfur dioxide and carbon dioxide may be correlated with oxidation exotherms recorded on D.T.A. charts. However, hydrocarbon absorption bands near 3.5 microns are weak and broad and will require further study.

Individual particles of gold have not been observed in organic matter. Poorly resolved clusters of submicron gold occasionally may appear in dolomitic siltstones of moderate organic content, where the gold is associated with clays or quartz. The textural features in most unoxidized ore samples indicate that organic matter was mobilized and redistributed possibly during mineralization.

### Role of Sulfur in Deposition

The importance of sulfur in the transportation and precipitation of metals in hydrothermal systems has been repeatedly investigated. Sulfides of gold, in common with Hg, Bi, Sb, As, and Te, are soluble in dilute aqueous solutions at different temperatures. The solubility of gold and mercury alkali-sulfide solutions was demonstrated long ago (Becker, 1). This was subsequently expanded into the "alkali sulphide theory of gold deposition" by F. Gordon Smith (13), which has gained moderate acceptance. He precipitated native gold from alkali-sulfide solutions.

Since the solubility of gold in aqueous solutions depends on free sulfur and high pH, it follows that precipitation of gold depends on lowering the sulfur content and the pH of the system. Such a mechanism for gold precipitation was proposed by Ransome (2) for the Goldfield deposits, and by Joralemon (19) for the Getchell deposit. The loss of sulfur as H₂S from hot springs is well documented. Discharge gases from Steamboat Springs are principally H₂S, CO₂, nitrogen, and argon (15). According to White (20), hydrogen sulfide is evolved from the water table wherever active thermal systems exist and, under favorable

conditions, is oxidized in part to sulfuric acid. He reports a pH of 7.9 for flowing saline water, while sinter only a few inches above the water table may have a pH of less than 2.

Sulfide minerals associated with gold deposition at Carlin include pyrite, stibnite, realgar, and cinnabar. The depth of the water table at the time of mineralization is not known, and only minor amounts of alunite and jarosite have been identified. Some degassing of thermal solutions may have occurred and possibly contributed in places to the supersaturation and ultimate precipitation of gold, arsenic, mercury, antimony, and iron.

## METALLIC MINERAL ASSOCIATES OF GOLD

Microscopic particles of gold have been observed in mutual contact with quartz, dolomite, barite, illitic clays, pyrite, and carbonaceous matter. Also a number of metallic minerals occur that have not been observed in mutual contact with gold; cinnabar, native arsenic, realgar, orpiment, and stibnite.

Base metal sulfides and sulfarsenides, including galena, sphalerite, jordanite, and possibly tennantite, have been detected in barite veins and quartz porphyry dikes quite low in gold. Metallic minerals that have been identified within or adjacent to ore zones at the Carlin mine are described in the following sections.

### Native Arsenic

Small, nearly spherical inclusions of native arsenic, 2 to 30 μ thick, occur in argillaceous siltstones and barite veins. Arsenic granules occur locally in the interstices between grains in argillaceous siltstones and in solution cavities in barite. Visible gold has been detected in polished mounts spatially associated with native arsenic but not in mutual contact.

### Pyrite

Pyrite is distributed throughout portions of unoxidized ore, ranging in concentration mostly from about 0.7 to 3 per cent; occasional samples may contain up to 10 per cent pyrite but are largely limited to the easterly extension of the main ore body. Pyrite crystals range predominantly from one to 20 μ in diameter and occur as minute inclusions in quartz crystals even where ore is highly oxidized. Pyrite is finely disseminated, similarly to gold, but, in some instances, may occur as replacements of microfossils.

In reflected light, pyrite is pale brass yellow and quite hard compared with associated gold. Morphologically, most pyrite crystals have been partially replaced and only occasionally appear as euhedral crystals, displaying pyritohedral and cubic outlines.

Growth rings and inclusions are common in pyrite and most crystals show microfracturing with cavity defects. Some grains are nearly spherical, a shape probably caused by crystal growth under conditions in which particles of colloidal size were precipitated initially, followed by some continued crystal growth directly from solution. Much of the gold appears to be slightly later than the pyrite, occurring along accessible cavities or external defects in the pyrite.

Occasionally gold is barely discernible microscopically as inclusions in pyrite. Such inclusions are liberated from the pyrite by oxidation. Gold may be more easily observed in polished sections of oxidized ores surrounded by hematite, goethite, and amorphous limonite.

X-ray diffraction patterns from flotation concentrates from the east ore body show major pyrite with minor amounts of quartz and clays as contaminants. Arsenopyrite or marcasite were not detected, but significant amounts of arsenic (1 to 2 per cent As) were noted by x-ray fluorescence analyses. Small amounts of copper and zinc (0.01 per cent) were also detected in the pyrite concentrates.

### Cinnabar

Small amounts of cinnabar are widely distributed, occurring as spotty disseminations along cavities in silicified beds. Fine, bright red crystals usually may be separated from most ore samples by panning and range from a few microns up to several millimeters long. Reddish brown crystals of cinnabar, up to nearly a centimeter long, were found near the northwest fault on the 6380-foot bench of the main ore body. Traces of schuetteite, $HgSO_4 \cdot 2H_2O$ were identified as occasional coatings on oxidized surfaces of the cinnabar (22). In thin section, crystals of cinnabar appear sparsely distributed in interstitial cavities along zones of porous silification and occasionally are surrounded by late fillings of calcite.

### Arsenic Sulfides

Realgar and orpiment have been identified in the Roberts Mountains and "Popovich" limestones and in mineralized seams of siltstone in the main ore, body. Realgar alters to orpiment and arsenolite, which appear as yellow coatings along exposed surfaces of barite veins and adjacent carbonaceous zones.

The distribution of arsenic in the ore body is somewhat parallel to gold. Higher gold values are usually accompanied by higher arsenic, as indicated by comparison of metal values across a high grade ore zone on the 6400-foot bench (Table VI). Gold content of from about 1 to 5 ounces per ton may be accompanied by from 0.1 to 0.4 per cent As. Some arsenic samples range up to several per cent. Partially oxidized ore in the main pit may contain arsenolite and locally scorodite.

### Stibnite

Antimony, like mercury and arsenic, is widely disseminated through the ore body and locally is concentrated in restricted pockets as finely crystalline stibnite. Stibnite crystals are usually highly prismatic and hairlike in appearance. Microscopically, crystals are opaque in transmitted light, light gray to white in reflected light, and strongly anisotropic. Stibnite is usually associated with realgar along silicified portions of barite veins.

Antimony commonly occurs in concentrations from 0.01 to 0.02 per cent in the ore and locally increases to several tenths of a per cent along barite veins mineralized with arsenic.

### Galena

Galena occurs sparsely with sphalerite in barite. Polished specimens show triangular pits typical of the cubic cleavage pattern of galena. Galena has been partially replaced by jordanite as the result of reaction with arsenic during mineralization.

Amounts of lead in the ore predominantly range from 0.001 to 0.003 per cent but may increase to nearly 0.01 per cent near quartz porphyry dikes and barite veins in the Roberts Mountains and "Popovich" formations. Sphalerite and galena are reported by R. A. Hardy (10) at Getchell, Nevada.

### Sphalerite

Sphalerite is associated with galena in barite replacements. Crystals are highly embayed by barite and microcrystalline quartz. Jordanite is locally associated with quartz veinlets that penetrate sphalerite. Sphalerite is medium gray in incident light and translucent amber yellow in transmitted light.

Alteration of sphalerite to an opaque phase

resembling tennantite has occurred along the external boundaries of some crystals.

## Jordanite, $Pb_{14}As_7S_{24}$

Borders of galena crystals are rimmed by microcrystalline replacements by jordanite. Jordanite is microcrystalline and similar in reflectance to galena, but the jordanite is anisotropic. At some barite localities, galena has been completely replaced by jordanite. Some pods of jordanite in barite range up to several millimeters across.

## Tennantite $(Zn,Cu,Fe)_{12}As_4S_{13}$

An isotropic gray, metallic mineral locally replaces sphalerite in barite veins where mineralized by late arsenic-bearing solutions. This opaque mineral may be zincian tennantite, since the mineral assemblage and chemical environment are compatible. However, as yet insufficient amounts of the mineral have been available for identification.

## Secondary Supergene Minerals

Several supergene products have been tentatively identified in weathered portions of the ore; schuetteite, $HgSO_4 \cdot 2HgO$, arsenolite, $As_2O_3$, carminite, $PbFe_2(AsO_4)_2(OH)_2$, and scorodite, $FeAsO_4 \cdot 2H_2O$. These minerals reflect the association and local reactions of lead, arsenic, mercury, and iron in an oxidizing environment. Oxidation products are usually fine grained and difficult to identify without combined microscopic and x-ray techniques.

## Paragenesis of Metallic Minerals

Although only a few of the metallic minerals described from the Carlin ore are found in any one specimen, there is a sufficient overlap of mineral assemblages to permit in outline, a probable sequence of metallic deposition (Figure 20).

Two main sequences of mineralization appear: (1) earlier base metal mineralization associated with barite and Cretaceous intrusives and (2) later low temperature Au-Hg-Sb-As mineralization possibly related to epithermal activity during the Tertiary.

Field and microscopic evidence indicate that cinnabar was deposited after the gold along porous silicified channels of the deposit. Crystals of cinnabar are locally surrounded by late calcite that fills porous interstices of the ore.

The gold has not been noticeably affected by recent weathering. There is no field or laboratory evidence to indicate that supergene redistribution of gold has occurred in significant amounts at the Carlin deposit.

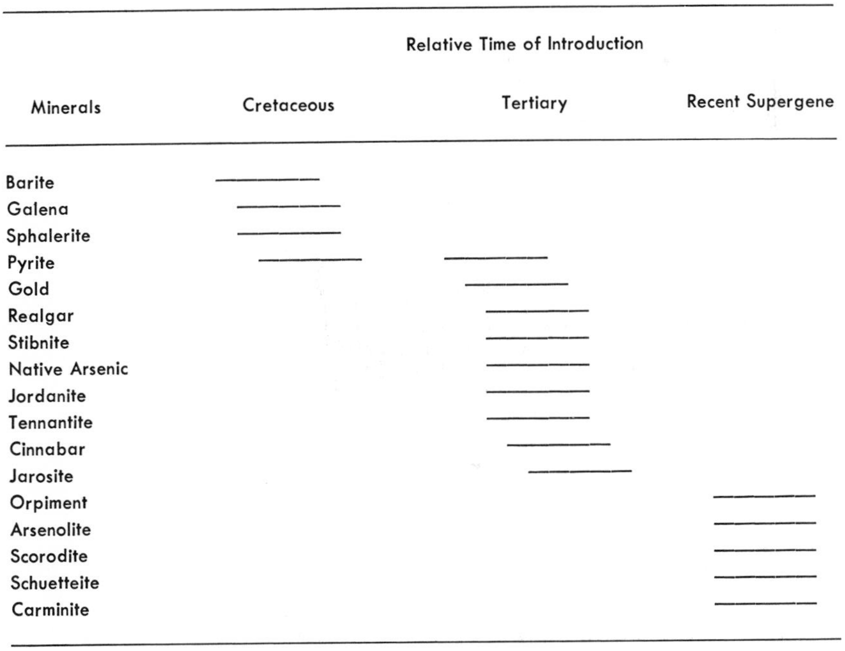

### Relative Time of Introduction

| Minerals | Cretaceous | Tertiary | Recent Supergene |
|---|---|---|---|
| Barite | ——— | | |
| Galena | ——— | | |
| Sphalerite | ——— | | |
| Pyrite | ——— | ——— | |
| Gold | | ——— | |
| Realgar | | ——— | |
| Stibnite | | ——— | |
| Native Arsenic | | ——— | |
| Jordanite | | ——— | |
| Tennantite | | ——— | |
| Cinnabar | | ——— | |
| Jarosite | | ——— | |
| Orpiment | | | ——— |
| Arsenolite | | | ——— |
| Scorodite | | | ——— |
| Schuetteite | | | ——— |
| Carminite | | | ——— |

FIG. 20. *Paragenesis of Metallic Minerals in Main Ore Body*

## ALTERATION FEATURES

### General Description

Among the most striking alteration features exposed in the Carlin pit are rounded white to light-gray areas caused by extensive silicification of the Roberts Mountains Formation along the east face of the main ore body. Areas are elliptical in outline with long axes 20 to 150 feet in length and elongate in the direction of bedding. Ellipses are isolated but occur in broadly spaced clusters shown in at least three pit areas. The elliptical forms may cut directly across bedding and also occur at multiple horizons in the deposit. They appear to represent ancient conduits, perhaps indicative of the direction of flow of late solutions causing silicification through the ore deposit. The conduit

areas are low in gold and do not appear to yield ore but may bear a zonal relationship to encircling ore zones. A siliceous-pyritic zone several feet thick frequently becomes brown on oxidation and surrounds the light gray conduits. The downward projection of the conduits is yet to be established.

In general, alteration at Carlin may be described in terms of four processes of replacement which have resulted in textural and compositional changes: (1) decarbonatization, (2) argillitization, (3) silicification, and (4) calcification.

### Decarbonatization

Decarbonatization represents the earliest stage of alteration in which carbonate minerals were removed selectively and redistributed. Se-

TABLE XI.   Chemical Composition of Miscellaneous Rock Types at Carlin

| Chemical Constituents | (1) Silicified Roberts Mts. Per Cent | (2) Silicified Roberts Mts. Per Cent | (3) Altered Porphyry Dike Per Cent | (4) Quartz Diorite Per Cent | (5) Altered Quartz Diorite Per Cent | Method of Analysis |
|---|---|---|---|---|---|---|
| $SiO_2$ | 91.22 | 74.14 | 63.11 | 59.94 | 53.46 | Wet Chemical |
| $Al_2O_3$ | 6.28 | 13.52 | 10.80 | 13.12 | 9.90 | " |
| $Fe_2O_3$ | 2.26 | 0.92 | 11.74 | 5.58 | 6.45 | " |
| CaO | 0.18 | 0.64 | 0.92 | | | " |
| MgO | 0.34 | 0.85 | 0.15 | | | " |
| $K_2O$ | <0.5 | 5.0 | <0.5 | 6.0 | 4.0 | Spectrographic |
| $Na_2O$ | <0.05 | 0.05 | <0.05 | 4.0 | 2.5 | " |
| $TiO_2$ | 0.25 | 0.7 | 1.0 | 0.75 | 0.85 | " |
| $MnO_2$ | 0.02 | 0.003 | 0.03 | 0.08 | 0.12 | " |
| BaO | 0.04 | 0.1 | 0.35 | 0.25 | 0.25 | " |
| $SrO_2$ | 0.005 | 0.008 | 0.05 | 0.08 | 0.05 | " |
| ZnO | 0.002 | <0.002 | 0.4 | 0.003 | 0.005 | " |
| CuO | 0.015 | 0.002 | 0.03 | 0.020 | 0.025 | " |
| PbO | <0.001 | <0.008 | 0.003 | 0.003 | 0.002 | " |
| $As_2O_5$ | 0.04 | 0.03 | 0.075 | 0.01 | 0.03 | " |
| HgO | <0.01 | <0.01 | <0.01 | <0.01 | <0.01 | " |
| $Sb_2O_5$ | 0.05 | <0.01 | <0.01 | <0.01 | <0.01 | " |
| NiO | 0.004 | <0.001 | 0.06 | 0.005 | 0.002 | " |
| $ZrO_2$ | 0.01 | 0.08 | 0.04 | 0.05 | 0.05 | " |
| CoO | <0.001 | <0.001 | 0.01 | 0.003 | 0.002 | " |
| $Cr_2O_3$ | 0.02 | 0.02 | 0.15 | 0.015 | 0.01 | " |
| $V_2O_5$ | 0.06 | 0.1 | 0.07 | 0.04 | 0.04 | " |
| $B_2O_5$ | 0.02 | 0.12 | 0.03 | <0.01 | 0.01 | " |
| Y | — | 0.007 | 0.015 | 0.005 | 0.005 | " |
| Au | 0.005 oz/t | 0.015 oz/t | 0.05 oz/t | Nil | 0.02 oz/t | Fire Assay |

(1) Silicified Roberts Mts. from Top Bench of West Ore Body.
(2) Silicified Roberts Mts. from 6460 Bench shown.
(3) Composite sample of altered porphyry dikes from 6380' and 6400' Benches of Main Ore Body.
(4) Quartz diorite from drill hole in intrusive 3 miles north of Main Pit.
(5) Altered sample from same drill hole as fresh quartz diorite.

lective removal of calcite by hydrothermal solutions occurred both in the Roberts Mountains and in the overlying "Popovich." Dolomite being less soluble commonly remained as relic rhomb-shaped crystals in a matrix of porous clays.

Calcite and dolomite were removed along a zone subjected to later silicification. There, relic sand grains may be observed distributed through a matrix of porous clays. However, later veinlets of calcite were locally re-introduced into some of the silicified zones. Largely decalcified rocks that include significant amounts of dolomite (5 to 30 per cent) often contain gold in larger amounts.

Fine-grained pyrite is disseminated through most of the altered rock but is not detected along zones of intense silicification. Such zones appear to have been largely stripped of their metal content, including iron, base metals, and gold (Table XI).

## Argillization

Argillic alteration apparently accompanied decarbonatization. Illite is the predominant clay mineral, while montmorillonite and kaolinite are less abundant. In most ore samples, clays constitute 20 to 60 per cent of the rock. However, the clay content of the rock prior to alteration is largely a matter of inference. Probably a portion of the clay in altered rocks was derived from original clays, concentrated during the removal of carbonates. However, both recrystallization and direct crystallization of clay minerals occurred during hydrothermal alteration. Recrystallization, at least, and even direct crystallization could account for the major portion of clays in argillized rocks of the Roberts Mountains Formation.

In thin sections of ore samples, illite occurs as finely crystalline clay flakes. Individual clay flakes are submicron in size but lie together in aggregates that show parallel extinction under crossed nicols. Oriented slides have been prepared from clay concentrates and analyzed by x-ray diffraction. X-ray diffraction patterns indicate that the illite is uncommonly well crystallized and usually does not contain interstratified layers of montmorillonite. Montmorillonite normally occurs as a minor but distinct mineral phase in most ore samples and increases locally in the vicinity of quartz porphyry dikes to become a major component within altered dikes and along contacts. Kaolinite also is locally prominent while minor dickite has been identified from altered dike samples along the northwest fault in the main ore body.

Consideration of polytype stabilities of clay mineral structures by Bailey (36), or polymorphs as determined by Güven and Burnham (35), suggests that low-grade metamorphic or hydrothermal illites usually are of the metastable 1Md and 1M structures. These metastable micas crystallize early and may persist indefinitely at low temperatures. Illite as used herein probably includes hydromica (which is often a synonym) and even fine sericite.

Irregularly crenulated and even vein-like clay seams of possible epigenetic origin follow the trend of lineation. Seams are lenticular, ranging from less than a millimeter to a centimeter or more in thickness, and do not show regularly bedded features. Illite is the dominant component forming as much as 80 per cent of some seams. Gold has been found to concentrate in the illite fractions of the rock as indicated by analyses of selected lithologic layers (Figure 21) and by clay concentrates (Table IX). Silicified beds found in portions of the ore body have been depleted in clays that may have been removed or destroyed.

FIG. 21. *Crenulated Clay Seams in Ore, as shown beneath a glass slide. (A) silicified argillaceous siltstone, 1 in. thick, 2.70 oz/ton Au; (B) carbonaceous silty clay, 0.6 in. thick, 2.98 oz/ton Au; (C) white bleached illitic clay seamlet, 0.3 in. thick, 3.02 oz/ton Au; and (D) siltstone-sandstone, 0.5 in. thick, 0.82 oz/ton Au.*

## Silicification

Silica in several types of field occurrence may be recognized at Carlin: (1) original detrital quartz grains in the Roberts Mountains Formation; (2) thin dark sedimentary chert bands sparsley distributed through the original

FIG. 24. *Drusy Dark-Gray Chalcedonic Seam from Close Up of Chimney on 6420-foot Bench.*

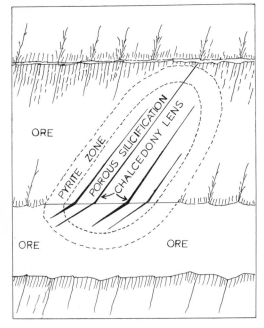

FIG. 22. *Schematic Diagram of a Chimney-like Form in the Main Ore Body. An ellipitical core of porous silification is cut by more or less parallel chalcedonic lenses and surrounded by a pyrite-bearing zone. Ore surrounds the chimney but the chimney itself is non-ore-bearing.*

FIG. 23. *Close-up View of Chimney Structure, showing dark chalcedonic tongues of drusy silica pinching out upward along bleached beds.*

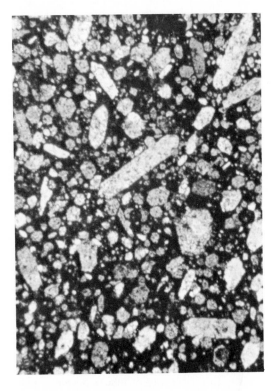

FIG. 25. *Photomicrograph of Recrystallized Siltstone Bed from 6380-foot Bench, showing euhedral doubly terminated crystals of quartz. ×26, partially crossed nicols.*

formation; (3) porous chimney-like ellipsoidal masses of fine recrystallized quartz (Figure 22); (4) chalcedonic tongues of gray silica, porous and drusy to hard and massive, that follow stratification but either pinch out in places or may swell to masses as much as 10 feet thick (Figures 23, 24); (5) recrystallized (often euhedral) quartz in the ore zone (Figure 25); and (6) small white quartz veins possibly accompanying surface induration that probably were formed by supergene recrystallization.

Lenses of cherty silicification in the original strata are thin dark gray and sinous, and intercalated with carbonate (calcite-bearing) strata, as would be expected in a syngenetic origin. Such dense dark-gray lenses of silica consist mostly of microcrystalline quartz and probably persist along less altered stratigraphic horizons

in the deposit, locally grading into fine drusy quartz between upper and lower layers of dark-gray chalcedony. In thin sections, the inner zone of drusy silica displays remarkable networks of interconnecting micro-channels ranging from about $30\,\mu$ up to a millimeter or more in diameter. The silica framework consists of a mozaic of anhedral microcrystalline quartz that ranges considerably in size and shape. The permeability of the drusy silica to water flow must be high compared to adjacent solid chalcedonic lenses.

Thin sections of finely crystalline silicification show a maze of open channels through a recrystallized framework of quartz crystals. Quartz crystals often display euhedral outlines (Figure 25) and range from about 30 to $100\,\mu$ long. Euhedral quartz crystals often are doubly terminated but show little or no tendency to

**TABLE XII.**  *Gold Content of Bleached, Silicified, and Related Rock Types in the Roberts Mountains Formation*

|  | Au Oz/Ton |
|---|---|
| **(A) Cross Section Through Light Gray Silicified Zone in Ore Body on 6380-foot Bench near Center of Main Pit (Fig. 64)** | |
| Dol. Arg. Siltstone above "white"[1] zone | 0.43 |
| Silicified chalcedonic seam at upper "white" contact | 0.08 |
| Arg. siltstone with drusy silica seams in "white" zone | 0.03 |
| Silicified chalcedonic seam at lower "white" contact | 0.22 |
| Arg. siltstone below "white" zone | 0.54 |
| Recrystallized siltstone-sandstone bed below "white" zone | 0.03 |
| **(B) Altered silicified Zone in Sparsely Mineralized Beds on 6460-foot Bench at Northeast End of Main Pit** | |
| Argillaceous siltstone above bleached | 0.065 |
| Argillaceous siltstone up-dip from "white" area | 0.04 |
| Argillaceous siltstone at contact of "white" area | 0.03 |
| Argillaceous siltstone near center of "white" area | 0.015 |
| Drusy and chalcedonic silica in "white" area | 0.02 |
| Drusy and chalcedonic silica at lower contact of "white" ellipse | 0.03 |
| Chalcedonic silica seam below "white" zone | 0.14 |
| Argillaceous siltstone below "white" zone | 0.055 |
| **(C) Silicified Beds Below Ore Zone in West Ore Body** | |
| Silicified outcrop with quartz veins at top of southwest ore body | <0.005 |
| Quartz vein in silicified beds exposed on top bench | <0.005 |
| Chalcedonic silica adjacent to quartz vein | 0.02 |
| Footwall adjacent to silicified beds | 0.01 |
| Argillaceous siltstones 50' below silicified zone in footwall | 0.02 |
| Argillaceous siltstones 50' above silicified zone in hanging wall | <0.005 |
| **(D) Miscellaneous Silicified Zones in East Face of Main Ore Body** | |
| Silicified Arg. siltstones on 6440-foot bench | <0.002 |
| Buff arg. siltstone above "white" zone on 6440-foot bench | 0.005 |
| Chalcedonic silica seams on 6600-foot bench | 0.03 |
| Chalcedonic silica in bleached zone on 6420-foot bench | <0.002 |

[1] "White" refers to ellipsoidal and other masses of finely crystalline silica (Figures 22, 23).

develop comb structures. Cavities in recrystallized siltstones range mostly between 10 and 150 $\mu$. The largest crystals rarely measure as much as several millimeters long.

Clays, carbonates, pyrite, and gold occur in minor amounts as minute inclusions in crystals of quartz. Possibly the gold inclusions in quartz are a final form of sparse precipitation. Interstitial cavities between crystals have been swept clean of residual clays but may be coated or filled locally with later epigenetic minerals, e.g., calcite, montmorillonite, realgar, and cinnabar. Gold usually is sparse in fine crystalline silicification and drusy silica lenses, being confined to minute inclusions in quartz crystals.

Drusy chalcedony (Figure 24) grades sharply into friable recrystallized portions of silicification ellipses that in turn pass transitionally into unsilicified argillaceous siltstones. Argillaceous siltstones adjacent to ellipsoidal zones are variably dolomitic and frequently ore bearing. Ellipses or "chimneys" show intermediate stages of silicification ranging from incipient quartz overgrowths to advanced stages of quartz authigenesis. Interstices are filled largely with leached clays, mostly illite, and the resulting rock is loosely coherent, friable and granular.

Friable white silicified areas are spatially associated and genetically related to lenses of drusy silica (Figure 24). A typical "chimney" observed in the main ore body (Figure 22) displays finger-like projections caused by bleaching solutions. Encircling the "chimney" is a brownish pyrite-bearing zone about 5 feet thick. Gold increases beyond the perimeter of the brownish zone. Another zone of fine crystalline silicification was examined in the east face above the 6400-foot bench, where arcuate upper and lower elliptical boundaries were in the direction of stratification. Numerous "chimneys" are grouped in the silicified areas designated in the east pit face (Figure 5).

Silicification also occurs as siliceous impregnations that both transect or follow bedding. Permeable decarbonatized rocks above the 6440-foot bench along the east face appear literally to have "soaked up" silica that forms irregular pockets of silicification with sharp boundaries (Figure 14). The bulk density of porous, slightly mineralized beds of argillaceous siltstone was increased from 1.90 to 2.58 in silicified pockets adjacent to unsilicified rock. The gold content apparently is greater in unsilicified rock than in silicified rock. This local form of silicification seemingly accompanied early stages of mineralization and diluted the amounts of gold in the ore body.

One large lens of chalcedonic silicification was observed to expand to 10 feet thick from a thin bed in the west ore body. At the surface not far distant, quartz veins, presumably supergene, locally are associated with chalcedonic lenses. White quartz veins exposed along the top of the hill south of the top bench follow the strike and dip of the bedding, but disappear down-dip, grading into dark-gray chalcedonic silica and siliceous impregnations.

The amounts of gold in various bleached and silicified portions of the main ore body are compared in Table XII. In Table XII (A), ore appears in argillic material both above and below the chimney.

## Calcification

Carbonate veinlets locally transect silicified zones locally replacing microcrystalline quartz. Chalcedonic to microcrystalline quartz also has been replaced by late calcite introduced along vertical fractures and interstitial vugs or channels. Permeable seams and beds in the ore may be impregnated with late calcite that fills interstices around euhedral quartz veins and permeates adjacent beds.

Decarbonated portions of the main ore body have been re-impregnated locally with calcite to form pseudo-limestones of higher bulk densities and lower amounts of gold. A sample of ore partially impregnated with calcite was separated into calcified and uncalcified fractions. A specific gravity of the uncalcified portion was 2.27 and that in the calcified portion 2.53. Gold, however, decreased from 0.49 to 0.20 oz/ton in the calcified material, possibly as a result of dilution from added carbonates.

## ORE GENESIS

The development at Carlin has not yet reached the point at which observations appear adequate to justify comparisons with other gold deposits in the Great Basin area. Thus, only brief reference to selected occurrences concerning which published data are available will be made.

The Carlin deposit appears to show some genetic similarities to the Getchell deposit. It may also be low-temperature epithermal bordering on telethermal (6). Feeders in the Carlin deposit lack many such common epithermal features as crustification, comb structure, and banding and even show some resemblance to siliceous sinters.

Vein development seldom is recognizable,

except occasionally in outcrop exposures. Permeable zones and feeders in the deposit often require microscopic examination for detection.

The similarity of the Getchell deposit to quicksilver deposits in the Great Basin has been discussed by Joralemon (19), who also suggests a close genetic relationship to Steamboat Springs.

The Carlin deposit in Nevada, together with Bootstrap and Getchell, and the Mercur deposit in western Utah and others (5), may comprise a genetically similar type, but considerably more study is needed to establish this relationship.

Recent studies of drill cores at Steamboat Springs (34) indicate that hot spring waters deposit small amounts of Hg, Sb, Au, and Ag and hydrothermally alter the rocks through which they flow. The most intense alteration is immediately adjacent to fractures where feldspars, carbonates, and clays are replaced by illite, quartz, and pyrite. This type of alteration also is prevalent at Carlin.

Hydrothermal alteration, coupled with the proven ability of such waters to deposit metals, indicate to Schoen and White (34) that spring waters are similar to some ore-forming fluids. It is well to remember, however, that the late silicification centers at Carlin, the features most suggestive of rising thermal centers, are deficient in gold. The genetic relationship of thermal waters at depth to epithermal ore solutions should receive further study.

Gold mineralization at Carlin is interpreted as epigenetic, of low-temperature origin, and formed under low-pressure conditions perhaps along the deeper roots of thermal spring activity.

## CONCLUSIONS

The cumulative evidence at Carlin points strongly to the epigenetic origin of the gold. Among other features, argillic replacement of dolomitic host rock, major deposition of colloidal-sized gold in clay aggregates, chemical changes from rock to ore, nearby igneous activity, and elliptical chimneys of recrystallized silicification, all point to low-temperature epithermal origin. Major stages in the creation of the gold deposits as now found include:

(1) Deposition of the Vinini, Roberts Mountains and "Popovich" formations.

(2) Overthrusting as represented by the Roberts Mountains fault.

(3) Cretaceous igneous invasion and subsequent deformation.

(4) Base-metal emplacement in veins and replacements.

(5) Hydrothermal alteration and gold deposition.

(6) Epithermal silicification.

(7) Supergene action, weathering, and oxidation.

Distribution of gold depends largely upon the permeability of decalcified strata of the Roberts Mountains Formation to ore-forming fluids and precipitation in favorable hosts such as illitic clays, carbonaceous matter, and pyrite. Quite near surface emplacement by the release of $H_2S$ may have contributed to the precipitation of gold, realgar, cinnabar, stibnite, and pyrite.

The fine gold occurrence at Carlin may be grouped along with other deposits at Getchell, Bootstrap, and Manhattan in Nevada, and at Mercur in Utah, as a shallow low-temperature epithermal deposit where vein development is hardly recognizable and ore minerals are finely disseminated in replacement bodies in carbonate sediments.

## REFERENCES CITED

1. Becker, G. F., 1887, Natural solution of cinnabar, gold and associated sulphides: Amer. Jour. Sci., 3d. ser., v. 33, p. 199–210.
2. Ransome, F. L., 1909, The geology and ore deposits of Goldfield, Nevada: U.S. Geol. Surv. Prof. Paper 66, 245 p.
3. Lincoln, F. L., 1911, Certain natural associations of gold: Econ. Geol., v. 6, p. 247–309.
4. Butler, B. S., *et al.*, 1920, The ore deposits of Utah: U.S. Geol. Surv. Prof. Paper 111, p. 391–395.
5. Ferguson, H. G., 1924, Geology and ore deposits of the Manhattan district, Nevada. U.S. Geol. Surv. Bull. 723, 163 p.
6. Graton, L. C., 1933, The depth zones in ore deposition: Econ. Geol., v. 28, p. 513–555.
7. Haycock, M. H., 1937, The role of the microscope in the study of gold ores: Canadian Inst. Min. and Met., Tr., v. 40, p. 405–414.
8. Mawdsley, J. B., 1938, Late gold and some of its implications: Econ. Geol., v. 33. p. 194–210.
9. Van Aubel, R., 1939, Sur l'importance dans les minerais d'or du calibre des particules, Annales des Mines, v. 16, p. 155–161.
10. Hardy, R. A., 1940, Geology of the Getchell mine: A.I.M.E. Tr., v. 144, p. 147–150.
11. Johnston, W. D., Jr., 1940, The gold quartz veins of Grass Valley, California: U.S. Geol. Surv. Prof. Paper 194, 101 p.
12. Gianella, V. P., 1941, Barite deposits in northern Nevada: A.I.M.E. Tr., v. 144, p. 294–299.
13. Smith, F. G., 1943, The alkali sulphide theory of gold deposition: Econ. Geol., v. 38, p. 561–589.
14. White, W. H., 1943, The mechanism and en-

vironment of gold deposition in veins: Econ. Geol., v. 38, p. 512–532.

15. Brannock, W. W., *et al.,* 1948, Preliminary geochemical results at Steamboat Springs, Nevada: Amer. Geophys. Union Tr., v. 29, no. 2, p. 211–226.

16. Pardee, J. T. and Park, C. F., Jr., 1948, Gold deposits of the southern Piedmont: U.S. Geol. Surv. Prof. Paper 66, 245 p.

17. Stillwell, F. L. and Edwards, A. B., 1949, An occurrence of sub-microscopic gold in the Dolphin East Load, Fiji: Aust. Inst. Min. and Met. Pr. no. 154–155, p. 31–46.

18. Chase, F. M., 1949, Origin of the Bendigo saddle reefs with comments on the formation of ribbon quartz: Econ. Geol., v. 44, p. 561–596.

19. Joralemon, P., 1951, The occurrence of gold at the Getchell mine, Nevada: Econ. Geol., v. 46, p. 267–309.

20. White, D. E., 1955, Thermal springs and epithermal ore deposits: Econ. Geol., 50th Ann. Vol., pt. 1, p. 155–161.

21. Roberts, R. J., *et al.,* 1958, Paleozoic rocks of north-central Nevada: Amer. Assoc. Petrol. Geols. Bull., v. 42, p. 2813–2857.

22. Bailey, E. H., *et al.,* 1959, Schuetteite, a new supergene mercury mineral: Amer. Mineral, v. 44, p. 1026–1038.

23. Roberts, R. J., 1960, Alinement of mining districts in north-central Nevada: U.S. Geol. Surv. Prof. Paper 400B, p. B17–B19.

24. Erickson, R. L., *et al.,* 1961, Geochemical anomalies in the upper plate of the Roberts thrust near Cortez, Nevada: U.S. Geol. Surv. Prof. Paper 424-D, p. D316–D320.

25. Molloy, W. and Kerr, P. F., 1961, Diffractometer patterns of A.P.I. reference clay minerals: Amer. Mineral., v. 46, p. 583–605.

26. Kerr, J. W., 1962, Paleozoic sequences and thrust slices of the Seetoya Mountains, Independence Range, Elko County, Nevada: Geol. Soc. Amer. Bull., v. 73, p. 439–460.

27. Armstrong, R. L., 1963, Geochronology and geology of the eastern Great Basin: Unpublished Ph.D. thesis, Yale Univ.

28. Van Olphen, H., 1963, An introduction to clay colloid chemistry: Interscience Publishers, N.Y., p. 301.

29. Erickson, R. L., *et al.,* 1964, Geochemical exploration near the Getchell mine, Humboldt County, Nevada: U.S. Geol. Surv. Bull. 1198-A, p. 26.

30. Kay, M. and Crawford, J. P., 1964, Paleozoic facies from the miogeosynclinal to the eugeosynclinal belt in thrust slices, central Nevada: Geol. Soc. Amer. Bull., v. 75, p. 425–454.

31. Hardie, B., 1965, Carlin gold, Lynn District: A.I.M.E. Fall Mtg., Reno, Nev.

32. McQuiston, F. W., Jr. and Hernlund, R. W., 1965, Newmont's Carlin gold project: Min. Cong. Jour., v. 52, no. 11, p. 26–30, 32, 38–39.

33. Schilling, J. H., 1965, Isotopic age determination of Nevada rocks: Nev. Bur. Mines Rept. 10, 79 p.

34. Schoen, R. and White, D. E., 1965, Hydrothermal alteration in GS-4 drill holes, Main Terrace, Steamboat Springs, Nevada: Econ. Geol., v. 60, 1411–1421.

35. Güven, N. and Burnham, C. W., 1965–1966, The crystal structure of 3T muscovite: in *Ann. Rept. Dir. Geophys. Lab.,* Carnegie Inst., Washington Year Book 65, p. 290–293.

36. Bailey, S. W., 1966, The status of clay mineral structures: 14th Nat. Cong. on Clays and Clay Minerals, Pr., Pergamon Press, p. 1–24.

37. Erickson, R. L., *et al.,* 1966, Gold geochemical anomaly in the Cortez district, Nevada: U.S. Geol. Surv. Circ. 534, 9 p.

# 47. Geology and Ore Deposits of the East Tintic Mining District, Utah[*]

W. M. SHEPARD,[†] H. T. MORRIS,[‡]
D. R. COOK[§]

## Contents

[*] Publication authorized by Kennecott Copper Corp. and the Director, U.S. Geological Survey.
[†] Kennecott Copper Corporation, Dividend, Utah.
[‡] U.S. Geological Survey, Menlo Park, California.
[§] Bear Creek Mining Company, Spokane, Washington.

## Illustrations

## Tables

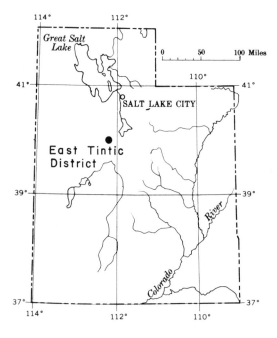

## ABSTRACT

The East Tintic district in central Utah has produced ores of gold, silver, copper, lead, and zinc valued at more than $120,000,000. All of this ore has been produced from blind ore bodies in Paleozoic sedimentary rocks that are concealed beneath hydrothermally altered volcanic rocks and cut by intrusive bodies of Eocene age. The Paleozoic rocks, which range in age from Early Cambrian to Mississippian, form the core and limbs of a large, north-trending asymmetric anticline that is cut by low-angle thrust faults, high-angle transcurrent faults, mineralized fissures and faults, and normal faults that are both older and younger than the mineralized fissures. The ore bodies generally may be grouped into two classes: (1) massive replacement bodies that are rich in silver, lead, zinc and manganese and (2) fissure veins that are valuable primarily for their content of gold, copper, and silver. The replacement ore bodies are localized chiefly in the Middle Limestone Member of the Ophir Formation of Middle Cambrian age where it has been thrust-faulted against older or younger rocks and cut by northeast-trending mineralized fissure zones. The fissure veins primarily are productive only in the Lower Cambrian Tintic Quartzite. Extensive low-grade deposits of lead-zinc-silver ore recently have been discovered in Devonian carbonate rocks in the Lower plate of the East Tintic thrust fault but as yet are unevaluated. The principal guides to ore include: (1) intersections of northeast-trending fissures and cross-breaking faults that brecciate the carbonate rocks; (2) zones of late-stage pyritic, calcitic, and sericitic alteration in the lavas; (3) primary geochemical anomalies in the altered rocks; and (4) mineral zoning patterns of the known ore bodies. The present development of the Burgin mine would seem to assure continued productivity of the district for at least a decade.

## INTRODUCTION

The East Tintic mining district in central Utah has achieved world prominence from its production of rich silver and lead ores from blind ore bodies. The recent development of the Burgin mine and the discovery of mineralized areas nearby, all concealed by thick flows of barren lava, give promise of the discovery of new ore bodies by the continued application of geological, geochemical, and geophysical techniques.

The district is in the east-central part of the East Tintic Mountains, a north-trending fault-block mountain range near the east-central margin of the Great Basin. It is approximately bounded by Meridians 112° and 112°5' west and Parallels 39°55' and 40° north and covers about 20 square miles. Officially, it constitutes the northeastern part of the Tintic mining district, but it has been popularly regarded as a separate district since the turn of the century. Although the ores of the two districts are mineralogically similar, the ore bodies of the East Tintic district are localized more obviously by structural features and generally do not form the extensive linear ore zones that characterize the replacement ore bodies of the Main Tintic district.

The topography in the area of the East Tintic district is rolling to moderately rugged,

ranging in elevation from about 5100 feet to 8100 feet. The steepest slopes are found on inliers of sedimentary rocks that project through the extensive lava flows and pyroclastic deposits that cover the greater part of the district. These steep slopes largely represent an ancient topography that is being exhumed through the erosional stripping of the less resistant, commonly altered volcanic rocks.

## PRODUCTION

The mines of the East Tintic district have yielded about 3.6 million tons of ore valued at $120,000,000 (21, p. 97). As shown in Table I, more than two-thirds of this ore was produced from the Tintic Standard mine and the balance from only eight other operating properties. Silver and base metal ores were first produced from the district in 1909, following the discovery of narrow replacement veins by shallow workings driven from the Eureka Lilly shaft. The discovery of these blind ore bodies stimulated exploration and development of adjacent properties, and the district rocketed into prominence with the discovery of the rich, totally concealed Tintic Standard ore bodies in 1916. By 1924, these deposits were yielding 150,000 tons of argentiferous lead ore per year, making the Tintic Standard the most productive silver mine in the United States, if

TABLE I.    Mine Production—East Tintic District, Utah, 1909–1966

| Mine | Interval | Tons | Gold (Ounces) | Silver (Ounces) | Copper (Pounds) | Lead (Pounds) | Zinc (Pounds) |
|------|----------|------|---------------|-----------------|-----------------|---------------|---------------|
| Apex Standard[1] | 1928–1937 | 13,728 | 1,373 | 188,074 | 109,824 | 741,312 | — |
| Burgin[2] | 1955–1966 | 150,600 | 22 | 2,137,400 | — | 52,551,700 | 19,699,800 |
| Eureka Lilly | 1909–1952 | 227,610 | 53,254 | 1,227,754 | 3,747,328 | 21,750,597 | 18,950 |
| Eureka Standard | 1928–1952 | 362,375 | 242,903 | 3,430,277 | 2,715,748 | 11,209,798 | 3,496,852 |
| Iron King[1] | | 14,000 | 1,400 | 19,601 | 28,000 | — | — |
| North Lily Group[3] (Includes: North Lily, Baltimore, Tintic Bullion, Eureka Bullion, Hannibal, Provo, and other properties operated through North Lily shaft) | 1927–1949 | 375,000 | 148,000 | 3,554,000 | 2,482,000 | 101,682,000 | 4,270,000 |
| Tintic Standard (Includes ore from Harold Mill dump 1943–1952) | 1913–1952 | 2,469,722 | 90,005 | 52,239,832 | 18,502,917 | 554,689,732 | 954,748 |
| 20th Century | 1943–1947 | 1,419 | 40 | 7,320 | 59,716 | 1,338 | — |
| Zuma[1] | 1928–1944 | 2,208 | 442 | 3,754 | 8,832 | 343,608 | — |
| Totals | | 3,616,662 | 537,439 | 62,808,012 | 27,654,365 | 742,970,085 | 28,440,350 |

[1] Data modified from Cook, D. R., editor, 1957, pl. 3.

[2] This report.

[3] Data modified from Kildale, M. B., in Cook, D. R., editor, 1957, p. 105.

All other data from U.S. Bureau of Mines.

*TABLE II.   Stratigraphic column, East Tintic District, Utah*

| System | Series | Formation or Unit | Lithology and Average Thickness |
|---|---|---|---|
| Quaternary | Recent | Younger Alluvium | Fanglomerate, gravel, sand, and silt; 0–100 ft. |
| Quaternary | Pleistocene | Bonneville Formation | Lacustrine gravel and sand; 20 ft. |
| Quaternary | Pleistocene | Alpine Formation | Lacustrine sand and silt; 25 ft. |
| Quaternary | Pleistocene | Older alluvium | Fanglomerate, colluvium, and stream gravels. |
| | | —Unconformity— | |
| Tertiary | Eocene | Andesite or latite dikes and related intrusion breccias | Purple porphyritic dikes, locally altered to kaolinite; probably contemporaneous with ore deposition. |
| | | —Intrusive contact— | |
| Tertiary | Eocene | Quartz monzonite porphyry | Greenish-gray, coarsely porphyritic dikes and plugs. |
| | | —Intrusive contacts— | |
| Tertiary | Eocene | Pebble dikes | Narrow dikes of intrusion breccia. |
| Tertiary | Eocene | Monzonite of Silver City stock and associated biotite monzonite porphyry | Greenish-gray, granitic to coarsely porphyritic monzonite; altered near veins. |
| Tertiary | Eocene | Monzonite porphyry of Sunrise Peak stock and associated hornblende monzonite porphyry | Medium- to dark-gray coarsely porphyritic monzonite; altered near veins. |
| | | —Intrusive contacts— | |
| Tertiary | Eocene | Laguna Springs Latite | Reddish-gray flows, tuffs and agglomerate; 0–2500 ft. |
| | | —Intrusive contact— | |
| Tertiary | Eocene | Swansea Quartz Monzonite | Granitic intrusive rock chiefly altered and bleached. |
| | | —Intrusive contact— | |
| Tertiary | Eocene | Packard Quartz Latite | Purplish-gray contorted flows and white tuff; 0–2700 ft. |
| Tertiary (?) | Eocene (?) | Apex Conglomerate | Brick-red conglomerate and sandy shale; 0–500 ft. |
| | | —Unconformity— | |
| Carboniferous Mississippian | Upper | Humbug Formation | Alternating blue limestone and buff sandstone; 650 ft. |
| Carboniferous Mississippian | Upper | Deseret Limestone | Blue-gray cherty and coquinoid limestone; 1000 ft. |
| Carboniferous Mississippian | Lower | Gardison Limestone | Blue-gray distinctly bedded cherty limestone; 500 ft. |
| Carboniferous Mississippian | Lower | Fitchville Formation | Seven distinctive units of limestone and cherty dolomite; curly laminated bed near top; 300 ft. |
| Devonian and Mississippian | | Pinyon Peak Limestone | Blue-gray shaly limestone, sandy at base 80 ft. |
| | | —Disconformity (?)— | |
| Devonian | Upper | Victoria Formation | Gray dolomite and buff quartzite; locally some lenses of penecontemporaneous breccia; 280 ft. |

TABLE II.   *Stratigraphic column, East Tintic District, Utah (Continued)*

| System | Series | Formation or Unit | Lithology and Average Thickness |
|---|---|---|---|
| Devonian, Silurian and Ordovician | | Bluebell Dolomite | Dusky-gray coarse-grained dolomite with some beds of sublithographic creamy white dolomite. Curly laminated marker beds near middle; 350–600 ft. |
| Ordovician | Upper | Fish Haven Dolomite | Mottled gray cherty dolomite; 300 ft. |
| | | —————————Disconformity————— | |
| | Lower | Opohonga Limestone | Blue-gray thin-bedded shaly limestone; 300–850 ft. |
| Cambrian | Upper | Ajax Dolomite | Dusky-gray cherty dolomite; 650 ft. |
| | | Opex Formation | Thin-bedded sandy limestone and shale; 250 ft. |
| | Middle | Cole Canyon Dolomite | Dusky-gray and creamy white dolomite; 850 ft. |
| | | Bluebird Dolomite | Dusky-gray dolomite with white markings; 190 ft. |
| | | Herkimer Limestone | Blue shaly limestone and green shale; 400 ft. |
| | | Dagmar Dolomite | Creamy-white laminated dolomite; 80 ft. |
| | | Teutonic Limestone | Blue shaly limestone with pisolitic zones; 420 ft. |
| | | Ophir Formation | Gray-green shale and blue oolitic limestone; 410 ft. |
| | Lower | Tintic Quartzite | Buff quartzite; gray-green phyllite beds near top, conglomerate zone near base; 3000 ft. |

## Structure

The thick section of Paleozoic rocks of the East Tintic Mountains is compressed into broad, north-trending folds and is cut by many faults including low-angle thrust faults, high-angle transcurrent faults, post-compressive normal faults, and, locally near the centers of intrusion, by moderately persistent mineralized fissures (Figure 1). The compressive structural features are all pre-lava in age and apparently resulted from the regional deformation of one of a series of large thrust sheets that moved eastward across central Utah in Cretaceous time (11, p. 377–400; 34, p. 1944–1951).

The dominant structures in the district are the north-trending, asymmetric East Tintic anticline and the East Tintic thrust fault that cuts its eastern limb. These structures are concealed largely beneath the lavas and are known chiefly from exposures in mine workings and from drill hole data. The anticline is one of the folds of the Tintic-Oquirrh fold system, which apparently is limited to the upper plate of the Midas thrust fault (34, p. 1948). A minimum amplitude of 10,000 feet is estimated for this fold on the basis of detailed measurements of the stratigraphic units that are exposed from the core of the anticline to the trough of the adjacent Tintic syncline.

Cross-sections prepared from subsurface data and limited surface exposures show the west limb of the East Tintic anticline dipping more or less uniformly to the west about 30° and the east limb overturned and sharply crumpled above the East Tintic thrust zone.

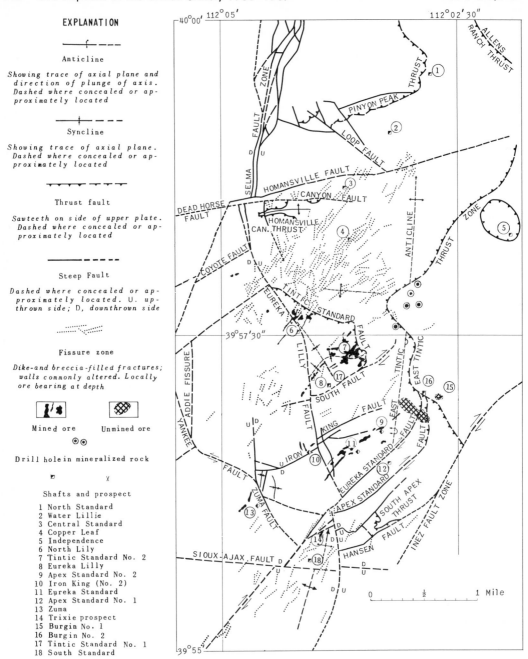

EXPLANATION

——┌——  ———
Anticline

*Showing trace of axial plane and
direction of plunge of axis.
Dashed where concealed or approximately located*

——┼——  ———
Syncline

*Showing trace of axial plane.
Dashed where concealed or approximately located*

▬▼▼▼▬ ▼ ▼ ▼ ▬
Thrust fault

*Sawteeth on side of upper plate.
Dashed where concealed or approximately located*

▬▬▬▬  ▬ ▬ ▬
Steep Fault

*Dashed where concealed or ap-
proximately located. U. upthrown side; D, downthrown side*

⋯⋯⋯⋯
Fissure zone

*Dike-and breccia-filled fractures;
walls commonly altered. Locally
ore bearing at depth*

Mined ore          Unmined ore

⊙⊙
Drill hole in mineralized rock

▪          χ

Shafts and prospect

1  North Standard
2  Water Lillie
3  Central Standard
4  Copper Leaf
5  Independence
6  North Lily
7  Tintic Standard No. 2
8  Eureka Lilly
9  Apex Standard No. 2
10 Iron King (No. 2)
11 Eureka Standard
12 Apex Standard No. 1
13 Zuma
14 Trixie prospect
15 Burgin No. 1
16 Burgin No. 2
17 Tintic Standard No. 1
18 South Standard

FIG. 1. *Subsurface Structure Map of East Tintic District, Utah, showing mined and unmined ore bodies.*

The crest of the fold, which plunges gently to the north and south from a point a few hundred feet east of the Tintic Standard mine, is cut by several transverse faults, some of which developed concurrently with the fold and thrust fault during the compressive stages of orogenic activity.

The East Tintic thrust fault, which localizes the principal ore bodies of the Burgin mine, cuts the concealed east limb of the anticline, displacing it 5000 feet or more along a plane dipping moderately to the west. In the Burgin mine, the rocks of the upper plate, which moved relatively eastward, include brecciated and deformed beds of the Ophir Formation that have been overturned and dragged under a wedge of Tintic Quartzite that rides above one or more subsidiary thrust strands. The rocks of the footwall plate of the main East Tintic thrust fault in the most fully developed parts of the mine are sheared and contorted beds extending from the lower part of the Opohonga Limestone to the Pinyon Peak and Fitchville Formations. The sublava position of the thrust zone has been traced by drill holes from the Burgin mine generally northward for about 2 miles and southward for approximately half a mile, where the thrust zone apparently terminates against the inferred Inez fault, which is believed to be a concealed tear fault of large displacement (31).

Other thrust faults, with somewhat less displacement than the East Tintic thrust but with the same general strike, dip, and sense of displacement, also are recognized in the East Tintic district. The best known minor thrust is a decollement near the contact of the Tintic Quartzite and the Ophir Formation in the upper plate of the East Tintic thrust. It is exposed only in the workings of the Tintic Standard and North Lily mines and has been named the Tintic Standard thrust fault by Lovering and his co-workers (15, p. 14). According to Lovering (15), the late asymmetric development of the East Tintic anticline crumpled the upper and lower plates of a zone of low-angle faults near the top of the Tintic Quartzite into a complex, northwest-trending asymmetric trough that concurrently was cross-folded near its southern end, producing a curving northeasterly- to easterly-striking cross-trough ma⁻ked by much broken and highly mineralized ground that is known locally as the Tintic Standard "pot hole." A re-evaluation of this structure, based on observations of the Eureka Standard and Apex Standard faults in the Burgin mine, suggests that the Tintic Standard thrust fault may terminate against a northeast-striking tear fault, which Lovering and his co-workers (15) had assumed to be part of the thrust plane. This tear fault, locally termed the South fault, appears to have formed a buttress against which the rocks of the upper plate were crumpled and drag folded, producing the irregular, northeast-striking cross-trough or pot hole. The great size and richness of the Tintic Standard ore body is due in large part to the unique structural form of the pot hole and to the large volume of brecciated limestone—prepared ground—that was contained within it.

The northwestern end of the main trough of the folded Tintic Standard thrust fault narrows rapidly at the fissure zone that passes near the North Lily mine, where the western side is vertical or overturned. In this part of the folded thrust, the formations in the constricted trough of the fold also are highly broken producing the so-called North Lily pot hole, which localizes the North Lily and Eureka Lilly replacement ore bodies.

The Pinyon Peak thrust, which is exposed near the North Standard shaft, is similar to the East Tintic thrust but has a stratigraphic throw ranging from only a few hundred to about 1500 feet. It is not known to be an ore-bearing structure. The displacement on this thrust is greatest on the lower east slopes of Pinyon Peak where the Middle Cambrian Cole Canyon Dolomite is thrust over the base of the Upper Ordovician Fish Haven Dolomite.

The transcurrent faults are part of a conjugate system of northeast- and northwest-trending fractures that cut the axes of the major folds at angles of 25° to 55°. In the East Tintic district, the best-known faults of this group are the Apex Standard and Eureka Standard faults, both of which are right-lateral tear faults that develop a progressively lower dip and more northerly strike as they approach and merge with the East Tintic thrust. The combined displacement on these two tear faults is 3000 to 4000 feet, as indicated by the horizontal separation of the axis of the East Tintic anticline. Small bodies of high-grade gold-telluride and argentiferous enargite ores, as well as somewhat larger but less profitable replacement deposits of copper, lead, and zinc ores, were mined in the Eurkea Standard fault zone in the Eureka Standard and Apex Standard mines. The Apex Standard fault zone, in contrast, contains only small ore shoots in the Apex Standard mine, but it has not been explored fully below the 900-foot level in that mine or southwest of the Apex Standard No. 1 shaft. The intersection of both of these faults with the East Tintic thrust may be one of the principal localizing features of the Burgin ore bodies.

The inferred Inez fault is believed to be a right-lateral, strike-slip fault similar to the Eureka Standard and Apex Standard faults; however it has much greater displacement. Regional studies (34) suggest that it may be the delimiting tear fault marking the southern edge of the Midas thrust fault, which extends northward to the Bingham district. The Inez fault has been bracketed by drill holes, but it is entirely unexplored and as yet cannot be evaluated as a possible ore-bearing structure.

The Yankee fault, which is poorly exposed and consequently is not well known, is believed to be a northwest-trending transcurrent fault. Unlike the northeast-trending shear faults, it apparently did not localize ore.

The post-compressive normal faults may be subdivided into two groups: (1) an east-trending set that is entirely pre-lava in age and (2) a north-trending set that has some post-lava displacement. The east-trending normal faults in the East Tintic district include the Homansville and Sioux-Ajax faults. Both of these faults seem to cut the transcurrent and thrust faults, but many of the critical intersections of these structures are concealed by the volcanic rocks in areas not exposed in mine openings. The Sioux-Ajax fault localizes ore bodies in several mines in the Tintic district.

The only north-trending normal fault recognized in the East Tintic district is the Eureka Lilly-Selma fault zone. This fault has a complex history including: (1) important displacement prior to the eruption of the lavas; (2) reactivation of the fault during the eruptive episode but prior to the deposition of ore; and (3) probable reactivation of the Selma fault segment of this fault zone as a Basin Range fault during the Pliocene and Pleistocene Epochs. The Eureka Lilly segment of the fault localizes small ore shoots in the North Lily mine, and ore also occurs along related fractures close to it in the Iron King No. 2 mine.

The youngest structures recognized in the area of ore deposition are the short, north-northeast-trending fissures that cut the lavas and intrusive rocks as well as the underlying sedimentary rocks. Most of these fissures dip steeply west and range from a few feet to 200 or 300 feet in displacement. They are the principal structures that localize monzonite plugs and dikes and the associated pebble dikes, and many of them provided channelways followed by the ore solutions. They are most readily recognized in the lavas and the larger monzonite intrusives by linear zones of hydrothermal alteration that extend outward from the fissure selvages. At depth, particularly where these fissures cut the Tintic Quartzite below the major lead-silver replacement ore bodies, they contain zones of breccia several feet wide that localize the shoots of pyritic auriferous and argentiferous copper ores.

Basin and Range faults are not known with certainty in the near vicinity of the East Tintic district, although, as stated above, the Selma fault may have been reactivated during the Late Tertiary and Quaternary Periods. The general straightness of the eastern edge of the East Tintic Mountains is suggestive of a Basin and Range fault zone, but geophysical studies do not indicate faulting of large displacement (25, p. 53).

## AGE OF MINERALIZATION

As noted by Lindgren and Loughlin (3, p. 104), the ore deposits of the East Tintic Mountains were formed shortly after the intrusion and consolidation of the Silver City monzonite stock and the associated bodies of biotite-monzonite porphyry. The dating of zircons from these intrusions by the lead-alpha method indicates an apparent absolute age of 38 to 46.5 million years (25, p. 30). The general accuracy of these numerical ages is confirmed by the intertonguing of agglomerate members of the Laguna Springs Latite and units equivalent to the Eocene Green River Formation in the southern part of the East Tintic volcanic field, 25 miles south-southeast of the East Tintic district. In this locality, the agglomerate also includes a lens of limestone that contains plant fossils of middle Eocene age (17, p. 234).

## ORE DEPOSITS—GENERAL STRATIGRAPHIC AND STRUCTURAL RELATIONS

In a private report written by Paul Billingsley in 1956, in which he analyzed a compilation of field notes taken in the North Lily and Eureka Lilly mines 30 years before, he stated:

"The immediate impression made by these maps and sections is one of extraordinary complexity of geology. This is in fact the case, and in the development of the two mines we were constantly taken by suprise and driven to purely empirical procedures in following the ore. A coherent pattern emerged only with time."

This statement succinctly summarizes the problem of East Tintic mine geology. This is an area of extreme geologic complexity with a masking blanket of Tertiary volcanic rocks covering the larger part of the area. Therefore, this complexity can be seen and studied only in mine workings and from the commonly incomplete or often frustrating data obtained from drill holes.

The ore bodies of the district generally may be divided into two categories: (1) the massive replacement bodies of the Tintic Standard, North Lily, Eureka Lilly, and Burgin mines that are rich in silver, lead, zinc, and manganese and (2) the fissure ores of the Eureka Standard, Eureka Lilly, and Apex Standard mines that are valuable primarly for their content of gold, copper, and silver.

Most of the ore discovered to date has been found in the lower part of the stratigraphic section, primarily in the Tintic Quartzite and the limy members of the Ophir Formation. This may be attributed, at least in part, to the concentration of exploration in certain parts of the district prior to the past 10 years and to the complex structural setting of the ore bodies. Before the postulation of the East Tintic thrust fault by Lovering and his coworkers in 1950 (16), it was generally believed that the subvolcanic structure at East Tintic was dominated by the East Tintic anticline and that the Tintic Quartzite and successively younger formations would be encountered in the core and east flank of this fold east of the area of the Tintic Standard and Eureka Standard mines. The small amount of exploration accomplished in drill holes from the surface and in the workings of the Apex Standard mine seemed to substantiate this premise. Only in the workings from the Independence shaft of the Silver Shield Mining and Milling Company in the northeast part of the district were much younger rocks encountered beneath the Tintic Quartzite; these later proved to be in the footwall of the East Tintic thrust fault. These younger—Mississippian—rocks were recognized and identified only by a few geologists during the short time this mine was in operation, and their significance was not generally appreciated. Consequently, after the major ore discoveries in the Tintic Standard mine and the subsequent interpretation of the complex structure by Billingsley and others, exploration was directed towards finding the Middle Ophir Limestone Member in structural situations similar to that at the Tintic Standard mine. This work was rewarded by the discovtry of the North Lily and the Eureka Lilly mines and the largely accidental discovery of the small, though rich, gold ore shoots of the Eureka Standard mine. Exploration along the East Tintic thrust fault during the past 10 years, however, has brought to light mineralized areas in the rocks much younger than the Ophir in the footwall block of the thrust fault. Thus far, these mineralized areas seem to be concentrated in the Devonian Victoria Formation, a host rock that has not been notably productive in the Main Tintic district.

In the overall picture, structural features are more important than specific host rocks in the localization of ore at East Tintic, and, in suitable structural environments, it seems probable that ore will be found in many formations that have not been deemed favorable to ore deposition in the district.

The principal productive area of the East Tintic district is divided into structural blocks with by far the largest share of the production having come from an area bounded on the east by the East Tintic thrust fault, on the west by the Eureka Lilly fault, on the north by the Tintic Standard fault and the East Tintic barrier, and on the south by the northeast-striking Apex Standard fault (Figure 1). Within this block, northeast-striking faults break the main block into a series of horsts and grabens. Ore is intimately related to minor folds and crenulations within these larger grossly deformed structural units.

The ore-bearing Ophir Formation is subdivided into three members: the Lower Shale Member, the Middle Limestone Member, and the Upper Shale Member. The average thickness of the Lower Shale Member in the East Tintic district is about 175 feet; this member contains a single carbonate bed about 10 feet thick that forms an excellent marker 70 to 90 feet above the base. The Middle Limestone Member, which averages about 145 feet in thickness, consists of several limestone beds interlayered with lenses and beds of green to light bluish-green shale. These limestone beds are the host rocks for the massive replacement ore bodies of the Tintic Standard, North Lily, and Eureka Lilly mines and probably also for the larger share of the ore developed to date in the Burgin mine. The Upper Shale Member of the Ophir, ranging in thickness from 70 to 90 feet, is a light greenish-gray fissile shale.

The stratigraphic position of the Ophir Formation, between the massive, underlying Tintic Quartzite and the massive limestone and dolomite section above it, has made the Ophir a locus of deformation in the East Tintic district. During the folding and subsequent faulting, deformation commonly was localized in the Ophir, and in those areas in which the Lower Shale Member was squeezed or faulted out, the limestone beds of the Middle Member were converted to ore where they were cut locally by ore-feeding fissures.

Ore occurring in limestone beds in stratigraphic units other than the Ophir Formation doubtless has been mined in the district, but the highly altered character of the rocks adjacent to ore as well as the general lack of diagnostic characteristics in the carbonate rocks of the Middle Cambrian section of the Tintic area makes positive identification of many of the ore-bearing rocks virtually impossible.

## FORM AND CHARACTER OF THE ORE BODIES

As previously stated, the ore deposits of the East Tintic district are notable for their complex structural environments. The ore bodies exploited in the five most important producing mines of the district will be discussed briefly in the light of their complex structural settings, noting particularly the stratigraphic relations of each deposit.

### Tintic Standard Mine

The Tintic Standard mine is the largest of the mines in the district, having a recorded production of over 2 million tons of high-grade silver-lead ore. The Tintic Standard "pot hole" is a unique structural node that localized massive replacement ore bodies (Figure 2). The deposits of this mine are associated with the low-angle Tintic Standard thrust fault located on the west side and near the crest of the East Tintic anticline. Movement on this minor thrust fault was localized in the shales and shaly limestone of the Ophir Formation between the underlying Tintic Quartzite and the overlying section of massive Cambrian and younger limestones. This movement served to slice out the Lower Shale Member of the Ophir in some areas, placing the limestone beds of the Middle Member directly against the quartzite footwall (Figure 3). At the places where through-breaking, north-northeast fissures in the quartzite fed directly into the overlying Middle Ophir limestones, large replacement bodies of silver-lead ore were formed, some of which lie directly against the quartzite footwall rocks.

The Tintic Standard structure has been interpreted as: (1) a folded thrust fault; (2) a downfaulted structure, bounded by a series of normal faults curving to form the pot-hole structure; and (3) a thrust fault folded into the complex structure by a tear fault along its southeast side. Regardless of origin, the pot hole structure has created a large volume of prepared ground that was highly amenable to replacement.

In the great Central ore body at the elevation of the 1100-foot level, the limestone beds of the Middle Ophir Member were replaced by ore nearly continuously from the South fault to the Tintic Standard thrust fault, a distance of nearly 600 feet. These beds dip gently

to the east in this sector of the mine, into the narrowing pot-hole structure. Down-dip, the beds are intersected by the main northeast-striking Tintic Standard fissure zone through which the ore-forming solutions moved. These solutions spread upward and outward through the limy beds, and, in this part of the mine, a mass of limestone as much as 200 feet in thickness was altered and mineralized between the 1100-foot and 1400-foot levels. Within this great mass, a large part of the limestone was replaced sufficiently to form ore, the gently-dipping beds locally being replaced by alternating bands of high-grade, silver-lead ore and low-grade, siliceous silver ore.

Above the 1200-foot level, the ore was largely oxidized, consisting of an earthy mixture of altered sanded dolomite, iron oxides, vuggy quartz and jasperoid, cerussite, argentojarosite, and other oxidized silver and lead minerals. The sanded dolomite above the ore bodies is iron-stained, leached, and locally brecciated owing to oxidation of disseminated pyrite and slumping of the altered rocks that lie above the leached and oxidized ore bodies. Heavy timbering was necessary throughout the ore body, and rock temperatures were exceptionally high.

Extending outward and upward from the Central ore body were other replacement ore bodies that occurred in fault blocks in, and adjacent to, the brecciated pot-hole structure. These ore bodies were similar in character to the Central ore body and form narrow pipe-like bodies that Billingsley and Crane (10) described as the fingertip extensions of the ore cluster.

On the lower levels of the Tintic Standard mine, the larger pebble-dike fissure veins were mined for their gold-copper-silver content where they attained sufficient grade and width to constitute ore.

### North Lily Mine

The North Lily mine was discovered in 1926 by drifting to the northwest on the Tintic Standard 700-foot level toward a target recommended by Paul Billingsley (6). This ore center is localized within a zone of northeast-striking monzonite porphyry dikes and persistent northeast-trending fissures that cut this and the adjacent Eureka Lilly ground. In a structural situation somewhat similar to that at Tintic Standard, the limestone beds of the Middle Ophir Member had been faulted against Tintic Quartzite of the East Tintic Barrier by the Tintic Standard thrust fault. These beds

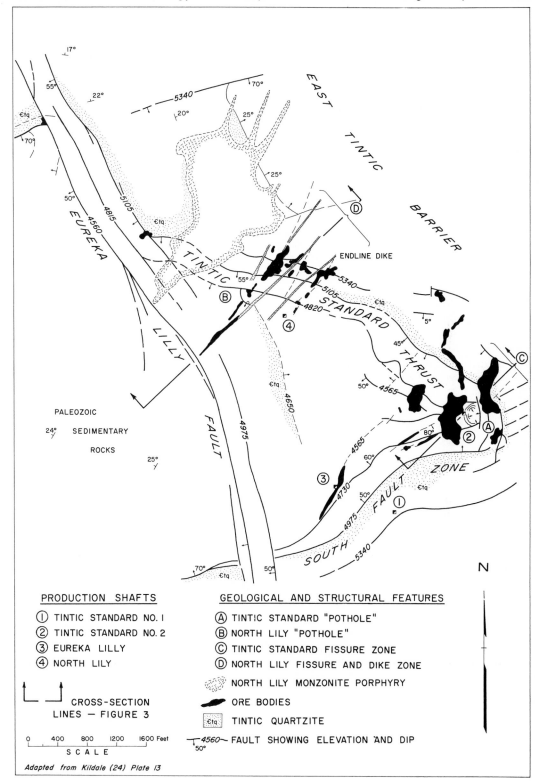

**PRODUCTION SHAFTS**

① TINTIC STANDARD NO. I
② TINTIC STANDARD NO. 2
③ EUREKA LILLY
④ NORTH LILY

CROSS-SECTION
LINES — FIGURE 3

0   400   800   1200   1600 Feet
S C A L E

Adapted from Kildale (24) Plate 13

**GEOLOGICAL AND STRUCTURAL FEATURES**

Ⓐ TINTIC STANDARD "POTHOLE"
Ⓑ NORTH LILY "POTHOLE"
Ⓒ TINTIC STANDARD FISSURE ZONE
Ⓓ NORTH LILY FISSURE AND DIKE ZONE

NORTH LILY MONZONITE PORPHYRY

ORE BODIES

€tq TINTIC QUARTZITE

4560 FAULT SHOWING ELEVATION AND DIP
50°

FIG. 2. *Composite Plan of Structure and Ore Bodies, Tintic Standard and North Lily Mine Area.*

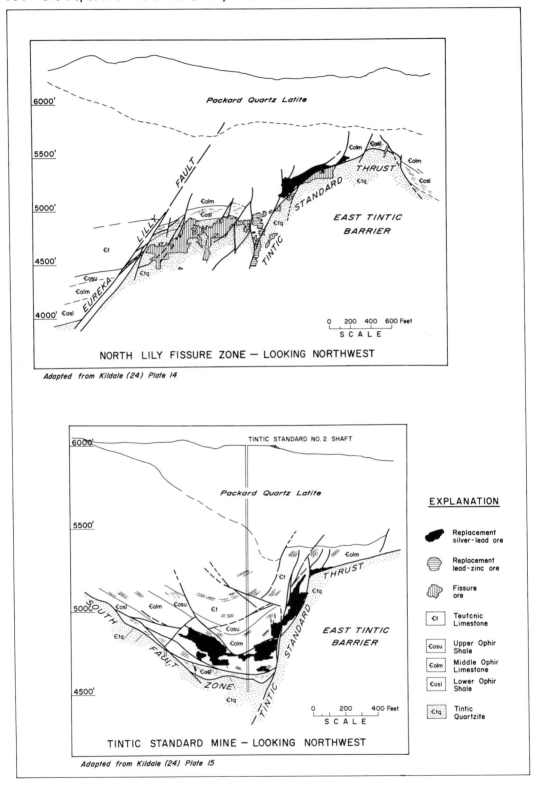

Adapted from Kildale (24) Plate 14

Adapted from Kildale (24) Plate 15

Fɪɢ. 3. *Generalized Cross-Sections through North Lily and Tintic Standard Mines.*

were then mineralized by solutions moving through fissures that cut the quartzite of the East Tintic Barrier as well as the quartzite beds southwest of the mine (Figure 2).

The main lead ore body of the North Lily mine was mined continuously from the 900-foot to the 600-foot level. A "keel" of pyritic gold ore extends down into the Tintic Quartzite along fissures below the upper part of the ore body, and lead ore also spreads easterly along the Tintic Standard thrust fault to the Endline Dike fissure zone. A replacement ore shoot in the Ophir Formation also was mined along the east side of the Endline Dike in the Eureka Lilly mine. Between the 1000-foot and 1200-foot levels along the North Lily fissure zone, the limestone beds of the Middle Ophir Member were down-faulted or down-folded in the hanging-wall block of the Tintic Standard thrust fault to form a large lead-zinc ore body above the North Lily "pot hole" area (Figure 3). This ore consisted of an incomplete replacement of the beds by galena, sphalerite, and pyrite. Below this ore body, in the pot hole itself, low-grade, lead-zinc-gold mineralization occurred in a mass of brecciated Lower Ophir Shale and monzonite.

At the south end of the North Lily mine, high-grade gold and gold-copper ore shoots occur along both the North Lily and Endline Dike fissures within the block of Tintic Quartzite south of the pot-hole area. These fissures converge toward the south and probably are not recognizable as two distinct zones at a distance of half a mile southwest of the North Lily mine. In both of these fissure zones, the ore occurs in part as fissure filling along strong steep fissures but also in part as cementing material enclosing fragments of quartzite or monzonite that make up rubbly breccias between the fissure walls. The richer parts of the ore shoots locally are controlled by small cross-fractures, and the southerly rake of the ore body as a whole indicates some localization by the bedding of the quartzite. This southerly rake also is indicative of deep conduits for ore-forming solutions that rose along the intersection of the northeast fissuring and the Eureka Lilly fault. The high-grade native gold-enargite ore becomes low grade and pyritic on the lower levels.

## Eureka Lilly Mine

The Eureka Lilly mine includes the area between the Tintic Standard and North Lily mines and also the property to the south and west of these two great mines. The ore from the Eureka Lilly has been produced from several different areas and geologic environments. Approximately 60 per cent of the ore came from lead-silver replacement ore bodies and siliceous precious-metal veins along the east side of the Endline Dike and the area immediately southeast of the North Lily mine. Including the production from the North Lily mine, the total tonnage of ore from the combined mines slightly exceeds 0.5 million tons.

Of somewhat lesser importance were siliceous gold-silver-copper ores from fissures along the South fault southwest of the Tintic Standard No. 1 shaft. These ores occurred both as massive fissure fillings and as cementing material for breccias within the fissure zones. The ore solutions appear to have been most strongly concentrated in local zones of open fissuring or unusually strong brecciation, and, in the Eureka Lilly mine fissure, ore of commercial grade is limited to a distance of a few 100 feet vertically below the intersection of the fissures with the South fault.

A small additional tonnage of oxidized lead-zinc replacement ore along fissures was mined from the area immediately northeast of the Eureka Lilly shaft from the 70-foot to 300-foot levels. This was the first production from the East Tintic district and was reported in considerable detail by Lindgren and Loughlin (3, p. 247–248).

## Eureka Standard Mine

The Eureka Standard mine was discovered in the late 1920's following the geologic recommendations of Paul Billingsley. The original premise was to explore for a structure similar to that of the great Tintic Standard pot hole in the graben-like block between the steep Iron King fault on the north, and the Eureka Standard fault on the south. However, in 1928, gold mineralization was discovered fortuitously in steep northeast-trending fissures on the footwall side of the strands of the Eureka Standard fault.

The valuable metals in this mine were primarily the gold, silver, and copper contained in narrow veins bearing pyrite, quartz, barite, tetrahedrite, enargite, and gold tellurides that cut the Tintic Quartzite. A small amount of silver-lead mineralization was discovered in the limestones of the Ophir Formation early in the exploration of the mine, but these discoveries were not developed further after the discovery of the gold-bearing ores.

The Eureka Standard mine was developed by a 1400-foot vertical shaft and six main levels driven along the strike of the Eureka Standard fault. Three ore shoots were mined, the southwest, main, and northeast shoots, each plunging at about 20° to the southwest and each having a plunge length of some 600 feet. Gold content diminished down dip in the individual fissures, and the fissures terminated upward against strands of the flatter Eureka Standard fault.

## Burgin Mine

The Burgin mine of Kennecott Copper Corporation is the newest addition to the ranks of the producing mines of the East Tintic district. This ore body was discovered as a result of exploration initiated by the Bear Creek Mining Company in 1956 along the East Tintic thrust fault. The studies leading up to the sinking of the exploration shaft and the subsequent underground work have been documented ably by Bush, Cook, Lovering, and Morris (26).

The principal ore body discovered to date is a complex replacement deposit that is localized in the hanging wall of the East Tintic thrust fault where it is intersected by the northeast-striking Eureka Standard and Apex Standard faults (Figure 4). Movement on the thrust fault has placed rocks of the Ophir Formation in contact with beds in the middle and lower parts of the Opohonga Formation in the area of the mine workings. The stratigraphic displacement across the fault is about 3000 to 3500 feet, with total displacement along the structure believed to be approximately 1 mile. In the mine workings, rocks in the footwall of the thrust fault range from the Opohonga Formation of early Ordovician age to the Pinyon Peak Formation of Devonian and Mississippian age. The attitudes of these rocks range from gently dipping to the east to completely overturned, nearly flat-lying beds that are exposed on the 1050-foot level near the No. 1 shaft. The rocks of the hanging wall are broken completely by faults, and the major mineralization discovered to date is in these brecciated hanging wall rocks lying directly on the footwall of the East Tintic thrust fault.

The unoxidized ore is an intimate mixture of lead and zinc sulfides with various amounts of argentite, rhodochrosite, barite, jasperoid, and quartz. The oxidized portions of the ore body contain cerussite, anglesite, smithsonite, cerargyrite, pyrolusite, chalcophanite, and unreplaced masses of the primary minerals. On the 1050-foot, or exploration, level of the mine, lenses of manganese oxides containing narrow lenses and stringers of oxidized base-metal ores are the only expression of the ore zone below. Development of the ore body, through 1966, has shown that its northeastern edge lies directly against the footwall rocks of the thrust fault, but its southwest edge is a complex steep fault contact with sanded dolomite. This contact is marked in some areas by jasperoid and manganese oxides and in other areas by a thin selvage of galena-sphalerite-rhodochrosite ore. At the present stage of development of the mine, the relationships of the various ore types are not understood fully and a more complete understanding doubtless will be achieved only as development and mining continue.

An ore body of particular interest currently is being mined above the 1050-foot level of the mine. This body has provided a major share of the ore produced during the period of de-watering and development of the lower levels of the mine. It occurs as a replacement of a highly brecciated limestone bed of the Middle Ophir Limestone Member that became the locus of pre-mineral caverns of considerable size. Prior to ore deposition, these caverns were filled with fine-bedded sediment, which also sifted into the interstices of the brecciated rocks below, and adjacent to, the cavern floors. During the mineralizing epoch, the entire mass of cave rubble and breccia was replaced by ore. The cave-fill portions of this ore body have yielded some ores of excellent grade: overall, the zone has averaged about 8 ounces per ton of silver, 10 per cent of lead, and 5.5 per cent of zinc. Surprisingly, this ore, which is located well above the permanent water table, is only slightly oxidized, averaging about 0.25 per cent of non-sulfide lead and zinc. An explanation for this phenomenon may be found in the comparative "tightness" of the entire ore zone caused by the enclosing shale beds.

The larger share of the ore discovered in the Burgin mine to date apparently occurs as replacements of the highly brecciated Ophir Formation immediately above the East Tintic thrust fault. This structure has been prospected for a distance of approximately 1500 feet along the strike in the mine and by drill holes for short distances north and south of the mine. In the ore-bearing interval of the thrust fault in the Burgin mine, the mineralizing solutions apparently rose along the Eureka Standard and other northeast-bearing faults parallel to it, as well as upward along the thrust fault zone, and spread laterally in the brecciated rocks of the hanging-wall block. Exploration

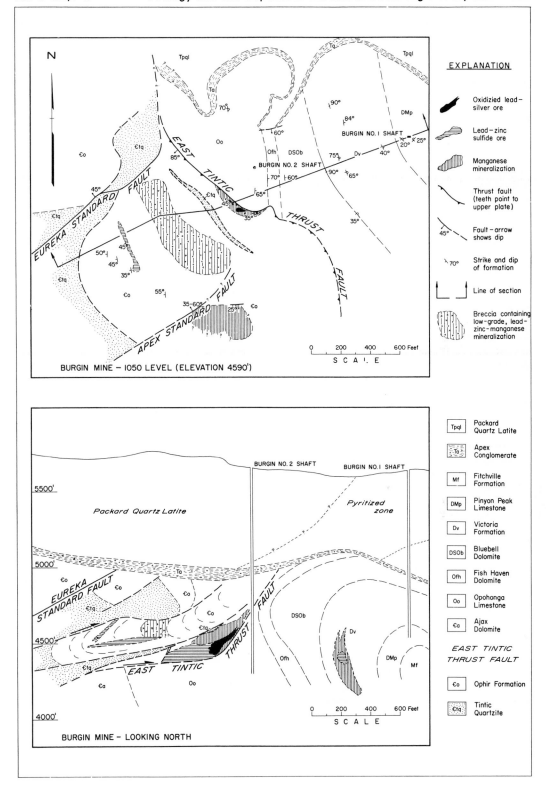

FIG. 4. *Plan and Generalized Cross-Section of Burgin Mine.*

drill holes and mine workings to date have not penetrated the thrust fault north of the Eureka Standard fault in the mine workings, although widespread surface drilling has bracketed this structure over a broad area.

It is noteworthy that the Burgin operation is the first to attempt the extraction of large tonnages of ore from beneath the water table at East Tintic. While workings of limited extent were opened below the water table in both the Tintic Standard and Eureka Standard mines, the present operation constitutes the first attempt to handle large quantities of the hot, corrosive water that is characteristic of the district.

## PRIMARY ORE AND GANGUE MINERALS

### Composition

The primary ore minerals of the replacement deposits are argentiferous galena and sphalerite with some of the silver probably present in blebs of argentite and tennantite in the galena and in small amounts of other silver minerals such as proustite, pearceite, and polybasite. Barite, rhodochrosite, manganosiderite, calcite, and the ubiquitous quartz, jasperoid,* and pyrite are the principal gangue minerals. Native gold, enargite (including the variety luzonite), and tetrahedrite are the principal ore minerals of the auriferous copper veins in Tintic Quartzite; and various gold and silver tellurides and tetrahedrite are the chief ore minerals in the gold-telluride veins in Tintic Quartzite. Quartz, pyrite, barite, and clay minerals are the main gangue minerals of the vein deposits.

Other primary minerals reported from the East Tintic mines, which generally are found in very small amounts, include: altaite, bismuthinite, bornite, chalcopyrite, chalcocite, jamesonite, marcasite, and tetradymite.

### Textures

The replacement silver, lead, and zinc ores generally are composed of fine-grained to coarsely crystalline, massive, galena and sphalerite. Some vugs are present, but good crystalline ore mineral specimens are not common.

* A rock consisting essentially of cryptocrystalline, chalcedonic, or phenocrystalline silica, which has formed by the replacement of some other material, ordinarily calcite or dolomite (Spurr, U.S. Geol. Survey Mon. 31, p. 219).

In some parts of the North Lily and Burgin mines, specimens of well-banded, fine-grained galena formed by metasomatic replacement of thin-bedded limy or shaly sediments have been found. As noted above, some of this material in the Burgin mine is a replacement of a stratified cave fill. Development of octahedral faces on galena crystals from the Burgin mine is quite common. Barite commonly occurs as thin platy crystals and as aggregates of platy crystals in the ores. In the sulfide ore of the Burgin mine, galena and sphalerite are closely intergrown, and some sphalerite is overgrown by rims of galena about 10 $\mu$ thick.

### Paragenesis

The sequence of hydrothermal alteration at East Tintic has been extensively studied by Lovering and his co-workers (15), and the paragenesis of the alteration has been painstakingly worked out. The minerals of the alteration sequence, which are of principal interest in the study of the ore deposits, are the hydrothermal dolomite of the early barren stage, the clay minerals and the "sanded" dolomite of the mid barren stage, and the quartz or jasperoid of the late barren stage. These were followed by pyrite of the early productive and productive stages that was commonly deposited in open spaces in the fractured and vuggy jasperoid. Lovering's (15, p. 17) general paragenetic sequence for the ore minerals shows an early deposition of minor amounts of gold, enargite, sphalerite, and galena during the early productive stage with later major deposition of sphalerite, galena, enargite, tetrahedrite, proustite, hessite, gold, and both primary and secondary chalcopyrite, and secondary chalcocite during the productive stage and subsequent weathering cycles. Recent work in the Burgin mine indicates at least two major stages of galena deposition, with an intervening episode of sphalerite. The relationship of the manganese and iron carbonates to the sulfides in the Burgin ores has not yet been studied in detail, but several stages of manganese carbonate, with intervening galena, deposition seem to be indicated.

## SECONDARY ORE AND GANGUE MINERALS

### Composition

Oxidation extends to considerable depths in the East Tintic district, the present permanent water table ranging in depth from less than 1100 to more than 1400 feet in the major producing mines. This has led to the formation of a great variety of secondary minerals, a few of which have been important as ores.

Important secondary minerals of the replacement deposits are cerussite and anglesite with the rare mineral argentojarosite occurring in sufficient quantity to constitute ore in the Tintic Standard mine. Secondary gangue minerals include oxides of iron, and, in the Burgin mine, various complex oxides of manganese including pyrolusite, chalcophanite, hetaerolite, crednerite, and quenselite.

A great variety of secondary minerals has been identified from the East Tintic mines including: antlerite, bindheimite, brochantite, copiapite, halloysite, epsomite, hydrocerussite, jarosite, lanarkite, malachite, massicot, zincian melanterite, mimetite, minium, native silver, plumbojarosite, smithsonite, and szomolnokite.

## Oxidation and Enrichment Phenomena

Oxidation and enrichment of silver ores have been important economically in several of the East Tintic mines with some small tonnages of extremely rich secondary silver ore having been mined, particularly in the Tintic Standard mine. Some secondary enrichment in silver has been encountered at or below the present water table and at the boundaries of different types of ore in the Burgin mine.

## EXPLORATION AND DEVELOPMENT

The East Tintic district long has been considered by prospectors and mining geologists alike to be an intriguing and baffling area for exploration. The presence of possible ore centers in carbonate rocks beneath a rather thin, nonproductive lava cap provides excellent exploration opportunities adjacent to the known mineralized areas.

The work of Crane (2), Lindgren and Loughlin (3) and later studies by Billingsley (9,10) and Kildale (14) established the basic stratigraphic and structural framework of the district. An understanding of the subsurface structure in the Paleozoic sedimentary rocks resulted in the recognition of favorable areas for ore deposition by many eminent geologists working in the district. More recently, the detailed stratigraphic and structural studies of the U.S. Geological Survey under the direction of Lovering and Morris (15,27,29) provided the basis for predicting a major thrust fault beneath the lavas. This interpretation led to

a new concept for ore localization and resulted in successful exploration for base-metal replacement deposits.

Concealed mineralization is suggested by various "ore guides" in the overlying volcanic rocks including: hydrothermal alteration, weakly mineralized fissures, and pebble dike intrusion breccias. Generalized relationships between hydrothermal alteration and mineralization in the district were established by the work of Lindgren (3) and Billingsley and Smith (7). Later, more specific alteration guides were developed by the U.S. Geological Survey (15), and Bear Creek Mining Company (23).

## Historic Sequence of Exploration Activities

Prior to 1907, a number of shallow shafts and prospect pits were sunk on surface indications of fissure vein mineralization in the East Tintic district, but no important discoveries of ore resulted from this activity. A small amount of oxidized replacement ore was shipped in 1909 from the Provo shaft, which is located on a small fault zone in carbonate rocks. The first significant discovery in the district was made in 1913 by the persistent efforts of E. J. Raddatz who sank the original Tintic Standard shaft and drifted beneath outcrops of jasperoidized limestone. These efforts resulted in the discovery of a northeasterly-trending mineralized fissure that led him to explore along strike into an area covered by lava. Fortunately for Raddatz, this lava concealed the Tintic Standard "pot hole," the host structure for the high-grade replacement mineralization of the famous Tintic Standard ore body. This exploration effort was the first success in a number of important blind ore-body discoveries. During its nearly 40 years of operations, the Tintic Standard mine produced more than 2,400,000 tons of high-grade, lead-silver ore from an area of a few acres; approximately $20,000,000 was paid in dividends from ore that averaged 24 ounces per ton of silver and 12 per cent of lead. The successful discovery and development of the Tintic Standard is attributed to Raddatz's dogged persistence, his uncanny "nose for ore," and good fortune. His subsequent exploration efforts in other areas of the Tintic districts were, unfortunately, not so rewarding.

The discovery of the Tintic Standard bonanza beneath pyritized volcanic rocks stimulated exploration activity in the district where similar relationships were thought to be present. A large number of shafts were sunk

other criteria are needed to locate target areas for exploration.

Argillic alteration characterizes the mid barren stage. It is most extensive in igneous rocks, but it is also locally prominent in carbonate rocks near hydrothermal conduits. Argillization is concentrated around centers of intrusion, and, in many of the resulting alteration halos, a zonation within the altered rocks is recognized. In dolomite, the "argillized" zone is generally represented by "sanded" areas that have been leached severely and converted locally to solution breccia. High porosity induced by acid leaching also is characteristic of the argillized igneous rocks; this porosity doubtless facilitated the movement of successive altering and ore depositing solutions.

Later barren-stage alteration zones containing pyrite, calcite, and barite are localized chiefly near major channels of mineralization. Cubic pyrite is disseminated widely in the lava more or less up-dip from fractures that have guided pre-ore altering solutions in the underlying Paleozoic rocks. Calcite of replacement origin may be present in the lavas for hundreds of feet laterally beyond the pyritic zone; the calcite is commonly up-rake from fissures in underlying Paleozoic rocks and spreads out on the footwall side of the projection of the fissure. This may be calcite displaced during silicification of nearby dolomite underlying the lava, and, although spread through a quite large volume of rock, such calcite suggests jasperoid in concealed carbonate rocks and, therefore, is an important guide to concealed ore bodies.

The areas altered during the early productive stage contain sericite, minor clear quartz, and some pyritohedral pyrite. Although this stage was somewhat earlier than the ore stage, pyritohedral pyrite is in large part contemporaneous with ore. Early productive stage alteration may follow fissures above an ore body for several hundred feet, but it does not extend laterally into the wall rocks for more than a short distance.

The productive stage is distinguished by the deposition of sulfides of the base metals, of sulfantimonides and sulfarsenides of copper and silver, and of gold and gold-silver tellurides, all accompanied by minor amounts of barite and terminated clear quartz crystals.

Weathering has masked the hydrothermal alteration effects in many areas. The zones of weathered pyritic alteration are indicated generally by brown, iron-stained areas surrounded by slightly argillized rock that has been bleached by the small volumes of sulfuric

acid generated by the oxidation of the disseminated pyrite. In many places, a band of manganese oxides separates the limonite-stained cap rock and the supergene bleached zone, where the outward moving acids were neutralized beyond the pyritized zone.

The most favorable combination of alteration zones in the lava is a large pyritized zone (marked by heavy limonite stain to a depth of 30 to 100 feet) with calcitized lava on one side (which suggests that the footwall side of a mineralized fracture is present at depth) and weakly argillized lava on the other side of the pyritized zone. Alteration halos of this type generally are up-rake from pre-volcanic, ore-bearing structures, and, therefore, are not necessarily directly above ore. A structural interpretation of sub-volcanic structure and stratigraphy is thus of prime importance in the evaluation of alteration halos in volcanic rocks. A good correlation seems to exist between jasperoidized dolomite or limestone at depth and calcitized lava at the surface. Calcitic alteration does not appear to have as great a vertical range in the lava cap as does pyritic alteration. The transition from altered to completely unaltered lava has been noted to occur above well-mineralized intercepts in drill holes, indicating rather thin alteration halos in some of the areas tested.

The occurrence of manganese oxides in altered sediments is an important ore guide because of the intimate association of manganese mineralization with some of the ore bodies. Anderson (30) has recognized mineralogical and trace-element criteria to distinguish the manganese oxides derived from manganese minerals associated with base-metal ores and those manganese minerals derived from other sources.

GEOCHEMICAL GUIDES Trace concentrations of ore stage heavy metals—geochemical halos—in the altered rocks provide a valuable guide to the location of ore-bearing structures. A technique to select the iron oxides from fractures in the lavas is a necessary sampling method for geochemical work. It is believed the residual ore fluids used the permeability of the fractured lavas for their final deposition of the ore metals. Geological control is needed in the interpretation of geochemical anomalies because the distribution of metals in the volcanic rocks has been found to be highly dependent on the geological structures in the underlying sedimentary rocks.

The identification of "productive jasperoid" on the 1050-foot level of the Burgin mine by megascopic and geochemical criteria was a critical guide to the discovery of the Burgin ore bodies.

GEOPHYSICAL GUIDES Several geophysical techniques have been attempted during the last 20 years in the East Tintic district but generally with disappointing results. However, gravity measurements taken underground in the Burgin mine have shown considerable promise in the search for nearby massive sulfides. Aeromagnetic surveys have been found to be only of questionable value in the direct detection of small intrusives related to ore centers; however, these surveys may be found more effective in the interpretation of broad features of the subsurface geology. To the present time, electromagnetic and induced polarization techniques have been found ineffective in penetrating the volcanic cover or the high-resistivity limestone units.

An extensive geothermal investigation has been conducted by the U.S. Geological Survey (35), and infrared measurements have been made by various private and government organizations. No practical ore guide has yet resulted from this work, although useful information has been obtained on the distribution of abnormal temperature gradients in the subsurface rocks.

DRILLING Reliable exploration information from drilling has been particularly difficult to obtain at East Tintic because of poor core recovery and unusual drilling difficulties. However, the employment of rotary drilling in the volcanic rocks; wireline equipment in the fractured, altered, and mineralized sedimentary rocks; and the use of special drilling muds and lost circulation materials have resulted in a dramatic improvement in core recovery and a lowering of drilling costs.

## Water and Ground Support Problems

Several unusual problems with geologic implications are recognized in the East Tintic district and at least brief mention should be made of these in any discussion of the district.

First of these problems is the presence of extremely hot water in the mines of the district. East of the Eureka Lilly fault, each of the mines that has reached the water table has encountered flows of thermal waters, and rock temperatures that are correspondingly high. These temperatures previously were thought to be due to heat generated by the oxidation of sulfides, however, work during the past

five years, primarily with water samples collected from the Burgin mine, revealed the striking similarity in trace-element content and isotopic ratios with waters from hot springs in western Utah. This similarity has led to the present hypothesis that this water is fed into the groundwater system from hot springs that discharge at the water table and nowhere reach the surface. Water temperatures as high as 147°F have been measured in the Burgin mine. This water contains a high percentage of dissolved material, principally as sulfates and carbonates, as well as a high concentration of sodium chloride. Although the pH of the water is near neutral, it is a strongly corrosive brine that requires stainless steel or suitably protected pumping and water transmission equipment. Lovering and Morris (35) present a detailed discussion of the hot water and its relationship to geothermal gradients in the district.

Second of the geologic-related problems to be considered here is the presence of "sanded dolomite" surrounding many of the silicified ore bodies of the district. This rock type consists of hydrothermal dolomite that was partly leached during the mid barren (argillic) stage of alteration, leaving an unconsolidated mesh of dolomite crystals that crumble and cave readily into underground openings. The tendency of this material to sluff is especially evident in areas where the sanded dolomite is wet, and it essentially flows as a slurry into mine workings or in chutes and bins. This slurry readily works its way into and through the narrowest of openings, and, once flowing, it is extremely difficult to control.

## SUMMARY—OUTLOOK FOR THE FUTURE

For more than 50 years, the East Tintic district has been a major producer of silver, lead, gold, copper, and zinc. The discovery of the Tintic Standard mine ushered in a new era in the exploration for blind ore bodies in the district. Continued exploration has led to the discovery of other mines, the most recent being the Burgin mine of Kennecott Copper Corporation. The search for blind ore bodies also has led to studies of alteration, trace element halos, and structure, all of which have added to the understanding of the district and to the body of knowledge concerning the utilization of these criteria as exploration tools.

The most significant development in the district in recent years has been the discovery of the East Tintic thrust fault and the subsequent re-interpretation of the geology of the district. This development has opened a large area in the eastern part of the district for exploration, an area that was previously believed to be underlain chiefly by Tintic Quartzite beneath the lava capping, an environment that was inhospitable to the deposition of large replacement ore bodies. Present knowledge leads to the belief that a large area along the strike of the thrust fault presents the best target for prospecting in the district, both in the massive breccia zones along the thrust structure and in the favorable rocks of the footwall block of the thrust fault. The present development of the Burgin mine would seem to assure the continued life of the district for at least a decade.

## REFERENCES CITED

1. Tower, G. W., Jr. and Smith, G. O., 1899, Geology and mining industry of the Tintic district, Utah: U.S. Gol. Surv. 19th Ann. Rept., pt. 3, p. 601–767.

2. Crane, G. W., 1917, Geology of the ore deposits of the Tintic mining district, Utah: A.I.M.E. Tr., v. 54, p. 342–355.

3. Lindgren, W. and Loughlin, G. F., 1919, Geology and ore deposits of the Tintic mining district, Utah: U.S. Geol. Surv. Prof. Paper 107, 282 p.

4. Parsons, A. B., 1925, The story of Tintic Standard: Eng. and Min. Jour., v. 120, no. 6, p. 205–210.

5. Parsons, A. B., 1925, The Tintic Standard Mine: Eng. and Min. Jour., v. 120, no. 17, p. 645–652.

6. Billingsley, P., 1927, North Lily development in East Tintic: Min. and Met., v. 8, no. 4 p. 182–183.

7. Billingsley, P., and Smith, N., 1927, Rhyolite alteration in the East Tintic district: Private rept., 21 p.

8. Walker, R. T., 1928, Deposition of ore in pre-existing limestone caves: A.I.M.E., Tech. Pub. no. 154, 43 p.

9. Billingsley, P., 1933, The utilization of geology in Tintic Utah: *in Ore deposits of the Western States* (Lindgren volume), A.I.ME., p. 716–722.

10. Billingsley, P. and Crane, G. W., 1933. The Tintic mining district: *in the Salt Lake Region,* 16th Int. Geol. Cong. Guidebook 17, p. 101–124.

11. Eardley, A. J., 1934, Structure and physiography of the Southern Wasatch Mountains, Utah: Mich. Acad. Sci., Arts, and Letters (1933 Papers), v. 19, p. 377–400.

12. Farmin, R., 1934, "Pebble dikes" and associated mineralization at Tintic, Utah: Econ. Geol., v. 29, p. 342–355.

13. Kransdorff, D., 1935, The geology of the

Eureka Standard mine, Tintic, Utah: Unpublished Ph. D. thesis, Harvard Univ.

14. Kildale, M. B., 1938, Structure and ore deposits of the Tintic district, Utah: Unpublished Ph.D. thesis, Stanford Univ.

15. Lovering, T. S., and others, 1949, Rock alteration as a guide to ore—East Tintic district, Utah: Econ. Geol. Mono. 1, 65 p.

16. Lovering, T. S., 1950, East Tintic district (Utah) geologic picture changed (abs.): Min. Cong. Jour., v. 36, no. 10, p. 78.

17. Muessig, S., 1951, Eocene volcanism in central Utah: Science, v. 114, no. 2957, p. 234.

18. Lovering, T. S., 1951, Structure of East Tintic district modified, *in* Hawkes, H. E., Jr., *Geochemistry, a symposium on the prospector's newest tool:* Min. Cong. Jour., v. 37, no. 9, p. 61, 84.

19. Morris, H. T., 1956, Exploration program in the Trixie area, East Tintic mining district, Utah County, Utah: U.S. Geol. Surv. Open File Rept.

20. Proctor, P. D., *et al.,* 1956, Preliminary geologic map of the Allens Ranch quadrangle, Utah: U.S. Geol. Surv. Mineral Investigations, Field Studies Map MF 45, 1:12,000.

21. Bush, J. B., 1957, Introduction to the geology and ore deposits of the East Tintic mining district, *in* Cook, D. R., *Editor, Geology of the East Tintic Mountains and ore deposits of the Tintic mining districts:* Utah Geol. Soc. Guidebook to the Geology of Utah no. 12, p. 97–102.

22. Bush, J. B., 1957, Ore deposits of the Eureka Standard, Apex Standard, and Iron King mines: *in* Cook, D. R., *Editor, Geology of the East Tintic Mountains and ore deposits of the Tintic mining districts:* Utah Geol. Soc. Guidebook to the Geology of Utah, no. 12, p. 120–123.

23. Howd, F. H., 1957, Hydrothermal alteration in the East Tintic mining district: *in* Cook, D. R., *Editor, Geology of the East Tintic Mountains and ore deposits of the Tintic mining districts:* Utah Geol. Guidebook to the Geology of Utah, No. 12, p. 124–134.

24. Kildale, M. B., 1957, Ore deposits of the Tintic Standard, North Lily, and Eureka Lilly mines: *in* Cook, D. R., *Editor, Geology of the East Tintic Mountains and ore deposits of the Tintic mining districts,* Utah Geol. Soc. Guidebook to the Geology of Utah No. 12, p. 103–119.

25. Morris, H. T., 1957, General geology of the East Tintic Mountains, Utah: *in* Cook, D. R., *Editor, Geology of the East Tintic Mountains and ore deposits of the Tintic mining districts:* Utah Geol. Soc. Guidebook to the Geology of Utah, No. 12, p. 1–56.

26. Bush, J. B., *et al.,* 1960, The Chief Oxide—Burgin area discoveries, East Tintic district, Utah; a case history: Econ. Geology, v. 55, p. 1116–1147, 1507–1540.

27. Lovering, T. S., *et al.*, 1960, Geologic and alteration maps of the East Tintic district, Utah: U.S. Geol. Surv. Mineral Investigations, Field Studies Map MF 230, 1:9600 (two sheets).

28. Anon., 1961, Million ton ore body disclosed: Min. Cong. Jour., v. 47, no. 4, p. 108.

29. Morris, H. T., and Lovering, T. S., 1961, Stratigraphy of the East Tintic Mountains, Utah: U.S. Geol. Surv. Prof. Paper 361, 145 p.

30. Anderson, J. A., 1964, Geochemistry of manganese oxides, a guide to sulfide ore deposits: Unpublished Ph. D. thesis, Harvard Univ.

31. Morris, H. T. and Shepard, W. M., 1964, Evidence for a concealed tear fault of large displacement in the central East Tintic Mountains, Utah: U.S. Geol. Surv. Prof. Paper 501-C, p. C19–C21.

32. Morris, H. T., 1964, Geology of the Eureka quadrangle, Utah and Juab Counties, Utah: U.S. Geol. Surv. Bull. 1142-K, p. K1–K29.

33. Proctor, P. D., 1964, Fringe zone alteration in carbonate rocks, North Tintic district, Utah: Econ. Geol. v. 59, p. 1564–1587.

34. Roberts, R. J., *et al.*, 1965, Pennsylvanian and Permian basins in northwestern Utah, northeastern Nevada and south-central Idaho: Bull. Amer. Assoc. Petrol. Geols., v. 49, p. 1926–1956.

35. Lovering, T. S., and Morris, H. T., 1965, Underground temperatures and heat flow in the East Tintic district, Utah: U.S. Geol. Surv. Prof. Paper 504-F, p. F1–F28.

# 48. The Eureka Mining District, Nevada*

T. B. NOLAN,† R. N. HUNT‡

## Contents

† U.S. Geological Survey, Washington, D.C.
‡ 2684 Hillsden Drive, Salt Lake City, Utah.
* T. B. Nolan is responsible for the section on history and general geology; R. N. Hunt for those on economic geology. Publication of data contributed by T. B. Nolan authorized by the Director, U.S. Geological Survey.

# Illustrations

# Tables

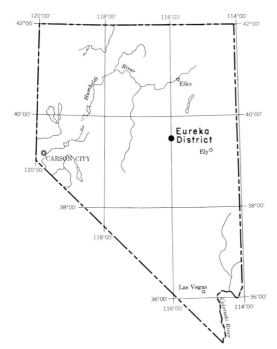

## ABSTRACT

In terms of present metal prices, analysis of extant records of the Eureka district indicate past production of the magnitude of $200,000,000 in recovered silver, lead, and gold.

Production to date has been thoroughly oxidized ore from massive replacements, fissures, manto, and pipe-like bodies—in large part from the massive Eldorado (middle Cambrian) and Hamburg (middle to late Cambrian) dolomite horizons. In the productive area of about 6 square miles carbonate rocks comprise 80 per cent of the section, which includes early Cambrian to late Ordovician horizons.

Four-fifths or more of the district's production came from a block of Eldorado dolomite in Ruby Hill that was moved in over younger horizons on strands of a regional thrust fault. The lower 500 to 600 feet of the block was much fractured and brecciated and became host to the ores. The Ruby Hill fault, a late normal fault, limits the Ruby Hill block of dolomite on the north for its entire length of nearly a mile and terminates the downward extent and exploitation of its ores. It introduced an hiatus of some 40 years in the district's development. Drilling of the past 20-odd years, followed by underground work, has now exposed the down-faulted extension of the Ruby Hill block of Eldorado dolomite and its contained ores, dropped 1800 feet in a dip shift to depths beneath the surface of 2000 to 3000 feet. The ores at depth are the sulfide counterpart of those mined in early years on the footwall side of the fault.

The ores of the district may be attributed to the same source as the intrusive quartz diorite outcropping in a limited way south of Ruby Hill but found at depth elsewhere in drill holes. The long north-south axis of the district, its anticlinal structure, its major faults, and the distribution of the intrusives and, in their relation to the surface, the ores themselves, all suggest that erosion has not cut deep enough to expose the extent of intrusives and the ores consequent upon them.

## HISTORY

The Eureka district, which celebrated its centenary in 1964, has a record of continued productivity that sets it apart from most of the older mining districts in the West. Although the bulk of its production occurred during the first 25 years or so of its history, it has enjoyed recurrent periods of activity that have made it a center of mining interest up to the present time.

Continued interest in the district, however, has been only in part due to its record of productivity. An important factor has been the degree to which new concepts of mining law and new techniques of prospecting, mining, and beneficiation have been applied in the exploitation of Eureka ores. The irregular replacement deposits found at Eureka, for example, differed greatly from the California gold-quartz veins whose form had influenced early mining law. Litigation attending the early mining on Ruby Hill resulted in Appellate and Supreme Court decisions confirming significant innovations in western mining law.

Similarly, in the early days the oxidized lead-silver-gold ores of the district required new and improved smelting methods, and Eureka was characterized by Hahn (16) as the "cradle of modern lead blast furnace smelting." Less well known, perhaps is the fact that some of the earliest experiments in both geophysical and geochemical prospecting were

conducted in the mines on Ruby Hill. Both were surprisingly modern in their general approach, including the formulation of the basic concept that "the indications of the existence of an orebody occupy a space greatly in excess of the size of the orebody itself." At Eureka, too, the "leasing" or "tribute" system was introduced as early as 1878. It was one of the first districts in the United States to utilize this method.

And even today, ingenious techniques employed to discover and develop the sulfide ores at depth in Ruby Hill despite heavy flows of water under high static head, interest the mining public.

## Mining Operations in the Area

The Eureka district was discovered in 1864 by prospectors from Austin, then a booming camp some 68 miles to the west. The original discovery, in New York Canyon, was small and never became a source of large production, but the discovery of rich oxidized lead-silver-gold ore bodies on Ruby Hill a few miles to the northwest led to rapid development and large production.

This highly productive period was concentrated in the years 1870 to 1893, during which the bulk of the district's production was made. A number of smelters were constructed to treat the oxidized arsenical ores, and the slag piles of the two largest producers, the Richmond Mining Company and the Eureka Consolidated, now mark the southern and northern entrances to the town of Eureka.

Company operations at Ruby Hill ceased about 1893, when the price of silver dropped below $1.00 an ounce. Much of the production since then has been from individual leases in and around old stopes. Since the period of peak production, there have been several revivals of activity leading to small but appreciable production, though none has approached the bonanza output of the earlier years. The Diamond-Excelsior and adjoining properties were active and productive in the 1890's and again in the 1950's, the Windfall in the late 1900's, Ruby Hill itself from 1905–1912 following the consolidation of many of the older mines as the Richmond-Eureka Mining Company, and the Holly and the Croesus in the early 1920's.

The last half of the district's history, however, is dominated first, by efforts to determine whether or not faulted extensions of the Ruby Hill ore zone existed and, later, by underground work to explore and develop the ore found at depth, as described in some detail below.

Currently the Ruby Hill Mining Company, jointly owned by the Richmond-Eureka Mining Company, a subsidiary of United States Smelting Refining and Mining Company, and by the Silver-Eureka Corporation, is engaged in detailed exploration of deep sulfide ore in the hanging wall of the Ruby Hill fault.

## Statistics of Mine Production

Table I gives recorded production for the Eureka district by decades from 1866 to the present. The figures for the early years, however, are subject to considerable doubt for there are large and unexplained differences between the Assessors' figures and production data for portions of this period from sources that are believed to be dependable. Analysis of these discrepancies has led to the suggestion that the value of the total production of the Eureka district is probably somewhat in excess of $120,000,000 (14, p. 56–59) rather than the $58,000,000 indicated in Table I. This total tonnage of ore mined appears to have been in the neighborhood of 2,000,000 tons and is estimated to have contained about 1,650,000 ounces of gold, 39,000,000 ounces of silver, and 625,000,000 pounds of lead. A few million pounds of copper and zinc are also reported to have been produced. At present prices* for gold, silver, and lead, the gross value of the ore produced would be

* $35.00 per ounce for gold; $1.29 per ounce for silver; and 16 cents per pound for lead.

something in excess of $200,000,000, a figure that perhaps more graphically expresses the relative importance of the district.

## PHYSIOGRAPHIC HISTORY AND PRESENT TOPOGRAPHY

The Eureka district lies at the north end of Fish Creek Range, an east-of-north-trending Basin Range located in east-central Nevada. This range ends not by a gradual decrease in altitude and disappearance beneath recent alluvium but by splitting into peneparallel ranges that extend northward from Eureka in a west-of-north direction. These are separated by Diamond Valley. The easterly of the two ranges, the Diamond Range, exhibits a more regular trend; the westerly is more irregular and is made up of the Mahogany Hills and Mountain Boy Range to the south and Whistler Mountain and the Sulphur Springs Range to the north.

The elevation of these linear mountain masses relative to the depression of the intervening Diamond Valley is believed to have been caused ultimately by movement along faults or fault-zones of normal displacement. But the present topographic character of the ranges indicates strongly that this Basin Range faulting has not been confined to a single brief period of time but rather has continued during a considerable part of Tertiary and Quaternary time and in many places been influenced by, or has utilized, fractures that were originally formed during earlier structural episodes. Essentially all of the mines in the district are

*TABLE I. Recorded Production of the Eureka District, by Decades, 1866–1964*

| Years | Tons Ore | Gold | | Silver | | Copper | | Lead | | Zinc | | Total Value |
|---|---|---|---|---|---|---|---|---|---|---|---|---|
| | | Ounces | Value | Ounces | Value | Pounds | Value | Pounds | Value | Pounds | Value | |
| 1866-1875 | 316,158 | -- | -- | -- | $19,530,994(a) | -- | -- | -- | -- | -- | -- | $12,310,490 |
| 1876-1885 | 733,509 | 27,913.00 | $3,169,483 | 923,432 | 21,367,641(a) | -- | -- | 250,307,000 | $1,985,034 | -- | -- | 28,619,794 |
| 1886-1895 | 227,507 | -- | -- | -- | 1,880,000(a) | -- | -- | 31,052,000 | -- | -- | -- | 4,696,443 |
| 1896-1905 | 57,073 | 9,924.27 | 205,153 | 272,996 | 150,572 | -- | -- | 26,608,237 | 91,763 | -- | -- | 1,367,944 |
| 1906-1915 | 200,777 | 56,011.58 | 1,157,861 | 655,174 | 378,515 | 298,968 | $47,936 | 12,352,636 | 574,220 | -- | -- | 2,041,524 |
| 1916-1925 | 86,773 | 19,662.65 | 406,463 | 642,816 | 561,042 | 1,405,101 | 219,802 | 15,491,521 | 1,164,238 | -- | -- | 1,511,208 |
| 1926-1935 | 164,153 | 15,501.48 | 381,023 | 380,423 | 224,106 | 199,522 | 26,111 | 10,940,032 | 745,619 | -- | -- | 1,925,951 |
| 1936-1945 | 81,812 | 17,597.04 | 111,336 | 287,644 | 47,113 | 99,297 | 515 | 1,037,728 | 13,056 | 2,878,800 | -- | 1,240,654 |
| 1946-1955 | 79,027 | 3,010.38 | 78,537 | 241,726 | 128,013 | 68,972 | -- | 3,130,283 | 248,583 | 9,467,800 | -- | 1,895,907 |
| 1956-1964 | 43,123 | 22,345.32 | 255,735 | 626,604 | 191,799 | 12,500 | 3,944 | 14,925,388 | 382,250 | 2,041,191 | $21,086 | 3,331,201 |
| Totals 1866-1964 | 1,989,912 | 171,965.72 | $5,765,591 | 4,030,815 | $44,459,795 | 2,084,360 | $298,308 | 365,844,825 | $5,204,763 | 14,387,791 | $21,086 | $58,941,116 |

Figures for the years 1866-1940 taken from the Records of Lander and Eureka County Assessors as published by the Nevada State Bureau of Mines in Nevada University Bulletin v. 37, no. 4, Geology and Mining Series No. 38, 1943, and for the years after 1940 from the records of the U. S. Bureau of Mines.

(a) Combined value of gold and silver.

found along Prospect Ridge,* the north-trending termination of the Fish Creek Range. The Ridge is the result of uplift relative to Spring Valley along the Spring Valley fault zone that outcrops along the west flank of the Ridge. The zone is composed of at least three en echelon faults, each striking nearly due north and dipping 55° to 60°W, the middle being known locally as the Sharp fault.

The displacement along this fault zone, at a maximum of about 2500 feet, caused an eastward tilting of both the hanging and foot-wall blocks. Displacement of remnants of a tilted older topographic surface on either side of the fault zone is in harmony with the stratigraphic evidence of displacement and indicates that movement occurred in fairly recent time. The Prospect Ridge block as a prominent topographic feature terminates against the line of the older Ruby Hill normal fault. Renewed movement along this northwest-striking, northeast-dipping fault amounted to about 400 feet, was probably contemporaneous with movement along the Spring Valley fault and is responsible for the lower ground found at the north end of the Ridge.

## GEOLOGIC HISTORY

### Sedimentary Rock Units

The Eureka district was one of the first in the Great Basin to be given detailed geologic study, and the stratigraphic section that was proposed as a result of this study by Hague and Walcott (1,3) has long been regarded as a standard in subsequent work in the region. Considering the state of knowledge and the physical difficulties encountered, the rock units and the relationships between them have undergone relatively little change in the intervening years. Such revisions as have been made may be found in papers by Walcott (4,5,6,7) Wheeler and Lemmon (9), Gianella (10), Sharp (11), Easton, *et al.* (12), and Nolan, *et al.* (14).

The 25 units that have been distinguished during recent geologic mapping in the district—in large part on and adjacent to Prospect Ridge—are listed in Table II, which also includes summary information on the age, thickness, and lithologic character of the rocks.

* Prospect Ridge is here used as the general term for the geographic feature extending from Prospect Peak northward through Mineral Hill, Ruby Hill, and Adams Hill to Mineral Point.

Seven of the formations are of special signifiance in relation to the ore bodies; six of these serve as host rocks at one place or another in the district, and the seventh is regarded as being genetically related to the ore solutions.

Three of the host-rock formations are made up in large part of dolomite, two of which have yielded by far the greater part of the production from the district. They are the Eldorado dolomite, the Hamburg dolomite, and the much less productive Hanson Creek formation.

The *Eldorado dolomite* of Middle Cambrian age is the oldest of the three and has also been the most productive, since it encloses the ore bodies of the mines on Ruby Hill. It is made up chiefly of massive thick-bedded gray dolomite that generally forms steep rough slopes. There are a few beds of limestone near the base of the formation, but these are unmineralized.

Two varieties of dolomite can be recognized. One is well-bedded, rather fine-grained, and is commonly dark gray to black in color. It shows a variety of sedimentary textures, including a local fine lamination, and is regarded as having formed through normal sedimentary processes. The other and more abundant variety is lighter-gray, coarser-grained, locally vuggy, and essentially textureless. Its distribution is independent of bedding and, hence, is considered to be the product of the recrystallization of the dark dolomite caused by circulating hydrothermal solutions. As is true of all dolomites in the district, the Eldorado dolomite is normally severely fractured.

The Eldorado dolomite is estimated to be 2500 feet thick. A much smaller thickness is found in the Ruby Hill mines, where faulting has cut out the lower part of the formation. On Prospect Ridge to the south, however, the thickness is apparently considerably greater, but this is believed to be the result of duplication by steep thrust faults.

The *Hamburg dolomite* of Middle and Late Cambrian age has been the next most productive formation. It closely resembles the Eldorado dolomite lithologically and, in many places, can be distinguished from it only by its stratigraphic position relative to the Secret Canyon or Dunderberg shales or to the Geddes limestone. As is true of the Eldorado, it is commonly altered to more coarsely crystalline dolomite and is thoroughly fractured, especially is the vicinity of ore bodies. Seemingly it is more susceptible to silicification than the

TABLE II.    *Geologic Formations Present in the Eureka Mining District*

| Age | Name | | Stratigraphic thickness (in feet) | Lithologic character |
|---|---|---|---|---|
| Quaternary | Alluvium | | 0–500± | Stream and slope alluvium, terrace gravels, and mine and smelter dumps. |
| | ——————Unconformity—————— | | | |
| Late Tertiary or Quaternary | Pyroxene andesite and basalt | | 700+ | Lava flows; a few dikes and small plugs. |
| | —Intrusive contact and unconformity— | | | |
| Oligocene or Miocene | Rhyolite tuff | | 400± | White, layered tuff. |
| | Rhyolite | | 100± of flows exposed | Chiefly intrusive plug, dikes, and breccia pipes; vitrophyre sill; and local lava flows. |
| Eocene | Hornblende andesite | | 300± of flows exposed | Dike and lava flows. |
| Late Cretaceous | Quartz porphyry | | -------------- | Sills and dikes. |
| | Quartz diorite | | -------------- | Intrusive plug south of Ruby Hill. |
| | ——————Intrusive contact—————— | | | |
| Early Cretaceous | Newark Canyon formation | | 200± | Fresh-water conglomerate, sandstone, grit, shale, and limestone. |
| | ——————Unconformity—————— | | | |
| Permian | Carbon Ridge formation | | 1,000± | Thin-bedded sandy and silty limestone; some included sandstone and dark shale. |
| | —Unconformity—Ely limestone absent— | | | |
| Late Mississippian | Diamond Peak formation | | 0–300 | Conglomerate, limestone, and sandstone. |
| | Chainman shale | | 500± exposed | Black shale with thin interbedded sandstone. |
| | ——————Break in section—————— | | | |
| Middle and Late Devonian | Devils Gate limestone | | 500± exposed | Thick-bedded limestone, locally dolomitized. |
| | —Break in section—Nevada, Lone Mountain, and Roberts Mountains formations not recognized in mapped area | | | |
| Late Ordovician | Hanson Creek formation | | 300± exposed | Dark-gray to black dolomite. |
| | ——————Unconformity?—————— | | | |
| Middle to Late(?) Ordovician | Eureka quartzite | | 300 | Thick-bedded vitreous quartzite. |
| | ——————Unconformity—————— | | | |
| Early and Middle Ordovician | Pogonip group | | 1,600–1,830 | Chiefly cherty thick-bedded limestone at top and bottom; thinner bedded shaly limestone in middle. |
| Late Cambrian | Windfall formation | Bullwhacker member | 400 | Thin-bedded sandy limestone. |
| | | Catlin member | 250 | Interbedded massive limestone, some cherty, and thin sandy limestone. |
| | Dunderberg shale | | 265 | Fissile brown shale with interbedded thin nodular limestone. |
| Middle and Late Cambrian | Hamburg dolomite | | 1,000 | Massively bedded dolomite; some limestone at base. |
| | Secret Canyon shale | Clarks Spring member | 425–450 | Thin-bedded platy and silty limestone, with yellow or red argillaceous partings. |
| | | Lower shale member | 200–225 | Fissile shale at surface; green siltstone underground. |
| Middle Cambrian | Geddes limestone | | 330 | Dark-blue to black limestone; beds 3–12 in. thick; some black chert. |
| | Eldorado dolomite | | 2,500± | Massive gray to dark dolomite; some limestone at or near base. |
| Early Cambrian | Pioche shale | | 400–500 | Micaceous khaki-colored shale; some interbedded sandstone and limestone. |
| | Prospect Mountain quartzite (base not exposed) | | 1,700+ | Fractured gray quartzite weathering pink or brown; a few thin interbeds of shale. |

Eldorado, particularly close beneath the Dunderberg shale.

The Hamburg dolomite is approximately 1000 feet thick. The mines on Adams Hill, those in New York Canyon (including the Diamond-Excelsior and adjoining properties), and those in a belt extending from the Windfall mine on the south to the Dunderberg mine on the north, are all located within the formation.

The remaining dolomite formation, the Hanson Creek, is of interest chiefly because it forms the wall rocks of the relatively unproductive Seventy Six mine in lower New York Canyon and is reported to be the site of the first discovery of ore in the Eureka district. The Hanson Creek is of Late Ordovician age. About 300 feet are exposed within the district.

Small amounts of ore have been found in limestone beds of the *Windfall Formation* and *Pogonip Group*. In both, the thickness of individual beds is less than that of the larger dolomite units, and the ore bodies are correspondingly smaller. The Windfall formation is of Late Cambrian age; it is the host rock for the Bullwhacker, Holly, and adjoining properties at the north end of Adams Hill. The Pogonip Group is locally mineralized south of Ruby Hill. It is of Early and Middle Ordovician age.

One quartzite formation, the *Eureka quartzite* of Middle to Late (?) Ordovician age, has been the site of ore deposition in the Hoosac Mine, at the south end of the district. This property was productive in the early days of the district, but little is known about the magnitude of the production or of the nature of the ore bodies.

### Intrusive Rocks

A small intrusive plug of quartz diorite crops out just south of Ruby Hill. Recent drilling, however, has shown that in depth the quartz diorite extends at least 3000 feet farther northeast than its outcrops (as shown in drill holes #606 and #608 of the Ruby Hill Mining Company). The plug, therefore, is believed to be the surface representation of a larger intrusive mass. To judge from the coincidence in space, it is also likely that the mineralizing solutions that formed the hypogene ore bodies emanated from depths beneath the larger mass.

The quartz diorite is poorly exposed; it is best seen in underground workings or in drill cores. The rock is rather variable in composition and texture, and much of it is altered. Quartz and andesine are the two most abundant minerals, and hornblende and biotite are the two ferromagnesian minerals. Small amounts of orthoclase and such accessory minerals as sphene, iron oxides, apatite, zircon, and allanite are also present. The common alteration products are calcite, epidote, chlorite, fine-grained quartz, pyrite, barite, and sericite. In the drill holes mentioned above, the quartz diorite is considerably altered; it is also cut by quartz veinlets carrying sulfides, largely pyrite, and flecks of molybdenite. The sedimentary rocks above the intrusive in the drill holes have been altered for distances of 125 to 150 feet to silicates, with abundant magnetite and some pyrite.

Sills and dikes of quartz porphyry are found on the north and east sides of Adams Hill in the vicinity of the Bowman, T. L., and Bullwhacker mines, and have been cut by drill holes in this area. The sill-like masses are found at or near the contacts between the Dunderberg shale and the Hamburg dolomite and the overlying Windfall limestone; the dikes tend to follow fault zones. The rock is highly altered but is probably closely related to the quartz diorite intrusive mass.

The quartz diorite has been dated by Jaffe and others (15, p. 73) by the lead-alpha method as being $62 \pm 12$ m.y. old. Armstrong (17, p. 164) obtained a date of 64 m.y. by the K/Ar method.

### Structure

The rocks underlying Prospect Ridge, to which most of the mining activity has been confined, have been considerably deformed by folding and faulting.

Superficially, the Ridge appears to be underlain by a monocline of near vertical Cambrian and Ordovician rocks. More detailed mapping, however, has shown that these rocks are cut by three zones of thrust faulting, by three major normal faults, and by one large transverse fault. One of the thrust faults has associated with it a recumbent anticline of some magnitude, and the branches of this thrust are themselves gently folded in a way that indicates the folding was contemporaneous with the movement along the thrust. Finally, mapping east of Prospect Ridge suggests that the Ridge itself rests on still another thrust fault, along which the lower Paleozoic rocks of Prospect Ridge have moved eastward over Mississippian and Permian sedimentary rocks.

The location and relationships of these structures are shown on the generalized geologic map (Figure 1). The innumerable smaller

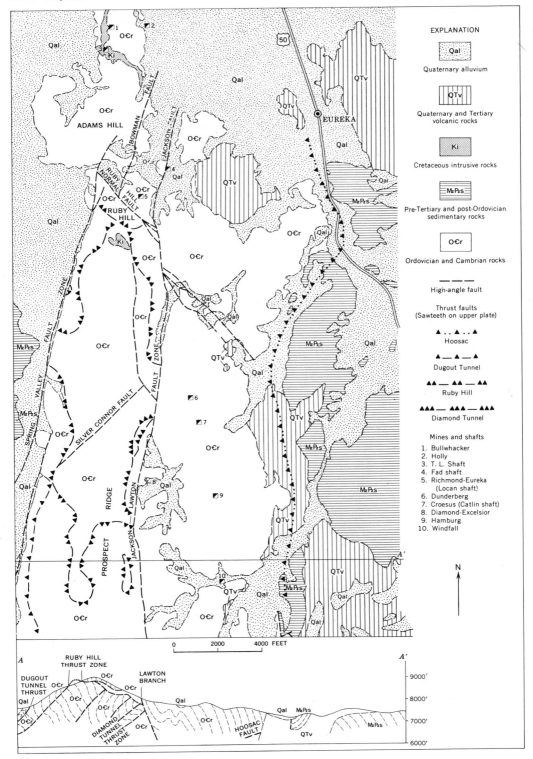

FIG. 1. *Generalized Geologic Map and Section of the Eureka District, Nevada.*

faults and minor flexures have been omitted, as have most of the individual formation boundaries.

The structurally highest of the thrust zones—the Dugout Tunnel thrust—is exposed on the southwest flank of Prospect Ridge. It has brought Ordovician and Devonian rocks over the Lower and Middle Cambrian rocks that form the bulk of Prospect Ridge. In this region, the rocks adjacent to the thrust zone contain no significant ore bodies.

The next lower thrust zone is extensively exposed in the mine workings under Ruby Hill, as well as on many other places along Prospect Ridge. It brings Lower and Middle Cambrian rocks over younger ones of late Cambrian age. The thrust zone is folded into a broad anticline whose axis extends somewhat west of north beneath Ruby Hill. In several places, notably on Ruby Hill, the fault has two or more major branches. In Ruby Hill, the lower of the two is folded more sharply than the upper, with the result that the intervening mass of rock made up of Lower Cambrian quartzite has been thinned to less than 50 feet. The rocks beneath the Ruby Hill thrust are overturned and form the upper limb of a recumbent syncline that is exposed in Zulu Canyon, southeast of Ruby Hill. The folding is believed to be contemporaneous with the thrust.

The dolomitic rocks above the Ruby Hill thrust zone are the host rocks for a large proportion of the district ore bodies. All the ore bodies on Ruby and Adams hills, as well as those in the dolomite belt that extends from the Dunderberg mine to the Windfall mine, lie in this thrust plate. In fact, some of the shallower ore bodies on Ruby Hill had the upper branch of the thrust as a footwall.

The Diamond Tunnel thrust zone is the lowest, and probably the oldest, of the three thrusts. It crops out only on the steep slopes at the head of New York Canyon and brings Middle and Upper Cambrian Hamburg dolomite over the uppermost Cambrian and Lower Ordovician rocks. The ore bodies of the Diamond-Excelsior and adjoining properties lie in the dolomite rocks above the thrust zone, which is here made up of several branches.

Because of its economic importance, the northwest-striking Ruby Hill normal fault is perhaps the best known structural feature of the mining district, since it terminated in depth the block of Eldorado dolomite in which the Ruby Hill ore bodies were found. The total displacement along the fault appears to be about 2000 feet.

The transverse Silver Connor fault and the compound Jackson-Lawton-Bowman fault system, both with displacements of 1000 feet or more have profoundly affected the distribution of the favorable dolomites in which most of the ore bodies are found. In addition, the location of the more intense mineralization in the vicinity of these faults suggests that they may have been influential in guiding the circulation of the mineralizing solutions.

The poorly exposed Hoosac fault, which limits the district on the east, was originally interpreted by Hague (3, p. 15–17) as a normal fault. Recent work, however, suggests that it may be a thrust fault, along which the rocks of Prospect Ridge have overridden much younger rocks to the east. If this interpretation is correct, the Hoosac fault may have acted as a sole to the three minor thrust-fault zones that are exposed to the west. The Spring Valley fault zone, bounding the district on the west is, as noted on a previous page, of quite recent origin.

The relationship of the thrust faults in the Eureka district and the associated structures to the major structural feature of central Nevada—the Roberts Mountains thrust with a minimum displacement of 50 miles (13)—is uncertain. Exposures are not conclusive as to whether the Eureka district was overriden by the Roberts Mountains thrust plate or whether the eastern limit of the thrust plate lies just west of the mining district. The former hypothesis appears less likely, because the character of the Eureka structural features suggests that they formed under a lighter load than that of the thick sedimentary plate that presumably overlay the Roberts Mountains thrust. If, on the other hand, the Roberts Mountains thrust failed to reach Eureka, the Eureka structural features might represent the near-surface disturbances in front of an advancing major thrust plate. The fact that the Hoosac fault involves Carboniferous and Permian rocks indicates that it postdated the Roberts Mountains thrust since this has been dated by J. F. Smith, in the Pine Valley region north of Eureka, as of late Devonian or very early Mississippian age. But its relationship to the thrusts on Prospect Ridge is unclear.

## THE ORES AND THEIR OCCURRENCE

The mines of the district are at scattered points at elevations of 6500 to 8500 feet within a north-south strip a mile or less in width beginning at Mineral Point immediately west of the town of Eureka and extending south

6 miles along the generally ascending divide of Mineral Hill and Prospect Ridge. As has been indicated, the area is one of close folding, complexly faulted but essentially anticlinal in structure. In the southern half, beneath planes of the Ruby Hill thrust zone, the closely folded, even overturned, section includes formations from the Prospect Mountain quartzite upward to and including the Pogonip limestone. In the Ruby Hill and northern half of the district, this sequence is duplicated in a thrust plate moved in over the highly folded formations of the lower plate by the Ruby Hill thrust zone. This upper plate and these horizons are warped into an open anticline plunging gently northward. The underlying Ruby Hill thrust zone itself is involved in this warping. As is the Cambrian-Ordovician section above it, it is gently arched in an open anticline plunging at a low angle to the north. Beneath the thrust the same horizons are closely folded even slightly overturned. To superimpose an unwarped section upon a closely folded sector must have required very considerable movement. The gentle folding of the fault itself and the section above it presumably occurred during a late stage of regional folding and after most of the movement upon the Ruby Hill thrust zone (Figures 1, 2).

A little north of center within the north-south strip, and now included in the property of the Ruby Hill Mining Company, are the old mines of Ruby Hill from which came the greatly preponderant portion of the district's production. This production came from three-dimensional replacement ore *bodies and smaller bodies of fissure-vein type in the Eldorado dolomite of the upper Ruby Hill thrust plate. To the north are small mines in the Hamburg dolomite and Windfall limestone on the crest and east flank of the anticlinally warped upper thrust plate. Their ores were in small, irregular replacements and still smaller bodies of fissure and manto types. In the small mines to the south of Ruby Hill were fissure and pipe-like ore bodies, most of them in the Hamburg dolomite on the east slope of Prospect Ridge.

That the ores of the district occur in these thick, massive carbonate formations, rather than in the underlying Prospect Mountain quartzite or the overlying Eureka quartzite or in two intercalated shale formations, may be ascribed, first of all, to their great volume in proportion to other rocks. The odds were four to one that in their travels the mineralizing agents passed through in carbonate rocks. Their distribution within the carbonate en-

vironment seems to have been determined by permeabilities created by stresses incident to the regional thrust faults and numerous later large normal faults, which resulted in the intense brecciation of large volumes of dolomite, particularly in the Eldorado dolomite of Ruby Hill resting upon the upper Ruby Hill thrust plane. As in most districts, from whatever source the mineralizers issued, an existing pattern of permeable features determined the loci and geometry of individual ore bodies. The chemical receptivity of the carbonate rocks may also be assumed to have been a factor in entrapping the metals at no great distance from a possible source in or beneath quartz diorite intrusives that, as will be indicated, probably have a greater distribution north and south at depth on the eastern flank of the anticline than their limited outcrops indicate. The silication of carbonate and argillaceous rocks above, and abutting against, these intrusives suggests that they were so chemically potent as may make it reasonable to associate the ores of the district genetically with them.

## RUBY HILL MINES

Ruby Hill is an erosional eminence carved out of a long northwest-southeast wedge of Eldorado dolomite, 1500 feet wide at its west end and tapering to a point nearly a mile to the east. North and south normal faults terminate it, the Sharp fault on the west and the Jackson fault on the east. These structures displace whatever extensions the dolomite wedge may have had in those directions at least several hundred feet downdip. For its entire length this wedge is flanked on the north by the Ruby Hill fault which in a dip shift has dropped its northerly extension approximately 1800 feet, bringing down opposite it the stratigraphically higher Geddes limestone, Secret Canyon shale and Hamburg dolomite. (Figures 2, 3). To a ton, the entire Ruby Hill production, 80 to 90 per cent of that of the district, came out of this block of Eldorado dolomite in the footwall of the Ruby Hill fault.

This Eldorado block rests upon the stratigraphically lower and older Prospect Mountain quartzite. But the interface is not stratigraphic. It is the upper of two planes of the Ruby Hill thrust fault, and it dips 25° to 35° northwest and northerly. Where visible, the bedding in the dolomite dips contrary-wise, to the east, impinging on the thrust fault plane at steep angles. The underlying Prospect Mountain quartzite itself rode into place on the lower strand of the Ruby Hill thrust zone over the

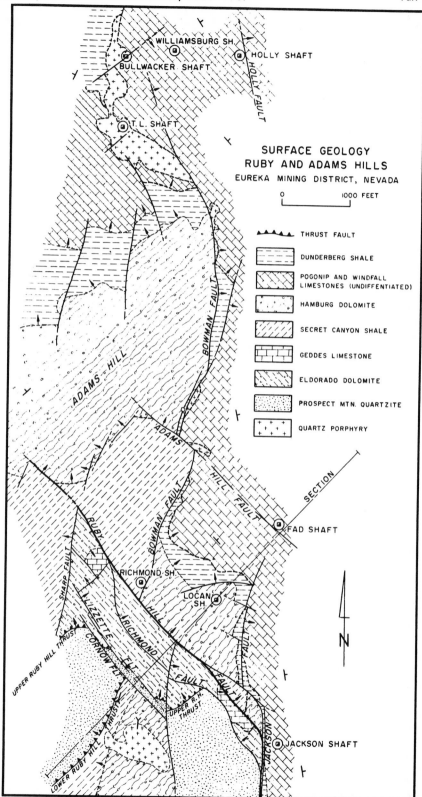

FIG. 2. *Surface Geology of Ruby and Adams Hills. Eureka District, Nevada.*

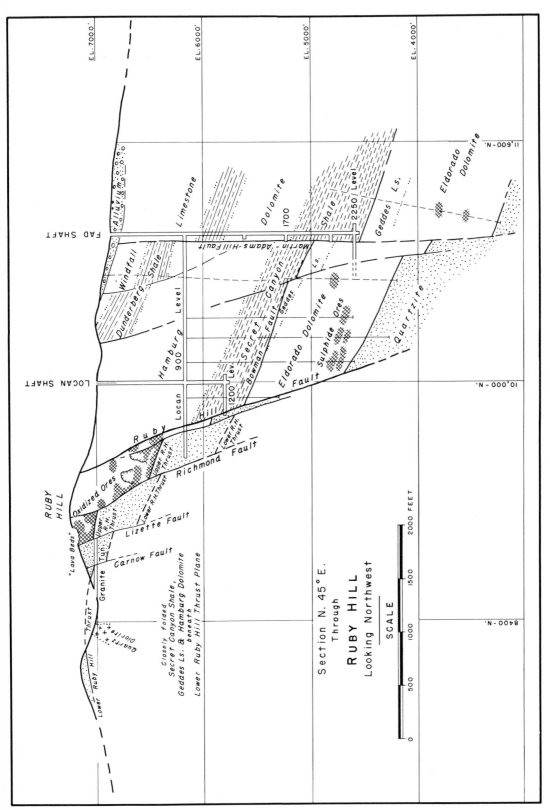

FIG. 3. *Section N45°E through Ruby Hill.*

younger, closely folded Eldorado, Secret Canyon, Geddes, and Hamburg formations (Figures 2, 3). In displacement, the lower strand appears to be much the greater of the two thrust planes. In the westerly third of Ruby Hill, in the old Richmond mine, the upper thrust plane divides into two parts, which in vertical section diverge westerly. The upper split continues as the undersurface of the Eldorado dolomite and the lower as the upper surface of the Prospect Mountain quartzite. Between them is a wedge which resembles what a mile and a half to the south has been mapped as Pioche formation. Here in the Richmond mine it consists of a thin stratum of highly sheared micaceous sandstone overlying dense light-gray limestone in which exploration to date has been without result.

The fracturing and intensive brecciation of large volumes of Eldorado dolomite resting upon or within a few hundred feet above the quartzite, as well as of the quartzite itself, may be attributed to stresses of the thrust movement. The irregularities of the breccias, the mingling in them of unlike fragments, and their appearance of being post-marmorization in age probably rules out their being formational breccias. Physical effects of stresses incident to the thrust movement appear to have been a major factor in limiting the replacement ore bodies of Ruby Hill to fractures and brecciated volumes within a vertical distance of a few hundred feet above the upper Ruby Hill thrust plane. Many formed directly above and upon the thrust plane.

Much of the Ruby Hill ore was in massive, somewhat tabular bodies resting upon or within a short distance above two faulted segments of the upper thrust plane known in early days as the "Alpha" and "Beta" faults. Ores penetrated upward from these structures into the massive dolomite through fissures and possibly through obscure textural permeabilities in the bedding, which is often completely obliterated in the brecciation and marmorization of the dolomite. Only locally is the bedding sometimes evident as gray and white bandings dipping easterly, usually at angles above 45°.

Smaller ore bodies formed in and along three lesser faults nearly parallel in strike and dip with the Ruby Hill fault—the Potts, Richmond, and Lizette. Ore bodies formed also along the West Lawton fault, which is one of two north-south normal faults of moderate displacement that displace the Richmond fault but themselves are displaced by, or terminate against, the Ruby Hill fault (Figures 2, 3 and 4). These ore bodies are localized on steeply

dipping fault structures well up within the Eldorado dolomite. These and other minor normal faults displace the underlying thrust plane wherever they intersect it. Two productive fissures of low westerly dip, well up in the dolomite, are nearly parallel and possibly sympathetic with the underlying thrust plane but show no evident displacement. The intersecting structures of this lattice within the Ruby Hill block of Eldorado dolomite terminate on dip or strike against the Ruby Hill fault. No structure has been found to pass through or displace that fault, unless it be the Jackson fault on the east which has been shown, by underground workings upon it, to curve to the south and merge tangentially with the Ruby Hill fault.

## Ruby Hill Fault

Mining in Ruby Hill began in 1869 on its south slope in dark brown to black outcrops of oxidized ore, mistakenly termed the "lava beds", and followed ramifying, coalescing runs that extended downward, generally northwesterly and northward direction, on or above the upper thrust plane and quartzite floor, to depths of about 1200 feet, where the ores terminated, if they had not sooner, against the Ruby Hill fault. To the miner, the Ruby Hill fault was like a great frustrating wall running more than 4000 feet for the entire length of the mines in Ruby Hill (Figures 3, 5).

"Unveinlike" irregularities, junctions, and ore bodies that lack upward connections with anything in the way of apices on the surface led to disputes as to extralateral rights and ownership among some of the six companies then owning Ruby Hill. In 1873, in a precedent extralateral right case, Richmond Mining Company vs Eureka Consolidated Mining Company, the trial lawyers described the Ruby Hill fault as a "clay and shale wall." They made it one of the definite boundaries legally necessary in describing the geometry of their "lode." The trial court held with them that the entire volume of Eldorado dolomite between the quartzite footwall and the "clay and shale wall" was the lode and vein. Affirmation by the United States Supreme Court in 1881 established in principle the concept of the "broad lode" in the exercise of extralateral rights under Federal mining law. Clarence King, then between completion of his field work on the 40th Parallel Survey and his appointment in 1879 as the first director of the United States Geological Survey, was an eloquent witness in this famous case. In this case, it was also

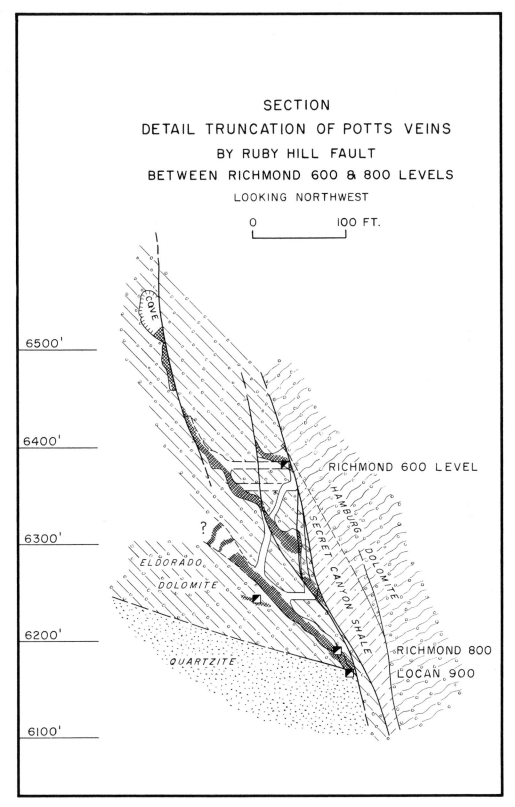

SECTION
DETAIL TRUNCATION OF POTTS VEINS
BY RUBY HILL FAULT
BETWEEN RICHMOND 600 & 800 LEVELS
LOOKING NORTHWEST

0          100 FT.

6500'

CAVE

6400'

RICHMOND 600 LEVEL

HAMBURG DOLOMITE

6300'

?

ELDORADO

DOLOMITE

SECRET CANYON SHALE

6200'

RICHMOND 800
LOGAN 900

QUARTZITE

6100'

FIG. 4. *Section: Detail of Truncation of Potts Veins by Ruby Hill Fault between Richmond 600 to 800 Levels.*

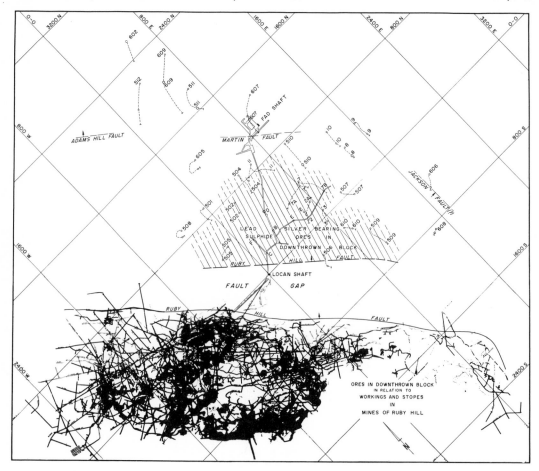

FIG. 5. *Ores in Down-thrown Block, in relation to workings and stopes in the mines of Ruby Hill.*

held and affirmed that a claim with divergent endlines located under the law of 1862 could have extralateral rights even in the direction of their divergence.

In accordance with prevailing views, in their further search for ores at depth the early miners sought the "true fissure vein" up which the ore depositing solutions must needs have risen. Though totally barren itself in all exposures of that day, the Ruby Hill fault seemed the obvious duct. Altogether, on various levels, 76 drifts totaling 4800 feet were driven upon the Ruby Hill fault itself, 28 crosscuts were driven through it into the shale and carbonate rocks of the hanging wall, five of 200 to 300 feet each, one of 1700 feet. A 1230 foot shaft, the Locan, was sunk in the hanging wall country and on its 1200-foot level, 145 feet lower than any workings opposite upon the footwall

side, a crosscut was driven from the shaft back through the fault. All this work disclosed no significant mineralization in the hanging wall of the Ruby Hill fault itself. The "clay wall" did in fact limit the "broad lode."

Later, however, in the early twenties, some 300 feet of drifting on the Locan 1200-foot level along a footwall strand of the fault did disclose considerable lead-bearing sulfide material. The Locan 1200-foot level is well below static water level. In his weekly reports, the superintendent referred to this strand as "the vein." Waldemar Lindgren, then under retainer, described these sulfides in the fault as crushed drag material.

The sharp juxtaposition of numerous ore bodies and a highly mineralized block of dolomite in the footwall against a totally barren hanging wall, the clean truncation and termination of ores, the details of brecciation, slicing, and displacement of ores by planes of the Ruby Hill fault, and the termination of ore-bearing structures themselves against that fault, appeared to indicate a very considerable post-mineral movement upon it (Figures 3, 4). But, as is true of major faults in many districts, the Ruby Hill fault has probably had a long history, possibly antedating in some part both the intrusives and the mineralization consequent upon them. In its eastern portion, the fault is occupied by a small rhyolite dike.

Its barrenness and lack of mineralization—gangue or ore minerals or wall-rock alteration—make it difficult to regard the Ruby Hill fault as the major mineralizing structure the oldtimers sought.

As will be suggested later, it is easier to consider the permeable zone of the upper Ruby Hill thrust plane, before its disruption by the Jackson, Ruby Hill, and other faults, as the major entry and distributing structure.

## Oxidized Ores

Except for occasional masses of incompletely oxidized ore at lowest points in the old mines, all past production was of thoroughly oxidized ore. The dollar value of the ores, often high, was provided by silver, lead, and gold. Silver and gold in any form were rarely, if ever, visible, although argentite has been recognized in the laboratory. Lead was present in oxidized derivates from galena, such as cerussite, anglesite, and plumbojarosite, and infrequently in relict galena. The lead arsenate, mimetite, and an oxide of lead and antimony, bindheimite, have been recognized. Zinc, when present, was a minor, nonpaying constituent, largely as smithsonite and calamine. As is often true under conditions of thorough oxidation, zinc was almost completely leached and in part redeposited in irregular usually low-grade disseminations and pockets in the carbonate walls. A few small lots of zinc carbonate ores were shipped in times of high metal prices.

The valuable minerals were in an iron-rich gangue of such minerals as limonite and goethite, with calcite, various quantities of unreplaced dolomite and very little if any quartz.

The "insoluble" content reported in smelter settlements was 10 to 12 per cent or less. In fact, the United States Smelting Refining and Mining Company acquired its original Ruby Hill properties in the days when its smelter at Midvale, Utah, received much siliceous custom ore, "dry ore," and required a low-silica, low-sulfur, base flux.

In early days, arsenic in small, but troublesome, amounts was removed in the local smelters as speiss. The 44,600 tons of speiss shipped 1918 to 1924 from the Richmond Mining Company's old smelter site assayed:

| Arsenic | Iron | Copper | Lead | Gold | Silver |
|---------|------|--------|------|------|--------|
| 32% | 56% | 1.66% | 2.06% | 0.196 oz | 2.28 oz |

In this century at Midvale, the arsenic in this and various ores was fumed off, collected, and sold as sodium arsenide for weed killers and as arsenic trioxide to chemical companies. During both world wars, arsenic was sold to the United States Army.

## Grade

(1) The ores mined in the past century were valued for their precious metals. Lead was a by-product and the collector of the silver. A compilation of recorded production (14, p. 58) reports 1,317,388 tons mined during the years 1869–1901 to have approximated the following grade:

| Gold | Silver | Lead |
|------|--------|------|
| 1.1 oz | 27 oz | 17% |

As an average, this amount of gold may be high. But it, as well as the figures for silver lead, are in accord with certain tonnages mined in virgin situations in later years.

(2) Below are figures from an analysis given by Curtis (2, p. 60–61) of a composite sample of all the ores from Ruby Hill treated by the Richmond Mining Company's smelter in the year 1877.

| Gold | Silver | Copper | Lead | Iron | Zinc | Arsenic | Antimony | Sulfur | Silica |
|------|--------|--------|------|------|------|---------|----------|--------|--------|
| 1.59 oz | 27.55 oz | 0.12% | 33.12% | 24.07% | 1.89% | 4.13% | 0.25% | 1.67% | 2.95% |

(3) 666,981 tons reported during the period 1902 to 1959 approximated:

| Gold | Silver | Lead |
|------|--------|------|
| 0.21 oz | 4.6 oz | 4.75% |

(4) 57,000 tons of company and leaser ores shipped in 1918 to 1927 and for which there are grade figures averaged:

| Gold | Silver | Copper | Lead |
|------|--------|--------|------|
| 0.293 oz | 6.81 oz | 0.41% | 9.32% |

(5) Complete assays of 2511 tons shipped in small lots from Ruby Hill in 1920 to 1925 averaged:

| Gold | Silver | Copper | Lead | Insol. | Iron | Zinc | Sulfur | Lime | Arsenic |
|------|--------|--------|------|--------|------|------|--------|------|---------|
| 0.305 oz | 4.21 oz | 0.32% | 6.63% | 12.3% | 34.0% | 5.32% | 0.51% | 2.34% | 2.90% |

The averages (3), (4), and (5) are of ores that largely were gleanings from in and around old slopes.

## Caves

The thorough oxidation of the original sulfide ores of Ruby Hill and other mines of the district developed sizeable caves immediately above the ores (Sections 3, 6, Figure 6). The process was in part one of gravity stoping once the competency of the sulfide ores gave way under oxidation to slumping and compaction. In such engineering records as remain of the stopes of early years, it is often impossible to determine the top of the ore as mined and the bottom of the cave. The roofs carry more or less stalactitic calcite. The undisturbed cave floors consist of tumbled slabs and rubble fallen from the roof and walls, partially cemented with calcite, in places with some gypsum. Beneath this floor of caved debris was the oxidized ore. In these features, Eureka is similar to other limestone districts of deep oxidation in the Southwest and Mexico. In Mexico centuries ago, the miner learned to prospect caves by sinking down through "saro" floors of sheeted calcite and gypsum, that had the appearance of so much ice.

## Exploration at Depth—Sulfide Ores in the Down-thrown Segment of the Ruby Hill-Eldorado Block

In 1919, a Canadian financed company, the Ruby Hill Development Company, became interested in the Ruby Hill fault problem. But it appears to have lost interest during the difficult years, 1920 to 1921 when the Pittman silver act (dollar silver) expired, and post-war metal prices reached a temporary low.

In 1937, a Canadian company, Ventures, Ltd., sought and obtained a lease upon the Richmond Eureka Mining Company's holdings. As its operating vehicle in Eureka, it formed The Eureka Corporation. After an initial unsuccessful hole, five successive holes, 1500 to 2000 feet away from the first, penetrated vertical intervals of 15 to 65 feet of heavy sulfide ore at depths of 2000 to 3000 feet in an area downdip from the most productive area on the footwall side of the Ruby Hill fault (Figures 2, 3). With this encouragement The Eureka Corporation sank the Fad shaft of four compartments to a depth of 2415 feet.

In March, 1948, at a point 175 feet from the shaft, a drift of the 2250 level, intended to expose the ore discovered in these drill holes, encountered a sudden flow of water beyond pump capacity. The water rose some 1200 feet, and the shaft was lost. After efforts to recover it, the shaft was abandoned. Substantial expenditures followed in further drilling from the surface of Ruby Hill and then on the easterly and northern slopes of Adams Hill. Erratic intervals of good ore encountered in holes on the north slope of Adams Hill lead to the sinking by The Eureka Corporation of a second shaft to a depth of 1127 feet, the TL shaft, 6000 feet northeast of the Fad shaft, from which a small tonnage of oxidized ore was mined and shipped.

In 1962, the Ruby Hill Mining Company was formed, 75 per cent owned by Richmond Eureka Mining Company, a subsidiary of United States Smelting Refining and Mining Company, and 25 per cent by the Silver Eureka Corporation, successor to The Eureka Corporation. In that year, the following companies united as lessees under the management

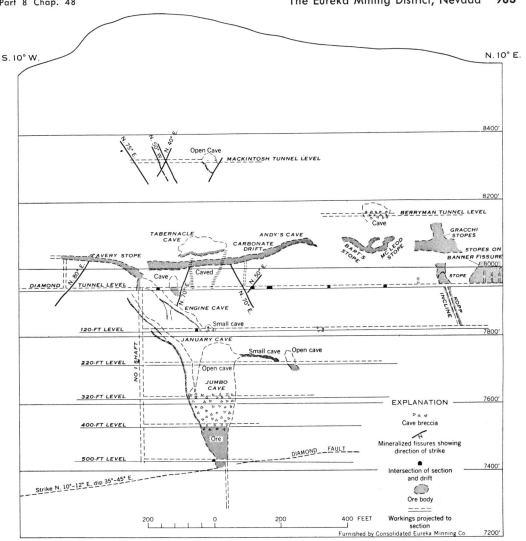

FIG. 6. *Section showing longitudinal projection in a plane N10°E of ore bodies associated with open caves, Diamond mine, Eureka district.*

of the Hecla Mining Company to continue the development of the sulfide ores in the downfaulted block:

Newmont Mining Corporation
Cyprus Mines Corporation
Hecla Mining Company
Richmond Eureka Mining Company
Silver Eureka Corporation

After further drilling from the surface, these companies under Hecla's management in 1964 undertook the recovery of the Fad shaft. A directed hole from the surface was successfully drilled to intercept, a few feet back from its face, the drift on the 2250 level that encoun-

tered the insurge of water in 1948. A water and pressure tight concrete plug was placed in the Fad shaft just above static water level, approximately 1030 feet below the shaft collar; 24,000 sacks of cement went into a slurry that was flowed down through the drill hole into the 2250 drift. At the same time, water (under sufficient pressure and in sufficient quantity to set up a reverse flow back into the 2250 face) was pumped down pipes through the shaft plug. This reverse flow carried the cement slurry into the face, sealing the water fissure by a cumulative deposit on its walls. Apparently a seal was effected before injection of the slurry ceased, and a considerable quantity of slurry flowed back into the drift, forming a partial plug in it and in the shaft station.

The work was sufficiently successful to permit, 16 years after it was abandoned, the unwatering of the Fad shaft at a pumping rate which at no time exceeded 2500 gpm. This compares with 9000 gpm pumped when recovery efforts ceased in December, 1948. Drifting toward the ore bodies penetrated in the drilling of the previous 20-odd years was resumed under cover of systematic cementation.

### Sulfide Ores

Sulfides were found in the old mines only in the lowest workings where pyrite and partially oxidized galena appeared mixed with valuable and worthless oxide minerals.

During the period 1939 to 1961, three successive campaigns of drilling, diamond drilling underground from the Locan 900 level and drilling from the surface with oil-well equip-

downward flow of ground water undepleted in oxygen and enroute to outlets to the north in the deep alluvium of Diamond Valley rather than to oscillations in static water level in Ruby Hill. Corroborating this inference is the fact that, in studies of the water situation in Ruby Hill, made before the unwatering of the Fad shaft was attempted, a slow definite flow downward at valley level in the shaft was detected and measured.

In detail, valuable metals in the sulfide ores are not disseminated with any uniformity but are irregular, coarse blotches in a heavy gangue of pyrite or in incompletely unreplaced dolomite. Quartz is not visibly present. The insoluble content is probably well under 10 per cent.

Estimates of grade are dependent upon sludges, cuttings, and core from diamond and rotary drill holes. The drilling indicates the probability of very substantial tonnages of a grade-in-place in the range of:

| Gold | Silver | Copper | Lead | Zinc | Iron |
|------|--------|--------|------|------|------|
| 0.15–0.2 oz | 7.0–9.0 oz | 0.10–0.20% | 5.0–7.0% | 9.0–12.0% | 20.0–30.0% |

ment, have located at depths of 2000 to 3000 feet the downthrown segments of the Eldorado block and of the early miners' "broad lode." Drilling has indicated a substantial tonnage of sulfide ore. (Figure 5) In these holes, vertical intervals in sulfides range from a few feet to many tens of feet, in one case 175 feet. As in the old mines, iron is the heavily preponderant metal, here in pyrite instead of oxides. Next quantitatively are zinc in sphalerite, much of it marmatitic, and lead in galena. Silver is not visible. In tests, it concentrates largely with galena, some with sphalerite. Gold is intergrown with and concentrates largely with pyrite and to some extent with arsenopyrite. A few hundredths to a few tenths of a per cent of copper is present but is invisible in the heavy pyrite gangue.

Even at depths of 1300 to 1500 feet below static water level, a minor but erratic and metallurgically significant amount of oxidation has occurred. It is visible as stains and coatings of oxides and carbonates, in small pockets and narrow seams of limonite, and in minute fractures within sphalerite crystals filled with smithsonite, cerussite, limonite, and other oxided metal-bearing minerals. That such oxidation is erratic rather than general suggests that it is attributable more to channels of

Arsenic is present in quantities that range from 1.0 to 3.0 per cent.

As will be detailed later, the lead-silver-bearing sulfide ore such as those for which the assay-range is given above, appear to occupy a definite area or belt distinct from a parallel one to the east that is composed dominantly of zinc ore.

At this writing, our knowledge of the sulfide ores is largely confined to information from 25 drill holes 840 to 4990 feet in depth, six drilled from the Locan 900 level, the rest from the surface. Necessarily holes were spotted blindly with no preconceptions as to controlling structures. Until underground development progresses further, it cannot be known what structural controls, other than the vertical limitation imposed by the physical effects of the underlying Ruby Hill thrust zone, may have determined the loci, geometry, and extent of the ores. As in the old mines, the sulfide ores in the down-faulted area are in large part within 500 feet vertically above the quartzite and upper Ruby Hill thrust plane.

## SILICATION AND SILICIFICATION

The walls of the ore bodies of Ruby Hill are unsilicated and unsilicified, but in general they are marmorized, a condition caused by their proximity to intrusives or by dynamic metamorphism or both. In the walls of the workings of the old mines accessible in 1962, there is no silicification nor any development of silicates. In the vicinity and to the west and south of the Mineral Hill stock, however, Secret Canyon shales and shaly limestones were converted to hornfels, Hamburg dolomite to garnet and diopside rock, and large volumes of Hamburg dolomite to coarse-grained marble. To the west and south of the stock and underground in the easterly end of Ruby Hill, pyrite, pyrrhotite, and magnetite are in small seams and disseminations in dense silicated rock. As yet, no commercial accumulations of metal-bearing minerals have been discovered in these silicated areas. There may be some possibility of ore farther away from the Mineral Hill stock, to the west in Eldorado dolomite beneath Secret Canyon shale in the floor of Spring Valley, dropped there by the Sharp fault or other branch of the Spring Valley fault. Similarly to the east in the Eldorado on the downthrow side of the Jackson fault.

The quartz diorite stock of Mineral Hill itself is not altered greatly, though feldspars are clouded and in places biotite is present and somewhat chloritized. Quartz diorite penetrated at depth by the two holes mentioned, 3000 feet to the northeast of the Mineral Hill stock (Figure 2), is of granitoid texture, highly altered, biotized, pyritized, and laced with quartz veinlets carrying sulfides including flakes of molybdenite. It is overlain by 125 to 150 feet of highly silicated sediments containing biotite, chlorite, epidote, magnetite, and disseminated pyrite.

On Adams Hill, 3000 to 5000 feet northwest of Ruby Hill, the Hamburg dolomite, stratigraphically just beneath the Dunderberg shale and in areas from which that formation has been stripped by erosion, is extensively silicified. Much of this siliceous material contained sufficient valuable mineral material to have been prospected and mined in early days. Also north and south of the Windfall mine, two and one-half miles south of Ruby Hill, uppermost beds of Hamburg dolomite adjacent to the Dunderberg shale are similarly silicified and at the Windfall mine carried gold with little or no base metal minerals.

## AGE OF MINERALIZATION

If the ores of Ruby Hill and the north half of the district are to be attributed to solutions accompanying or following upon the intrusion of the quartz diorite, they may be of pre-Eocene age. Geologic relations in the district indicate that extrusive and intrusive andesite, rhyolite, and other rocks of volcanic association are younger than the quartz diorite. Age relations determined by the Larsen method using zircon concentrations from the igneous rocks indicate that the quartz diorite antedates the andesite and rhyolite and may be of a sufficiently greater age to place the quartz diorite and consequent mineralization in late Cretaceous (16, p. 15).

## MINES NORTH OF RUBY HILL

### Mines of Adams Hill in Hamburg Dolomite

Northeast of Ruby Hill, Adams Hill is pock-marked with small openings driven in the Hamburg dolomite in the early days. These workings exploited small showings of oxidized gold-silver-lead ore in fissures and along bedding planes and in runs seemingly independent of visible structures.

Small tonnages of early day ore from Adams Hill, were high in gold and silver, ranging up to 2 to 3 ounces of gold and 150 ounces of silver. No doubt screening and sorting raised the grade above that of the ore as it was broken.

The ore from rabbit warrens of small interconnecting surface workings in the areas of silicified dolomite described near the top of the Hamburg probably was valuable only for silver and gold. There may have been an increment of supergene silver. In these openings, there is little evidence of base metals.

### T.L. Mine and Shaft

North of Adams Hill is the T.L. shaft and mine. In 1956 to 1957 from on and above the three levels, 850, 950, 1050, The Eureka Corporation mined and shipped 31,374 tons of thoroughly oxidized ore averaging:

| Gold | Silver | Lead | Zinc |
|------|--------|------|------|
| 0.426 oz | 11.5 oz | 17.2% | 3.0% |

This ore came from loosely connected bodies of small cross-section in the lower third of the Hamburg dolomite. Collectively the ore bodies rake downward to the northeast, their rake roughly paralleling the Bullwhacker fault, which strikes northeast, dips 35° to the south-

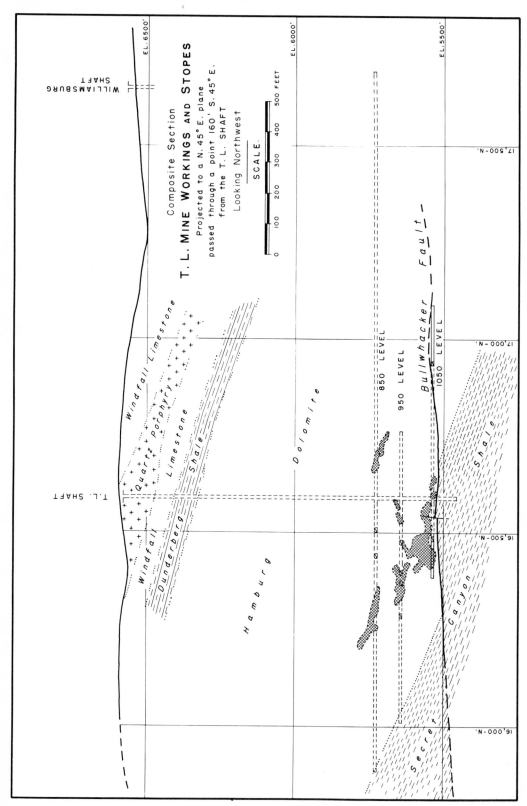

FIG. 7. *Composite Section, showing T.L. mine workings and stopes.*

east, and has a normal throw of about 225 feet. With two exceptions, these ore bodies are out in the hanging wall and in detail neither in nor along the plane of the fault. On their upward rake, the exceptional bodies approach the fault from the hanging wall side at considerable angles and terminate abruptly against it. Their geometry suggests displacement by the fault, which may be the fact, since two small ore bodies in the footwall of the Bullwhacker fault also approach the fault at wide angles and terminate against it. Figure 7 is a composite profile of the T.L. ore bodies projected to a vertical plane paralleling the Bullwhacker fault on strike but divergent upward from it.

The T.L. mine ores consist of various combinations of cerussite, anglesite, and plumbojarosite in a heavy iron-oxide gangue. Sizable pockets occur of clean cerussite. The ores are not materially different from the oxidized ores in the Eldorado dolomite of Ruby Hill.

## Bullwhacker, Williamsburg, and Holly Mines

Immediately to the north and northeast of the T.L. mine, in the order named, lie the Bullwhacker, Williamsburg, and Holly mines. Like those in the T.L. mine, their ores were in the hanging wall of the Bullwhacker fault. They were lead-silver-bearing ores of various grades in small fissures and in manto and pipelike runs in the bedding of the Windfall limestone (Figure 8). Much of Holly ore was composed of only partially oxidized sulfides that yielded variable, rather low recoveries in a small flotation mill operated in the mid-twenties. Many of the ore bodies in the Bullwhacker mine bordered or were not far from the upper and lower contacts of the Bullwhacker quartz-porphyry sill.

74,700 tons of recorded production from stopes in these northmost mines of the district averaged:

| Gold | Silver | Lead |
| --- | --- | --- |
| 0.235 oz | 28.7 oz | 37.5% |

Probably included were moderately low-grade concentrates from the Holly mill.

## MINES SOUTH OF RUBY HILL

A dozen small mines lie at scattered points 1 to 4 miles due south of Ruby Hill. As far

as is known, their ores were thoroughly oxidized, largely in the Hamburg and to a lesser extent in the Eldorado dolomite. Of these, seven are credited with gross production exceeding $200,000. Only one, the Diamond mine of the Consolidated Eureka Mining Company, has been active in recent years.

### Diamond Mine

Properties now in the holdings of the Consolidated Eureka Mining Company are credited with production during the period 1873 to 1939 of 84,772 tons having a gross value of $2,134,116. Ore produced during the years 1954 to 1956 averaged: 0.751 oz gold, 39.7 oz silver, 28.1 per cent lead, grossed $125.04 per ton, and was shipped directly to a smelter. These ores came in part from small bodies along a steeply dipping north-south fissure but more largely from steeply raking pipe of small diameter. The steeply dipping bedding of the Hamburg dolomite may be a factor in determining their attitude, possibly also in some degree, obscure fissures. The structural relations of such pipelike runs of ore in limestone in other districts are often obscure.

### Silver Connor Mine

The Silver Connor mine is credited with production during the years 1877 to 1921 of 5,937 tons having a gross value of $167,132 (16). The valuable metal was largely silver. This ore was in the Silver Connor fault, a steeply dipping fissure of considerable displacement between dolomite walls.

### Dunderberg Mine

A mile and three quarters south of Ruby Hill is the Dunderberg mine which is credited (16) with 14,293 tons grossing $488,000 mined during the period 1873 to 1893. According to Curtis (2, p. 63) Dunderberg ores closely resembled those of Ruby Hill.

129 shipments are reported to have averaged:

| Gold | Silver | Lead |
| --- | --- | --- |
| 1.29 oz | 31.27 oz | 30% |

These ores occur near the stratigraphic top of the Hamburg in a silicified area near a rhyolite dike.

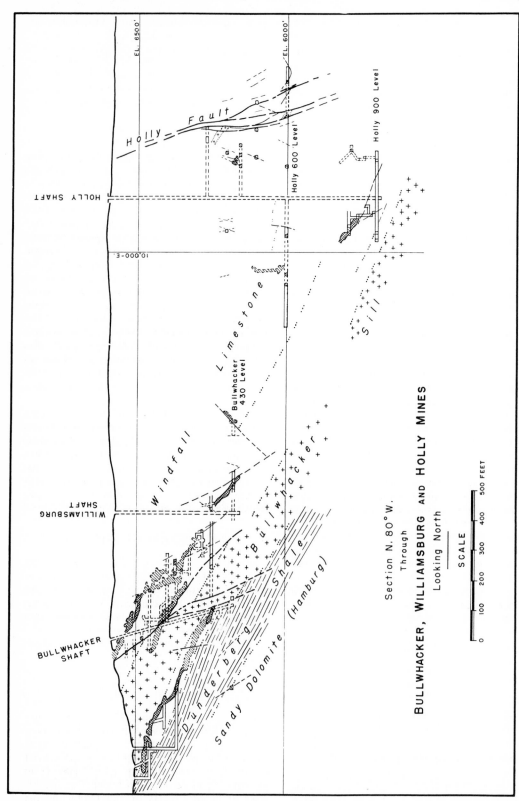

FIG. 8. *Section N80°W through Bullwhacker, Williamsburg, and Holly mines.*

## Windfall Mine

About a mile and a half south of the Dunderberg and also in the stratigraphic top of the Hamburg is the Windfall mine credited with 65,000 tons that grossed $349,000 and were mined during the period 1908 to 1916 (16). This was rather low-grade gold ore with little or no base metals. A few small lots, only, carried appreciable amounts of lead and silver. The ores in both the Dunderberg and Windfall are in close proximity to the Dunderberg shale. In this they are similar to the siliceous ores described in Adams Hill. In the vicinity of the Windfall mine, the Hamburg dolomite has been extensively altered to a friable, "sandy" material, possibly by hydrothermal action, seemingly without the introduction of foreign material. This alteration is not confined to the immediate vicinity of the gold ores. The known ore bodies apparently were five elliptical ore shoots 50 to 150 feet in diameter. They are believed to have terminated or at least to have followed to shallow depths possibly of less than 300 feet.

## ZONING

From the standpoint of such metalliferous zoning as characterizes in various degrees of completeness so many of our western districts of probable Tertiary or late Mesozoic mineralization, the distribution of metals in the Eureka district thus far appears somewhat anomalous if all the ores are attributed to agencies emanating only from the area of the small quartz diorite stock south of Ruby Hill.

Insofar as it is developed, the picture is that of a belt of silver-lead bearing mineralization beginning in Ruby Hill and extending a mile and a half northward upon the crest and easterly side of the north-south broad anticline in which the north half of the district lies. This belt embraces Ruby Hill, the little mines of Adams Hill and the T.L. mine, all beneath the Dunderberg shale in the Hamburg dolomite, and, farther north and above the Dunderberg shale, the Bullwhacker, Williamsburg, and Holly mines in the Windfall limestone.

The last 10 holes drilled in recent years from the surface northeast of Ruby Hill have indicated a considerable area of iron-zinc mineralization with erratic but low lead and silver content and occasional minute showings of molybdenum. This area is adjacent to, easterly and northeasterly of, that of the lead-silver-bearing ores of Ruby Hill. In part, the zinc ores lie between the lead-zinc-bearing ores of Ruby Hill and the quartz diorite discovered

to the east at depth in drill holes 606 and 608 (Figure 9).

In camps such as Bingham, Utah, Ely, Nevada, and the Santa Rita-Hanover-Fierro area, New Mexico, but less clearly evident in Leadville and in some other districts, lead-silver-bearing ores occur in the outer reaches of the spread of mineralization around stock-like intrusives. In Bingham, Ely, and the Santa Rita area, in and adjacent to stocks, are large areas of predominantly pyritic mineralization with sufficient copper and molybdenum to make them profitable to mine. At Bingham, farther outward, but adjacent to the disseminated cupriferous ores, were massive replacement copper ores of 2 to 4 per cent copper. Still farther out, a mile wide belt of lead-zinc-silver ore encircles the cupriferous areas. Silver content increases toward the periphery of this outer belt. Similarly in the Santa Rita area, a definite belt of zinc ore with very low or no lead and silver lies between the intrusives with their cupriferous mineralization and outlying areas of lead-silver ores, the Ground Hog mine area and that of Central City on one side and Georgetown and Shingle Canyon on the other. No two districts are strictly alike geologically or in their zonal aspects. But similar and consistent radial changes in the character of the mineralization are found in other districts, notably Butte, Bisbee, and even in the San Juan region of Colorado. Zoning in the distribution of metals, and in some districts also in attendant silicate alterations, is especially notable in districts having volcanic features and associations. At Eureka, erosion appears not to have cut deep enough to expose the complete pattern of intrusives and their relations to mineralization probably consequent upon them.

Development, however, has proceeded far enough to indicate the probability of a certain degree of zoning, such as that suggested in Figure 8. Small dikes in Ruby Hill and small sill-like sheets to the northward within, just below, or just above the Dunderberg shale (encountered in drill holes in and along the Bowman fault), and, farther north, the Bullwhacker sill, may well be apophyses of intrusives at depth easterly of and paralleling the known lead-silver ores and possibly at the extreme north even beneath them. This is in accord with the north-south axis of the district along which are aligned its folding, many major faults, its known intrusives, both small and larger bodies of andesite, rhyolite, quartz diorite, and quartz porphyry and the district's ore occurrences themselves (Figure 9).

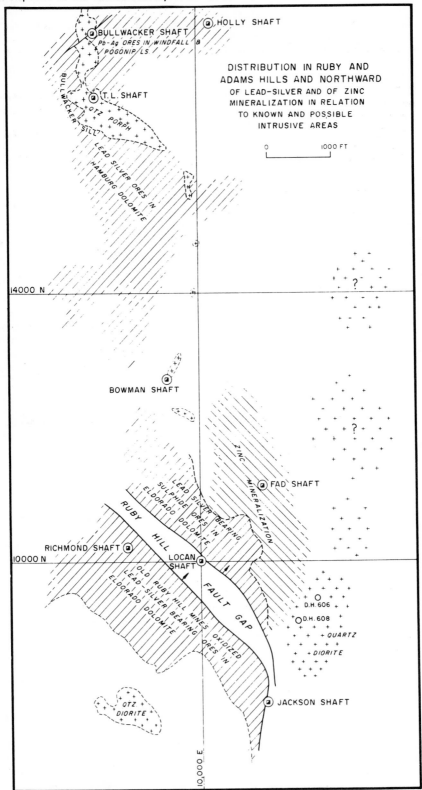

FIG. 9. *Distribution in Ruby and Adams Hills and Northward of Lead-Silver Ores and of Zinc-low Lead-Low Silver Mineralization, in relation to known and possible intrusive areas.*

In Eureka, mineralized areas and ore bodies may be present at depths below economic reach. The district has been handicapped by its distance from sources of power. Because of the hiatus in exploitation occasioned by the Ruby Hill fault, Eureka remains the only major mining district in the West that has not received the full benefit of modern underground exploration and development.

## REFERENCES CITED

1. Hague, A., 1883, Abstract of report on the geology of the Eureka district, Nevada: U.S. Geol. Surv. 3d Ann. Rept. p. 237–272.
2. Curtis, J. S., 1884, Silver-lead deposits of Eureka, Nev.: U.S. Geol. Surv. Mono. 7, 200 p.
3. Hague, A., 1892, Geology of the Eureka district, Nevada: U.S. Geol. Surv. Mono. 20 (with atlas), 419 p.
4. Walcott, C. D., 1908, Nomenclature of some Cambrian Cordilleran formations: Smithsonian Misc. Coll. v. 53, pub. no. 1804, p. 1–12.
5. ——— 1908, Cambrian sections of the Cordilleran area: Smithsonian Misc. Coll. v. 53, pub. no. 1812, p. 166–230.
6. ———1923, Nomenclature of some post Cambrian and Cambrian Cordilleran formations: Smithsonian Misc. Coll. v. 67, no. 8, p. 457–476.
7. ——— 1925, Cambrian and Ozarkian trilobites: Smithsonian Misc. Coll. v. 75, no. 3, p. 61–146.
8. Dwight, A. S., 1936, A brief history of blast-furnace lead smelting in America; Metallurgy of lead and zinc: A.I.M.E. Tr., v. 121, p. 10–12.
9. Wheeler, H. E. and Lemmon, D. M., 1939, Cambrian formations of the Eureka and Pioche districts, Nevada: Univ. Nev. Bull. v. 33, no. 3, 60 p. (Geol. and Min. ser. no. 31).
10. Gianella, V. P., 1946, Igneous fusion of tuff at Eureka, Nevada (abs.): Geol. Soc. Amer. Bull., v. 57, p. 1251–1252.
11. Sharp, W., 1947, The story of Eureka: A.I.M.E. Tr., v. 178, p. 206–217 (Tech. Pub. 2196, 12 p.).
12. Easton, W. H., et al., 1953, Revision of straitigraphic units in Great Basin: Amer. Assoc. Petrol. Geols. Bull., v. 37, p. 143–151.
13. Gilluly, J., 1954, Further light on the Roberts thrust, north-central Nevada: Science, v. 119, p. 423.
14. Nolan, T. B., et al., 1956, Stratigraphy in the vicinity of Eureka, Nevada: U.S. Geol. Surv. Prof. Paper 276, 77 p.
15. Jaffe, H. W., et al., 1959, Lead-alpha determinations of accessory minerals of igneous rocks (1953–1957): U.S. Geol. Surv. Bull. 1097-B, p. 65–148.
16. Nolan, T. B., 1962, The Eureka mining district, Nevada: U.S. Geol. Surv. Prof. Paper 406, 78 p.
17. Armstrong, R. L., 1963, Geochronology and geology of the eastern Great Basin in Nevada and Utah: Unpublished Ph.D. Thesis, Yale Univ.